Woodhead Publishing Series in Biomedicine: Number 73

Medicinal Plants

Chemistry, Biology and Omics

Authored by

Da Cheng Hao

Xiao-Jie Gu

Pei Gen Xiao

ELSEVIER

AMSTERDAM • BOSTON • CAMBRIDGE • HEIDELBERG
LONDON • NEW YORK • OXFORD • PARIS • SAN DIEGO
SAN FRANCISCO • SINGAPORE • SYDNEY • TOKYO

Woodhead Publishing is an imprint of Elsevier

WP WOODHEAD PUBLISHING

Woodhead Publishing Limited is an imprint of Elsevier
80 High Street, Sawston, Cambridge, CB22 3HJ, UK
225 Wyman Street, Waltham, MA 02451, USA
Langford Lane, Kidlington, OX5 1GB, UK

ISBN: 978-0-08-100085-4 (print)
ISBN: 978-0-08-100103-5 (online)

British Library Cataloguing in Publication Data
A catalogue record for this book is available from the British Library

Library of Congress Cataloging-in-Publication Data
A catalog record for this book is available from the Library of Congress

Library of Congress Number: 2015942335

For Information on all Woodhead Publishing publications visit our website at http://store.elsevier.com/

Contents

Preface

Medicinal plants provide myriad pharmaceutically active components, which has been commonly used in traditional Chinese medicine (TCM) and worldwide ethnomedicine for thousands of years. Increasing interest in plant-based medicinal resources has led to additional discoveries of many novel compounds, such as steroidal alkaloids, saponins, terpenoids, glycosides, in various angiosperm and gymnosperm species, and to investigations on their chemotaxonomy, molecular phylogeny, and pharmacology. In continuation with our studies on pharmacophylogeny, in this book we review the phytochemistry, chemotaxonomy, molecular biology, and phylogeny of selected medicinal plant tribes and genera and their relevance to drug efficacy. Literature search is used to characterize the global scientific effort in the flexible technologies being applied. The interrelationship within traditional Chinese medicinal plant groups and between Chinese species and species outside of China is clarified by the molecular phylogenetic inferences based on nuclear and chloroplast DNA sequences. The incongruence between chemotaxonomy and molecular phylogeny is revealed and discussed. It is indispensable to study more species, according to the principles of pharmacophylogeny, for both the sustainable utilization of medicinal resources and finding novel compounds with potential clinical utility. Systems biology and omics technologies (genomics, transcriptomics, proteomics, metabolomics, etc.) will play an increasingly important role in future pharmaceutical research involving bioactive compounds of land plants.

Biodiversity represents an endless source for the discovery of pharmacological active compounds in the development of new medicinal drugs. Bioactive compounds occur in a broad diversity of organisms ranging from bacteria to flowering plants. Discoveries are made through systematic taxonomical investigations and the analysis of herbal medicines used for thousands of years in TCM, Ayurvedic medicine, and worldwide ethnomedicine. The medicinal drugs represent a broad spectrum of chemical structures such as steroidal alkaloids, saponins, terpenoids, polyphenol, and glycosides. Promising compounds are further investigated and developed for improved analog drugs through approaches such as chemotaxonomy, molecular phylogeny, and poly-pharmacology. Evolutionary biology and chemical ecology, especially when approached by high-throughput genomic technology, are closely related to the above fields and would significantly facilitate both, a more systematic approach toward the conservation of medicinal plant diversity as well as drug discovery and development. Some researchers would argue that the discovery of many compounds is passé—the chemical side has been looked at for many years. Evolutionary approaches could thus provide new twist and add new dimension in medicinal plant studies. For example, approved and clinical-trial natural product drugs are obtained from clustered and disjunct taxonomic clades and similarly are herbal indigenous pharmacopeias biased toward the selection of certain phylogenetic clusters.

Evolutionary analyses have been performed at different levels, including genes involved in the biosynthesis of secondary metabolites, its pathways and networks, population dynamics and the molecular interaction of species with the ecosystem or parts thereof (chemical ecology). Also, have studies focusing on the evolution of biosynthesis pathways provided novel insights and directed approaches in synthetic biology and metabolic engineering. Evolutionary approaches offer a rich science-based method to prospect plant diversity having revealed predictive power for bio-prospecting traditional medicines. The trend of integrating evolution into studies of medicinal plants is perceivable and therefore it is time to summarize the current progress in the relevant fields in order to make full use of evolutionary biology and revolutionize the roadmap of medicinal plant research.

This book wish to reflect the current progress in phylogeny, chemotaxonomy, molecular biology, and phytochemistry of selected medicinal plant tribes and genera in the light of evolutionary biology and genomics. In the context of evolution, each chapter of this book is a kind of fusion of commentary, perspective, and review, which aims to characterize the global scientific effort as well as the flexible technologies and methods applied, and also illustrate how evolutionary biology could further our understanding of a number of aspects of medicinal plant research.

Five features of this book

1. reviews and summarizes best practice and essential developments in medicinal plant chemistry and biology;
2. discusses the principles and applications of various chemical, biology, and omics techniques used to discover medicinal compounds, bioactivities, and underlying evolutionary relationship;
3. explores the analysis and classification of novel plant-based medicinal compounds;
4. includes case studies on pharmacophylogeny;
5. compares and integrates traditional knowledge and current perception of worldwide medicinal plants.

The book is designed for use by senior undergraduate and graduate students, researchers, and professionals in medicinal plant, phytochemistry, pharmacognosy, molecular biology, biotechnology, agriculture, and pharmacy working in the academic and industrial sectors. Students and researchers in pharmacology, medicinal chemistry, plant systematics, food and nutrition, clinical medicine, evolution and ecology, as well as professionals in pharmaceutical industries might also be interested in plants included in this book.

Chapter authorship

Chapters 1, 3, 6, 10, 11, and 15, Da Cheng Hao (DH) and Xiao Jie Gu (XG); Chapter 2, DH; Chapters 4, 5, 7–9, 12–14, DH, Pei Gen Xiao and XG.

This book is supported by Academic Publication Fund of Dalian Jiaotong University. Friends and colleagues in many parts of the world lent support to this book. We would like to thank all those who have published their findings that we cite in the chapters. Special thanks go to the project editor Dr. Glyn Jones and the project manager Mr. Harriet Clayton from Elsevier UK (Woodhead) and other team members for their interest, support and encouragement.

About the authors

1. Dr. Da Cheng Hao, associate professor/principle investigator, School of Environment and Chemical Engineering/Biotechnology Institute, Dalian Jiaotong University, Dalian, China

   Dr. Hao got his Bachelor's degree in Medicine, Master's degree in Science, and PhD degree in Biotechnology from Xi'an Jiaotong University, National University of Singapore and Chinese Academy of Sciences, respectively. He had the post-doc training in Institute of Medicinal Plant Development (IMPLAD), Chinese Academy of Medical Sciences (CAMS), under the supervision of Prof. Pei Gen Xiao and Prof. Shi Lin Chen. He was a visiting scholar of John Innes Centre, UK for 1 year (2012–2013), supported by Ministry of Education, China.
2. Dr. Xiao Jie Gu, lecturer, School of Environment and Chemical Engineering/Biotechnology Institute, Dalian Jiaotong University, Dalian, China

   Dr. Gu got her Bachelor's degree and PhD degree in pharmaceutical science from Liaoning University of Traditional Chinese Medicine and China Pharmaceutical University, respectively. Her major research interest is medicinal plant and pharmacognosy.
3. Prof. Pei Gen Xiao, the founder of pharmacophylogenetics and the leading scientist of Chinese medicinal plant and Chinese Materia Medica studies, and also a well-known ethnopharmacologist.

   Prof. Xiao graduated in 1953 from Xiamen University majoring in Biology. After graduation, he served at the Institute of Materia Medica, CAMS. Since 1983 he has been a professor and was designated as the director of IMPLAD, CAMS. Starting from 1996 he has been designated as honorary director of IMPLAD, CAMS, and the head of Key Laboratory on Resource Utilization and Conservation of Chinese Materia Medica, State Administration of Traditional Chinese Medicine. Due to his outstanding scientific achievements, Prof. Xiao has been elected as member (academician) of Chinese Academy of Engineering, Division of Medicine and Health Engineering since 1994 and was also elected as president of the International Society on Ethnopharmacology in 1994.

Chemotaxonomy: a phylogeny-based approach

1

1.1 Introduction

Chemotaxonomy, also called chemosystematics, is used to classify and identify organisms (mainly plants), according to perceptible differences and similarities in their biochemical compositions. The compounds studied in all cases are either primary metabolites or secondary metabolites (SMs). Examples of chemotaxonomic markers used in recent years are summarized below. Chemotaxonomy contributes to the classification of plants when uncertainty exists using classical botanical methods. Chemosystematics can be regarded as a fusion science that complements available morphological and molecular data to improve plant systematics and to facilitate pharmaceutical resource discovery.

1.2 Chemotaxonomic marker

1.2.1 Primary metabolite

1.2.1.1 Fatty acid

Among the various biochemical markers, fatty acids (FAs) or lipid profiles represent a chemically relatively inert class of compounds that is easy to isolate from biological material. FA (Figure 1.1) profiles are chemotaxonomic markers that define groups of various taxonomic ranks in flowering plants, trees, and other embryophytes. The FA profiles of 2076 microalgal strains from the Culture Collection of Algae at Göttingen University (SAG) were determined in the stationary phase (Lang et al., 2011). Seventy-six different FAs and 10 other lipophilic substances were identified and quantified. The FA profiles were added into a database. FA distribution patterns were found to reflect phylogenetic relationships at the level of phyla and classes. At lower taxonomic levels, for example, between closely related species and among multiple isolates of the same species, FA contents may be rather variable. FA distribution patterns are suitable chemotaxonomic markers to define taxa of higher rank in algae. Due to their extensive variation at the species level, it is difficult to make predictions about the FA profile in a novel isolate.

The distribution of FAs in 13 species of macroalgae (Chlorophyta, Ochrophyta, and Rhodophyta) and one sea grass (*Spartina* sp.), collected on the Rio de Janeiro state coast, was determined (Fleury et al., 2011). Statistical analyses showed the

Medicinal plants: chemistry, biology and omics. http://dx.doi.org/10.1016/B978-0-08-100085-4.00001-3

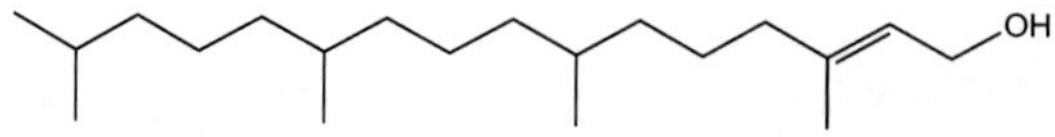

(*E*)-phytol

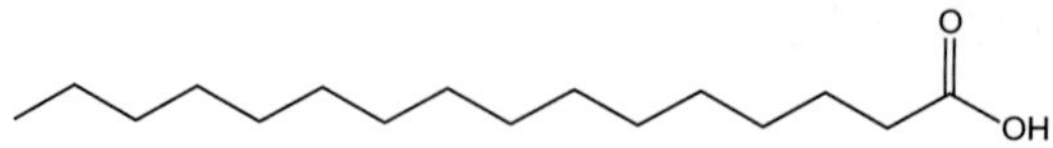

Hexadecanoic acid (palmitic acid)

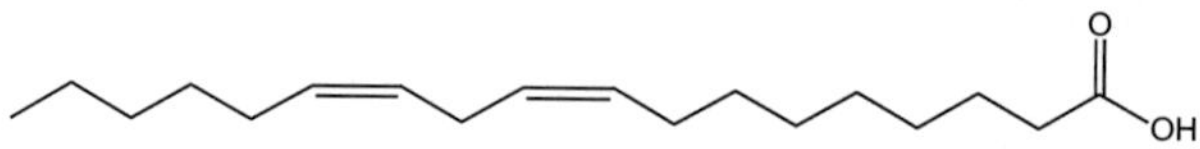

Linoleic acid

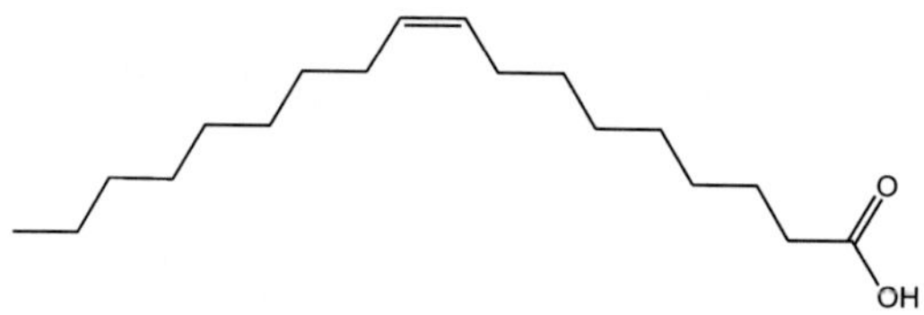

Oleic acid

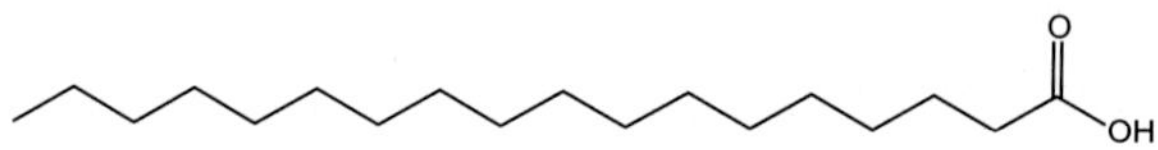

Stearic acid

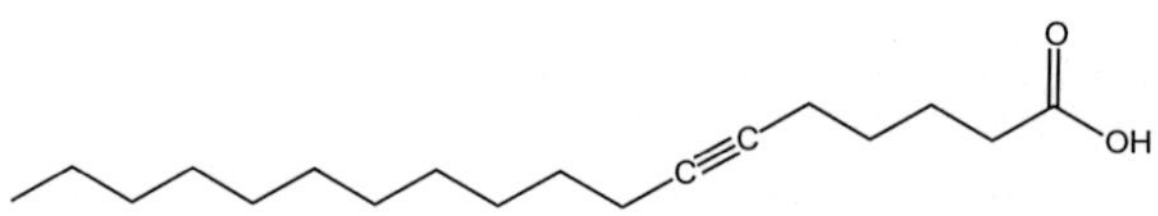

6-Octadecynoic acid

Figure 1.1 Fatty acids.

effectiveness as taxonomic and phylogenetic markers of the distribution of the methyl FA esters in these macrophytes.

In *Geranium* (Geraniaceae) and highly related *Erodium* taxa from Serbia and Macedonia, the investigated essential oils consisted mainly of FAs and FA-derived compounds (45.4–81.3%), with hexadecanoic acid and (*E*)-phytol as the major components (Radulović and Dekić, 2013). *Geranium* and *Erodium* taxa are phylogenetically closely related, and there is no great intergeneric oil-composition variability.

The FA composition of 12 *Brassica* species (Brassicaceae) was analyzed by GC-FID and confirmed by gas chromatography–mass spectrometry (GC–MS) (Barthet, 2008). According to the C18:1 $(n-7)/(n-9)$ ratios for chemotaxonomy, the surveyed species could be arranged into three groups. The first group includes *Brassica napus*, *B. rapa*, and *B. tournefortii* with *Eruca sativa* branching only related to *B. napus*. The second group includes *B. tournefortii*, *Raphanus sativus*, and *Sinapis alba*. The last group includes *B. juncea*, *B. carinata*, and *B. nigra* with no similarity/relationship between them and between the other species.

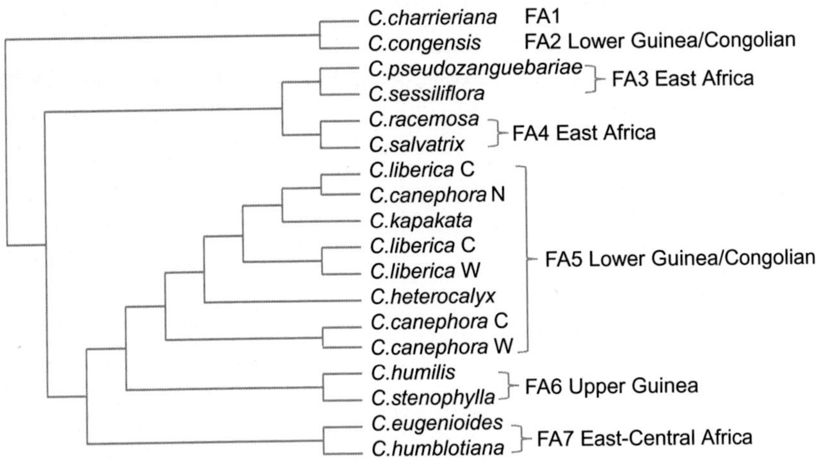

Figure 1.2 A simplified scheme showing the hierarchical clustering analysis of the seed fatty acid composition of 59 *Coffea* genotypes (according to Dussert et al., 2008). Correspondence between groups of species obtained through HCA of seed FA data and clades inferred from DNA sequences (Maurin et al., 2007) is shown.

The FA composition of the seed oil of 23 *Stachys* (Labiatae) taxa was analyzed by GC–MS (Gören et al., 2012). The main compounds were linoleic (27.1–64.3%), oleic (20.25–48.1%), palmitic (4.3–9.1%), stearic (trace to 5.2%), and 6-octadecynoic (2.2–34.1%) acids. The latter compound could be a chemotaxonomic marker of the genus *Stachys*.

FAs and sterols were determined in 59 genotypes of 17 distinct *Coffea* species (Rubiaceae) (Dussert et al., 2008). Interestingly, while groupings based on seed FA composition showed remarkable ecological and geographic coherence (Figure 1.2), no phylogeographic explanation was found for the clusters retrieved from sterol data. When compared with previous phylogenetic studies, the groups deduced from seed FA composition were remarkably congruent with the clades inferred from nuclear and plastid DNA sequences (Table 1.1). Leaf FA composition is useful in chemotaxonomy of Rubiaceae (Mongrand et al., 2005). Principal component analysis (PCA) allowed a clear-cut separation of Coffeae, Psychotrieae, and Rubieae.

1.2.1.2 Protein, amino acid, and carbohydrate

The complete amino acid sequence of [2Fe–2S] ferredoxin from *Panax ginseng* (Araliaceae) was determined (Mino, 2006). Phylogenetic analysis based on the amino acid sequence of ferredoxin suggests that *P. ginseng* is related taxonomically to umbelliferous plants.

Eighteen species of the genus *Euphorbia* (Euphorbiaceae) have proteolytic enzymes in their lattices, nine of them are characterized by the type of endopeptidases (cysteine endopeptidase, serine endopeptidase, metallo-endopeptidase, and aspartic endopeptidase), which are responsible for the activity (Domsalla et al., 2010), and

Table 1.1 Distribution of phenylethanoid glycoside in Gesneriaceae species

Species	Acteoside	Paraboside B	Isonuomioside A	Paraboside II	Paraboside III
Beccarinda tonkinensis	*	–	–	–	–
Hemiboea subcapitata	*	–	–	–	–
Hemiboea flaccida	*	–	*	–	–
Chirita macrodonta	*	–	–	–	–
Chirita pumila	*	–	–	–	–
Chiritopsis repanda	*	–	–	–	–
Paraboea peltifolia	*	–	–	–	–
Paraboea nutans	*	–	*	–	–
Paraboea rufescens	*	*	–	–	–
Rhabdothamnopsis sinensis	*	*	–	–	–
Lysionotus pauciflorus	*	–	–	–	*

*, present; _, not detected.

all nine are serine endopeptidases. The lattices of 64 different species were examined concerning proteolytic activity and serine protease activity, five of which are mentioned in the literature to be proteolytic active and four are known to contain at least one serine endopeptidase. All tested samples were able to degrade labeled casein; the activity of six lattices was completely inhibited by specific serine protease inhibitors; 15 samples were not influenced; and in 43 lattices, a remaining activity was measured, indicating that other types of endopeptidases seem to be involved.

Differences in cell-wall composition and structure, corresponding to the carbohydrate fingerprint region (1200–800 cm^{-1}) of the FT-IR spectrum, can provide the basis for chemotaxonomy of flowering plants (Kim et al., 2004).

1.2.1.3 Alkanes

The PCA of the contents of nine *n*-alkanes showed a clear separation of the Serbian spruce populations from those of the two investigated pine species, which partially overlapped (Nikolić et al., 2013). The separation of the species was due to high contents of the *n*-alkanes C29 and C31 (*Picea omorika*); C19, C20, C21, C22, C23, and C24 (*Pinus heldreichii*); and C28 (*Pinus peuce*).

Samples of 195 *Pinus nigra* trees from seven populations belonging to several infraspecific taxa (*Pinus nigra* ssp. *nigra*, *Pinus nigra* var. *gocensis*, *Pinus nigra* ssp. *pallasiana*, and *Pinus nigra* var. *banatica*) were analyzed (Bojović et al., 2012). The size of the *n*-alkanes ranged from C16 to C33, with the exception of *Pinus nigra* ssp. *nigra*, for which it ranged from C18 to C33. The most abundant were C23, C25, C27, and C29 alkanes. The needle waxes of populations I–III and V were characterized by a higher content of C23, C25, and C27 alkanes and a lower content of C24, C26, C28, and C30 alkanes, compared to the other populations, and the trees of these populations could be assigned to *Pinus nigra* ssp. *nigra*. The samples of population VI were characterized by higher amounts of C22, C24, C30, and C32 alkanes and lower amounts of C25 and C27 alkanes, and the trees could be *Pinus nigra* ssp. *pallasiana*. The samples of population VII, consisting of trees belonging to *Pinus nigra* var. *banatica*, were richer in C29, C31, and C33 alkanes. The wax compositions of populations IV and V, both composed of trees previously determined as *Pinus nigra* var. *gocensis*, showed a tendency of splitting. The alkane composition of population IV was closer to that of *Pinus nigra* ssp. *pallasiana* pines, while that of population V was more similar to that of *Pinus nigra* ssp. *nigra* pines. In the central part of the Balkan Peninsula, significant diversification and differentiation of the populations of black pine exist, and these populations could be defined as different intraspecific taxa. *n*-Alkanes are valid as chemotaxonomic characters within this aggregate.

Analyses by GC and GC–MS of an essential oil sample obtained from dry fruits of *Scandix balansae* (Apiaceae) allowed the identification of 81 components (Radulović and Denić, 2013), comprising 91.4% of the total oil composition. The major identified volatile compounds were medium-chain length *n*-alkanes, that is, tridecane (6.7%; Figure 1.3), pentadecane (13.4%), and heptadecane (19.3%), and a long-chain homologue nonacosane (7.6%). A number of minor oil constituents, among them tetradecyl 3-methylbutanoate, and octadecyl 2-methylpropanoate, 3-methylbutanoate, and pentanoate, have a restricted natural occurrence not only in umbellifers but also in the plant kingdom, whereas the last ester is a new natural compound in general. The identity of these rare plant constituents that present excellent chemotaxonomic marker candidates for *Scandix* was unambiguously confirmed by coinjection of the oil sample with appropriate standards. These samples and additional 58 oils obtained from Scandiceae were compared using multivariate statistical analyses (MVAs), which demonstrated that the evolution of the volatiles' metabolism of Scandiceae taxa neither was genus-specific nor follows their morphological evolution.

The leaf cuticular *n*-alkane chain length distribution pattern was used as an alternative taxonomic marker for eggplant (Solanaceae) and related species (Haliński et al., 2011). The results are in good agreement with current knowledge of the systematics of these plants.

1.2.1.4 Alkynes

Among 106 and 81 constituents, *S*-containing polyacetylene (Figure 1.4) compounds and triquinane sesquiterpenoids made up ~80% of *Echinops bannaticus* (Compositae) and *E. sphaerocephalus* oils, respectively (Radulović and Denić, 2013). A multivariate statistical comparison of the essential oil composition data for these two and additional six taxa of this genus available from the literature permitted an examination of the mutual relationships of the taxa within this morphologically highly uniform genus. PCA and agglomerative hierarchical clustering revealed a grouping of *E. bannaticus*

Tridecane

Pentadecane

Heptadecane

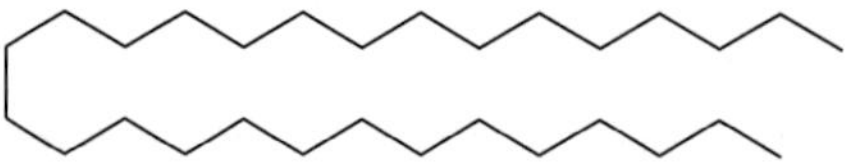
Nonacosane

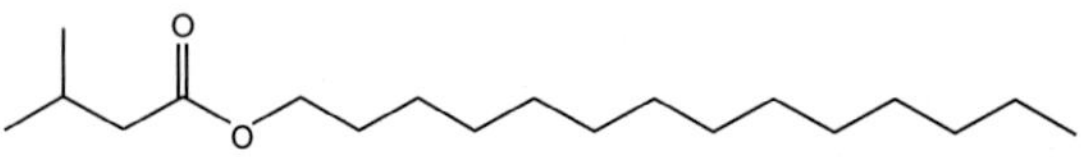

Tetradecyl 3-methylbutanoate

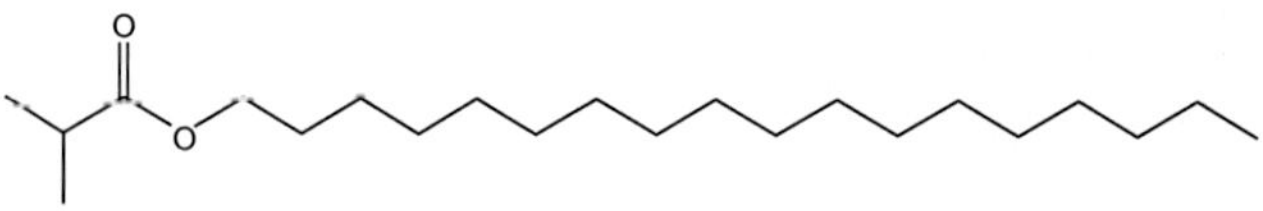

Octadecyl 2-methylpropanoate

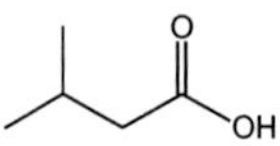

3-Methylbutanoate

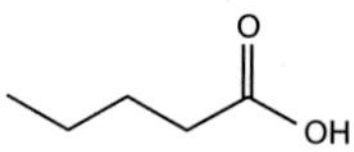

Pentanoate

Figure 1.3 Some alkanes as chemotaxonomic markers mentioned in the text.

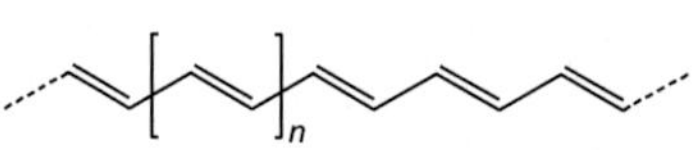

Polyacetylene

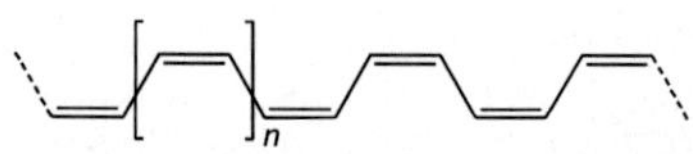

Figure 1.4 Alkynes.

and *E. sphaerocephalus* (section *Echinops*), and their close relationship with *E. grijsii*, suggesting a circumscription of this Chinese taxon to the section *Echinops*. PCA correlation matrix offered valuable insight into the biosynthetic links between essential oil constituents, and these agreed excellently with the currently proposed ones for the polyacetylene *S*-containing compounds, triquinanes, and monoterpenes.

1.2.1.5 Carotenoid

Berries and leaves from six varieties of Carpathians' sea buckthorn (*Hippophae rhamnoides* L., ssp. *carpatica*) were analyzed for their carotenoid composition (free and esterified) using HPLC-PAD, GC–MS, and UHPLC–PAD-ESI-MS techniques

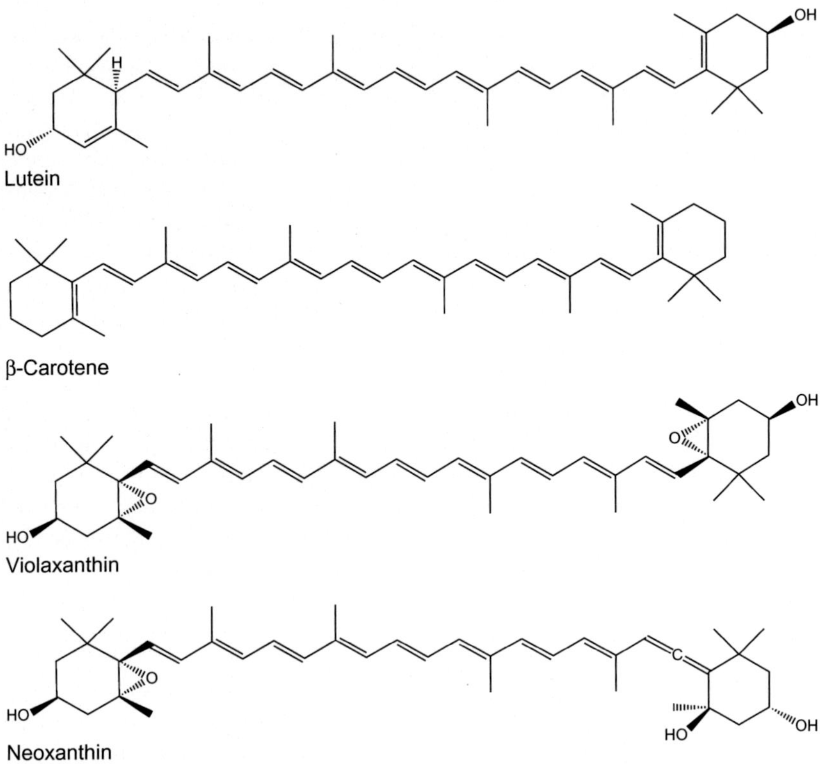

Figure 1.5 Carotenoids.

(Pop et al., 2014). GC–MS revealed the FA profile specific for each berry variety, while targeted UHPLC–MS identified the FAs involved in carotenoid esterification: palmitic (C16:0), myristic (C14:0), and stearic (C18:0). Total carotenoid content varied between 53 and 97 mg/100 g dry weight in berries and between 3.5 and 4.2 mg/100 g DW in leaves. The carotenoid diesters were the main fraction among berry varieties having zeaxanthin dipalmitate as major compound, while leaves contained only free carotenoids like lutein (Figure 1.5), β-carotene, violaxanthin, and neoxanthin. PCA identified the suitable carotenoid biomarkers characteristic for the Carpathians' sea buckthorn from Romania with contribution to their taxonomic classification and authenticity recognition.

1.2.2 Secondary metabolite

The bryophytes contain the Marchantiophyta (liverworts), Bryophyta (mosses), and Anthocerotophyta (hornworts). The Marchantiophyta have a cellular oil body that produces various mono-, sesqui-, and diterpenoids; aromatic compounds like bibenzyl; bisbibenzyls; and acetogenins (Asakawa et al., 2013). Most sesqui- and diterpenoids obtained from liverworts are enantiomers of those found in higher plants. Many of these compounds display a characteristic odor and have interesting biological activities, including antimicrobial, antifungal, antiviral, cytotoxic, insecticidal, insect antifeedant, NO production-inhibitory, antioxidant, piscicidal, neurotrophic, and muscle-relaxing activities, and are involved in allergenic contact dermatitis and in the release of superoxide anion radicals, 5-lipoxygenase, calmodulin, hyaluronidase, cyclooxygenase, DNA polymerase β, and α-glucosidase. Each liverwort synthesizes unique components, which are valuable for their chemotaxonomic classification.

Initial collection of Cameroon herbs was composed of 3742 phytochemicals previously isolated from 67 families, along with 319 hemisynthetic products, giving a total of 4061 chemical structures (Ntie-Kang et al., 2013). Removal of duplicates gave 2770 pure compounds. Emphasis was laid on those plant families from which at least 2.5% of the SMs have been isolated, which include Leguminosae (13.9%), Moraceae (10.6%), Guttiferae (10.1%), Rutaceae (6.5%), Meliaceae (4.5%), Euphorbiaceae (4.4%), Compositae (3.9%), Zingiberaceae (3.4%), Ochnaceae (3.2%), Bignoniaceae (3.1%), Sapotaceae (3.1%), and Apocynaceae (2.8%). Terpenoids were most abundant in Cameroon medicinal plants (26.0% of the isolated compounds). This was followed by flavonoids (19.6%), alkaloids (11.8%), xanthones (5.4%), quinones (5.0%), and glycosides (4.9%), showing a similar trend with a previous analysis of 1859 metabolites.

1.2.2.1 Essential oil and volatile terpene

The structures of some monoterpenoids, sesquiterpenoids, aromatic compounds, aliphatic hydrocarbons, and other compounds included in essential oil are shown in Figure 1.6a–e.

It might be possible to use chemical analysis of SMs emitted from the trees to differentiate clones growing in various diverse environments, as their terpenoid emissions are directly influenced by the environmental conditions in which they grow (Niogret et al., 2013). The endogenous factors are related to anatomical and physiological characteristics of the plants and to the biosynthetic pathways of the volatiles, which not only might change either in the different tissues of the plants or in different seasons but also could be influenced by DNA adaptation (Barra, 2009). Those factors lead to ecotypes or chemotypes in the same plant species. In recent years, chemotaxonomy has been widely used to classify plants with essential oils characterized by intraspecific chemical polymorphism.

Considerable intra- and interspecific essential oil component variation was detected in six subspecies of *Phebalium squamulosum* (Rutaceae: Boronieae), suggesting the existence of distinct chemotypes and supporting previously observed segregate species based on morphological evidence (Sadgrove et al., 2014). GC–MS identified 145 (including 64 tentatively identified) volatile compounds from peel oils of *Citrus*, *Poncirus*, and *Fortunella* (Rutaceae) (Liu et al., 2013a). The chemotaxonomic results based on peel oils are congruent with the Swingle taxonomy system, and *Citrus*, *Poncirus*, and *Fortunella* were almost completely separated. *Citrophorum*, *Cephalocitrus*, and *Sinocitrus*, which belong to the subgenus *Citrus*, can be differentiated by chemotaxonomy.

Mangshanyegan (*Citrus nobilis* Lauriro), a wild germplasm in the citrus family, contains volatile compounds similar to those from pomelo.

The major monoterpenes among the volatiles, that is, β-phellandrene (4), limonene (6), and γ-terpinene (5), and phenylpropanoids, that is, estragole (3), (*E*)-anethole (7), and myristicin (1), showed to be useful chemotaxonomic markers of six *Gingidia* (Umbelliferae) species from New Zealand and Australia (Sansom et al., 2013).

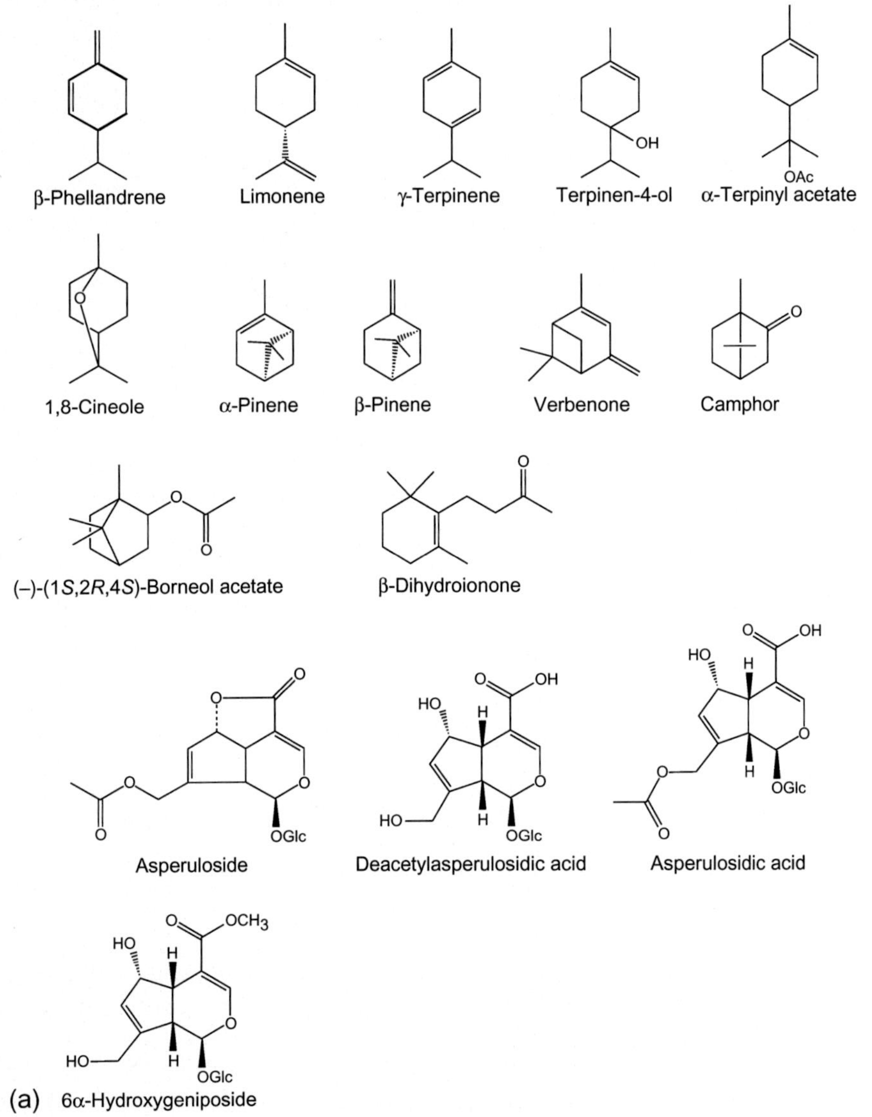

Figure 1.6 Essential oil: (a) monoterpenoids;

(Continued)

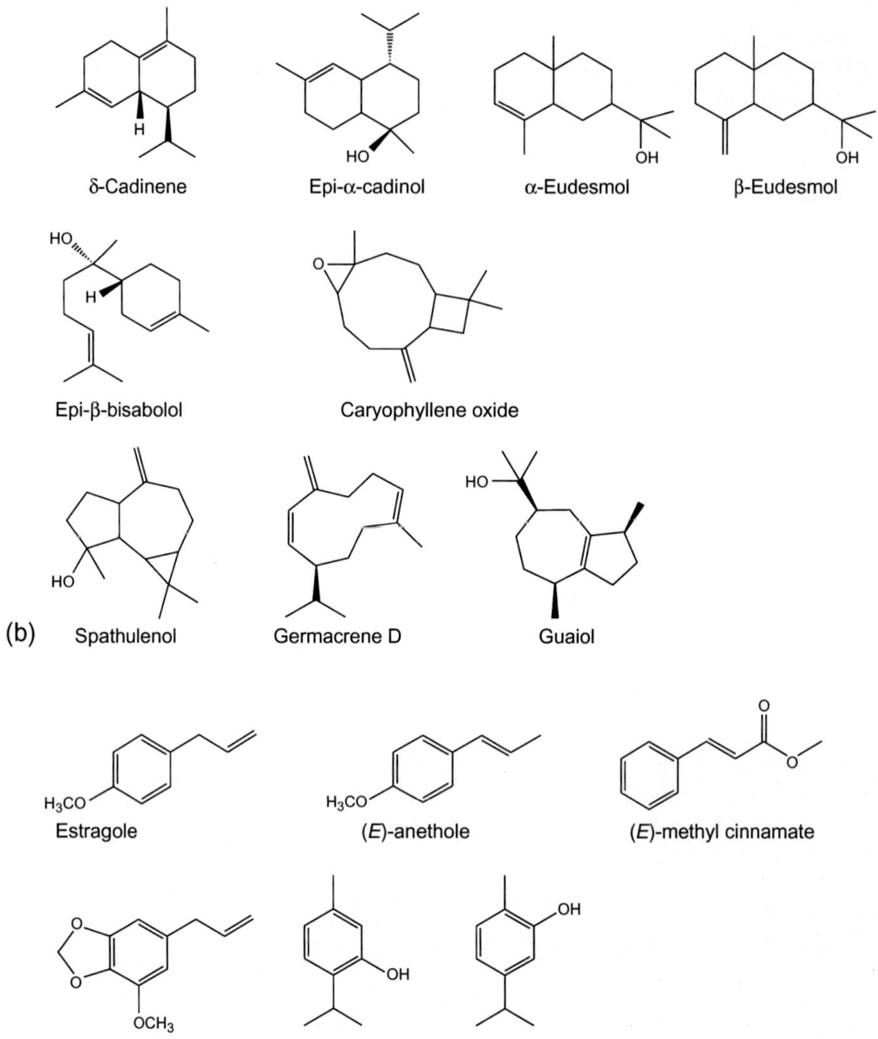

Figure 1.6 Continued. (b) sesquiterpenoids; (c) aromatic compounds;

(Continued)

Essential oils, as chemotaxonomic markers, could be useful to classify *Artemisia* (Compositae) species and to characterize biodiversity in the different populations (Maggio et al., 2012).

The hydrodistilled essential oils obtained from aerial flowering parts of *Teucrium stocksianum* ssp. *stocksianum* (TSS) and *T. stocksianum* ssp. *gabrielae* (TSG) from Iran were analyzed by capillary GC and GC–MS (Sonboli et al., 2013). The oil analysis of two subspecies led to the identification of 65 compounds that accounted for

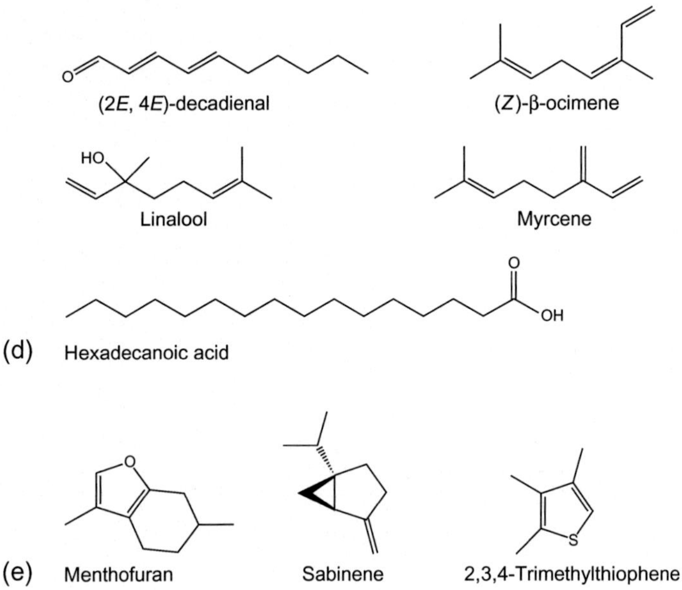

Figure 1.6 Continued. (d) aliphatic hydrocarbons; (e) others.

93.3% and 95.1% of the total oil compositions, respectively. Sesquiterpenoids (52.9%) constituted the main compounds in the essential oil of TSS represented mainly by *cis*-sesquisabinene hydrate (12.0%), epi-β-bisabolol (6.6%), guaiol (5.4%), and β-eudesmol (4.4%), while monoterpenoids (61.2%) were found to be the major components of the oil of TSG, represented by α-pinene (23.0%), β-pinene (13.0%), myrcene (6.3%), and sabinene (6.3%). The principal component in both subspecies was α-pinene (22.0 and 23.0%, respectively) and β-pinene (6.5 and 13.0%, respectively). epi-α-Cadinol, myrcene, and sabinene, detected as principal compounds of TSG, were characterized in lower amounts (<1.5%) in the oil of TSS. Seven components were identified in the oil of TSS corresponding to 25.9% of total oil, which were absent in TSG, in which *cis*-sesquisabinene hydrate (12.0%), guaiol (5.4%), and β-eudesmol (4.4%) were abundant.

The oils of *Teucrium polium* (Lamiaceae) and *T. montanum* consisted mainly of sesquiterpenes (64.3 and 72.7%, respectively), with germacrene D (4; 31.0%) and δ-cadinene (10; 8.1%) as the main constituents, respectively (Radulović et al., 2012). In contrast, the monoterpene menthofuran (1; 11.9%) predominated in the oil of *T. scordium* ssp. *scordioides*, which clearly distinguished this species from the other *Teucrium* taxa.

Hydrodistilled essential oils of 21 accessions of *Ocimum basilicum* belonging to two different varieties (*Ocimum basilicum* var. *purpurascens* and *Ocimum basilicum* var. *dianatnejadii*) from Iran were characterized by GC-FID and GC–MS analyses (Pirmoradi et al., 2013). The oil yield was found to be between 0.6 and 1.1% (v/w). Forty-nine compounds, accounting for 96.6–99.7% of the oil compositions,

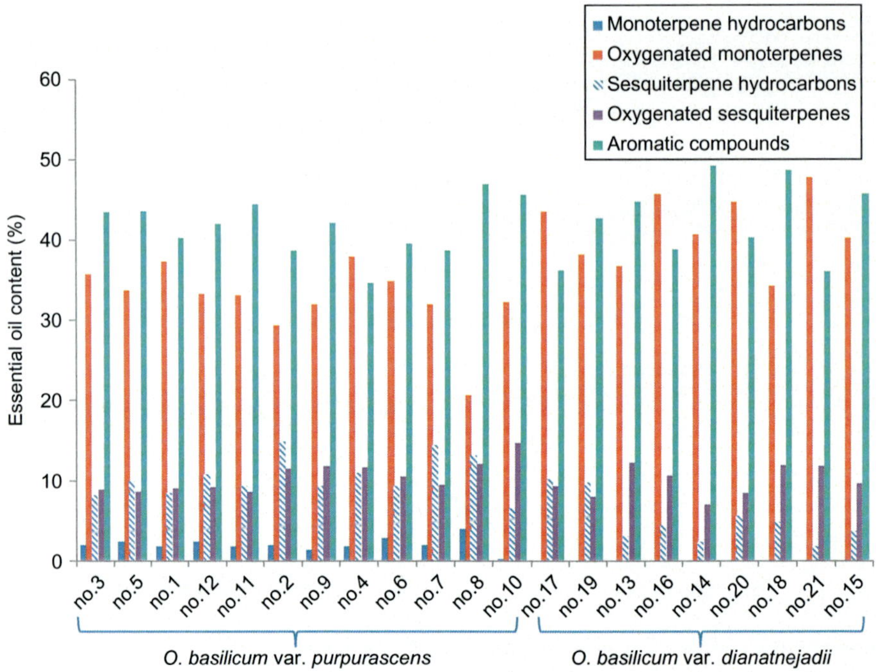

Figure 1.7 Essential oil content (percentage) and composition of 21 *O. basilicum* accessions.

were identified. Aromatic compounds, represented mainly by methyl chavicol (33.6–49.1%), and oxygenated monoterpenes, represented by linalool (14.4–39.3%), were the main components in all essential oils (Figure 1.7). Monoterpene hydrocarbons were present in the essential oils of all accessions of the *Ocimum basilicum* var. *purpurascens*, whereas they were completely absent in those of the *Ocimum basilicum* var. *dianatnejadii*, indicating that monoterpene hydrocarbons might be considered as marker constituents of the *Ocimum basilicum* var. *purpurascens*. The cluster analysis (CA) showed a clear separation of the *Ocimum basilicum* var. *purpurascens* accessions and the *Ocimum basilicum* var. *dianatnejadii* accessions, although the data showed no major chemotype variation between the studied varieties. The CA revealed only one principal chemotype (methyl chavicol/linalool) for both varieties.

The essential oil variability in seven native populations belonging to different infraspecific taxa of *Pinus nigra* (*Pinus nigra* ssp. *nigra*, *Pinus nigra* var. *gocensis*, *Pinus nigra* ssp. *pallasiana*, and *Pinus nigra* var. *banatica*) growing wild in Serbia was analyzed (Sarac et al., 2013). In the needles of 195 trees from seven populations, 58 essential oil components were identified. The major components were α-pinene (43.6%) and germacrene D (29.8%), composing 73.4% of the total oil composition. Based on the average chemical profile of the main terpene components (with contents >5%), the studied populations were found to be most similar to populations from central Italy and Greece (*Pinus nigra* ssp. *nigra*). CA showed the division of the populations into three principal groups: the first group consisted of populations I, II, III, IV, and V (considered as *Pinus nigra* ssp. *nigra* group); the second of population VI (*Pinus nigra* ssp. *pallasiana* group); and the third of population VII, which had the most distinct oil composition (*Pinus nigra* ssp. *banatica* group).

Three relict conifers are clearly separated according to terpene profile with 22 common compounds (Nikolić et al., 2011). In addition, *Picea omorika* has the most abundant *O*-containing monoterpenes and sesquiterpenes; *Pinus heldreichii* and *Pinus peuce* have the largest abundance of sesquiterpene and monoterpene hydrocarbons. The chemosystematic value of the total ketone content, especially of thujone isomers and fenchone, is confirmed (Tsiri et al., 2009), as oil analysis for *Thuja* genus (Cupressaceae) has been proved as a reliable chemosystematic tool in previous studies on different species and subspecies.

The essential oil compositions of leaves, flowers, and rhizomes of *Alpinia galanga* (Zingiberaceae), *A. calcarata*, *A. speciosa*, and *A. allughas* were examined and compared by capillary GC and GC–MS (Padalia et al., 2010). Monoterpenoids were the major oil constituents. 1,8-Cineole, alpha-terpineol, (*E*)-methyl cinnamate, camphor, terpinen-4-ol, and α- and β-pinenes were the major constituents commonly distributed in leaf and flower essential oils. The presence of endo-fenchyl acetate, exo-fenchyl acetate, and endo-fenchol was the unique feature of rhizome essential oils of *A. galanga*, *A. calcarata*, and *A. speciosa*. The rhizome oil of *A. allughas* was dominated by β-pinene. Significant qualitative and quantitative variations were observed in essential oil compositions of the different parts of *Alpinia* species growing in subtemperate and subtropical regions of northern India. CA was performed to find similarities and differences in essential oil compositions based on representative molecular skeletons. 1,8-Cineole, terpinen-4-ol, camphor, pinenes, (*E*)-methyl cinnamate, and fenchyl derivatives were used as chemotaxonomic markers.

To evaluate the chemotaxonomic significance of the essential oils of 23 populations of 18 Iranian *Ferula* (Umbelliferae) species, the chemical composition of the oils was investigated by GC-FID and GC–MS (Kanani et al., 2011). Eighty-four constituents, representing 81.3–99.7% of the total composition of the oils, have been identified. The main constituents were α-terpinyl acetate (73.3%), 2,3,4-trimethylthiophene (2; 49.0%), sabinene (75.3%), verbenone (5; 69.4%), β-pinene (59.0–66.3%), and (*Z*)-β-ocimene (41.7%). CA of the percentage content of the essential oil components of the *Ferula* species resulted in the characterization of four groups, that is, taxa containing either (i) monoterpene hydrocarbons, (ii) oxygenated monoterpenes, (iii) organosulfur compounds, or (iv) monoterpene, sesquiterpene, and aliphatic hydrocarbons (Figure 1.6d) as the principal classes of compounds. The chemical independence of *F. hirtella* from *F. szowitsiana* and of *F. galbaniflua* from *F. gummosa* at the specific level was concluded and their positions as distinct species were confirmed.

The essential oil analysis is also useful in chemotaxonomy of *Hypericum* (Guttiferae; Yuce and Bagci, 2012).

Iridoid, derivative of monoterpene, is useful in chemotaxonomy of *Linaria* (Scrophulariaceae; Guiso et al., 2007) and *Veronica* L. (Plantaginaceae; Saracoglu et al., 2011). From a qualitative point of view, the iridoidic pattern of the two accessions of *Crucianella maritima* was similar (Venditti et al., 2014), since the same compounds (asperuloside, asperulosidic acid, and deacetyl asperulosidic acid) were isolated. Asperuloside was the main compound in both accessions. Asperulosidic acid was the second most abundant compound in the accession from *Sardinia*, while the accession from *Latium* exhibited a similar amount of asperulosidic acid and deacetyl asperulosidic acid. These iridoids can be chemotaxonomic markers for Rubiaceae family, especially for the Rubioideae subfamily to which *C. maritima* belongs.

Several roots or rhizomes of rubiaceous species are used as the emetic and antiamoebic drug ipecac. True ipecac (*Carapichea ipecacuanha*) is chemically well characterized, in contrast to striated or false ipecac derived from the rhizomes of *Ronabea emetica* (syn. *Psychotria emetica*; Rubiaceae). Besides its previous use as substitute of ipecac, the latter species is applied in traditional medicine of Panama, and fruits of its relative *Ronabea latifolia* are reported as curare additives from Colombia. Compounds of *R. emetica* were isolated using standard chromatographic techniques (Berger et al., 2011) and structurally characterized by NMR spectroscopy and MS. Organ-specific distribution in *R. emetica* as well as in *R. latifolia* was assessed by comparative HPLC analysis. Four iridoid glucosides, asperuloside, 6α-hydroxygeniposide, deacetylasperulosidic acid, and asperulosidic acid, were extracted from leaves of *R. emetica*. Rhizomes, used in traditional medicine, were dominated by deacetylasperulosidic acid. HPLC profiles of *R. latifolia* were largely corresponding. These results contrast to the general tendency of producing emetine-type and indole alkaloids in species of *Psychotria* and closely related genera and merit chemotaxonomic significance, characterizing the newly delimited genus *Ronabea*. Chemotaxonomy resolves the historic problem of adulteration of ipecac by establishing the chemical profile of *R. emetica*, the false ipecac, as one of its less known sources.

Licorice (*Glycyrrhiza glabra*, Leguminosae) is a plant of considerable commercial importance in traditional medicine and for the flavor and sweets industry. *Glycyrrhiza* species are very competitive targets for phytochemical studies, and knowledge about the volatile composition is important for understanding the olfactory and taste properties. Volatile constituents from *G. glabra*, *G. inflata*, and *G. echinata* roots were profiled using steam distillation and solid-phase microextraction (Farag and Wessjohann, 2012). Two phenols, thymol and carvacrol, were found exclusively in essential oil and headspace samples of *G. glabra* and with highest amounts for samples that originated from Egypt. In *G. echinata* oil, (2*E*, 4*E*)-decadienal (21%) and β-caryophyllene oxide (24%) were main constituents, whereas 1α, 10α-epoxyamorpha-4-ene (13%) and β-dihydroionone (8%) predominated *G. inflata*. Principal component and hierarchical cluster analyses clearly separated *G. echinata* and *G. inflata* from *G. glabra*, with phenolics and aliphatic aldehydes contributing mostly for species segregation. Thymol and carvacrol, exclusively in *G. glabra*, could serve as chemotaxonomic markers and might be considered as potentially relevant for taste.

Vibrational spectroscopy can be used to discriminate between different essential oil profiles from individual oil plants of the same species (chemotypes) (Baranska et al., 2005). The spectroscopic data correlate very well with those found by GC analysis. Electronic-nose (e-nose) instruments, derived from numerous types of aromasensor technologies, have been used in wood chemotaxonomy (Wilson, 2013). The volatile compound BinBase mass spectral database is well suited for between-study comparisons of chemotaxonomy investigations (Skogerson et al., 2011). Together, oil analysis can be of considerable help, providing basic information needed for the chemosystematic approach of a genus.

1.2.2.2 Sesquiterpene

The sesquiterpene dialdehyde contents can be used to differentiate *Pseudowintera* (Winteraceae) species (Wayman et al., 2010). *P. insperata* individuals had high levels (3.0–6.9% of leaf dry wt.) of the coumarate, *P. axillaris* had high levels (2.2–6.9%) of paxidal, and *P. colorata* from different areas of New Zealand contained varying levels of polygodial (1.4–2.9%) and 9-deoxymuzigadial (0–2.9%).

The New Caledonian endemic *Treubia isignensis* var. *isignensis*, which is a morphologically primitive liverwort, was extracted with diethyl ether, and the crude extract analyzed by TLC and GC–MS (Coulerie et al., 2014). The species is chemically very primitive since it produces only maaliane, eudesmane, aristolane, and gorgonane sesquiterpene hydrocarbons, which are significant chemical markers of the species; neither oxygenated terpenoids nor aromatic compounds were detected.

α-Bisabolol (Figure 1.8) is a commercially important aroma chemical currently obtained from the candeia tree (*Vanillosmopsis erythropappa*; Asteraceae). Continuous overharvesting of the candeia tree has prompted the urgent need to identify alternative crops as a source of this sesquiterpene alcohol. A chemotaxonomic assessment of two *Salvia* species indigenous to South Africa recommended them as a potential source of α-bisabolol (Sandasi et al., 2012). The essential oil obtained by hydrodistillation of the aerial parts was analyzed by GC–MS and mid-infrared spectroscopy (MIRS). Orthogonal projections to latent structures discriminant analysis (OPLS-DA) were used for multivariate classification of the oils based on GC–MS and MIRS data. Partial least squares (PLS) calibration models were developed on the MIRS data for the quantification of α-bisabolol using GC–MS as the reference method. A clear distinction between *Salvia stenophylla* and *S. runcinata* oils was observed using

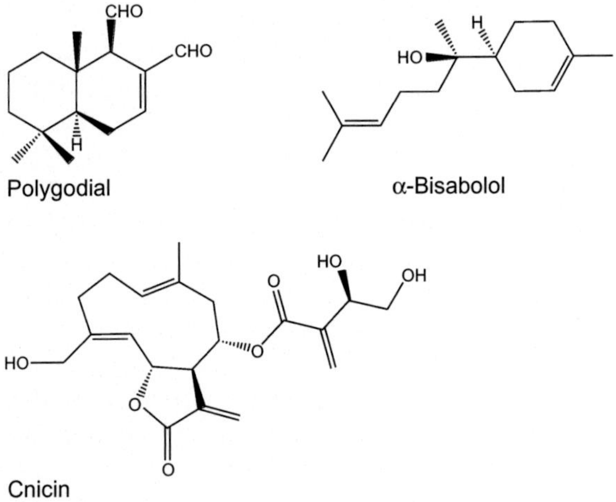

Figure 1.8 Sesquiterpene.

OPLS-DA on both GC–MS and MIRS data. The MIR calibration model showed high coefficient of determination (0.999) and low error of prediction (RMSEP 0.54%) for α-bisabolol content.

Asteraceae, one of the largest families among angiosperms, is chemically characterized by the production of sesquiterpene lactones (SLs). Except for *Gonospermum* species collected on the island of Tenerife, those collected on the island of El Hierro and, in a previous study those from La Gomera, contain SLs that can be used as chemotaxonomic markers confirming the inclusion of *Gonospermum*, *Lugoa*, and species of *Tanacetum* endemic to the Canary Islands in a genus that does not support the monophyly of Gonosperminae (Triana et al., 2010).

The presence of the phytotoxic sesquiterpene (−)-hamanasic acid A {(−)HAA; 7-carboxy-8-hydroxy-1(2), 12(13)-dien-bisabolene} isolated from *Flourensia campestris* (FC; Asteraceae) was investigated in the South American species of the genus (López et al., 2014), together with the evaluation of the phytotoxic activity of their leaf aqueous extracts. (−)HAA was identified and isolated from *F. fiebrigii* (FF) and *F. oolepis* (FO), being chemically (GC–MS and NMR) and biologically (bioassayed on lettuce) indistinguishable from that of FC, while no (−)HAA was found in *F. hirta* (FH), *F. riparia* (FR), and *F. niederleinii* (FN). Its leaf content in FF was similar to that found in FC (~15 mg/g) and significantly higher than in FO (0.8 mg/g). The screening for the presence of (−)HAA in other species showed that its natural occurrence is restricted only to *Flourensia* species. No (−)HAA could be detected in any of the 37 most representative species of the communities (26 natives and 11 exotics), despite many of them belong to the same family and tribe as *Flourensia* spp. Leaf aqueous extracts of all *Flourensia* species exhibited strong inhibitory effects on lettuce germination and on root and shoot growth, regardless of the presence and content of (−)HAA, suggesting the existence of other powerful phytotoxic compounds in those *Flourensia* spp. lacking (−)HAA. Relative to previous exomorphological groupings of the genus, the chemotaxonomic data would give support to the close link described between FC and FF, but not with FR. The fact that (−)HAA was found in FO, which belongs to a second different line, points out that species position in this lineage would deserve to be revisited. The restricted production of (−)HAA by *Flourensia* in their communities sustains its supposed allelochemical role.

Morphological characters and molecular analyses of *Cichorium calvum* (Compositae) and *C. pumilum* do not allow clear discrimination between these closely related wild species. Chemical markers can be selected from the SMs of *C. calvum*, which are unique to this species. From roots of *C. calvum*, ten sesquiterpene lactones were isolated (Michalska et al., 2014), including seven lactucin-type guaianolides reported earlier from *C. pumilum*. Aerial parts also afforded SMs common to both species, along with the megastigmane glucosides staphylionoside D, saussureoside B, and komaroveside A. These norisoprenoids occur in *Cichorium* species, and chemical discrimination of *C. calvum* is possible based on its norisoprenoid composition.

Sesquiterpene lactones related to nerolidol could be used as chemotaxonomic markers of genus *Tanacetum* (Compositae) (Triana et al., 2013). Sesquiterpene lactones from *Lactuca canadensis* (Compositae) have chemotaxonomic significance (Michalska et al., 2013). The main sesquiterpene lactones of *Centaurea zuccariniana*

(Compositae) were malacitenolide, cnicin, and 4′-*O*-acetylcnicin (Ćirić et al., 2012), which agrees with previous study of Greek *Centaurea* sp. belonging to the section *Acrolophus*, and could be of chemotaxonomic significance for the genus *Centaurea*.

Pericarps of *Illicium* (Illiciaceae) species were chemically characterized (Hu et al., 2010). Twenty-two samples from 17 *Illicium* species were subject to HPLC–MS. The chromatographic data were analyzed by CA using SAS software. *Illicium* can be divided into five chemical sections. The distribution of pseudoanisatin, 6-deoxypseudoanisatin, and pseudomajucin was evaluated in 22 samples. LC–MS chromatograms can be used to identify the Chinese star anise.

1.2.2.3 Diterpene

Sideritis (Lamiaceae) species from the Mediterranean region can be classified into four groups (Fraga, 2012). The first group is formed by taxa containing triterpenes, but not diterpenes. A second group is constituted by species having bicyclic diterpenes of the labdane type and not diterpenes. The third group is characterized by its content in tetracyclic diterpenes of the ent-kaurene type. A fourth group is composed of plants with tetracyclic diterpenes of the ent-beyer-15-ene and/or ent-atis-13-ene class. With an UPLC–QTOF-MS/MS method, 21 diterpenoids from different *Salvia* species were separated within 10 min and were unequivocally or tentatively identified via comparisons with authentic standards and literature (Zhou et al., 2009), which is useful in chemotaxonomy.

Caldesia grandis is a rare and endangered aquatic plant of the Alismataceae family. Forty-three chemical components are present in *C. grandis* and closely related genera such as *Alisma*, *Sagittaria*, and *Echinodorus* (Zheng et al., 2007). Their common components are diterpenoids. Kaurane diterpenoids are found in *Caldesia*, *Alisma*, and *Sagittaria*; clerodane diterpenoids in *Sagittaria* and *Echinodorus*; and pimarene and abietane diterpenoids in *Sagittaria*. Kaurane and abietane diterpenoids represent evolutionarily derived compounds, clerodane diterpenoids are primordial, and pimarene diterpenoids are intermediate. The chemotaxonomy, karyotypic analysis, and fossil records of those genera showed that *Caldesia* was evolutionarily closer to *Alisma* than to *Sagittaria* and *Echinodorus*. Possible chronological order of evolution is *Echinodorus*, *Sagittaria*, *Alisma*, and *Caldesia*.

Acid compounds in the extract of *Pinus thunbergii* needles comprised mainly labdane-type diterpenoids (*trans*-communic acid) (Shpatov et al., 2013), while in the extracts of defoliated twigs and outer bark, the acids were represented predominantly by abietane-type compounds (neoabietic, dehydroabietic, abietic, levopimaric, and palustric acids; Figure 1.9). The major neutral components of the needle extract were 10-nonacosanol, labdanoids (18-hydroxy-13-epi-manoyl oxide and *trans*-communol), and beta-sitosterol. In the extract of defoliated twigs, labdanoids (18-hydroxy-13-epi-manoyl oxide, *trans*-communol, and 13-epi-torulosol), serratane triterpenoids (3β-methoxyserrat-14-en-21-one), and beta-sitosterol were the main neutral constituents, while serratanoids (3β-methoxyserrat-14-en-21-one) alone dominated among the neutral compounds of the outer bark extract. The distribution of lipophilic metabolites in the studied parts of *P. thunbergii* shoot system may be applied for chemotaxonomy purposes.

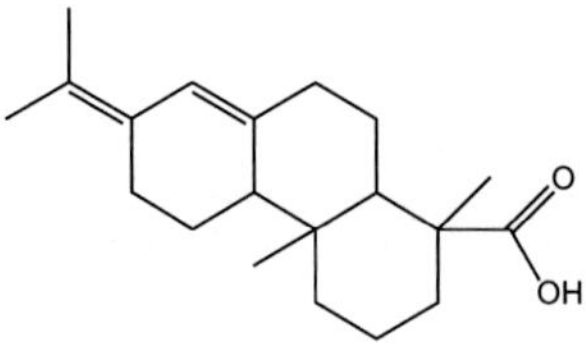

Neoabietic acid

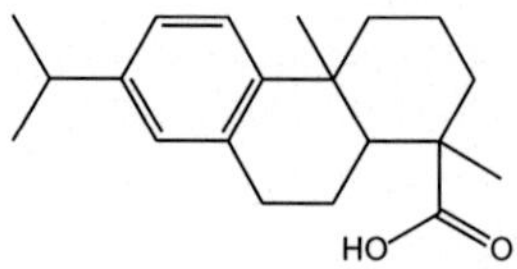

Dehydroabietic acid

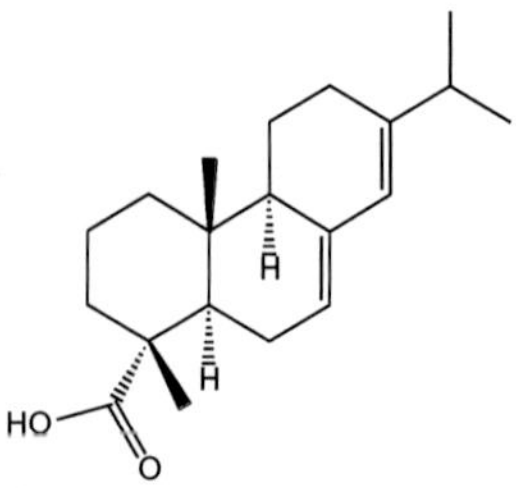

Abietic acid

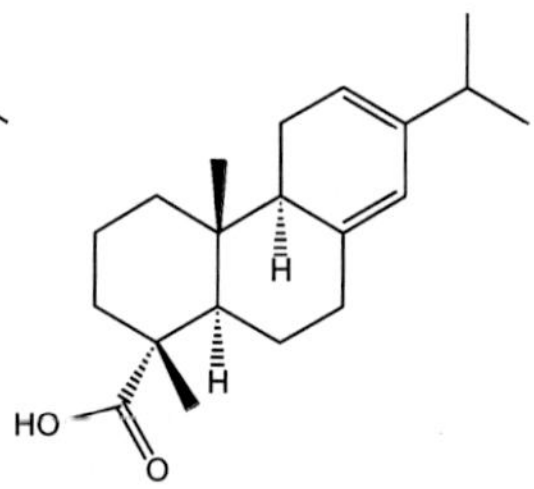

Levopimaric acid

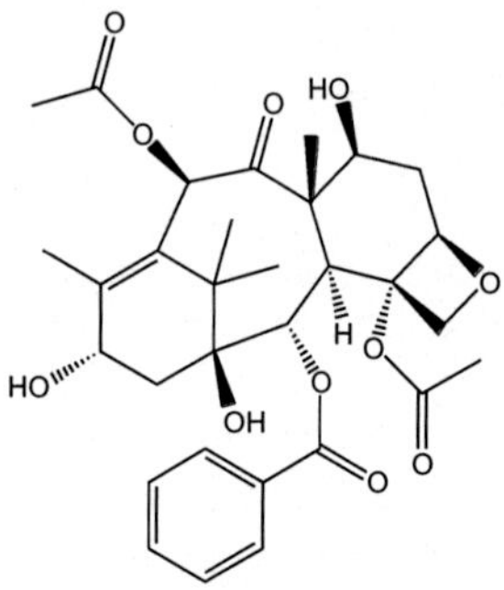

Baccatin III

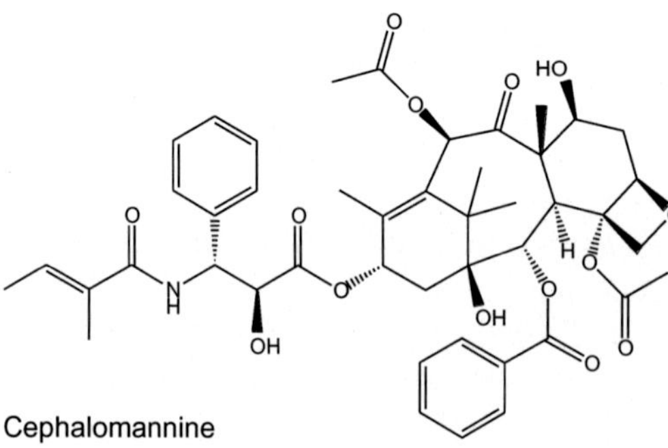

Cephalomannine

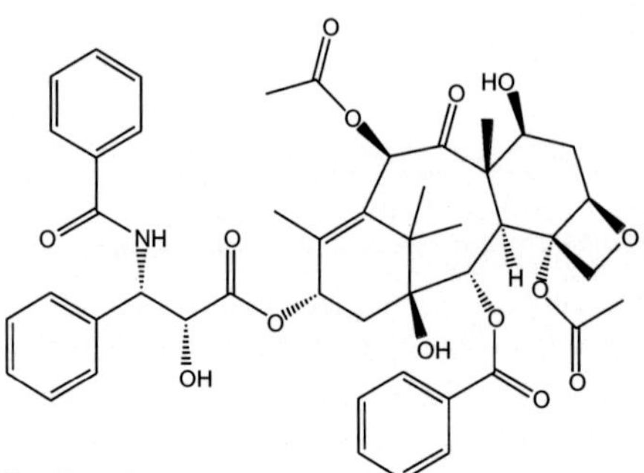

Paclitaxel

Figure 1.9 Diterpene.

The composition of the varieties of *Taxus* growing in Estonia was analyzed by capillary electrophoresis with diode array detection (CE-DAD) for the separation of phenolic compounds and by HPLC–MS for the determination of toxoids (Truus et al., 2012). Fingerprints scanned at 214 nm on the basis of CE separation at pH 9.3 were used to characterize seven varieties of yew. The contents of four key taxoids (10-deacetylbaccatin, baccatin III, cephalomannine, and paclitaxel; Figure 1.9) in six *Taxus* varieties were comparatively determined by HPLC–MS. The set of electropherograms/chromatograms were subject to PCA, using the peak areas of 16 phenolic compounds and 14 taxoids as characteristics. The formation of distinct clusters in accordance with botanical classification proves the suitability of PCA for differentiating varieties of *Taxus*.

1.2.2.4 Triterpene saponin

Five Chinese *Pulsatilla* (Ranunculaceae) species showed distinct HPLC fingerprints of triterpene saponin, although they share 10 peaks (Li et al., 2011). Fourteen accessions were divided into four groups: all accessions from *P. koreana* were in group I, *P. ambigua* in group II, *P. dahurica* and *P. turczaninovii* in group III, and *P. chinensis* in group IV. The significant differences between *P. koreana* and *P. dahurica* and between *P. turczaninovii* and *P. ambigua* were observed. The chemotaxonomic results were in agreement with the traditional taxonomic study. Since DNA sequences of *P. koreana* and *P. ambigua* are not available, it is difficult to compare molecular phylogeny (Figure 1.10) and chemotaxonomic results.

Fifteen cytotoxic polyhydroxyoleanene saponins, aesculiosides C1–C15, were isolated from husks of *Aesculus californica* (Yuan et al., 2013). The triterpenoid saponins from *A. californica* have greater structural diversity than those from any other investigated species thus far in the genus *Aesculus* (Hippocastanaceae). The chemotaxonomic characteristic of aesculiosides C1–C15 is that the unit attached to the C3 of the aglycone is a glucopyranosyl moiety, instead of a glucuronopyranosyl group in the saponins that have been isolated from other *Aesculus* species. The saponins isolated from *A. californica* then provide important evolutionary and chemotaxonomic knowledge of the *Aesculus* genus, a well-known intercontinental disjunct genus in the Northern Hemisphere.

Five azukisapogenol glycosides were isolated from the aerial parts of alsike clover (*Trifolium hybridum*; Leguminosae), three of which were identified as 3-*O*-[−α-L-arabinopyranosyl(1→2)]-β-D-glucuronopyranosyl azukisapogenol, 3-*O*-[−β-D-glucuronopyranosyl(1→2)-β-D-glucuronopyranosyl]-29-*O*-β-D-glucopyranosyl azukisapogenol, and 3-*O*-[−α-L-arabinopyranosyl(1→2)-β-D-glucuronopyranosyl]-29-*O*-β-D-glucopyranosyl azukisapogenol (Pérez et al., 2013). These saponins have chemotaxonomic features that may be recognized as specific of *Trifolium* species.

1.2.2.5 Phenolic compound

Paeonol, paeoniflorin, and their analogs were analyzed in the roots of 14 species and 2 subspecies of *Paeonia* (Guo et al., 2008b). The existence and content of these compounds were discussed in three sections, sect. *Moutan*, sect. *Paeonia*, and sect.

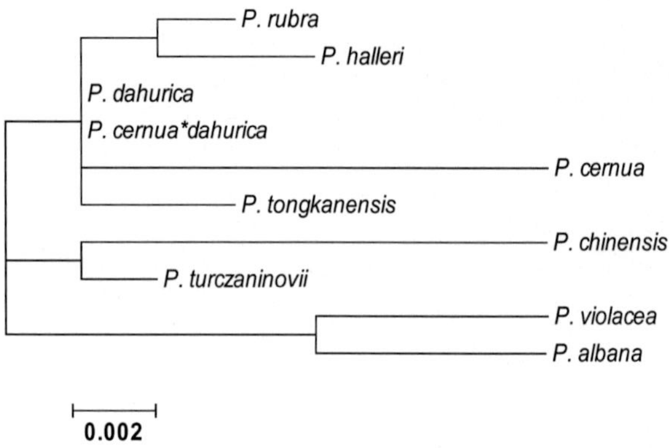

Figure 1.10 *Pulsatilla* ITS phylogenetic tree inferred with maximum likelihood (ML) method. The evolutionary history was inferred based on the general time-reversible (GTR) model. The tree with the highest log likelihood (−957.6810) is shown. Initial trees for the heuristic search were obtained automatically by applying neighbor-joining and BioNJ algorithms to a matrix of pairwise distances estimated using the maximum composite likelihood (MLC) approach and then selecting the topology with superior log-likelihood value. The tree is drawn to scale, with branch lengths measured in the number of substitutions per site. The analysis involved 10 nucleotide sequences. All positions with less than 95% site coverage were eliminated. That is, fewer than 5% alignment gaps, missing data, and ambiguous bases were allowed at any position. There were 548 positions in the final dataset. Evolutionary analyses were conducted in MEGA6 (Tamura et al., 2013).

Onaepia. In sect. *Moutan*, paeonol and its analogs were abundant in all species. In sect. *Paeonia*, low content of paeonol and its analogs was found in *P. lactiflora*, *P. anomala* ssp. *veitchii*, *P. mairei*, and *P. intermedia*. None of these compounds were found in sect. *Onaepia*. Paeonol has a simple structure and is distributed widely in plant; low abundance and absence may be the result of evolution. The relationship among the three sections of *Paeonia* might be that woody sect. *Moutan* is more primitive and derived from the ancestor of *Paeonia* first. The herbaceous sect. *Paeonia* is more closely related to sect. *Moutan* than to sect. *Onaepia*. In sect. *Moutan*, there are less paeonol and its analogs in the species of subsect. *Vaginatae* than in those of subsect. *Delavayanae*. Thus, the former may be more advanced. In sect. *Paeonia*, the taxa with minor content of paeonol and its analogs are diploid except *P. mairei*. Among them, *P. lactiflora* and *P. anomala* ssp. *veitchii* are relatively primitive in morphology. None of paeonol and its analogs were detected in the species with specialized form.

Eighteen SMs were isolated from the heartwood of *Bagassa guianensis*, including six moracins, eight stilbenoids, and three known flavonoids (Royer et al., 2010), suggesting that *B. guianensis* is closely related to *Morus* sp. in phylogeny and should be included in the Moreae sensu stricto tribe of the Moraceae family. Eight chemotaxonomic markers (phenylpropanoids, phenylethanol derivatives, flavonoids, and phenolic acids) were determined to chemically classify *Rhodiola* (Crassulaceae) samples of seven species (Liu et al., 2013b). The chemotaxonomy agrees well with the ITS-based molecular phylogeny.

Salvia (sage) is the largest plant genus in the family Lamiaceae, embracing ~900 species. Conjugated (i.e., binary) chromatographic fingerprints were introduced for 20 *Salvia* species that are grown and cultivated in Poland (Ciesla et al., 2010). Apart from videoscans traditionally used for a comparison of the high-performance thin-layer chromatography (HPTLC) fingerprints, digital scanning profiles and images obtained with the use of the image processing program were employed. Polyphenolic standards are shown in Figure 1.11. The proposed procedure is rapid when compared with the similar ones presented in the literature, and moreover, it is easy to handle. The proposed method offers a possibility to discern the investigated species. It can be applied not only for chemotaxonomic purposes but also for finding new plant species that can be used as medical herbs (as it has been proposed, with *S. triloba* having a similar profile to *S. officinalis*).

The polyphenols of tea leaves as chemotaxonomic markers were examined to investigate the phenetic relationship between 89 wild (the small-leaved *C. sinensis* var. *sinensis* and large-leaved *C. sinensis* var. *assamica*), hybrid, and cultivated tea trees from China and Japan (Li et al., 2010). (−)-Epigallocatechin 3-*O*-gallate (EGCG) (1); (−)-epigallocatechin (EGC) (2); (−)-epicatechin 3-*O*-gallate (ECG) (3); (−)-epicatechin (EC) (4); and (+)-catechin, strictinin, and gallic acid were used as polyphenolic markers. Of the 13 polyphenol patterns observed, PCA indicated that

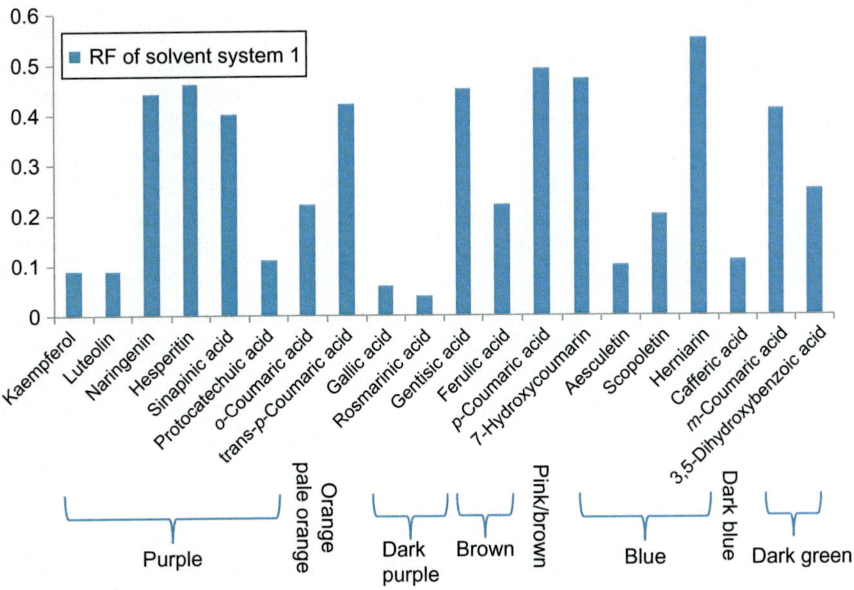

Figure 1.11 Polyphenolic standards used in high-performance thin-layer chromatography of 20 *Salvia* species. The retardation factor (Rf) is defined as the ratio of the distance traveled by the center of a spot to the distance traveled by the solvent front. Color after spraying with H_2SO_4 ($\lambda = 366$ nm) is shown below the compound name. Polyphenolic compounds such as rutin, chlorogenic acid, hyperoside, and quercetin have an Rf of zero (Ciesla et al., 2010), while vanillic acid, cinnamic acid, coumarin, acacetin, and apigenin are not detected at $\lambda = 366$ nm.

the structure types of the flavonoid B-rings, such as the pyrogallol-(1 and 2) and catechol-(3 and 4) types, greatly influenced the classification. Ward's minimum-variance cluster analysis was used to produce a dendrogram that consisted of three subclusters. One subcluster (A) consisted of old tea trees "Gushu" cha (*C. sinensis* var. *assamica*) and "Taidi" cha, suggesting that relatively primitive tea trees contain greater amounts of compounds 3 and 4 and lower amounts of compounds 1 and 2. The subclusters B and C, made up of Chinese hybrids (subcluster B) and Japanese and Taiwanese tea trees (subcluster C), had lower contents of 3 and 4 than subcluster A. The more the amount of 1 and 2 (and the less of 3 and 4), the younger the tea line. Based on morphological characteristics, geographic information, and the historical information on tea trees, these results show good agreement with the current theory of tea tree origins, suggesting that the Xishuangbanna district and Pu'er City of Yunnan province are among the origin sites of the tea tree species.

Phenylethanoid glycosides are useful in the chemotaxonomy of the genus *Phlomis* (Lamiaceae) (Kirmizibekmez et al., 2005). Phenylethanoid glycosides were the predominant group of polyphenols in the studied samples contributing 60% of the total phenolic content for *Teucrium polium* (Labiatac) and *T. scordium* and around 90% for *T. montanum* and *T. chamaedrys* (Mitreski et al., 2014). The systematic analysis for identification and quantification of all present phenolic compounds contributes to the chemotaxonomy of *Teucrium* species and to the valorization based on their phenolic profiles and content.

Five phenylethanoid glycosides, acteoside, paraboside B, isonuomioside A, paraboside II, and paraboside III, were quantitatively determined in 11 species of Gesneriaceae by HPLC (Bai et al., 2013). Phenylethanoid glycosides were found in most Gesneriaceae plants, but the type of phenylethanoid glycosides varied in different species. Acteoside was distributed in most plants (Table 1.1), while paraboside B, isonuomioside A, paraboside II, and paraboside III were sporadically distributed in those plants. On ITS phylogenetic tree, most trib. Didymocarpeae sequences are basal to those of trib. Trichosporeae (Figure 1.12). The chemotaxonomic results support morphological viewpoint that trib. Trichosporeae is more advanced evolutionarily than trib. Didymocarpeae.

The esters of cinnamic acid derivatives with iridoid and phenylethanoid glycosides and an unusually high concentration of verminoside were the most distinctive chemotaxonomic characters of the sun hebes (*Veronica*; Plantaginaceae) (Taskova et al., 2012). The chemical profiles of the species were compared and used to assess the phylogenetic relationships in the group.

Xanthones are not universally present in Gentianaceae, but about 100 different compounds have been reported from 121 species in 21 genera (Jensen and Schripsema, 2002). A coherent theory for the biosynthesis of xanthones, based partly on published biosynthetic results and partly on biosynthetic reasoning, is postulated and used to group the compounds into biosynthetic categories. Arranging the genera according to the xanthones present gives rise to four groups, which is well correlated with molecular data (trnL intron and matK sequences).

Six coumarin compounds, nodakenin (1), oxypeucedanin (2), bisabolangelone (3), notopterol (4), imperatorin (5), and isoimperatorin, of the dried roots of *Ostericum koreanum* (kangwhoal; Umbelliferae), were simultaneously determined

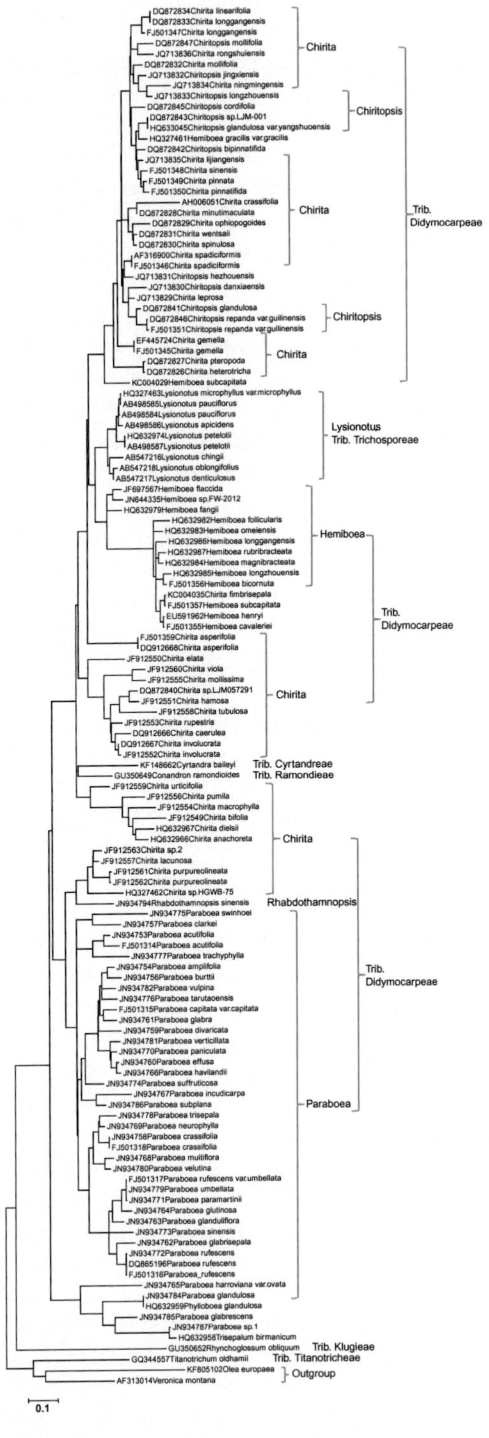

Figure 1.12 Phylogenetic relationship of Gesneriaceae ITS sequences inferred with ML method and GTR+I+G model. The tree is drawn to scale, with branch lengths measured in the number of substitutions per site. The analysis involved 130 nucleotide sequences. There were a total of 1073 positions in the final dataset. Evolutionary analyses were conducted in MEGA6. NCBI GenBank accession number is shown in front of taxon name.

by HPLC (Kim et al., 2012a), which can be used to unambiguously identify 38 samples of different origins.

A rapid reversed-phase (RP) HPLC method was developed and applied for simultaneous separation and the determination of flavonoids and phenolic acids in eight *Plantago* (Plantaginaceae) taxa (*P. altissima*, *P. argentea*, *P. coronopus*, *P. holosteum* ssp. *depauperata*, *P. holosteum* ssp. *holosteum*, *P. holosteum* ssp. *scopulorum*, *P. lagopus*, and *P. maritima*) growing in Croatia (Jurišić Grubešić et al., 2013). The contents of the analyzed phenolic compounds (% of the dry weight of the leaves, dw) varied among species: rutin (max. 0.024%, *P. argentea*), hyperoside (max. 0.020%, *P. lagopus*), quercitrin (max. 0.013%, *P. holosteum* ssp. *holosteum*), quercetin (max. 0.028%, *P. holosteum* ssp. *scopulorum*), chlorogenic acid (max. 0.115%, *P. lagopus*), and caffeic acid (max. 0.046%, *P. coronopus*). Isoquercitrin was detected only in *P. argentea* (0.020%), while isochlorogenic acid content was below the limit of quantification in all investigated species. Multivariate analyses (UPGMA and PCA) showed significant differences in contents of polyphenolic compounds between different *Plantago* taxa, which might be employed as chemotaxonomic markers in the study of the complex genus *Plantago*.

Tannic, gallic, caffeic, vanillic, ferulic, chlorogenic, and cinnamic acids were detected in varying amounts in different parts of 20 varieties of ber (*Ziziphus mauritiana*; Rhamnaceae), which are useful chemotaxonomic markers (Singh et al., 2007). Phenolic acids and flavonoids are of chemotaxonomic significance, especially for the distinction of the diploid taxa of the *Achillea millefolium* L. aggregate (Asteraceae; Benedek et al., 2007). Metabolite profiles indicate considerable phytochemical diversity in the genus *Urtica* (Urticaceae), which largely falls into a group characterized by high contents of hydroxy FAs (e.g., most Andean-American taxa) and another group characterized by high contents of phenolic acids (especially the *U. dioica* clade) (Farag et al., 2013). However, most highly supported phylogenetic clades were not retrieved in the metabolite cluster analyses.

Mercurialis annua and *M. perennis* are used in complementary medicine. Analytic methods to allow a chemotaxonomic differentiation of these species by means of chemical marker compounds were established (Lorenz et al., 2012). The exclusive presence of pyridine-3-carbonitrile and nicotinamide in CH_2Cl_2 extracts obtained from the herbal parts of *M. annua* was demonstrated by GC–MS. Further chromatographic separation of the CH_2Cl_2 extracts via polyamide yielded a MeOH fraction exhibiting a broad spectrum of side-chain saturated *n*-alkylresorcinols. While the *n*-alkylresorcinol pattern was similar for both plant species, some specific differences were observed for particular *n*-alkylresorcinol homologues. The investigation of H_2O extracts by LC–MS/MS revealed the presence of depside constituents. In *M. perennis*, a mixture of mercurialis acid (2*R*-(*E*-caffeoyl)-2-oxoglutarate) and phaselic acid (*E*-caffeoyl-2-malate) could be detected; in *M. annua*, only phaselic acid was found. The configuration of the depside could be 2*S* in *M. annua* and 2*R* in *M. perennis*.

2-(2-Phenylethyl) chromones and dibenzophenones were the characteristic components of the genus *Aquilaria* (Thymelaeaceae; Huang et al., 2013) or even subfamily Aquilarioideae. Flavonoids, diterpenes, and triterpenes are also useful in chemotaxonomy of *Aquilaria*.

Cannabinoids are terpenophenolic compounds unique to *Cannabis* (Cannabaceae). The proportion of high THC/CBD chemotype plants in most accessions assigned to *C. sativa* was <25% and in most accessions assigned to *C. indica* was >25% (Hillig and Mahlberg, 2004). Plants with relatively high levels of tetrahydrocannabivarin (THCV) and/or cannabidivarin (CBDV) were common only in *C. indica*. These results support a two-species concept of *Cannabis*.

1.2.2.6 Flavonoid

Six major flavonoids, including sophoricoside, genistin, genistein, rutin, quercetin, and kaempferol, in *Styphnolobium japonicum* (Leguminosae) were simultaneously determined by LC–ESI-MS/MS (Chang et al., 2013). The quantitative difference in content of six active compounds was useful for chemotaxonomy of many samples from different sources and the standardization and differentiation of many similar samples. Twelve flavonoid compounds were used to differentiate 34 sea buckthorn berries samples (Chen et al., 2007). No obvious difference between *Hippophae rhamnoides* ssp. *sinensis* (Elaeagnaceae) and *H. rhamnoindes* ssp. *yunnanensis* suggested that the two subspecies might have a very close relationship in terms of chemotaxonomy. Flavonoid glycosides were used to authenticate Unani herbal drug chamomile (*Matricaria chamomilla*) from its adulterants, that is, *Anthemis nobilis*, *Matricaria aurea*, and *Inula vestita* (Ahmad et al., 2009).

Pharmacologically active isoflavone aglycones genistein, daidzein, formononetin, and biochanin A were used to classify 13 *Trifolium* (clover; Leguminosae) species, native to Poland (Zgórka, 2009). Naphthodianthrones (e.g., hypericin and pseudohypericin), flavonol glycosides (e.g., isoquercitrin and hyperoside), biflavonoids (e.g., amentoflavone), phloroglucinol derivatives (e.g., hyperforin and adhyperforin), and xanthones may serve as chemotaxonomic markers at various taxonomic levels (i.e., family to species) (Crockett and Robson, 2011), indicating that particular biosynthetic pathways have been conserved within a taxon or, alternatively, have arisen two or more times within a taxon through evolutionary convergence. Flavonoids are useful chemotaxonomic markers of the genus *Iris* (Iridaceae; Wang et al., 2010).

7-Methoxylated flavonoids are a chemotaxonomic trait frequently found in the family Anacardiaceae (Feuereisen et al., 2014).

Sea buckthorn (*Hippophae rhamnoides*) is rich in many bioactive compounds (e.g., vitamins, phenolics, and carotenoids) important for human health and nutrition. Among the phenolics, berries and leaves contain a wide range of flavonols that are good quality and authenticity biomarkers. Six varieties of cultivated sea buckthorn (*Hippophae rhamnoides* ssp. *carpatica*) berries and leaves were analyzed by UHPLC–PDA-ESI-MS (Pop et al., 2013). Berries and leaves contained mainly isorhamnetin (I) glycosides in different ratios. Whereas I-3-neohesperidoside, I-3-glucoside, I-3-rhamnosylglucoside, I-3-sophoroside-7-rhamnoside, and free isorhamnetin were predominant for berries (out of 17 compounds identified), I-3-rhamnosylglucoside, I-3-neohesperidoside, I-3-glucoside, quercetin-3-pentoside, kaempferol-3-rutinoside, and quercetin-3-glucoside were predominant in leaves (out of 19 compounds identified). Berries contained, on average, 917 mg/100 g DW flavonol glycosides. Leaves had higher content of flavonol glycosides than berries, on average 1118 mg/100 g DW. The variation of the quantitative dataset analyzed using PCA accounted for 91% of the total variance in the case of berries and 73% in case of leaves, demonstrating a good discrimination among samples. The flavonol derivatives can be biomarkers to discriminate among varieties and to recognize specifically the berry versus leaf composition.

Dasymaschalon and *Desmos* are two independent genera of family Annonaceae, which is supported by gross morphology, leaf anatomy, and molecular phylogeny. These genera contain formyl-substituted flavonoids with substituted A-ring and unsubstituted B-ring, which could be the chemotaxonomic markers (Zhou et al., 2012).

Flavonoid glycoconjugates from roots and leaves of eight North American lupine species (*Lupinus elegans*, *L. exaltatus*, *L. hintonii*, *L. mexicanus*, *L. montanus*, *L. rotundiflorus*, *L. stipulatus*, and *Lupinus* sp.), three Mediterranean species (*L. albus*, *L. angustifolius*, and *L. luteus*), and one species from South America domesticated in Europe (*L. mutabilis*) were analyzed using two LC–MS systems (Wojakowska et al., 2013). As a result of the LC–MS profiling using the CID/MSn experiments, structures of 175 flavonoid glycoconjugates found in 12 lupine species were identified at three confidence levels according to the Metabolomics Standards Initiative, mainly at levels 2 and 3. Among the flavonoid derivatives recognized in the plant extracts were isomeric or isobaric compounds, differing in the degree of hydroxylation of the aglycones and the presence of glycosidic, acyl, or alkyl groups in the molecules. The elemental composition of the glycoconjugate molecules was established from the exact m/z values of the protonated/deprotonated molecules ($[M+H]^+/[M-H]^-$) measured with the accuracy better than 5 ppm. Information concerning structures of the aglycones, the type of sugar moieties (hexose, deoxyhexose, or pentose), and, in some cases, their placement on the aglycones as well as the acyl substituents of the flavonoid glycoconjugates was achieved. Information obtained from the flavonoid conjugate profiling was used for the chemotaxonomic comparison of the studied lupine species. A clear-cut discrimination of the Mediterranean and North American lupines was obtained.

It is necessary to establish the HPLC fingerprint of flavonoids of six frequently used Chinese materia medica for regulating Qi flow, including *Citri grandis* (Mao Ju Hong), *C. grands* (Guang Ju Hong), *Citri Reticulatae Pericarpium* (Chen Pi), *Citri Reticulatae Pericarpium Viride* (Qing Pi), *Aurantii Fructus* (Zhi Ke), and *Aurantii Fructus Immaturus* (Zhi Shi) from *Citrus* (Chen and Lin, 2011). HPLC was performed on a C18 column with methanol–water (with acetic acid). The six herbal drugs were divided into naringin type and hesperidin type. *C. grandis* and *C. grands* had fifteen common peaks; *Citri Reticulatae Pericarpium*, *Citri Reticulatae Pericarpium Viride*, *Aurantii Fructus*, and *Aurantii Fructus Immaturus* had ten common peaks. All herbs had five common peaks. The holistic similarity of chromatograms of *C. grandis* and *C. grands* was in the range of 0.928–0.996. For *Citri Reticulatae Pericarpium*, *Citri Reticulatae Pericarpium Viride*, and *Aurantii Fructus Immaturus*, it was in the range of 0.922–0.997. But the similarity between *Aurantii Fructus* and the mutual model was only 0.454–0.773. The established fingerprints of flavonoids can be used to compare the differences intuitively. The peak height and peak areas of characteristic peaks are distinct, but whether it is connected with the different function of regulating Qi flow of the six medical materials is awaiting further study.

129 leaf samples from 35 species and one variety of the Chinese *Epimedium* (Berberidaceae), most of which were placed under subgen. *Epimedium* and sect. *Diphyllon*, were analyzed by HPLC method (Guo et al., 2008a). The HPLC profiles of all samples for icariin and similar compounds were achieved, sorted, and analyzed. According to the second peak group ("ABCI" peak group) characters, chromatograms were divided into four main types and nine subtypes. By correlation analysis with flower morphology, II-3 was suggested to be the most primitive type; II-1, IV, and I-3 were primitive and closely related to II-3; I-1 was basic type; and I-2, I-4, III, and II-2 were derived types. The HPLC chromatogram-type division corresponds to W. T. Stearn's classification on sect. *Diphyllon* with four series in 2002.

1.2.2.7 Quinone

1,4-Naphthoquinone derivatives 7-methyljuglone and plumbagin possess a diverse and well-documented array of biological activities, and the chemotaxonomic distribution of naphthoquinones (NQs) among *Drosera* (Droseraceae) species is of phytopharmaceutical interest. These two NQs are ubiquitously coproduced in species-specific ratios (Egan and van der Kooy, 2012), and that 7-methyljuglone appears negatively associated with the occurrence of pigmentation in sundews. The prospective antifeedant function of 7-methyljuglone was evaluated in relation to allocation in various organs and ontogenetic phases of *D. capensis*, revealing that much higher levels were accumulated in young and reproductive organs, most likely for defense. Investigation into the relationship between the biosynthesis of NQs and carnivory showed that the production of 7-methyljuglone is optimally induced and localized in leaves in response to capture of insect prey. This SM is important in ecological interactions and holds implication in chemotaxonomy of the genus.

1.2.2.8 Alkaloid

N-Methylcytisine, cytisine, and jussiaeiines A, C, and D, belonging to quinolizidine alkaloids, are recognized as markers of the genus *Ulex* (Leguminosae) in Portugal (Máximo et al., 2006). Isoquinoline alkaloids are abundant in the family Menispermaceae and have utility in chemotaxonomy (de Wet et al., 2011). Pyrimidine-beta-carboline-type alkaloid is seldom reported and is particularly important for *Annona* genus (Annonaceae) chemotaxonomy (Costa et al., 2006). Tropane alkaloids are used in chemotaxonomy of the pantropical genus *Merremia* (Convolvulaceae) (Jenett-Siems et al., 2005). Several glycoalkaloids (solasonine, α-solamargine, β-solamargine, and α-solanine) and their aglycones (solasodine and solanidine) are a valuable tool to resolve the taxonomic controversy of *Solanum nigrum* complex (Solanaceae) based on morphological characters (Mohy-Ud-Din et al., 2010). *S. retroflexum* did not show marked chemical difference and hence might be regarded as a variety or subspecies of *S. nigrum*.

The alkaloid pattern of *Lapiedra martinezii* (Amaryllidaceae) comprises 49 compounds of homolycorine, lycorine, tazettine, haemantamine, and narciclasine types (Ríos et al., 2013). The populations located in the north and south margins of the distribution area displayed alkaloid patterns different from those of the central area. The variations in alkaloid content could be interpreted in a phylogenetic sense. Steroidal alkaloids are useful for chemotaxonomy of *Fritillaria* (Liliaceae; Li et al., 2009). Chemotaxonomic study indicates that not all *Sceletium* species (Aizoaceae) contain the mesembrine-type alkaloids usually associated with *Sceletium* (Patnala and Kanfer, 2013). It is thus important to identify the correct *Sceletium* species to ensure correct alkaloidal content for the manufacture and quality control of products containing this plant material.

Colombian coca farmers have traditionally cultivated three varieties of coca for cocaine production (*Erythroxylum novogranatense* var. *novogranatense*, *E. novogranatense* var. *truxillense*, and *E. coca* var. *ipadu*). Within the past 13 years, 15 new cultigens of cocaine-bearing *Erythroxylum* have been propagated by Colombian coca farmers, each with differing physical characteristics yet producing cocaine alkaloids at similar levels found in the historical and native varieties (Casale et al., 2014). Five plants per cultigen were randomly selected and examined for alkaloid content to determine their varietal characteristics when compared to the three known varieties. Ten cultigens gave classic *E. coca* var. *ipadu* alkaloid profiles, four cultigens produced alkaloid profiles consistent with a hybridization of *E. novogranatense* and *E. coca* var. *ipadu*, while one cultigen gave heterogeneous alkaloid profiles that could not be characterized.

Chemotaxonomic results support the circumscription of the family proposed by Wu et al., who considered that the Berberidaceae should be treated as four independent families: Nandinaceae, Berberidaceae (s.s.), Podophyllaceae, and Leonticaceae (Peng et al., 2006). Phytochemically, the monotypic family Nandinaceae is characterized by a rich spectrum of benzylisoquinoline alkaloids (BIAs), such as berberine, palmatine, jatrorrhi zine, coptisine, magnoflorine, domesticine, nandinine, and protopine. The existence of the cyanogenic compound nandinin, biflavonoid amentoflavone, and benzaldehyde-4-*O*-glucoside in this family indicates its relatively distant relation with other three families. *Nandina indica*, the only species of Nandinaceae, has been used for clearing heat and counteracting toxins or as antitussive. Berberidaceae (s.s.), which consists of *Berberis* L. and *Mahonia* Nutt., contains mainly BIAs, for example, berberine, palmatine, jatrorrhizine, columbamine, and magnoflorine, particularly a higher content of biisobenzylquinoline alkaloids represented by berbamine and oxyacanthine. The plants in this family have been used for clearing heat and counteracting toxins. Plants in both *Berberis* and *Mahonia* have long been used as the main sources of the drugs berberine and berbamine. Podophyllaceae can be divided into two tribes. The tribe Podophylleae, consisting of *Podophyllum* (including *Sinopodophyllum* and *Dysosma*) and *Diphylleia*, contains various podophyllotoxin lignans, and the plants in this tribe have been used as the most important source for the manufacture of the anticancer drugs, that is, podophyllotoxin's derivatives. The plants have been used for activating blood, revolving stasis, relieving swelling, removing toxin, and clearing heat. The tribe Epimedieae, consisting of *Epimedium*, *Vancouveria*, *Achlys*, *Jeffersonia* (*Plagiorhegma*), and *Ranzania*, has diversified chemical constituents. Both *Epimedium* and *Vancouveria* contain predominately bioactive icariin flavonoids, the characteristic constituents of this group. The plants in *Epimedium* have been used as a male sexual tonic and as medicines for dispelling wind and removing dampness. The phytochemistry of the remaining three genera *Achlys*, *Jeffersonia*, and *Ranzania* has not been thoroughly investigated. Leonticaceae, including *Gymnospermium*, *Leontice*, *Caulophyllum*, and *Bongardia*, contains β-amyrin triterpenoids and quinolizidine alkaloids and has been used for activating blood, revolving stasis, dispelling wind, and removing dampness.

1.2.2.9 Lignan

The family Schisandraceae (Magnoliidae) contains approximately 60 species that are disjunctly distributed in the southeast of Asia and North America. It was divided into two genera, *Schisandra* and *Kadsura*, represented by 29 species in China, 19 in *Schisandra* and 10 in *Kadsura* (Xu et al., 2008). Dibenzocyclooctadiene lignans are the main chemical components of the family. Besides their traditionally recognized hepatoprotective function, they also exhibit antioxidant, anticancer, and anti-HIV potential. Those dibenzocyclooctadiene lignans possessing hydroxyl or angeloyloxy groups at C6 or C9 in the ethylidenecyclooctane ring tend to exhibit a higher anticancer activity. Spirobenzofuranoid dibenzocyclooctadienes, mostly present in *Kadsura*, contain a special tetrahydrofuran ring spanning the biphenyl linkage and demonstrate anti-PAF activities, which support the traditional use of *Kadsura* to improve blood circulation and "remove dampness." Spirobenzofu ranoid dibenzocyclooctadienes could be the bioactive marker compounds in *Kadsura* and markers for quality control. The distribution of all known lignans in the family showed that *Kadsura* is relatively advanced in evolution. Cycloartanone triterpenes occur in both *Schisandra* and *Kadsura*. Those with the A-ring open tend to exhibit greater anticancer and anti-HIV activity. 7/7/5/6 triterpene lactones, showing strong cytotoxicity, were discovered in *Kadsura longipedunculata* and have potential as anticancer agents. Nortriterpenoids possessing a unique skeleton were found in *S. lancifolia* and *S. micrantha*; some exhibited anticancer or anti-HIV activity.

1.2.2.10 Glucosinolate

Glucosinolates (GLs) were characterized in the seed and root of *Aurinia leucadia* (Brassicaceae) and *A. sinuata* and quantified based on the HPLC analysis of desulfo-GLs (Blazevic et al., 2013). Glucoalyssin (GAL, 1), glucobrassicanapin (GBN, 2), and glucoberteroin (GBE, 3) were the major GLs in *A. leucadia* and *A. sinuata*. GC–MS analysis of the volatile fractions obtained after enzyme hydrolysis showed that they mostly contain isothiocyanates (ITCs) originating from the parent GLs. C5 alkyl GLs 1, 2, and 3 can be chemotaxonomic markers of the *Aurinia* genus.

1.2.2.11 Glycoside

Chemical investigation of the glandular trichome exudate from *Geranium carolinianum* (Geraniaceae) led to the characterization of unique disaccharide derivatives

(Asai et al., 2011), *n*-octyl 4-*O*-isobutyryl-α-L-rhamnopyranosyl-(1 → 2)-6-*O*-isobutyryl-β-D-glucopyranoside, *n*-octyl 4-*O*-isobutyryl-α-L-rhamnopyranosyl-(1 → 2)-6-*O*-(2-methylbutyryl)-β-D-glucopyranoside, and *n*-octyl 4-*O*-(2-methylbutyryl)-α-L-rhamnopyranosyl-(1 → 2)-6-*O*-isobutyryl-β-D-glucopyranoside, named caroliniasides A–C, respectively. *n*-Alkyl glycoside derivatives, the rare type of SMs, could be used in chemotaxonomy as they are found in glandular trichome exudates of *Geranium* plants.

1.2.3 Macroelement and trace element

Satureja montana (Lamiaceae) and *S. subspicata* are used as spice and pepper substitute, for preparing tea and juice, and as a medicine. Fourteen populations (seven per species) of *S. montana* and *S. subspicata* growing in Croatia were examined to determine the chemical composition of the essential oil (analyzed by GC-FID and GC–MS), the content of macroelements (Na, K, Ca, and Mg) and trace elements (B, Fe, Cu, Mn, Zn, Al, Pb, Cr, Cd, Ni, Hg, and As) analyzed by ICP-AES (Dunkić et al., 2012) and antioxidant compounds (by UV/VIS spectrophotometer), and the types and distribution of trichomes (by scanning electron microscopy). The main constituents of the essential oil were carvacrol and thymol in *S. montana* and all populations belong to one phenol chemotype, while α-eudesmol, β-eudesmol, and spathulenol dominated in *S. subspicata* and three chemotypes could be distinguished. Both species possess considerably higher quantities of Ca and Mg and moderate concentrations of K and Na, while Hg and As levels were below the limit of quantification.

1.3 Metabolomics

Metabolomics is an omics approach that aims to comprehensively analyze all metabolites in a biological sample and has great potential for chemotaxonomy. For example, ultraperformance liquid chromatography–quadrupole time-of-flight high-definition mass spectrometry (UPLC–QTOF-HDMS) was used to detect 22 metabolites shared by the mother root of *Aconitum carmichaelii* (CHW; Ranunculaceae) and lateral root of *A. carmichaelii* (SFZ) and 13 metabolites shared by the CHW and root of *A. kusnezoffii* (CW) (Sun et al., 2013). Of note, songorine, carmichaeline, and isotalatizidine were not identified in CW but are present in the SFZ and CHW.

1.3.1 Asterids of core eudicot

1D- and 2D-NMR-based metabolomics classified 11 South American *Ilex* (Aquifoliaceae) species into four groups (Kim et al., 2010). Group A (*I. paraguariensis*) was metabolically characterized with a higher amount of xanthines and phenolics including phenylpropanoids and flavonoids; group B (*I. dumosa* var. *dumosa* and *I. dumosa* var. *guaranina*) with oleanane-type saponins; group C (*I. brasiliensis*, *I. integerrima*, *I. pseudobuxus*, and *I. theezans*) with arbutin and dicaffeoylquinic acids; and group D

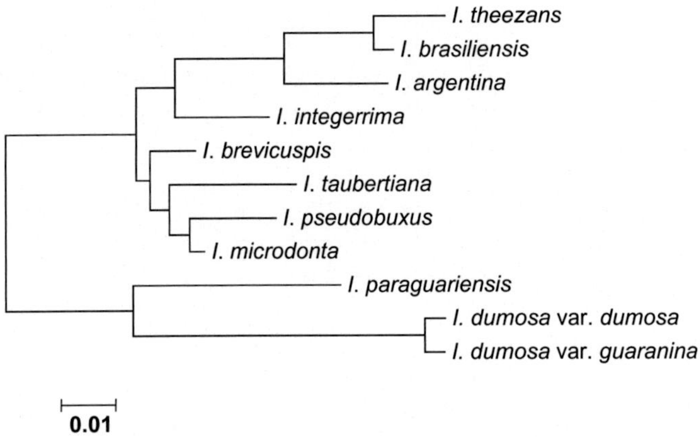

Figure 1.13 *Ilex* ITS phylogenetic tree inferred with ML method. The evolutionary history was inferred based on the GTR model. The tree with the highest log likelihood (−1654.0571) is shown. The tree is drawn to scale, with branch lengths measured in the number of substitutions per site. The analysis involved 11 nucleotide sequences. All positions with less than 95% site coverage were eliminated. That is, fewer than 5% alignment gaps, missing data, and ambiguous bases were allowed at any position. There were 544 positions in the final dataset. Evolutionary analyses were conducted in MEGA6 (Tamura et al., 2013).

(*I. argentina*, *I. brevicuspis*, *I. microdonta*, and *I. taubertiana*) with the highest level of ursane-type saponins. These results are largely congruent with ITS phylogenetic tree (Figure 1.13) and illustrate the utility of metabolomics in chemotaxonomy.

Chemical differences between 19 *Salvia* species (Lamiaceae) were thoroughly investigated, and 86 constituents were characterized by a sensitive HPLC–DAD-ESI-MSn method (Qiao et al., 2009). Ser. *Brachylomae*, ser. *Digitaloidites*, ser. *Castaneae*, and ser. *Miltiorrhizae* were the sibling series and that ser. *Brachylomae*, ser. *Maximowicziana*, and ser. *Campanulatae* were unique with regard to their constituents in addition to having componential affinity (Figure 1.14). This study provides a simple chemotaxonomy approach and a reasonable bioactivity interpretation of *Salvia* species. Twenty-eight different iridoid glucosides and 10 caffeoyl phenylethanoid glucosides (CPGs), as well as salidroside and arbutin, were isolated from 14 species of tribe Veroniceae (Plantaginaceae) and characterized by NMR (Taskova et al., 2006), which are useful in chemotaxonomy of Veroniceae and its allies in the Plantaginaceae. Four steroidal saponins, two xanthone glycosides, two isoflavonoids, and one anthraquinone in *Anemarrhena asphodeloides* (Liliaceae) can be simultaneously quantified by HPLC–ESI-MS (Sun et al., 2012), making it feasible to chemically differentiate and classify the *Anemarrhena* samples. Metabolomics has grown greatly as a functional genomics tool and has become an invaluable diagnostic tool for biochemical phenotyping of medicinal plants.

The possible applicability of (un)targeted metabolomics (volatile metabolites) for revealing taxonomic/evolutionary relationships among *Senecio* species (Asteraceae; tribe Senecioneae) was explored (Radulović et al., 2014). Essential oil composition

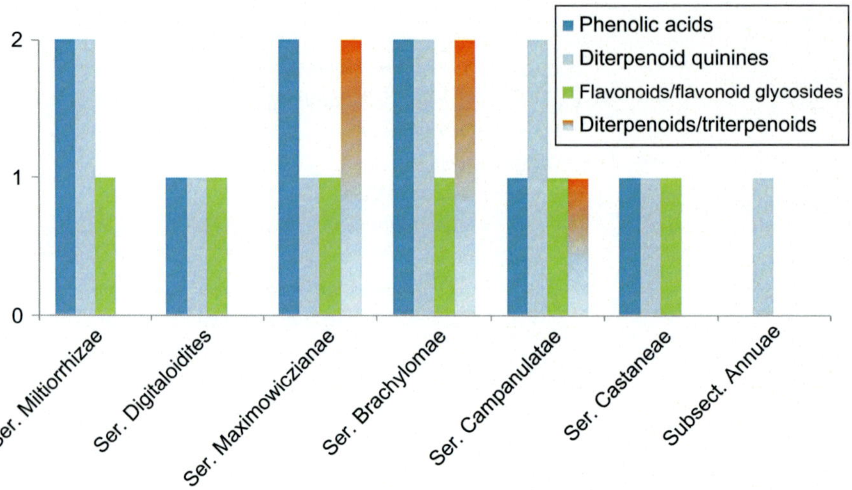

Figure 1.14 Comparison of *Salvia* chemical components. 2 of *y* axis, major component; 1, detected; 0, absent/rarely detected.

data of selected *Senecio*/Senecioneae/Asteraceae taxa (93 samples) were mutually compared by MVA, that is, agglomerative hierarchical clustering and PCA. The MVA input dataset included the very first composition data on the essential oil extracted from the aerial parts of *S. viscosus* as well as on four different Serbian populations of *S. vernalis* (oils from aerial parts and roots; eight samples). This metabolomic screening suggested short-chain alk-1-enes (e.g., oct-1-ene, non-1-ene, and undec-1-ene), with restricted general occurrence in Plantae, as chemotaxonomic markers/targets for future metabolomic studies of *Senecio*/Senecioneae taxa. The MVA showed that the evolution of the terpene metabolism (volatile mono- and sesquiterpenoids) within the tribe Senecioneae was not genus-specific but confirmed plant organ-specific production/accumulation of volatiles within *S. vernalis* and suggested the existence of at least two volatile chemotypes for this species.

Chemotaxonomic study of 25 *Eurya* taxa (24 species and 1 variety; Theaceae), a total of 28 samples, was carried out with HPLC method (Shi et al., 2014). Twenty-three chemical components were identified in terms of the relative retention time of the chromatograms. The chromatograms were kept constant in the same species and varied among different species, which could be used to distinguish species from each other. In order to reveal the interspecific relationships and the taxonomic positions of these species, chromatogram data were subject to CA using UPGMA and NJ methods. HPLC method was rapid, intuitive, and repeatable and can help classify the *Eurya* plants and the resource utilization study.

Plants belonging to the *Lippia* genus have been widely used in ethnobotany throughout South and Central America and in tropical Africa as foods, medicines, and sweeteners and in beverage flavoring. Various taxonomic problems involving some genera from Verbenaceae, including *Lippia*, have been reported. The metabolite profiling of 15 extracts of various organs of six *Lippia* species was performed and compared using UHPLC–PDA-TOF-MS (Funari et al., 2012). Fourteen phenolic compounds that were previously isolated from *L. salviaefolia* and *L. lupulina* were used as references. The annotation of the remaining LC peaks was based on concomitant online high-mass-accuracy measurements and subsequent molecular formula assignments, which is generic and time-saving, avoids isolation/purification procedures, enables an efficient LC peak annotation of most of the studied compounds, and is well adapted for plant chemotaxonomic studies. The interconversion of four flavanone glucoside isomers was additionally highlighted by analytic HPLC isolation and immediate analysis using fast UHPLC gradients. Dereplication results and hierarchical data analysis demonstrated that *L. salviaefolia*, *L. balansae*, *L. velutina*, and *L. sidoides* displayed significant chemical similarities, while the compositions of *L. lasiocalicyna* and *L. lupulina* differed substantially.

1.3.2 Rosids of core eudicot

An LC–ESI-MS-based metabolite profiling was performed to classify the phenotypes of *Lespedeza* (Leguminosae) species (Kim et al., 2012b). MVAs such as PCA and hierarchical clustering analysis (HCA) were used for the clustering pattern analysis and distance analysis between species, respectively. Leaves were classified into four phenotypes according to species. In both the genetic and the chemotaxonomic classification methods, the distance between *L. cyrtobotrya* and *L. bicolor* was the closest, and *L. cuneata* was the farthest away from the other three species. Orthogonal partial least squares discriminant analysis was employed to identify significantly different phytochemicals between species. *L. cyrtobotrya* and *L. bicolor* were classified by identifying significantly different phytochemicals. Leaves and stems showed different phenotypic classifications based on the chemotaxonomic classification. Stem samples of the other three species could not be differentiated, whereas *L. cyrtobotrya* stem samples were clustered within species. Four wild-type *Lespedeza* spp. were classified by analyzing the combined nrITS and cp trnL–trnF sequences. The phenotypic classification of leaves coincided more with the genotypic classification than that of stems.

1.3.3 Other eudicots

The section *Moutan* of the genus *Paeonia* consists of eight species that are distributed in a small area of China. A wide range of metabolites, including monoterpenoid glucosides, flavonoids, tannins, stilbenes, triterpenoids, steroids, paeonols, and phenols, have been found in this section. HPLC-DAD-based metabolic fingerprinting was applied to the classification of eight species: *Paeonia suffruticosa* (Mu Dan), *P. qiui*, *P. ostii*, *P. rockii*, *P. jishanensis*, *P. decomposita*, *P. delavayi*, and *P. ludlowii* (He et al., 2014). Forty-seven peaks exhibited an occurrence frequency of 75% in all 23 tree peony samples, 43 of which were identified according to their retention times and UV absorption spectra, together with combined HPLC–QTOF-MS. The 43 isolated compounds included 17 monoterpenoid glucosides, 11 galloyl glucoses, 5 flavonoids, 6 paeonols, and 4 phenols. PCA and hierarchical CA (HCA) showed a clear separation between the species based on metabolomic similarities and four groups were identified, which agrees well with the classical classification based on the morphological characteristics and geographic distributions of the subsections *Vaginatae* F.C. Stern and *Delavayanae* F.C. Stern with the exception of *P. decomposita*, a transition species between these subsections. *P. ostii* and *P. suffruticosa* could be considered one species, which is consistent with the viewpoint of medicinal botanists but different from that of classical morphological processing. Significant variations were obtained in the metabolic profiles of *P. delavayi*, and no significant difference was found between *P. delavayi* and *P. ludlowii*, which indicates that these species have a close genetic relationship. The combination of HPLC-DAD and multivariate analyses has great potential for guiding future chemotaxonomic studies to examine the potential pharmaceutical value of tree peony species and appears to help clarify the confusion and skepticism associated with the reported morphology- and molecular marker-based taxonomy of tree peonies.

The metabolomic analysis of three *Cimicifuga* species was performed using H NMR spectroscopy and pattern recognition (PR) techniques (Shen et al., 2013). A broad range of metabolites could be detected by ^{1}H NMR spectroscopy without any chromatographic separation. The analysis using PCA and discriminant partial least squares (DPLS) of the ^{1}H NMR spectrum showed a clear discrimination between *C. foetida* and the other two species. The major metabolites responsible for the discrimination were triterpenoid saponins and saccharides. The combination of ^{1}H NMR and PR provides a useful tool for chemotaxonomic analysis and authentication of *Cimicifuga* species and could be used for the quality control of plant materials.

1.3.4 Gymnosperm

Three *Chamaecyparis* species (*C. formosensis*, *C. obtusa*, and *C. obtusa* var. *formosana*; Cupressaceae) are difficult to distinguish by the naked eye. It would be valuable to find a simple and rapid method to differentiate these three species (Lin et al., 2011). The chemical compositions of biogenic volatile organic compounds (BVOCs) from mature leaves were analyzed using solid-phase microextraction (SPME)-GC–MS. CA and PCA were conducted for the BVOC constituents to reveal the differences among species. The compositions of BVOCs from the three species were distinct. These species were clearly differentiated according to the results of CA and PCA.

1.3.5 Bryophyte

Bryophytes are the second largest taxonomic group in the plant kingdom; yet, studies conducted to better understand their chemical composition are rare. The chemical composition of bryophytes common in northern Europe was characterized by elemental, spectral, and nondestructive analytic methods, such as Fourier transform IR spectrometry (FT-IR), solid-phase ^{13}C NMR spectrometry, and pyrolysis-gas chromatography-mass spectrometry (Py-GC–MS) (Maksimova et al., 2013). Bryophytes consist mainly of carbohydrates. Judging by FT-IR spectra, the OH groups in combination of C–O groups were the most abundant. The ^{13}C NMR spectra provided information on the presence of such compounds as phenolics and lipids. The amount of phenolic compounds in bryophytes is relatively small. Lignin is absent in the studied bryophytes. CA was used to better understand differences in the chemical composition of bryophyte samples and to evaluate possible usage of these methods in the chemotaxonomy of bryophytes.

1.4 Cheminformatics and database

No many DNA marker sequences are available for *Glycyrrhiza*, making it challenging to infer the phylogenetic relationship (Figure 1.15). In order to study the chemical components of licorice deeply and systematically, a licorice compounds database was constructed based on the comprehensive review of the compounds found in *Glycyrrhiza* (Xiang et al., 2012). The database was used to classify the licorice components in order to quantitatively analyze the distribution of each type of compound and the compounds in the medicinal *Glycyrrhiza* plants. 422 compounds have been

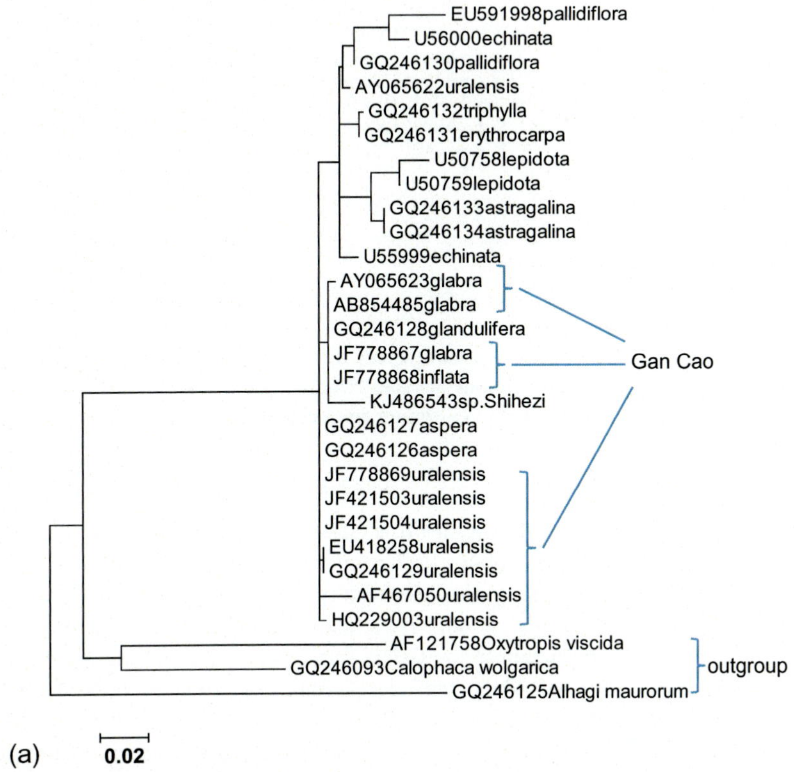

Figure 1.15 *Glycyrrhiza* phylogenetic tree inferred with ML method. (a) ITS;

(Continued)

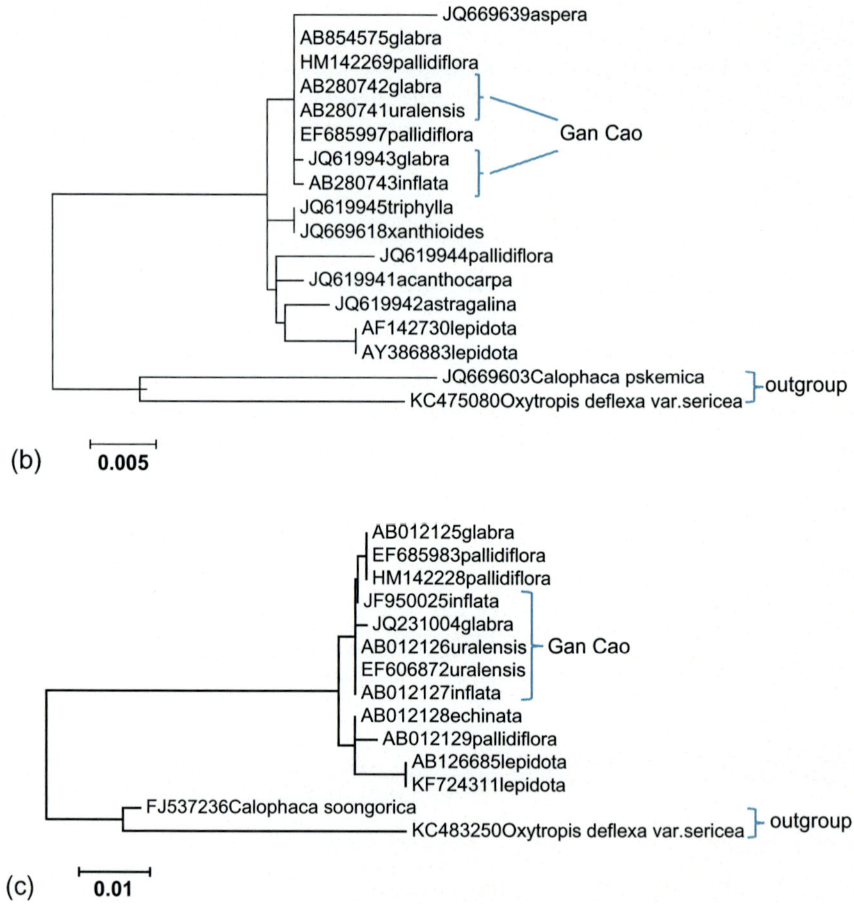

Figure 1.15 Continued. (b) matK; (c) rbcL. Trees are drawn to scale, with branch lengths measured in the number of substitutions per site. Evolutionary analyses were conducted in MEGA6. *G. uralensis*, *G. glabra*, and *G. inflata* are recorded in Chinese Pharmacopoeia as the source plant of "Gan Cao."

reported in *Glycyrrhiza*, which fall into five classes, flavonoids, coumarins, triterpenoids, stilbenoids, and others (Table 1.2). To date, 170 compounds have been isolated from *G. uralensis*, 134 from *G. glabra*, 52 from *G. inflata*, and 31 from *G. yunnanensis*. The category "stilbenoids" should be added, and "dibenzoylmethanes" should belong to chalcones. The database facilitates further chemotaxonomic study.

The most commonly used ANN (artificial neural network) architecture for PR and classification is the self-organizing map (SOM). SOMs of molecular descriptors for SLs show evident similarities among the Heliantheae, Helenieae, and Eupatorieae tribes as well as between the Anthemideae and the Inuleae tribes (Scotti et al., 2012). These observations are in agreement with systematic classifications that were

Table 1.2 **Distribution of various compounds in *Glycyrrhiza* species**

Compound	***G. aspera***	***G. glabra***	***G. inflata***	***G. pallidiflora***	***G. squamulosa***	***G. uralensis***	***G. yunnanensis***	***G. kanscensis***	***G. triphylla***	**North America *Glycyrrhiza***	***G. acanthocarpa***
Flavones	*	*	*	*	*	*	*	*	*	*	*
Flavonols	*	*	*	*	*	**	*	*	*	*	*
Flavanones	*	**	**	*	*	**	*	*	*	*	*
Flavanonols	–	*	–	–	–	*	–	–	–	–	–
Chalcones	*	*	*	*	*	*	*	*	*	*	*
Dibenzoylmethanes	–	–	*	–	–	–	–	–	–	–	–
Isoflavones	*	*	*	*	*	*	*	*	*	*	*
Isoflavanones	*	–	–	–	–	*	–	–	–	–	–
Isoflavans	*	**	*	*	*	**	*	*	*	*	*
Isoflav-3-enes	*	*	–	*	–	–	–	–	–	–	–
Pterocarpans	**	*	–	–	–	*	–	–	–	–	–
Pterocarpenes	–	–	–	–	–	*	–	–	–	–	–
Dihydrochalcones	–	*	–	*	–	*	–	–	–	–	–
Homoisoflavanone	–	–	–	*	–	–	–	–	–	–	–
Aurones	–	*	–	–	–	–	–	–	–	–	–
Simple coumarins	*	*	–	–	–	*	–	–	–	–	–
3-Arylcoumarins	*	*	–	–	–	*	–	–	–	–	–
Coumestans	*	–	–	–	–	*	–	–	–	–	–
4-Arylcoumarin	–	–	*	–	–	–	–	–	–	–	–

Continued

Table 1.2 **Continued**

Compound	*G. aspera*	*G. glabra*	*G. inflata*	*G. pallidiflora*	*G. squamulosa*	*G. uralensis*	*G. yunnanensis*	*G. kanscensis*	*G. triphylla*	North America *Glycyrrhiza*	*G. acanthocarpa*
11-Oxo-olean-12-en-29/30-oic acids	*	**	**	_	_	**	_	*	_	_	_
Olean-12-ene-29/30-oic acids	_	*	*	_	_	**	**	_	*	_	_
Olean-11,13(18)-diene-29/30-oic acids	*	*	*	*	*	*	**	*	*	*	*
Olean-9 (11), 12-diens	_	_	_	_	_	_	*	_	_	_	_
Friedelane	_	_	_	_	_	_	*	_	_	_	_
Lupanes	*	*	*	*	*	*	*	*	*	*	*
Dihydrostilbenes	_	*	_	_	_	*	_	_	_	*	_
2-Arylbenzofurans	*	*	*	_	_	*	_	_	_	_	_
Prenylated dihydrophenanthrenes	_	_	_	_	_	*	_	_	_	_	_
Diphenylethylketones	_	_	_	_	_	*	_	_	_	_	_
Resorcinols	_	_	_	_	_	_	_	_	_	_	*
Phenolic acid	*	*	*	*	*	*	*	*	*	*	*

**, abundant; *, present; _, absent/rarely detected.

proposed by Bremer, which use mainly morphological and molecular data. Descriptors obtained through fragments or by the two-dimensional representation of the SL structures were adequate to obtain significant results, and better results were not achieved by using descriptors derived from three-dimensional representations of SLs.

1.5 Chemotype

Definitions of classification position, classification standard, nomenclature, and naming methods of chemotype are suggested based on discussing the significance of chemotype of medicinal plants (Hua et al., 2009). Chemotype should be established in infraspecific categories of "forma." Chemotype identification mainly has two aspects. One is that the main constituents are distinct or one or two components are half or more than half of the total chemical content. The main constituents come from the same biosynthetic pathway. The other is that the chemical variation is heritable. The chemotype-based chemotaxonomy of medicinal plants has important theoretical and practical value for quality assessment, resource development, and the genuine medicinal research. It also ensures the safe and effective use of clinical medicine.

1.5.1 Flower

The medicinal use of *Cannabis* is increasing as countries worldwide are setting up official programs to provide patients with access to safe sources of medicinal-grade *Cannabis*. An important question is which variety of *Cannabis* should be made available for medicinal use (Hazekamp and Fischedick, 2012). Drug varieties of *Cannabis* are commonly distinguished through the use of popular names, *Cannabis indica* and *Cannabis sativa*. Although more than 700 different cultivars have already been described, it is unclear whether such classification reflects any relevant differences in chemical composition. Some attempts have been made to classify *Cannabis* varieties based on chemical composition, but they have mainly been useful for forensic applications, distinguishing drug varieties, with high THC content, from the nondrug hemp varieties. The biologically active terpenoids should be included in these approaches. A better classification system, based on a range of potentially active constituents, is needed. The cannabinoids and terpenoids, present in high concentrations in *Cannabis* flowers, are the main candidates. Cultivars obtained from multiple sources were compared. Based on the analysis of 28 major compounds, followed by PCA of the quantitative data, constituents that defined the samples into distinct chemovar groups were identified. PCA approach is useful for chemotaxonomic classification of *Cannabis* varieties.

The volatile compositions of hydrodistilled essential oils in the flower heads of *Chrysanthemum indicum* (Compositae) from eight populations in China were analyzed by GC–MS (Zhang et al., 2010). 169 compounds representing 88.79–99.53% of the oils were identified, and some remarkable differences were found in the constituent percentages of the eight populations. The predominant components of the essential oils were 1,8-cineole (0.62–7.34%), (+)-(1*R*,4*R*)-camphor (0.17–27.56%), caryophyllene oxide (0.54–5.8%), β-phellandrene (0.72–1.87%), (−)-(1*S*,2*R*,4*S*)-borneol acetate (0.33–8.46%), 2-methyl-6-(*p*-tolyl)hept-2-ene (0.3–8.6%), 4,6,6-trimethylbicyclo[3.1.1]hept-3-en-2-yl acetate (0.17–26.48%), and hexadecanoic acid (0.72–15.97%). The chemotaxonomic value of the essential oil compositions was discussed according to the results of CA and PCA. The eight populations were divided into five chemotypes, and the scores together with the loadings revealed clearly different chemical properties of each population.

1.5.2 Fruit

"Long-storage" tomato (*Solanum lycopersicum*) is traditionally cultivated under no water supply, the fruits of which combine a good taste with excellent nutritional properties. HPLC–DAD/ESI-MS was used to identify the phenolic profile in 10 landraces of long-storage tomato (Siracusa et al., 2012), grown under a typical semiarid climate, as compared to a processing tomato hybrid cultivated in the same environment, under both well-irrigated and unirrigated conditions. Sixteen SMs, belonging to the classes of cinnamoylquinic acids and flavonoids, were identified. Quantitative analyses were performed to monitor the changes in the phenolic content along the batch. Landraces originating from the same area exhibit different fruit morphologies but own a similar biochemical profile. The two controls (well irrigated and unirrigated) are placed into the same cluster, suggesting that these SMs in tomato fruits may be more genetics-dependent than environment-dependent. The analysis of phenols represents a useful tool to assess the genetic variability in tomato, and these compounds could be chemotaxonomic markers in the traceability of this niche product.

Tomatoes, the second most important vegetable crop worldwide, are a key component in the so-called Mediterranean diet, which is strongly associated with a reduced risk of chronic degenerative diseases. The differences in the total and individual polyphenol content and hydrophilic antioxidant capacity of seven varieties of tomato cultivated in Vegas Bajas del Guadiana, Badajoz (Spain), were evaluated (Vallverdú-Queralt et al., 2011), which were collected from two consecutive harvests (2008–2009). Hydrophilic antioxidant capacity was evaluated using the TEAC assay, while the Folin–Ciocalteu assay with a previous cleanup was used to establish total polyphenol content. Individual polyphenols were quantified using LC–ESI-MS/MS on a triple quadrupole. All compounds were found to be significantly different when the analysis of variance was performed. PCA results show that phenolic compounds and hydrophilic antioxidant capacity were responsible for the differences among tomato samples according to variety.

1.5.3 Root

The ^{1}H NMR fingerprints of fractionated nonpolar and polar extracts (control substance for plant drug [CSPD] A and B) from the roots of 12 specimens of *Saposhnikovia divaricata* (Fang Feng; Umbelliferae) were achieved with FT-NMR spectrometer and assigned by comparison to each other and to the ^{1}H NMR spectra of the isolated individual compounds (Xin et al., 2010). These fingerprints were found to be uniform in terms of the specificity for the implication of all 12 specimens being systematically of the same origin. The uniformity was affirmed by HPLC, which revealed exactly identical specificity for the identified *S. divaricata* with the NMR appearances of the corresponding CSPD on the part of the composition of characteristic constituents when comparing to corresponding individual compounds. The specific signals from the chemotaxonomically significant compounds of chromones and coumarins in *S. divaricata* are exhibited distinctively in the composite features of both NMR fingerprints and HPLC profiles.

1.5.4 Bulb

"Cipolla di Giarratana," a locally cultivated white onion (Liliaceae) landrace, is an item in the "List of Traditional Agro-food Products" of the Italian Department for Agriculture and listed as "slow food presidium" by the Slow Food Foundation for Biodiversity. Ten local accessions were investigated for their biomorphological and biochemical characteristics in five experimental locations (Riggi et al., 2013). HPLC–DAD/ESI-MS was used to identify the phenolic profile and quantify phenolic content in bulbs: quercetin, quercetin 3,4′ di-*O*-glucoside, and quercetin 4′-*O*-glucoside were detected as major components. The "Cipolla di Giarratana" landrace is characterized by a high bulb weight (436 g) and high diameter (11 cm). The total flavonol content ranged between 68 and 408 mg/kg fresh weight in nine of the 10 accessions. Flavonol patterns could be chemotaxonomic descriptors to characterize onion germplasm.

1.6 Conclusions

The formation of some SMs, such as pyrrolizidine alkaloids, in unrelated plant families could be the result of convergent evolution and thus not useful as a taxonomic marker at the family level. Actually, it is nonsense to discuss the evolutionary relationship solely based on the chemotaxonomic data. The use of molecular markers in plant systematics is the mainstream, complemented by chemotaxonomy and morphology. Chemotaxonomy provides a valuable framework that allows the comparison and placement of many other experimental data in a pharmacophylogenetic or chemotaxonomic context. SM-based genus-level chemotaxonomy is useful in finding alternative source plant of specific SM. Intraspecific chemotaxonomy is essential for defining chemotype, quality assessment, pharmaceutical resource development, and the genuine medicine research.

References

Ahmad, M., Ajabkhan, M., Zafar, M., et al., 2009. Chemotaxonomic authentication of herbal drug chamomile. Asian J. Chem. 21, 3395–3410.

Asai, T., Sakai, T., Ohyama, K., Fujimoto, Y., 2011. n-Octyl α-L-rhamnopyranosyl-(1→2)-β-D-glucopyranoside derivatives from the glandular trichome exudate of *Geranium carolinianum*. Chem. Pharm. Bull. 59 (6), 747–752.

Asakawa, Y., Ludwiczuk, A., Nagashima, F., 2013. Phytochemical and biological studies of bryophytes. Phytochemistry 91, 52–80.

Bai, Z.F., Wang, X.Q., Xiao, P.G., Liu, Y., 2013. Phenylethanoid glycosides distribution in medicinal plants of Gesneriaceae. China J. Chin. Mat. Med. 38 (24), 4267–4270.

Baranska, M., Schulz, H., Krüger, H., Quilitzsch, R., 2005. Chemotaxonomy of aromatic plants of the genus Origanum via vibrational spectroscopy. Anal. Bioanal. Chem. 381 (6), 1241–1247.

Barra, A., 2009. Factors affecting chemical variability of essential oils: a review of recent developments. Nat. Prod. Commun. 4 (8), 1147–1154.

Barthet, V.J., 2008. (n-7) and (n-9) cis-Monounsaturated fatty acid contents of 12 *Brassica* species. Phytochemistry 69 (2), 411–417.

Benedek, B., Gjoncaj, N., Saukel, J., Kopp, B., 2007. Distribution of phenolic compounds in Middle European taxa of the *Achillea millefolium* L. aggregate. Chem. Biodivers. 4 (5), 849–857.

Berger, A., Fasshuber, H., Schinnerl, J., et al., 2011. Iridoids as chemical markers of false ipecac (*Ronabea emetica*), a previously confused medicinal plant. J. Ethnopharmacol. 138 (3), 756–761.

Blazevic, I., De Nicola, G.R., Montaut, S., Rollin, P., 2013. Glucosinolates in two endemic plants of the *Aurinia* genus and their chemotaxonomic significance. Nat. Prod. Commun. 8 (10), 1463–1466.

Bojović, S., Sarac, Z., Nikolić, B., et al., 2012. Composition of n-alkanes in natural populations of *Pinus nigra* from Serbia – chemotaxonomic implications. Chem. Biodivers. 9 (12), 2761–2774.

Casale, J.F., Mallette, J.R., Jones, L.M., 2014. Chemosystematic identification of fifteen new cocaine-bearing *Erythroxylum* cultigens grown in Colombia for illicit cocaine production. Forensic Sci. Int. 237, 30–39.

Chang, L., Zhang, X.X., Ren, Y.P., et al., 2013. Simultaneous quantification of six major flavonoids from *Fructus sophorae* by LC-ESI-MS/MS and statistical analysis. Indian J. Pharm. Sci. 75 (3), 330–338.

Chen, Y.G., Lin, L., 2011. Study and comparison on HPLC fingerprints of flavonoids of frequently used Chinese materia medica in *Citrus*. China J. Chin. Mat. Med. 36 (19), 2660–2665.

Chen, C., Zhang, H., Xiao, W., Yong, Z.P., Bai, N., 2007. High-performance liquid chromatographic fingerprint analysis for different origins of sea buckthorn berries. J. Chromatogr. A 1154 (1–2), 250–259.

Ciesla, Ł., Hajnos, M., Staszek, D., et al., 2010. Validated binary high-performance thin-layer chromatographic fingerprints of polyphenolics for distinguishing different *Salvia* species. J. Chromatogr. Sci. 48 (6), 421–427.

Ćirić, A., Karioti, A., Koukoulitsa, C., et al., 2012. Sesquiterpene lactones from *Centaurea zuccariniana* and their antimicrobial activity. Chem. Biodivers. 9 (12), 2843–2853.

Costa, E.V., Pinheiro, M.L., Xavier, C.M., et al., 2006. A pyrimidine-beta-carboline and other alkaloids from *Annona foetida* with antileishmanial activity. J. Nat. Prod. 69 (2), 292–294.

Coulerie, P., Nour, M., Thouvenot, L., Asakaw, Y., 2014. Sesquiterpene hydrocarbons from the liverwort *Treubia isignensis* var. *isignensis* with chemotaxonomic significance. Nat. Prod. Commun. 9 (8), 1059–1060.

Crockett, S.L., Robson, N.K., 2011. Taxonomy and chemotaxonomy of the genus *Hypericum*. Med. Aromat. Plant Sci. Biotechnol. 5 (1), 1–13.

de Wet, H., van Heerden, F.R., van Wyk, B.E., 2011. Alkaloidal variation in cissampelos capensis (Menispermaceae). Molecules 16 (4), 3001–3009.

Domsalla, A., Görick, C., Melzig, M.F., 2010. Proteolytic activity in latex of the genus Euphorbia – a chemotaxonomic marker? Pharmazie 65 (3), 227–230.

Dunkić, V., Kremer, D., Dragojević Müller, I., et al., 2012. Chemotaxonomic and micromorphological traits of *Satureja montana* L. and *S. subspicata* Vis. (Lamiaceae). Chem. Biodivers. 9 (12), 2825–2842.

Dussert, S., Laffargue, A., de Kochko, A., Joët, T., 2008. Effectiveness of the fatty acid and sterol composition of seeds for the chemotaxonomy of *Coffea* subgenus *Coffea*. Phytochemistry 69 (17), 2950–2960.

Egan, P.A., van der Kooy, F., 2012. Coproduction and ecological significance of naphthoquinones in carnivorous sundews (*Drosera*). Chem. Biodivers. 9 (6), 1033–1044.

Farag, M.A., Wessjohann, L.A., 2012. Volatiles profiling in medicinal licorice roots using steam distillation and solid-phase microextraction (SPME) coupled to chemometrics. J. Food Sci. 77 (11), C1179–C1184.

Farag, M.A., Weigend, M., Luebert, F., et al., 2013. Phytochemical, phylogenetic, and anti-inflammatory evaluation of 43 *Urtica* accessions (stinging nettle) based on UPLC-Q-TOF-MS metabolomic profiles. Phytochemistry 96, 170–183.

Feuereisen, M.M., Hoppe, J., Zimmermann, B.F., et al., 2014. Characterization of phenolic compounds in Brazilian pepper (*Schinus terebinthifolius* Raddi) Exocarp. J. Agric. Food Chem. 62 (26), 6219–6226.

Fleury, B.G., Figueiredo, L., Marcoli, M.I., et al., 2011. Fatty acids as chemotaxonomic markers of marine macrophytes from Rio de Janeiro state, Brazil. Nat. Prod. Commun. 6 (5), 667–672.

Fraga, B.M., 2012. Phytochemistry and chemotaxonomy of *Sideritis* species from the Mediterranean region. Phytochemistry 76, 7–24.

Funari, C.S., Eugster, P.J., Martel, S., et al., 2012. High resolution ultra high pressure liquid chromatography-time-of-flight mass spectrometry dereplication strategy for the metabolite profiling of Brazilian *Lippia* species. J. Chromatogr. A 1259, 167–178.

Gören, A.C., Akçicek, E., Dirmenci, T., et al., 2012. Fatty acid composition and chemotaxonomic evaluation of species of *Stachys*. Nat. Prod. Res. 26 (1), 84–90.

Guiso, M., Tassone, G., Nicoletti, M., et al., 2007. Chemotaxonomy of iridoids in *Linaria vulgaris*. Nat. Prod. Res. 21 (13), 1212–1216.

Guo, B.L., Hong, D.Y., Xiao, P.G., 2008a. Further research on chemotaxonomy of paeonol and analogs in *Paeonia* (Ranunculaceae). Acta Phytotax. Sin. 46 (5), 724–729.

Guo, B.L., Pei, L., Xiao, P.G., 2008b. Further research on taxonomic significance of flavonoids in *Epimedium* (Berberidaceae). Acta Phytotax. Sin. 46 (6), 874–885.

Haliński, L.P., Szafranek, J., Stepnowski, P., 2011. Leaf cuticular n-alkanes as markers in the chemotaxonomy of the eggplant (*Solanum melongena* L.) and related species. Plant Biol. 13 (6), 932–939.

Hazekamp, A., Fischedick, J.T., 2012. *Cannabis* – from cultivar to chemovar. Drug Test. Anal. 4 (7–8), 660–667.

He, C., Peng, B., Dan, Y., Peng, Y., Xiao, P.G., 2014. Chemical taxonomy of tree peony species from China based on root cortex metabolic fingerprinting. Phytochemistry 107, 69–79.

Hillig, K.W., Mahlberg, P.G., 2004. A chemotaxonomic analysis of cannabinoid variation in *Cannabis* (Cannabaceae). Am. J. Bot. 91 (6), 966–975.

Hu, Y., Duan, T.X., Cao, F., et al., 2010. Evaluation of LC-MS chromatograms of pericarps from *Illium* species. China J. Chin. Mat. Med. 35 (14), 1836–1839.

Hua, Y.L., Huang, L.Q., Chen, M.L., Xiao, P.G., 2009. Discussing on significance, position and classification standard of chemotype of medicinal plants. China J. Chin. Mat. Med. 34, 924–928.

Huang, S.Z., Zhao, Y.X., Mei, W.L., et al., 2013. Chemotaxonomy studies on the genus *Aquilaria*. J. Trop. Subtrop. Bot. 21 (3), 273–280.

Jenett-Siems, K., Weigl, R., Böhm, A., et al., 2005. Chemotaxonomy of the pantropical genus *Merremia* (Convolvulaceae) based on the distribution of tropane alkaloids. Phytochemistry 66 (12), 1448–1464.

Jensen, S.R., Schripsema, J., 2002. Chemotaxonomy and pharmacology of Gentianaceae. In: Struwe, L., Albert, V. (Eds.), Gentianaceae – Systematics and Natural History. Cambridge University Press, Cambridge, pp. 573–631.

Jurišić Grubešić, R., Srečnik, G., Kremer, D., et al., 2013. Simultaneous RP-HPLC-DAD separation, and determination of flavonoids and phenolic acids in *Plantago* L. species. Chem. Biodivers. 10 (7), 1305–1316.

Kanani, M.R., Rahiminejad, M.R., Sonboli, A., et al., 2011. Chemotaxonomic significance of the essential oils of 18 *Ferula* species (Apiaceae) from Iran. Chem. Biodivers. 8 (3), 503–517.

Kim, S.W., Ban, S.H., Chung, H., et al., 2004. Taxonomic discrimination of flowering plants by multivariate analysis of Fourier transform infrared spectroscopy data. Plant Cell Rep. 23 (4), 246–250.

Kim, H.K., Saifullah, S., Khan, S., et al., 2010. Metabolic classification of South American *Ilex* species by NMR-based metabolomics. Phytochemistry 71 (7), 773–784.

Kim, S., Kim, K.Y., Han, C.S., et al., 2012a. Simultaneous analysis of six major compounds in Osterici Radix and Notopterygii Rhizoma et Radix by HPLC and discrimination of their origins from chemical fingerprint analysis. Arch. Pharm. Res. 35 (4), 691–699.

Kim, Y.M., Lee, J., Park, S.H., et al., 2012b. LC-MS-based chemotaxonomic classification of wild-type *Lespedeza* sp. and its correlation with genotype. Plant Cell Rep. 31 (11), 2085–2097.

Kirmizibekmez, H., Montoro, P., Piacente, S., et al., 2005. Identification by HPLC-PAD-MS and quantification by HPLC-PAD of phenylethanoid glycosides of five *Phlomis* species. Phytochem. Anal. 16 (1), 1–6.

Lang, I., Hodac, L., Friedl, T., Feussner, I., 2011. Fatty acid profiles and their distribution patterns in microalgae: a comprehensive analysis of more than 2000 strains from the SAG culture collection. BMC Plant. Biol. 11, 124.

Li, H.J., Jiang, Y., Li, P., 2009. Characterizing distribution of steroidal alkaloids in *Fritillaria* spp. and related compound formulas by liquid chromatography-mass spectrometry combined with hierarchical cluster analysis. J. Chromatogr. A 1216 (11), 2142–2149.

Li, J.H., Nesumi, A., Shimizu, K., et al., 2010. Chemosystematics of tea trees based on tea leaf polyphenols as phenetic markers. Phytochemistry 71 (11–12), 1342–1349.

Li, H.Y., Li, H.B., Hao, N., et al., 2011. Study on HPLC fingerprint characteristics and chemotaxonomy of *Pulsatilla* medicinal plants. Zhongguo Zhong Yao Za Zhi 36 (11), 1478–1482.

Lin, C.Y., Chen, Y.J., Cheng, S.S., Chang, S.T., 2011. Rapid differentiation of three *Chamaecyparis* species (Cupressaceae) grown in Taiwan using solid-phase microextraction-gas chromatography/mass spectrometry, cluster analysis, and principal component analysis. J. Agric. Food Chem. 59 (20), 10854–10859.

Liu, C., Jiang, D., Cheng, Y., et al., 2013a. Chemotaxonomic study of *Citrus*, *Poncirus* and *Fortunella* genotypes based on peel oil volatile compounds – deciphering the genetic origin of Mangshanyegan (*Citrus nobilis* Lauriro). PLoS One 8 (3), e58411.

Liu, Z., Liu, Y., Liu, C., et al., 2013b. The chemotaxonomic classification of *Rhodiola* plants and its correlation with morphological characteristics and genetic taxonomy. Chem. Cent. J. 7 (1), 118.

López, D., Piazza, L.A., Silva, M.P., et al., 2014. Distribution of (−)-hamanasic acid A in South American species of *Flourensia* and phytotoxic effects of leaf aqueous extracts. Nat. Prod. Commun. 9 (3), 341–345.

Lorenz, P., Duckstein, S., Conrad, J., et al., 2012. An approach to the chemotaxonomic differentiation of two European dog's mercury species: *Mercurialis annua* L. and *M. perennis* L. Chem. Biodivers. 9 (2), 282–297.

Maggio, A., Rosselli, S., Bruno, M., et al., 2012. Chemical composition of essential oil from Italian populations of *Artemisia alba* Turra (Asteraceae). Molecules 17 (9), 10232–10241.

Maksimova, V., Klavina, L., Bikovens, O., et al., 2013. Structural characterization and chemical classification of some bryophytes found in Latvia. Chem. Biodivers. 10 (7), 1284–1294.

Maurin, O., Davis, A.P., Chester, M., et al., 2007. Toward a phylogeny for *Coffea* (Rubiaceae): identifying well supported lineages based on nuclear and plastid DNA sequences. Ann. Bot. 100, 1565–1583.

Máximo, P., Lourenço, A., Tei, A., Wink, M., 2006. Chemotaxonomy of Portuguese *Ulex*: quinolizidine alkaloids as taxonomical markers. Phytochemistry 67 (17), 1943–1949.

Michalska, K., Szneler, E., Kisiel, W., 2013. Sesquiterpene lactones from *Lactuca canadensis* and their chemotaxonomic significance. Phytochemistry 90, 90–94.

Michalska, K., Beharav, A., Kisiel, W., 2014. Chemotaxonomic value of magastigmane glucosides of *Cichorium calvum*. Nat. Prod. Commun. 9 (3), 311–312.

Mino, Y., 2006. Protein chemotaxonomy. XIII. Amino acid sequence of ferredoxin from Panax ginseng. Biol. Pharm. Bull. 29 (8), 1771–1774.

Mitreski, I., Stanoeva, J.P., Stefova, M., et al., 2014. Polyphenols in representative *Teucrium* species in the flora of R. Macedonia: LC/DAD/ESI-MS(n) profile and content. Nat. Prod. Commun. 9 (2), 175–180.

Mohy-Ud-Din, A., Khan, Z., Ahmad, M., Kashmiri, M.A., 2010. Chemotaxonomic value of alkaloids in *Solanum nigrum* complex. Pak. J. Bot. 42 (1), 653–660.

Mongrand, S., Badoc, A., Patouille, B., et al., 2005. Chemotaxonomy of the Rubiaceae family based on leaf fatty acid composition. Phytochemistry 66 (5), 549–559.

Nikolić, B., Ristić, M., Tešević, V., et al., 2011. Terpene chemodiversity of relict conifers *Picea omorika, Pinus heldreichii,* and *Pinus peuce*, endemic to Balkan. Chem. Biodivers. 8 (12), 2247–2260.

Nikolić, B., Tešević, V., Bojović, S., Marin, P.D., 2013. Chemotaxonomic implications of the n-alkane composition and the nonacosan-10-ol content in *Picea omorika, Pinus heldreichii*, and *Pinus peuce*. Chem. Biodivers. 10 (4), 677–686.

Niogret, J., Epsky, N.D., Schnell, R.J., et al., 2013. Terpenoid variations within and among half-sibling avocado trees, *Persea americana* Mill. (Lauraceae). PLoS One 8 (9), e73601.

Ntie-Kang, F., Lifongo, L.L., Mbaze, L.M., et al., 2013. Cameroonian medicinal plants: a bioactivity versus ethnobotanical survey and chemotaxonomic classification. BMC Complement. Altern. Med. 13, 147.

Padalia, R.C., Verma, R.S., Sundaresan, V., Chanotiya, C.S., 2010. Chemical diversity in the genus *Alpinia* (Zingiberaceae): comparative composition of four *Alpinia* species grown in Northern India. Chem. Biodivers. 7 (8), 2076–2087.

Patnala, S., Kanfer, I., 2013. Chemotaxonomic studies of mesembrine-type alkaloids in *Sceletium* plant species. South Africa J. Sci. 109, 1–5.

Peng, Y., Chen, S.B., Liu, Y., Chen, S., Xiao, P.G., 2006. A pharmacophylogenetic study of the Berberidaceae (s.l.). Acta Phytotax. Sin. 44 (3), 241–257.

Pérez, A.J., Kowalczyk, M., Simonet, A.M., et al., 2013. Isolation and structural determination of triterpenoid glycosides from the aerial parts of alsike clover (*Trifolium hybridum* L.). J. Agric. Food Chem. 61 (11), 2631–2637.

Pirmoradi, M.R., Moghaddam, M., Farhadi, N., 2013. Chemotaxonomic analysis of the aroma compounds in essential oils of two different *Ocimum basilicum* L. varieties from Iran. Chem. Biodivers. 10 (7), 1361–1371.

Pop, R.M., Socaciu, C., Pintea, A., et al., 2013. UHPLC/PDA-ESI/MS analysis of the main berry and leaf flavonol glycosides from different Carpathian *Hippophaë rhamnoides* L. varieties. Phytochem. Anal. 24 (5), 484–492.

Pop, R.M., Weesepoel, Y., Socaciu, C., et al., 2014. Carotenoid composition of berries and leaves from six Romanian sea buckthorn (*Hippophae rhamnoides* L.) varieties. Food Chem. 147, 1–9.

Qiao, X., Zhang, Y.T., Ye, M., et al., 2009. Analysis of chemical constituents and taxonomic similarity of *Salvia* species in China using LC/MS. Planta Med. 75 (15), 1613–1617.

Radulović, N.S., Dekić, M.S., 2013. Volatiles of *Geranium purpureum* Vill. and *Geranium phaeum* L.: chemotaxonomy of balkan *Geranium* and *Erodium* species (Geraniaceae). Chem. Biodivers. 10 (11), 2042–2052.

Radulović, N.S., Denić, M.S., 2013. Essential oils from the roots of *Echinops bannaticus* Rochel ex Schrad. and *Echinops sphaerocephalus* L. (Asteraceae): chemotaxonomic and biosynthetic aspects. Chem. Biodivers. 10 (4), 658–676.

Radulović, N., Dekić, M., Joksović, M., Vukićević, R., 2012. Chemotaxonomy of Serbian *Teucrium* species inferred from essential oil chemical composition: the case of *Teucrium scordium* L. ssp. scordioides. Chem. Biodivers. 9 (1), 106–122.

Radulović, N.S., Mladenović, M.Z., Blagojević, P.D., 2014. (Un)targeted metabolomics in Asteraceae: probing the applicability of essential-oil profiles of senecio L. (Senecioneae) taxa in chemotaxonomy. Chem. Biodivers. 11 (9), 1330–1353.

Riggi, E., Avola, G., Siracusa, L., Ruberto, G., 2013. Flavonol content and biometrical traits as a tool for the characterization of "Cipolla di Giarratana": a traditional Sicilian onion landrace. Food Chem. 140 (4), 810–816.

Ríos, S., Berkov, S., Martínez-Francés, V., Bastida, J., 2013. Biogeographical patterns and phenological changes in *Lapiedra martinezii* LAG related to its alkaloid diversity. Chem. Biodivers. 10 (7), 1220–1238.

Royer, M., Herbette, G., Eparvier, V., et al., 2010. Secondary metabolites of *Bagassa guianensis* Aubl. wood: a study of the chemotaxonomy of the Moraceae family. Phytochemistry 71 (14–15), 1708–1713.

Sadgrove, N.J., Telford, I.R., Greatrex, B.W., Jones, G.L., 2014. Composition and antimicrobial activity of the essential oils from the *Phebalium squamulosum* species complex (Rutaceae) in New South Wales, Australia. Phytochemistry 97, 38–45.

Sandasi, M., Kamatou, G.P., Viljoen, A.M., 2012. An untargeted metabolomic approach in the chemotaxonomic assessment of two *Salvia* species as a potential source of α-bisabolol. Phytochemistry 84, 94–101.

Sansom, C.E., Heenan, P.B., Perry, N.B., et al., 2013. Chemosystematic analyses of gingidia volatiles. Chem. Biodivers. 10 (12), 2226–2234.

Sarac, Z., Bojović, S., Nikolić, B., et al., 2013. Chemotaxonomic significance of the terpene composition in natural populations of *Pinus nigra* J.F. Arnold from Serbia. Chem. Biodivers. 10 (8), 1507–1520.

Saracoglu, I., Oztunca, F.H., Nagatsu, A., Harput, U.S., 2011. Iridoid content and biological activities of *Veronica cuneifolia* subsp. *cuneifolia* and *V. cymbalaria*. Pharm. Biol. 49 (11), 1150–1157.

Scotti, M.T., Emerenciano, V., Ferreira, M.J., et al., 2012. Self-organizing maps of molecular descriptors for sesquiterpene lactones and their application to the chemotaxonomy of the Asteraceae family. Molecules 17 (4), 4684–4702.

Shen, L., Zhao, Y.Y., Xie, H.P., Liu, W.H., 2013. Classification of *Cimicifuga* species based on ^{1}H-NMR fingerprint combined with pattern recognition technique. Zhongguo Zhong Yao Za Zhi 38 (2), 217–222.

Shi, X., Li, C., Li, K., Ye, C., 2014. Chemotaxonomy of *Eurya* by HPLC chromatograms. Acta Sci. Nat. Univ. Sunyatseni 53 (2), 101–107.

Shpatov, A.V., Popov, S.A., Salnikova, O.I., et al., 2013. Low-volatile lipophilic compounds in needles, defoliated twigs, and outer bark of *Pinus thunbergii*. Nat. Prod. Commun. 8 (12), 1759–1762.

Singh, D.P., Maurya, S., Singh, S.P., et al., 2007. Chemotaxonomic variability in *Zizyphus mouritiana* varieties and its pharmacological properties in relation to human health. J. Herb. Pharmacother. 7 (3–4), 229–237.

Siracusa, L., Patanè, C., Avola, G., Ruberto, G., 2012. Polyphenols as chemotaxonomic markers in Italian "long-storage" tomato genotypes. J. Agric. Food Chem. 60 (1), 309–314.

Skogerson, K., Wohlgemuth, G., Barupal, D.K., Fiehn, O., 2011. The volatile compound BinBase mass spectral database. BMC Bioinformatics 12, 321.

Sonboli, A., Bahadori, M.B., Dehghan, H., et al., 2013. Chemotaxonomic importance of the essential-oil composition in two subspecies of *Teucrium stocksianum* Boiss. from Iran. Chem. Biodivers. 10 (4), 687–694.

Sun, Y., Du, Y., Liu, Y., et al., 2012. Simultaneous determination of nine components in Anemarrhena asphodeloides by liquid chromatography-tandem mass spectrometry combined with chemometric techniques. J. Sep. Sci. 35 (14), 1796–1807.

Sun, H., Wang, M., Zhang, A., et al., 2013. UPLC-Q-TOF-HDMS analysis of constituents in the root of two kinds of Aconitum using a metabolomics approach. Phytochem. Anal. 24 (3), 263–276.

Tamura, K., Stecher, G., Peterson, D., et al., 2013. MEGA6: molecular evolutionary genetics analysis version 6.0. Mol. Biol. Evol. 30, 2725–2729.

Taskova, R.M., Gotfredsen, C.H., Jensen, S.R., 2006. Chemotaxonomy of Veroniceae and its allies in the Plantaginaceae. Phytochemistry 67 (3), 286–301.

Taskova, R.M., Kokubun, T., Garnock-Jones, P.J., Jensen, S.R., 2012. Iridoid and phenylethanoid glycosides in the New Zealand sun hebes (*Veronica*; Plantaginaceae). Phytochemistry 77, 209–217.

Triana, J., Eiroa, J.L., Ortega, J.J., et al., 2010. Chemotaxonomy of *Gonospermum* and related genera. Phytochemistry 71 (5–6), 627–634.

Triana, J., Eiroa, J.L., Morales, M., et al., 2013. A chemotaxonomic study of endemic species of genus *Tanacetum* from the Canary Islands. Phytochemistry 92, 87–104.

Truus, K., Vaher, M., Borissova, M., et al., 2012. Characterization of yew tree (*Taxus*) varieties by fingerprint and principal component analyses. Nat. Prod. Commun. 7 (9), 1143–1146.

Tsiri, D., Graikou, K., Pobłocka-Olech, L., et al., 2009. Chemosystematic value of the essential oil composition of *Thuja* species cultivated in Poland—antimicrobial activity. Molecules 14 (11), 4707–4715.

Vallverdú-Queralt, A., Medina-Remón, A., Martínez-Huélamo, M., et al., 2011. Phenolic profile and hydrophilic antioxidant capacity as chemotaxonomic markers of tomato varieties. J. Agric. Food Chem. 59 (8), 3994–4001.

Venditti, A., Altieri, A., Bianco, A., 2014. Monoterpenoids glycosides content from two Mediterranean populations of *Crucianella maritima* L. Nat. Prod. Res. 28 (8), 586–588.

Wang, H., Cui, Y., Zhao, C., 2010. Flavonoids of the genus *Iris* (Iridaceae). Mini Rev. Med. Chem. 10 (7), 643–661.

Wayman, K.A., de Lange, P.J., Larsen, L., et al., 2010. Chemotaxonomy of *Pseudowintera*: sesquiterpene dialdehyde variants are species markers. Phytochemistry 71 (7), 766–772.

Wilson, A.D., 2013. Diverse applications of electronic-nose technologies in agriculture and forestry. Sensors 13 (2), 2295–2348.

Wojakowska, A., Piasecka, A., García-López, P.M., et al., 2013. Structural analysis and profiling of phenolic secondary metabolites of Mexican lupine species using LC-MS techniques. Phytochemistry 92, 71–86.

Xiang, C., Qiao, X., Ye, M., Guo, D., 2012. Classification and distribution analysis of components in *Glycyrrhiza* using licorice compounds database. Acta Pharm. Sin. 47 (8), 1023–1030.

Xin, Y.Y., Deng, A.J., Du, G.H., et al., 2010. Fingerprinting analysis of *Saposhnikovia divaricata* using 1H nuclear magnetic resonance spectroscopy and high performance liquid chromatography. J. Integr. Plant Biol. 52 (9), 782–792.

Xu, L.J., Liu, H.T., Peng, Y., Xiao, P.G., 2008. A preliminary pharmacophylogenetic investigation in Schisandraceae. Acta Phytotax. Sin. 46 (5), 692–723.

Yuan, W., Wang, P., Su, Z., et al., 2013. Cytotoxic triterpenoid saponins from husks of *Aesculus californica* (Spach) Nutt. Phytochemistry 90, 95–105.

Yuce, E., Bagci, E., 2012. The essential oils of the aerial parts of two *Hypericum* taxa (*Hypericum triquetrifolium* and *Hypericum aviculariifolium* subsp. *depilatum* var. *depilatum* (Clusiaceae)) from Turkey. Nat. Prod. Res. 26 (21), 1985–1990.

Zgórka, G., 2009. Ultrasound-assisted solid-phase extraction coupled with photodiode-array and fluorescence detection for chemotaxonomy of isoflavone phytoestrogens in *Trifolium* L. (Clover) species. J. Sep. Sci. 32 (7), 965–972.

Zhang, C., Qin, M.J., Shu, P., et al., 2010. Chemical variations of the essential oils in flower heads of *Chrysanthemum indicum* L. from China. Chem. Biodivers. 7 (12), 2951–2962.

Zheng, X.W., Pan, H.Y., Wang, Y.G., Chen, J.K., 2007. Chemical analysis of diterpenoids in *Caldesia grandis* and its chemotaxonomic implication. Acta Phytotax. Sin. 45 (4), 570–575.

Zhou, Y., Xu, G., Choi, F.F., et al., 2009. Qualitative and quantitative analysis of diterpenoids in *Salvia* species by liquid chromatography coupled with electrospray ionization quadrupole time-of-flight tandem mass spectrometry. J. Chromatogr. A 1216 (24), 4847–4858.

Zhou, X.L., Shi, N., Bai, J., et al., 2012. Phylogenetic relationship between *Dasymaschalon* (Hook. f. et Thoms.) Dalle Torre et Harms and *Desmos* Lour. as well as study on their constituents. Chin. Trad. Herb. Drugs 43 (9), 1852–1857.

High-throughput sequencing in medicinal plant transcriptome studies

2

2.1 Introduction

Research on plant-derived products employed in medicine has always been led by chemists and phytochemists; however, as the association between the secondary metabolites and the active genes that encode them is elucidated, the means by which genetics and genomics are applied will become more efficient in advancing natural product discoveries. Plants synthesize a myriad of secondary metabolites, the biochemical screening of which is viewed as indispensable for the discovery of novel chemicals that can be developed as drugs. Studies of their biosynthesis and the relevant genetic mechanism can facilitate the large-scale production of drugs via molecular breeding, metabolic engineering, and transgenic plants. RNA sequencing, which is more cost-effective and more feasible than the whole genome sequencing, is becoming a powerful tool in medicinal plant research and has accelerated the investigation of the plant gene expression. Roche 454 pyrosequencing and Illumina high-throughput sequencing are popular sequencing platforms in the medicinal plant transcriptome studies. Advances in the sequencing workflow, from sample preparation to data analysis, enable rapid profiling and deep investigation of the medicinal plant transcriptome. Good sequencing randomness is obtained in the high-throughput transcriptome sequencing (Figure 2.1), as the distribution of reads in the assembled unigenes is largely homogeneous. Protein-coding sequence (CDS) prediction, a necessary step in the functional annotation of genes based on the transcriptome data, can be performed based on the assembled unigenes (e.g., Figure 2.2). This chapter summarizes the recent progress in the application of high-throughput sequencing in the medicinal plant transcriptome studies.

2.2 Metabolic pathway analysis

2.2.1 *Terpenoid and saponin*

Traditional Chinese medicinal plants have been used in disease prevention and treatment for thousands of years and currently constitute a rich source of medicinal compounds and drug candidates. The species-specific knowledge of plant metabolism has been obtained by determining the expression of the transcriptomes from some traditional Chinese medicinal plants. *Artemisia annua* (sweet wormwood or Qing Hao) has

Medicinal plants: chemistry, biology and omics. http://dx.doi.org/10.1016/B978-0-08-100085-4.00002-5

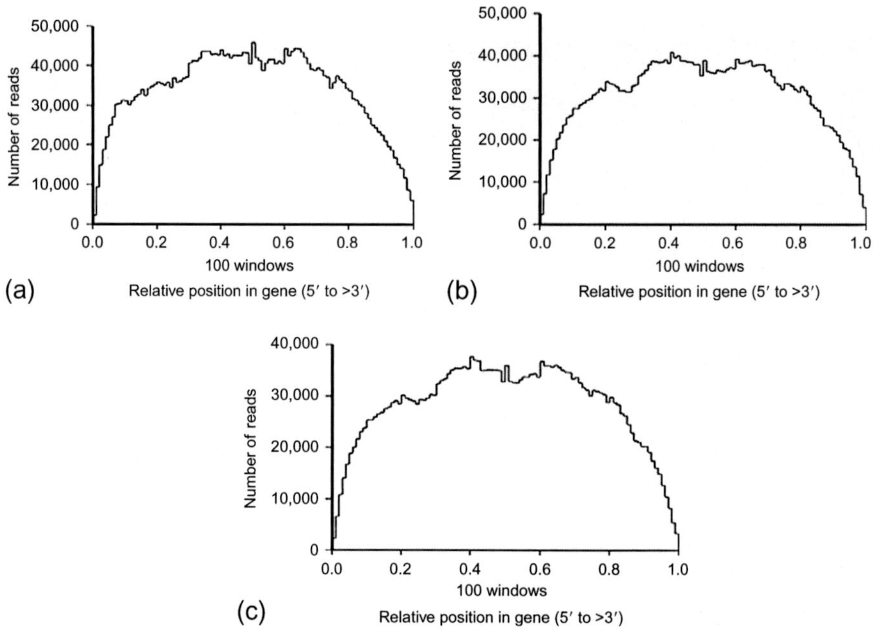

Figure 2.1 Sequencing randomness inferred from the reads distribution of the *Salvia sclarea* leaf transcriptome (Hao et al., 2015). (a) 0 h of 7.5 mM MeJA treatment; (b) 10 h; (c) 26 h. We sum the numbers of the reads aligned to different positions of the reference genes. Because genes have different lengths, we normalize the positions covered by the reads in the reference genes to relevant positions (i.e., ratio between the positions of reads on the reference genes and length of the genes). If randomness of mRNA fragmentation is ideal, there should be a roughly even distribution of reads in the reference genes. The horizontal coordinate is the relevant position from the 5′ end to the 3′ end, and the vertical coordinate is the corresponding reads number.

been used as a remedy by Chinese herbalists for more than 2000 years (Maude et al., 2010), but it was not subjected to scientific scrutiny until the 1970s. The most prominent antimalarial drug artemisinin, a sesquiterpene lactone, is produced in the glandular trichomes of *A. annua*. However, only limited genomic information was available in this nonmodel plant species. The *A. annua* glandular trichome transcriptome has been globally characterized using 454 pyrosequencing (Wang et al., 2009). By BLAST search against the NCBI nonredundant protein database, putative functions were assigned to over 28,573 unigenes, including previously undescribed enzymes likely involved in sesquiterpene biosynthesis. In higher plants, the terpenoid precursor isopentenyl diphosphate (IPP) can be produced from both MVA and MEP (methyl-D-erythritol 4-phosphate) pathway routes, which is then converted to its isomer dimethylallyl pyrophosphate (DMAPP). Unigenes encoding the MEP and MVA pathway enzymes and all the sesquiterpene artemisinin pathway enzymes were found in this pyrosequencing dataset. Unigenes corresponding to MEP pathway enzymes

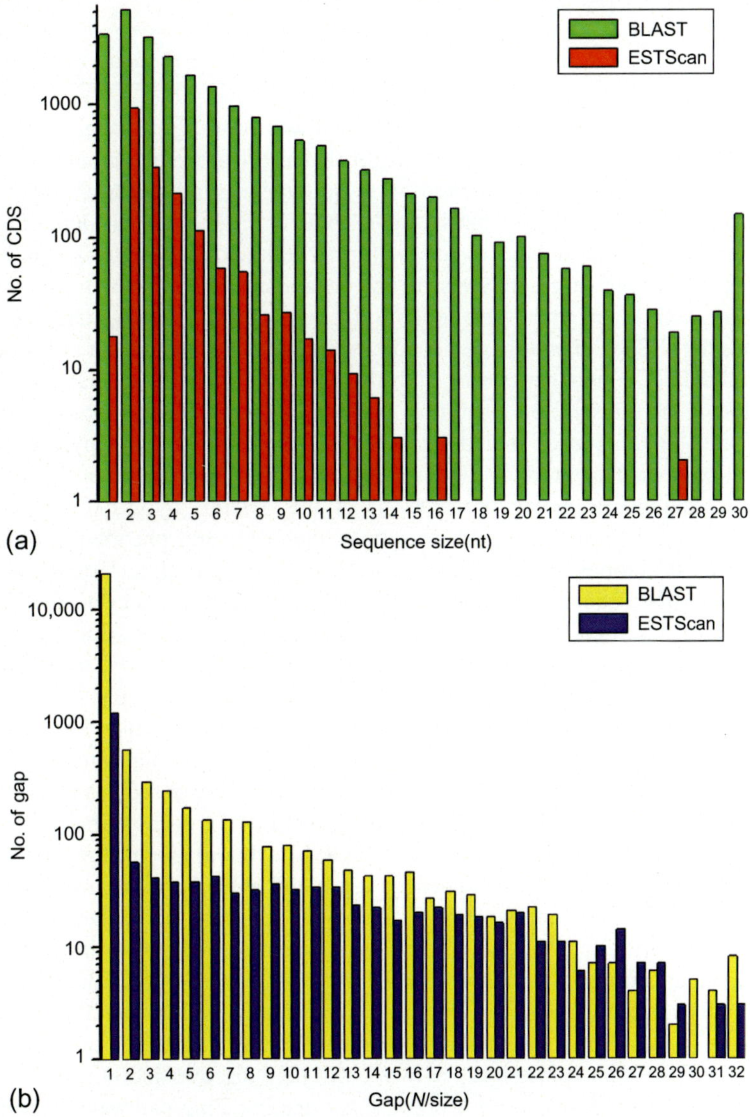

Figure 2.2 Prediction of protein-coding sequence (CDS) from the assembled *T. mairei* unigenes. Unigenes are firstly aligned by BLASTX (E-value < 0.00001) to protein databases in the priority order of nr, Swiss-Prot, KEGG, and COG. Unigenes aligned to databases with higher priority will not enter the next circle. The alignments end when all circles are finished. Proteins with the highest ranks in BLAST results are taken to decide the coding region sequences of unigenes; then, the coding region sequences are translated into amino acid sequences with the standard codon table. Thus, both the nucleotide sequences (5′–3′) and amino acid sequences of the unigene-coding region are acquired. Unigenes that cannot be aligned to any database are scanned by ESTScan (http://www.ch.embnet.org/software/ESTScan.html) to get the nucleotide sequence (5′–3′) and amino acid sequence of the coding regions. (a) Length distribution of CDS predicted from BLAST results and by ESTScan. 1, 200; 2, 300; 3, 400; 4, 500; 5, 600; 6, 700; 7, 800; 8, 900; 9, 1000; 10, 1100; 11, 1200; 12, 1300; 13, 1400; 14, 1500; 15, 1600; 16, 1700; 17, 1800; 18, 1900; 19, 2000; 20, 2100; 21, 2200; 22, 2300; 23, 2400; 24, 2500; 25, 2600; 26, 2700; 27, 2800; 28, 2900; 29, 3000; 30, >3000. (b) Gap (N) distribution of CDS predicted from BLAST results and by ESTScan. 1, 0; 2, 0.01; 3, 0.02; 4, 0.03; 5, 0.04; 6, 0.05; 7, 0.06; 8, 0.07; 9, 0.08; 10, 0.09; 11, 0.1; 12, 0.11; 13, 0.12; 14, 0.13; 15, 0.14; 16, 0.15; 17, 0.16; 18, 0.17; 19, 0.18; 20, 0.19; 21, 0.2; 22, 0.21; 23, 0.22; 24, 0.23; 25, 0.24; 26, 0.25; 27, 0.26; 28, 0.27; 29, 0.28; 30, 0.29; 31, 0.3; 32, >0.3.

were two folds more abundant than MVA pathway transcripts, suggesting that the MEP pathway may serve as a major route for DMAPP/IPP production in the *A. annua* trichomes. Three unigenes annotated as sesquiterpene synthases were selected for RACE (rapid amplification of cDNA ends)-PCR to retrieve the full-length cDNAs, which provide the raw sequences for the further functional characterization of these enzymes. Moreover, large amounts of unigenes annotated as phenylpropanoid and flavonoid pathway enzymes were found in the assembled pyrosequencing expressed sequence tag (EST) collection. Almost all medicinal plants have little or no genomic data available. The new-generation high-throughput sequencing offers rapid characterization of the transcriptome and thus provides a comprehensive tool for gene discovery and elucidation of metabolic pathways. In 2010, the 454 pyrosequencing platform was used to produce EST databases from cDNA libraries derived from enriched glandular trichome preparations of the Artemis hybrid (Graham et al., 2010). Key genes associated with metabolic pathways and phenotypic traits such as trichome development and plant architecture that could affect artemisinin yield are found in the sequencing dataset, and their relative abundance in the different libraries are quantified. The CAP3 software was used in the transcriptome sequence clustering and *de novo* assembly of both *A. annua* and another Chinese medicinal plant *Epimedium sagittatum* (Yin Yang Huo in Chinese) (Zeng et al., 2010). Flavonoids are the major medicinal compounds of *Epimedium* species. Twenty-nine EST consensus sequences relating to the secondary metabolic process were found, including genes encoding key enzymes in the flavonoid biosynthetic pathway such as phenylalanine ammonia lyase (PAL), cinnamate 4-hydroxylase (C4H), 4-coumarate-CoA ligase (4CL), and uridine diphosphate glucose (UDPG)-flavonoid glucosyltransferase. Furthermore, more than 300 EST sequences representing transcription regulators were annotated. Functional studies of these genes will facilitate molecular modification, gene transformation, and metabolic engineering to enhance the target drug production.

The nonmevalonate pathway or MEP/DOXP pathway of isoprenoid biosynthesis is an alternative metabolic pathway leading to the formation of IPP and DMAPP that has not been elucidated in many medicinal plants. The transcriptome study of the MEP pathway and the related metabolic network will help discover more natural products for developing new drugs and is a premise for manipulating pathways in plants and reconstituting plant pathways in microbial hosts (e.g., Ajikumar et al., 2010). High-throughput sequencing, as a cost-effective approach of sequence determination, has dramatically improved the efficiency and speed of gene discovery. Paclitaxel and its analogs and derivatives have been popular in the treatment of many types of clinical cancer (Kageyama, 2008). In the taxane-producing *T. cuspidata*, all genes encoding the seven enzymes in the plastidial MEP pathway were identified in the 454 datasets (Wu et al., 2011). The sequences encoding 4-hydroxy-3-methylbut-2-enyl diphosphate reductase (HDR), with 69 ESTs, were substantially abundant. In another important taxane source plant, *Taxus mairei*, the congener of *T. cuspidata*, all genes encoding the seven enzymes in the plastidial MEP pathway were identified in the Illumina datasets (Hao et al., 2011). HDR is the most abundantly expressed (Figure 2.3a), followed by 4-hydroxy-3-methylbut 2-enyl diphosphate synthase (HDS) and 1-deoxy-D-xylulose 5-phosphate synthase (DXS), the first enzyme of the MEP pathway,

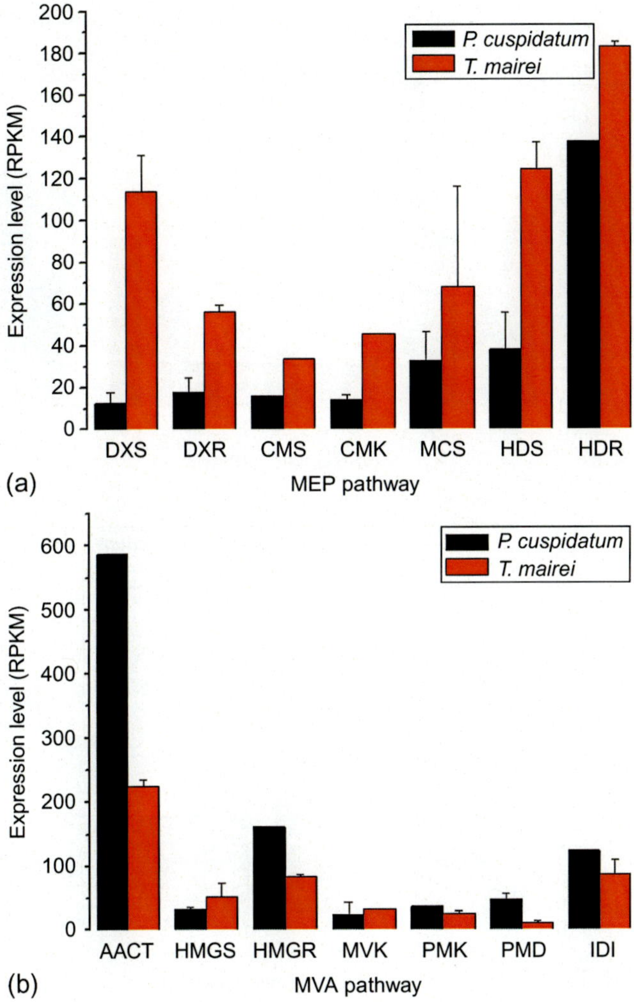

Figure 2.3 Unigene expression level (RPKM value) of four metabolic pathways in *Taxus mairei* and *Polygonum cuspidatum*. (a) MEP pathway, DXS, 1-deoxyxylulose 5-phosphate synthase; DXR, DXP reductoisomerase; CMS, MEP cytidylyltransferase; CMK, 4-(cytidine-5′-diphospho)-2-C-methyl-D-erythritol kinase; MCS, 2-C-methyl-D-erythritol 2,4-cyclodiphosphate synthase; HDS, 4-hydroxy-3-methylbut 2-en-yl-diphosphate synthase; HDR, 1-hydroxy-2-methyl-butenyl 4-diphosphate reductase; (b) MVA pathway, AACT, acetoacetyl CoA thiolase; HMGS, HMG-CoA synthase; HMGR, HMG-CoA reductase; MVK, MVA kinase; PMK, MVP kinase; PMD, MVPP decarboxylase; IDI, IPP isomerase;

(Continued)

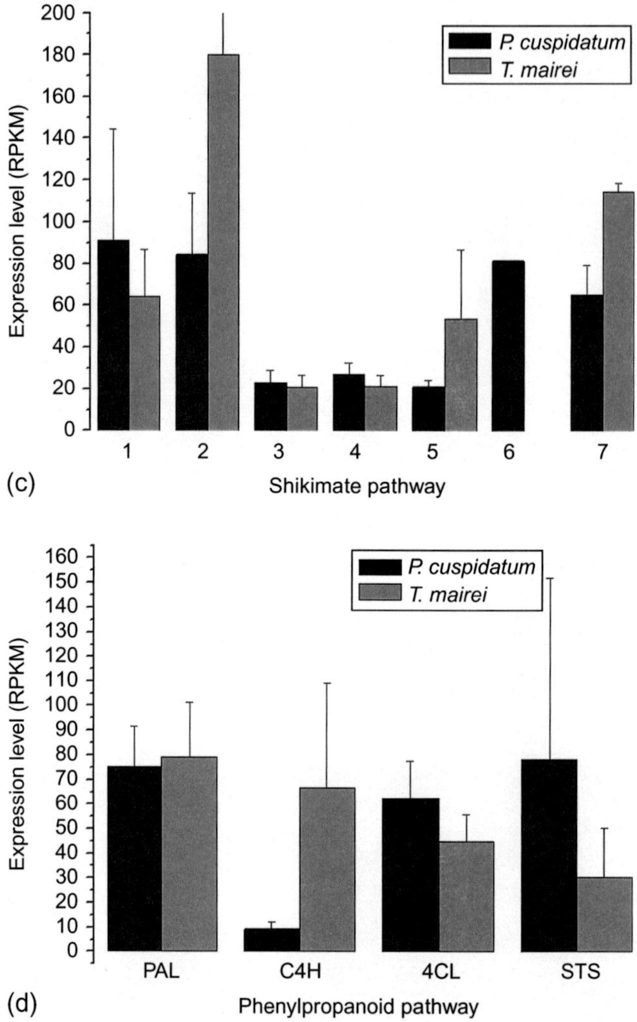

Figure 2.3 Continued. (c) shikimate pathway, 1, DAHP synthase; 2, 3-dehydroquinate synthase; 3, shikimate 5-dehydrogenase; 4, shikimate: NADP oxidoreductase; 5, shikimate kinase; 6, EPSP synthase; 7, chorismate synthase; (d) phenylpropanoid pathway, PAL, phenylalanine ammonia lyase; C4H, cinnamic acid 4-hydroxylase; 4CL, 4-coumarate: CoA ligase; STS, stilbene synthase. Bars represent the standard error of the average.

implying the conserved transcription regulation between the two *Taxus* species. Based on the same sequencing platform, the activity of other related pathways can be compared between different species (e.g., Figure 2.3b–d). Hundreds of novel genes in the MEP pathway and other closely related metabolic pathways have been found in *Camellia sinensis* (tea plant) (Shi et al., 2011), *S. miltiorrhiza* (Wenping et al.,

2011), *Euphorbia fischeriana* (Lang Du) (Barrero et al., 2011), and other medicinal plants mentioned above.

Plant natural products have been chosen for millennia by humans for various uses such as flavor, fragrances, and medicines. These compounds often are only produced in low amounts and are difficult to chemically synthesize. Elucidation of the underlying biosynthetic processes might help alleviate these issues (e.g., via metabolic engineering), but investigation of this is hampered by the low levels of relevant gene expression and expansion of the corresponding enzymatic gene families. The often inducible nature of such metabolic processes enables the selection of those genes whose expression pattern suggests a role in the production of the targeted natural product. Metabolomics and transcriptomics were combined to investigate the inducible biosynthesis of the bioactive diterpenoid tanshinones from *Salvia miltiorrhiza* (Dan Shen) (Gao et al., 2014b). Untargeted metabolomics investigations of elicited hairy root cultures indicated that tanshinone production was a dominant component of the metabolic response, increasing at later time points. A transcriptomic approach was applied to define not only a comprehensive transcriptome (20,972 nonredundant genes) but also its response to induction, revealing 6358 genes that exhibited differential expression, with significant enrichment for the upregulation of genes involved in stress, stimulus, and immune response processes. Consistent with metabolomics analysis, there appears to be a slower but more sustained increase in transcript levels of known genes from diterpenoid and tanshinone biosyntheses. Among the coregulated genes were 70 transcription factors (TFs) and eight cytochromes P450 (CYPs), providing targets for future investigation. Danshen terpenoid metabolism showed a biphasic response to elicitation, with early induction of sesqui- and triterpenoid biosyntheses, followed by later and more sustained production of the diterpenoid tanshinones.

KEGG analysis identified 501 and 952 transcripts from transcriptomes of *Ocimum sanctum* (family Lamiaceae) and *O. basilicum* (Rastogi et al., 2014), respectively, related to secondary metabolism with higher percentage of transcripts for the biosynthesis of terpenoids in *O. sanctum* and phenylpropanoids in *O. basilicum*. Higher DGE in *O. basilicum* was validated through qPCR and correlated to higher essential oil content and chromosome number (*O. sanctum*, $2n=16$ and *O. basilicum*, $2n=48$).

2.2.2 Saponin

Saponins are mainly amphipathic glycosides that possess many biological activities and confer potential health benefits to humans. Since the major steroidal components are present in the roots of *Asparagus racemosus*, comparative *de novo* transcriptome analysis of root versus leaf tissue was performed, and some root-specific transcripts involved in steroidal saponin biosynthesis were identified using high-throughput transcriptome sequencing (Upadhyay et al., 2014). 126,861 unigenes were generated with an average length of 1200 bp. Differentially expressed genes (DEGs) in root were identified using the RPKM method using digital subtraction between root and leaf. Twenty-seven putative secondary metabolite (SM)-related transcripts were validated for their expression in root or leaf tissue using qRT-PCR analysis. Most of the selected transcripts showed preferential expression in root as compared with leaf supporting the digitally subtracted result obtained. The methyl jasmonate (MeJA) application induces the SM-related gene transcripts leading to their increased accumulation in plants. Therefore, the identified transcripts related to saponin biosynthesis were further analyzed for their induced expression after 3, 5, and 12 h of exogenous application of MeJA in a tissue-specific manner.

Saponins are derived from geranyl pyrophosphate (GPP). GPP is synthesized by sequential head-to-tail addition of IPP and its allelic isomer DMAPP. IPP and DMAPP are synthesized via the cross talks between the cytosolic MVA pathway and plastidial MEP pathway. The MVA pathway starts from the condensation of acetyl-CoA, whereas the MEP pathway needs pyruvate and glyceraldehyde 3-phosphate. In the *Chlorophytum borivilianum* transcriptome dataset (Kalra et al., 2013), multiple transcripts encoding almost all known enzymes involved in the MVA pathway, MEP pathway, and saponin biosynthesis pathway were identified. In almost all the cases, more than one unique sequence was annotated as same enzyme. These unique sequences may represent either different fragments of a single transcript or different members of a gene family or both. IPP was converted to squalene by a series of enzymes that included IPP isomerase (six unigenes), geranylgeranyl diphosphate synthase (five unigenes), farnesyl diphosphate synthase (14 unigenes), and squalene synthase (nine unigenes). Squalene monooxygenase (14 unigenes) catalyzes the conversion of squalene to 2,3-oxidosqualene. The cyclization event of 2,3-oxidosqualene is catalyzed by a class of enzymes, OSCs. The cyclization of 2,3-oxidosqualene is a rate-limiting step, and this event is also the branch point for sterol and triterpenoid biosynthesis in many plants. Three OSC genes of *C. borivilianum*, cycloartenol synthase (16 unigenes), β-amyrin synthase (one unigene), and dammarenediol II synthase (one unigene) exist in this dataset. Steroidal saponins, borivilianosides, are the major type of saponins present in *C. borivilianum*. To date, no reports reveal the presence of triterpenoid saponins in *C. borivilianum*. However, two singleton sequences (Transcript 664097 and 798016) matched with β-amyrin synthase and dammarenediol II synthase of *V. vinifera* and *R. communis*, respectively.

2.2.3 Flavonoids and phenolics

Tea is a popular natural nonalcoholic beverage consumed worldwide due to its bioactive ingredients, particularly catechins (flavan-3-ols). Catechins not only contribute to tea quality but also serve important functions in the antistress regulation of secondary metabolic pathways. Catechins vary greatly among tea plant (*Camellia sinensis*) cultivars. Transcriptomes from leaf tissues of four tea plant cultivars, "Yunnanshilixiang," "Chawansanhao," "Ruchengmaoyecha," and "Anjibaicha," were sequenced using Illumina HiSeq™ 2000 (Wu et al., 2014c). Catechin contents were measured through reversed-phase high-performance liquid chromatography (RP-HPLC). A unified unigene database was constructed. 146,342 pairs of putative orthologs from the four tea plant cultivars were generated. 68,890 unigenes (47.1%) were aligned to the sequences of seven public databases with a cut-off E-value of 1e-05. 217 DEGs were found through RPKM values, and 150 unigenes were assigned to the flavonoid biosynthetic pathway using the integrated function annotation. The (−)-EGC and (−)-EC contents were significantly lower, and the (+)-GC and (+)-C contents were abnormally higher in "Ruchengmaoyecha" than in "Yunnanshilixiang," "Chawansanhao," and "Anjibaicha." The proportion of catechins was confirmed by selecting critical genes (ANS, ANR, and LAR) for qRT-PCR analysis.

Flavonoids make a very important contribution to the organoleptic qualities of grapes and wines. Flavonoids of Cabernet Sauvignon grown in Changli, Hebei of east China, and Gaotai, Gansu of west China, were analyzed (Li et al., 2014b). These regions have distinct climates contributing to their different "terroir." RNA sequencing was performed to trace transcriptome changes in Cabernet Sauvignon berries at pea size, veraison, and ripening, corresponding to E-L 31, 35, and 38, respectively. The accumulation of flavonols, flavan-3-ols, and anthocyanins together with the expression of relevant genes was analyzed and compared between the two regions. The biosynthesis patterns were similar between two regions, but more flavonols, anthocyanins, and trihydroxylated flavonoids accumulated in grapes from Gaotai before berry harvest, possibly due to the higher transcript levels of the genes that encode biosynthetic enzymes and their potential candidate TFs. The lower levels of flavan-3-ols, mainly (−)-epigallocatechin, in the preveraison grapes from Changli, might be due to limited flow of carbon to the F3′5′H branch pathway, as the ratio of F3′5′H to F3′H was lower in these berries. The combination of climatic factors profoundly affects the flavonoid pathway in grapes of China, providing regionally specific metabolism patterns.

Magnolia sprengeri is one of the most highly valuable medicinal and ornamental plants of the Magnoliaceae family. The natural color of *M. sprengeri* is variable. The transcriptome of white and red petals of *M. sprengeri* was sequenced using Illumina technology (Shi et al., 2014). The resulting reads were assembled into 77,048 unique sequences, of which 28,243 could be annotated by Gene Ontology (GO) analysis. The main enzymes involved in the flavonoid biosynthesis, such as phenylalanine ammonia lyase, cinnamate 4-hydroxylase, dihydroflavonol 4-reductase, flavanone 3-hydroxylase, flavonoid 3′-hydroxylase, flavonol synthase, chalcone synthase, and anthocyanidin synthase, were identified in the transcriptome. 270 TFs were sorted into three families, including MYB, bHLH, and WD40 types. Among these TFs, eight showed fourfold or greater changes in transcript abundance in red petals compared with white petals. HPLC analysis of anthocyanin compositions showed that the main anthocyanin in the petals of *M. sprengeri* is cyanidin-3-O-glucoside chloride, and its content in red petals was 26-fold higher than that in white petals. Thus, the metabolic pathways involved in the biosynthesis and catabolism of *M. sprengeri* flavonoids were reconstructed.

Herbaceous peony (*Paeonia lactiflora*) is a traditional flower and medicinal plant in China and an attractive wedding flower worldwide. Yellow is the rarest and its price is ten times that of the other colors. Two cDNA libraries from *P. lactiflora* chimera with red outer petals and yellow inner petals were sequenced using an Illumina HiSeq™ 2000 platform (Zhao et al., 2014b). 66,179,398 and 65,481,444 raw reads from red outer petal and yellow inner petal cDNA libraries were generated, which were assembled into 61,431, and 70,359 unigenes with an average length of 628 and 617 bp, respectively. 61,408 nonredundant all-unigenes were obtained, with 37,511 all-unigenes (61.08%) annotated in public databases. 6345 all-unigenes were differentially expressed between the red outer petals and yellow inner petals, with 3899 upregulated and 2446 downregulated all-unigenes, and the flavonoid metabolic pathway related to color development was identified using the Kyoto Encyclopedia of Genes and Genomes (KEGG) database. The expression patterns of 10 candidate DEGs involved in the flavonoid metabolic pathway were examined, and flavonoids were qualitatively and quantitatively analyzed. Numerous anthoxanthins (flavone and flavonol) and a few anthocyanins were detected in the yellow inner petal, which were all lower than those in the red outer petal due to the low expression levels of the phenylalanine ammonia lyase (PlPAL) gene, flavonol synthase (PlFLS) gene, dihydroflavonol 4-reductase (PlDFR) gene, anthocyanidin synthase (PlANS) gene, anthocyanidin 3-*O*-glucosyltransferase (Pl3GT) gene, and anthocyanidin 5-*O*-glucosyltransferase (Pl5GT) gene. RNA-Seq analysis is an efficient approach to identify critical genes in *P. lactiflora* and other nonmodel plants. The flavonoid metabolic pathway and glucide metabolic pathway were identified to be involved in the yellow formation in *P. lactiflora*. PlPAL, PlFLS, PlDFR, PlANS, Pl3GT, and Pl5GT were selected as potential candidates involved in the flavonoid metabolic pathway, which induces inhibition of anthocyanin biosynthetic genes mediating yellow formation in *P. lactiflora*.

Scutellarin (a flavone) and chlorogenic acids are the primary active components in *Erigeron breviscapus*. Using Illumina sequencing on GAIIx platform, 64,605,972 raw sequencing reads were generated and assembled into 73,092 nonredundant unigenes (Jiang et al., 2014), 44,855 of which (61.37%) were annotated in the public databases Nr, Swiss-Prot, KEGG, and COG. The transcripts encoding the known enzymes involved in flavonoid and chlorogenic acid biosyntheses were discovered in the Illumina dataset. Three candidate cytochrome P450 (CYP) genes were discovered that might encode flavone 6-hydroxylase, converting apigenin to scutellarin. Four unigenes encoding the homologues of maize P1 (R2R3 MYB transcription factors) were defined, which might regulate the biosynthesis of scutellarin. 11,077 simple sequence repeats (SSRs) were identified from 9255 unigenes, of which trinucleotide motifs were the most abundant motif.

2.2.4 Alkaloids

The purported presence of alkaloids in *Podophyllum* species has been enigmatic since the nineteenth century, remaining unresolved until now. *P. hexandrum* and *P. peltatum* are sources of podophyllotoxin (a lignan), extensively used as a chemical scaffold for various anticancer drugs. Integrated omics technologies (including advanced mass spectrometry/metabolomics, transcriptome sequencing/gene assemblies, and bioinformatics) gave unequivocal evidence that both species possess a previously unknown aporphine alkaloid metabolic pathway (Marques et al., 2014). RNA-Seq transcriptome sequencing and bioinformatics-guided gene assemblies/analyses suggested the presence of transcripts homologous to genes encoding all known steps in aporphine alkaloid biosynthesis. A comprehensive metabolomics analysis, including UPLC–TOF-MS and MALDI-MS imaging, enabled the detection, identification, localization, and quantification of the aporphine alkaloids, magnoflorine, corytuberine, and muricinine, in the underground and aerial tissues. These findings have evolutionary and phylogenetic implications.

The major medicinal alkaloids isolated from *Uncaria rhynchophylla* (Gou Teng) capsules are rhynchophylline (RIN) and isorhynchophylline (IRN). Extracts containing these terpenoid indole alkaloids (TIAs) can inhibit the formation and destabilize preformed fibrils of amyloid β-protein (a pathological marker of Alzheimer's disease) and have been shown to improve the cognitive function of mice with Alzheimer's-like symptoms. RNA sequencing of pooled *Uncaria* capsule RNA samples taken at three developmental stages that accumulate different amounts of RIN and IRN was performed (Guo et al., 2014). More than 50 million high-quality reads from a cDNA library were generated and *de novo* assembled. Sequences for all known enzymes involved in TIA synthesis were identified. 193 CYP, 280 methyltransferase, and 144 isomerase genes were identified, which are potential candidates for enzymes involved in RIN and IRN synthesis. DGE profile analysis was performed on the three-capsule developmental stages and based on genes possessing expression profiles consistent with RIN and IRN levels; four CYPs, three methyltransferases, and three isomerases were identified as the candidates most likely to be involved in the later steps of RIN and IRN biosynthesis. The transcriptome data provide an important resource for understanding the formation of major bioactive constituents in the capsule extract from *Uncaria* and provide information that may aid in metabolic engineering to increase yields of these important alkaloids.

The medicinal plant, *Catharanthus roseus*, accumulates a wide range of TIAs, which are well-documented therapeutic agents. Deep transcriptome sequencing of *C. roseus* was carried out to identify the pathways and enzymes (genes) involved in the biosynthesis of these compounds (Verma et al., 2014). About 343 million reads were generated from the leaf, flower, and root of *C. roseus* using the Illumina platform. Optimization of *de novo* assembly involving a two-step process resulted in 59,220 unique transcripts with an average length of 1284 bp. Sixty-five percent of *C. roseus* transcripts showed homology with sequences available in various public repositories, while the remaining 35% of unigenes might be *C. roseus*-specific. In silico analysis revealed the presence of 11,620 genic SSRs (excluding mononucleotide repeats) and 1820 TF-encoding genes in the *C. roseus* transcriptome. Expression analysis showed roots and leaves to be actively participating in bisindole alkaloid production with a clear indication that enzymes involved in the pathway of vindoline and vinblastine biosyntheses are restricted to aerial tissues. Such large-scale transcriptome study provides a rich source for understanding plant-specialized metabolism and is expected to promote research towards the production of plant-derived pharmaceuticals.

Galantamine is an Amaryllidaceae alkaloid used to treat the symptoms of Alzheimer's disease, which is primarily isolated from daffodil (*Narcissus* spp.), snowdrop (*Galanthus* spp.), and summer snowflake (*Leucojum aestivum*). The absence of genetic information on biosynthetic pathways is a limiting factor in the development of synthetic biology platforms for many important botanical medicines. A new bioinformatic approach using several recent technological improvements was applied to search for genes in the proposed galantamine biosynthetic pathway (Kilgore et al., 2014), first targeting methyltransferases (MTs) due to strong signature amino acid sequences in the proteins. Using Illumina sequencing, a *de novo* transcriptome assembly was constructed for daffodil. BLAST was used to identify sequences that contain signatures for plant OMTs in this transcriptome. The program Haystack was used to identify MTs that fit a model for galantamine biosynthesis in leaf, bulb, and inflorescence. One candidate gene for the methylation of norbelladine to 4′-*O*-methylnorbelladine in the proposed galantamine biosynthetic pathway was identified. This MT cDNA was expressed in *E. coli*, and the resulting protein was found to be a norbelladine 4′-*O*-MT of the proposed galantamine biosynthetic pathway.

Atropa belladonna is a medicinal plant and main commercial source of tropane alkaloids (TAs) including scopolamine and hyoscyamine, which are anticholinergic drugs widely used clinically. Based on the high-throughput transcriptome sequencing results (http://medicinalplantgenomics.msu.edu/), the digital expression patterns of unigenes representing nine structural genes (ODC, ADC, AIH, CPA, SPDS, PMT, CYP80F1, H6H, and TRII) involved in TA biosynthesis were constructed (Qiang et al., 2014), and simultaneously, the expression analysis of four released genes in NCBI (PMT, CYP80F1, H6H, and TRII) for verification was performed using qPCR, as well as the TA content detection in eight different tissues. Four genes (ODC, ADC, AIH, and CPA) in the upstream pathway and the two branch pathway genes (SPDS and TRII) were found to be expressed in all examined tissues with high expression level in the secondary root, while the three TA pathway-specific genes (PMT, CYP80F1, and H6H) were only expressed in the secondary roots and primary roots, mainly in the secondary roots. The qPCR detection results of PMT, CYP80F1, and H6H were consistent with the digital expression patterns, but their expression levels in the primary root were too low to be detected. The highest content of hyoscyamine was found in tender stems (3.364 mg/g), followed by tender leaves (1.526 mg/g), roots (1.598 mg/g), young fruits (1.271 mg/g), and fruit sepals (1.413 mg/g). The highest content of scopolamine was detected in fruit sepals (1.003 mg/g), followed by tender stems (0.600 mg/g) and tender leaves (0.601 mg/g). Both old stems and old leaves had the lowest content of hyoscyamine and scopolamine. The gene expression profile and TA accumulation indicated that TAs in *A. belladonna* were mainly biosynthesized in the secondary root and then transported and deposited in tender aerial parts. Screening the *A. belladonna* secondary root transcriptome database will facilitate unveiling the unknown enzymatic reactions and the mechanisms of transcriptional control.

In *C. borivilianum* transcriptome dataset, in addition to saponins and flavonoids, genes involved in the biosynthesis of benzylisoquinoline alkaloids (BIAs) were also discovered (Kalra et al., 2013). BIA biosynthesis begins with a metabolic lattice of decarboxylations, ortho-hydroxylations, and deaminations that convert tyrosine to both dopamine and 4-hydroxyphenylacetaldehyde. This reaction is catalyzed by tyrosine decarboxylase (TYDC; eight unigenes). The induction of TYDC mRNAs in parsley and *Arabidopsis* suggests that, in addition to BIAs, tyramine serves as precursor to a ubiquitous class of defense-responsive metabolites. Dopamine and 4-hydroxyphenylacetaldehyde are condensed by norcoclaurine synthase (NCS) (13 unigenes) to yield the trihydroxybenzylisoquinoline alkaloid (*S*)-norcoclaurine, which is the central precursor to all BIAs in plants. (*S*)-Norcoclaurine is converted to (*S*)-reticuline by a 6-*O*-methyltransferase (six unigenes), an *N*-methyltransferase, a CYP (EC 1.14.13.71; two unigenes), and a 4′-O-MT. (*S*)-Reticuline is the central intermediate that can undergo various rearrangements and modifications to yield different structural classes of benzylisoquinolines. Berberine bridge enzyme (BBE) (reticuline oxidase; one unigene) that catalyzes the conversion of reticuline to (*S*)-scoulerine was also found, but other enzymes of this pathway were not.

2.2.5 Glucosinolate

Broccoli (*Brassica oleracea* var. *italica*), a member of Cruciferae, is an important vegetable containing high concentrations of various nutritive and functional molecules especially the anticarcinogenic glucosinolates. The sprouts of broccoli contain 10–100 times higher level of glucoraphanin, the main contributor of anticarcinogenesis, than the edible florets. Around 85 million 251 bp reads were obtained by transcriptome sequencing (Gao et al., 2014a,b). After *de novo* assembly and searching the assembled transcripts against the *Arabidopsis thaliana* and NCBI nr databases, 19,441 top-hit transcripts were clustered as unigenes with an average length of 2133 bp. Twenty-five putative glucosinolate metabolism genes sharing 62.04–89.72% nucleotide sequence identity with the *Arabidopsis* orthologs were identified. Broccoli glucosinolate metabolic pathway has high colinearity to *Arabidopsis*. Many biosynthetic and degradation genes showed higher expression after germination than in seeds, especially the myrosinase TGG2 (20–130 times higher expression). These results along with the previous reports about these genes' studies in *Arabidopsis* and the glucosinolate concentration in broccoli sprouts indicate that the breakdown products of glucosinolates may play important roles in seed germination and sprout development.

2.2.6 Polyketide

Bitter acids (e.g., humulone) are prenylated polyketides synthesized in the lupulin glands of the hop plant (*Humulus lupulus*), which are important contributors to the bitter flavor and stability of beer. Bitter acids are formed from acyl-CoA precursors derived from branched-chain amino acid (BCAA) degradation and C5 prenyl diphosphates from the MEP pathway (Clark et al., 2013). RNA-Seq was used to obtain the transcriptomes of isolated lupulin glands, cones with glands removed and leaves from high α-acid hop cultivars (Figure 2.4). Levels of BCAAs, acyl-CoA intermediates, and bitter acids in glands, cones, and leaves were determined. Transcripts encoding all the enzymes of BCAA metabolism were significantly more abundant in lupulin glands, indicating that BCAA biosynthesis and subsequent degradation occur in these specialized cells. Branched-chain acyl-CoAs and bitter acids were present at higher levels in glands compared with leaves and cones. RNA-Seq analysis showed the gland-specific expression of the MEP pathway, enzymes of sucrose degradation, and several TFs that may regulate bitter acid biosynthesis in glands. Two branched-chain aminotransferase (BCAT) enzymes, HlBCAT1 and HlBCAT2, were abundant, with gene expression

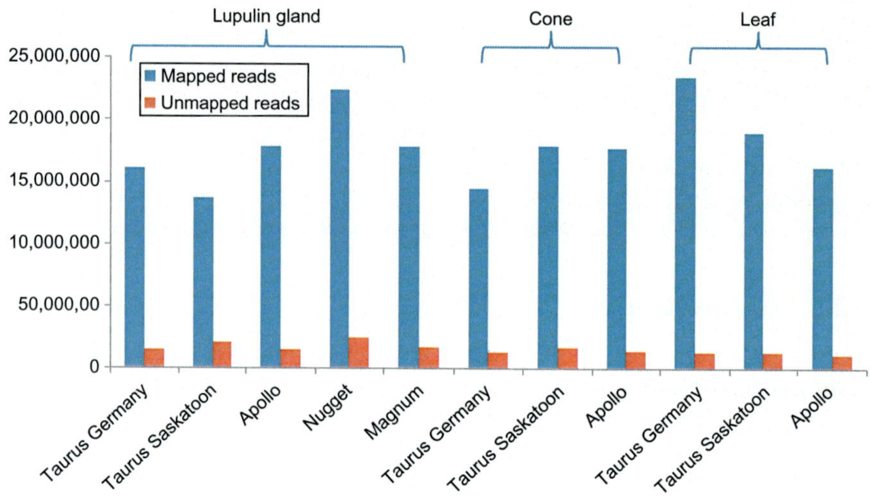

Figure 2.4 *Humulus lupulus* sequencing reads for each cultivar RNA sample. RNA purified from 11 samples (five glands, three cones, and three leaves) was reverse-transcribed and the cDNAs sequenced on a single lane of an Illumina HiSeq 2000 instrument as a multiplexed sample (Clark et al., 2013). Each cDNA produced 15.7–24.7 million reads. At least 87% of the reads from each sample were mapped to the reference transcriptome for each tissue.

quantification by RNA-Seq and qRT-PCR indicating that HlBCAT1 was specific to glands, while HlBCAT2 was present in glands, cones, and leaves. Recombinant HlBCAT1 and HlBCAT2 catalyzed forward (biosynthetic) and reverse (catabolic) reactions with similar kinetic parameters. HlBCAT1 is targeted to the mitochondria where it likely plays a role in BCAA catabolism. HlBCAT2 is a plastidial enzyme likely involved in BCAA biosynthesis. Phylogenetic analysis of the hop BCATs and those from other plants showed that they group into distinct biosynthetic (plastidial) and catabolic (mitochondrial) clades.

2.2.7 CYPs

Using 454 pyrosequencing, the research group led by Shilin Chen has obtained the transcriptome datasets of 10 medicinal plants (Tables 2.1 and 2.2), including *Taxus cuspidata* (yew) (Wu et al., 2011), *Ginkgo biloba* (Yin Xing) (Lin et al., 2011), *Huperzia serrata* (Qian Ceng Ta/Jin Bu Huan), *Phlegmariurus carinatus* (Luo et al., 2010), *Panax quinquefolius* (American ginseng) (Sun et al., 2010), *Panax ginseng* (Chen et al., 2011b), *Panax notoginseng* (San Qi/Tian Qi) (Luo et al., 2011), *Salvia miltiorrhiza* (Chinese sage/Dan Shen) (Li et al., 2010b), *Camptotheca acuminata* (happy tree) (Sun et al., 2011), and *Glycyrrhiza uralensis* (Chinese licorice) (Li et al., 2010a). These transcriptome sequencing studies provide a wealth of unigene sequences that are involved in various secondary metabolite biosynthesis processes, especially those of CYP and glycosyltransferase (GT). In plants, heme containing

Table 2.1 Examples of metabolic pathway studies using high-throughput transcriptome sequencing in medicinal plants

Species	Family	Sequencing platform	Tissue	No. of raw reads	No. of unique transcripts	No. of transcription factor	No. of CYPs	No. of GTs	Medicinal compounds	Reference
Brassica oleracea	Cruciferae	Illumina	Sprout/seed	85 million	19,441	1633	10	2	Glucosinolate	Gao et al. (2014a,b)
Erigeron breviscapus	Compositae	Illumina	Leaf/flower	64,605,972	73,092	846	1200	NS	Scutellarin/ chlorogenic acid	Jiang et al. (2014)
Stevia rebaudiana	Asteraceae	Illumina	Leaf	191,590,282	80,160	NS	NS	143	Diterpenoid steviol glycoside	Chen et al. (2014a)
Ocimum sanctum/ *O. basilicum*	Lamiaceae	Illumina	Leaf	45.97 million (*O. sanctum*)/50.84 million (*O. basilicum*)	69,117 (*O. sanctum*)/ 130,043 (*O. basilicum*)	3489/6074	386/ 801	NS	Terpenoids (*O. sanctum*)/ phenylpropanoids (*O. basilicum*)	Rastogi et al. (2014)
Narcissus pseudonarcissus	Amaryllidaceae	Illumina	Leaf/bulb/ inflorescence	65 million	106,450	NS	NS	NS	Galantamine	Kilgore et al. (2014)
Catharanthus roseus	Apocynaceae	Illumina	Leaf/flower/ root	343 million	59,220	1820	NS	NS	Vindoline/vinblastine	Verma et al. (2014)
Uncaria rhynchophylla	Rubiaceae	Illumina	Capsule	51 million	100,940	NS	193	NS	Rhynchophylline/ isorhynchophylline	Guo et al. (2014)
Paeonia lactiflora	Paeoniaceae	Illumina	Red outer petal/yellow inner petal	66,179,398 (red)/ 65,481,444 (yellow)	61,431 (red)/ 70,359 (yellow)	NS	NS	NS	Anthocyanin/ flavonoid	Zhao et al. (2014b)
Magnolia sprengeri	Magnoliaceae	Illumina	Petal	39,652,898 (red petal)/68,698,774 (white petal)	77,048	270	NS	NS	Flavonoid	Shi et al. (2014)
Podophyllum hexandrum/ *P. peltatum*	Berberidaceae	Illumina	Rhizome	NS	227,885 (*P. hexandrum*)/ 147,960 (*P. peltatum*)	NS	13	NS	Aporphine alkaloid/ lignin	Marques et al. (2014)
Sinopodophyllum hexandrum	Berberidaceae	Illumina	Rhizome	125,957,408 (15 °C)/ 44,346,624 (25 °C)	60,089	16,473	116	35	Aporphine alkaloid/ lignin	Kumari et al. (2014)
Asparagus racemosus	Asparagaceae	Illumina/454 pyrosequencing/ SOLiD	Leaf/root	54,893,366 (leaf)/ 59,911,356 (root)	126,861	1640	NS	NS	Steroidal saponin	Upadhyay et al. (2014)

Continued

Table 2.1 Continued

Species	Family	Sequencing platform	Tissue	No. of raw reads	No. of unique transcripts	No. of transcription factor	No. of CYPs	No. of GTs	Medicinal compounds	Reference
Camellia sinensis	Theaceae	Illumina	Leaf	95 million	86,523	NS	NS	NS	Catechins	Wu et al., 2014a, 2014b, 2014c
Chlorophytum borivilianum	Agavaceae	Illumina	Leaf	22,595,634	101,589	8369	490	72	Steroidal saponin/ flavonoid/alkaloid	Kalra et al. (2013)
Humulus lupulus	Cannabaceae	Illumina	Lupulin gland/cone/ leaf	~222.2 million	~170 000	48	NS	NS	Prenylated polyketides	Clark et al. (2013)
Siraitia grosvenorii	Cucurbitaceae	Illumina	Fruit	48,755,516	43,891	NS	85	90	Mogrosides (triterpenoid saponins)	Tang et al. (2011)
Ricinus communis	Euphorbiaceae	Illumina	Seed, leaf, flower	NS	75,090	1060	NS	NS	Lipids	Brown et al. (2012)
Salvia miltiorrhiza	Labiatae	Illumina	Flower, stem, leaf, root	8,353,971	56,774	1341	NS	NS	Tanshinone, salvianolic acid	Wenping et al. (2011)
Punica granatum	Punicaceae	Illumina	Fruit peel	103,000,000	9839	38	NS	32	Phenolic, flavonoids	Ono et al. (2011)
Armoracia rusticana	Cruciferae	454 Pyrosequencing	Leaf/root/ sprout/stem	592,507	14,871	NS	NS	NS	Peroxidases/volatile compound	Näätsaari et al. (2014)
Solanum lycopersicum	Solanaceae	454 Pyrosequencing	Stem trichome	979,076	27,195	743	NS	NS	Lycopene/terpene	Spyropoulou et al. (2014)
Bupleurum chinense	Apiaceae	454 Pyrosequencing	Root	195,088	24,037	NS	246	102	Saikosaponins	Sui et al. (2011)
Panax ginseng	Araliaceae	454 Pyrosequencing	Root	217,529	31,741	NS	133	235	Ginsenosides (triterpenoid saponins)	Chen et al. (2011b)
Panax notoginseng	Araliaceae	454 Pyrosequencing	Root	188,185	30,852	906	174	242	Triterpenoid saponins	Luo et al. (2011)

Panax quinquefolius	Araliaceae	454 Pyrosequencing	Root	209,747	31,088	NS	150	235	Ginsenosides (triterpenoid saponins)	Sun et al. (2010)
Epimedium sagittatum	Berberidaceae	454 Pyrosequencing	Leaf	228,768	76,459	300	NS	NS	Flavonoids	Zeng et al. (2010)
Glycyrrhiza uralensis	Fabaceae	454 Pyrosequencing	Root, stem, leaf	59,219	27,229	NS	125	172	Glycyrrhizin	Li et al. (2010a)
Ginkgo biloba	Ginkgoaceae	454 Pyrosequencing	Leaf	64,057	22,304	446	89	NS	Flavonoid, ginkgolides	Lin et al. (2011)
Salvia miltiorrhiza	Labiatae	454 Pyrosequencing	Root	46,722	18,235	577	70	NS	Tanshinone, salvianolic acid	Li et al. (2010b)
		454 Pyrosequencing/ Illumina	Hairy root	1,061,065/24.6 million	20,972	1162	125	NS		Gao et al. (2014a,b)
Salvia sclarea	Labiatae	454 Pyrosequencing	Calyx	190,148	45,822	986	NS	NS	NS	Legrand et al. (2010)
Huperzia serrata	Lycopodiaceae	454 Pyrosequencing	Leaf, root	140,930	36,763	504	20	NS	Lycopodium alkaloids	Luo et al. (2010)
Phlegmariurus carinatus	Lycopodiaceae	454 Pyrosequencing	Leaf, root	79,920	31,812	469	NS	NS	Lycopodium alkaloids	Luo et al. (2010)
Camptotheca acuminata	Nyssaceae	454 Pyrosequencing	Leaf	74,858	30,358	NS	99	NS	Terpenoid, indole alkaloid, camptothecin	Sun et al. (2011)
Fraxinus	Oleaceae	454 Pyrosequencing	Phloem plug	203,718	58,673	NS	192	NS	NS	Bai et al. (2011)
Digitalis purpurea	Plantaginaceae	454 Pyrosequencing	Leaf, stem, flower, root	66,103	23,532	NS	36	56	Cardiac glycosides	Wu et al. (2012)
Paeonia suffruticosa	Paeoniaceae	454 Pyrosequencing	Flower buds	625,342	23,652	484	NS	NS	Various	Gai et al. (2012)
Taxus cuspidata	Taxaceae	454 Pyrosequencing	Leaf	81,148	20,557	291	NS	NS	Paclitaxel, taxanes	Wu et al. (2011)

CYP, cytochrome P450; GT, glycosyltransferase; NS, data not shown in the original studies.

Table 2.2 Examples of molecular markers from high-throughput transcriptome sequencing in medicinal plants

Species	Family	Sequencing platform	Tissue	Mean read length (bp)	Average length of unigene (bp)	No. of transcripts matching to plants	No. of SSRs	No. of SNPs	Medicinal compounds	Reference
Catharanthus roseus	Apocynaceae	Illumina	Seedling	101	997	15,417	11,004	NS	Vindoline/vinblastine	Kumar et al. (2014)
Hippophae rhamnoides	Elaeagnaceae	Illumina	Leaf/root	NS	610	NS	6790	NS	Essential oil/ flavonoid	Jain et al. (2014)
Brassica juncea	Cruciferae	Illumina	Inflorescence/ pod/seedling	101	NS	NS	NS	135,693	Glucosinolate/ phenolics	Paritosh et al. (2014)
Vernicia fordii	Euphorbiaceae	Illumina	Seed	NS	945	NS	6366	NS	Fatty acid	Zhang et al. (2014)
Isatis indigotica	Brassicaceae	Illumina	Leaf/root	101	974	28,184	6400	NS	Indole alkaloid	Tang et al. (2014)
Lilium "Sorbonne"	Liliaceae	Illumina	Flower	90	NS	30,986	2762	NS	Flavonoid	Zhang et al. (2015a)
Daucus carota	Apiaceae	Illumina	Root	75	216	NS	NS	11,369	Carotenoid	Rong et al. (2014)
Ipomoea batatas	Convolvulaceae	Illumina	Root	75	581	35,051 (62.02%)	4114	NS	Alkaloids	Wang et al. (2010)
Sesamum indicum	Pedaliaceae	Illumina	Root, leaf, flower, seed, shoot tip	90	629	46,584 (54.03%)	7702	NS	Lipids	Wei et al. (2011)
Punica granatum	Punicaceae	Illumina	Fruit peel	NS	NS	NS	115	NS	Phenolic, flavonoids	Ono et al. (2011)
Sonneratia alba	Sonneratiaceae	Illumina	Root	75	581	22,242 (72.6%)	2358	NS	Polyphenol, flavonoids, polysaccharides	Chen et al. (2011a)
Diospyros kaki	Ebenaceae	454 Pyrosequencing	Flower	314	579	NS	42,711	NS	Triterpenoid/ polyphenol	Luo et al. (2014)
Cicer arietinum/*C. reticulatum*	Fabaceae	454 Pyrosequencing	Seedling/flower/ pod	472	NS	NS	1415	51,632	Polyphenol/flavonoid	Deokar et al. (2014)

Panax notoginseng	Araliaceae	454 Pyrosequencing	Root	410	581	21,658 (70.20%)	2772	NS	Triterpenoid saponins	Luo et al. (2011)
Artemisia annua	Asteraceae	454 Pyrosequencing	Leaf, flower bud, cotyledon	230	NS	NS	49	34,419	Sesquiterpene (artemisinin), monoterpene	Graham et al. (2010)
Artemisia tridentata	Asteraceae	454 Pyrosequencing, Illumina	Leaf	403.9 (454)	716 (contig)	21,436 (72.6%)	119	20,952	Terpenoids	Bajgain et al. (2011)
Epimedium sagittatum	Berberidaceae	454 Pyrosequencing	Leaf	224.9	375.9 (contig)	29,466 (38.5%)	2810	NS	Flavonoids	Zeng et al. (2010)
Citrullus lanatus	Cucurbitaceae	454 Pyrosequencing	Fruit flesh	302.8	540.3 (contig)	41,212 (54.90%)	5000	NS	NS	Guo et al. (2011)
Scabiosa columbaria	Dipsacaceae	454 Pyrosequencing, Illumina	Root, leaf, flower bud	212(454), 75 (Illumina)	270 (contig)	48,740 (44.5%)	4320	75,054	NS	Angeloni et al. (2011)
Ginkgo biloba	Ginkgoaceae	454 Pyrosequencing	Leaf	NS	NS	14,388 (64.5%)	204	NS	Flavonoid, ginkgolides	Lin et al. (2011)
Ribes nigrum	Grossulariaceae	454 Pyrosequencing	Leaf bud	230	407	21,451 (64%)	3000	7000	Polyphenol, lipids	Russell et al. (2011)
Salvia miltiorrhiza	Labiatae	454 Pyrosequencing	Root	414	NS	13,308 (73%)	223	NS	Tanshinone, salvianolic acid	Li et al. (2010b)
Huperzia serrata	Lycopodiaceae	454 Pyrosequencing	Leaf, root	405	608 (contig)	16,274 (44.3%)	2729	NS	Lycopodium alkaloids	Luo et al., 2010
Phlegmariurus carinatus	Lycopodiaceae	454 Pyrosequencing	Leaf, root	382	532 (contig)	14,070 (44.2%)	1573	NS	Lycopodium alkaloids	Luo et al. (2010)
Fraxinus	Oleaceae	454 Pyrosequencing	Phloem plug	NS	649 (contig)	32,270 (55%)	980	1272	NS	Bai et al. (2011)
Paeonia suffruticosa	Ranunculaceae	454 Pyrosequencing	Flower buds	358	NS	12,345 (52.2%)	2253	NS	Various	Gai et al. (2012)
Capsicum annuum	Solanaceae	454 Pyrosequencing	Fruit	375	521 (contig)	18,664 (54%)	758	1536	NS	Lu et al. (2012)
Taxus cuspidata	Taxaceae	454 Pyrosequencing	Leaf	389	456.6	14,095 (68.6%)	753	NS	Paclitaxel, taxanes	Wu et al. (2011)

NS, data not shown in the original studies.

CYP is a superfamily of monooxygenases that catalyze the addition of one oxygen atom from O_2 into a substrate, with a substantial reduction of the other atom to water. The function of CYP families is attributed to chemical defense mechanisms under terrestrial environmental conditions, and many of them might be involved in the secondary metabolism. *P. notoginseng* and *P. ginseng* are important medicinal plants of the Araliaceae family and are the major components of many traditional Chinese medicinal empirical formulas (Hu et al., 2008). Triterpenoid saponins are the bioactive constituents in these species and their relative *P. quinquefolius*. In *P. notoginseng*, the candidate CYP genes most likely to be involved in the hydroxylation of aglycones for triterpenoid saponin biosynthesis were found from 174 CYPs by phylogenetic analysis (Luo et al., 2011); in *P. quinquefolius*, a total of 150 CYP unique sequences were found in the 454 cDNA library, from which one CYP was selected as the candidate most likely to be involved in ginsenoside biosynthesis through a MeJA inducibility experiment and tissue-specific expression pattern analysis based on a real-time PCR assay (Sun et al., 2010); in *P. ginseng*, one of the most highly valued medicinal plant of traditional Chinese medicine, 133 CYP unique sequences were found (Chen et al., 2011b), illustrating the high efficiency of the 454 pyrosequencing compared with the traditional Sanger sequencing. CYPs are known to be involved in a wide range of biosynthetic pathways in medicinal plants, including those leading to the synthesis of glycyrrhizin (Li et al., 2010a), camptothecin (Sun et al., 2011), ginkgolide and flavonoid (Lin et al., 2011), tanshinone and salvianolic acid (Li et al., 2010b), and lycopodium alkaloids (Luo et al., 2010; Table 2.1), many of which are defense-related phytoalexins (Morant et al., 2003). The discovery of novel CYP genes is accelerated by the immense capacity of both 454 pyrosequencing and Illumina high-throughput sequencing. In *Bupleurum chinense* (Chai Hu), 246 CYP unique sequences were found in the 454 dataset, two of which were identified as the most likely candidates involved in saikosaponin biosynthesis (Sui et al., 2011); 192 and 36 CYP genes were recovered from *Fraxinus* spp. (Bai et al., 2011) and *Digitalis purpurea* (common foxglove) (Wu et al., 2012) 454 transcriptomic databases, respectively. In *Siraitia grosvenorii* (Buddha fruit/Luo Han Guo), 85 CYP unigenes were identified from the Illumina transcriptome dataset, seven of which were selected as the candidates most likely to be involved in mogroside biosynthesis (Tang et al., 2011).

2.2.8 Glycosyltransferase

Glycosyltransferases (GTs) are another large multigene family in plants. In general, glycosylation is the last step in the biosynthesis of secondary metabolites, and sugar conjugation results in both increased stability and water solubility. Using BLAST search, 172 unigenes (1205 ESTs) in the 454 transcriptome study of *G. uralensis* showed sequence similarities to GTs in the KEGG database (Li et al., 2010a). Among these unigenes, 27 (83 ESTs) encoded UDP-glycosyltransferases (UGTs), which are involved in the biosynthesis of secondary metabolites. Eleven unigenes (33 ESTs) that encoded glucuronosyltransferases were found, which might be involved in the last steps of glycyrrhizin biosynthesis. From these two categories, 17 contigs were chosen for organ-specific expression pattern analysis by real-time PCR. The expression patterns of six GT unigenes were similar to that of CYP88D6, and they were regarded as candidate GTs that encode the enzymes responsible for glycyrrhizin biosynthesis and will be the subject of further study. Among 235 GTs, four UGTs were selected as candidate genes involved in ginsenoside biosynthesis based on their MeJA-inducible and tissue-specific expression patterns in *P. quinquefolius* (Sun et al., 2010). According to the sequence similarity search for the 235 *P. ginseng* GTs, six transcripts showed sequence similarities to *P. quinquefolius* candidate UGTs (Chen et al., 2011b). Among 242 GTs, 16 UGTs that contain more than 10 reads in the *P. notoginseng* root cDNA library were the candidate enzymes involved in triterpenoid saponin biosynthesis (Luo et al., 2011). Plants tend to have far more GT genes than any other organism sequenced to date, and this can be partially explained by the numerous glycosylated secondary metabolites. Candidate UGTs responsible for the formation of saikosaponin, cardiac glycoside, mogroside, and hydrolysable tannin were found in *B. chinense* (Sui et al., 2011), *D. purpurea* (Wu et al., 2012), *S. grosvenorii* (Tang et al., 2011), and *Punica granatum* (pomegranate) (Ono et al., 2011), respectively, with either 454 pyrosequencing or Illumina high-throughput sequencing (Table 2.1).

One hundred forty-three UGT unigenes were identified from the transcriptome of *Stevia rebaudiana*, some of which might be involved in diterpenoid steviol glycoside (SG) biosynthesis (Chen et al., 2014a). In the lignin biosynthesis pathway, lignin monomers like coumaryl alcohol, coniferyl alcohol, and sinapyl alcohol are translocated in the form of glucosides from cytosol to cell wall. UGTs are involved in the biosynthesis of lignin (Kumari et al., 2014). Enzyme activity is expected to be essential for vacuolar storage of otherwise toxic lignans and is shown to be correlated with lignan glucoside accumulation. A podophyllotoxin-glucose glucosidase has been isolated from *Podophyllum peltatum*. BLAST search identified 35 unigenes encoding GTs. Ten UGTs exhibited modulation by twofold and above at the two temperatures as analyzed through FPKM, of which eight were downregulated, while two showed upregulation at 15 °C as compared with 25 °C. In-depth analysis of these modulated UGTs could identify the possible candidates associated with podophyllotoxin biosynthesis.

2.2.9 *Other enzymes*

MTs are transferases that participate in the transfer of a methyl group from a donor to an acceptor. O-methylation plays an important role in the biosynthesis of lignans including podophyllotoxin (Kumari et al., 2014). Ferulic acid and sinapic acid are methylated compounds and are precursors of monolignols (coniferyl and sinapyl alcohols), the moieties involved in lignin biosynthesis. MTs are essential in the PP pathway: the activity of CCoAOMT is essential for coniferyl alcohol and sinapyl alcohol biosyntheses; COMT is the key enzyme involved in methylation using hydroxycinnamates as substrates. The conversion of matairesinol to yatein involves four steps that include hydroxylation, dual methylation, and methylenedioxy bridge formation. Of 63 MTs identified from *Sinopodophyllum hexandrum* (*Podophyllum hexandrum*) transcriptome sequencing, 18 MTs exhibited differential modulation by twofold and above, in which two MTs showed upregulation and 16 showed downregulation at

15 °C as compared with that at 25 °C. Deeper analysis of these MTs could be promising to identify the genes associated with podophyllotoxin biosynthesis.

Horseradish peroxidases (HRPs) from *Armoracia rusticana* have long been utilized as reporters in various diagnostic assays and histochemical stainings. Regardless of their increasing importance in the field of life sciences and traditional uses in folk medicine, chemical synthesis, and other industrial applications, the HRP isoenzymes, their substrate specificities, and enzymatic properties are poorly characterized. A normalized, size-selected *A. rusticana* transcriptome library was sequenced using 454 Titanium technology (Näätsaari et al., 2014), resulting in 14,871 isotigs with an average length of 1133 bp. Sequence databases, ORF finding, and ORF characterization were utilized to identify peroxidase genes from the isotigs. The sequences were manually reviewed and verified with Sanger sequencing of PCR-amplified genomic fragments, resulting in the discovery of 28 secretory peroxidases, including 23 novel ones. Twenty-two isoenzymes including allelic variants were successfully expressed in *Pichia pastoris* and showed peroxidase activity with at least one of the substrates tested, thus enabling their development into commercial pure isoenzymes. Transcriptome sequencing combined with sequence motif search is a powerful concept for the discovery and quick supply of new enzymes and isoenzymes from any plant or other eukaryotic organisms.

2.2.10 Regulation

TFs, the sequence-specific DNA-binding proteins, play important roles in the regulation of gene expressions in response to developmental programs and environmental alterations in plants (Singh et al., 2002). Genetic manipulation of the TF might increase the target drug production and/or divert the metabolic flux to the biosynthesis of the useful medicinal compounds. Predictions using InterPro found 906 *P. notoginseng* unique sequences representing putative homologues belonging to different TF families, covering the ARF, AUX/IAA, B3, MYB, basic helix-loop-helix (bHLH), bZIP, homeobox, homeodomain-like/related, pathogenesis-related/ERF, WRKY, and zinc finger family proteins (Luo et al., 2011). Abundant TFs were also found in the transcriptome datasets of many other medicinal plants (Table 2.1).

Glandular trichomes are the production and storage organs of specialized metabolites such as terpenes, which play a role in the plant's defense system. Identification of TFs that control the expression of terpene synthases could shed light on the regulation of terpene biosynthesis in *Solanum lycopersicum* trichomes (Spyropoulou et al., 2014). A trichome transcriptome database was created with 27,195 contigs that contained 743 annotated TFs. A quantitative expression database was obtained of jasmonic acid-treated trichomes. Sixteen candidate TFs were selected for further analysis. One TF of the MYC bHLH class and one of the WRKY class were able to transiently transactivate *S. lycopersicum* terpene synthase promoters in *Nicotiana benthamiana* leaves. Strikingly, SlMYC1 was shown to act synergistically with a previously identified zinc finger-like TF, expression of terpenoids 1 (SlEOT1) in transactivating the SlTPS5 promoter. High-throughput sequencing of tomato stem trichomes led to the discovery of two TFs that activated several terpene synthase promoters.

2.2.11 Integrated use of omics platform

Omics and bioinformatics are essential to understanding the molecular systems that underlie various plant functions, including those related to drug discovery and development. Combining various omics technologies, for example, genome sequencing, transcriptome sequencing, and metabolomics, is becoming a popular research paradigm in medicinal plant studies. *Papaver somniferum* (opium poppy) is the source for the pharmaceutical BIAs including morphine, codeine, and sanguinarine. Opioid receptor agonists are one of the most effective classes of analgesic drugs available for clinical use (Beedle and Zamponi, 2001). Treatment with the fungal elicitor could increase the biosynthesis and accumulation of sanguinarine and induce other defense responses in opium poppy cell cultures (Desgagné-Penix et al., 2010). Using 454 pyrosequencing, 427,369 ESTs with an average length of 462 bp were generated from the cDNA library prepared for opium poppy cell cultures treated with a fungal elicitor for 10 h. Transcripts encoding all known sanguinarine biosynthetic enzymes were identified in the EST database, five of which were represented among the 50 most abundant transcripts. Liquid chromatography–tandem mass spectrometry (LC–MS/MS) of total protein extracts from cell cultures treated with a fungal elicitor for 50 h facilitated the identification of 1004 proteins. Query of the opium poppy-specific EST database substantially enhanced peptide identification. Eight out of 10 known sanguinarine biosynthetic enzymes and many relevant primary metabolic enzymes were represented in the peptide database. The integration of deep transcriptome and proteome analyses provides an effective platform to catalog the components of secondary metabolism and to identify genes encoding uncharacterized enzymes. *Cannabis sativa* is another plant cultivated worldwide for its medicinal and intoxicating properties. The molecular pharmacology of cannabinoids has been under intense study (Bertalovitz et al., 2010). Selective breeding has produced cannabis plants for specific uses, including high-potency marijuana strains and hemp cultivars for fiber and seed production. For increasing the target drug yield via molecular breeding, the molecular biology underlying cannabinoid biosynthesis and other traits of interest must be explored. The genomic DNA and RNA from the marijuana strain Purple Kush have been sequenced using short-read approaches (van Bakel et al., 2011). The draft haploid genome sequence is 534 Mb in size and the transcriptome consists of 30,000 genes. Comparing the transcriptome of Purple Kush with that of the hemp cultivar "Finola" revealed that many genes of the cannabinoid and precursor pathways are more highly expressed in the former. The exclusive occurrence of Δ9-tetrahydrocannabinolic acid synthase in the Purple Kush transcriptome, and its replacement by cannabidiolic acid synthase in "Finola," may explain why the psychoactive cannabinoid Δ9-tetrahydrocannabinol (THC) is produced in marijuana but not in hemp. The whole genome and transcriptome sequences will aid the development of therapeutic marijuana strains with modified cannabinoid profiles and provide a basis for the breeding of the plant with better agronomic characteristics. The integration of deep genome and transcriptome analyses has also been applied to the study of the insect pathogenic fungus *Cordyceps militaris* (Zheng et al., 2011), a unique traditional medicine that contains pharmacologically active components including cordycepin, cordycepic acids, polysaccharides and macrolides. In addition, the integration of transcriptome sequencing and metabolomics has been used in the studies of *T. cuspidata* (Lee et al., 2010), *T. mairei* (Hao et al., 2011), *A. annua* (Graham et al., 2010), and *Ricinus communis* (castor oil plant) (Brown et al., 2012).

2.3 Molecular marker mining

2.3.1 Simple sequence repeats

Simple sequence repeats (SSRs) (microsatellites) are repeating DNA sequences of 1–6 nucleotides that are present in tandem arrays in all eukaryotic genomes. Compared with other types of molecular markers, SSR marker has many advantages including high abundance, random distribution in the entire genome, high information content, codominant inheritance, and reproducibility. Almost all wild plant species used in traditional Chinese medicine are endangered due to commercial overexploitation, while sustainable utilization, conservation genctics, systematics, and marker-assisted selection of these medicinal plants in supply crisis are less studied due to the lack of molecular markers. High-throughput sequencing facilitates the large-scale discovery of SSRs that could be used in the linkage map development, quantitative trait loci (QTL) mapping, marker-assisted selection, parentage analysis, cultivar fingerprinting, and so on. cDNAs of *E. sagittatum* are sequenced using 454 pyrosequencing (Zeng et al., 2010). A total of 2810 EST-SSRs is identified from the 76,459 consensus sequences of *Epimedium* transcriptome dataset. Trinucleotide SSR is the dominant repeat type (55.2%), followed by dinucleotide (30.4%), tetranuleotide (7.3%), hexanucleotide (4.9%), and pentanucleotide (2.2%) SSR. The dominant repeat motif is AAG/CTT (23.6%) followed by AG/CT (19.3%), ACC/GGT (11.1%), AT/AT (7.5%), and AAC/GTT (5.9%). Thirty-two SSRs are randomly selected, and primer pairs are synthesized for testing the transferability across 52 *Epimedium* species. Eighteen primer pairs could be successfully transferred to *Epimedium* species and 16 of these show high genetic diversity with 0.35 of observed heterozygosity and a high number of alleles per locus (11.9). *E. sagittatum* EST-SSR transferability to the major *Epimedium* germplasm is up to 85.7%. Trinucleotide SSR is also the dominant repeat type found in the transcriptome datasets of *Artemisia tridentata* (common sagebrush) (Bajgain et al., 2011), *Capsicum annuum* (red pepper) (Lu et al., 2012), *Citrullus lanatus* (watermelon) (Guo et al., 2011), *Scabiosa columbaria* (pincushion flower) (Angeloni et al., 2011), *Fraxinus* (Bai et al., 2011; Table 2.2), *T. mairei* (Figure 2.5), and *Polygonum cuspidatum* (Japanese knotweed/Hu Zhang; Hao et al., 2012), while dinucleotide SSR is most abundant in *Vernicia fordii* (Zhang et al., 2014), *Lilium* "Sorbonne" (AG/CT; Zhang et al., 2015a), *Sonneratia alba* (a kind of mangrove plant) (Chen et al., 2011a), *Sesamum indicum* (sesame) (Wei et al., 2011), *S. miltiorrhiza* (Li et al., 2010b), *P. notoginseng* (Luo et al., 2011), and *G. biloba* (Lin et al., 2011). Mononucleotide SSR is predominant in *Isatis indigotica* (Ban Lan Gen; Tang et al., 2014). These transcriptome datasets and SSRs thereof will be a powerful resource for further studies such as taxonomy, molecular breeding, genetics, and genomics in the respective medicinal plant.

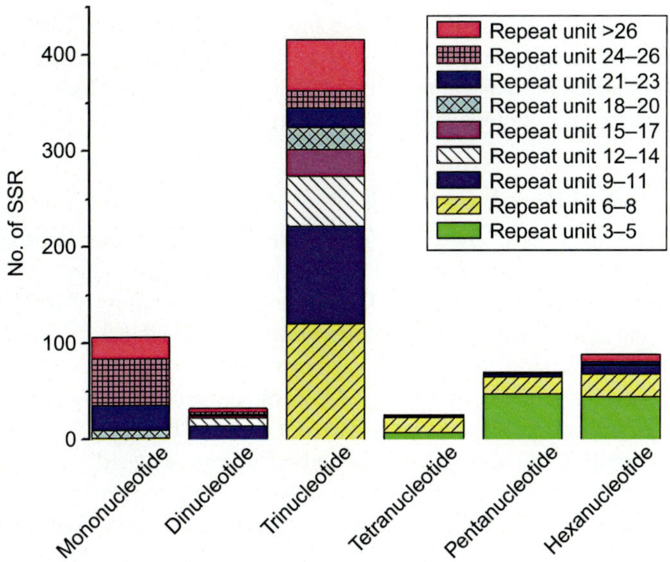

Figure 2.5 Distribution of repeat sizes for SSR markers of *Taxus mairei*. A number of SSR markers from dinucleotide, trinucleotide, and other (mono-, tetra-, penta-, and hexanucleotide) categories with different numbers of the repeat unit are shown.

2.3.1.1 Flower

Oriental persimmon (*Diospyros kaki*) ($2n=6x=90$) is a major commercial and deciduous fruit tree that is believed to originate in China. Using Roche 454 sequencing technology, the transcriptome from RNA of the flowers of *D. kaki* was analyzed (Luo et al., 2014). 1,250,893 reads were generated and 83,898 unigenes were assembled. 42,711 SSR loci were identified from 23,494 unigenes and 289 PCR primer pairs were designed. Of these primers, 155 (53.6%) showed robust PCR amplification, and 98 revealed polymorphism between 15 persimmon genotypes, indicating a polymorphic rate of 63.23% of the productive primers for characterization and genotyping of the genus *Diospyros*.

2.3.1.2 Fruit

Gene-based microsatellite markers are becoming more popular as compared with traditional random genomic microsatellite markers due to a rapid and inexpensive method of isolation and their cross species portability. Microsatellites were found in the transcriptome of sea buckthorn (*Hippophae rhamnoides*), a plant with immense medicinal, nutritional, and ecological value (Jain et al., 2014). *De novo* assembly of over 80 million high-quality short reads yielded 88,297 putative unigenes, 7.69% of which harbored microsatellite repeats with an average of one microsatellite per 6.704 Kb transcriptome. Dinucleotide repeats were most abundant followed by trinucleotide repeats. Microsatellites were densely populated in coding regions, followed by 3′ and 5′ untranslated regions. AG and AAG type repeats were the most frequently represented. Of the microsatellite positive unigenes, 48.81% could be assigned GO terms. The utility of unigene specific microsatellites was assessed on the basis of polymorphism(s) detected in 18 sea buckthorn collections from Leh (India) using a set of randomly selected 25 unigene specific microsatellites.

2.3.1.3 Seedling

Transcriptomic data of *C. roseus* offer ample sequence resources for providing better insights into gene diversity and accelerating genomic studies and breeding in *Catharanthus* (Kumar et al., 2014). Transcriptome sequencing of a 26-day-old *C. roseus* seedling tissue using Illumina GAIIx platform resulted in 3.37 Gb of nucleotide sequence data comprising 29,964,104 reads that were *de novo* assembled into 26,581 unigenes. Based on similarity searches, 58% of the unigenes were annotated of which 13,580 unique transcripts were assigned 5016 GO terms. 7687 unigenes have cluster of orthologous group (COG) classifications, and 4006 were assigned to 289 KEGG pathways. 5221 (19.64%) of transcripts were distributed to 81 known TF families. 11,004 SSRs were identified in 26.62% transcripts from which 2520 SSR markers were designed, which exhibited a nonrandom pattern of distribution. The most abundant was the trinucleotide repeats (AAG/CTT) followed by the dinucleotide repeats (AG/CT). SSRs were preferentially associated with the 5′UTRs with a predicted role in the regulation of gene expression. A PCR validation of a set of 48 primers revealed 97.9% successful amplification, and 76.6% of them showed polymorphism across different *Catharanthus* species as well as accessions of *C. roseus*.

2.3.2 Single nucleotide polymorphisms

A single nucleotide polymorphism (SNP) is a DNA sequence variation occurring when a single nucleotide in the genome differs between members of a biological species and paired chromosomes in an individual. SNPs can be utilized to determine genetic variation, construct genetic linkage maps, and associate with phenotypic variants, for example, traits for breeding in medicinal plants. SNP markers are increasingly becoming the marker system of choice, although they are less studied via the high-throughput transcriptome sequencing than SSRs. For robust supply of artemisinin from *A. annua*, Graham et al. (2010) identified 34,419 SNPs from DNA sequences contained in the five 454 transcriptome databases derived from the Artemis F1 hybrid material, the market leader for artemisinin production. This polymorphism was confirmed experimentally with 19 amplified fragment length polymorphism (AFLP) primer combinations that revealed 322 polymorphic markers. The SNP markers were used for fast-track breeding. Extensive genetic variation was used to build a comprehensive genetic map with nine linkage groups, which was then used to obtain a QTL map that accounts for a significant amount of the variation in key traits controlling artemisinin yield in the replicated field trials. Enrichment for positive QTLs in parents of new high-yielding hybrids verifies that it is possible to convert *A. annua* into a robust crop. Over 700,000 reads were generated in the 454 transcriptome sequencing of *Ribes nigrum* (Russell et al., 2011), from which 7000 SNPs were found. The selected SNPs were confirmed across various mapping populations and other selected *Ribes* species. SNP-based maps were built up from two blackcurrant mapping populations, incorporating 48% and 27% of analyzed SNPs, respectively. A high proportion of visually monomorphic SNPs were investigated further by quantitative trait mapping of theta score outputs from BeadStudio analysis, which enabled more SNPs to be placed on the two maps. SNPs were also found in the transcriptome datasets of *A. tridentata* (Bajgain et al., 2011), *C. annuum* (Lu et al., 2012), *S. columbaria* (Angeloni et al., 2011), and *Fraxinus* (Bai et al., 2011; Table 2.2). The use of high-throughput sequencing technology for the development of markers is superior to previously described methods, in both numbers of markers and their biological informativeness.

2.3.2.1 Root

Understanding the molecular basis of domestication can provide insights into the processes of rapid evolution and crop improvement. The processes of carrot domestication were demonstrated, and genes under selection were identified based on transcriptome analyses (Rong et al., 2014). The root transcriptomes of widely differing cultivated and wild carrots were sequenced. A method accounting for sequencing errors was introduced to optimize SNP discovery. 11,369 SNPs were identified, of which 622 (out of 1000 tested SNPs) were validated and used to genotype a large set of cultivated carrots, wild carrots, and other wild *Daucus carota* subspecies, primarily of European origin. Phylogenetic analysis indicated that eastern carrots may originate from Western Asia and western carrots may be selected from eastern carrots. Different wild *D. carota* subspecies may have contributed to the domestication of cultivated carrots. Genetic diversity was significantly reduced in western cultivars, probably through bottlenecks and selection. A high genetic diversity (>85% of the genetic diversity in wild populations) is retained in western cultivars. Model simulation indicated high and asymmetric gene flow from wild to cultivated carrots, spontaneously and/or by introgression breeding. Nevertheless, high genetic differentiation exists between cultivated and wild carrots (Fst 0.295) showing strong selection. Expression patterns differed radically for some genes between cultivated and wild carrot roots, which may be related to changes in root traits. The upregulation of water-channel-protein gene expression in cultivars might be involved in changing water content and transport in roots. The activated expression of carotenoid-binding protein genes in cultivars could be related to the high carotenoid accumulation in roots. The silencing of allergen-protein-like genes in cultivated carrot roots suggested strong human selection to reduce allergy, suggesting that regulatory changes of gene expressions may have played a predominant role in domestication. The reduction in genetic diversity in western cultivars due to domestication bottleneck/selection may have been offset by introgression from wild carrots. Differential gene expression patterns between cultivated and wild carrot roots may indicate strong selection for favorable cultivation traits.

2.3.2.2 Seed

In the whole genome sequencing, the genetic map provides an essential framework for accurate and efficient genome assembly and validation. A high-density genetic map was developed using RAD-Seq (restriction-site associated DNA sequencing) genotyping by sequencing (RAD-Seq GBS) and Illumina GoldenGate assays (Deokar et al., 2014), and the alignment of the current map with the kabuli chickpea genome assembly was examined. 51,632 genic SNPs were identified by 454 transcriptome sequencing of *Cicer arietinum* and *C. reticulatum* genotypes. An Illumina GoldenGate assay for 1536 SNPs was developed. 1519 SNPs were successfully assayed across 92 recombinant inbred lines (RILs), of which 761 SNPs were polymorphic between the two parents. The next-generation sequencing(NGS)-based GBS was applied to the same population generating 29,464 high-quality SNPs, which were clustered into 626 recombination bins based on common segregation patterns. Data from the two approaches were used for the construction of a genetic map using a population derived from an intraspecific cross. The map consisted of 1336 SNPs including 604 RAD recombination bins and 732 SNPs from Illumina GoldenGate assay. The map covered 653 cM of the chickpea genome with an average distance between adjacent markers of 0.5 cM. The alignment of the map with the CDC Frontier genome assembly revealed an overall conserved marker order; however, a few local inconsistencies within the *C. arietinum* pseudochromosome 1 (Ca1), Ca5, and Ca8 were detected. The map enabled the alignment of 215 unplaced scaffolds from the CDC Frontier draft genome assembly. The alignment also revealed varying degrees of recombination rates and hot spots across the chickpea genome.

2.3.2.3 Inflorescence

Brassica juncea (AABB) is an allotetraploid species containing genomes of *B. rapa* (AA) and *B. nigra* (BB). It is a major oilseed crop and medicinal plant in South Asia and China. *B. juncea* has two well-defined gene pools—Indian and east European. Hybrids between the two gene pools are heterotic for yield. A large number of qualitative and quantitative traits need to be introgressed from one gene pool to the other. SNPs are available from RNA-Seq-generated contigs and are useful for general mapping, fine mapping of selected regions, and comparative arrangement of gene blocks on *B. juncea* A and B genomes (Paritosh et al., 2014). RNA isolated from two lines of *B. juncea*—Varuna (Indian type) and Heera (east European type)—was sequenced using Illumina paired-end sequencing technology and assembled using the Velvet *de novo* program. A and B genome-specific contigs were identified in two steps. 135,693 SNPs were recorded in the assembled partial gene models of Varuna and Heera, 85,473 in the A genome, and 50,236 in the B. Using KASPar technology, 999 markers were added to an earlier intron polymorphism marker-based map of a *B. juncea* Varuna × Heera DH population. Many new gene blocks were identified in the B genome. A number of SNP markers covered single-copy homoeologues of the A and B genomes, and these were used to identify homoeologous blocks between the two genomes. Comparison of the block architecture of A and B genomes revealed extensive differences in gene block associations and block fragmentation patterns. Sufficient SNP markers are available for general- and specific-region fine mapping of crosses between lines of two diverse *B. juncea* gene pools, supporting the hypothesis that the two genomes evolved from independent hexaploid events.

2.4 Adaptation and plant development

2.4.1 Flower

Rheum nobile (Ta Huang) is an alpine plant with translucent bracts concealing the inflorescence, which produce a "glasshouse" effect promoting the development of fertile pollen grains in such conditions. Its rhizome and root are famous drugs in TCM and Tibetan medicine. The current understanding of the adaptation of such bracts to alpine environments mainly focuses on the phenotypic and physiological changes. The upper bract and the lower rosulate leaf from the same *R. nobile* stem were subject to RNA sequencing (Wang et al., 2014a), which identified candidate genes that may be involved in the alpine adaption of the translucent bract in "glasshouse" plants and illustrated the changes in gene expression underlying the adaptive and complex evolution of the bract phenotype. 174.2 million paired-end reads from each transcriptome were assembled into 25,249 unigenes. By comparing the gene expression profiles, 1063 and 786 genes upregulated in the upper bract and the lower leaf, respectively, were identified. Functional enrichment analyses of these genes recovered a number of important pathways, including flavonoid biosynthesis, mismatch repair, and photosynthesis-related pathways. These pathways are mainly involved in three types of functions: nine genes in the UV protective process, nine mismatch repair-related genes, and 88 genes associated with photosynthesis.

Pollen grains of *Lilium longiflorum* are a long-established model system for pollen germination and tube tip growth. Due to their size, protein content, and almost synchronous germination in synthetic media, they provide a simple system for physiological measurements and sufficient material for biochemical studies like protein purifications, enzyme assays, organelle isolation or determination of metabolites during germination, and pollen tube elongation. Despite recent progresses in molecular biology techniques, sequence information of expressed proteins or transcripts in lily pollen is still scarce. RNA-Seq was used to investigate the lily pollen transcriptome resulting in more than 50 million high-quality reads with a length of 90 bp (Lang et al., 2015). Sequenced transcripts were assembled and annotated and finally visualized with MapMan software tools and compared with other RNA-Seq or genome data including *Arabidopsis* pollen, *Lilium* vegetative tissues, and the *Amborella trichopoda* genome.

Seedlessness is a desirable character in lemons and other citrus species. Seedless fruit can be induced in many ways, including through self-incompatibility (SI). SI is widely used as an intraspecific reproductive barrier that prevents self-fertilization in flowering plants. "Xiangshui" lemon is an important seedless cultivar whose seedlessness has been caused by SI. Candidate genes associated with SI were identified using

RNA-Seq (Zhang et al., 2015b). 61,224 unigenes were obtained (average, 948 bp; N50 1457 bp), among which 47,260 unigenes were annotated by comparison to six public databases (Nr, Nt, Swiss-Prot, KEGG, COG, and GO). DEGs were identified by comparing the transcriptomes of non-, self-, and cross-pollinated stigmas with styles of the "Xiangshui" lemon. Several DEGs that might be associated with SI were identified, such as those involved in pollen tube growth, programmed cell death, signal transduction, and transcription. NADPH oxidase genes associated with apoptosis were highly upregulated in the self-pollinated transcriptome. The expression pattern of 12 genes was analyzed by qRT-PCR. A putative S-RNase gene was identified that had not been previously associated with self-pollen rejection in lemon or citrus.

RNA-Seq was used to study the response of the reproductive tissues of almond (*Prunus dulcis*) to frost stress (Mousavi et al., 2014). RNA sequencing resulted in more than 20 million reads from the anther and ovary of almond. Around 40,000 contigs were assembled and annotated *de novo* in each tissue. The profile of gene expression in ovary showed significant alterations in 5112 genes, whereas in anther, 6926 genes were affected by freezing stress. Around 2000 of these genes were shared in both ovary and anther libraries. GO indicated the involvement of DEGs, responding to freezing stress, in metabolic and cellular processes. qRT-PCR analysis verified the expression pattern of eight genes randomly selected from the DEGs. These results add to the limited available information on almond and Rosaceae.

Zantedeschia aethiopica is an evergreen perennial plant cultivated worldwide and commonly used for ornamental and medicinal purposes including the treatment of bacterial infections. RNA-Seq technology was used for transcriptome assembly and characterization to improve understanding of its biology (Cândido Ede et al., 2014). Following *Z. aethiopica* spathe tissue RNA extraction, high-throughput RNA sequencing was performed to obtain both abundant and rare transcript data. Functional profiling based on KEGG Orthology (KO) analysis highlighted contigs that were involved predominantly in genetic information (37%) and metabolism (34%) processes. Predicted proteins involved in the plant circadian system, hormone signal transduction, secondary metabolism, and basal immunity are highlighted. In silico screening of the transcriptome dataset for antimicrobial peptide(AMP)-encoding sequences identified three lipid transfer proteins (LTP) as potential AMPs involved in plant defense. Spathe predicted protein maps suggest that major plant efforts are expended in guaranteeing the maintenance of cell homeostasis, characterized by high investment in carbohydrate, amino acid, and energy metabolism and in genetic information.

2.4.2 Stem

Ramie (*Boehmeria nivea* L.) is one of the oldest crops in China and the second most important fiber crop in terms of area sown. Ramie root and leaf have therapeutic use in TCM. Ramie fiber, extracted from the plant bast, is important in the textile industry. A whole sequencing run was performed on the 454 GS FLX+platform using four separately pooled parts of ramie bast (Chen et al., 2014c), which generated 1,030,057 reads with an average length of 457 bp. Among the 58,369 unigenes (13,386 contigs and 44,983 isotigs) that were generated through *de novo* assembly, 780 were differentially expressed. Thirteen genes that belong to the cellulose synthase gene family (four), the expansin gene family (three), and the xyloglucan endotransglucosylase/hydrolase (XTH) gene family (six) were upregulated at the top part of the bast, which was in contrast to the other three parts, which indicated that the early stage (represented by the top part of the bast) might be important for the molecular regulation of ramie fiber development. Four of the 13 unigenes from the expansin (two) and XTH (two) families shared a coincident expression pattern during the whole growth season, implying that they were more relevant to ramie fiber development during the different seasons than the other genes.

Transcriptome sequencing and gene expression profiling of ramie suggest that Pol II and Pol III subunits and the genes of the galactose metabolism pathway had higher expressions in phloem compared with xylem (Chen et al., 2014b). The expression of fatty acid metabolism pathway genes is more abundant in xylem than phloem. High activities of RNA synthesis and galactose metabolism pathway guarantee fiber synthesis in phloem.

2.4.3 Seed

Sand rice (*Agriophyllum squarrosum*) is an annual desert plant adapted to mobile sand dunes in arid and semiarid regions of Central Asia. The sand rice seeds are used in TCM and Mongolian medicine and have excellent nutrition value. Sand rice is a potential food crop resilient to ongoing climate change. This species has undergone only little agronomic modifications through classical breeding in recent years. A deep transcriptomic sequencing of sand rice was performed, which uncovers 67,741 unigenes (Zhao et al., 2014a). Phylogenetic analysis based on 221 single-copy genes a showed close relationship between sand rice and the recently domesticated crop sugar beet. Transcriptomic comparisons showed a high level of global sequence conservation between these two species. Conservation of sand rice and sugar beet orthologs assigned to response to salt stress GO term suggests that sand rice is also a potential salt-tolerant plant. Sand rice is far more tolerant to high temperature. A set of genes likely relevant for resistance to heat stress was functionally annotated according to expression levels, sequence annotation, and comparisons corresponding to transcriptome profiling results in *Arabidopsis*. Future screening the genetic variation among different ecotypes and constructing a draft genome sequence will further facilitate agronomic trait improvement and final domestication of sand rice.

Coconut palm (*Cocos nucifera*) is a symbol of the tropics and a source of numerous edible and nonedible products of economic value. *De novo* transcript assembly from RNA-Seq data and analysis of gene expression in seed tissues (embryo and endosperm) and leaves of a dwarf coconut variety were performed (Huang et al., 2014b). Assembly of 10 Gb sequencing data for each tissue resulted in 58,211 total unigenes in embryo, 61,152 in endosperm, and 33,446 in leaf. Within each unigene pool, 24,857 could be annotated in embryo, 29,731 in endosperm, and 26,064 in leaf. A KEGG analysis identified 138, 138, and 139 pathways, respectively, in transcriptomes of embryo, endosperm, and leaf tissues. Coconut seeds are of the extraordinarily large size.

Homology searches were used to identify putative homologues of factors required for RNA-directed DNA methylation, which is important during coconut seed development, particularly in maturing endosperm.

The potential biodiesel plant castor bean (*Ricinus communis*) has been the focus for bioenergy research due to the availability of its genome that raises the bar for genome-wide studies claiming advances that impact the "genome-phenome challenge." Phytohormone ABA was applied as an exogenous factor for the improvement of storage reserve accumulation with a focus on the complex interaction of pathways associated with seed filling (Chandrasekaran et al., 2014). After the application of exogenous ABA, increased ABA levels were measured in the developing seeds cultured *in vitro* using the ELISA technique, and the content of major biomolecules (total lipids, sugars, and protein) was quantified. Exogenous ABA (10 μM) enhanced the accumulation of soluble sugar content (6.3%), followed by the deposition of total lipid content (4.9%). Differential gene expression was analyzed using Illumina RNA sequencing technology, and 2568 (1507 upregulated/1061 downregulated) DEGs were identified, which were involved in sugar metabolism (e.g., glucose 6-phosphate, fructose 1,6-bisphosphate, glycerol 3-phosphate, and pyruvate kinase), lipid biosynthesis (e.g., ACS, ACBP, GPAT2, GPAT3, FAD2, FAD3, SAD1, and DGAT1), storage protein synthesis (e.g., SGP1, zinc finger protein, RING-H2 protein, nodulin 55, and CYP), and ABA biosynthesis (e.g., NCED1, NCED3, and beta-carotene). Metabolite measurements supported by genes and pathway expression results provide new insights to understand the ABA signaling mechanism towards seed storage filling and contribute useful information for facilitating oilseed crop functional genomics on an aim for utilizing castor bean agricultural and bioenergy use.

2.4.4 Fruit

Fruit ripening is a complex, genetically programmed process that occurs in conjunction with the differentiation of chloroplasts into chromoplasts and involves changes to the organoleptic properties of the fruit. An integrative analysis of the transcriptome and proteome was performed to identify important regulators and pathways involved in fruit ripening in a spontaneous late-ripening mutant ("Fengwan" orange, *Citrus sinensis*) and its wild type ("Fengjie 72-1") (Wu et al., 2014a,b,c). At the transcript level, 628 genes showed a twofold or more expression difference between the mutant and wild type as detected by RNA sequencing. At the protein level, 130 proteins differed by 1.5-fold or more in their relative abundance, as indicated by iTRAQ (isobaric tags for relative and absolute quantitation) analysis. A comparison of the transcriptome and proteome data revealed some aspects of the regulation of metabolism during orange fruit ripening. First, a large number of differential genes were found to belong to the plant hormone pathways and cell wall-related metabolism. Second, there is a correlation between ripening-associated transcripts and sugar metabolites, suggesting the importance of these metabolic pathways during fruit ripening. Third, a number of genes showed inconsistency between the transcript and protein level, implying post-transcriptional events.

Fruit cracking has long been a topic of great concern for growers and researchers of *Litchi chinensis* Sonn. RNA-Seq was used for *de novo* assembly and characterization of the transcriptome of cracking pericarp of lychee (Li et al., 2014a). Comparative transcriptomic analyses were performed on noncracking and cracking fruits. Approximately 26 million and 29 million high-quality reads were obtained from the two groups of samples and were assembled into 46,641 unigenes with an average length of 993 bp. Four unigenes (LcAQP, PIP, NIP, and SIP) involved in water transport, five unigenes (e.g., LcKS, GA2ox, and GID1) involved in GA metabolism, 21 unigenes (e.g., LcCYP707A, GT, β-Glu, PP2C, ABI1, and ABI5) involved in ABA metabolism, 13 unigenes (e.g., LcTPC, Ca^{2+}/H^{+} exchanger, Ca^{2+}-ATPase, CDPK, and CBL) involved in Ca transport, and 24 unigenes (e.g., LcPG, EG, PE, EXP, β-Gal, and XET) involved in cell wall metabolism were identified as genes that are differentially expressed in cracked fruits compared with noncracked fruits. These results help to understand the molecular mechanisms behind fruit cracking in lychee and other fruits, especially Sapindaceae plants.

2.4.5 Leaf

Avicennia marina is a widely distributed mangrove species that thrives in high-salinity habitats. It plays a significant role in supporting the coastal ecosystem and holds unique potential for studying molecular mechanisms underlying ecological adaptation. Because of its numerous merits, including medicinal value, this species is facing increasing pressure of exploitation and deforestation. Illumina sequencing of an *A. marina* foliar cDNA library was performed to generate a transcriptome dataset for gene and marker discovery (Huang et al., 2014a,b). Forty million high-quality reads were assembled into 91,125 unigenes with a mean length of 463 bp, which covered most of the publicly available *A. marina* ESTs and greatly extended the repertoire of transcripts for this species. 54,497 and 32,637 unigenes were annotated based on homology to sequences in the NCBI nonredundant (nr) and the Swiss-Prot protein databases, respectively. Both GO analysis and KEGG pathway analysis revealed some transcriptomic signatures of stress adaptation for this halophytic species. An extraordinary amount of transcripts derived from fungal endophytes were detected, demonstrating the utility of transcriptome sequencing in surveying endophyte diversity without isolating them. 3423 candidate SSRs were identified from 3141 unigenes with a density of one SSR locus every 8.25 kb sequence. These data provide valuable resources for ecological, genetic, and evolutionary studies in *A. marina*.

Citrullus colocynthis is a very drought-tolerant medicinal species, closely related to watermelon (*C. lanatus* var. *lanatus*). mRNA Illumina sequencing technology and bioinformatic strategies were used to analyze the *C. colocynthis* leaf transcriptome under drought treatment (Wang et al., 2014a,b,c). Leaf samples at four different time points (0, 24, 36, or 48 h of withholding water) were used for RNA extraction and Illumina sequencing. qRT-PCR of several drought-responsive genes was performed to confirm the accuracy of RNA sequencing. 5038 full-length cDNAs were detected, with 2545 genes showing significant changes during drought stress. Principal component analysis (PCA) indicated that drought was the major contributing factor regulating transcriptome changes. Upregulation of many TFs, stress signaling factors, detoxification genes, and genes involved in phytohormone signaling and citrulline metabolism occurred under the water-deficit conditions. These data provide a valuable resource for the future functional analysis of candidate genes in defense of drought stress.

Digitalis purpurea produces cardiac glycosides and is well known in the treatment of heart failure. A strand-specific RNA-Seq library of mixed leaf, root, stem, and flower was constructed and sequenced using Illumina HiSeq platforms to characterize the transcriptome of *D. purpurea* with a focus on alternative splicing (AS) events and the effect of AS on protein domains (Wu et al., 2014a,b,c). *De novo* RNA-Seq assembly resulted in 48,475 genes. 3265 AS genes were detected, including 5408 AS events. Both GT and monooxygenase, which were involved in the biosynthesis of cardiac glycosides, are regulated by AS. 2422 AS events occurred in coding regions, and 959 AS events were located in the regions of 882 unique protein domains, which could affect protein function.

The perennial species *Rhazya stricta* grows in arid zones and carries out typical C3 photosynthesis under daily extremes of heat, light intensity, and low humidity. In order to identify processes attributable to its adaptation to this harsh environment, the foliar transcriptome of apical and mature leaves harvested from the field at three time periods of the same day was profiled (Yates et al., 2014). 28,018 full-length transcript sequences were recovered and 45.4% were differentially expressed throughout the day. Microarray experiments in *A. thaliana* and other desert species were compared to identify trends in circadian and stress response profiles between species. 34% of the DEGs were homologous to *Arabidopsis* circadian-regulated genes. Independent of circadian control, significant overlaps with *Arabidopsis* genes were observed only with heat and salinity/high light stress-responsive genes. Groups of DEGs common to other desert plant species were identified. There are protein families specific to *R. stricta* that were found to have diverged from their homologues in other species and were overexpressed at midday. Temporal profiling is essential to assess the significance of genes apparently responsive to abiotic stress. In *R. stricta*, the circadian clock is a major regulator of DEGs, even of those annotated as stress-responsive in other species, which may be an important feature of the adaptation of *R. stricta* to its extreme but predictable environment. The majority of DEGs were not circadian-regulated. Of these, some were common to other desert species, and others were distinct to *R. stricta*, suggesting that they are important for the adaptation of such plants to arid environments.

2.4.6 Root

Lupinus albus serves as model plant for the root-induced mobilization of sparingly soluble soil phosphates via the formation of cluster roots (CRs) that mediate the secretion of protons, citrate, phenolics, and acid phosphatases (APases). Next-generation sequencing (NGS) was employed to investigate the molecular mechanisms behind these complex adaptive responses at the transcriptome level (Wang et al., 2014a,b,c). Different stages of CR development, including the preemergent (PE), juvenile (JU), and mature (MA) stages, were compared. The primary metabolism underwent significant modifications during CR maturation, promoting the biosynthesis of organic acids, as had been deduced from physiological studies. Citrate catabolism was downregulated, associated with citrate accumulation in MA clusters. Upregulation of the phenylpropanoid pathway reflected the accumulation of phenolics. Specific transcript expression of ALMT and MATE transporter genes correlated with the exudation of citrate and flavonoids. The expression of transcripts related to nucleotide degradation and APases in MA clusters coincided with the remobilization and hydrolysis of organic phosphate resources. Hormone-related gene expression suggested a central role of ethylene during CR maturation, which was associated with the upregulation of the iron deficiency regulated network that mediates the ethylene-induced expression of Fe-deficiency responses in other species. Transcripts related to abscisic acid and jasmonic acid were upregulated in MA clusters, while auxin- and brassinosteroid-related genes and cytokinin receptors were most strongly expressed during CR initiation. Key regulations proposed by the RNA-Seq data were confirmed by qRT-PCR and physiological analyses.

The peanut (*Arachis hypogaea*), one of the most important oil legumes in the world, is heavily damaged by white grubs. Tissue-specific promoters are needed to incorporate insect resistance genes into the peanut by genetic transformation to control the subterranean pests. Transcriptome sequencing is the most effective way to analyze differential gene expression in this nonmodel species and contribute to promoter cloning (Geng et al., 2014). The transcriptomes of the roots, seeds, and leaves of the peanut were sequenced using Illumina technology. A simple digital expression profile was established based on a number of transcripts per million clean tags (TPM) from various tissues, from which 584 root-specific candidate transcript assembly contigs (TACs) and 316 seed-specific candidate TACs were identified. Among these candidate TACs, 55.3% were root-specific and 64.6% were seed-specific by semiquantitative RT-PCR analysis. The consistency of semiquantitative RT-PCR with the simple digital expression profile was correlated with the length and TPM value of TACs. Some root-specific TACs are involved in stress resistance and respond to auxin stimulus, while seed-specific candidate TACs are involved in embryo development, lipid storage, and long-chain fatty acid biosynthesis. One root-specific promoter was cloned and characterized. This high-yield screening system can be used for other nonmodel plants to explore tissue-specific or spatially specific promoters.

Stress acclimation is an effective mechanism that plants acquired for adaptation to dynamic environments. Cassava (*Manihot esculenta*), a major tropical crop and medicinal plant, can be tolerant to much lower temperature after chilling acclimation. The transcriptome and microRNAome of Cassava root and leaf were profiled and analyzed, using high-throughput deep sequencing, across the normal condition, a moderate chilling stress (14 °C), a harsh stress (4 °C) after chilling acclimation (14 °C), and a chilling shock from 24 to 4 °C (Zeng et al., 2014). Moderate stress and chilling shock triggered comparable degrees of transcriptional perturbation, and about two-thirds of DEGs reversed their expression from upregulation to downregulation or vice versa in response to harsh stress after experiencing moderate stress. MicroRNAs played important roles in the process of this massive genetic circuitry rewiring. Chilling acclimation helped the plant develop immunity to further harsh stress by exclusively inducing genes with a function for nutrient reservation, therefore providing protection, while chilling shock induced genes with a function for viral reproduction, therefore causing damage.

2.4.7 Turion

Higher plants exhibit a remarkable phenotypic plasticity to adapt to adverse environmental changes. The greater duckweed *Spirodela*, an aquatic plant, presents exceptional tolerance to cold winters through its dormant structure of turions in place of seeds. Abundant starch in turions permits them to sink and escape the freezing surface of waters. Due to their clonal propagation, they are the fastest growing biomass on earth, providing yet an untapped source for industrial applications (Wang et al., 2014d). Next-generation sequencing was used to examine the transcriptome of turion development triggered by exogenous ABA. 208 Genes showed more than a fourfold increase compared with 154 downregulated genes in developing turions. The analysis of upregulated DEGs in response to dormancy revealed an enriched interplay among various pathways: signal transduction, seed dehydration, carbohydrate and secondary metabolism, and senescence. The genes responsible for rapid growth and biomass accumulation through DNA assembly, protein synthesis, and carbon fixation are repressed. Three members of late-embryogenesis-abundant protein family are exclusively expressed during turion formation. Key genes in starch synthesis such as APS1, APL3, and GBSSI are highly expressed, which could artificially be reduced for redirecting carbon flow from photosynthesis to create a higher energy biomass.

2.5 Comparative transcriptomics and phylogeny

2.5.1 Bioinformatics

Hyb-Seq, the combination of target enrichment and genome skimming, allows simultaneous data collection for low-copy nuclear genes and high-copy genomic targets for plant systematics and evolution studies. Genome and transcriptome assemblies for milkweed (*Asclepias syriaca*) were used to design enrichment probes for 3385 exons from 768 genes (>1.6 Mb), followed by Illumina sequencing of enriched libraries (Weitemier et al., 2014). Hyb-Seq of 12 individuals (10 *Asclepias* species and two related genera) resulted in at least partial assembly of 92.6% of exons and 99.7% of genes and an average assembly length >2 Mb. Complete plastomes and nuclear ribosomal DNA cistrons were assembled using off-target reads. Phylogenomic analyses demonstrated signal conflict between genomes. The Hyb-Seq approach enables targeted sequencing of thousands of low-copy nuclear exons and flanking regions, as well as genome skimming of high-copy repeats and organellar genomes, to efficiently produce genome-scale datasets for phylogenomics.

Orthology inference is central to phylogenomic analyses. Phylogenomic datasets commonly include transcriptomes and low-coverage genomes that are incomplete and contain errors and isoforms, which could severely violate the underlying assumptions of orthology inference with existing heuristics (Yang and Smith, 2014). A procedure that uses phylogenies for both homology and orthology assignment has been presented.

The transcriptome is the readout of the genome. Identifying common features in it across distant species can reveal fundamental principles. The ENCODE and modENCODE consortia have generated large amounts of matched RNA sequencing data for human, worm, and fly (Gerstein et al., 2014). Uniform processing and comprehensive annotation of these data allow comparison across metazoan phyla, extending beyond earlier within-phylum transcriptome comparisons and revealing ancient, conserved features. By similar approach, coexpression modules shared across medicinal plants could be discovered, many of which could be enriched in developmental genes. Expression patterns can be used to align the development stages in distant species, and novel pairing could be revealed. The extent of the noncanonical, noncoding transcription of distant species could be compared. The gene expression levels, both coding and noncoding, could be quantitatively predicted from chromatin features at the promoter using a "universal model" based on a single set of organism-independent parameters. Phylotranscriptomics could be helpful in upgrading pharmacophylogenetics to pharmacophylogenomics (Hao et al., 2014).

2.5.2 Asterids of eudicot

The eggplant (*Solanum melongena*) and turkey berry (*S. torvum*), a wild ally of eggplant with promising multidisease resistance traits, have great economic, medicinal, and genetic importance. Comprehensive, high-quality *de novo* transcriptome assemblies of the two *Leptostemonum* clade species were built from short-read RNA sequencing data (Yang et al., 2014), from which 34,174 unigenes for eggplant and 38,185 unigenes for turkey berry were obtained. Functional annotations based on sequence similarity to known plant datasets revealed a distribution of functional categories for both species very similar to that of tomato. Comparison of eggplant, turkey berry, and another 11 plant proteomes resulted in 276 high-confidence single-copy orthologous groups, reasonable phylogenetic tree inferences (Figure 2.6), and reliable divergence time estimations. Eggplant and its wild *Leptostemonum* clade relative turkey berry split from each other in the Late Miocene, ~6.66 million years ago (mya), and that *Leptostemonum* split from the potato clade in the Middle Miocene, ~15.75 mya. 621 and 815 plant resistance genes were identified in eggplant and turkey berry, respectively, indicating the variation of disease resistance genes between them. These results provide a comprehensive transcriptome resource for two *Solanum* species and insight into their evolutionary history and biological characteristics. RNA-Seq is a fast, reliable, and cost-effective method for assessing genome evolution in nonmodel species.

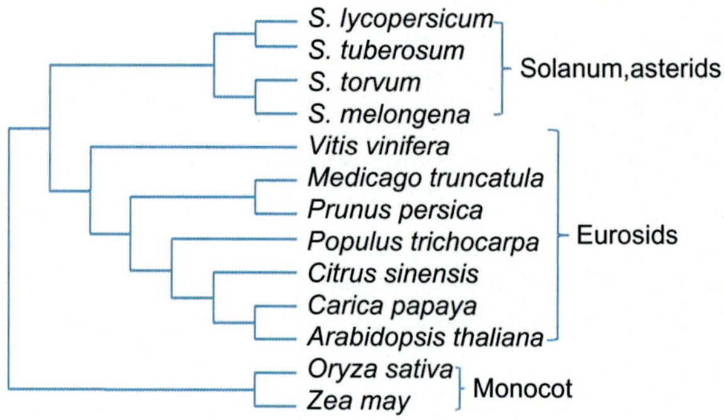

Figure 2.6 Maximum likelihood (ML) unrooted tree based on the second-codon positions of 276 single-copy genes. Quite different substitution rates are commonly observed for the three-codon positions (Yang et al., 2014), with the third position being especially variable as a result of the degeneracy of the genetic code. Third-position substitutions are likely to be saturated and may accumulate mutational bias, which may affect the accuracy of phylogeny estimations.

2.5.3 Rosids of eudicot

Previous phylogenetic studies of the grape family (Vitaceae) yielded poorly resolved deep relationships. NGS now offers access to protein-coding sequences easily, quickly, and cost-effectively. 417 orthologous single-copy nuclear genes were extracted from the transcriptomes of 15 species of the Vitaceae (Wen et al. 2013; Figure 2.7a), covering its phylogenetic diversity. The resulting transcriptome phylogeny provides robust support for the deep relationships (Figure 2.7b), showing the phylogenetic utility of transcriptome data for plants over a timescale at least since the mid-Cretaceous.

Large phylogenetic data matrices can be assembled accurately from even short (50 bp average) transcript sequences, so even nonoptimal plant material, for example, that was preserved in "RNAlater" could be used for transcriptome data generation. The transcriptomes can yield resolution for previously difficult to resolve radiations, especially at the family level. It should be kept in mind that the third positions of coding sequences and the 5′ and 3′ untranslated regions evolve relatively rapidly, which may affect the phylogenetic inference. Transcriptome data can effectively lead to the identification of truly single-copy transcripts and offer the conserved sequences necessary to generate primer pairs that can be used to amplify and sequence rapidly evolving intron regions for studies at and below the species level, generally without cloning steps. As the RNA-Seq approach is still relatively costly, extensive taxon sampling is not currently feasible. More samplings in other plant families will show what we can accomplish using transcriptomes and steps towards resolving the deep phylogenetic relationships of the plant family.

2.5.4 Other eudicots

Reconstructing species-level phylogenies for nonmodel groups remains a challenge. The use of a number of independent genes is required to resolve phylogenetic relationships, especially for groups displaying low polymorphism. Low-copy nuclear exons

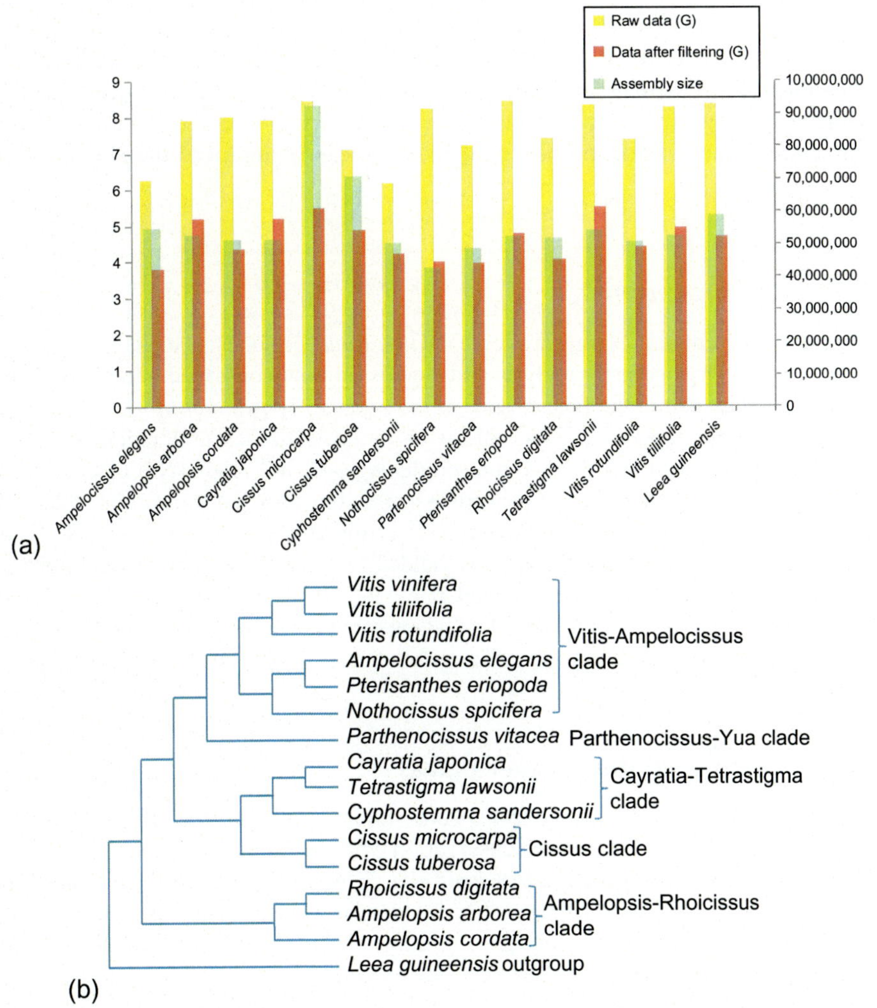

Figure 2.7 (a) Transcriptome sequencing of 15 species of Vitaceae and Leeaceae. Ninety bp paired-end DNA sequence reads from non-normalized cDNA libraries were obtained for each of the 15 species using an Illumina HiSeq 2000, which were assembled *de novo* into contigs for further analysis. (b) ML tree of Vitaceae using nucleotide sequences of 229 genes from the 15 transcriptomes of Vitaceae (Wen et al., 2013). The same topology was recovered from the 417 gene dataset.

and noncoding regions, such as 3′ untranslated regions (3′UTRs) or introns, constitute a potentially interesting source of nuclear (nr) DNA variation. New nr orthologous markers can be identified using both public nucleotide databases and transcriptomic data generated for the group of interest by using NGS technology (Tonnabel et al., 2014). To identify PCR primers for the genus *Leucadendron* (Proteaceae), a framework aimed at minimizing the probability of paralogy and maximizing polymorphism was adopted. When possible, the right-hand primer was anchored into the 3′UTR and the left-hand primer into the coding region. Seven new nr markers were selected, three of which included 3′UTRs. The sequenced 3′UTRs yielded higher polymorphism rates than the ITS (internal transcribed spacer) region did. Strong incongruences with the phylogenetic signal contained in the ITS region and the new designed markers were not found, and new markers significantly improved the phylogeny of *Leucadendron*.

2.5.5 Gymnosperm

The increasing amount of transcriptome data on *Pinus* provides an excellent resource for multigene phylogenetic analysis and studies on how conserved genes and functions are maintained in the face of species divergence. The *P. tabuliformis* transcriptome from a normalized cDNA library of multiple tissues and individuals was sequenced in a full 454 GS-FLX run, producing 911,302 reads (Niu et al., 2013). The high-quality overlapping ESTs were assembled into 46,584 transcripts, and more than 700 SSRs and 92,000 SNPs/InDels were characterized. Comparative analysis of the transcriptome of six conifer species yielded 191 orthologs, from which a phylogenetic tree, evolutionary patterns, and rates of gene diversion were inferred. 938 Fast-evolving sequences were identified, some of which perhaps evolved in response to positive selection and might be responsible for speciation in the *Pinus* lineage.

2.5.6 Moss and other lower plants

The *Marchantia polymorpha* (Di Qian) transcriptome was analyzed using TBLASTX to assess similarity and sequence conservation with the transcript datasets for related sequenced species, specifically *Physcomitrella*, *Selaginella*, and the algae *Chlamydomonas* (Sharma et al., 2014). 10,949 (23.5%) transcripts were shared among three species. Consistent with the phylogenetic position of these species with respect to *M. polymorpha*, *Physcomitrella* showed the greatest number of transcript hits (19,114), that is, bryophyte-specific genes, followed by *Selaginella*, which had hits with 5340 more transcripts than *Chlamydomonas*. The sequence conservation of *M. polymorpha* transcripts with proteomes from the model dicot *A. thaliana* and the model monocot *Oryza sativa* was analyzed using a BLASTX search with an E-value cut-off of 1e-05. The percentage of transcripts showing significant homology to monocots is higher than those showing homology to dicots, as 16,590 transcripts (35.6%) showed hits with rice proteins compared with 9801 transcripts (21%) that showed hits with proteins from dicots.

Using the OrthoMCL tool, 20,072 *M. polymorpha* transcripts were assigned to 8468 orthogroup MCL clusters. The proportion of orthogroups/protein number for each plant is different: 25.2% for *Physcomitrella*, 42.1% for *M. polymorpha*, 41.1% for *O. sativa*, 37.7% for *Arabidopsis*, and 57.25% for *Chlamydomonas*.

2.6 Digital gene expression

Two alternative approaches to gene expression analysis are sequence-based and have become increasingly popular due to rapid developments in sequencing technologies. The first is the direct sequencing of cDNA, termed RNA-Seq, and its utility in medicinal plant transcriptome research is summarized and exemplified above. The second approach is based on sequencing serial analysis of gene expression (SAGE) libraries (digital gene expression (DGE)), a method that generates a digital output proportional to the number of transcripts per mRNA (Anisimov, 2008). DGE, with the transcriptome dataset as the reference genome in the nonmodel species, has the benefit of not requiring presynthesized oligonucleotide probes (as in microarrays), allowing the direct enumeration of transcript molecules, that is, digital quantification, which is directly comparable across different experiments. The *T. cuspidata* transcriptome was determined with 454 pyrosequencing, and then DGE tag profiling was performed to compare gene expression in prospective CMCs (innately undifferentiated cambial meristematic cells) with that in DDCs (dedifferentiated plant cells) (Lee et al., 2010). It was found that 563 genes were differentially expressed in CMCs, with 296 upregulated and 267 downregulated. Both stress and biotic defense response genes were prominently overrepresented in CMCs, which is consistent with a stem cell identity for these cultured cells. Genes encoding key enzymes integral to the biosynthesis of paclitaxel were induced more strongly in CMCs than in DDCs. These stem cells may provide a cost-effective and environmentally friendly platform for sustainable production of a variety of important plant natural products. Hao et al. (2011) investigated the transcriptome difference of three *T. mairei* tissues using a tag-based DGE system. A sequencing depth of over 3.15 million tags per sample was obtained, and a large number of genes associated with tissue-specific functions and a taxane biosynthetic pathway were identified. The expression of the taxane biosynthetic genes is significantly higher in the root than in the leaf and the stem, while high activity of the taxane-producing pathway in the root was also revealed via metabolomics analyses. Roots have relatively simple chemical profiles and possess high yields of valuable taxanes such as paclitaxel, cephalomannine, 10-deacetylpaclitaxel, and 7-xylosyltaxanes. Rational exploitation of the taxanes that the root contains is of great help for alleviating the taxane supply crisis. Illumina DGE has also been used in a sweet orange red-flesh mutant (MT), compared with the wild type (WT), to study the molecular process regulating lycopene, a secondary metabolite, accumulation (Xu et al., 2009). Comparing a spontaneously early flowering trifoliate orange MT (*Poncirus trifoliata*) with the WT tree suggests that newly initiated transcription occurs in the MT and 2735 genes had more than twofold expression difference between the MT and the WT (Zhang et al., 2011). DGE analysis is extremely sensitive for detecting gene expression differences, revealing that in *S. grosvenorii*, watermelon, sweet orange, and *C. militaris*, a far greater number of genes are development stage-regulated than had previously been identified (Tang et al., 2011; Guo et al., 2011; Yu et al., 2012; Zheng et al., 2011). More importantly, DGE analysis could reveal the existence of the differentially expressed gene clusters that might be regulated by the same set of TFs, a phenomenon that is hard to be observed in microarray studies. How the knowledge regarding the function and regulation of the gene clusters could be used in the drug discovery and development warrants further study.

The salinization and alkalization of soil are widespread environmental problems, and alkaline-salt stress is more destructive than neutral salt stress. Understanding the mechanism of plant tolerance to saline-alkaline stress has become a major challenge. Gene expression profiling of flax (*Linum usitatissimum*) was analyzed under alkaline-salt stress (AS2), neutral salt stress (NSS), and alkaline stress (AS) by digital gene expression (Yu et al., 2014). Three-week-old flax seedlings were placed in 25 mM Na_2CO_3 (pH 11.6) (AS2), 50 mM NaCl (NSS), and NaOH (pH 11.6) (AS) for 18 h. There were 7736, 1566, and 454 DEGs in AS2, NSS, and AS compared with CK, respectively. The GO category gene enrichment analysis revealed that photosynthesis was particularly affected in AS2, carbohydrate metabolism was particularly affected in NSS, and the response to biotic stimulus was most affected in AS. The expression pattern of five categories of genes, including transcription factors, signaling transduction proteins, phytohormones, reactive oxygen species proteins, and transporters, was analyzed under three stress conditions. Some key regulatory gene families involved in abiotic stress, such as WRKY, MAPKKK, ABA, PrxR, and ion channels, were differentially expressed. Compared with NSS and AS, AS2 triggered more DEGs and special pathways, indicating that the mechanism of AS2 was more complex than NSS and AS. These data provide novel insights into the molecular mechanisms of plant saline-alkaline tolerance and offer a number of candidate genes as potential markers of tolerance to abiotic stress.

2.7 Conclusion

The application of the high-throughput sequencing in the medicinal plant transcriptome studies has just begun, although we have witnessed the huge success in this field. Many more important medicinal plants, especially those in supply crisis or near extinction, have to be sequenced at the transcriptomic level for biodiversity conservation and sustainable utilization. To date, all medicinal plant transcriptome studies used either Roche 454 pyrosequencing or Illumina high-throughput sequencing. As costs and capabilities of these technologies continue to improve, we are just beginning to see the possible utility of these platforms. The high-throughput sequencing methods can be grouped into three types: sequencing by synthesis (e.g., 454 and Illumina), sequencing by ligation (e.g., SOLiD), and single-molecule sequencing (e.g., Helicos and Pacific Biosciences). SOLiD and Helicos have not been used to sequence the medicinal plant transcriptome, and their competitiveness compared with 454 and Illumina is thus unknown. In addition, every cell in an individual plant may contain a genetic variation that affects cellular function, and such genomic heterogeneity could be relevant to the complex traits that might be useful in drug development and production. Hopefully, the single-molecule sequencing would be used to measure the single-cell transcriptome and the findings thereof would be of great value for future drug research and development.

References

Ajikumar, P.K., Xiao, W.H., Tyo, K.E., et al., 2010. Isoprenoid pathway optimization for Taxol precursor overproduction in *Escherichia coli*. Science 330, 70–74.

Angeloni, F., Wagemaker, C.A., Jetten, M.S., et al., 2011. De novo transcriptome characterization and development of genomic tools for Scabiosa columbaria L. using next-generation sequencing techniques. Mol. Ecol. Resour. 11, 662–674.

Anisimov, S.V., 2008. Serial Analysis of Gene Expression (SAGE): 13 years of application in research. Curr. Pharm. Biotechnol. 9, 338–350.

Bai, X., Rivera-Vega, L., Mamidala, P., et al., 2011. Transcriptomic signatures of ash (*Fraxinus* spp.) phloem. PLoS One 6, e16368.

Bajgain, P., Richardson, B.A., Price, J.C., et al., 2011. Transcriptome characterization and polymorphism detection between subspecies of big sagebrush (*Artemisia tridentata*). BMC Genomics 12, 370.

Barrero, R.A., Chapman, B., Yang, Y., et al., 2011. De novo assembly of *Euphorbia fischeriana* root transcriptome identifies prostratin pathway related genes. BMC Genomics 12, 600.

Beedle, A.M., Zamponi, G.W., 2001. Molecular determinants of opioid analgesia: modulation of presynaptic calcium channels. Drug Dev. Res. 54, 118–128.

Bertalovitz, A.C., Ahn, K.H., Kendall, D.A., 2010. Ligand binding sensitivity of the extracellular loop two of the cannabinoid receptor 1. Drug Dev. Res. 71, 404–411.

Brown, A.P., Kroon, J.T., Swarbreck, D., et al., 2012. Tissue-specific whole transcriptome sequencing in castor, directed at understanding triacylglycerol lipid biosynthetic pathways. PLoS One 7, e30100.

Cândido Ede, S., Fernandes Gda, R., de Alencar, S.A., et al., 2014. Shedding some light over the floral metabolism by arum lily (*Zantedeschia aethiopica*) spathe de novo transcriptome assembly. PLoS One 9 (3), e90487.

Chandrasekaran, U., Xu, W., Liu, A., 2014. Transcriptome profiling identifies ABA mediated regulatory changes towards storage filling in developing seeds of castor bean (*Ricinus communis* L.). Cell Biosci. 4, 33.

Chen, S., Zhou, R., Huang, Y., et al., 2011a. Transcriptome sequencing of a highly salt tolerant mangrove species *Sonneratia alba* using Illumina platform. Mar. Genomics 4, 129–136.

Chen, S.L., Luo, H., Li, Y., et al., 2011b. 454 EST analysis detects genes putatively involved in ginsenoside biosynthesis in *Panax ginseng*. Plant Cell Rep. 30, 1593–1601.

Chen, J., Hou, K., Qin, P., et al., 2014a. RNA-Seq for gene identification and transcript profiling of three *Stevia rebaudiana* genotypes. BMC Genomics 15, 571.

Chen, J., Liu, F., Tang, Y., et al., 2014b. Transcriptome sequencing and profiling of expressed genes in phloem and xylem of ramie (*Boehmeria nivea* L. Gaud). PLoS One 9 (10), e110623.

Chen, J., Pei, Z., Dai, L., et al., 2014c. Transcriptome profiling using pyrosequencing shows genes associated with bast fiber development in ramie (*Boehmeria nivea* L.). BMC Genomics 15 (1), 919.

Clark, S.M., Vaitheeswaran, V., Ambrose, S.J., et al., 2013. Transcriptome analysis of bitter acid biosynthesis and precursor pathways in hop (*Humulus lupulus*). BMC Plant Biol. 13, 12.

Deokar, A.A., Ramsay, L., Sharpe, A.G., et al., 2014. Genome wide SNP identification in chickpea for use in development of a high density genetic map and improvement of chickpea reference genome assembly. BMC Genomics 15, 708.

Desgagné-Penix, I., Khan, M.F., Schriemer, D.C., et al., 2010. Integration of deep transcriptome and proteome analyses reveals the components of alkaloid metabolism in opium poppy cell cultures. BMC Plant Biol. 10, 252.

Gai, S., Zhang, Y., Mu, P., et al., 2012. Transcriptome analysis of tree peony during chilling requirement fulfillment: assembling, annotation and markers discovering. Gene 497 (2), 256–262.

Gao, J., Yu, X., Ma, F., Li, J., 2014a. RNA-seq analysis of transcriptome and glucosinolate metabolism in seeds and sprouts of broccoli (*Brassica oleracea var. italic*). PLoS One 9 (2), e88804.

Gao, W., Sun, H.X., Xiao, H., et al., 2014b. Combining metabolomics and transcriptomics to characterize tanshinone biosynthesis in *Salvia miltiorrhiza*. BMC Genomics 15, 73.

Geng, L., Duan, X., Liang, C., et al., 2014. Mining tissue-specific contigs from peanut (*Arachis hypogaea* L.) for promoter cloning by deep transcriptome sequencing. Plant Cell Physiol. 55 (10), 1793–1801.

Gerstein, M.B., Rozowsky, J., Yan, K.K., 2014. Comparative analysis of the transcriptome across distant species. Nature 512 (7515), 445–448.

Graham, I.A., Besser, K., Blumer, S., et al., 2010. The genetic map of *Artemisia annua* L. identifies loci affecting yield of the antimalarial drug artemisinin. Science 327, 328–331.

Guo, S., Liu, J., Zheng, Y., et al., 2011. Characterization of transcriptome dynamics during watermelon fruit development: sequencing, assembly, annotation and gene expression profiles. BMC Genomics 12, 454.

Guo, Q., Ma, X., Wei, S., et al., 2014. De novo transcriptome sequencing and digital gene expression analysis predict biosynthetic pathway of rhynchophylline and isorhynchophylline from *Uncaria rhynchophylla*, a non-model plant with potent anti-alzheimer's properties. BMC Genomics 15, 676.

Hao, D.C., Ge, G.B., Xiao, P.G., et al., 2011. The first insight into the tissue specific taxus transcriptome via Illumina second generation sequencing. PLoS One 6 (6), e21220.

Hao, D.C., Ma, P., Mu, J., Chen, S., Xiao, P.G., et al., 2012. De novo characterization of the root transcriptome of a traditional Chinese medicinal plant *Polygonum cuspidatum*. Sci. China Life Sci. 55 (5), 452–466.

Hao, D.C., Chen, S.L., Osbourn, A., et al., 2015. Temporal transcriptome changes induced by methyl jasmonate in *Salvia sclarea*. Gene 558, 41–53. http://dx.doi.org/10.1016/j.gene.2014.12.043.

Hao, D.C., Xiao, P.G., Liu, M., et al., 2014. Pharmaphylogeny vs. pharmacophylogenomics: molecular phylogeny, evolution and drug discovery. Acta Pharm. Sin. 49 (10), 1387–1394.

Hu, Y., Liu, P., Guo, D.H., et al., 2008. Behavioral and biochemical effects of Kaixin-San, a traditional Chinese medicinal empirical formula. Drug Dev. Res. 69, 267–271.

Huang, J., Lu, X., Zhang, W., et al., 2014a. Transcriptome sequencing and analysis of leaf tissue of *Avicennia marina* using the Illumina platform. PLoS One 9 (9), e108785.

Huang, Y.Y., Lee, C.P., Fu, J.L., et al., 2014b. De novo transcriptome sequence assembly from coconut leaves and seeds with a focus on factors involved in RNA-directed DNA methylation. G3 (Bethesda). 4, 2147–2157. http://dx.doi.org/10.1534/g3.114.013409.

Jain, A., Chaudhary, S., Sharma, P.C., 2014. Mining of microsatellites using next generation sequencing of seabuckthorn (Hippophae rhamnoides L.) transcriptome. Physiol. Mol. Biol. Plants 20 (1), 115–123.

Jiang, N.H., Zhang, G.H., Zhang, J.J., et al., 2014. Analysis of the transcriptome of *Erigeron breviscapus* uncovers putative scutellarin and chlorogenic acids biosynthetic genes and genetic markers. PLoS One 9 (6), e100357.

Kageyama, Y., 2008. Development of drugs against hormone-refractory prostate cancer. Drug Dev. Res. 69, 431–450.

Kalra, S., Puniya, B.L., Kulshreshtha, D., et al., 2013. De Novo transcriptome sequencing reveals important molecular networks and metabolic pathways of the plant, *Chlorophytum borivilianum*. PLoS One 8 (12), e83336.

Kilgore, M.B., Augustin, M.M., Starks, C.M., et al., 2014. Cloning and characterization of a norbelladine 4′-*O*-methyltransferase involved in the biosynthesis of the Alzheimer's drug galanthamine in *Narcissus sp. aff. pseudonarcissus*. PLoS One 9 (7), e103223.

Kumar, S., Shah, N., Garg, V., Bhatia, S., 2014. Large scale in-silico identification and characterization of simple sequence repeats (SSRs) from de novo assembled transcriptome of *Catharanthus roseus* (L.) G. Don. Plant Cell Rep. 33 (6), 905–918.

Kumari, A., Singh, H.R., Jha, A., et al., 2014. Transcriptome sequencing of rhizome tissue of *Sinopodophyllum hexandrum* at two temperatures. BMC Genomics 15, 871.

Lang, V., Usadel, B., Obermeyer, G., 2015. De novo sequencing and analysis of the lily pollen transcriptome: an open access data source for an orphan plant species. Plant Mol Biol. 87 (1–2), 69–80.

Lee, E.K., Jin, Y.W., Park, J.H., et al., 2010. Cultured cambial meristematic cells as a source of plant natural products. Nat. Biotechnol. 28, 1213–1217.

Legrand, S., Valot, N., Nicolé, F., et al., 2010. One-step identification of conserved miRNAs, their targets, potential transcription factors and effector genes of complete secondary metabolism pathways after 454 pyrosequencing of calyx cDNAs from the Labiate *Salvia sclarea* L. Gene 450, 55–62.

Li, Y., Luo, H.M., Sun, C., et al., 2010a. EST analysis reveals putative genes involved in glycyrrhizin biosynthesis. BMC Genomics 11, 268.

Li, Y., Sun, C., Luo, H.M., et al., 2010b. Transcriptome characterization for *Salvia miltiorrhiza* using 454 GS FLX. Acta Pharm. Sin. 45, 524–529.

Li, Q., He, F., Zhu, B.Q., et al., 2014a. Comparison of distinct transcriptional expression patterns of flavonoid biosynthesis in Cabernet Sauvignon grapes from east and west China. Plant Physiol. Biochem. 84C, 45–56.

Li, W.C., Wu, J.Y., Zhang, H.N., et al., 2014b. De Novo assembly and characterization of pericarp transcriptome and identification of candidate genes mediating fruit cracking in *Litchi chinensis* Sonn. Int. J. Mol. Sci. 15 (10), 17667–17685.

Lin, X., Zhang, J., Li, Y., et al., 2011. Functional genomics of a living fossil tree *Ginkgo* based on next generation sequencing technology. Physiol. Plant. 143, 207–218.

Lu, F.H., Cho, M.C., Park, Y.J., 2012. Transcriptome profiling and molecular marker discovery in red pepper, *Capsicum annuum* L. TF68. Mol. Biol. Rep. 39, 3327–3335.

Luo, H., Li, Y., Sun, C., et al., 2010. Comparison of 454-ESTs from *Huperzia serrata* and *Phlegmariurus carinatus* reveals putative genes involved in lycopodium alkaloid biosynthesis and developmental regulation. BMC Plant Biol. 10, 209.

Luo, H., Sun, C., Sun, Y., et al., 2011. Analysis of the transcriptome of *Panax notoginseng* root uncovers putative triterpene saponin-biosynthetic genes and genetic markers. BMC Genomics 12 (S5), S5.

Luo, C., Zhang, Q.L., Luo, Z.R., 2014. Development and characterization of novel transcriptome-derived microsatellites for genetic analysis of persimmon. Genet. Mol. Res. 13 (2), 3013–3024.

Marques, J.V., Dalisay, D.S., Yang, H., et al., 2014. A multi-omics strategy resolves the elusive nature of alkaloids in *Podophyllum* species. Mol. Biosyst. 10 (11), 2838–2849.

Maude, R.J., Woodrow, C.J., White, L.J., 2010. Artemisinin antimalarials: preserving the "magic bullet" Drug Dev. Res. 71, 12–19.

Morant, M., Bak, S., Moller, B.L., Werck-Reichhart, D., 2003. Plant cytochromes P450: tools for pharmacology, plant protection and phytoremediation. Curr. Opin. Biotechnol. 14, 151–162.

Mousavi, S., Alisoltani, A., Shiran, B., et al., 2014. De novo transcriptome assembly and comparative analysis of differentially expressed genes in *Prunus dulcis* Mill. in response to freezing stress. PLoS One 9 (8), e104541.

Näätsaari, L., Krainer, F.W., Schubert, M., et al., 2014. Peroxidase gene discovery from the horseradish transcriptome. BMC Genomics 15, 227.

Niu, S.H., Li, Z.X., Yuan, H.W., et al., 2013. Transcriptome characterisation of *Pinus tabuliformis* and evolution of genes in the *Pinus* phylogeny. BMC Genomics 14, 263.

Ono, N.N., Britton, M.T., Fass, J.N., et al., 2011. Exploring the transcriptome landscape of pomegranate fruit peel for natural product biosynthetic gene and SSR marker discovery. J. Integr. Plant Biol. 53, 800–813.

Paritosh, K., Gupta, V., Yadava, S.K., et al., 2014. RNA-seq based SNPs for mapping in *Brassica juncea* (AABB): synteny analysis between the two constituent genomes A (from *B. rapa*) and B (from *B. nigra*) shows highly divergent gene block arrangement and unique block fragmentation patterns. BMC Genomics 15, 396.

Qiang, W., Wang, Y.X., Zhang, Q.Z., et al., 2014. Expression pattern of genes involved in tropane alkaloids biosynthesis and tropane alkaloids accumulation in *Atropa belladonna*. Zhongguo Zhong Yao Za Zhi 39 (1), 52–58.

Rastogi, S., Meena, S., Bhattacharya, A., et al., 2014. De novo sequencing and comparative analysis of holy and sweet basil transcriptomes. BMC Genomics 15, 588.

Rong, J., Lammers, Y., Strasburg, J.L., et al., 2014. New insights into domestication of carrot from root transcriptome analyses. BMC Genomics 15, 895.

Russell, J.R., Bayer, M., Booth, C., et al., 2011. Identification, utilisation and mapping of novel transcriptome-based markers from blackcurrant (*Ribes nigrum*). BMC Plant Biol. 11, 147.

Sharma, N., Jung, C.H., Bhalla, P.L., Singh, M.B., 2014. RNA sequencing analysis of the gametophyte transcriptome from the liverwort, *Marchantia polymorpha*. PLoS One 9 (5), e97497.

Shi, C.Y., Yang, H., Wei, C.L., et al., 2011. Deep sequencing of the *Camellia sinensis* transcriptome revealed candidate genes for major metabolic pathways of tea-specific compounds. BMC Genomics 12, 131.

Shi, S.G., Yang, M., Zhang, M., et al., 2014. Genome-wide transcriptome analysis of genes involved in flavonoid biosynthesis between red and white strains of *Magnolia sprengeri* pamp. BMC Genomics 15, 706.

Singh, K., Foley, R.C., Onate-Sanchez, L., 2002. Transcription factors in plant defense and stress responses. Curr. Opin. Plant Biol. 5, 430–436.

Spyropoulou, E.A., Haring, M.A., Schuurink, R.C., 2014. RNA sequencing on *Solanum lycopersicum* trichomes identifies transcription factors that activate terpene synthase promoters. BMC Genomics 15, 402.

Sui, C., Zhang, J., Wei, J., et al., 2011. Transcriptome analysis of *Bupleurum chinense* focusing on genes involved in the biosynthesis of saikosaponins. BMC Genomics 12, 539.

Sun, C., Li, Y., Wu, Q., et al., 2010. De novo sequencing and analysis of the American ginseng root transcriptome using a GS FLX Titanium platform to discover putative genes involved in ginsenoside biosynthesis. BMC Genomics 11, 262.

Sun, Y., Luo, H., Li, Y., et al., 2011. Pyrosequencing of the *Camptotheca acuminata* transcriptome reveals putative genes involved in camptothecin biosynthesis and transport. BMC Genomics 12, 533.

Tang, Q., Ma, X., Mo, C., et al., 2011. An efficient approach to finding *Siraitia grosvenorii* triterpene biosynthetic genes by RNA-seq and digital gene expression analysis. BMC Genomics 12, 343.

Tang, X., Xiao, Y., Lv, T., et al., 2014. High-throughput sequencing and de novo assembly of the *Isatis indigotica* transcriptome. PLoS One 9 (9), e102963.

Tonnabel, J., Olivieri, I., Mignot, A., et al., 2014. Developing nuclear DNA phylogenetic markers in the angiosperm genus *Leucadendron* (Proteaceae): a next-generation sequencing transcriptomic approach. Mol. Phylogenet. Evol. 70, 37–46.

Upadhyay, S., Phukan, U.J., Mishra, S., Shukla, R.K., 2014. De novo leaf and root transcriptome analysis identified novel genes involved in steroidal sapogenin biosynthesis in *Asparagus racemosus*. BMC Genomics 15, 746.

van Bakel, H., Stout, J.M., Cote, A.G., et al., 2011. The draft genome and transcriptome of *Cannabis sativa*. Genome Biol. 12, R102.

Verma, M., Ghangal, R., Sharma, R., et al., 2014. Transcriptome analysis of *Catharanthus roseus* for gene discovery and expression profiling. PLoS One 9 (7), e103583.

Wang, W., Wang, Y., Zhang, Q., et al., 2009. Global characterization of *Artemisia annua* glandular trichome transcriptome using 454 pyrosequencing. BMC Genomics 10, 465.

Wang, Z., Fang, B., Chen, J., et al., 2010. De novo assembly and characterization of root transcriptome using Illumina paired-end sequencing and development of cSSR markers in sweet potato (Ipomoea batatas). BMC Genomics 11, 726.

Wang, L., Zhou, H., Han, J., et al., 2014a. Genome-scale transcriptome analysis of the alpine "glasshouse" plant *Rheum nobile* (Polygonaceae) with special translucent bracts. PLoS One 9 (10), e110712.

Wang, W., Wu, Y., Messing, J., 2014b. RNA-Seq transcriptome analysis of *Spirodela* dormancy without reproduction. BMC Genomics 15, 60.

Wang, Z., Hu, H., Goertzen, L.R., et al., 2014c. Analysis of the *Citrullus colocynthis* transcriptome during water deficit stress. PLoS One 9 (8), e104657.

Wang, Z., Straub, D., Yang, H., et al., 2014d. The regulatory network of cluster-root function and development in phosphate-deficient white lupin (*Lupinus albus*) identified by transcriptome sequencing. Physiol. Plant. 151 (3), 323–338.

Wei, W., Qi, X., Wang, L., et al., 2011. Characterization of the sesame (*Sesamum indicum* L.) global transcriptome using Illumina paired-end sequencing and development of EST-SSR markers. BMC Genomics 12, 451.

Weitemier, K., Straub, S.C., Cronn, R.C., et al., 2014. Hyb-Seq: combining target enrichment and genome skimming for plant phylogenomics. Appl. Plant Sci. 2 (9), 1400042.

Wen, J., Xiong, Z., Nie, Z.L., et al., 2013. Transcriptome sequences resolve deep relationships of the grape family. PLoS One 8 (9), e74394.

Wenping, H., Yuan, Z., Jie, S., et al., 2011. De novo transcriptome sequencing in *Salvia miltiorrhiza* to identify genes involved in the biosynthesis of active ingredients. Genomics 98, 272–279.

Wu, Q., Sun, C., Luo, H., et al., 2011. Transcriptome analysis of *Taxus cuspidata* needles based on 454 pyrosequencing. Planta Med. 77, 394–400.

Wu, B., Li, Y., Yan, H., et al., 2012. Comprehensive transcriptome analysis reveals novel genes involved in cardiac glycoside biosynthesis and mlncRNAs associated with secondary metabolism and stress response in *Digitalis purpurea*. BMC Genomics 13, 15.

Wu, B., Suo, F., Lei, W., Gu, L., 2014a. Comprehensive analysis of alternative splicing in *Digitalis purpurea* by strand-specific RNA-Seq. PLoS One 9 (8), e106001.

Wu, J., Xu, Z., Zhang, Y., et al., 2014b. An integrative analysis of the transcriptome and proteome of the pulp of a spontaneous late-ripening sweet orange mutant and its wild type improves our understanding of fruit ripening in citrus. J. Exp. Bot. 65 (6), 1651–1671.

Wu, Z.J., Li, X.H., Liu, Z.W., et al., 2014c. De novo assembly and transcriptome characterization: novel insights into catechins biosynthesis in *Camellia sinensis*. BMC Plant Biol. 14 (1), 277.

Xu, Q., Yu, K., Zhu, A., et al., 2009. Comparative transcripts profiling reveals new insight into molecular processes regulating lycopene accumulation in a sweet orange (*Citrus sinensis*) red-flesh mutant. BMC Genomics 10, 540.

Yang, Y., Smith, S.A., 2014. Orthology inference in nonmodel organisms using transcriptomes and low-coverage genomes: improving accuracy and matrix occupancy for phylogenomics. Mol. Biol. Evol. 31 (11), 3081–3092.

Yang, X., Cheng, Y.F., Deng, C., et al., 2014. Comparative transcriptome analysis of eggplant (*Solanum melongena* L.) and turkey berry (*Solanum torvum* Sw.): phylogenomics and disease resistance analysis. BMC Genomics 15, 412.

Yates, S.A., Chernukhin, I., Alvarez-Fernandez, R., et al., 2014. The temporal foliar transcriptome of the perennial C3 desert plant *Rhazya stricta* in its natural environment. BMC Plant Biol. 14, 2.

Yu, K., Xu, Q., Da, X., et al., 2012. Transcriptome changes during fruit development and ripening of sweet orange (*Citrus sinensis*). BMC Genomics 13, 10.

Yu, Y., Huang, W., Chen, H., et al., 2014. Identification of differentially expressed genes in flax (Linum usitatissimum L.) under saline-alkaline stress by digital gene expression. Gene 549 (1), 113–122.

Zeng, S., Xiao, G., Guo, J., et al., 2010. Development of an EST dataset and characterization of EST-SSRs in a traditional Chinese medicinal plant, *Epimedium sagittatum* (Sieb. Et Zucc.) Maxim. BMC Genomics 11, 94.

Zeng, C., Chen, Z., Xia, J., et al., 2014. Chilling acclimation provides immunity to stress by altering regulatory networks and inducing genes with protective functions in cassava. BMC Plant Biol. 14, 207.

Zhang, J.Z., Ai, X.Y., Sun, L.M., et al., 2011. Transcriptome profile analysis of flowering molecular processes of early flowering trifoliate orange mutant and the wild-type [*Poncirus trifoliata* (L.) Raf.] by massively parallel signature sequencing. BMC Genomics 12, 63.

Zhang, L., Jia, B., Tan, X., et al., 2014. Fatty acid profile and unigene-derived simple sequence repeat markers in tung tree (*Vernicia fordii*). PLoS One 9 (8), e105298.

Zhang, M.F., Jiang, L.M., Zhang, D.M., Jia, G.X., 2015a. De novo transcriptome characterization of *Lilium* 'Sorbonne' and key enzymes related to the flavonoid biosynthesis. Mol. Genet. Genomics 290, 399–412.

Zhang, S., Ding, F., He, X., et al., 2015b. Characterization of the 'Xiangshui' lemon transcriptome by de novo assembly to discover genes associated with self-incompatibility. Mol. Genet. Genomics 290, 365–375.

Zhao, D., Jiang, Y., Ning, C., et al., 2014a. Transcriptome sequencing of a chimaera reveals coordinated expression of anthocyanin biosynthetic genes mediating yellow formation in herbaceous peony (*Paeonia lactiflora* Pall.). BMC Genomics 15, 689.

Zhao, P., Capella-Gutíerrez, S., Shi, Y., et al., 2014b. Transcriptomic analysis of a psammophyte food crop, sand rice (*Agriophyllum squarrosum*) and identification of candidate genes essential for sand dune adaptation. BMC Genomics 15 (1), 872.

Zheng, P., Xia, Y., Xiao, G., et al., 2011. Genome sequence of the insect pathogenic fungus *Cordyceps militaris*, a valued traditional Chinese medicine. Genome Biol. 12, R116.

Taxus medicinal resources: a comprehensive study

3

3.1 Introduction

The genus *Taxus*, commonly known as yew, a gymnosperm in the family Taxaceae, includes 24 species (Spjut, 2007) of small trees or shrubs that are a major source of medicinal compounds, particularly the anticancer/cardiovascular drug Taxol (paclitaxel). Taxol was first isolated from *T. brevifolia* Nutt. (Wani et al., 1971), a species found in the Pacific Northwest of North America (Spjut, 2007). Native species in China include the more widely distributed *T. biternata* Spjut, *T. celebica* (Warb.) H. L. Li, *T. contorta* Griff., *T. chinensis* (Pilg.) Rehder, *T. kingstonii* Spjut, *T. mairei* (Lemée & Lév) S. Y. Hu ex T. S. Liu, *T. sumatrana* (Miq.) de Laub., *T. umbraculifera* (Sieb. ex Endl.) C. Lawson, *T. wallichiana* Zucc., *T. yunnanensis* W. C. Cheng & L. K. Fu, and other localized species (Spjut, 2007; www.worldbotanical.com/nomenclature), all of which have been referred to as Chinese yew under a limited species concept for yew in China (e.g., Rehder, 1940). Cultivated species such as *T.* × *media*, which include many cultivars, are largely hybrids involving wild species in Japan that include *T. cuspidata* Sieb. & Zucc. and European, Southwest Asian, and North African species, *T. baccata* L., *T. fastigiata* Lindley, and *T. recurvata* Hort. ex C. Lawson, respectively (Spjut, www.worldbotanical.com/introduction). These species and their cultivars provide economic and ecological benefits to China.

A major concern is the attachment of the scientific findings on the genus *Taxus* to species names that lack standards. It continues to be a problem in the pharmacological literature as well as in the botanical taxonomy literature. Botanists in China generally followed Cheng and Fu's (1978) taxonomy, which *Taxus yunnanensis* was used for what others today call *T. wallichiana*. Botanists outside China would refer to *T. yunnanensis* and *T. mairei* as *T. sumatrana*. Before 1978 they may have also been called *T. chinensis*. The West Himalayan yew has been referred to as *T. baccata*, *T. wallichiana*, *T. fuana*, and *T. contorta*. Without having knowledge of what sources were used for the species determination, one really does not know for sure what species of *Taxus* are really being reported. Recently, Richard W. Spjut of World Botanical Associates saw an image on Wikipedia reported to be *T. sumatrana* from a botanical garden. Upon checking the source, and a database on places where *T. sumatrana* was reportedly being cultivated, about 10 or more botanical gardens, the source of the garden material was several places in The Philippines. The image, somewhat blurry, appeared to be *T. obscura* which Spjut classifies in the Chinensis Subgroup. Thus in a

Medicinal plants: chemistry, biology and omics. http://dx.doi.org/10.1016/B978-0-08-100085-4.00003-7

DNA phylogeny study it might not group with *T. sumatrana* or *T. mairei*, or *T. kingstonii* (personal communication).

Farjon has often been cited as an authority based on his checklist prior to 2010, and his 2010 handbook on conifers. However, his classification has changed in the last decade without really providing a taxonomic justification for his decisions. For example, in one checklist he considered *T. fuana* as endemic to the type locality, no doubt based on Spjut's annotation of specimens at Kew in 1997 (personal communication with Spjut), and since Spjut did not see a *Taxus contorta* at Kew from China at that time, Farjon's conclusion in a 1998 and 2001 checklists were unjustified. Other taxonomic reviews have been limited to a particular geographic region, while not considering the earliest applicable name that may actually apply from a yew collected outside the region. This is what happened with *T. yunnanensis*, *T. fuana*, *T. celebica*, *T. mairei*. Also, conifer taxonomists give too much weight to defining species by geographical boundaries.

Historically, the Chinese yew has additional names (Li et al., 2011) discovered in the excavation and classification of the historical documents in the Siku Quanshu (Guy, 1987), the largest collection of books in Chinese history from which the property and flavor, the channel tropism, and the effects and functions of the Chinese yew were summarized. The use of *Taxus* for treating cancer has been traced back to the Tang dynasty—nearly 1200 years earlier than what has been reported for species in western countries (Li et al., 2011). Nevertheless, western science research on *Taxus* has led to a greater understanding of its chemistry (e.g., Wani et al., 1971; Kong et al., 2007; Chen and Liang, 2008; Wu et al., 2010; Feng et al., 2011; Watchueng et al., 2011; Figures 3.1 and 3.2), biotechnological applications (Yukimune et al., 1996; Gao et al., 2003; Lee et al., 2010; Hao et al., 2010a), genomics (Trapp and Croteau, 2001; Hao et al., 2011a, b; Onrubia et al., 2011),

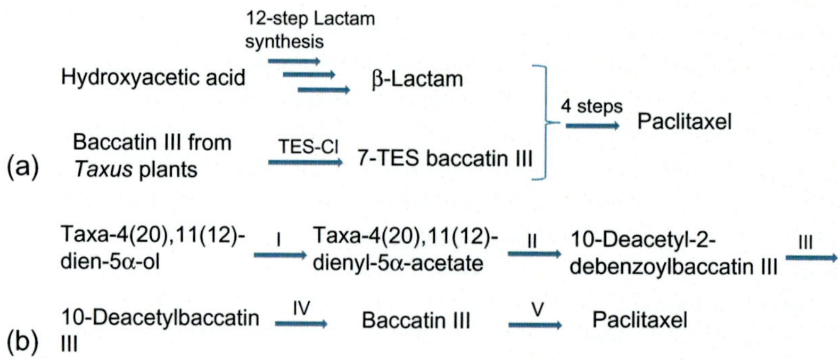

Figure 3.1 Semisynthesis (a) and biosynthesis (b) of paclitaxel. I. 5-*O*-Acetylation by a taxadien-5a-ol acetyltransferase; II. several oxidation reactions; III. 2-*O*-benzoylation by a 2-*O*-debenzoylbaccatin III benzoyltransferase; IV. 10-*O*-acetylation by a 10-deacetylbaccatin III acetyltransferase; V. 13-*O*-acylation by a phenylpropanoyltransferase (including biosynthesis of CoA thioester substrates), followed by *N*-benzoylation by an *N*-benzoyltransferase. Ac, Acetyl; Ph, phenyl; TES, triethylsilyl.

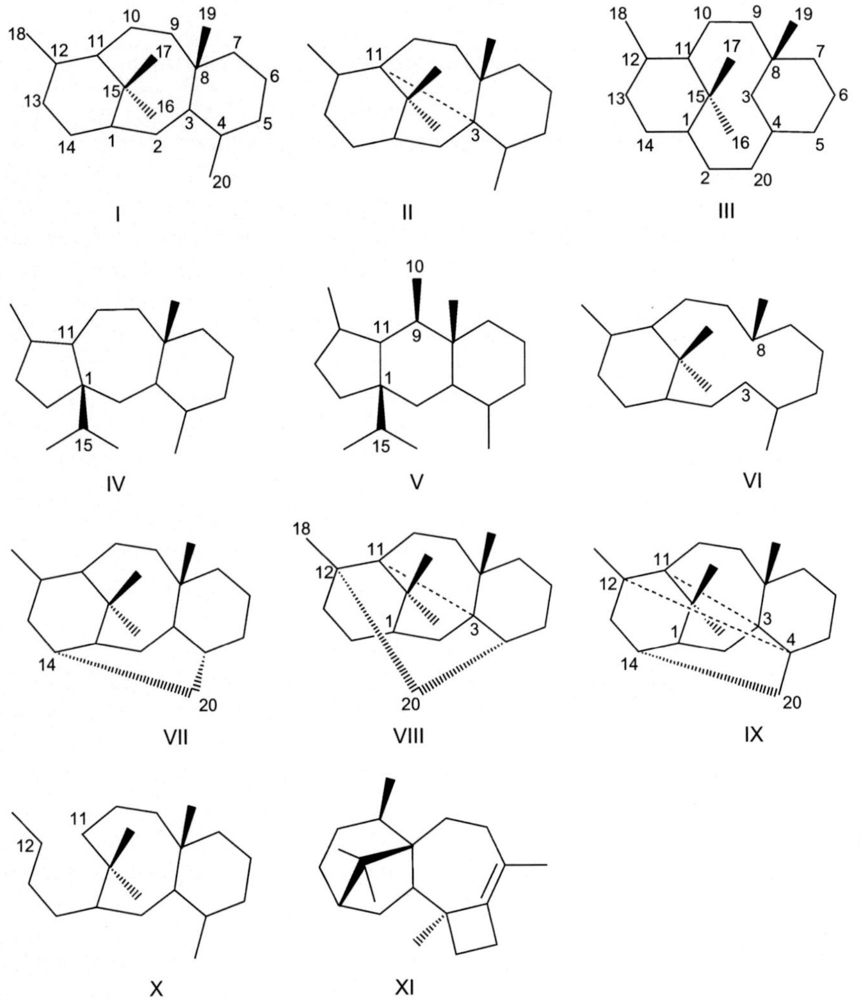

Figure 3.2 Taxane skeletons.

pharmacology (Ojima et al., 1999; Zhang et al., 2008, 2009), microbial symbionts (Stierle et al., 1993; Zhan et al., 2003; Tan and Cuo, 2006; Chen et al., 2009; Soca-Chafre et al., 2011), toxicology (Brown and Hull, 1951; Gausterer et al., 2012), and taxonomy (Spjut, 2007). This chapter, however, does not cover the full range of such studies; instead, we focus on recent progress in molecular biology and genomics of *Taxus*, as well as its chemistry and metabolomics. The bibliometric method is used to quantify and characterize the global scientific production of *Taxus*-related research.

3.2 From molecular biology to genomics

3.2.1 Molecular phylogeny, taxonomy, and evolution

Plants of the genus *Taxus* are sources of a number of physiologically and pharmacologically active compounds of different classes, especially the anticancer paclitaxel and many other taxane derivatives. There are at least 10 species of *Taxus*. The species of *Taxus* are more geographically than morphologically separable. There is a large variation in taxane content between the different species and cultivars (van Rozendaal et al., 2000; Hao et al., 2011a). It is essential to find suitable plants for the various production protocols of paclitaxel. The correct identification of the *Taxus* species is not only a prerequisite for the relevant plant breeding and selection, and good agricultural practice (GAP), but also a precondition for the chemical and pharmacological investigations of the respective *Taxus* species, and good manufacturing practice (GMP) (Khan, 2006; Hao et al., 2008b). *Taxus mairei*, an endemic *Taxus* species of southern China, is the most important plant source for 7-xylosyltaxanes that can be converted to paclitaxel and other useful taxanes via chemical transformation and biotransformation (Hao et al., 2008a). Since it is difficult to authenticate and differentiate one species from other plant sources of taxanes by morphology, molecular studies are definitely required to solve the problem. The molecular studies with the limited taxon sampling have provided only few detailed insights into relationships within *Taxus* (Li et al., 2001; Collins et al., 2003; Shah et al., 2008). In Li et al.'s (2001) study, the controversial *T. sumatrana*, *T. yunnanensis*, and *T. wallichiana* were not included, and only internal transcribed spacer (ITS) sequences were used to infer the phylogenetic relationship. Collins et al. (2003) noted that *Taxus* species delimitation remained a problem, but their RAPD and chloroplast (cp) DNA data focused on two hybrids and their parental species. Shah et al. (2008) utilized the sequence data of the nuclear (nr) ribosomal DNA ITS region and the cp trnL-F region to delimit *T. baccata*, *T. wallichiana*, and *T. fuana*. Hao et al. (2008b) substantially increased the taxonomic sampling of nrDNA and cpDNA for *Taxus* and provided a comprehensive picture of their phylogeny. One cp (trnS-trnQ spacer) and three nr (taxadiene synthase (TS), 10-deacetylbaccatin III-10β-*O*-acetyltransferase (DBAT), and 18S rDNA) molecular markers were combined to infer the interspecific relationship. Three of the four New World species (*T. brevifolia*, *T. floridana*, and *T. globosa*) form a well-supported clade, whereas *T. canadensis* initially branches, appearing distantly related to both Old World taxa and New World species. In Asia, *Taxus chinensis*, *T. mairei*, *T. sumatrana*, and *T. wallichiana* cluster together and are sisters to a clade containing *T. baccata* and *T. contorta* (i.e., *T. fuana*). *Taxus yunnanensis* is more closely related to *T. wallichiana* than to the four other *Taxus* species from China; *T. contorta* is closer to the Euro-Mediterranean *T. baccata* than to the Asian species. This study provides a genetic method for the authentication of economically important *Taxus* species and proposes a robust phylogenetic hypothesis for the genus. Using trnS-trnQ spacer sequences, one is able to distinguish *T. mairei* from all other species of *Taxus*. Schirone et al. (2010) then used trnS-trnQ in the origin analysis of *T. baccata* in the Azores and found that the Azorean population represents a different evolutionary line within *Taxus*. In addition, other cp markers, including matK, rbcL, and psbA-trnH, were combined with trnL-F and ITS to infer the phylogenetic relationship between and within the families Taxaceae and Cephalotaxaceae (Hao et al., 2008c; Figure 3.3), and the results were largely congruent with those inferred by trnS-trnQ, TS, DBAT, and 18S rDNA (Hao et al., 2008b). Based on these results, Christenhusz et al. (2011) proposed that the small group *Amentotaxus* be within the family Taxaceae. A novel alignment-free DNA bar code sequence identification algorithm, BRONX, which accounts for observed within taxon variability and hierarchic relationships among taxa, was proposed based on these DNA marker sequences (Little, 2011).

DNA bar coding is a taxonomic method that uses a short genetic marker in an organism's DNA to identify it as belonging to a particular species. It differs from molecular phylogeny in that the main goal is not to determine classification but to identify an unknown sample in terms of a known classification (Kress et al., 2005). The use of nucleotide sequence variations to investigate evolutionary relationships is not a new concept. Molecular markers have been successfully used in molecular systematics for decades (Hao et al., 2009a, 2010a). DNA bar coding provides a

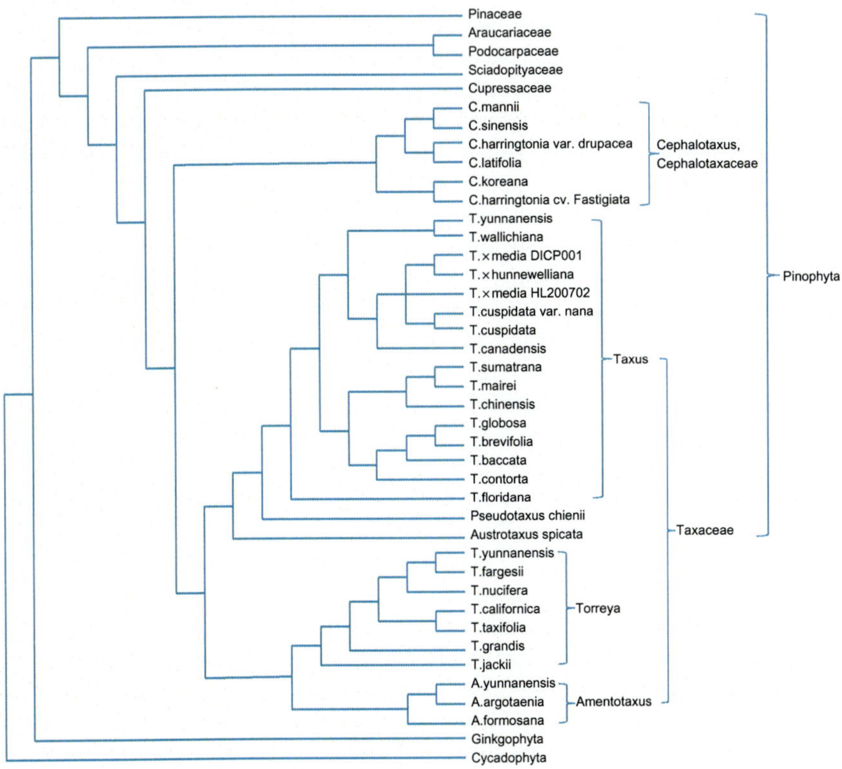

Figure 3.3 Phylogenetic relationship of *Taxus* and *Cephalotaxus*, modified from Hao et al. (2008c).

standardized method for this process via the use of a short DNA sequence from a particular region of the genome to provide a "bar code" for identifying species (Zhu et al., 2010; Little, 2011). Liu et al. (2011) evaluated the utility of five candidate plant DNA bar coding regions (rbcL, matK, trnH-psbA, trnL-F, and ITS) in Eurasian yews. As single loci, trnL-F and ITS showed the highest species discriminatory power, each resolving 11 of 11 lineages (i.e., bar code taxa). The proposed CBOL core bar code (rbcL + matK) resolved eight of 11 lineages. Based on overall performance, trnL-F and ITS, separately or combined, are proposed as the bar code for Eurasian *Taxus*, which can be used to rapidly and reliably identify *Taxus* species in Eurasia for conservation protection and for monitoring illegal trade. In plants, establishing a standardized DNA bar coding system has been more challenging than in animals. Positive selection and repetitive sequences might influence the discriminatory power of the commonly used DNA bar codes (Hao et al., 2010b,c,d,e). Notwithstanding, DNA bar coding would be routinely used in the identification and authentication of the medicinal plant species during drug development.

Plants synthesize an enormous number of secondary metabolites that provide an increasingly exploited reservoir for the generation of pharmaceutically active agents, and many more await discovery. In *Taxus*, paclitaxel (Taxol), a well-known anticancer agent, and related taxane compounds are major components in the mixture of secondary metabolites that play an important ecological role in plant defense. Over the past 20 years, major advances have been made in the identification of genes responsible for paclitaxel biosynthesis (Croteau et al., 2006), a process requiring more than 20 enzymatic reactions involving the construction of the tetracyclic skeleton and the addition of the various oxygen and acyl functional groups. Among the intermediate steps, the cyclization of geranylgeranyl diphosphate (GGPP) to taxadiene is catalyzed by TS (Koepp et al., 1995), and the acetylation of 10-deacetylbaccatin III (10-DAB) to baccatin III is catalyzed by 10-deacetylbaccatin III-10-*O*-acetyltransferase (DBAT; Figure 3.1). Positive, diversifying selection is an important evolutionary force that accelerates the divergence between homologous proteins (Swanson et al., 2001). Among the proteins identified to be under positive selection are immune response- and defense-related proteins (Bishop, 2005; Nielsen et al., 2005) and toxin proteins (Liu et al., 2005). Since paclitaxel biosynthetic enzymes catalyze the formation of an important defense molecule paclitaxel and the related taxanes, it is intriguing to study whether the adaptive evolution affects any sites of any enzymes. Evolutionary patterns of sequence divergence were analyzed in genes TS and DBAT (Hao et al., 2009b). N-terminal fragments of TS, full-length DBAT, and ITS were PCR amplified from 15 closely related *Taxus* species and sequenced. While both TS and DBAT are, overall, under purifying selection, a number of amino acids of TS under positive selection are identified based on the inference using maximum likelihood models. Positively selected amino acids in the N-terminal region of TS suggest that this region might be more important for enzyme function than previously thought. Interestingly, in the X-ray crystal structure of TS, the carboxy-terminal catalytic domain is a class I terpenoid cyclase that binds and activates substrate GGPP with a three-metal ion cluster (Köksal et al., 2011). The N-terminal domain and a third "insertion" domain together adopt the fold of a vestigial class II terpenoid cyclase. A class II cyclase activates the isoprenoid substrate by protonation instead of ionization, and the TS structure reveals a definitive connection between the two distinct cyclase classes in the evolution of terpenoid biosynthesis. Moreover, lineages, that is, *T. yunnanensis* and *T. mairei*, with significantly elevated rates of amino acid substitution, that is, undergoing adaptive evolution, are identified using a genetic algorithm (Hao et al., 2009b). These findings demonstrate that the pattern of adaptive paclitaxel biosynthetic enzyme evolution can be documented between closely related *Taxus* species, where species-specific taxane metabolism has evolved recently. Knowledge about codons that are under positive selection and purifying selection is important for studies of plant secondary metabolism and phylogenetics and could facilitate the development of more broadly applicable enzymes for the biotransformation of taxanes. Biocatalytic methods offer the potential to produce metabolites that are difficult to synthesize by traditional medicinal chemistry (Hunter et al., 2011). The substrate selectivity, regioselectivity, and other properties of the taxane biosynthetic enzymes could be altered by directed evolution to produce the desired metabolites and drug candidates.

3.2.2 Genomics and transcriptomics

3.2.2.1 Genomics

Conifers have dominated forests for more than 200 million years and are of huge ecological and economic importance. The draft assembly of the 20-Gb genome of the Norway spruce (*Picea abies*) has been presented (Nystedt et al., 2013), which is the first available for any gymnosperm. The number of well-supported genes (28,354) is similar to the >100 times smaller genome of *Arabidopsis thaliana*, and there is no evidence of a recent whole-genome duplication in the gymnosperm lineage. The large genome size seems to result from the slow and steady accumulation of a diverse set of long terminal repeat (LTR) transposable elements (TEs), possibly due to the lack of an efficient elimination mechanism. Comparative sequencing of *Pinus sylvestris*, *Abies sibirica*, *Juniperus communis*, *T. baccata*, and *Gnetum gnemon* reveals that the TE diversity is shared among extant conifers. The expression of 24-nucleotide small RNAs, previously implicated in TE silencing, is tissue-specific and much lower than in other plants. Numerous long (>10,000 bp) introns, gene-like fragments, uncharacterized long noncoding RNAs, and short RNAs were identified, which opens up new genomic avenues for conifer forestry and breeding.

The *T. mairei* cp genome is 129,513 bp in length, with 113 single-copy genes and two duplicated genes (trnI-CAU and trnQ-UUG) (Zhang et al., 2014a). Among the 113 single-copy genes, 9 are intron-containing. Compared with other land plant cp genomes, the *T. mairei* cp genome has lost one of the large inverted repeats (IRs) found in angiosperms, ferns, liverworts, and gymnosperms such as *Cycas revoluta* and *Ginkgo biloba* L. Compared with related species, the gene order of *T. mairei* has a large inversion of ~110 kb including 91 genes (from rps18 to accD) with gene contents unarranged. Repeat analysis identified 48 direct and 2 IRs 30 bp long or longer with a sequence identity greater than 90%. Repeated short segments were found in genes rps18, rps19, and clpP. Analysis also revealed 22 microsatellite (simple sequence repeat (SSR)) loci and almost all are composed of A or T.

Plastid-to-nucleus DNA transfer provides a rich genetic resource to the complexity of plant nuclear genome architecture. The complete plastomes of two yews, *Amentotaxus formosana* and *T. mairei* (Taxaceae of Coniferales), were sequenced (Hsu et al., 2014). Comparative genomic analyses recovered an evolutionary scenario for plastomic reorganization from ancestral to extant plastomes in the three sampled Taxaceae genera, *Amentotaxus*, *Cephalotaxus*, and *Taxus*. Specific primers were designed to amplify nonsyntenic regions between ancestral and extant plastomes, and 12.6 kb of nupts (nuclear plastid DNA) were identified based on phylogenetic analyses. These nupts have significantly accumulated GC-to-AT mutations, reflecting a nuclear mutational environment shaped by the spontaneous deamination of 5-methylcytosine. The ancestral initial codon of rps8 is retained in the T. nupts, but its corresponding extant codon is mutated and requires C-to-U RNA editing, suggesting that nupts can help recover scenarios of the nucleotide mutation process. The Taxaceae nupts may have been retained since the Cretaceous, and they carry information of both ancestral genomic organization and nucleotide composition, which offer clues for understanding the plastome evolution in conifers.

3.2.2.2 Transcriptomics

Paclitaxel is a chemotherapeutic drug used in the treatment of many types of cancer and coronary heart diseases, but the increasing demands have created a supply crisis and raised serious environmental concerns (Kingston, 2001). Presently, the mainstay solution is semisynthesis from several precursors that have the same core skeleton and can be isolated from renewable *Taxus* resources (Ge et al., 2008a,b). Paclitaxel and its precursors belong to a group of typical secondary metabolites named the taxane diterpenoids or taxoids, and their distribution and composition are highly variable in different species and different tissues. Therefore, choosing suitable *Taxus* species and screening the constituents in each tissue are essential to the cost-effective production of taxane drugs. Previous studies mainly focused on *Taxus* needles from various species, which displayed distinct chemical distributions (van Rozendaal et al., 2000; Ge et al., 2008b). In recent years, the root constituents from various *Taxus* species are systematically investigated, and it is found that roots have relatively simple chemical profiles and possess high yields of valuable taxanes such as paclitaxel (P), cephalomannine (C), 10-deacetyl paclitaxel (10-DAT), and 7-xylosyltaxanes (Ge et al., 2008a). Rational exploitation of the taxanes in the root is of great help for alleviating the taxane supply crisis. However, the biosynthetic pathway of paclitaxel and other taxanes is not fully elucidated, and the underlying molecular mechanism of the metabolic difference between different *Taxus* tissues was not studied, which hamper improvements in taxane drug production.

During the past few years, the high demand for low-cost sequencing has driven the development of high-throughput second-generation sequencing technologies that parallelize the sequencing process, producing thousands or millions of sequences at once (Hall, 2007; Schuster, 2008). Roche 454 pyrosequencing and Illumina high-throughput sequencing are based on sequencing by synthesis, SOLiD and Polonator are based on sequencing by ligation, and Helicos and Pacific Biosciences utilize the single-molecule sequencing (Egan et al., 2012). The inexpensive production of large volumes of sequence data via second-generation sequencing is the primary advantage over conventional methods (Metzker, 2010). Collins et al. (2008) used a combination of mapping and *de novo* assembly tools in the analysis of data from Illumina sequencing of the polyploid plant *Pachycladon enysii* and demonstrated that transcriptome analysis using high-throughput short-read sequencing need not be restricted to the genomes of model organisms. The *de novo* transcriptome sequencing and characterization based on Illumina second-generation sequencing has been performed successfully for many medicinal plants (e.g., Ono et al., 2011; Brown et al., 2012). Illumina produces orders of magnitude more sequences at a fraction of the cost of 454 pyrosequencing.

Besides transcriptome sequencing, another approach to gene expression analysis, digital gene expression (DGE) profiling, can be performed on the Illumina Genome Analyzer and HiSeq 2000 sequencing platforms. DGE tag sequencing is an implementation of the LongSAGE (serial analysis of gene expression) protocol on the Illumina sequencing platform that increases utility while reducing both the cost and time required to generate gene expression profiles. The ultrahigh-throughput sequencing capability of the Illumina platform allows the cost-effective generation of libraries containing an average of 20 million tags, a 200-fold improvement over classical LongSAGE (Morrissy et al., 2010). Illumina DGE has been used in a sweet orange red-flesh mutant to study the molecular process regulating lycopene, a secondary metabolite, accumulation (Xu et al., 2009). Lee et al. (2010) compared the expression profiles of cultured dedifferentiated *Taxus cuspidata* cells and undifferentiated cambial meristematic cells via DGE and identified marker genes and transcriptional programs in the latter that are consistent with a stem cell identity. These stem cells may provide a lucrative and environmentally friendly platform for the sustainable production of a variety of important plant natural products.

The estimated genome size of *Taxus* is 10,000 Mb (Leitch and Hanson, 2001), and 20.8% of the fosmid end sequences of the *T. mairei* genomic library are repetitive elements (Hao et al., 2011b), which are composed of retroelements, DNA transposons, satellites, simple repeats, and low-complexity sequences. Since the whole-genome sequences of *Taxus* are not available presently, it is realistic to first perform the transcriptome sequencing for collecting key information of the large and complex *Taxus* genome. In order to create a backbone for the *Taxus* and medicinal plant research community, including a large part of the transcriptome and expression patterns in different tissues, as well as to look at phenotypes and try to look at the correlation between gene expression and taxane metabolites, the *de novo* assembly of *Taxus mairei* transcriptome using Illumina paired-end sequencing technology is performed (Hao et al., 2011a). In a single run, 13,737,528 sequencing reads corresponding to 2.03 Gb total nucleotides are produced. These reads were assembled into 36,493 unique sequences. Based on a similarity search with known proteins, 23,515 unigenes were identified to have the BLAST hit with a cutoff *E*-value above 10^{-5}. Furthermore, the transcriptome difference of three tissues is investigated with the tag-based DGE system. A sequencing depth of over 3.15 million tags per sample is obtained, and a large number of genes associated with tissue-specific functions and the taxane biosynthetic pathway are identified. The expression of the taxane biosynthetic genes is significantly higher in the root than in the leaf and the stem, while high activity of the taxane-producing pathway in the root is also revealed via metabolomic analyses. In addition, many antisense transcripts and novel transcripts are found; clusters with similar differential expression patterns, enriched gene ontology (GO) terms, and enriched metabolic pathways with regard to the differentially expressed genes are revealed for the first time. Compared with two *T. cuspidata* transcriptome data sets generated on the 454 sequencing platform (Lee et al., 2010; Wu et al., 2011), this study provides the most comprehensive sequence resource available for *Taxus* study and will help define mechanisms of tissue-specific functions and secondary metabolism in nonmodel plant organisms.

The Illumina sequencing platform was also used to identify microRNAs (miRNAs) from *T. chinensis* cells and to investigate the effect of the taxoid elicitor methyl jasmonate (MeJA) on miRNA expression (Qiu et al., 2009). Plant miRNAs have an impact on the regulation of several biological processes such as development, growth, and metabolism by negatively controlling the gene expression at the post-transcriptional level. To elucidate the role of miRNAs in *Taxus*, we used a deep sequencing approach to analyze the small RNA and degradome sequence tags of *T. mairei* leaves (Hao et al., 2012b). For miRNAs, the sequencing library generated 14.9 million short sequences, resulting in 13.1 million clean reads. The library contains predominantly small RNAs with 21 nucleotides in length, followed by 19 and 20 nt small RNAs. Around 29% of total small RNAs are matched to the *T. mairei* transcriptome. By sequence alignment, we identified 871 mature miRNAs, 15 miRNAs*, and 869 miRNA precursors representing known plant miRNA families. There are 547 unique small RNAs matching the miRNA precursors. We predict 37 candidate novel miRNAs from the unannotated small RNAs that could be mapped to the reference transcriptome. The expression of the selected candidates was for the first time quantified by real-time reverse transcription PCR. The novel miRNA m0034 turns out to be from the intron sequence of the paclitaxel biosynthetic gene TS. The 21 potential targets of the nine novel miRNAs are also predicted. Additionally, 56 targets for known miRNA families and 15 targets for novel candidate miRNA families were identified by a high-throughput degradome sequencing approach. It is found that two paclitaxel biosynthetic genes, taxane 13α-hydroxylase and taxane 2α-*O*-benzoyltransferase, are the cleavage targets of miR164 and miR171, respectively. This study represents the first transcriptome-based analysis of miRNAs and degradome in gymnosperms.

Deep sequencing reveals transcriptome reprogramming of *T.* × *media* cells to the elicitation with MeJA (Sun et al., 2013).

Several important genes in Taxol biosynthesis are currently still unknown and have been shown to be difficult to isolate directly from *Taxus*, including the gene encoding taxoid 9α-hydroxylase. *Ginkgo biloba* suspension cells exhibit taxoid hydroxylation activity and provides an alternate means of identifying genes encoding enzymes with taxoid 9α-hydroxylation activity (Zhang et al., 2014b). Through the analysis of high-throughput RNA sequencing data from *G. biloba*, two candidate genes with high similarity to *Taxus* CYPs were identified. *In vitro* cell-free protein synthesis assays and LC–MS analysis show that one candidate that belongs to the CYP716B, a subfamily whose biochemical functions have not been previously studied, possessed 9α-hydroxylation activity. This work will aid the future identification of the taxoid 9α-hydroxylase gene from *Taxus*.

3.2.2.3 Molecular marker

753 microsatellite motifs were identified from the *T. cuspidata* transcriptome sequencing data set (Wu et al., 2012). PCR primers were obtained for 519 EST-SSRs, while randomly selected cloning sequencing revealed that 87.5% of ESTs were the same as the results of Sanger sequencing.

Fifteen microsatellite loci were targeted in *Pseudotaxus chienii* using the fast isolation by AFLP of sequences containing repeats (FIASCO) protocol (Deng et al., 2013). Polymorphism was evaluated in five populations of *P. chienii* and five populations of *T. mairei*. Of these loci, 13 were polymorphic in *P. chienii*, whereas 15 were polymorphic in *T. mairei*.

Ten highly polymorphic microsatellite markers were obtained from 454 DNA sequencing of *T. wallichiana* (Gajurel et al., 2013). Characterization of the new microsatellite loci was done in 99 individuals collected from three valleys with different climatic regimes. The number of alleles per locus varied from 4 to 12. The observed heterozygosity of populations, averaged across loci, ranged from 0.30 to 0.59. These new markers could substantially increase the resolution for detailed studies in phylogeography, population genetics, and parentage analysis.

It is little known how the profoundly complex topography and habitat heterogeneity generated by the uplift of the Qinghai-Tibetan Plateau (QTP) during the late Tertiary affected the population genetic structure of the endangered *T. yunnanensis*, which is an ancient tree/shrub distributed in southwest China. Recently, the species has suffered a sharp decline due to excessive logging for its famous anticancer metabolite Taxol, resulting in smaller and more isolated populations. Eleven polymorphic microsatellites were used to genotype 288 individuals of 14 populations from a range-wide sampling in China (Miao et al., 2014). Two different population groups that were once isolated have persisted *in situ* during glacial periods in both areas and have not merged since. Habitat fragmentation has led to significant genetic bottlenecks, high inbreeding, and population divergence in this species. The two different population groups of *T. yunnanensis* could be attributed to restricted gene flow caused via isolation by geographic barriers and by habitat heterogeneity during uplift of the QTP or the existence of two separate glacial refugia during the Pleistocene. *In situ* and *ex situ* conservation of the two evolutionarily significant units (ESUs), artificial gene flow between populations, and a comprehensive understanding of the pollination system in this endangered species are suggested.

3.2.2.4 Microbe

Metagenomics is the study of metagenomes, genetic material recovered directly from environmental samples. Basic findings at the metagenomic levels could be applied to drug discovery and development. The regional variability of the bacterial community composition and diversity was studied by the comparative analysis of three large 16S rRNA gene clone libraries from the *Taxus* rhizosphere in different regions of China (subtropical and temperate regions). One hundred and forty-six clones were screened for three libraries. Phylogenetic analysis of 16S rRNA gene sequences demonstrated that the abundance of sequences affiliated with γ-proteobacteria, β-proteobacteria, and Actinobacteria was higher in the library from the *T.* × *media* rhizosphere of the temperate region compared with the subtropical *T. mairei* rhizosphere. On the other hand, Acidobacteria was more abundant in libraries from the subtropical *T. mairei* rhizosphere. Richness estimates and diversity indexes of three libraries revealed major differences, indicating a higher richness in the *Taxus* rhizosphere bacterial communities of the subtropical region and considerable variability in the bacterial community composition within this region. By enrichment culture, a novel Actinobacteria strain DICP16 was isolated from the *T.* × *media* rhizosphere and was identified as *Leifsonia shinshuensis* sp. via the 16S rRNA gene and gyrase B sequence analyses. DICP16 was able to remove the xylosyl group from 7-xylosyl-10-DAB and 7-xylosyl-10-DAT, thereby making the xylosyltaxanes available as sources of 10-DAB and paclitaxel. A xylosidase isolated from DICP16 could hydrolyze 20-C, β-(1 → 6)-xyloside of ginsenoside Rb3 (G-Rb3) into ginsenoside Rd but did not hydrolyze the other β-D-glucosidic bonds of G-Rb3 (Luan et al., 2008). Moreover, two fungal strains (*Mucor spinosus* AS 3.345 0 and *Cunninghamella echinulata* AS 3.340 0) and a bacterial strain (*Proteus vulgaris* AS 1.120 8) were chosen to transform sinenxan A, an abundant taxane in *Taxus* (Zhan et al., 2003). Three products were obtained and identified as 10-deacetylsinenxan A1, 6α-hydroxy-10-deacetylsinenxan A2, and 9α-hydroxy-10-deacetylsinenxan A3, respectively. These studies illustrate the utility of metagenomics and microbial transformation in drug development. In the near future, both the culture-dependent and culture-independent approaches should be integrated in the metagenomic, metatranscriptomic, and metaproteomic studies of the *Taxus* rhizosphere and phyllosphere.

The ability of the endophytic fungus of hazel *Penicillium aurantiogriseum* NRRL 62431 to independently synthesize paclitaxel was established by liquid chromatography (LC)–mass spectrometry (MS) and proton nuclear magnetic resonance (NMR) (Yang et al., 2014). The genome of *P. aurantiogriseum* was sequenced, and gene candidates that may be involved in paclitaxel biosynthesis were identified by comparison with the 13 known paclitaxel biosynthetic genes in *Taxus*. Paclitaxel biosynthetic gene candidates in NRRL 62431 have evolved independently and that horizontal gene transfer between this endophytic fungus and its plant host is unlikely, which sheds new light on how paclitaxel-producing endophytic fungi synthesize paclitaxel and will facilitate metabolic engineering for the industrial production of paclitaxel from fungi.

3.3 Bioactivity, pharmacology, and therapeutic use

T. wallichiana Zucc., known as Himalayan yew, is a medium-sized, temperate, Himalayan forest tree of medicinal importance. In India, this evergreen tree is found at altitudes between 1800 and 3300 m above mean sea level (Juyal et al., 2014). It has been used by the native populations for treating the common cold, cough, fever, and pain. Its uses are described in Ayurveda and Unani medicine.

Three species of yews *T. contorta* Griff., *T. mairei* (Lemée & Lév.) S.Y. Hu ex T.S. Liu, and *T. wallichiana* Zucc. distributed in the Hindu Kush Himalayan (HKH) region have been commercially exploited in recent decades to extract taxane. Indigenous people of this region are using yews for several other purposes including gastrointestinal disorders, respiratory problems, and skeletal system disorders and as edible fruit, fodder, fish poison, traditional veterinary medicine, etc. Ethnobotanical knowledge from 10 major ethnic/caste groups of Mongol and Caucasian origins in the Nepal Himalayas was documented in 2010 and 2011 from 27 sites covering the extant distribution range of the three species (Poudel et al., 2013). Seventy-two key informants, recommended by the majority of people in informal group discussions at each study site, were interviewed to collect information on the importance of yews. The key informants cited 45 uses under 21 categories. A greater use diversity and high consensus value for use types were recorded for medicinal uses (gastrointestinal ailments, cough and cold, skeletomuscular system problems, and others), followed by fruit consumption, household tools, agriculture implements, and timber. A decline of yew populations and associated traditional knowledge among the younger generations of indigenous people was found. A strong agreement of ethnobotanical knowledge on yews between communities of Mongol and Caucasian origins was shown. The potential for additional therapeutic applications in yews of the HKH region, besides cancer treatment, was revealed. To compensate the low yield of paclitaxel and the fact that three yew species are involved, the reported species-specific curative properties need to be validated scientifically and evaluated clinically. Initiatives should be taken immediately to stop the further degradation of yew populations and the associated indigenous knowledge in the HKH region.

3.3.1 Anticancer activity

Paclitaxel (Taxol) inhibits the depolymerization of microtubules. Aqueous extract of *T. chinensis* (Pilger) Rehd (AETC), which is free of paclitaxel, has been used as a TCM formula for thousands of years. It is bitter in flavor and nature with tropism to the heart meridian. AETC is usually used to treat cancer, kidney disease, and rheumatism alone or in combination with other herbs. AETC has inhibitory effects on lung carcinoma A549 cells and induces cell cycle arrest and apoptosis. AETC inhibits the A549 cells through the epidermal growth factor receptor (EGFR)/mitogen-activated protein kinase (MAPK) pathway *in vitro* and *in vivo* (Shu et al., 2014).

Aqueous extract of *T. mairei* has no significant effect on angiogenesis (Cui et al., 2013) but might inhibit A549 xenograft growth by inhibiting the survivin protein and EGFR phosphorylation.

Water decoctions of *T. cuspidata*, which have hydrophilic paclitaxel derivatives, showed antitumor effects on pancreatic cancer (Qu and Chen, 2014). The extract of *T. cuspidata* (TC) needles and twigs showed inhibition rates of 70–90% in different human cancer cell lines (HL-60, BGC-823, KB, Bel-7402, and HeLa) but only 5–7% in normal mouse T/B lymphocytes (Shang et al., 2011), demonstrating the broad-spectrum anticancer activity and low toxicity to normal cells. TC extract inhibited cancer cell growth by inducing apoptosis and G2/M cell cycle arrest. TC extract and 5-FU, combined as a cocktail, synergistically inhibited the growth of cancer cells *in vitro*, with combination index (CI) values ranging from 0.90 to 0.26 in MCF-7 cells, CI ranging from 0.93 to 0.13 for IC40 to IC90 in PC-3M-1E8 cells, and CI < 1 in A549 cells. The cocktail had lower cytotoxicity in the normal human cell (HEL) than 5-FU used alone. TC extract did not affect the pharmacokinetics of 5-FU in rats.

Paclitaxel activated the Raf-1/extracellular signal-regulated kinase (ERK) pathway, leading to an activation of ribosomal S6 kinase (RSK)/YB-1 signaling (Shiota et al., 2014). The activated Raf-1/ERK pathway was blunted by YB-1 knockdown in prostate cancer cells, indicating regulation between Raf-1/ERK signaling and YB-1. ERK or RSK was activated in taxane-resistant prostate cancer cells, resulting in YB-1 activation. YB-1 knockdown and RSK inhibition using RSK-specific siRNA or the small-molecule inhibitor SL0101 successfully blocked the activation of YB-1, leading to suppression of prostate cancer growth and sensitization to paclitaxel. RSK/YB-1 signaling contributes to taxane resistance and implicates the therapeutics targeting RSK/YB-1 signaling such as RSK inhibitor as a promising novel therapy against prostate cancer, especially in combination with taxane.

Aqueous extract from *T. baccata* inhibits adenosine deaminase (ADA) activity significantly in cancerous and noncancerous human gastric and colon tissues (Durak et al., 2014). In addition to other proposed mechanisms, accumulated adenosine due to the inhibition of ADA enzyme might also play a part in the anticancer properties of *Taxus* species.

7,7″-Dimethoxyagastisflavone, isolated from the needles of *T.* × *media* cv. *Hicksii*, induced apoptotic or autophagic cell death in different cancer cells (Hwang et al., 2012).

A polysaccharide isolated from *T. yunnanensis* suppressed tumor cell proliferation. A complex, water-soluble polysaccharide, PSY-1, was isolated from the leaves of *T. mairei* and exhibited antineoplastic effects (Zheng et al., 2014). PSY-1 effectively suppressed the migration and invasion ability of the melanoma cancer cell line B16-F10 and caused the downregulation of MMP-2 and MMP-9, and the NF-κB pathway was involved in the antimetastatic effects imposed by PSY-1. TMP70W, a polysaccharide isolated from *T. yunnanensis*, displayed mild cytotoxicity against K562 cells with the IC50 value of 39.63 ± 2.37 μg/ml and inhibitory activity against MCF-7 cells ($32.08 \pm 0.39\%$ at the concentration of 400 μg/ml) in a concentration-dependent manner (Yan et al., 2013).

The endophytic fungus *Fusarium solani* isolated from *T. celebica* produced Taxol and its precursor baccatin III in liquid-grown culture (Chakravarthi et al., 2013). Both fungal Taxol and baccatin III inhibited cell proliferation of a number of cancer cell lines with IC50 ranging from 0.005 to 0.2 μM for fungal Taxol and 2–5 μM for fungal baccatin III. They also induced apoptosis in JR4-Jurkat cells with a possible involvement of antiapoptotic Bcl-2 and loss in mitochondrial membrane potential, which was unaffected by inhibitors of caspase-9,-2, or -3 but was prevented in presence of caspase-10 inhibitor. DNA fragmentation was observed in cells treated with fungal Taxol and baccatin III.

3.3.2 Effects on the cardiovascular system

The *syn*ergy between percutaneous coronary intervention with *Taxus* and cardiac surgery (SYNTAX) trial demonstrated that in patients with 3-vessel or left main coronary artery disease, coronary artery bypass graft surgery (CABG) was associated with a lower rate of cardiovascular death, myocardial infarction, stroke, or repeat revascularization compared with percutaneous coronary revascularization with drug (paclitaxel)-eluting stents (DES-PCI) (Cohen et al., 2014). Among patients with less complex diseases, DES-PCI may be preferred on both clinical and economic grounds.

For patients with complex chronic total occlusions in native coronary arteries, the use of the paclitaxel-coated balloon after bare-metal stenting was associated with similar clinical results and a nonsignificantly higher in-stent late loss compared with a matched population with paclitaxel-eluting stent implantation (Wöhrle and Werner, 2013).

The leaves of *T. mairei* are used traditionally to fill pillows in some rural areas of China. Its volatile substances could improve sleep quality, stabilize blood pressure, and have a diuretic capacity as recorded in ancient Chinese materia medica. Volatile components from leaves of *T. mairei* (VCLT) prevented the increase of systolic blood pressure (SBP) and plasma angiotensin II in L-NNA (injection of *N*ω-nitro-L-arginine)-treated rats (Yang et al., 2012). Although VCLT do not significantly reduce blood triglycerides (TGs), high-density lipoprotein cholesterol (HDL-C), and low-density lipoprotein cholesterol (LDL-C), they decrease total cholesterol (TC) while increasing plasma NO levels in a dose-dependent manner.

Seven *T. cuspidata* compounds showed stronger inhibitory effects than acetylsalicylic acid (ASA) on platelet aggregation induced by arachidonic acid (AA) (IC50 14.4, 64.5, 35.5, 16.0, 21.9, 28.6, and 48.2 vs. 63.0 μM) or U46619 (IC50 34.8, 24.9, 36.2, 35.0, 46.9, 71.9, and 68.7 vs. 340 μM) (Kim and Yun-Choi, 2010). Taxinine, taxinine B, 2-deacetoxytaxinine B, and taxacin, with a cinnamoyl group at the C5 position, showed strong inhibitory effects against AA-induced aggregation compared with taxinine A (with an OH group at C5) or compounds with an oxetane ring at C4,5, such as taxchinin B and paclitaxel. All seven compounds were 5–13-fold more strongly inhibitory than ASA against U46619-induced aggregation.

3.3.3 Effects on the nervous system

Sciadopitysin, a taxane diterpenoid isolated from the 95% ethanol extract of *T. chinensis*, exhibited the potency against Aβ aggregation and the formation of fibrils (Gu et al., 2013). Cellular assay indicated that sciadopitysin increased the cell viability of the SH-SY5Y cell and demonstrated neuroprotection against Aβ protein-induced insult in the primary cortical neurons.

Lignans of *T. baccata* exhibited a moderate inhibition against both butyrylcholinesterase and lipoxygenase (LOX) (Kucukboyaci et al., 2010), which play a role in the pathogenesis of Alzheimer's disease. They were inactive towards acetylcholinesterase. The compounds displayed a great scavenging activity against DPPH especially at 500 and 1000 μg/ml, and they exert noteworthy ferric reducing antioxidant power

(FRAP). The FRAP of taxiresinol (3.552 ± 0.02), isolariciresinol (3.021 ± 0.71), and 3-demethylisolariciresinol (3.533 ± 0.01) was as high as that of the reference chlorogenic acid (3.618 ± 0.01) at 1000 μg/ml. None of the compounds exhibited chelating ability against ferrous ions.

GSK (glycogen synthase kinase)-3β plays an important physiological role in the regulation of numerous signaling pathways involved in cell differentiation and the morphological development of neurons. Three *T. yunnanensis* lignans (3*S*, 4*R*-4′-hydroxy-6,3′-dimethoxyisoflavan-4-ol, 7*R*-7-hydroxytaxiresinol, and tanegool) highly stimulated the autophosphorylation of GSK-3β and the GSK-3β-mediated phosphorylation of two basic brain proteins (bMBP, pI 11.3 and rhTP, pI8.2) (Ohtsuki et al., 2012) but inhibited dose-dependently the phosphorylation of an acidic protein (rCRMP-2, pI 6.0) by the kinase. These lignans showed binding affinities with bMBP (bovine myelin basic protein) and rhTP (recombinant human tau protein) but had low affinities with rCRMP (rat collapsin response mediator protein)-2. The binding of tanegool and 7*R*-7-hydroxytaxiresinol to these two basic proteins induced their novel potent phosphorylation sites for GSK-3β. These lignans, but not EGCG, induced Tyr phosphorylation of GSK-3β *in vitro*, suggesting that these lignans act as novel effective activators for GSK-3β and the GSK-3β-mediated phosphorylation of their binding basic proteins. Tanegool (IC50 1 μM) is an effective inhibitor for the phosphorylation of rCRMP-2 by the kinase *in vitro*.

3.3.4 Anti-inflammatory and analgesic activities

Tasumatrol B, isolated from the bark of *T. wallichiana*, revealed significant analgesic activity in comparison to a saline group based on an acetic acid-induced model (Qayum et al., 2012). Tasumatrol B showed significant anti-inflammatory activity. However, tasumatrol B, 1,13-diacetyl-10-deacetylbaccatin III, and 4-deacetylbaccatin III failed to exhibit any considerable activity in the hot plate test and the *in vitro* LOX inhibitory assay.

Taxusabietane A, isolated from the bark of *T. wallichiana*, revealed considerable LOX inhibitory activity with the IC50 value 57 ± 0.31 (Khan et al., 2011). Standard compound baicalein showed the IC50 value 22.1 ± 0.03 μM. Taxusabietane A also showed significant anti-inflammatory activity induced by carrageenan.

Alcoholic extract of *T. baccata* not only has bronchodilating activity but also decreases bronchial hyperreactivity by decreasing the infiltration of inflammatory cells in the airway and inhibiting the release of histamine-like mediators from the mast cell by stabilizing it (Patel et al., 2011). 95% ethanol extract of *T. baccata* exhibits potent anti-inflammatory activity at 200 mg/kg 4 h after administration in comparison with ether extract and the reference standard, aspirin (Dutta et al., 2010), which provides a scientific basis for the folklore use of the plant in treating acute inflammation.

3.3.5 Other effects

T. wallichiana has anticonvulsant and antipyretic activities and antibacterial and antifungal activities (Juyal et al., 2014).

Polyprenols from *T. mairei* prevent the development of CCl_4-induced liver fibrosis in rats (Yu et al., 2012), which might be related to the reduction of oxidative damage, the inhibition of hepatic stellate cell activation, the downregulation of profibrogenic stimuli, and the protection of hepatocytes.

Paclitaxel has adjuvant property, although not as strong as synthetic docetaxel, which is more soluble in water and easier to manipulate in medication (Chen et al., 2012).

Crude polysaccharides of *T. cuspidata* displayed antidiabetic activity in streptozotocin-induced diabetic mice (Zhang et al., 2012).

3.3.6 Toxicity and side effects

One of the rare side effects caused by taxanes is a bilateral cystoid macular edema (CME; Kuznetcova et al., 2012). The particularity of this type of CME is that it is angiographically silent showing no leakage or pooling on fluorescein angiography (FA). The content of the CME seems to be made up of viscous fluid. As the origin of the CME is not inflammatory, classical treatments have no effect, and only discontinuation of the taxane drug allows reversal.

Toxic effects of the yew have been known since ancient times. Yew toxicity is due to the content of cyanogenic glycosides and a mixture of alkaloids known as taxines. Taxine B is probably responsible for the most part of the adverse effects in poisoned organisms. This particular taxoid is common in body fluids of the yew-poisoned. Laboratory examination was performed to confirm substances that lead to fatality of a pair of olive baboons (*Papio anubis*) following ingestion of yew seeds (Kominkova et al., 2013). When both cagemates (male and female) died suddenly, poisoning was suspected because many berries had fallen into the cage from a nearby fruiting *T. baccata* during the windy night before. MS analysis of taxoids confirmed poisoning by taxanes. The presence of taxin B/isotaxin B was confirmed in all investigated samples. Apparently, in urine and bile, there were concentrations ranging 150–220 ng/ml and in serum concentrations ranging 25–30 ng/ml.

Taxus plants are common ornamental shrubs that contain cardiotoxic alkaloids. Histologic lesions in the calves included multifocal cardiac myocyte hypereosinophilia (Sula et al., 2013), sarcolemma fragmentation, pyknosis, karyolysis, myocyte loss, and a mild interstitial lymphoplasmacytic infiltrate with edema. Moderate fibrinosuppurative interstitial pneumonia was the only other significant finding. Cardiac changes were attributed to damage from the initial exposure to *T. cuspidata* 6 days prior to death. Myocardial fibrosis associated with previous ingestion of yew in a Holstein heifer is evidence for chronic yew toxicity in cattle (Burcham et al., 2013).

3.4 From chemistry to metabolomics

3.4.1 Taxane diterpenoids

Taxane skeletons are shown in Figure 3.2, and examples of recently isolated taxane diterpenoids are summarized in Figure 3.4 and Table 3.1. Representative naturally occurring taxanes are shown in Figure 3.5.

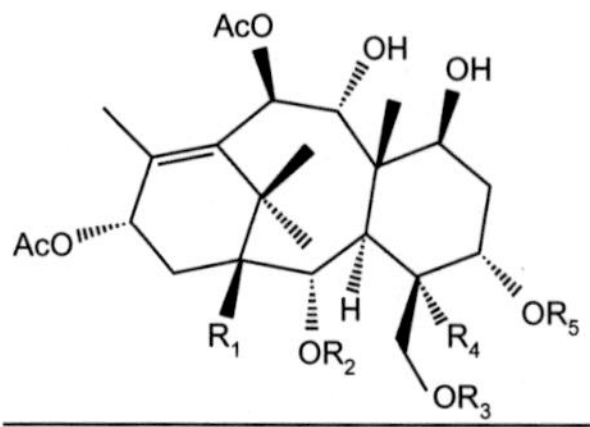

	R_1	R_2	R_3	R_4	R_5
1	OH	Bz	Ac	OH	H
2	H	H	H	H	Ac

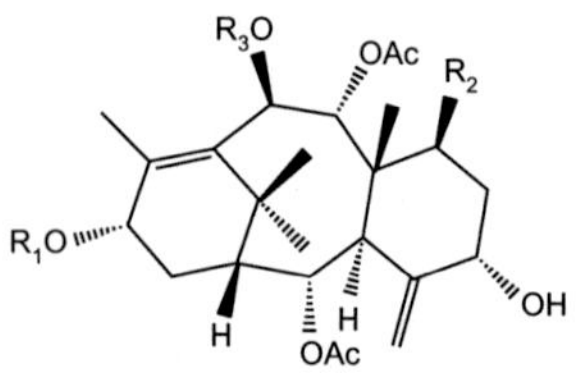

	R_1	R_2	R_3
3	Glc	H	Ac
4	Ac	OAc	H

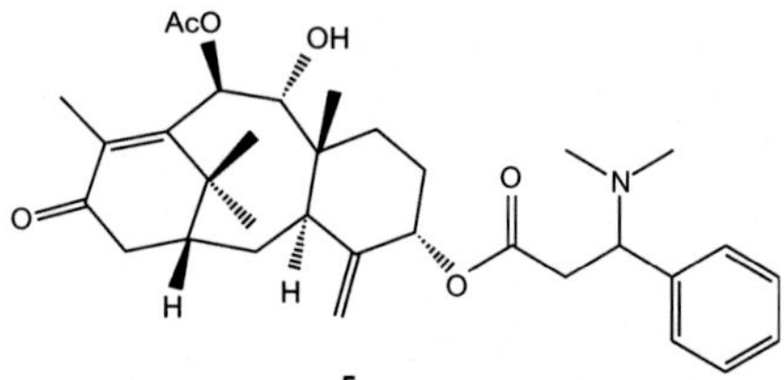

5

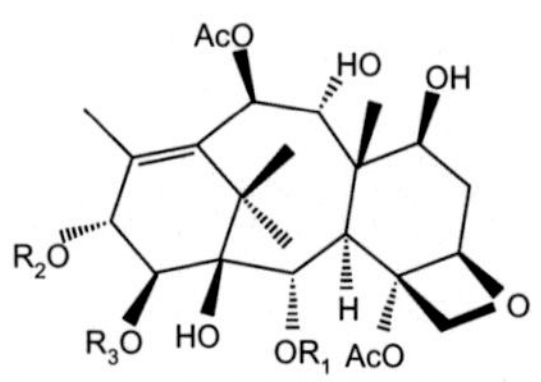

	R_1	R_2	R_3
6	Bz	Ac	H
7	Bz	H	H
8	H	H	Bz

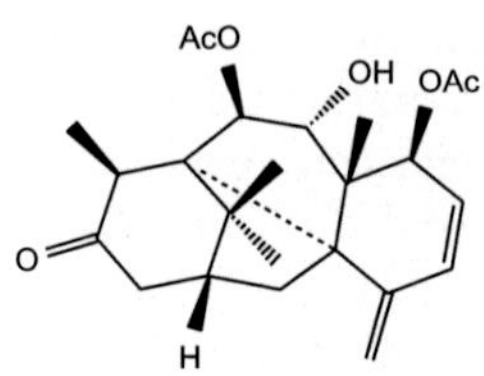

9

Figure 3.4 Examples of taxane compounds found in *Taxus*.

(*Continued*)

Ninety-three new taxanes have been isolated, together with 37 known taxoids including Taxol® (paclitaxel) and cephalomannine, from the Canadian yew, *T. canadensis*, in the past 30 years (Li et al., 2013). These new taxoids possess various skeletons containing 5/7/6, 6/10/6, 6/5/5/6, 6/8/6, and 6/12 ring systems, and six new taxanes with four novel skeletons, that is, a taxane with a 6/6/8/6 ring system, a taxane with a [3.3.3]

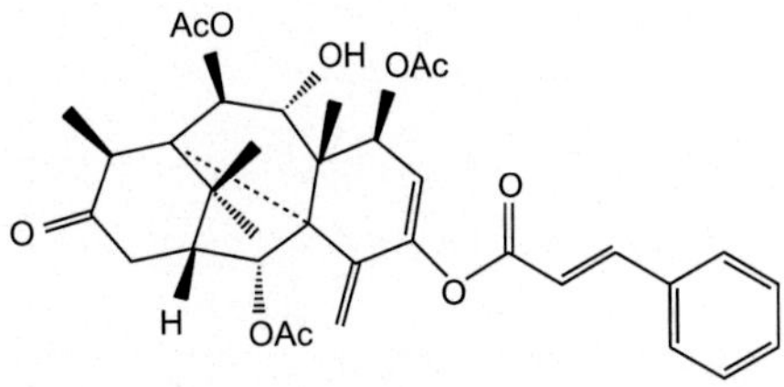

10

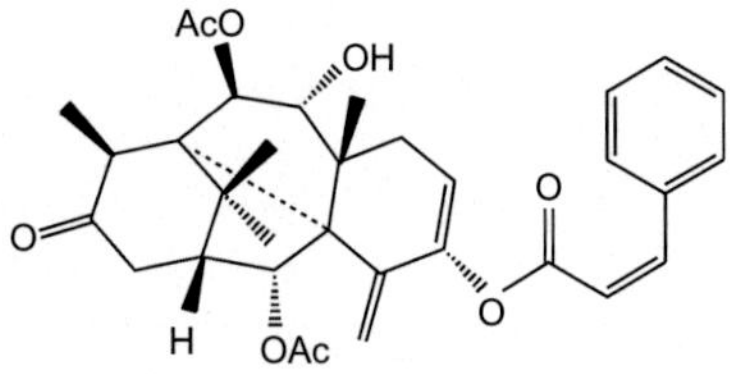

11

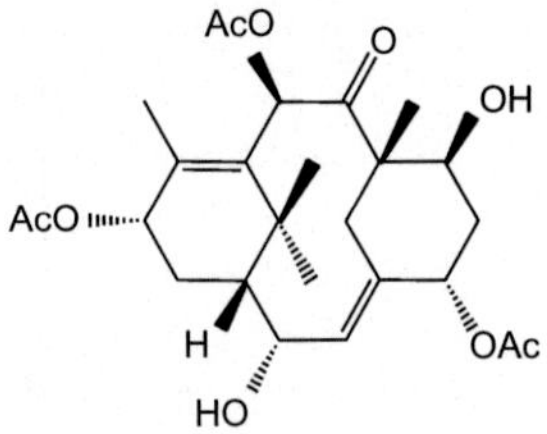

12

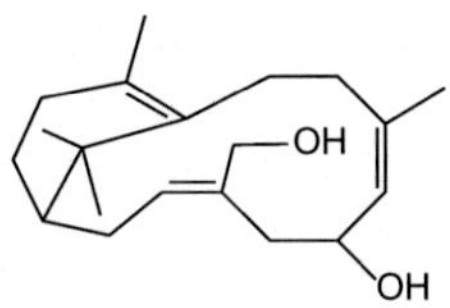

13

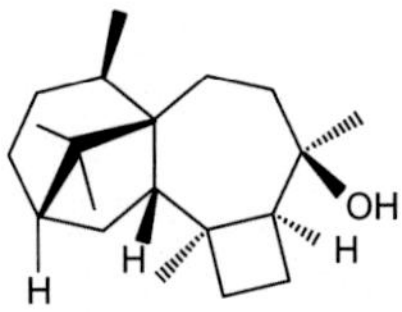

14

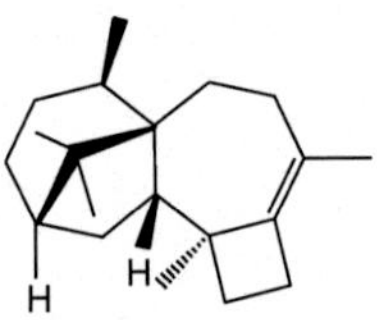

15

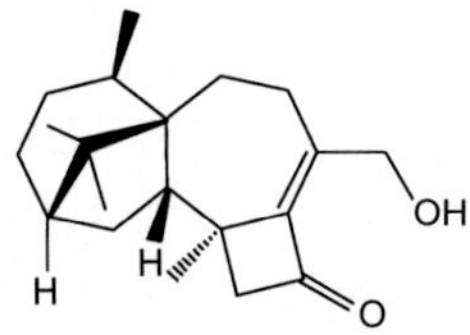

16

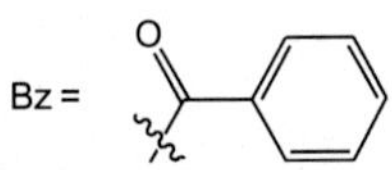

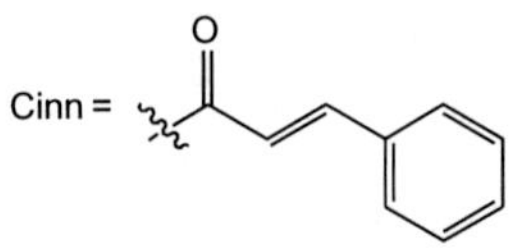

Figure 3.4 Continued.

propellane skeleton, three taxanes with [3.3.3] [3.4.5] dipropellane systems, and a novel taxane with a unique 5/5/4/6/6/6 hexacyclic skeleton, containing a unique [3.3.2] propellane, were isolated for the first time from natural sources. 13-Acetyl-9-dihydrobaccatin III, a very useful starting material for the semisynthesis of Taxol and Taxotere, represents the most abundant taxane in the needles of this species, which establishes *T. canadensis* as distinct from the remaining species. Some chemical modifications on the taxanes isolated from this species have been carried out.

3.4.2 Alkaloids

LC–MS/MS allows the simultaneous identification and quantification of the commercially available yew alkaloid taxoids: paclitaxel, 10-deacetyltaxol (10-DAT), baccatin III (BAC III), 10-deacetylbaccatin III (10-DAB III), cephalomannine (Taxol B), and

Table 3.1 **Taxanes isolated from *Taxus* in recent years**

	Compounds	Types	Species	Tissues	References
1	(2α,5α,7β,9α,10β,13α)-10,13,20-Tris(acetyloxy)-1,4,5,7,9-pentahydroxytax-11-en-2-yl benzoate	I	*Taxus cuspidata*	Needles	Cheng et al. (2010)
2	5α,10β,13α-Triacetoxytax-11-ene-2α,7β,9α,20-tetraol	I	*T. cuspidata*	Leaves	Ni et al. (2011)
3	2α,9α,10β-Triacetoxy-13α-(β-D-glucopyranosyloxy)taxa-4(20)-11-dien-5α-ol	I			
4	Taxadiene	I	*T. chinensis* var. *mairei*	Whole seedlings	Zhou et al. (2014)
5	10β-Acetoxy-5α-[(3′-dimethylamino-3′-phenyl)-propionyloxy] moiety-9α-hydroxy-tax-4(20)11-dien-13-one	I	*T. cuspidata*	Seeds	Wang et al. (2013)
6	Baccatin VIII	I	*T. yunnanensis*	Twigs and leaves	Hai et al. (2014)
7	Baccatin IX				
8	Baccatin X				
9	7β,10β-Diacetoxy-9α-hydroxy-3α,11α-cyclotaxa-4(20),5-dien-13-one	II	*T. cuspidata*	Seeds	Wang et al. (2013)
10	2α,7β,10β-Triacetoxy-5-cinnamyloxy-9α-hydroxy-3α,11α-cyclotaxa-4(20),5-dien-13-one	II			
11	2,10-Diacetyl-5(*Z*)-cinnamoylphototaxicin II	II	*T. canadensis*	Needles	Yang et al. (2009)
12	5α,10β,13β-Triacetoxy-2α,7β-dihydroxy-2(3 → 20)abeotaxa-4(20),11-dien-9-one	III	*T. cuspidata*	Leaves	Ni et al. (2011)
13	Compound 1	XI	*T. baccata*	Endophytic fungus	Adelin et al. (2014)
14	Compound 2				
15	Compound 3				
16	Compound 4				

I, normal 6/8/6-ring taxanes; II, 3,11-cyclotaxanes; III, 2(3 → 20)abeotaxanes; IV, 11(15 → 1)abeotaxanes; V, 11(15 → 1),11(10 → 9)diabeotaxanes; VI, 3,8-secotaxanes (bicyclic taxanes); VII, 14,20-cyclotaxane; VIII, 3,11:12,20-dicyclotaxane; IX, 3,11:4,12,14:20-tricyclotaxane; X, 11,12-secotaxane; XI, harziane type.

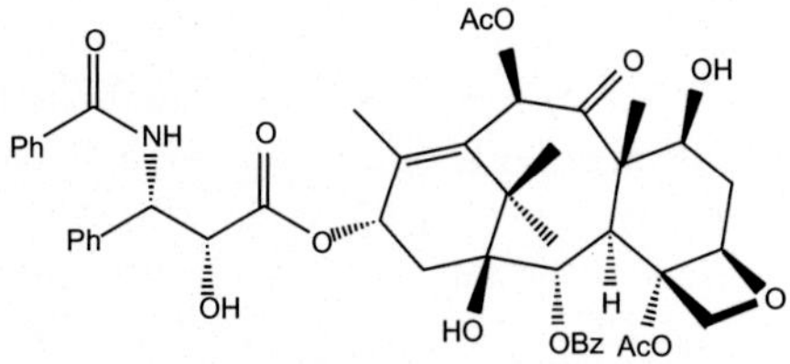

Taxol (paclitaxel, taxol A)

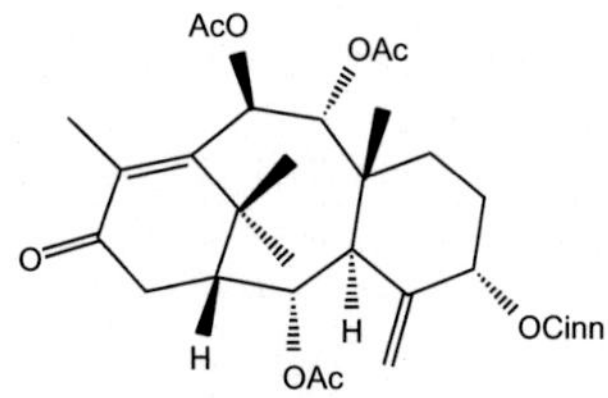

Taxinine

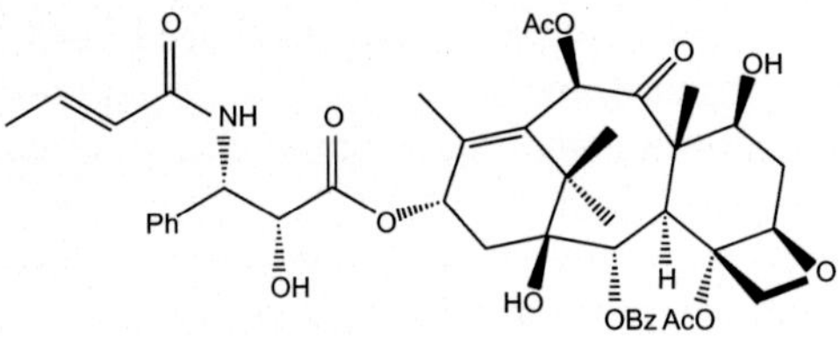

Cephalomannine (taxol B)

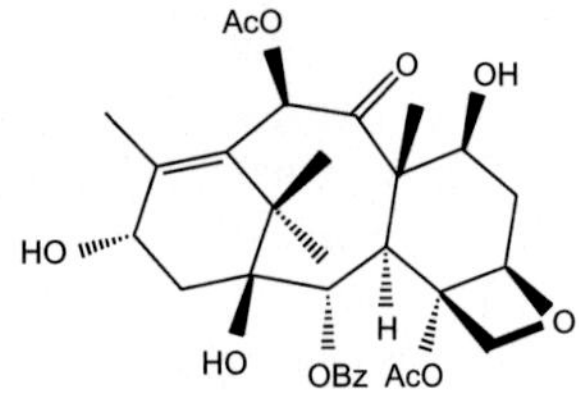

Baccatin III

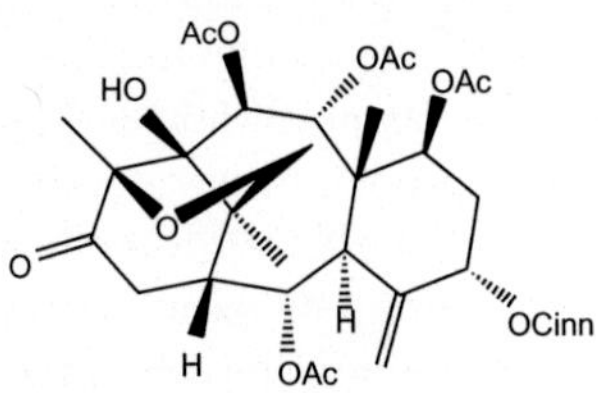

Taxagifine

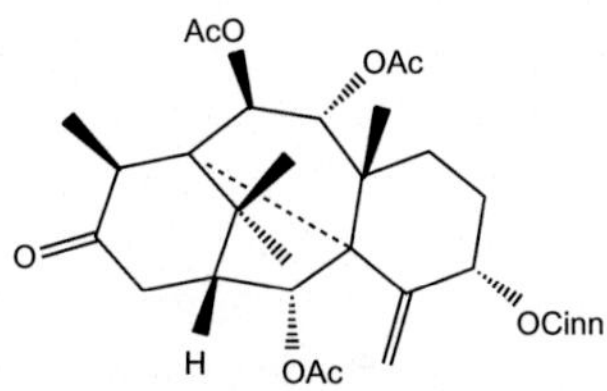

Taxuspine C

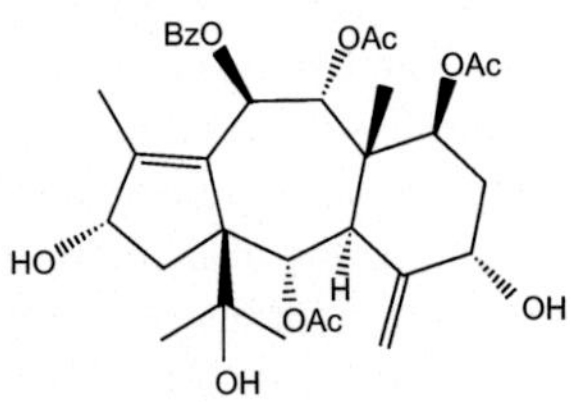

Wallifoliol

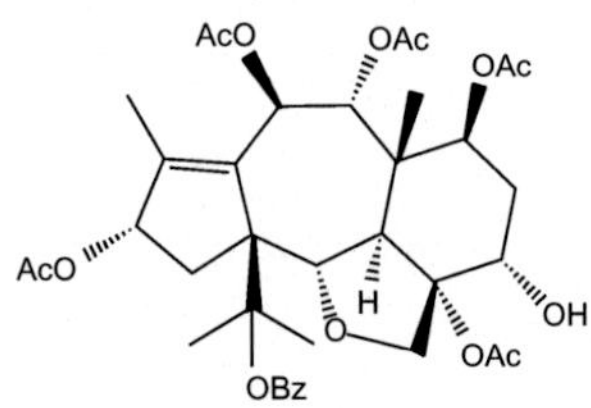

Taxuspine B

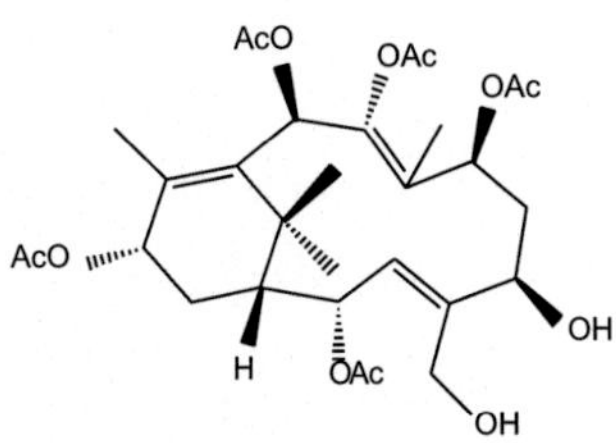

Canadensene

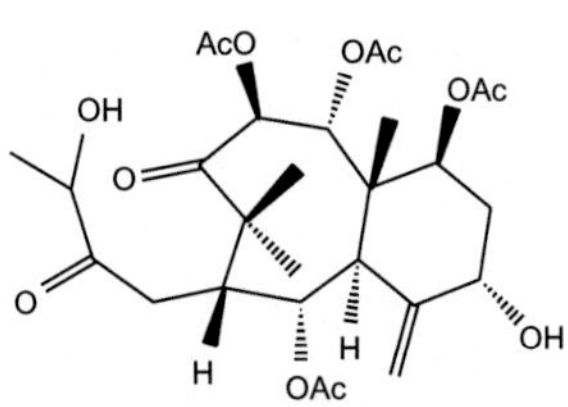

Taxusecone

Figure 3.5 Structures of representative naturally occurring taxanes.

3,5-dimethoxyphenol (3,5-DMP), also encompassing the qualitative analysis of the alkaloidal diterpenoids (Grobosch et al., 2013); reference mass spectra obtained from a yew leaf extract: monoacetyltaxine (MAT: 568.4), taxine B (584.2), monohydroxydiacetyltaxine (MHDAT: 626.4), triacetyltaxine (TAT: 652.4), and monohydroxytriacetyltaxine (MHTAT: 668.4). In fatality cases, paclitaxel, 10-DAT, and cephalomannine were not identified in urine and cardiac and femoral blood, but all taxoids and 3,5-DMP were present in the stomach content and excreted into the bile. In urine, the highest 3,5-DMP concentration was 7500 and 23,000 μg/l after enzymatic hydrolysis. In intentional and accidental poisonings, when electrocardiogram (ECG) examinations revealed ventricular tachycardia and/or prolonged QRS intervals, taxines were identified in plasma/serum, even after the ingestion of a few number of yew leaves, when 3,5-DMP was not even found. According to the data from one near-fatal intentional poisoning, elimination half-life of MAT, taxine B, MHDAT, and MHTAT in serum was calculated with 11–13 h, and taxines were detected up to $t = +122$ h postingestion of approximately two handfuls of yew leaves.

The *T. baccata* bark is known to contain 2'β-deacetoxyaustrospicatine. Initial examination of heartwood extracts for 2'β-deacetoxyaustrospicatine by LC–MS revealed the presence of this basic taxoid at about 0.0007% dry weight, using a standard isolated from bark (Kite et al., 2013). Analyses for taxine B, however, proved negative at the extract concentration analyzed. Observing other basic taxoids within the heartwood extracts was facilitated by developing generic LC–MS methods that utilized a fragment arising from the *N*-containing acyl group of basic taxoids as a reporter ion. Combining all ion collisions with high-resolution ion filtering by the orbitrap was most effective, in terms of both the number of basic taxoids detected and sensitivity. Numerous basic taxoids, in addition to 2'β-deacetoxyaustrospicatine, were revealed by this method in heartwood extracts of *T. baccata*. Red wine readily extracted the basic taxoids from heartwood, while coffee extracted them less efficiently. Contamination with basic taxoids could be detected in soft cheese that had been spread onto wood. The generic LC–MS method for detecting basic taxoids complements specific methods for detecting taxine B when investigating yew poisoning cases in which the analysis of complex extracts may be required or taxine B has not been detected.

3.4.3 Other compounds

Examples of flavonoids and lignans are shown in Figure 3.6 and Table 3.2. Thirty-three kinds of fatty acids were identified from leaves of *T. mairei* by GC–MS (Tang et al., 2013).

A novel heteropolysaccharide (TMP70W) was isolated and purified from leaves of *T. yunnanensis* by anion-exchange chromatography and gel permeation chromatography (Yan et al., 2013). Its molecular weight was 36.94 kDa and structural features were elucidated by partial acid hydrolysis, periodate oxidation-Smith degradation, methylation analysis, GC–MS, HPAEC-PAD, FT-IR, and NMR. The repeating unit of TMP70W had a backbone composed of (1→5)-linked α-L-Araf, (1→2,5)-linked α-L-Araf, and (1→6)-linked-β-D-Galp with a branch at the position of C-2 of arabinose.

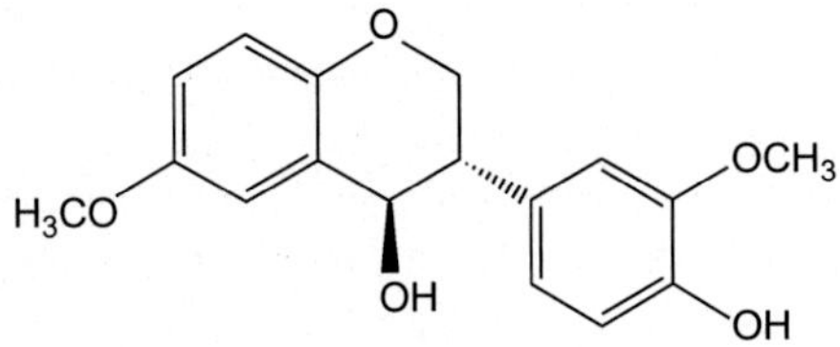

17

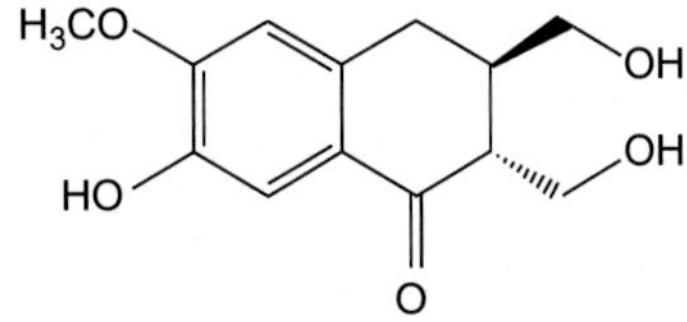

18

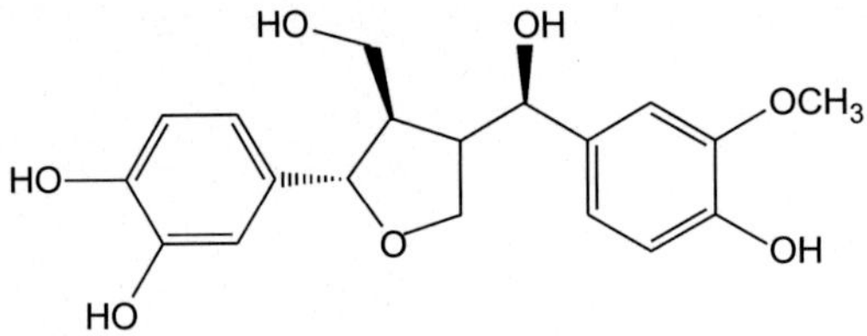

19

Figure 3.6 Examples of other compounds found in *Taxus*.

Table 3.2 Other compounds isolated from *Taxus* in recent years (Tezuka et al., 2011)

	Compounds	Types	Species	Tissues
17	(3*S*,4*R*)-4′-Hydroxy-6,3′-dimethoxyisoflavan-4-ol	A	*Taxus yunnanensis*	Wood
18	2,3-Bis(hydroxymethyl)-7-hydroxy-6-dimethoxy-1-tetralone	B		
19	(7*R*)-7-Hydroxytaxiresinol	B		

A, flavonoid; B, lignan.

Taximin was identified as a gene encoding a plant-specific, small, cysteine-rich signaling peptide, through a transcriptome survey of jasmonate-elicited *T. baccata* suspension cells grown in two media cultures (Onrubia et al., 2014). Taximin expression increased in a coordinated manner with that of paclitaxel biosynthetic genes. Tagged Taximin peptides were shown to enter the secretory system and localize to the plasma membrane. In agreement with this, the exogenous application of synthetic Taximin peptide variants could transiently modulate the biosynthesis of taxanes in *T. baccata* cell suspension cultures. Taximin peptide is widely conserved in the higher plant kingdom with a high degree of sequence conservation. Taximin overexpression could stimulate the production of nicotinic alkaloids in *Nicotiana tabacum* hairy root cultures in a synergistic manner with jasmonates. In contrast, no pronounced effects of Taximin overexpression on the specialized metabolism in *Medicago truncatula* roots were observed. This study increases our understanding of the regulation of *Taxus* diterpene biosynthesis in particular and plant metabolism in general. Taximin might increase the practical potential of metabolic engineering of medicinal plants.

3.4.4 Chemotaxonomy

Chemotaxonomy, also called chemosystematics, is the attempt to classify and identify organisms (originally plants) according to confirmable differences and similarities in their biochemical compositions. Chemotaxonomy-based plant selection is a prerequisite for the successful natural product research. Due to difficulty in PCR amplification, molecular markers are very often inapplicable for yew extracts. More importantly, the gene variations cannot represent the variations at the metabolite level that are closely related to the manufacturing process of taxanes. Novel classifications based on metabolic analysis are thus highly desirable. During the past 20 years, some substantial quantitative and qualitative variations among different *Taxus* species have been found using modern analytic techniques (van Rozendaal et al., 2000; Wang et al., 2011). However, there was no practical chemical classification that can be applied to yew species identification. The systemic analysis of yew constituents is a big challenge, due to the numerous constituents from different classes and varying metabolite levels caused by many nongenetic factors such as developmental stage, climate, elevation, and slope exposure. As a rapid, cost-efficient, and popular analysis method, HPLC (high-performance liquid chromatography) fingerprinting has been regarded as the first choice for medicinal plant identification and quality control (Lu et al., 2005; Xie et al., 2006). A holistic approach of fingerprint analysis, profile similarity-based clustering, and choice of taxonomic markers capable of capturing the greatest chemical variations is proposed for *Taxus* classification (Ge et al., 2008b). Thirty samples representing eight *Taxus* species are collected and analyzed, and the fingerprint-based data are extracted and processed by hierarchical cluster analysis (HCA) and principal component analysis (PCA). Based on the PCA loadings, 12 chemical constituents, identified by LC–mass spectrometry (MS), are selected as the chemotaxonomic markers that can be used to establish a more sensible classification. Eight studied species are divided into six well-supported groups, and most samples can be assigned to the correct species. Traditional chemotaxonomic and chemosystematic studies are frequently used to infer relationships among plant taxa, by using the average concentration of several preselected compounds (van Rozendaal et al., 1999). However, they could not be used to examine the variations within species and may result in wrong conclusions in cases where the intraspecific variation is large (Becerra, 2003; Wink, 2003). In contrast, profile-based classification can investigate variations within and among species by comparison of fingerprints (Vieira et al., 2003). Moreover, the fingerprint similarity-based taxonomy, which relies on the ratio of selected constituents, can improve the misclassifications caused by large quantitative differences.

3.4.5 Metabolomics and functional genomics

Metabolomics is the systematic study of the unique chemical fingerprints that specific cellular processes leave behind (Daviss, 2005), that is, the study of their small-molecule metabolite profiles. Gas chromatography (GC) coupled with time-of-flight (TOF) MS was used to profile metabolite changes of *T. cuspidata* cells under laminar shear stress (Han and Yuan, 2009a). Yet, LC–MS has been recognized as the most appropriate analytic technique of metabolomics for the structural assignment of taxanes in complex samples due to its specificity and structure characterization ability (Guitton et al., 2005; Tanaka et al., 2011; Wang et al., 2011). To understand the responses of phospholipids in shear stress-induced mechanotransduction, a lipidomic approach using LC/electrospray ionization (ESI)/MS was employed to profile phospholipid species of *T. cuspidata* cells under laminar shear stress (Han and Yuan, 2009b). However, rapid metabolic profiling of yew materials is still a challenge, since there are numerous trace ingredients from different classes and many analogs with similar chromatographic behavior and many regio- and stereoisomers (Vivekanandan et al., 2006). In recent years, ultraperformance liquid chromatography (UPLC), which employs sub-2 mm stationary phase particles to achieve superior theoretical plates and extremely high resolution in very short analytic times, has attracted the wide attention of pharmaceutical and biochemical analysts (Plumb et al., 2004; O'Connor et al., 2006). UPLC and its hyphenated system also showed excellent sensitivity and improved peak capability (Churchwell et al., 2005; Novakova et al., 2006). These advantages make it possible to rapidly detect trace ingredients from crude yew samples. In 2008, an efficient and sensitive profiling approach was developed for the complex yew samples, using UPLC/ESI/MS (Ge et al., 2008a). The appropriate in-source collision-induced dissociation (CID) energy was employed to produce informative characteristic ions that could be used for the stereochemical and substructural assignment of yew constituents. The method was successfully applied in the rapid screening of yew hair roots from various species, and 53 constituents including 47 taxoids were detected from the partially purified root extract. C-7 hydroxytaxane stereoisomers could be identified based on their different fragment ions under the optimal profiling conditions. It was also observed that hair roots from different *Taxus* species exhibited nearly identical chemical distribution, indicating that they had similar metabolic framework (Hao et al., 2011a). *Taxus* root resources display benign medicinal prospects because they have relatively simple chemical profiles and possess high yields of valuable taxanes such as paclitaxel, cephalomannine (Figure 3.5), 10-DAT, and 7-xylosyltaxanes. Furthermore, these pharmaceutically important taxanes compose over 55% of the peak areas in the entire chromatogram (0–19 min), which is quite different from the needles that usually have major divergent pathways apart from the paclitaxel biosynthesis such as the formation of abundant taxine B and taxinine M (Figure 3.7) (Ketchum et al., 2003). These detected taxanes and their connectivity facilitate further studies on the taxane drug discovery and development from the *Taxus* root.

To date, more than 400 natural taxanes and hundreds of synthetic analogs have been discovered (Ojima et al., 2005; Liu et al., 2008; Chen and Liang, 2008; Wu et al., 2010; Wang et al., 2011). The core structure of paclitaxel and its analogs consists of four rings with substituents at different carbon sites. Most taxanes have a β-hydroxyl at the C-7 site, and 7α stereoisomers often coexist because the conversion of 7-hydroxyl from β-equatorial orientation into α-axial orientation is very facile (Tian and Stella, 2008). The C-7 epimerization readily takes place in solvent under heating or under mildly basic conditions such as physiological pH (MacEachern-Keith et al., 1997). 7-Epi-paclitaxel and other 7α-hydroxyltaxanes have been detected from natural sources, in the manufacturing process for paclitaxel-active pharmaceutical

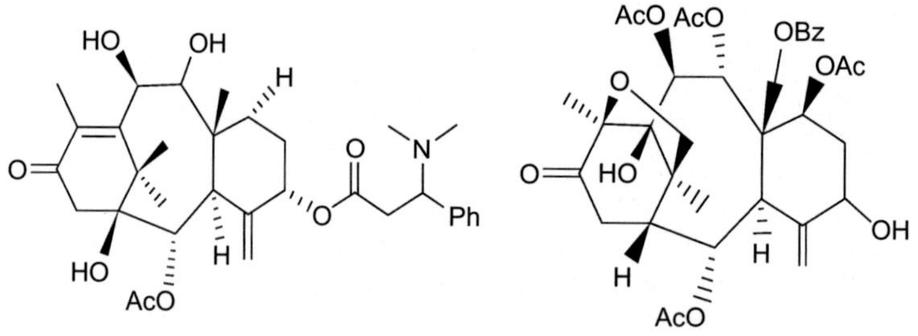

Figure 3.7 Structure of taxine B and taxinine M, two prominent side-route metabolites of *Taxus*. Highly abundant taxine B was found in the needles of *T. baccata* and *T. cuspidata*, while abundant taxinine M was detected in the needles of *T. mairei*, *T. yunnanensis*, and *T. chinensis*.

ingredients, and from taxane-related metabolic samples (Ge et al., 2009). As a common impurity and metabolite of taxane drugs, the monitoring of 7-epi-paclitaxel and other 7α-hydroxyltaxanes is important for quality control and pharmacological research of taxane drugs. However, the identification and characterization of these 7α-hydroxyltaxanes from realistic samples were difficult due to their presence in trace amounts and the coexistence of the corresponding 7β-hydroxyl isomers. In 2009, different ESI-MS methods were utilized to analyze several pairs of taxane stereoisomers including paclitaxel and 7-epi-paclitaxel (Ge et al., 2009). Both ESI-MS and tandem MS provided stereochemically dependent mass spectra in negative ion mode, and all studied stereoisomers could be easily discriminated based on their feature ions or distinct fragmentation patterns. MS/MS experiments for several taxane analogs at various collision energies were performed to elucidate potential dissociation pathways. The gas-phase deprotonation potentials were calculated to estimate the most thermodynamically favorable deprotonation site. The results of the theoretical studies agreed well with the fragmentation patterns of paclitaxel and 7-epi-paclitaxel observed from MS/MS experiments. LC/ESI/MS was a useful and sensitive technique for the assignment of C-7 taxane stereoisomers from realistic samples.

Several solutions have been utilized to overcome the supply crisis and to produce taxane drugs, including plant cell culture (Aoyagi et al., 2002; Baebler et al., 2002; Zhao and Yu, 2005) and semisynthesis from common natural precursors that can be isolated from renewable yew resources such as needles and roots (Figure 3.1a and 3.8) (Ge et al., 2008a,b; Hao et al., 2011a). Several natural taxanes including 10-DAT, 10-DAB, baccatin III, 9-dihydro-13-acetylbaccatin III, cephalomannine, and 10-deacetyl-7-xylosylpaclitaxel can be converted to paclitaxel or docetaxel via a few steps, and they are recognized as the paclitaxel equivalents or valuable taxanes (Ge et al., 2008a,b; Hao et al., 2011a). Paclitaxel and these precursors are found in most *Taxus* species, but their distribution and level are highly variable with species and tissues. Therefore, the screening of the valuable yew constituents from different species and tissues is crucial to the cultivation of yew trees and cost-effective manufacturing of taxane drugs. The rapid identification and accurate quantification of target compounds are critical for the quality control of medicinal plants. LC-based

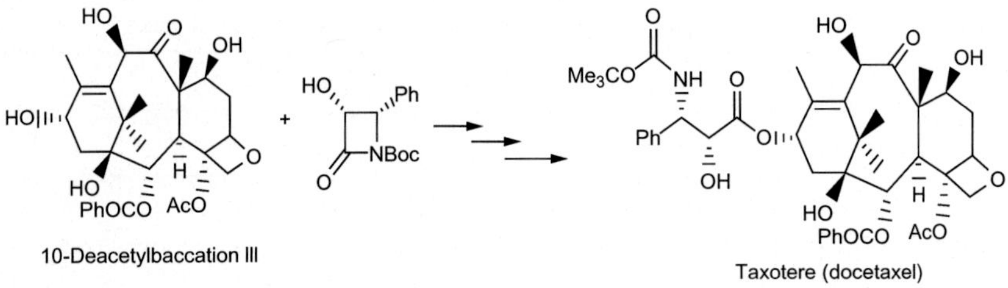

Figure 3.8 Semisynthesis of docetaxel, the analog of the anticancer drug paclitaxel.

techniques have been widely used for the assignment and quantification of taxanes in biological or botanical samples (Ge et al., 2008a,b). However, the rapid, sensitive, and simultaneous determination of target taxanes from yew materials was difficult due to the chemical complexity of the crude yew extract that contains trace amounts of taxanes. UPLC and ultrafast liquid chromatography (UFLC), which employ fine stationary-phase particles to achieve extreme high resolution with short analytic time and good sensitivity, have attracted the wide attention of pharmaceutical analysts for its rapid determination of trace constituents from complex samples (Dong et al., 2009). These advantages facilitate the large-scale screening of trace constituents from crude yew samples. A rapid and valid method incorporating UFLC with MS and UV detection has been developed for the simultaneous determination of paclitaxel and its six semisynthesis precursors in needles and hair roots from various *Taxus* species (Ge et al., 2010). All target analytes could be identified by comparing their retention times and UV and MS spectra with authentic standards, while seven valuable taxanes in botanical samples can be rapidly determined by UFLC-diode array detection (DAD) with excellent sensitivity. Analysis of more than one hundred yew samples from nine species showed significant variations in distribution and content of seven evaluated taxanes (Tables 3.3 and 3.4) (also see Hao et al., 2011a). Thus, different development strategies should be used for the sustainable utilization of various yew resources.

Metabolomic and transcriptomic profiling data were integrated to elucidate the crucial network controls on Taxol and its precursor biosynthesis during the taxane core functionalization within methyl jasmonate (MeJA)-induced *T. chinensis* cells (Song et al., 2014). Twelve metabolites were identified using LC/ESI/MS that contain Taxol (paclitaxel), baccatin III (BAC III) and its analogs, a group structurally bearing multiple free hydroxyls (TAX), and another group of multiple acyl taxanes (MAT), including taxuyunnanine C (TC) and its analogs. The metabolomic profile showed a higher increase in TAX than in MAT. The ratio of BAC III and Taxol to the total taxane content increased more significantly in TAX than in MAT. The MAT proportion did not significantly change, although they are predominant components in cell cultures compared with TAX. Quantitative real-time polymerase chain reaction (qRT-PCR) was used to determine the transcription level of 20 genes, among which 11 were reported responsible for Taxol biosynthesis and 9 were obtained from previous transcriptomic data. The total expression levels of hydroxylase after 24 h and 6 days were higher than those of acylase. The principal component analysis (PCA) validated the

Table 3.3 Quantitative difference in the average concentration of six valuable taxanes in needles of *T. mairei* (different ages)

	Age of trees (years)							
Taxane	**Three**	**Four**	**Five**	**Six**	**Seven**	**Average**	**SD**	**RSD (%)**
10-DAB	168.2	104.6	87.9	70.5	85.4	103.3	34.2	33.1
Baccatin	25.9	21.6	14.0	10.6	9.9	16.4	6.3	38.3
10-DAXT	341.3	576.7	558.4	513.1	539.5	505.8	84.9	16.8
10-DAT	157.7	170.3	158.7	66.2	125.9	135.7	37.8	27.8
Cephalomannine	73.0	84.5	170.2	35.7	106.8	81.4	26.4	32.4
Paclitaxel	52.7	82.1	84.5	22.3	53.7	59.1	22.8	38.6
Total (μg/g dry needle)	818.8	1039.8	1010.7	718.4	921.2	901.7		

Three samples were collected from three individual yews at each age. DHB (9-dihydro-13-acetylbaccatin III) was not detected in *T. mairei* and was uniquely found in *T. canadensis* (Ge et al., 2010; Hao et al., 2011a). 10-DAB, 10-Deacetylbaccatin III; 10-DAXT, 10-deacetyl-7-xylosylpaclitaxel; 10-DAT, 10-deacetylpaclitaxel. SD, Standard deviation; RSD, relative standard deviation.

Table 3.4 Quantitative difference in the average concentration of six valuable taxanes in roots of *T. mairei* (different age)

	Age of trees (years)							
Taxane	**Three**	**Four**	**Five**	**Six**	**Seven**	**Average**	**SD**	**RSD (%)**
10-DAB	125.0	177.6	56.4	87.4	195.6	128.4	52.6	40.9
Baccatin	200.6	290.2	165.8	248.2	150.5	211.1	51.9	24.6
10-DAXT	243.0	544.6	209.5	519.0	802.5	463.7	218.0	47.0
10-DAT	226.6	681.4	121.4	361.6	280.3	334.3	190.3	56.9
Cephalomannine	231.3	361.6	155.9	182.0	246.1	235.4	71.0	30.2
Paclitaxel	329.9	480.6	103.3	360.0	406.9	336.1	127.1	37.8
Total (μg/g dry powder)	1356.5	2056.0	709.3	1398.2	1675.1	1439.0		

Three samples were collected from three individual yews at each age. DHB (9-dihydro-13-acetylbaccatin III) was not detected in *T. mairei* and was uniquely found in *T. canadensis* (Ge et al., 2010; Hao et al., 2011a). 10-DAB, 10-Deacetylbaccatin III; 10-DAXT, 10-deacetyl-7-xylosylpaclitaxel; 10-DAT, 10-deacetylpaclitaxel. SD, Standard deviation; RSD, relative standard deviation.

metabolomic data, indicating that hydroxylation was more crucial than acylation for controlling the flux towards TAX biosynthesis. PCA contribution comparison showed that two undefined genes of OHX1 and ACX3 might have good potential in TAX upregulation and MAT downregulation. This study provides the first experimental evidence on the contribution of total hydroxylation to taxane biosynthesis.

3.5 Proteomics

Pinus massoniana (AR-sensitive) and *T. mairei* (AR-resistant) are widely distributed in southern China. Under acid rain (AR) stress, significant necrosis and collapsed lesions were found in *P. massoniana* needles with remarkable yellowing and wilting tips, whereas *T. mairei* did not exhibit chlorosis and visible damage (Hu et al., 2014). Due to the activation of a large number of stress-related genes and the synthesis of various functional proteins to counteract AR stress, it is important to study the differences in AR tolerance mechanisms by comparative proteomic analysis of tolerant and sensitive species. Sixty-five and 26 differentially expressed proteins were identified in *P. massoniana* and *T. mairei*, respectively. Among them, proteins involved in metabolism, photosynthesis, signal transduction, and transcription were radically downregulated in *P. massoniana*, whereas most of the proteins participating in metabolism, cell structure, photosynthesis, and transcription were increased in *T. mairei*. These results suggest the distinct patterns of protein expression in the two woody species in response to AR, allowing a deeper understanding of diversity on AR tolerance in forest tree species.

3.6 Bibliometric analysis of *Taxus* research

A bibliometric analysis is performed to evaluate the global scientific production of *Taxus* research, characterize *Taxus* research activities, and identify patterns, tendencies, and regularities of *Taxus*-related articles. Data are based on the Science Citation Index Expanded (SCI Expanded) from the Web of Science database. Articles referring to *Taxus* are assessed by the trend of publication output during 1991–2010. The records are downloaded into Microsoft Excel 2007, and additional coding is manually performed for all data analysis. Globally, 2916 papers were published during the 20-year study period. The most productive countries, institutions, and Web of Science subject categories and journals, as well as the most cited articles, are identified. The mainstream research on *Taxus* was in the plant sciences, biochemistry and molecular biology, cardiac and cardiovascular systems, biotechnology, and applied microbiology. The G7 industrial countries and China and India held the majority of total world production. Research on the various economically important *Taxus* species remained the hot spot during the 20-year study period, whereas that on the related topic "paclitaxel-eluting stents" increased dramatically since 2002 (Chen and Zhan, 2011). The analysis of single words in abstracts is performed to make specific inferences about the scientific literature and identify the subjective focus and emphasis specified by authors. Similar to the distribution of paper titles, "Taxol," "cell(s)," and "baccata" were among the most frequently used single words during 1991–2010. Interestingly, "isolated," "analysis," and "species" were also emphasized in abstracts. Isolation, purification, and phytochemical analyses of paclitaxel and taxoids from the respective *Taxus* species and endophytic fungi are the hot issues (Guenard et al., 1993; Baloglu and Kingston, 1999). The analysis of the distribution of the author keywords suggests that, among various *Taxus* species, *T. cuspidata* was the most commonly studied (Ge et al., 2008b), followed by *T. baccata* (Parmar et al., 1999), *T. chinensis*

(Eisenreich et al., 1996), *T. mairei* (Hao et al., 2011a,b), *T. yunnanensis* (Parmar et al., 1999; Ge et al., 2008a,b), and *T. canadensis* (Ketchum et al., 1999), while *T. wallichiana* (Himalayan yew) (Parmar et al., 1999; Ge et al., 2008b), *T. brevifolia* (Trapp and Croteau, 2001), and *T.* × *media* (Hao et al., 2008a,b,c) were less commonly studied. Among various *Taxus* tissues, the stem bark was the most frequently studied (Wani et al., 1971; Eldridge et al., 2002), followed by needles (Ge et al., 2008b; Hao et al., 2011a,b), pollen (O'Leary et al., 2007; Zimmermann, 2010), seeds (Li et al., 2005; Huo et al., 2007), and roots (Onrubia et al., 2011; Hao et al., 2011a). With the synthetic analysis of words in article title, author keyword, abstract, and KeyWords Plus (Hao et al., 2012a), some *Taxus* research hot spots have been noticed. *Taxus* cell culture systems, which are commercially successful, allow for the sustainable production of *Taxus* secondary metabolites, which is not limited by the low yields associated with natural harvest or the high cost associated with complex chemical synthesis (Wilson and Roberts, 2012). Semisynthesis from the precursor baccatin III or 10-DAB, which can be extracted from *Taxus* leaves and roots, proved to be another valid method for the commercial production of paclitaxel (Fu et al., 2009). Encouraging findings with the endophytic fungi of *Taxus* resulted in much interest in the prospect of using endophytes as the producer of paclitaxel and other taxanes (Miller et al., 2008; Soca-Chafre et al., 2011). The genome size of fungi is much smaller than that of *Taxus*, making it easier to perform the whole-genome sequencing, assembly, and bioinformatic prediction of the endophytes, which would lay a solid foundation for biosynthesis studies, metabolic engineering, and high-yield strain breeding. The application of compounds derived from *Taxus* in clinical cardiology, pharmacology, and oncology and studies related to *Taxus* chemistry, metabolism, cytology, and microbiology are the ongoing Taxus-related research in the twenty-first century. Gaps are still present in knowledge about the genomics, epigenomics, transcriptomics, proteomics, metabolomics, and bioinformatics of *Taxus* and their endophytic fungi. For example, only the conventional two-dimensional gel electrophoresis was used to study the responses of *T. cuspidata* cells to local microenvironments in different zones of immobilized support matrices (Cheng and Yuan, 2006). The proteomic study of *Taxus* is still in its infancy. Epigenomics is the study of the complete set of epigenetic modifications on the genetic material of a cell, known as the epigenome. The field is analogous to genomics and proteomics, which are the study of the genome and proteome of a cell (Russell, 2010), respectively. It is unknown how the epigenomic mechanisms regulate the biosynthesis of secondary metabolites in *Taxus*.

3.7 Conclusion and prospects

Plant-derived natural products hold great promise for the discovery and development of new pharmaceuticals. Careful consideration of the entire process of discovery and development, a "systems" approach, will be required to realize this great promise effectively. The biological and chemical studies of *Taxus* with respect to the emerging use of omics technologies provide a paradigm of the active integration of various state-of-the-art methodologies into the early stage of drug research and development. Both aspects are complementary and indispensable in the taxonomy and authentication of *Taxus* and in the elucidation of the correlation between the genotype and the metabolic phenotype. It is no doubt that *Taxus* research is an exciting and revolutionary multidisciplinary field at the center of many current key scientific issues. Although semisynthesis and subsequent plant cell culture-based production efforts have decreased the need for harvesting the endangered yew tree, production still depends on plant-based processes. Recent developments in metabolic engineering and synthetic biology offer new pathways for the overproduction of complex natural products by optimizing more technically amenable microbial hosts (e.g., *E. coli*) (Ajikumar et al., 2010; Morrone et al., 2010). Moreover, the crystal structure of the TS, the enzyme that catalyzes the first committed step of paclitaxel biosynthesis, and the evolution of modular architecture in terpene biosynthesis have been elucidated (Köksal et al., 2011), which will accelerate the study of *Taxus* metabolism and physiology and promote the research and development of the microbial cell factory. Confronting the environmental challenge, the regulation of *Taxus* biological processes has to be of multiple levels, that is, the genomic level, epigenomic level, transcriptional and posttranscriptional levels, and translational and posttranslational levels. Knowledge about the regulation of almost all these levels is lacking, although it is essential for the sustainable development and utilization of the *Taxus* medicinal resources. Systems biology and various omics technologies will play an increasingly important role in the coming decades.

Acknowledgments

Richard W. Spjut of World Botanical Associates helps edit and polish the manuscript. We thank him for his constructive comments.

References

Adelin, E., Servy, C., Martin, M., et al., 2014. Bicyclic and tetracyclic diterpenes from a *Trichoderma* symbiont of *Taxus baccata*. Phytochemistry 97, 55–61.

Ajikumar, P.K., Xiao, W.H., Tyo, K.E., et al., 2010. Isoprenoid pathway optimization for Taxol precursor overproduction in *Escherichia coli*. Science 330, 70–74.

Aoyagi, H., Dicosmo, F., Tanaka, H., 2002. Efficient paclitaxel production using protoplasts isolated from cultured cell of *Taxus cuspidata*. Planta Med. 68, 420–424.

Baebler, S., Camloh, M., Kovac, M., et al., 2002. Jasmonic acid stimulates taxane production in cell suspension culture of yew (*Taxus* × *media*). Planta Med. 68, 475–476.

Baloglu, E., Kingston, D.G.I., 1999. The taxane diterpenoids. J. Nat. Prod. 62, 1448–1472.

Becerra, J.X., 2003. Synchronous coadaptation in an ancient case of herbivory. Proc. Natl. Acad. Sci. U. S. A. 100, 12804–12807.

Bishop, J.G., 2005. Directed mutagenesis confirms the functional importance of positively selected sites in polygalacturonase inhibitor protein. Mol. Biol. Evol. 22, 1531–1534.

Brown, R.G., Hull, F.E., 1951. *Taxus* (yew) poisoning of cattle. J. Am. Vet. Med. Assoc. 118, 398–399.

Brown, A.P., Kroon, J.T., Swarbreck, D., et al., 2012. Tissue-specific whole transcriptome sequencing in castor, directed at understanding triacylglycerol lipid biosynthetic pathways. PLoS One 7, e30100.

Burcham, G.N., Becker, K.J., Tahara, J.M., et al., 2013. Myocardial fibrosis associated with previous ingestion of yew (*Taxus* sp.) in a Holstein heifer: evidence for chronic yew toxicity in cattle. J. Vet. Diagn. Investig. 25 (1), 147–152.

Chakravarthi, B.V., Sujay, R., Kuriakose, G.C., et al., 2013. Inhibition of cancer cell proliferation and apoptosis-inducing activity of fungal taxol and its precursor baccatin III purified from endophytic *Fusarium solani*. Cancer Cell Int. 13 (1), 105.

Chen, X.Y., Liang, J.Y., 2008. Studies on chemical constituents of planted *Taxus mairei* (III). Acta Pharm. Sin. 43, 1205–1207.

Chen, W.P., Zhan, H.B., 2011. Recent developments of drug eluting stent coatings. Acta Pharm. Sin. 46, 1301–1307.

Chen, S., Liu, J., Gong, H., Yang, D., 2009. Identification and antibacterial activity of secondary metabolites from *Taxus* endophytic fungus. Chin. J. Biotechnol. 25, 368–374.

Chen, J., Yuan, L., Fan, Q., et al., 2012. Adjuvant effect of docetaxel on the immune responses to influenza A H1N1 vaccine in mice. BMC Immunol. 13, 36.

Cheng, J.S., Yuan, Y.J., 2006. Proteomic analysis reveals the spatial heterogeneity of immobilized *Taxus cuspidata* cells in support matrices. Proteomics 6, 2199–2207.

Cheng, M., Dong, M., Li, Y., et al., 2010. A new taxane with an opened oxetane ring from the needles of *Taxus cuspidata*. Chem. Biodivers. 7, 2035–2039.

Christenhusz, M., Reveal, J., Farjon, A., et al., 2011. A new classification and linear sequence of extant gymnosperms. Phytotaxa 19, 55–70.

Churchwell, M.I., Twaddle, N.C., Meeker, L.R., Doerge, D.R., 2005. Improving LC-MS sensitivity through increases in chromatographic performance: comparisons of UPLC-ES/MS/MS to HPLC-ES/MS/MS. J. Chromatogr. B 825, 134–143.

Cohen, D.J., Osnabrugge, R.L., Magnuson, E.A., et al., 2014. Cost-effectiveness of percutaneous coronary intervention with drug-eluting stents versus bypass surgery for patients with 3-vessel or left main coronary artery disease: final results from the Synergy Between Percutaneous Coronary Intervention With TAXUS and Cardiac Surgery (SYNTAX) trial. Circulation 130 (14), 1146–1157.

Collins, D., Mill, R.R., Moller, M., 2003. Species separation of *Taxus baccata*, *T. canadensis*, and *T. cuspidata* (Taxaceae) and origins of their reputed hybrids inferred from RAPD and cpDNA data. Am. J. Bot. 90, 175–182.

Collins, L.J., Biggs, P.J., Voelckel, C., Joly, S., 2008. An approach to transcriptome analysis of non-model organisms using short-read sequences. Genome Inform. 21, 3–14.

Croteau, R., Ketchum, R.E.B., Long, R.M., et al., 2006. Taxol biosynthesis and molecular genetics. Phytochem. Rev. 5, 75–97.

Cui, Q.L., Shao, M., Shu, Q.J., 2013. Study on inhibitory effect of aqueous extract of *Taxus chinensis* var. *mairei* on growth of A549 lung cancer xenografts in nude mice and its mechanism. Zhongguo Zhong Yao Za Zhi 38 (20), 3549–3553.

Daviss, B., 2005. Growing pains for metabolomics. The Scientist 19, 25–28.

Deng, Q., Su, Y.J., Wang, T., 2013. Microsatellite loci for an old rare species, *Pseudotaxus chienii*, and transferability in *Taxus wallichiana* var. *mairei* (Taxaceae). Appl. Plant Sci. 1 (5), 1200456.

Dong, P.P., Ge, G.B., Zhang, Y.Y., et al., 2009. Quantitative structure-retention relationship studies for taxanes including epimers and isomeric metabolites in ultra fast liquid chromatography. J. Chromatogr. A 1216, 7055–7062.

Durak, Z.E., Büber, S., Devrim, E., et al., 2014. Aqueous extract from taxus baccata inhibits adenosine deaminase activity significantly in cancerous and noncancerous human gastric and colon tissues. Pharmacogn. Mag. 10 (S2), S214–S216.

Dutta, S., Mariappan, G., Sarkar, D., Sarkar, P., 2010. Assessment of anti-inflammatory activity of *Taxus baccata* Linn. bark extract. Anc. Sci. Life 29 (3), 19–21.

Egan, A.N., Schlueter, J., Spooner, D.M., 2012. Applications of next-generation sequencing in plant biology. Am. J. Bot. 99, 175–185.

Eisenreich, W., Menhard, B., Hylands, P.J., et al., 1996. Studies on the biosynthesis of taxol: the taxane carbon skeleton is not of mevalonoid origin. Proc. Natl. Acad. Sci. U. S. A. 93, 6431–6436.

Eldridge, G.R., Vervoort, H.C., Lee, C.M., et al., 2002. High-throughput method for the production and analysis of large natural product libraries for drug discovery. Anal. Chem. 74, 3963–3971.

Feng, L., Wanninayake, U., Strom, S., et al., 2011. Mechanistic, mutational, and structural evaluation of a *Taxus* phenylalanine aminomutase. Biochemistry 50, 2919–2930.

Fu, Y., Li, S., Zu, Y., et al., 2009. Medicinal chemistry of paclitaxel and its analogues. Curr. Med. Chem. 16, 3966–3985.

Gajurel, J.P., Cornejo, C., Werth, S., et al., 2013. Development and characterization of microsatellite loci in the endangered species *Taxus wallichiana* (Taxaceae). Appl. Plant Sci. 1 (3), 1200281.

Gao, W.Y., Jia, W., Duan, H.Q., Xiao, P.G., 2003. Industrialization of medicinal plant tissue culture. China J. Chin. Mat. Med. 28, 385–390.

Gausterer, C., Stein, C., Stimpfl, T., 2012. Application of direct PCR in a forensic case of yew poisoning. Int. J. Legal Med. 126, 315–319.

Ge, G.B., Luan, H.W., Zhang, Y.Y., et al., 2008a. Profiling of yew hair roots from various species using ultra-performance liquid chromatography/electrospray ionization mass spectrometry. Rapid Commun. Mass Spectrom. 22, 2315–2323.

Ge, G.B., Zhang, Y.Y., Hao, D.C., et al., 2008b. Chemotaxonomic study of medicinal *Taxus* species with fingerprint and multivariate analysis. Planta Med. 74, 773–779.

Ge, G.B., Zhang, R., Ai, C.Z., et al., 2009. Stereochemical differentiation of C-7 hydroxyltaxane isomers by electrospray ionization mass spectrometry. Rapid Commun. Mass Spectrom. 23, 425–432.

Ge, G.B., Liang, S.C., Hu, Y., et al., 2010. Rapid qualitative and quantitative determination of seven valuable taxanes from various *Taxus* species by UFLC-ESI-MS and UFLC-DAD. Planta Med. 76, 1773–1777.

Grobosch, T., Schwarze, B., Felgenhauer, N., et al., 2013. Eight cases of fatal and non-fatal poisoning with *Taxus baccata*. Forensic Sci. Int. 227 (1–3), 118–126.

Gu, Q., Li, Y., Chen, Y., et al., 2013. Sciadopitysin: active component from *Taxus chinensis* for anti-Alzheimer's disease. Nat. Prod. Res. 27 (22), 2157–2160.

Guenard, D., Gueritte-Voegelein, F., Potier, P., 1993. Taxol and taxotere—discovery, chemistry and structure-activity relationships. Acc. Chem. Res. 26, 160–167.

Guitton, J., Cohen, S., Tranchand, B., et al., 2005. Quantification of docetaxel and its main metabolites in human plasma by liquid chromatography/tandem mass spectrometry. Rapid Commun. Mass Spectrom. 19, 2419–2426.

Guy, K.R., 1987. The Emperor's Four Treasuries: Scholars and the State in the Late Ch'ien-lung era (Harvard East Asian Monographs 129). Harvard University Press, Cambridge, MA.

Hai, P., Wen, S., Li, Y., et al., 2014. New taxane diterpenoids from *Taxus yunnanensis*. Nat. Prod. Bioprospect. 4, 47–51.

Hall, N., 2007. Advanced sequencing technologies and their wider impact in microbiology. J. Exp. Biol. 210, 1518–1525.

Han, P.P., Yuan, Y.J., 2009a. Metabolic profiling as a tool for understanding defense response of *Taxus cuspidata* cells to shear stress. Biotechnol. Prog. 25, 1244–1253.

Han, P.P., Yuan, Y.J., 2009b. Lipidomic analysis reveals activation of phospholipid signaling in mechanotransduction of *Taxus cuspidata* cells in response to shear stress. FASEB J. 23, 623–630.

Hao, D.C., Ge, G.B., Yang, L., 2008a. Bacterial diversity of *Taxus* rhizosphere: culture-independent and culture-dependent approaches. FEMS Microbiol. Lett. 284, 204–212.

Hao, D.C., Huang, B.L., Yang, L., 2008b. Phylogenetic relationship within the genus *Taxus* inferred from protein-coding genes and non-protein-coding DNA sequences. Biol. Pharm. Bull. 31, 260–265.

Hao, D.C., Xiao, P.G., Huang, B.L., et al., 2008c. Interspecific relationships and origins of Taxaceae and Cephalotaxaceae revealed by partitioned Bayesian analyses of chloroplast and nuclear DNA sequences. Plant Syst. Evol. 276, 89–104.

Hao, D.C., Huang, B., Chen, S.L., Mu, J., 2009a. Molecular evolution and phylogenetic analysis of the chloroplast trnL-trnF region in Taxaceae and Cephalotaxaceae. Biochem. Genet. 47, 351–369.

Hao, D.C., Yang, L., Huang, B.L., 2009b. Molecular evolution of paclitaxel biosynthetic genes TS and DBAT of *Taxus* species. Genetica 135, 123–135.

Hao, D.C., Chen, S.L., Xiao, P.G., Peng, Y., 2010a. Authentication of medicinal plants by DNA based markers and genomics. Chin. Herb. Med. 2, 250–261.

Hao, D.C., Chen, S.L., Xiao, P.G., 2010b. Molecular evolution and positive Darwinian selection of the chloroplast maturase matK. J. Plant Res. 123, 241–247.

Hao, D.C., Chen, S.L., Xiao, P.G., 2010c. Sequence characteristics and divergent evolution of the chloroplast psbA-trnH noncoding region in gymnosperms. J. Appl. Genet. 51, 259–273.

Hao, D.C., Mu, J., Chen, S.L., Xiao, P.G., 2010d. Physicochemical evolution and positive selection of the gymnosperm matK proteins. J. Genet. 89, 81–89.

Hao, D.C., Mu, J., Xiao, P.G., 2010e. Molecular evolution and positive Darwinian selection of the gymnosperm photosynthetic Rubisco enzyme. Bot. Stud. 51, 491–510.

Hao, D.C., Ge, G.B., Xiao, P.G., et al., 2011a. The first insight into the tissue specific taxus transcriptome via Illumina second generation sequencing. PLoS One 6, e21220.

Hao, D.C., Yang, L., Xiao, P.G., 2011b. The first insight into the *Taxus* genome via fosmid library construction and end sequencing. Mol. Genet. Genomics 285, 197–205.

Hao, D.C., Xiao, P.G., Peng, Y., et al., 2012a. Research progress and trend analysis of biology and chemistry of *Taxus* medicinal resources. Yao Xue Xue Bao 47 (7), 827–835.

Hao, D.C., Yang, L., Xiao, P.G., Liu, M., 2012b. Identification of *Taxus* microRNAs and their targets with high-throughput sequencing and degradome analysis. Physiol. Plant. 146 (4), 388–403.

Hsu, C.Y., Wu, C.S., Chaw, S.M., 2014. Ancient nuclear plastid DNA in the yew family (taxaceae). Genome Biol. Evol. 6 (8), 2111–2121.

Hu, W.J., Chen, J., Liu, T.W., et al., 2014. Comparative proteomic analysis of differential responses of *Pinus massoniana* and *Taxus wallichiana* var. *mairei* to simulated acid rain. Int. J. Mol. Sci. 15 (3), 4333–4355.

Hunter, D.J., Behrendorff, J.B., Johnston, W.A., et al., 2011. Facile production of minor metabolites for drug development using a CYP3A shuffled library. Metab. Eng. 13, 682–693.

Huo, C.H., Wang, Y.F., Zhang, X.P., et al., 2007. A new metabolite with a new substitution pattern from the seeds of the Chinese Yew, *Taxus mairei*. Chem. Biodivers. 4, 84–88.

Hwang, C.H., Lin, Y.L., Liu, Y.K., et al., 2012. 7,7″-Dimethoxyagastisflavone-induced apoptotic or autophagic cell death in different cancer cells. Phytother. Res. 26 (4), 528–534.

Juyal, D., Thawani, V., Thaledi, S., Joshi, M., 2014. Ethnomedical properties of taxus wallichiana zucc. (Himalayan yew). J. Tradit. Complement. Med. 4 (3), 159–161.

Ketchum, R.E., Gibson, D.M., Croteau, R.B., Shuler, M.L., 1999. The kinetics of taxoid accumulation in cell suspension cultures of *Taxus* following elicitation with methyl jasmonate. Biotechnol. Bioeng. 62, 97–105.

Ketchum, R.E., Rithner, C.D., Qiu, D., et al., 2003. *Taxus* metabolomics: methyl jasmonate preferentially induces production of taxoids oxygenated at C-13 in *Taxus* × *media* cell cultures. Phytochemistry 62, 901–909.

Khan, I.A., 2006. Issues to botanicals. Life Sci. 78, 2033–2038.

Khan, I., Nisar, M., Shah, M.R., et al., 2011. Anti-inflammatory activities of Taxusabietane A isolated from *Taxus wallichiana* Zucc. Fitoterapia 82 (7), 1003–1007.

Kim, S.Y., Yun-Choi, H.S., 2010. A comparative optical aggregometry study of antiplatelet activity of taxanes from *Taxus cuspidata*. Thromb. Res. 125 (6), e281–e284.

Kingston, D.G.I., 2001. Taxol, a molecule for all seasons. Chem. Commun. 10, 867–880.

Kite, G.C., Rowe, E.R., Veitch, N.C., et al., 2013. Generic detection of basic taxoids in wood of European Yew (*Taxus baccata*) by liquid chromatography-ion trap mass spectrometry. J. Chromatogr. B Anal. Technol. Biomed. Life Sci. 915–916, 21–27.

Koepp, E., Hezari, M., Zajicek, J., et al., 1995. Cyclization of geranylgeranyl diphosphate to taxa-4(5),11(12)-diene is the committed step of Taxol biosynthesis in Pacific yew. J. Biol. Chem. 270, 8686–8690.

Köksal, M., Jin, Y., Coates, R.M., et al., 2011. Taxadiene synthase structure and evolution of modular architecture in terpene biosynthesis. Nature 469, 116–120.

Kominkova, M., Stanisavljevic, M., Zitka, O., et al., 2013. Yew poisoning of olive baboons (*Papio anubis*) in captivity: laboratory diagnosis. Neuro Endocrinol. Lett. 34 (S2), 130–133.

Kong, J.Q., Wang, W., Zhu, P., Cheng, K.D., 2007. Recent advances in the biosynthesis of Taxol. Acta Pharm. Sin. 42, 358–365.

Kress, W.J., Wurdack, K.J., Zimmer, E.A., et al., 2005. Use of DNA barcodes to identify flowering plants. Proc. Natl. Acad. Sci. U. S. A. 102, 8369–8374.

Kucukboyaci, N., Orhan, I., Sener, B., et al., 2010. Assessment of enzyme inhibitory and antioxidant activities of lignans from *Taxus baccata* L. Z. Naturforsch. C 65 (3–4), 187–194.

Kuznetcova, T.I., Cech, P., Herbort, C.P., 2012. The mystery of angiographically silent macular oedema due to taxanes. Int. Ophthalmol. 32 (3), 299–304.

Lee, E.K., Jin, Y.W., Park, J.H., et al., 2010. Cultured cambial meristematic cells as a source of plant natural products. Nat. Biotechnol. 28, 1213–1217.

Leitch, I.J., Hanson, L., 2001. Nuclear DNA C-values complete familial representation in gymnosperms. Ann. Bot. 88, 843–849.

Li, J., Davis, C.C., Tredici, P.D., Donoghue, M.J., 2001. Phylogeny and biogeography of *Taxus* (Taxaceae) inferred from sequences of the internal transcribed spacer region of nuclear ribosomal DNA. Harv. Pap. Bot. 6, 267–274.

Li, Z.P., Wang, C.L., Gu, J.S., Shi, Q.W., 2005. Studies on chemical constituents in seeds of *Taxus mairei* II. China J. Chin. Mat. Med. 30, 1260–1263.

Li, L., Feng, Z., Liu, D., 2011. Comprehensive textual research on concept, content and function of Chinese yew. China J. Chin. Mat. Med. 36, 1682–1685.

Li, Y., Qin, F., Wang, S.M., et al., 2013. Chemical studies on *Taxus canadensis*. Chem. Biodivers. 10 (10), 1729–1753.

Little, D.P., 2011. DNA barcode sequence identification incorporating taxonomic hierarchy and within taxon variability. PLoS One 6, e20552.

Liu, Z., Bos, J.I.B., Armstrong, M., et al., 2005. Patterns of diversifying selection in the phytotoxin-like scr74 gene family of *Phytophthora infestans*. Mol. Biol. Evol. 22, 659–672.

Liu, X.Q., Zhang, X.D., Zhu, Y.L., et al., 2008. Structure identification of biflavones and determination of Taxol from *Taxus media*. J. Chin. Med. Mat. 31, 1498–1501.

Liu, J., Möller, M., Gao, L.M., et al., 2011. DNA barcoding for the discrimination of Eurasian yews (*Taxus* L., Taxaceae) and the discovery of cryptic species. Mol. Ecol. Resour. 11, 89–100.

Lu, G.H., Chan, K., Liang, Y.Z., et al., 2005. Development of high-performance liquid chromatographic fingerprints for distinguishing Chinese Angelica from related Umbelliferae herbs. J. Chromatogr. A 1073, 383–392.

Luan, H.W., Hu, Y., Liu, X.B., Hao, D.C., Yang, L., 2008. Purification and characterization of a β-D-xylosidase from *Leifsonia Shinshuensis* DICP16. Chin. J. Biotechnol. 24, 867–873.

MacEachern-Keith, G.J., Wagner Butterfield, L.J., Incorvia Mattina, M.J., 1997. Paclitaxel stability in solution. Anal. Chem. 69, 72–77.

Metzker, M.L., 2010. Sequencing technologies – the next generation. Nat. Rev. Genet. 11, 31–46.

Miao, Y.C., Lang, X.D., Zhang, Z.Z., Su, J.R., 2014. Phylogeography and genetic effects of habitat fragmentation on endangered *Taxus yunnanensis* in southwest China as revealed by microsatellite data. Plant Biol. 16 (2), 365–374.

Miller, K., Neilan, B., Sze, D.M., 2008. Development of Taxol and other endophyte produced anti-cancer agents. Recent Pat. Anticancer Drug Discov. 3, 14–19.

Morrissy, S., Zhao, Y., Delaney, A., et al., 2010. Digital gene expression by tag sequencing on the illumina genome analyzer. Curr. Protoc. Hum. Genet. 11, 1–36 (Chapter 11: Unit 11).

Morrone, D., Lowry, L., Determan, M.K., et al., 2010. Increasing diterpene yield with a modular metabolic engineering system in *E. coli*: comparison of MEV and MEP isoprenoid precursor pathway engineering. Appl. Microbiol. Biotechnol. 85, 1893–1906.

Ni, Z., Dong, M., Cong, B., et al., 2011. A novel taxane 13-glucoside and other taxanes from the leaves of *Taxus cuspidata*. Planta Med. 77 (3), 281–283.

Nielsen, R., Bustamante, C., Clark, A.G., et al., 2005. A scan for positively selected genes in the genomes of Humans and Chimpanzees. PLoS Biol. 3, 976–985.

Novakova, L., Matysova, L., Solich, P., 2006. Advantages of application of UPLC in pharmaceutical analysis. Talanta 68, 908–918.

Nystedt, B., Street, N.R., Wetterbom, A., et al., 2013. The Norway spruce genome sequence and conifer genome evolution. Nature 497 (7451), 579–584.

O'Connor, D., Mortishire-Smith, R., Morrison, D., et al., 2006. Ultra-performance liquid chromatography coupled to time-of-flight mass spectrometry for robust, high-throughput quantitative analysis of an automated metabolic stability assay, with simultaneous determination of metabolic data. Rapid Commun. Mass Spectrom. 20, 851–857.

Ohtsuki, K., Miyai, S., Yamaguchi, A., et al., 2012. Biochemical characterization of novel lignans isolated from the wood of *Taxus yunnanensis* as effective stimulators for glycogen synthase kinase-3β and the phosphorylation of basic brain proteins by the kinase in vitro. Biol. Pharm. Bull. 35 (3), 385–393.

Ojima, I., Chakravarty, S., Inoue, T., et al., 1999. A common pharmacophore for cytotoxic natural products that stabilize microtubules. Proc. Natl. Acad. Sci. U. S. A. 96, 4256–4261.

Ojima, I., Borella, C.P., Wu, X., et al., 2005. Design, synthesis and structure-activity relationships of novel taxane-based multidrug resistance reversal agents. J. Med. Chem. 48, 2218–2228.

O'Leary, S.J., Poulis, B.A., von Aderkas, P., 2007. Identification of two thaumatin-like proteins (TLPs) in the pollination drop of hybrid yew that may play a role in pathogen defence during pollen collection. Tree Physiol. 27, 1649–1659.

Ono, N.N., Britton, M.T., Fass, J.N., et al., 2011. Exploring the transcriptome landscape of pomegranate fruit peel for natural product biosynthetic gene and SSR marker discovery. J. Integr. Plant Biol. 53 (10), 800–813, http://dx.doi.org/10.1111/j.1744-7909.2011.01073.x.

Onrubia, M., Moyano, E., Bonfill, M., et al., 2011. The relationship between TXS, DBAT, BAPT and DBTNBT gene expression and taxane production during the development of *Taxus baccata* plantlets. Plant Sci. 181, 282–287.

Onrubia, M., Pollier, J., Vanden Bossche, R., et al., 2014. Taximin, a conserved plant-specific peptide is involved in the modulation of plant-specialized metabolism. Plant Biotechnol. J. 12 (7), 971–983.

Parmar, V.S., Jha, A., Bisht, K.S., et al., 1999. Constituents of the yew trees. Phytochemistry 50, 1267–1304.

Patel, P., Patel, K., Gandhi, T., 2011. Evaluation of effect of *Taxus baccata* leaves extract on bronchoconstriction and bronchial hyperreactivity in experimental animals. J. Young Pharm. 3 (1), 41–47.

Plumb, R., Castro-Perez, J., Granger, J., et al., 2004. Ultra-performance liquid chromatography coupled to quadrupole-orthogonal time-of-flight mass spectrometry. Rapid Commun. Mass Spectrom. 18, 2331–2337.

Poudel, R.C., Gao, L.M., Möller, M., et al., 2013. Yews (*Taxus*) along the Hindu Kush-Himalayan region: exploring the ethnopharmacological relevance among communities of Mongol and Caucasian origins. J. Ethnopharmacol. 147 (1), 190–203.

Qayum, M., Nisar, M., Shah, M.R., et al., 2012. Analgesic and antiinflammatory activities of taxoids from *Taxus wallichiana* Zucc. Phytother. Res. 26 (4), 552–556.

Qiu, D., Pan, X., Wilson, I.W., et al., 2009. High throughput sequencing technology reveals that the taxoid elicitor methyl jasmonate regulates microRNA expression in Chinese yew (*Taxus chinensis*). Gene 436, 37–44.

Qu, C., Chen, Z., 2014. Antitumor effect of water decoctions of taxus cuspidate on pancreatic cancer. Evid. Based Complement. Alternat. Med. 2014, 291675.

Rehder, A., 1940. Manual of Cultivated Trees and Shrubs, second ed. MacMillan Co., New York.

Russell, P.J., 2010. iGenetics, third ed. Pearson Benjamin Cummings, San Francisco.

Schirone, B., Ferreira, R.C., Vessella, F., et al., 2010. *Taxus baccata* in the Azores: a relict form at risk of imminent extinction. Biodivers. Conserv. 19, 1547–1565.

Schuster, S.C., 2008. Next-generation sequencing transforms today's biology. Nat. Methods 5, 16–18.

Shah, A., Li, D.Z., Möller, M., et al., 2008. Delimitation of *Taxus fuana* Nan Li & R.R. Mill (Taxaceae) based on morphological and molecular data. Taxon 57, 211–222.

Shang, W., Qiao, J., Gu, C., et al., 2011. Anticancer activity of an extract from needles and twigs of *Taxus cuspidata* and its synergistic effect as a cocktail with 5-fluorouracil. BMC Complement. Altern. Med. 11, 123.

Shiota, M., Itsumi, M., Yokomizo, A., et al., 2014. Targeting ribosomal S6 kinases/Y-box binding protein-1 signaling improves cellular sensitivity to taxane in prostate cancer. Prostate 74 (8), 829–838.

Shu, Q., Shen, M., Wang, B., et al., 2014. Aqueous extract of *Taxus chinensis* (Pilger) Rehd inhibits lung carcinoma A549 cells through the epidermal growth factor receptor/mitogen-activated protein kinase pathway in vitro and in vivo. J. Tradit. Chin. Med. 34 (3), 293–301.

Soca-Chafre, G., Rivera-Orduña, F.N., Hidalgo-Lara, M.E., et al., 2011. Molecular phylogeny and paclitaxel screening of fungal endophytes from *Taxus globosa*. Fungal Biol. 115, 143–156.

Song, G.H., Zhao, C.F., Zhang, M., et al., 2014. Correlation analysis of the taxane core functional group modification, enzyme expression, and metabolite accumulation profiles under methyl jasmonate treatment. Biotechnol. Prog. 30 (2), 269–280.

Spjut, R.W., 2007. Taxonomy and nomenclature of *Taxus* (Taxaceae). J. Bot. Res. Inst. Texas 1, 203–289.

Stierle, A., Strobel, G., Stierle, D., 1993. Taxol and taxane production by *Taxomyces andreanae*, an endophytic fungus of Pacific yew. Science 260, 214–216.

Sula, M.J., Morgan, S., Bailey, K.L., et al., 2013. Characterization of cardiac lesions in calves after ingestion of Japanese yew (*Taxus cuspidata*). J. Vet. Diagn. Investig. 25 (4), 522–526.

Sun, G., Yang, Y., Xie, F., et al., 2013. Deep sequencing reveals transcriptome re-programming of *Taxus* × *media* cells to the elicitation with methyl jasmonate. PLoS One 8 (4), e62865.

Swanson, W.J., Yang, Z., Wolfner, M.F., Aquadro, C.F., 2001. Positive Darwinian selection in the evolution of mammalian female reproductive proteins. Proc. Natl. Acad. Sci. U. S. A. 98, 2509–2514.

Tan, X.M., Cuo, S.X., 2006. Root microstructure and distribution of the endophytic fungi in *Taxus chinensis* var. *mairei*. Acta Acad. Med. Sin. 28, 372–374.

Tanaka, K., Li, F., Morikawa, K., et al., 2011. Analysis of biosynthetic fluctuations of cultured *Taxus* seedlings using a metabolomic approach. Phytochemistry 72, 1760–1766.

Tang, B., Zhang, F.S., Li, X., et al., 2013. Supercritical CO_2 fluid extraction and component analysis of leaves oil from *Taxus chinensis* var. *mairei*. Zhong Yao Cai 36 (12), 2023–2027.

Tezuka, Y., Morikawa, K., Li, F., et al., 2011. Cytochrome P450 3A4 inhibitory constituents of the wood of *Taxus yunnanensis*. J. Nat. Prod. 74 (1), 102–105.

Tian, J., Stella, V.J., 2008. Degradation of paclitaxel and related compounds in aqueous solutions I: epimerization. J. Pharm. Sci. 97, 1224–1235.

Trapp, S.C., Croteau, R.B., 2001. Genomic organization of plant terpene synthases and molecular evolutionary implications. Genetics 158, 811–832.

van Rozendaal, E., Kurstjens, S., van Beek, T.A., van den Berg, R.G., 1999. Chemotaxonomy of *Taxus*. Phytochemistry 52, 427–433.

van Rozendaal, E.L., Lelyveld, G.P., van Beek, T.A., 2000. Screening of the needles of different yew species and cultivars for paclitaxel and related taxoids. Phytochemistry 53, 383–389.

Vieira, R.F., Grayer, R.J., Paton, A.J., 2003. Chemical profiling of *Ocimum americanum* using external flavonoids. Phytochemistry 63, 555–567.

Vivekanandan, K., Swamy, M.G., Prasad, S., et al., 2006. Identification of isocephalomannine in the presence of cephalomannine isomers and alkali metal ion adducts in a paclitaxel active pharmaceutical ingredient using electrospray tandem mass spectrometry. Rapid Commun. Mass Spectrom. 20, 1731–1735.

Wang, Y.F., Shi, Q.W., Dong, M., et al., 2011. Natural taxanes: developments since 1828. Chem. Rev. 111, 7652–7709.

Wang, L., Ding, L., Huo, S., et al., 2013. Three new taxoids from the seed of *Taxus cuspidata*. J. Nat. Med. 67 (4), 827–832.

Wani, M.C., Taylor, H.L., Wall, M.E., et al., 1971. Plant antitumor agents. 6. Isolation and structure of taxol, a novel antileukemic and antitumor agent from *Taxus brevifolia*. J. Am. Chem. Soc. 93, 2325–2327.

Watchueng, J., Kamnaing, P., Gao, J.M., et al., 2011. Efficient purification of paclitaxel from yews using high-performance displacement chromatography technique. J. Chromatogr. A 1218, 2929–2935.

Wilson, S.A., Roberts, S.C., 2012. Recent advances towards development and commercialization of plant cell culture processes for the synthesis of biomolecules. Plant Biotechnol. J. 10, 249–268.

Wink, M., 2003. Evolution of secondary metabolites from an ecological and molecular phylogenetic perspective. Phytochemistry 64, 3–19.

Wöhrle, J., Werner, G.S., 2013. Paclitaxel-coated balloon with bare-metal stenting in patients with chronic total occlusions in native coronary arteries. Catheter. Cardiovasc. Interv. 81 (5), 793–799.

Wu, D., Li, Y.N., Wu, L.J., Gao, H.Y., 2010. Isolation and identification of chemical constituents from *Taxus cuspidata* Sibe et Zucc. Acta Pharm. Sin. 45, 1398–1401.

Wu, Q., Sun, C., Luo, H., et al., 2011. Transcriptome analysis of *Taxus cuspidata* needles based on 454 pyrosequencing. Planta Med. 77, 394–400.

Wu, Q., Duan, X.Q., Chen, X., Xiao, P.G., 2012. Identification of EST-SSRs in *Taxus cuspidata* based on high-throughput sequencing. Zhongguo Zhong Yao Za Zhi 37 (24), 3728–3733.

Xie, P., Chen, S., Liang, Y., et al., 2006. Chromatographic fingerprint analysis – a rational approach for quality assessment of traditional Chinese herbal medicine. J. Chromatogr. A 1112, 171–180.

Xu, Q., Yu, K., Zhu, A., et al., 2009. Comparative transcripts profiling reveals new insight into molecular processes regulating lycopene accumulation in a sweet orange (*Citrus sinensis*) red-flesh mutant. BMC Genomics 10, 540.

Yan, C., Yin, Y., Zhang, D., et al., 2013. Structural characterization and in vitro antitumor activity of a novel polysaccharide from *Taxus yunnanensis*. Carbohydr. Polym. 96 (2), 389–395.

Yang, C., Wang, J., Luo, J., Kong, L., 2009. J. Asian Nat. Prod. Res. 11 (5–6), 533–537.

Yang, W.X., Zhao, Z.G., Wang, L.H., et al., 2012. Control of hypertension in rats using volatile components of leaves of *Taxus chinensis* var. *mairei*. J. Ethnopharmacol. 141 (1), 309–313.

Yang, Y., Zhao, H., Barrero, R.A., et al., 2014. Genome sequencing and analysis of the paclitaxel-producing endophytic fungus *Penicillium aurantiogriseum* NRRL 62431. BMC Genomics 15, 69.

Yu, J., Wang, Y., Qian, H., et al., 2012. Polyprenols from *Taxus chinensis var. mairei* prevent the development of CCl_4-induced liver fibrosis in rats. J. Ethnopharmacol. 142 (1), 151–160.

Yukimune, Y., Tabata, H., Higashi, Y., Hara, Y., 1996. Methyl jasmonate-induced overproduction of paclitaxel and baccatin III in *Taxus* cell suspension cultures. Nat. Biotechnol. 14, 1129–1132.

Zhan, J.X., Zhong, J.J., Dai, J.G., et al., 2003. Microbial transformation of sinenxan A, a rich constituent in callus cultures of *Taxus*. Acta Pharm. Sin. 38, 555–558.

Zhang, J.W., Ge, G.B., Liu, Y., et al., 2008. Taxane's substituents at C3' affect its regioselective metabolism: different in vitro metabolism of cephalomannine and paclitaxel. Drug Metab. Dispos. 36, 418–426.

Zhang, Y.Y., Liu, Y., Zhang, J.W., et al., 2009. C-7 configuration as one of determinants in taxanes metabolism by human cytochrome P450 enzymes. Xenobiotica 39, 903–914.

Zhang, D., Meng, H., Yang, H.S., 2012. Antidiabetic activity of *Taxus cuspidata* polysaccharides in streptozotocin-induced diabetic mice. Int. J. Biol. Macromol. 50 (3), 720–724.

Zhang, N., Han, Z., Sun, G., et al., 2014a. Molecular cloning and characterization of a cytochrome P450 taxoid 9á-hydroxylase in *Ginkgo biloba* cells. Biochem. Biophys. Res. Commun. 443 (3), 938–943.

Zhang, Y., Ma, J., Yang, B., et al., 2014b. The complete chloroplast genome sequence of *Taxus chinensis* var. *mairei* (Taxaceae): loss of an inverted repeat region and comparative analysis with related species. Gene 540 (2), 201–209.

Zhao, C.F., Yu, L.J., 2005. LC-ESI-MS metabolic profiling analysis of taxanes from the extracts of *Taxus chinensis* cell cultures. Acta Pharm. Sin. 40, 734–739.

Zheng, Z.Q., Fu, Y.Y., Li, B.H., et al., 2014. PSY-1, a *Taxus chinensis* var. *mairei* extract, inhibits cancer cell metastasis by interfering with MMPs. Nat. Prod. Commun. 9 (2), 241–245.

Zhou, Z., Qu, W., Liu, M., et al., 2014. A new taxane diterpenoid from *Taxus chinensis* var. *mairei*. Nat. Prod. Res. 28 (8), 530–533.

Zhu, Y.J., Chen, S.L., Yao, H., et al., 2010. DNA barcoding the medicinal plants of the genus *Paris*. Acta Pharm. Sin. 45, 376–382.

Zimmermann, B., 2010. Characterization of pollen by vibrational spectroscopy. Appl. Spectrosc. 64, 1364–1373.

Phytochemical and biological research of *Fritillaria* medicinal resources

4

4.1 Introduction

Fritillaria is a genus of about 130–165 species (Rix, 2001; Xiao et al., 2007) within the monocot family Liliaceae, native to temperate regions of the Northern Hemisphere. *Fritillaria* is a botanical source of various pharmaceutically active components, which have been used in traditional Chinese medicine for thousands of years. Many species (such as *F. cirrhosa*, *F. thunbergii*, and *F. verticillata*) are used in traditional Chinese cough remedies. In China Pharmacopoeia (http://www.chp.org.cn/cms/home/), they are listed as Chuan Bei Mu, Zhe Bei Mu, Yi Bei Mu, Hubei Bei Mu, and Ping Bei Mu, respectively, and are often in formulations combined with extracts of loquat (*Eriobotrya japonica*). *F. verticillata* bulbs are also traded as Bei Mu in Japan. Chuan Bei Mu and Zhe Bei Mu are the most famous Chinese herbal medicines obtained from *Fritillaria* species. The former is from *F. cirrhosa*, *F. unibracteata*, *F. przewalskii*, *F. delavayi*, *F. taipaiensis*, and *F. wabuensis*, while the latter from *F. thunbergii* and *F. thunbergii* var. *chekiangensis*. According to traditional descriptions, *Fritillaria* is slightly cold and affects the lungs (to clear heat and moisten dryness, used for hot-type bronchitis with dry cough) and the heart (to calm heart fire). *Fritillaria* is also used for treating lumps beneath the skin, such as scrofulous swellings and breast lumps. Zhe Bei Mu is often used for the treatment of lumps and the moistening property attributed to Chuan Bei Mu is not needed; it has been adopted into some Chinese herb formulas for treating cancers. Different *Fritillaria* species possess different chemical profiles and may have different pharmacological effects. However, it is quite difficult to authenticate *Fritillaria* species only by morphology.

To date, at least 30 *Fritillaria* species have been characterized for their chemical components (Li et al., 2006a). Advances in analytical chemistry and molecular biology techniques facilitate the in-depth studies of *Fritillaria* pharmaceutical resources. However, it is essential to study more species for both the sustainable utilization of *Fritillaria* medicinal resources and finding novel compounds with potential clinical utility. In this chapter, we focus on recent progress in phytochemistry and chemotaxonomy of *Fritillaria*, as well as molecular taxonomy and phylogeny. To date, very few studies have attempted to correlate DNA-based phylogeny of medicinal plants with phytochemistry or with medicinal properties (Rønsted et al., 2008). Here, we reconstruct the molecular phylogeny of worldwide *Fritillaria* species and compare the results with those of chemotaxonomy. Phylogeny has great explanatory power and offers a unique perspective to complement chemotaxonomy.

Medicinal plants: chemistry, biology and omics. http://dx.doi.org/10.1016/B978-0-08-100085-4.00004-9

4.2 Chemical components and bioactivity

4.2.1 *Steroidal alkaloids*

Alkaloids are a group of naturally occurring chemical compounds that contain mostly basic nitrogen atoms. The steroidal alkaloids from *Fritillaria* can be classified into two groups based on the carbon framework: isosteroidal alkaloids and steroidal alkaloids (Figure 4.1). According to the linkage patterns between rings E and F, isosteroidal alkaloids can be divided into cevanine (A), jervine (B), and veratramine (C) types (Figures 4.1 and 4.2). Steroidal alkaloids can be divided into solanidine (D)

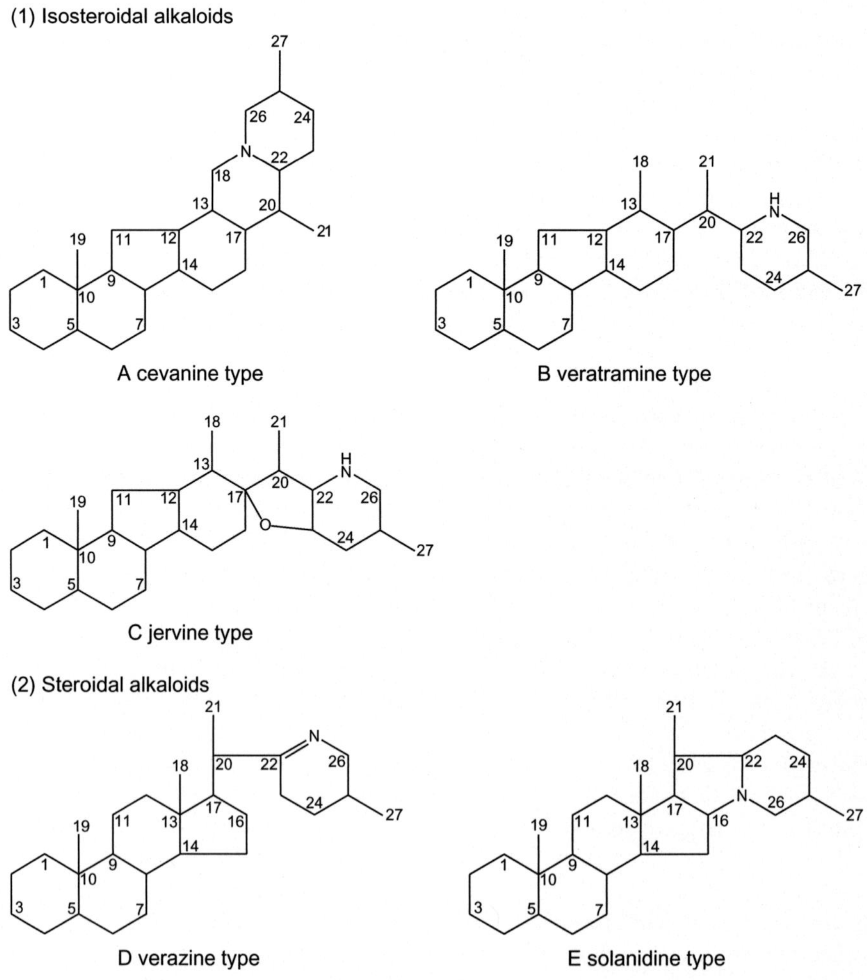

Figure 4.1 Types of steroidal alkaloids in *Fritillaria*. (1) Isosteroidal alkaloids; (2) steroidal alkaloids. Refer to Table 4.1 for more details.

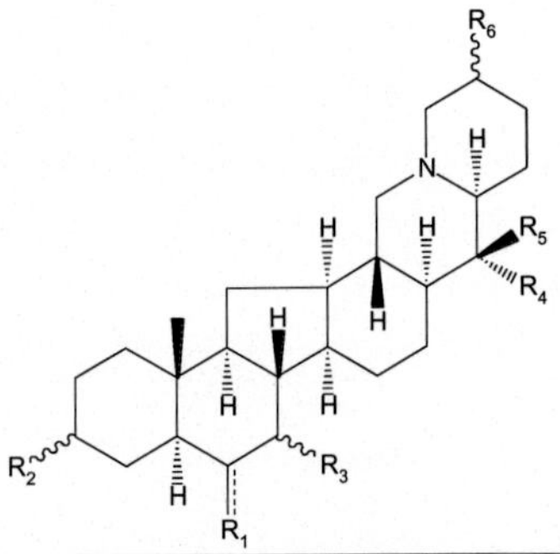

1	$R_1 = R_2 = \beta$-OH; $R_3 = H_2$; R_4 = Me; R_5 = H; $R_6 = \alpha$-Me
2	$R_1 = \beta$-OH; $R_2 = R_3 = \alpha$-OH; R_4 = Me; R_5 = H; $R_6 = \alpha$-Me
7	R_1 = O; $R_2 = \beta$-OAc; $R_3 = H_2$; R_4 = Me; R_5 = OH; $R_6 = \beta$-Me
8	$R_1 = \alpha$-OH; $R_2 = \beta$-OAc; $R_3 = H_2$; R_4 = Me; R_5 = OH; $R_6 = \beta$-Me
11	R_1 = O; $R_2 = \beta$-OGlc; $R_3 = H_2$; R_4 = Me; R_5 = H; $R_6 = \alpha$-Me
13	R_1 = O; $R_2 = \beta$-OGlc; $R_3 = H_2$; R_4 = Me; R_5 = H; $R_6 = \beta$-Me
14	$R_1 = R_2 = \beta$-OH; $R_3 = H_2$; R_4 = H; R_5 = Me; $R_6 = \alpha$-Me

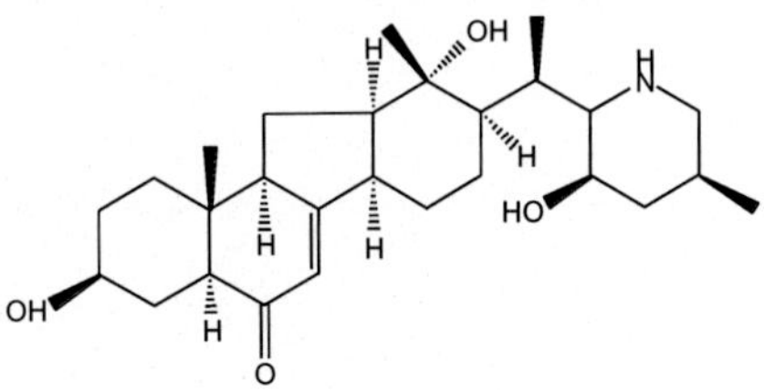

3

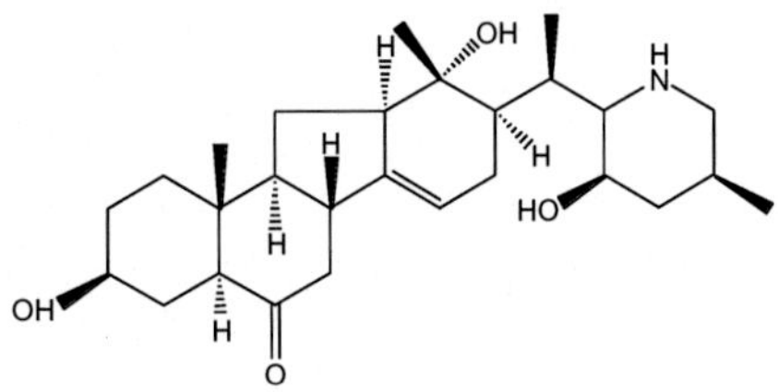

4

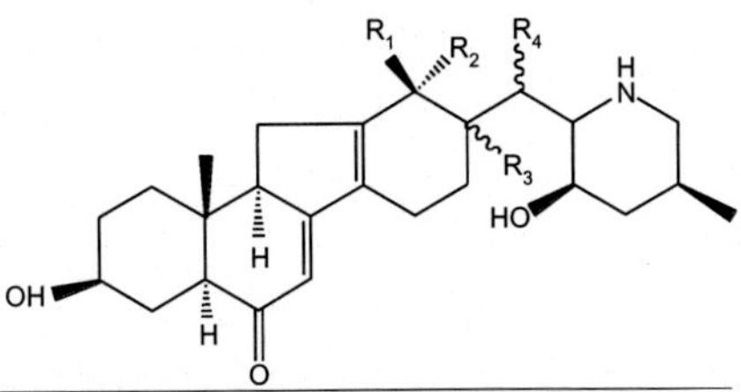

5	R_1 = Me; R_2 = H; $R_3 = \alpha$-H; $R_4 = \beta$-Me
9	R_1 = H; R_2 = Me; $R_3 = \beta$-H; $R_4 = \alpha$-Me
10	R_1 = OH; R_2 = Me; $R_3 = \beta$-H; $R_4 = \alpha$-Me

Figure 4.2 Steroidal alkaloids of *Fritillaria*. D/E configurations of cevanine group steroidal alkaloids of 1, 2, 7, 8, 11, 13, 14, and 15 are *trans*-form. Refer to Tables 4.1 and 4.2 for more details.

(Continued)

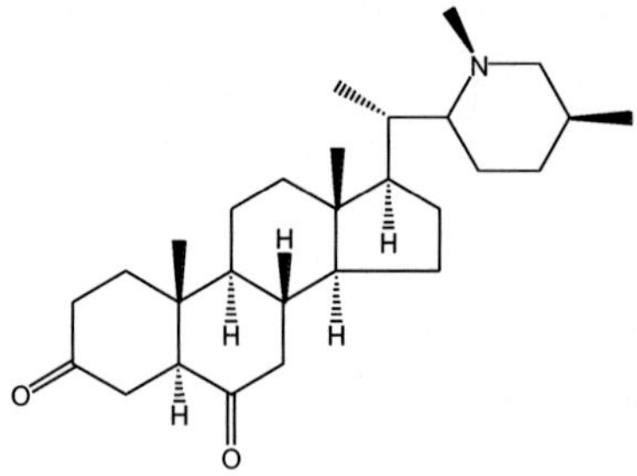

6

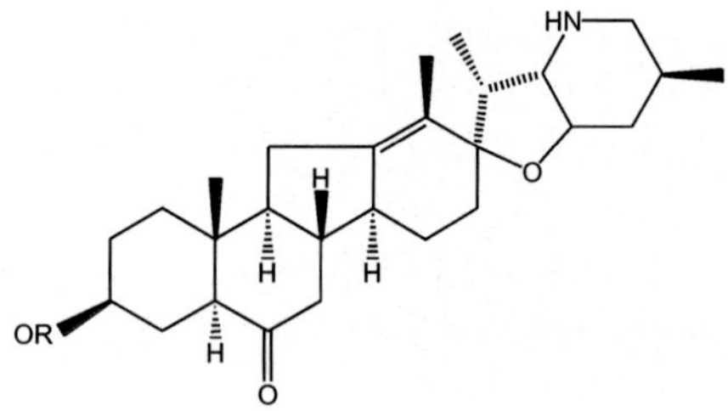

12 R = S_1

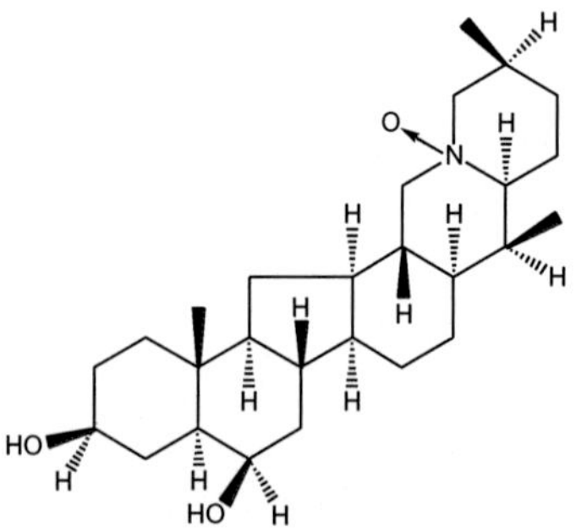

15

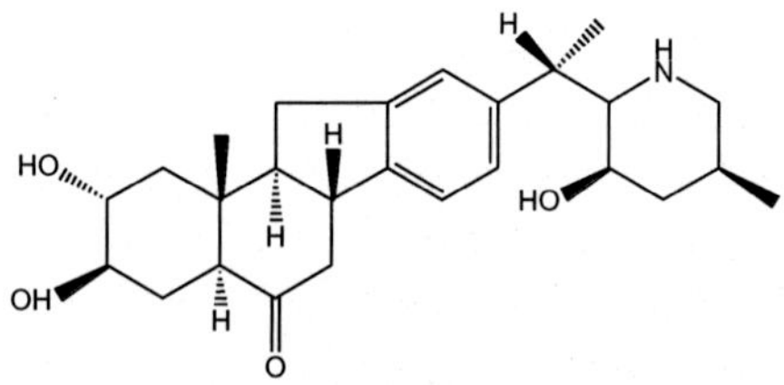

16

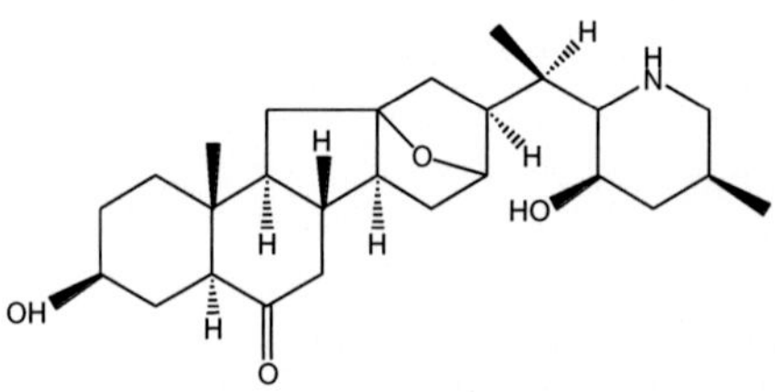

17

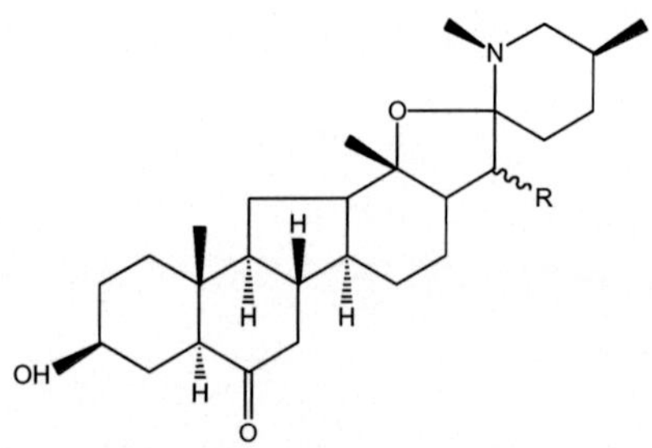

18 R = α-Me
19 R = β-Me

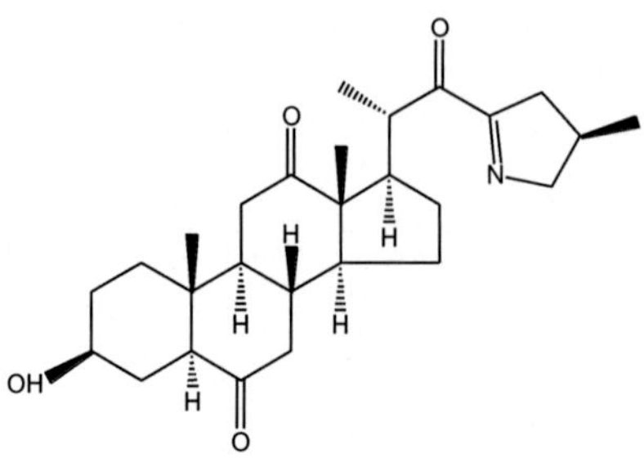

20

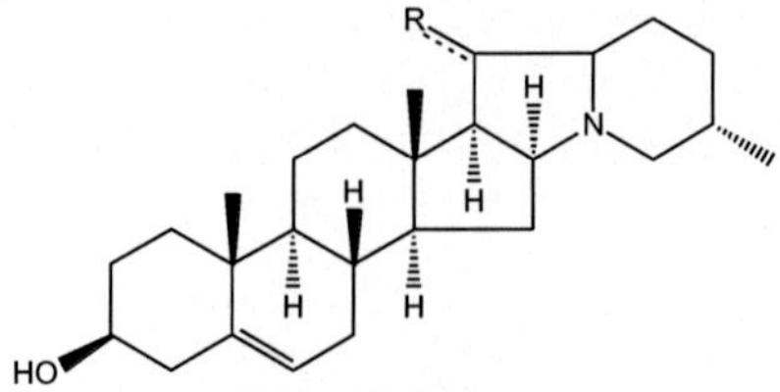

21 R = α-Me
22 R = CH_2

Figure 4.2 Continued.

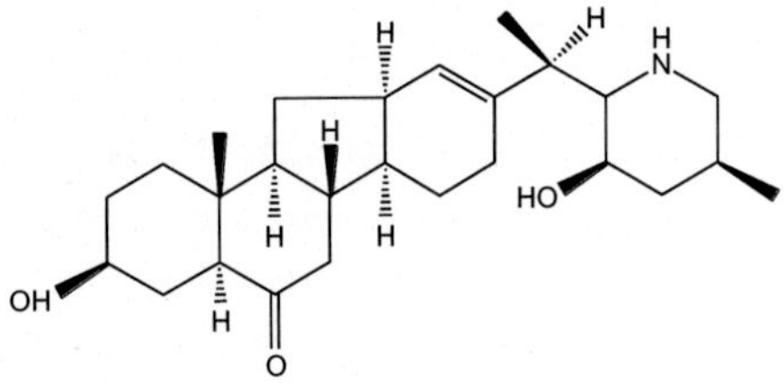

23

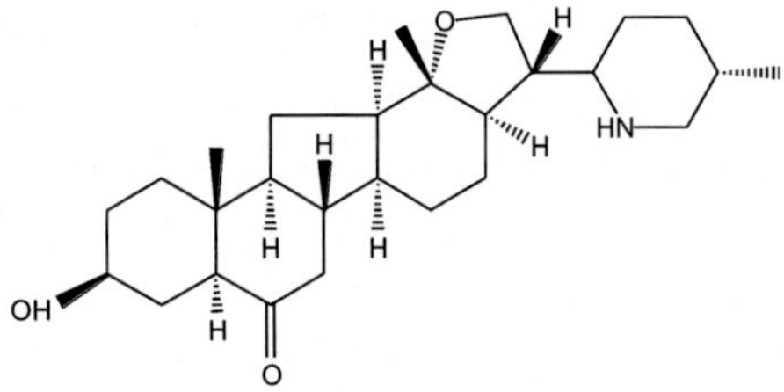

24

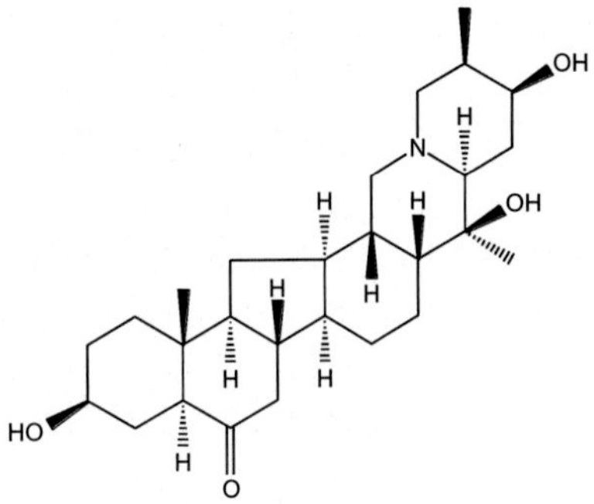

25

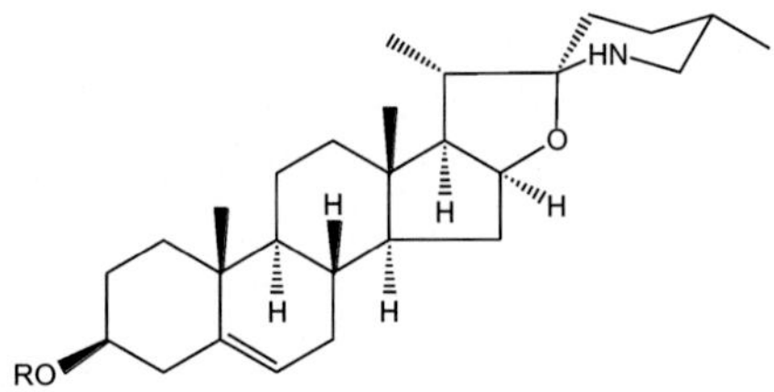

26 R = S_5

Figure 4.2 Continued.

and secosolanidine (E) types (Xiao et al., 2007), depending on whether the nitrogen atom is incorporated into an indolizidine ring or a piperidine ring (Figure 4.1). Steroidal alkaloids found in *Fritillaria* before 2005 have been summarized (Li et al., 2006a). Table 4.1 summarizes the steroidal alkaloids found since 2006.

Puqiedine 1 and 3α-puqiedin-7-ol 2 were isolated from *F. puqiensis*, their structures being determined on the basis of NMR spectroscopy and MS (Jiang et al., 2006). The new cevanine-type alkaloids, (3β, 5α)-20-hydroxy-6-oxocevan-3-yl acetate 7 and (3β, 5α, 6α)-6, 20-dihydroxycevan-3-yl acetate 8, were isolated from bulbs of *F. hupehensis* and characterized by spectral analysis (Zhang et al., 2008). Bulbs of *F. unibracteata* in Sichuan, China, contained the new glycoalkaloid puqiedinone-3-*O*-β-D-glucopyranoside 11, whose structure was elucidated by spectroscopic and chemical methods (Zhang et al., 2011). Phytochemical investigation of bulbs of *F. lichuanensis* yielded a new isosteroidal alkaloid hupehenizioiside 13 (Pi et al., 2006a). A further investigation of bulbs of *F. lichuanensis* has resulted in the isolation of two unique cevanine-type alkaloids, lichuanine ((20S, 25R)-5α, 14α-cevanine-3β, 6β-diol) 14 and lichuanisinine ((20S, 25S)-5α,14α-cevanine-3β, 6β-diol-*N*-oxide) 15 (Pi et al., 2006b). Extraction of bulbs of *F. puqiensis* yielded three new veratramine-type alkaloids, namely, puqienine C 3, puqienine D 4, and puqienine E 5 (Jiang et al., 2006). Compounds 9 and 10 have been obtained from *F. hupehensis*. Their structures were deduced to be (3β,5α,13α,23β)-7,8,12,14-tetradehydro-5,6,12,13-tetrahydro-3,23-dihydroxyveratraman-6-one and (3β,5α,13α,23β)-7,8,12,14-tetradehydro-5,6,12,13-tetrahydro-3,13,23-trihydroxyveratraman-6-one, respectively, on the basis of chemical

Table 4.1 Steroidal alkaloids found in *Fritillaria* bulbs in recent years

Taxon	Steroidal alkaloids	Type	References
F. puqiensis	Puqiedine 1	A	Jiang et al. (2006)
	3α-Puqiedin-7-ol **2**	A	Jiang et al. (2006)
	Puqienine C **3**	B	Jiang et al. (2006)
	Puqienine D **4**	B	Jiang et al. (2006)
	Puqienine E **5**	B	Jiang et al. (2006)
	Puqietinedione **6**	D	Jiang et al. (2006)
F. hupehensis	(3β,5α)-20-Hydroxy-6-oxocevan-3-yl acetate **7**	A	Zhang et al. (2008)
	(3β,5α,6α)-6,20-Dihydroxycevan-3-yl acetate **8**	A	Zhang et al. (2008)
	(3β,5α,13α,23β)-7,8,12,14-Tetradehydro-5,6,12,13-tetrahydro-3,23-dihydroxyveratraman-6-one **9**	B	Zhang et al. (2008)
	(3β,5α,13α,23β)-7,8,12,14-Tetradehydro-5,6,12,13-tetrahydro-3,13,23-trihydroxyveratraman-6-one **10**	B	Zhang et al. (2008)
F. unibracteata	Puqiedinone-3-*O*-β-D-glucopyranoside **11**	A	Zhang et al. (2011)
	Peimisine-3-*O*-β-D-glucopyranoside **12**	C	Zhang et al. (2011)
F. lichuanensis	Hupehenizioiside **13**	A	Pi et al. (2006a)
F. lichuanensis	Lichuanine ((20S,25R)-5α,14α-cevanine-3β,6β-diol) **14**	A	Pi et al. (2006b)
	Lichuanisinine ((20S,25S)-5α,14α-cevanine-3β,6β-diol-*N*-oxide) **15**	A	Pi et al. (2006b)
F. ussuriensis	Pingbeimunone A **16**	B	Yang and Duan (2012)
F. puqiensis	Puqienine F **17**	B	Li et al. (2006b)
F. monantha	Pengbeimine B **18**	C	Liu et al. (2007a)
	Pengbeimine D **19**	C	Liu et al. (2007a)
F. delavayi	Delavidine **20**	D	Cao et al. (2008)
F. anhuiensis	(22S,25S)-Solanid-5-en-3β-ol **21**	E	Shou et al. (2010)
	(22S,25S)-Solanid-5,20(21)-dien-3β-ol **22**	E	Shou et al. (2010)
F. pallidiflora	(20R,22R,23R,25R)-3β,23-Dihydroxy-*N*-methyl-veratram-13(17)-en-6-one **23**	B	Shen et al. (2012b)
F. shuchengensis	Suchengbeisine **24**	B	Huang et al. (2013)
F. pallidiflora	5α,14α,17β-Cevanin-6-oxo-3β,20β,24β-triol **25**	A	Xu et al. (2014)
F. meleagris	(22S,25S)-Spirosol-5-en-3β-ylO-β-D-glucopyranosyl-(1→4)-*O*-[α-L-rhamnopyranosyl-(1→2)]-β-D-glucopyranoside **26**	D	Matsuo et al. (2013)

Isosteroidal alkaloids (Figures 4.1 and 4.2): A cevanine type; B veratramine type; C jervine type.
Steroidal alkaloids (Figures 4.1 and 4.2): D verazine type; E solanidine type.
D/E configurations of cevanine group steroidal alkaloids of 1, 2, 7, 8, 11, 13, 14, and 15 are *trans*-form.

shift comparisons and 1H, 1H-COSY, HMBC correlations (Zhang et al., 2008). Bulbs of *F. ussuriensis* have yielded a new alkaloid pingbeimunone A 16 with an aromatized D-ring, whose structure was elucidated on the basis of spectral analysis. Chemical investigation of the bulbs of *F. puqiensis* afforded puqienine F 17. This novel veratramine alkaloid possessed a 12, 16-epoxy ring (Li et al., 2006b). Peimisine-3-*O*-β-D-glucopyranoside 12, possessing the furan ring (E) fused onto a piperidine ring system forming an ether bridge between carbon atoms C17 and C23, was isolated from *F. unibracteata*, which might be identical with the known jervine-type compound peimisine on the basis of NMR data (Zhang et al., 2011). Pengbeimine B 18 and pengbeimine D 19 have a novel skeleton wherein C13 and C22 are connected with oxygen ring from *F. monantha*. Their absolute structures were proved by X-ray diffraction (Liu et al., 2007a).

Extract of the dried ground bulbs of *F. puqiensis* afforded verazine-type alkaloid puqietinedione 6. Its structure was formulated as (22R, 25S)-*N*-methyl-22, 26-epiminocholest-3, 6-dione by comparison of the NMR data with those of puqietinone (Jiang et al., 2006). Delavidine 20, an unusual 6, 12, 22-triketo verazine-type steroidal alkaloid with a pyrrolidine side chain, was isolated from the methanol extraction of the bulbs of *F. delavayi* (Cao et al., 2008). The fresh bulbs of *F. anhuiensis* contain alkaloids (22S, 25S)-solanid-5-en-3β-ol 21 and (22S, 25S)-solanid-5, 20(21)-dien-3β-ol 22, which are the first solanidine-type alkaloids with 22-S configuration discovered from nature. Their structures were elucidated by NMR spectroscopy (Shou et al., 2010).

4.2.2 Pharmacology and therapeutic use

4.2.2.1 Effects on respiratory system

The alkaloids imperialine, 3β-acetylimperialine, and sinpeinine A, depending on their concentration, relax carbachol-induced contraction of guinea pig tracheal rings (Lin et al., 2006a). The relaxant action of imperialine and sinpeinine A, the *cis*-D/E cevanine alkaloids, is due to their selective inhibitory effects on muscarinic M2 receptors. Verticine, verticinone, imperialine, imperialine-3beta-D-glucoside, and puqietinone, that is, both *trans*- and *cis*-D/E cevanine alkaloids, significantly elevate the cAMP concentration in the HEK cells transfected with muscarinic M2 receptor plasmid (Zhou et al., 2006). Imperialine, chuanbeinone, verticinone, and verticine from *F. cirrhosa*, one of the six source plants of Chuan Bei Mu, showed antitussive, expectorant, and anti-inflammatory activities (Wang et al., 2011). The four alkaloids imperialine, imperialine-β-*N*-oxide, isoverticine, and isoverticine-β-*N*-oxide isolated from *F. wabuensis*, another source plant of Chuan Bei Mu, significantly inhibited cough frequency and increased latent period of cough in mice induced by ammonia (Wang et al., 2012). They may be the active ingredients of Chuan Bei Mu.

4.2.2.2 Effects on cardiovascular system

Puqienine E is an angiotensin-converting enzyme inhibitory steroidal alkaloid from *F. puqiensis*, which showed the antihypertensive effect *in vitro* (An et al., 2010). Puqienine B and puqienine A of *F. puqiensis* also exhibited moderate antihypertensive effect.

N-Demethylpuqietinone, hupeheninoside, ebeiedinone, yibeinoside A, and chuanbeinone inhibited the bioactivity of human whole blood acetylcholinesterase and plasma butyrylcholinesterase (Lin et al., 2006b).

4.2.2.3 Anticancer activity

Four new steroidal penta- and hexacyclic veratraman- and cevan-based alkaloids were obtained from *F. hupehensis* (Zhang et al., 2008). Compounds 1 and 2 showed significant inhibitory effects against HeLa and HepG2 tumor cell lines. Verticinone was shown to induce differentiation in human leukemia cells (Yun et al., 2008), which also induces cell cycle arrest and apoptosis in immortalized and malignant human oral keratinocytes.

4.2.2.4 Other effects

The alkaloid verticinone could exert a good antinociceptive effect on inflammatory pain and cancer-related neuropathic pain probably through both peripheral and central mechanisms, and it might be partly involved with some sedation effects (Xu et al., 2011c). The newly found peimisine-3-*O*-β-D-glucopyranoside of *F. unibracteata*, one of the six source plants of Chuan Bei Mu, showed moderate protection effect on neurotoxicity of PC12 cell lines induced by rotenone (Zhang et al., 2011).

4.2.3 Saponins and terpenoids

Plant saponins are widely distributed among plants and have a wide range of biological properties. Steroidal saponins consist of a steroidal aglycone, a C27 spirostane skeleton, generally comprising of a six-ring structure (Figure 4.3). Since 2006, new steroidal saponins have been isolated from the dry bulbs of *F. pallidiflora* (Shen et al., 2011, 2012a) and *F. meleagris* (Table 4.2). Four saponins showed cytotoxicity against C6 and HeLa cell lines with IC_{50} values 5.1–75.8 μM (Shen et al., 2012a).

5β-Spirostanol glycoside and a cholestane derivative of *P. meleagris* induced apoptotic cell death in HL-60 cells through different mechanisms of action (Matsuo et al., 2013). The 22R-spirosolanol glycoside selectively induced apoptosis in A549 cells without affecting the caspase-3 activity level.

Ten terpenoid derivatives have been reported since 2004. Compounds 44–50 and 53 belong to diterpenes, and compounds 51 and 52 belong to triterpenoids (Table 4.3, Figure 4.4). The new kaurane diterpene (Table 4.3), ent-3β-butanoyloxykaur-15-en-17-ol, and four known kaurane diterpenes, isolated from the bulbs of *F. ebeiensis*, showed neuroprotective effects against MPP(+)-induced neuronal cell death in human dopaminergic neuroblastoma SH-SY5Y cell (Xu et al., 2011b). Two labdane diterpenes of *F. ebeiensis*, 6α, 7β-dihydroxy-labda-8(17), 12(E), 14-triene and 6-oxo-2α-hydroxy-labda-7, 12(E), 14-triene, also showed similar effects (Xu et al., 2011a). A new diterpene isopimara-7, 15-dien-19-oic acid from *F. imperialis* showed prolyl endopeptidase inhibitory activity (Atta-ur-Rahman et al., 2005).

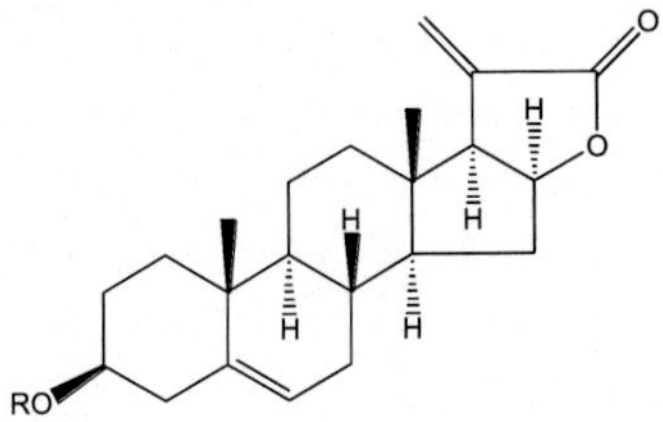

27

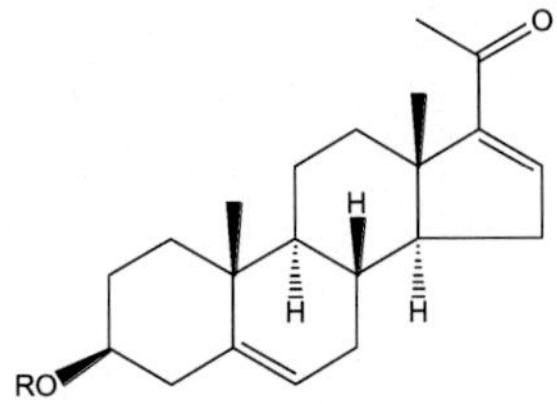

28

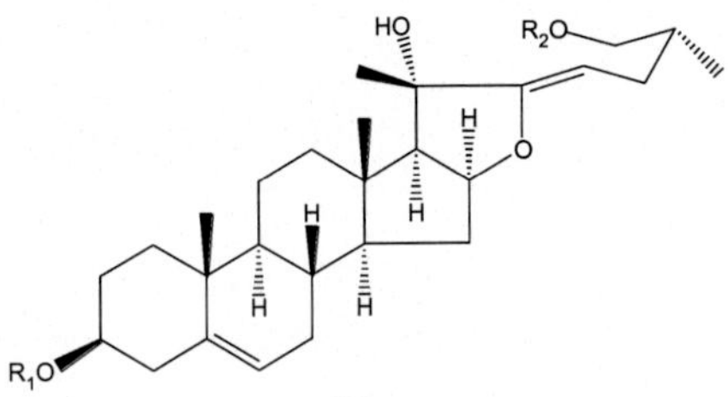

29

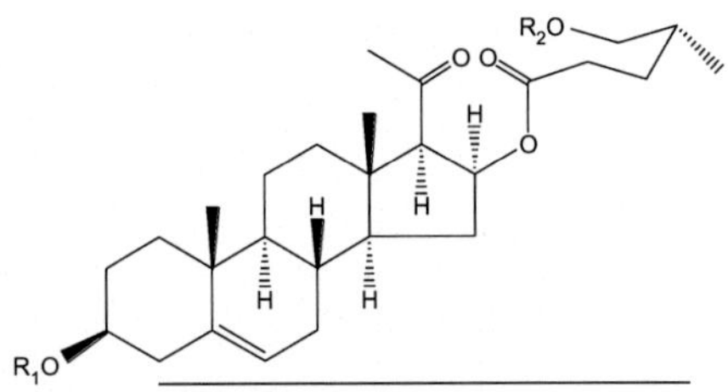

30 $R_1 = S_1$; $R_2 = S_4$
33 $R_1 = S_2$; $R_2 = S_4$

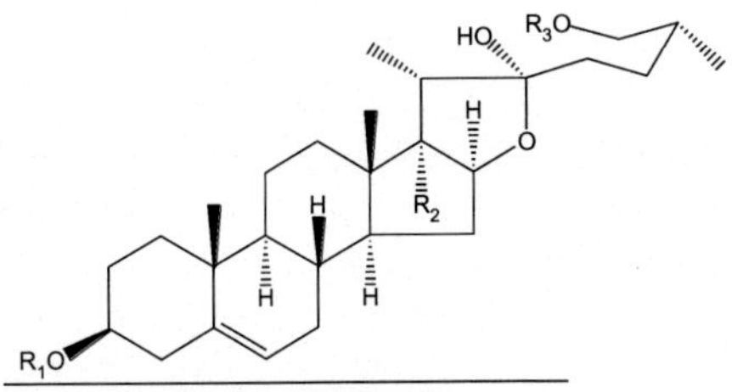

31 $R_1 = S_2$; $R_2 = H$; $R_3 = S_1$
39 $R_1 = S_5$; $R_2 = OH$; $R_3 = S_1$
40 $R_1 = S_7$; $R_2 = H$; $R_3 = S_1$
41 $R_1 = S_7$; $R_2 = OH$; $R_3 = S_1$

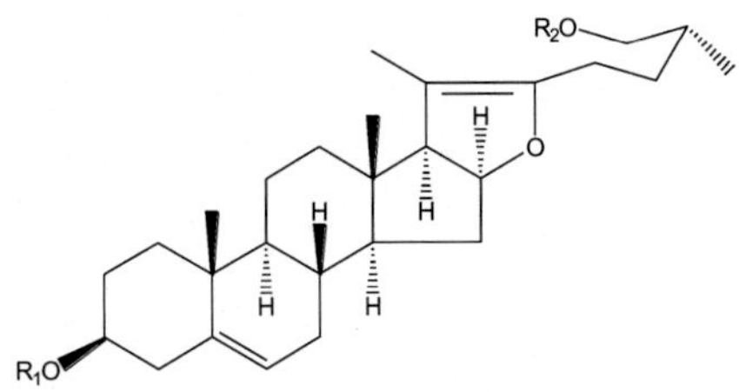

32

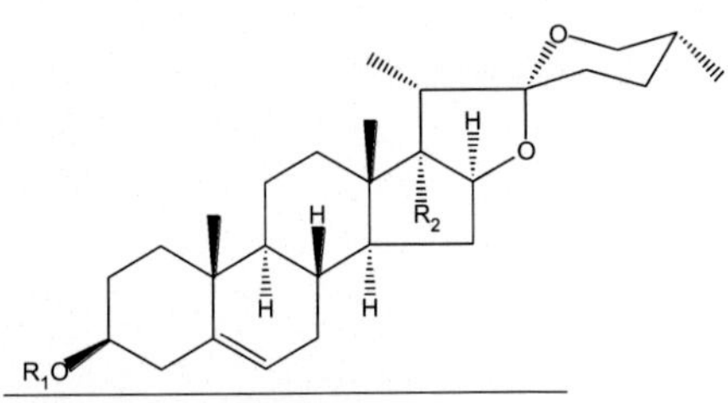

34 $R_1 = S_4$; $R_2 = H$
36 $R_1 = S_6$; $R_2 = OH$

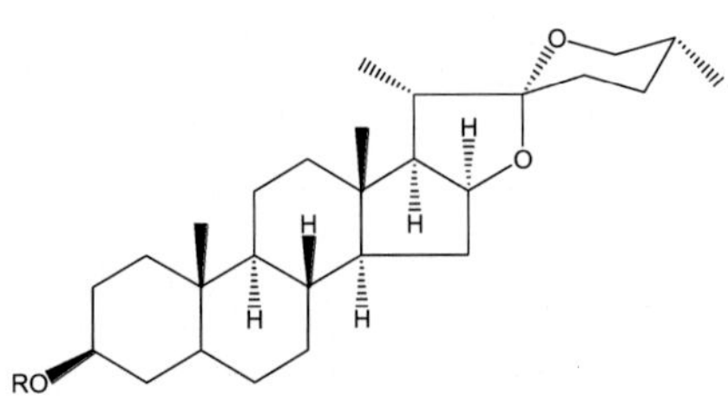

35

Figure 4.3 Steroidal saponins of *Fritillaria*. Refer to Table 4.2 for more details.

(Continued)

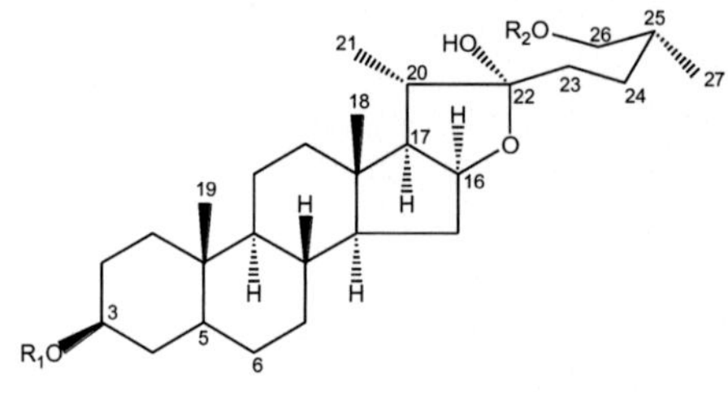

37

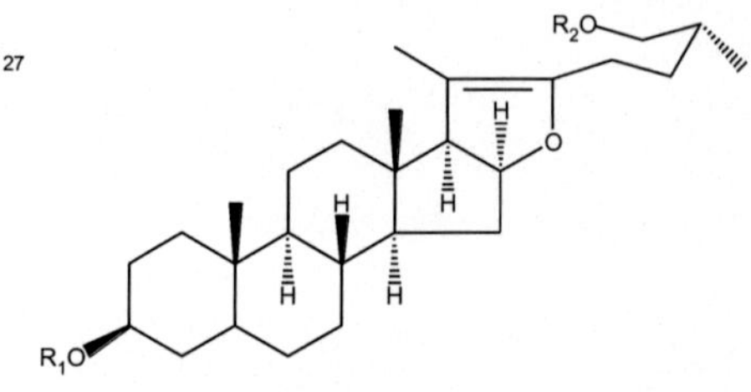

38

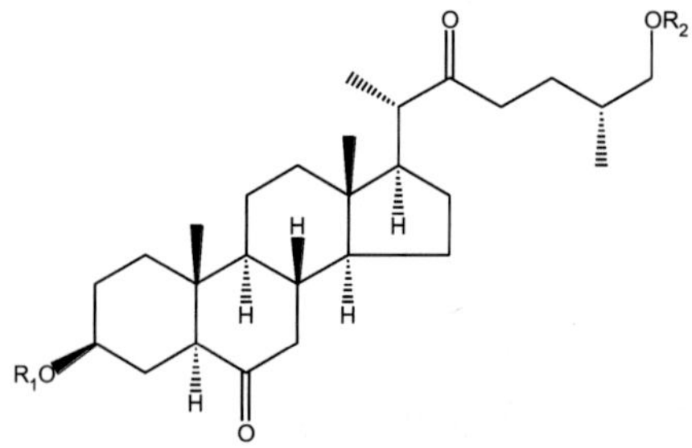

42

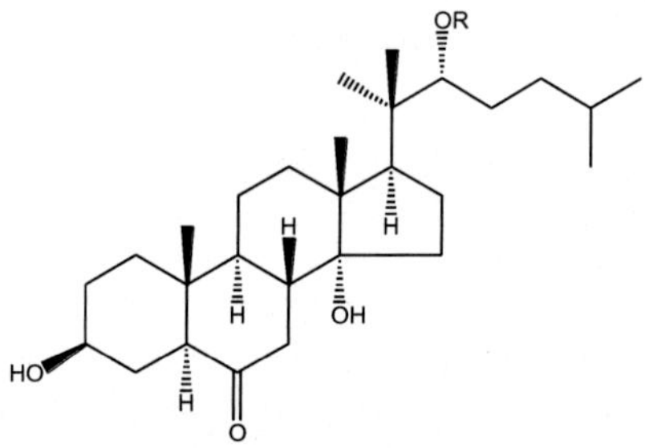

43

S_1 =

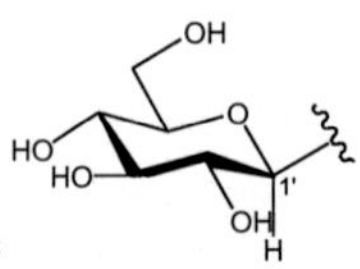

S_2 =

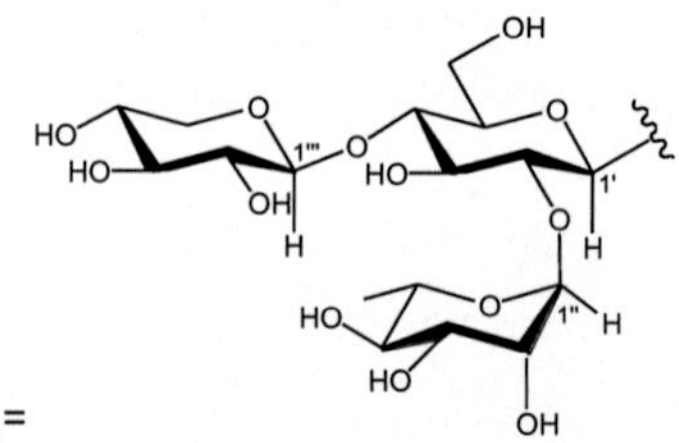

S_3 =

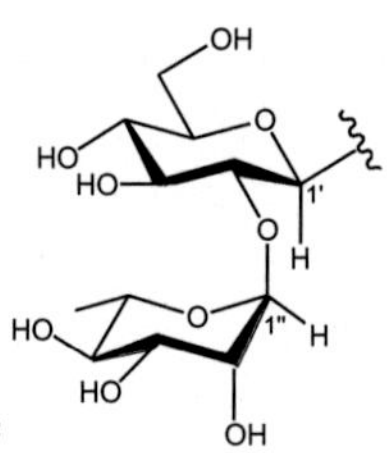

S_4 =

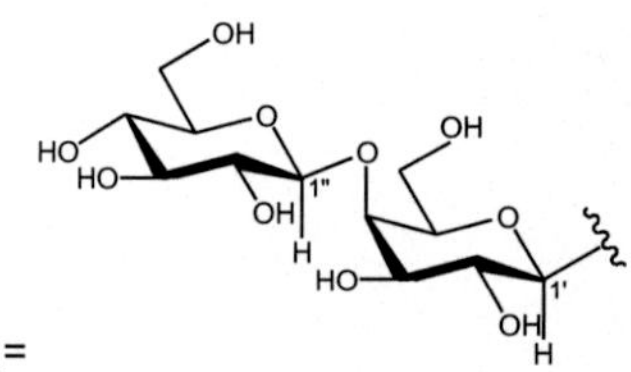

S_5 =

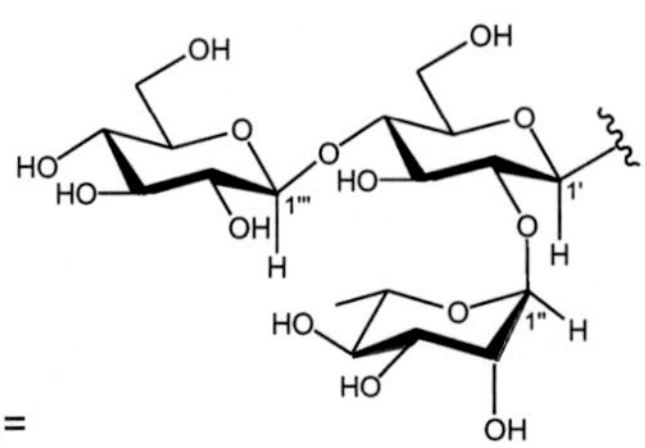

S_6 =

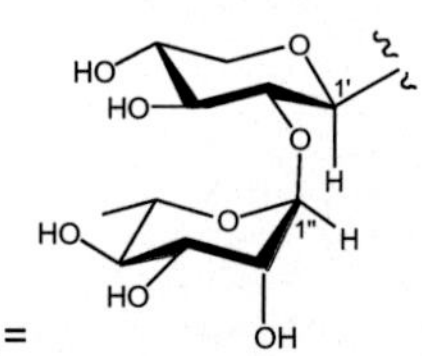

Figure 4.3 Continued.

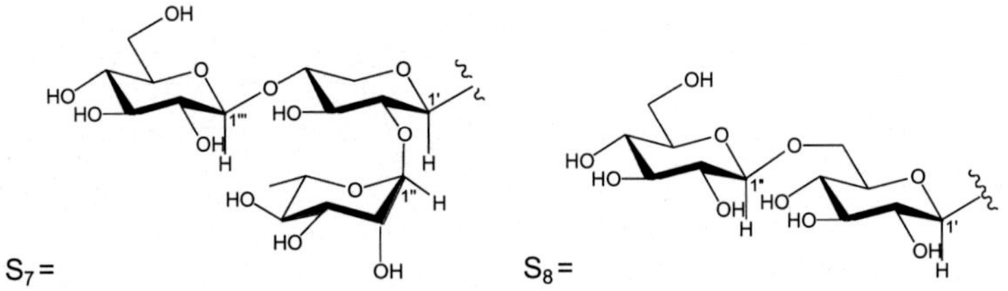

Figure 4.3 Continued.

Table 4.2 **Steroidal saponins found in *Fritillaria* bulbs in recent years**

Taxon	Saponins	Aglycone	Oligosaccharide chain	References
F. pallidiflora	Pallidifloside D **27**	A	$R=S_3$	Shen et al. (2012a)
	Pallidifloside E **28**	B	$R=S_2$	Shen et al. (2012a)
	Pallidifloside G **29**	C	$R_1=S_2$, $R_2=S_1$	Shen et al. (2012a)
	Pallidifloside H **30**	C	$R_1=S_2$, $R_2=S_1$	Shen et al. (2012a)
	Pallidifloside I **31**	C	$R_1=S_2$, $R_5=S_1$	Shen et al. (2012a)
F. pallidiflora	Pallidifloside A **32**	C	$R_1=S_2$, $R_2=S_1$	Shen et al. (2011)
	Pallidifloside B **33**	C	$R_1=S_3$, $R_2=S_1$	Shen et al. (2011)
	Pallidifloside C **34**	D	$R_1=S_4$	Shen et al. (2011)
F. meleagris	(25R)-5β-Spirostan-3β-ylO-β-D-glucopyranosyl-(1→4)-*O*-[α-L-rhamnopyranosyl-(1→2)]-β-D-glucopyranoside **35**	D	$R=S_5$	Matsuo et al. (2013)
	(25R)-17α-Hydroxyspirost-5-en-3β-ylO-α-L-rhamnopyranosyl-(1→2)-β-D-xylopyranoside **36**	D	$R_1=S_6$	Matsuo et al. (2013)
	(25R)-26-[(β-D-Glucopyranosyl)]oxy-22α-hydroxy-5β-furostan-3β-ylO-β-D-glucopyranosyl-(1→4)-*O*-[α-L-rhamnopyranosyl-(1→2)]-β-D-glucopyranoside **37**	C	$R_1=S_5$, $R_2=S_1$	Matsuo et al. (2013)

Continued

Table 4.2 Continued

Taxon	Saponins	Aglycone	Oligosaccharide chain	References
	(25R)-26-[(β-D-Glucopyranosyl)]oxy-5β-furost-20(22)-en-3β-ylO-β-D-glucopyranosyl-(1→4)-*O*-[α-L-rhamnopyranosyl-(1→2)]-β-D-glucopyranoside **38**	C	$R_1=S_5$, $R_2=S_1$	Matsuo et al. (2013)
	(25R)-26-[(β-D-Glucopyranosyl)oxy]-17α,22α-dihydroxyfurost-5-en-3β-ylO-β-D-glucopyranosyl-(1→4)-*O*-[α-L-rhamnopyranosyl-(1→2)]-β-D-glucopyranoside **39**	C	$R_1=S_5$, $R_5=S_1$	Matsuo et al. (2013)
	(25R)-26-[(β-D-Glucopyranosyl)oxy]-22α-hydroxyfurost-5-en-3β-ylO-β-D-glucopyranosyl-(1→4)-*O*-[α-L-rhamnopyranosyl-(1→2)]-β-D-xylopyranoside **40**	C	$R_1=S_7$, $R_5=S_1$	Matsuo et al. (2013)
	(25R)-26-[(β-D-Glucopyranosyl)oxy]-17α,22α-dihydroxyfurost-5-en-3β-ylO-β-D-glucopyranosyl-(1→4)-*O*-[α-L-rhamnopyranosyl-(1→2)]-β-D-xylopyranoside **41**	C	$R_1=S_7$, $R_5=S_1$	Matsuo et al. (2013)
	(25R)-3β-[(*O*-β-D-Glucopyranosyl-(1→6)-β-D-glucopyranosyl)oxy]-26-[(β-D-	E	$R_1=S_8$, $R_2=S_1$	Matsuo et al. (2013)

Continued

Table 4.2 Continued

Taxon	Saponins	Aglycone	Oligosaccharide chain	References
	glucopyranosyl) oxy]-5α-cholestane-6,22-dione **42**			
	(20R,22R)-22-[(β-D-Glucopyranosyl) oxy]-3β,14α,20-trihydroxy-5α-cholestane-6-one **43**	E	R = S_1	Matsuo et al. (2013)

Aglycone: A, C-22 steroidal lactone type; B, pregnane type; C, furostanol type; D, spirostanol type; E, cholestane type.

S_1 = –Glc; S_2 = –Glc^2—Rha (4|Xyl); S_3 = –Glc^2—Rha; S_4 = –Gal^4—Glc; S_5 = –Glc^4—Glc (2|Rha);

S_6 = –Xyl^2—Rha; S_7 = –Xyl^4—Glc (2|Rha); S_8 = –Glc^6—Glc.

Table 4.3 Terpenoids found in *Fritillaria* in recent years

Taxon	Type	Terpenes	Tissue	References
F. ebeiensis	A	6α,7β-Dihydroxy-labda-8(17),12 (E),14-triene **44**	Bulbs	Xu et al. (2011a)
	A	6-oxo-2α-Hydroxy-labda-7,12 (E),14-triene **45**	Bulbs	Xu et al. (2011a)
F. anhuiensis	A	12,15-Epoxy-8(17),13-labdadien-19-ol **46**	Bulbs	Kang et al. (2007)
F. anhuiensis	A	14(RS),15-Dihydroxy-8(17),12 (E)-labdadien-19-oic acid **47**	Bulbs	Shou et al. (2012)
F. ebeiensis	B	Ent-3β-butanoyloxykaur-15-en-17-ol **48**	Bulbs	Xu et al. (2011b)
F. monantha	B	Ent-16,17-epoxy-kauran-3α-ol **49**	Bulbs	Liu et al. (2007b)
F. hupehensis	B	Fritillahupehin **50**	Bulbs	Zhang et al. (2004)
F. hupehensis	C	25-Hydroxy-9,19-cycloart-22-ene-3-one **51**	Leaves, stems	Pi et al. (2009)
	C	(23Z)-9,19-Cycloart-23-ene-3α,25-diol **52**	Leaves, stems	Pi et al. (2009)
F. imperialis	D	Isopimara-7,15-dien-19-oic acid **53**	Bulbs	Atta-ur-Rahman et al. (2005)

Aglycone: A labdane type; B ent-kaurane type; C cycloartane type; D isopimarane type.

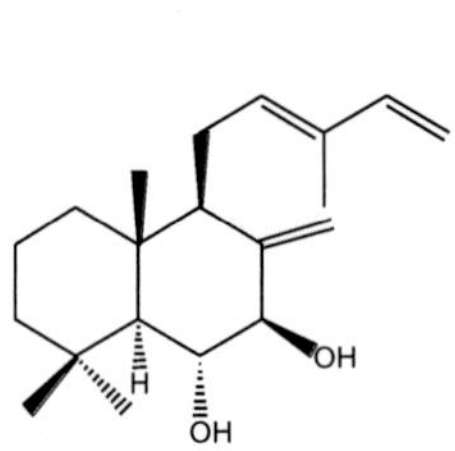

44

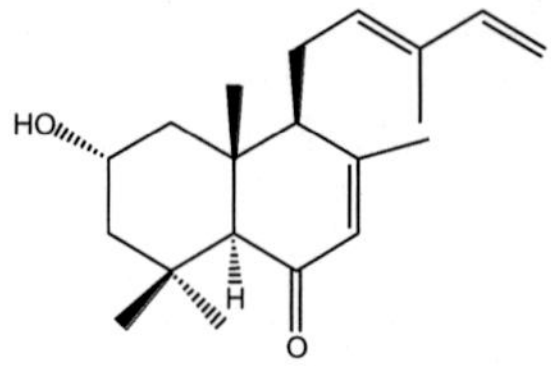

45

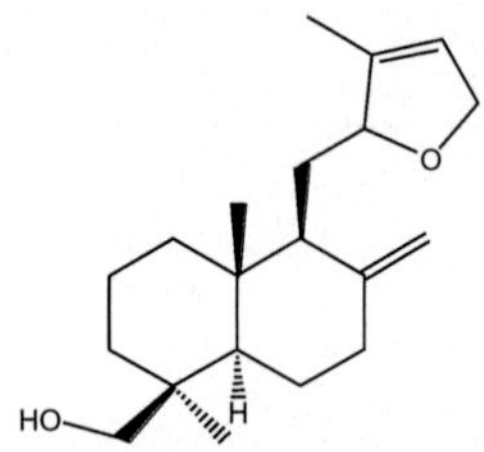

46

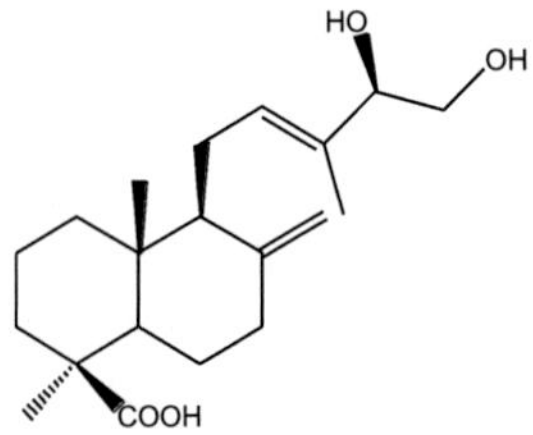

47

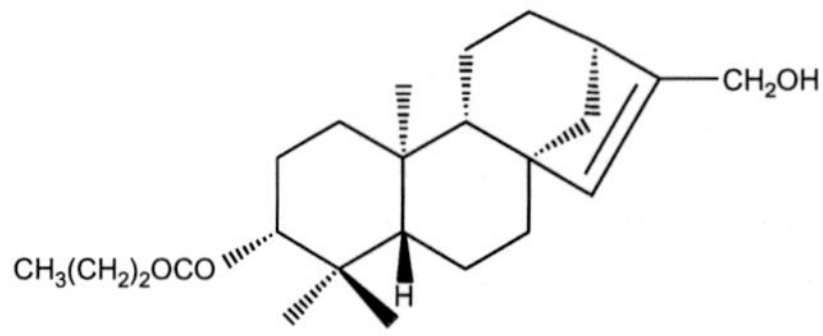

48

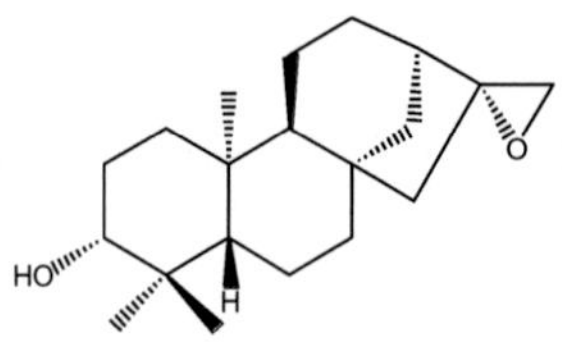

49

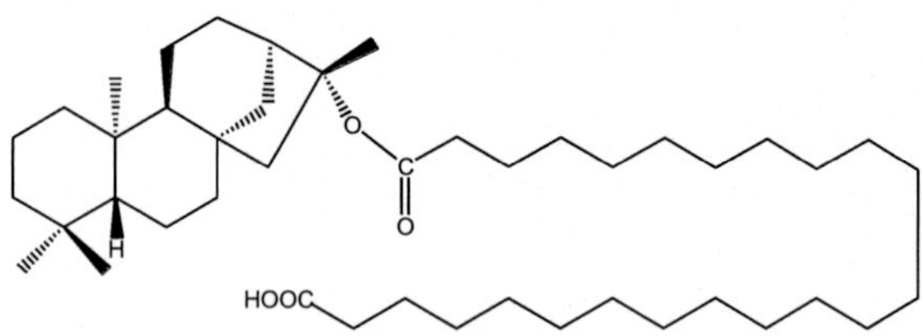

50

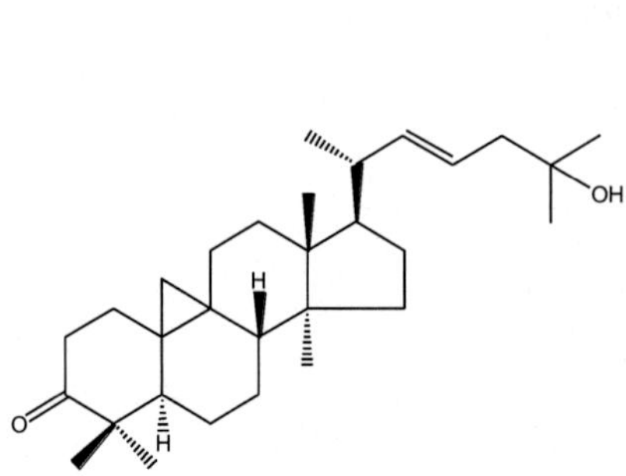

51

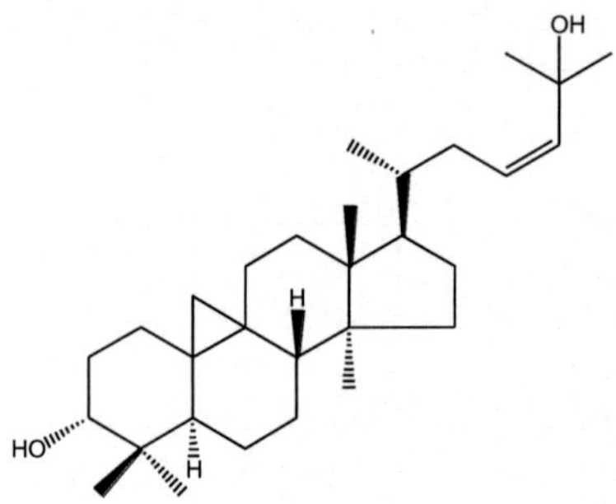

52

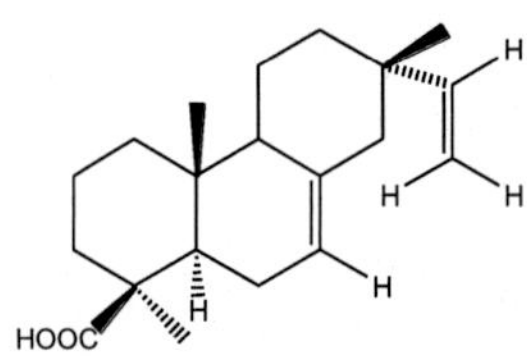

53

Figure 4.4 Terpenoids of *Fritillaria*. Refer to Table 4.3 for more details.

4.2.4 Other compounds

FUP-1, a water-soluble polysaccharide obtained from *F. ussuriensis*, exerts antioxidant activity through its own radical-scavenging activity and by boosting the enzymatic and nonenzymatic antioxidant defense system of the host (Liu et al., 2012). Three nucleosides and three bases were isolated from *F. puqiensis* and identified as 2′-*O*-methyladenosine (I), uridine (II), adenosine (III), uracil (IV), thymine (V), and adenine (VI) (Zhou et al., 2008). Compounds I, II, and IV–VI were isolated from *Fritillaria* for the first time. Compound I is a rare methylated ribonucleoside. Adenosine is involved in decreasing the blood pressure, slowing the heart rate, and relaxing the smooth muscle and has sedative effects (Zhang et al., 2010b). Adenosine, uridine, adenine, guanosine, and thymidine were quantified with reverse-phase (RP) high-performance liquid chromatography (HPLC). The order of contents in different *Fritillaria* bulbs was *F. hupehensis* > *F. thunbergii* > *F. cirrhosa* approximately *F. ussuriensis*. A sensitive and reliable HPLC diode array detector (DAD) method was developed to simultaneously determine nine nucleosides and nucleobases including uracil, cytidine, guanine, uridine, thymine, inosine, guanosine, thymidine, and adenosine in 13 *Fritillaria* species (Cao et al., 2010), which could be helpful to control the quality of *Fritillaria* bulbs. Seven nucleosides and nucleobases (uracil, cytidine, uridine, guanosine, thymidine, adenosine, and adenine) in *F. taipaiensis*, one of the six source plants of Chuan Bei Mu, can be rapidly and accurately quantified by HPLC-DAD (Huang et al., 2011). Bulbus of *F. delavayi*, one of the six source plants of Chuan Bei Mu, is the most commonly used antitussive and apophlegmatic in China. It contains uracil, cytidine, inosine, uridine, guanosine, thymidine, adenosine, hypoxanthine, adenine, and 2-deoxyadenosine that can be quantified with HPLC, which is an efficient way to evaluate the quality consistency of *F. delavayi* (Duan et al., 2012). Thirty-nine volatile components were identified from *F. thunbergii* flowers with gas chromatography-time-of-flight mass spectrometry (GC-TOF-MS), including the octadecatrienoic acid methyl esters and some aromatic aldehydes and ketones, such as benzeneacetaldehyde and 1-(2-hydroxy-5-methylphenyl)-ethanone (Liang et al., 2011). The volatile component 3-methyl-2-butene-1-thiol, causing the foxy odor, was identified from the headspace of *F. imperialis* flower bulbs (Helsper et al., 2006). Element fingerprints were deciphered for *F. thunbergii* from 10 major producing regions of China (Yuan et al., 2010). Analysis by inductively coupled plasma optical emission spectrometry allowed simultaneous determination of 18 elements (Al, B, Ca, Cd, Co, Cr, Cu, Fe, K, Mg, Na, Mn, Mo, S, Ni, P, Pb, and Zn) in *F. thunbergii*. Among the many elements of *F. ussuriensis*, Cu, Zn, Fe, Mg, and Mn are the dominant chemicals (Yao et al., 2008). In addition, a glucosylsterol, β-sitosterol-3-*O*-glucopyranoside, was isolated from *F. verticillata*, which inhibits sortase, a bacterial surface protein anchoring transpeptidase (Kim et al., 2003). Many other compounds were also isolated from various *Fritillaria* species (Li et al., 2005). Their functions and pharmacological effects await further studies.

4.2.5 Bioactivity of compound mixture

The compatible application of *Aconiti Radix Cocta* and *F. cirrhosa* could enhance the analgesic effect of *Aconiti Radix Cocta* and reduce the expectorant and antitussive effects of *F. cirrhosa* (Tan et al., 2013), which vary in terms of different matching ratio and dose. The compatible application of *Aconiti Radix Cocta* and *F. thunbergii* shows no effect on the antitussive effect of *F. thunbergii*. These results provide experimental basis for in-depth studies on the combined effect of *Aconiti Radix Cocta* and *Fritillaria*—two of 18 incompatible pairs in TCM.

F. cirrhosa (FC) are widely used in Chinese medicine to treat several aliments and as an adjuvant to chemotherapy of lung cancer. Cell growth responses to FC were compared in ovarian and endometrial cancer cell lines (Kavandi et al., 2013). Dose-dependent cell growth inhibition was observed following higher doses in all cell lines, while lower doses stimulated growth in only endometrial cell lines. Higher doses of FC significantly decreased cell growth on soft agar and decreased the invasive potential of cancer cells. Treatment of cells with FC resulted in activation of caspase-3, G0/G1 phase cell cycle arrest, downregulation of cyclins D1 and D3, and induction of p27. FC decreased NF-κB DNA binding, reduced expression of phosphorylated IκBα, abrogated NF-κB activation, and downregulated NF-κB-regulated metastasis-promoting proteins in cancer cells. Knockdown of NF-κB attenuated FC-induced cell growth inhibition. These results suggest that inhibition of NF-κB activation may be an important mechanism for growth suppression by FC.

Compound granule prescription of thunberg fritillary bulb can relieve the bone marrow suppression in refractory acute leukemia patients caused by the chemotherapy (Li et al., 2012), which is mainly reflected by the slowdown of reduction in white blood cells.

4.3 Chemotaxonomy

All plants produce secondary metabolites in the coevolution of host and pathogen/herbivore; however, the structural types are often specific and restricted to taxonomically related plant groups. This observation was the basis of the development of chemotaxonomy. Much evidence suggests that 5α-cevanine isosteroidal alkaloids are the characteristic constituents of *Fritillaria* (Yu and Xiao, 1992a). The C-13 and C-17 of the molecules may be rational positions uniting a nitrogenous group in their biosyntheses that generate two kinds of 5α-cevanine isosteroidal alkaloids. The two hydrogen atoms of C-13 and C-17 are at the state of *trans*-configuration (e.g., verticine and verticinone; Figure 4.2) and *cis*-configuration (e.g., delavine, chuanbeinone, and songbeinine), respectively. In *F. cirrhosa*, *F. cirrhosa* var. *ecirrhosa*, *F. delavayi* (source plants of Chuan Bei Mu), and *F. pallidiflora* (source plant of Yi Bei Mu), the *trans*-configuration 5α-cevanine isosteroidal alkaloids were not detected in thin layer chromatography (TLC), and only *cis*-configuration ones were abundant (Yu and Xiao, 1992a). Only *cis*-configuration 5α-cevanine isosteroidal alkaloids were obtained by ingredient separation of *F. unibracteata* (source plant of Chuan Bei Mu). In contrast, the *trans*-configuration 5α-cevanine isosteroidal alkaloids were detected in *F. thunbergii* (source plant of Zhe Bei Mu), *F. anhuiensis*, *F. ebeiensis*, and *F. wuyangensis* (source plants of Anhui Bei Mu), while *cis*-configuration ones were not detected. *F. hupehensis* (source plant of Hubei Bei Mu), *F. ussuriensis* (source plant of Ping Bei Mu), and *F. karelinii* (source plant of Yi Bei Mu) contained both *trans*- and *cis*-configuration 5α-cevanine isosteroidal alkaloids. *Fritillaria* samples were prederivatized by trimethylsilylimidazole and six isosteroidal alkaloids (ebeiedine, ebeiedinone, verticine, isoverticine, verticinone, and imperialine) were analyzed by gas chromatography (Li et al., 1999), which is more sensitive than TLC. Sixteen species were classified into four groups. *F. thunbergii*, *F. thunbergii* var. *chekiangensis*, *F. anhuiensis*, *F. ebeiensis*, *F. ebeiensis* var. *purpurea*, and *F. hupehensis* contained only *trans*-configuration 5α-cevanine isosteroidal alkaloids (verticine, verticinone, and ebeiedine); *F. pallidiflora*, *F. walujewii*, and *F. yuminensis* (source plants of Yi Bei Mu) contained only *cis*-configuration ones (e.g., imperialine); *F. cirrhosa*, *F. cirrhosa* var. *viridiflora*, *F. unibracteata*, *F. delavayi*, and *F. ussuriensis* contained both *trans*- and *cis*-configuration ones; *F. meleagroides* and *F. maximowiczii* contained neither *trans*- nor *cis*-configuration ones.

Bei Mu is one of the most commonly used Chinese herbal medicines and is derived from the bulbs of some fritillaries. During the last decades, many new species and infraspecific taxa of *Fritillaria* have been described in China, including 80 species, 52 varieties, and six forms. The significant increase of new taxa has influenced the investigation, application, quality control, and commodity circulation of Bei Mu. Xiao et al. (2007) performed a benchmark study of the morphological, geographic, phytochemical, and historical aspects of Chinese medicinal fritillaries, with regard to their current application. The source plants of Bei Mu, from a pharmacophylogenetic point of view, were classified into six groups. (1) Zhe Bei Mu is mainly derived from the cultivated *F. thunbergii* and *F. thunbergii* var. *chekiangensis* in Zhejiang and Jiangsu provinces and contains D/E *trans*-cevanine group alkaloids, predominantly verticine, isoverticine, and verticinone. No D/E *cis*-cevanine group alkaloids were detected (Xiao et al., 2007). (2) Yi Bei Mu consists of *F. walujewii*, *F. tortifolia*, *F. yuminensis*, *F. karelinii*, and *F. pallidiflora*, which are distributed in Xinjiang, China, and Central Asia and contain D/E *cis*-cevanine group alkaloids, for example, imperialine; no D/E *trans*-forms of cevanine group alkaloids were found. (3) Ping Bei Mu is obtained from cultivated *F. ussuriensis* in northeast China and Korea and contains both D/E *trans*- and *cis*-forms of cevanine group alkaloids. (4) Chuan Bei Mu is the most important medicinal Bei Mu, which is derived mainly from *F. cirrhosa*, *F. cirrhosa* var. *viridiflora*, *F. przewalskii*, *F. delavayi*, *F. wabuensis*, *F. taipaiensis*, *F. taipaiensis* var. *ningxiaensis*, and *F. unibracteata* in the Hengduan Mountains and their adjacent regions. *F. cirrhosa*, *F. przewalskii*, and *F. unibracteata* have smaller bulbs than the other species and are the main source of the commodity Qingbei, which is regarded as Bei Mu of the highest quality. *F. cirrhosa*, distributed in Tibet, Yunnan, Sichuan, and Qinghai, shows many morphological variations and, towards its northwestern limit, is replaced by a very closely related species, *F. taipaiensis*. The bulbs of *F. delavayi*, called Lubei, are in the group of Chuan Bei Mu. All kinds of Chuan Bei Mu contain both D/E *trans*-cevanine group and *cis*-cevanine group alkaloids. (5) Hubei Bei Mu is mainly from bulbs of cultivated *F. hupehensis* (Xiao, 2002), which is very closely related to *F. monantha* and *F. egregia*. It contains both D/E *trans*- and *cis*-forms of cevanine group alkaloids; verticine and isoverticine are representative of the D/E *trans*-form, while hupehenine, hupeheninoside, hupehenrine, and hupehenizine are representative of the D/E *cis*-form. (6) Anhui Bei Mu is derived from *F. anhuiensis*, *F. ebeiensis*, *F. ebeiensis* var. *purpurea*, and *F. wuyangensis* and contains only D/E *trans*-cevanine group alkaloids. The six Bei Mu groups have six “Dao-di” (genuine) production areas, respectively. This classification is in line with current clinical application and drug marketing. All except Anhui Bei Mu are recorded in the latest edition of China Pharmacopoeia (2010). Xiao et al. (2007) suggested that the publication and naming of new taxa of the most important economic plants, not limited to Bei Mu, should be more cautious. The chemosystematics study is greatly helpful for not only the sustainable utilization of *Fritillaria* medicinal resources but also the conservation of critically endangered *Fritillaria* species such as those in western China.

Due to the complexity of Bei Mu botanical origin in the herbal markets, it is urgent to develop a reliable method for species identification. Morphological and histological techniques have respective limitations. To find an applicable authentication method, 27 steroidal alkaloids (16 cevanine-type, six veratramine-type, one jervine-type, and four secosolanidine-type alkaloids) in 17 *Fritillaria* species and 12 Bei Mu-containing compound formulas were identified and characterized by an HPLC–MS method (Li et al., 2009a). The estimated relative composition of steroidal alkaloids was used in a chemotaxonomic study of *Fritillaria* based on hierarchal cluster analysis. The content and composition of steroidal alkaloids are controlled more by genetic than by environmental factors. *F. puqiensis* is basal to and far away from other species, suggesting its chemical uniqueness. The presence of multiple types of steroidal skeletons, especially of large amount of veratramine-type (e.g., puqienine B) and secosolanidine-type (e.g., puqietinone) alkaloids in one species, is rare in *Fritillaria* (Jiang et al., 2006; Li et al., 2009a,b), which suggests that *F. puqiensis* is not a subspecies of *F. thunbergii* as proposed previously (Yu and Xiao, 1992b). The remaining 16 species fell into two groups. Group I, mainly containing the *trans*-D/E cevanine alkaloids such as verticine, verticinone, isoverticine, and ebeiedinone, includes the source plants of Zhe Bei Mu, Ping Bei Mu, Hubei Bei Mu, and Anhui Bei Mu and *F. maximowiczii*, an adulterant of Chuan Bei Mu (Luo et al., 2012). The chemical profile of Zhe Bei Mu is closer to that of Ping Bei Mu than to those of Hubei Bei Mu and Anhui Bei Mu, which is geographically closer to Zhe Bei Mu. Group II, containing highly abundant *cis*-D/E cevanine alkaloids (e.g., imperialine and sinpeinine A), includes the source plants of Chuan Bei Mu and Yi Bei Mu. The chemical profile of *F. meleagroides*, which is geographically closer to Yi Bei Mu, is closer to that of Chuan Bei Mu and renders *F. meleagroides* a useful medicinal resource. The previous classification agrees with the early TLC studies (Yu and Xiao, 1992a) and is congruent with the pharmacophylogenetic treatment of Chinese medicinal Bei Mu proposed by Xiao et al. (2007). The characteristic distribution patterns of the examined compounds were utilized to determine the botanical origin of Bei Mu-containing compound formulas. Ten out of 12 products meet their label claims in terms of the Bei Mu raw material. The qualitative and quantitative differences in steroidal alkaloids were useful not only for chemotaxonomy of medicinal *Fritillaria* but also for species identification in compound formulas. Due to its price, most of the Chuan Bei Mu on the market is not pure; quite often, the high-yield *Fritillaria* product Ping Bei Mu is mixed with it to a greater or lesser extent (Luo et al., 2012). This kind of adulteration mitigates the therapeutic efficiency. The HPLC–MS method is important in quality control of Bei Mu-containing herbal medicine preparations, allowing for preventing Bei Mu confusion and detecting the possible adulteration.

LC coupled with electrospray ionization (ESI) quadrupole time-of-flight (QTOF)-MS/MS was performed to study the fragmentation behaviors of steroidal alkaloids of *Fritillaria* (Zhou et al., 2010). Forty-one steroidal alkaloids (29 cevanine-type, one jervine-type, six veratramine-type, and five secosolanidine-type alkaloids) were selectively identified in eight medicinal *Fritillaria* species. Twenty-six compounds were unambiguously identified by comparing with the reference compounds, and 15 compounds were tentatively identified or inferred according to their MS/MS data. The LC/QTOF-MS total ion chromatograms of eight species confirmed the chemotaxonomy results of previous study (Li et al., 2009a,b). For example, Ping Bei Mu shared more peaks with Zhe Bei Mu than with Chuan Bei Mu; Yi Bei Mu shared more peaks with Chuan Bei Mu than with Zhe Bei Mu. Confirming the previous findings, the major chemical differences between source plants of Yi Bei Mu and those of Zhe Bei Mu, Hubei Bei Mu, and Anhui Bei Mu were revealed by the analysis of *F. pallidiflora* and *F. walujewii* with ultraperformance LC with evaporative light-scattering detection (UPLC-ELSD) (Duan et al., 2010). Peimisine and sipeimine are highly abundant in Yi Bei Mu, while peimine and peiminine, which are highly abundant in Zhe Bei Mu, Hubei Bei Mu, and Anhui Bei Mu, are of low abundance or not detected in Yi Bei Mu.

The major alkaloids from *F. hupehensis* bulbs (FHB) are analyzed by the combined use of the following two methods: the simultaneous quantitation of three alkaloids by using HPLC-ELSD and the simultaneous characterization of seven alkaloids by using HPLC–ESI-MS(n) (Zhang et al., 2014). The other four congeneric species were also analyzed. Both correlation coefficients of similarity in chromatograms were calculated for quantitation of HPLC profiles. The chromatogram profile combining similarity evaluation could efficiently identify and distinguish FHB from other *Fritillaria* species. Differentiating the bulbs of *Fritillaria* is also possible by wooden-tip ESI-MS (Xin et al., 2014a).

Besides steroidal alkaloids, chemical elements were also analyzed to differentiate *Fritillaria* samples. Flame atomic absorption spectrometry (FAAS) and graphite furnace atomic absorption spectrometry were used to analyze Zn, Fe, Mn, Cu, Na, Mg, K, Ca, As, Pb, Cd, Co, Cr, Sr, and Al of *Fritillaria* samples from four production regions (Yu et al., 2007). Zn, Mn, Cd, and Sr were highly abundant in *F. thunbergii* and Cu and Ca highly abundant in Xiang Bei (a kind of commercial Zhe Bei Mu originated in Xiangshan, Zhejiang province), whereas Fe, Mg, K, Co, and Al were highly abundant in *F. thunbergii* var. *chekiangensis* and Na, As, Pb, and Cr highly abundant in *F. cirrhosa*. Principal component analysis unequivocally grouped the *Fritillaria* samples. Eighteen elements in *F. thunbergii* were analyzed with inductively coupled plasma optical emission spectrometry to differentiate samples from 10 major production regions (Yuan et al., 2010). Despite these studies, the relationships among the *Fritillaria* species have not been totally resolved. These studies did not include *Fritillaria* species outside of China, although the classification of medicinal

Fritillaria should be determined in the context of worldwide distribution. The distribution of secondary metabolites and other chemical entities apparently is of value for taxonomy, but it has to be analyzed cautiously, as any other adaptive trait. Molecular methods have to be combined with morphology and chemotaxonomy in the elucidation of the pharmacophylogeny of *Fritillaria*.

4.4 Molecular taxonomy, molecular phylogeny, and genomics

4.4.1 *Molecular biology*

RAPD (random amplification of polymorphic DNA) is a kind of PCR reaction, but the DNA fragments that are amplified are random (Hao et al., 2010). The arbitrary and short primers (8–12 nucleotides) are used in the PCR, which uses a large template of genomic DNA. By resolving the resulting patterns, a semiunique profile can be garnered from a RAPD reaction. RAPD has been used to discriminate Chuan Bei Mu, Hubei Bei Mu, and other source plants of Bei Mu (Li et al., 2006c; Yin et al., 2007). RAPD was also used to study the genetic diversity of four *F. thunbergii* cultivars, for example, Duozi (many seeds), broad-leaf, narrow-leaf, and Lingxia, and five *Fritillaria* species (*F. thunbergii*, *F. cirrhosa*, *F. anhuiensis*, *F. thunbergii* var. *chekiangensis*, and *F. pallidiflora*) (Lu et al., 2009). Nine polymorphic primers were selected for further RAPD amplification. Fifty-seven amplicons with 48 polymorphic bands, accounting for 84.21% of all the bands, were generated. Among four *F. thunbergii* cultivars, Duozi differed from the other three. Lingxia was closer to narrow-leaf than to broad-leaf. *F. thunbergii* var. *chekiangensis* and *F. thunbergii* had close relationship, while *F. cirrhosa* was closer to Zhe Bei Mu than to *F. pallidiflora*. Screened RAPD markers can be applied to identify different *F. thunbergii* cultivars and *Fritillaria* species. RAPD was also used to determine whether *F. wabuensis* could be used as the source plant of Chuan Bei Mu (Li et al., 2010). All taxa could be divided into four groups. *F. thunbergii* was basal to other taxa; *F. sichuanica* was basal to the source plants of Chuan Bei Mu; the group consisting of *F. taipaiensis*, *F. delavayi*, one *F. cirrhosa* var. *ecirrhosa* sample, and one *F. cirrhosa* sample is sister to the group consisting of *F. wabuensis*, *F. cirrhosa*, *F. cirrhosa* var. *ecirrhosa*, *F. unibracteata*, and *F. przewalskii*. The taxonomic status of 42 taxa of *Fritillaria* in Turkey was examined by RAPD-PCR and seed protein analysis (Celebi et al., 2008). *F. imperialis* and *F. persica* are close relatives. *F. acmopetala* subsp. *acmopetala* and *F. sororum* are very close relatives according to morphological, RAPD, and protein results. Therefore, these two taxa can be considered as synonyms. *F. zagrica*, *F. caucasica*, *F. baskilensis*, *F. armena*, and *F. pinardii* are grouped together. All these separate taxa could be treated as synonyms based on morphological and molecular data. However, no East Asia species were included in this study.

Eleven species of *Fritillaria* are recorded as *Bulbus fritillariae* in the Chinese Pharmacopoeia. *Bulbus fritillariae cirrhosae* (BFC) is a group of six *Fritillaria* species with higher efficiency and lower toxicity derived mainly from wild sources. Because of their higher market price, five other *Fritillaria* species are often sold deceptively as BFC in the herbal market. To ensure the efficacy and safety of medicinal herbs, a DNA-based identification method was developed to authenticate the commercial sources of BFC (Xin et al., 2014b). A putative DNA marker (0.65 kb) specific for BFC was identified using the RAPD technique. A DNA marker representing a sequence characterized amplified region (SCAR) was developed from an RAPD amplicon. The SCAR marker was successfully applied to differentiate BFC from other species of *Fritillaria*. The SCAR marker was also useful in identifying the commercial samples of BFC.

F. tubiformis subsp. *moggridgei* is a rare alpine geophyte with shiny yellow flowers, which is sporadically distributed across the southwestern Alps where it is biogeographically close to *F. tubiformis* var. *burnatii*. The latter has dark purple flowers and ranges in the majority of the Western and Central Alps. An RAPD-based analysis was conducted to study the genetic status of these taxa and the distribution of genetic variability of the subspecies by sampling seven populations distributed across the subspecies' range (Mucciarelli et al., 2014). Four populations of *F. tubiformis* var. *burnatii* were chosen within this range and included in the genetic analysis. 264 individuals were analyzed and 201 polymorphic loci were scored. Genetic diversity scored in the subspecies was in line with expectations for endemic species ($H_e = 0.194$). *F. tubiformis* var. *burnatii* showed lower intraspecific diversity ($H_e = 0.173$), notwithstanding a wider range than the subspecies. Most phenotypic variation (about 83%) was allocated within populations. PCoA analysis and Bayesian clustering separated populations into two genetically differentiated groups corresponding with the subspecific taxa. Three populations ascribed to the *F. tubiformis* subsp. *moggridgei* repeatedly showed genetic admixture with *F. tubiformis* var. *burnatii* populations. Although the different flower color, the two taxa are genetically very similar and share a consistent part of their gene pool; the majority of genetic variability is allocated within populations rather than among them; a representative amount of genetic diversity can be preserved by sampling from a restricted number of populations. RAPD markers are efficient in analyzing genetic variation, and the results contribute to the preservation of biodiversity of the species.

ISSR (intersimple sequence repeat) is a general term for a genome region between microsatellite loci. The complementary sequences to two adjacent microsatellites are used as PCR primers; the variable region between them is amplified. ISSR was used to study the relationship and genetic polymorphism of 19 *Fritillaria* populations in Sichuan province (Li et al., 2009a,b). Eleven primers were selected from 35 ISSR primers and 179 DNA fragments were amplified from 19 populations. Samples were classified into four groups. ISSR was also used to classify 22 ornamental *Fritillaria* taxa, including species from both East Asia and the Mediterranean region (Wu et al., 2010). Thirteen primers were used for ISSR amplification and 160 polymorphic bands were generated. Samples were classified into two major groups. In the small group, *F. davidii* of subgenus *Davidii* is closer to *F. anhuiensis* than to *F. camtschatcensis* of subgenus *Liliorhiza*. In the large group, there are three subgroups: (1) the subgroup consisting of Mediterranean taxa *F. acmopetala* subsp. *wendelboi*, *F. acmopetala*,

F. elwesii, and *F. uva-vulpis*; (2) the subgroup consisting of source plants of Yi Bei Mu, Zhe Bei Mu, and Chuan Bei Mu, which belong to subgenus *Fritillaria*; (3) the subgroup consisting of Mediterranean taxa *F. imperialis*, *F. raddeana* (subgenus *Petilium*), and *F. persica* (subgenus *Theresia*). Subgroup 1, belonging to subgenus *Fritillaria*, is closer to subgroup 2 than to subgroup 3, which does not contradict with the RAPD results (Celebi et al., 2008) and phylogeny results inferred from ITS and matK sequences (see below). Genetic diversity and differentiation of four landraces of *F. thunbergii* and two congeners *F. cirrhosa* and *F. anhuiensis* were evaluated using ISSR markers (Li et al., 2011). The genetic diversity of *F. thunbergii* was high at the specie level but relatively low at the landrace level. *F. anhuiensis* was closer to *F. cirrhosa* than to *F. thunbergii*, which agrees with the previous ISSR study (Liu et al., 2010) and RAPD study (Lu et al., 2009). Interestingly, in UPGMA tree, *F. thunbergii* cultivar Duozi was basal to other taxa including other *F. thunbergii* cultivars, *F. thunbergii* var. *chekiangensis*, *F. anhuiensis*, and *F. przewalskii* (Liu et al., 2010). It is possible that Duozi is a hybrid taxon selected during long-term cultivation. Low levels of genetic diversity were observed in all seven populations of *F. thunbergii* from the major production area Pan'an, Zhejiang province (Liu et al., 2010), which were lower than that of fritillary populations from other production areas. However, the high genetic diversity and heterogeneity of *F. thunbergii* in Pan'an, inferred from ISSR markers, was also reported (Zhou and Wang, 2012).

AFLP (amplified fragment length polymorphism) uses restriction enzymes to cut genomic DNA, followed by ligation of adaptors to the sticky ends of the restriction fragments. A subset of the restriction fragments are then amplified using primers complementary to the adaptor and part of the restriction site fragments. The amplified fragments are visualized on denaturing polyacrylamide gels through autoradiography (Zhang et al., 2010a,b,c) or fluorescence (Xu et al., 2010b) methodologies. AFLP was used to investigate the genetic diversity of six representational populations of *F. thunbergii* including 32 individuals (Xu et al., 2010a,b). High intrapopulation genetic diversity of cultivated *F. thunbergii* was revealed, while genetic differentiation among the populations was not significant. AFLP was also used to study the genetic diversity and structure of *F. cirrhosa* in southwest China (Zhang et al., 2010a,b,c). Two primer combinations produced 148 reproducible, unambiguous, and polymorphic bands, ranging from 120 to 750 bp. Populations from Hengduan Mountains exhibited higher level of genetic diversity than those from the east margin of the distribution range. The majority of variation (72.29%) was distributed within populations, and there is a relatively high gene migration rate between populations. As an important medicinal plant, wild *F. cirrhosa* has been harvested for thousands of years, with the greatest pressure in the past 30 years. Overharvesting might result in loss of genetic variation in wild populations. As a perennial outcrossing herb, seed dispersal by wind or water plays an important role in maintaining gene flow for *F. cirrhosa*. For such an endangered and overharvested plant, it is unusual for *F. cirrhosa* to possess relatively high genetic diversity at population and species levels. It is inferred that the major diversification of modern *Fritillaria* occurred in Himalaya-Hengduan region during periods of the uplift of the Qinghai-Tibetan Plateau and the global geologic and ecological changes during the late Tertiary. Many plant groups had adaptive radiation, differentiation, and speciation after migrating to this region, for example, *Pedicularis* (Yang et al., 2003); *Aconitum*, *Corydalis*, *Primula*, *Silene*, and *Rhododendron* (Luo et al., 2005); and *Clematis* (Xie et al., 2011). Relationships among some closely related species of Hengduan Mountains and adjacent regions, especially the source plants of Chuan Bei Mu and other taxa of subgenus *Fritillaria*, are not resolved (Luo and Chen, 1996; Li et al., 2009a,b, 2010), suggesting that these taxa may have diverged recently. Geologic and climatic changes in the late Tertiary to Quaternary (2.588 my-present) might be important for the speciation of *Fritillaria*, especially in eastern Asia. Long-distance dispersal of the fruits by wind, water, or animals and strong environmental adaptability might explain the present worldwide distribution and high species diversity. Although *F. cirrhosa* maintains relatively high genetic diversity, sharply decreasing numbers of wild populations and individuals remain a threat. Zhang et al. (2010a,b,c) suggested a method of sustainable collection, as protecting a proportion of mature individuals is necessary and important for maintaining effective population size and evolutionary potential. Even at the wholesale price in China, good quality Chuan Bei Mu can cost over $200/kg. As the most valued traditional medicinal Bei Mu, domestication and cultivation of *F. cirrhosa* and other source plants of Chuan Bei Mu are vital both satisfying market demand and protecting the wild resource (Zhang et al., 2010a,b,c).

4.4.2 Phylogeny

With the rapid development of DNA sequencing technology, some nuclear DNA and chloroplast DNA markers have been used in *Fritillaria* taxonomy and phylogenetic study. PCR primers were designed to specifically amplify 5s rDNA sequences from two traditional source plants of Chuan Bei Mu, *F. przewalskii* and *F. unibracteata* (Tan et al., 2011), with which no products were amplified from the new Chuan Bei Mu source plant *F. wabuensis* and other medicinal *Fritillaria*. The internal transcribed spacer 1 (ITS1) regions of the nuclear ribosomal DNA (nrDNA) of nine species and one variety of *Fritillaria* were sequenced (Wang et al., 2007). A mutation site in the ITS1 region shared by *F. cirrhosa* and other source species of Chuan Bei Mu was found and can be recognized by the restriction endonuclease SmaI. PCR-RFLP (restriction fragment length polymorphism) was used to differentiate four Chuan Bei Mu species, recorded in China Pharmacopoeia of 2005 version, from other medicinal *Fritillaria* species and is a successful authentication method. This method is also applicable to *F. wabuensis* and *F. taipaiensis* that are recorded in China Pharmacopoeia of 2010 version as additional source plants of Chuan Bei Mu (Xu et al., 2010a,b). ITS2 sequence was used as a DNA bar code to discriminate raw plants of Chuan Bei Mu and its adulterants (Luo et al., 2012). Different Chuan Bei Mu plants clustered together and could be distinguished from their adulterants on the neighbor-joining (NJ) phylogenetic tree inferred from ITS2 sequences. Intriguingly, different samples of *F. cirrhosa* failed to form a single cluster on the NJ tree, so did samples of other Chuan Bei Mu source species. It is inferred that these taxa are evolutionarily young and still in the stage of rapid species radiation and speciation. Macroscopic morphology, microscopic morphology, and flower characters have diverged considerably, and therefore, different species are delimited, but the molecular sequences and chemical profiles have not diverged significantly. The maximum parsimony (MP) phylogenetic tree based on ITS sequences (ITS1 + 5.8s rDNA + ITS2) suggests that Chuan Bei Mu source species are closer to *F. thunbergii* than to *F. yuminensis*, a source species of Yi Bei Mu (Gao et al., 2012). However, only 10 *Fritillaria* ITS sequences were used in this study. Rønsted et al. (2005) performed phylogenetic analyses of 37 taxa of *Fritillaria*, 15 species of *Lilium*, and several outgroup taxa from Liliaceae to explore the generic delimitation of *Fritillaria* in relation to *Lilium* and infrageneric relationships within *Fritillaria*. DNA sequences from the maturase-coding plastid matK gene and the trnK intron, the intron of the ribosomal protein-coding rpl16 plastid gene and the nr ITS, were PCR amplified and sequenced. *Fritillaria* and *Lilium* are sister taxa. *Fritillaria* subgenera *Fritillaria* and *Liliorhiza* are supported in part, and some enigmatic species usually included in *Fritillaria* (subgenera *Petilium* and *Theresia* and the monotypic subgenus *Korolkowia*) are closely related. The results support the new classification of *Fritillaria* proposed by Rix (2001), who suggested eight subgenera, that is, *Fritillaria*, *Rhinopetalum*, *Japonica*, *Theresia*, *Petilium*, *Liliorhiza*, *Davidii*, and *Korolkowia*. The independent origins of the underground bulbils found in *F. davidii*, an adulterant of Chuan Bei Mu (Luo et al., 2012), and subgenus *Liliorhiza* were proposed (Rønsted et al., 2005). Regretfully, this study included few Chinese species and the results were not cross validated by NJ and/or other tree reconstruction methods, while Li (2009) only studied species distributed in Sichuan, China.

We retrieve matK and ITS sequences of *Fritillaria* and reconstruct phylogenetic trees based on matK, ITS, and the combined dataset. NJ, MP, and maximum likelihood (ML) phylogenetic relationships are inferred from 57 matK, 69 ITS, and 56 matK + ITS sequences. On the matK + ITS tree inferred by ML method (Hao et al., 2013), there are two major clades, one (I) consisting of subgenus *Fritillaria* taxa collected from Sichuan, China (Li et al., 2009b; Li, 2009), and the other (II) comprising other taxa. Two *F. cirrhosa* taxa are basal to other taxa in clade I, but other *F. cirrhosa* taxa intermingle with Sichuan species and their interrelationship is actually not resolved. *F. thunbergii* and *F. hupehensis* that Li (2009) claimed might not be bona fide ones, which were collected in Sichuan and were not from the native distribution areas. The so-called *F. thunbergii* might be a misidentification and the putative "*F. hupehensis*" could be a hybrid between *F. cirrhosa* and other Sichuan species, which is inferred from the ITS tree, matK tree (Hao et al., 2013), and ISSR results (Li et al., 2009b). In clade II, the Mediterranean species of subgenus *Fritillaria* are basal to the remaining taxa, which is supported by the NJ inference but contradicts with the MP result (Rønsted et al., 2005). *F. gibbosa* of subgenus *Rhinopetalum* and *F. japonica* of subgenus *Japonica* are basal to two sister subclades, one consisting of subgenus *Liliorhiza* species of northeast Asia and North America and the other comprising species of subgenus *Davidii*, *Fritillaria*, *Theresia*, *Korolkowia*, and *Petilium*. *F. thunbergii* is basal to *Korolkowia* and *Petilium*, while *F. pallidiflora* and *F. persica* are sisters to the branch containing *F. thunbergii*. Interestingly, in the ITS tree, *F. thunbergii* (NCBI GenBank accession no. HQ448863) is sister to *F. hupehensis* (Hubei Bei Mu), *F. anhuiensis* (Anhui Bei Mu), and *F. monantha* (Figure 4.5a), but their interrelationship is not resolved.

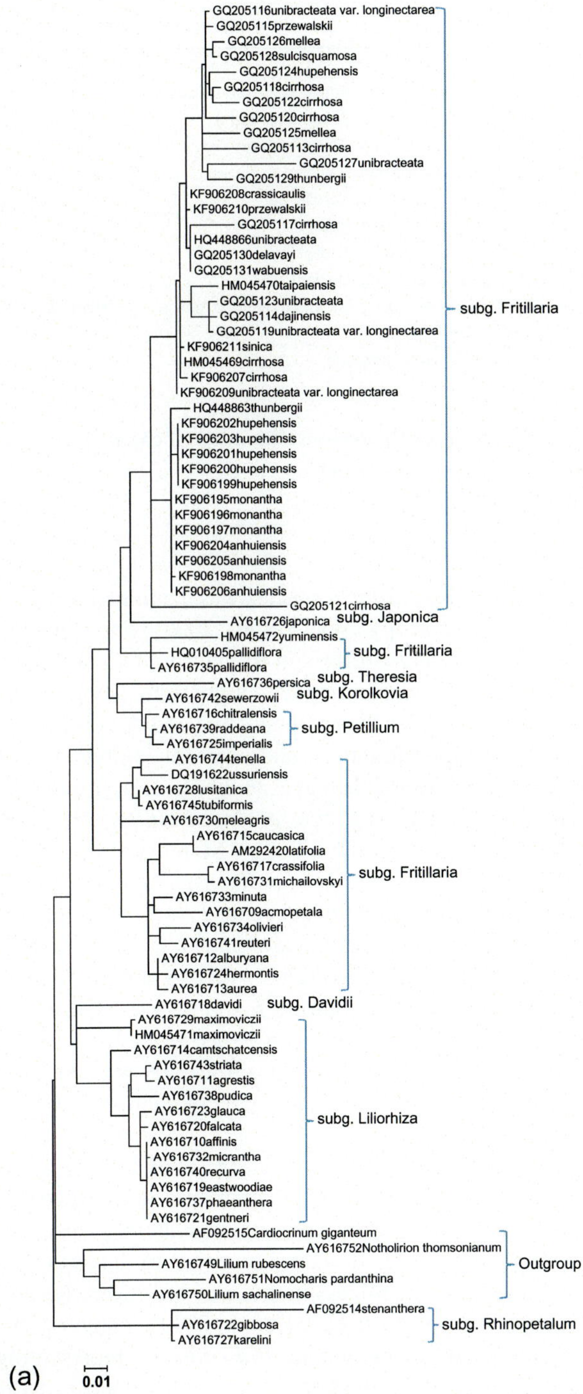

Figure 4.5 Molecular phylogenetic analysis of *Fritillaria*. (a) The evolutionary relationship of 88 ITS sequences was inferred by ML method based on the GTR+I+G model. The tree with the highest log likelihood (−5593.9445) is shown. The tree is drawn to scale, with branch lengths measured in the number of substitutions per site. There were a total of 846 positions in the final dataset. Evolutionary analyses were conducted in MEGA6.

(Continued)

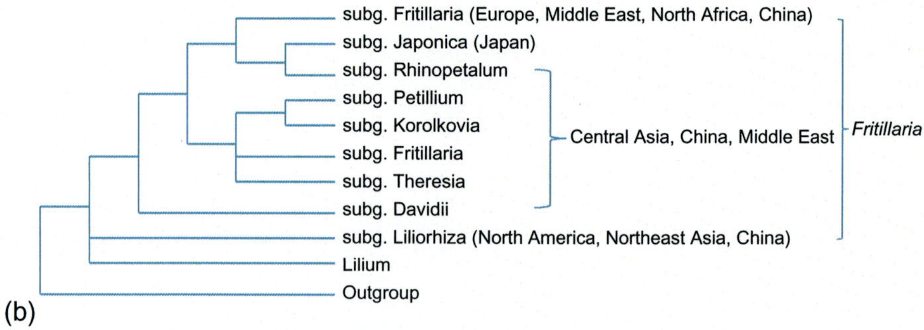

Figure 4.5 Continued. (b) The evolutionary relationship of chloroplast matK+rbcL+rpl16 sequences was inferred by Bayesian method.
According to Day et al. (2014).

The leaf epidermis of 16 *Fritillaria* species was examined using light microscopy and scanning electron microscopy (Wang et al., 2009). The stomatal characteristics support the origin of subgenus *Fritillaria* in China from two floristic elements. Some species spread from Central Asia, joining the Turgai flora, along the Altai-Sayan route from the south of Siberia to the Sayan Mountains and Altai Mountains. Others are distributed in the Hengduan Mountains and adjacent regions of China and belong to the Phava flora. It is not impossible that *F. thunbergii* originated from hybridization between Sichuan species and *F. pallidiflora*, since *Fritillaria* hybrids are not rare in the Sino-Japanese floristic region (Hill, 2011). Moreover, we have some intriguing findings on the ITS tree. *F. ussuriensis* of northeast China and North Korea is sister to *F. tenella*, a Mediterranean species of subgenus *Fritillaria*, although its twin-scaled bulb is similar to that of Japanese endemic *Fritillaria*. *F. karelinii*, a source species of Yi Bei Mu (Xiao et al., 2007), is sister to *F. gibbosa* and *F. stenanthera*, and these three are basal to the other taxa. In contrast, *F. yuminensis* is sister to *F. pallidiflora*. Since other molecular marker sequences of *F. ussuriensis* and *F. karelinii* are not available or very few, their phylogenetic positions are still elusive. The molecular marker sequences of the unique *F. puqiensis* are not available; thus, its phylogenetic position remains intangible.

Fritillaria is a genus of approximately 140 species of bulbous perennial plants that includes taxa of both horticultural and medicinal importance. Aside from being commercially valuable, *Fritillaria* species have attracted attention because of their exceptionally large genome sizes, with all values recorded to date in excess of 30 Gb. A phylogenetic reconstruction encompassing most of the currently recognized species diversity in the genus was performed (Day et al., 2014). Three regions of the plastid genome were sequenced in 117 individuals of *Fritillaria*, representing 92 species (c. 66% of the genus), and in representatives of nine other genera of Liliaceae. Eleven low-copy nuclear gene regions were also screened in selected species for their potential utility. Phylogenetic analysis of a combined plastid dataset using maximum parsimony and Bayesian inference provided support for the monophyly of the majority of currently recognized subgenera. However, subgenus *Fritillaria*, which is by far the largest of the subgenera and includes the most important species used in traditional Chinese medicine (TCM), is found to be polyphyletic (Figure 4.5b). Clade containing the source plants of Chuan Bei Mu, Hubei Bei Mu, and Anhui Bei Mu might be treated as a separate subgenus. The Japanese endemic subgenus *Japonica*, which contains the species with the largest recorded genome size for any diploid plant, is resolved as sister to the predominantly Middle Eastern and Central Asian subgenus *Rhinopetalum*, which is significantly incongruent with ITS tree (Figure 4.5a). Convergent or parallel evolution of phenotypic traits may be a common cause of incongruence between morphology-based classifications and the results of molecular phylogeny. While relationships between most of the major *Fritillaria* lineages can be resolved, these results also highlight the need for data from additional independently evolving loci, which is particularly challenging in light of the huge nuclear genomes found in these plants.

The discovery that some of the most important species used in TCM are closely related to widely cultivated members of subgenus *Petilium* and *Theresia* suggests that substituting more commonly commercially available bulbs for the rare species, which are currently collected directly from the wild, might be possible. Species used for traditional medicines in independent cultures are significantly clustered in phylogenetic trees (Saslis-Lagoudakis et al., 2012), implying that traditional knowledge can be effective in identifying plants with bioactive compounds. Future testing of species such as *F. imperialis* and *F. eduardii*, which are being commercially bred, could establish whether these bulbs contain the same bioactive compounds that have been identified from some of the Chinese species. The ability to substitute commercial varieties suitable for cultivation for native Chinese species has the potential to reduce the burden on wild populations by alleviating collecting pressure on traditional bulb sources.

The molecular phylogeny results are not congruent with those inferred from chemotaxonomy (Li et al., 2009a; Zhou et al., 2010). Many secondary metabolites have a taxonomically restricted distribution. However, the same secondary metabolites (e.g., several groups of alkaloids and cardiac glycosides) in other plant taxa that are not closely related in a phylogenetic context are also found (Rønsted et al., 2008; Wink et al., 2010). The patchy distribution might be explained as follows: First, the occurrence of the same set of steroidal alkaloids in distantly related taxa, for example, *F. cirrhosa* and *F. pallidiflora*, *F. thunbergii*, *F. ussuriensis*, and *F. maximowiczii*, might be due to rampant parallelism or convergent evolution. Second, the genes encoding the enzymes of steroidal alkaloid biosynthesis might be regulated transcriptionally, that is, they are switched on or off in a certain context. However, little is known about the *Fritillaria* transcriptome and the regulation mechanism at the transcription level. Third, a patchy distribution can be due to the presence of endophytic fungi, which could produce secondary metabolites (e.g., ergot alkaloids in Convolvulaceae and peimisine and peiminine in *F. unibracteata* var. *wabensis*) (Wink et al., 2010; Pan et al., 2014). Lateral gene transfer from symbiotic fungi is also possible. Fourth, as the rhizosphere bacteria and the soil properties have a close relationship with the geoherbalism of *F. thunbergii* (Shi et al., 2011), it is intriguing to investigate how these biotic and abiotic factors shape the chemical profiles of various *Fritillaria* species.

4.4.3 Genomics

The genomic and transcriptomic studies may not only help elucidate the secondary metabolism pathways in *Fritillaria* but also facilitate taxonomic and phylogenetic investigations. *Fritillaria* species have exceptionally large genomes (1C=30,000–127,000 Mb) and a bicontinental distribution. Most North American species (subgenus *Liliorhiza*) differ from Eurasian species by their discrete phylogenetic position and increased amounts of heterochromatin. Fosmid libraries are powerful tools in dissecting the huge genome (Hao et al., 2011b) and were constructed from genomic DNAs of *F. affinis* of North America and *F. imperialis* of Eurasia and used to explore highly repeated sequences (Ambrozová et al., 2011). Repeats corresponding to 6.7% and 4.7% of the *F. affinis* and *F. imperialis* genome, respectively, were identified. Chromoviruses and the Tat lineage of Ty3/gypsy group long terminal repeat retrotransposons were the predominant components of the highly repeated fractions in the *F. affinis* and *F. imperialis* genomes, respectively. A heterogeneous, extremely AT-rich satellite repeat was isolated from *F. affinis*. The FriSAT1 repeat localized in heterochromatic bands makes up approximately 26% of the *F. affinis* genome and substantial genomic fractions in several other *Liliorhiza* species. The giant *Fritillaria* genomes are composed of many diversified families of transposable elements, which are useful in phylogenetic studies. The genome obesity may be partly determined by the failure of removal mechanisms to counterbalance effectively the retrotransposon amplification (Ambrozová et al., 2011). In the future, it is necessary to characterize the genome of Chinese species.

The decreasing sequencing cost makes it feasible to perform the large-scale transcriptome analysis of medicinal plants (Hao et al., 2012a). Two thousand one hundred and fifty-eight high-quality expressed sequence tags (ESTs) were generated from a cDNA library of *F. cirrhosa* bulbs and 1343 unique transcripts assembled (Sun et al., 2011). After removing ribosomal RNA sequences, 1330 putative protein-coding sequences were obtained, among which 765 (57.5%) had at least one significant match to the Swiss-Prot protein database via a BLASTX similarity search. The 1330 unique transcripts were further functionally classified for gene discovery, using the Gene Ontology (GO) and Kyoto Encyclopedia of Genes and Genomes (KEGG) databases. More than ten transcripts that are likely involved in the biosynthesis of *F. cirrhosa* alkaloids and corresponding regulatory activities were discovered, including HMGR, FPSs, CYP450s, and aminotransferases. This report describes the first example of transcriptome analysis from *Fritillaria* and is useful for further cloning and identification of candidate genes related to steroidal alkaloid biosynthesis. EST sequences can be used to find orthologs that are conserved among species of different evolutionary levels and are potential phylogenetic markers (Hao et al., 2012b).

4.5 Conclusions

Fritillaria plants contain many pharmaceutically active constituents, which has been commonly used in traditional Chinese medicine for thousands of years. Increasing interest in *Fritillaria* medicinal resources has led to additional discoveries of alkaloids, terpenoids, saponins, and many other compounds in various *Fritillaria* species and to studies on their chemotaxonomy, molecular phylogeny, and pharmacology. However, around 80% of all 165 species have not been explored in phytochemistry, molecular biology, and pharmacology, which hampers the quality control and therapeutic use of *Fritillaria*-related products. The potential for finding new medicinal *Fritillaria* species and lead compounds for drug development is huge. Chemosystematics tries to classify and identify plants by spotting differences and similarities in their biochemical compositions, which is of great help in pharmaceutical resource discovery. Yet, the results of chemotaxonomy have to be compared with those of molecular phylogeny and traditional morphology-based classification, in order to rationalize the quality control and guarantee the authenticity of plant materials used in the clinical setting, research laboratory, and pharmaceutical industry. Molecular phylogeny studies help researchers gain deeper insights into the known and potential medicinal *Fritillaria* species and are vital to the conservation and sustainable utilization of *Fritillaria* resources. More importantly, chemical and biological research of *Fritillaria* medicinal resources should not be restricted in the above aspects, for example, the biosynthetic pathway of secondary metabolites is little studied; the regulation of *Fritillaria* biological processes at genomic level, epigenomic level, transcriptional and posttranscriptional levels, and translational and posttranslational levels is totally unknown, although it is essential for the sustainable development and utilization of the *Fritillaria* medicinal resources. The biological and chemical studies of *Taxus* (Hao et al., 2011b) and *Polygonum* (Hao et al., 2012b) with the use of omics technologies provide a paradigm of the active integration of various state-of-the-art methodologies into the early stage of the drug research and development. Systems biology and omics techniques will play a noteworthy role in future *Fritillaria* pharmaceutical studies.

References

Ambrozová, K., Mandáková, T., Bures, P., et al., 2011. Diverse retrotransposon families and an AT-rich satellite DNA revealed in giant genomes of *Fritillaria lilies*. Ann. Bot. 107 (2), 255–268.

An, J.J., Zhou, J.L., Li, H.J., et al., 2010. Puqienine E: an angiotensin converting enzyme inhibitory steroidal alkaloid from *Fritillaria puqiensis*. Fitoterapia 81 (3), 149–152.

Atta-ur-Rahman, A.M.N., Choudhary, M.I., et al., 2005. New diterpene isopimara-7,15-dien-19-oic acid and its prolyl endopeptidase inhibitory activity. Nat. Prod. Res. 19 (1), 13–22.

Cao, X.W., Chen, S.B., Li, J., et al., 2008. Steroidal alkaloids from the bulbs of *Fritillaria delavayi* Franch. (Liliaceae). Biochem. Syst. Ecol. 36 (8), 665–668.

Cao, X.W., Li, J., Chen, S.B., et al., 2010. Simultaneous determination of nine nucleosides and nucleobases in different *Fritillaria* species by HPLC-diode array detector. J. Sep. Sci. 33 (11), 1587–1594.

Celebi, A., Teksen, M., Acik, L., et al., 2008. Taxonomic relationships in genus *Fritillaria* (Liliaceae): evidence from RAPD-PCR and SDS-PAGE of seed proteins. Acta Bot. Hun. 50 (3–4), 325–343.

Day, P.D., Berger, M., Hill, L., et al., 2014. Evolutionary relationships in the medicinally important genus *Fritillaria* L. (Liliaceae). Mol. Phylogenet. Evol. 80C, 11–19.

Duan, B.Z., Huang, L.F., Chen, S.L., 2010. Simultaneous determination of peimisine and sipeimine in *Fritillaria walujewii* regel and *Fritillaria pallidiflora* Schrenk by UPLC-ELSD. Acta Pharm. Sin. 45 (12), 1541–1544.

Duan, B., Huang, L., Chen, S., 2012. Chemical fingerprint analysis of *Fritillaria delavayi* Franch. by high-performance liquid chromatography. J. Sep. Sci. 35 (4), 513–518.

Gao, Y.D., Zhou, S.D., He, X.J., Wan, J., 2012. Chromosome diversity and evolution in tribe Lilieae (Liliaceae) with emphasis on Chinese species. J. Plant Res. 125 (1), 55–69.

Hao, D.C., Chen, S.L., Xiao, P.G., et al., 2010. Authentication of medicinal plants by DNA-based markers and genomics. Chin. Herb. Med. 2 (4), 250–261.

Hao, D.C., Ge, G.B., Xiao, P.G., et al., 2011a. The first insight into the tissue specific taxus transcriptome via Illumina second generation sequencing. PLoS One 6 (6), e21220.

Hao, D.C., Yang, L., Xiao, P.G., 2011b. The first insight into the *Taxus* genome via fosmid library construction and end sequencing. Mol. Genet. Genomics 285 (3), 197–205.

Hao, D.C., Chen, S.L., Xiao, P.G., Liu, M., 2012a. Application of high-throughput sequencing in the medicinal plant transcriptome studies. Drug Dev. Res. 73 (8), 487–498.

Hao, D.C., Ma, P., Mu, J., Chen, S., Xiao, P.G., et al., 2012b. De novo characterization of the root transcriptome of a traditional Chinese medicinal plant *Polygonum cuspidatum*. Sci. China Life Sci. 55 (5), 452–466.

Hao, D.C., Gu, X.J., Xiao, P.G., Peng, Y., 2013. Phytochemical and biological research of *Fritillaria* medicinal resources. Chin. J. Nat. Med. 11 (4), 330–344.

Helsper, J.P., Bücking, M., Muresan, S., et al., 2006. Identification of the volatile component(s) causing the characteristic foxy odor in various cultivars of *Fritillaria imperialis* L. (Liliaceae). J. Agric. Food Chem. 54 (14), 5087–5091.

Hill, L., 2011. A taxonomic history of Japanese endemic *Fritillaria* (Liliaceae). Kew Bull. 66 (2), 227–240.

Huang, L., Duan, B., Wang, L., et al., 2011. Simultaneous determination of 7 nucleosides and nucleobases in aqueous extracts of *Fritillaria taipaiensis* by HPLC-DAD. China J. Chin. Mater. Med. 36 (5), 585–588.

Huang, S., Zhou, X.L., Wen, J., et al., 2013. A novel steroidal alkaloid from *Fritillaria shuchengensis*. J. Nat. Med. 67 (3), 647–651.

Jiang, Y., Li, P., Li, H.J., Yu, H., 2006. New steroidal alkaloids from the bulbs of *Fritillaria puqiensis*. Steroids 71 (9), 843–848.

Kang, L., Zhou, J.X., Shen, Z.W., 2007. A new diterpenoid from *Fritillaria anhuiensis*. Acta Pharm. Sin. 42 (1), 58–60.

Kavandi, L., Lee, L.R., Bokhari, A.A., et al., 2013. The Chinese herbs *Scutellaria baicalensis* and *Fritillaria cirrhosa* target NFκB to inhibit proliferation of ovarian and endometrial cancer cells. Mol. Carcinog. 54 (5), 368–378, http://dx.doi.org/10.1002/mc.22107.

Kim, S.H., Shin, D.S., Oh, M.N., et al., 2003. Inhibition of sortase, a bacterial surface protein anchoring transpeptidase, by beta-sitosterol-3-*O*-glucopyranoside from *Fritillaria verticillata*. Biosci. Biotechnol. Biochem. 67 (11), 2477–2479.

Li, K.Q., 2009. Studies on Genetic Diversity of Genus *Fritillaria* from Sichuan Province and Effects of Temperature and Light on the Growth and Development of *Fritillaria cirrhosa*. Master Degree Thesis, Sichuan Agricultural University.

Li, S.L., Li, P., Lin, G., et al., 1999. Existence of 5α-cevanine isosteroidal alkaloids in bulbs of *Fritillaria* L. Acta Pharm. Sin. 34 (11), 842–847.

Li, F., Yan, D.M., Liu, H.N., et al., 2005. Review on chemical components research on domestic *Fritillaria* plants of China. World Sci. Technol.: Modern Trad. Chin. Med. Materia. Med. 7 (2), 26–31.

Li, H.J., Jiang, Y., Li, P., 2006a. Chemistry, bioactivity and geographical diversity of steroidal alkaloids from the Liliaceae family. Nat. Prod. Rep. 23 (5), 735–752.

Li, H.J., Jiang, Y., Li, P., Ye, W.C., 2006b. Puqienine F, a novel veratramine alkaloid from the bulbs of *Fritillaria puqiensis*. Chem. Pharm. Bull. 54 (5), 722–724.

Li, Y.F., Tang, L., Chen, F., 2006c. RAPD analysis of eight *Fritillaria* species. Chin. Tradit. Patent. Med. 28 (10), 1527–1529.

Li, H.J., Jiang, Y., Li, P., 2009a. Characterizing distribution of steroidal alkaloids in Fritillaria spp. and related compound formulas by liquid chromatography-mass spectrometry combined with hierarchical cluster analysis. J. Chromatogr. A 1216 (11), 2142–2149.

Li, K.Q., Wu, W., Zheng, Y., et al., 2009b. Genetic diversity of *Fritillaria* from Sichuan province based on ISSR. China J. Chin. Mater. Med. 34 (17), 2149–2154.

Li, Q., Chen, X., Wang, S., 2010. Study on the relative in molecular biology among *Fritillariae* of Sichuan species. West China J. Pharm. Sci. 25 (2), 140–143.

Li, S., Hu, K., Guo, J., et al., 2011. Genetic diversity and relationship of *Fritillaria thunbergii* Miq. landraces and related taxa. Biochem. Syst. Ecol. 39 (4–6), 725–731.

Li, D.Y., Yu, T.T., Bai, H., Chen, X.Y., 2012. Clinical study on effect of compound granule prescription of thunberg fritillary bulb in relieving post-chemotherapy bone marrow suppression in RAL patients. Zhongguo Zhong Yao Za Zhi 37 (20), 3155–3157.

Liang, J., Cao, X., Li, J., et al., 2011. Analysis of volatile components of flowers of *Fritillaria thunbergii* by GC-TOF-MS. China J. Chin. Mater. Med. 36 (19), 2689–2692.

Lin, B.Q., Ji, H., Li, P., et al., 2006a. Selective antagonism activity of alkaloids from bulbs *Fritillariae* at muscarinic receptors: functional studies. Eur. J. Pharmacol. 551 (1–3), 125–130.

Lin, B.Q., Ji, H., Li, P., et al., 2006b. Inhibitors of acetylcholine esterase in vitro – screening of steroidal alkaloids from *Fritillaria* species. Planta Med. 72 (9), 814–818.

Liu, H.N., Li, F., Luo, Y.M., Zhu, W.F., 2007a. Two novel isosteroid alkaloids from *Fritillaria monatha*. J. Asian Nat. Prod. Res. 9 (6–8), 563–567.

Liu, H.N., Li, F., Luo, Y.M., et al., 2007b. Diterpenoids from bulbus of *Fritillaria monanth*. Acta Pharm. Sin. 42 (11), 1152–1154.

Liu, X.X., Chen, C., Pan, L.L., et al., 2010. Genetic diversity of bulbus *Fritillariae thunbergii* Miq. varieties based on ISSR markers. J. Zhejiang Univ. (Agric. Life Sci.) 36 (3), 246–254.

Liu, C., Chang, J., Zhang, L., et al., 2012. Purification and antioxidant activity of a polysaccharide from bulbs of *Fritillaria ussuriensis* Maxim. Int. J. Biol. Macromol. 50 (4), 1075–1080.

Lu, H., Zhu, S.H., Zhou, S.J., et al., 2009. Specific RAPD screening of breeds in *Fritillaria thunbergii* and genetic diversity analysis of five species of *Fritillaria*. J. Ningbo Univ. 22 (1), 44–47.

Luo, Y.B., Chen, X.Q., 1996. A revision of *Fritillaria* L. (Liliaceae) in the Hengduan Mountains and adjacent regions, China (II). Acta Phytotax. Sin. 34 (5), 547–553.

Luo, Y., Zhang, F.M., Yang, Q.E., 2005. Phylogeny of Aconitum subgenus *Aconitum* (Ranunculaceae) inferred from ITS sequences. Plant Syst. Evol. 252 (1–2), 11–25.

Luo, K., Ma, P., Yao, H., et al., 2012. Molecular identification of *Fritillariae Cirrhosae Bulbus* and its adulterants. World Sci. Technol.: Modern Trad. Chin. Med. Materia. Med. 14 (1), 1–6.

Matsuo, Y., Shinoda, D., Nakamaru, A., Mimaki, Y., 2013. Steroidal glycosides from the bulbs of *Fritillaria meleagris* and their cytotoxic activities. Steroids 78 (7), 670–682.

Mucciarelli, M., Ferrazzini, D., Belletti, P., 2014. Genetic variability and population divergence in the rare *Fritillaria tubiformis subsp. moggridgei* Rix (Liliaceae) as revealed by RAPD analysis. PLoS One 9 (7), e101967.

Pan, F., Hou, K., Gao, F., et al., 2014. Peimisine and peiminine production by endophytic fungus Fusarium sp. isolated from Fritillaria unibracteata var. wabensis. Phytomedicine 21 (8–9), 1104–1109.

Pi, H.F., Ruan, H.L., Zhang, Y.H., et al., 2006a. Steroidal alkaloids from bulbs of *Fritillaria lichuanensis*. J. Asian Nat. Prod. Res. 8 (1–2), 133–136.

Pi, H.F., Ruan, H.L., Zhan, Y.H., Wu, J.Z., 2006b. Two new steroidal alkaloids from bulbs of *Fritillaria lichuanensis*. J. Asian Nat. Prod. Res. 8 (3), 253–257.

Pi, H.F., Zhang, P., Ruan, H.L., et al., 2009. Two new triterpenoids from the leaves and stems of *Fritillaria hupehensis*. J. Asian Nat. Prod. Res. 11 (9), 779–782.

Rix, E.M., 2001. *Fritillaria*: A Revised Classification. The Fritillaria Group of the Alpine Garden Society, Edinburgh.

Rønsted, N., Law, S., Thornton, H., et al., 2005. Molecular phylogenetic evidence for the monophyly of *Fritillaria* and *Lilium* (Liliaceae; Liliales) and the infrageneric classification of *Fritillaria*. Mol. Phylogenet. Evol. 35 (3), 509–527.

Rønsted, N., Savolainen, V., Molgaard, P., et al., 2008. Phylogenetic selection of *Narcissus* species for drug discovery. Biochem. Syst. Ecol. 36 (5–6), 417–422.

Saslis-Lagoudakis, C.H., Savolainen, V., Williamson, E.M., et al., 2012. Phylogenies reveal predictive power of traditional medicine in bioprospecting. Proc. Natl. Acad. Sci. U. S. A. 109 (39), 15835–15840.

Shen, S., Chen, C.J., Bu, R., et al., 2011. Three new steroidal saponins from *Fritillaria pallidiflora*. J. Asian Nat. Prod. Res. 13 (11), 1014–1022.

Shen, S., Li, G., Huang, J., et al., 2012a. Steroidal saponins from *Fritillaria pallidiflora* Schrenk. Fitoterapia 83 (4), 785–794.

Shen, S., Li, G., Huang, J., et al., 2012b. Chemical constituents from *Fritillaria pallidiflora* Schrenk. Biochem. Syst. Ecol. 45, 183–187.

Shi, J.Y., Yuan, X.F., Lin, H.R., et al., 2011. Differences in soil properties and bacterial communities between the rhizosphere and bulk soil and among different production areas of the medicinal plant *Fritillaria thunbergii*. Int. J. Mol. Sci. 12 (6), 3770–3785.

Shou, Q.Y., Tan, Q., Wu, S.Z., 2010. Two 22S-solanidine-type steroidal alkaloids from *Fritillaria anhuiensis*. Fitoterapia 81 (2), 81–84.

Shou, Q.Y., Wohlmuth, H., He, X., et al., 2012. Chemical constituents from *Fritillaria anhuiensis*. Biochem. Syst. Ecol. 45, 16–19.

Sun, C., Sun, Y.Q., Song, J.Y., et al., 2011. Discovery of genes related to steroidal alkaloid biosynthesis in *Fritillaria cirrhosa* by generating and mining a dataset of expressed sequence tags (ESTs). J. Med. Plant Res. 5 (21), 5307–5314.

Tan, Y., Zhang, L.H., Li, M.C., et al., 2011. Identification of Chinese herb *Fritillaria cirrhosa* D. Don by DNA fingerprint. Chin. Pharm. J. 46 (1), 14–16.

Tan, S.F., Liu, C.F., Wang, C.S., et al., 2013. Valuation on analgesic, expectorant and antitussive effects of compatible use of *Aconiti radix cocta* and *Fritillaria cirrhosa* or *Fritillaria thunbergii*. Zhongguo Zhong Yao Za Zhi 38 (16), 2706–2713.

Wang, C.Z., Li, P., Ding, J.Y., et al., 2007. Simultaneous identification of *Bulbus Fritillariae cirrhosae* using PCR-RFLP analysis. Phytomedicine 14 (9), 628–632.

Wang, Q., Zhou, S.D., Deng, X.Y., et al., 2009. Comparative morphology of the leaf epidermis in *Fritillaria* (Liliaceae) from China. Bot. J. Linn. Soc. 160 (1), 93–109.

Wang, D., Zhu, J., Wang, S., et al., 2011. Antitussive, expectorant and anti-inflammatory alkaloids from *Bulbus Fritillariae Cirrhosae*. Fitoterapia 82 (8), 1290–1294.

Wang, D., Wang, S., Chen, X., et al., 2012. Antitussive, expectorant and anti-inflammatory activities of four alkaloids isolated from Bulbus of *Fritillaria wabuensis*. J. Ethnopharmacol. 139 (1), 189–193.

Wink, M., Botschen, F., Gosmann, C., et al., 2010. Chemotaxonomy seen from a phylogenetic perspective and evolution of secondary metabolism. In: Wink, M. (Ed.), Biochemistry of Plant Secondary Metabolism. second ed. Annual Plant Reviews, vol. 40. Wiley-Blackwell, New York.

Wu, X.Q., Cheng, H.Z., Zhou, Y.Z., et al., 2010. Analysis of genetic relationship ornamental *Fritillaria* germplasm resources by ISSR markers. Northern Horti. 18, 157–160.

Xiao, P.G., 2002. A review on the study of Hu Bei Bei Mu. China J. Chin. Mater. Med. 27 (10), 726–728.

Xiao, P.G., Jiang, Y., Li, P., et al., 2007. The botanical origin and pharmacophylogenetic treatment of Chinese material medica Beimu. Acta Phytotax. Sin. 45 (4), 473–487.

Xie, L., Wen, J., Li, L.Q., 2011. Phylogenetic analyses of *Clematis* (Ranunculaceae) based on sequences of nuclear ribosomal ITS and three plastid regions. Syst. Bot. 36 (4), 907–921.

Xin, G.Z., Hu, B., Shi, Z.Q., et al., 2014a. Rapid identification of plant materials by wooden-tip electrospray ionization mass spectrometry and a strategy to differentiate the bulbs of *Fritillaria*. Anal. Chim. Acta 820, 84–91.

Xin, G.Z., Lam, Y.C., Maiwulanjiang, M., et al., 2014b. Authentication of *Bulbus Fritillariae Cirrhosae* by RAPD-derived DNA markers. Molecules 19 (3), 3450–3459.

Xu, C.L., Li, H.J., Li, P., et al., 2010a. Molecular method for the identification of *Bulbus Fritillariae Cirrhosae*. J. Chin. Pharm. Univ. 41 (3), 226–230.

Xu, J.Z., Zhang, H.Y., Ma, X.Y., et al., 2010b. AFLP analysis on genetic diversity for germplasm resources of *Fritillaria thunbergii* cultivated in Zhejiang Province. Chin. Tradit. Herb. Drugs 41 (1), 109–113.

Xu, F., Xu, S., Wang, L., et al., 2011a. Antinociceptive efficacy of verticinone in murine models of inflammatory pain and paclitaxel induced neuropathic pain. Biol. Pharm. Bull. 34 (9), 1377–1382.

Xu, J., Guo, P., Liu, C., et al., 2011b. Neuroprotective kaurane diterpenes from *Fritillaria ebeiensis*. Biosci. Biotechnol. Biochem. 75 (7), 1386–1388.

Xu, J., Liu, C., Guo, P., et al., 2011c. Neuroprotective labdane diterpenes from *Fritillaria ebeiensis*. Fitoterapia 82 (5), 772–776.

Xu, W., Liu, M., Chen, D., Wang, J., 2014. Chemical constituents from bulbs of *Fritillaria pallidiflora* Schrenk. Biochem. Syst. Ecol. 57, 198–202.

Yang, Z.D., Duan, D.Z., 2012. A new alkaloid from *Fritillaria ussuriensis* Maxim. Fitoterapia 83 (1), 137–141.

Yang, F.S., Wang, X.Q., Hong, D.Y., 2003. Unexpected high divergence in nrDNA ITS and extensive parallelism in floral morphology of *Pedicularis* (Orobanchaceae). Plant Syst. Evol. 240 (1–4), 91–105.

Yao, Y.H., Xu, G.H., Zhang, J.D., et al., 2008. Determination of trace elements in traditional Chinese medicine from Changbai Mountain by ICP-MS. Spectrosc. Spectr. Anal. 28 (5), 1165–1167.

Yin, C.P., Liu, W.T., Xu, S.Q., et al., 2007. Discriminating Chuan Bei Mu and Hu Bei Bei Mu by RAPD. Herald Med. 26 (4), 359–361.

Yu, S.C., Xiao, P.G., 1992a. The existence of isosteroidal alkaloids in Fritillaria L. (Liliaceae) and its taxonomical significance. Acta Phytotax. Sin. 30 (5), 450–459.

Yu, S.C., Xiao, P.G., 1992b. On the taxonomical state of *Fritillaria sulcisquamosa* and *F. puqiensis*. Acta Phytotax. Sin. 30 (3), 277–278.

Yu, R.P., Cheng, Z.F., Gong, X.Q., 2007. Determination and comparison of trace elements in four sorts of *Bulbus Fritillariae* by microwave digestion-atomic absorption spectrometry. Spectrosc. Spectr. Anal. 27 (12), 2591–2594.

Yuan, X., Shi, J., Yang, Y., et al., 2010. Establishment of element fingerprint and multielement analysis of *Fritillaria thunbergii* by inductively coupled plasma optical emission spectrometry. Biol. Trace Elem. Res. 135 (1–3), 304–313.

Yun, Y.G., Jeon, B.H., Lee, J.H., et al., 2008. Verticinone induces cell cycle arrest and apoptosis in immortalized and malignant human oral keratinocytes. Phytother. Res. 22 (3), 416–423.

Zhang, Y.H., Ruan, H.L., Pi, H.F., et al., 2004. Structural elucidation of fritillahupehin from bulbs of *Fritillaria hupehensis* Hsiao et K.C. Hsia. J. Asian Nat. Prod. Res. 6 (1), 29–34.
Zhang, Y.H., Yang, X.L., Zhang, P., et al., 2008. Cytotoxic alkaloids from the bulbs of *Fritillaria hupehensis*. Chem. Biodivers. 5 (2), 259–266.
Zhang, D.Q., Gao, L.M., Yang, Y.P., 2010a. Genetic diversity and structure of a traditional Chinese medicinal plant species, *Fritillaria cirrhosa* (Liliaceae) in southwest China and implications for its conservation. Biochem. Syst. Ecol. 38 (2), 236–242.
Zhang, J., Song, C., Chen, B., et al., 2010b. Simultaneous determination of 5 nucleotides in *Bulbus Fritillariae* by RP-HPLC. China J. Chin. Mater. Med. 35 (1), 67–70.
Zhang, S., Wei, J., Chen, S.L., et al., 2010c. Floral dynamic and pollination habit of *Fritillaria cirrhosa*. China J. Chin. Mater. Med. 35 (1), 27–29.
Zhang, Q.J., Zheng, Z.F., Yu, D.Q., 2011. Steroidal alkaloids from the bulbs of *Fritillaria unibracteata*. J. Asian Nat. Prod. Res. 13 (12), 1098–1103.
Zhang, Z.F., Lu, L.Y., Liu, Y., 2014. Comparing major alkaloids of *Fritillariae Hupehensis* Bulbs (FHB) and congeneric plants by HPLC-ELSD and HPLC-ESI-MS(n). Nat. Prod. Res. 28 (15), 1171–1175.
Zhou, J., Wang, J., 2012. ISSR analysis of genetic diversity of *F. thunbergii*. J. Zhejiang Agric. Sci. 2, 156–159.
Zhou, Y., Ji, H., Lin, B.Q., et al., 2006. The effects of five alkaloids from *Bulbus Fritillariae* on the concentration of cAMP in HEK cells transfected with muscarinic M(2) receptor plasmid. Am. J. Chin. Med. 34 (5), 901–910.
Zhou, J.L., Jiang, Y., Bi, Z.M., et al., 2008. Study on nucleosides from *Fritillaria puqiensis*. Chin. Pharm. J. 43 (12), 894–896.
Zhou, J.L., Xin, G.Z., Shi, Z.Q., et al., 2010. Characterization and identification of steroidal alkaloids in *Fritillaria* species using liquid chromatography coupled with electrospray ionization quadrupole time-of-flight tandem mass spectrometry. J. Chromatogr. A 1217 (45), 7109–7122.

Phytochemical and biological research of Chelidonieae pharmaceutical resources

5

5.1 Introduction

Chelidonieae, native to temperate and subtropical regions of the Northern Hemisphere, is a eudicot tribe of the family Papaveraceae, Ranunculales, and includes 23 species of eight genera, that is, *Sanguinaria* (one species/none in China), *Bocconia* (ten species/none in China), *Eomecon* (one species/in China), *Stylophorum* (three species/two in China), *Hylomecon* (one species/in China), *Dicranostigma* (four species/in China), *Chelidonium* (one species/in China), and *Macleaya* (two species/ in China) (Feng et al., 1985). Chelidonieae is a botanical source of various pharmaceutically active components, which has been used in traditional Chinese medicine and other folk medicine for many centuries. In Chinese medicine, *Chelidonium majus* is used as an analgesic, antitussive, anti-inflammatory, and detoxicant. Traditionally, celandine was also used as a remedy for jaundice, scurvy, scrofula, gout, toothache, peptic ulcers, and piles and most notably as a topical to treat abnormal growths, probably owing to the antimitotic properties of the benzylisoquinoline alkaloids, sanguinarine and chelerythrine (Gilca et al., 2010). *Macleaya cordata* (plume poppy) is used in traditional Chinese medicine for its anti-inflammatory and antibacterial activities (Vrba et al., 2012). *Sanguinaria canadensis* (bloodroot) was used historically by Native Americans for its curative properties as an emetic, respiratory aid, and other treatments. More medicinal studies have been performed for *Chelidonium*, *Sanguinaria*, and *Macleaya* than for other genera, although the latter also contains bioactive ingredients and is used in folk medicine.

Different Chelidonieae species possess different chemical profiles and may have different pharmacological effects. Various Chelidonieae species have been characterized for their chemical components (Feng et al., 1985; Huang and Du, 2002; Gilca et al., 2010; Gong et al., 2010). Advances in analytical chemistry and molecular biology techniques facilitate the in-depth studies of Chelidonieae pharmaceutical resources. However, it is essential to integrate systems biology and omics techniques into further inquiry for both the sustainable utilization of Chelidonieae medicinal resources and finding novel compounds with potential therapeutic utility. In this brief chapter, we summarize recent progress in phytochemistry and pharmacology of Chelidonieae, as well as chemotaxonomy and phylogeny. To date, very few studies have attempted to correlate DNA-based phylogeny of medicinal plants with phytochemistry or with medicinal properties (Hao et al., 2012a). Here, we reconstruct the molecular phylogeny of Chelidonieae genera and compare the results with those of

Medicinal plants: chemistry, biology and omics. http://dx.doi.org/10.1016/B978-0-08-100085-4.00005-0

chemotaxonomy. Phylogeny has great explanatory power and offers a unique perspective to complement chemotaxonomy.

5.2 Alkaloids

Alkaloids are a diverse group of about 12,000 low-molecular-weight, nitrogen-containing compounds found in around 20% of plant species (Alcantara et al., 2005). Chelidonine and sanguinarine are members of the large and diverse group of benzylisoquinoline alkaloids, which contains more than 2500 different structures identified in plants. The Chelidonieae alkaloids can be classified into four major groups based on the carbon framework: benzophenanthridine-type, protopine-type, protoberberine-type, and aporphine-type alkaloids (Table 5.1, Figures 5.1 and 5.2). Benzophenanthridine-type alkaloids are a group of natural products with potential therapeutic utility and include four structure categories: (1) hexahydrobenzophenanthridine, including chelidonine (7), chelamine (8), chelamidine (9), α-homochelidonine (10), oxychelidonine (11), and methoxychelidonine (12); (2) dihydrobenzophenanthridine, including methyl 2′-(7,8-dihydrosanguinarine-8-yl)acetate (1), maclekarpine A–E (2-6), dihydrochelerythrine (13), dihydrosanguinarine (14), dihydrochelirubine (15), dihydrochelilutine (16), N-demethyl-9, 10-dihydroxysanguinarine (17), 6-methoxydihydrochelerythrine (18), 6-methoxydihydrosanguinarine (19), N-demethyldihydroxysanguinarine (20), and oxysanguinarine (21); (3) quaternary benzophenanthridine, including chelerythrine (22), sanguinarine (23), chelirubine (24), and chelilutine (25); and (4) dimer of dihydrobenzophenanthridine, including chelidimerine (26), chelerythridimerine (27), sanguidimerine (28), and rhoeadine (29).

Total synthesis of 12-methoxydihydrochelerythrine has been performed (Watanabe et al., 2003). Introduction of a methoxy group into the 12-position of the benzophenanthridine skeleton could cause enhanced activity against MDA-MB-231 cancer cells. Total syntheses of chelidonine and norchelidonine featuring an enamide-benzyne-[2+2] cycloaddition initiated cascade have been successful (Ma et al., 2012). The cascade includes a pericyclic ring opening and intramolecular Diels–Alder reaction. A Pd II-catalyzed ring-opening strategy was adopted in the concise enantioselective total syntheses of (+)-homochelidonine, (+)-chelamidine, (+)-chelidonine, (+)-chelamine, and (+)-norchelidonine (Fleming et al., 2008).

Protopine-type alkaloids include protopine (30), allocryptopine (31), and cryptopine (32) (Figure 5.2). Protoberberine-type alkaloids include berberine (33), canadine (34), scoulerine (35), coptisine (36), stylopine (37), columbamine (38), and corysamine (39). Aporphine-type alkaloids include menisperine (40), magnoflorine (41), corydine (42), and isocorydine (43). Alkaloids found in Chelidonieae before 2002 have been summarized (Huang and Du, 2002). Table 5.1 summarizes the compounds found since 2003. According to the medical importance and the resource utilization of China Chelidonieae, much attention should be paid to the following alkaloids and taxa: chelidonine (*Chelidonium majus*), isocorydine (*Dicranostigma*), tetrahydrocoptisine (*Stylophorum*), chelerythrine (whole tribe, especially *Macleaya*), and sanguinarine (whole tribe, especially *Macleaya*) (Feng et al., 1985; Gilca et al., 2010).

Table 5.1 Alkaloids isolated from tribe Chelidonieae

No.	Compounds	Types	Plants	Tissues	References
1	(2′,6′-Epoxy-1′α,2′α,3′β,4′α,5′α-pentahydroxy) hexane-(1′→6)-dihydrochelerythrin	I_A	a	Roots	Lei et al. (2014)
2	Maclekarpine D	I_A	a	Roots	Yu et al. (2014)
3	(5′*R*)-3′-Methyl-2′(5′*H*)furanone-(5′→6)-(6*S*)-dihydrochelerythrin = maclekarpine A	I_A	a	Roots	Lei et al. (2014) and Yu et al. (2014)
4	(5′*R*)-3′-Methyl-2′(5′*H*)furanone-(5′→6)-(6*S*)-dihydrosanguinarine = maclekarpine B	I_A	a	Roots	Lei et al. (2014) and Yu et al. (2014)
5	(5′*R*)-3′-Methyl-2′(5′*H*)furanone-(5′→6)-(6*R*)-dihydrosanguinarine = maclekarpine C	I_A	a	Roots	Lei et al. (2014) and Yu et al. (2014)
6	6-(1′-Hydroxyethyl)dihydrochelerythrine	I_A	a	Roots	Lei et al. (2014)
7	(1′*E*)-5′-Methoxy-6′-hydroxy-cinnamenyl-(1′→6)-dihydrosanguinarine = maclekarpine E	I_A	a	Roots	Lei et al. (2014) and Yu et al. (2014)
8	11-Acetonyldihydrochelerythrine	I_A	b	Barks	Lei et al. (2014)
9	6-Acetonyl-dihydrosanguinarine	I_A	a	Whole plants	Lei et al. (2014) and Yu et al. (2014)
			b	Aerial parts	
			c	Whole plants	
			d	Fruits, stems	
			l	Leaves	
10	6-Acetonyl-dihyrochelerythrine	I_A	b	Barks, aerial parts	Lei et al. (2014) and Yu et al. (2014)
			d	Fruits	
			l	Leaves	

Continued

Table 5.1 Continued

No.	Compounds	Types	Plants	Tissues	References
11	6-Butoxydihydrochelerythrine	I_A	a	Roots	Lei et al. (2014) and Yu et al. (2014)
			e	Roots	
12	6-Butoxy-dihydrosanguinarine	I_A	e	Roots	Lei et al. (2014)
13	6-Ethoxydihydrosanguinarine	I_A	a	Whole plants	Lei et al. (2014) and Yu et al. (2014)
			d	–	
14	6-Methoxydihydrochelerythrine = angoline	I_A	a	Roots, stems, aerial parts	Lei et al. (2014) and Yu et al. (2014)
			b	Barks, stems, leaves	
			d	Stems	
			f	Whole plants	
15	6-Methoxydihydrosanguinarine	I_A	a	Roots, stems	Lei et al. (2014) and Yu et al. (2014)
			d	Fruits	
			f	Whole plants	
16	8-Hydroxydihydrochelerythrine	I_A	d	Seeds	Lei et al. (2014), Yu et al. (2014), and Baek et al. (2013)
			f	Whole plants	
17	8-Hydroxy-dihydrosanguinarine	I_A	d	Seeds	Lei et al. (2014) and Baek et al. (2013)
			f	Whole plants	
18	Bocconoline	I_A	d	Fruits	Lei et al. (2014) and Yu et al. (2014)
			m	–	
19	Dihydrochelerythrine	I_A	a	Whole plants, roots, stems	Lei et al. (2014), Yu et al. (2014), and Qing et al. (2014)
			b	Aerial parts, barks	
			d	Fruits	
			g	Fruits	
			n	Leaves	

20	Dihydrochelirubine	I_A	g	Fruits	Lei et al. (2014)
			n	Leaves	
21	Dihydrosanguinarine	I_A	a	Whole plant, roots, stems, leaves aerial parts, barks	Lei et al. (2014), Yu et al. (2014), Sun et al. (2014), and Qing et al. (2014)
			b	Whole plants	
			c	Fruits	
			d	Stems, fruits	
			g	Whole plants	
			f	Leaves	
			n		
22	Methyl 2′-(7,8-dihydrosanguinarine-8-yl) acetate	I_A	f	Whole plants	Lei et al. (2014)
23	Oxychelerythrine	I_A	d	Fruits	Lei et al. (2014) and Yu et al. (2014)
			g	Fruits	
24	Oxysanguinarine	I_A	b	Barks	Lei et al. (2014) and Yu et al. (2014)
			d	Fruits, stems	
			g	Fruits	
			h	Rhizomes	
			i	Rhizomes	
			o	Leaves, roots, seeds	
25	Spallidamine	I_A	a	Roots	Lei et al. (2014) and Yu et al. (2014)
			d	Whole plants	
26	12-Methoxydihydrochelerythrine	I_A	b	–	Yu et al. (2014)
27	6,12-Dimethoxydihydrocheleritrine	I_A	b	–	Yu et al. (2014)
28	8-Methoxydihydrosanguinarine	I_A	d	Seeds	Yu et al. (2014)
29	6-Methoxydihydrochelirubine	I_A	b	–	Yu et al. (2014)

Continued

Table 5.1 Continued

No.	Compounds	Types	Plants	Tissues	References
30	6-Carboxymethyldihydrochelerythrine	I_A	d	Whole plants	Yu et al. (2014)
31	6α-Isobutanonyldihydrochelerythrine	I_A	d	Fruits	Yu et al. (2014)
32	6α-Isobutanonyldihydrosanguinarine	I_A	d	Fruits	Yu et al. (2014)
33	*R*-6-((*R*)-1-Hydroxyethyl)-dihydrochelerythrie	I_A	d	Whole plants	Yu et al. (2014)
34	(+)-Chelamine	I_B	f	Roots	Lei et al. (2014)
35	Chelidonine	I_B	f	Whole plants	Lei et al. (2014) and Yu et al. (2014)
			j	Roots	
			k	Rhizomes	
			l	Roots	
36	(±)-Homochelidonine	I_B	d	–	Lei et al. (2014) and Yu et al. (2014)
			f	Roots	
37	(±)-Norchelidonine	I_B	f	Roots, whole plants	Lei et al. (2014)
38	Isochelidonine	I_B	f	Whole plants	Lei et al. (2014)
39	12-Methoxynorchelerythrine	I_C	g	Fruits	Lei et al. (2014) and Yu et al. (2014)
40	Norchelerythrine	I_C	d	Whole plants	Lei et al. (2014) and Yu et al. (2014)
			g	Fruits	
41	Norsanguinarine	I_C	d	Fruits, stems	Lei et al. (2014) and Yu et al. (2014)
42	Pancorine	I_C	a	Roots	Lei et al. (2014) and Yu et al. (2014)

43	6-Ethoxychelerythrine	I_D	d	–	Lei et al. (2014) and Yu et al. (2014)
			i	Rhizomes	
44	Chelerythrine	I_D	a	–	Lei et al. (2014), Yu et al. (2014), and Qing et al. (2014)
			b	Barks	
			c	Roots	
			d	Fruits	
			e	Roots	
			h	Rhizomes	
			i	Rhizomes	
			k	Rhizomes	
			l	Roots, stalks, leaves	
			o	Barks, stems, seeds	
45	Chelilutine	I_D	a	–	Lei et al. (2014) and Yu et al. (2014)
			d	Roots	
			f	Whole plants	
			k	Rhizomes	
46	8-*O*-Demethylchelerythrine	I_D	d	–	Yu et al. (2014)
47	Chelirubine	I_D	a	–	Lei et al. (2014) and Yu et al. (2014)
			d	Fruits	
			k	Rhizomes	
			m	–	
48	Macarpine	I_D	a	–	Lei et al. (2014) and Yu et al. (2014)
			d	Callus tissues	
			f	Whole plants	

Continued

Table 5.1 **Continued**

No.	Compounds	Types	Plants	Tissues	References
49	Sanguinarine	I_D	a	Aerial parts	Lei et al. (2014), Yu et al. (2014), and Qing et al. (2014)
			c	Roots, whole plants	
			d	Fruits, stems	
			e	Roots	
			h	Rhizomes	
			i	Rhizomes	
			j	Roots	
			k	Rhizomes	
			l	Roots, stalks, leaves	
			m	Leaves	
			o	Barks, stems	
50	Sanguilutine	I_D	d	Fruits	Lei et al. (2014)
51	Sanguirubine	I_D	a	Whole plants	Lei et al. (2014)
52	bis[6-(5,6-Dihydrochelerythrinyl)]ether	I_E	a	Roots	Lei et al. (2014) and Yu et al. (2014)
53	Bocconarborine A	I_E	a	Whole plants	Lei et al. (2014) and Yu et al. (2014)
			b	Aerial parts	
			d	Fruits, leaves	
54	Bocconarborine B	I_E	b	Aerial parts	Lei et al. (2014) and Yu et al. (2014)
			d	Fruits, leaves	
55	Chelerythridimerine	I_E	b	Barks	Lei et al. (2014) and Yu et al. (2014)

56	Chelidimerine	I_E	b	Aerial parts	Lei et al. (2014) and Yu et al. (2014)
			d	Fruits, leaves	
			f	Whole plants	
			i	Rhizomes	
57	Sanguidimerine	I_E	b	Aerial parts	Lei et al. (2014) and Yu et al. (2014)
			d	Fruits, leaves	
58	1,3-bis(1-Hydrochelerythriny1)acetone	I_E	b	–	Lei et al. (2014)
59	Arnottianamide	I_F	a	Roots	Lei et al. (2014)
60	1-Canadine	II	f	Whole plants	Lei et al. (2014) and Yu et al. (2014)
			l	Roots, stalks, leaves	
61	Chelanthifoline	II	d	Fruits	Lei et al. (2014)
62	Stylopine	II	f	Leaves, whole plants	Lei et al. (2014)
			j	Roots	
			k	Rhizomes	
63	Tetrahydroberberine	II	k	Rhizomes	Lei et al. (2014)
64	Tetrahydrocoptisine	II	f	Whole plants	Lei et al. (2014)
65	Tetrahydropalmatrubine	II	d	Cultured cells	Yu et al. (2014)
66	Corytenchine	II	d	Cultured cells	Yu et al. (2014)
67	(−)-Isocorypalmine	II	l	Leaves, roots	Yu et al. (2014)
68	(−)-Scoulerine	II	l	Roots, leaves	Yu et al. (2014)
69	Dehydrocicanthifoline	II	d	Fruits	Lei et al. (2014) and Yu et al. (2014)
70	8-Oxycoptisine	II	f	Whole plants	Lei et al. (2014)

Continued

Table 5.1 Continued

No.	Compounds	Types	Plants	Tissues	References
71	Coptisine	II	a	Roots	Lei et al. (2014), Yu et al. (2014), and Qing et al. (2014)
			d	Fruits	
			e	Roots	
			f	Whole plants	
			k	Rhizomes	
			l	Roots, stalks, leaves	
72	Berberine	II	a	Roots	Lei et al. (2014), Yu et al. (2014), and Qing et al. (2014)
			d	Fruits	
			e	Roots	
			f	Whole plants	
			k	Rhizomes	
			l	Roots, stalks, leaves	
73	Berberrubine	II	a	Whole plants	Lei et al. (2014) and Yu et al. (2014)
74	Columbamine	II	l	Roots, stalks, leaves	Yu et al. (2014)
75	Dehydrocorytenchine	II	d	Cultured cells	Yu et al. (2014)
76	Dehydrocheilanthifoline	II	d	Fruits	Lei et al. (2014) and Yu et al. (2014)
77	Corysamine	II	l	–	Yu et al. (2014)
78	5-Hydroxy-coptisine	II	c	Whole plants	Zhong et al. (2014)
79	(−)-*cis*-*N*-Methylcanadinium	II	l	Roots	Yu et al. (2014)
80	*cis*-Protopinium	II	c	Whole plants	Lei et al. (2014)
81	*trans*-Protopinium	II	c	Whole plants	Lei et al. (2014)

82	α-Allocryptopine	III	a	Whole plants	Lei et al. (2014), Yu et al. (2014), and Qing et al. (2014)
			d	Fruits	
			m	Leaves	
			o	Roots	
83	Cryptopine	III	a	Aerial parts	Lei et al. (2014) and Yu et al. (2014)
			c	Rhizomes	
			d	Fruits	
			k	Rhizomes	
84	Protopine	III	a	Aerial parts, whole plants	Lei et al. (2014), Yu et al. (2014), Sun et al. (2014), and Qing et al. (2014)
			c	Whole plants	
			d	Fruits	
			e	Roots	
			f	Whole plants	
			h	Rhizomes	
			i	Rhizomes	
			k	Rhizomes	
85	β-Allocryptopine	III	a	Aerial parts, whole plants	Lei et al. (2014), Yu et al. (2014), and Qing et al. (2014)
			d	Fruits	
86	Protopine *N*-oxide	III	d	Whole plants	Yu et al. (2014)
87	Dihydrocryptopine	III	c	Whole plants	Sun et al. (2014)
88	Corydine	IV	c	Whole plants	Lei et al. (2014) and Sun et al. (2014)
			f	Roots	
89	*N*-Norcorydine	IV	f	Roots	Lei et al. (2014)
90	*N*-Methylhernovine	IV	c	Whole plants	Lei et al. (2014)
91	Glaucine	IV	c	Whole plants	Lei et al. (2014)

Continued

Table 5.1 Continued

No.	Compounds	Types	Plants	Tissues	References
92	Isocorydine	IV	c	Whole plants	Lei et al. (2014) and Sun et al. (2014)
93	Dicranostigmine	IV	c	Whole plants	Lei et al. (2014)
94	Magnoflorine	IV	c	Whole plants	Lei et al. (2014)
95	10-*O*-Methylhernovine	IV	c	Whole plants	Sun et al. (2014)
96	Corytuberine	IV	c	Whole plants	Sun et al. (2014)
97	Lagesianine A	IV	c	Whole plants	Sun et al. (2014)
98	Nantenine	IV	c	Whole plants	Sun et al. (2014)
99	Papaverine	V	d	Cultured cells	Yu et al. (2014)
100	Laudanine	V	d	Cultured cells	Yu et al. (2014)
101	Codamine	V	d	Cultured cells	Yu et al. (2014)
102	Pseudocodamine	V	d	Cultured cells	Yu et al. (2014)
103	Rhoeadine	VI	l	Leaves, stalks	Yu et al. (2014)
104	Sinoacutine	VII	c	Whole plants	Lei et al. (2014)
105	(−)-Turkiyenine	VIII	f	Whole plants	Lei et al. (2014)
106	Corysamine	VIII	d	Fruits	Lei et al. (2014)
107	Sparteine	VIII	f	Whole plants	Lei et al. (2014)

I: Benzophenanthridine (BPA); I_A: dihydrobenzophenanthridine; I_B: QBPA; I_C: *N*-demethylation BPA; I_D: hexahydrobenzophenanthridine; I_E: dimeric BPA; I_F: seco-BPA. II: proberberine type; III: protopine type; IV: aporphine type; V: benzylisoquinoline type; VI: rhoeadine type; VII: promorphinane type; VIII: other alkaloids.
a, *M. microcarpa*; b, *B. arborea*; c, *D. leptopodum*; d, *M. cordata*; e, *D. lactucoides*; f, *C. majus*; g, *B. pearcei*; h, *S. canadensis*; i, *E. chionantha*; j, *S. diphyllum*; k, *H. japonica*; l, *B. frutescens*; m, *B. cordata*; n, *B. integrifolia*; o, *B. Latisepala*; p, *H. vernalis*

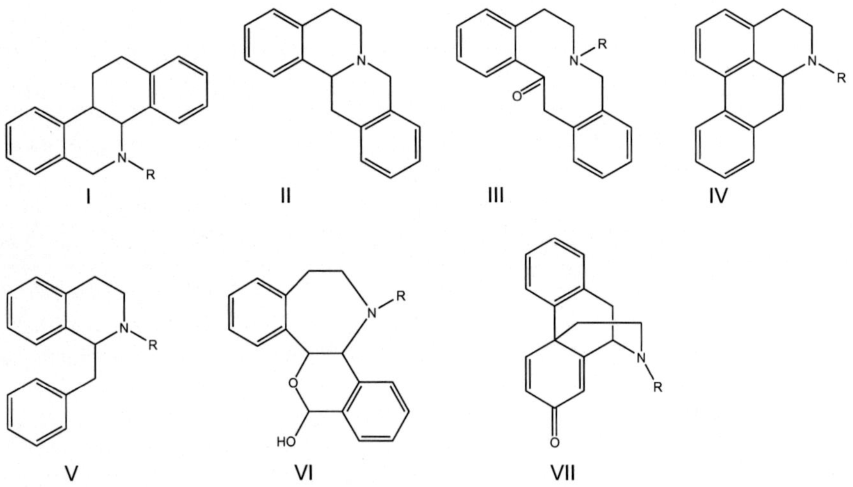

Figure 5.1 Types of alkaloids in Chelidonieae. I: Benzophenanthridine (BPA); II: proberberine type; III: protopine type; IV: aporphine type; V: benzylisoquinoline type; VI: rhoeadine type; VII: promorphinane type.

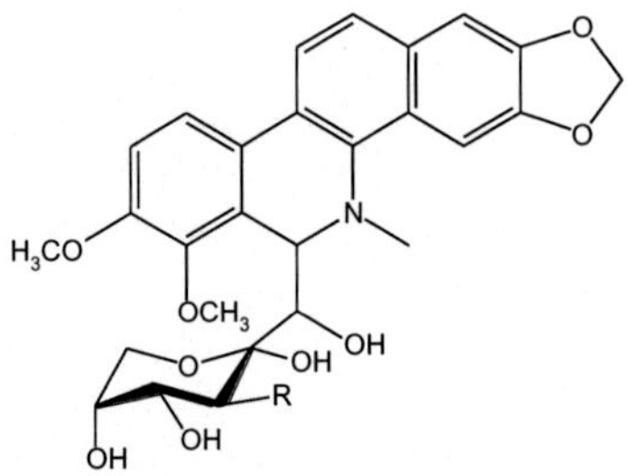

1	R=CH_3
2	R=OH

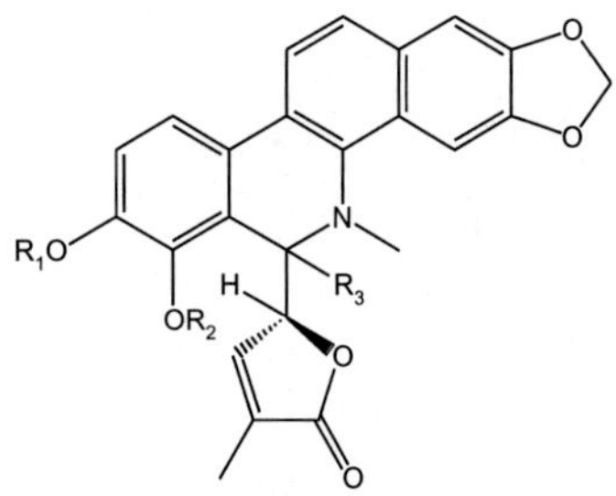

3	$R_1=R_2=CH_3$; $R_3=\beta$-H
4	$R_1+R_2=3CH_2$; $R_3=\beta$-H
5	$R_1+R_2=CH2$; $R_3=\alpha$-H

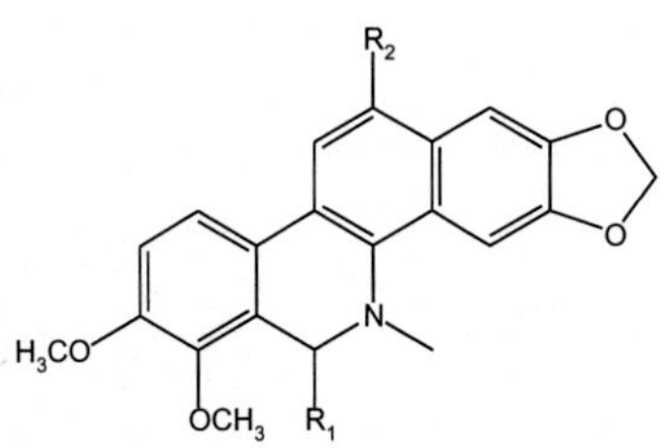

Figure 5.2 Structures of alkaloids isolated from tribe Chelidonieae. Refer to the text and Table 5.1 for more details.

(Continued)

6	$R_1=CH(OH)CH_3$; $R_2=H$
10	$R_1=CH_2COCH_3$; $R_2=H$
11	$R_1=OCH_2(CH_2)_2CH_3$; $R_2=H$
14	$R_1=OCH_3$; $R_2=H$
16	$R_1=OH$; $R_2=H$
18	$R_1=CH_2OH$; $R_2=H$
19	$R_1=H$; $R_2=H$
23	$R_1==O$; $R_2=H$
26	$R_1=H$; $R_2=OCH$
27	$R_1=R_2=OCH_3$
30	$R_1=CH_2COOH$; $R_2=H$

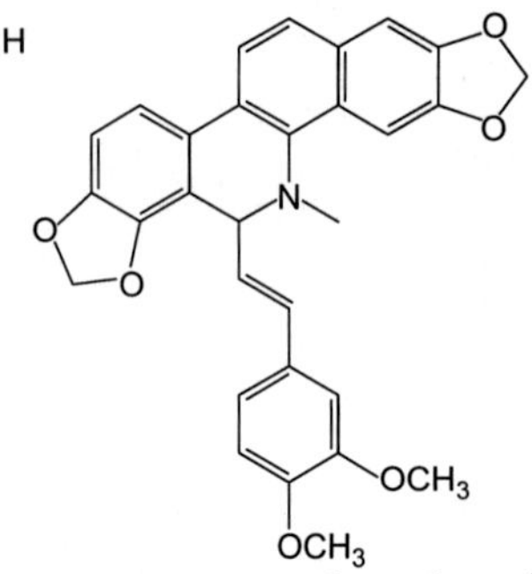

7

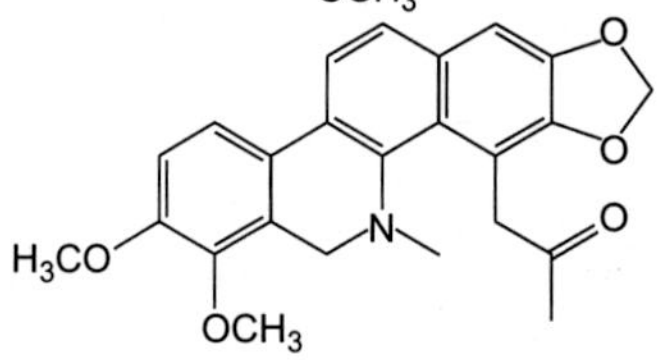

8

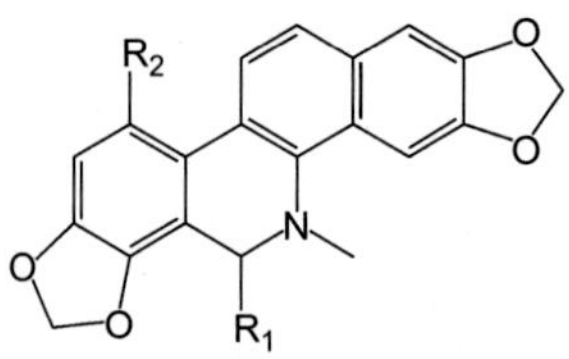

9	$R_1=CH_2COCH_3$; $R_2=H$
12	$R_1=OCH_2(CH_2)_2CH_3$; $R_2=H$
13	$R_1=OCH_2CH_3$;$R_2=H$
15	$R_1=OCH_3$; $R_2=H$
17	$R_1=OH$; $R_2=H$
20	$R_1=H$; $R_2=OCH_3$
21	$R_1=R_2=H$
22	$R_1=CH_2COOCH_3$; $R_2=H$

Figure 5.2 Continued.

24 $R_1 = =O$; $R_2 = H$

25 $R_1 = CH_2COOH$; $R_2 = H$

28 $R_1 = OCH_3$; $R_2 = H$

29 $R_1 = R_2 = OCH_3$

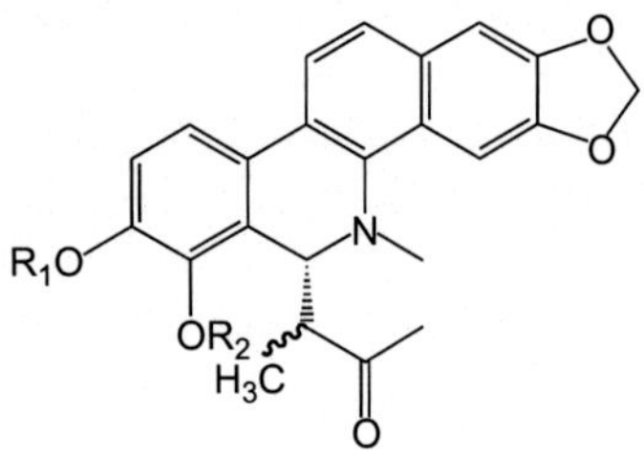

33 6R, 1′R

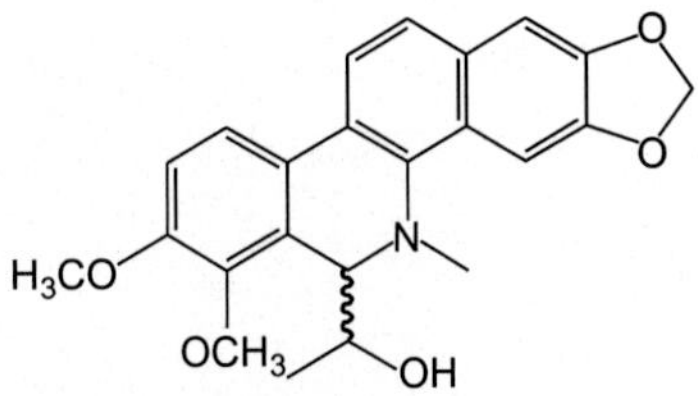

31 $R_1 = R_2 = CH_3$

32 $R_1 + R_2 = CH_2$

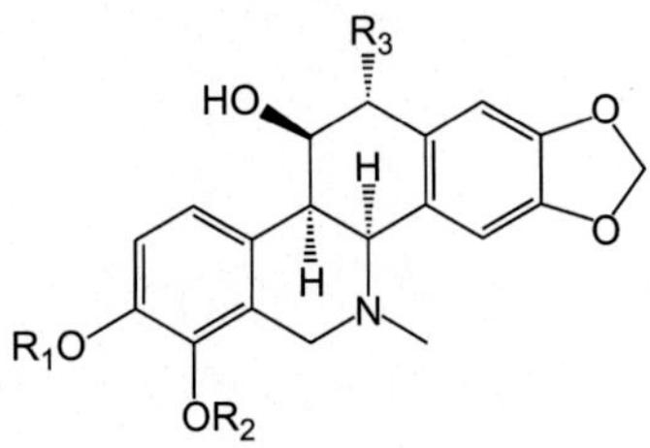

34 $R_1 + R_2 = CH_2$; $R_3 = OH$

35 $R_1 + R_2 = CH_2$; $R_3 = H$

36 $R_1 = R_2 = CH_3$; $R_3 = H$

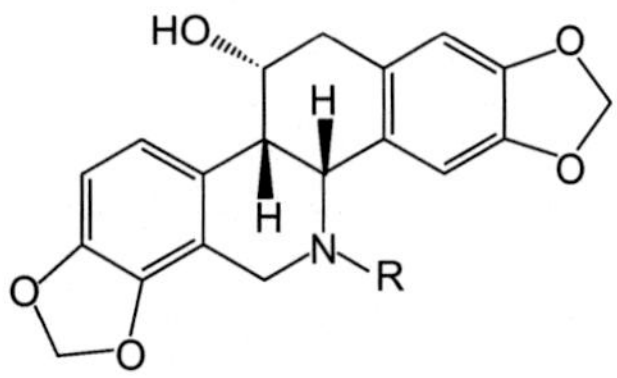

37 R=H

38 $R = CH_3$

Figure 5.2 Continued.

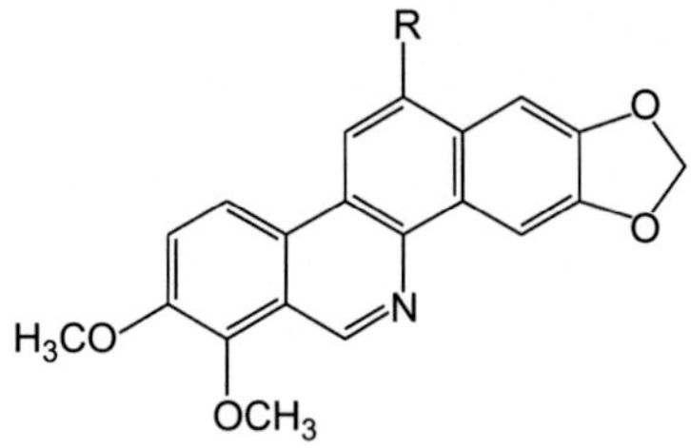

39 R=OCH_3

40 R=H

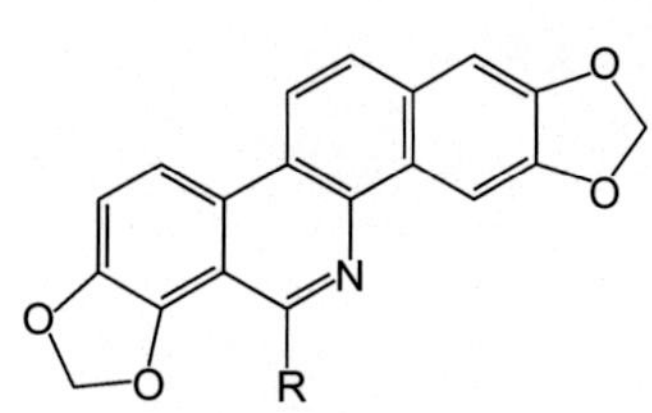

41 R=H

42 R=OCH_3

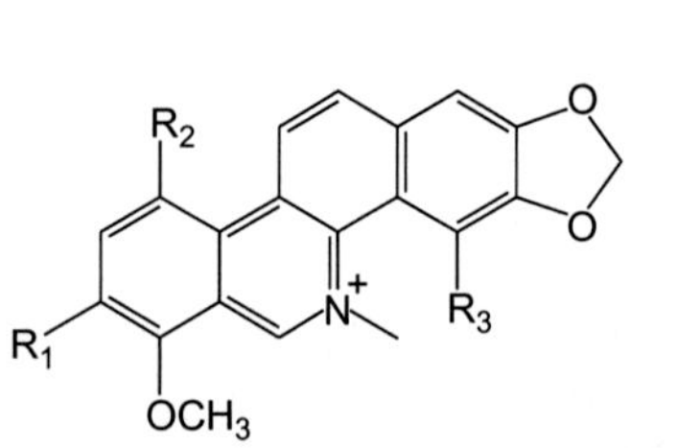

43 $R_1=OCH_3$; $R_2=H$; $R_3=OCH_2CH_3$

44 $R_1=OCH_3$; $R_2=R_3=H$

45 $R_1=R_2=OCH_3$; $R_3=H$

46 $R_1=OH$; $R_2=R_3=H$

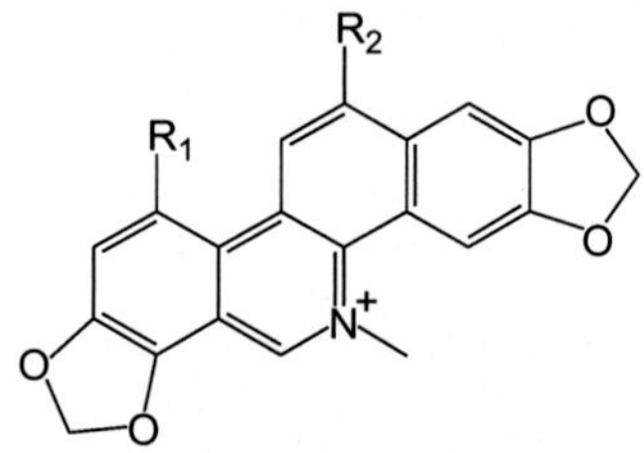

47 $R_1=OCH_3$; $R_2=H$

48 $R_1=R_2=OCH_3$

49 $R_1=R_2=H$

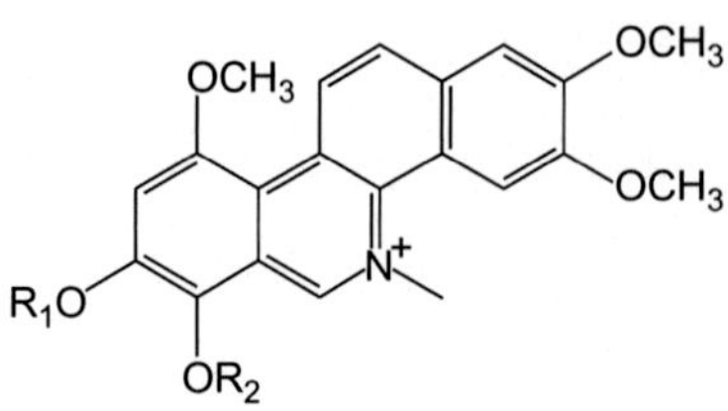

50 $R_1=R_2=CH_3$

51 $R_1+R_2=CH_2$

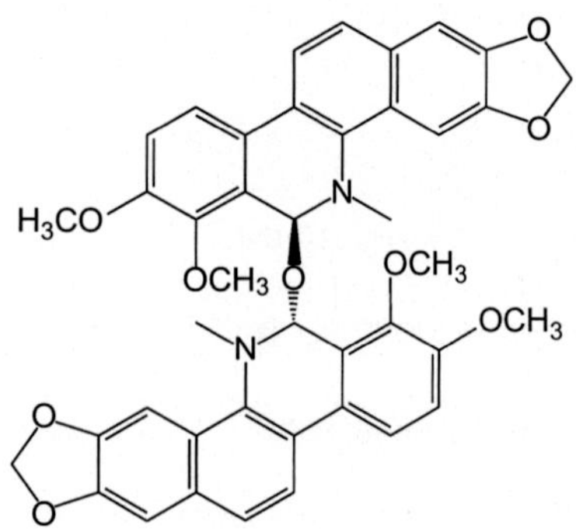

52

Figure 5.2 Continued.

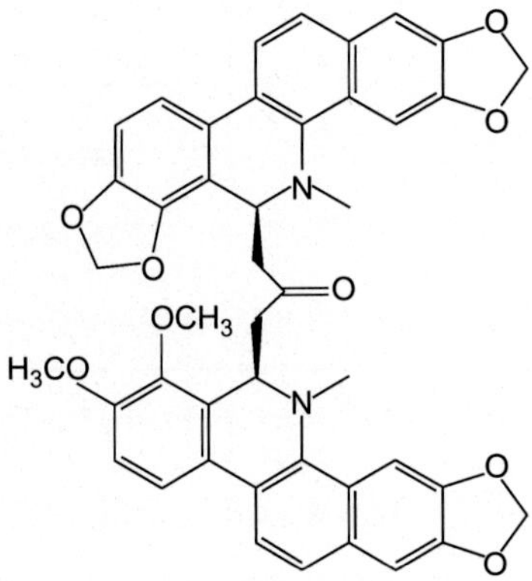

53

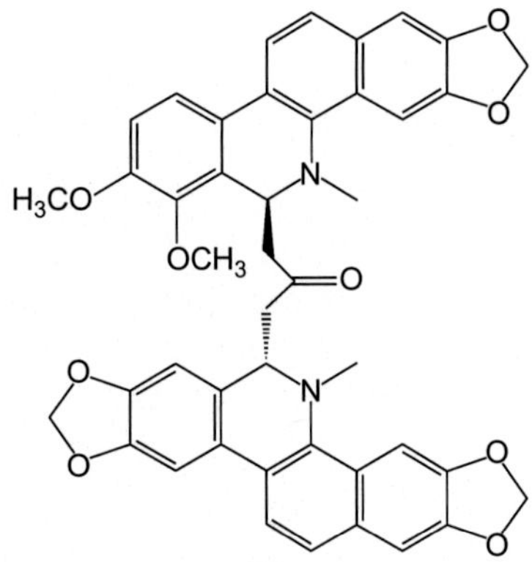

54

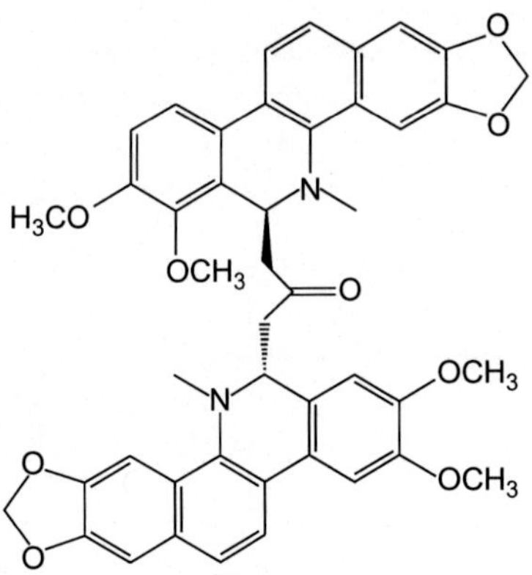

55

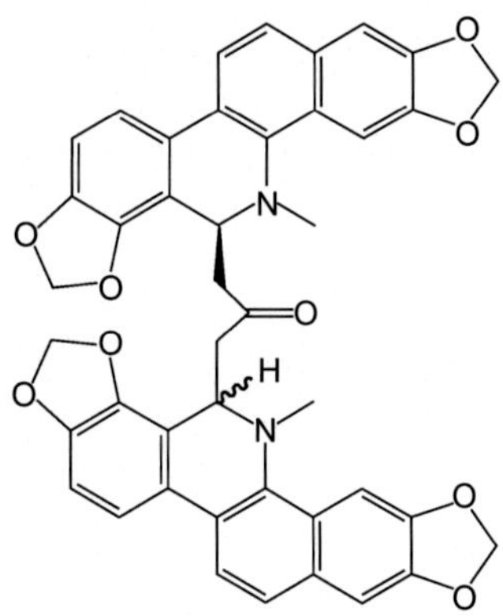

56 α **57** β

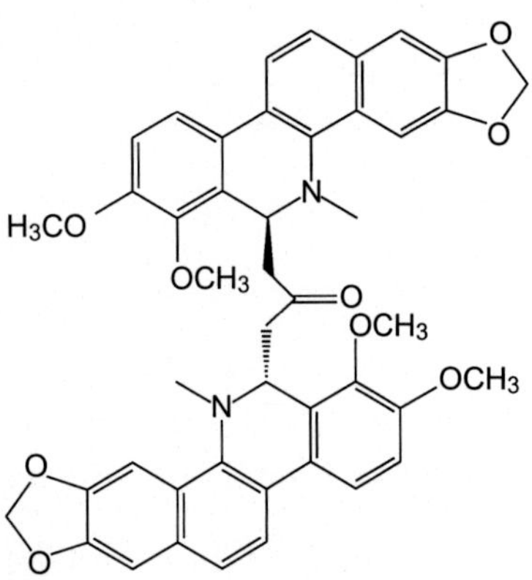

58

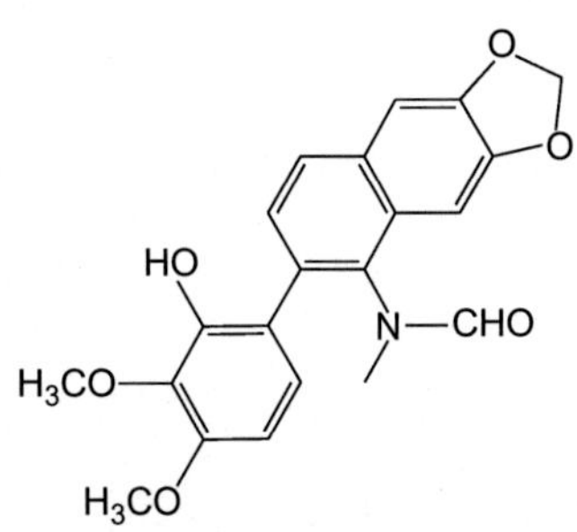

59

Figure 5.2 Continued.

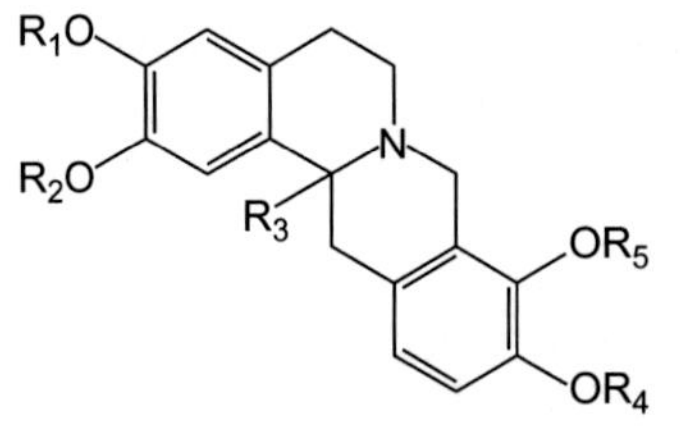

60 $R_1+R_2=CH_2$; $R_3=\beta$-H; $R_4=R_5=CH_3$

61 $R_1=H$; $R_2=CH_3$; $R_3=\alpha$-H; $R_4+R_5=CH_2$

62 $R_1+R_2=CH_2$; $R_3=\beta$-H; $R_4+R_5=CH_2$;

63 $R_1+R_2=CH_2$; $R_3=\alpha$-H; $R_4=R_5=CH_3$

64 $R_1+R_2=CH_2$; $R_3=\alpha$-H; $R_4+R_5=CH_2$

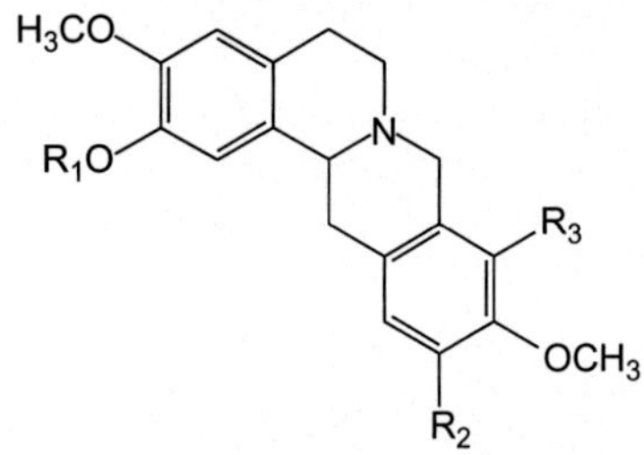

65 $R_1=CH_3$; $R_2=H$; $R_3=OH$

66 $R_1=CH_3$; $R_2=OH$; $R_3=H$

67 $R_1=R_2=H$; $R_3=CH_3$

68 $R_1=R_2=R_3=H$

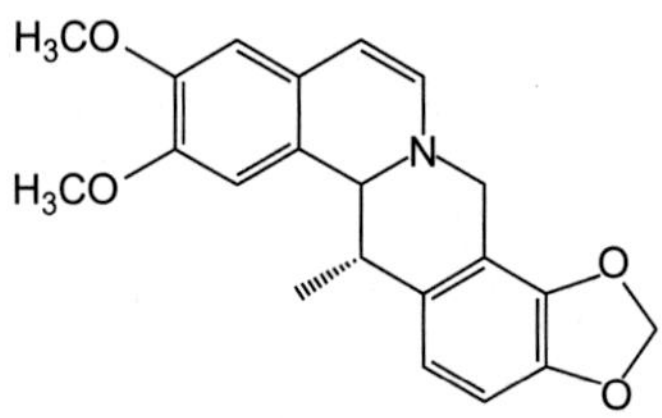

69

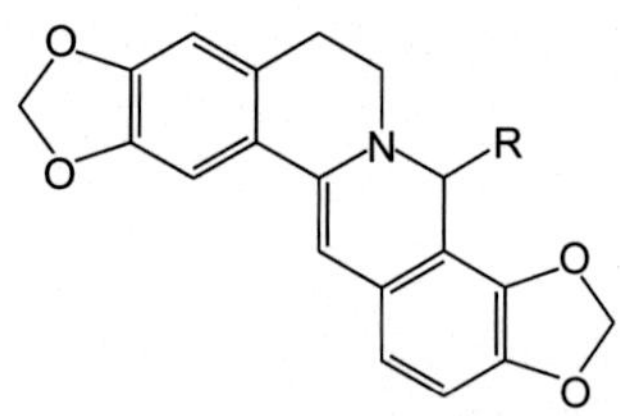

70 R==O

71 R=H

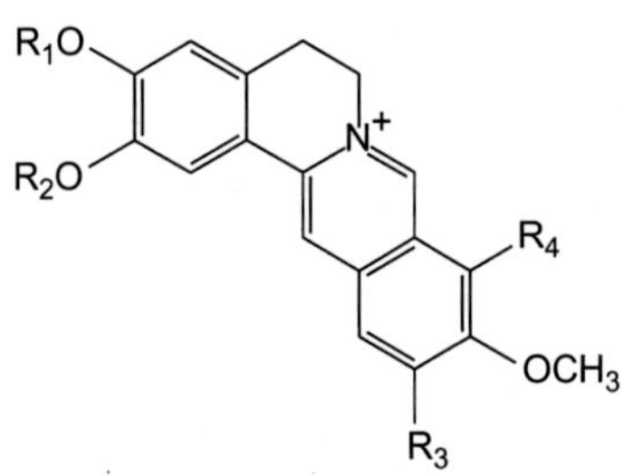

72 $R_1+R_2=CH_2$; $R_3=H$; $R_4=OCH_3$

73 $R_1+R_2=CH_2$; $R_3=H$; $R_4=OH$

Figure 5.2 Continued.

74 $R_1=CH_3$; $R_2=R_3=H$; $R_4=OCH_3$

75 $R_1=R_2=CH_3$; $R_3=OH$; $R_4=H$

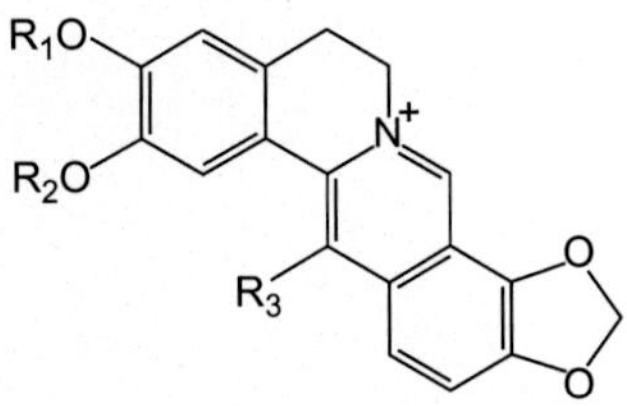

76 $R_1=H$; $R_2=CH_3$; $R_3=H$

77 $R_1+R_2=CH_2$; $R_3=CH_3$

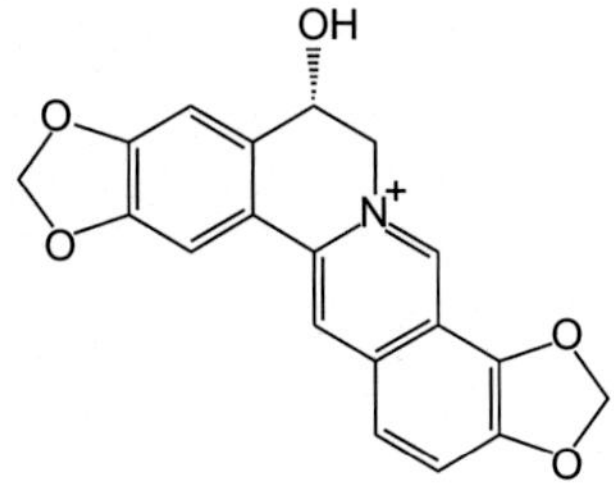

78

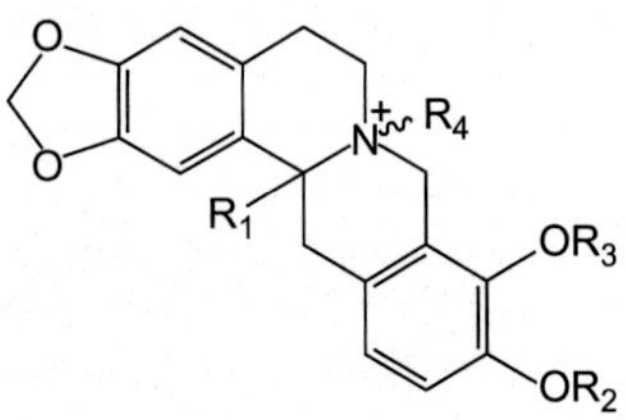

79 $R_1=H$; $R_2=R_3=CH_3$; $R4=CH_3$

80 $R_1=OH$; $R_2+R_3=CH_2$; $R_4=\alpha\text{-}CH_3$

81 $R_1=OH$; $R_2+R_3=CH_2$; $R_4=\beta\text{-}CH_3$

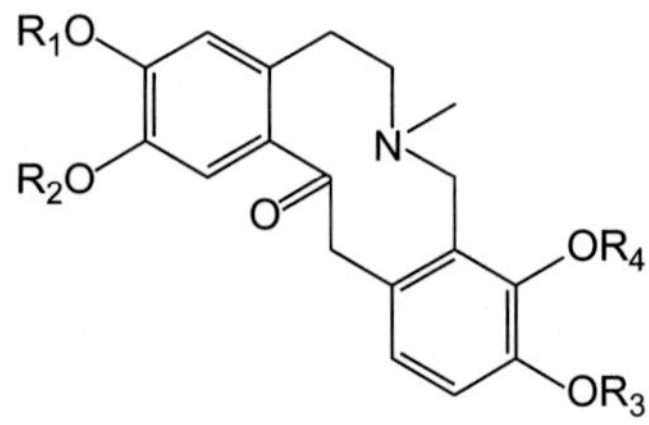

82 $R_1+R_2=CH_2$; $R_3=R_4=CH_3$

83 $R_1=R_2=CH_3$; $R_3+R_4=CH_2$

84 $R_1+R_2=R_3+R_4=CH_2$

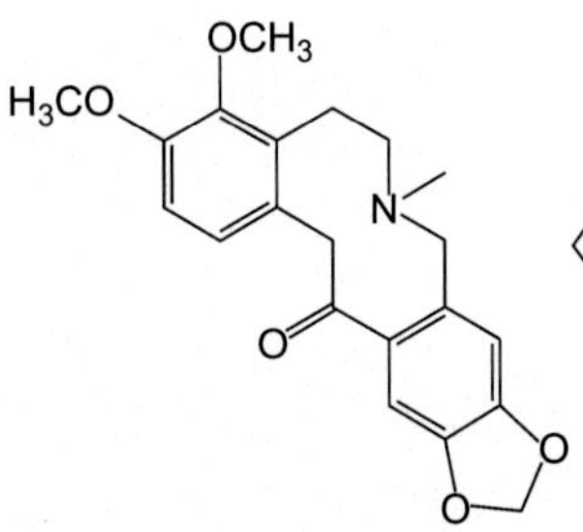

85

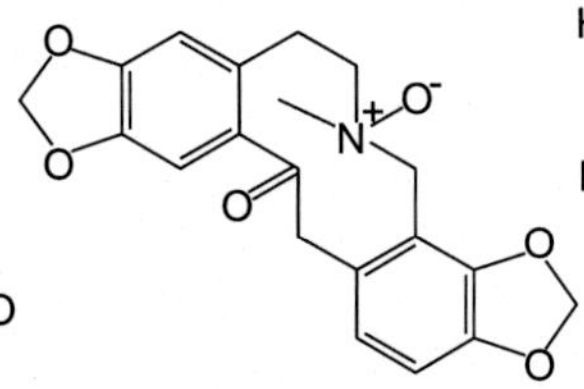

86

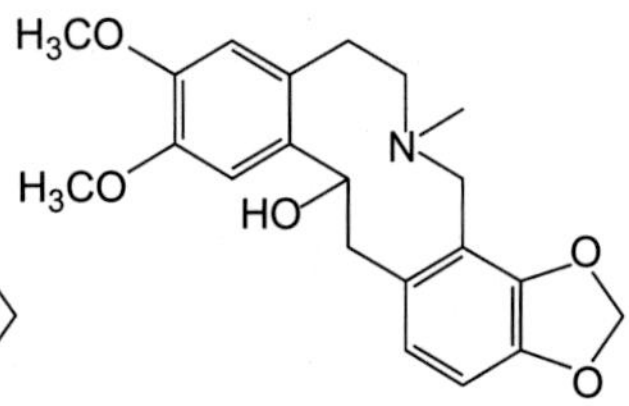

87

Figure 5.2 Continued.

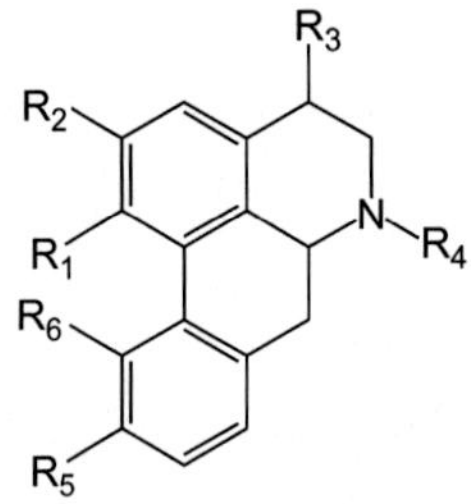

	R_1	R_2	R_3	R_4	R_5	R_6
88	OCH_3	OH	H	H	OCH_3	OCH_3
89	OH	OCH_3	H	CH_3	OCH_3	OH
90	OCH_3	OCH_3	OH	CH_3	OCH_3	OH
95	OH	OCH_3	H	H	OCH_3	OCH_3
96	OH	OCH_3	H	CH_3	OCH_3	OCH_3
97	OCH_3	OH	H	CH_3	OH	OCH_3

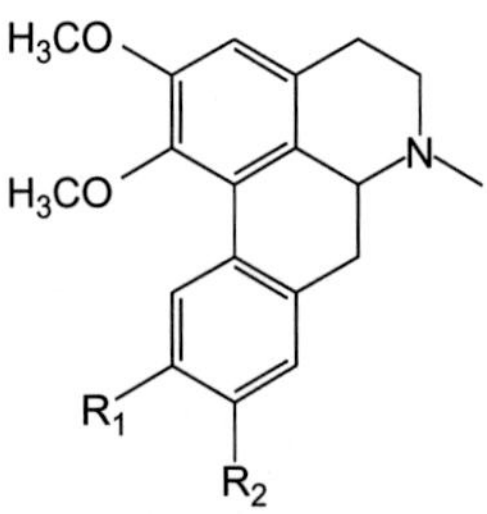

91	R_1=α-CH_3; R_2=OCH_3
92	R_1=β-CH_3; R_2=OCH_3
98	R_1+R_2=O-CH_2-O

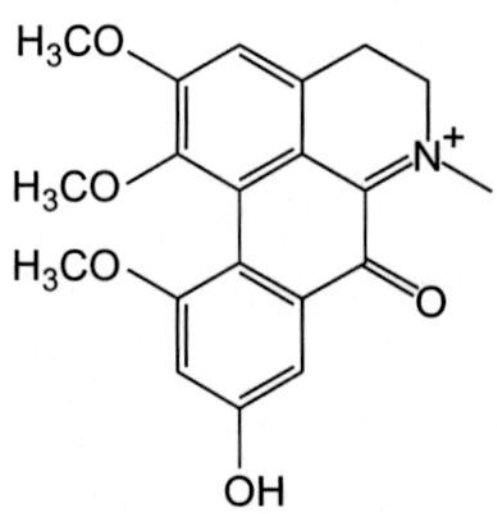

93

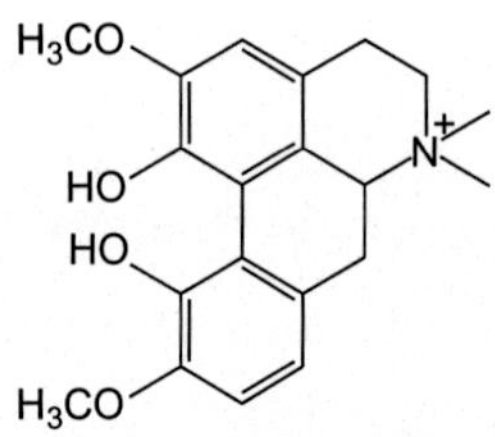

94

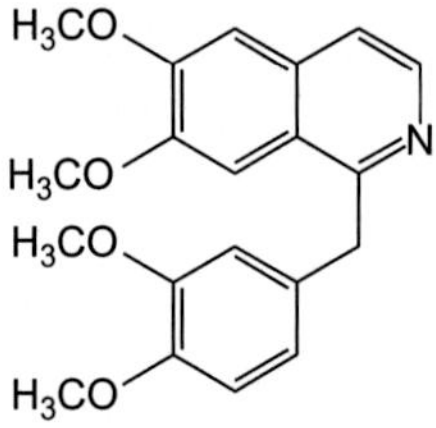

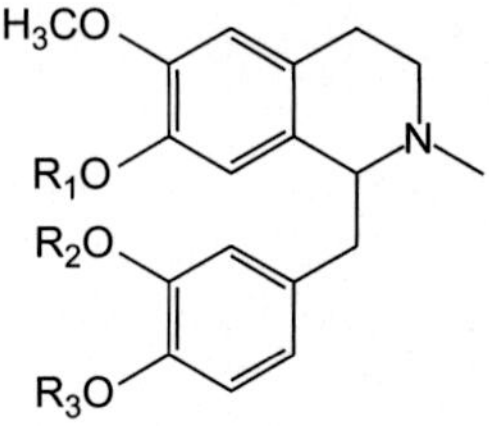

Figure 5.2 Continued.

99

100 $R_1=R_3=CH_3$; $R_2=H$

101 $R_1=H$; $R_2=R_3=CH_3$

102 $R_1=R_2=CH_3$; $R_3=H$

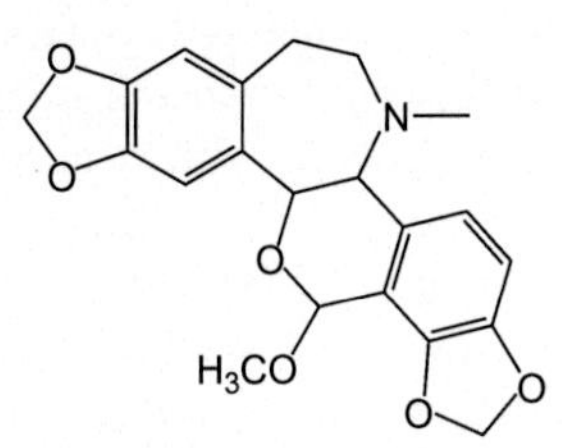

103

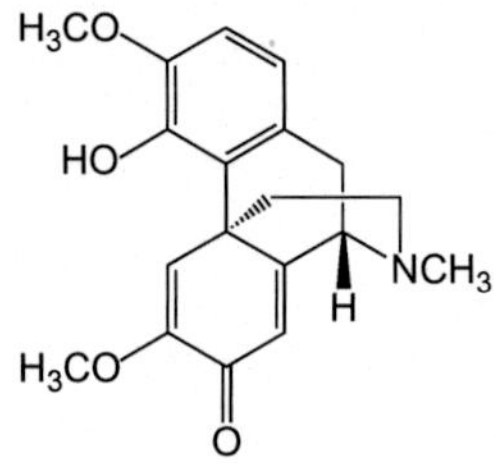

104

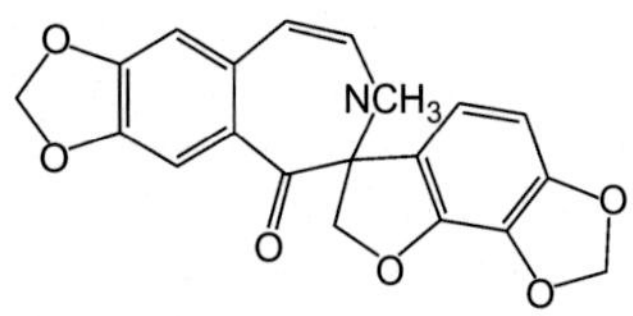

105

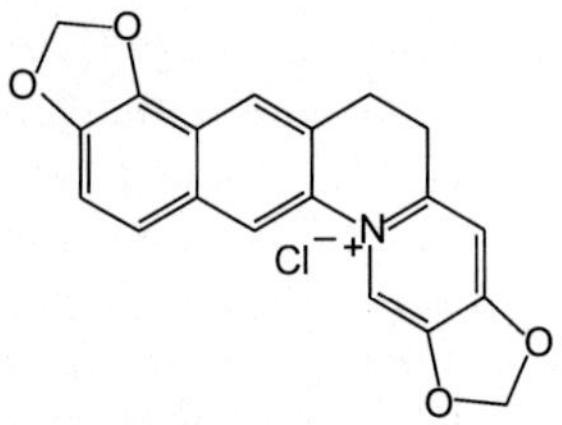

106

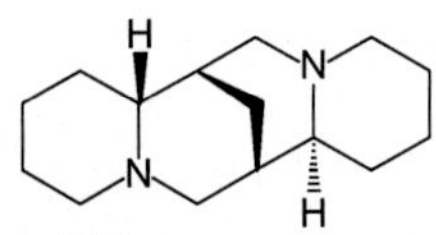

107

Figure 5.2 Continued.

Benzophenanthridines (BPAs) are predominant in *Bocconia* and *Macleaya*, followed by protoberberines, protopines, and other types (Yu et al., 2014). The BPA type of alkaloids, derived from protoberberines via N-C6 bond cleavage and the formation of C6–C13 bonds, is one of the characteristic and chemotaxonomic components in the family Papaveraceae.

Four types of alkaloids of *Dicranostigma leptopodum*, including two aporphines (isocorydine and corydine), two protopines (protopine and allocryptopine), a morphine (sinoacutine), and three quaternary protoberberine alkaloids (berberrubine, 5-hydroxycoptisine, and berberine), were separated well on a SinoChrom ODS-BP column (Chen et al., 2014).

5.3 Pharmacology and therapeutic use of alkaloids

5.3.1 Anticancer activity: multiple mechanisms

Chelidonine produces *in vivo* a significant antiproliferative effect on planarian stem cells in a dose-dependent fashion (Isolani et al., 2012). Chelidonine treatment resulted in mitotic abnormalities and the number of cells able to proceed to anaphase/telophase significantly reduced when compared to the controls. Sanguinarine causes cell cycle blockade and apoptosis of human prostate carcinoma cells via modulation of cyclin kinase inhibitor-cyclin-cyclin-dependent kinase machinery (Adhami et al., 2004). Modified alkaloids from *Chelidonium majus* L. induce G2/M arrest, caspase-3 activation, and apoptosis in human acute lymphoblastic leukemia MT-4 cells (Fil'chenkov et al., 2006). Isocorydine inhibits cell proliferation in hepatocellular carcinoma cell lines by inducing G2/M cell cycle arrest and apoptosis (Sun et al., 2012). Sanguinarine, chelerythrine, chelidonine, sanguilutine, and chelilutine exhibit strong antiproliferative activity in malignant melanoma cells regardless of their p53 status (Hammerová et al., 2011). Despite the significant similarity of benzophenanthridine alkaloids molecular structures, the mechanism of cell death induction is different for each alkaloid.

Chelidonine reduces telomerase activity through downregulation of the telomerase catalytic subunit (hTERT) expression in HepG2 cells (Noureini and Wink, 2009). Senescence induction might not be directly caused by reducing telomerase activity as it occurs after a few population doublings. G-quadruplex (G4) structures can be formed at the single-stranded overhang of telomeric DNA, and ligands able to stabilize this structure have recently been identified as potential anticancer drugs (Bessi et al., 2012). Human telomere d[(TTAGGG)4] undergoes a conformational transition to the Na^+-form upon binding with sanguinarine in the presence of K^+ (Pradhan et al., 2011). Sanguinarine and berberine induce the formation of G4 as well as increase its stability (Ji et al., 2012). Five DNA-interacting compounds, that is, coptisine, berberine, allocryptopine, sanguinarine, and chelerythrine, were identified from *Macleaya cordata* and *Chelidonium majus*, and the proliferation of four types of human solid cancer cell lines was markedly inhibited by these compounds (Zhang et al., 2011).

Sanguinarine, chelerythrine, and chelidonine possess prominent apoptotic effects toward cancer cells (Kemény-Beke et al., 2006; Kaminskyy et al., 2008a). Sanguinarine induces apoptosis in primary effusion lymphoma cells (Hussain et al., 2007) and apoptosis of HT-29 human colon cancer cells via the regulation of Bax/Bcl-2 ratio and caspase-9-dependent pathway (Lee et al., 2012a). Sanguinarine activates prodeath Bcl-2 family proteins and mitochondrial apoptosis pathway in immortalized human HaCaT keratinocytes (Adhami et al., 2003). Sanguinarine sensitizes human gastric adenocarcinoma AGS cells to TRAIL (tumor necrosis factor-related apoptosis-inducing ligand)-mediated apoptosis via downregulation of AKT and activation of caspase-3 (Choi et al., 2009). Sanguinarine induced apoptosis in human leukemia U937 cells via Bcl-2 downregulation and caspase-3 activation (Han et al., 2008). Combinatory treatment of sanguinarine and TRAIL may overcome the resistance of breast cancer cells due to overexpression of Akt or Bcl-2 (Kim et al., 2008). Rapid cytochrome c release from mitochondria in CEM T-leukemia cells exposed to sanguinarine or chelerythrine was not accompanied by changes in Bax, Bcl-2, and Bcl-X (L/S) proteins in the mitochondrial fraction and preceded activation of the initiator caspase-8 (Kaminskyy et al., 2008a). Mitochondria play a decisive role in defining rate and intensity of apoptosis induction by different alkaloids. Chelerythrine and sanguinarine blocked absorption and accumulation of calcium cations in mitochondria and inhibited oxidative phosphorylation, while the coptisine significantly diminished those indices (Kamins'kyĭ et al., 2006). Chelidonine, colchicines, and colchamine had no influence on the studied characteristics. The effect of alkaloids upon mitochondria functional state correlated tightly with their DNA intercalating properties: Chelerythrine and sanguinarine were strong intercalators, while coptisine was a weak one, and chelidonine, colchicine, and colchamine did not interact with DNA and caused no changes in its melting point. Proapoptotic activity of Ukrain, a semisynthetic thiophosphoric acid derivative of chelidonine, is based on *Chelidonium majus* alkaloids and mediated via a mitochondrial death pathway (Habermehl et al., 2006). Induction of necrosis and apoptosis to KB cancer cells by sanguinarine is associated with reactive oxygen species production and mitochondrial membrane depolarization (Chang et al., 2007). 6-Methoxydihydrosanguinarine from *Hylomecon* induced apoptosis in HepG2 cells, which is mediated by reactive oxygen species (Yin et al., 2005). 6-Methoxydihydrosanguinarine also induced apoptosis in HT29 colon carcinoma cells (Lee et al., 2004).

Capillary electrophoretic study of the synergistic biological effects of alkaloids from *C. majus* showed a differential ability of celandine alkaloids to penetrate into the normal and cancer cell interior (Kulp and Bragina, 2013), which was inversely proportional to their cytotoxic activity. While the most effective transport of celandine alkaloids from the cell medium to the cell interior was observed for normal murine fibroblast NIH/3T3 cells (about 55% of total content), cytotoxicity tests demonstrated selective and profound apoptotic effects of a five-alkaloid combination in the mouse melanoma B16F10 cell line.

Sanguinarine potently suppressed blood vessel formation *in vivo* in mouse Matrigel plugs and the chorioallantoic membrane of chick embryos (Eun and Koh, 2004). Sanguinarine inhibits vascular endothelial growth factor (VEGF)-induced tumor angiogenesis in a fibrin gel matrix, which could be mediated by blocking the VEGF-induced Akt activation (Basini et al., 2007a,b).

Sanguinarine and chelerythrine produce a dose-dependent increase in DNA damage and cytotoxicity in both primary mouse spleen cells and L1210 mouse lymphocytic leukemic cells (Kaminskyy et al., 2008b). Chelidonine did not show a significant cytotoxicity or damage DNA in both cell types, but completely arrested growth of L1210 cells. Examination of nuclear morphology revealed more cells with apoptotic features upon treatment with chelerythrine and sanguinarine, but not chelidonine. Chelidonine and sanguinarine induced apoptosis in human acute T-lymphoblastic leukemia MT-4 cells, which does not correlate with their DNA-damaging effects (Philchenkov et al., 2008). Chelidonine and sanguinarine differed drastically in their cell cycle phase-specific effects, since only the former arrested MT-4 cells in G2/M phase. Glioblastoma is a highly malignant brain tumor with a highly invasive phenotype. Ukrain caused a significant, dose-related decrease of glioblastoma cell proliferation and a tendency to downregulation of cell–matrix interaction regulator SPARC (secreted protein acidic and rich in cysteine) at the protein level (Gagliano et al., 2006). Ukrain also upregulates glial fibrillary acidic protein, which prevents tumor invasion, but not the gap junction protein connexin 43 expression, and induces apoptosis in human cultured glioblastoma cells (Gagliano et al., 2007). Induction of apoptosis by sanguinarine in C6 rat glioblastoma cells is associated with the modulation of the Bcl-2 family and activation of caspases through downregulation of extracellular signal-regulated kinase and Akt (Han et al., 2007). These results suggest the drug may be a useful therapeutic tool for brain tumors.

Five dihydrobenzophenanthridine alkaloids, named as maclekarpines A–E, were isolated from the roots of *Macleaya microcarpa*, together with 10 known benzophenanthridine/dihydrobenzophenanthridine derivatives and a known amide (Deng and Qin, 2010; Table 5.1 and Figure 5.2). Some of these compounds showed cytotoxicity against five human tumor cell lines.

Cancer cells often develop multidrug resistance (MDR), which is a multidimensional problem involving several mechanisms and targets. Chelidonine and an alkaloid extract from *C. majus*, which contains protoberberine and benzo[c] phenanthridine alkaloids, have the ability to overcome MDR of different cancer cell lines through interaction with ABC transporters (El-Readi et al., 2013), CYP3A4 and GST, through induction of apoptosis, and through their cytotoxic effects. Chelidonine and the alkaloid extract inhibited P-gp/MDR1 activity in a concentration-dependent manner in Caco-2 and CEM/ADR5000 and reversed their doxorubicin resistance. Chelidonine and the alkaloid extract inhibited the activity of the drug-modifying enzymes CYP3A4 and GST in a dose-dependent manner. The alkaloids induced apoptosis in MDR cells, which was accompanied by an activation of caspase-3, caspase-8, and caspase-6/caspase-9, and phosphatidylserine (PS) exposure. cDNA arrays were applied to identify differentially expressed genes after treatment with chelidonine and the alkaloid extract. The expression analysis identified a common set of regulated genes related to apoptosis, cell cycle, and drug metabolism. Treatment of Caco-2 cells with 50 μg/ml alkaloid extract and 50 μM chelidonine for up to 48 h resulted in a significant decrease in mRNA levels of P-gp/MDR1, MRP1, BCRP, CYP3A4, GST, and hPXR and in a significant increase in caspase-3 and caspase-8 mRNA. Chelidonine is a promising model compound for overcoming MDR and for enhancing cytotoxicity of chemotherapeutics, especially against leukemia cells. Its efficacy needs to be confirmed in animal models.

5.3.2 Acetylcholinesterase and butyrylcholinesterase inhibitory activity

One Alzheimer's disease treatment strategy to enhance cholinergic function is the use of acetylcholinesterase (AChE, EC 3.1.1.7) inhibitors to increase the amount of acetylcholine present in the synapses between cholinergic neurons (Cho et al., 2006). Chelidonine belonged to reversible AChE inhibitors of a competitive type, all other examined alkaloids are inhibitors of a mixed competitive–noncompetitive type, and a greater contribution to the inhibition was made by the competitive constituent (Kuznetsova et al., 2005). Among all examined agents, berberine, sanguinarine, and "sanguirythrine" were the strongest AChE inhibitors and chelidonine and Ukrain were much weaker. Two *Chelidonium majus* isolation artifacts 6-ethoxydihydrosanguinarine and 6-ethoxydihydrochelerythrine exhibited the highest activity against human blood AChE and human plasma butyrylcholinesterase (HuBuChE) (Cahlíková et al., 2010).

The most active of the naturally occurring alkaloids was chelidonine, which inhibited both AChE and HuBuChE in a dose-dependent manner. Inhibiting BuChE might cause hepatotoxic side effects and thus the selective inhibitors are preferred. Alkaloids 8-hydroxydihydrochelerythrine, 8-hydroxydihydrosanguinarine, and berberine from C. majus showed a potent AChE inhibitory activity (Cho et al., 2006). 8-Hydroxydihydrochelerythrine showed competitive and selective inhibition for AChE.

5.3.3 Monoamine oxidase inhibitory activity

Ukrain and chelidonine are the strongest inhibitors of rat liver mitochondrial monoamine oxidase (MAO) (Iagodina et al., 2003) and strongly inhibit liver MAOs of minks (Iagodina, 2010). The agent Sanguirythrin and alkaloids berberine and sanguinarine produce the weaker MAO inhibitory effect (Kuznetsova et al., 2005). MAO inhibitors are pronounced antidepressants. The combination of malignotoxicity and antidepressive activity in drug Ukrain seems to be favorable for its clinical applications (Iagodina et al., 2003).

5.3.4 Anti-inflammatory, antioxidant, and immunomodulatory activities

Chelidonine and 8-hydroxydihydrosanguinarine, isolated from *Chelidonium majus*, showed strong inhibitory activities toward the LPS-induced NO production in macrophage RAW264.7 cells with IC50 values of 7.3 and 4.5 μM, respectively (Park et al., 2011). Stylopine from *C. majus* inhibits LPS-induced inflammatory mediators in RAW 264.7 cells (Jang et al., 2004). Sanguinarine may, under appropriate conditions, increase the capacity of the enzymatic antioxidant defense system via activation of the p38 MAPK/Nrf2 pathway (Vrba et al., 2012).

Tetrahydrocoptisine, one of the main active components of *C. majus*, protects rats from LPS-induced acute lung injury (Li et al., 2014). 6-Acetonyl-5, 6-dihydrosanguinarine (ADS) from *C. majus* triggers proinflammatory cytokine production via ROS-JNK/ERK-NFκB signaling pathway (Kim et al., 2013).

5.3.5 Antimicrobial activity

The antibacterial activity of chelerythrine on *Streptococcus mutans* was significant, while no inhibition zone in each concentration of chelidonine (Cheng et al., 2006). There is some degree of inhibitory effect of chelerythrine on the cell surface hydrophobicity and adherence of *S. mutans* (Cheng et al., 2007). Chelerythrine possesses powerful anticariogenic potential and can be used for prevention of dental caries. Sanguinarine kills methicillin-resistant *Staphylococcus aureus* (MRSA) by compromising the cytoplasmic membrane (Obiang-Obounou et al., 2011). Sanguinarine (+) binding site on Na^+/K^+-ATPase has been identified (Janovská et al., 2010). Benzophenanthridine-type alkaloids 8-hydroxydihydrosanguinarine and 8-hydroxydihydrochelerythrine were potently active against MRSA strains, which has created nosocomial problem worldwide

(Zuo et al., 2008). 6-Methoxydihydrosanguinarine, 6-acetonylhydrosanguinarine, and dihydrosanguinarine, isolated from *Hylomecon hylomeconoides*, were very active against MRSA strains (Choi et al., 2010). Two potent isoquinoline alkaloids, 8-hydroxydihydrosanguinarine and 8-hydroxydihydrochelerythrine, were identified as the major active principles against extended-spectrum β-lactamase-producing strains (Zuo et al., 2011). Sanguinarine and chelerythrine inhibited the growth of the bacterium Helicobacter pylori that causes gastrointestinal ailments (Mahady et al., 2003). Sanguinarine has also been incorporated into expectorant mixtures and has a strong bactericidal effect upon gram-positive bacteria, particularly *Bacillus anthracis* and staphylococci (García et al., 2006). These medicinal properties are due to the interaction of sanguinarine with DNA. Microarrays were used to analyze the *Mycobacterium tuberculosis* genome-wide transcriptional changes triggered by treatment with subinhibitory concentrations of chelerythrine (Liang et al., 2011). A total of 759 genes were differentially regulated by chelerythrine. Some important genes that were significantly regulated are related to different pathways, such as urease, methoxy-mycolic acid synthase, surface-exposed lipids, heat shock response, and protein synthesis.

Dihydrosanguinarine and dihydrochelerythrine showed inhibitory activity against phytopathogenic fungi (Feng et al., 2011). Sanguinarine and chelerythrine demonstrated a significant antifungal activity against the six test fungi (Liu et al., 2009), and they also demonstrated strong antibacterial activity. Alkaloids 8-hydroxydihydrosanguinarine and 8-hydroxydihydrochelerythrine demonstrated potent activity against resistant clinical yeast isolates (Meng et al., 2009), while dihydrosanguinarine, dihydrochelerythrine, sanguinarine, and chelerythrine had some degree of antifungal activity.

5.3.6 Antiparasitic, insecticidal, and anthelmintic activities

Seven benzophenanthridine compounds were isolated from the methanolic extracts of *Bocconia pearcei* (Fuchino et al., 2010). Among them, dihydrosanguinarine showed the most potent leishmanicidal activities. Chelidonine exhibited significant activity against *Dactylogyrus intermedius* and might be potential sources of new antiparasitic drugs (Yao et al., 2011b). Sanguinarine, chelerythrine, cryptopine, beta-allocryptopine, protopine, and 6-methoxyl-dihydro-chelerythrine might also be potential plant-based medicines for the treatment of *D. intermedius* infection (Wang et al., 2010; Li et al., 2011). Sanguinarine from the leaves of *Macleaya cordata* inhibits *Ichthyophthirius multifiliis* in grass carp (Yao et al., 2010). Dihydrosanguinarine and dihydrochelerythrine from *Macleaya microcarpa* showed antiparasitic efficacy against *I. multifiliis* in richadsin (Yao et al., 2011a,b). Sanguinarine can destroy liver functions of *Oncomelania hupensis* through decreasing glucogen content and changing activities of some important enzymes in snail liver (Sun et al., 2011). Sanguinarine could alter the expression of proteins in livers of Oncomelania snails (Liu et al., 2010).

8-Hydroxydihydrochelerythrine and 8-methoxydihydrosanguinarine, isolated from the seeds of *Macleaya cordata*, decreased the survival of the cotton aphid by $76.1 \pm 7.9\%$ and $73.6 \pm 14.6\%$ at 100 ppm, respectively (Baek et al., 2013).

5.3.7 Effects on nervous system

Sanguinarine is isolated from the roots of *Macleaya cordata* and *M. microcarpa* and possesses several biological activities such as anti-inflammatory and antioxidant effects. Sanguinarine markedly induces the expression of HO (heme oxygenase)-1, which leads to a neuroprotective response in mouse hippocampus-derived neuronal HT22 cells from apoptotic cell death induced by glutamate (Park et al., 2014). Sanguinarine significantly attenuated the loss of mitochondrial function and membrane integrity associated with glutamate-induced neurotoxicity and inhibited HT22 cell apoptosis. JC-1 staining, which is a well-established measure of mitochondrial damage, was decreased after treatment with sanguinarine in glutamate-challenged HT22cells. Sanguinarine diminished the intracellular accumulation of ROS and Ca^{2+}. Sanguinarine induced HO-1 and NQO-1 expression via activation of Nrf2. siRNA-mediated knockdown of Nrf2 or HO-1 significantly inhibited sanguinarine-induced neuroprotective response. These findings revealed the therapeutic potential of sanguinarine in preventing the neurodegenerative diseases.

Glycine transporter inhibitors modulate the transmission of pain signals. Chelerythrine and sanguinarine selectively inhibit the glycine transporter GlyT1 with comparable potency in the low micromolar range, while berberine shows no inhibition at all (Jursky and Baliova, 2011).

5.3.8 Effects on cardiovascular system

Sanguinarine and chelerythrine, isolated from the root of *Bocconia frutescens*, were significant inhibitors of binding of specific ligands to the human angiotensin II AT1 receptor (Caballero-George et al., 2002), which provide some justification for the traditional use of *B. frutescens* L. to control hypertension. Chelidonine and some protoberberine alkaloids exhibited no affinity for the human angiotensin II AT1 and endothelin 1 ET(A) receptors. Sanguinarine inhibits binding of candesartan to the human angiotensin AT1 receptor (Caballero-George et al., 2003). Sanguinarine interacts with the receptor in a slow, nearly irreversible, and noncompetitive manner. Antiplatelet effect of sanguinarine is correlated with calcium mobilization and thromboxane and cAMP production (Jeng et al., 2007).

5.3.9 Other effects

Chelerythrine, sanguinarine, and an alkaloid extract from *Macleaya cordata*—sanguiritrin—were found to be inhibitors of aminopeptidase A and dipeptidyl peptidase IV (Sedo et al., 2002). Reduced glutathione interacts with thiophosphoric derivatives of alkaloids of *Chelidonium majus in vitro* (Glazev and Nefedov, 2009). Similar interaction of thiophosphoric derivatives of alkaloids may involve SH groups of amino acid residues, at the active sites of some metabolic enzymes.

A significant initial increase in uterine activity was visible at each dosage of celandine (Kuenzel et al., 2013). Inhibition of uterine activity was seen over longer periods of 5 and 10 min, particularly for a medium-dose range of 1–2 mg/ml. At a dosage of 2 mg/ml in particular, celandine almost always led to significant values. Following intra-arterial administration in a swine uterus perfusion model, celandine initially causes a significant increase in contractility, which is followed over time by a relaxation phase. *C. majus* might be used to promote targeted sperm transport.

5.4 Pharmacology and therapeutic use of Chelidonieae extracts

5.4.1 Anticancer and immunomodulatory activities

Among 18 plant extracts, *Sanguinaria canadensis* extract ranked second in the tumoricidal potency (Mazzio and Soliman, 2009). Alcohol tinctures and water infusions were generated from bloodroot flowers, leaves, rhizomes, and roots (Senchina et al., 2009a). Infusions demonstrated greater immunomodulatory capabilities than tinctures, and flower- and root-based extracts showed greater immunomodulatory properties than leaf- or rhizome-based extracts. Several extracts were able to augment peripheral blood mononuclear cell (PBMC) proliferation and diminish K562 proliferation, suggesting a selective anticarcinogenic activity. The rhizome alcohol tincture had a markedly stronger effect against K562 cells than other extracts. Bloodroot extracts may have potential as therapeutic immunomodulators. Bloodroot extracts significantly increased cytokine production compared to other stimulants or controls and could offset exercise-associated effects on immune activity (Senchina et al., 2009b). The homeopathic *Chelidonium majus* 30C and 200C exhibited antitumor and antioxidative stress potential against artificially induced hepatic tumors and hepatotoxicity in rats (Banerjee et al., 2010). Antitumor, antigenotoxic, and hepatoprotective effects of the *C. majus* extract were also observed in mice (Biswas et al., 2008). The extract of *C. majus* L. had a strong antioxidant potential and exerted the antiproliferative activity via apoptosis on leukemia cells (Nadova et al., 2008). The isoquinoline alkaloids and the flavonoid components may play an important role in both cancer chemoprevention through its antioxidant activity and modern cancer chemotherapy as cytotoxic and apoptosis-inducing agent.

5.4.2 Antiparasitic and antimicrobial activities

The combination of two homeopathic drugs Chelidonium 30 and Nosode 30 demonstrated considerable *in vivo* antimalarial activity with chemosuppression of 91.45% on day 7 (Bagai and Walter, 2014). The combination also significantly enhanced the mean survival time of mice (22.5 ± 6.31 days vs. 8.55 ± 0.83 days in *Plasmodium berghei*-infected control). The increase in levels of the liver function marker enzymes tested in serum of treated mice were significantly less than those observed in control on day 10. The serum urea and creatinine used for assessment of renal sufficiency were slightly elevated above normal but were statistically significant as compared to infected control.

The extracts of *B. frutescens* showed antimycobacterial activity (Cruz-Vega et al., 2008) and antitrichomonal activity (Calzada et al., 2007).

5.4.3 Anti-inflammatory activity

C. majus methanol extract significantly suppressed the progression of collagen-induced arthritis, and this action was characterized by the decreased production of tumor necrosis factor (TNF)-α, IL-6, IFN-γ, B cells, and T cells (in the spleen) and increased proportion of $CD4^+CD25^+$ regulatory T cells *in vivo* (Lee et al., 2007). Water extract from *C. majus* enhances nitric oxide (NO) and TNF-α production via nuclear factor (NF)-κB activation in mouse peritoneal macrophages (Chung et al., 2004).

MCE (*Macleaya cordata* extract) had a significant linear effect on body weight of chickens on days 21 and 35 (Khadem et al., 2014), and the gain/feed ratio was improved only over the whole period, whereas feed intake was not different. Only MCE but not OTC (oxytetracycline) decreased the percentage of abdominal fat. Plasma α1-AG concentration increased from days 21 to 35, with the values being lower in the treatment groups. Both OTC and MCE significantly reduced the jejunal mucosal expression of inducible NO synthase. For most parameters measured, there was a clear linear dose-response to MCE treatment. The results are consistent with the anti-inflammatory theory of growth promotion in production animals.

Hylomecon hylomeconoides ethanol extract (HHE) inhibited LPS-induced NO and IL-6 production (Chae et al., 2012). HHE suppressed the phosphorylation of ERK1/2 and p38 in LPS-induced RAW 264.7 in a dose-dependent manner. Major constituents of the chloroform-soluble extract are dihydrosanguinarine and 6-methoxydihydrosanguinarine.

5.4.4 Other effects

The extracts of *Bocconia frutescens* induced concentration-dependent contraction of rat aortic rings, suggesting that this plant has potential health benefits for the treatment of ailments such as venous insufficiency (Ibarra-Alvarado et al., 2010). The aqueous extract of *B. frutescens* showed the high antisecretory activity and is used in the treatment of gastrointestinal disorders such as diarrhea (Velázquez et al., 2006). *Chelidonium majus* leaves methanol extract and its chelidonine ingredient reduced cadmium-induced nephrotoxicity in rats (Koriem et al., 2013). An extract from *C. majus* showed radiation protective effect and might be useful in reducing the time needed for reconstitution of hematopoietic cells after irradiation treatment (Song et al., 2003). The dentifrice containing both NaF and *Sanguinaria* showed a significantly greater effect on the remineralization of the enamel lesion than the other test dentifrices (Hong et al., 2005). *C. majus* extract showed inhibitory effects on atopic dermatitis-like skin lesions in NC/Nga mice (Yang et al., 2011).

C. majus is traditionally used for disorders with symptoms like pain, bloating, and abdominal cramp after meals. Suppressed glycine-induced response and elevated glutamate-induced response by *C. majus* may increase neuronal excitability and activation of descending pain control system, and this mechanism can be suggested as one of the analgesic actions of *C. majus* (Shin et al., 2003).

A natural feed additive (*Macleaya cordata*), containing sanguinarine, improved performance and health status of weaning pigs (Kantas et al., 2014). Blood analysis from the Sang groups and especially the Sang 50 (50 g Sangrovit®/t of feed) group revealed low values of haptoglobin and serum amyloid A.

5.5 Toxicity

Sanguinarine is a powerful escharotic contained in the root of *Sanguinaria canadensis*. Herbalists prescribe bloodroot for multiple conditions including skin lesions and sore throats. Lack of regulation of information on the Internet allows alternative therapies to be promoted without full consideration of potential toxicity (Cienki and Zaret, 2010). The use of escharotic agents carries risk of incomplete removal of tumor, damage of surrounding healthy tissues, and marked scarring with poor cosmetic outcome (Moran and Helm, 2008). A possible link between the use of *S. canadensis*-containing products and the preneoplastic lesion, leukoplakia, has been suggested (Vlachojannis et al., 2012). The expression levels of Bcl-2 and Bax proteins in myocardial cells in poisoning groups were much greater than those in the control groups (Zhang et al., 2009). The detection of Bcl-2 and Bax proteins level by immunohistochemistry still could be an ancillary method. Pathological changes induced by *Macleaya cordata* total alkaloids could be found through the apoptosis detection (Zhang et al., 2006). There is striking evidence for *Chelidonium*-induced liver injury with high causality gradings (Teschke et al., 2012). *Chelidonium* hepatotoxicity is caused by an idiosyncratic reaction of the metabolic form, but there is uncertainty with respect to its culprit. The alkaloids from *C. majus* L. that had a significant inhibitory effect in mitochondrial respiration were those that contain a positive charge due to a quaternary nitrogen atom, that is, chelerythrine, sanguinarine, berberine, and coptisine, with malate + glutamate or succinate as substrates (Barreto et al., 2003). In submitochondrial particles, berberine and coptisine had a marked inhibitory effect on NADH dehydrogenase activity but practically no effect on succinate dehydrogenase activity, whereas chelerythrine and sanguinarine inhibited succinate dehydrogenase more strongly than NADH dehydrogenase, which is in agreement with the results found for mitochondrial respiration.

C. majus does not potentiate the hepatic effect of a subtoxic dose of acetaminophen in rats (Mazzanti et al., 2013).

5.6 Other compounds

Nonalkaloids include chelidonic acid (CA), chelidoniol, malate, citric acid, tyramine, choline, saponin, flavonol glycoside (Lee et al., 2012b), terpenoid (Table 5.2), lupenyl acetate, volatile oil, peptides, vitamin, etc. CA, a constituent of *Chelidonium majus* (greater celandine), has many pharmacological effects, including mild analgesic,

Table 5.2 **Other compounds isolated from tribe Chelidonieae**

No.	Compounds	Types	Plants	Tissues
108	Adenosine	I	a	Roots
109	Arnottianamide	I	a	Roots
110	*n*-*p*-Coumaroyltyramine	I	a	Roots
111	*n*-Methyl-4,5-methylene-di-ol-succinimide	I	d	Stems
112	*m*-Hydroxybenzoic acid	II	d	Aerial parts, seeds
113	*p*-Hydroxybenzoic acid	II	d	Aerial parts, seeds
114	Gallic acid	II	d	Aerial parts, seeds
115	Gentisic acid	II	d	Aerial parts, seeds
116	Protocatechuic acid	II	d	Aerial parts, seeds
117	Caffeic acid	II	d	Aerial parts, seeds
			f	Whole plants
118	(+)-(*E*)-Caffeoyl-L-malic acid	II	f	Aerial parts
119	(−)-4-(*E*)-Caffeoyl-L-threonic acid	II	f	Aerial parts
120	(−)-2-(*E*)-Caffeoyl-L-threonic acid lactone	II	f	Aerial parts
121	2-(−)-Caffeoyl-D-glyceric acid	II	f	Aerial parts
122	Ferulic acid	II	d	Aerial parts, seeds
			f	Aerial parts
123	4-*O*-β-D-Glucoside of (*E*)-ferulic acid	II	a	Roots
124	Coumaric acid	II	d	Aerial parts, seeds
			f	Aerial parts
125	Sinapic acid	II	d	Aerial parts, seeds
126	3,4′,6′-Trihydroxy-5-methoxy-2′-methyl-biphenyl-2-carboxylic acid	II	d	Stems
127	3-*O*-Feruloylquinic acid	II	a	Roots
128	Methyl 3-*O*-feruloylquinate	II	a	Roots
129	Chelidonic acid	II	f	Aerial parts
			j	Roots
130	15-Nonacosanol *E. chionantha*	III	i	Aerial parts
131	Citric acid	III	f	Whole plants
132	Malic acid	III	f	Whole plants
133	Succinic acid	III	d	Aerial parts, seeds
134	Butyl stearate	III	i	Aerial parts

Continued

Table 5.2 Continued

No.	Compounds	Types	Plants	Tissues
135	Butyl palmitate	III	i	Aerial parts
136	Monopalmilin	III	a	Roots
137	*n*-Dotriacontanol	III	i	Aerial parts
138	*n*-Nonanoic acid	III	i	Aerial parts
139	Butyric acid	III	d	Seed oil
140	Capronic acid	III	d	Seed oil
141	Caprylic acid	III	d	Seed oil
142	Capric acid	III	d	Seed oil
143	Undecanoic acid	III	d	Seed oil
144	Tridecanoic acid	III	d	Seed oil
145	Myristic acid	III	d	Seed oil
146	*cis*-10-Pentadecanoic acid	III	d	Seed oil
147	Palmitoleic acid	III	d	Seed oil
148	All-*cis*-7,10,13-Hexadecatrienoic acid	III	d	Seed oil
149	Heptadecanoic acid	III	d	Seed oil
150	*cis*-10-Heptadecanoic acid	III	d	Seed oil
151	Stearic acid	III	d	Seed oil
152	Elaidic acid	III	d	Seed oil
153	Oleic acid	III	d	Seed oil
154	*cis*-11-Octadecenoic acid	III	d	Seed oil
155	Linolelaidic acid	III	d	Seed oil
156	Linoleic acid	III	d	Seed oil
157	Linolenic acid	III	d	Seed oil
158	Arachidic acid	III	d	Seed oil
159	Eicosenoic acid	III	d	Seed oil
160	Palmitic acid	III	d	Seed oil
			i	Aerial parts
161	4-Hydroxy-3-methoxy-cinnamaldehyde	IV	d	Stems
162	4-Hydroxy-3-methoxybenzaldehyde	IV	d	Stems
163	3,4,5-Trimethoxy-phenol	IV	d	Stems
164	Dyphylline	IV	j	Roots
165	*n*-Pentadecane	V	i	Aerial parts
166	*n*-Hexadecane	V	i	Aerial parts
167	*n*-Heptadecane	V	i	Aerial parts
168	*n*-Octadecane	V	i	Aerial parts
169	*n*-Nonadecane	V	i	Aerial parts
170	*n*-Eicosane	V	i	Aerial parts
171	*n*-Heneicosane	V	i	Aerial parts
172	*n*-Docosane	V	i	Aerial parts
173	*n*-Tricosane	V	i	Aerial parts
174	*n*-Tetracosane	V	i	Aerial parts

Table 5.2 Continued

No.	Compounds	Types	Plants	Tissues
175	*n*-Pentacosane	V	i	Aerial parts
176	*n*-Hexacosane	V	i	Aerial parts
177	*n*-Heptacosane	V	i	Aerial parts
178	*n*-Octacosane	V	i	Aerial parts
179	*n*-Nonacosane	V	i	Aerial parts
180	*n*-Triacontane	V	i	Aerial parts
181	Z-14-Nonacosene	V	i	Aerial parts
182	3α,22α-Dihydroxy-olean-12(13)-en-30-oic acid	VI	i	Aerial parts
183	3α-Hydroxy-oleanan-12(13)-ene-30-oic acid	VI	d	Stems
184	3β-Hydroxy-oleanan-12(13)-ene-30-oic acid	VI	a	Roots
185	3-Oxoolean-12(13)-en-30-oic acid	VI	a	Roots
186	β-Amyrin acetate	VI	i	Aerial parts
187	Oleanolic acid	VI	a	Stems
188	Lupenyl acetate	VI	i	Whole plants
189	1-Oxohop-2,22(30)-dien-29-oic	VI	a	Roots
190	Dicranostigmone	VI	c	Whole plants
191	β-Sitosterol	VII	d	Stems
			i	Aerial parts
192	β-Daucosterol	VII	a	Roots
			i	Aerial parts
193	Stigmasterol	VII	a	Roots
194	(6*R*,9*R*)-3-Oxo-α-ionyl-9-*O*-α-L-rhamnopyranosyl-(1″→2′)-β-D-glucopyranoside	VIII	p	Aerial parts
195	(6*R*,9*R*)-9-Hydroxymegastigman-4-en-3-one 9-*O*-α-L-Rhamnopyranosyl-(1″→2′)-β-D-glucopyranoside	VIII	p	Aerial parts
196	3-Hydroxy-5,6-epoxy-β-ionol-9-*O*-β-D-glucopyranoside	VIII	p	Aerial parts
197	Megastigmane-7-ene-3,5,6,9-tetraol-9-*O*-β-D-glucopyranoside	VIII	p	Aerial parts
198	Megastigmane-7-en-3,6-epoxy-5,9-diol 9*R*-*O*-β-D-glucopyranoside	VIII	p	Aerial parts
199	Megastigmane-7-en-3,6-epoxy-5,9-diol 9*R*-*O*-β-D-Xylopyranosyl-(1″→6′)-β-D-glucopyranoside	VIII	p	Aerial parts
200	Megastigmane-7-en-3,6-epoxy-5,9-diol 9*R*-*O*-α-L-Arabinopyranosyl-(1″→6′)-β-D-glucopyranoside	VIII	p	Aerial parts

I: Amides; II: aromatic acid and their derivatives; III: aliphatic acids and their derivatives; IV: other aromatics; V: alkanes and alkenes; VI: triterpenoids; VII: sterols; VIII: megastigmane glycosides.
108–193 are from Lei et al. (2014); 194–200 are from Lee et al. (2011).

anti-inflammatory, and antimicrobial (Kim et al., 2012). CA showed inhibitory effects on interleukin-6 production by blocking NF-κB and caspase-1 in HMC-1 cells (Shin et al., 2011). Milky sap isolated from *C. majus* L. serves as a rich source of various biologically active substances such as proteins, enzymes, alkaloids, flavonoids, and phenolic acids.

Fifteen phenolic compounds were identified from the aerial parts of *C. majus* (Grosso et al., 2014), including nine new flavonoids and three new hydroxycinnamic acids. Only phenolic compounds were quantified by a validated HPLC-DAD method, the pair quercetin-3-*O*-rutinoside + quercetin-3-*O*-glucoside dominating all the 29 extracts. The phenolic profile could be used in the quality control of *C. majus* samples.

Phenolic acids, essential oils (2-methoxy-4-vinylphenol), polysaccharides, flavonoids, and steroids were found from *M. cordata*, and a triterpene (3α-hydroxyolean-12-en-30-oic acid) was present in *B. arborea* (Yu et al., 2014).

Nucleases CMN1 and CMN2 isolated from *C. majus* milky sap exhibit apoptotic activity in HeLa tumor cell line, but not in CHO cells, without inflammatory reaction (Nawrot et al., 2008). A novel extracellular peroxidase/DNase from *C. majus* combined with other proteins is probably involved in the development and differentiation of the plant and defense against different pathogens (Nawrot et al., 2007a). The biological activity of *C. majus* whole plants and extracts may depend not only on its alkaloidal content but also on the presence of biologically active proteins. Proteomic analysis of *C. majus* milky sap using two-dimensional gel electrophoresis and tandem mass spectrometry (MS) identified 21 proteins, including disease-/defense-related, signaling, Krebs cycle, nucleic acid binding, and other proteins (Nawrot et al., 2007b). The majority of the identified proteins can be linked to direct and indirect stress and defense reactions, for example, against different pathogens. A low-sulfated poly-glycosaminoglycan moiety of 3800 Da, isolated from the aqueous extract of *C. majus*, showed antiretroviral activity (Gerencer et al., 2006). The protein-bound polysaccharide extracted from *C. majus* showed immunomodulatory and antitumor activity (Song et al., 2002).

5.7 Chemotaxonomy

A chemotaxonomic study of Chelidonieae suggests that this tribe is linked to the other tribes of Papaveraceae in the presence of protopine-type alkaloids, and the ubiquitous occurrence of quaternary benzophenanthridine-type alkaloids (e.g., chelerythrine and sanguinarine) is a chemical character of Chelidonieae (Feng et al., 1985; Figures 5.1 and 5.2). According to the thin-layer chromatography (TLC) results, Chelidonieae can be divided into two groups. One group, including Sanguinaria, Eomecon, Macleaya, and Bocconia, is characterized by the absence of TLC-detected aporphine-, tetrahydroberberine-, and reduced benzophenanthridine-type alkaloids; the other group, including Stylophorum, Hylomecon, Dicranostigma, and Chelidonium, is characterized by the abundant tetrahydroberberine- and reduced benzophenanthridine-type alkaloids. Interestingly, plants of the former group have undulatedly or palmately incised leaves, as well as short capsules, whereas those of the latter group have deeply pinnatifid leaves and long and slender capsules. In Chelidonieae, only Dicranostigma plants contain abundant aporphine-type alkaloids, such as isocorydine, corydine, isocorypalmine, and magnoflorine (Gong et al., 2010), while these are also abundantly present in the neighboring tribe Papavereae.

Recently, ultraperformance liquid chromatography (UPLC), which employs sub-2 mm stationary phase particles to achieve superior theoretical plates and extremely high resolution in very short analytic times, has attracted the wide attention of pharmaceutical and biochemical analysts (Hao et al., 2012b). UPLC and its hyphenated system showed excellent sensitivity and improved peak capability. These advantages make it possible to rapidly detect trace ingredients from crude plant samples. UPLC has been used to study the chemical profiles of *Fritillaria* (Hao et al., 2013), *Taxus* (Hao et al., 2012b), and *Ilex* (Li et al., 2012) and provides abundant information for chemotaxonomy. UPLC has been used to quantify eight alkaloids in the roots, stems, leaves, and fruits of Macleaya cordata (Zhong et al., 2011). Chelidonieae should be analyzed with the state-of-the-art metabolomic techniques to obtain more objective and accurate chemotaxonomic results. The distribution of secondary metabolites and other chemical entities apparently is useful for taxonomy, but it has to be analyzed carefully, as any other adaptive trait. Molecular methods have to be combined with morphology and chemotaxonomy in the elucidation of the pharmacophylogeny of Chelidonieae.

5.8 Molecular phylogeny and omics

5.8.1 Phylogeny

The chloroplast atpB and rbcL sequences, trnK restriction sites, and morphological characters were combined to elucidate Papaveraceae phylogeny (Hoot et al., 1997). *Sanguinaria* is basal to other Chelidonieae genera; *Macleaya* is basal to *Stylophorum*, while *Dicranostigma* is closer to *Glaucium* than to *Stylophorum*. Only four genera of Chelidonieae were included in this analysis and no nuclear markers were used. The chloroplasts rbcL, matK, and trnL-F and nuclear ribosomal 26S rDNA were used to reconstruct phylogenetic relationships within the order Ranunculales (Wang et al., 2009). The sister relationships between *Eomecon* and *Sanguinaria* and between *Hylomecon* and *Macleaya* were revealed. These four genera formed a cluster. However, other four Chelidonieae genera were not studied and 26S rDNA marker is of low resolving power.

In the present study, we retrieve rbcL, matK, trnL-F, nuclear 18S rRNA, and ITS sequences of Chelidonieae and reconstruct phylogenetic trees based on the respective markers and the combined dataset. Only rbcL and ITS sequences are available for all genera. On the rbcL+ITS tree (Figure 5.3), there are two main clades: In one clade, *Chelidonium* is closer to *Stylophorum*, and *Hylomecon* is basal to them; in another clade, *Dicranostigma* is basal to the cluster of four genera, and the sister relationships between *Eomecon* and *Sanguinaria* and between *Bocconia* and *Macleaya* were

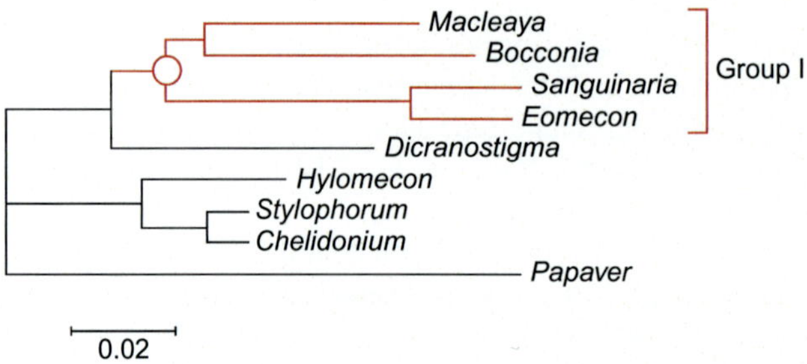

Figure 5.3 Phylogenetic relationship of Chelidonieae inferred from nuclear ITS and chloroplast rbcL sequences. The evolutionary history was inferred by using the maximum likelihood method based on the GTR+G model. The tree with the highest log likelihood (−6202.0940) is shown. Initial trees for the heuristic search were obtained automatically as follows: When the number of common sites was <100 or less than one fourth of the total number of sites, the maximum parsimony method was used; otherwise, BIONJ method with MCL distance matrix was used. A discrete γ distribution was used to model evolutionary rate differences among sites (five categories, parameter = 0.1613). The tree is drawn to scale, with branch lengths measured in the number of substitutions per site. The bootstrap support values are on the branch. The analysis involved 18 nucleotide sequences (ITS+rbcL) retrieved from NCBI GenBank, which were used in the previous phylogenetic studies (Hoot et al., 1997; Wang et al., 2009; Lee et al., 2010). There were a total of 2213 positions in the final dataset. Evolutionary analyses were conducted in MEGA5 (Tamura et al., 2011).

revealed. This classification does not contradict with the chemotaxonomic results (Feng et al., 1985). Phylogenetic relationship revealed by other molecular markers agrees with this inference (data not shown). The chloroplast and nuclear DNA markers can be used in the identification of illicit plants such as *Papaver somniferum* L. (opium poppy; Lee et al., 2010), as well as Chelidonieae plants with medicinal use. In addition, RAPD (random amplification of polymorphic DNA) markers have been used to study genetic diversity of *Eomecon chionantha* (Hu et al., 2011). In the future, these molecular markers should be used to elucidate the intrageneric relationship of *Stylophorum*, *Dicranostigma*, and *Bocconia*.

The pharmacologically important benzylisoquinoline alkaloids have a scattered distribution among angiosperm families. Within the order Ranunculales, benzylisoquinoline alkaloids are distributed in Papaveraceae, Menispermaceae, Berberidaceae, and Ranunculaceae (Zhu and Xiao, 1991a,b; Peng et al., 2006), suggesting that the biosynthetic pathway evolved from their common ancestor. No/trace amount benzylisoquinoline alkaloids were detected in the families Lardizabalaceae and Circaeasteraceae and the genus *Glaucidium* of Ranunculaceae (Wang, 2007). The relevant biosynthetic pathway might be lost in these groups; alternatively, the gene transcriptional regulation mechanisms might be distinct in these plants. On the other hand, abundant benzylisoquinoline alkaloids were also found in Annonaceae (Magnoliales), Lauraceae (Laurales), and Colchicaceae (Liliales) (Zhang et al., 2011). Whether benzylisoquinoline alkaloid biosynthesis evolved independently in these taxonomically distant families is worthy of further investigations. Many other types of plant secondary metabolites have scattered distributions in the plant kingdom, and an independent origin of biosynthesis in different lineages is not rare. Plant tropane alkaloid biosynthesis evolved independently in the Solanaceae and Erythroxylaceae (Jirschitzka et al., 2012). The pyrrolizidine alkaloids are found in Orchidaceae, Fabaceae, Boraginaceae, and Asteraceae. The analysis of the first biosynthetic step, homospermidine synthase, indicated that this enzyme has been recruited from a primary metabolic enzyme, deoxyhypusine synthase, at least four times over the course of evolution (Reimann et al., 2004). The cardenolides, a type of glycosylated steroid found in the Liliaceae, Ranunculaceae, Brassicaceae, and Apocynaceae, have also been proposed to have originated polyphyletically (Wink, 2003). Therefore, the scattered distribution in chemotaxonomy could be partially explained by the continual gain (and loss) of biosynthetic pathways throughout evolution. As in *Fritillaria* (Hao et al., 2013), convergent evolution, transcriptional regulation, endophytic fungi (Wang et al., 2014), and/or rhizosphere microbes might also contribute to the chemical profiles of various Chelidonieae species.

5.8.2 Transcriptomics and metabolomics

Macleaya cordata and *M. microcarpa* are traditional antiviral, anti-inflammatory, and insecticidal herb medicines for their isoquinoline alkaloids. They are also the basis of the popular natural animal food addictive in Europe. *Macleaya* spp. transcriptome was sequenced and integrated with iTRAQ proteome analysis to identify potential genes involved in alkaloids biosynthesis (Zeng et al., 2013). Ten samples of both species were investigated. 69,367 and 78,255 unigenes were obtained for *M. cordata* and *M. microcarpa*, respectively, in which about two-thirds of them were similar to sequences in public databases. By metabolism profiling, reverse patterns for alkaloids sanguinarine, chelerythrine, protopine, and allocryptopine were observed in different organs of two species. More than 1000 proteins were identified from iTRAQ proteome data. The root might be the organ for major alkaloids biosynthesis. The alkaloids' storage and transport were also important for their accumulation. The ultrastructure of laticifers by SEM helps to prove the alkaloids could accumulate in the mature roots.

5.8.3 Proteomics

C. majus and *Corydalis cava* are phylogenetically closely related (Papaveraceae family). The medicinal and pharmaceutical interest in these plants is based on their synthesis of pharmaceutically important compounds, such as alkaloids, flavonoids, phenolic acids, and proteins. *C. majus* shoot and *C. cava* tuber extracts have been used in traditional folk medicine to treat many diseases, such as fungal, bacterial, and viral infections; liver disorders; fever; posttraumatic, colic, abdominal, and menstrual pains; and even cancer. A global comparative proteomic analysis of pharmacologically important extracts from these two closely related unsequenced plant species was performed to gain insights into the protein basis of these plant organs and to compare their common and specific proteomic compositions (Nawrot et al., 2014). A shotgun proteomic approach was adopted, combined with label-free protein quantitation according to the exponentially modified protein abundance index (emPAI). A mean number of 228 protein identification results were recorded in *C. cava* tuber extracts and about 1240 in *C. majus* shoot extracts. Comparative analysis revealed a similar stress- and defense-related protein composition of pharmacologically active plant species and showed the presence of different pathogenesis-related and low-molecular-weight inducible antimicrobial peptides. These findings could be the basis for further elucidation of the mechanism of the strong pharmacological activities of these medicinal plant extracts.

5.9 Conclusions

Chelidonieae plants contain many pharmaceutically active ingredients, which have been commonly used in traditional Chinese medicine for many centuries. Increasing interest in Chelidonieae pharmaceutical resources has led to additional discoveries of alkaloids, terpenoids, flavonoids, and many other compounds in various Chelidonieae species and to studies on their chemotaxonomy, molecular phylogeny, and pharmacology. However, only *Chelidonium* and *Sanguinaria* have been explored in phytochemistry and pharmacology frequently, while few studies focus on other genera, which hamper the quality control and therapeutic use of Chelidonieae-related products. The potential for finding new medicinal lead compounds for drug development is huge. Chemotaxonomy tries to classify and identify plants by spotting differences and similarities in their biochemical compositions, which is of great help in pharmaceutical resource discovery. However, the results of chemotaxonomy have to be compared with those of molecular phylogeny and traditional morphology-based classification, in order to rationalize the quality control and guarantee the authenticity of plant materials used in the clinical setting, research laboratory, and pharmaceutical industry. Molecular phylogeny studies help researchers gain deeper insights into the physiological and biochemical mechanisms of Chelidonieae species and are essential to the conservation and sustainable utilization of Chelidonieae resources. Moreover, chemical and biological research of Chelidonieae pharmaceutical resources should not be limited to the above aspects, for example, the biosynthetic pathway of secondary metabolites has not been studied; the regulation of Chelidonieae biological processes at genomic level, epigenomic level, transcriptional and posttranscriptional levels, and translational and posttranslational levels is completely unknown, although it is essential for the sustainable development and utilization of the Chelidonieae resources. The biological and chemical studies of *Papaver* (Desgagné-Penix et al., 2010), *Taxus* (Hao et al., 2011) and *Polygonum* (Hao et al., 2012a) with the use of omics technologies provide a paradigm of the active integration of various state-of-the-art methodologies into the early stage of drug research and development. Systems biology and omics techniques will play a determinant role in future Chelidonieae pharmaceutical studies.

References

Adhami, V.M., Aziz, M.H., Mukhtar, H., Ahmad, N., 2003. Activation of prodeath Bcl-2 family proteins and mitochondrial apoptosis pathway by sanguinarine in immortalized human HaCaT keratinocytes. Clin. Cancer Res. 9, 3176–3182.

Adhami, V.M., Aziz, M.H., Reagan-Shaw, S.R., et al., 2004. Sanguinarine causes cell cycle blockade and apoptosis of human prostate carcinoma cells via modulation of cyclin kinase inhibitor-cyclin-cyclin-dependent kinase machinery. Mol. Cancer Ther. 3, 933–940.

Alcantara, J., Bird, D.A., Franceschi, V.R., Facchini, P.J., 2005. Sanguinarine biosynthesis is associated with the endoplasmic reticulum in cultured opium poppy cells after elicitor treatment. Plant Physiol. 138, 173–183.

Baek, M.Y., Park, H.J., Kim, G.M., et al., 2013. Insecticidal alkaloids from the seeds of *Macleaya cordata* on cotton aphid (*Aphis gossypii*). J. Korean Soc. Appl. Biol. Chem. 56 (2), 135–140.

Bagai, U., Walter, N.S., 2014. Antiplasmodial potential of homeopathic drugs *Chelidonium* and nosode against *Plasmodium berghei* infection. J. Complement. Integr. Med. 11 (3), 195–201.

Banerjee, A., Pathak, S., Biswas, S.J., et al., 2010. *Chelidonium majus* 30C and 200C in induced hepato-toxicity in rats. Homeopathy 99, 167–176.

Barreto, M.C., Pinto, R.E., Arrabaça, J.D., Pavão, M.L., 2003. Inhibition of mouse liver respiration by *Chelidonium majus* isoquinoline alkaloids. Toxicol. Lett. 146, 37–47.

Basini, G., Bussolati, S., Santini, S.E., Grasselli, F., 2007a. Sanguinarine inhibits VEGF-induced angiogenesis in a fibrin gel matrix. Biofactors 29, 11–18.

Basini, G., Santini, S.E., Bussolati, S., Grasselli, F., 2007b. Sanguinarine inhibits VEGF-induced Akt phosphorylation. Ann. N. Y. Acad. Sci. 1095, 371–376.

Bessi, I., Bazzicalupi, C., Richter, C., et al., 2012. Spectroscopic, molecular modeling, and NMR-spectroscopic investigation of the binding mode of the natural alkaloids berberine and sanguinarine to human telomeric G-quadruplex DNA. ACS Chem. Biol. 7, 1109–1119.

Biswas, S.J., Bhattacharjee, N., Khuda-Bukhsh, A.R., 2008. Efficacy of a plant extract (*Chelidonium majus* L.) in combating induced hepatocarcinogenesis in mice. Food Chem. Toxicol. 46, 1474–1487.

Caballero-George, C., Vanderheyden, P.M., Apers, S., et al., 2002. Inhibitory activity on binding of specific ligands to the human angiotensin II AT(1) and endothelin 1 ET(A) receptors: bioactive benzo[c]phenanthridine alkaloids from the root of *Bocconia frutescens*. Planta Med. 68, 770–775.

Caballero-George, C., Vanderheyden, P.M., Solis, P.N., et al., 2003. In vitro effect of sanguinarine alkaloid on binding of [3H]candesartan to the human angiotensin AT1 receptor. Eur. J. Pharmacol. 458, 257–262.

Cahlíková, L., Opletal, L., Kurfürst, M., et al., 2010. Acetylcholinesterase and butyrylcholinesterase inhibitory compounds from *Chelidonium majus* (Papaveraceae). Nat. Prod. Commun. 5, 1751–1754.

Calzada, F., Yépez-Mulia, L., Tapia-Contreras, A., 2007. Effect of Mexican medicinal plant used to treat trichomoniasis on *Trichomonas vaginalis* trophozoites. J. Ethnopharmacol. 113, 248–251.

Chae, H.S., Kang, O.H., Keum, J.H., et al., 2012. Anti-inflammatory effects of *Hylomecon hylomeconoides* in RAW 264.7 cells. Eur. Rev. Med. Pharmacol. Sci. 16 (S3), 121–125.

Chang, M.C., Chan, C.P., Wang, Y.J., et al., 2007. Induction of necrosis and apoptosis to KB cancer cells by sanguinarine is associated with reactive oxygen species production and mitochondrial membrane depolarization. Toxicol. Appl. Pharmacol. 218, 143–151.

Chen, Y., Li, M., Liu, J., et al., 2014. Simultaneous determination of the content of isoquinoline alkaloids in *Dicranostigma leptopodum* (Maxim) Fedde and the effective fractionation of the alkaloids by high-performance liquid chromatography with diode array detection. J. Sep. Sci. 38 (1), 9–17. http://dx.doi.org/10.1002/jssc.201400905.

Cheng, R.B., Chen, X., Liu, S.J., et al., 2006. Experimental study of the inhibitory effects of *Chelidonium majus* L. extractive on *Streptococcus mutans* in vitro. Shanghai J. Stomatol. 15, 318–320.

Cheng, R.B., Chen, X., Liu, S.J., Zhang, X.F., 2007. Effect of Chelerythrine on cell surface hydrophobicity and adherence of *Streptococcus mutans*. Shanghai J. Stomatol. 16, 68–72.

Cho, K.M., Yoo, I.D., Kim, W.G., 2006. 8-Hydroxydihydrochelerythrine and 8-hydroxydihydrosanguinarine with a potent acetylcholinesterase inhibitory activity from *Chelidonium majus* L. Biol. Pharm. Bull. 29, 2317–2320.

Choi, W.Y., Jin, C.Y., Han, M.H., et al., 2009. Sanguinarine sensitizes human gastric adenocarcinoma AGS cells to TRAIL-mediated apoptosis via down-regulation of AKT and activation of caspase-3. Anticancer Res. 29, 4457–4465.

Choi, J.G., Kang, O.H., Chae, H.S., et al., 2010. Antibacterial activity of *Hylomecon hylomeconoides* against methicillin-resistant *Staphylococcus aureus*. Appl. Biochem. Biotechnol. 160, 2467–2474.

Chung, H.S., An, H.J., Jeong, H.J., et al., 2004. Water extract isolated from *Chelidonium majus* enhances nitric oxide and tumour necrosis factor-alpha production via nuclear factor-kappaB activation in mouse peritoneal macrophages. J. Pharm. Pharmacol. 56, 129–134.

Cienki, J.J., Zaret, L., 2010. An Internet misadventure: bloodroot salve toxicity. J. Altern. Complement. Med. 16, 1125–1127.

Cruz-Vega, D.E., Verde-Star, M.J., Salinas-González, N., et al., 2008. Antimycobacterial activity of *Juglans regia*, *Juglans mollis*, *Carya illinoensis* and *Bocconia frutescens*. Phytother. Res. 22, 557–559.

Deng, A.J., Qin, H.L., 2010. Cytotoxic dihydrobenzophenanthridine alkaloids from the roots of *Macleaya microcarpa*. Phytochemistry 71, 816–822.

Desgagné-Penix, I., Khan, M.F., Schriemer, D.C., et al., 2010. Integration of deep transcriptome and proteome analyses reveals the components of alkaloid metabolism in opium poppy cell cultures. BMC Plant Biol. 10, 252.

El-Readi, M.Z., Eid, S., Ashour, M.L., et al., 2013. Modulation of multidrug resistance in cancer cells by chelidonine and *Chelidonium majus* alkaloids. Phytomedicine 20 (3–4), 282–294.

Eun, J.P., Koh, G.Y., 2004. Suppression of angiogenesis by the plant alkaloid, sanguinarine. Biochem. Biophys. Res. Commun. 317, 618–624.

Feng, R.Z., Lian, W.Y., Fu, G.X., Xiao, P.G., 1985. Chemotaxonomy and resource utilization of the tribe Chelidonieae (Papaveraceae). Acta Phytotax. Sin. 23, 36–42.

Feng, G., Zhang, J., Liu, Y.Q., 2011. Inhibitory activity of dihydrosanguinarine and dihydrochelerythrine against phytopathogenic fungi. Nat. Prod. Res. 25, 1082–1089.

Fil'chenkov, O.O., Zavelevych, M.P., Khranovs'ka, N.M., et al., 2006. Modified alkaloids from *Chelidonium majus* L. induce G2/M arrest, caspase-3 activation, and apoptosis in human acute lymphoblastic leukemia MT-4 cells. Ukr. Biokhim. Zh. 78, 81–87.

Fleming, M.J., McManus, H.A., Rudolph, A., et al., 2008. Concise enantioselective total syntheses of (+)-homochelidonine, (+)-chelamidine, (+)-chelidonine, (+)-chelamine and (+)-norchelidonine by a Pd II-catalyzed ring-opening strategy. Chemistry 14, 2112–2124.

Fuchino, H., Kawano, M., Mori-Yasumoto, K., et al., 2010. In vitro leishmanicidal activity of benzophenanthridine alkaloids from *Bocconia pearcei* and related compounds. Chem. Pharm. Bull. 58, 1047–1050.

Gagliano, N., Moscheni, C., Torri, C., et al., 2006. Effect of Ukrain on matrix metalloproteinase-2 and Secreted Protein Acidic and Rich in Cysteine (SPARC) expression in human glioblastoma cells. Anticancer Drugs 17, 189–194.

Gagliano, N., Moscheni, C., Torri, C., et al., 2007. Ukrain modulates glial fibrillary acidic protein, but not connexin 43 expression, and induces apoptosis in human cultured glioblastoma cells. Anticancer Drugs 18, 669–676.

García, V.P., Valdés, F., Martín, R., Luis, J.C., Afonso, A.M., Ayala, J.H., 2006. Biosynthesis of antitumoral and bactericidal sanguinarine. J. Biomed. Biotechnol. 2006 (2), 63518.

Gerencer, M., Turecek, P.L., Kistner, O., et al., 2006. In vitro and in vivo anti-retroviral activity of the substance purified from the aqueous extract of *Chelidonium majus* L. Antivir. Res. 72, 153–156.

Gilca, M., Gaman, L., Panait, E., et al., 2010. *Chelidonium majus*—an integrative review: traditional knowledge versus modern findings. Forsch. Komplementmed. 17, 241–248.

Glazev, A.A., Nefedov, L.I., 2009. Reduced glutathione interaction with thiophosphoric derivatives of alkaloids of *Chelidonium majus* L. in vitro. Biomed. Khim. 55, 114–120.

Gong, J., Luo, R.F., Ni, S.F., et al., 2010. Pharmaceutical studies of endangered traditional Chinese medicine *Dicranostigma*. Northwest Pharm. J. 25, 153–155.

Grosso, C., Ferreres, F., Gil-Izquierdo, A., et al., 2014. Box-Behnken factorial design to obtain a phenolic-rich extract from the aerial parts of *Chelidonium majus* L. Talanta 130, 128–136.

Habermehl, D., Kammerer, B., Handrick, R., et al., 2006. Proapoptotic activity of Ukrain is based on *Chelidonium majus* L. alkaloids and mediated via a mitochondrial death pathway. BMC Cancer 6, 14.

Hammerová, J., Uldrijan, S., Táborská, E., Slaninová, I., 2011. Benzo[c]phenanthridine alkaloids exhibit strong anti-proliferative activity in malignant melanoma cells regardless of their p53 status. J. Dermatol. Sci. 62, 22–35.

Han, M.H., Kim, S.O., Kim, G.Y., et al., 2007. Induction of apoptosis by sanguinarine in C6 rat glioblastoma cells is associated with the modulation of the Bcl-2 family and activation of caspases through downregulation of extracellular signal-regulated kinase and Akt. Anticancer Drugs 18, 913–921.

Han, M.H., Yoo, Y.H., Choi, Y.H., 2008. Sanguinarine-induced apoptosis in human leukemia U937 cells via Bcl-2 downregulation and caspase-3 activation. Chemotherapy 54, 157–165.

Hao, D.C., Ge, G.B., Xiao, P.G., et al., 2011. The first insight into the tissue specific taxus transcriptome via Illumina second generation sequencing. PLoS One 6 (6), e21220.

Hao, D.C., Xiao, P.G., Ge, G.B., Liu, M., 2012a. Biological, chemical and omics research of *Taxus* medicinal resources. Drug Dev. Res. 73 (8), 477–486.

Hao, D.C., Ma, P., Mu, J., Chen, S.L., Xiao, P.G., et al., 2012b. De novo characterization of the root transcriptome of a traditional Chinese medicinal plant *Polygonum cuspidatum*. Sci. China Life Sci. 55, 452–466.

Hao, D.C., Gu, X.J., Xiao, P.G., Peng, Y., 2013. Phytochemical and biological research of *Fritillaria* medicinal resources. Chin. J. Nat. Med. 11 (4), 330–344.

Hong, S.J., Jeong, S.S., Song, K.B., 2005. Effects of sanguinaria in fluoride-containing dentifrices on the remineralisation of subsurface carious lesion in vitro. Int. Dent. J. 55, 128–132.

Hoot, S.B., Kadereit, J.W., Blattner, F.R., et al., 1997. Data congruence and phylogeny of the Papaveraceae s.l. based on four data sets: atpB and rbcL sequences, trnK restriction sites, and morphological characters. Syst. Bot. 22, 575–590.

Hu, X.H., Chen, X., Xiao, Y.A., et al., 2011. Optimization of RAPD reaction system for *Eomecon chionantha* Hance, an endemic herbaceous of China. Bull. Bot. Res. 31, 89–94.

Huang, S., Du, F.L., 2002. Research progress on Papaveraceae Chelidonieae's chemical component and phytochemistry classification basis. Hunan Guiding J. TCMP 8, 582–584.

Hussain, A.R., Al-Jomah, N.A., Siraj, A.K., et al., 2007. Sanguinarine-dependent induction of apoptosis in primary effusion lymphoma cells. Cancer Res. 67, 3888–3897.

Iagodina, O.V., 2010. Comparative substrate-inhibitor analysis of liver monoamine oxidases of minks. Zh. Evol. Biokhim. Fiziol. 46, 380–386.

Iagodina, O.V., Nikol'skaia, E.B., Faddeeva, M.D., 2003. Inhibition of liver mitochondrial monoamine oxidase activity by alkaloids isolated from *Chelidonium* and *Macleaya* and by their derivative drugs. Tsitologiia 45, 1032–1037.

Ibarra-Alvarado, C., Rojas, A., Mendoza, S., et al., 2010. Vasoactive and antioxidant activities of plants used in Mexican traditional medicine for the treatment of cardiovascular diseases. Pharm. Biol. 48, 732–739.

Isolani, M.E., Pietra, D., Balestrini, L., et al., 2012. The in vivo effect of chelidonine on the stem cell system of planarians. Eur. J. Pharmacol. 686, 1–7.

Jang, S.I., Kim, B.H., Lee, W.Y., et al., 2004. Stylopine from *Chelidonium majus* inhibits LPS-induced inflammatory mediators in RAW 264.7 cells. Arch. Pharm. Res. 27, 923–929.

Janovská, M., Kubala, M., Simánek, V., Ulrichová, J., 2010. Fluorescence of sanguinarine: spectral changes on interaction with amino acids. Phys. Chem. Chem. Phys. 12, 11335–11341.

Jeng, J.H., Wu, H.L., Lin, B.R., et al., 2007. Antiplatelet effect of sanguinarine is correlated to calcium mobilization, thromboxane and cAMP production. Atherosclerosis 191, 250–258.

Ji, X., Sun, H., Zhou, H., et al., 2012. The interaction of telomeric DNA and C-myc22 G-quadruplex with 11 natural alkaloids. Nucleic Acid Ther. 22, 127–136.

Jirschitzka, J., Schmidt, G.W., Reichelt, M., et al., 2012. Plant tropane alkaloid biosynthesis evolved independently in the Solanaceae and Erythroxylaceae. Proc. Natl. Acad. Sci. U. S. A. 109, 10304–10309.

Jursky, F., Baliova, M., 2011. Differential effect of the benzophenanthridine alkaloids sanguinarine and chelerythrine on glycine transporters. Neurochem. Int. 58, 641–647.

Kamins'kyĭ, V.O., Kryv'iak, N.V., Lutsyk, M.D., Stoĭka, R.S., 2006. Effect of alkaloids from celandine on calcium accumulation and oxidative phosphorylation in mitochondria depending on their DNA intercalating properties. Ukr. Biokhim. Zh. 78, 73–78.

Kaminskyy, V., Kulachkovskyy, O., Stoika, R., 2008a. A decisive role of mitochondria in defining rate and intensity of apoptosis induction by different alkaloids. Toxicol. Lett. 177, 168–181.

Kaminskyy, V., Lin, K.W., Filyak, Y., Stoika, R., 2008b. Differential effect of sanguinarine, chelerythrine and chelidonine on DNA damage and cell viability in primary mouse spleen cells and mouse leukemic cells. Cell Biol. Int. 32, 271–277.

Kantas, D., Papatsiros, V.G., Tassis, P.D., et al., 2014. The effect of a natural feed additive (*Macleaya cordata*), containing sanguinarine, on the performance and health status of weaning pigs. Anim. Sci. J. 86, 92–98. http://dx.doi.org/10.1111/asj.12240.

Kemény-Beke, A., Aradi, J., Damjanovich, J., et al., 2006. Apoptotic response of uveal melanoma cells upon treatment with chelidonine, sanguinarine and chelerythrine. Cancer Lett. 237, 67–75.

Khadem, A., Soler, L., Everaert, N., Niewold, T.A., 2014. Growth promotion in broilers by both oxytetracycline and *Macleaya cordata* extract is based on their anti-inflammatory properties. Br. J. Nutr. 112 (7), 1110–1118.

Kim, S., Lee, T.J., Leem, J., et al., 2008. Sanguinarine-induced apoptosis: generation of ROS, down-regulation of Bcl-2, c-FLIP, and synergy with TRAIL. J. Cell. Biochem. 104, 895–907.

Kim, D.S., Kim, S.J., Kim, M.C., et al., 2012. The therapeutic effect of chelidonic acid on ulcerative colitis. Biol. Pharm. Bull. 35, 666–671.

Kim, D.H., Lee, J.H., Park, S., et al., 2013. 6-Acetonyl-5,6-dihydrosanguinarine (ADS) from *Chelidonium majus* L. triggers proinflammatory cytokine production via ROS-JNK/ERK-NFκB signaling pathway. Food Chem. Toxicol. 58, 273–279.

Koriem, K.M., Arbid, M.S., Asaad, G.F., 2013. *Chelidonium majus* leaves methanol extract and its chelidonine alkaloid ingredient reduce cadmium-induced nephrotoxicity in rats. J. Nat. Med. 67 (1), 159–167.

Kuenzel, J., Geisler, K., Strahl, O., et al., 2013. *Chelidonium majus* and its effects on uterine contractility in a perfusion model. Eur. J. Obstet. Gynecol. Reprod. Biol. 169 (2), 213–217.

Kulp, M., Bragina, O., 2013. Capillary electrophoretic study of the synergistic biological effects of alkaloids from *Chelidonium majus* L. in normal and cancer cells. Anal. Bioanal. Chem. 405 (10), 3391–3397.

Kuznetsova, L.P., Sochilina, E.E., Faddeeva, M.D., Iagodina, O.V., 2005. Effect of some isoquinoline alkaloids on enzymatic activity of acetylcholinesterase and monoamine oxidase. Ukr. Biokhim. Zh. 77, 147–153.

Lee, Y.J., Yin, H.Q., Kim, Y.H., et al., 2004. Apoptosis inducing effects of 6-methoxydihydrosanguinarine in HT29 colon carcinoma cells. Arch. Pharm. Res. 27, 1253–1257.

Lee, Y.C., Kim, S.H., Roh, S.S., et al., 2007. Suppressive effects of *Chelidonium majus* methanol extract in knee joint, regional lymph nodes, and spleen on collagen-induced arthritis in mice. J. Ethnopharmacol. 112, 40–48.

Lee, E.J., Hwang, I.K., Kim, N.Y., et al., 2010. An assessment of the utility of universal and specific genetic markers for opium poppy identification. J. Forensic Sci. 55, 1202–1208.

Lee, S.Y., Choi, S.U., Lee, K.R., 2011. Three new megastigmane glycosides from *Hylomecon vernalis*. Bull. Kor. Chem. Soc. 32 (10), 3813–3816.

Lee, J.S., Jung, W.K., Jeong, M.H., et al., 2012a. Sanguinarine induces apoptosis of HT-29 human colon cancer cells via the regulation of Bax/Bcl-2 ratio and caspase-9-dependent pathway. Int. J. Toxicol. 31, 70–77.

Lee, S.Y., Kim, K.H., Lee, I.K., et al., 2012b. A new flavonol glycoside from *Hylomecon vernalis*. Arch. Pharm. Res. 35, 415–421.

Lei, Q.F., Zhao, X., Xu, L., Peng, Y., Xiao, P.G., 2014. Chemical constituents of plants from tribe Chelidonieae and their bioactivities. Chin. Herb. Med. 6 (1), 1–21.

Li, X.L., Yao, J.Y., Zhou, Z.M., et al., 2011. Activity of the chelerythrine, a quaternary benzo[c]phenanthridine alkaloid from *Chelidonium majus* L. on *Dactylogyrus intermedius*. Parasitol. Res. 109, 247–252.

Li, L., Peng, Y., Ma, G., He, C., Feng, Y., Lei, Q., Xiao, P.G., 2012. Quantitative analysis of five kudinosides in the large-leaved Kudingcha and related species from the genus *Ilex* by UPLC-ELSD. Phytochem. Anal. 23 (6), 677–683.

Li, W., Huang, H., Niu, X., et al., 2014. Tetrahydrocoptisine protects rats from LPS-induced acute lung injury. Inflammation 37, 2106–2115 (Epub ahead of print).

Liang, J., Zeng, F., Guo, A., et al., 2011. Microarray analysis of the chelerythrine-induced transcriptome of *Mycobacterium tuberculosis*. Curr. Microbiol. 62, 1200–1208.

Liu, H., Wang, J., Zhao, J., et al., 2009. Isoquinoline alkaloids from *Macleaya cordata* active against plant microbial pathogens. Nat. Prod. Commun. 4, 1557–1560.

Liu, M., Peng, L., Liu, J.J., et al., 2010. Identification of differentially expressed proteins in the liver of Oncomelania snails induced by *Eomecon chinanthe* sanguinarine. Chin. J. Prev. Med. 44, 490–493.

Ma, Z.X., Feltenberger, J.B., Hsung, R.P., 2012. Total syntheses of chelidonine and norchelidonine via an enamide-benzyne-[2+2] cycloaddition cascade. Org. Lett. 14, 2742–2745.
Mahady, G.B., Pendland, S.L., Stoia, A., Chadwick, L.R., 2003. In vitro susceptibility of Helicobacter pylori to isoquinoline alkaloids from *Sanguinaria canadensis* and *Hydrastis canadensis*. Phytother. Res. 17, 217–221.
Mazzanti, G., Di Sotto, A., Di Giacomo, S., et al., 2013. *Chelidonium majus* L. does not potentiate the hepatic effect of acetaminophen. Exp. Toxicol. Pathol. 65 (7–8), 1117–1120.
Mazzio, E.A., Soliman, K.F., 2009. In vitro screening for the tumoricidal properties of international medicinal herbs. Phytother. Res. 23, 385–398.
Meng, F., Zuo, G., Hao, X., et al., 2009. Antifungal activity of the benzo[c]phenanthridine alkaloids from *Chelidonium majus* Linn against resistant clinical yeast isolates. J. Ethnopharmacol. 125, 494–496.
Moran, A.M., Helm, K.F., 2008. Histopathologic findings and diagnostic difficulties posed with use of escharotics agents for treatment of skin lesions: a case report and review of the literature. J. Cutan. Pathol. 35, 404–406.
Nadova, S., Miadokova, E., Alfoldiova, L., et al., 2008. Potential antioxidant activity, cytotoxic and apoptosis-inducing effects of *Chelidonium majus* L. extract on leukemia cells. Neuro Endocrinol. Lett. 29, 649–652.
Nawrot, R., Kalinowski, A., Gozdzicka-Jozefiak, A., 2007a. Proteomic analysis of *Chelidonium majus* milky sap using two-dimensional gel electrophoresis and tandem mass spectrometry. Phytochemistry 68, 1612–1622.
Nawrot, R., Lesniewicz, K., Pienkowska, J., Gozdzicka-Jozefiak, A., 2007b. A novel extracellular peroxidase and nucleases from a milky sap of *Chelidonium majus*. Fitoterapia 78, 496–501.
Nawrot, R., Wołuń-Cholewa, M., Goździcka-Józefiak, A., 2008. Nucleases isolated from *Chelidonium majus* L. milky sap can induce apoptosis in human cervical carcinoma HeLa cells but not in Chinese Hamster Ovary CHO cells. Folia Histochem. Cytobiol. 46, 79–83.
Nawrot, R., Zauber, H., Schulze, W.X., 2014. Global proteomic analysis of *Chelidonium majus* and *Corydalis cava* (Papaveraceae) extracts revealed similar defense-related protein compositions. Fitoterapia 94, 77–87.
Noureini, S.K., Wink, M., 2009. Transcriptional down regulation of hTERT and senescence induction in HepG2 cells by chelidonine. World J. Gastroenterol. 15, 3603–3610.
Obiang-Obounou, B.W., Kang, O.H., Choi, J.G., et al., 2011. The mechanism of action of sanguinarine against methicillin-resistant *Staphylococcus aureus*. J. Toxicol. Sci. 36, 277–283.
Park, J.E., Cuong, T.D., Hung, T.M., et al., 2011. Alkaloids from *Chelidonium majus* and their inhibitory effects on LPS-induced NO production in RAW264.7 cells. Bioorg. Med. Chem. Lett. 21, 6960–6963.
Park, S.Y., Jin, M.L., Kim, Y.H., et al., 2014. Involvement of heme oxygenase-1 in neuroprotection by sanguinarine against glutamate-triggered apoptosis in HT22 neuronal cells. Environ. Toxicol. Pharmacol. 38 (3), 701–710.
Peng, Y., Chen, S.B., Chen, S.L., Xiao, P.G., 2006. Preliminary pharmacophylogenetic study on Ranunculaceae. China J. Chin. Mat. Med. 31, 1124–1128.
Philchenkov, A., Kaminskyy, V., Zavelevich, M., Stoika, R., 2008. Apoptogenic activity of two benzophenanthridine alkaloids from *Chelidonium majus* L. does not correlate with their DNA damaging effects. Toxicol. In Vitro 22, 287–295.
Pradhan, S.K., Dasgupta, D., Basu, G., 2011. Human telomere d[(TTAGGG)4] undergoes a conformational transition to the Na^+-form upon binding with sanguinarine in presence of K^+. Biochem. Biophys. Res. Commun. 404, 139–142.
Qing, Z., Cheng, P., Liu, X., et al., 2014. Structural speculation and identification of alkaloids in *Macleaya cordata* fruits by high-performance liquid chromatography/quadrupole-time-of-

flight mass spectrometry combined with a screening procedure. Rapid Commun. Mass Spectrom. 28 (9), 1033–1044.

Reimann, A., Nurhayati, N., Backenköhler, A., Ober, D., 2004. Repeated evolution of the pyrrolizidine alkaloid-mediated defense system in separate angiosperm lineages. Plant Cell 16, 2772–2784.

Sedo, A., Vlasicová, K., Barták, P., et al., 2002. Quaternary benzo[c]phenanthridine alkaloids as inhibitors of aminopeptidase N and dipeptidyl peptidase IV. Phytother. Res. 16, 84–87.

Senchina, D.S., Flinn, G.N., McCann, D.A., et al., 2009a. Bloodroot (*Sanguinaria canadensis* L., Papaveraceae) enhances proliferation and cytokine production by human peripheral blood mononuclear cells in an in vitro model. J. Herbs Spices Med. Plants 15, 45.

Senchina, D.S., Hallam, J.E., Dias, A.S., Perera, M.A., 2009b. Human blood mononuclear cell in vitro cytokine response before and after two different strenuous exercise bouts in the presence of bloodroot and *Echinacea* extracts. Blood Cells Mol. Dis. 43, 298–303.

Shin, M.C., Jang, M.H., Chang, H.K., et al., 2003. Modulation of Chelidonii herba on glycine-activated and glutamate-activated ion currents in rat periaqueductal gray neurons. Clin. Chim. Acta 337, 93–101.

Shin, H.J., Kim, H.L., Kim, S.J., et al., 2011. Inhibitory effects of chelidonic acid on IL-6 production by blocking NF-κB and caspase-1 in HMC-1 cells. Immunopharmacol. Immunotoxicol. 33, 614–619.

Song, J.Y., Yang, H.O., Pyo, S.N., et al., 2002. Immunomodulatory activity of protein-bound polysaccharide extracted from *Chelidonium majus*. Arch. Pharm. Res. 25, 158–164.

Song, J.Y., Yang, H.O., Shim, J.Y., et al., 2003. Radiation protective effect of an extract from *Chelidonium majus*. Int. J. Hematol. 78, 226–232.

Sun, W.X., Yuan, S.S., Huang, Q.Y., et al., 2011. Study on liver injury of *Oncomelania hupensis* caused by *Eomecon chinanthe* sanguinarine. Chin. J. Schisto. Control 23, 82–84.

Sun, H., Hou, H., Lu, P., et al., 2012. Isocorydine inhibits cell proliferation in hepatocellular carcinoma cell lines by inducing g2/m cell cycle arrest and apoptosis. PLoS One 7, e36808.

Sun, R., Jiang, H., Zhang, W., et al., 2014. Cytotoxicity of aporphine, protoberberine, and protopine alkaloids from *Dicranostigma leptopodum* (Maxim.) Fedde. Evid. Based Complement. Alternat. Med. 2014, 580483.

Tamura, K., Peterson, D., Peterson, N., et al., 2011. MEGA5: molecular evolutionary genetics analysis using maximum likelihood, evolutionary distance, and maximum parsimony methods. Mol. Biol. Evol. 28, 2731–2739.

Teschke, R., Glass, X., Schulze, J., Eickhoff, A., 2012. Suspected Greater Celandine hepatotoxicity: liver-specific causality evaluation of published case reports from Europe. Eur. J. Gastroenterol. Hepatol. 24, 270–280.

Velázquez, C., Calzada, F., Torres, J., et al., 2006. Antisecretory activity of plants used to treat gastrointestinal disorders in Mexico. J. Ethnopharmacol. 103, 66–70.

Vlachojannis, C., Magora, F., Chrubasik, S., 2012. Rise and fall of oral health products with Canadian bloodroot extract. Phytother. Res. 26 (10), 1423–1426.

Vrba, J., Orolinova, E., Ulrichova, J., 2012. Induction of heme oxygenase-1 by *Macleaya cordata* extract and its constituent sanguinarine in RAW264.7 cells. Fitoterapia 83, 329–335.

Wang, W., 2007. Phylogenetic Analysis of Ranunculales: Molecular and Morphological Evidence. PhD Dissertation, Institute of Botany, Chinese Academy of Sciences, Beijing, China.

Wang, W., Lu, A.M., Ren, Y., et al., 2009. Phylogeny and classification of Ranunculales: evidence from four molecular loci and morphological data. Perspect. Plant Ecol. Evol. Syst. 11, 81–110.

Wang, G.X., Zhou, Z., Jiang, D.X., et al., 2010. In vivo anthelmintic activity of five alkaloids from *Macleaya microcarpa* (Maxim) Fedde against *Dactylogyrus intermedius* in *Carassius auratus*. Vet. Parasitol. 171, 305–313.

Wang, X.J., Min, C.L., Ge, M., Zuo, R.H., 2014. An endophytic sanguinarine-producing fungus from *Macleaya cordata, Fusarium proliferatum* BLH51. Curr. Microbiol. 68 (3), 336–341.

Watanabe, T., Ohashi, Y., Yoshino, R., et al., 2003. Total synthesis of 12-methoxydihydrochelerythrine and anti-tumour activity of its quaternary base: toward an efficient synthetic route for 12-alkoxybenzo[c]phenanthridine bases via naphthoquinone monooxime from 2-benzofuranyl-1-tetralone derivative. Org. Biomol. Chem. 1, 3024–3032.

Wink, M., 2003. Evolution of secondary metabolites from an ecological and molecular phylogenetic perspective. Phytochemistry 64, 3–19.

Yang, G., Lee, K., Lee, M.H., et al., 2011. Inhibitory effects of *Chelidonium majus* extract on atopic dermatitis-like skin lesions in NC/Nga mice. J. Ethnopharmacol. 138, 398–403.

Yao, J.Y., Shen, J.Y., Li, X.L., et al., 2010. Effect of sanguinarine from the leaves of *Macleaya cordata* against Ichthyophthirius multifiliis in grass carp (*Ctenopharyngodon idella*). Parasitol. Res. 107, 1035–1042.

Yao, J.Y., Zhou, Z.M., Li, X.L., et al., 2011a. Antiparasitic efficacy of dihydrosanguinarine and dihydrochelerythrine from *Macleaya microcarpa* against *Ichthyophthirius multifiliis* in richadsin (*Squaliobarbus curriculus*). Vet. Parasitol. 183, 8–13.

Yao, J.Y., Zhou, Z.M., Pan, X.Y., et al., 2011b. In vivo anthelmintic activity of chelidonine from *Chelidonium majus* L. against *Dactylogyrus intermedius* in *Carassius auratus*. Parasitol. Res. 109, 1465–1469.

Yin, H.Q., Kim, Y.H., Moon, C.K., Lee, B.H., 2005. Reactive oxygen species-mediated induction of apoptosis by a plant alkaloid 6-methoxydihydrosanguinarine in HepG2 cells. Biochem. Pharmacol. 70, 242–248.

Yu, X., Gao, X., Zhu, Z., et al., 2014. Alkaloids from the tribe Bocconieae (papaveraceae): a chemical and biological review. Molecules 19 (9), 13042–13060.

Zeng, J., Liu, Y., Liu, W., et al., 2013. Integration of transcriptome, proteome and metabolism data reveals the alkaloids biosynthesis in *Macleaya cordata* and *Macleaya microcarpa*. PLoS One 8 (1), e53409.

Zhang, D.Y., Zhu, J.H., Zhao, W.Z., et al., 2006. Apoptosis in myocardial cells after macleaya cordata total alkaloids poisoning in rats. J. Forensic Med. 22, 330–332.

Zhang, D.Y., Kong, Q.Q., Zhang, C.B., et al., 2009. Expression of Bcl-2 and Bax proteins in myocardial cells after *Macleaya cordata* alkaloids poisoning in rats. J. Forensic Med. 25, 246–248.

Zhang, Z.F., Guo, Y., Wei, X.H., 2011. Screen for natural benzylisoquinoline alkaloids against tumor. China J. Chin. Mat. Med. 36, 2684–2688.

Zhong, M., Huang, K.L., Zeng, J.G., et al., 2011. Rapid determination of main alkaloids from organs of *Macleaya cordata* (Willd) R. Br. by UPLC-ESI-MS. J. Central South Univ. (Sci. Technol.) 42, 1899–1905.

Zhong, M., Ma, Y.X., Liu, J.X., Di, D.L., 2014. A new quaternary protoberberine alkaloid isolated from *Dicranostigma leptopodum* (Maxim) Fedde. Nat. Prod. Res. 28 (8), 507–510.

Zhu, M., Xiao, P.G., 1991a. Distribution of benzylisoquinoline alkaloids in Berberidaceae. Chin. Trad. Herb. Drugs 22, 207–208.

Zhu, M., Xiao, P.G., 1991b. Chemosystematic studies on *Thalictrum* L. in China. J. GUCAS 29, 358–369.

Zuo, G.Y., Meng, F.Y., Hao, X.Y., et al., 2008. Antibacterial alkaloids from *Chelidonium majus* linn (papaveraceae) against clinical isolates of methicillin-resistant *Staphylococcus aureus*. J. Pharm. Pharm. Sci. 11, 90–94.

Zuo, G.Y., Meng, F.Y., Han, J., et al., 2011. In vitro activity of plant extracts and alkaloids against clinical isolates of extended-spectrum β-lactamase (ESBL)-producing strains. Molecules 16, 5453–5459.

Phytochemical and biological research of *Papaver* pharmaceutical resources

6

6.1 Introduction

Opium poppy (*Papaver somniferum*, Papaveraceae) is one of the world's oldest medicinal plants and remains the only commercial source for the narcotic analgesics such as morphine and codeine and semisynthetic derivatives such as oxycodone and naltrexone. The plant also produces several other benzylisoquinoline alkaloids (BIAs) with potent pharmacological properties including the vasodilator papaverine, the cough suppressant and potential anticancer drug noscapine, and the antimicrobial agent sanguinarine. Opium poppy has served as a model system to investigate the biosynthesis of BIAs in plants. The application of biochemical and functional genomics has resulted in the discovery of biosynthetic genes involved in the formation of major BIAs in opium poppy. The availability of extensive biochemical genetic tools and information pertaining to benzylisoquinoline alkaloid metabolism is facilitating the study of a wide range of phenomena including the structural biology of novel catalysts, the genomic organization of biosynthetic genes, the cellular and subcellular localization of biosynthetic enzymes, and a variety of biotechnological applications. In this chapter, we highlight recent developments and summarize the frontiers of knowledge regarding the biochemistry, bioactivity, and omics of benzylisoquinoline alkaloid and other medicinal compounds in opium poppy and other *Papaver* species.

6.2 Chemical constituents

6.2.1 *Benzylisoquinoline alkaloid and other alkaloids*

BIAs are a large, diverse group of ~2500 specialized plant metabolites. Types of BIAs found in *Papaver* plants are shown in Figure 6.1 and Table 6.1. Examples of alkaloids found in *Papaver* plants are shown in Figure 6.2.

Papaver alkaloids play a major role in medicine and pharmacy. [ring-(13)C(6)]-tyramine as a biogenetic precursor of these alkaloids was fed to *P. somniferum* seedlings (Schmidt et al., 2007). The alkaloid pattern was elucidated by both direct-infusion high-resolution ESI-FT-ICR mass spectrometry (MS) and liquid chromatography (LC)/electrospray tandem MS. The structure of about 20 alkaloids displaying an incorporation of the labeled tyramine could be elucidated, which belong

Medicinal plants: chemistry, biology and omics. http://dx.doi.org/10.1016/B978-0-08-100085-4.00006-2

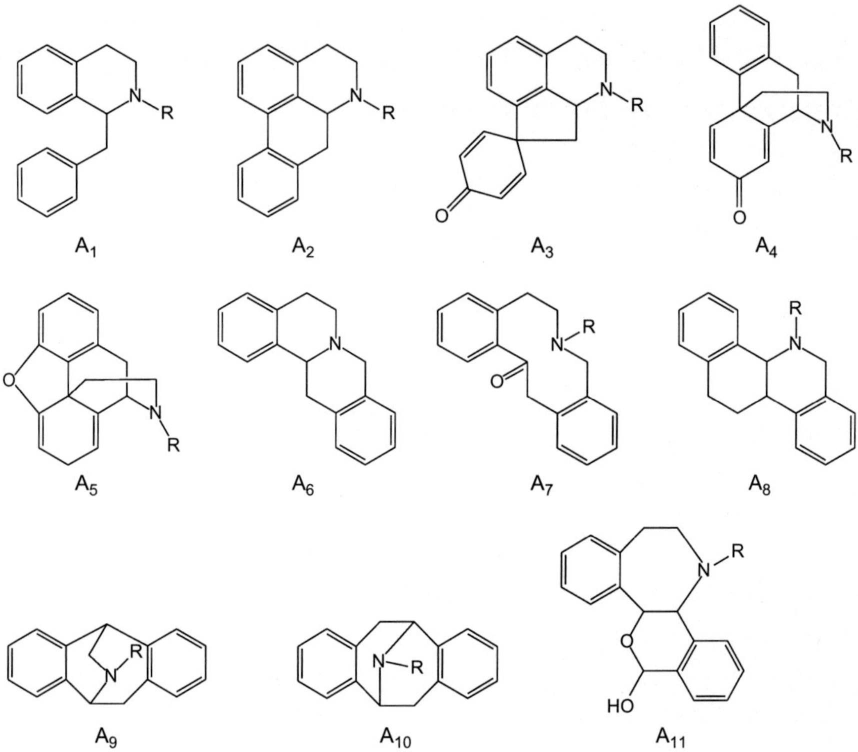

Figure 6.1 BIA structural subgroups derived from the basic benzylisoquinoline subunit.

to different classes, for example, morphinan, benzylisoquinoline, protoberberine, benzo[c]phenanthridine, phthalide isoquinoline, and protopine.

Morphinan, tetrahydrobenzylisoquinoline, benzo[c]phenanthridine, and phthalideisoquinoline alkaloids were determined qualitatively and quantitatively by HPLC and LC–MS analysis in tissues of the Tasmanian *P. somniferum* elite cultivar C048-6-14-64 (Frick et al., 2005). The data were compared with the results from the low-morphine cultivar "Marianne." In the elite cultivar, 91.2% of the latex alkaloids consist of the three pharmaceutically most valuable alkaloids: morphine, codeine, and thebaine. In the root system, the major alkaloids are sanguinarine/10-hydroxysanguinarine and dihydrosanguinarine/10-hydroxydihydrosanguinarine. In the stems, a gradient in relative total alkaloid content from the top downward toward the roots was observed. The concentration of morphine was decreasing toward the roots, whereas an increasing gradient from the upper to the lower stem parts was detected for codeine. The relative total alkaloid concentration in leaves remained constant. The cultivar "Marianne" displayed a shifted pattern of alkaloid accumulation and reduced levels of total alkaloid. In the condiment cultivar, 80.5% of the alkaloids of the latex consisted of the two phthalide isoquinoline alkaloids narcotoline and noscapine. Only 18.8% of the relative total alkaloid content was morphinan alkaloids.

Table 6.1 **Various types of compounds found *Papaver* plants in recent years**

No.	Compounds	Types	Species	Tissues	References
1	Miltanhoridine	A_1	a	Aerial parts	Sari et al. (2006)
2	Miltanthoridinone	A_1	a	Aerial parts	Sari et al. (2006)
3	(+)-*O*-Methylarmepavine	A_1	b	Aerial parts	Istatkova et al. (2012)
4	(−)-6a,7-Dehydrofloripavidine	A_2	c	Aerial parts	Sari et al. (2004)
5	Magnoflorine	A_2	d	Aerial parts	Liu et al. (2014a,b)
6	(+)-Amuronine	A_3	e	Aerial parts	Istatkova et al. (2008)
7	(−)-Dihydroamuronine	A_3	e	Aerial parts	Istatkova et al. (2008)
8	(−)-8,14-Dihydroflavinantine	A_4	b,e	Aerial parts	Istatkova et al. (2008,2012), Philipov et al. (2007)
9	Racem. 8,14-dihydroamurine	A_4	b,e	Aerial parts	Istatkova et al. (2012), Philipov et al. (2007)
10	(+)-Amurine	A_4	d,e	Aerial parts	Liu et al. (2014a,b), Philipov et al. (2007)
11	(−)-Flavinantine	A_4	b,e,f	Aerial parts	Istatkova et al. (2012), Philipov et al. (2007), Cheng et al. (2011)
12	*O*-Methylflavinantine	A_4	d	Aerial parts	Liu et al. (2014a,b)
13	6-Methoxy-17-methyl-2,3-[methylenebis (oxy)]-morphinan-5-en-7-one	A_4	d	Aerial parts	Liu et al. (2014a,b)
14	Bismorphine B	A_5	g	Capsules	Morimoto et al. (2003)
15	(−)-Mecambridine	A_6	b	Aerial parts	Istatkova et al. (2012)
16	Alborine	A_6	b,d	Aerial parts	Istatkova et al. (2012), Liu et al. (2014a,b)
17	(−)-Mecambridine methohydroxide	A_6	b	Aerial parts	Istatkova et al. (2012)

Continued

Table 6.1 Continued

No.	Compounds	Types	Species	Tissues	References
18	Simplicifolianine	A_6	d,j	Aerial parts	Liu et al. (2014a,b), Wangchuk et al. (2013a)
19	Scoulerine	A_6	d	Aerial parts	Liu et al. (2014a,b)
20	Sinactine	A_6	d	Aerial parts	Liu et al. (2014a,b)
21	Stylopine	A_6	d	Aerial parts	Liu et al. (2014a,b)
22	Coptisine	A_6	d	Aerial parts	Liu et al. (2014a,b)
23	Papavorientine	A_6	h	Roots	Ma et al. (2000)
24	Orientalidine	A_6	h	Roots	Ma et al. (2000)
25	Pangrenine	A_6	i	Aerial parts	Siddikov et al. (2005)
26	Cryptopine	A_7	d,k	Aerial parts	Liu et al. (2014a,b), Lee et al. (2011a)
27	Protopine	A_7	d,j,k	Aerial parts	Liu et al. (2014a,b), Wangchuk et al. (2013a), Lee et al. (2011a)
28	Pseudoprotopine	A_7	e	Aerial parts	Istatkova et al. (2008)
29	Allocryptopine	A_7	e,k	Aerial parts	Istatkova et al. (2008), Lee et al. (2011a)
30	Papavorientine	A_8	h	Roots	Ma et al. (2000)
31	Norsanguinarine	A_8	d,j	Aerial parts	Wangchuk et al. (2013a), Guo et al. (2014b)
32	Dihydrosanguinarine	A_8	j	Aerial parts	Wangchuk et al. (2013a)
33	6-Methoxydihydrosanguinarine	A_8	j	Aerial parts	Wangchuk et al. (2013a)
34	Oxysanguinarine	A_8	j	Aerial parts	Wangchuk et al. (2013a)
35	Cavidilinine	A_8	d	Aerial parts	Guo et al. (2014b)

36	(−)-Amurensinine	A_9	b,e	Aerial parts	Istatkova et al. (2012), Philipov et al. (2007)
37	Amurensinine *N*-oxide A	A_9	d,e	Aerial parts	Liu et al. (2014a,b), Istatkova et al. (2008)
38	Amurensinine *N*-oxide B	A_9	d,e	Aerial parts	Liu et al. (2014a,b), Istatkova et al. (2008)
39	(−)-*O*-Methylthalisopavine	A_9	e	Aerial parts	Philipov et al. (2007)
40	Amurensine	A_9	e	Aerial parts	Philipov et al. (2007)
41	Nudicaulin I	A_{12}	e,l,m	Yellow petals	Tatsis et al. (2013)
42	Nudicaulin II	A_{12}	e,l,m	Yellow petals	Tatsis et al. (2013)
43	Nudicaulin III	A_{12}	e,l	Yellow petals	Tatsis et al. (2013)
44	Nudicaulin IV	A_{12}	e,l	Yellow petals	Tatsis et al. (2013)
45	Nudicaulin V	A_{12}	e,l,m	Yellow petals	Tatsis et al. (2013)
46	Nudicaulin VI	A_{12}	e,l,m	Yellow petals	Tatsis et al. (2013)
47	Nudicaulin VII	A_{12}	e,l	Yellow petals	Schliemann et al. (2006), Tatsis et al. (2013)
48	Nudicaulin VIII	A_{12}	e,l	Yellow petals	Schliemann et al. (2006), Tatsis et al. (2013)
49	Nudicaulin IX	A_{12}	m	Yellow petals	Tatsis et al. (2013)
50	Nudicaulin X	A_{12}	m	Yellow petals	Tatsis et al. (2013)
51	Oleracein E	A_{12}	d	Aerial parts	Guo et al. (2014a)
52	*N*-*p*-Hydroxyl-trans-coumaroyltyramine	A_{12}	d	Aerial parts	Liu et al. (2014a,b), Guo et al. (2014a)
53	Tricin-7-*O*-β-D-glucopyranoside	B	d	Aerial parts	Liu et al. (2014a,b)
54	Tricin	B	d	Aerial parts	Liu et al. (2014a,b)
55	Luteolin-7-*O*-β-D-glucopyranoside	B	d	Aerial parts	Liu et al. (2014a,b)

Continued

Table 6.1 Continued

No.	Compounds	Types	Species	Tissues	References
56	Luteolin	B	d	Aerial parts	Liu et al. (2014a,b)
57	Apigenin	B	d,e	Aerial parts	Liu et al. (2014a,b), Guo et al. (2014a)
58	Chrysoeriol	B	e	Aerial parts	Guo et al. (2014a)
59	Quercetin	B	d	Aerial parts	Liu et al. (2014a,b)
60	Quercetin-3-*O*-gentiobioside	B	d	Aerial parts	Liu et al. (2014a,b)
61	Quercetin-3-*O*-[α-L-rhamnopyranosyl(1→2)]-β-D-glucopyranoside	B	d	Aerial parts	Liu et al. (2014a,b)
62	Quercetin-3-*O*-[β-D-galactopyranosyl(1→2)]-β-D-glucopyranoside	B	d	Aerial parts	Liu et al. (2014a,b)
63	Hyperoside	B	d	Aerial parts	Liu et al. (2014a,b)
64	Isoquercetin	B	d	Aerial parts	Liu et al. (2014a,b)
65	Kaempferol-3-*O*-[β-D-galactopyranosyl (1→2)]-β-D-glucopyranoside	B	d,e,l,m	Aerial parts, petals	Liu et al. (2014a,b)
66	Kaempferol-3-*O*-β-D-glucopyranoside	B	d	Aerial parts	Liu et al. (2014a,b)
67	Kaempferol	B	d	Aerial parts	Liu et al. (2014a,b)
68	Kaempferide	B	d	Aerial parts	Liu et al. (2014a,b)
69	Hydnocarpin	B	d,e	Aerial parts	Liu et al. (2014a,b)
70	3-(Kaempferol-8-yl)-2,3-epoxyflavanone	B	d	Aerial parts	Liu et al. (2014a,b)
71	Kaempferol-3-*O*-β-D-xylopyranosyl(1→2)-β-D-glucopyranoside	B	d	Aerial parts	Guo et al. (2014b)
72	Kaempferol-3-*O*-β-sophoroside-7-*O*-β-glucoside	B	e,l,m	Petals	Tatsis et al. (2013)

73	Kaempferol-3-*O*-β-[(6-malonyl)-sophoroside]-7-*O*-β-glucoside	B	e,l,m	Petals	Tatsis et al. (2013)
74	Kaempferol-3-*O*-β-(sophoroside)-7-*O*-β-[(6-malonyl)-glucoside]	B	e,l,m	Petals	Tatsis et al. (2013)
75	Kaempferol-3-*O*-β-[(6-malonyl)-sophoroside]-7-*O*-β-[(6-malonyl)-glucoside]	B	e,l,m	Petals	Tatsis et al. (2013)
76	Gossypitrin	B	e,l,m	Petals	Tatsis et al. (2013)
77	Pelargonidin 3-*O*-β-sophoroside-7-*O*-β-glucoside	B	e,l,m	Petals	Tatsis et al. (2013)
78	Pelargonidin 3-*O*-β-[(6-malonyl)-sophoroside]-7-*O*-β-glucoside	B	e,l,m	Petals	Tatsis et al. (2013)
79	Pelargonidin 3-*O*-β-[(6-malonyl)-sophoroside]-7-*O*-β-[(6-malonyl)-glucoside]	B	e,l,m	Petals	Tatsis et al. (2013)
80	Pelargonidin 3-*O*-β-sophoroside	B	e,l,m	Petals	Tatsis et al. (2013)
81	Pelargonidin 3-*O*-β-[(6-malonyl)-sophoroside]	B	e,l,m	Petals	Tatsis et al. (2013)
82	Anthocyanin	B	n	Blue petals	Tanaka et al. (2001)
83	Cycloart-23-ene-3β, 25-diol	C	d	Aerial parts	Guo et al. (2014b)
84	3α-Hydroxyolean-12-en-30-oic acid	C	d	Aerial parts	Guo et al. (2014b)
85	3β-Hydroxyolean-12-en-30-oic acid	C	d	Aerial parts	Guo et al. (2014b)
86	β-Sitosterol	D	e	Aerial parts	Guo et al. (2014a)
87	Stigmast-5-ene-3β-yl formate	D	e	Aerial parts	Guo et al. (2014a)
88	3β-Hydroxy-7α-ethoxy-24β-ethylcholest-5-ene	D	e	Aerial parts	Guo et al. (2014a)
89	Cinnamic acid	E	d	Aerial parts	Liu et al. (2014a,b)
90	*p*-Hydroxycinnamic acid esters	E	d	Aerial parts	Liu et al. (2014a,b)
91	*p*-Hydroxycinnamic acid	E	d	Aerial parts	Liu et al. (2014a,b)
92	Caffeic acid glucoside	E	d	Aerial parts	Liu et al. (2014a,b)

Continued

Table 6.1 Continued

No.	Compounds	Types	Species	Tissues	References
93	Caffeic acid	E	d	Aerial parts	Liu et al. (2014a,b)
94	6,7-Dihydroxy-coumarin	E	d	Aerial parts	Liu et al. (2014a,b)
95	Cinnamamide	E	d	Aerial parts	Liu et al. (2014a,b)
96	Citric acid	F	d	Aerial parts	Liu et al. (2014a,b)
97	Methyl linoleate	F	f	Aerial parts	Cheng et al. (2011)
98	5-Hydroxygeranyllinalol	F	d	Aerial parts	Guo et al. (2014b)
99	5, 15-*O*-Diacetyl-3-*O*-phenyl-6(17)-epoxylathyrol	F	d	Aerial parts	Guo et al. (2014b)

A, BIA structural subgroups: A_1, benzylisoquinoline type; A_2, aporphine type; A_3, proaporphine type; A_4, promorphinane type; A_5, morphinane type; A_6, proberberine type; A_7, protopine type; A_8, benzophenanthridine type; A_9, isopavine type; A_{10}, pavine type (Figure 6.1); A_{11}, rhoeadine type (Figure 6.1); A_{12}, other alkaloids.
B, flavonoids; C, triterpenoids; D, steroids; E, phenylpropanoids; F, other compounds.
a, *P. triniifolium* Boiss.; b, *P. pseudocanescens* M. Pop.; c, *P. fugax* Poir.; d, *M. horridula* Hook.f. Thoms; e, *P. nudicaule* L.; f, *M. quintuplinervia*; g, *P. somniferum* L.; h, *P. orientale* L.; i, *P. angrenicum* P.; j, *M simplicifoli*a (D. Don); k, *P. coreanum* Nakai; l, *P. alpinum*; m, *M. cambrica*; n, *M. grandis*.
41 and 42, 43 and 44, 45 and 46, 47 and 48, and 49 and 50 are pairs of diastereomers.

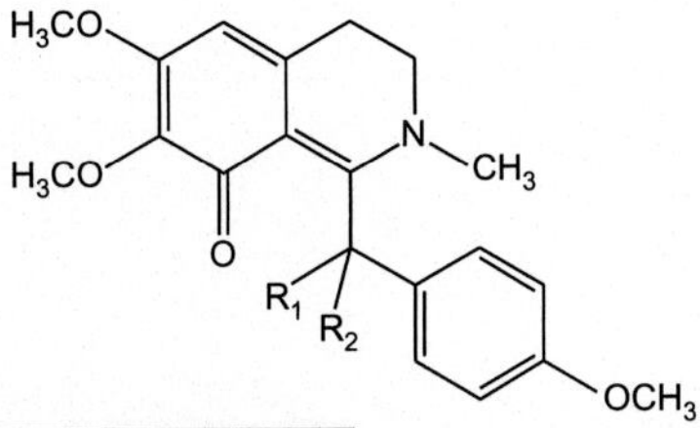

1 $R_1 = R_2 = H$
2 $R_1 + R_2 = O$

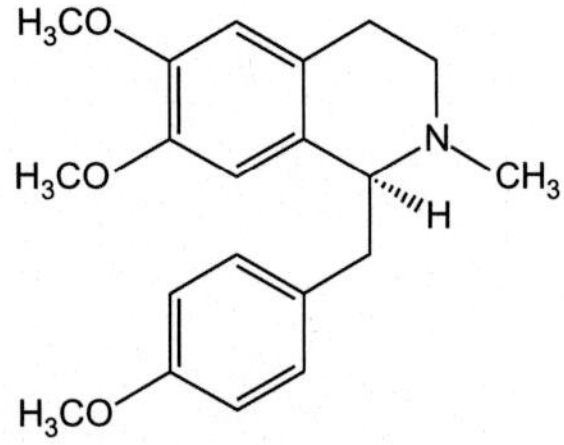

3

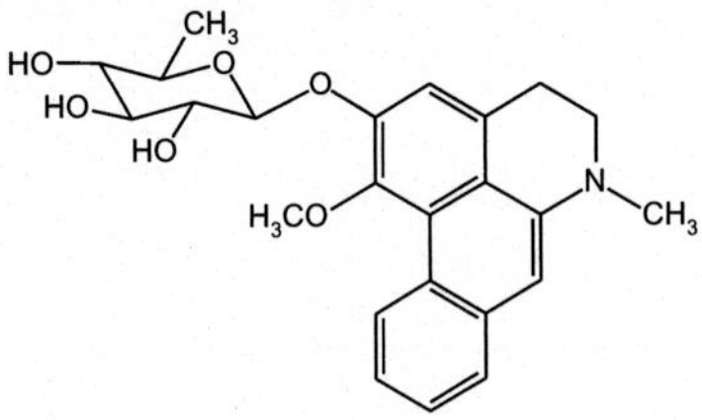

4

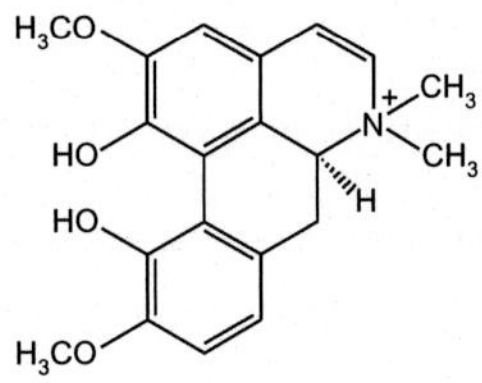

5

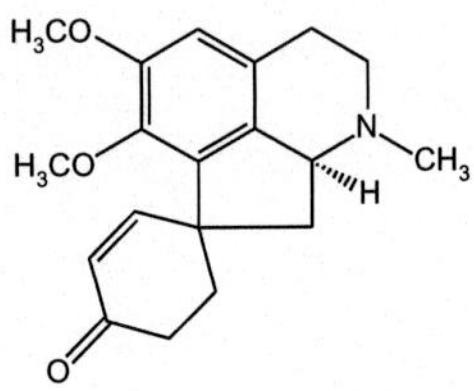

6

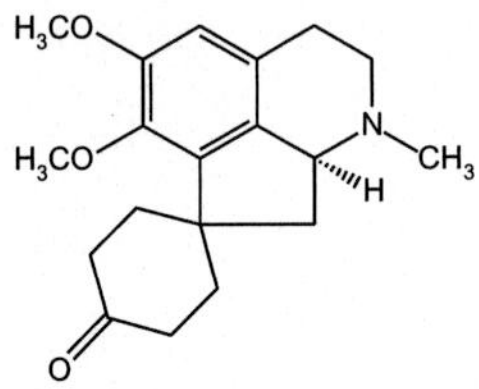

7

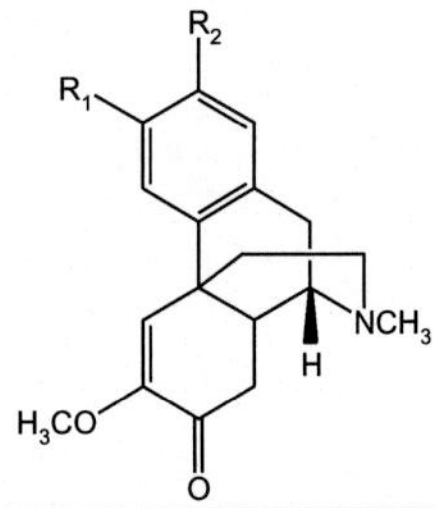

8 $R_1 = OH; R_2 = OCH_3$
9 $R_1 + R_2 = O\text{-}CH_2\text{-}O$

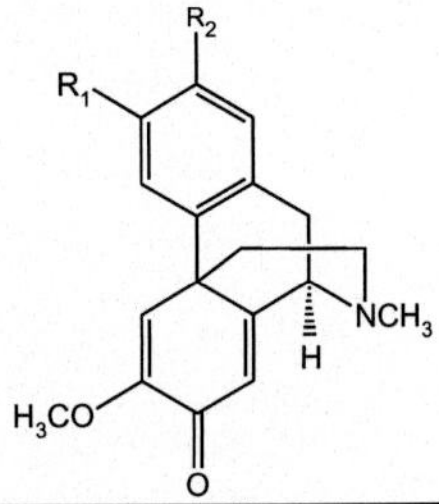

10 $R_1 + R_2 = O\text{-}CH_2\text{-}O$
11 $R_1 = OH; R_2 = OCH_3$
12 $R_1 = R_2 = OCH_3$

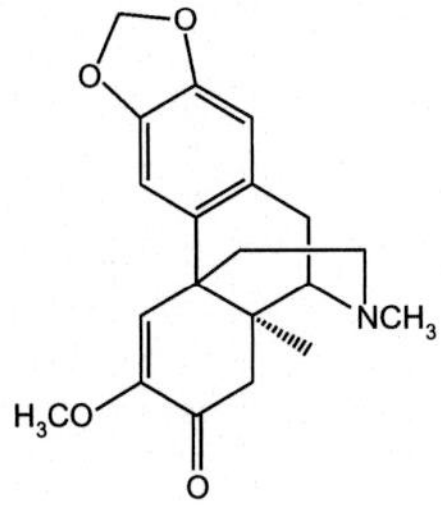

13

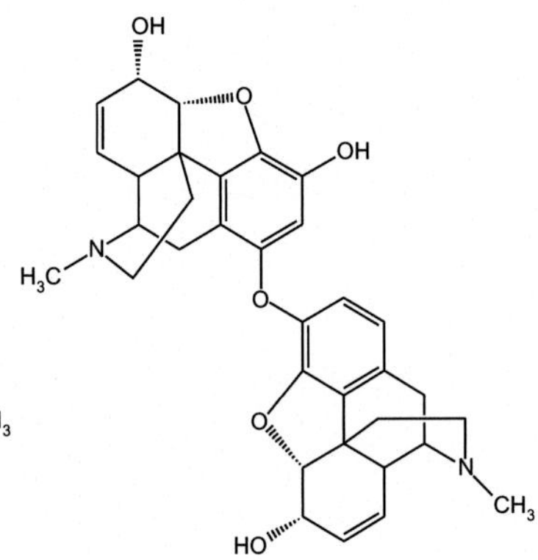

14

Figure 6.2 Examples of alkaloids found in *Papaver* plants.

(Continued)

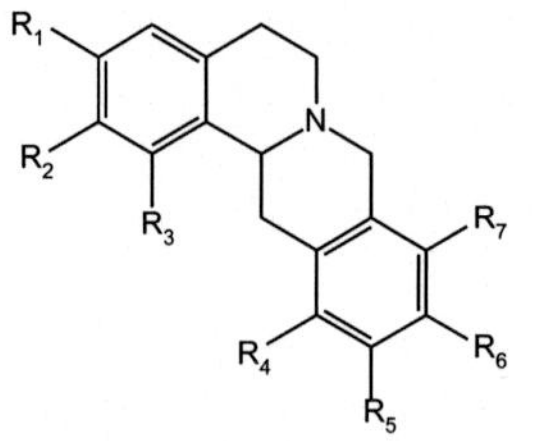

15	$R_1 + R_2 = O\text{-}CH_2\text{-}O$; $R_3 = R_5 = R_6 = OCH_3$; $R_4 = OCH_2OH$; $R_7 = H$
19	$R_1 = R_6 = OCH_3$; $R_2 = R_7 = OH$; $R_3 = R_4 = R_5 = H$
20	$R_1 = R_2 = OCH_3$; $R_6 + R_7 = O\text{-}CH2\text{-}O$; $R_3 = R_4 = R_5 = H$
21	$R_1 + R_2 = R_6 + R_7 = O\text{-}CH2\text{-}O$; $R_3 + R_4 = R_5 = H$
23	$R_1 + R_3 = R_4 = R_6 = H$; $R_2 = R_5 = OCH_3$; $R_7 = OH$

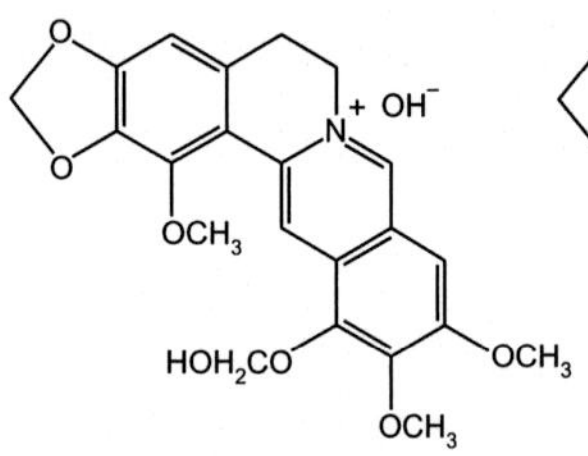

16

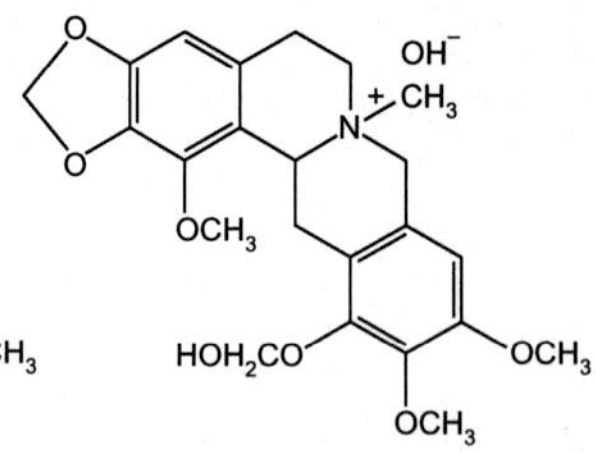

17

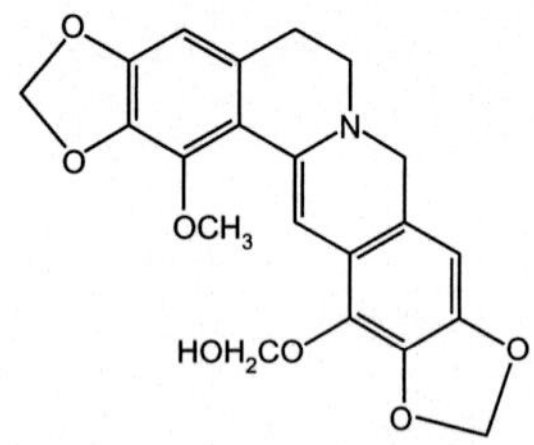

18

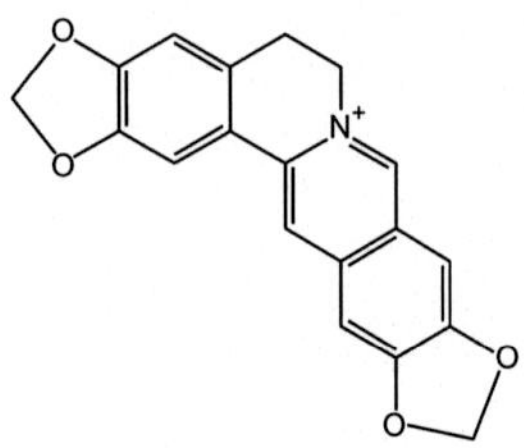

22

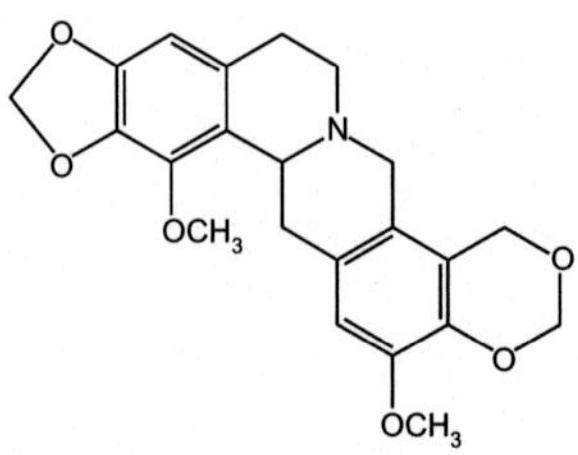

24

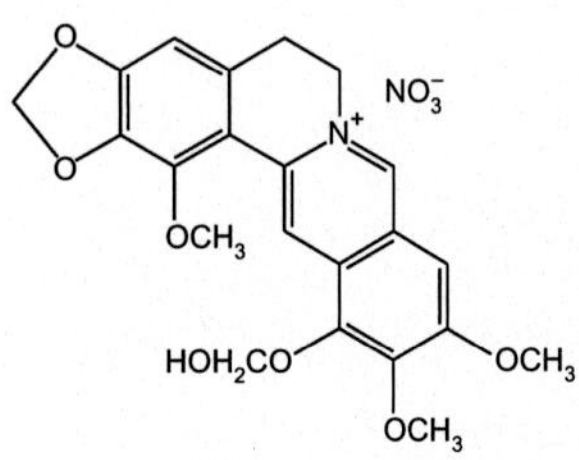

25

Figure 6.2 Continued.

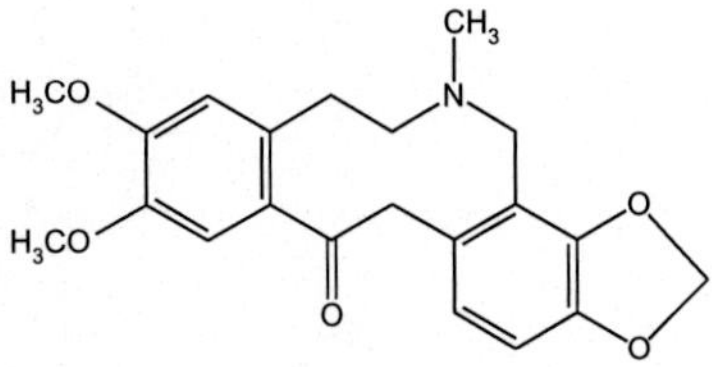

26

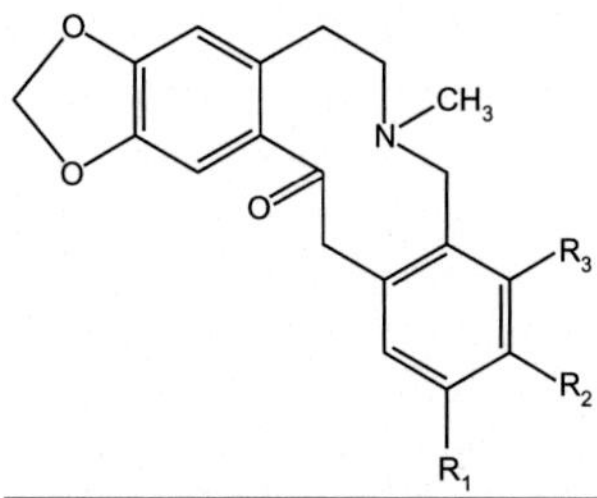

27 $R_1 = H; R_2 + R_3 = O\text{-}CH_2\text{-}O$

28 $R_1 + R_2 = O\text{-}CH_2\text{-}O; R_3 = H$

29 $R_1 = H; R_2 = R_3 = OCH_3$

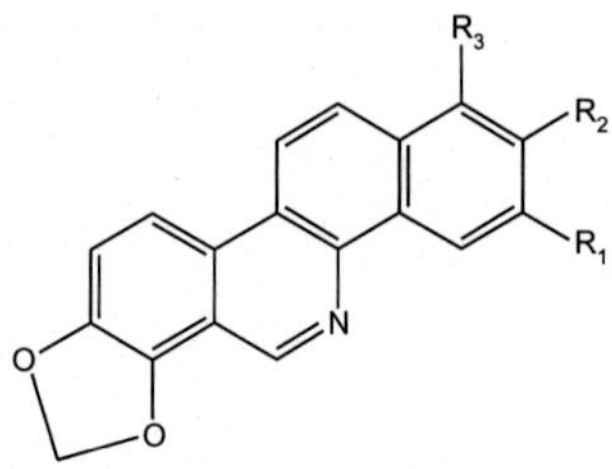

30 $R_1 = O\text{-}CH_3; R_2 + R_3 = O\text{-}CH_2\text{-}O;$

31 $R_2 + R_3 = O\text{-}CH_2\text{-}O$

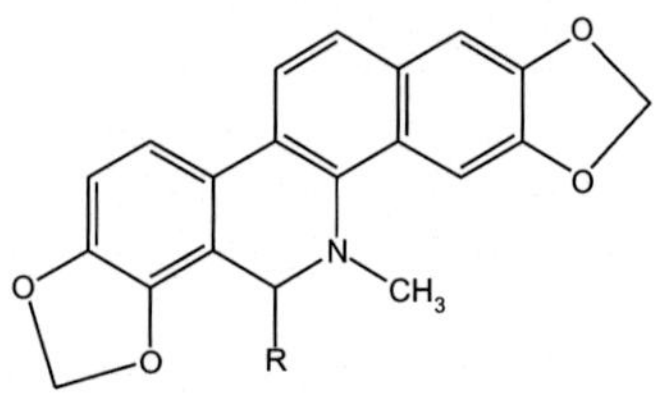

32 $R = H$

33 $R = OCH_3$

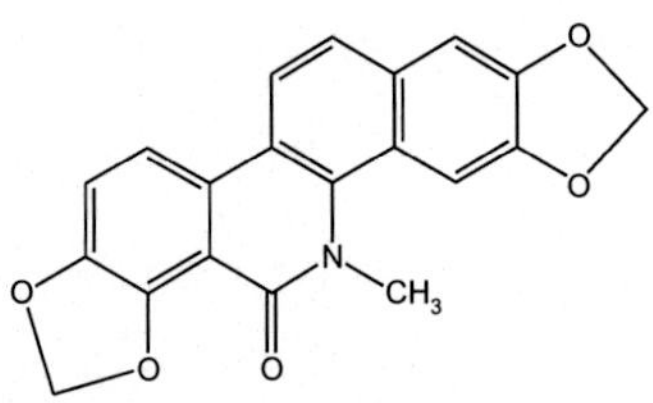

34

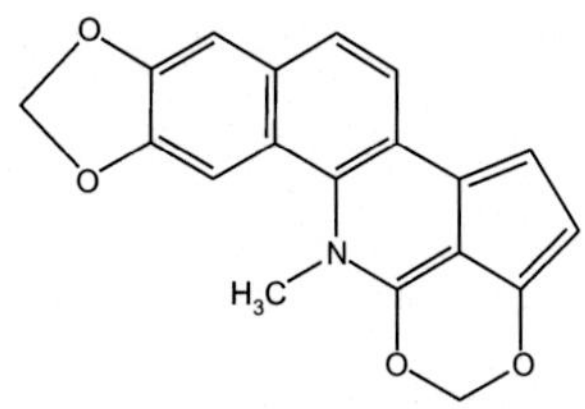

35

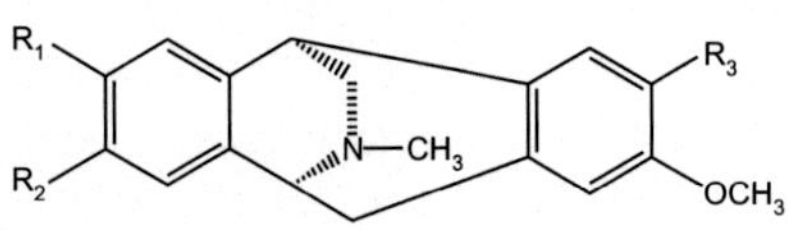

36 $R_1 + R_2 = O\text{-}CH_2\text{-}O; R_3 = OCH_3$

39 $R_1 = R_2 = R_3 = OCH_3$

Figure 6.2 Continued.

40	$R_1 + R_2$ = O-CH_2-O; R_3 = OH

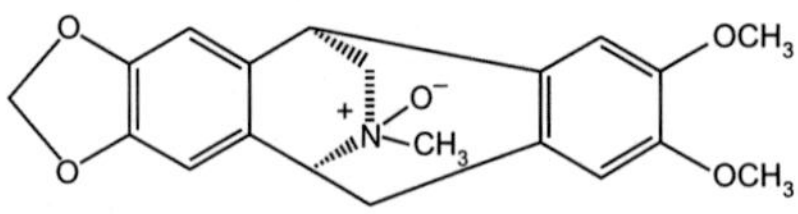

37

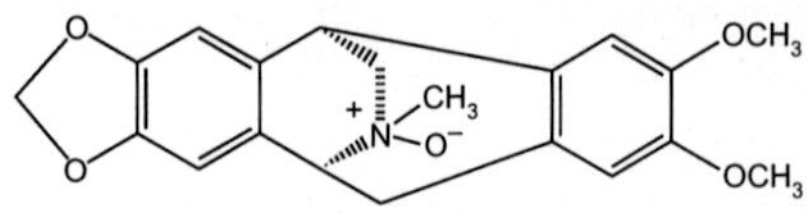

38

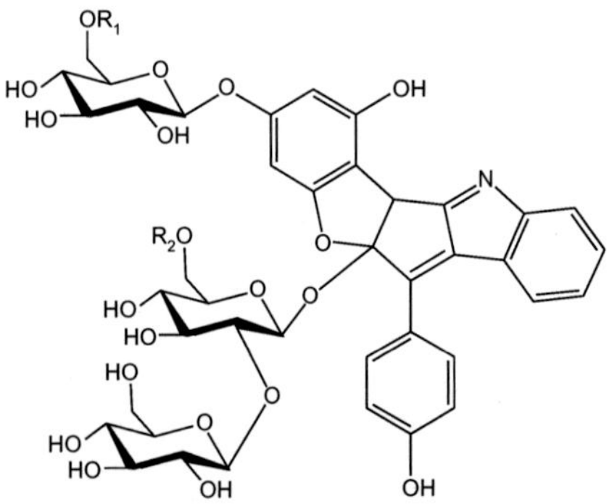

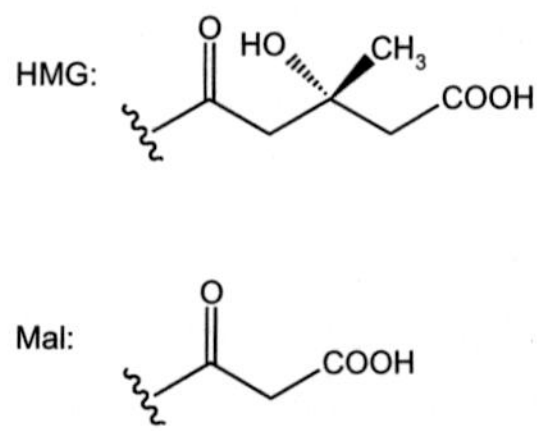

41, 42	$R_1 = R_2$ = H
43, 44	R_1 = H; R_2 = Mal
45, 46	R_1 = Mal; R_2 = H
47, 48	R_1 = Mal; R_2 = Mal
49, 50	R_1 = HMG; R_2 = H

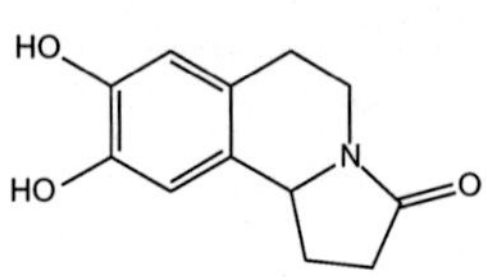

51

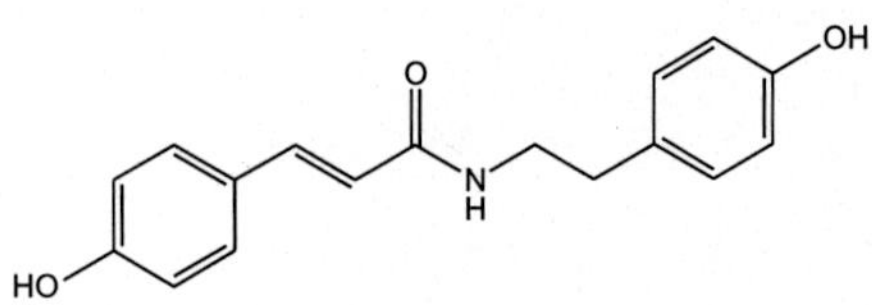

52

Figure 6.2 Continued.

In contrast to the narcotic cultivar, in which the benzo[c]phenanthridines in roots dominated over the morphinan and tetrahydrobenzylisoquinoline alkaloids, the concentration of benzo[c]phenanthridines in "Marianne" was similar to that of morphinan and tetrahydrobenzylisoquinoline alkaloids. These data suggest a differential alkaloid regulation in each cultivar of *P. somniferum*.

Eight alkaloids of the isoquinoline and promorphinane types were isolated from aerial parts of *P. pseudocanescens* of Mongolian origin (Istatkova et al., 2012). 8,14-Dihydroamurine, 8,14-dihydroflavinantine, and flavinantine are promorphinanes. Alborine, mecambridine, and mecambridine methohydroxide are retroprotoberberines. Amurensinine is an isopavine alkaloid, and *O*-methylarmepavine is a benzylisoquinoline alkaloid. *O*-Methylarmepavine is a new alkaloid for the genus *Papaver*. Promorphinane-type alkaloids have been found for the first time in the species.

Protopine and allocryptopine represent the main components of *P. coreanum* (Lee et al., 2011a). Four native *Papaver* species of Iran, that is, *P. glaucum*, *P. tenuifolium*, *P. dubium*, and *P. fugax*, were collected from their natural habitat and subjected to HPLC analysis for determination of their morphine (Salehi et al., 2007), codeine, and thebaine content. *P. dubium* and *P. glaucum* contained all three narcotic alkaloids, while morphine was not found in *P. fugax*, and *P. tenuifolium* was free from codeine.

6.2.2 Other compounds

Examples of other compounds found in *Papaver* are shown in Figure 6.3. Poppy and safflower oils were rich in linoleic acid at 74.5% and 70.5% level, respectively (Bozan and Temelli, 2008). The amount of total tocols was 11.0 mg/100 g poppy and 12.1 mg/100 g safflower seed. Poppy oil was rich in γ-tocopherol (30.9 mg/100 g), while α-tocopherol (44.1 g/100 g) was dominant in safflower oil. Only α-tocotrienol and γ-tocotrienol were found in the oils. Oxidative stability of oils was measured at 110 °C at the rate of 20 l/h air flow rate, and poppy oil (5.56 h) was the most stabile oil followed by safflower oil (2.87 h) and flax oil (1.57 h). There were no correlation between oxidative stability and unsaturation degree of fatty acids and tocol levels of the oils. All seeds investigated provide a healthy oil profile and may have potential as a source of specialty oils on a commercial scale.

1-Pentanol (3.3–4.9%), 1-hexanal (10.9–30.9%), 1-hexanol (5.3–33.7%), 2-pentylfuran (7.2–10.0%), and caproic acid (2.9–11.5%) were identified as the main volatile compounds in poppy seed oil samples (Krist et al., 2005). The triglyceride (TAG) composition of these oils was analyzed by MALDI-ReTOF-MS and ESI-IT-MS/MS. The predominant TAG components were composed of linoleic, oleic, and palmitic acids, comprising approximately 70% of the oils. TAG patterns of the different poppy varieties were very homogeneous, in spite of the different pressing methods of the plant seeds.

Two depsides, *p*-hydroxybenzoic acid, and its methyl ester, protocatechuic acid; 2-(4-hydroxyphenyl)-ethanol; 2-(3,4-dihydroxyphenyl)-ethanol; the flavonoids kaempferol, quercetin, luteolin, and hypolaetin; and the glycosides 3-*O*-β-D-glucopyranosylquercetin (isoquercitrin), 3-*O*-β-D-glucopyranosylkaempferol

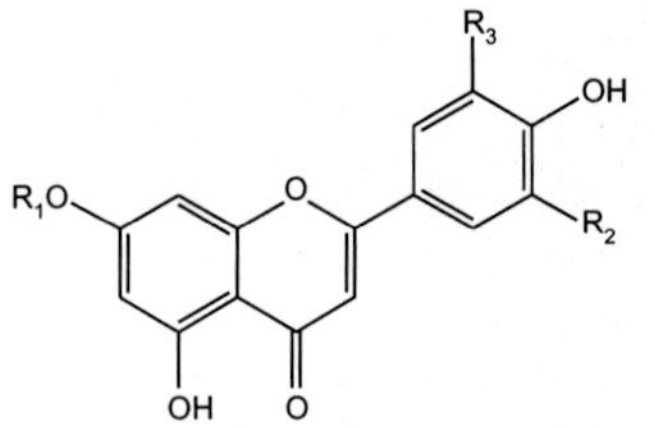

53 R_1 = β-D-Glc; R_2 = R_3 = OCH_3

54 R_1 = H; R_2 = R_3 = OCH_3

55 R_1 = β-D-Glc; R_2 = OH; R_3 = H

56 R_1 = R_3 = H; R_2 = OH

57 R_1 = R_2 = R_3 = H

58 R_1 = OH; R_2 = H; R_3 = OCH_3

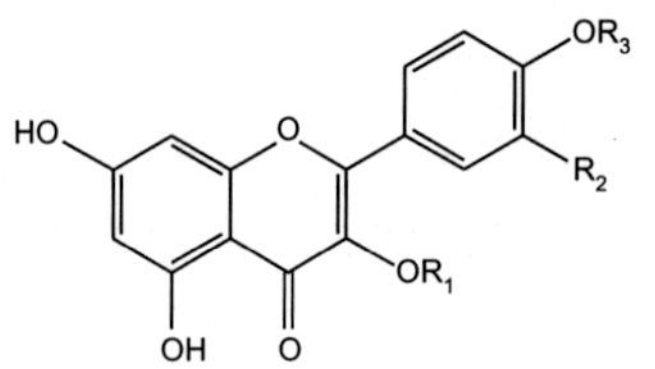

59 R_1 = R_3 = H; R_2 = OH

60 R_1 = β-D-Glc-(1→6)-β-D-Glc; R_2 = OH; R_3 = H

61 R_1 = α-L-Rha-(1→2)-β-D-Glc; R_2 = OH; R_3 = H

62 R_1 = β-D-Gal-(1→2)-β-D-Glc; R_2 = OH; R_3 = H

63 R_1 = β-D-Gal; R_2 = OH; R_3 = H

64 R_1 = β-D-Glc; R_2 = OH; R_3 = H

65 R_1 = β-D-Glc-(1→2)-β-D-Glc; R_2 = R_3 = H

66 R_1 = β-D-Glc; R_2 = R_3 = H

67 R_1 = R_2 = R_3 = H

Figure 6.3 Other compounds found in *Papaver*.

(Continued)

68	$R_1 = R_2 = H$; $R_3 = CH_3$
71	R_1 = β-D-xyl-(1→2)-β-D-Glc; $R_2 = R_3 = H$

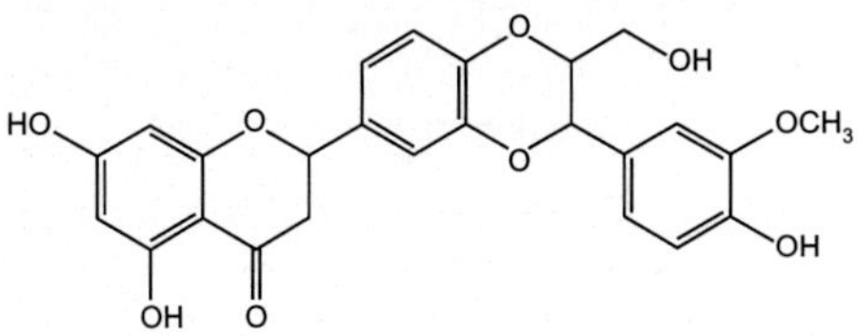

69

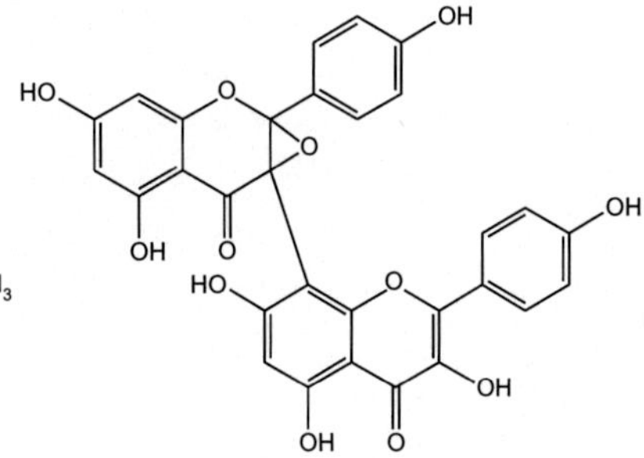

70

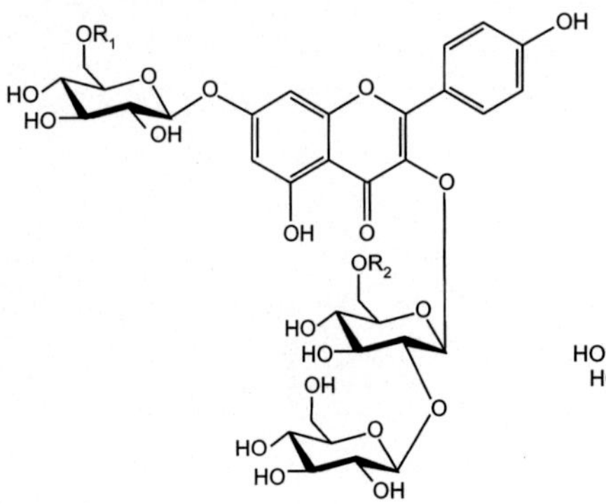

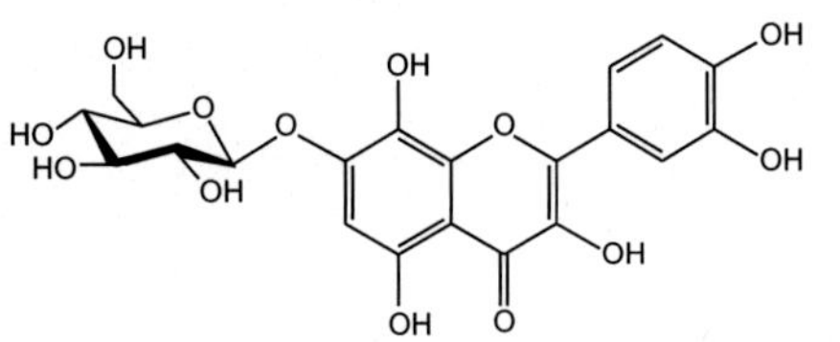

76

72	$R_1 = R_2 = H$
73	$R_1 = H$; R_3 = Mal
74	R_1 = Mal; $R_2 = H$
75	R_1 = Mal; R_3 = Mal

Figure 6.3 Continued.

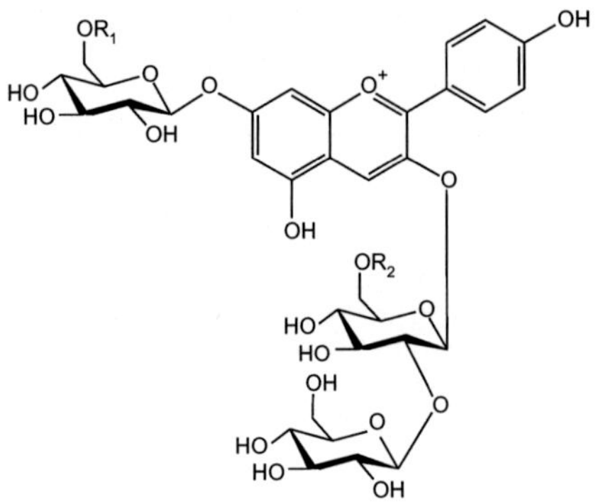

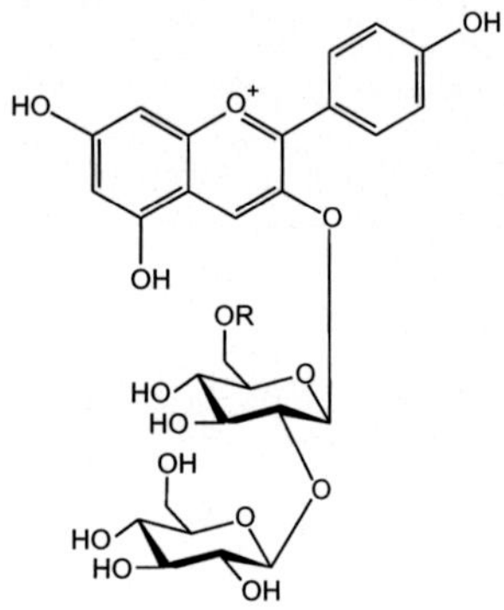

77	$R_1 = R_2 = H$
78	$R_1 = H$; $R_2 = Mal$
79	$R_1 = Mal$; $R_2 = Mal$

80	R = H
81	R = Mal

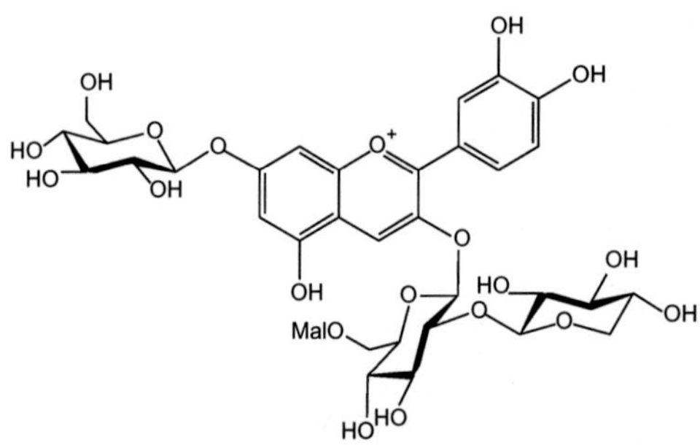

82

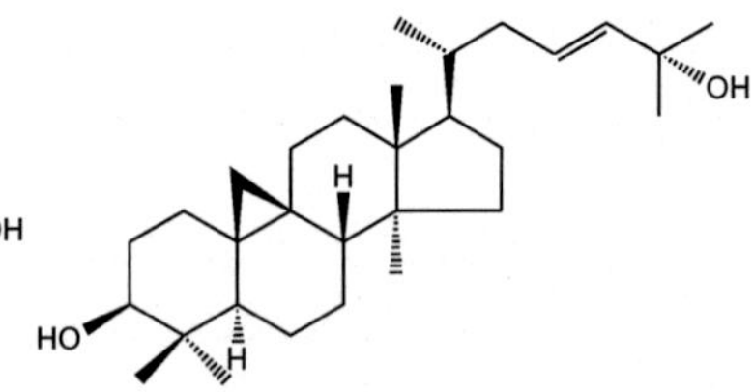

83

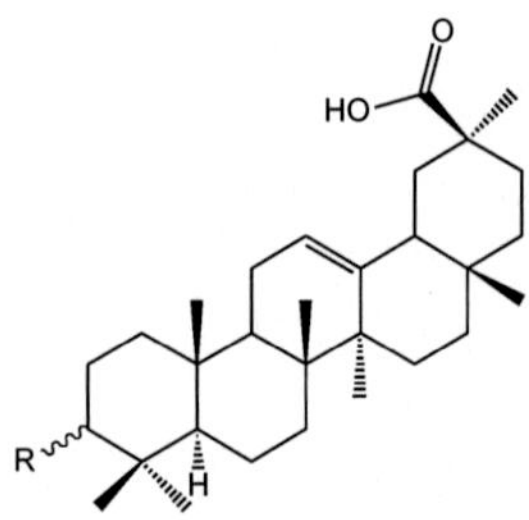

84	R = α-OH

Figure 6.3 Continued.

85 R = β-OH

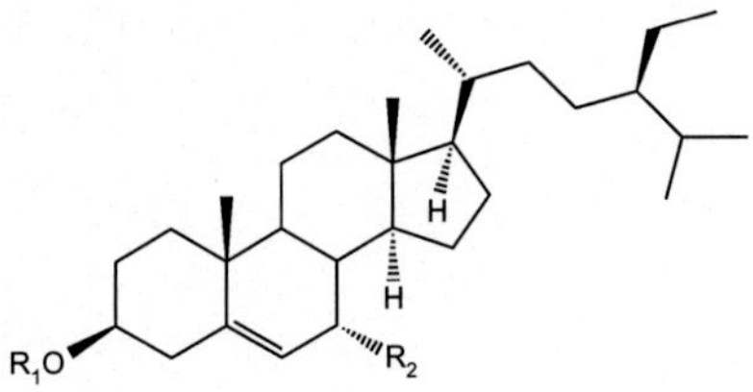

86 $R_1 = R_2 = H$

87 $R_1 = CHO; R_2 = H$

88 $R_1 = H; R_2 = OEt$

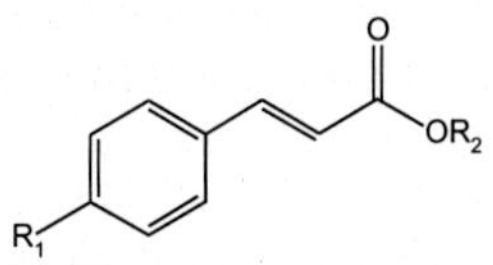

89 $R_1 = R_2 = H$

90 $R_1 = OH; R_2 =$ β-D-Glc

91 $R_1 = OH; R_2 = H$

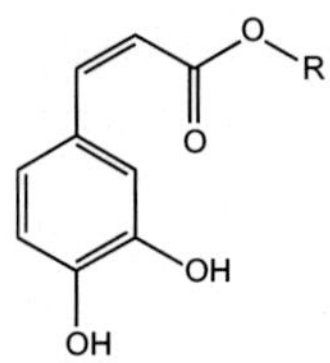

92 R = β-D-Glc

93 R = H

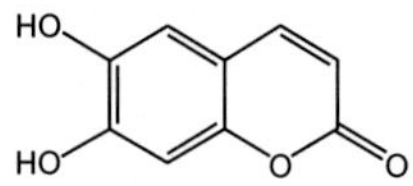

94

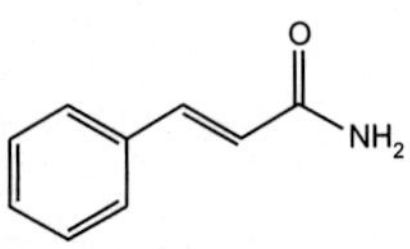

95

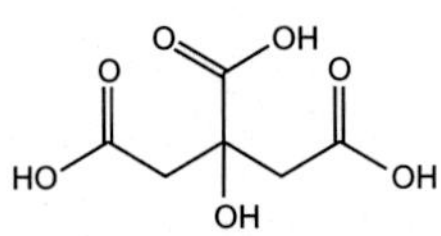

96

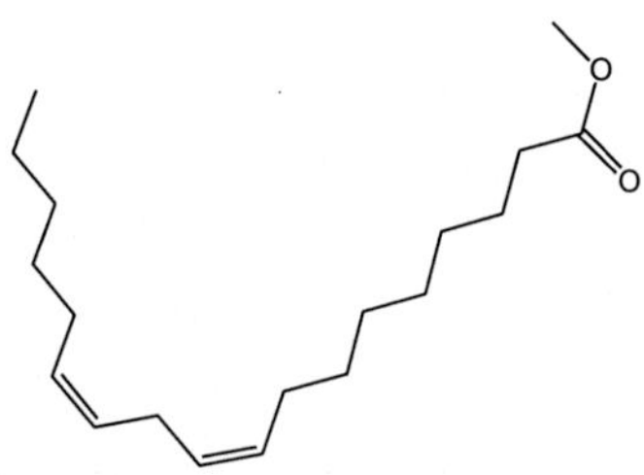

97

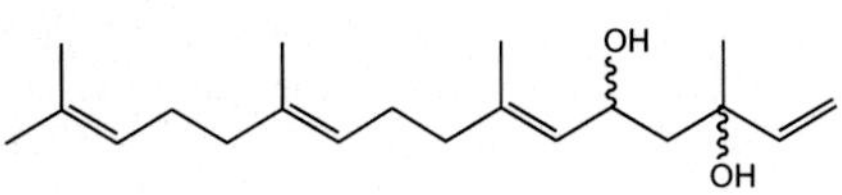

98

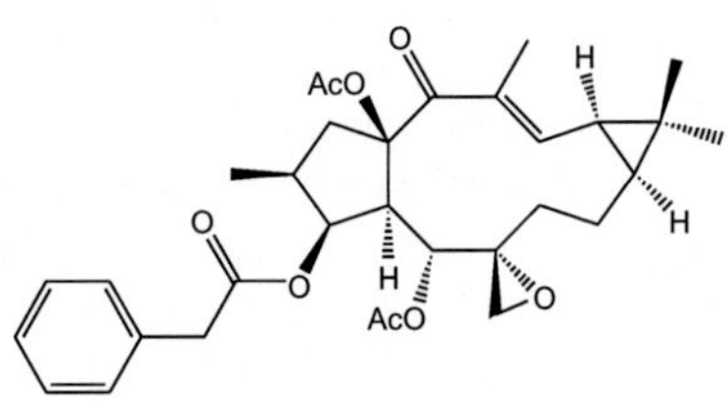

99

Figure 6.3 Continued.

(astragaline), and 3-*O*-β-D-galactopyranosylquercetin (hyperoside) were isolated from the methanol extract of the petals of *P. rhoeas* (corn poppy) (Hillenbrand et al., 2004).

n-Hexadecanoic acid (27.65%) and 6,10,14-trimethyl-2-pentadecanone (16.33%) are predominant among 47 essential oil compounds of *Meconopsis oliverana* (Gao et al., 2013).

Among Papaveraceae plants, meconic acid is found only in *Papaver*, *Meconopsis*, *Stylomecon*, and *Roemeria* (Carolan et al., 2006).

6.3 Pharmacology and therapeutic use

6.3.1 Antioxidant, antimutagenic, and anticarcinogenic effects

Concentration of *P. rhoeas* extract corresponding to IC50 in the DPPH assay possessed the highest antioxidant activity in the anti-ROS biological assay (Todorova et al., 2014). A concentration with 50% scavenging activity expressed the most pronounced antimutagenic properties decreasing Zeo (radiomimetic zeocin)-induced gene conversion twofold, reverse mutation fivefold, and total aberrations fourfold. The same concentration possessed well-expressed anticarcinogenic properties measured as reduction of MMS-induced Ty1 transposition rate fivefold and fourfold when Zeo was used as an inductor. The *P. rhoeas* extracts are worthy of further investigations and could be used as a food additive.

The possible cytotoxicity, genotoxicity, and antioxidant effect of extract from flowers of *P. rhoeas* were investigated in human lymphoblastoid cell line (TK6) (Hasplova et al., 2011). Antioxidant activity was determined using the DPPH assay. The plant extract exhibited dose-dependent free radical scavenging ability. The growth activity assay was used for determination of cytotoxicity. To assess potential genotoxicity, the comet assay was used. The lower extract concentrations (0.25 and 0.5 mg/ml) exerted neither cytotoxic nor genotoxic effects in TK6 cells, but they stimulated cell proliferation. The concentration 25 mg/ml scavenged almost 85% of DPPH free radical. This concentration also had strong cytotoxic and genotoxic effects on TK6 cells. The balance between beneficial and harmful effects should be always considered when choosing the effective dose.

Oleracein E of *Meconopsis horridula* showed significant cytotoxicity against HepG2, with inhibitory rate of 52.2% at 10 μmol/l (Guo et al., 2014a).

Meconopsis integrifolia is a high-mountain endemic species used as a traditional Tibetan and Mongolian herb to treat hepatitis, pneumonia, and edema. Ethanolic extract of *M. integrifolia* exhibited strong antioxidant ability *in vitro* (Zhou et al., 2013). In the rats with CCl_4-induced liver injury, the groups treated with *M. integrifolia* and silymarin showed significantly lower levels of ALT, AST, ALP, and TB. *M. integrifolia* demonstrated good antioxidant activities in both the liver and kidney of the rats *in vivo*. The ethanolic extract from *M. quintuplinervia* exhibited *in vitro* and *in vivo* antioxidant activity (He et al., 2012).

The ability of scavenging OH and DPPH radicals of the *M. oliverana* essential oils is stronger than that of BHT (Gao et al., 2013).

6.3.2 Antimicrobial and antiparasitic activities

Aqueous, ethanolic, and hydroalcoholic extracts of *P. somniferum* (Ying Su) seeds were active against the two acne-causing bacteria, that is, *Propionibacterium acnes* and *Staphylococcus epidermidis* (Chaudhary et al., 2013). Against *P. acne*, aqueous and ethanolic extracts of Tukhm Khashkhash (*P. somniferum* seeds as Unani drugs) showed zones of inhibition of 13 ± 1.04 and 14 ± 1.8 mm, respectively. Against *S. epidermidis*, hydroalcoholic and ethanolic extracts of Tukhm Khashkhash showed zones of inhibition of 15 ± 1.09 and 13 ± 1.6 mm, respectively.

The antiviral effect of *P. pseudocanescens* alkaloids was tested against the *in vitro* replication of several viruses, which belong to different taxonomic groups and represent significant human pathogens (Istatkova et al., 2012). Particular alkaloids possess selective antiviral effects against the replication of poliovirus 1 and human rhinovirus 14, two viruses from the *Enterovirus* genus of the Picornaviridae family.

Among the five *Meconopsis simplicifolia* alkaloids tested, simplicifolianine showed the most potent antiplasmodial activities against the *Plasmodium falciparum* strains (Wangchuk et al., 2013a), TM4/8.2 (chloroquine-antifolate-sensitive strain) and K1CB1 (multidrug-resistant strain), with IC50 values of 0.78 and 1.29 μg/ml, respectively. The compounds tested did not show any significant cytotoxicity activities against human oral carcinoma KB cells and normal Vero cells of African kidney epithelial cells.

6.3.3 Effects on nervous system

Behavioral and pharmacological studies of ethanolic and aqueous extracts showed that *P. rhoeas* extract reduced locomotory, exploratory, and postural behavior of mice (Soulimani et al., 2001). This was evaluated through two specific behavioral tests: the staircase test and free exploratory test. These behavioral and pharmacological effects are more pronounced when the solvent used for extraction is 10% ethanol and is not antagonized by benzodiazepines, opioids, and dopaminergic and cholinergic compounds. The plant extract did not induce sleep in mice after treatment with an infrahypnotic dose of pentobarbital and has a sedative effect at a 400 mg/kg dosage.

Poppy seed oil protected the hippocampus after cerebral ischemia and reperfusion in rats, although it did not alter significantly the frequency of cell death (Cevik-Demirkan et al., 2012).

Administration of the extract of *P. rhoeas* reduced the acquisition and expression of morphine-induced behavioral sensitization in mice (Sahraei et al., 2006a). *P. rhoeas* reduced the acquisition but not the expression of morphine-induced conditioned place preference (Sahraei et al., 2006b). The extract of *P. rhoeas* can ameliorate the withdrawal syndrome in morphine-dependent mice (Pourmotabbed et al., 2004).

6.3.4 Anti-inflammatory activity

P. rhoeas extract could inhibit acute phase of formalin test in mice by opioidergic, glutamatergic, and nitricergic mechanisms (Saeed-Abadi et al., 2012) and can induce corticosterone plasma level, which may be responsible for inhibition of inflammation and chronic phase of pain induced by formalin.

The crude extracts of *Meconopsis simplicifolia* inhibited the production of proinflammatory cytokine and TNF-α in LPS-activated THP-1 monocytic cells (Wangchuk et al., 2013b), which rationalize the traditional use of this plant in the Bhutanese traditional medicine as an individual plant or in combination with other ingredients for the treatment of disorders bearing relevance to the inflammatory conditions.

6.3.5 Other effects

Stress could reduce food and water intake of male NMRI mice and increase delay to eating times, which were inhibited by the *P. rhoeas* extract pretreatment (Ranjbaran et al., 2013). Administration of the extract of *P. rhoeas* can reduce the side effects of stress but increases plasma corticosterone level, which may be due to its effects on the adrenal gland.

6.3.6 Toxicity and safety

Exposure to opium in diabetic patients may be positively associated with the risk of coronary artery disease (CAD) and with the angiographically determined severity and extent of the disease (Hosseini et al., 2011). Dosage of opium consumption may correlate with severity of CAD.

The use of poppy seed tea (PST) as the major source of opiates could be considered favorably within "harm reduction" philosophies, because of its low cost, legal availability, and oral route of administration (Braye et al., 2007). On the other hand, PST could act as a "gateway drug" by inducing opioid dependence and introducing people to the culture of drug abuse.

Poppy seed-containing foods are popular dishes in Hungary and some other Central European countries. Poppy seeds used as food may contain considerable amounts of alkaloids, which raises the question of food safety. Morphine, codeine, thebaine, and noscapine concentrations of poppy seed samples from the period 2001–2010 and consumption data from two Hungarian surveys, carried out in 2003 and 2009, were evaluated (Zentai et al., 2012). Exposure calculations were made for morphine intake by both point estimate and probabilistic methods, and the uncertainty of the calculated values was estimated. The point estimate for the acute consumer exposure, calculated using the 97.5th percentiles of morphine concentration and of poppy seed consumption and taking into account the reduction of morphine content by processing, was 78.64 μg $(kg\ bw)^{-1}\ day^{-1}$ for adults and 116.90 μg $(kg\ bw)^{-1}\ day^{-1}$ for children. Based on probabilistic estimations, the 97.5th and 99th percentile exposures ranged between 18.3–25.4 and 25.6–47.4 μg $(kg\ bw)^{-1}\ day^{-1}$ for adults and between 32.9 and 66.4 μg $(kg\ bw)^{-1}\ day^{-1}$ for children, respectively. As a no observed effect level (NOEL) had not been established, the significance of exposure could not be assessed.

6.4 Molecular biology, phylogeny, and omics

6.4.1 Expressed sequence tag

An integrated approach using targeted metabolite profiles and modest expressed sequence tag (EST) libraries each containing approximately 3500 unigenes was developed to discover and functionally characterize novel genes involved in plant-specialized metabolism (Liscombe et al., 2009). EST databases were established for benzylisoquinoline alkaloid-producing cell cultures of *Eschscholzia californica*, *P. bracteatum*, and *Thalictrum flavum* and are a rich repository of alkaloid biosynthetic genes. ESI-FTICR-MS and ESI-MS/MS analyses facilitated unambiguous identification and relative quantification of the alkaloids in each system. Manual integration of known and candidate biosynthetic genes in each EST library with benzylisoquinoline alkaloid biosynthetic networks assembled from empirical metabolite profiles allowed identification and functional characterization of four *N*-methyltransferases (NMTs). One cDNA from *T. flavum* encoded pavine *N*-methyltransferase (TfPavNMT), which showed a unique preference for (+/−)-pavine and represents the first isolated enzyme involved in the pavine alkaloid branch pathway. The correlation of the occurrence of specific alkaloids, the complement of ESTs encoding known benzylisoquinoline alkaloid biosynthetic genes, and the differential substrate range of characterized NMTs demonstrated the feasibility of bilaterally predicting enzyme function and species-dependent specialized metabolite profiles.

SSR (simple sequence repeat, or microsatellite) markers in the opium poppy were identified from an EST database composed of 20,340 sequences (Lee et al., 2011a,b). In total, 2780 SSR-containing sequences were identified. The most frequent microsatellite had an AT/TA motif (37%). Twenty-two opium poppy EST-SSR markers were developed, and polymorphisms of six markers (psom 2, 4, 12, 13, 17, and 22) were utilized in 135 individuals under narcotic control investigation. An average of three alleles per locus (range: 2–5 alleles) with a mean heterozygosity of 0.167 was detected. Six loci identified 29 unique profiles in 135 individuals. The EST-SSR markers exhibited small degrees of genetic differentiation (fixation index $=0.727$, $p<0.001$). Other variable markers will be needed to facilitate the forensic identification of the opium poppy. To determine the potential for cross species amplification, six markers were tested in five *Papaver* species and two *Eschscholzia* genera. The psom 4 and psom 17 primer pair was transferable.

Micro-RNAs (miRNAs) are regulatory noncoding class of small RNAs functioning in many organisms. Computational approaches identified 20 conserved opium poppy miRNAs belonging to 16 miRNA families in EST database (Unver et al., 2010). Lengths of mature miRNAs varied from 20 to 23 nucleotides located at the different positions of precursor RNAs. Uracil was found to be a dominant nucleotide both in poppy pre-miRNA sequences ($31.28 \pm 7.06\%$ of total nucleotide composition) and in the first position at the $5'$ end of the mature poppy miRNAs. Quantitative real-time PCR (qRT-PCR) assays were applied to compare and validate expression levels of selected *P. somniferum* miRNAs and their target transcripts. Some of the predicted miRNAs and their target genes were found to be differentially expressed in leaf and root tissues. A meaningful correlation between three of the four analyzed pairs of miRNAs and their target transcript expression levels was detected. Additionally, using these predicted miRNAs as queries, 41 potential target mRNAs were found in NCBI protein-coding nucleotide (mRNA) database of all plant species. Some target mRNAs were found to be transcription factors regulating plant development, morphology, and flowering time. Other target mRNAs of identified conserved miRNAs involve in metabolic processes, signal transduction, and stress responses.

6.4.2 Transcriptomics and proteomics

Noscapine is an antitumor alkaloid from opium poppy that binds tubulin, arrests metaphase, and induces apoptosis in dividing human cells. Elucidation of the biosynthetic pathway will enable improvement in the commercial production of noscapine and related bioactive molecules. Transcriptomic analysis revealed the exclusive expression of 10 genes encoding five distinct enzyme classes in a high-noscapine-producing poppy variety, HN1 (Winzer et al., 2012). The analysis of an F2 mapping population indicated that these genes are tightly linked in HN1, and bacterial artificial chromosome (BAC) sequencing confirmed that they exist as a complex gene cluster for plant alkaloids. Virus-induced gene silencing resulted in accumulation of pathway intermediates, allowing gene function to be linked to noscapine synthesis and a novel biosynthetic pathway to be proposed.

Stem cDNA libraries from each of the eight opium poppy cultivars were subjected to 454 pyrosequencing, and searchable EST databases were created from the assembled reads (Desgagné-Penix et al., 2012). These deep and integrated metabolite and transcript databases provide a nearly complete representation of the genetic and metabolic variances responsible for the differential occurrence of specific BIAs in each cultivar as demonstrated using the biochemically well-characterized pathway from tyrosine to morphine. Similar correlations between the occurrence of specific transcripts and alkaloids effectively reveal candidate genes encoding uncharacterized biosynthetic enzymes as shown using cytochromes P450 (CYP) potentially involved in the formation of papaverine and noscapine.

P. rhoeas, an annual plant species in the Papaveraceae family, is part of the biodiversity of agricultural ecosystems and also a noxious agronomic weed. SSR markers were developed to study the genetic diversity of *P. rhoeas*, using an enriched microsatellite library coupled with 454 next-generation sequencing (Kati et al., 2012). A total of 13,825 sequences were obtained that yielded 1795 microsatellite loci, and automated primer design was successful for 598 loci. 74 loci were tested for amplification with 97 primer pairs. Thirty loci passed the tests and were subsequently tested for polymorphism using 384 *P. rhoeas* plants originating from 12 populations from France. Eleven loci showed reliable polymorphism not affected by the presence of null alleles. The number of alleles and the expected heterozygosity ranged from 3 to 7.4 and from 0.27 to 0.73, respectively. A low but significant genetic differentiation among populations was observed. The 11 validated polymorphic microsatellite markers developed will be useful in studies of genetic diversity and population structure of *P. rhoeas*, assisting in designing management strategies for the control or the conservation of this species.

Comparison of the plastid rpl16 gene and the rpl16-rpl14 spacer sequences allowed distinguishing five species, that is, *P. bracteatum*, *P. orientale*, *P. pseudo-orientale*, *P. rhoeas*, *P. setigerum*, and *P. somniferum* from one another (Hosokawa et al., 2004), but sequences from *P. setigerum* and *P. somniferum* were identical. It is difficult to distinguish between *P. bracteatum*, *P. orientale*, and *P. pseudo-orientale* at the vegetative stage of growth. Furthermore, *P. rhoeas* was clearly discriminated from *P. setigerum* and *P. somniferum* by the sequence of the rpl16 exon using young seedlings.

Comparative transcriptome analysis of opium poppy and several other *Papaver* species that do not accumulate morphinan alkaloids showed that known genes encoding BIA biosynthetic enzymes are expressed at higher levels in *P. somniferum* (Ziegler et al., 2009). The transcriptional induction of alkaloid metabolism in cultured cells provides an opportunity to identify components of this process via the integration of deep transcriptome and proteome databases generated using next-generation technologies (Desgagné-Penix et al., 2010). A cDNA library was prepared for opium poppy cell cultures treated with a fungal elicitor for 10 h. Using 454 GS-FLX Titanium pyrosequencing, 427,369 ESTs with an average length of 462 bp were generated. Assembly of these sequences yielded 93,723 unigenes, of which 23,753 were assigned Gene Ontology (GO) annotations. Transcripts encoding all known sanguinarine biosynthetic enzymes were identified in the EST database, five of which were represented among the 50 most abundant transcripts. Liquid chromatography-tandem mass spectrometry (LC–MS/MS) of total protein extracts from cell cultures treated with a fungal elicitor for 50 h facilitated the identification of 1004 proteins. Proteins were fractionated by one-dimensional SDS-PAGE and digested with trypsin prior to LC–MS/MS analysis. Query of an opium poppy-specific EST database substantially enhanced peptide identification. Eight out of 10 known sanguinarine biosynthetic enzymes (Figure 6.4) and many relevant primary metabolic enzymes were represented in the peptide database. The integration of deep transcriptome and proteome analyses provides an effective platform to catalog the components of secondary metabolism and to identify genes encoding uncharacterized enzymes. The proteome database represents the most relevant alkaloid-producing enzymes, compared with the much deeper and more complete transcriptome library. The transcript database contained full-length mRNAs encoding most alkaloid biosynthetic enzymes, which is a key requirement for the functional characterization of novel gene candidates.

The BIA papaverine, synthesized in low amount in most varieties of *P. somniferum*, is used as a vasodilator muscle relaxant and antispasmodic. A comparative transcriptome analysis of high papaverine mutant (pap1) and normal cultivar (BR086) of *P. somniferum* was performed (Pathak et al., 2013). This natural mutant synthesizes more than 12-fold papaverine in comparison to BR086. More than 238 Mb transcriptome data were obtained separately for pap1 and BR086. Assembly of reads generated 127,342 and 106,128 unigenes in pap1 and BR086, respectively. Digital gene expression (DGE) analysis of transcriptomes revealed 3336 differentially expressing unigenes. Enhanced expression of (S)-norcoclaurine-6-*O*-methyltransferase (6OMT),

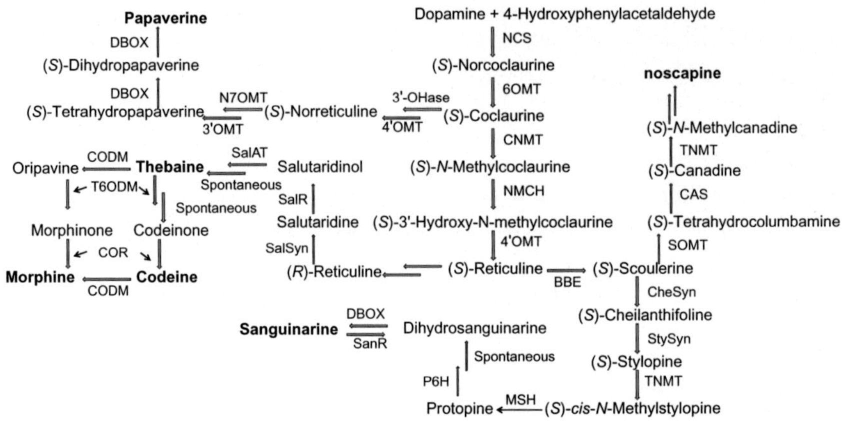

Figure 6.4 Biosynthetic pathway of BIAs in opium poppy. TYDC, tyrosine/dopa decarboxylase; NCS, norcoclaurine synthase; 6OMT, (S)-norcoclaurine 6-*O*-methyltransferase; CNMT, (S)-coclaurine *N*-methyltransferase; NMCH, (S)-*N*-methylcoclaurine 3′-hydroxylase; 4′OMT, (S)-3′-hdroxy-*N*-methylcoclaurine 4′-*O*-methyltransferase; BBE, berberine bridge enzyme; CheSyn, cheilanthifoline synthase; StySyn, stylopine synthase; TNMT, tetrahydroprotoberberine *N*-methyltransferase; MSH, methylstylopine hydroxylase; P6H, protopine 6-hydroxylase; DBOX, dihydrobenzophenanthridine oxidase; DRS, 1,2-dehydroreticuline synthase; DRS, 1,2-dehydroreticuline reductase; SalSyn, salutaridine synthase; SalR, salutaridine reductase; SalAT, salutaridinol 7-*O*-acetyltransferase; THS, thebaine synthase; T6ODM, thebaine 6-*O*-demethylase; COR1, codeinone reductase 1; CODM, codeine *O*-demethylase; 7OMT, (R,S)-reticuline 7-*O*-methyltransferase; N7OMT, (R,S)-norreticuline 7-*O*-methyltransferase.

(S)-3′-hydroxy-*N*-methylcoclaurine 4′-*O*-methyltransferase (4′OMT), and norreticuline 7-*O*-methyltransferase (N7OMT) and downregulation of reticuline 7-*O*-methyltransferase (7OMT) in pap1 in comparison to BR086 suggest (S)-coclaurine as the route for papaverine biosynthesis. Several methyltransferases and dehydrogenases with enhanced expression in pap1 were identified. (S)-coclaurine is the branch-point intermediate and the preferred route for papaverine biosynthesis (Figure 6.4). Differentially expressing methyltransferases and dehydrogenases identified will help in elucidating complete biosynthetic pathway of papaverine. The information generated will be helpful in developing strategies for enhanced biosynthesis of papaverine through biotechnological approaches.

Immunofluorescence labeling and shotgun proteomics were used to establish the cell type-specific localization of morphine biosynthesis in *P. somniferum* (Onoyovwe et al., 2013). Polyclonal antibodies for each of six enzymes involved in converting (R)-reticuline to morphine detected corresponding antigens in sieve elements of the phloem, as described previously for all upstream enzymes transforming (S)-norcoclaurine to (S)-reticuline. Validated shotgun proteomics performed on whole-stem and latex total protein extracts generated 2031 and 830 distinct protein families, respectively. Proteins corresponding to nine morphine biosynthetic enzymes were represented in the whole stem, whereas only four of the final five pathway enzymes were detected in the latex. Salutaridine synthase was detected in the whole stem, but not in the latex subproteome. The final three enzymes converting thebaine to morphine were among the most abundant active latex proteins despite a limited occurrence in laticifers. Multiple charge isoforms of two key *O*-demethylases in the latex were revealed by two-dimensional immunoblot analysis. Salutaridine biosynthesis appears to occur only in sieve elements, whereas conversion of thebaine to morphine is predominant in adjacent laticifers, which contain morphine-rich latex. Complementary use of immunofluorescence labeling and shotgun proteomics has substantially resolved the cellular localization of morphine biosynthesis in opium poppy.

In contrast to the plant, cell cultures of opium poppy do not accumulate alkaloids constitutively but produce sanguinarine in response to treatment with certain fungal-derived elicitors. The induction of sanguinarine biosynthesis provides a model platform to characterize the regulation of BIA pathways and other defense responses. Proteome analysis of elicitor-treated opium poppy cell cultures by two-dimensional denaturing-polyacrylamide gel electrophoresis coupled with LC-tandem MS facilitated the identification of 219 of 340 protein spots based on peptide fragment fingerprint searches of a combination of databases (Zulak et al., 2009). Of the 219 hits, 129 were identified through preexisting plant proteome databases, 63 were identified by matching predicted translation products in opium poppy EST databases, and the remainder shared evidence from both databases. Metabolic enzymes represented the largest category of proteins and included S-adenosylmethionine synthetase, several glycolytic enzymes, and a nearly complete set of tricarboxylic acid cycle enzymes, one alkaloid, and several other secondary metabolic enzymes. The abundance of chaperones, heat shock proteins, protein degradation factors, and pathogenesis-related proteins provided a comprehensive proteomics view on the coordination of plant defense responses. Qualitative comparison of protein abundance in control and elicitor-treated cell cultures allowed the separation of induced and constitutive or suppressed proteins. DNA microarrays were used to corroborate increases in protein abundance with a corresponding induction in cognate transcript levels.

6.4.3 Phylogeny

The internal transcribed spacer (ITS) of 18S–26S nuclear ribosomal (nr) DNA, trnL intron, and the trnL-trnF intergenic spacer of plastid DNA were used in phylogenetic analysis, which demonstrated that the genus *Papaver* is not monophyletic unless the closely related *Roemeria*, *Stylomecon*, and *Meconopsis cambrica* are included in a revised circumscription of *Papaver* (Carolan et al., 2006). Three distinct clades are resolved in a combined ITS and trnL-F analysis. Clade 1 consists of *Papaver* sect. Meconella and Asian *Meconopsis*. Clade 2 contains a group identified as *Papaver* s.s., comprising sections Carinatae, Meconidium, Oxytona, Papaver, Pilosa, Pseudopilosa, and Rhoeadium. Clade 3 consists of *Papaver* sect. Argemonidium and *Roemeria refracta*. A number of diagnostic indels support these groupings. In clade 2,

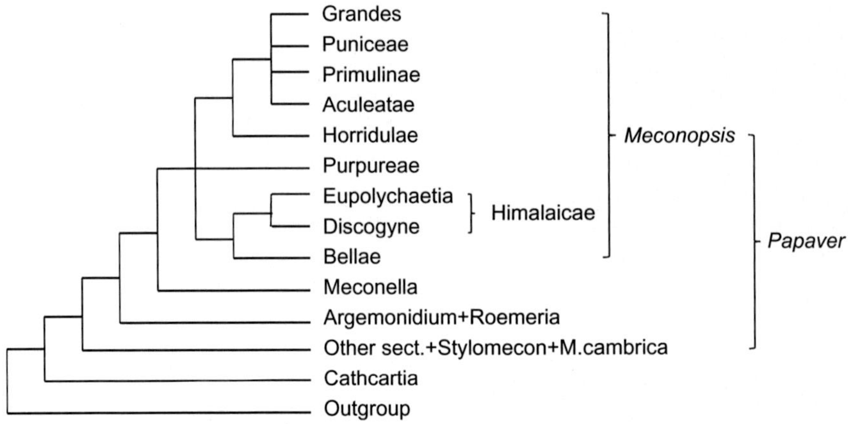

Figure 6.5 Phylogenetic relationship of *Papaver* ITS sequences.

sects. Papaver and Rhoeadium are not monophyletic. A number of morphological characters such as valvate capsule dehiscence, dark or light filaments, and sessile stigmatic disks have arisen in parallel. The phylogenetic trees are incongruent with the existing taxonomy of *Papaver*. *Roemeria*, *Stylomecon heterophylla*, and *Meconopsis cambrica* should be included in *Papaver* (Figure 6.5).

ITS and the chloroplast DNA (cpDNA) trnL-F region were used for phylogenetic reconstruction of *Meconopsis* (Himalayan blue poppy) and its close relatives *Papaver*, *Roemeria*, and *Stylomecon* (Liu et al., 2014b). Five main clades were identified. 41 species of Asian *Meconopsis* fall into clades 1 and 5, which are separated in the phylogenetic tree by clades 2, 3, and 4 of *Papaver* and its allies. There are seven *Papaver* species in China. *P. nudicaule* and *P. radicatum* belong to the section Meconella (clade 2) and are more closely related to Eumeconopsis than other clades; *P. pavoninum* is in clade 2, Argemonidium; *P. orientale*, *P. somniferum*, and *P. rhoeas* are in clade 4 (Figure 6.5).

Clade 5 includes only four Asian *Meconopsis* species (*M. chelidonifolia*, *M. oliverana*, *M. smithiana*, and *M. villosa*), with the remaining Asian species included in clade 1. In clade 1, five subclades (1a-1e) were recognized in the nr ITS tree. Three species (*M. discigera*, *M. pinnatifolia*, and *M. torquata*) of subgenus Discogyne were imbedded in subclade 1a, indicating that the present definition of subgenera in *Meconopsis* should be rejected. Series Superbae and Robustae should be united, with subgenus Discogyne included in this group.

These subclades are inconsistent with any series or sections of the present classifications, suggesting that classifications of the genus should be completely revised based on molecular, morphological, and biogeographic evidence. The *Meconopsis* subgenus Eumeconopsis should belong to *Papaver*; alternatively, *Papaver* could be split into three genera, that is, *Meconella*, *Argemonidium*, and *Papaver*. Obviously, the latter is not parsimonious. Clade 5, basal to other four clades, should be advanced to generic level and be named as *Cathcartia* (Figure 6.5).

Meconopsis is an iconic genus of alpine forbs that includes medicinal and ornamental species. ITS sequences were used in phylogenetic analyses of *Meconopsis* (Xie et al., 2014). *Meconopsis* evolved as a ≥ octaploid clade, with considerable interior structure reflecting further changes in ploidy levels and phenotypic differentiation. *Meconopsis* is a Tibetan region clade, which diverged from the *Meconella* clade of *Papaver* ($2n=14$) approximately 16.6 Ma. This is soon after the "hard" collision of the Indian continent with Asia caused uplift of the Himalaya and Hengduan ranges, greatly extended the Tibetan plateau, and initiated monsoonal climates. Eight major clades within *Meconopsis* are well supported, and these correspond closely to previously recognized subgenus groups. However, the relationship among the clades is poorly resolved, probably because they diverged rapidly ~15–11 Ma during Tibetan plateau uplift.

Two clades, Grandes ($2n=76$) and Puniceae ($2n=84$), are ~ dodecaploid but might originate independently. Other clades are octaploid ($2n=56$). The transition to tetraploidy ($2n=4x=28$) or octaploidy possibly enabled *Meconopsis* to invade cold monsoonal environments (Xie et al., 2014). The eight clades have distinct distributions, variously associated with the Himalaya, the Eastern Plateau, and Hengduan ranges. The dodecaploid Grandes clade occurs in the Himalaya and Hengduan ranges, but *M. integrifolia* and *M. pseudointegrifolia* have adapted to drier plateau habitats to the north. The Purpureae and dodecaploid Puniceae clades spread across the Eastern Plateau and Hengduan ranges. Some *Meconopsis* species were not monophyletic, suggesting that a combination of multilocus molecular and phenotypic traits is required when defining and revising species.

The complex tectonic events and climatic oscillations in the Qinghai-Tibetan Plateau (QTP), the largest and highest plateau in the world, are thought to have had great effects on the evolutionary history of the native plants. A range-wide phylogeographic analysis of *M. integrifolia* based on trnL-trnF and trnfM-trnS regions was performed (Yang et al., 2012), and 26 haplotypes that were phylogenetically divided into six clades dated to the late Tertiary were defined. The six clades correspond, respectively, to highly differentiated population groups that do not overlap in geographic distribution, implying that the mountain ranges acting as corridors or barriers greatly affected the evolutionary history of the QTP plants. The older clade of *M. integrifolia* only occurs in the southwest of the species' range, while the distributions of younger clades extend northeastward in the eastern QTP, suggesting that climatic divergence resulting from the uplift of the QTP triggered the initial divergence of *M. integrifolia* native to the plateau. The nrDNA ITS region was used to clarify the unexpected phylogenetic relationships of cpDNA haplotypes between *M. integrifolia* ($2n=76$) and *M. betonicifolia* ($2n=76$). The topological incongruence between the cp and nr phylogenies suggests an ancestral hybridization between two species. Our study indicates that geographic isolation and hybridization are two important mechanisms responsible for the population differentiation and speciation of *Meconopsis*, a species-rich genus with complex polyploids.

The perennial herb *Meconopsis cambrica*, a western European endemic, was thought to be the only European species of the otherwise Himalayan genus *Meconopsis* and has been interpreted as a tertiary relict species. By analyzing rbcL and ITS sequence variation, the split between *M. cambrica* and its sister clade *Papaver* was dated to the Middle to Upper Miocene (12.8 My, 6.4–19.2 My HPD) (Valtueña et al., 2012). Within *M. cambrica*, cpDNA sequence variation reveals the existence of two groups of populations with a comparable level of genetic variation: a northern group from Great Britain, the Massif Central, the western Pyrenees and the Iberian System, and a southern group from the central and eastern Pyrenees. Populations from the Cantabrian Mountains were placed in both groups. Based on ITS sequence variation, the divergence between these two groups can be dated to 1.5 My (0.4–2.8 My), and the age of the British populations is estimated as 0.37 My (0.0–0.9 My). Amplified fragment length polymorphism (AFLP) results confirm the distinctive nature of the populations from Britain, the Massif Central, and the central and eastern Pyrenees. These patterns of latitudinal variation of *M. cambrica* differ from patterns of longitudinal differentiation found in many other temperate species and imply glacial survival of the northern populations in northerly refugia. The primary differentiation into northern and southern cpDNA groups dates to near the onset of the Quaternary and suggests that an ancient phylogeographic pattern has survived through several glacial periods. This species has persisted for a long period with a highly fragmented and probably highly localized distribution.

6.4.4 Molecular biology

The relative nuclear DNA content at the vegetative stage was measured by flow cytometric (FCM) analysis in three closely related species of the genus *Papaver* section Oxytona, that is, *P. orientale*, *P. pseudo-orientale*, and *P. bracteatum*, based on the difference of the chromosome numbers (Aragane et al., 2014). The nr DNA content differed between three species, and most of the commercially distributed plants examined could be identified as *P. bracteatum*. The remaining plants were *P. pseudo-orientale*, a nonnarcotic plant. The FCM results agreed with the results obtained by the morphological analysis, the intergenic spacer sequence of rpl16-rpl14 of cp DNA, and the presence of thebaine.

Of the 110 *Papaver* species, only *P. somniferum* and *P. setigerum* are controlled poppies in Korea. It is important to check if the morphologically similar poppy samples contain controlled substances such as morphine and codeine for forensic purpose. Since the alkaloid content of *Papaver* plants varies according to their growing stage, chemical components analysis alone is not enough to identify exact species. Hundreds of poppy plants suspected to be *P. somniferum* were found in Jeju Island, South Korea (Choe et al., 2012). They had a slightly different but overall similar appearance to *P. somniferum*. Using GC–MS analysis, codeine, rhoeadine, papaverine, protopine, noscapine, setigeridine, and trace amounts of morphine were detected in these samples. Although their chemical components were different from what has been described in literatures for *P. setigerum*, they could be assumed to be *P. setigerum* based on their morphological features and GC–MS results. Chromosome numbers using their seeds showed $2n=44$, and the numbers were in accordance with those of *P. setigerum*. Nucleotide substitution or insertion/deletion of ITS, 18S rRNA, rbcL (large subunit of ribulose 1,5-bisphosphate carboxylase), trnL-trnF IGS (intergenic spacer), trnL intron, and psbA-trnH was assessed as universal genetic markers for *P. setigerum*. Six target genes involved in the biosynthesis of BIAs, including TYDC (tyrosine/dopa decarboxylase), SAT (salutaridinol-7-*O*-acetyltransferase), BBE (berberine bridge enzyme), COR (codeinone reductase), CYP80B1 ((S)-*N*-methylcoclaurine 3′-hydroxylase), and NCS (norcoclaurine synthase), were tested as *Papaver*-specific genetic markers by the existence of their PCR products. The sequences of the six universal genetic markers and six *Papaver*-specific markers for *P. setigerum* were identified. The trnL intron and psbA-trnH sequences of seven *Papaver* species were also identified.

The identification of opium poppy was tested using 10 genetic markers that are universal for all plants or specific to a few poppy plants (Lee et al., 2010). The genetic distances of universal markers such as ITS, 18S rRNA, rbcL, and trnL-trnF intergenic spacer of 14 species included in the Papaveraceae and Fumariaceae families were acquired by sequence comparisons. Both the ITS region and trnL-trnF IGS showed high levels of interspecific divergence. Six *Papaver* genus-specific markers were developed from coding regions involved in morphine biosynthesis. Three markers (TYDC, NCS, and BBE) produced amplicons only in opium poppy, providing a presence/absence test for opium poppy, while three additional markers (CYP80B1, SAT, and COR) were genus-specific. These 10 markers might be useful for the forensic DNA analysis of opium poppy.

AFLP markers were developed to identify polymorphism between DM (downy mildew)-resistant and DM-susceptible opium poppy genotypes (Dubey et al., 2010). These markers could be used in future genetic studies for analysis of linkage to the DM resistance trait. Comparative sequence analysis of the plastid-encoded rpoC1 gene from both genotypes revealed four single-nucleotide substitutions at four positions in DM-resistant genotypes that caused three amino acid changes in the protein sequence (Dubey et al., 2009)—(1) A to C transversion at position 825 (Glu275Asp), (2) A to G transition at position 1203 (Ile401Met), and (3) T to C transition at position 1422 and G to A transition at position 1423 both in the same codon of the reading frame (Ala475Thr). This finding may lead to implication of possible role of RNA polymerase β′ subunit in resistance to DM caused by *Peronospora arborescens*.

Meconopsis horridula is one of the eight most famous flowers in Yunnan, China. A modified biotin–streptavidin capture method was used to detect 13 microsatellite markers in the genome of *M. horridula* (Zhao et al., 2010). The polymorphism of each locus was assessed in 24 samples collected from four populations. The number of alleles per locus ranged from 2 to 7 (mean: 3.2). The observed and expected heterozygosities ranged from 0.0833 to 0.9167 and from 0.0816 to 0.8050, respectively. Nine of the 13 microsatellite markers were successfully amplified in three other congeneric species. These polymorphic SSR markers could be useful for studying the population genetics of *M. horridula* and for assessing genetic variation in conservation programs of *Meconopsis*.

6.4.5 Metabolomics

A method that uses metabolite profiling for species discrimination between *P. somniferum*, *Papaver rhoeas*, and *P. setigerum* was developed (Choe et al.,

2011), and the gas chromatography/mass spectrometry (GC–MS) data of the alkaline extract were processed with in-house Microsoft Visual Basic modules and the chemical information was analyzed through multivariate statistical analyses such as hierarchical cluster analysis (HCA), principal component analysis (PCA), and discriminant analysis (DA). The GC–MS results combined with multivariate analysis demonstrated that the metabolite profiling was an efficient technique for the classification and this method will provide a powerful tool for the identification of Korea *Papaver* species.

Variations in plant secondary metabolism as a result of genetic diversity are often associated with perturbations in other metabolic pathways. ^{1}H nuclear magnetic resonance (NMR) metabolite profiling, as part of a functional genomics platform, was used for the analysis of primary and secondary metabolism in opium poppy (Hagel et al., 2008). Aqueous and chloroform extracts of six different opium poppy cultivars were subjected to chemometric analysis. PCA of the ^{1}H NMR spectra for latex extracts clearly distinguished two varieties, including a low-alkaloid variety and a high-thebaine, low-morphine cultivar. Distinction was also made between pharmaceutical-grade opium poppy cultivars and a condiment variety. Such phenotypic differences were not observed in root extracts. Loading plots confirmed that morphinan alkaloids contributed predominantly to the variance in latex extracts. Quantification of 34 root and 21 latex metabolites showed major differences in the accumulation of specific alkaloids in the latex of the low-alkaloid and high-thebaine, low-morphine varieties. Relatively few differences were found in the levels of other metabolites, indicating that the variation was specific for alkaloid metabolism. Exceptions in the low-alkaloid cultivar included an increased accumulation of the alkaloid precursor tyramine and reduced levels of sucrose, some amino acids, and malate. Real-time PCR analysis of 42 genes involved in primary and secondary metabolism showed differential gene expression mainly associated with alkaloid biosynthesis. Reduced alkaloid levels in the condiment variety were associated with the reduced abundance of transcripts encoding several alkaloid biosynthetic enzymes.

Metabolite fingerprinting and compound-specific profiling showed the extensive reprogramming of primary metabolic pathways in association with the induction of alkaloid biosynthesis in response to elicitor treatment (Zulak et al., 2008). Using Chenomx NMR Suite v. 4.6, a software package capable of identifying and quantifying individual compounds based on their respective signature spectra, the levels of 42 diverse metabolites were monitored over a 100 h time course in control and elicitor-treated opium poppy cell cultures. Overall, detectable and dynamic changes in the metabolome of elicitor-treated cells, especially in cellular pools of carbohydrates, organic acids, and nonprotein amino acids, were detected within 5 h after elicitor treatment. The metabolome of control cultures also showed substantial modulations 80 h after the start of the time course, particularly in the levels of amino acids and phospholipid pathway intermediates. Specific flux modulations were detected throughout primary metabolism, including glycolysis, the tricarboxylic acid cycle, nitrogen assimilation, phospholipid/fatty acid synthesis, and the shikimate pathway, all of which generate secondary metabolic precursors. The response of cell cultures to elicitor treatment involves the extensive reprogramming of primary and secondary metabolism and associated cofactor biosynthetic pathways.

Tasmanian opium accounts for 25% of the world's legal supply of opium straw, and in 1998–1999, sufficient numbers of flower pods (66,013) to manufacture ca 500 kg of heroin were stolen. The heroin signature program was developed to determine the origin of heroin from Tasmania (Odell et al., 2008). Tasmanian poppies contain a unique alkaloid, oripavine, which is the source of "marker" impurities in illicit heroin produced from Tasmanian poppy straw. Treatment of oripavine (500 mg) under Thiboumery and Mohr heroin processing conditions, followed by simple evaporative workup, afforded 613 mg of a dark orange residue (1), which upon extensive chromatographic purification yielded oripavine 3-acetate (2), 22 mg; 3-acetyl-*N*-acetyldesthebaine (3), 35 mg; 3-acetyl-6-methoxy-4,5-epoxyphenanthrene (4), 5.8 mg; 3,4-diacetyl-6-methoxyphenanthrene (5), 27 mg; and 3,4,6-methoxy-5-[2(*N*-methylacetamido)]ethylphenanthrene (6), 52 mg. Compounds (2–6) are derived from oripavine and are unique to heroin derived from the Tasmanian poppy. The analysis of illicit heroin samples seized from Turkey, Pakistan, Columbia, and Myanmar did not reveal any of these marker compounds. Four marker compounds (3–6) were identified in seized heroin samples from Australia suggesting that they are of Tasmanian origin.

6.5 Conclusions

Recent advances in transcriptomics, proteomics, and metabolomics have created unprecedented opportunities for isolating and characterizing novel BIA biosynthetic genes. The application of next-generation sequencing and cDNA microarrays for selecting gene candidates based on comparative transcriptome analysis is becoming a new paradigm in medicinal plant research. The establishment of corresponding transcript and protein databases generated by next-generation technologies in a system with a well-defined metabolite profile facilitates an improved linkage between genes, enzymes, and pathway components. The advanced mass spectrometric techniques to perform deep proteome and targeted metabolite analyses on BIA-producing plant tissues should be highlighted. Methodologies for functionally characterizing biosynthetic gene candidates through *in vitro* enzyme assays and transient gene silencing in plants should be updated continually.

References

Aragane, M., Watanabe, D., Nakajima, J., et al., 2014. Rapid identification of a narcotic plant *Papaver bracteatum* using flow cytometry. J. Nat. Med. 68 (4), 677–685.

Bozan, B., Temelli, F., 2008. Chemical composition and oxidative stability of flax, safflower and poppy seed and seed oils. Bioresour. Technol. 99 (14), 6354–6359.

Braye, K., Harwood, T., Inder, R., et al., 2007. Poppy seed tea and opiate abuse in New Zealand. Drug Alcohol Rev. 26 (2), 215–219.

Carolan, J.C., Hook, I.L., Chase, M.W., et al., 2006. Phylogenetics of *Papaver* and related genera based on DNA sequences from ITS nuclear ribosomal DNA and plastid trnL intron and trnL-F intergenic spacers. Ann. Bot. 98 (1), 141–155.

Cevik-Demirkan, A., Oztaşan, N., Oguzhan, E.O., et al., 2012. Poppy seed oil protection of the hippocampus after cerebral ischemia and re-perfusion in rats. Biotech. Histochem. 87 (8), 499–505.

Chaudhary, S.S., Tariq, M., Zaman, R., Imtiyaz, S., 2013. The in vitro anti-acne activity of two unani drugs. Anc. Sci. Life 33 (1), 35–38.

Cheng, F., Guo, M., Zhang, Y., et al., 2011. Tracking the active component of Tibetan medicine *Meconopsis quintuplinervia* from Gansu. Zhong Yao Cai 34 (1), 69–71.

Choe, S., Kim, S., Lee, C., et al., 2011. Species identification of *Papaver* by metabolite profiling. Forensic Sci. Int. 211 (1–3), 51–60.

Choe, S., Lee, E., Jin, G.N., et al., 2012. Genetic and chemical components analysis of *Papaver setigerum* naturalized in Korea. Forensic Sci. Int. 222 (1–3), 387–393.

Desgagné-Penix, I., Khan, M.F., Schriemer, D.C., et al., 2010. Integration of deep transcriptome and proteome analyses reveals the components of alkaloid metabolism in opium poppy cell cultures. BMC Plant Biol. 10, 252.

Desgagné-Penix, I., Farrow, S.C., Cram, D., et al., 2012. Integration of deep transcript and targeted metabolite profiles for eight cultivars of opium poppy. Plant Mol. Biol. 79 (3), 295–313.

Dubey, M.K., Shasany, A.K., Dhawan, O.P., et al., 2009. Genetic variation revealed in the chloroplast-encoded RNA polymerase beta′ subunit of downy mildew-resistant genotype of opium poppy. J. Hered. 100 (1), 76–85.

Dubey, M.K., Shasany, A.K., Dhawan, O.P., et al., 2010. AFLP studies on downy-mildew-resistant and downy-mildew-susceptible genotypes of opium poppy. J. Genet. 89 (1), 9–19.

Frick, S., Kramell, R., Schmidt, J., et al., 2005. Comparative qualitative and quantitative determination of alkaloids in narcotic and condiment *Papaver somniferum* cultivars. J. Nat. Prod. 68 (5), 666–673.

Gao, A., Zhao, B., Gong, J., et al., 2013. Chemical components of essential oils from *Meconopsis oliverana* and their antioxidant activity. Zhongguo Zhong Yao Za Zhi 38 (2), 284–288.

Guo, Z.Q., Guo, Q., Zhang, S., et al., 2014a. Chemical constituents from the aerial parts of *Meconopsis horridula* (Papaveraceae). Biochem. Syst. Ecol. 55, 329–332.

Guo, Z.Q., Guo, Q., Zhu, Z.X., et al., 2014b. Chemical constituents from a Tibetan medicine *Meconopsis horridula*. Zhongguo Zhong Yao Za Zhi 39 (7), 1152–1156.

Hagel, J.M., Weljie, A.M., Vogel, H.J., Facchini, P.J., 2008. Quantitative 1H nuclear magnetic resonance metabolite profiling as a functional genomics platform to investigate alkaloid biosynthesis in opium poppy. Plant Physiol. 147 (4), 1805–1821.

Hasplova, K., Hudecova, A., Miadokova, E., et al., 2011. Biological activity of plant extract isolated from *Papaver rhoeas* on human lymfoblastoid cell line. Neoplasma 58 (5), 386–391.

He, J., Huang, B., Ban, X., et al., 2012. In vitro and in vivo antioxidant activity of the ethanolic extract from *Meconopsis quintuplinervia*. J. Ethnopharmacol. 141 (1), 104–110.

Hillenbrand, M., Zapp, J., Becker, H., 2004. Depsides from the petals of *Papaver rhoeas*. Planta Med. 70 (4), 380–382.

Hosokawa, K., Shibata, T., Nakamura, I., Hishida, A., 2004. Discrimination among species of *Papaver* based on the plastid rpl16 gene and the rpl16-rpl14 spacer sequence. Forensic Sci. Int. 139 (2–3), 195–199.

Hosseini, S.K., Masoudkabir, F., Vasheghani-Farahani, A., et al., 2011. Opium consumption and coronary atherosclerosis in diabetic patients: a propensity score-matched study. Planta Med. 77 (17), 1870–1875.

Istatkova, R., Philipov, S., Yadamsurenghiin, G.O., et al., 2008. Alkaloids from *Papaver nudicaule* L. Nat. Prod. Res. 22 (7), 607–611.

Istatkova, R., Nikolaeva-Glomb, L., Galabov, A., et al., 2012. Chemical and antiviral study on alkaloids from *Papaver pseudocanescens* M. Pop. Z. Naturforsch. C 67 (1–2), 22–28.

Kati, V., Corre, V.L., Michel, S., et al., 2012. Isolation and characterisation of 11 polymorphic microsatellite markers in *Papaver rhoeas* L. (Corn Poppy), a major annual plant species from cultivated areas. Int. J. Mol. Sci. 14 (1), 470–479.

Krist, S., Stuebiger, G., Unterweger, H., et al., 2005. Analysis of volatile compounds and triglycerides of seed oils extracted from different poppy varieties (*Papaver somniferum* L.). J. Agric. Food Chem. 53 (21), 8310–8316.

Lee, E.J., Hwang, I.K., Kim, N.Y., et al., 2010. An assessment of the utility of universal and specific genetic markers for opium poppy identification. J. Forensic Sci. 55 (5), 1202–1208.

Lee, D.U., Park, J.H., Wessjohann, L., Schmidt, J., 2011a. Alkaloids from *Papaver coreanum.* Nat. Prod. Commun. 6 (11), 1593–1594.

Lee, E.J., Jin, G.N., Lee, K.L., et al., 2011b. Exploiting expressed sequence tag databases for the development and characterization of gene-derived simple sequence repeat markers in the opium poppy (*Papaver somniferum* L.) for forensic applications. J. Forensic Sci. 56 (5), 1131–1135.

Liscombe, D.K., Ziegler, J., Schmidt, J., et al., 2009. Targeted metabolite and transcript profiling for elucidating enzyme function: isolation of novel *N*-methyltransferases from three benzylisoquinoline alkaloid-producing species. Plant J. 60 (4), 729–743.

Liu, J., Wu, H., Zheng, F., et al., 2014a. Chemical constituents of *Meconopsis horridula* and their simultaneous quantification by high-performance liquid chromatography coupled with tandem mass spectrometry. J. Sep. Sci. 37 (18), 2513–2522.

Liu, Y.C., Liu, Y.N., Yang, F.S., Wang, X.Q., 2014b. Molecular phylogeny of Asian *Meconopsis* based on nuclear ribosomal and chloroplast DNA sequence data. PLoS One 9 (8), e104823.

Ma, W.G., Fukushi, Y., Tahara, S., Osawa, T., 2000. Fungitoxic alkaloids from Hokkaido Papaveraceae. Fitoterapia 71 (5), 527–534.

Morimoto, S., Suemori, K., Taura, F., Shoyama, Y., 2003. New dimeric morphine from opium poppy (*Papaver somuniferum*) and its physiological function. J. Nat. Prod. 66 (7), 987–989.

Odell, L.R., Skopec, J., McCluskey, A., 2008. Isolation and identification of unique marker compounds from the Tasmanian poppy *Papaver somniferum* N. Implications for the identification of illicit heroin of Tasmanian origin. Forensic Sci. Int. 175 (2–3), 202–208.

Onoyovwe, A., Hagel, J.M., Chen, X., et al., 2013. Morphine biosynthesis in opium poppy involves two cell types: sieve elements and laticifers. Plant Cell 25 (10), 4110–4122.

Pathak, S., Lakhwani, D., Gupta, P., et al., 2013. Comparative transcriptome analysis using high papaverine mutant of *Papaver somniferum* reveals pathway and uncharacterized steps of papaverine biosynthesis. PLoS One 8 (5), e65622.

Philipov, S., Istatkova, R., Yadamsurenghiin, G.O., et al., 2007. A new 8, 14-dihydropromorphinane alkaloid from *Papaver nudicaule* L. Nat. Prod. Res. 21 (9), 852–856.

Pourmotabbed, A., Rostamian, B., Manouchehri, G., et al., 2004. Effects of *Papaver rhoeas* extract on the expression and development of morphine-dependence in mice. J. Ethnopharmacol. 95 (2–3), 431–435.

Ranjbaran, M., Mirzaei, P., Lotfi, F., et al., 2013. Reduction of metabolic signs of acute stress in male mice by *Papaver rhoaes* hydro-alcoholic extract. Pak. J. Biol. Sci. 16 (19), 1016–1021.

Saeed-Abadi, S., Ranjbaran, M., Jafari, F., et al., 2012. Effects of *Papaver rhoeas* (L.) extract on formalin-induced pain and inflammation in mice. Pak. J. Biol. Sci. 15 (21), 1041–1044.

Sahraei, H., Faghih-Monzavi, Z., Fatemi, S.M., et al., 2006a. Effects of *Papaver rhoeas* extract on the acquisition and expression of morphine-induced behavioral sensitization in mice. Phytother. Res. 20 (9), 737–741.

Sahraei, H., Fatemi, S.M., Pashaei-Rad, S., et al., 2006b. Effects of *Papaver rhoeas* extract on the acquisition and expression of morphine-induced conditioned place preference in mice. J. Ethnopharmacol. 103 (3), 420–424.

Salehi, P., Sonboli, A., Zavareh, A.F., et al., 2007. Narcotic alkaloids of four papaver species from Iran. Z. Naturforsch. C 62 (1–2), 16–18.

Sari, A., Gray, A.I., Sariyar, G., 2004. A new dehydroaporphine alkaloid from *Papaver fugax*. Nat. Prod. Res. 18 (3), 265–268.

Sari, A., Gray, A.I., Sariyar, G., 2006. Two new benzylisoquinoline alkaloids from *Papaver triniifolium*. Nat. Prod. Res. 20 (5), 493–496.

Schliemann, W., Schneider, B., Wray, V., et al., 2006. Flavonols and an indole alkaloid skeleton bearing identical acylated glycosidic groups from yellow petals of *Papaver nudicaule*. Phytochemistry 67 (2), 191–201.

Schmidt, J., Boettcher, C., Kuhnt, C., et al., 2007. Poppy alkaloid profiling by electrospray tandem mass spectrometry and electrospray FT-ICR mass spectrometry after [ring-13C6]-tyramine feeding. Phytochemistry 68 (2), 189–202.

Siddikov, D., Zhalolov, I., Tashkhodzhaev, B., et al., 2005. Alkaloids from *Papaver angrenicum*. II. Crystal and molecular structure of the new diisoquinoline alkaloid pangrenine. Chem. Nat. Compd. 41 (4), 442–445.

Soulimani, R., Younos, C., Jarmouni-Idrissi, S., et al., 2001. Behavioral and pharmacotoxicological study of Papaver rhoeas L. in mice. J. Ethnopharmacol. 74 (3), 265–274.

Tanaka, M., Fujimori, T., Uchida, I., et al., 2001. A malonylated anthocyanin and flavonols in blue *Meconopsis* flowers. Phytochemistry 56 (4), 373–376.

Tatsis, E.C., Böhm, H., Schneider, B., 2013. Occurrence of nudicaulin structural variants in flowers of papaveraceous species. Phytochemistry 92, 105–112.

Todorova, T., Pesheva, M., Gregan, F., Chankova, S., 2014. Antioxidant, antimutagenic, and anticarcinogenic effects of papaver rhoeas L. extract on *Saccharomyces cerevisiae*. J. Med. Food. 18 (4), 460–467.

Unver, T., Parmaksiz, I., Dündar, E., 2010. Identification of conserved micro-RNAs and their target transcripts in opium poppy (*Papaver somniferum* L.). Plant Cell Rep. 29 (7), 757–769.

Valtueña, F.J., Preston, C.D., Kadereit, J.W., 2012. Phylogeography of a Tertiary relict plant, *Meconopsis cambrica* (Papaveraceae), implies the existence of northern refugia for a temperate herb. Mol. Ecol. 21 (6), 1423–1437.

Wangchuk, P., Keller, P.A., Pyne, S.G., et al., 2013a. A new protoberberine alkaloid from *Meconopsis simplicifolia* (D. Don) Walpers with potent antimalarial activity against a multidrug resistant *Plasmodium falciparum* strain. J. Ethnopharmacol. 150 (3), 953–959.

Wangchuk, P., Keller, P.A., Pyne, S.G., Taweechotipatr, M., 2013b. Inhibition of TNF-α production in LPS-activated THP-1 monocytic cells by the crude extracts of seven Bhutanese medicinal plants. J. Ethnopharmacol. 148 (3), 1013–1017.

Winzer, T., Gazda, V., He, Z., et al., 2012. A *Papaver somniferum* 10-gene cluster for synthesis of the anticancer alkaloid noscapine. Science 336 (6089), 1704–1708.

Xie, H., Ash, J.E., Linde, C.C., et al., 2014. Himalayan-Tibetan plateau uplift drives divergence of polyploid poppies: *Meconopsis* Viguier (Papaveraceae). PLoS One 9 (6), e99177.

Yang, F.S., Qin, A.L., Li, Y.F., Wang, X.Q., 2012. Great genetic differentiation among populations of *Meconopsis integrifolia* and its implication for plant speciation in the Qinghai-Tibetan Plateau. PLoS One 7 (5), e37196.

Zentai, A., Sali, J., Szeitzné-Szabó, M., et al., 2012. Exposure of consumers to morphine from poppy seeds in Hungary. Food Addit. Contam. Part A Chem. Anal. Control Expo. Risk Assess. 29 (3), 403–414.

Zhao, Y., Zhang, S.B., Yang, J., Zhang, L., 2010. Characterization of 13 microsatellite loci developed from *Meconopsis horridula*. Genet. Mol. Biol. 33 (3), 539–541.

Zhou, G., Chen, Y., Liu, S., et al., 2013. In vitro and in vivo hepatoprotective and antioxidant activity of ethanolic extract from *Meconopsis integrifolia* (Maxim.) Franch. J. Ethnopharmacol. 148 (2), 664–670.

Ziegler, J., Facchini, P.J., Geissler, R., et al., 2009. Evolution of morphine biosynthesis in opium poppy. Phytochemistry 70 (15–16), 1696–1707.

Zulak, K.G., Weljie, A.M., Vogel, H.J., Facchini, P.J., 2008. Quantitative ^{1}H NMR metabolomics reveals extensive metabolic reprogramming of primary and secondary metabolism in elicitor-treated opium poppy cell cultures. BMC Plant Biol. 8, 5.

Zulak, K.G., Khan, M.F., Alcantara, J., et al., 2009. Plant defense responses in opium poppy cell cultures revealed by liquid chromatography-tandem mass spectrometry proteomics. Mol. Cell. Proteomics 8 (1), 86–98.

Chemical and biological studies of *Aconitum* pharmaceutical resources

7

7.1 Introduction

Aconitum L. (Ranunculaceae) is a large genus of about 300 species distributed in the temperate regions of the Northern Hemisphere (Xiao et al., 2006; Hao et al., 2013a). This genus consists of three well-defined subgenera, that is, *Lycoctonum* (DC.) Peterm., *Aconitum*, and *Gymnaconitum* (Stapf) Rapaics (Xiao et al., 2006; Jabbour and Renner, 2012). Southwestern China, particularly its Hengduan Mountains region, where at least 166 species are found, is the most important center of diversity and speciation of the genus (Jabbour and Renner, 2012; Luo et al., 2005; Luo and Yang, 2005). Many species in this genus are poisonous and have been used as medicinal plants. In China, at least 76 *Aconitum* species have been medicinally used (Xiao et al., 2006). They are mainly used for the treatment of plaque, sepsis, intoxication, cold- and immunosuppression-induced ailments, rheumatoid arthritis, and various types of pain, including migraine, swelling induced by trauma and fracture, and facial paralysis. Pharmacologically, they can be developed as analgesic, antirheumatic, and antiarrhythmic agents.

The most well-known medicinal compound of *Aconitum* is diterpenoid alkaloid. For conservation and sustainable exploitation of *Aconitum* pharmaceutical resources, it is essential to study the chemical profile of various species and pharmacological effects of multiple compounds therein, to evaluate the taxonomic value of diterpenoid alkaloids mainly at subgeneric and serial levels, and to infer their biosynthetic pathways. The correlation between molecular phylogeny, chemical ingredients, and pharmaceutical uses in the genus *Aconitum* is also discussed based on the data of the toxicity and therapeutic value of the species in the genus, which falls into the research field of pharmacophylogenetics (Xiao et al., 2006).

7.2 Diterpenoid alkaloids

In a chemotaxonomic viewpoint, plant diterpenoid alkaloids have four skeletal types: C18, C19, C20, and bisditerpenoid alkaloids. They can be subdivided into 14 subgroups (Figure 7.1), that is, C18: appaconine-type (1) and ranaconine-type (2); C19: aconitine-type (3), lycoctonine-type (4), 7,17-seco-type (5), and lactone-type (6); C20: atisine-type (7), denudatine-type (8), hetidine-type (9), hetisine-type (10),

Medicinal plants: chemistry, biology and omics. http://dx.doi.org/10.1016/B978-0-08-100085-4.00007-4

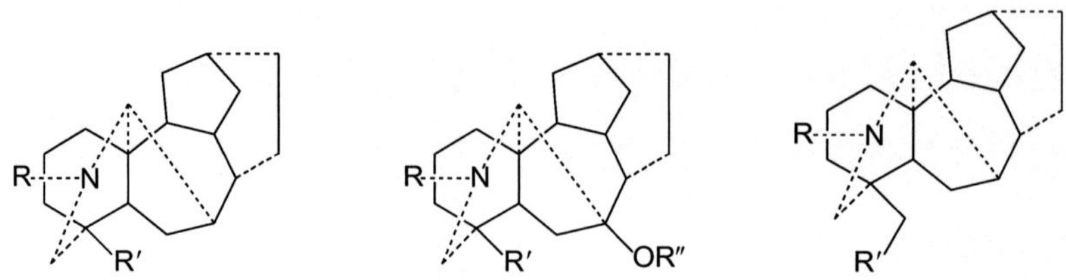

Lappaconines (I) Ranaconines (II) Aconitine (III)

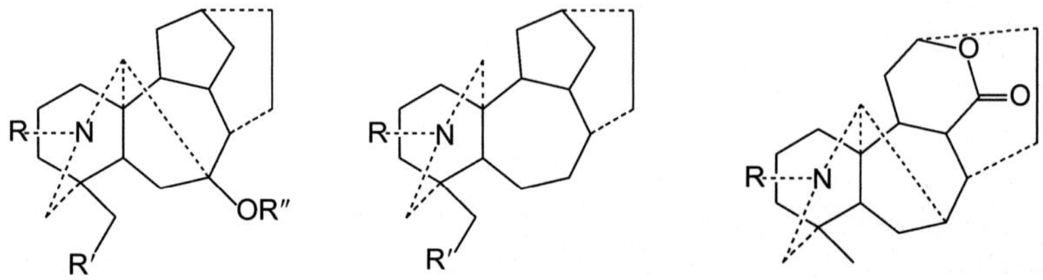

Lycoctonine (IV) 7, 17-Seco type (V) Lactones (VI)

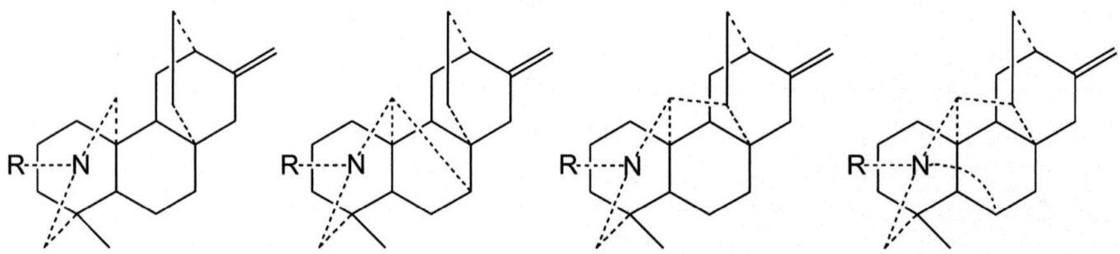

Atisine (VII) Denudatine (VIII) Hetidine (IX) Hetisine (X)

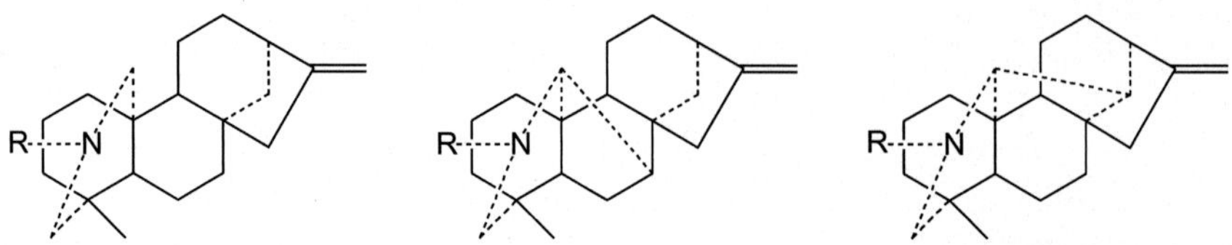

Veatchine (XI) Napelline (XII) Anopterine (XIII)

Figure 7.1 Type of diterpenoid alkaloids found in *Aconitum*. C18, I–II; C19, III–VI; C20, VII–XIII; bisditerpenoid alkaloid, XIV.
Modified from Xiao et al. (2006).

(Continued)

veatihine-type (11) (not present in *Aconitum* species), napelline-type (12), and anopterine-type (13) (none in *Aconitum*); and bisditerpenoid alkaloids (14). The newly found diterpenoid alkaloids are summarized in Figure 7.2 and Table 7.1.

Atisine content ranged from 0.14% to 0.37% and total alkaloids (aconites) from 0.20% to 2.49%, among 14 accessions of *A. heterophyllum* (Malhotra et al., 2014). Two accessions contained the highest atisine content with 0.30% and 0.37% and the highest alkaloids content with 2.22% and 2.49%, respectively. No atisine was detected in leaves and shoots of *A. heterophyllum*.

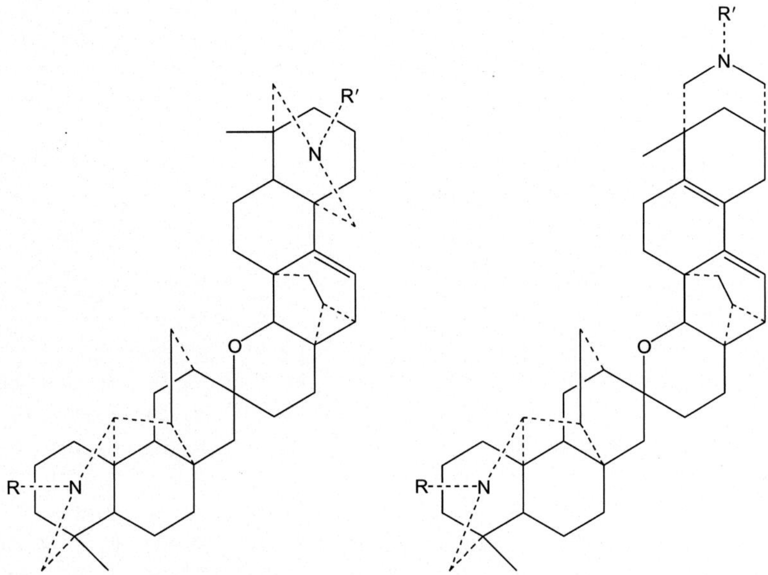

Bisditerpenoid alkaloid (XIV)

Figure 7.1 Continued.

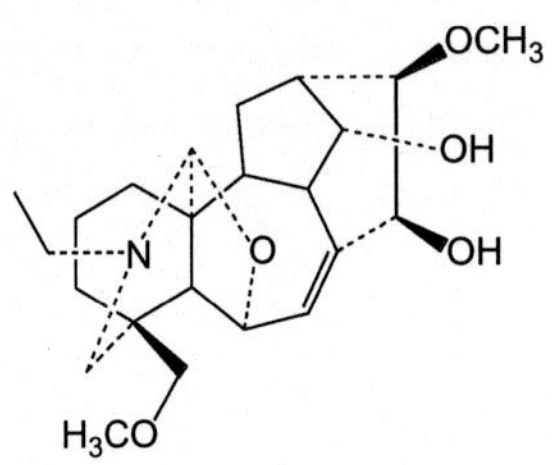

1

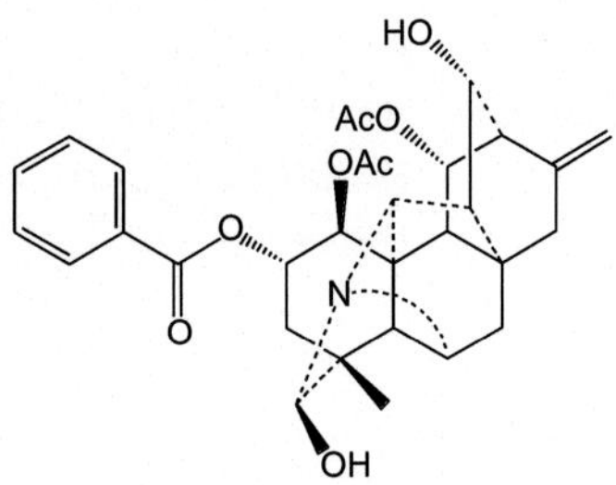

2

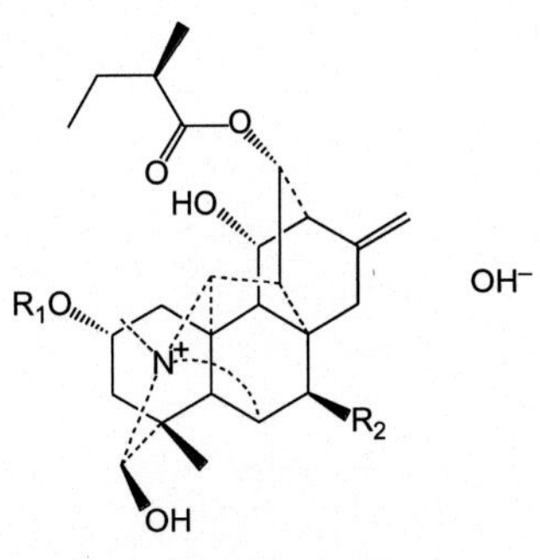

3	$R_1=COCH_2CH_3$	$R_2=H$
4	$R_1=COCH_2CH_3$	$R_2=OH$
5	$R_1=COCH(CH_3)_2$	$R_2=OH$

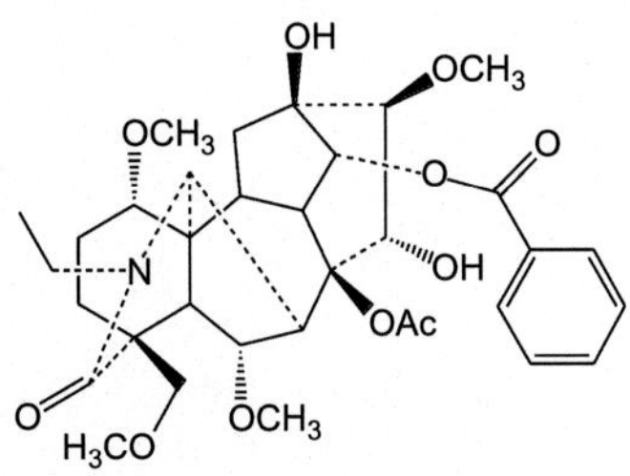

6

Figure 7.2 Diterpenoid alkaloids found in *Aconitum* in recent years. Refer to Table 7.1 for more details. Cin, cinnamoyl; Dim, 3, 4-dimethoxybenzoyl. *ring A is in chair conformation.

(Continued)

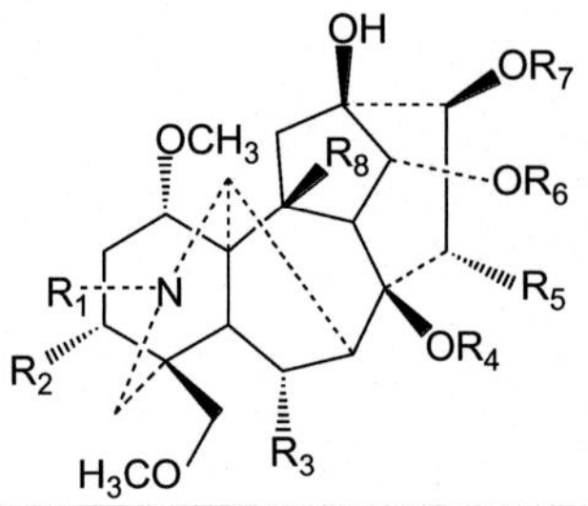

7	R_1=CH_2CH_3	R_2=OH	R_3=OCH_3	R_4=Ac	R_5=OH	R_6=Bz	R_7=CH_3	R_8=OH
8	R_1=CH_3	R_2=OH	R_3=OCH_3	R_4=CH_3	R_5=OH	R_6=Bz	R_7=CH_3	R_8=OH
9*	R_1=CH_3	R_2=OH	R_3=OCH_3	R_4=CH_3	R_5=OH	R_6=Bz	R_7=CH_3	R_8=OH
10	R_1=CH_2CH_3	R_2=OH	R_3=OCH_3	R_4=CH_3	R_5=OH	R_6=Bz	R_7=CH_3	R_8=OH
11	R_1=CH_3	R_2=OH	R_3=OCH_3	R_4=H	R_5=OH	R_6=Bz	R_7=CH_3	R_8=OH
12	R_1=CH_2CH_3	R_2=OH	R_3=OCH_3	R_4=Ac	R_5=H	R_6=Bz	R_7=CH_3	R_8=OH
13	R_1=CH_3	R_2=H	R_3=OCH_3	R_4=Ac	R_5=OH	R_6=Bz	R_7=CH_3	R_8=OH
14	R_1=CH_2CH_3	R_2=H	R_3=OCH_3	R_4=Ac	R_5=OH	R_6=Bz	R_7=CH_3	R_8=H
15	R_1=CH_2CH_3	R_2=H	R_3=OCH_3	R_4=H	R_5=H	R_6=Bz	R_7=CH_3	R_8=H
16	R_1=CH_2CH_3	R_2=OH	R_3=OCH_3	R_4=H	R_5=OH	R_6=Bz	R_7=CH_3	R_8=H
17	R_1=CH_3	R_2=OH	R_3=OCH_3	R_4=Bz	R_5=OH	R_6=Bz	R_7=CH_3	R_8=H
18	R_1=CH_2CH_3	R_2=H	R_3=OCH_3	R_4=CH_3	R_5=OH	R_6=Bz	R_7=CH_3	R_8=H
19	R_1=CH_2CH_3	R_2=H	R_3=H	R_4=H	R_5=OH	R_6=Bz	R_7=CH3	R_8=H
28	R_1=H	R_2=OH	R_3=OCH_3	R_4=H	R_5=OH	R_6=H	R_7=CH3	R_8=H
29	R_1=CH_3	R_2=OH	R_3=OCH_3	R_4=H	R_5=OH	R_6=H	R_7=H	R_8=OH

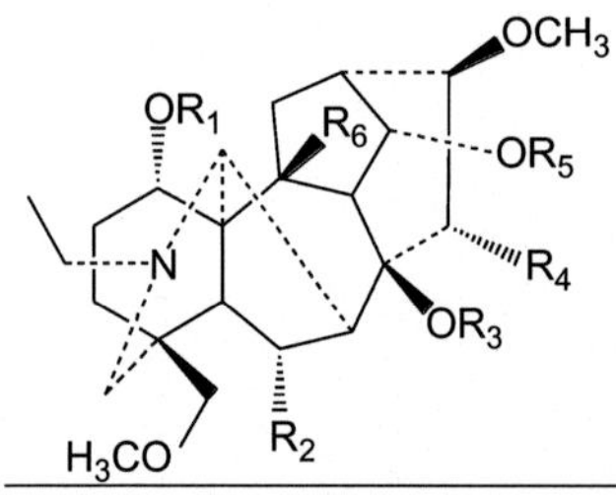

20	R_1=CH_3	R_2=OH	R_3=CH_3	R_4=OH	R_5=Bz	R_6= H
21	R_1=CH_3	R_2=OH	R_3=H	R_4=OH	R_5=Bz	R_6=H
22	R_1=CH_3	R_2=OH	R_3=CH_2CH_3	R_4=OH	R_5=Bz	R_6=H
23	R_1=CH_3	R_2=H	R_3=H	R_4=OH	R_5=Bz	R_6=H
24	R_1=H	R_2=H	R_3=H	R_4=OH	R_5=Bz	R_6=H
25	R_1=H	R_2=OCH_3	R_3=H	R_4=OH	R_5=Bz	R_6=H
26	R_1=H	R_2=OCH_3	R_3=H	R_4=OH	R_5=Cin	R_6=H
27	R_1=CH_3	R_2=OCH_3	R_3=Ac	R_4=OH	R_5=Bz	R_6=H
33	R_1=CH_3	R_2=OCH_3	R_3=As	R_4=H	R_5=H	R_6=H
42	R_1=CH_3	R_2=H	R_3=CH_3	R_4=H	R_5=H	R_6=OH
51	R_1=H	R_2=H	R_3=Dim	R_4=H	R_5=H	R_6=H

Figure 7.2 Continued.

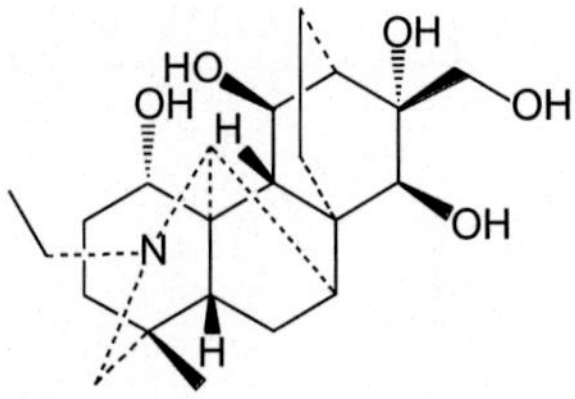

30

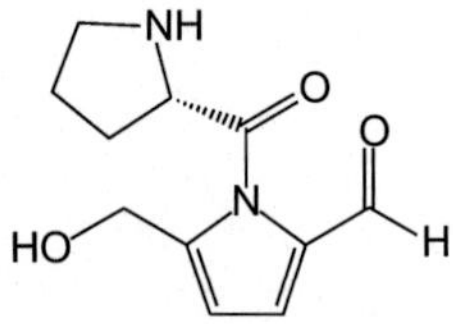

31

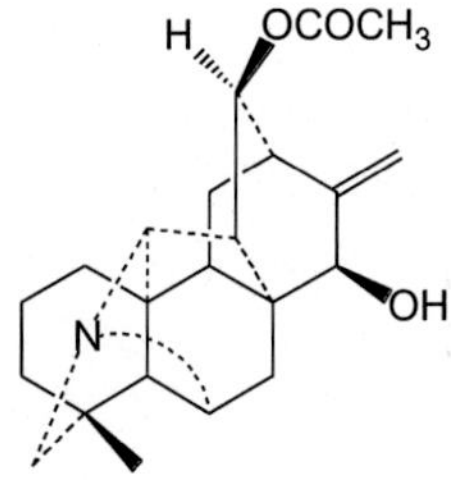

32

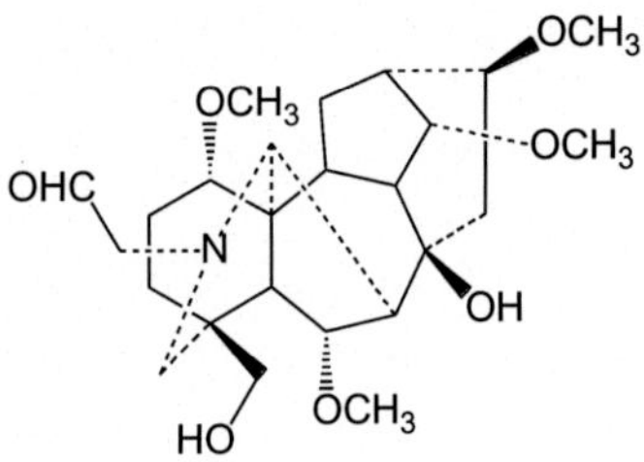

35

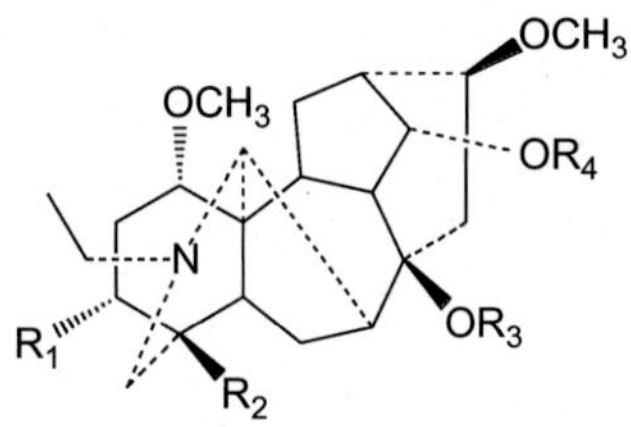

33	R_1=H	R_2=CH_3	R_3=Bz	R_4=Ac
55	R_1=OH	R_2=H	R_3=H	R_4=H

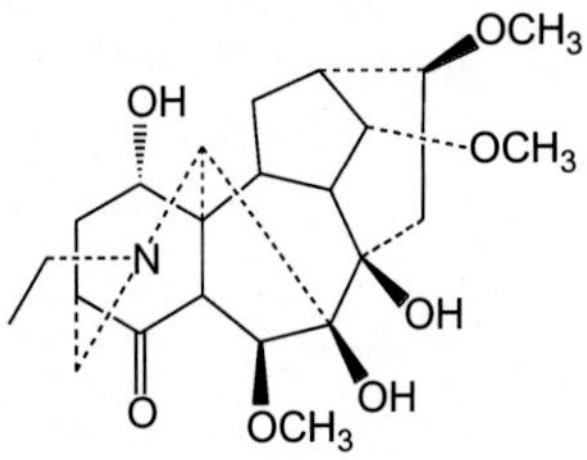

36

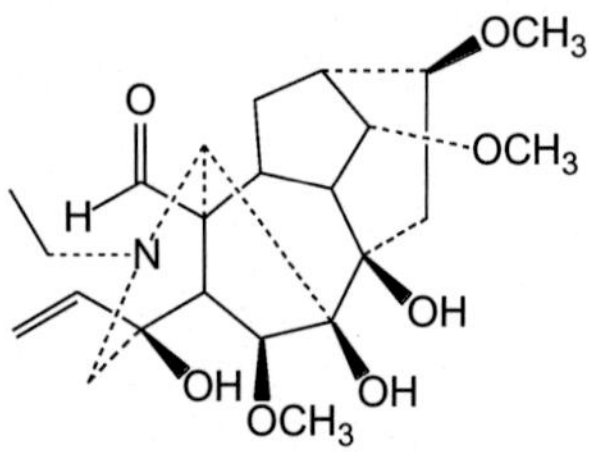

37

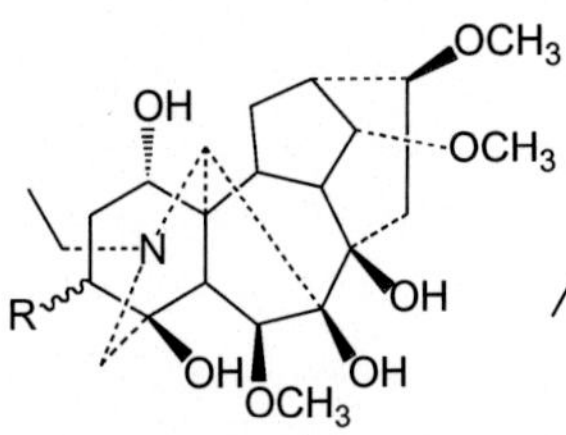

38	R=α-OH
39	R=β-OH
40	R=β-Cl

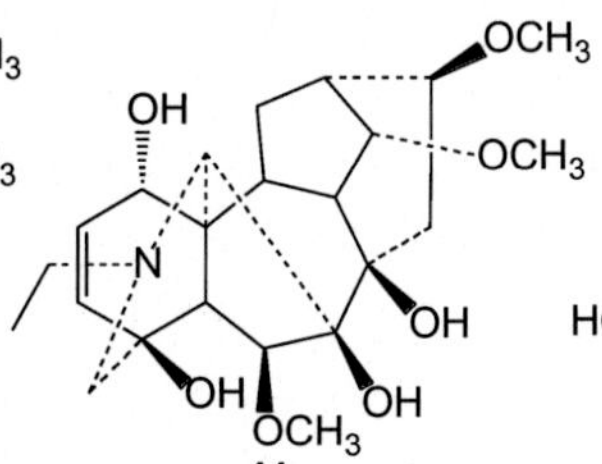

41

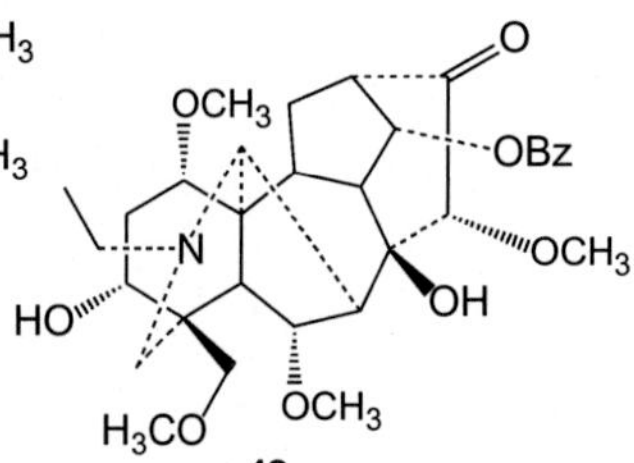

42

Figure 7.2 Continued.

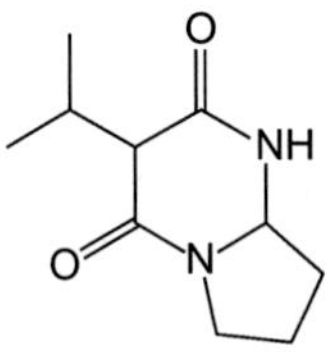

44

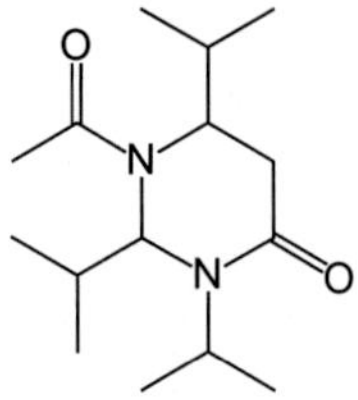

45

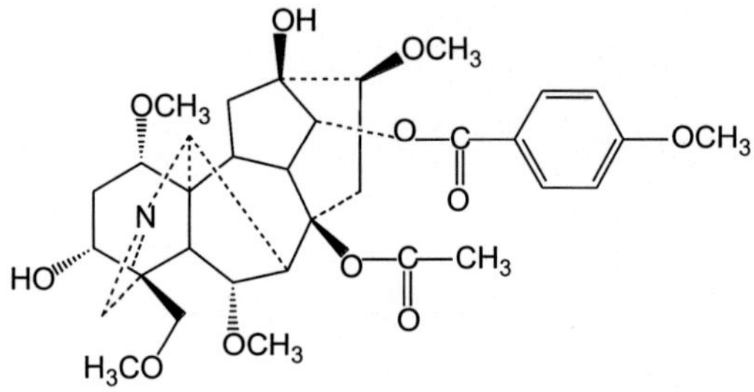

46

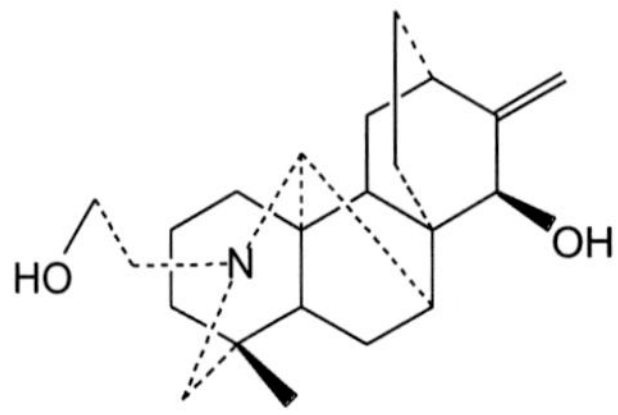

47

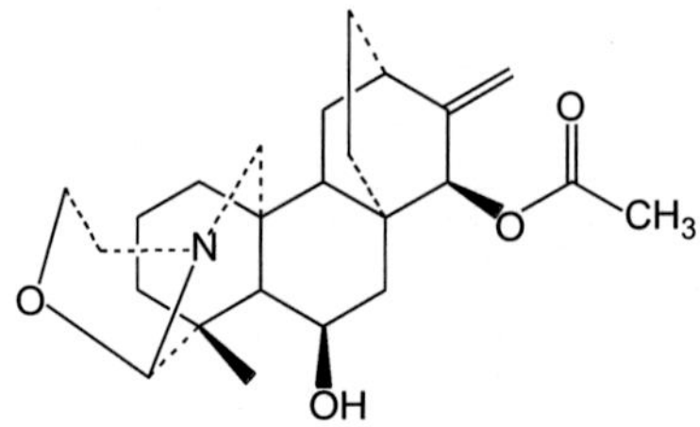

48

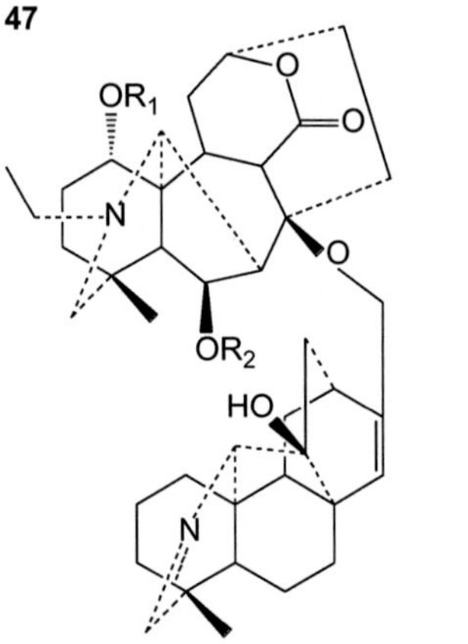

49	R_1=H	R_2=Ac
50	R_1=R_2=H	

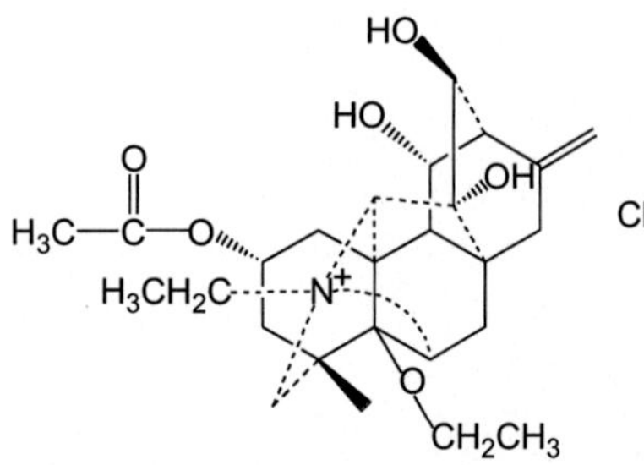

51

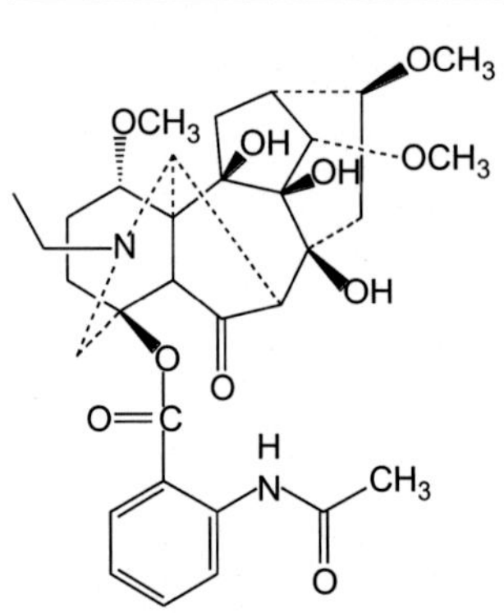

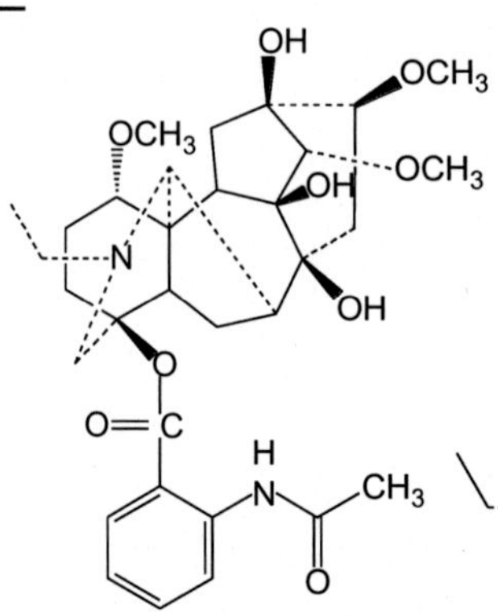

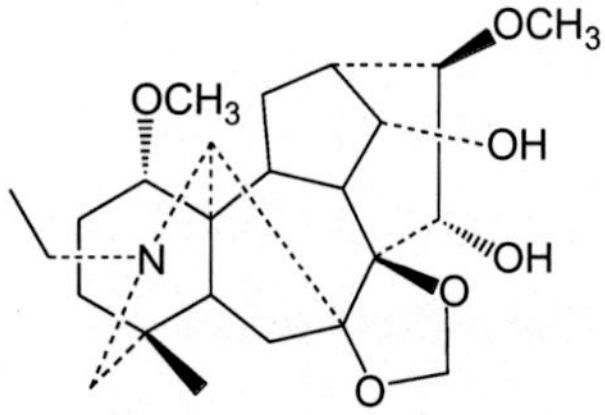

Figure 7.2 Continued.

53 **54** **56**

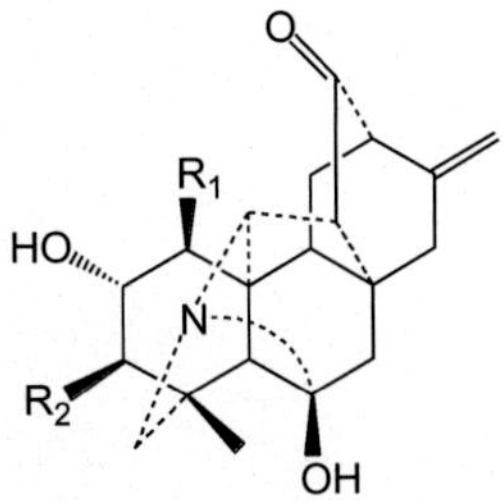

57	R_1=OH	R_2=H
58	R_1=H	R_2=OH

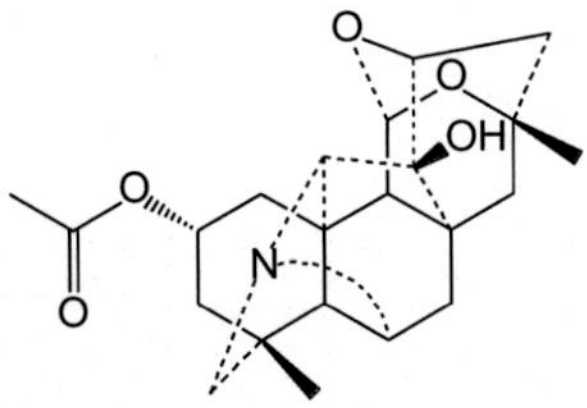

59

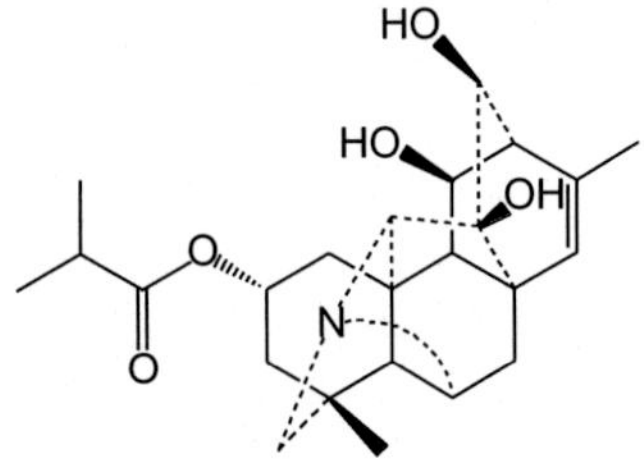

60

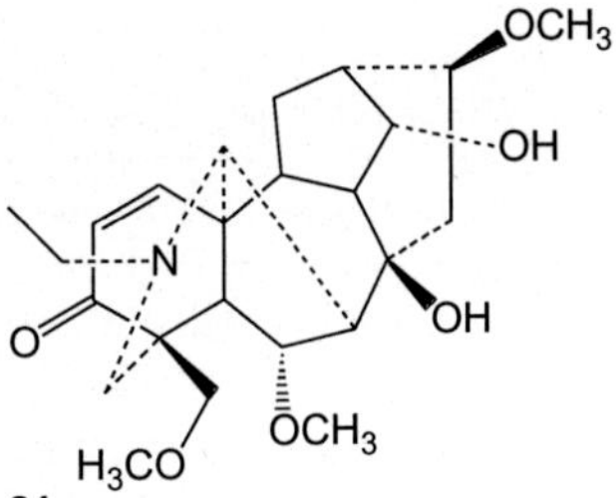

61

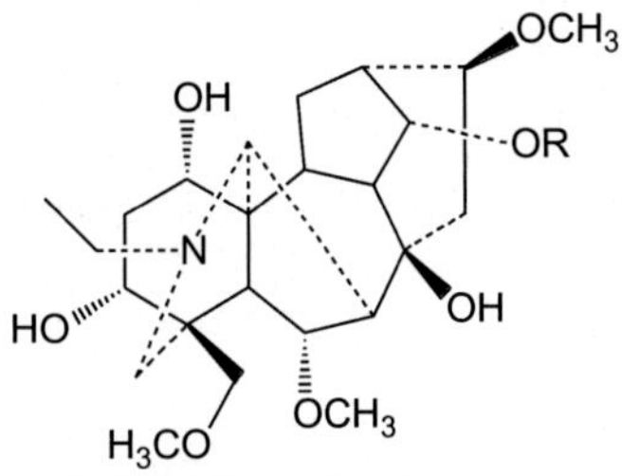

62	R=H
63	R=Ac

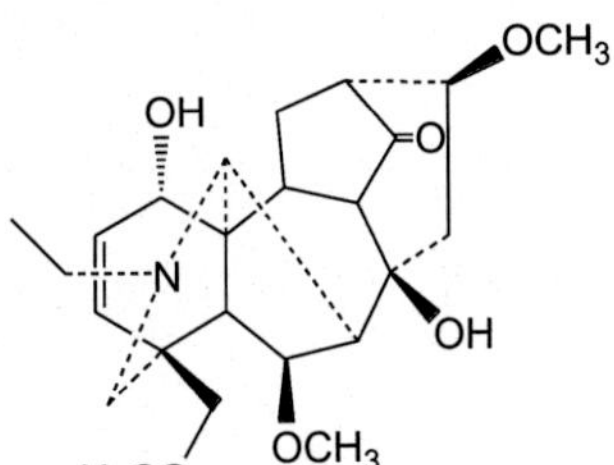

64

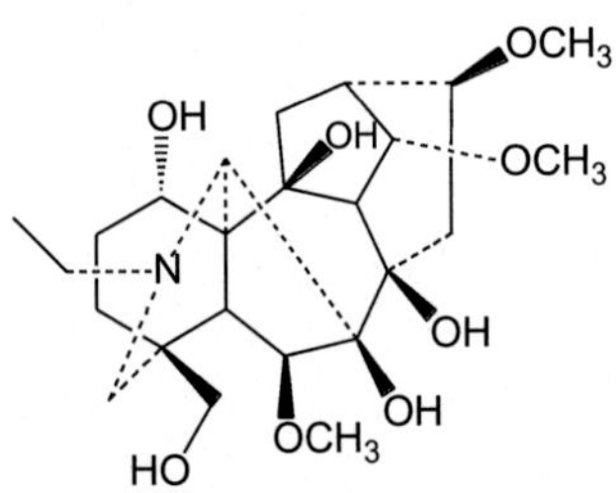

65

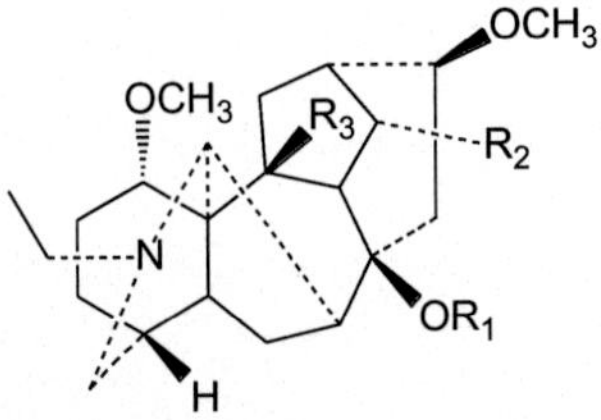

66	R_1=CH_2CH_3	R_2=OAc	R_3=H
67	R_1=H	R_2=OH	R_3=OH
68	R_1=H	R_2==O	R_3=H
69	R_1=CH_2CH_3	R_2=OH	R_3=H

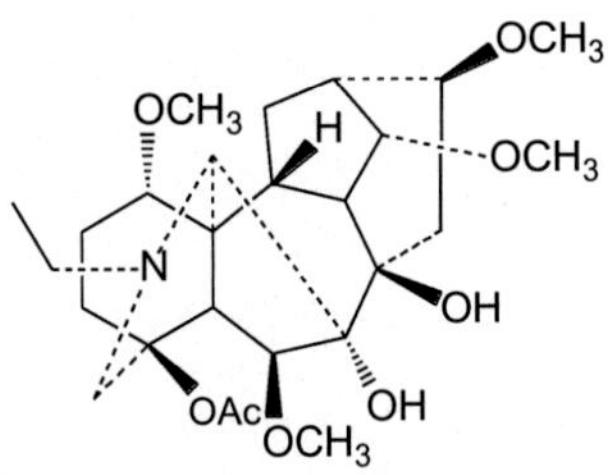

70

Figure 7.2 Continued.

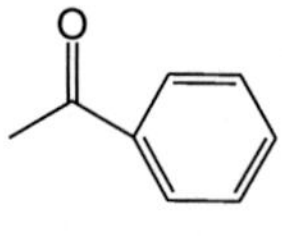

Bz

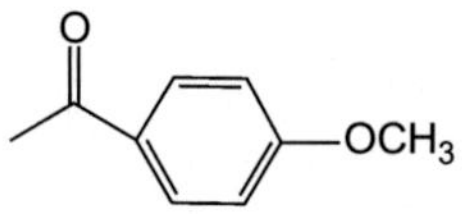

As

Ac

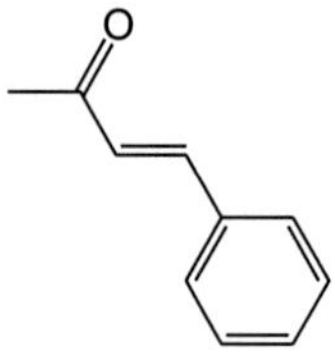

Cin

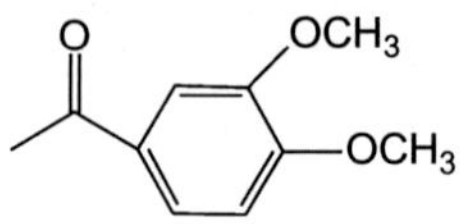

Dim

Figure 7.2 Continued.

Table 7.1 Diterpenoid alkaloids found in *Aconitum* in recent years

Taxon	Diterpenoid alkaloid	Type	Tissue	Reference
A. carmichaeli	Guiwuline **1**	V	Root	Wang et al. (2012a,b)
A. carmichaeli	(+)-(13R,19S)-1β,11α-Diacetoxy-2α-benzoyloxy-13,19-dihydroxyhetisan **2**	X	Lateral Root	Jiang et al. (2012)
	(−)-(13R,19S)-11α,19-Dihydroxy-*N*-methyl-13-(S-2-methylbutyryloxy)-2α-propionyloxyhetisanium hydroxide **3**	X		
	(−)-(13R,19S)-7β,11α,19-Trihydroxy-*N*-methyl-13-(S-2-methylbutyryloxy)-2α-propionyloxyhetisanium hydroxide **4**	X		
	(+)-(13R,19S)-2α-Isobutyryloxy-7β,11α,19-trihydroxy-*N*-methyl-13-(S-2-methylbutyryloxy) hetisanium hydroxide **5**	X		
	(−)-(A-c)-8β-Acetoxy-14α-benzoyloxy-*N*-ethyl-13β,15α-dihydroxy-1α,6α,16β,18-tetramethoxy-19-oxo-aconitane **6**	III		
	(−)-(A-b)-8β-Acetoxy-14α-benzoyloxy-*N*-ethyl-3α,10β,13β,15α-tetrahydroxy-1α,6α,16β,18-tetramethoxyaconitane **7**	III		

Continued

Table 7.1 **Continued**

Taxon	Diterpenoid alkaloid	Type	Tissue	Reference
	(−)-(A-b)-14α-Benzoyloxy-3α,10β,13β,15α-tetrahydroxy-1α,6α,8β,16β,18-pentamethoxy-*N*-methylaconitane **8**	III		
	(−)-(A-c)-14α-Benzoyloxy-3α,10β,13β,15α-tetrahydroxy-1α,6α,8β,16β,18-pentamethoxy-*N*-methylaconitane **9**	III		
	(−)-(A-b)-14α-Benzoyloxy-*N*-ethyl-3α,10β,13β,15α-tetrahydroxy-1α,6α,8β,16β,18-pentamethoxyaconitane **10**	III		
	(−)-(A-b)-14α-Benzoyloxy-3α,8β,10β,13β,15α-pentahydroxy-1α,6α,16β,18-tetramethoxy-*N*-methylaconitane **11**	III		
	(−)-(A-b)-8β-Acetoxy-14α-benzoyloxy-*N*-ethyl-3α,10β,13β-trihydroxy-1α,6α,16β,18-tetramethoxyaconitane **12**	III		
	(−)-(A-b)-8β-Acetoxy-14α-benzoyloxy-10β,13β,15α-trihydroxy-1α,6α,16β,18-tetramethoxy-*N*-methylaconitane **13**	III		
	(−)-(A-b)-8β-Acetoxy-14α-benzoyloxy-*N*-ethyl-13β,15α-dihydroxy-1α,6α,16β,18-tetramethoxyaconitane **14**	III		
	(−)-(A-b)-14α-Benzoyloxy-*N*-ethyl-8β,13β-dihydroxy-1α,6α,16β,18-tetramethoxyaconitane **15**	III		
	(−)-(A-b)-14α-Benzoyloxy-*N*-ethyl-3α,8β,13β,15α-tetrahydroxy-1α,6α,16β,18-tetramethoxyaconitane **16**	III		
	(−)-(A-b)-8β,14α-Dibenzoyloxy-*N*-ethyl-3α,13β,15α-trihydroxy-1α,6α,16β,18-tetramethoxyaconitane **17**	III		
	(−)-(A-b)-14α-Benzoyloxy-*N*-ethyl-13β,15α-dihydroxy-1α,6α,8β,16β,18-pentamethoxyaconitane **18**	III		

Continued

Table 7.1 **Continued**

Taxon	Diterpenoid alkaloid	Type	Tissue	Reference
	(−)-(A-b)-14α-Benzoyloxy-*N*-ethyl-8β,13β,15α-trihydroxy-1α,16β,18-trimethoxyaconitane **19**	III		
	(−)-(A-b)-14α-Benzoyloxy-*N*-ethyl-6α,15α-dihydroxy-1α,8β,16β,18-tetramethoxyaconitane **20**	III		
	(−)-(A-b)-14α-Benzoyloxy-*N*-ethyl-6α,8β,15α-trihydroxy-1α,16β,18- trimethoxyaconitane **21**	III		
	(−)-(A-b)-14α-Benzoyloxy-8β-ethoxy-*N*-ethyl-6α,15α-dihydroxy-1α,16β,18-trimethoxyaconitane **22**	III		
	(−)-(A-b)-14α-Benzoyloxy-*N*-ethyl-8β,15α-dihydroxy-1α,16β,18-trimethoxyaconitane **23**	III		
	(−)-(A-b)-14α-Benzoyloxy-*N*-ethyl-1α,8β,15α-trihydroxy-16β,18-dimethoxyaconitane **24**	III		
	(−)-(A-b)-14α-Benzoyloxy-*N*-ethyl-1α,8β,15α-trihydroxy-6α,16β,18-trimethoxyaconitane **25**	III		
	(−)-(A-b)-14α-Cinnamoyloxy-*N*-ethyl-1α,8β,15α-trihydroxy-6α,16β,18-trimethoxyaconitane **26**	III		
	(−)-(A-b)-8β-Acetoxy-14α-benzoyloxy-*N*-ethyl-15α-hydroxy-1α,6α,16β,18-tetramethoxyaconitane **27**	III		
A. carmichaeli	*N*-Deethylaconine **28**	III	Lateral root	Liu et al. (2012)
	Beiwutinine **29**	III		
A. carmichaeli	Aconicarmine **30**	VIII	Lateral root	Xiong et al. (2012)
	Aconicaramide **31**	A		
A. carmichaeli	Carmichaeline A **32**	X	Root	Li et al. (2013d)
A. transsectum	Aconitramines D **33**	I	Root	Shen et al. (2012)
	Aconitramines E **34**	III		

Continued

Table 7.1 Continued

Taxon	Diterpenoid alkaloid	Type	Tissue	Reference
A. vulparia	Alexhumboldtine **35**	III	Root	Kurtoğlu et al. (2012)
A. barbatum var. puberulum	Puberunine **36**	II	Root	Mu et al. (2012)
	Puberudine **37**	II		
	Puberumine A **38**	II		
	Puberumine B **39**	II		
	Puberumine C **40**	II		
	Puberumine D **41**	II		
A. kusnezoffii	1,15-Dimethoxy-3-hydroxy-14-benzoyl-16-ketoneoline **42**	III	Root	Xu et al. (2011)
A. anthora	10-Hydroxy-8-*O*-methyltalatizamine **43**	III	Whole plant	Forgo et al. (2011)
A. taipeicum	3-Isopropyl-tetrahydropyrrolo[1,2-a]pyrimidine-2,4(1H,3H)-dione **44**	A	Root	Xu et al. (2010b)
	1-Acetyl-2,3,6-triisopropyl-tetrahydropyrimidin-4(1H)-one **45**	A		
A. macrorhynchum	Macrorhynine C **46**	III	Root	Yang et al. (2009)
A. heterophyllum	Heterophyllinine A **47**	VIII	Root	Nisar et al. (2009)
	Heterophyllinine B **48**	VII		
A. tanguticum var. trichocarpum	Trichocarpine A **49**	XIV	Whole plant	Lin et al. (2009)
	Trichocarpine B **50**	XIV		
A. coreanum	Guanfu base R **51**	X	Stem, leaf	Li and Liang (2009)
A. longzhoushanense	Longzhoushansine **52**	III	Root	He et al. (2009)
A. heterophyllum	6-Dehydroacetylsepaconitine **53**	III	Root	Ahmad et al. (2008)
	13-Hydroxylappaconitine **54**	III		
A. toxicum	Acotoxicine **55**	I	Root	Csupor et al. (2008)
A. taipaicum	Isodelelatine **56**	II	Root	He et al. (2008)
A. coreanum	Guanfu base T **57**	X	Root	Tang et al. (2008)
	Guanfu base U **58**	X		
A. coreanum	Guanfu base J **59**	X	Root	Xing et al. (2014)
	Guanfu base N **60**	X		

Continued

Table 7.1 Continued

Taxon	Diterpenoid alkaloid	Type	Tissue	Reference
A. duclouxii	Ducloudine C **61**	III	Root	Yin et al. (2014a)
	Ducloudine D **62**	III		
	Ducloudine E **63**	III		
A. duclouxii	Ducloudine F **64**	IV	Root	Yin et al. (2014b)
A. moldavicum	1-*O*-Demethylswatinine **65**	III	Root	Borcsa et al. (2014)
A. weixiense	Weisaconitine A **66**	I	Root	Zhao et al. (2013)
	Weisaconitine B **67**	I		
	Weisaconitine C **68**	I		
	Weisaconitine D **69**	I		
A. kirinense	Kirinenine A **70**	II	Root	Zhang et al. (2013)

C_{18}: I, lappaconines; II, ranaconines; C_{19}: III, aconitine; IV, lycoctonine; V, 7, 17-seco type; VI, lactones; C_{20}: VII, atisine; VIII, denudatine; IX, hetidine; X, hetisine; XI, veatchine; XII, napelline; XIII, anopterine; XIV, bisditerpenoid alkaloid. A, amide alkaloid; Cin, cinnamoyl; Dim, 3,4-dimethoxybenzoyl.
The ring A of compound 9 is of chair conformation, while the ring A of compounds 6–8 and 10–27 is of boat conformation. In ref. Jiang et al. (2012), whether the ring A of every compound is of chair conformation (A–c) or of boat conformation (A–b) is defined, which facilitate the differentiation between compounds 8 and 9. These two alkaloids have the same skeleton and substituents on the two-dimensional figure; however, their A rings are distinct in three-dimensional structure. Other references did not mention the chair/boat conformation of the ring A. Refer to Figures 7.1 and 7.2 for more details.

7.3 Pharmacology and toxicology of *Aconitum* alkaloids and extracts

7.3.1 Anticancer activity

Aconitine, hypaconitine, mesaconitine, and oxonitine, isolated from *Aconitum carmichaelii*, were found to strongly inhibit the growth of the HepG2 cell line, suggesting that the existence and quantity of the ester groups have a major influence on the cytotoxicity of the diterpenoid alkaloids (Gao et al., 2012). Aconitine-containing agent BC1 resulted in 77.3% growth inhibition of solid form of Ehrlich's carcinoma (Dasyukevich and Solyanik, 2007). The actively proliferating murine aorta endothelial cells are highly sensitive to the action of BC1 (Garmanchouk et al., 2005). BC1 induced cell apoptosis and caused inversion of surface charge of endothelial cells. *A. vaginatum* can inhibit the proliferation, invasion, and metastasis of human lung adenocarcinoma A549 cell lines, and the mechanism might be inhibiting matrix metalloproteinase −2 and −9 activity (Xu et al., 2010a). In the occurrence of cytotoxic effects of atisine-type alkaloids, replacement by an acyl group at both C-11 and C-15 resulted in the enhancement of activity of the parent alkaloids compared to those having hydroxy groups at these positions (Wada et al., 2011), and the presence of a hydroxy group at the C-6 position was required for the cytotoxic effects. These acylated alkaloid derivatives inhibit A549 cell growth through G1 arrest. Two new amide alkaloids of *A. taipeicum* exhibited more significant cell growth inhibitory activities against HL-60 leukemia cells than Adriamycin (Xu et al., 2010b), and they also showed antitumor activities against K562 cells. Some norditerpenoid alkaloids, for example, neoline, 8-*O*-methylcolumbianine, 1,14-diacetylcardiopetaline, 18-*O*-demethylpubescenine, 14-deacetylpubescenine, pubescenine, 14-deacetylajadine, lycoctonine, browniine, delphatine, dehydrotakaosamine, and ajadelphinine, exhibited selective cytotoxicity to cancerous versus noncancerous cells (de Inés et al., 2006), which could be related to the inhibition of ATP production. *Aconitum* alkaloids inhibit growth of A172 human malignant glioma cells (Wada et al., 2007).

Taipeinine A (JNQ2), a C19 diterpenoid alkaloid from the roots of *A. taipeicum*, inhibited the proliferation of HepG2 cells in a dose- and time-dependent manner and blocked the cell cycle at the G1/S phase (Zhang et al., 2014). High dosage of JNQ2 induced significant apoptosis of tumor cells. The invasiveness of HepG2 cells was inhibited by JNQ2. The molecular mechanism of JNQ2 antitumor effect might be the upregulation of the protein expression of Bax and caspase-3 and the downregulation of the protein expression of Bcl-2 and CCND1.

Two *Aconitum* C20 diterpenoid alkaloid derivatives, 11-m-trifluorometylbenzoyl-pseudokobuisne and 11-anisoyl-pseudokobusine, showed significant suppressive effects against non-Hodgkin's lymphoma Raji cells (Hazawa et al., 2011). The atisine-type structure characteristic of C20 diterpenoid alkaloids plays a very important role in the pharmacological properties. The C-11 residues play an important role for the antitumor properties and the lower toxicity to hematopoiesis. The novel compounds derived from C20 diterpenoid alkaloids showed a significantly suppressive effect in four cancer cell lines (Hazawa et al., 2009). In contrast, natural C19 norditerpenoid alkaloids and their derivatives showed either no effect or only a slight effect. One derivative compound also showed radiosensitizing properties on A549 cells. These effects are not related to either the cell cycle distribution, the enhancement of apoptosis, or the γ-H2AX expression. Distribution of different types of diterpenoid alkaloids in plants is summarized in Table 7.2.

However, Hazawa et al. (2012) showed that C6 derivatives of *Aconitum* C20 diterpenoid alkaloid additively or synergistically sustained proliferation of CD34+ hematopoietic stem/progenitor cells derived from the placental/umbilical cord blood. It seems that identical structures belonging to C20 diterpenoid alkaloids surprisingly have opposing proliferative effects depending on the derivatization sites (C6 versus C11).

7.3.2 Effects on cardiovascular system and cardiotoxicity

Higenamine (HG) is a potent cardioactive benzylisoquinoline alkaloid isolated from Aconiti tuber, which has long been used as a cardiotonic in traditional Chinese medicine (Zhou and Du, 2003). HG exerts various effects on the cardiocirculatory system, inotropic and chronotropic in isolated rat atria. It also relaxes isolated rat aorta. It

Table 7.2 Distribution of diterpenoid alkaloids in plants

Taxon	Diterpenoid alkaloid		
	C_{18}	C_{19}	C_{20}
1. Ranunculaceae			
Aconitum, subgenus *Aconitum*	I	III, IV, V, VI	VII, VIII, IX, X, XII
Aconitum, subgenus *Lycoctonum*	I, II	III, IV, V	
Aconitum, subgenus *Gymnaconitum*		III, IV	VII, VIII
Delphinium	Sporadic	III, IV	X, XIV
Consolida		IV	VIII, X
Thalictrum			VII, IX
2. Rosaceae			
Spiraea			VII, IX, X
3. Garryaceae			
Garrya			XI
4. Escalloniaceae			
Anopterus			XIII
5. Polygonaceae			
Rumex			X

C18: I, lappaconines; II, ranaconines; C19: III, aconitine; IV, lycoctonine; V, 7, 17-seco type; VI, lactones; C20: VII, atisine; VIII, denudatine; IX, hetidine; X, hetisine; XI, veatchine; XII, napelline; XIII, anopterine; bisditerpenoid alkaloid: XIV.
Modified from Xiao et al. (2006).

inhibits epinephrine, ADP, or collagen-induced platelet aggregation in platelet-rich plasma. HG inhibits lipopolysaccharide (LPS)-induced nitrate accumulation and the expression of iNOS (inducible nitric oxide synthase) mRNA in RAW 264.7 cells. HG lowers blood pressure in rats and increases the recovery rates in acute thrombosis model of mice and lowers the weight of thrombus formed in the arteriovenous shunt model of rats. Higenamine also has ameliorative effects in the LPS-induced disseminated intravascular coagulation (DIC) model.

Both raw and processed aconite roots improved contractile force significantly on isolated toad cordis (Wang et al., 2009c). The most amplification of contractile force was $91.1 \pm 87.6\%$ ($P < 0.01$), while the effect on isolated heart rate was not significant. The alteration of LVSP (left ventricular systolic peak pressure), $\pm$ dp/dtmax, and LVEDP (left ventricular end-diastolic pressure) was noteworthy, and the heart rate increased dramatically by processed aconite roots. Mesaconine, hypaconine, and beiwutinine, isolated from *A. carmichaeli*, showed the strongest cardiac actions on the isolated perfused bullfrog heart (Liu et al., 2012). Mesaconine has protective effects, including improved inotropic effect and left ventricular diastolic function, on myocardial ischemia-reperfusion injury in rat at a dose of 10^{-9} mol/l. However, mesaconine has almost no effect on heart rate. Acehytisine, isolated from the root of *A. coreanum*, showed blocking effects on pacemaker currents (If) in sinoatrial node cells and human HCN4 channels expressed in *Xenopus laevis* oocytes (Fan et al., 2012), which might account for its antiarrhythmic actions. The potassium channel encoded by the human ether-à-go-go (hERG)-related gene plays an important role in repolarization of the cardiac action potential. Both Guanfu base A (GFA) and Guanfu base G (GFG) of *A. coreanum* blocked hERG channel current (Huang et al., 2012). This effect was stronger after GFG treatment. This blockade was dependent on open and inactivated channel states. GFA could be a rather promising antiarrhythmic drug without severe side effects, whereas GFG could cause QT prolongation and requires further research. Compared to acute coronary occlusion micropigs treated with saline, acehytisine hydrochloride (AHH) prolongs the P-wave duration and PR interval, shortens QTc interval, prolongs ARP and AEP, also prolongs V-A reverse conduction time and VARC-ERP, but shortens VFRP (Liang et al., 2006). No proarrhythmia effect was found in AHH-treated groups.

Significant blocking activity on GIRK channel was exerted by aconitine (45% at 10 μM), but no blocking activities of the other investigated *A. napellus* subsp. *firmum* alkaloids were detected (Kiss et al., 2013). The tested compounds were inactive on hERG channel in the tested concentration.

Pyroaconitine, ajacine, septentriodine, and delectinine, isolated from *A. moldavicum*, demonstrated significant Nav 1.2 channel inhibition (42–57%) at 10 μM concentration (Borcsa et al., 2014); several other compounds (acovulparine, acotoxicine, hetisinone, 14-benzoylaconine-8-*O*-palmitate, aconitine, and lycoctonine) exerted moderate inhibitory activity (22–30%).

Blocking specific K^+ channels has been proposed as a promising strategy for the treatment of neurodegenerative diseases (Song et al., 2008). Talatisamine is a novel specific blocker for the delayed rectifier K^+ channels in rat hippocampal neurons; thus, it is a leading compound worthy of further investigations.

The high dose of Zhi Fu Zi (*Radix Aconiti lateralis preparata*) exerted more portal hypotensive effects than propranolol and simultaneously ameliorated the systemic arterial hypotension in portal vein ligation rats (Lin et al., 2007). Aconiti tuber increases plasma nitrite and nitrate levels and might affect peripheral vascular function via increased NO production in humans (Yamada et al., 2005).

The cardiotoxicity and neurotoxicity of aconitine and related alkaloids are due to their actions on the voltage-sensitive sodium channels of the cell membranes of excitable tissues, including the myocardium, nerves, and muscles (Chan, 2009). Aconitine and mesaconitine bind with high affinity to the open state of the voltage-sensitive sodium channels at site two, thereby causing a persistent activation of the sodium channels, which become refractory to excitation. The electrophysiological mechanism of arrhythmia induction is triggered activity due to delayed after depolarization and early after depolarization. The arrhythmogenic properties of aconitine are in part due to its cholinolytic (anticholinergic) effects mediated by the vagus nerve. Aconitine has a positive inotropic effect by prolonging sodium influx during the action potential. It has hypotensive and bradycardic actions due to activation of the ventromedial nucleus of the hypothalamus. At 10 μM, aconitine, 14-benzoylaconine 8-*O*-palmitate, songoramine, gigactonine, and neolinine of *A. anthora* demonstrated significant hERG K+ channel inhibition (Forgo et al., 2011); all other compounds exerted only low (6–21%) inhibitory activity. Aconitine blocks hERG and Kv1.5 potassium channels, contributes to cardiotoxic effects of *Aconitum*, and has been known to induce arrhythmia in animal models (Li et al., 2010). The frequency of arrhythmia remarkably decreased with time and repeated administration of aconitine (Wada et al., 2005).

Aconitine-induced Ca^{2+} overload causes arrhythmia and triggers apoptosis through p38 MAPK signaling pathway in rats (Sun et al., 2014).

The cardiotoxicity of Fu Zi was dose-dependent, and the high-dose group obviously manifested heart damage in histology and a certain degree of arrhythmia (Cai et al., 2013). Significant changes of 14 lipid metabolites, which are involved in phospholipid metabolism, sphingolipid metabolism, saturated fatty acid oxidation, and unsaturated fatty acid peroxidation, were identified and considered as the potential biomarkers of Fu Zi toxicity.

Thirty-three C19 diterpenoid alkaloids, 22 prepared from known C19 diterpenoid alkaloids and 11 isolated from *Aconitum* and *Delphinium* species, were evaluated for their cardiac activity in the isolated bullfrog heart assay (Jian et al., 2012). Among them, 11 compounds exhibited cardiac activity, with average rate of amplitude increase in the range of 16–118%. Compound 7, mesaconine (17), hypaconine (25), and beiwutinine (26) exhibited strong cardiac activities relative to the reference drug. The structure–activity relationship (SAR) data indicated that an α-hydroxyl group at C-15, a hydroxyl group at C-8, an α-methoxyl or hydroxyl group at C-1, and a secondary amine or *N*-methyl group in ring A are important structure features necessary for the cardiac activities of the aconitine-type C19 diterpenoid alkaloids without any ester groups. An α-hydroxyl group at C-3 is also helpful for the cardiac activity of these alkaloids.

7.3.3 Effects on nervous system, analgesic activity, and neurotoxicity

The analgesic effect of crude Fu Zi and its processed products was decreased in opioid mu-receptor knockout mice, but the effect remained unchanged in mice with opioid mu-receptors, indicating that the analgesic effect of Fu Zi is centrally mediated (Liou et al., 2005). The processed Aconiti tuber (PAT) has antinociceptive effects and inhibitory effects on morphine tolerance by activation of kappa-opioid receptor (Wu et al., 2011a). PAT relieves neuropathic pain in the rat chronic constriction injury model (Xu et al., 2006a). High doses of PAT inhibit the acute but potentiate the chronic antinociception of morphine (Shu et al., 2008a). PAT at subanalgesic doses could inhibit or reverse the antinociceptive tolerance to morphine (Shu et al., 2008b). PAT might be superior to some *N*-methyl-D-aspartate (NMDA) receptor antagonists, which do not reverse already-developed morphine tolerance. Bulleyaconitine A isolated from *A. bulleyanum* displays long-acting local anesthetic properties *in vitro* and *in vivo* (Wang et al., 2007) and has been used as an adjuvant for prolonged cutaneous analgesia in the rat (Wang et al., 2008). A new franchetine-type C19 diterpenoid alkaloid, guiwuline, isolated from *A. carmichaeli*, exhibited potential analgesic activity (ED50, 15 mg/kg) and showed little toxicity to mice (LD50, 500 mg/kg) (Wang et al., 2012a). Among 26 new diterpenoid alkaloids isolated from *A. carmichaeli*, the isomer with ring A in the chair conformation was found to be more active than that with ring A in the boat conformation (Jiang et al., 2012). Twenty-five semisynthetic C18 or C19 diterpenoid alkaloids were prepared from lappaconitine, crassicauline A, or yunaconitine (Wang et al., 2009a,b,c). Four crassicauline A analogs and three yunaconitine analogs exhibited good analgesic activities. Among them, 8-*O*-deacetyl-8-*O*-ethylcrassicauline A (ED50 0.0972 mg/kg) and 8-*O*-ethylyunaconitine (ED50 0.0591 mg/kg) were the most potent analgesics relative to the reference drugs. A tertiary amine in ring A, an acetoxyl or an ethoxyl group at C-8, an aromatic ester at C-14, and the saturation state of the ring D are important structural features necessary to the analgesic activity of the C19 diterpenoid alkaloids. The local anesthetic activity of *Aconitum* alkaloids can be explained by their structural properties (Bello-Ramírez and Nava-Ocampo, 2004). The alkaloids with the highest anesthetic activity had an aroyl/aroyloxy group at R14 position, while the weaker anesthetic alkaloids had the aroyloxy group at R4.

Through its action on voltage-sensitive sodium channels in the axons, aconitine blocks neuromuscular transmission by decreasing the evoked quantal release of acetylcholine (Chan, 2009). Aconitine, mesaconitine, and hypaconitine can induce strong contractions of the ileum through acetylcholine release from the postganglionic cholinergic nerves. *Aconitum* significantly evoked dopamine release from cultured PC12 neuron cells and from the nucleus accumbens of mice (Zhao et al., 2010). *Aconitum* exposure induced reactive oxygen species formation with the decrease of superoxide dismutase and glutathione peroxidase. Excessive extracellular dopamine can create stresses on cellular antioxidant systems and induce neuron apoptosis. Some diterpenoid alkaloids of the genera *Aconitum* and *Delphinium* act at neuronal nicotinic acetylcholine receptors (nAChRs) and exhibit potent *N*-cholinolytic activity (Turabekova et al., 2010). Quantitative structure–activity relationship (QSAR) analysis and molecular docking of alkaloids into AChBP (homologue of the ligand binding domain of nAChRs) are helpful in elucidating the mechanism of *Aconitum*-induced neurotoxicity.

The detected analgesic activity of diterpene alkaloids from *A. baikalensis* was comparable to that of sodium metamizole (Nesterova et al., 2014). The mechanisms of analgesia were different in diterpene alkaloids of different structure. The antinociceptive effect of atisine alkaloids (12-epinapelline *N*-oxide, songorine) was naloxone-dependent and realized via opioid receptor modulation.

8-*O*-Ethylaconosine of *A. weixiense* showed writhing inhibitions of 24% (50 mg/kg), 26% (100 mg/kg) and 34% (200 mg/kg), respectively (Zhao et al., 2013), as compared to the reference drug aspirin (63%) at a dose of 200 mg/kg.

Among 26 new diterpenoid alkaloids isolated from *A. carmichaeli*, 15, 16, and 19 showed neuroprotective activity (Jiang et al., 2012). Diterpene alkaloids of *A. baicalensis* exhibit antidepressant properties, possibly due to modulation of sensitivity to serotonin (Nesterova et al., 2011). Bullatine A, a diterpenoid alkaloid of the genus *Aconitum*, could attenuate ATP-induced BV-2 microglia death/apoptosis via P2X receptor pathways (Li et al., 2013a).

Heishunpian total alkaloids have a certain role of antidiarrheal and its mechanism may be related to antagonizing muscarinic receptors (Shao et al., 2013).

7.3.4 Anti-inflammatory activity

It is known that the traditional processing of aconite drugs increases the amount of lipo-alkaloids. It was found that compounds esterified with unsaturated fatty acids demonstrated significant cyclooxygenase (COX)-2 inhibitory effects, while in the COX-1 assay, only 14-benzoylaconine-8-*O*-eicosapentaenoate exerted remarkable activity (Borcsa et al., 2011). The inhibition of leukotriene B4 formation was pronounced in cases of long-chain fatty acid derivatives. The water–alcohol extracts from *A. baikalense* and *A. septentrionale* inhibited the development of main symptoms of inflammation, that is, exudation, pain, and fever, to the same extent as nonsteroidal anti-inflammatory agents (Nesterova et al., 2009). The ethanol extracts from the aboveground parts of *A. baicalensis* and the aboveground parts and roots of *A. septentrionale* L. suppressed the exudative and proliferative stages of model chronic inflammation (Pashinskiĭ et al., 2006). Aconite cake-separated moxibustion has an ideal therapeutic effect on primary knee osteoarthritis of liver and kidney deficiency type, and the therapeutic effect at anaphase is better than that of sodium diclofenate slow-released tablet (Sun et al., 2008). *Radix Aconiti Lateralis* could significantly decrease the contents of iNOS and NO and increase content of superoxide dismutase (SOD) in the heart, lung, and kidney tissues of mice with endotoxemia (Xu et al., 2006a,b).

A. tanguticum has been widely used as a remedy for infectious diseases in traditional Tibetan medicine in China. The total alkaloids of *A. tanguticum* exhibited a potent protective effect on LPS-induced acute lung injury in rats through its anti-inflammatory activity (Wu et al., 2014).

Detoxicated Fu Zi (d-Fu Zi) was demonstrated to possess chondroprotective activity on MIA-induced OA model by *in vivo* preventing the articular degeneration and the reduction of bone density and Mankin score (Tong et al., 2014) and by *in vitro* promoting the chondrocyte proliferation and inhibiting the MIA-induced chondrocyte damage. A total of 23 compounds were identified in d-Fu Zi, most of which were deduced as the nontoxic derivatives of aconite alkaloids.

A. pseudo-laeve var. *erectum* (APE) has been widely shown in herbal medicine to have a therapeutic effect on inflammatory conditions. The extract of APE is involved in the biological bone metabolism process, particularly osteoclast-mediated bone resorption (Baek et al., 2014). APE inhibits receptor activator of nuclear factor (NF) kappa-B ligand-induced osteoclastogenesis via the c-Fos/NF of activated T cells and cytoplasmic 1 signaling pathway and prevents LPS-induced bone loss in mice.

Huangcaowu (*A. vilmorinianum*) is locally used for analgesia and anti-inflammation in Yunnan province and significantly improved allodynia, swelling, hyperemia, and vascular permeability in arthritic knee joints (Li et al., 2013b). It showed the higher anti-arthritic effects than *Aconiti Radix* and *Aconiti Kusnezoffii Radix*.

7.3.5 Effects on energy metabolism

Aconite root could improve the energy metabolism in rats, by influencing the metabolic process of sugar, lipid, and amino acid (Yu et al., 2011), which may be the main molecular mechanism of warming yang and dispelling cold for the treatment of the cold syndrome according to Chinese medicine theory. *A. napellus* administered around the clock induced hyperthermia overall and in a time-dependent manner (de la Peña et al., 2011), with greatest effects during the resting span. Thus, time of day may significantly affect the outcome of *A. napellus* and other homeopathic treatments and should be considered in determining optimal dosing and treatment time in order to increase the desired outcome and decrease undesired effects. Fu Zi can increase and maintain the dogs' body heat for at least 6 h (Chen et al., 2009). The body weight gain in aconitine-administered mice was less than that of the control group until day 22 (Wada et al., 2006). Transient rectal hypothermia occurred within 30 min after the last administration of aconitine. Then, the rectal temperature gradually increased to normal level in respect to time. The drug metabolism of aconitine increased and the toxicity of aconitine decreased due to long-term administrations of aconitine.

Fingerprints of Fu Zi, Yanfuzi, Heishunpian, and Paofupian were established by ultrahigh-performance liquid chromatography (UPLC), and effects of Fu Zi and its processed products on rat liver mitochondrial metabolism were studied by microcalorimetry (Zheng et al., 2014). Spectrum-effect relationships between UPLC fingerprints and energy metabolism of mitochondria were investigated using canonical correlation analysis (CCA). Due to inherent differences in chemical compositions, the energy metabolism of mitochondria was differentially affected by Fu Zi and its processed products. The bioactivity sequence of the tested products was Fu Zi > Heishunpian > Paofupian > Yanfuzi. Results of CCA showed that the compounds mesaconitine, benzoylaconitine, and benzoylhypacoitine might be the principal active components.

7.3.6 Antimicrobial and pesticidal activities

Atisinium chloride, the major alkaloid from *A. orochryseum*, has moderate antiplasmodial activities against the TM4 strain and the K1 strain of *Plasmodium falciparum* (Wangchuk et al., 2010). Demethylenedelcorine, 18-*O*-methylgigactonine, and lepenine of *A. sinomontanum* exhibited bactericidal and fungicidal activities (Yuan and Wang, 2012). Demethylenedelcorine and 18-*O*-methylgigactonine exhibited better pesticidal activities against *Mythimna separata*, while lepenine had the induce-feeding activity against *M. separata*. Chasmanine, crassicauline A, karacoline, sachaconitine, talatisamine, and yunaconitine from *A. episcopale* exhibited feeding deterrent activity against the red flour beetle, *Tribolium castaneum* (Liu et al., 2011). Two new aconitine-type norditerpenoid alkaloids 6-dehydroacetylsepaconitine and 13-hydroxylappaconitine, along with three known norditerpenoid alkaloids of *A. heterophyllum*, exhibited significant antibacterial activity (Ahmad et al., 2008).

Fu Zi and its processed products could be distinguished according to their toxic effects on *E. coli* (Zhao et al., 2014). The IC50 values of 14.6 mg/ml for Fu Zi, 59.2 mg/ml for Yanfupian, 118.3 mg/ml for Heifupian, and 182.7 mg/ml for Danfupian illustrated that the sequence of toxicity on *E. coli* was Fu Zi > Yanfupian > Heifupian > Danfupian.

7.3.7 Other effects

Activation of functions of hemopoietic progenitor cells along with increased feeder capacity of the stromal elements in the hemopoiesis-inducing microenvironment with decreased levels mesenchymal progenitor cells in the bone marrow is the mechanisms underlying the hemostimulating action of napelline of *A. baicalense* (Zyuz'kov et al., 2013).

Hair growth-promoting effect of *Aconiti Ciliare Tuber* extract is mediated by the activation of Wnt/β-catenin signaling (Park et al., 2012).

Sachaconitine and 14-acetylsachaconitine of *A. recemulosum* moderately inhibit PAF-induced platelet aggregation (Ge et al., 2009). Higenamine, a β2-adrenergic receptor agonist in the extract of *Radix Aconiti Lateralis Preparata*, contributes to the successful relief of bronchoconstriction (Bai et al., 2008). Heterophyllinine A and heterophyllinine B of *A. heterophyllum* are selective butyrylcholinesterase inhibitors (Nisar et al., 2009). Heishunpian, the fire-processed product of the root of *A. carmichaeli*, lowers the plasma glucose concentrations of streptozotocin-induced diabetic rats through activation of opioid mu-receptors of peripheral tissues, resulting in enhanced glucose utilization (Liou et al., 2006).

The significant hypolipidemic effect of *A. heterophyllum* may be linked to its ability to inhibit HMGR activity and block intestinal fat absorption (Subash and Augustine, 2012). The increase in HDL-c may be linked to its ability to activate LCAT enzyme.

Based on drug–target relationships of FDA approved drugs, a model for predicting targets was established by random forest algorithm (Wu et al., 2011a,b). The predicted targets of 22 compounds of *Aconiti Lateralis Radix Praeparata* (Fu Zi) are validated by literature. Each compound in the established network was correlated with 16.3 targets on average, while each target was correlated with 4.77 compounds on average, which reflects the "multicompound and multitarget" characteristic of Chinese medicine (Hao and Xiao, 2014).

Both Cao Wu (*A. kusnezoffii*) and aconitine had direct embryotoxic effects during the rat organogenetic period and should be treated with caution in pregnant women (Xiao et al., 2007, 2008).

7.4 Polysaccharide

The crude monkshood (also known as *Radix Aconiti Lateralis Preparata* or aconite) polysaccharide had inhibitory activity towards *Staphylococcus aureus* and *Escherichia coli* (Lin et al., 2011). The water-soluble polysaccharides from *A. kusnezoffii* showed antioxidant and immunological activities (Gao et al., 2011). A hot alkali-extracted polysaccharide from *A. kusnezoffii* had significant ferrous ion-chelating ability, reducing power and scavenging effects on DPPH (1,1-diphenyl-2-picrylhydrazyl) radical, hydroxyl radical, and superoxide anion and H_2O_2 and self-oxidation of 1,2,3-phentriol, suggesting that it could be a novel natural antioxidant (Gao et al., 2010b).

A RG-II-type polysaccharide purified from *A. coreanum* alleviates LPS-induced inflammation by inhibiting the NF-κB signal pathway (Li et al., 2014a).

The contents of polysaccharides in parent root, daughter root, and rootlet of *A. carmichaeli* were 22.02%, 33.53%, and 6.10%, respectively (Lu et al., 2011). Parent root, daughter root, and rootlet mainly contained glucose and small amount of galacturonic acid, galactose, and arabinose. Daughter root also contained mannose, and rootlet also contained mannose, rhamnose, and xylose. The non-starch-type polysaccharide fractions, extracted from *Radix Aconiti*, *Radix Aconiti Lateralis*, and *Radix Aconiti Kusnezoffii*, had good antitumor activities (Gao et al., 2010a). The non-starch-type polysaccharides, especially the pectic polysaccharide fractions, not only had significant immunostimulating activities including nonspecific immunity, cellular immunity, and humoral immunity but also could restore the antitumor drug-suppressed immune function. Therefore, the polysaccharides from *Aconitum* species might be good immunostimulating modifiers and could be used in the cancer therapy.

FPS-1 is an α-(1→6)-D-glucan isolated from *A. carmichaeli* (Fu Zi), with an average molecular weight of about 14,000 Da (Zhao et al., 2006). FPS-1showed potent stimulating effects on murine lymphocyte proliferation induced by concanavalin A or LPS both *in vitro* and *in vivo* and on splenocyte antibody production. Monkshood (*A. carmichaeli*) polysaccharide combined with Adriamycin long-circulating temperature-sensitive liposome can enhance killer activity of NK cells and transformation of T cells, supporting their synergic antitumor effect (Dong et al., 2006).

ACP-a1, a *A. coreanum* polysaccharide with a molecular weight of 3.2×10^5 Da, significantly inhibited the growth of hepatoma H22 transplanted in mice and prolonged the survival time of H22 tumor-bearing mice (Li et al., 2013c). The body weight, peripheral white blood cell (WBC) count, thymus index, and spleen index of H22 tumor bearing were also improved after ACP-a1 treatment. ACP-a1 could promote the secretion of serum cytokines in H22 tumor-bearing mice, such as IL-2, TNF-α, and IFN-γ.

Autophagy attenuates starvation-induced cardiomyocyte death, and polysaccharides from Fu Zi (FPS) increases autophagy activity to protect against starvation-induced cytotoxicity in H9c2 cells, likely through AMPK/mTOR pathway activation (Liao et al., 2013).

The cholesterol-lowering effect of Fu Zi polysaccharide in hypercholesteremic rats is partially caused by increased hepatic LDL-R (low-density lipoprotein receptor) and cytochrome P450 (CYP) 7α-1 expression and decreased 3-hydroxy-3-methylglutaryl (HMG)-CoA (Huang et al., 2010). The neurogenic and antidepressant effects of Fu Zi polysaccharide-1 may involve BDNF (brain-derived neurotrophic factor) signaling in the hippocampus (Yan et al., 2010).

7.5 Other compounds

Examples of other compounds are shown in Table 7.3 and Figure 7.3.

Four caffeoyl derivatives of *A. koreanum*, caffeic acid, 4,5-dicaffeoylquinic acid, 3,5-dicaffeoylquinic acid, and 3,5-dicaffeoylquinic acid methyl ester, showed potent

Table 7.3 **Other compounds found in *Aconitum* in recent years**

Taxon	Type	Compounds	Tissue	Reference
A. burnatii Gáyer	A	Quercetin 3-*O*-β-D-glucopyranoside-7-*O*-(6-E-p-coumaroyl)-β-D-glucopyranosyl-(1→3)-α-L-rhamnopyranoside **71**	Aerial part	Vitalini et al. (2010)
	A	Quercetin 3-*O*-β-D-glucopyranoside-7-*O*-β-D-glucopyranosyl-(1→3)-α-L-rhamnopyranoside **72**		Vitalini et al. (2010)
A. variegatum	A	Kaempferol 3-*O*-β-D-galactopyranoside-7-*O*-α-L-arabinopyranoside **73**		Vitalini et al. (2010)
A. anthora	A	Quercetin 3-*O*-((β-D-glucopyranosyl-(1→3)-(4-*O*-(E-p-coumaroyl))- α-L-rhamnopyranosyl-(1→6)-β-D-galactopyranoside))-7-*O*-α-L-rhamnopyranoside **74**	Aerial part	Mariani et al. (2008)
	A	Kaempferol 3-*O*-((β-D-glucopyranosyl-(1→3)-(4-*O*-(E-p-coumaroyl))- α-L-rhamnopyranosyl-(1→6)-β-D-galactopyranoside))-7-*O*-α-L-rhamnopyranoside **75**		
A. tanguticum	A	Quercetin 3-*O*-α-L-rhamnopyranosyl-(1→2)-[β-D-glucopyranosyl-(1→3)-α-L-(4-*O*-*trans*-p-coumaroylrhamnopyranosyl)-(1→6)]-β-D-galactopyranoside-7-*O*-α-L-rhamnopyranoside **76**	Whole plant	Xu et al. (2013)
	A	Kaempferol 3-*O*-α-L-rhamnopyranosyl-(1→2)-[β-D-glucopyranosyl-(1→3)-α-L-(4-*O*-trans-p-coumaroylrhamnopyranosyl)-(1→6)]-β-D-galactopyranoside-7-*O*-α-L-rhamnopyranoside **77**		
A. tanguticum	B	4-Hydroxyphenethoxy-8-*O*-β-D-[6-*O*-(4-*O*-β-D-glucopyranosyl)-sinapoyl]-glucopyranoside **78**	Whole plant	Li et al. (2014b)
	B	3,4-Dimethoxy-trans-cinnamic acid-9-*O*-β-D-allopyranoside **79**		
A. carmichaelii	C	Methyl (±)-3-(4′-hydroxy-3′-methoxyphenyl)-3-sulfopropionate **80**	Lateral root	Jiang et al. (2014)
	C	4-Methyl (−)-(R)-hydroxyeucomate **81**		
	C	4-Butyl (−)-(R)-hydroxyeucomate **82**		
	C	4-Butyl-1-methyl (+)-(R)-2-*O*-(4′-hydroxy-3′-methoxybenzoyl)malate **83**		
	C	1-Butyl-4-methyl (+)-(R)-2-*O*-(4′-hydroxy-3′-methoxybenzoyl)malate **84**		
	C	Dimethyl (+)-(R)-2-*O*-(4′-hydroxy-3′-methoxybenzoyl)malate **85**		
	C	Dimethyl (+)-(R)-2-*O*-(4′-hydroxybenzoyl)malate **86**		

A, Flavonoids; B, phenylpropanoid glycosides; C, aromatic acid derivatives. Refer to Figure 7.3 for more details.

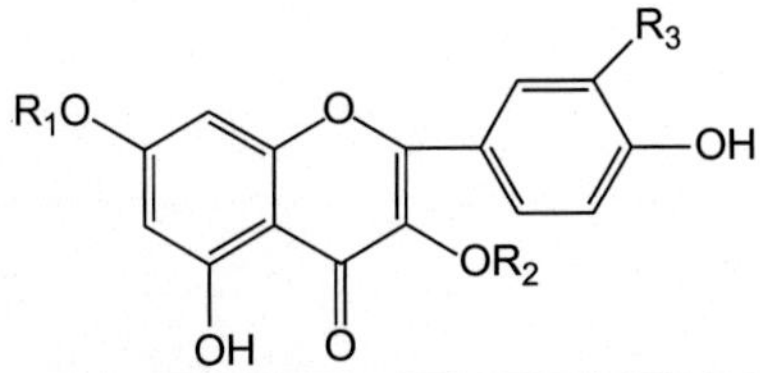

71	R_1=S1	R_2=Glc	R_3=OH
72	R_1=Glc(1→3)-Rha	R_2=Glc	R_3=OH
73	R_1=Ara	R_2=Gal	R_3=H
74	R_1=Rha	R_2=S_2	R_3=OH
75	R_1=Rha	R_2=S_2	R_3=H
76	R_1=Rha	R_2=Rha(1→2)-[Glc(1→3)-(4-*O*-coumaroyl)Rha-(1→6)]-Gal	R_3=OH
77	R_1= Rha	R_2=Rha(1→2)-[Glc(1→3)-(4-*O*-coumaroyl)Rha-(1→6)]-Gal	R_3=H

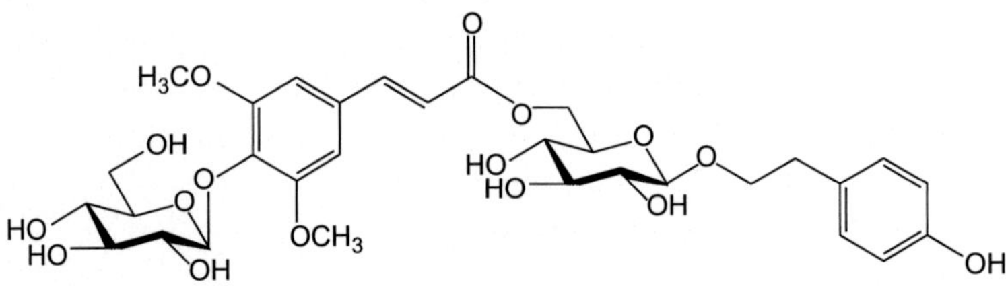

78

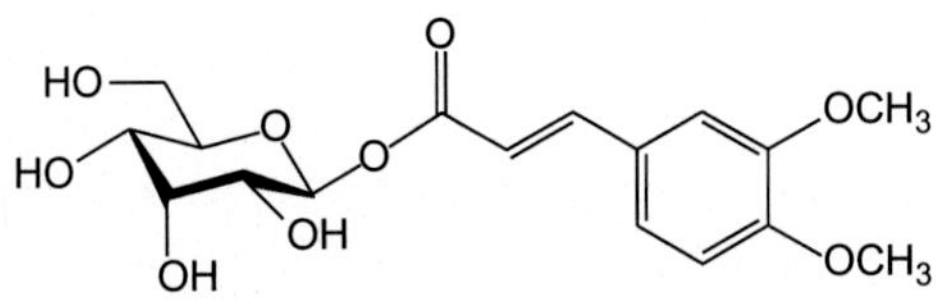

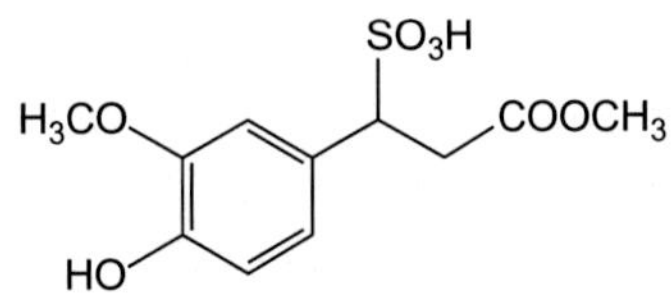

Figure 7.3 Other compounds.

(Continued)

DPPH radical scavenging and NO inhibitory activities as compared with positive controls (Park et al., 2009). These compounds dose-dependently inhibited the expressions of iNOS and COX-2 and their mRNA levels.

New flavonol glycosides are isolated from *A. naviculare* (Shrestha et al., 2006), *A. napellus* sp. *lusitanicum* (Luis et al., 2006), *A. burnatii*, *A. variegatum* (Vitalini et al., 2010), and *A. anthora* (Mariani et al., 2008), some of which showed antioxidant activities. The splenic NK-cell activity of cold-stressed mice was significantly reduced, and the reduction was dose-dependently recovered by PA (processed aconite root) administration (Makino et al., 2009). PA exhibited a thermogenic effect on hypothermia induced by cold stress in mice by additional upregulation of uncoupling protein −1 level in brown adipose tissue, which was already enhanced by hypothermia, and that

79 **80**

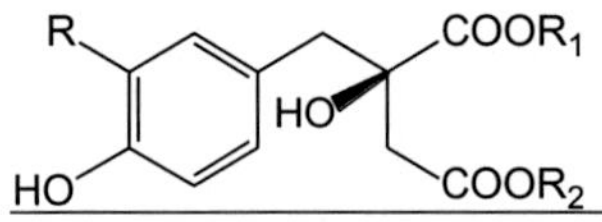

81	R=OH	R_1=H	R_2=CH_3
82	R=OH	R_1=H	R_2=*n*-But

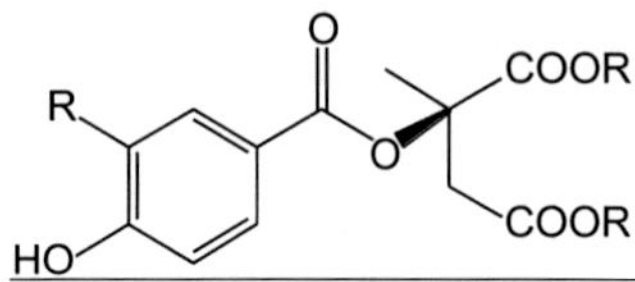

83	R=OCH_3	R_1=CH_3	R_2=CH_3
84	R=OCH_3	R_1=n-But	R_2=CH_3
85	R=OCH_3	R_1=CH_3	R_2=CH_3
86	R=H	R_1=CH_3	R_2=CH_3

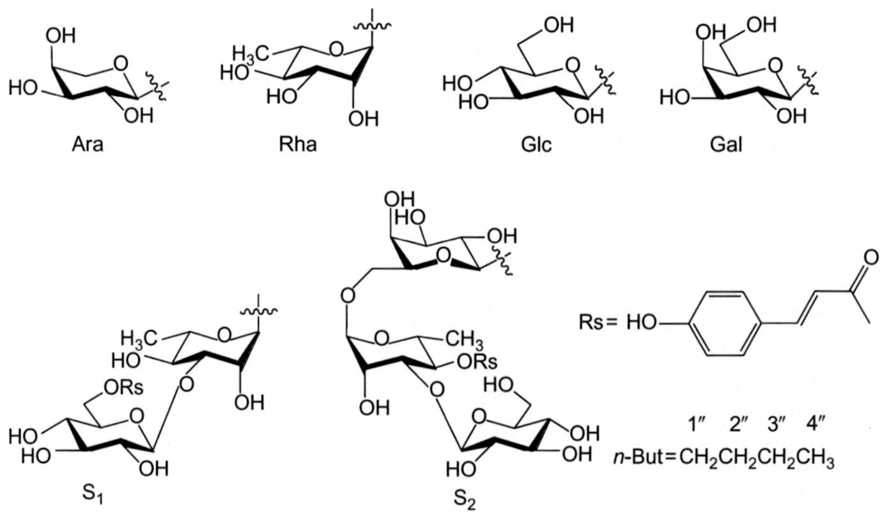

Figure 7.3 Continued.

the active ingredients present in PA are nonalkaloidal low-molecular-weight compounds. The major fatty acids in *A. szechenyianum* are linoleic acid, palmitic acid, linolenic acid, oleic acid, and stearic acid, and the unsaturated fatty acids occupy 66.4% of the total fatty acids (Qu et al., 2011).

Twenty-eight compounds were separated and identified from petroleum fraction of the flower, stem, and leaf of *A. taipeicum* by GC–MS (Xu et al., 2012). Most of them were alkane, fat acids, and their esters and alkenes. The nonacosane covered 13.05%

of the total peaks, while 19-methyl-18, 21-hexatriacontanediether 8.18%, ethylene glycol monooctadecyl ether 7.85%, 3,7,11,15-tetramethyl-2-hexadecen-1-ol 7.80%, and methyl palmitate 6.67%.

7.6 Chemotaxonomy

The monotypic subgenus *Gymnaconitum* has only one species, *A. gymnandrum*, which is primitive phytochemically (Xiao et al., 1984), as only C19 (aconitine and lycoctonine) and C20 (atisine and denudatine) diterpenoid alkaloids are present and no C18 ones found. The whole plant of this species is used as insecticide. The species in *Aconitum* subgenus *Lycoctonum* contain mainly the C18 (lappaconine-type and ranaconine-type) and C19 diterpenoid alkaloids (lycoctonine-type). Lycoctonum is relatively primitive in phytochemistry, which supports the previous morphological evidence (Wang, 1979). Roots of the plants in this subgenus show a relatively lower toxicity (Xiao et al., 2006) and have been used for the treatment of rheumatism, pains, irregular menstruation, etc. Because of the lower toxicity, the species in this subgenus are worthy of a more detailed phytochemical investigation for the development of new medicines.

Subgenus *Aconitum* has about 250 species that are typically divided into two sections, section *Sinaconitum*, with only *A. polycarpum*, and section *Aconitum* including the remaining species (Figure 7.4a). The species in China in the latter section are morphologically separated into 10 series. (1) Series *Tangutica* and series *Rotundifolia* are dwarf alpine plants, generally less than 35 cm tall. Phytochemically, they contain mainly the lactone-type C19 diterpenoid alkaloids and sporadically C20 diterpenoid alkaloids. The lactone-type alkaloids have been found to occur exclusively within these two series and thus can be considered as the characteristic chemical constituents of these series. Their roots show a relatively lower toxicity, comparable to those of *Lycoctonum*. The whole plants of the species have been traditionally used in China's Tibetan, Mongolian, and Uygur regions for the treatment of high fever. (2) Series *Bullatifolia* contain mainly the denudatine-type and the napelline-type C20 diterpenoid alkaloids and the aconitine-type C19 diterpenoid alkaloids. This series may occupy an intermediate position in subgenus *Aconitum* from a chemotaxonomic view. Their toxicity is higher than that of the abovementioned series. The roots in this series are used for the treatment of pains and rheumatism. (3) Series *Brachypoda* includes *A. brachypodum*, *A. pendulum*, *A. polyschistum*, and several other morphologically close species. Their roots are recorded as "Xue Shang Yi Zhi Hao" in the Chinese Pharmacopoeia and have been used as antirheumatic and analgesic remedies. Phytochemically, they contain mainly the aconitine-type C19 diterpenoid alkaloids, an evolutionarily advanced type. Their toxicity is comparable to that of *Bullatifolia*. It is noteworthy that *A. coreanum* contains mainly the C20 diterpenoid alkaloids and has a much lower toxicity (comparable to *Tangutica* and *Rotundifolia*) than the others of this series. From a chemotaxonomic view, it is reasonable to segregate *A. coreanum* and *A. anthoroideum* from series *Brachypoda* (Xiao et al., 2006). (4) Series *Stylosa*,

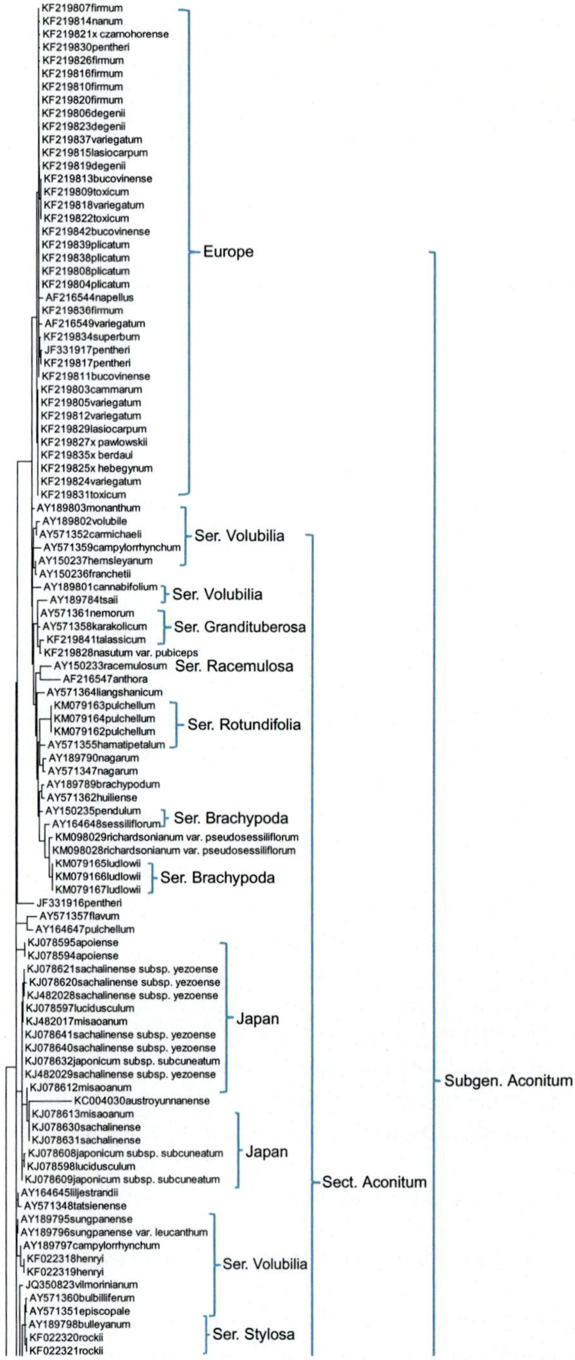

Figure 7.4 Phylogenetic relationship of *Aconitum* and related taxa. The evolutionary history was inferred by using the ML method based on the GTR+I+G model. The trees are drawn to scale, with branch lengths measured in the number of substitutions per site. Evolutionary analyses were conducted in MEGA6. (a) 212 ITS sequences; there were 887 positions in the final dataset.

(Continued)

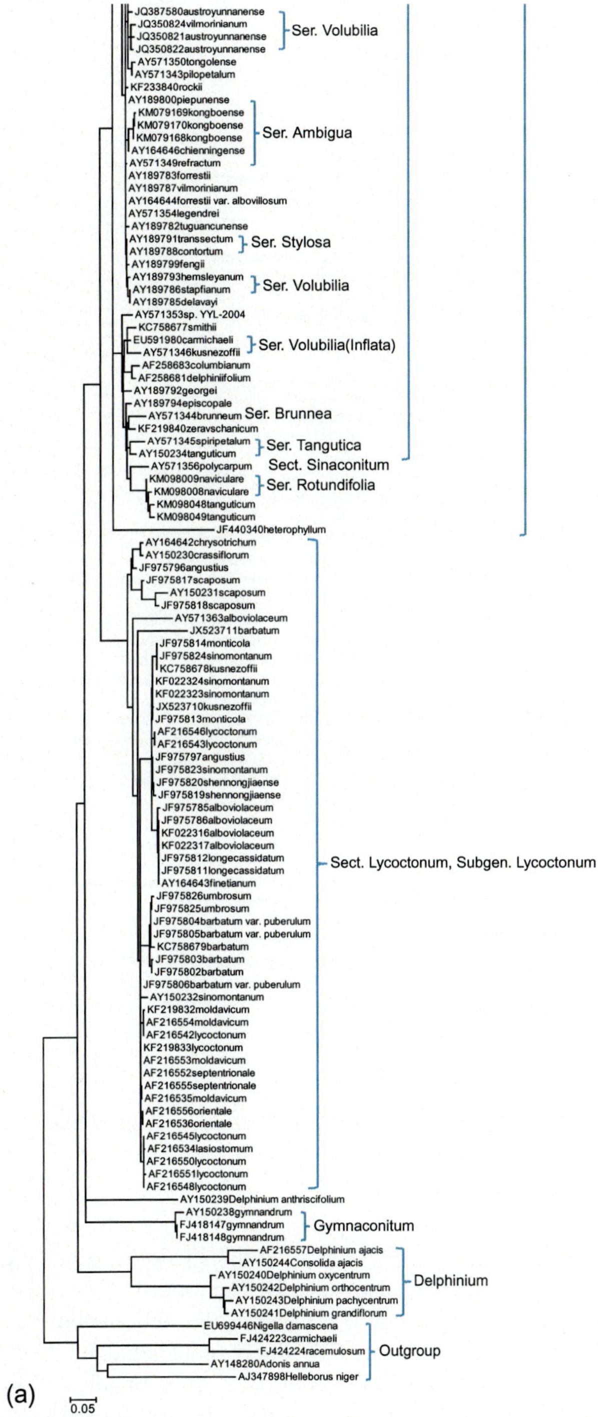

Figure 7.4 Continued.

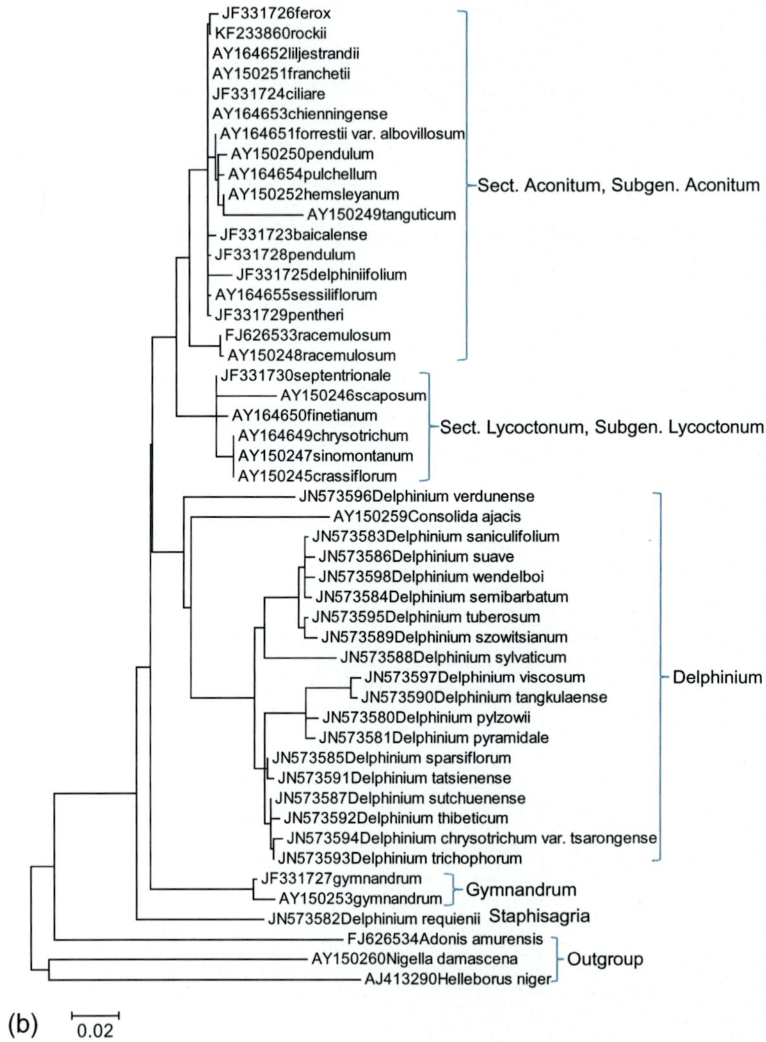

Figure 7.4 Continued. (b) 49 trnL-trnF sequences; there were 1229 positions in the final dataset.

with plants of larger roots, are the major source of "Da Wu Tou." The series contain largely the aconitine-type diester C19 diterpenoid alkaloids, which are characterized by the presence of anisoyloxy residues at C-14. The roots display a high toxicity. From a chemotaxonomic view, *A. contortum*, having the hetidine-type C20 diterpenoid alkaloids and the dranaconine-type C18 diterpenoid alkaloids but no anisoyloxy residues, is a unique species in series *Stylosa*. Its systematic position needs to be reevaluated. (5) Series *Ambigua* contains mainly the aconitine-type C19 diterpenoid alkaloids with anisoyloxy residues, which is similar to series *Stylosa*. (6) Series

Volubilia, characterized by twining stems, contains the aconitine-type diester C19 diterpenoid alkaloids with the presence of an anisoyl or a benzoyl group at C-14. Several species in this series, such as *A. sczukinii* and *A. volubile*, contain the highly advanced 15-hydroxyl aconitine-type C19 diterpenoid alkaloids, indicating its possible affinity to series *Inflata*. Their toxicity is comparable to that of *Brachypoda*. The roots of many species in this series are used as folklore drugs for the treatment of trauma and rheumatic pains. In this series, *A. hemsleyanum*, a polymorphic species in gross morphology, exhibits also a great interpopulational phytochemical variation. (7) Series *Inflata* includes two most well-known medicinal species, *A. carmichaeli* (Wu Tou, Chuan Wu, and Fu Zi) and *A. kusnezoffii* (Bei Wu Tou and Cao Wu). Both are listed in the Pharmacopoeia of the People's Republic of China. They contain the aconitine-type 15-hydroxyl monoester or diester C19 diterpenoid alkaloids, which are evolutionarily young chemical ingredients in the genus *Aconitum*. Their roots are highly toxic. (8) Series *Grandituberosa*, morphologically characterized by having several chain-like arranged tubers, contains the highly advanced aconitine-type C19 diterpenoid alkaloids. The roots, with the LD50 to mice being 29 mg/kg (i.v.), have a high toxicity. (9) Series *Racemulosa* contains mainly the lycoctonine-type and the aconitine-type amino alcohol C19 diterpenoid and C20 diterpenoid alkaloids. The roots of *A. racemulosum* are known as "Xue Wu" and have the effects of activating blood circulation and removing stasis. Chemotaxonomically, this series show some primitive features. (10) Series *Brunnea* is phytochemically not well studied. This series contains both C20 diterpenoid alkaloids, for example, denudatine and songorine, and the highly advanced aconitine-type alkaloids, indicating its transitional position between *Bullatifolia* and *Brachypoda*.

Matrix-assisted laser desorption/ionization time-of-flight mass spectrometry (MALDI-TOF-MS) was used to distinguish raw *Aconitum* roots (high toxicity) and processed roots (low toxicity) (Pan and Liu, 2005), while a rapid resolution liquid chromatographic method for fingerprint analysis of raw and processed Cao Wu (*A. kusnezoffii*) was also reported (Qiao et al., 2009). Hypaconitine (HA) could be used as an indicator to monitor the quality of raw and processed Chuan Wu and Cao Wu (Sun et al., 2012). In addition, the raw and processed aconite can be distinguished by monitoring the ratio of aconitine and mesaconitine to HA. High-performance liquid chromatography (HPLC) was used to quantify alkaloids of various medicinal *Aconitum* species (Wang et al., 2006; Kang et al., 2010). The contents of alkaloids in different species varied significantly. HA, aconitine, and benzoylmesaconine are most abundant in *A. carmichaeli* (a source plant of Chuan Wu), *A. pendulum* (a source plant of Xue Shang Yi Zhi Hao), and *A. transsectum* (a source plant of Cao Wu), respectively, which are distinct from the chemical profile of *A. hemsleyanum* (a source plant of Cao Wu) (Wang et al., 2006). HPLC fingerprint and hierarchical cluster analysis could efficiently identify and distinguish *A. kusnezoffii* (a source plant of Cao Wu) from its related species *A. karacolicum* (used as Cao Wu in Xinjiang, China), *A. austroyunnanense* (used as Cao Wu in Yunnan, China), and *A. contortum* (Zhao et al., 2009). The metabolomic characteristics of the crude aconite root and its processed products were explored by UPLC (ultra-performance liquid chromatography)-TOF-MS combined with pattern recognition methods and ingenuity pathway analysis

(Wang et al., 2012a,b). UPLC has been used in the chemotaxonomic investigations of *Taxus* (Hao et al., 2012) and *Fritillaria* (Hao et al., 2013b) and would contribute to the pharmacophylogenetic study of *Aconitum*.

7.7 Molecular phylogeny and genomics

7.7.1 Phylogeny

The phylogenetic relationship of *Aconitum* is elusive, and various molecular markers were used to tackle this issue. Zhang et al. (2003a) used nuclear ITS (internal transcribed spacer) to investigate the phylogeny of the *A. delavayi* complex of Hengduan Mountains, Southwest China. The phylogenetic relationship of 27 taxa, including eight taxa of the *A. delavayi* complex, is reconstructed. The morphological series *Volubilia* is not a monophyletic group, and series *Stylosa* cannot be distinguished from series *Ambigua*, which suggest that the morphological serial delimitation of the subgenus *Aconitum* proposed by Wang (1979) is problematic. The close systematic relationship among some endemic species of Hengduan Mountains might be due to the recent speciation events. *A. delavayi* complex is not a natural lineage, as three clades were seen on the ITS tree. Luo et al. (2005) used ITS to study the phylogeny of 51 species and one variety of the subgenus *Aconitum*. Again, traditional circumscription of some infrasubgeneric taxa is not supported by the molecular data. Species from northern China show relatively close relationship with those from Europe, North America, Japan, and Korea. High morphological variation and low genetic diversity found in Hengduan species imply a recent species radiation. Seed and petal morphology, instead of other traditional morphology characters, are better suitable for elucidating the evolutionary relationship of *Aconitum*. The chloroplast (cp) psbA-trnH was used to identify medicinal *Aconitum* (He et al., 2010). The sequence inversions were found in the two most commonly used species, *A. carmichaeli* and *A. kusnezoffii*. The 19 taxa were grouped into 10 clusters according to the sequence alignment, and most species can be authenticated by psbA-trnH. Population studies on *Aconitum* species, using either ITS (Wang et al., 2009a,b,c), RAPD (Zhang et al., 2005), and ISSR markers (Luo et al., 2006) or allozyme variation (Zhang et al., 2003a,b), confirm a rather high population diversity, and the polymorphism observed in these studies indicates that more investigations are needed at population or species level to ensure a better characterization of *Aconitum* species.

Jabbour and Renner (2012) utilized chloroplast trnL-F spacer, trnL intron, and nuclear ITS sequences to study the phylogenetic relationship of 185 species of the tribe Delphinieae, including 57 *Aconitum* species. Three short-lived Mediterranean species belonging to *Delphinium* subgenus *Staphisagria* are the sister clade to all other Delphinieae (Figure 7.4b), suggesting that *Staphisagria* should be regarded as a genus if *Delphinium* and *Aconitum* are to keep mutually monophyletic. Delphinieae originated in the early Oligocene (c. 32.3 Ma) and expanded to North America around 2.9–3.3 Ma ago. Late Miocene bursts in diversification rates in the Himalayan and southwestern Chinese clades of *Aconitum* and *Delphinium* might be associated with transitions from short-lived to long-lived life histories. On the phylogenetic tree based on ITS and cp sequences, the subgenus *Gymnaconitum* is outside of other two subgenera (Figure 7.4), while the subgenus *Lycoctonum* is basal to the subgenus *Aconitum*. Series *Rotundifolia* and *Brachypoda*, as well as *Tangutica* and *Bullatifolia*, are not monophyletic groups and cluster together. The series *Ambigua*, *Stylosa*, *Volubilia*, and *Inflata* are also not monophyletic but intermingled on the phylogenetic tree, suggesting extensive morphological convergence. Series *Grandituberosa* is closer to *Volubilia* than to other series. *A. brunneum* and *A. racemulosum* are distinct species. Branches leading to taxa of the subgenus *Aconitum* are all extremely short, implying that there were bursts of species radiation during evolution, especially shortly after the uplift of Qinghai–Tibet Plateau. Since the number of species and sequence is limited in the above studies, we retrieve more marker sequences from NCBI GenBank to reconstruct the phylogenetic tree (Figure 7.4). The topology of ITS tree, psbA-trnH tree (not shown), and trnL-F tree is largely congruent and confirms the findings of Luo et al. (2005) and Jabbour and Renner (2012). On the ITS tree, *A. liangshanicum* of series *Ambigua* is basal to the clade consisted of series *Ambigua*, *Bullatifolia*, *Rotundifolia*, and *Brachypoda*. On the psbA-trnH tree, *A. coreanum* of series *Brachypoda* and *A. contortum* of series *Stylosa* are far away from other members of the same series, which are reminiscent of their distinct alkaloid profiles (Xiao et al., 2006).

7.7.2 Transcriptomics

A. heterophyllum, popularly known as Atis or Patis, is an important medicinal herb of Northwestern and Eastern Himalayas. Quantitative expression analysis of 15 genes of MVA/MEP pathways in *A. heterophyllum* roots versus shoots, differing for atisine content (0–2.2-fold), showed 11–100-fold increase in transcript amounts of four genes of MVA pathway (HMGS, HMGR, PMK, and IPPI) and four genes of MEP pathway (DXPS, ISPD, HDS, and GDPS), respectively (Malhotra et al., 2014). The overall expression of eight genes decreased to 5–12-fold after comparative expression analysis between roots of high (0.37%)- versus low (0.14%)-atisine content accessions, but their relative transcript amounts remained higher in high-content accessions, thereby implying their role in atisine biosynthesis and accumulation. PCA analysis revealed a positive correlation between MVA/MEP pathways genes and alkaloids content. The future high-throughput transcriptome sequencing could of MVA/MEP pathway genes, which are a prerequisite for the genetic improvement of medicinal *Aconitum*.

7.7.3 Functional genomics and metabolomics

Processed *A. carmichaelii* Debx (Baifupian in Chinese, BFP) elicits differential toxic responses in rats under various health conditions. Sensitive and reliable biomarkers of toxicity were identified, with the corresponding metabolic pathways being unveiled. Thirty male Sprague-Dawley rats were divided into five groups ($n=6$) and received oral administration of BFP extract (0.32, 0.64, 1.28, or 2.56 g/kg per day) or an equal volume of drinking water (control) for 15 days (Tan et al., 2014a). The metabolomic profiles of rat serum were analyzed by LC–Q-TOF-MS. Linear regression analysis and Ingenuity Pathway Analysis (IPA) were used to elucidate the differentially altered metabolites and associated network relationships. BFP could induce prominent toxicity in the heart, liver, and kidneys at a dose of 2.56 g/kg per day. Betaine upregulation and phosphatidylcholine downregulation were detected in the serum samples of drug-treated groups in a dose-dependent manner, which could be sensitive biomarkers for the toxic responses of BFP. Perturbations of RhoA signaling, choline metabolism, and free radical scavenging were partly responsible for the toxic effects of BFP.

Five therapeutic biomarkers (phosphate, betaine, (4-hydroxyphenyl) acetaldehyde, 5-hydroxyindol-3-acetic acid, and 5′-phosphoribosyl-*N*-formylglycinamide) and two corresponding canonical pathways (amino acid metabolism and purine nucleotide metabolism) were identified to be involved in the therapeutic mechanism of BFP treating the kidney-yang deficiency syndrome (KYDS) (Tan et al., 2014b).

7.7.4 Molecular marker

ISSR was used to investigate the genetic diversity and variations of *A. leucostomum* and to supply essential characteristics for identifying *Aconitum* crude drugs (Gao et al., 2014).

nrDNA ITS sequence-based SCAR marker was developed to authenticate *A. heterophyllum* in ayurvedic raw drug source and prepared herbal products (Seethapathy et al., 2014). *A. vilmorinianum* and *A. austroyunnanense*, distributed and used as herbal drug in Yunnan, China, can be identified by their ITS sequences (Zhang et al., 2012), and there are more variable sites in ITS1 than in ITS2 sequences.

7.8 Conclusions

Aconitum plants hold great potential for in-depth investigation for various biological activities, especially on anticancer therapy and their effects on cardiovascular and central nervous systems. The key barrier for the wide medical utilization of *Aconitum* species is their usually extremely high toxicity. From a pharmacophylogenetic point of view, therefore, the less toxic species, that is, those in subgenus *Lycoctonum* and in series *Tangutica*, *Rotundifolia*, and *Racemulosa* and two species (*A. coreanum* and *A. anthoroideum*) in series *Brachypoda* in subgenus *Aconitum*, should be preferably studied with various omics tools. The thorough metabolomic studies should be performed for these species, including their roots, stems, and leaves. The metabolic phenotypes should be linked with their genomic structure and functions and tissue-specific transcriptomic alterations (Hao et al., 2011a,b). Further comprehensive studies on these species will boost their better medicinal utilization and conservation.

References

Ahmad, M., Ahmad, W., Ahmad, M., Zeeshan, M., Obaidullah, Shaheen, F., 2008. Norditerpenoid alkaloids from the roots of *Aconitum heterophyllum* Wall with antibacterial activity. J. Enzyme Inhib. Med. Chem. 23 (6), 1018–1022.

Baek, J.M., Kim, J.Y., Cheon, Y.H., et al., 2014. *Aconitum pseudo-laeve var. erectum* inhibits receptor activator of nuclear factor kappa-B ligand-induced osteoclastogenesis via the c-Fos/nuclear factor of activated T-cells, cytoplasmic 1 signaling pathway and prevents lipopolysaccharide-induced bone loss in mice. Molecules 19 (8), 11628–11644.

Bai, G., Yang, Y., Shi, Q., et al., 2008. Identification of higenamine in *Radix Aconiti Lateralis Preparata* as a beta2-adrenergic receptor agonist. Acta Pharmacol. Sin. 29 (10), 1187–1194.

Bello-Ramírez, A.M., Nava-Ocampo, A.A., 2004. The local anesthetic activity of *Aconitum* alkaloids can be explained by their structural properties: a QSAR analysis. Fundam. Clin. Pharmacol. 18 (2), 157–161.

Borcsa, B., Widowitz, U., Csupor, D., et al., 2011. Semisynthesis and pharmacological investigation of lipo-alkaloids prepared from aconitine. Fitoterapia 82 (3), 365–368.

Borcsa, B., Fodor, L., Csupor, D., et al., 2014. Diterpene alkaloids from the roots of *Aconitum moldavicum* and assessment of Nav 1.2 sodium channel activity of aconitum alkaloids. Planta Med. 80 (2–3), 231–236.

Cai, Y., Gao, Y., Tan, G., et al., 2013. Myocardial lipidomics profiling delineate the toxicity of traditional Chinese medicine *Aconiti Lateralis radix praeparata*. J. Ethnopharmacol. 147 (2), 349–356.

Chan, T.Y., 2009. Aconite poisoning. Clin. Toxicol. 47 (4), 279–285.

Chen, T.T., Qi, C., Guo, H., et al., 2009. The effects of Fu Zi on changes in the body heat of dogs. J. Acupunct. Meridian Stud. 2 (1), 71–74.

Csupor, D., Forgo, P., Wenzig, E.M., Bauer, R., Hohmann, J., 2008. Bisnorditerpene, norditerpene, and lipo-alkaloids from *Aconitum toxicum*. J. Nat. Prod. 71 (10), 1779–1782.

Dasyukevich, O.I., Solyanik, G.I., 2007. Comparative study of anticancer efficacy of aconitine-containing agent BC1 against ascite and solid forms of Ehrlich's carcinoma. Exp. Oncol. 29 (4), 317–319.

de Inés, C., Reina, M., Gavín, J.A., González-Coloma, A., 2006. In vitro cytotoxicity of norditerpenoid alkaloids. Z. Naturforsch. C 61 (1–2), 11–18.

de la Peña, S.S., Sothern, R.B., López, F.S., et al., 2011. Circadian aspects of hyperthermia in mice induced by *Aconitum napellus*. Pharmacogn. Mag. 7 (27), 234–242.

Dong, L.F., Zhang, Y.J., Liu, J.S., et al., 2006. Anti-tumor effect of monkshood polysaccharide with adriamycin long circulating temperature-sensitive liposome and its mechanism. Xi Bao Yu Fen Zi Mian Yi Xue Za Zhi 22 (4), 458–462.

Fan, X., Chen, Y., Xing, J., et al., 2012. Blocking effects of acehytisine on pacemaker currents (If) in sinoatrial node cells and human HCN4 channels expressed in *Xenopus laevis* oocytes. J. Ethnopharmacol. 139 (1), 42–51.

Forgo, P., Borcsa, B., Csupor, D., et al., 2011. Diterpene alkaloids from *Aconitum anthora* and assessment of the hERG-inhibiting ability of *Aconitum* alkaloids. Planta Med. 77 (4), 368–373.

Gao, T., Bi, H., Ma, S., Lu, J., 2010a. The antitumor and immunostimulating activities of water soluble polysaccharides from *Radix Aconiti*, *Radix Aconiti Lateralis* and *Radix Aconiti Kusnezoffii*. Nat. Prod. Commun. 5 (3), 447–455.

Gao, T., Bi, H., Ma, S., Lu, J., 2010b. Structure elucidation and antioxidant activity of a novel alpha-(1 → 3), (1 → 4)-D-glucan from *Aconitum kusnezoffii* Reichb. Int. J. Biol. Macromol. 46 (1), 85–90.

Gao, T., Ma, S., Song, J., et al., 2011. Antioxidant and immunological activities of water-soluble polysaccharides from *Aconitum kusnezoffii* Reichb. Int. J. Biol. Macromol. 49 (4), 580–586.

Gao, F., Li, Y.Y., Wang, D., et al., 2012. Diterpenoid alkaloids from the chinese traditional herbal "fuzi" and their cytotoxic activity. Molecules 17 (5), 5187–5194.

Gao, F.C., Sun, Y., Zhang, J., Zhang, F., 2014. ISSR analysis for genetic polymorphism of *Aconitum leucostomum* from different habitats. Zhong Yao Cai 37 (1), 26–29.

Garmanchouk, L.V., Pyaskovskaya, O.N., Yanish, Y.V., et al., 2005. Influence of aconitine-containing herbal extract BC1 on proliferative and electrokinetic characteristics of endothelial cells. Exp. Oncol. 27 (4), 262–266.

Ge, Y., Mu, S., Zhang, J., et al., 2009. Diterpenoid alkaloids from roots of *Aconitum recemulosum* and their inhibitory effects on PAF-induced platelet aggregation. Zhongguo Zhong Yao Za Zhi 34 (15), 1935–1937.

Hao, D.C., Xiao, P.G., 2014. Network pharmacology: a Rosetta stone for traditional Chinese medicine. Drug Dev. Res. 75 (5), 299–312.

Hao, D.C., Ge, G.B., Xiao, P.G., et al., 2011a. The first insight into the tissue specific taxus transcriptome via Illumina second generation sequencing. PLoS One 6 (6), e21220.

Hao, D.C., Yang, L., Xiao, P.G., 2011b. The first insight into the Taxus genome via fosmid library construction and end sequencing. Mol. Genet. Genomics 285 (3), 197–205.

Hao, D.C., Xiao, P.G., Peng, Y., et al., 2012. Research progress and trend analysis of biology and chemistry of Taxus medicinal resources. Acta Pharm. Sin. 47, 827–835.

Hao, D.C., Gu, X.J., Xiao, P.G., et al., 2013a. Recent advances in the chemical and biological studies of *Aconitum* pharmaceutical resources. J. Chin. Pharm. Sci. 22 (3), 209–221.

Hao, D.C., Gu, X.J., Xiao, P.G., Peng, Y., 2013b. Phytochemical and biological research of *Fritillaria* medicinal resources. Chin. J. Nat. Med. 11 (4), 330–344.

Hazawa, M., Wada, K., Takahashi, K., et al., 2009. Suppressive effects of novel derivatives prepared from *Aconitum* alkaloids on tumor growth. Invest. New Drugs 27 (2), 111–119.

Hazawa, M., Takahashi, K., Wada, K., et al., 2011. Structure-activity relationships between the Aconitum C20-diterpenoid alkaloid derivatives and the growth suppressive activities of non-Hodgkin's lymphoma Raji cells and human hematopoietic stem/progenitor cells. Invest. New Drugs 29 (1), 1–8.

Hazawa, M., Wada, K., Kashiwakura, I., 2012. *Aconitum* C(20)-diterpenoid alkaloid derivatives regulate the growth and differentiation of cord blood derived human hematopoietic stem/progenitor cells. Phytother. Res. 26 (8), 1262–1264.

He, Y.Q., Ma, Z.Y., Yang, Q., et al., 2008. A new norditerpenoid alkaloid from *Aconitum taipaicum*. Acta Pharm. Sin. 43 (9), 934–937.

He, P., Jian, X.X., Chen, D.L., Wang, F.P., 2009. Diterpenoid alkaloids from *Aconitum longzhoushanense*. Nat. Prod. Commun. 4 (1), 19–22.

He, J., Wong, K.L., Shaw, P.C., et al., 2010. Identification of the medicinal plants in *Aconitum* L. by DNA barcoding technique. Planta Med. 76 (14), 1622–1628.

Huang, X., Tang, J., Zhou, Q., et al., 2010. Polysaccharide from fuzi (FPS) prevents hypercholesterolemia in rats. Lipids Health Dis. 9, 9.

Huang, X., Yang, Y., Zhu, J., et al., 2012. Comparative effects of Guanfu base A and Guanfu base G on HERG K+ channel. J. Cardiovasc. Pharmacol. 59 (1), 77–83.

Jabbour, F., Renner, S.S., 2012. A phylogeny of Delphinieae (Ranunculaceae) shows that *Aconitum* is nested within *Delphinium* and that Late Miocene transitions to long life cycles in the Himalayas and Southwest China coincide with bursts in diversification. Mol. Phylogenet. Evol. 62 (3), 928–942.

Jian, X.X., Tang, P., Liu, X.X., et al., 2012. Structure-cardiac activity relationship of C19-diterpenoid alkaloids. Nat. Prod. Commun. 7 (6), 713–720.

Jiang, B., Lin, S., Zhu, C., et al., 2012. Diterpenoid alkaloids from the lateral root of *Aconitum carmichaelii*. J. Nat. Prod. 75 (6), 1145–1159.

Jiang, Z., Jiang, B., Zhu, C., et al., 2014. Aromatic acid derivatives from the lateral roots of *Aconitum carmichaelii*. J. Asian Nat. Prod. Res. 16 (9), 891–900.

Kang, X.Q., Fan, Z.C., Zhang, Z.Q., 2010. Simultaneous determination of three *Aconitum* alkaloids in six herbal medicines by high-performance liquid chromatography. J. Chromatogr. Sci. 48 (10), 860–865.

Kiss, T., Orvos, P., Bánsághi, S., et al., 2013. Identification of diterpene alkaloids from Aconitum napellus subsp. firmum and GIRK channel activities of some *Aconitum* alkaloids. Fitoterapia 90, 85–93.

Kurtoğlu, S., Sen, B., Melikoğlu, G., et al., 2012. Diterpenoid alkaloids of *Aconitum vulparia* Rchb. Z. Naturforsch. C 67 (3–4), 103–107.

Li, Y., Liang, S., 2009. Studies on chemical constituents from leaves and stems of *Aconitum coreanum*. Zhongguo Zhong Yao Za Zhi 34 (10), 1220–1222.

Li, Y., Tu, D., Xiao, H., et al., 2010. Aconitine blocks HERG and Kv1.5 potassium channels. J. Ethnopharmacol. 131 (1), 187–195.

Li, H., Sun, M., Xu, J., et al., 2013a. Immunological response in H22 transplanted mice undergoing *Aconitum coreanum* polysaccharide treatment. Int. J. Biol. Macromol. 55, 295–300.

Li, J., Ren, W., Huang, X.J., et al., 2013b. Bullatine A, a diterpenoid alkaloid of the genus *Aconitum*, could attenuate ATP-induced BV-2 microglia death/apoptosis via P2X receptor pathways. Brain Res. Bull. 97, 81–85.

Li, M., He, J., Jiang, L.L., et al., 2013c. The anti-arthritic effects of *Aconitum vilmorinianum*, a folk herbal medicine in Southwestern China. J. Ethnopharmacol. 147 (1), 122–127.

Li, S., Xiong, J., Zhang, Y., et al., 2013d. Carmichaeline A: a new C20-diterpenoid alkaloid from *Aconitum carmichaeli*. Nat. Prod. Commun. 8 (2), 155–156.

Li, X., Jiang, J., Shi, S., et al., 2014a. A RG-II type polysaccharide purified from *Aconitum coreanum* alleviates lipopolysaccharide-induced inflammation by inhibiting the NF-κB signal pathway. PLoS One 9 (6), e99697.

Li, Y.R., Xu, L., Li, C., et al., 2014b. Two new compounds from *Aconitum tanguticum*. J. Asian Nat. Prod. Res. 16 (7), 730–734.

Liang, Y., Zhu, J., Yang, Y.M., et al., 2006. Electrophysiological effects of acehytisine hydrochloride in a porcine model of acute coronary occlusion. Zhonghua Xin Xue Guan Bing Za Zhi 34 (11), 1035–1039.

Liao, L.Z., Chen, Y.L., Lu, L.H., et al., 2013. Polysaccharide from Fuzi likely protects against starvation-induced cytotoxicity in H9c2 cells by increasing autophagy through activation of the AMPK/mTOR pathway. Am. J. Chin. Med. 41 (2), 353–367.

Lin, J.S., Chan, C.Y., Yang, C., et al., 2007. Zhi-fuzi, a cardiotonic Chinese herb, a new medical treatment choice for portal hypertension? Exp. Biol. Med. 232 (4), 557–564.

Lin, L., Chen, D.L., Liu, X.Y., et al., 2009. Bis-diterpenoid alkaloids from *Aconitum tanguticum var. trichocarpum*. Nat. Prod. Commun. 4 (7), 897–901.

Lin, G.M., Liu, L., Shao, W., 2011. Microcalorimetry studies on the antibacterial effect of crude monkshood polysaccharide. J. Zhejiang Univ. Sci. B 12 (7), 563–567.

Liou, S.S., Liu, I.M., Lai, M.C., Cheng, J.T., 2005. Comparison of the antinociceptive action of crude Fuzei, the root of *Aconitum*, and its processed products. J. Ethnopharmacol. 99 (3), 379–383.

Liou, S.S., Liu, I.M., Lai, M.C., 2006. The plasma glucose lowering action of Hei-Shug-Pian, the fire-processed product of the root of Aconitum (*Aconitum carmichaelii*), in streptozotocin-induced diabetic rats. J. Ethnopharmacol. 106 (2), 256–262.

Liu, Z.L., Cao, J., Zhang, H.M., et al., 2011. Feeding deterrents from *Aconitum episcopale* roots against the red flour beetle, *Tribolium castaneum*. J. Agric. Food Chem. 59 (8), 3701–3706.

Liu, X.X., Jian, X.X., Cai, X.F., et al., 2012. Cardioactive C_{19}-diterpenoid alkaloids from the lateral roots of *Aconitum carmichaeli* "Fu Zi" Chem. Pharm. Bull. 60 (1), 144–149.

Lu, Y., Bu, H., Yang, L., et al., 2011. Comparasion of polysaccharides in parent root, daughter root and rootlet of *Aconitum carmichaeli*. Zhongguo Zhong Yao Za Zhi 36 (9), 1154–1157.

Luis, J.C., Valdés, F., Martín, R., et al., 2006. DPPH radical scavenging activity of two flavonol glycosides from *Aconitum napellus* sp. lusitanicum. Fitoterapia 77 (6), 469–471.

Luo, Y., Yang, Q.E., 2005. Taxonomic revision of Aconitum (Ranunculaceae) from Sichuan, China. Acta Phytotax. Sin. 43, 289–386.

Luo, Y., Zhang, F.M., Yang, Q.E., 2005. Phylogeny of Aconitum subgenus *Aconitum* (Ranunculaceae) inferred from ITS sequences. Plant Syst. Evol. 252, 11–25.

Luo, Q., Ma, D.W., Wang, Y.H., 2006. ISSR identification of genetic diversity in *Aconitum carmichaelii*. Chin. Tradit. Herb. Drugs 37, 1554–1557.

Makino, T., Kato, K., Mizukami, H., 2009. Processed aconite root prevents cold-stress-induced hypothermia and immuno-suppression in mice. Biol. Pharm. Bull. 32 (10), 1741–1748.

Malhotra, N., Kumar, V., Sood, H., et al., 2014. Multiple genes of mevalonate and non-mevalonate pathways contribute to high aconites content in an endangered medicinal herb, *Aconitum heterophyllum* Wall. Phytochemistry 108, 26–34. http://dx.doi.org/10.1016/j.phytochem.2014.08.025.

Mariani, C., Braca, A., Vitalini, S., et al., 2008. Flavonoid characterization and in vitro antioxidant activity of *Aconitum anthora* L. (Ranunculaceae). Phytochemistry 69 (5), 1220–1226.

Mu, Z.Q., Gao, H., Huang, Z.Y., et al., 2012. Puberunine and puberudine, two new C(18)-diterpenoid alkaloids from *Aconitum barbatum var. puberulum*. Org. Lett. 14 (11), 2758–2761.

Nesterova, I.V., Povet'eva, T.N., Aksinenko, S.G., et al., 2009. Evaluation of anti-inflammatory activity of extracts from Siberian plants. Vestn. Ross. Akad. Med. Nauk 11, 30–34.

Nesterova, Y.V., Povetieva, T.N., Suslov, N.I., et al., 2011. Antidepressant activity of diterpene alkaloids of *Aconitum baicalense* Turcz. Bull. Exp. Biol. Med. 151 (4), 425–428.

Nesterova, Y.V., Povet'yeva, T.N., Suslov, N.I., et al., 2014. Analgesic activity of diterpene alkaloids from *Aconitum baikalensis*. Bull. Exp. Biol. Med. 157 (4), 488–491.

Nisar, M., Ahmad, M., Wadood, N., et al., 2009. New diterpenoid alkaloids from *Aconitum heterophyllum* Wall: selective butyrylcholinestrase inhibitors. J. Enzyme Inhib. Med. Chem. 24 (1), 47–51.

Pan, A.H., Liu, J., 2005. Study on identification of aconiti by matrix-assisted laser desorption/ionization time-of-flight mass spectrometry. Fa Yi Xue Za Zhi 21 (2), 118–120.

Park, K.H., Park, M., Choi, S.E., et al., 2009. The anti-oxidative and anti-inflammatory effects of caffeoyl derivatives from the roots of *Aconitum koreanum* R. RAYMOND. Biol. Pharm. Bull. 32 (12), 2029–2033.

Park, P.J., Moon, B.S., Lee, S.H., et al., 2012. Hair growth-promoting effect of *Aconiti Ciliare Tuber* extract mediated by the activation of Wnt/β-catenin signaling. Life Sci. 91 (19–20), 935–943.

Pashinskiĭ, V.G., Gaĭdamovich, N.N., Povet'eva, T.N., et al., 2006. The influence of aconitum preparations on the development of chronic inflammation. Eksp. Klin. Farmakol. 69 (2), 55–57.

Qiao, Y.L., Zhang, Y.H., Zhang, W., Zhang, J.L., 2009. A rapid resolution liquid chromatographic method for fingerprint analysis of raw and processed caowu (*Aconitum kusnezoffii*). J. AOAC Int. 92 (2), 653–662.

Qu, W.X., Mou, Z.L., Cui, H.Y., Zhang, Z.Q., 2011. Analysis of fatty acids in A. szechenyianum Gay. by microwave-assisted extraction and gas chromatography–mass spectrometry. Phytochem. Anal. 22 (3), 199–204.

Seethapathy, G.S., Balasubramani, S.P., Venkatasubramanian, P., 2014. nrDNA ITS sequence based SCAR marker to authenticate *Aconitum heterophyllum* and *Cyperus rotundus* in Ayurvedic raw drug source and prepared herbal products. Food Chem. 145, 1015–1020.

Shao, F., Chen, H.J., Liu, R.H., et al., 2013. Effects of heishunpian total alkaloids on *Cassia acutifolia* induced mice diarrhea and contraction of isolated intestinal smooth muscle in rats. Zhong Yao Cai 36 (11), 1805–1809.

Shen, Y., Ai, H.L., Zi, S.H., et al., 2012. Two new C19-diterpenoid alkaloids from *Aconitum transsectum*. J. Asian Nat. Prod. Res. 14 (3), 244–248.

Shrestha, B.B., Dall'Acqua, S., Gewali, M.B., et al., 2006. New flavonoid glycosides from *Aconitum naviculare* (Brühl) Stapf, a medicinal herb from the trans-Himalayan region of Nepal. Carbohydr. Res. 341 (12), 2161–2165.

Shu, H., Hayashida, M., Arita, H., et al., 2008a. High doses of processed Aconiti tuber inhibit the acute but potentiate the chronic antinociception of morphine. J. Ethnopharmacol. 119 (2), 276–283.

Shu, H., Hayashida, M., Huang, W., et al., 2008b. The comparison of effects of processed Aconiti tuber, U50488H and MK-801 on the antinociceptive tolerance to morphine. J. Ethnopharmacol. 117 (1), 158–165.

Song, M.K., Liu, H., Jiang, H.L., et al., 2008. Discovery of talatisamine as a novel specific blocker for the delayed rectifier K^+ channels in rat hippocampal neurons. Neuroscience 155 (2), 469–475.

Subash, A.K., Augustine, A., 2012. Hypolipidemic effect of methanol fraction of *Aconitum heterophyllum* wall ex Royle and the mechanism of action in diet-induced obese rats. J. Adv. Pharm. Technol. Res. 3 (4), 224–228.

Sun, K., Yang, J., Shen, D.K., 2008. Clinical observation on treatment of primary knee osteoarthritis of liver and kidney deficiency type with Aconite cake-separated moxibustion. Zhongguo Zhen Jiu 28 (2), 87–90.

Sun, A., Gao, B., Ding, X., et al., 2012. Quantitative and qualitative analysis of aconitum alkaloids in raw and processed Chuanwu and Caowu by HPLC in combination with automated analytical system and ESI/MS/MS. J. Anal. Methods Chem. 2012, 936131.

Sun, G.B., Sun, H., Meng, X.B., et al., 2014. Aconitine-induced Ca^{2+} overload causes arrhythmia and triggers apoptosis through p38 MAPK signaling pathway in rats. Toxicol. Appl. Pharmacol. 279 (1), 8–22.

Tan, Y., Ko, J., Liu, X., et al., 2014a. Serum metabolomics reveals betaine and phosphatidylcholine as potential biomarkers for the toxic responses of processed *Aconitum carmichaelii* Debx. Mol. Biosyst. 10 (9), 2305–2316.

Tan, Y., Liu, X., Lu, C., et al., 2014b. Metabolic profiling reveals therapeutic biomarkers of processed *Aconitum carmichaeli* Debx in treating hydrocortisone induced kidney-yang deficiency syndrome rats. J. Ethnopharmacol. 152 (3), 585–593.

Tang, Q., Liu, J., Xue, J., et al., 2008. Preparative isolation and purification of two new isomeric diterpenoid alkaloids from *Aconitum coreanum* by high-speed counter-current chromatography. J. Chromatogr. B Analyt. Technol. Biomed. Life Sci. 872 (1–2), 181–185.

Tong, P., Xu, S., Cao, G., et al., 2014. Chondroprotective activity of a detoxicated traditional Chinese medicine (Fuzi) of *Aconitum carmichaeli* Debx against severe-stage osteoarthritis model induced by mono-iodoacetate. J. Ethnopharmacol. 151 (1), 740–744.

Turabekova, M.A., Rasulev, B.F., Dzhakhangirov, F.N., et al., 2010. *Aconitum* and *Delphinium* alkaloids of curare-like activity. QSAR analysis and molecular docking of alkaloids into AChBP. Eur. J. Med. Chem. 45 (9), 3885–3894.

Vitalini, S., Braca, A., Passarella, D., Fico, G., 2010. New flavonol glycosides from *Aconitum burnatii* Gáyer and *Aconitum variegatum* L. Fitoterapia 81 (7), 940–947.

Wada, K., Nihira, M., Hayakawa, H., et al., 2005. Effects of long-term administrations of aconitine on electrocardiogram and tissue concentrations of aconitine and its metabolites in mice. Forensic Sci. Int. 148 (1), 21–29.

Wada, K., Nihira, M., Ohno, Y., 2006. Effects of chronic administrations of aconitine on body weight and rectal temperature in mice. J. Ethnopharmacol. 105 (1–2), 89–94.

Wada, K., Hazawa, M., Takahashi, K., et al., 2007. Inhibitory effects of diterpenoid alkaloids on the growth of A172 human malignant cells. J. Nat. Prod. 70 (12), 1854–1858.

Wada, K., Hazawa, M., Takahashi, K., et al., 2011. Structure-activity relationships and the cytotoxic effects of novel diterpenoid alkaloid derivatives against A549 human lung carcinoma cells. J. Nat. Med. 65 (1), 43–49.

Wang, W.T., 1979. *Aconitum* L. In: Flora Reipublicae Popularis Sinicae 27. Science Press, Beijing, pp. 113–362.

Wang, Z., Wen, J., Xing, J., He, Y., 2006. Quantitative determination of diterpenoid alkaloids in four species of *Aconitum* by HPLC. J. Pharm. Biomed. Anal. 40 (4), 1031–1034.

Wang, C.F., Gerner, P., Wang, S.Y., Wang, G.K., 2007. Bulleyaconitine A isolated from aconitum plant displays long-acting local anesthetic properties in vitro and in vivo. Anesthesiology 107 (1), 82–90.

Wang, C.F., Gerner, P., Schmidt, B., et al., 2008. Use of bulleyaconitine A as an adjuvant for prolonged cutaneous analgesia in the rat. Anesth. Analg. 107 (4), 1397–1405.

Wang, J.L., Shen, X.L., Chen, Q.H., et al., 2009a. Structure-analgesic activity relationship studies on the C(18)- and C(19)-diterpenoid alkaloids. Chem. Pharm. Bull. 57 (8), 801–807.

Wang, L., Abbott, R.J., Zheng, W., et al., 2009b. History and evolution of alpine plants endemic to the Qinghai-Tibetan Plateau: *Aconitum gymnandrum* (Ranunculaceae). Subgen. Gymnaconitum. Mol. Ecol. 18 (4), 709–721.

Wang, L., Zhang, D., Qu, X., et al., 2009c. Experimental research on cardiotonic effect of active compound from raw and processing aconite roots. Zhongguo Zhong Yao Za Zhi 34 (5), 596–599.

Wang, D.P., Lou, H.Y., Huang, L., et al., 2012a. A novel franchetine type norditerpenoid isolated from the roots of *Aconitum carmichaeli* Debx. with potential analgesic activity and less toxicity. Bioorg. Med. Chem. Lett. 22 (13), 4444–4446.

Wang, X., Wang, H., Zhang, A., et al., 2012b. Metabolomics study on the toxicity of aconite root and its processed products using ultraperformance liquid-chromatography/electrospray-ionization synapt high-definition mass spectrometry coupled with pattern recognition approach and ingenuity pathways analysis. J. Proteome Res. 11 (2), 1284–1301.

Wangchuk, P., Bremner, J.B., Samten, N.V., et al., 2010. Antiplasmodial activity of atisinium chloride from the Bhutanese medicinal plant, *Aconitum orochryseum*. J. Ethnopharmacol. 130 (3), 559–562.

Wu, G., Huang, W., Zhang, H., et al., 2011a. Inhibitory effects of processed Aconiti tuber on morphine-induced conditioned place preference in rats. J. Ethnopharmacol. 136 (1), 254–259.

Wu, L., Gao, X., Wang, L., et al., 2011b. Prediction of multi-target of *Aconiti Lateralis Radix Praeparata* and its network pharmacology. Zhongguo Zhong Yao Za Zhi 36 (21), 2907–2910.

Wu, G., Du, L., Zhao, L., et al., 2014. The total alkaloids of *Aconitum tanguticum* protect against lipopolysaccharide-induced acute lung injury in rats. J. Ethnopharmacol. 155 (3), 1483–1491.

Xiao, P.G., Chen, D.H., Song, W.L., 1984. Phytochemical evidence related to the phylogeny of *Aconitum gymnandrum* Maxim. Acta Phytotax. Sin. 22, 53–56.

Xiao, P.G., Wang, F.P., Gao, F., et al., 2006. A pharmacophylogenetic study of *Aconitum* L. (Ranunculaceae) from China. Acta Phytotax. Sin. 44, 1–46.

Xiao, K., Wang, L., Liu, Y., et al., 2007. Study of aconitine toxicity in rat embryos in vitro. Birth Defects Res. B Dev. Reprod. Toxicol. 80 (3), 208–212.

Xiao, K., Yan, G.Y., Wang, L., et al., 2008. Toxicity of Caowu on in vitro cultured rat embryos. Sichuan Da Xue Xue Bao Yi Xue Ban 39 (3), 441–444.

Xing, B., Jin, S., Wang, H., et al., 2014. New diterpenoid alkaloids from *Aconitum coreanum* and their anti-arrhythmic effects on cardiac sodium current. Fitoterapia 94, 120–126.

Xiong, L., Peng, C., Xie, X., et al., 2012. Alkaloids isolated from the lateral root of *Aconitum carmichaelii*. Molecules 17, 9939–9946.

Xu, H., Arita, H., Hayashida, M., et al., 2006a. Pain-relieving effects of processed Aconiti tuber in CCI-neuropathic rats. J. Ethnopharmacol. 103 (3), 392–397.

Xu, L.L., Wang, L., Wang, Y.Q., 2006b. Effects of radix salviae miltiorrhiae, radix aconiti lateralis preparata and rhizoma anemarrhena on nitric oxide systems of endotoxemia mice. Zhongguo Zhong Xi Yi Jie He Za Zhi. 26 (Suppl.), 102–107.

Xu, X.F., Chen, Y.H., Wang, J., et al., 2010a. Effect and mechanism of *Aconitum vaginatum* on the proliferation, invasion and metastasis in human A549 lung carcinoma cells. Zhong Yao Cai 33 (12), 1909–1912.

Xu, Y., Guo, Z.J., Wu, N., 2010b. Two new amide alkaloids with anti-leukaemia activities from *Aconitum taipeicum*. Fitoterapia 81 (8), 1091–1093.

Xu, N., Zhao, D.F., Liang, X.M., et al., 2011. Identification of diterpenoid alkaloids from the roots of *Aconitum kusnezoffii* Reihcb. Molecules 16 (4), 3345–3350.

Xu, W.M., Guo, Z.J., Han, L., et al., 2012. Analysis of petroleum fraction from flower, stem and leaf of *Aconitum taipeicum* by GC-MS. Zhong Yao Cai 35 (5), 726–728.

Xu, L., Zhang, X., Lin, L., et al., 2013. Two new flavonol glycosides from the Tibetan medicinal plant *Aconitum tanguticum*. J. Asian Nat. Prod. Res. 15 (7), 737–742.

Yamada, K., Suzuki, E., Nakaki, T., et al., 2005. Aconiti tuber increases plasma nitrite and nitrate levels in humans. J. Ethnopharmacol. 96 (1–2), 165–169.

Yan, H.C., Qu, H.D., Sun, L.R., et al., 2010. Fuzi polysaccharide-1 produces antidepressant-like effects in mice. Int. J. Neuropsychopharmacol. 13 (5), 623–633.

Yang, L.J., Luo, Y.P., Chen, W., et al., 2009. A novel norditerpenoid alkaloid from *Aconitum macrorhynchum*. J. Asian Nat. Prod. Res. 11 (8), 728–731.

Yin, T., Cai, L., He, J., et al., 2014a. Three new diterpenoid alkaloids from the roots of *Aconitum duclouxii*. J. Asian Nat. Prod. Res. 16 (4), 345–350.

Yin, T., Cai, L., Zhou, H., et al., 2014b. A new C_{19}-diterpenoid alkaloid from the roots of *Aconitum duclouxii*. Nat. Prod. Res. 28 (19), 1649–1654.

Yu, H., Ji, X., Wu, Z., Wang, S., 2011. Effects of aconite root on energy metabolism and expression of related genes in rats. Zhongguo Zhong Yao Za Zhi 36 (18), 2535–2538.

Yuan, C.L., Wang, X.L., 2012. Isolation of active substances and bioactivity of *Aconitum sinomontanum* Nakai. Nat. Prod. Res. 26 (22), 2099–2102.

Zhang, F.M., Ge, S., Chen, W.L., 2003a. Phylogeny of the *Aconitum delavayi* complex (Ranunculaceae) based on evidence from nuclear ribosomal ITS sequences. Acta Phytotax. Sin. 41, 220–228.

Zhang, F.M., Kong, H.Z., Ge, S., 2003b. Allozyme variation and population differentiation of the *Aconitum delavayi* complex (Ranunculaceae) in the Hengduan Mountains of China. Biochem. Genet. 41, 47–55.

Zhang, F.M., Chen, W.L., Yang, Q.E., Ge, S., 2005. Genetic differentiation and relationship of populations in the *Aconitum delavayi* complex (Ranunculaceae) and their taxonomic implications. Plant Syst. Evol. 254, 39–48.

Zhang, X.N., Du, C.H., Fu, D.H., et al., 2012. Study on ITS sequences of *Aconitum vilmorinianum* and its medicinal adulterant. Zhong Yao Cai 35 (9), 1410–1413.

Zhang, S., Jiang, Y., Bi, Y., et al., 2013. Diterpenoid alkaloids from *Aconitum kirinense*. J. Asian Nat. Prod. Res. 15 (1), 78–83.

Zhang, H., Guo, Z., Han, L., et al., 2014. The antitumor effect and mechanism of taipeinine A, a new C19-diterpenoid alkaloid from *Aconitum taipeicum*, on the HepG2 human hepatocellular carcinoma cell line. J. BUON 19 (3), 705–712.

Zhao, C., Li, M., Luo, Y., Wu, W., 2006. Isolation and structural characterization of an immunostimulating polysaccharide from fuzi, *Aconitum carmichaeli*. Carbohydr. Res. 341 (4), 485–491.

Zhao, Y.Y., Zhang, Y., Lin, R.C., Sun, W.J., 2009. An expeditious HPLC method to distinguish *Aconitum kusnezoffii* from related species. Fitoterapia 80 (6), 333–338.

Zhao, Y., Bu, Q., Zhou, Y., et al., 2010. Mechanism study of Aconitum-induced neurotoxicity in PC12 cells: involvement of dopamine release and oxidative damage. Neurotoxicology 31 (6), 752–757.

Zhao, D.K., Ai, H.L., Zi, S.H., et al., 2013. Four new C18-diterpenoid alkaloids with analgesic activity from *Aconitum weixiense*. Fitoterapia 91, 280–283.

Zhao, Y., Wang, J., Sun, X., et al., 2014. Microcalorimetry coupled with chemometric techniques for toxicity evaluation of *Radix Aconiti Lateralis Preparata* (Fuzi) and its processed products on *Escherichia coli*. Appl. Microbiol. Biotechnol. 98 (1), 437–444.

Zheng, Q., Zhao, Y., Wang, J., et al., 2014. Spectrum-effect relationships between UPLC fingerprints and bioactivities of crude secondary roots of *Aconitum carmichaelii* Debeaux (Fuzi) and its three processed products on mitochondrial growth coupled with canonical correlation analysis. J. Ethnopharmacol. 153 (3), 615–623.

Zhou, S.J., Du, G.Y., 2003. Effects of higenamine on the cardio-circulatory system. Zhongguo Zhong Yao Za Zhi 28 (10), 910–913.

Zyuz'kov, G.N., Zhdanov, V.V., Udut, E.V., et al., 2013. Mechanisms of napelline action stimulating the regeneration of hemopoietic tissue in cytostatic myelosuppression. Bull. Exp. Biol. Med. 155 (4), 439–442.

Chemical and biological studies of Cimicifugeae pharmaceutical resources

8

8.1 Introduction

Cimicifugeae plants belong to subfamilies Helleboroideae and Ranunculaceae and are traditionally composed of five genera: *Cimicifuga*, *Actaea*, *Anemonopsis*, *Souliea*, and *Beesia*. There are more than 40 species in the tribe Cimicifugeae, which are native to temperate regions of the Northern Hemisphere. The genus *Actaea* is closely related to *Cimicifuga* and *Souliea*, and based on combined DNA sequence data, similarity in biochemical constituents, and morphology, many botanists include those genera within *Actaea* (Compton and Culham, 2002). If not merged, *Cimicifuga* has 28 species and eight are distributed in China. Among them, *C. foetida*, *C. dahurica*, and *C. heracleifolia* are officially listed in Chinese Pharmacopoeia as "Sheng Ma," which have been used since ancient times to treat wind-heat headache, sore throat, toothache, and uterine prolapse. Other *Cimicifuga* species are also used in folk medicine to treat physical wounds, promote eruption, and detoxify (Gao et al., 2008). There are eight species in the genus *Actaea*, six of which are distributed in North America. Among them, *A. racemosa*, also called black cohosh, has long been used by Native Americans to tackle rheumatism and menopausal and nervous problems. Presently, black cohosh products are among the most popular health-promoting medicines in the United States and Europe (Jiang et al., 2006). Two *Actaea* species are used in traditional Chinese medicine and their therapeutic use is distinct from *Cimicifuga* (Gao et al., 2008). *Souliea* medicine is antipyretic and antitoxic, which is also used to eliminate dampness, clear heart fire, and relieve anxiety. *Beesia* is endemic in China and is used to treat cold caused by exterior heat, joint pain of rheumatism, diarrhea, sore throat, headache, and snake bite. On the phylogenetic tree based on chloroplast and nuclear DNA sequences, *Beesia* and the monotypic genus *Anemonopsis* group together and are basal to the cluster formed by *Cimicifuga*, *Actaea*, and *Souliea* (Wang et al., 2009). However, little is known about the chemical constituents and bioactivities of these two genera. *C. foetida* and *A. racemosa* are most frequently studied. In this chapter, we summarize the recent progress in the phytochemical and biological studies of Cimicifugeae.

8.2 Triterpenoid saponins

9,19-Cycloartane-type triterpenoid saponins are mainly distributed in Cimicifugeae and *Thalictrum* of Ranunculaceae and *Astragalus* of Leguminosae (Tian et al., 2006a; Liu et al., 2011). This unique type of saponin is the major pharmaceutical

Medicinal plants: chemistry, biology and omics. http://dx.doi.org/10.1016/B978-0-08-100085-4.00008-6

component of the well-known Chinese medicines Huang Qi and Sheng Ma. Triatomic ring is formed at C9 and C19. The conformations of rings A, B, C, and D are chair, twist chair, twist boat, and envelope, respectively (Figures 8.1 and 8.2). A/B and C/D are of trans-condensation, while B/C is of *cis*-condensation.

The structures of rings A–D are not quite different in diverse cycloartane-type saponins, whereas the side chains tied with ring D vary greatly, due to difference in oxygenation and the approach of ring formation (Table 8.1 and Figures 8.1 and 8.2). C-3β is frequently glycosylated and usually links with one or more sugars. The most common is pentose (usually xylose and arabinose), while hexose (e.g., glucose and galactose) is occasional. Sometimes, C-15 is also glycosylated. Chlorination diversifies *Cimicifuga racemosa* triterpene glycosides (Chen et al., 2007).

8.3 Bioactivities and adverse effects of saponins and extracts from Cimicifugeae

8.3.1 Anticancer activity

Novel natural product anticancer compounds, due to their safety and efficiency, provide a gorgeous substitute to synthetic chemicals. Cimigenol from *C. foetida* (Sheng Ma) exerted potent cytotoxic activity against SMMC-7721 (7.87 μM) and A-549 (12.16 μM), while cimiacerin B also showed obvious cytotoxicity against the A-549 cell line (Lu et al., 2012). The research group led by Prof. Xiao PG found that *C. foetida* extract inhibits proliferation of hepatocellular cells via induction of cell cycle arrest and apoptosis (Tian et al., 2007a). *Cimicifuga racemosa* (black cohosh; CR) reduced Ki-67 and cyclin D1 protein expression in fibroadenomas and may have chemopreventive potential for breast cancer (Einbond et al., 2012). Cimiside E from *C. heracleifolia*, one of source plants of Sheng Ma, arrests cell cycle and induces cell apoptosis in gastric cancer cells (Guo et al., 2009). Total glycosides from the aerial part (TGA) of *C. dahurica*, another source plant of Sheng Ma, induced G0/G1 HepG2 cell cycle arrest at lower concentration and triggered G2/M arrest and apoptosis at higher concentrations (Tian et al., 2007b). An increase in the ratio of Bax/Bcl-2 was implicated in TGA-induced apoptosis. Besides, TGA dose-dependently inhibited the growth of the implanted mouse H22 tumor. Treatment with the triterpene glycoside actein induced a stress response and apoptosis in human breast cancer cells, suggesting that compounds from *Cimicifuga* species may be useful in the prevention and treatment of breast cancer (Einbond et al., 2008). Gene expression profiling reveals effects of CR on the estrogen receptor (ER)-positive breast cancer cell line MCF-7 (Gaube et al., 2007). No estrogenic but antiproliferative and proapoptotic gene expression was shown for CR in MCF-7 cells at the transcriptional level. The effects may be caused by the activation of different pathways. The cycloartane glycosides and their aglycones could be identified as the active components in CR. The methanolic extract of CR also activated genes that enhance apoptosis and repressed cell cycle genes in MDA-MB-453 human breast cancer cells (Einbond et al., 2007). The CR extract BNO 1055 inhibits proliferation of human prostate cancer-derived LNCaP cells (Seidlová-Wuttke et al., 2006).

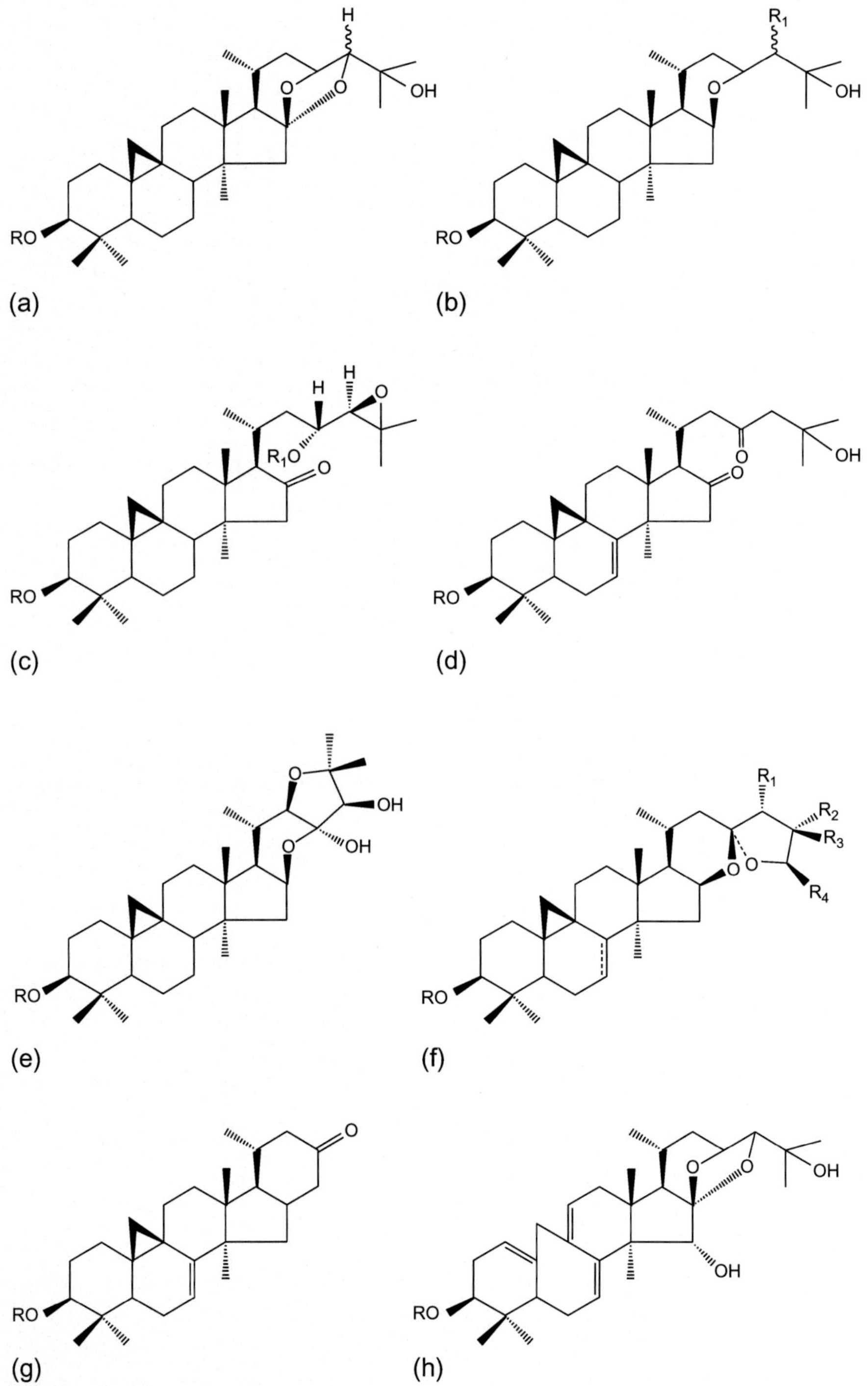

Figure 8.1 Chemical structure types of 9,19-cycloartane triterpenoids in tribe Cimicifugeae. A, Cimigenol type. B, Hydroshengmanol type. C, Shengmanol type. D, 16,23-Dione type. E, Cimiacerogenin type. F, Cimifugenin type. G, Side chain-seco type. H, Neocimicigenoside type.

(Continued)

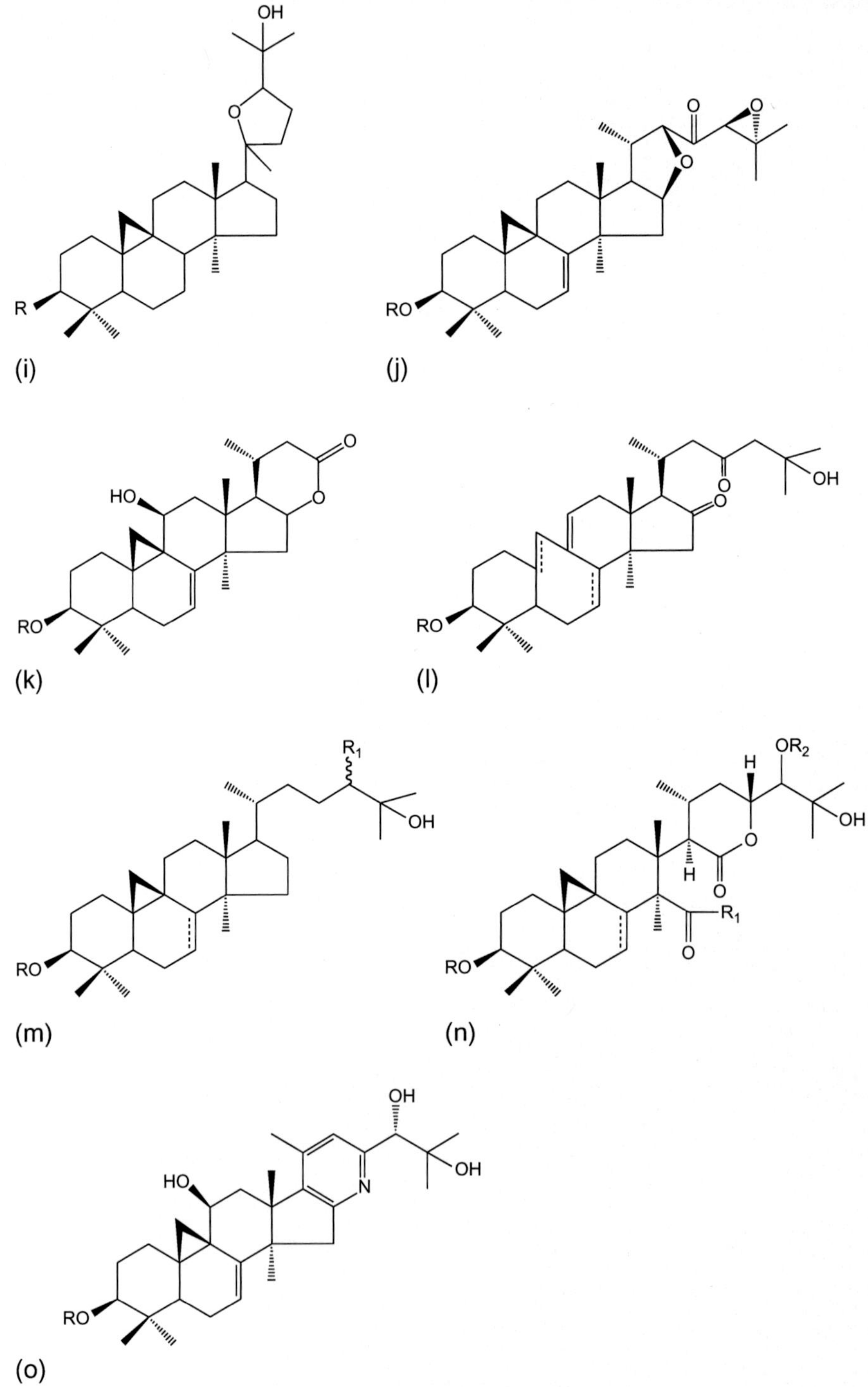

Figure 8.1 Continued. I, Beesioside type. J, Asiaticoside type. K, Ranunculane type. L, 9,10-Seco-9,19-cycloartane type. M, 9,19-Cycloartane type. N, Actaeaepoxide type. O, Tetranor type. P, 15,16-Seco-cycloartane type. Q, Cycloartane triterpene alkaloid type. R, 16,22-Seco-cycloartane type. A-O types belong to 9,19-cycloartane. P–R types are shown in Figure 8.2.

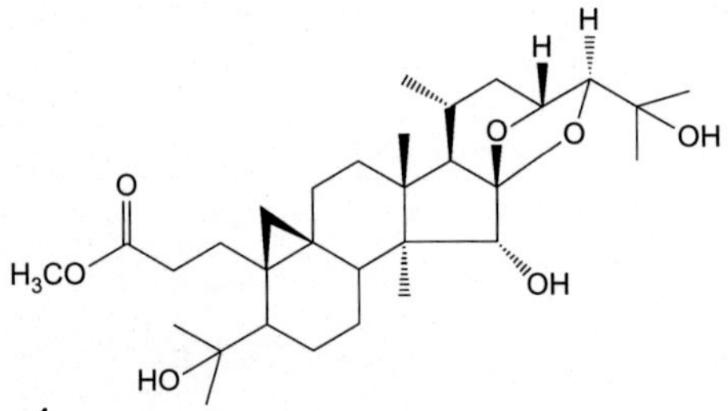

1

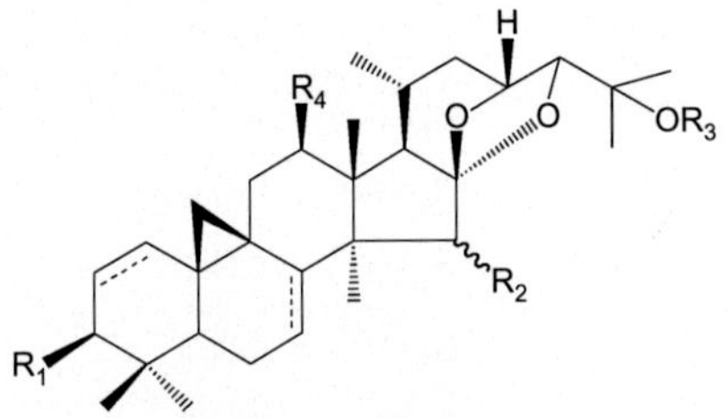

2	R_1 = *O*-Ara(2′,4′-*O*-diacetyl)	R_2 = α-OH	R_3 = H	R_4 = H	C24-S
3	R_1 = *O*-Ara(3′,4′-*O*-diacetyl)	R_2 = α-OH	R_3 = H	R_4 = H	C24-S
4	R_1 = *O*-Ara(4′-*O*-acetyl)	R_2 = α-OH	R_3 = H	R_4 = H	C24-S
6	R_1 = =*O*	R_2 = α-OH	R_3 = H	R_4 = H	C24-R
7	R_1 = *O*-Gal	R_2 = α-OH	R_3 = Ac	R_4 = H	C24-R,$\Delta^{7(8)}$
8	R_1 = *O*-Gal	R_2 = α-OH	R_3 = H	R_4 = H	C24-R,$\Delta^{7(8)}$
10	R_1 = OH	R_2 = α-OH	R_3 = H	R_4 = =O	C24-S
11	R_1 = OH	R_2 = α-H	R_3 = H	R_4 = OH	C24-S
12	R_1 = *O*-Ara(2′-*O*-acetyl)	R_2 = α-OH	R_3 = H	R_4 = H	C24-S
13	R_1 = *O*-Xyl(2′-*O*-acetyl)	R_2 = α-OH	R_3 = H	R_4 = H	C24-S
15	R_1 =*O*-Xyl	R_2 = β-D-Glc	R_3 = H	R_4 = H	
16	R_1 =*O*-Xyl	R_2 = β-D-Glc	R_3 = Ac	R_4 = H	
17	R_1 = *O*-Glc(1→3)-Xyl	R_2 = α-OH	R_3 = H	R_4 = H	C24-S
18	R_1 = *O*-Glc(1→2)-Xyl	R_2 = α-OH	R_3 = H	R_4 = H	C24-S
19	R_1 = *O*-Xyl(3′-O-acetyl)	R_2 = α-OH	R_3 = H	R_4 = OAc	C24-S

Figure 8.2 Examples of triterpenoids and saponins found in tribe Cimicifugeae.

(Continued)

20	R_1 = *O*-Xyl(4′-*O*-acetyl)	R_2 = α-OH	R_3 = Ac	R_4 = H	C24-S
21	R_1 = *O*-Xyl(2′-*O*-acetyl)	R_2 = α-OH	R_3 = CH_3	R_4 = H	C24-S
22	R_1 = *O*-Xyl(2′-*O*-acetyl)	R_2 = α-OH	R_3 = CH_2CH_3	R_4 = H	C24-S
23	R_1 = *O*-Xyl(1→3)-Xyl	R_2 = α-OH	R_3 = H	R_4 = OH	C24-S
24	R_1 = *O*-Xyl(1→3)-Xyl	R_2 = α-OH	R_3 = Ac	R_4 = H	C24-S
25	R_1 = OH	R_2 = α-OH	R_3 = H	R_4 = H	C24-S
26	R_1 = *O*-Ara(2′-*O*-acetyl)	R_2 = α-OH	R_3 = H	R_4 = H	C24-R
27	R_1 = *O*-Ara(2′-*O*-acetyl)	R_2 = α-OH	R_3 = Ac	R_4 = H	C24-R
28	R_1 = *O*-Xyl(2′-*O*-acetyl)(p)	R_2 = α-OH	R_3 = Ac	R_4 = H	C24-S
29	R_1 = *O*-Xyl(1′3)-Xyl	R_2 = α-OH	R_3 = H	R_4 = OH	C24-S
31	R_1 = *O*-Ara[2′-*O*-(*E*)-2-butenoyl]	R_2 = α-OH	R_3 = H	R_4 = H	C24-S
32	R_1 = *O*-Ara(4′-*O*-acetyl)	R_2 = α-OH	R_3 = Ac	R_4 = H	C24-S
33	R_1 = *O*-Ara(3′-*O*-acetyl)	R_2 = α-OH	R_3 = Ac	R_4 = H	C24-S
34	R_1 = OH	R_2 = α-OH	R_3 = H	R_4 = OH	C24-S,$\Delta^{7(8)}$
36	R_1 = = *O*	R_2 = α-OH	R_3 = H	R_4 = H	C24-S,$\Delta^{7(8)}$
37	R_1 = = *O*	R_2 = α-OH	R_3 = H	R_4 = H	C24-S,$\Delta^{1(2)}$,$\Delta^{7(8)}$

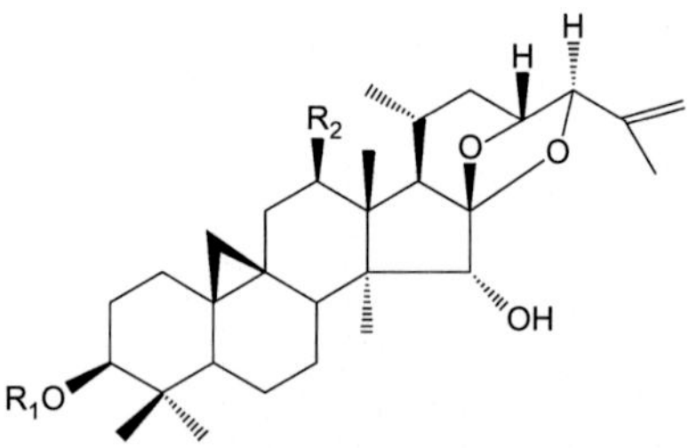

5	R_1 = Ara(3′-*O*-acetyl)	R_2 = H
9	R_1 = H	R_2 = OH
14	R_1 = Ara	R_2 = OH
30	R_1 = Xyl(2′-*O*-acetyl)	R_2 = H

Figure 8.2 Continued.

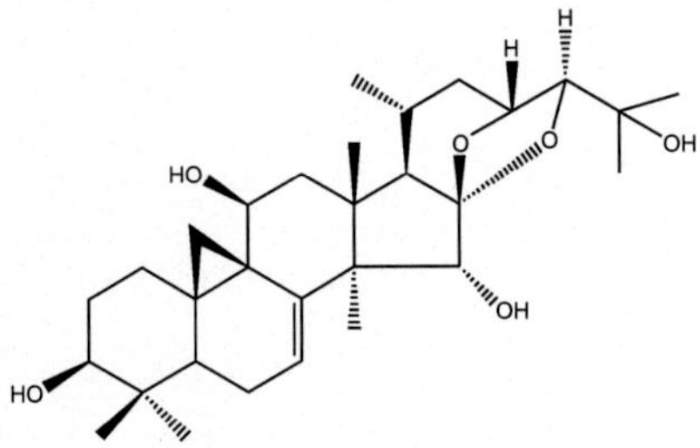

35

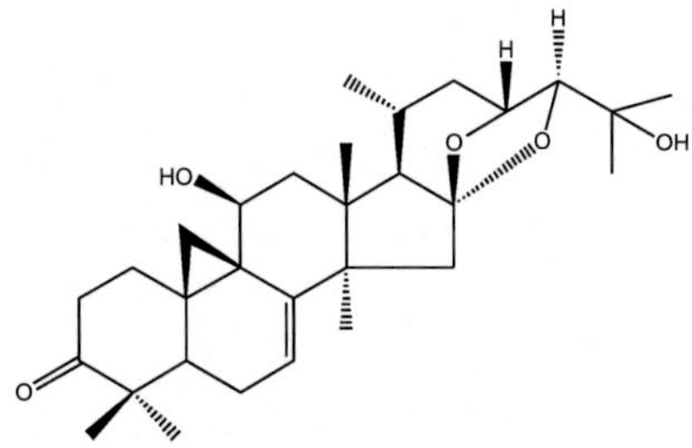

38

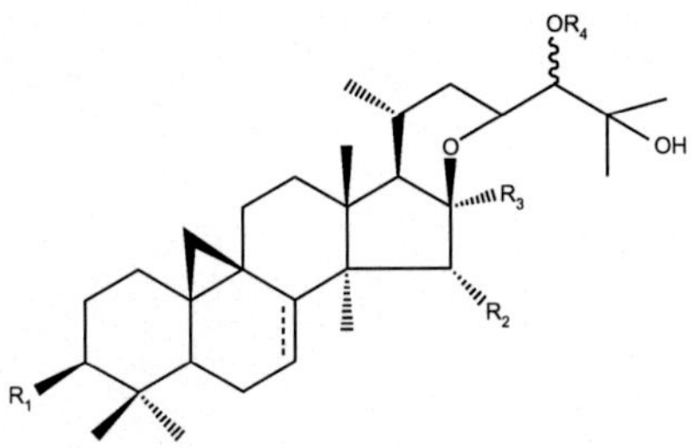

39	R_1 = O-Gal	R_2 = OH	R_3 = OH	R_4 = H	C24-S,$\Delta^{7(8)}$
40	R_1 = O-Gal	R_2 = OH	R_3 = OH	R_4 = H	C24-R,$\Delta^{7(8)}$
42	R_1 = O-Gal	R_2 = OH	R_3 = OH	R_4 = Ac	C24-S,$\Delta^{7(8)}$
43	R_1 = O-Ara(2′-O-acetyl)	R_2 = =O	R_3 = H	R_4 = Ac	C24-S
44	R_1 = O-Ara	R_2 = =O	R_3 = H	R_4 = Ac	C24-S
46	R_1 = O-Xyl(1→3)-Xyl	R_2 = OH	R_3 = OH	R_4 = Ac	C24-S
47	R_1 = O-Gal	R_2 = =O	R_3 = H	R_4 = Ac	C24-S
48	R_1 = O-Xyl(1→2)-Glc	R_2 = OH	R_3 = OH	R_4 = Ac	C24-R
51	R_1 = O-Xyl(2′-O-acetyl)	R_2 = =O	R_3 = H	R_4 = Ac	C24-S
52	R_1 = O-Xyl	R_2 = =O	R_3 = H	R_4 = Ac	C24-R
53	R_1 = H	R_2 = =O	R_3 = H	R_4 = Ac	C24-S
54	R_1 = H	R_2 = =O	R_3 = H	R_4 = Ac	C24-S,$\Delta^{7(8)}$
55	R_1 = O-Xyl	R_2 = OH	R_3 = OH	R_4 =Glc	C24-S,$\Delta^{7(8)}$
56	R_1 = O-Xyl	R_2 = OH	R_3 = OH	R_4 =Glc	C24-R,$\Delta^{7(8)}$
57	R_1 = O-Xyl	R_2 = OGlc	R_3 = OH	R_4 = Ac	C24-R,$\Delta^{7(8)}$

Figure 8.2 Continued.

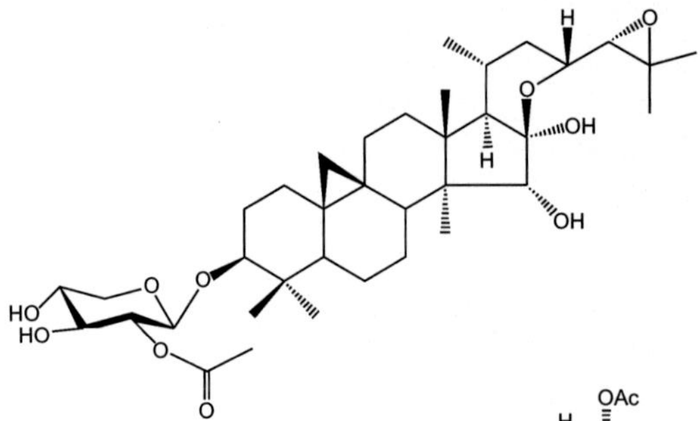

41

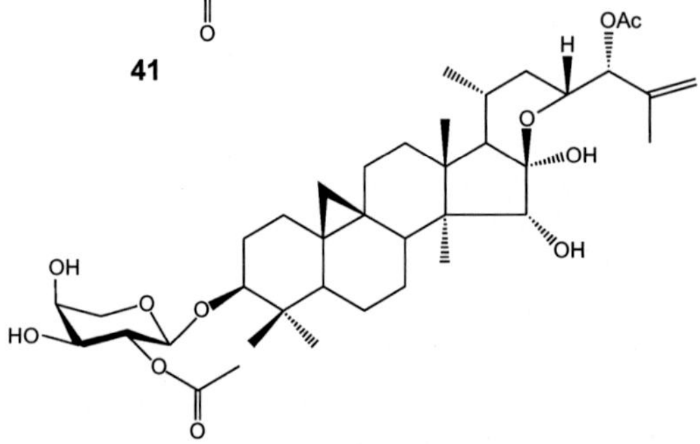

45

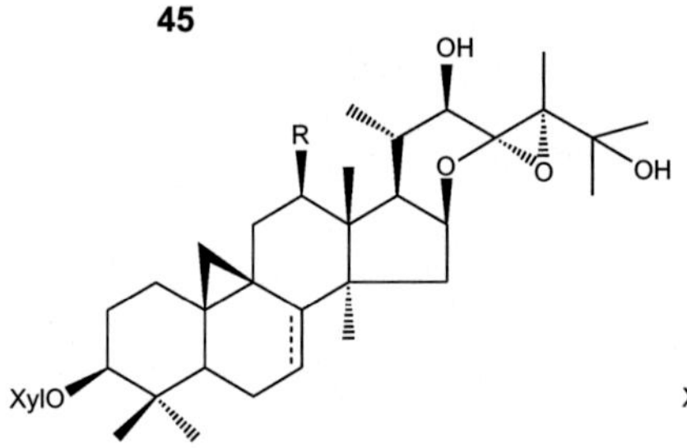

49	R = OAc
50	R = H,$\Delta^{7(8)}$

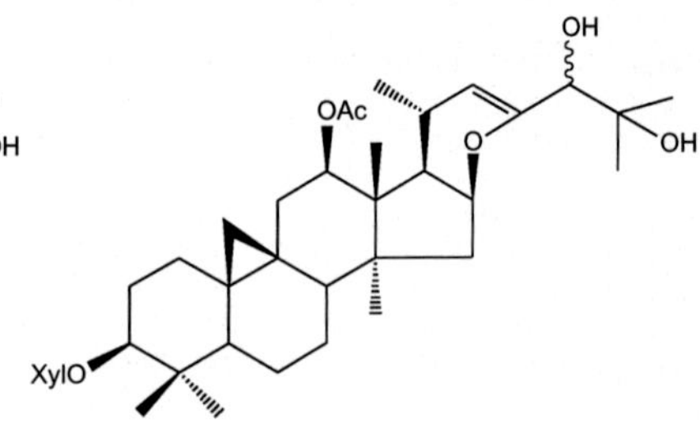

58

Figure 8.2 Continued.

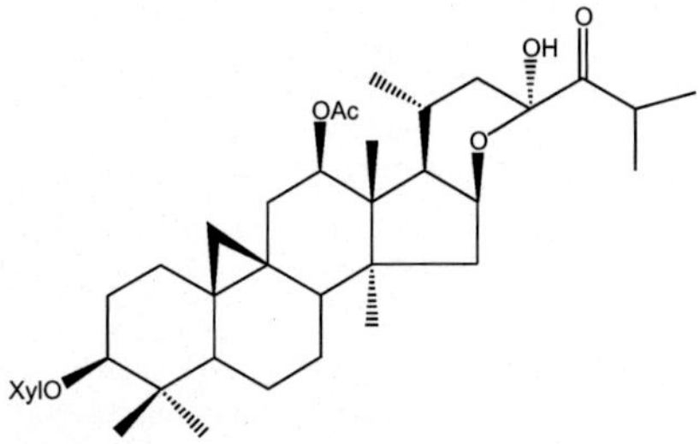

59

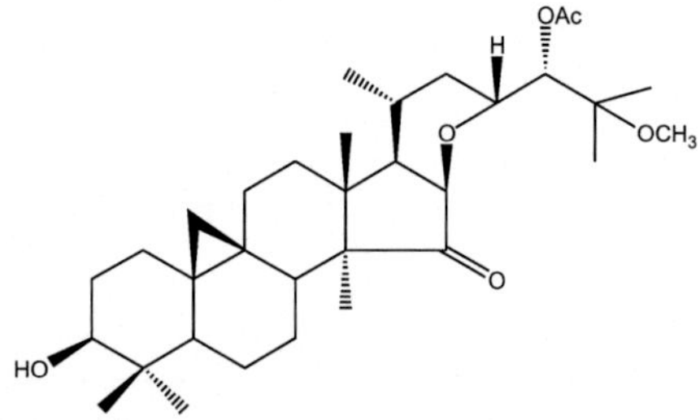

60

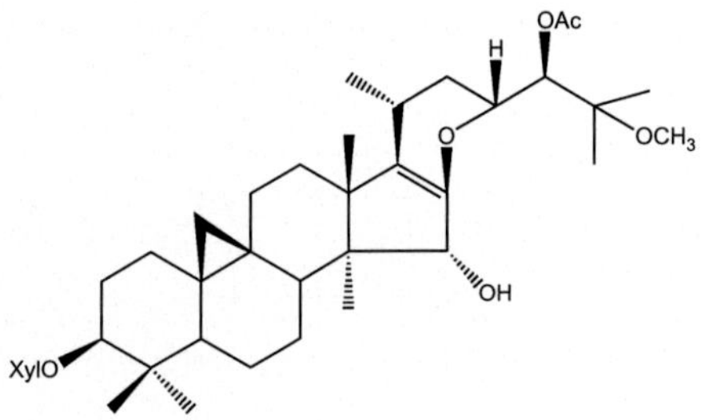

61

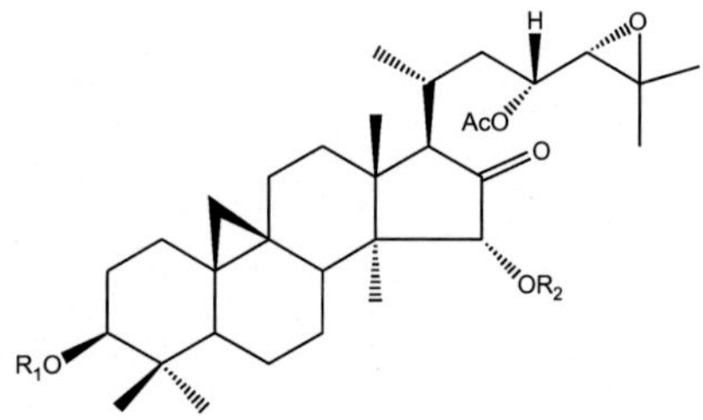

62	R_1 = Ara(2′-*O*-acetyl)	R_2 = H
63	R_1 = Ara	R_2 = Ac

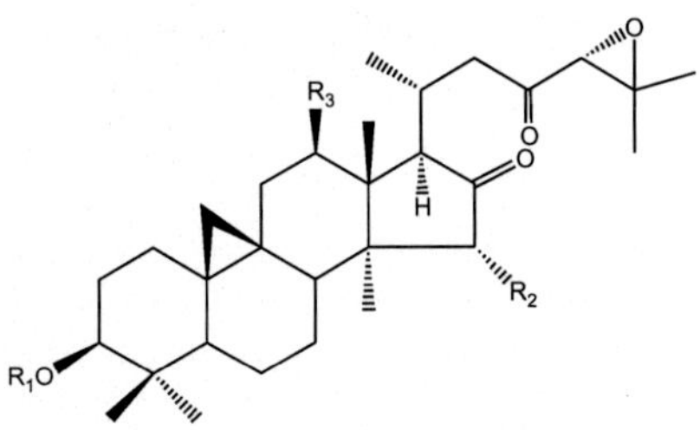

64	R_1 = H	R_2 = H	R_3 = H
72	R_1 = Ara(2′-*O*-acetyl)	R_2 = OH	R_3 = OAc

Figure 8.2 Continued.

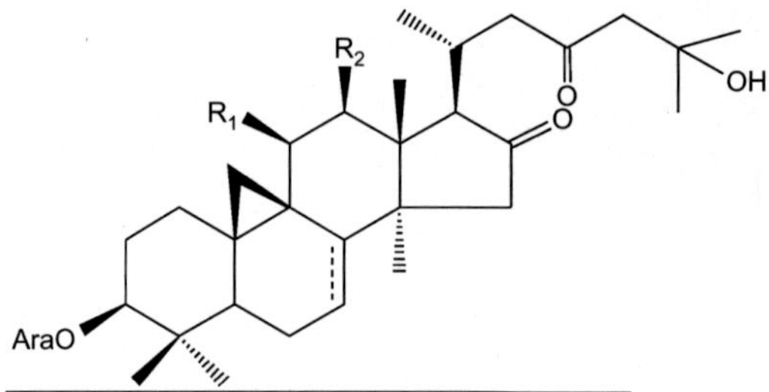

65	$R_1 = H$	$R_2 = OH, \Delta^{7(8)}$
66	$R_1 = H$	$R_2 = OAc, \Delta^{7(8)}$
67	$R_1 = OH$	$R_2 = H$

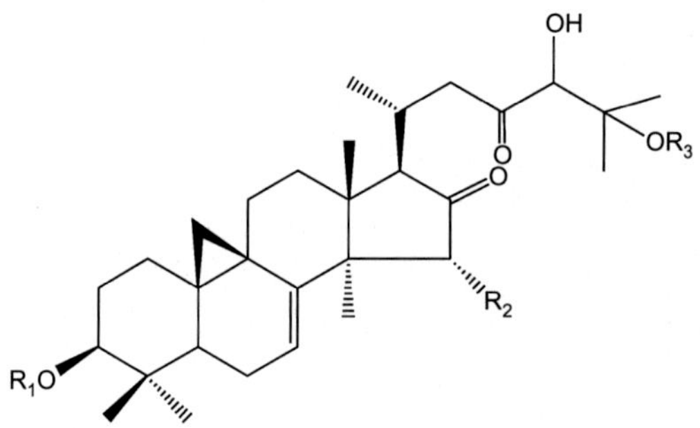

68	$R_1 = Xyl$	$R_2 = OH$	$R_3 = Glc$
69	$R_1 = Xyl$	$R_2 = H$	$R_3 = Glc$
70	$R_1 = OH$	$R_2 = OH$	$R_3 = Xyl(1{\rightarrow}6)\text{-}Glc$

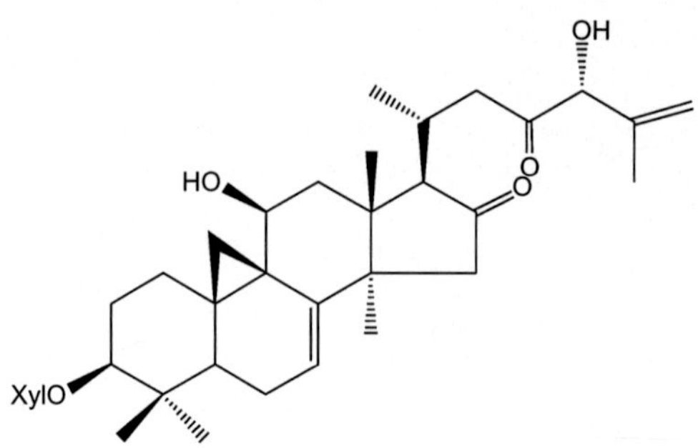

71

Figure 8.2 Continued.

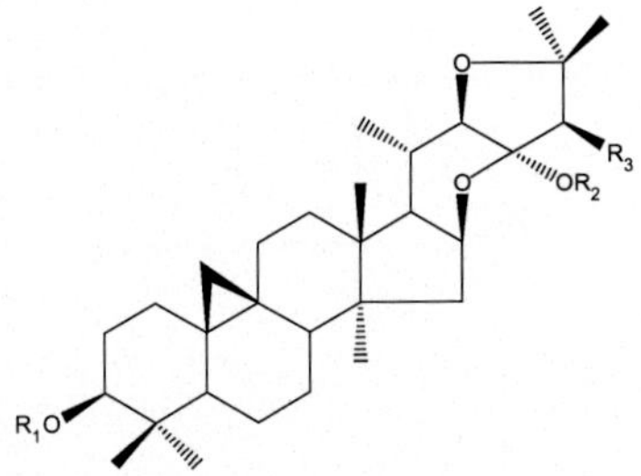

73	R_1 = Xyl(2′-*O*-3″-*O*-methylmalonyl)	$R_2 = CH_3$	R_3 = H
74	R_1 = Xyl(1→3)-Xyl	R_2 = H	R_3 = OH
75	R_1 = Xyl(4′-*O*-acetyl)	R_2 = H	R_3 = OH

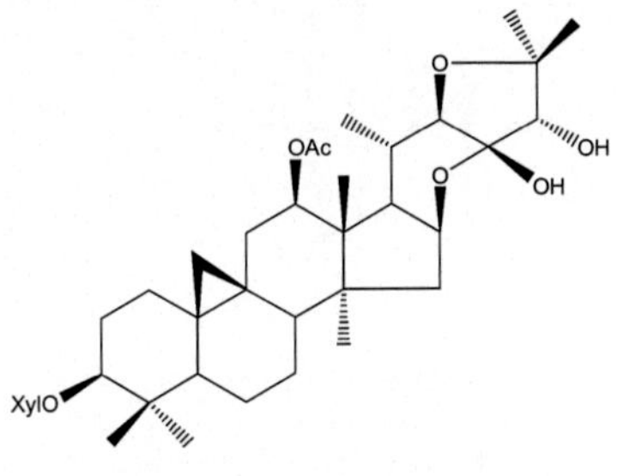

76

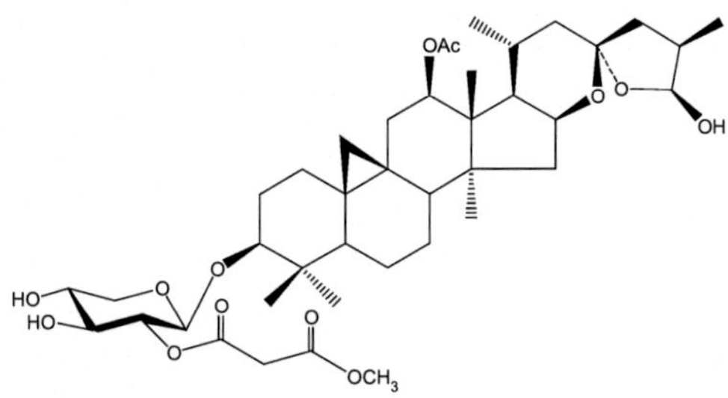

77

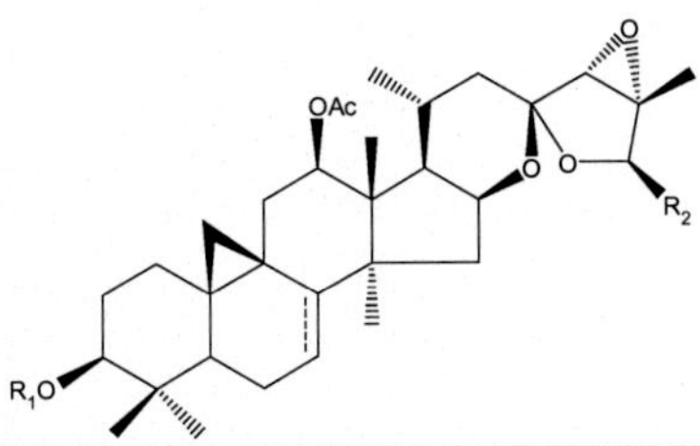

78	R_1 = Xyl(3′-*O*-acetyl)	R_2 = OH	C23-R,$\Delta^{7(8)}$
79	R_1 = Xyl(4′-*O*-acetyl)	R_2 = H	C23-S,$\Delta^{7(8)}$
80	R_1 = Xyl	R_2 = H	C23-S,$\Delta^{7(8)}$
81	R_1 = Xyl(1→3)-Xyl	R_2 = H	C23-S

Figure 8.2 Continued.

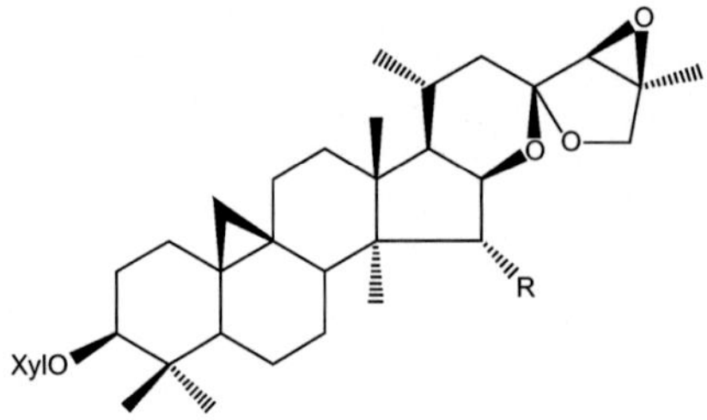

82	R = OH
83	R = H

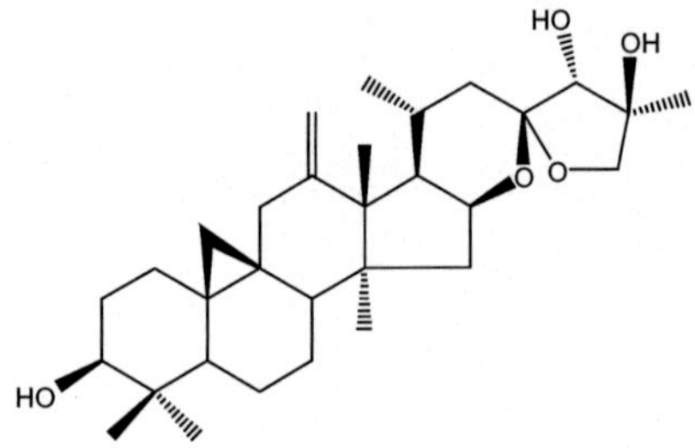

84

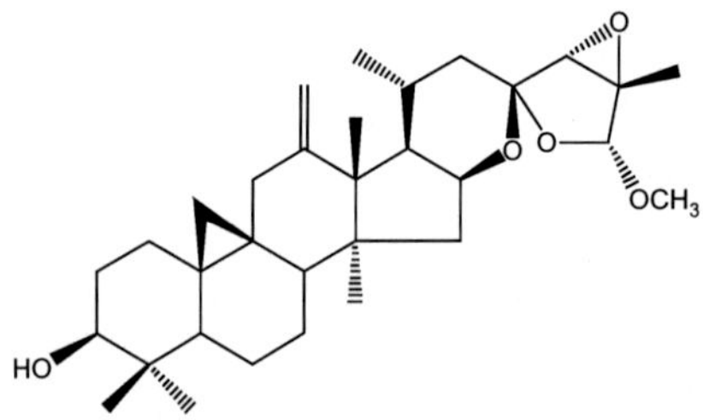

85

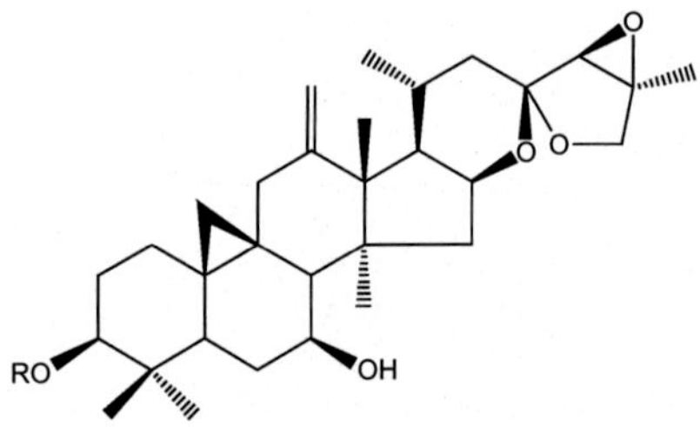

86	R = Ara
87	R = Xyl

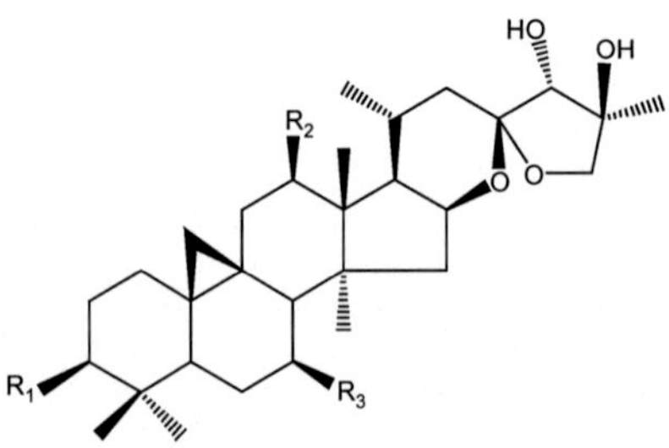

88	R_1 = = *O*	R_2 = OH	R_3 = H
89	R_1 = OH	R_2 = OH	R_3 = H
90	R_1 = OH	R_2 = OAc	R_3 = H
91	R_1 = *O*-Ara	R_2 = OH	R_3 = OH

Figure 8.2 Continued.

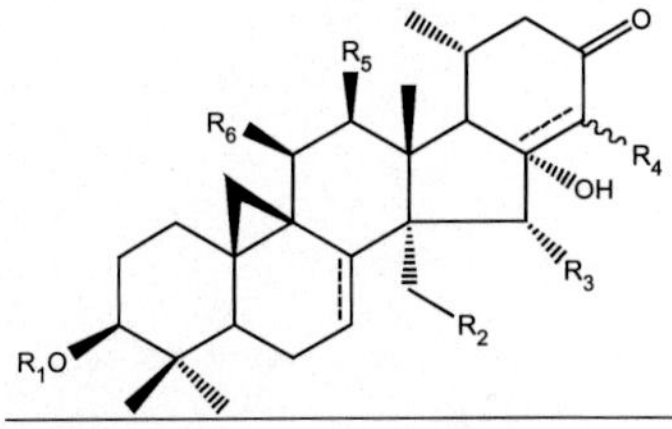

92	R_1 = Xyl	R_2 = H	R_3 = OH	R_4 = α-OH	R_5 = H	R_6 = H
93	R_1 = Xyl	R_2 = H	R_3 = OH	R_4 = α-OH	R_5 = H	R_6 = H,$\Delta^{7(8)}$
94	R_1 = Xyl	R_2 = H	R_3 = OH	R_4 = α-OH	R_5 = OAc	R_6 = H,$\Delta^{7(8)}$
95	R_1 = Xyl	R_2 = H	R_3 = OH	R_4 = β-OH	R_5 = H	R_6 = OH,$\Delta^{7(8)}$,$\Delta^{16(24)}$
96	R_1 = Xyl	R_2 = H	R_3 = H	R_4 = α-OH	R_5 = H	R_6 = OH,$\Delta^{7(8)}$
97	R_1 = Xyl(1→3)-Xyl	R_2 = OH	R_3 = H	R_4 = α-OH	R_5 = H	R_6 = OH,$\Delta^{7(8)}$

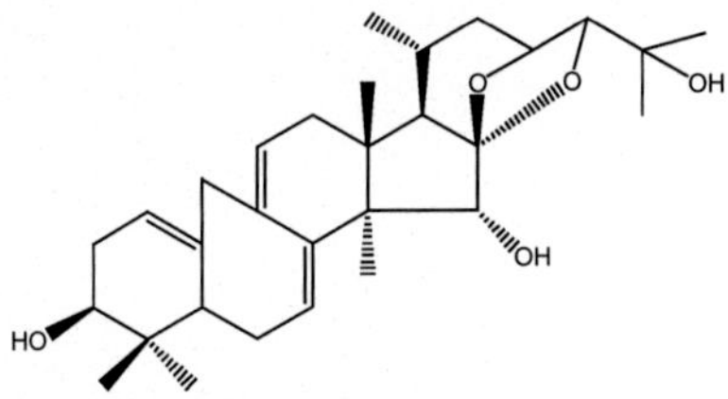

98

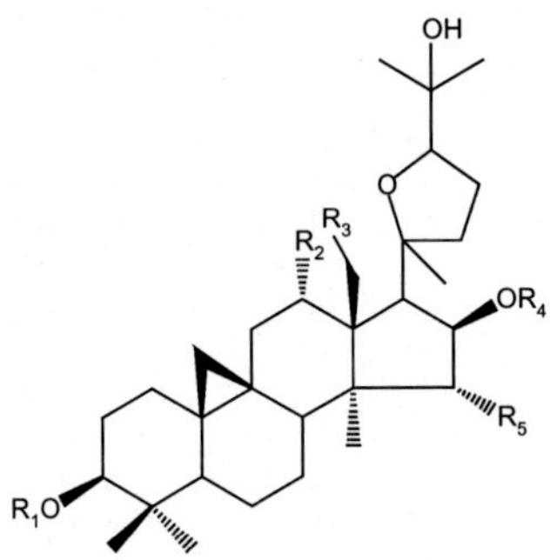

99	R_1 = Glc	R_2 = H	R_3 = OH	R_4 = H	R_5 = H
100	R_1 = Glc(1→6)-Glc	R_2 = H	R_3 = OH	R_4 = H	R_5 = H
101	R_1 = Xyl	R_2 = H	R_3 = OH	R_4 = Ac	R_5 = OAc
104	R_1 = Xyl	R_2 = H	R_3 = H	R_4 = H	R_5 = OAc

Figure 8.2 Continued.

105	R_1 = Xyl	R_2 = OH	R_3 = H	R_4 = H	R_5 = OH

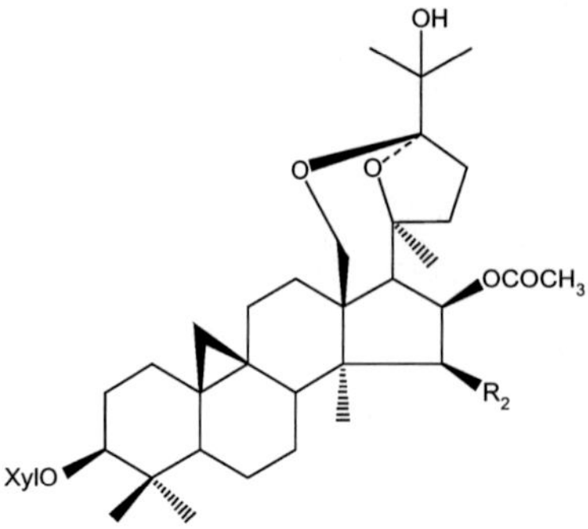

102	R = OH
103	R = H

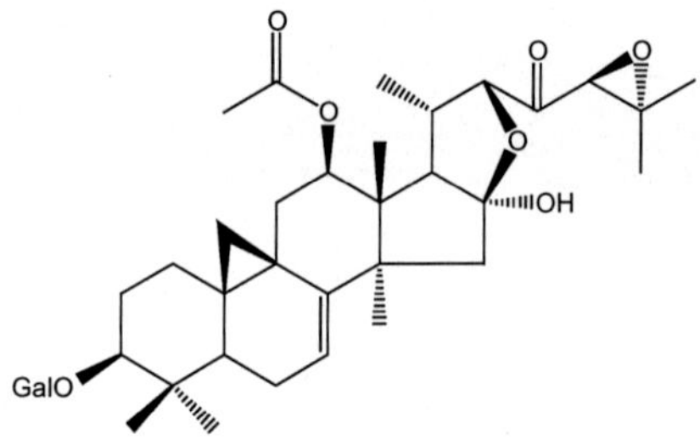

106

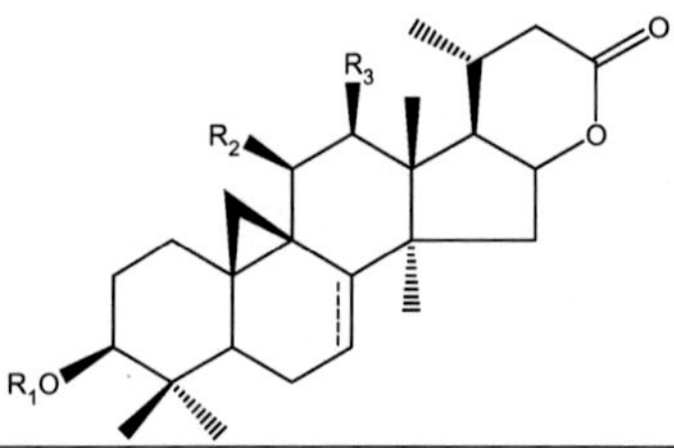

107	R_1 = Xyl	R_2 = OH	R_3 = H,$\Delta^{7(8)}$
108	R_1 = H	R_2 = H	R_3 = OAc

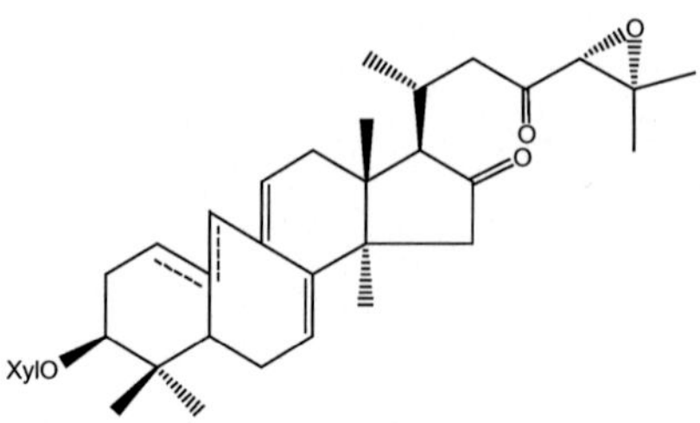

109	$\Delta^{1(10)}$
110	$\Delta^{10(19)}$

Figure 8.2 Continued.

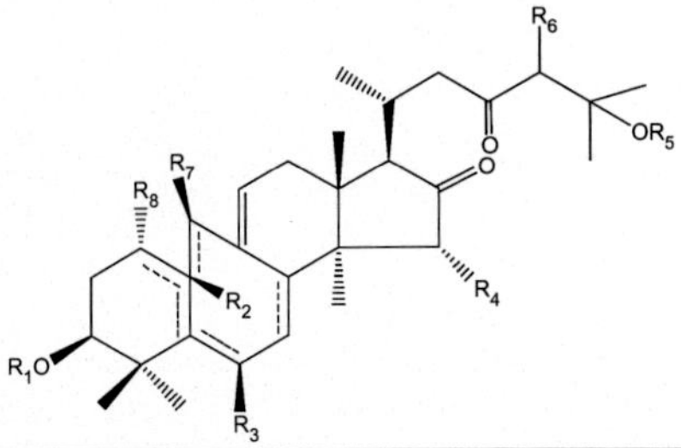

111	R_1=Xyl	R_2=H	R_3=H	R_4=H	R_5=Glc	R_6=OH	R_7=H	R_8=OH, $\Delta^{1(10)}$, $\Delta^{7(8)}$
112	R_1=Ara		R_3=H	R_4=H	R_5=H	R_6=H	R_7=H	R_8=OH, $\Delta^{5(10)}$
113	R_1=Ara	R_2=H	R_3=H	R_4=H	R_5=H	R_6=H	R_7=OH	R_8=H
114	R_1=Ara	R_2=H	R_3=H	R_4=OH	R_5=H	R_6=H	R_7=OH	R_8=H
115	R_1=Ara	R_2=OH	R_3=H	R_4=H	R_5=H	R_6=H	R_7=H	R_8=H
116	R_1=Ara		R_3=OH	R_4=OH	R_5=H	R_6=H	R_7=OH	R_8=H, $\Delta^{5(10)}$
117	R_1=Ara		R_3=H	R_4=OH	R_5=H	R_6=H	R_7=H	R_8=H, $\Delta^{10(19)}$
118	R_1=Ara		R_3=H	R_4=OH	R_5=H	R_6=H	R_7=H	R_8=H, $\Delta^{5(6)}$, $\Delta^{10(19)}$

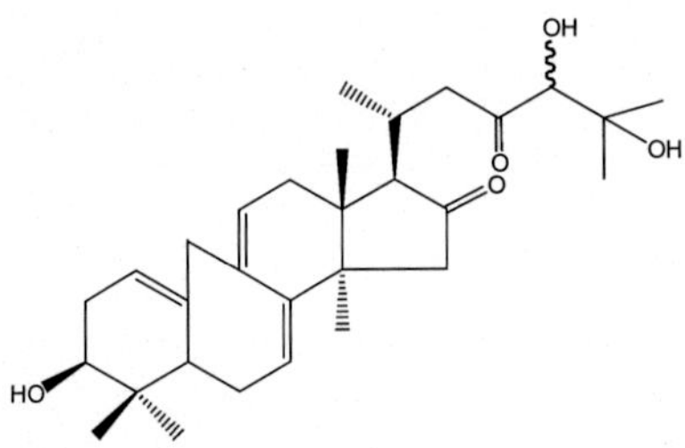

119	24α
120	24β

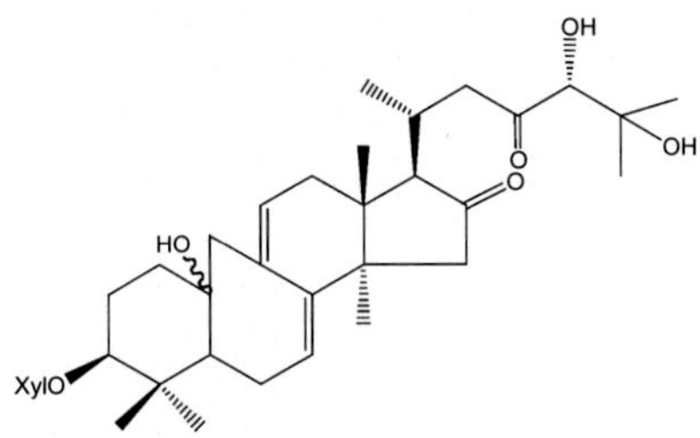

121	10β
122	10α

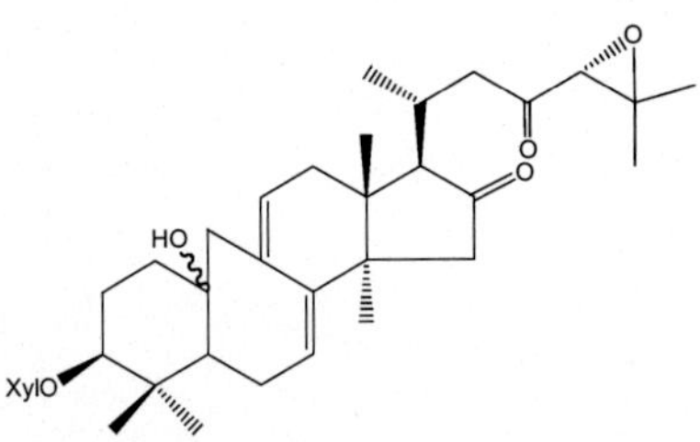

Figure 8.2 Continued.

123	10β
124	10α

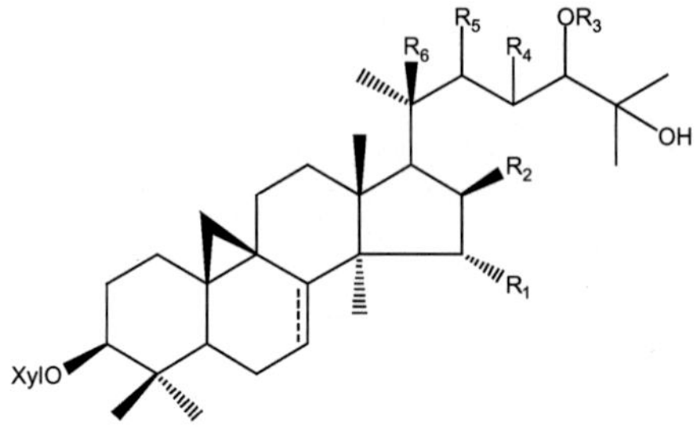

125	R_1 = H	R_2 = OH	R_3 = H	R_4 = H		C24-R
126	R_1 = OH	R_2 = = O	R_3 = Ac	R_4 = OAc	C23-R	C24-S,$\Delta^{7(8)}$
127	R_1 = OH	R_2 = = O	R_3 = H	R_4 = OAc	C23-R	C24-R,$\Delta^{7(8)}$
128	R_1 = H	R_2 = OGlc	R_3 = H	R_4 = OH	C23-R	C24-S
129	R_1 = H	R_2 = OGlc	R_3 = H	R_4 = H		C24-ζ
130	R_1 = H	R_2 = OH	R_3 = Ac	R_4 = OH	C23-ζ	C24-ζ

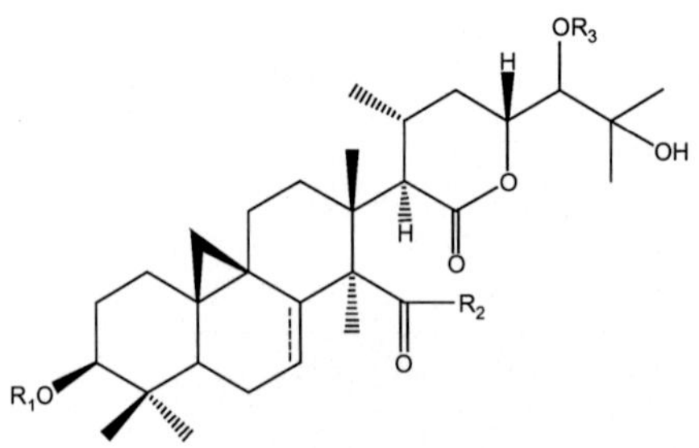

131	R_1 = H	R_2 = H	R_3 = H	C24-S,$\Delta^{7(8)}$
132	R_1 = Xyl	R_2 = OH	R_3 = Ac	C24-R
133	R_1 = Xyl	R_2 = OH	R_3 = H	C24-R
134	R_1 = Xyl	R_2 = OH	R_3 = H	C24-S
135	R_1 = H	R_2 = H	R_3 = H	C24-S
136	R_1 = Ara	R_2 = H	R_3 = H	C24-S
137	R_1 = Xyl	R_2 = H	R_3 = H	C24-S

Figure 8.2 Continued.

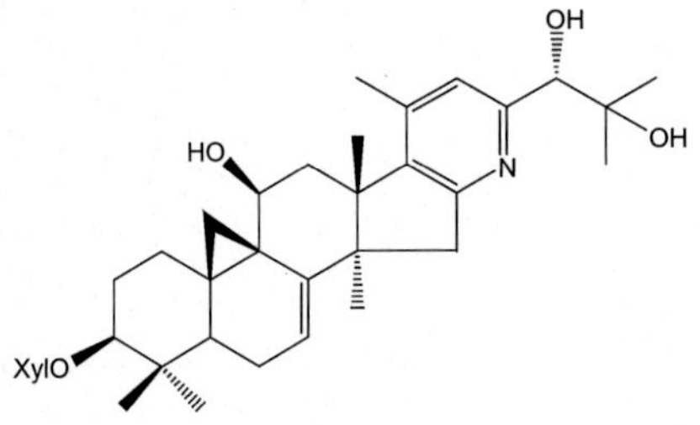

138

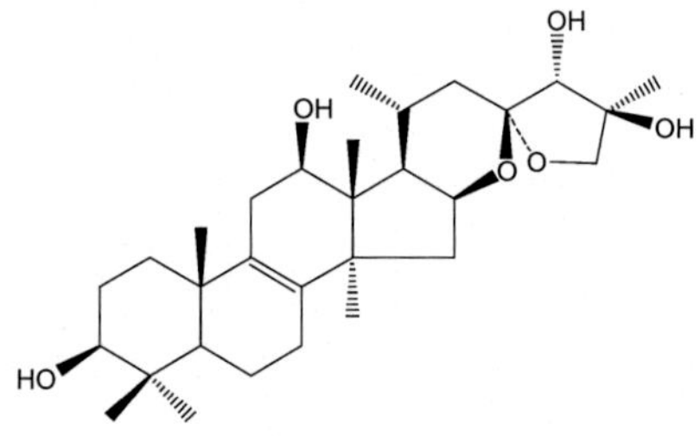

139

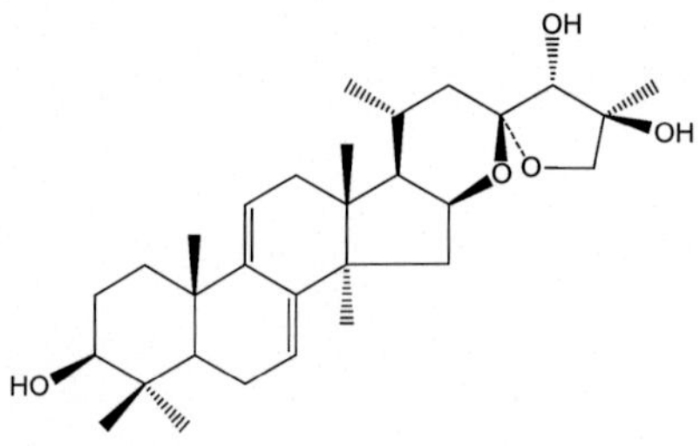

140

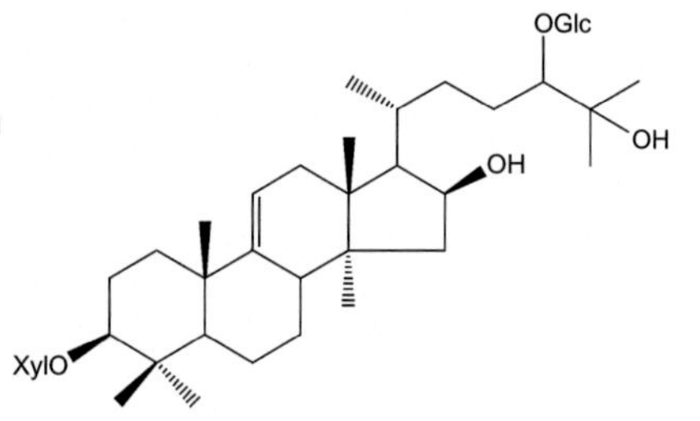

141

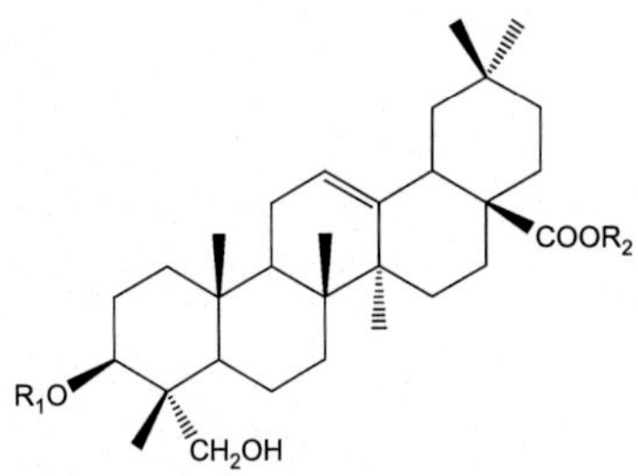

142	R_1 = All(1→3)-Rha(1→2)-[Glc(1→4)]-Ara	R_2 = Rha(1→4)-Glc(1→6)-Glc
143	R_1 = Glc(1→4)-[Rha(1→2)]-Ara	R_2 = Rha(1→4)-Glc(1→6)-Glc(1→4)-Rha(1→4)-Glc(1→6)-Glc

Figure 8.2 Continued.

Three cycloartane triterpenoids from the aerial parts of *C. foetida* exhibited broad-spectrum and moderate cytotoxic activities (Nian et al., 2011). Human heat shock protein (Hsp) 27 is increased in various human cancers and exhibits cytoprotective activity that affects tumorigenesis and the susceptibility of tumors to cancer therapy, but its expression at 2000 jig/L *C. foetida* extract is diminished (Soler et al., 2011). Three 9, 19-cycloartane triterpene glycosides exhibited significant cytotoxicity against human HepG2 cells (Nian et al., 2010). Cimicifugoside, a triterpenoid from *C. simplex*, is a specific nucleoside transport inhibitor that displays synergistic potentiation of methotrexate cytotoxicity (Yawata et al., 2009).

The triterpene glycoside actein from black cohosh activates stress- and statin-associated responses and is bioavailable in Sprague–Dawley rats (Einbond et al.,

Table 8.1 Triterpenoid saponins found in Cimicifugeae medicinal plants in recent years

	Compound	Aglycone	Species	Tissue	References
1	Methyl 3,4-seco-4-hydroxy-3-cimigenolate	A	*Cimicifuga foetida*	Aerial part	Nian et al. (2012)
2	Cimigenol-3-*O*-[2′,4′-*O*-diacetyl]-α-L-arabinopyranoside	A			
3	Cimigenol-3-*O*-[3′,4′-*O*-diacetyl]-α-L-arabinopyranoside	A			
4	Cimigenol-3-*O*-[4′-*O*-acetyl]-α-L-arabinopyranoside	A			
5	25-Anhydrocimigenol-3-*O*-[3′-*O*-acetyl]-α-L-arabinopyranoside	A			
6	24-Epi-cimigenol-3-one	A	*C. foetida*		Lu et al. (2012)
7	Shengmaxinside A	A	*C. simplex*	Root	Kuang et al. (2011)
8	Shengmaxinside B	A			
9	12β-Hydroxy-25-anhydrocimigenol	A	*C. foetida*	Aerial part	Nian et al. (2011)
10	Cimigenol-12-one	A			
11	12β-Hydroxy-15-deoxycimigenol	A			
12	2′-*O*-acetyl-24-epi-cimigenol-3-*O*-α-L-arabinopyranoside	A			
13	2′-*O*-Acetylcimigenol-3-*O*-β-D-xylopyranoside	A			
14	25-Anhydrocimigenol-3-*O*-α-L-arabinopyranoside	A			
15	3-*O*-β-D-Xylopyranosyl cimigenol 15-*O*-β-D-glucopyranoside	A	*C. foetida*	Rhizome	Shen (2010)
16	25-*O*-Acetylcimigenol-3-*O*-β-D-xylopyranosyl 15-*O*-β-D-galactopyranoside	A			
17	Cimifoetiside A	A	*C. foetida*	Aerial part	Pan et al. (2009)
18	Cimifoetiside B	A			
19	(3′,12β)-*O*-Diacetyl-cimigenol-3-*O*-β-D-xylopyranoside	A	*Actaea asiatica*	Rhizome	Fan et al. (2009)
20	(4′,25)-*O*-Diacetyl-cimigenol-3-*O*-β-D-xylopyranoside	A			

21	2′-*O*-Acetyl-25-*O*-methyl-cimigenol-3-*O*-β-D-xylopyranoside	A			
22	2′-*O*-Acetyl-25-*O*-ethyl-cimigenol-3-*O*-β-D-xylopyranoside	A			
23	Cimifoside A	A	*C. foetida*	Rhizome	Sun et al. (2008)
24	Cimifoside B	A			
25	12β-Acetoxycimigenol	A	*Actaea pachypoda*	Root	Ali et al. (2007a, b,c)
26	Cimicifoetiside A	A	*C. foetida*	Rhizome	Sun et al. (2007)
27	Cimicifoetiside B	A			
28	Soulieoside A	A	*Souliea vaginata*	Rhizome	Zhou et al. (2004)
29	12β-Hydroxycimigenol-3-*O*-β-D-xylopyranosyl-(1 → 3)-β-D-xylopyranoside	A	*Souliea vaginata*	Rhizome	Zhou et al. (2006)
30	25-Anhydrocimicigenol-3-*O*-β-D-(2′-*O*-acetyl)xylopyranoside	A			
31	Cimigenol-3-*O*-[2′-*O*-(*E*)-2-butenoyl]-α-L-arabinopyranoside	A	*C. dahurica*	Root	Nian et al. (2013a)
32	25-*O*-Acetylcimigenol-3-*O*-[4′-*O*-acetyl]-α-L-arabinopyranoside	A			
33	25-*O*-Acetylcimigenol-3-*O*-[3′-*O*-acetyl]-α-L-arabinopyranoside	A			
34	12β-Hydroxy-7(8)-en-cimigenol	A			
35	11β-Hydroxy-7(8)-en-cimigenol	A			
36	Cimigenol-7(8)-en-3-one	A			
37	Cimigenol-1(2),7(8)-dien-3-one	A			
38	11β-Hydroxy-15-deoxycimigenol-7(8)-en-3-one	A			
39	24-*O*-Hydroxy-7,8-didehydrohydroshengmanol 3-*O*-β-D-galactopyranoside	B	*C. simplex*		Kuang et al. (2012)
40	24-Epi-24-*O*-hydroxy-7,8-didehydrohydroshengmanol 3-*O*-β-D-galactopyranoside	B			

Continued

Table 8.1 Continued

	Compound	Aglycone	Species	Tissue	References
41	Shengmanol-3-*O*-[2′-*O*-acetyl]-α-L-arabinopyranoside	B	*C. foetida*	Aerial parts	Nian et al. (2012)
42	Shengmaxinside C	B	*C. simplex*	Root	Kuang et al. (2011)
43	2′,24-Di-*O*-acetylisodahurinol-3-*O*-α-L-arabinopyranoside	B	*C. foetida*	Aerial parts	Nian et al. (2011)
44	24-*O*-Acetylisodahurinol-3-*O*-α-L-arabinopyranoside	B			
45	2′,24-Di-*O*-acetyl-25-anhydroshengmanol-3-*O*-α-L-arabinopyranoside	B			
46	Cimifoside C	B	*C. foetida*	Rhizome	Sun et al. (2008)
47	Cimifoetiside VI	B	*C. foetida*	Aerial parts	Pan et al. (2007)
48	Cimifoetiside VII	B			
49	7,8-Dihydroactaeaepoxide 3-*O*-β-D-xylopyranoside	B	*Actaea pachypoda*	Root	Ali et al. (2007a, b,c)
50	12-Deacetoxyactaeaepoxide 3-*O*-β-D-xylopyranoside	B			
51	Soulieoside B	B	*Souliea vaginata*	Rhizome	Zhou et al. (2004)
52	24-*O*-Acetyldahurinol-3-*O*-β-D-xylopyranoside	B	*Souliea vaginata*	Rhizome	Zhou et al. (2006)
53	24-*O*-Acetylisodahurinol	B	*C. dahurica*	Root	Nian et al. (2013a)
54	24-*O*-Acetyl-7(8)-en-isodahurinol	B			
55	Heracleifolinoside D	B	*C. heracleifolia*	Rhizome	Liu et al. (2013)
56	Heracleifolinoside E	B			

57	Heracleifolinoside F	B			
58	Cimifoetidanoside G	B	*C. foetida*	Rhizome	Chen et al. (2014)
59	Cimifoetidanoside F	B			
60	25-Methoxy-24-*O*-acetylisohurinol	B	*C. yunnanensis*	Aerial part	Bao et al. (2014)
61	24-*O*-Acetylhydroshengmanol (Δ-16,17)-enol ether-3-*O*-β-D-xylopyranoside (24*S*)	B	*Actaea racemosa*	Root, rhizome	Findeis et al. (2012)
62	2′,23-Di-*O*-acetylshengmanol-3-*O*-α-L-arabinopyranoside	C	*C. foetida*	Aerial part	Nian et al. (2011)
63	15,23-*O*-Diacetyl-7(8)-en-shengmanol-3-*O*-α-L-arabinopyranoside	C	*C. dahurica*	Root	Nian et al. (2013a)
64	7,8-Dihydro-11-dehydroxycimicidanol	D	*C. foetida*	Aerial part	Nian et al. (2012)
65	Podocarpaside H	D	*Actaea podocarpa*	Root	Ali et al. (2007a, b,c)
66	Podocarpaside I	D			
67	Podocarpaside J	D			
68	Heracleifolinoside A	D	*C. heracleifolia*	Rhizome	Liu et al. (2013)
69	Heracleifolinoside B	D			
70	Heracleifolinoside C	D			
71	Cimifoetidanoside E	D	*C. foetida*	Rhizome	Chen et al. (2014)
72	(3β,12β,15α,24*R*)-12,2′-diacetoxy-24,25-epoxy-15-hydroxy-16,23-dione-3-*O*-α-L-arabinopyranoside	D	*C. foetida*	Root	Zhu et al. (2014)
73	23-*O*-methyl-24-deoxy-2′-*O*-(3″-methylmalonyl)-cimiaceroside B	E	*C. foetida*	Rhizome	Sun et al. (2011)
74	Cimiaceroside C	E	*C. foetida*	Rhizome	Sun et al. (2008)

Continued

Table 8.1 Continued

	Compound	Aglycone	Species	Tissue	References
75	Soulieoside C	E	*Souliea vaginata*	Rhizome	Zhou et al. (2004)
76	Cimifoetidanoside H	E	*C. foetida*	Rhizome	Chen et al. (2014)
77	2′-*O*-(3″-methylmalonyl)actein	F	*C. foetida*	Rhizome	Sun et al. (2011)
78	3′-*O*-acetyl-cimicifugoside	F	*Actaea asiatica*	Rhizome	Fan et al. (2009)
79	4′-*O*-acetyl-23-epi-26-deoxycimifugoside	F			
80	(23*R*)-26-deoxycimicifugoside	F	*Actaea asiatica*	Rhizome	Lu et al. (2008)
81	Cimifoside D	F	*C. foetida*	Rhizome	Sun et al. (2008)
82	12-Deacetyloxy-15α-hydroxy-23-epi-26-deoxyactein	F	*Souliea vaginata*	Rhizome	Zhou et al. (2006)
83	12-Deacetyloxy-23-epi-26-deoxyactein	F			
84	Yunnanterpene G	F	*C. yunnanensis*	Aerial part	Bao et al. (2014)
85	26-Methoxy-acteol-12(18)-en	F			
86	7β-Hydroxy-23-epi-acteol-3-*O*-α-L-arabinopyranoside	F			
87	7β-Hydroxy-23-epi-acteol-3-*O*-β-D-xylosepyranoside	F			
88	Yunnanterpene A	F	*C. yunnanensis*	Aerial part	Nian et al. (2013b)
89	Yunnanterpene B	F			
90	Yunnanterpene C	F			
91	Yunnanterpene F	F			
92	3β,15α,16α,24α-Tetrahydroxy-25,26,27-trinor-16,24-cyclo-cycloartane-23-one 3-*O*-β-D-xylopyranoside	G	*C. heracleifolia*	Rhizome	Nishida and Yoshimitsu (2011)

93	3β,15α,16α,-24α-Tetrahydroxy-25,26,27-trinor-16,24-cyclo-cycloart-7-en-23-one 3-*O*-β-D-xylopyranoside	G			
94	12β-Acetoxy-3β,15α,16α,-24α-tetrahydroxy-25,26,27-trinor-16,24-cyclo-cycloart-7-en-23-one 3-*O*-β-D-xylopyranoside	G			
95	15α-Hydroxy-16-dehydroxy-16(24)-en-foetidinol-3-*O*-β-D-xylopyranoside	G	*C. foetida*	Rhizome	Lu et al. (2009)
96	28-Hydroxy-foetidinol-3-*O*-β-D-xylopyranoside	G			
97	Foetidinol-3-*O*-β-D-xylopyranosyl-(1″→3′)-β-D-xylopyranoside	G			
98	9,10-Seco-1(10),7(8),9(11)-trien cimigenol	H	*C. dahurica*	Root	Nian et al. (2013a)
99	Beesioside G	I	*Beesia calthifolia*	Rhizome	Ju et al. (2002)
100	Beesioside H	I			
101	Beesioside J	I			
102	Beesioside K	I			
103	Beesioside L	I			
104	Beesioside M	I			
105	Beesioside N	I			
106	3β,16α-Dihydroxy-12-acetoxy-16,22-cyclo-23-ketone-24 *R*,25-epoxy-cycloartane-7-ene 3-*O*-β-D-galactopyranoside	J	*C. simplex*		Kuang et al. (2012)
107	3β,11β-Dihydroxy-24,25,26,27-tetranor-cycloart-7-en-23,16β-olide 3-*O*-β-D-xylopyranoside	K	*C. heracleifolia*	Rhizome	Nishida and Yoshimitsu (2011)
108	Cimilactone C	K	*C. yunnanensis*	Aerial part	Nian et al. (2013b)
109	Isocimipodocarpaside	L	*C. racemosa*		Jamróz et al. (2012)

Continued

Table 8.1 Continued

	Compound	Aglycone	Species	Tissue	References
110	Cimipodocarpaside	L	*C. racemosa*		Jamróz et al. (2011)
111	Foetidinoside B	L	*C. foetida*		Lu et al. (2010)
112	Podocarpaside A	L	*Actaea podocarpa*	Root	Ali et al. (2007a, b,c)
113	Podocarpaside B	L			
114	Podocarpaside C	L			
115	Podocarpaside D	L			
116	Podocarpaside E	L			
117	Podocarpaside F	L			
118	Podocarpaside G	L			
119	Cimifoetidanol A	L	*C. foetida*	Rhizome	Chen et al. (2014)
120	Cimifoetidanol B	L			
121	Cimifoetidanoside A	L			
122	Cimifoetidanoside B	L			
123	Cimifoetidanoside C	L			
124	Cimifoetidanoside D	L			
125	Foetinoside	M	*C. foetida*		Lu et al. (2012)
126	23*R*,24*S*-diacetoxy-3β,15α,25-trihydroxy-cycloart-7-en-16-one 3-*O*-β-D-xylopyranoside	M	*C. heracleifolia*	Rhizome	Nishida and Yoshimitsu (2011)
127	23*R*-acetoxy-3β,15α,24*R*,25-tetrahydroxy-cycloart-7-en-16-one 3-*O*-β-D-xylopyranoside	M			
128	Foetidinosides C	M	*C. foetida*		Lu et al. (2010)
129	Foetidinosides D	M			

130	Foetidinosides E	M			
131	15,16-Seco-7,8-didehydro-15-formyl-16-oxohydroshengmanol	N	*C. foetida*	Aerial parts	Nian et al. (2012)
132	24-Acetoxy-15,16-seco-cycloartane 3-*O*-β-D-xylopyranoside	N	*Cimicifuga*	Rhizome	Yoshimitsu et al. (2007)
133	24-Epi-hydro-15,16-seco-cycloartane 3-*O*-β-D-xylopyranoside	N			
134	24-Epimer-hydro-15,16-seco-cycloartane 3-*O*-β-D-xylopyranoside	N			
135	15,16-Seco-shengmanol C	N	*C. yunnanensis*	Aerial part	Bao et al. (2014)
136	15,16-Seco-cimiterpene A	N	*C. yunnanensis*	Aerial part	Nian et al. (2013b)
137	15,16-Seco-cimiterpene B	N			
138	Cimicifugadine	O	*C. foetida*	Root	Dan et al. (2007)
139	Yunnanterpene D	P	*C. yunnanensis*	Aerial part	Nian et al. (2013b)
140	Yunnanterpene E	P			
141	Foetidinosides A	Q	*C. foetida*		Lu et al. (2010)
141	Eranthisaponin A	R	*Eranthis cilicica*	Tuber	Watanabe et al. (2003)
142	Eranthisaponin B				

Chemical structure types of 9,19-cycloartane triterpenoids in tribe Cimicifugeae: A, cimigenol type; B, hydroshengmanol type; C, shengmanol type; D, 16,23-dione type; E, cimiacerogenin type; F, cimifugenin type; G, side chain-seco type; H, 9,10-seco-1(10),7(8),9(11)-trien cimigenol type; I, beesioside type; J, 16,22-seco-cycloartane type; K, tetranor type; L, 9,10-seco-9,19-cycloartane type; M, 9,19-cycloartane type; N, 15,16-seco-cycloartane type; O, cycloartane triterpene alkaloid type; P, 9,19-seco-cimifugenin type; Q, lanostane type; R, oleanane type.

2009). Actein reduced free fatty acid and cholesterol content in the liver and inhibited the growth of HepG2 liver cancer cells. Actein strongly inhibited the growth of human breast cancer cells and induced a dose-dependent release of calcium into the cytoplasm (Einbond et al., 2013). The ER IP3 receptor antagonist heparin blocked this release, indicating that the receptor is required for activity. Heparin partially blocked the growth inhibitory effect, while the MEK inhibitor U0126 enhanced it. Consistent with this, actein synergized with the ER mobilizer thapsigargin. Further, actein preferentially inhibited the growth of 293T (NF-κB) cells.

Acerinol, a cyclolanstane triterpenoid from *C. acerina*, reverses ABCB1-mediated multidrug resistance in HepG2/ADM and MCF-7/ADR cells (Liu et al., 2014). The antiproliferative effects of the ethanolic *C. racemosa* extract BNO-1055 result from hindered nucleoside uptake of prostate cancer cells due to impaired ENT nucleoside transporter activity (Dueregger et al., 2013).

25-Anhydrocimigenol-3-*O*-β-D-xylopyranoside isolated from *Souliea vaginata* showed the anticancer activity against hepatoma, as it might induce apoptosis and G0/G1 cell cycle arrest (Tian et al., 2006b). The anticancer activity of *Beesia* and *Anemonopsis* awaits further studies. In the future, the structure–activity relationship of various saponins and their potentials in the treatment of various human cancers should be studied for novel drug development.

8.3.2 Effects on menopausal symptoms and female reproductive system

C. racemosa (CR) has been used in Europe as a medicinal plant for more than a century, and its roots have been widely used for the treatment of menopausal symptoms. Recent randomized controlled studies have shown that CR consumption alleviates "hot flushes," and due to the lack of uterotropic effects, it can be a safe option to estrogen replacement therapy (Rachoń et al., 2008). The isopropanol extract of CR may act on the hypothalamic nuclei and have therapeutic effects on menopausal symptoms (Hui et al., 2012). Hot flashes are a disorder of thermoregulation due to the lack of estrogens and are the most common and characteristic climacteric complaint. The CR special extract BNO 1055 prevents hot flashes in ovariectomized rats (Kapur et al., 2010). The combination of CR with *Hypericum perforatum* (HP) had a positive effect on climacteric complaints, while CR monotherapy as well as HP and *Agnus castus* was not better than placebo (Laakmann et al., 2012). A standardized isopropanolic CR extract (remifemin) is safe and effective for menopausal symptoms (Ross, 2012). It was an effective agent to manage Chinese women with climacteric symptom (Bai et al., 2009). It has similar therapeutic effect and lower incidence of adverse effect when compared with tibolone. For the CR extract, the onset of affecting abnormal thermoregulation took longer than that of estradiol valerate in ovariectomized rats (Ma et al., 2011b). CR extract had a significant effect on day central body temperature (CBT) but did not affect night CBT of ovariectomized rats.

CR extract might be a reasonable treatment in tamoxifen-treated breast cancer patients with predominantly psychovegetative symptoms (Rostock et al., 2011).

It is hypothesized that the established positive allosteric modulation of gamma-aminobutyric acid (GABA) type A receptors may contribute to the beneficial effects of CR extracts in the treatment of climacteric symptoms (Cicek et al., 2010). Four cycloartane glycosides of CR significantly enhanced GABA-induced chloride currents. CR-containing preparations improved menopausal symptoms overall by 26% (95% confidence interval 11–40%; Shams et al., 2010). Naloxone blockade and positron emission tomography neuroimaging helped confirm the central opioid activity of CR in postmenopausal women (Reame et al., 2008).

CR extracts reduced endometrial proliferation in comparison to the placebo (Alves et al., 2008). CR decreases local estrogen formation in normal human breast tissue *in vitro*, which may contribute to the lack of hormonal effects of black cohosh in breast tissue observed in previous studies (Stute et al., 2007). Adding CR rhizome dry extract to clomiphene citrate induction can improve the pregnancy rate and cycle outcomes in couples with unexplained infertility and recurrent clomiphene citrate induction failure (Shahin et al., 2008). *C. racemosa* extract (Klimadynon), as a phytoestrogen, can be used as an alternative to clomiphene citrate for ovulation induction in women with polycystic ovarian syndrome (Kamel, 2013).

C. heracleifolia var. *bifida* ethanol extract does not activate or inactivate ER-α in a direct manner, but may affect factors in ER signal transduction pathway (Park et al., 2012).

8.3.3 Osteoprotective effects

Triterpenoids from *Cimicifuga* rhizome are a novel class of inhibitors on bone resorption and ovariectomy-induced bone loss (Li et al., 2007). CR extract has osteoprotective effects, and its triterpene saponins are responsible for the reduction of bone marrow fat and the secretion of proinflammatory cytokines (Seidlova-Wuttke et al., 2009, 2012a). CR facilitated metaphyseal fracture healing in the early stage of osteoporosis in ovariectomized rats (Kolios et al., 2010). Deoxyactein isolated from CR protects osteoblastic MC3T3-E1 cells against antimycin A-induced cytotoxicity (Choi, 2013a). Deoxyactein stimulates osteoblast function and inhibits bone-resorbing mediators in MC3T3-E1 cells (Choi, 2013b). *C. heracleifolia* significantly preserved trabecular bone mass, bone volume, trabecular number, trabecular thickness, structure model index, and bone mineral density of proximal tibia metaphysis or distal femur metaphysis, thus preventing ovariectomy-induced bone loss in mice (Ahn et al., 2012).

Cycloartane triterpenoids of *C. foetida*, such as cimigenol, actein, (23*R*, 24*S*) 25-*O*-acetyl-cimigenol-3-*O*-beta-D-xylopyranoside, and (23*R*, 24*S*) cimigenol-3-*O*-beta-D-xylopyranoside, exhibit osteoclast inhibition activity (Dan et al., 2009). A triterpene glycoside from CR inhibits osteoclastogenesis by modulating tumor necrosis factor (TNF)-α and RANKL (a member of the TNF superfamily) signaling pathways (Qiu et al., 2007). The special extract of CR BNO 1055 was shown to have bone protective effects without exerting estrogenic effects in the uterus or mammary gland, as well as other organs that also express ERs (Seidlova-Wuttke et al., 2009). Ethanolic extract of CR potentiates bone nodule formation in MC3T3-E1 preosteoblast cells (Chan et al., 2008). Postmenopausal women treated with an isopropanolic extract of CR underwent a decrease in the urinary concentration of *N*-telopeptides, a marker of bone resorption, and an increase in alkaline phosphatase, a marker of bone formation, at the third month of therapy (García-Pérez et al., 2009).

8.3.4 Anti-inflammatory and immunosuppressive activities

Cycloartane-type triterpene glycosides from the rhizomes of *C. heracleifolia* and *C. foetida* showed anticomplement activity (Lee et al., 2012; Qiu et al., 2006). Cyclolanostane triterpene diglycosides from the aerial part of *C. foetida* effectively inhibited the proliferation of murine splenocytes induced by concanavalin A (Pan et al., 2009). A nonphysiological accumulation of fat cells in abdomen and joints results in increased production of proinflammatory cytokines that have adverse effects on serum lipids, glucose, and joint cartilage. The special extract of CR BNO 1055 was shown to reduce the size of the fat depot. This extract and its saponin prevent the metabolic syndrome and deterioration of cartilage in the knee joint of ovariectomized rats (Seidlova-Wuttke et al., 2012a,b). Combined prescription of *Aralia cordata* and *C. heracleifolia* and its major compounds inhibit matrix proteinases and vascular endothelial growth factor through the regulation of mitogen-activated protein kinase (MAPK) pathway, thus providing a therapy for osteoarthritis (Huh et al., 2011).

Isoimperatorin, cimiside E, and 23-*O*-acetylshengmanol-3-xyloside from *Cimicifuga* rhizome inhibit TNF-α-induced vascular cell adhesion molecule (VCAM)-1 expression in human endothelial cells (ECs) (Moon et al., 2011). Peroxisome proliferator-activated receptor-γ (PPAR-γ) upregulation and PI3K, ERK 1/2, and protein kinase C (PKC) signal pathways are involved in their anti-inflammatory activities. Cimiracemate A from multiple *Cimicifuga* species suppressed the lipopolysaccharide-induced TNF-α production in the blood macrophages (Yang et al., 2009). It might modulate a signaling MAPK and transcription factor nuclear factor-κB activities. Among the triterpene glycosides, 23-epi-26-deoxyactein was identified as an active principle in CR extracts, which inhibits nitric oxide (NO) production by reducing iNOS expression without affecting activity of the enzyme, thus contributing to the anti-inflammatory activities of CR (Schmid et al., 2009b). Lipophilic constituents of CR bind to an as yet unidentified ER, likely inversely involved in inflammation (Bolle et al., 2007).

8.3.5 Effects on cardiovascular system

CR relaxes the isolated rat thoracic aorta through endothelium-dependent and endothelium-independent mechanisms (Kim et al., 2011). CR extract elicited the vasorelaxant effect via the NO/cGMP pathway. CR-induced endothelium-independent vasorelaxation appears to involve the inhibition of calcium influx mediated by the opening of inward rectifier potassium channels. 7,8-Didehydrocimigenol (DHC), a triterpenoid of *Cimicifuga* rhizome, increased PPAR-γ expression in ECs in a time- and dose-dependent manner (Mun et al., 2011). 7,8-DHC inhibits TNF-α-induced VCAM-1 but not ICAM-1 expression through upregulation of PPAR-γ in human ECs and can be used for the treatment of cardiovascular disorders such as atherosclerosis.

Black cohosh extract induces plasminogen activator inhibitor type-1 (PAI-1) protein expression in the VSMCs likely via an oxidant mechanism (Lee et al., 2013). It also stimulates the enzyme activity of PAI-1 and reduces that of free t-PA (tissue-type plasminogen activator), suggesting that black cohosh might exert a negative influence on fibrinolysis.

8.3.6 Antioxidant activity

Heracleifolinoside B, isolated from *C. heracleifolia*, is effectively resistant to hypoxia and reoxygenation-induced human umbilical vein EC injury (Liu et al., 2013), with cell viabilities of 61.95 ± 2.04%, 77.04 ± 4.44%, and 83.65 ± 3.29% at concentrations of 1, 10, and 100 μM, respectively.

8.3.7 Effects on nervous system

A triterpene monoglycoside 1, isolated from *Actaea racemosa*, has an IC50 of 100 nM for selectively reducing the production of amyloidogenic Aβ42 while having a much smaller effect on the production of Aβ40 (IC50 6.3 μM) in cultured cells overexpressing APP (Findeis et al., 2012). This compound was found to modulate the pool of total Aβ produced by reducing the proportion of Aβ42 while increasing the relative amounts of shorter and less amyloidogenic Aβ37 and Aβ39. Concentrations of 1 sufficient to lower levels of Aβ42 substantially (up to 10 μM) did not significantly affect the processing of Notch or other aspects of APP processing. When 1 (10 μg) was administered to CD-1 normal mice intracerebroventricularly, the level of Aβ42 in brain was reduced. Assays for off-target pharmacology and the absence of overt signs of toxicity in mice dosed with compound 1 suggest a comparatively selective pharmacology for this triterpenoid. Compound 1 represents a new lead for the development of potential treatments for Alzheimer's disease via modulation of γ-secretase.

C. foetida extract possesses antidepressant-like properties in rodents (Ye et al., 2012), which could be related to its serotonergic and noradrenergic activation and normalization of the hypothalamic–pituitary–adrenal axis.

23-*O*-Acetylshengmanol 3-*O*-β-D-xylopyranoside (Ac-SM) isolated from *A. racemosa*, an herbal remedy for the treatment of mild menopausal disorders, is an efficacious modulator of GABAA receptors composed of α1, β2, and γ2S subunits. Ac-SM shifted the GABA concentration-response curve toward higher GABA sensitivity (about threefold) and significantly increased the maximal GABA response by 44 ± 13%, indicating a pharmacological profile distinct from a pure allosteric GABAA receptor modulator (Strommer et al., 2014). In mice, Ac-SM significantly reduced anxiety-related behavior in the elevated plus maze test at a dose of 0.6 mg/kg, total ambulation in the open field test at doses ≥6 mg/kg, stress-induced hyperthermia at doses ≥0.6 mg/kg, and significantly elevated seizure threshold at doses ≥20 mg/kg body weight. High efficacy and long biologic half-life of Ac-SM suggest that potential cumulative sedative side effects upon repetitive intake of *A. racemosa* preparations might not be negligible.

8.3.8 Other positive effects

Oral administration of CR extract attenuates psychological and physiological stress responses (Nadaoka et al., 2012a). The CR extract can alleviate acute stress responses by suppressing the changes of amine-to-metabolite ratio in brain (Nadaoka et al., 2012b). The CR extract significantly prevented the development of water immersion stress-induced gastric mucosal ulcers in rats (Nadaoka et al., 2012c).

Dichloromethane fraction of *C. heracleifolia* decreases the level of melanin synthesis by activating the ERK or AKT signaling pathway in B16F10 cells (Jang et al., 2009). *C. heracleifolia* would be a useful therapeutic agent for treating hyperpigmentation and an effective component in whitening and/or lightening cosmetics. Isoferulic acid exhibited inhibitory effect on mushroom tyrosinase (Gong et al., 2013).

The *C. racemosa* extract Ze 450 (23-epi-26-deoxyactein, protopine, and cimiracemoside C) showed antidiabetic effects *in vitro* and *in vivo* in ob/ob mice (Moser et al., 2014).

8.3.9 Adverse effects and safety

Eight triterpene glycosides of CR were identified as competitive CYP (cytochrome P450) 3A4 inhibitors with IC50 values ranging from 2.3 to 5.1 μM, while the alkaloids protopine and allocryptopine were identified as competitive CYP2D6 inhibitors (Li et al., 2011). Coadministration of CR with tamoxifen might interfere with the clinical efficacy of this drug. CYPs 1A2, 2D6, 2C9, 2C19, and 3A4 were inhibited *in vitro* by CR extracts (Huang et al., 2010; Ho et al., 2011). Fukinolic acid derivatives and triterpene glycosides from CR inhibit CYP isozymes, but are not cytotoxic to HepG2 cells *in vitro*. However, Pang et al. (2011) suggested that the incidence of herb–drug interaction in patients administered CR may not be mediated by human pregnane X receptor and CYP3A4. With the exceptions of St. John's wort and goldenseal, the information currently available suggests that concomitant intake of the herbal drugs, for example, the well-known ginseng and CR, is not a major risk for drugs that are metabolized by CYPs (Zadoyan and Fuhr, 2012).

Compared to St John's wort and ginger root extract, black cohosh was the strongest inhibitor for both CYP and carboxylesterase-mediated biotransformation of tamoxifen and irinotecan, respectively, to their active metabolites (Gorman et al., 2013). The clinical significance of these results needs to be evaluated in a comprehensive clinical trial.

CR extract induced dose-dependent hematological changes, that is, a nonregenerative macrocytic anemia, and increased frequencies of peripheral micronucleated red blood cells in female B6C3F1/N mice and Wistar Han rats (Mercado-Feliciano et al., 2012).

Current evidence does not support an association between black cohosh and increased risk of breast cancer (Fritz et al., 2014). There is a lack of evidence supporting the efficacy of black cohosh for the reduction of hot flashes in breast cancer patients. Given conflicting but promising results, and apparent safety, further research is warranted. CR would not influence breast cancer risk if given to women before tumor formation. However, it increases metastatic mammary cancer in transgenic mice expressing c-erbB2 (Davis et al., 2008).

CR impairs fatty acid β-oxidation and induces oxidative stress in livers of ovariectomized rats with renovascular hypertension (Campos et al., 2012). Quality problems in a few CR products were evident that require additional regulatory quality specifications (Teschke et al., 2011). The pattern of pathological injury of liver cells in two patients, following consumption of black cohosh, is identical to troxis necrosis (Enbom et al., 2014), seen during autoimmune hepatitis. Recognition of the possibility of the acute hepatic injury by the herbal supplement black cohosh is essential for early accurate diagnosis and timely patient management.

Black cohosh extracts exposure in B6C3F1/N mice for 28 days at doses up to 1000 mg/kg had minimal immune effects, with the exception of an increased cytotoxic T cell response (Smith et al., 2013).

8.4 Bioactivity of other compounds

Examples of other compounds are shown in Table 8.2 and Figure 8.3.

8.4.1 Phenolic compounds

Petasiphenone, a phenol isolated from CR, *in vitro* inhibits proliferation of the human prostate cancer cell line LNCaP (Jarry et al., 2007). The radical scavenging activity of the *Actaea* extracts correlates to their polyphenolic composition (Nuntanakorn et al., 2007). Isoferulic acid (3-hydroxy-4-methoxycinnamic acid) from *Rhizoma Cimicifugae* is an effective natural antioxidant in both lipid and aqueous media (Wang et al., 2011). Actaealactone and a new phenylpropanoid ester derivative, cimicifugic acid G from CR, displayed antioxidant activity in the 1,1-diphenyl-2-picrylhydrazyl (DPPH) free radical assay (Nuntanakorn et al., 2006).

Actaea spicata, native to Europe and west Asia, has been traditionally used for the treatment of various ailments such as rheumatism, inflammation, nerve diseases, lumbago, scrofula, and chorea. A significant antioxidant activity, that is, mean percentage inhibition of DPPH radical, was observed in methanol extract and ethyl acetate fraction of *A. spicata* at the concentration of 10 μg/ml and 5 μg/ml, respectively (Madaan et al., 2011). Polyphenols are responsible for antioxidant activity of *A. spicata*.

Huang Qi (*Radix Astragali*) and Sheng Ma (*C. foetida*) demonstrate significantly better antioxidant effects when used in combination than when used alone. The combination of isoferulic acid (from Sheng Ma) and calycosin (from Huang Qi) at a dose ratio of 1:1 resulted in significant synergy in scavenging DPPH radicals, H_2O_2-induced HepG2 cells bioassay, and ferric reducing antioxidant power (FRAP) assay

Table 8.2 Other compounds found in Cimicifugeae medicinal plants in recent years

	Compound	Aglycone	Species	Tissue	References
144	8,11-Dihydro-5-hydroxy-2,9-dihydroxymethyl-4H-pyrano[2,3-g][1]benzoxepin-4-one	A	*Eranthis cilicica*	Tuber	Kuroda et al. (2009)
145	5,7-Dihydroxy-8-[(2*E*)-4-hydroxy-3-methylbut-2-enyl]-2-methyl-4H-1-benzopyran-4-one	A			
146	5,7-Dihydroxy-2-hydroxy-3-methyl-8-[(2*E*)-4-hydroxy-3-methylbut-2-enyl]-4H-1-benzopyran-4-one	A			
147	7-[(β-D-glucopyranosyl)oxy]-5-hydroxy-8-[(2*E*)-4-hydroxy-3-methylbut-2-enyl]-2-methyl-4H-1-benzopyran-4-one	A			
148	7-[(β-D-glucopyranosyl)oxy]-5-hydroxy-2-hydroxymethyl-8-[(2*E*)-4-hydroxy-3-methylbut-2-enyl]-4H-1-benzopyran-4-one	A			
149	9-[(*O*-β-D-glucopyranosyl-(1 → 6)-β-D-glucopyranosyl)oxy]methyl-8,11-dihydro-5,9-dihydroxy-2-methyl-4H-pyrano[2,3-g][1]benzoxepin-4-one	A			
150	8,11-Dihydro-5,9-dihydroxy-9-hydroxymethyl-2-methyl-4H-pyrano[2,3-g][1]benzoxepin-4-one	A			
151	7-[(*O*-β-D-glucopyranosyl-(1 → 6)-β-D-glucopyranosyl)oxy]methyl-4-hydroxy-5H-furo[3,2-g][1]benzopyran-5-one	A			
152	Norkhelloside	A	*Cimicifuga heracleifolia*	Rhizome	Liu et al. (2013)
153	Cimicifugamide A	B	*Cimicifuga dahurica*	Rhizome	Zhang et al. (2013)
154	Cimicifugolide A	C	*Actaea cimicifuga*	Root, rhizome	Ma et al. (2013)
155	Cimicifugolide B	C			
156	Cimicifugolide C	C			

A, Chromone; B, phenolic amide glycoside; C, monoterpene lactone.

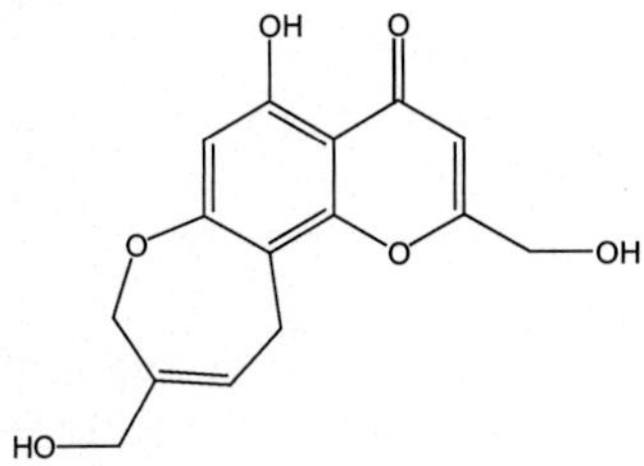

144

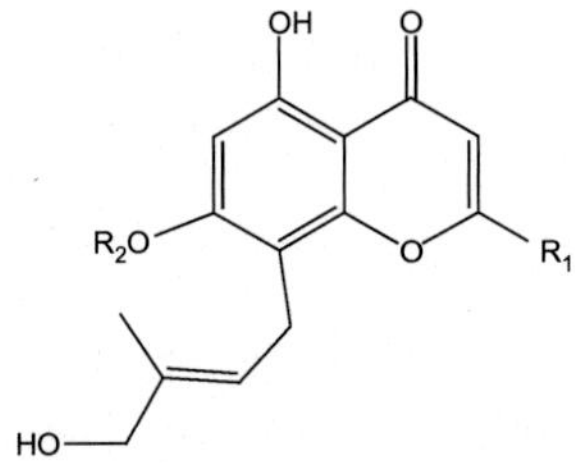

145	$R_1 = CH_3$	$R_2 = H$
146	$R_1 = CH_2OH$	$R_2 = H$
147	$R_1 = CH_3$	$R_2 = Glc$
148	$R_1 = CH_2OH$	$R_2 = Glc$

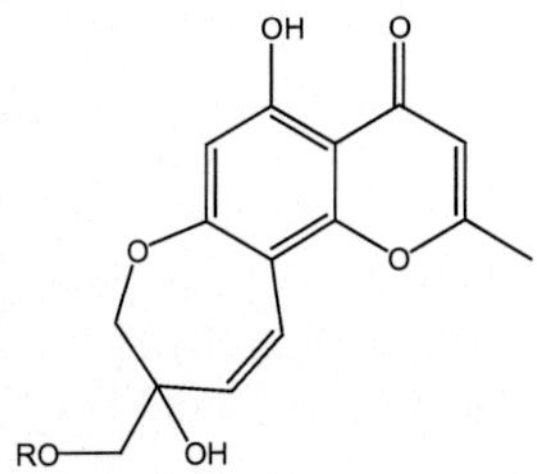

149	R = Glc(1→6)-Glc
150	R = H

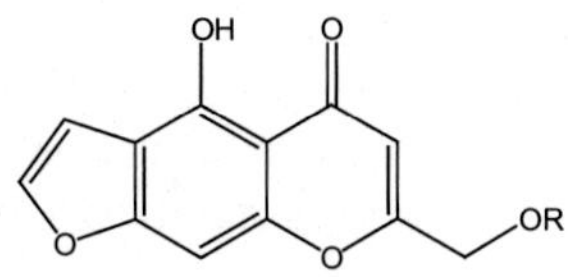

151	R = Glc(1→6)-Glc
152	R = Api(f)(1→6)-Glc

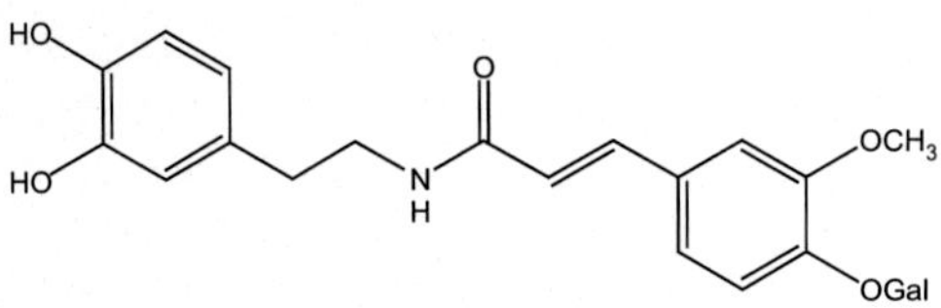

153

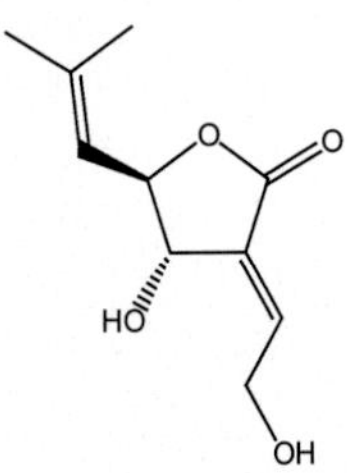

154

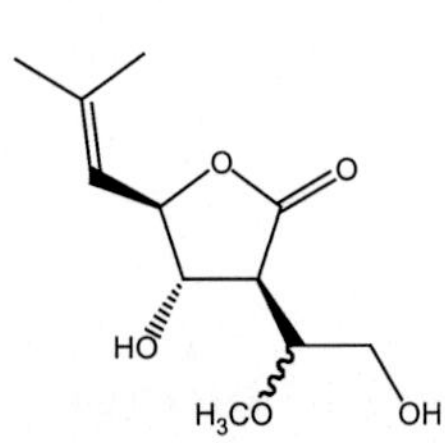

155

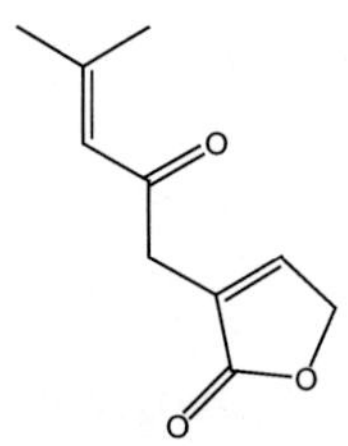

156

Figure 8.3 Examples of other compounds found in tribe Cimicifugeae.

(Wang et al., 2014). These findings support the synergistic efficiency of Huang Qi and Sheng Ma as functional foods, dietary supplements, and drugs.

Fukinolic acid and cimicifugic acids A–J, from a mixture of *C. dahurica* and *C. heracleifolia*, showed stronger hyaluronidase inhibitory activities than the positive control, rosmarinic acid (Iwanaga et al., 2010a). Cimicifugic acids K–N from *C. simplex* showed more potent hyaluronidase inhibitory activities than rosmarinic acid (Iwanaga et al., 2010b). Isoferulic acid was the major active principle in CR root extract, responsible for the observed inhibition of interleukin (IL)-6, TNF-α, and interferon-γ, but not for IL-8 stimulation (Schmid et al., 2009a,b). The effect of this compound may explain the anti-inflammatory activities of CR and its beneficial actions in rheumatism and other inflammatory diseases.

Carboxymethyl isoferulate, cimicifugic acid A, and cimicifugic acid B, isolated from *C. heracleifolia* (Yim et al., 2012), showed cytotoxic activity against cancer cells in a dose-dependent manner at the concentrations of 2.5–40 μM, and western blotting analysis showed that these compounds increased expression of cleaved poly ADP ribose polymerase (PARP), a critical apoptosis marker.

8.4.2 Others

Cimicifugin from *C. foetida* inhibits human respiratory syncytial virus (HRSV; Wang et al., 2012a). *C. foetida* dose-dependently inhibited viral attachment and could increase heparin effect on viral attachment (Wang et al., 2012b). Additionally, *C. foetida* time-dependently and dose-dependently inhibited HRSV internalization.

Serotonergic receptors and transporters are involved in thermoregulation. Various guanidine alkaloids and Pictet–Spengler adducts were detected in CR (Gödecke et al., 2009a). These strongly basic and frequently zwitterionic nitrogenous metabolites contribute considerable chemical diversity to the polar serotonergic fraction of CR, which is used to alleviate menopausal symptoms. New phytochemical methods and LC (liquid chromatography)–MS (mass spectrometry) led to the identification of *N*(ω)-methylserotonin as serotonergic active principle of CR (Gödecke et al., 2009a,b). *N*(ω)-methylserotonin showed 5-hydroxytryptamine (serotonin) 7 receptor binding induced cAMP and blocked serotonin reuptake, suggesting that *N*(ω)-methylserotonin may be responsible for the serotonergic activity of CR (Powell et al., 2008).

A new 4α-methyl sterol from the aerial parts of *C. foetida*, cimisterol A, exhibited broad-spectrum and potent cytotoxic activities against human HL-60, Jurkat, K562, U937, HepG-2, and SGC-7091 cell lines (Nian et al., 2012). The bioactivities of Cimicifugeae alkaloids await further studies.

8.5 Chemotaxonomy and authentication

Black cohosh (*Actaea racemosa*, syn. *C. racemosa*) is ranked among the top-selling herbs in the United States. There is a risk for adulteration with the similar-looking *C. americana*, which grows in the same habitats of the eastern United States. Other adulterants found in the current global market are the Asian

Cimicifuga species *C. foetida*, *C. heracleifolia*, and *C. dahurica*. Three of the 11 black cohosh products were found to contain the marker compound cimifugin and not cimiracemoside C, indicating that these products contain Asian *Actaea* instead of black cohosh (Jiang et al., 2006). One product contained both black cohosh and an Asian *Actaea* species. The chemotaxonomic distinctiveness of the HPLC fingerprints allows identification of 10 *Cimicifuga* species, including three North American species and seven Asian species (He et al., 2006). The triterpene glycosides cimigenol-3-*O*-arabinoside, cimifugin, and cimifugin-3-*O*-glucoside were suitable species-specific markers for the distinction of CR from the other *Cimicifuga* species. A rapid and reliable high-performance thin-layer chromatographic (HPTLC) method was developed for identification of CR and detection of its most common contaminants by fingerprint profiles (Ankli et al., 2008). The HPLC and LC–MS fingerprints for polyphenols and triterpene glycosides revealed distinct patterns that make CR clearly distinguishable from most other species of *Actaea* (Jiang et al., 2011). Cimifugin and cimiracemoside F were found to be important to distinguish CR from most Asian species of *Actaea*. CR products continue to be one of the most popular botanical supplements in the United States and world markets, and correct identification of different *Actaea* species is a key first step for their good manufacturing practice.

Fifteen chemical markers, including three cimifugin derivatives, 11 triterpene glycosides, and one alkaloid, were identified with HPLC-TOF-ESI–MS technique and principal component analysis (PCA), and the 16 *Actaea* species were divided into three groups, that is, Asian (seven species), North American (eight species), and *A. racemosa* groups (Ma et al., 2011a,b). The adulteration of botanical supplements is a major problem that efficacy of the original product can be diminished and even toxicity can arise. Therefore, identification of marker compounds may help in quality control and standardization of botanical supplements. Ultra performance liquid chromatography (UPLC) is a relatively new technique giving new possibilities in liquid chromatography, which is faster and consumes less solvent than HPLC. UPLC-UV/ELSD and UPLC–MS were successfully used to analyze different CR market products and to distinguish between two other *Actaea* species (Avula et al., 2009). Thin-layer chromatography (TLC) and combined TLC-bioluminescence (Bioluminex) are efficient, economical, and effective techniques that provide characteristic patterns and toxicity profiles for CR, *Actaea pachypoda* Ell. (white cohosh), *A. podocarpa* DC. (yellow cohosh), and other congeners (Verbitski et al., 2008). In addition to identification, the fingerprint method provided insight into chemical interconversion processes occurring between the diverse triterpene glycosides contained in Cimicifugeae.

The metabolic profiling of 25 black cohosh products was analyzed using LC–MS/MS (Masada-Atsumi et al., 2014). It was found that *C. dahurica* and possibly *C. foetida* are misused as sources of the black cohosh products and, in some cases, the extracts of black cohosh were adulterated with the plant materials of *C. dahurica*. These three species can be distinguished by three marker compounds, that is, actein, compound X ($C_{37}H_{56}O_{11}$), and compound Y (suggested as acetylacteol-3-*O*-arabinoside), in a specific mass range, which is helpful in establishing regulations for the safe use of the black cohosh products.

The metabolomic analysis of three source species of Sheng Ma was performed using H NMR spectroscopy and pattern recognition (PR) techniques (Shen et al., 2013). A broad range of metabolites could be detected by ^{1}H NMR spectroscopy without any chromatographic separation. The analysis using PCA and discriminant partial least square (DPLS) of the ^{1}H NMR spectrum showed a clear discrimination between *C. foetida* and the other two species. The major metabolites responsible for the discrimination were triterpenoid saponins and saccharides. The combination of ^{1}H NMR and PR provides a useful tool for chemotaxonomic analysis and authentification of *Cimicifuga* species and could be used for the quality control of plant materials.

8.6 Molecular phylogeny and genomics

8.6.1 Phylogeny

Sequence analysis of nuclear internal transcribed spacer (ITS) and fluorescence melting curve analysis of LightCycler real-time polymerase chain reaction (PCR) products were used to authenticate *C. foetida* from four substitutes: *C. heracleifolia*, *C. dahurica*, *C. acerina*, and *C. simplex* (Xue et al., 2009). This method is expensive and did not resolve the phylogenetic relationship within Cimicifugeae. We retrieve the ITS and chloroplast (cp) DNA sequences from NCBI GenBank and construct the phylogenetic tree. On the ITS tree (Figure 8.4a), *Beesia* and *Anemonopsis* are basal to other genera and *Eranthis* is basal to the intermingled *Actaea*, *Cimicifuga* and *Souliea* sequences. There are three major clades of *Cimicifuga*. The first clade includes *C. foetida*, *C. yunnanensis*, *C. brachycarpa*, *C. kashmiriana*, and *C. europaea*, which is supported by the cp trnL-F tree (Figure 8.4b). *C. heracleifolia* is closer to *C. dahurica* than to *C. simplex*. The second clade includes *C. acerina*, *C. japonica*, and *C. biternata*, which are distributed in Japan. *C. acerina* also has extensive distribution in China and its rhizome is traditionally used as herbal medicine and pesticide. The third clade includes four North American species.

The boundary between *Actaea* and *Cimicifuga* is not unambiguous (Figure 8.4). Surprisingly, *Souliea* is not basal to *Actaea* and *Cimicifuga*. *C. nanchuanensis*, an endemic endangered species in Sichuan, China, is closer to *Souliea* than to *Actaea*. More sequence data are needed to resolve its phylogenetic position. On the trnL-F tree, the group containing *C. foetida* and *C. yunnanensis* is closer to *C. heracleifolia* and *C. dahurica*, whereas *C. simplex* is closer to these Asia species. *C. yunnanensis* clusters with *C. foetida* (alternative names *C. mairei* and *C. frigida*) on both ITS and cp trees, implying that its medicinal value is worth further exploration. On both ITS and trnL-F trees, *A. racemosa* (black cohosh) is basal to other *Actaea* species.

To avoid confusion, every taxon should have only one species name; thus, all synonyms should be deleted.

8.6.2 Transcriptomics

A. racemosa cDNA libraries were constructed from young leaf, rhizome, and root tissues (Spiering et al., 2011). Expressed sequence tag (EST) sequencing was performed and 1590 unigenes were assembled. Seventy putative secondary metabolism genes

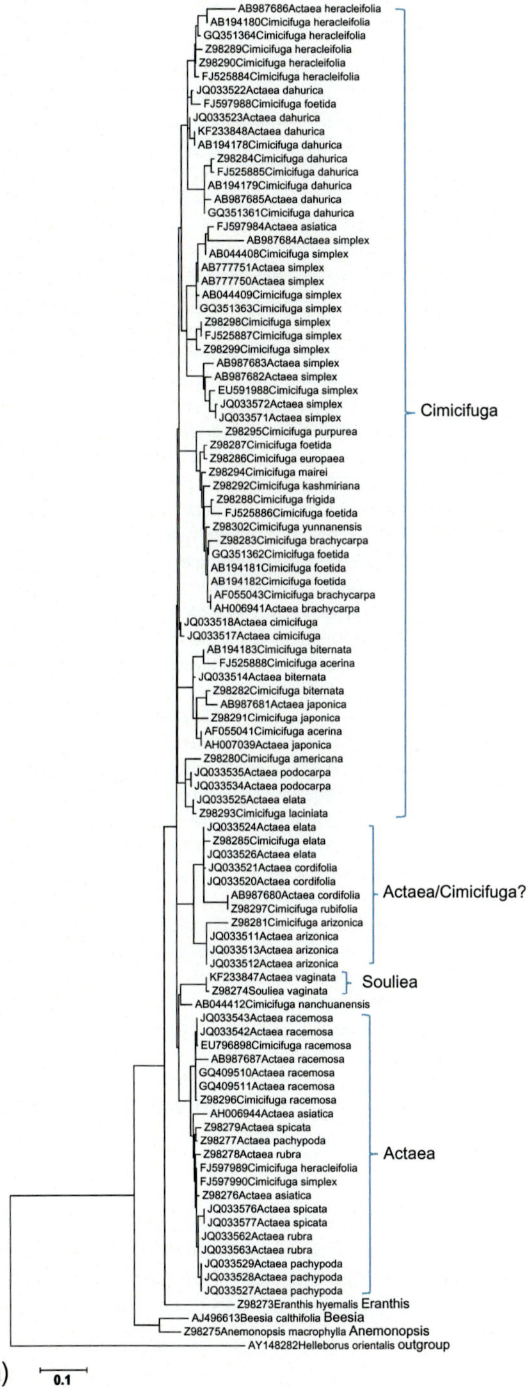

(a)

Figure 8.4 Evolutionary relationships of Cimicifugeae sequences. (a) ITS.

(Continued)

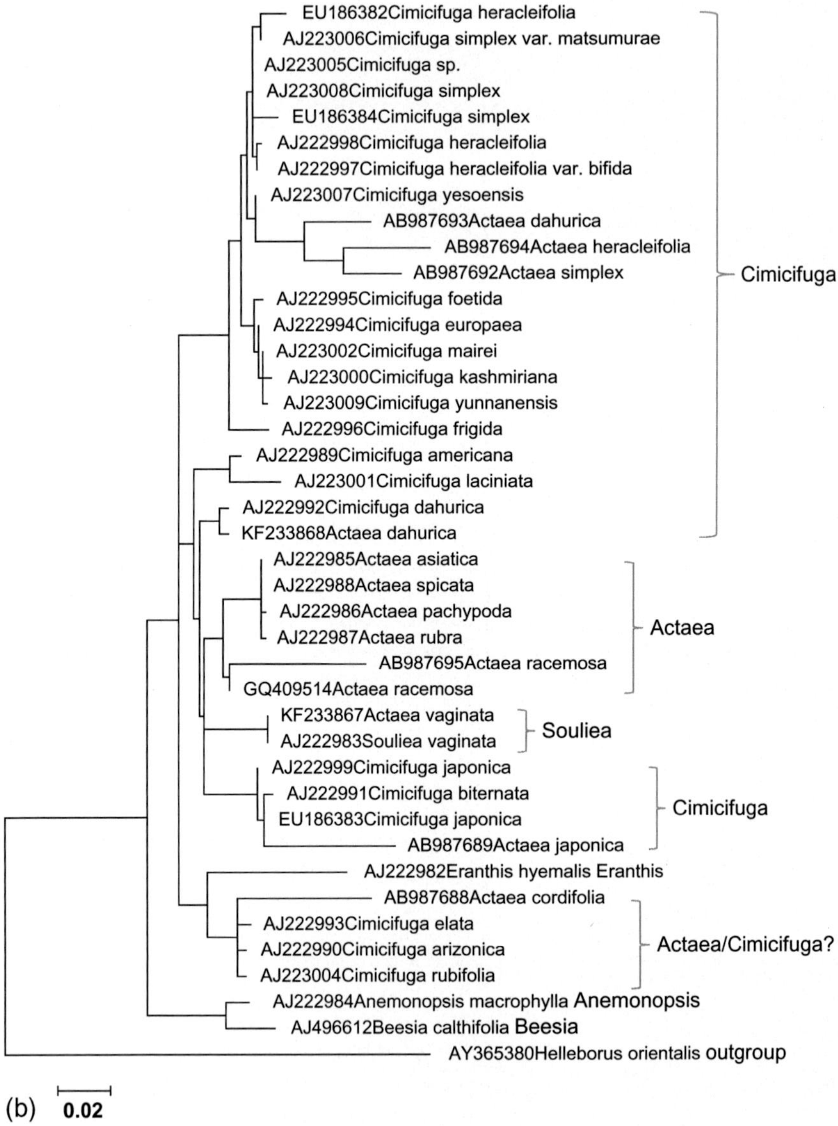

Figure 8.4 Continued. (b) trnL-trnF. The evolutionary history was inferred by using the ML method based on GTR+I+G model. The trees are drawn to scale, with branch lengths measured in the number of substitutions per site. The analysis involved 99 ITS and 41 trnL-F sequences. Evolutionary analyses were conducted in MEGA5 (Tamura et al., 2011). *C. mairei*=*C. frigida*=*C. foetida*.

were identified, including 2,3 oxidosqualene cyclase, BAHD-type acyltransferase, and tryptophan decarboxylase. This preliminary transcriptome study provides initial insight into gene content and diversity in black cohosh and generates tools and resources for detailed investigations of secondary metabolite genes and enzymes in this important medicinal plant. Seven microsatellite markers were found for *A. racemosa* (Pate et al., 2012). Most of the loci cross amplified in *A. pachypoda*, *A. podocarpa*, and *A. rubra*, indicating the utility of these markers for the genus. In the future, more microsatellite markers could be mined from the transcriptome and genomic survey datasets (Hao et al., 2011, 2012), which will provide tools for population genetics studies and molecular breeding.

8.6.3 Molecular marker

The lengths of ITS2 sequence of the three origin plants of *Rhizoma Cimicifugae C. heracleifolia*, *C. foetida*, and *C. dahurica* were 217, 219, and 219 bp, respectively (Ren et al., 2014). Their intraspecific genetic distance was much lower than the interspecific genetic distance with their closely related species. The NJ tree of ITS2 indicated that the three origin plants of *Rhizoma Cimicifugae* formed a monophyletic clade. ITS2 is suitable for the authentication of *Rhizoma Cimicifugae* and its adulterants.

Two matK nucleotides that consistently distinguish black cohosh from related species were identified (Baker et al., 2012). Using these nucleotides, an assay was able to correctly identify all of the black cohosh samples in the validation set. None of the other *Actaea* species were falsely identified as black cohosh. Of 36 dietary supplements sequenced, 27 (75%) had a sequence that exactly matched black cohosh. The remaining nine samples (25%) had a sequence identical to that of three Asian *Actaea* (*Cimicifuga*) species (*A. cimicifuga*, *A. dahurica*, and *A. simplex*). Manufacturers should routinely test plant material using a reliable assay to ensure accurate labeling.

8.7 Conclusion

The phylogeny, pharmacology, and chemical composition of the tribe Cimicifugeae of the family of Ranunculaceae are analyzed and compared. The main chemical constituents of this tribe are 9, 19-cycloartane triterpenoid glycosides. In ethnopharmacology, Cimicifugeae plants are commonly used as antipyretic, antitoxic, and anti-inflammation agents and utilized for the treatment of menopausal symptoms. Recent bioactivity research demonstrated that some species showed antiosteoporotic, antiviral, anticancer, antiallergic, and hormonal-like activities. *Actaea* and *Cimicifuga* have a close relationship in chemical constituents, phylogenetic position, and medicinal use. They could be considered as one clade of tribe Cimicifugeae. *Beesia* and *Anemonopsis* are evolutionarily ancient and beesioside-type triterpenoids had never been found in *Actaea* and *Cimicifuga*. Meanwhile, pentacyclic triterpenoids and beesioside-type triterpenoids were both isolated from *Beesia*, as well as *Souliea*, but they are absent in *Cimicifuga* and *Actaea*. Indolinone alkaloids were detected in either *Souliea* or *Cimicifuga* and *Actaea*, implying that the biosynthetic pathway has changed dramatically during evolution. There is a close relationship between the cycloartane triterpenoids and the antitoxic, antipyretic effects of Cimicifugeae. The cycloartane triterpenoids would be a new resource of antitumor and antiviral agents. Various omics techniques should be put into full play in the drug development research of Cimicifugeae.

References

Ahn, B.S., Yang, M., Jang, H., et al., 2012. Evaluation of the antiosteoporotic potential of *Cimicifuga heracleifolia* in female mice. Phytother. Res. 26 (5), 663–668.

Ali, Z., Khan, S.I., Fronczek, F.R., Khan, I.A., 2007a. 9, 10-seco-9, 19-Cyclolanostane arabinosides from the roots of *Actaea podocarpa*. Phytochemistry 68 (3), 373–382.

Ali, Z., Khan, S.I., Pawar, R.S., et al., 2007b. 9, 19-cyclolanostane derivatives from the roots of *Actaea pachypoda*. J. Nat. Prod. 70 (1), 107–110.

Ali, Z., Khan, S.I., Khan, I.A., 2007c. New cycloartane-type triterpene arabinosides from the roots of *Actaea podocarpa* and their biological study. Planta Med. 73 (7), 699–703.

Alves, D.L., Lima, S.M., da Silva, C.R., et al., 2008. Effects of *Trifolium pratense* and *Cimicifuga racemosa* on the endometrium of Wistar rats. Maturitas 61 (4), 364–370.

Ankli, A., Reich, E., Steiner, M., 2008. Rapid high-performance thin-layer chromatographic method for detection of 5% adulteration of black cohosh with *Cimicifuga foetida*, *C. heracleifolia*, *C. dahurica*, or *C. americana*. J. AOAC Int. 91 (6), 1257–1264.

Avula, B., Wang, Y.H., Smillie, T.J., Khan, I.A., 2009. Quantitative determination of triterpenoids and formononetin in rhizomes of black cohosh (*Actaea racemosa*) and dietary supplements by using UPLC-UV/ELS detection and identification by UPLC-MS. Planta Med. 75 (4), 381–386.

Bai, W.P., Wang, S.Y., Liu, J.L., et al., 2009. Efficacy and safety of remifemin compared to tibolone for controlling of perimenopausal symptoms. Zhonghua Fu Chan Ke Za Zhi 44 (8), 597–600.

Baker, D.A., Stevenson, D.W., Little, D.P., 2012. DNA barcode identification of black cohosh herbal dietary supplements. J. AOAC Int. 95 (4), 1023–1034.

Bao, N., Nian, Y., Zhu, G., et al., 2014. Cytotoxic 9,19-cycloartane triterpenes from the aerial parts of *Cimicifuga yunnanensis*. Fitoterapia 99, 191–197.

Bolle, P., Mastrangelo, S., Perrone, F., Evandri, M.G., 2007. Estrogen-like effect of a *Cimicifuga racemosa* extract sub-fraction as assessed by in vivo, ex vivo and in vitro assays. J. Steroid Biochem. Mol. Biol. 107 (3–5), 262–269.

Campos, L.B., Gilglioni, E.H., Garcia, R.F., et al., 2012. *Cimicifuga racemosa* impairs fatty acid β-oxidation and induces oxidative stress in livers of ovariectomized rats with renovascular hypertension. Free Radic. Biol. Med. 53 (4), 680–689.

Chan, B.Y., Lau, K.S., Jiang, B., et al., 2008. Ethanolic extract of *Actaea racemosa* (black cohosh) potentiates bone nodule formation in MC3T3-E1 preosteoblast cells. Bone 43 (3), 567–573.

Chen, S.N., Lankin, D.C., Nikolic, D., et al., 2007. Chlorination diversifies *Cimicifuga racemosa* triterpene glycosides. J. Nat. Prod. 70 (6), 1016–1023.

Chen, J., Li, P., Tang, X., et al., 2014. Cycloartane triterpenoids and their glycosides from the rhizomes of *Cimicifuga foetida*. J. Nat. Prod. 77, 1997–2005.

Choi, E.M., 2013a. Deoxyactein isolated from *Cimicifuga racemosa* protects osteoblastic MC3T3-E1 cells against antimycin A-induced cytotoxicity. J. Appl. Toxicol. 33 (6), 488–494.

Choi, E.M., 2013b. Deoxyactein stimulates osteoblast function and inhibits bone-resorbing mediators in MC3T3-E1 cells. J. Appl. Toxicol. 33 (3), 190–195.

Cicek, S.S., Khom, S., Taferner, B., et al., 2010. Bioactivity-guided isolation of GABA(A) receptor modulating constituents from the rhizomes of *Actaea racemosa*. J. Nat. Prod. 73 (12), 2024–2028.

Compton, J.A., Culham, A., 2002. Phylogeny and circumscription of tribe Actaeeae (Ranunculaceae). Syst. Bot. 27 (3), 502–511.

Dan, C., Zhou, Y., Ye, D., et al., 2007. Cimicifugadine from *Cimicifuga foetida*, a new class of triterpene alklaoids with novel reactivity. Org. Lett. 9 (9), 1813–1816.

Dan, C., Liang, J., Zhou, Y., Ding, L., 2009. Cycloartane triterpenoid of *Cimicifuga foetida*. Chin. J. Chin. Materia Med. 34 (15), 1930–1934.

Davis, V.L., Jayo, M.J., Ho, A., et al., 2008. Black cohosh increases metastatic mammary cancer in transgenic mice expressing c-erbB2. Cancer Res. 68 (20), 8377–8383.

Dueregger, A., Guggenberger, F., Barthelmes, J., et al., 2013. Attenuation of nucleoside and anti-cancer nucleoside analog drug uptake in prostate cancer cells by *Cimicifuga racemosa* extract BNO-1055. Phytomedicine 20 (14), 1306–1314.

Einbond, L.S., Su, T., Wu, H.A., et al., 2007. Gene expression analysis of the mechanisms whereby black cohosh inhibits human breast cancer cell growth. Anticancer Res. 27 (2), 697–712.

Einbond, L.S., Wen-Cai, Y., He, K., et al., 2008. Growth inhibitory activity of extracts and compounds from *Cimicifuga* species on human breast cancer cells. Phytomedicine 15 (6–7), 504–511.

Einbond, L.S., Soffritti, M., Esposti, D.D., et al., 2009. Actein activates stress- and statin-associated responses and is bioavailable in Sprague–Dawley rats. Fundam. Clin. Pharmacol. 23 (3), 311–321.

Einbond, L.S., Soffritti, M., Degli Esposti, D., et al., 2012. Chemopreventive potential of black cohosh on breast cancer in Sprague–Dawley rats. Anticancer Res. 32 (1), 21–30.

Einbond, L.S., Mighty, J., Redenti, S., Wu, H.A., 2013. Actein induces calcium release in human breast cancer cells. Fitoterapia 91, 28–38.

Enbom, E.T., Le, M.D., Oesterich, L., et al., 2014. Mechanism of hepatotoxicity due to black cohosh (*Cimicifuga racemosa*): histological, immunohistochemical and electron microscopy analysis of two liver biopsies with clinical correlation. Exp. Mol. Pathol. 96 (3), 279–283.

Fan, Y.S., Yao, Z., Zhang, Y.W., Duan, H.Q., 2009. Six new cycloartane triterpene glycosides from *Actaea asiatica*. J. Asian Nat. Prod. Res. 11 (7), 588–596.

Findeis, M.A., Schroeder, F., McKee, T.D., et al., 2012. Discovery of a novel pharmacological and structural class of gamma secretase modulators derived from the extract of *Actaea racemosa*. ACS Chem. Neurosci. 3 (11), 941–951.

Fritz, H., Seely, D., McGowan, J., et al., 2014. Black cohosh and breast cancer: a systematic review. Integr. Cancer Ther. 13 (1), 12–29.

Gao, J.C., Peng, Y., Yang, M.S., Xiao, P.G., 2008. A preliminary pharmacophylogenetic study of tribe Cimicifugeae (Ranunculaceae). J. Syst. Evol. 46, 516–536.

García-Pérez, M.A., Pineda, B., Hermenegildo, C., et al., 2009. Isopropanolic *Cimicifuga racemosa* is favorable on bone markers but neutral on an osteoblastic cell line. Fertil. Steril. 91 (4 Suppl.), 1347–1350.

Gaube, F., Wolfl, S., Pusch, L., et al., 2007. Gene expression profiling reveals effects of *Cimicifuga racemosa* (L.) NUTT. (black cohosh) on the estrogen receptor positive human breast cancer cell line MCF-7. BMC Pharmacol. 7, 11.

Gödecke, T., Lankin, D.C., Nikolic, D., et al., 2009a. Guanidine alkaloids and Pictet–Spengler adducts from black cohosh (*Cimicifuga racemosa*). J. Nat. Prod. 72 (3), 433–437.

Gödecke, T., Nikolic, D., Lankin, D.C., et al., 2009b. Phytochemistry of cimicifugic acids and associated bases in *Cimicifuga racemosa* root extracts. Phytochem. Anal. 20 (2), 120–133.

Gong, S., Yin, M., Yun, Z., 2013. Kinetics of inhibitory effect of isoferulic acid on mushroom tyrosinase. J. Cosmet. Sci. 64 (4), 235–241.

Gorman, G.S., Coward, L., Darby, A., Rasberry, B., 2013. Effects of herbal supplements on the bioactivation of chemotherapeutic agents. J. Pharm. Pharmacol. 65 (7), 1014–1025.

Guo, L.Y., Joo, E.J., Son, K.H., et al., 2009. Cimiside E arrests cell cycle and induces cell apoptosis in gastric cancer cells. Arch. Pharm. Res. 32 (10), 1385–1392.

Hao, D.C., Yang, L., Xiao, P.G., 2011. The first insight into the *Taxus* genome via fosmid library construction and end sequencing. Mol. Genet. Genomics 285 (3), 197–205.

Hao, D.C., Ma, P., Mu, J., Chen, S.L., Xiao, P.G., et al., 2012. De novo characterization of the root transcriptome of a traditional Chinese medicinal plant *Polygonum cuspidatum*. Sci. China Life Sci. 55 (5), 452–466.

He, K., Pauli, G.F., Zheng, B., et al., 2006. *Cimicifuga* species identification by high performance liquid chromatography-photodiode array/mass spectrometric/evaporative light scattering detection for quality control of black cohosh products. J. Chromatogr. A 1112 (1–2), 241–254.

Ho, S.H., Singh, M., Holloway, A.C., Crankshaw, D.J., 2011. The effects of commercial preparations of herbal supplements commonly used by women on the biotransformation of fluorogenic substrates by human cytochromes P450. Phytother. Res. 25 (7), 983–989.

Huang, Y., Jiang, B., Nuntanakorn, P., et al., 2010. Fukinolic acid derivatives and triterpene glycosides from black cohosh inhibit CYP isozymes, but are not cytotoxic to Hep-G2 cells in vitro. Curr. Drug Saf. 5 (2), 118–124.

Huh, J.E., Shin, Y.J., Baek, Y.H., et al., 2011. Combined prescription (OAH19T) of *Aralia cordata* Thunb and *Cimicifuga heracleifolia* Komar and its major compounds inhibit matrix proteinases and vascular endothelial growth factor through the regulation of mitogen-activated protein kinase pathway. J. Ethnopharmacol. 135 (2), 414–421.

Hui, Z., Xiaoyan, M., Mukun, Y., et al., 2012. Effects of black cohosh and estrogen on the hypothalamic nuclei of ovariectomized rats at different temperatures. J. Ethnopharmacol. 142 (3), 769–775.

Iwanaga, A., Kusano, G., Warashina, T., Miyase, T., 2010a. Hyaluronidase inhibitors from "*Cimicifugae Rhizoma*" (a mixture of the rhizomes of *Cimicifuga dahurica* and *C. heracleifolia*). J. Nat. Prod. 73 (4), 573–578.

Iwanaga, A., Kusano, G., Warashina, T., Miyase, T., 2010b. Phenolic constituents of the aerial parts of *Cimicifuga simplex* and *Cimicifuga japonica*. J. Nat. Prod. 73 (4), 609–612.

Jamróz, M.K., Jamróz, M.H., Dobrowolski, J.C., et al., 2011. Novel and unusual triterpene from Black Cohosh. Determination of structure of 9,10-seco-9,19-cyclolanostane xyloside (cimipodocarpaside) by NMR, IR and Raman spectroscopy and DFT calculations. Spectrochim. Acta A Mol. Biomol. Spectrosc. 78 (1), 107–112.

Jamróz, M.K., Jamróz, M.H., Cz Dobrowolski, J., et al., 2012. One new and six known triterpene xylosides from *Cimicifuga racemosa*: FT-IR, Raman and NMR studies and DFT calculations. Spectrochim. Acta A Mol. Biomol. Spectrosc. 93, 10–18.

Jang, J.Y., Lee, J.H., Kang, B.W., et al., 2009. Dichloromethane fraction of *Cimicifuga heracleifolia* decreases the level of melanin synthesis by activating the ERK or AKT signaling pathway in B16F10 cells. Exp. Dermatol. 18 (3), 232–237.

Jarry, H., Stromeier, S., Wuttke, W., Nahrstedt, A., 2007. Petasiphenone, a phenol isolated from *Cimicifuga racemosa*, in vitro inhibits proliferation of the human prostate cancer cell line LNCaP. Planta Med. 73 (2), 184–187.

Jiang, B., Kronenberg, F., Nuntanakorn, P., et al., 2006. Evaluation of the botanical authenticity and phytochemical profile of black cohosh products by high-performance liquid chromatography with selected ion monitoring liquid chromatography-mass spectrometry. J. Agric. Food Chem. 54 (9), 3242–3253.

Jiang, B., Ma, C., Motley, T., et al., 2011. Phytochemical fingerprinting to thwart black cohosh adulteration: a 15 *Actaea* species analysis. Phytochem. Anal. 22 (4), 339–351.

Ju, J.H., Liu, D., Lin, G., et al., 2002. Beesiosides G, H, and J–N, seven new cycloartane triterpene glycosides from *Beesia calthifolia*. J. Nat. Prod. 65 (2), 147–152.

Kamel, H.H., 2013. Role of phyto-oestrogens in ovulation induction in women with polycystic ovarian syndrome. Eur. J. Obstet. Gynecol. Reprod. Biol. 168 (1), 60–63.

Kapur, P., Wuttke, W., Seidlova-Wuttke, D., 2010. The *Cimicifuga racemosa* special extract BNO 1055 prevents hot flashes in ovariectomized rats. Phytomedicine 17 (11), 890–894.

Kim, E.Y., Lee, Y.J., Rhyu, M.R., 2011. Black cohosh (*Cimicifuga racemosa*) relaxes the isolated rat thoracic aorta through endothelium-dependent and -independent mechanisms. J. Ethnopharmacol. 138 (2), 537–542.

Kolios, L., Schumann, J., Sehmisch, S., et al., 2010. Effects of black cohosh (*Cimicifuga racemosa*) and estrogen on metaphyseal fracture healing in the early stage of osteoporosis in ovariectomized rats. Planta Med. 76 (9), 850–857.

Kuang, H., Su, Y., Yang, B., et al., 2011. Three new cycloartenol triterpenoid saponins from the roots of *Cimicifuga simplex* Wormsk. Molecules 16 (6), 4348–4357.

Kuang, H., Su, Y., Wang, Q., et al., 2012. Three new cycloartenol glycosides from the roots of *Cimicifuga simplex*. Planta Med. 78 (6), 622–625.

Kuroda, M., Uchida, S., Watanabe, K., et al., 2009. Chromones from the tubers of *Eranthis cilicica* and their antioxidant activity. Phytochemistry 70 (2), 288–293.

Laakmann, E., Grajecki, D., Doege, K., et al., 2012. Efficacy of *Cimicifuga racemosa*, *Hypericum perforatum* and *Agnus castus* in the treatment of climacteric complaints: a systematic review. Gynecol. Endocrinol. 28 (9), 703–709.

Lee, J.H., Cuong, T.D., Kwack, S.J., et al., 2012. Cycloartane-type triterpene glycosides from the rhizomes of *Cimicifuga heracleifolia* and their anticomplementary activity. Planta Med. 78 (12), 1391–1394.

Lee, D.Y., Roh, C.R., Kang, Y.H., et al., 2013. Effects of black cohosh on the plasminogen activator system in vascular smooth muscle cells. Maturitas 76 (1), 75–80.

Li, J.X., Liu, J., He, C.C., et al., 2007. Triterpenoids from *Cimicifugae rhizoma*, a novel class of inhibitors on bone resorption and ovariectomy-induced bone loss. Maturitas 58 (1), 59–69.

Li, J., Gödecke, T., Chen, S.N., et al., 2011. In vitro metabolic interactions between black cohosh (*Cimicifuga racemosa*) and tamoxifen via inhibition of cytochromes P450 2D6 and 3A4. Xenobiotica 41 (12), 1021–1030 [Epub ahead of print].

Liu, J., Zhao, Z.Z., Chen, H.B., 2011. Review of Astragali radix. Chin. Herb. Med. 3 (2), 90–105.

Liu, Y., Wu, Z., Li, C., et al., 2013. Heracleifolinosides A–F, new triterpene glycosides from *Cimicifuga heracleifolia*, and their inhibitory activities against hypoxia and reoxygenation. Planta Med. 79 (3–4), 301–307.

Liu, D.L., Li, Y.J., Yao, N., et al., 2014. Acerinol, a cyclolanstane triterpenoid from *Cimicifuga acerina*, reverses ABCB1-mediated multidrug resistance in HepG2/ADM and MCF-7/ADR cells. Eur. J. Pharmacol. 733, 34–44.

Lu, B., Fan, Y.S., Duan, H.Q., 2008. A new cycloartane triterpene saponin from rhizome of *Actaea asiatica*. Chin. J. Chin. Materia Med. 33 (13), 1558–1561.

Lu, L., Chen, J., Nian, Y., et al., 2009. Trinor-cycloartane glycosides from the rhizomes of *Cimicifuga foetida*. Molecules 14 (4), 1578–1584.

Lu, L., Chen, J.C., Song, H.J., et al., 2010. Five new triterpene bisglycosides with acyclic side chains from the rhizomes of *Cimicifuga foetida* L. Chem. Pharm. Bull. 58 (5), 729–733.

Lu, L., Chen, J.C., Li, Y., et al., 2012. Studies on the constituents of *Cimicifuga foetida* collected in Guizhou Province and their cytotoxic activities. Chem. Pharm. Bull. 60 (5), 571–577.

Ma, C., Kavalier, A.R., Jiang, B., Kennelly, E.J., 2011a. Metabolic profiling of *Actaea* species extracts using high performance liquid chromatography coupled with electrospray ionization time-of-flight mass spectrometry. J. Chromatogr. A 1218 (11), 1461–1476.

Ma, X., Zhang, H., Wang, K., et al., 2011b. Effects of an isopropanolic-aqueous black cohosh extract on central body temperature of ovariectomized rats. J. Ethnopharmacol. 138 (1), 156–161.

Ma, L., Wang, Y., Tang, G., et al., 2013. New monoterpene lactones from *Actaea cimicifuga*. Planta Med. 79 (3–4), 308–311.

Madaan, R., Bansal, G., Kumar, S., Sharma, A., 2011. Estimation of total phenols and flavonoids in extracts of *Actaea spicata* roots and antioxidant activity studies. Indian J. Pharm. Sci. 73 (6), 666–669.

Masada-Atsumi, S., Kumeta, Y., Takahashi, Y., et al., 2014. Evaluation of the botanical origin of black cohosh products by genetic and chemical analyses. Biol. Pharm. Bull. 37 (3), 454–460.

Mercado-Feliciano, M., Cora, M.C., Witt, K.L., et al., 2012. An ethanolic extract of black cohosh causes hematological changes but not estrogenic effects in female rodents. Toxicol. Appl. Pharmacol. 263 (2), 138–147.

Moon, L., Ha, Y.M., Jang, H.J., et al., 2011. Isoimperatorin, cimiside E and 23-O-acetylshengmanol-3-xyloside from Cimicifugae rhizome inhibit TNF-α-induced VCAM-1 expression in human endothelial cells: involvement of PPAR-γ upregulation and PI3K, ERK1/2, and PKC signal pathways. J. Ethnopharmacol. 133 (2), 336–344.

Moser, C., Vickers, S.P., Brammer, R., et al., 2014. Antidiabetic effects of the *Cimicifuga racemosa* extract Ze 450 in vitro and in vivo in ob/ob mice. Phytomedicine 21 (11), 1382–1389.

Mun, L., Jun, M.S., Kim, Y.M., et al., 2011. 7,8-didehydrocimigenol from *Cimicifugae rhizoma* inhibits TNF-α-induced VCAM-1 but not ICAM-1 expression through upregulation of PPAR-γ in human endothelial cells. Food Chem. Toxicol. 49 (1), 166–172.

Nadaoka, I., Yasue, M., Sami, M., Kitagawa, Y., 2012a. Oral administration of *Cimicifuga racemosa* extract affects immobilization stress-induced changes in murine cerebral monoamine metabolism. Biomed. Res. 33 (2), 133–137.

Nadaoka, I., Watanabe, K., Yasue, M., et al., 2012b. Oral administration of *Cimicifuga racemosa* extract attenuates immobilization stress-induced reactions. Nat. Prod. Commun. 7 (1), 15–18.

Nadaoka, I., Yasue, M., Kitagawa, Y., Koga, Y., 2012c. Oral administration of *Cimicifuga racemosa* extract attenuates psychological and physiological stress responses. Biomed. Res. 33 (3), 145–152.

Nian, Y., Zhang, Y.L., Chen, J.C., et al., 2010. Cytotoxic chemical constituents from the roots of *Cimicifuga foetida*. J. Nat. Prod. 73 (2), 93–98.

Nian, Y., Zhang, X.M., Li, Y., et al., 2011. Cycloartane triterpenoids from the aerial parts of *Cimicifuga foetida* Linnaeus. Phytochemistry 72 (11–12), 1473–1481.

Nian, Y., Wang, H.Y., Su, J., et al., 2012. A cytotoxic 4α-methyl steroid from the aerial parts of *Cimicifuga foetida* L. Fitoterapia 83 (2), 293–297.

Nian, Y., Wang, H.Y., Zhou, L., et al., 2013a. Cytotoxic cycloartane triterpenes of the traditional Chinese medicine "Shengma" (*Cimicifuga dahurica*). Planta Med. 79, 60–69.

Nian, Y., Zhu, H., Tang, W., et al., 2013b. Triterpenes from the aerial parts of *Cimicifuga yunnanensis* and their antiproliferative effects on p53(N236S) mouse embryonic fibroblasts. J. Nat. Prod. 76 (5), 896–902.

Nishida, M., Yoshimitsu, H., 2011. Six new cycloartane glycosides from cimicifuga rhizome. Chem. Pharm. Bull. 59 (10), 1243–1249.

Nuntanakorn, P., Jiang, B., Einbond, L.S., et al., 2006. Polyphenolic constituents of *Actaea racemosa*. J. Nat. Prod. 69 (3), 314–318.

Nuntanakorn, P., Jiang, B., Yang, H., et al., 2007. Analysis of polyphenolic compounds and radical scavenging activity of four American *Actaea* species. Phytochem. Anal. 18 (3), 219–228.

Pan, R.L., Chen, D.H., Si, J.Y., Zhao, X.H., 2007. Cimifoetisides VI and VII Two new cyclolanostanol triterpene glycosides from the aerial parts of *Cimicifuga foetida*. J. Asian Nat. Prod. Res. 9 (2), 97–102.

Pan, R.L., Chen, D.H., Si, J.Y., et al., 2009. Immunosuppressive effects of new cyclolanostane triterpene diglycosides from the aerial part of *Cimicifuga foetida*. Arch. Pharm. Res. 32 (2), 185–190.

Pang, X., Cheng, J., Krausz, K.W., et al., 2011. Pregnane X receptor-mediated induction of Cyp3a by black cohosh. Xenobiotica 41 (2), 112–123.

Park, J., Shim, M., Rhyu, M.R., Lee, Y., 2012. Estrogen receptor mediated effects of *Cimicifuga* extracts on human breast cancer cells. Pharmazie 67 (11), 947–950.

Pate, S.J., Clement, J.A., McCoy, J.A., et al., 2012. Development and characterization of microsatellite markers for *Actaea racemosa* (black cohosh, Ranunculaceae). Am. J. Bot. 99 (7), e274–e276.

Powell, S.L., Gödecke, T., Nikolic, D., et al., 2008. In vitro serotonergic activity of black cohosh and identification of N(omega)-methylserotonin as a potential active constituent. J. Agric. Food Chem. 56 (24), 11718–11726.

Qiu, M., Kim, J.H., Lee, H.K., Min, B.S., 2006. Anticomplement activity of cycloartane glycosides from the rhizome of *Cimicifuga foetida*. Phytother. Res. 20 (11), 945–948.

Qiu, S.X., Dan, C., Ding, L.S., et al., 2007. A triterpene glycoside from black cohosh that inhibits osteoclastogenesis by modulating RANKL and TNFalpha signaling pathways. Chem. Biol. 14 (7), 860–869.

Rachoń, D., Vortherms, T., Seidlová-Wuttke, D., Wuttke, W., 2008. Effects of black cohosh extract on body weight gain, intra-abdominal fat accumulation, plasma lipids and glucose tolerance in ovariectomized Sprague–Dawley rats. Maturitas 60 (3–4), 209–215.

Reame, N.E., Lukacs, J.L., Padmanabhan, V., et al., 2008. Black cohosh has central opioid activity in postmenopausal women: evidence from naloxone blockade and positron emission tomography neuroimaging. Menopause 15 (5), 832–840.

Ren, W.C., Ma, X.X., Yu, J.L., et al., 2014. Identification of *Cimicifugae rhizoma* and its adulterants using ITS2 sequence. Zhongguo Zhong Yao Za Zhi 39 (12), 2184–2188.

Ross, S.M., 2012. Menopause: a standardized isopropanolic black cohosh extract (remifemin) is found to be safe and effective for menopausal symptoms. Holist. Nurs. Pract. 26 (1), 58–61.

Rostock, M., Fischer, J., Mumm, A., et al., 2011. Black cohosh (*Cimicifuga racemosa*) in tamoxifen-treated breast cancer patients with climacteric complaints – a prospective observational study. Gynecol. Endocrinol. 27 (10), 844–848.

Schmid, D., Woehs, F., Svoboda, M., et al., 2009a. Aqueous extracts of *Cimicifuga racemosa* and phenolcarboxylic constituents inhibit production of proinflammatory cytokines in LPS-stimulated human whole blood. Can. J. Physiol. Pharmacol. 87 (11), 963–972.

Schmid, D., Gruber, M., Woehs, F., et al., 2009b. Inhibition of inducible nitric oxide synthesis by *Cimicifuga racemosa* (*Actaea racemosa*, black cohosh) extracts in LPS-stimulated RAW 264.7 macrophages. J. Pharm. Pharmacol. 61 (8), 1089–1096.

Seidlová-Wuttke, D., Thelen, P., Wuttke, W., 2006. Inhibitory effects of a black cohosh (*Cimicifuga racemosa*) extract on prostate cancer. Planta Med. 72 (6), 521–526.

Seidlova-Wuttke, D., Jarry, H., Wuttke, W., 2009. Effects of estradiol benzoate, raloxifen and an ethanolic extract of *Cimicifuga racemosa* in nonclassical estrogen regulated organs of ovariectomized rats. Planta Med. 75 (12), 1279–1285.

Seidlova-Wuttke, D., Stecher, G., Kammann, M., et al., 2012a. Osteoprotective effects of *Cimicifuga racemosa* and its triterpene-saponins are responsible for reduction of bone marrow fat. Phytomedicine 19 (10), 855–860.

Seidlova-Wuttke, D., Eder, N., Stahnke, V., et al., 2012b. *Cimicifuga racemosa* and its triterpene-saponins prevent the Metabolic Syndrome and deterioration of cartilage in the knee joint of ovariectomized rats by similar mechanisms. Phytomedicine 19 (8–9), 846–853.

Shahin, A.Y., Ismail, A.M., Zahran, K.M., Makhlouf, A.M., 2008. Adding phytoestrogens to clomiphene induction in unexplained infertility patients – a randomized trial. Reprod. Biomed. Online 16 (4), 580–588.

Shams, T., Setia, M.S., Hemmings, R., et al., 2010. Efficacy of black cohosh-containing preparations on menopausal symptoms: a meta-analysis. Altern. Ther. Health Med. 16 (1), 36–44.

Shen, J.W., 2010. Studies on the chemical constituents of two traditional medicine. PhD Dissertation, Northwest Institute of Plateau Biology, Chinese Academy of Sciences.

Shen, L., Zhao, Y.Y., Xie, H.P., Liu, W.H., 2013. Classification of *Cimicifuga* species based on ^{1}H-NMR fingerprint combined with pattern recognition technique. Zhongguo Zhong Yao Za Zhi 38 (2), 217–222.

Smith, M.J., Germolec, D.R., Frawley, R.P., White Jr., K.L., 2013. Immunomodulatory effects of black cohosh (*Actaea racemosa*) extract in female B6C3F1/N mice. Toxicology 308, 146–157.

Soler, M.C., Molina, J.L., Díaz, H.A., et al., 2011. Effect of the standardized *Cimicifuga foetida* extract on Hsp 27 expression in the MCF-7 cell line. Biol. Res. 44 (3), 243–249.

Spiering, M.J., Urban, L.A., Nuss, D.L., et al., 2011. Gene identification in black cohosh (*Actaea racemosa* L.): expressed sequence tag profiling and genetic screening yields candidate genes for production of bioactive secondary metabolites. Plant Cell Rep. 30 (4), 613–629.

Strommer, B., Khom, S., Kastenberger, I., et al., 2014. A cycloartane glycoside derived from *Actaea racemosa* L. modulates GABAA receptors and induces pronounced sedation in mice. J. Pharmacol. Exp. Ther. 351 (2), 234–242.

Stute, P., Nisslein, T., Götte, M., et al., 2007. Effects of black cohosh on estrogen biosynthesis in normal breast tissue in vitro. Maturitas 57 (4), 382–391.

Sun, L.R., Qing, C., Zhang, Y.L., et al., 2007. Cimicifoetisides A and B, two cytotoxic cycloartane triterpenoid glycosides from the rhizomes of *Cimicifuga foetida*, inhibit proliferation of cancer cells. Beilstein J. Org. Chem. 3, 3.

Sun, L., Yan, J., Nian, Y., et al., 2008. New triterpene diglycosides from the rhizome of *Cimifuga foetida*. Molecules 13 (8), 1712–1721.

Sun, L.R., Yan, J., Zhou, L., et al., 2011. Two new triterpene glycosides with monomethyl malonate groups from the rhizome of *Cimicifuga foetida* L. Molecules 16 (7), 5701–5708.

Tamura, K., Peterson, D., Peterson, N., et al., 2011. MEGA5: molecular evolutionary genetics analysis using maximum likelihood, evolutionary distance, and maximum parsimony methods. Mol. Biol. Evol. 28, 2731–2739.

Teschke, R., Schwarzenboeck, A., Schmidt-Taenzer, W., et al., 2011. Herb induced liver injury presumably caused by black cohosh: a survey of initially purported cases and herbal quality specifications. Ann. Hepatol. 10 (3), 249–259.

Tian, Z., Xiao, P.G., Wen, J., et al., 2006a. Review of bioactivities of natural cycloartane triterpenoids. Chin. J. Chin. Materia Med. 31, 625–629.

Tian, Z., Zhou, L., Huang, F., Chen, S., Yang, J., Wu, E., Xiao, P.G., Yang, M., 2006b. Anticancer activity and mechanisms of 25-anhydrocimigenol-3-O-beta-D-xylopyranoside isolated from *Souliea vaginata* on hepatomas. Anticancer Drugs 17 (5), 545–551.

Tian, Z., Pan, R., Chang, Q., Si, J., Xiao, P.G., Wu, E., 2007a. *Cimicifuga foetida* extract inhibits proliferation of hepatocellular cells via induction of cell cycle arrest and apoptosis. J. Ethnopharmacol. 114 (2), 227–233.

Tian, Z., Si, J.Y., Chang, Q., Zhou, L., Chen, S., Xiao, P.G., Wu, E., 2007b. Antitumor activity and mechanisms of action of total glycosides from aerial part of *Cimicifuga dahurica* targeted against hepatoma. BMC Cancer 7, 237.

Verbitski, S.M., Gourdin, G.T., Ikenouye, L.M., et al., 2008. Detection of *Actaea racemosa* adulteration by thin-layer chromatography and combined thin-layer chromatography-bioluminescence. J. AOAC Int. 91 (2), 268–275.

Wang, W., Lu, A.M., Ren, Y., et al., 2009. Phylogeny and classification of Ranunculales: evidence from four molecular loci and morphological data. Persp. Plant Ecol. Evol. Syst. 11 (2), 81–110.

Wang, X., Li, X., Chen, D., 2011. Evaluation of antioxidant activity of isoferulic acid in vitro. Nat. Prod. Commun. 6 (9), 1285–1288.

Wang, K.C., Chang, J.S., Chiang, L.C., Lin, C.C., 2012a. *Cimicifuga foetida* L. inhibited human respiratory syncytial virus in HEp-2 and A549 cell lines. Am. J. Chin. Med. 40 (1), 151–162.

Wang, K.C., Chang, J.S., Lin, L.T., et al., 2012b. Antiviral effect of Cimicifugin from *Cimicifuga foetida* against human respiratory syncytial virus. Am. J. Chin. Med. 40 (5), 1033–1045.

Wang, F., Zhao, S., Li, F., et al., 2014. Investigation of antioxidant interactions between *Radix Astragali* and *Cimicifuga foetida* and identification of synergistic antioxidant compounds. PLoS One 9 (1), e87221.

Watanabe, K., Mimaki, Y., Sakuma, C., Sashida, Y., 2003. Eranthisaponins A and B, two new bisdesmosidic triterpene saponins from the tubers of *Eranthis cilicica*. J. Nat. Prod. 66 (6), 879–882.

Xue, C.Y., Li, D.Z., Wang, Q.Z., 2009. Application of LightCycler polymerase chain reaction and melting curve analysis to the authentication of the traditional Chinese medicinal plant *Cimicifuga foetida*. Planta Med. 75 (8), 873–875.

Yang, C.L., Chik, S.C., Li, J.C., et al., 2009. Identification of the bioactive constituent and its mechanisms of action in mediating the anti-inflammatory effects of black cohosh and related *Cimicifuga* species on human primary blood macrophages. J. Med. Chem. 52 (21), 6707–6715.

Yawata, A., Matsuhashi, Y., Kato, H., et al., 2009. Inhibition of nucleoside transport and synergistic potentiation of methotrexate cytotoxicity by cimicifugoside, a triterpenoid from *Cimicifuga simplex*. Eur. J. Pharm. Sci. 38 (4), 355–361.

Ye, L., Hu, Z., Du, G., et al., 2012. Antidepressant-like effects of the extract from *Cimicifuga foetida* L. J. Ethnopharmacol. 144 (3), 683–691.

Yim, S.H., Kim, H.J., Park, S.H., et al., 2012. Cytotoxic caffeic acid derivatives from the rhizomes of *Cimicifuga heracleifolia*. Arch. Pharm. Res. 35 (9), 1559–1565.

Yoshimitsu, H., Nishida, M., Nohara, T., 2007. Three new 15,16-seco-cycloartane glycosides from *Cimicifuga* rhizome. Chem. Pharm. Bull. 55 (5), 789–792.

Zadoyan, G., Fuhr, U., 2012. Phenotyping studies to assess the effects of phytopharmaceuticals on in vivo activity of main human cytochrome p450 enzymes. Planta Med. 78 (13), 1428–1457.

Zhang, F., Han, L., Pan, G., et al., 2013. A new phenolic amide glycoside from *Cimicifuga dahurica*. Acta Pharm. Sin. 48 (8), 1281–1285.

Zhou, L., Yang, J., Zou, J., Tu, G., 2004. Three new cycloartane triterpene glycosides from *Souliea vaginata*. Chem. Pharm. Bull. 52 (5), 622–624.

Zhou, L., Yang, J., Tu, G., Zou, J., 2006. Cyclolanostane triterpene glycosides from *Souliea vaginata*. Chem. Pharm. Bull. 54 (6), 823–826.

Zhu, D., Nian, Y., Wang, H., et al., 2014. New 9, 19-cycloartane triterpenoid from the root of *Cimicifuga foetida*. Chin. J. Nat. Med. 12 (4), 294–296.

Chemical and biological research of *Clematis* medicinal resources

9

9.1 Introduction

Clematis is a genus of about 355 species (Wang and Li, 2005) within the buttercup family Ranunculaceae. The genus is composed of mostly vigorous, woody, climbing vines/lianas, which are mainly distributed in the temperate regions of the northern hemisphere. *Clematis* is a botanical source for various pharmaceutically active components, which has been used in traditional Chinese medicine for thousands of years. Roots, rhizomes, and stems of some *Clematis* species are used to dispel wind-damp, unblock channels, and relieve pain. Wei Ling Xian and Chuan Mu Tong, recorded in the Chinese Pharmacopoeia, are the most famous Chinese herbal medicines obtained from *Clematis* species. The former is from *Clematis chinensis*, *C. hexapetala*, and *C. mandshurica*, while the latter is from *C. armandii* and *C. montana*. Traditionally, *Clematis* medicine is used orally to treat syphilis, gout, rheumatism, bone disorders, and chronic skin conditions and as a diuretic. In folk medicine, *Clematis* is used topically for blisters and as a poultice to treat purulent wounds and ulcers. Different *Clematis* species may have different pharmacological effects. However, it is quite difficult to authenticate *Clematis* species only by morphology.

To date, at least 30 *Clematis* species have been characterized for their chemical components. However, it is essential to study more species for both the sustainable utilization of *Clematis* medicinal resources and finding novel compounds with potential clinical utility. In this brief chapter, we focus on recent progress in phytochemistry and chemotaxonomy of *Clematis*, as well as molecular taxonomy and pharmacology.

9.2 Chemical components

9.2.1 Triterpenoid saponins

Triterpenoid saponins are triterpenes that belong to the group of saponin compounds. Triterpenes belong to a large group of compounds arranged in a four- or five-ring configuration of 30 carbons with several oxygens attached. The aglycone of *Clematis* plants varies, such as oleanolic acid type (Figure 9.1, A), olean-3β, 28-diol type (B), hederagenin type (C), and hederagenin-11,13-dien type (D). The conjugated glycosyl groups include glucose (Glc, D configuration), rhamnose (Rha, L configuration), galactose (Gal, D configuration), arabinose (Ara, L configuration), xylose (Xyl, D configuration), and ribose (Rib, D configuration). Novel triterpenoid saponins isolated from *Clematis* plants since 2009 are shown in Table 9.1. Types A and C are

Medicinal plants: chemistry, biology and omics. http://dx.doi.org/10.1016/B978-0-08-100085-4.00009-8

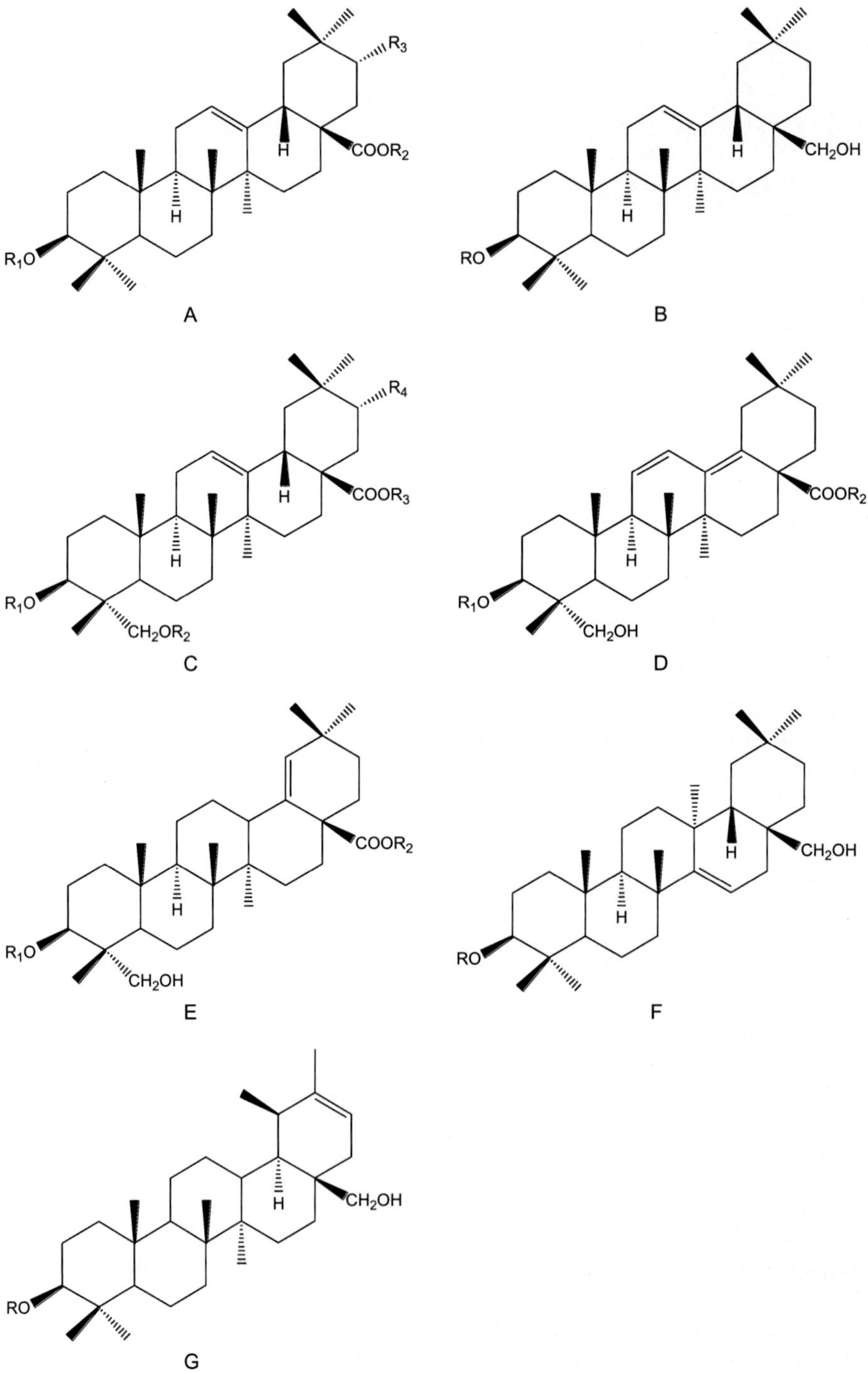

Figure 9.1 Aglycone and the substituent linked with glycosyl groups of triterpenoid saponins of *Clematis*. A, oleanolic acid type; B, olean-3β, 28-diol type; C, hederagenin type; D, hederagenin-11,13-dien type; E, olean-18-en type; F, taraxerane type; G, ursan-20-en type. CA, caffeoyl; IF, isoferuloyl; MC, p-methoxy cinnamyl; DMC, 3,4-dimethoxy cinnamyl.

(Continued)

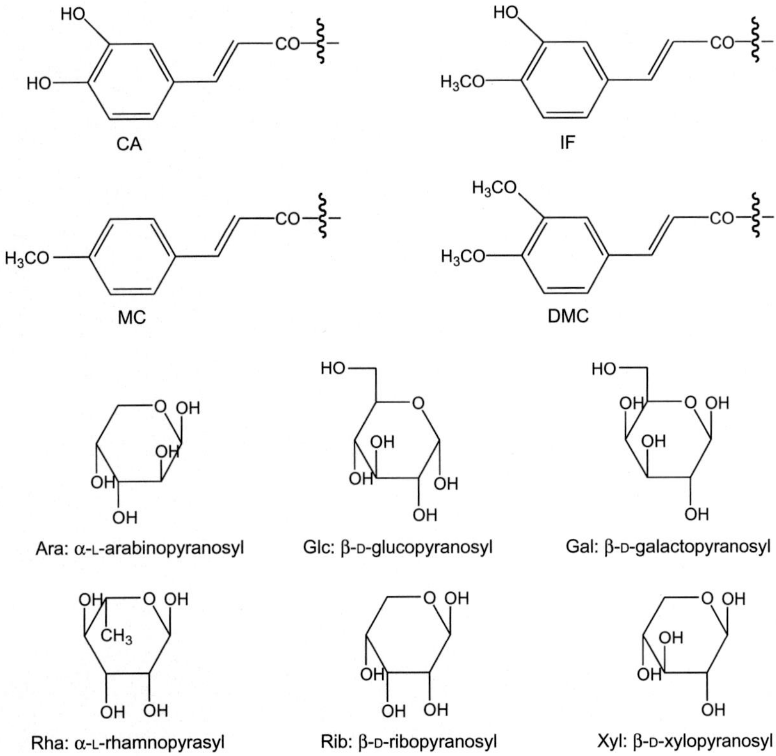

Figure 9.1 Continued.

predominant in *Clematis* plants (Sun and Yang, 2009; Hai et al., 2012). Most of these saponins are bidesmosidic, that is, C-3 and C-28 link with the oligosaccharide chains. The monodesmosidic saponin is less common. Some oligosaccharide chains are substituted with acetyl, caffeoyl (CA), isoferuloyl (IF), *p*-methoxy cinnamyl (MC), and 3,4-dimethoxy cinnamyl (DMC) groups (Figure 9.1). In addition, more than 20 secondary glycosides have been found (Sun and Yang, 2009), which lose their C-28 oligosaccharide chains after hydrolysis.

9.2.2 Flavonoids and anthocyanins

Flavonoids (or bioflavonoids) are a class of plant secondary metabolites. More than 50 flavonoid compounds have been isolated or detected from *Clematis* (Table 9.2 and Figure 9.2; Sun and Yang, 2009; Wang et al., 2009a,b; Du et al., 2010; Hashimoto et al., 2011; Sakaguchi et al., 2012), the aglycones of which are mainly apigenin, kaempferol, luteolin, and quercetin. Sugar moieties are connected to the aglycone through either the oxygen atom or the carbon atom. Flavonoids of *Clematis* are categorized, according to chemical structure, into flavonols, flavones, flavanones, isoflavones and their glycosides, xanthones, and anthocyanidins.

Table 9.1 Triterpenoid saponins found in *Clematis* medicinal plants in recent years

	Compound	Type	Oligosaccharide chain	Species	Tissue	References
1	3β-*O*-{β-D-Ribopyranosyl-(1→3)-α-L-rhamnopyranosyl-(1→2)-[β-D-glucopyranosyl-(1→4)]-β-D-xylopyranosyl} oleanolic acid	A	R_1 = Rib(1→3)-Rha(1→2)-Glc(1→4)-Xyl R_2 = H R_3 = H	a	Roots	Hai et al. (2012)
2	Mandshunoside B	A	R_1 = Glc(1→4)-Glc(1→4)-Rib(1→3)-Rha(1→2)-Ara R_2 = H R_3 = H	b	Roots, rhizomes	He et al. (2011)
3	Clematochinenoside F	A	R_1 = Glc(1→4)-Glc(1→4)-Rib(1→3)-Rha(1→2)-Ara R_2 = Rha(1→4)-Glc(1→6)-Glc R_3 = H	c	Roots, rhizomes	Fu et al. (2010)
4	3β-*O*-[(α-L-Arabinopyranosyl)-oxy]-olean-12-en-28-oic acid	A	R_1 = Ara R_2 = H R_3 = H	d	Roots, rhizomes	Ding et al. (2009)
5	3β-*O*-α-L-Rhamnopyranosyl-(1→2)-α-L-arabinopyranosyl oleanoic acid	A	R_1 = Rha(1→2)-Ara R_2 = H R_3 = H	d	Roots, rhizomes	Ding et al. (2009)
6	Clematomandshurica saponin E	A	R_1 = Rha(1→6)-Glc(1→4)-Glc(1→4)-Rib(1→3)-Rha(1→2)-Ara R_2 = Glc(1→6)-Glc R_3 = H	b	Roots, rhizomes	Dong et al. (2010)
7	Clematichinenoside AR_2	A	R_1 = Rha(1→6)-Glc(1→4)-Glc(1→4)-Rib(1→3)-Rha(1→2)-Ara R_2 = Rha(1→4)-Glc(1→6)-Glc R_3 = OH	c	Roots	Liu et al. (2009)

8	Clematomandshurica saponin F	A	R_1 = Glc(1 → 3)-Rha(1 → 2)-Glc(1 → 4)-Ara R_2 = Rha(1 → 2)-Rha(1 → 4)-Glc(1 → 6)-Glc R_3 = H	b	Roots, rhizomes	Gong et al. (2013)
9	Clematomandshurica saponin G	A	R_1 = Glc(1 → 4)-Rib(1 → 3)-Rha(1 → 2)-Rha (1 → 6)-Glc(1 → 4)-Ara R_2 = Rha(1 → 4)-Glc(1 → 6)-Glc R_3 = H	b	Roots, rhizomes	Gong et al. (2013)
10	Clematomandshurica saponin H	A	R_1 = Glc(1 → 4)-Rib(1 → 3)-Rha(1 → 2)-Glc (1 → 4)-Ara R_2 = Rha(1 → 4)-Glc(1 → 6)-Glc R_3 = H	b	Roots, rhizomes	Gong et al. (2013)
11	Clematomandshurica saponin I	A	R_1 = Rib(1 → 3)-Rha(1 → 2)-Glc(1 → 4)-Ara R_2 = Rha(1 → 4)-Glc(1 → 6)-Glc R_3 = H	b	Roots, rhizomes	Gong et al. (2013)
12	Clematomandshurica saponin J	A	R_1 = Rha(1 → 6)-IF(1 → 2)-Glc(1 → 4)-Rib (1 → 3)-Rha(1 → 2)-Glc(1 → 4)Ara R_2 = H R_3 = H	b	Roots, rhizomes	Gong et al. (2013)
13	Clematomandshurica saponin K	A	R_1 = Rha(1 → 6)-IF(1 → 2)-Glc(1 → 4)-Rib (1 → 3)-Rha(1 → 2)-Ara R_2 = H R_3 = H	b	Roots, rhizomes	Gong et al. (2013)
14	3β-*O*-β-D-Ribopyranosyl oleanolic acid 28-*O*-α-L-rhamnopyranosyl-(1 → 4)-β-D-glucopyranosyl-(1 → 6)-β-D-glucopyranoside	A	R_1 = Rib R_2 = Rha(1 → 4)-Glc(1 → 6)-Glc R_3 = H	f	Whole plants	Zhang et al. (2013d)
15	3-*O*-β-D-Ribopyranosyl-(1 → 3)-α-L-rhamnopyranosyl-(1 → 2)-β-D-xylopyranosyl oleanolic acid 28-*O*-β-D-glucopyranosyl ester	A	R_1 = Rib(1 → 3)-Rha(1 → 2)-Xyl R_2 = Glc R_3 = H	g	Whole plants	Tian et al. (2013)

Continued

Table 9.1 Continued

	Compound	Type	Oligosaccharide chain	Species	Tissue	References
16	3-*O*-β-D-Ribopyranosyl-(1→3)-α-L-rhamnopyranosyl-(1→2)-α-L-arabinopyranosyl oleanolic acid 28-*O*-β-D-Glucopyranosyl-(1→6)-β-D-glucopyranoside	A	R_1=Rib(1→3)-Rha(1→2)-Ara R_2=Glc(1→6)-Glc R_3=H	a	Roots	Zhao et al. (2014a)
17	3-*O*-β-D-Ribopyranosyl-(1→3)-α-L-rhamnopyranosyl-(1→2)-[β-D-glucopyranosyl-(1→4)]-α-L-arabinopyranosyl oleanolic acid 28-*O*-α-L-rhamnopyranosyl-(1→4)-β-D-glucopyranosyl-(1→6)-β-D-glucopyranoside	A	R_1=Rib(1→3)-Rha(1→2)-[Glc(1→4)]-Ara R_2=Rha(1→4)-Glc(1→6)-Glc R_3=H	a	Roots	Zhao et al. (2014a)
18	3-*O*-β-D-Ribopyranosyl-(1→3)-α-L-rhamnopyranosyl-(1→2)-β-D-xylopyranosyl-3β,21α-dihydroxyolean-12-en-28-oic acid	A	R_1=Rib(1→3)-Rha(1→2)-Xyl R_2=H R_3=OH	a	Roots	Zhao et al. (2014a)
19	3-*O*-β-D-Ribopyranosyl-(1→3)-α-L-rhamnopyranosyl-(1→2)-[β-D-glucopyranosyl-(1→4)]-α-L-arabinopyranosyl oleanolic acid	A	R_1=Rib(1→3)-Rha(1→2)-[Glc(1→4)]-Ara R_2=R_3=H	a	Roots	Zhao et al. (2014a)
20	Clematiunicinoside A	A	R_1=Rib(1→3)-Rha(1→2)-[Glc(1→4)]-Ara R_2=Rha(1→4)-Glc(1→6)-Glc R_3=H	h	Roots	Li et al. (2014)
21	Clematiunicinoside C	A	R_1=Rib(1→3)-Rha(1→2)-Ara R_2=Rha(1→4)-Glc(1→6)-Glc R_3=OH	h	Roots	Li et al. (2014)
22	Clematiunicinoside D	A	R_1=Glc(1→4)-Rib(1→3)-Rha(1→2)-Xyl R_2=Rha(1→4)-Glc(1→6)-Glc R_3=H	h	Roots	Li et al. (2014)

23	Clematiunicinoside E	A	R_1 = Glc(1 → 4)-Glc(1 → 4)-Rib(1 → 3)-Rha (1 → 2)-Ara R_2 = Rha(1 → 4)-Glc(1 → 6)-Glc R_3 = H	h	Roots	Li et al. (2014)
24	Clematiunicinoside F	A	R_1 = Glc(1 → 3)-Glc(1 → 4)-Glc(1 → 4)-Rib (1 → 3)-Rha(1 → 2)-Ara R_2 = Rha(1 → 4)-Glc(1 → 6)-Glc R_3 = H	h	Roots	Li et al. (2014)
25	Clematiunicinoside G	A	R_1 = Glc(1 → 3)-[Rha(1 → 6)]-Glc(1 → 4)-Glc(1 → 4)-Rib(1 → 3)-Rha(1 → 2)-Ara R_2 = Rha(1 → 4)-Glc(1 → 6)-Glc R_3 = H	h	Roots	Li et al. (2014)
26	Clematiunicinoside H	A	R_1 = Glc(1 → 3)-IF(1 → 2)-Glc(1 → 4)-Glc (1 → 4)-Rib(1 → 3)-Rha(1 → 2)-Ara R_2 = Rha(1 → 4)-Glc(1 → 6)-Glc R_3 = H	h	Roots	Li et al. (2014)
27	3β-*O*-β-D-Xylopyranosyl-(1 → 3)-α-L-rhamnopyranosyl-(1 → 2)-[β-D-glucopyranosyl]-12-oleanene-3β, 28-diol	B	R = Xyl(1 → 3)-Rha(1 → 2)-Glc	a	Roots	Zhao et al. (2012)
28	Mandshunosides A	C	R_1 = Glc(1 → 4)-Glc(1 → 4)-Rib(1 → 3)-Rha (1 → 2)-Ara R_2 = R_3 = R_4 = H	b	Roots, rhizomes	He et al. (2011)
29	Clematochinenoside A	C	R_1 = Glc(1 → 2)-Rha(1 → 6)-Glc(1 → 3)-Glc (1 → 4)-Glc(1 → 4)-Rib(1 → 3)-Rha(1 → 2)-Ara R_2 = R_4 = H R_3 = Rha(1 → 4)-Glc(1 → 6)-Glc	c	Roots, rhizomes	Fu et al. (2010)

Continued

Table 9.1 Continued

	Compound	Type	Oligosaccharide chain	Species	Tissue	References
30	Clematochinenoside B	C	R_1 = Glc(1 → 2)-Rha(1 → 6)-Glc(1 → 3)-Glc (1 → 4)-Glc(1 → 4)-Rib(1 → 3)-Rha(1 → 2)-Ara R_2 = R_4 = H R_3 = Rha(1 → 4)-Glc(1 → 6)-Glc	c	Roots, rhizomes	Fu et al. (2010)
31	Clematochinenoside C	C	R_1 = Rha(1 → 6)-Glc(1 → 4)-Glc(1 → 3)-Glc (1 → 4)-Glc(1 → 4)-Rib(1 → 3)-Rha(1 → 2)-Ara R_2 = R_4 = H R_3 = Rha(1 → 4)-Glc(1 → 6)-Glc	c	Roots, rhizomes	Fu et al. (2010)
32	Clematochinenoside D	C	R_1 = Glc(1 → 4)-Glc(1 → 4)-Rib(1 → 3)-Rha (1 → 2)-Ara R_2 = R_4 = H R_3 = Rha(1 → 4)-Glc(1 → 6)-Glc	c	Roots, rhizomes	Fu et al. (2010)
33	Clematochinenoside E	C	R_1 = Glc(1 → 4)-Glc(1 → 4)-Rib(1 → 3)-Rha (1 → 2)-Ara R_2 = R_4 = H R_3 = Rha(1 → 4)-Glc(1 → 6)-Glc	c	Roots, rhizomes	Fu et al. (2010)
34	Clematochinenoside G	C	R_1 = Glc(1 → 3)-Glc(1 → 4)-Glc(1 → 4)-Rib (1 → 3)-Rha(1 → 2)-Ara R_2 = R_4 = H R_3 = Rha(1 → 4)-Glc(1 → 6)-Glc	c	Roots, rhizomes	Fu et al. (2010)
35	Hederagenin 3β-*O*-α-L-arabinopyranoside	C	R_1 = Ara R_2 = R_3 = R_4 = H	d	Roots, rhizomes	Ding et al. (2009)
36	α-Hederin	C	R_1 = R_2 = R_3 = R_4 = H	d	Roots, rhizomes	Ding et al. (2009)

37	Parvilobaside A	C	R_1 = Rib(1 → 3)-Rha(1 → 2)-Ara R_2 = Ac R_3 = R_4 = H	e	Stems	Yan et al. (2009)
38	3-*O*-β-D-Xylopyranosyl-(1 → 2)-α-L-arabinopyranosyl hederagenin 28-*O*-α-L-rhamnopyranosyl-(1 → 4)-β-D-glucopyranosyl-(1 → 6)-β-D-glucopyranoside	C	R_1 = Xyl(1 → 2)-Ara R_2 = R_4 = H R_3 = Rha(1 → 4)-Glc(1 → 6)-Glc	f	Whole plants	Zhang et al. (2013d)
39	3-*O*-{β-D-Xylopyranosyl-(1 → 2)-[β-D-glucopyranosyl-(1 → 4)]-α-L-arabinopyranosyl} hederagenin 28-*O*-α-L-rhamnopyranosyl-(1 → 4)-β-D-glucopyranosyl-(1 → 6)-β-D-glucopyranoside	C	R_1 = Xyl(1 → 2)-Glc(1 → 4)-Ara R_2 = R_4 = H R_3 = Rha(1 → 4)-Glc(1 → 6)-Glc	f	Whole plants	Zhang et al. (2013d)
40	Clematangoside B	C	R_1 = Rha(1 → 2)-Ara R_2 = H R_3 = Rha(1 → 4)-Glc(1 → 6)-Glc R_4 = OH	f	Whole plants	Zhang et al. (2013b)
41	Clematangoside C	C	R_1 = R_2 = R_4 = H R_3 = Glc(1 → 6)-Glc(1 → 4)-Rha(1 → 4)-Glc (1 → 6)-Glc	f	Whole plants	Zhang et al. (2013b)
42	Clematangoside D	C	R_1 = R_2 = R_4 = H R_3 = Rha(1 → 4)-Glc(1 → 6)-Glc(1 → 4)-Rha (1 → 4)-Glc(1 → 6)-Glc	f	Whole plants	Zhang et al. (2013b)
43	3-*O*-β-D-Ribopyranosyl-(1 → 3)-α-L-rhamnopyranosyl-(1 → 2)-[β-D-glucopyranosyl-(1 → 4)]-β-D-xylopyranosyl hederagenin	C	R_1 = Rib(1 → 3)-Rha(1 → 2)-[Glc(1 → 4)]-Xyl R_2 = R_3 = R_4 = H	g	Whole plants	Tian et al. (2013)
44	3-*O*-β-D-Ribopyranosyl-(1 → 3)-α-L-rhamnopyranosyl-(1 → 2)-β-D-xylopyranosyl hederagenin	C	R_1 = Rib(1 → 3)-Rha(1 → 2)-Xyl R_2 = R_3 = R_4 = H	g	Whole plants	Tian et al. (2013)

Continued

Table 9.1 Continued

	Compound	Type	Oligosaccharide chain	Species	Tissue	References
45	3-*O*-β-D-Ribopyranosyl-(1→3)-α-L-rhamnopyranosyl-(1→2)-[β-D-glucopyranosyl-(1→4)]-α-L-arabinopyranosyl hederagenin	C	R_1=Rib(1→3)-Rha(1→2)-[Glc(1→4)]-Ara R_2=R_3=R_4=H	g	Whole plants	Tian et al. (2013)
46	Clematochinenoside H	C	R_1=Rha(1→6)-Glc(1→3)-Glc(1→4)-Glc(1→4)-Rib(1→3)-Rha(1→2)-Ara R_2=R_4=H R_3=Rha(1→4)-Glc(1→6)-Glc	c	Roots, rhizomes	Fu et al. (2013)
47	Clematochinenoside I	C	R_1=Glc(1→2)-Rha(1→6)-[Glc(1→4)]-Glc(1→3)-IF(1→2)-Glc(1→4)-Glc(1→4)-Rib(1→3)-Rha(1→2)-Ara R_2=R_4=H R_3=Rha(1→4)-Glc(1→6)-Glc	c	Roots, rhizomes	Fu et al. (2013)
48	Clematochinenoside J	C	R_1=Glc(1→2)-Glc(1→4)-[Rha(1→6)]-Glc(1→3)-IF(1→2)-Glc(1→4)-Glc(1→4)-Rib(1→3)-Rha(1→2)-Ara R_2=R_4=H R_3=Rha(1→4)-Glc(1→6)-Glc	c	Roots, rhizomes	Fu et al. (2013)
49	Clematiunicinoside B	C	R_1=Rib(1→3)-Rha(1→2)-[Glc(1→4)]-Ara R_2=R_4=H R_3=Rha(1→4)-Glc(1→6)-Glc	h	Roots	Li et al. (2014)
50	3β-*O*-[β-D-Ribopyranosyl-(1→3)-α-L-rhamnopyranosyl-(1→2)-α-L-arabinopyranosyl] hederagenin-11,13-dien-28-oic acid	D	R_1=Rib(1→3)-Rha(1→2)-Ara R_2=H	a	Roots	Hai et al. (2012)

51	Clematangoside A	E	R_1 = Rha(1 → 2)-Ara R_2 = Rha(1 → 4)-Glc(1 → 6)-Glc	f	Whole plants	Zhang et al. (2013b)
52	3β-*O*-[β-D-Xylopyranosyl-(1 → 3)-α-L-rhamnopyranosyl-(1 → 2)-β-D-glucopyranosyl]-28-hydroxytaraxeran-14-en	F	R = Xyl(1 → 3)-Rha(1 → 2)-Glc	a	Roots	Zhao et al. (2014b)
53	3β-*O*-[β-D-Xylopyranosyl-(1 → 3)-α-L-rhamnopyranosyl-(1 → 2)-β-D-glucopyranosyl]-28-hydroxy-18αH-ursan-20-en	G	R = Xyl(1 → 3)-Rha(1 → 2)-Glc	a	Roots	Zhao et al. (2014b)

Aglycone: A, oleanolic acid type; B, olean-3β, 28-diol type; C, hederagenin type; D, hederagenin-11,13-dien type; E, olean-18-en type; F, taraxerane type; G, ursan-20-en type.
a, *C. argentilucida*; b, *C. mandshurica*; c, *C. chinensis*; d, *C. ganpiniana*; e, *C. parviloba*; f, *C. tangutica*; g, *C. lasiandra*; h, *C. uncinata*.
The locations of R_1, R_2, and R_3 are shown in Figure 9.1.

Table 9.2 Other compounds found in *Clematis* medicinal plants in recent years

No.	Compound	Type	Species	Tissue	References
54	Kaempferol 3-*O*-α-L-rhamnopyranosyl-(1→6)-β-D-glucopyranoside-7,4′-di-*O*-β-D-glucopyranoside	Flavonol glycoside	Cultivars	Flowers	Sakaguchi et al. (2013)
55	(6-*O*-β-D-Glucopyranosyl-1H-indol-3-yl) carboxylic acid methyl ester	Indole alkaloid	*C. terniflora* DC.	Aerial parts	Li et al. (2013)
56	(7*R*,8*S*)-9-Acetyl-dehydrodiconiferyl alcohol	Lignan	*C. armandii*	Stems	Xiong et al. (2014)

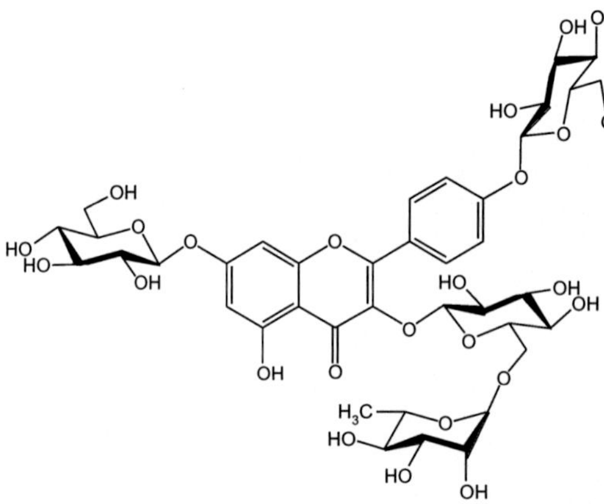

54

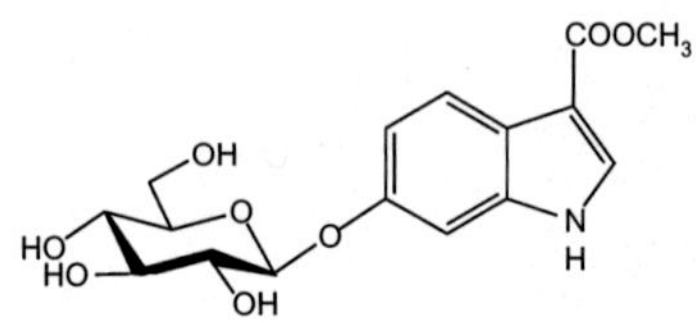

55

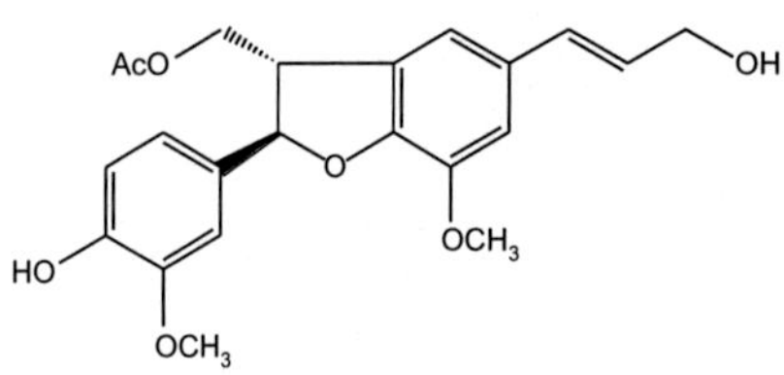

56

Figure 9.2 Other compounds isolated from *Clematis* species.

9.2.3 Others

Twenty-four lignans isolated from *Clematis* are mainly eupomatene lignans, cyclolignans, monoepoxylignans, bisepoxylignans, and lignanolides (Sun and Yang, 2009). Six coumarin compounds have been isolated (Figure 9.3; Sun and Yang, 2009; Li et al., 2009). Eleven alkaloids isolated from *Clematis* fall into two categories: aporphine and terpenoid alkaloid.

Protoanemonin concentrations varied between different *Clematis* species, and leaves contained higher concentrations than stems and roots (Jin et al., 2013).

Many other compounds were also isolated from various *Clematis* species, such as phenolic glycosides (Yan et al., 2010), volatile oils, triterpenes, steroids, organic acids and their derivatives (Yang et al., 2009), macrocyclic compounds, and others (Wang et al., 2009a,b). Their functions and pharmacological effects await further studies.

9.3 Bioactivity and therapeutic use

9.3.1 Anti-inflammatory activity

The triterpene saponin AR-6 from *C. chinensis* has a potential anti-inflammatory effect in rats with collagen-induced arthritis (CIA), which inhibits the expression of nuclear factor kappa B (NF-κB) p65 subunits, TNF-α, and COX-2 (Peng et al., 2012). Antiarthritic effects of AR-6 correlated with significant decrease of NO and TNF-α produced by peritoneal macrophages (Sun et al., 2010c). AR-6 also significantly decreased the proliferation of synoviocyte. Multiple triterpene saponins of *C. chinensis* showed the potential anti-inflammatory activity with inhibitory activities against COX-1 and COX-2 enzymes (Fu et al., 2010). Saponins might be responsible for the inhibitory effects of *C. chinensis* on the proinflammatory and degradative

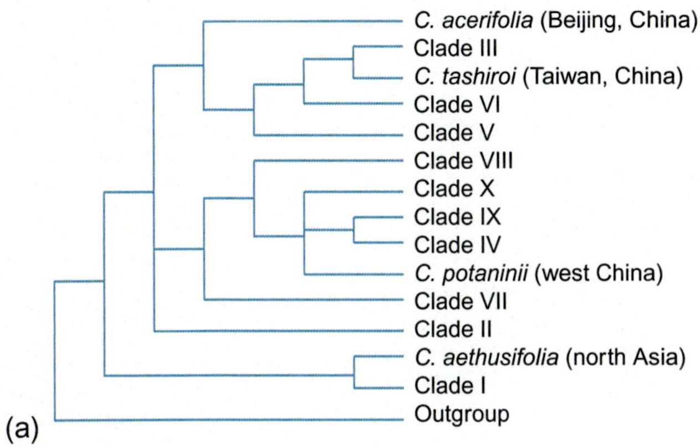

Figure 9.3 Phylogenetic relationship of *Clematis* inferred from (a) ITS + chloroplast atpB-rbcL + rpoB-trnC + trnQ-psbA (according to Xie et al., 2011).

(Continued)

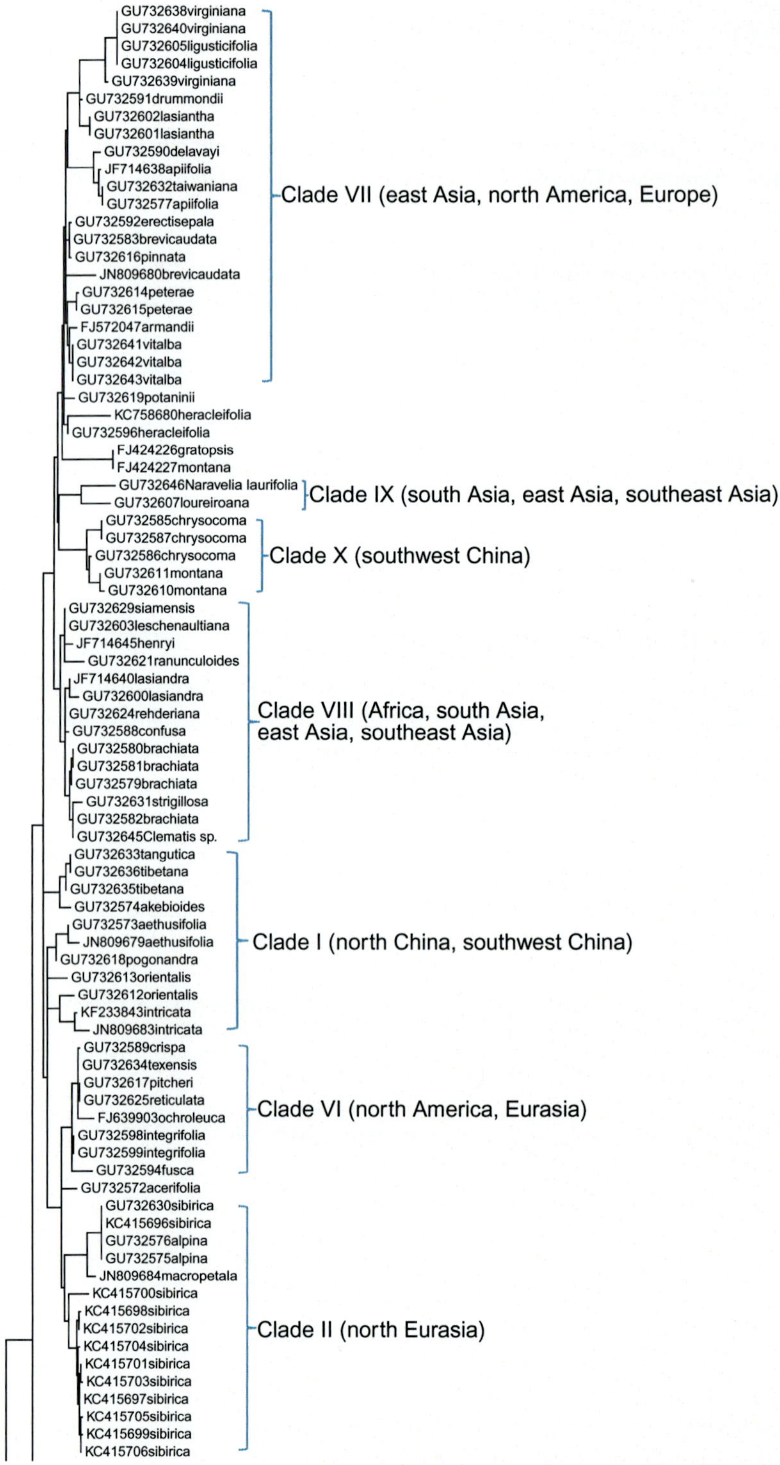

Figure 9.3 Continued.

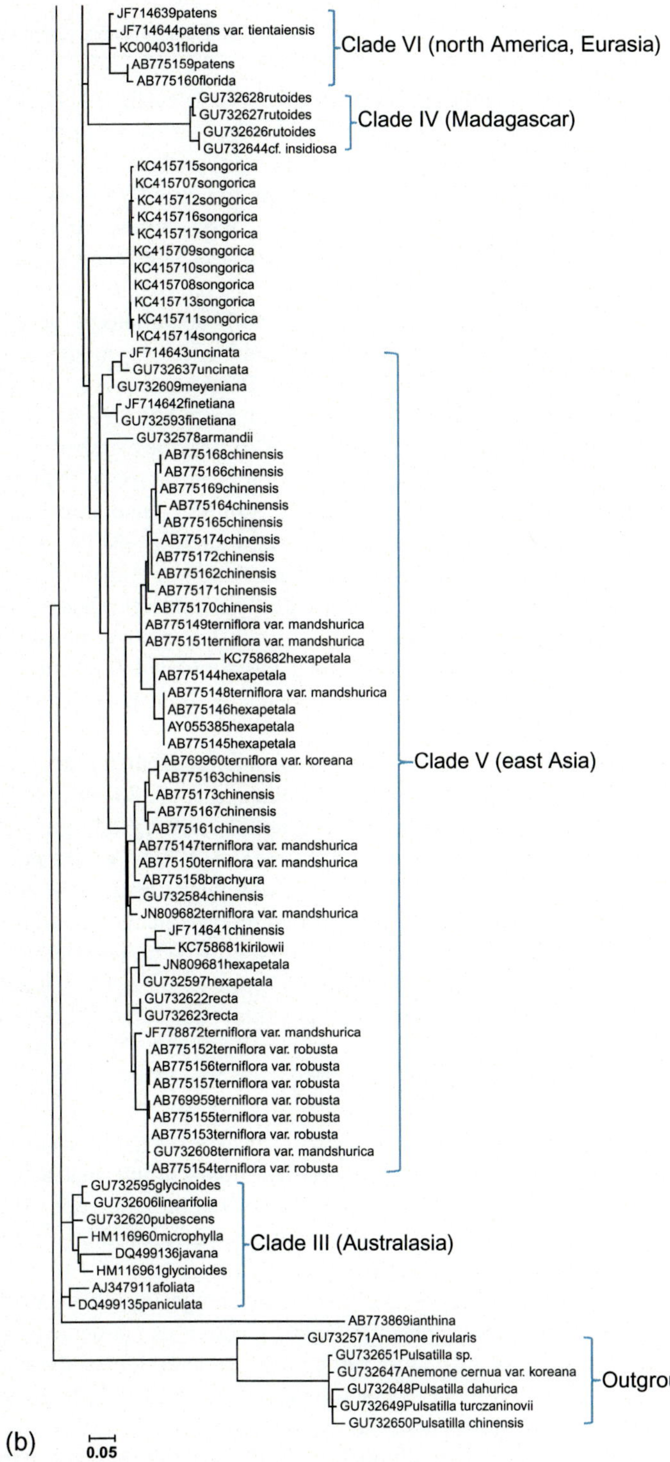

Figure 9.3 Continued. (b) ITS. The ML analysis involved 167 nucleotide sequences; ML phylogenetic tree is shown. NCBI accession number of each sequence is shown in front of species name.

mediators associated with inflammatory arthritis (Hsieh et al., 2011). The saponin fraction from this plant showed the therapeutic effect on osteoarthritis induced by monosodium iodoacetate through protecting articular cartilage (Wu et al., 2010).

Clematichinenoside (AR-6) is extracted from the dried root and rhizome of *C. chinensis* Osbeck, *C. hexapetala* Pall. or *C. manshurica*. Administration of AR-6 (32, 16, and 8 mg/kg), especially the high dose level of 32 mg/kg, significantly suppressed the swelling of hind paws of CIA rats ($p < 0.01$) and inhibited their body weight loss ($p < 0.01$) (Han et al., 2013). Based on histopathologic observation, all AR-6 groups showed great amelioration compared with the model group. AR-6 significantly reduced the production of TNF-α, PI3K, and p-Akt expression and decreased TNF-α mRNA, PI3K mRNA, and p-Akt mRNA in CIA rat synovium.

Treatment with α-viniferin, an oligostilbene of trimeric resveratrol, isolated from *C. mandshurica* decreased LPS-induced production of nitric oxide (NO) and prostaglandin E2 (PGE2) (Dilshara et al., 2014a). α-Viniferin also downregulated the LPS-induced expression of proinflammatory genes such as iNOS and COX-2 by suppressing the activity of NF-κB via dephosphorylation of Akt/PI3K.

α-l-Rhamnopyranosyl-(1 → 6)-β-D-glucopyranosyl-3-indolecarbonate (RG3I) of *C. mandshurica* suppresses the production of proinflammatory mediators such as NO and PGE2 and their regulatory genes in LPS-stimulated BV2 microglial cells by inhibiting the PI3K/Akt- and ROS-dependent NF-κB signaling pathway (Dilshara et al., 2014b).

Trans-isoferulic acid (TIA) of *C. mandshurica* suppresses the production of proinflammatory mediators such as NO and PGE2 (Dilshara et al., 2014c), as well as their regulatory genes, in LPS-stimulated BV2 microglial cells, by inhibiting PI3K/Akt-dependent NF-κB activity and enhancing Nrf2-mediated HO-1 expression.

The anti-neuroinflammatory bioassay showed that lignan (7*R*,8*S*)-9-acetyl-dehydrodiconiferyl alcohol, (7*R*,8*S*)-dehydrodiconiferyl alcohol, erythro-guaiacyl-glycerol-β-coniferyl ether, and threo-guaiacylglycerol-β-coniferyl ether displayed significant inhibitory effects on NO production (Xiong et al., 2014). Two neolignans exhibited more potent activities than the positive control (*N*(*G*)-monomethyl-L-arginine (L-NMMA)), with an IC50 value of 9.3 and 3.9 μM. These neolignans were also found to concentration-dependently suppress the TNF-α release in LPS-stimulated BV2 cells.

The saponin-rich fraction from *C. chinensis* Osbeck roots (SFC) could effectively alleviate experimental osteoarthritis induced by monosodium iodoacetate in rats through protecting articular cartilage and inhibiting local inflammation. SFC protects rabbit chondrocytes against NO-induced apoptosis via preventing mitochondrial impairment and caspase-3 activation (Wu et al., 2013).

C. mandshurica root is widely used in Asia as an analgesic and anti-inflammatory agent. The root extract exhibits anti-inflammatory activities via decreasing the production of proinflammatory mediators through the suppression of the pathways of NF-κB and MAPKs in LPS-induced macrophages (Lee et al., 2014). Results of the CA-induced rat hind paw edema assay show an antiedema effect of *C. mandshurica* root. These findings support the traditional use of *C. mandshurica* root in the inflammatory symptoms of rheumatic arthritis and acute icteric hepatitis.

SKI306X, a mixed extract of three herbs, *C. mandshurica* (CM), *Prunella vulgaris* (PV), and *Trichosanthes kirilowii* (TK), is chondroprotective in animal models of osteoarthritis (OA) (Choi et al., 2014). SKI306X inhibited IL-1β-induced glycosaminoglycan (GAG) release from cartilage explants, while SKI306X and CM inhibited IL-1β-induced matrix metalloproteinase (MMP) gene expression. Unexpectedly, SKI306X greatly stimulated IL-1β+oncostatin M-induced ADAMTS (a disintegrin and metalloproteinase with thrombospondin motifs)-4 gene expression. Some fractions of SKI306X also inhibited ADAMTS-4 activity.

Flavonoids were the main active compounds in four preparative extracts of *C. terniflora*, which might be responsible for antinociceptive activity and anti-inflammatory activity in rats with carrageenan-induced chronic nonbacterial prostatitis (Chen et al., 2011). Anti-inflammatory, antinociceptive, and antipyretic properties of the aqueous extract of *C. brachiata* leaf in male rats may be due in part to the flavonoids (Mostafa et al., 2010).

9.3.2 Antioxidant activity and effects on cardiovascular system

Oxidative stress and excessive NO production are thought to play considerable roles in ischemia/reperfusion (I/R) injury that impairs cardiac function. Clematichinenoside, a triterpenoid saponin isolated from the roots of *C. chinensis* (Wei Ling Xian), attenuated infarct size; decreased lactate dehydrogenase, creatine kinase, and malondialdehyde levels; and enhanced superoxide dismutase activity (Zhang et al., 2013a). Clematichinenoside improved hemodynamic indexes, such as left ventricular developed pressure, maximum left ventricular developed pressure, and increase/decrease rate in the isolated guinea pig heart after reperfusion. Clematichinenoside also inhibited excessive production of NO through downregulating inducible NOS and upregulating endothelial NOS during I/R injury. Clematichinenoside attenuates I/R injury *in vivo*, *ex vivo*, and *in vitro* via an antioxidant effect and restoration of the balance between inducible NOS and endothelial NOS.

Five triterpenoid saponins of *C. tangutica* exhibited antimyocardial ischemia activities with ED50 values in the range of 75.77–127.22 μM (Zhang et al., 2013b). Seven saponins of *C. tangutica* exhibited cardioprotective effects by decreasing the levels of creatine kinase-MB (CK-MB) and lactate dehydrogenase (LDH) (Zhang et al., 2013d).

Acidic proteoglycan of *C. huchouensis* showed excellent antioxidant activity in three *in vitro* assays (Zhang et al., 2014). Two flavone C-glycosides from *C. rehderiana* showed potent antioxidant activities (Du et al., 2010).

9.3.3 Anticancer activity

Seven monodesmosidic saponins exhibited cytotoxic activity against the human glioblastoma U251MG cells with IC50 values ranging from 6.95 to 38.51 μM (Zhao et al., 2014b). Eight monodesmosidic saponins exhibited cytotoxic activity against the HL-60, HepG-2, and SGC-7901 cell lines with IC50 values in the range of 0.87–19.48 μM (Zhao et al., 2014a). Three bidesmosidic triterpenoid saponins of *C. uncinata* exhibited inhibitory effect on CaSki cervical cancer cells (Li et al., 2014). Five triterpenoid saponins from the whole plant of *C. lasiandra* showed significant cytotoxicity to human tumor cell lines HL-60, Hep-G2, and SGC-7901 with IC50 1.40–9.50 μM (Tian et al., 2013). Three triterpene saponins of *C. mandshurica* exhibited antiproliferative effects against PC-3 human prostate cancer cells with GI50 values of 1.29, 1.50, and 0.71 μM, respectively (Gong et al., 2013).

Two new saponins from *C. argentilucida* exhibited significant cytotoxicity against human leukemia cells, hepatocellular carcinoma cells, and glioblastoma cells (Hai et al., 2012). Triterpene saponins from *C. mandshurica* inhibit human colon cancer cells (He et al., 2011). Four triterpene glycosides from *C. ganpiniana* showed cytotoxicity against cancer cells and antibacterial activity (Ding et al., 2009).

α-Hederin, a monodesmosidic triterpenoid saponin distributed in *Clematis*, *Hedera*, or *Nigella* species, displays many biological activities. α-Hederin of *C. ganpiniana* showed strong inhibitory activity on the growth of breast cancer cells and induced apoptosis in these cells (Cheng et al., 2014a). α-Hederin induced depolarization of mitochondrial membrane potential, which released Apaf-1 and cytochrome c from the intermembrane space into the cytosol, where they promoted caspase-3 and caspase-9 activation.

An oleanane-type triterpene saponin, D Rhamnose β-hederin (DRβ-H), was isolated from *C. ganpiniana*, a Chinese traditional antitumor herb (Cheng et al., 2014b). DRβ-H showed strong inhibitory activity on the growth of various breast cancer cells and induced apoptosis in these cells. DRβ-H inhibited PI3K/Akt and activated ERK signaling pathway. The PI3K inhibitor LY294002 synergistically enhanced DRβ-H-induced apoptosis, whereas the MEK inhibitor U0126 reduced the apoptosis rate. DRβ-H regulated the ratio of proapoptotic and antiapoptotic Bcl-2 family proteins. Furthermore, DRβ-H induced depolarization of mitochondrial membrane potential, which released Apaf-1 and cytochrome c from the intermembrane space into the cytosol, where they promoted caspase-9 and caspase-3 activation.

An indole alkaloidal glucoside and eleutheroside E, isolated from the aerial parts of *C. terniflora*, showed significant cytotoxicity against human ECA-109 (Li et al., 2013).

C. montana lectin (CML), a novel mannose-binding lectin purified from *C. montana* stem, has hemagglutinating activity in rabbit erythrocytes and apoptosis-inducing activity in tumor cells (Lu et al., 2014).

9.3.4 Other effects

C. longicauda Steud. ex A. Rich. and *C. burgensis* Engl. are used in southwest Ethiopia to treat otorrhea and eczema. Antimicrobial activity and MIC of crude extracts were determined by disk diffusion and broth dilution (Hawaze et al., 2012). The methanol and petroleum ether extracts of both plants showed antibacterial and antifungal activities. Sensitivity of reference strains was concentration-dependent. Methanol and petroleum ether extracts of *C. burgensis* leaves exerted greater inhibitory effects than *C. longicauda* extracts, whereas aqueous extracts of both plants were inactive. The MIC study revealed a concentration of 0.78 mg/ml on bacteria and 3.125 mg/ml on fungi for methanol extract and 1.56 mg/ml on both fungi and bacteria for petroleum ether extract. Phytochemical screening indicated the presence of proteins, fixed oils, carbohydrates, tannins, saponins, flavonoids, and steroids. Preliminary chromatographic investigation showed fluorescing spots with *R(f)* values that ranged from 0.05 to 0.96 for phenolic compounds and saponins.

Two aporphine alkaloids showed potent antifungal activities (Chen et al., 2009). A novel mannose-binding lectin was isolated from *C. montana* and showed antiviral and apoptosis-inducing activities (Peng et al., 2009).

The formulation obtained from Mediterranean dietary plant *C. vitalba* exhibited the strongest inhibitory effect on pancreatic lipase (IC50 0.99 mg/ml) and on α-amylase (IC50 31.52 μg/ml) (Anonymous, 2013). HPTLC analysis of the extracts revealed the predominance of (±)-catechin, caffeic acid, and chlorogenic acid in *C. vitalba* formulation at concentration of 23.18 ± 3.14, 13.63 ± 0.65, and 18.88 ± 0.76 mg/g, respectively. GC/MS analysis identified fatty acids and terpene composition.

9.4 Chemotaxonomy

The genus *Clematis* was established by Linnaeus (1753) and nine species were depicted. Further taxonomic studies increased the number of species: Johnson (1997) summarized 314 species and Grey-Wilson (2000) 297 species; Wang and Li (2005) described 355 species in the new classification of the genus, with ~150 species distributed in China. This system proposed four subgenera and ten sections based on some important morphological characteristics of vegetative and reproductive organs. The four subgenera are (1) *Cheiropsis* Peterm., section *Cheiropsis* DC.; (2) *Clematis* Keener & Dennis, sect. *Clematis* Tamura, *Meclatis* Baill., *Fruticella* Tamura, *Naraveliopsis* Hand.-Mazz, *Viticella* DC., and *Tubulosae* Decne.; (3) *Viorna* Gray, sect. *Viorna* Prantl and *Archiclemaits* Tamura; and (4) *Atragene* Torr. & Gray, sect. *Atragene* DC. There is no unified recognition of the taxonomy, derivation, and systematic position of this genus.

According to pharmacophylogeny, plants with close genetic relationship are similar in systematic classification and chemical constituents (Xiao et al., 2006, 2007). A Chemotaxonomic study of sect. *Viorna*, subsect. *Viornae* was performed using flavonoid distribution patterns (Dennis and Bierner, 1980). The supposed close link between *C. viorna* and *C. reticulata* from their morphology and distribution was validated by the similarities of their flavonoid profiles. The hypothesis that *C. glaucophylla* recently originated from *C. versicolor* was supported by their identical flavanoid profiles, and the diversity in flavonoid profiles of *C. pitcheri* specimens was found to parallel their morphological variations. Lebreton (1986) used flavonoids as chemical markers to discuss the taxonomy of the Ranunculaceae. Feng et al. (2011) compared the total flavonoid content from aerial parts of 11 *Clematis* species. *C. apiifolia* leaves had the highest (5.581%) total flavonoid, while *C. huchouensis* had only 0.195%. The amount of flavones in different species was quite different. The cluster analysis based on the flavonoid content differentiated the species of the subgenus *Clematis* and those of the subgenus *Viorna* into two groups, with the exceptions of *C. terniflora* and *C. huchouensis*. This study suggests that the aerial parts of *C. apiifolia* and *C. finetiana* could be used for flavonoid drug development.

In a taxonomic study, saponins were used as chemotaxonomic markers in differentiating various taxa of plants (Fay and Dale, 1993; Jiang et al., 2008; Kim et al., 2010). Although saponins have been isolated from different *Clematis* species (Sun and Yang, 2009; Wang et al., 2009a,b; Table 9.1 and Figure 9.2), only a limited number of them have been used for the chemotaxonomic study of *Clematis*. Ishtiaq et al. (2010a,b) performed a chemotaxonomic analysis of 15 populations of 12 species, belonging to sections *Rectae*, *Clematis*, *Meclatis*, *Tubulosae*, and *Viorna*, with high-performance liquid chromatography (HPLC) coupled with diode array detector and electrospray ionization (ESI)–mass spectrometry (MS) (Sun et al., 2007). Nine saponins were identified. A chemical cluster tree was constructed based on the presence or absence of saponins and their peak area size in different taxa. Three clades were identified. Clade I includes species of sect. *Rectae* of subgenus *Clematis*, *C. finetiana*, *C. armandii*, *C. chinensis*, and *C. terniflora*, which share the following five saponins: Huzhangoside B (HGB), Clematichinenoside C (CCC), Seiboldianoside A (SDA), Huzhangoside D (HGD), and Clematichinenoside B (CCB). Clade II consists of *C. apiifolia*, *C. argentilucida*, *C. ganpiniana*, and *C. peterae*, which belong to sect. *Clematis* and share saponins CCB, HGD, CCC, and HGB. *C. henryi* of subgenus *Viorna* shares HGD and Clemochinenoside A (CCA) with *C. heracleifolia* of subgenus *Clematis*, and they were grouped together in clade III. This paradox implies that more saponins and/or other secondary metabolites should be used in the chemotaxonomic analysis. *C. huchouensis* of subgenus *Clematis* was quite different in total flavonoid (Feng et al., 2011) and saponin composition, which may have distinct secondary metabolic pathway.

Many studies have investigated the chemical constituents and pharmacological effects of *Radix et Rhizoma Clematidis* (Wei Ling Xian), *Caulis Clematidis Armandii* (Chuan Mu Tong), and other *Clematis* species. Triterpenoid saponins are rich in many species (Table 9.1) and contribute multiple pharmacological effects. However, crude drugs derived from different species are also traded in the market without authentication of efficacy, and these may be misused clinically. Therefore, it is necessary to determine the distribution and the contents of the pharmaceutical ingredients in the approved *Clematis* species for quality control. Sun et al. (2008) developed a simple and accurate method involving HPLC with evaporative light-scattering detection for the simultaneous determination of five triterpenoid saponins in *Clematis*. Ten *Clematis* samples were analyzed under optimized conditions, and the chemical profiles were used to identify the botanical origin, the development of new medicinal resources, and the chemotaxonomic study. The contents of 2 (HGD) and 3 (3-*O*-α-L-ara(2-1)-α-L-rha(4-1)-β-D-rib pulsatiloside C) were much higher in *C. chinensis* than in other species, which corresponds to its effects in clinical use. However, these saponins were not detected in the aerial part of *C. manshurica*, suggesting that the aerial part of this plant should not be used as a source of the drug *Radix et Rhizoma Clematidis*. This study also suggested the necessity of the simultaneous determination of these components to improve the quality control of this traditional Chinese drug. The triterpenoid saponins were rich in *C. ganpiniana* and *C. apiifolia* var. *obtusidentata*, which could be developed as new medicinal resources. No triterpenoid saponins were detected in *C. armandii*, which agreed with previous chemical studies (Huang et al., 2003). Therefore, its pharmacological effects might be related to other components. The contents of the five compounds in the roots and stems of *C. ganpiniana*, *C. apiifolia*, and *C. apiifolia* var. *obtusidentata* were higher than those in the whole plants except for 1 (CCB) and 4 (CCC) in *C. apiifolia*, which implies that the root and stem might be a better medicinal part than the aerial part.

The research group led by Cheng Yi-Yu and Xiao Pei-Gen performed a series of *Clematis* studies. A simple and rapid gradient HPLC–MS method was developed and validated for concurrent determination of four triterpenoid saponins, that is, CCB, HGD, CCC, and HGB, in *Radix Clematidis* and related *Clematis* species (Sun et al., 2010a). Under the optimized conditions, the analytes were well separated in 25 min. The results showed that the samples contain a wide range (0.0652–41.7 mg/g) of the four analytes. *Radix Clematidis* (Tianmu Mountain, Zhejiang, China) has the highest saponin content (41.7 mg/g), which was about seven times that in *Radix Clematidis* (commodity of Chinese market) (5.80 mg/g). This might be due to different harvest times and places, suggesting that the source of *Radix Clematidis* should be restricted. This method has good linearity, high sensitivity, precision, and accuracy and consumes shorter time, which is helpful in controlling the quality of traditional Chinese medicine derived from *Clematis*. Triterpenoid saponins were detected by qualitative HPLC–MSn analysis of root and rhizome from 18 *Clematis* samples and the whole plant of *C. puberula* var. *ganpiniana* and *C. terniflora* (Sun et al., 2010b). The HPLC–MSn detected 17 oleanolic acid or hederagenin saponins. Saponin profiles were used for phylogenetic studies. All root and rhizome samples could be grouped into three branches based on the 17 peaks. The first group included *C. lasiandra* (S1; subgenus *Viorna*), *C. finetiana* (S5), *C. terniflora* (S21; whole plant), *C. grandidentata* (S9), and the outgroup *R. sieboldii* (S22). The second group included *C. puberula* var. *ganpiniana* (S2), *C. peterae* var. *trichocarpa* (S4), *C. courtoisii* (S7), *C. huchouensis* (S10), *C. chinensis* (S11), *C. hexapetala* (S13), *C. uncinata* (S12), and *C. mandshurica* (S14 and S15), which belong to subgenus *Clematis*. The third group included *C. uncinata* (S3), *C. terniflora* (S6), *C. chinensis* (S8), *C. puberula* var. *ganpiniana* (S19 and S20), *C. apiifolia* (S18), and *C. argentilucida* (S17), which also belong to subgenus *Clematis* and are closer to the second group than to the first group. The two *C. chinensis* taxa (S8 and S11) did not form one group, reflecting different harvest time and region. The similar reason is valid for two samples of *C. uncinata*. HGB could be unequivocally identified in all taxa. All 17 compounds were present in *C. puberula* var. *ganpiniana*, *C. apiifolia*, and *C. argentilucida*. Whether these saponins are suitable for chemotaxonomy of the whole genus needs to be studied thoroughly and more taxa should be included. Flavonoids and lignans are important secondary metabolites in *Clematis*, which should be examined in more detail to see if they reveal any more information about the relationships among *Clematis* species.

Proteins are essential parts of living organisms, since they are the main components of the physiological metabolic pathways of cells. Proteins can be biomarkers for identification of botanic drugs. Two-dimensional gel electrophoresis of leaf protein was used to explore the phylogenetic relationship among species of section *Clematis* (Ishtiaq et al., 2010a,b). A high level of polymorphism was present among the 1085 spots scored: only 255 spots were common to all. Nine proteins were solely present in *C. chinensis* and absent in *C. finetiana* and *C. armandii*, although they are closer to each other than to other taxa. *C. apiifolia*, *C. peterae*, *C. argentilucida*, and *C. ganpiniana* cluster together. The genetic distance 0.4–0.45 and 0.25–0.55 was observed at intra-subsection level in *Clematis* and *Rectae*. Proteomics is a promising taxonomic tool and the differentially expressed proteins provide clues for exploring the difference in metabolic pathway of different species. In addition, Shi et al. (2006) suggested that the macrocyclic glucosides from *C. chinensis*, *C. armandii*, *C. hexapetala*, and *C. mandshurica* provided a significant reference point for the chemotaxonomy of sect. *Clematis*. Despite these studies, the relationships among the *Clematis* species have not been resolved. Molecular methods have to be combined with morphology and chemotaxonomy in the elucidation of the pharmacophylogeny of *Clematis*.

9.5 Molecular taxonomy and molecular phylogeny

9.5.1 Molecular marker

Random amplified polymorphic DNA (RAPD) is a kind of PCR, but the DNA fragments that are amplified are haphazard (Hao et al., 2010). The arbitrary and short primers (8–12 nucleotides) are used in the PCR, which uses a large template of genomic DNA. By resolving the resulting patterns, a semiunique profile can be garnered from a RAPD reaction. There are 59 species and 24 varieties in 6 sections of 3 subgenera in Yunnan (Jiang et al., 2007), 56 of which are endemic to China and 16 of which are only found in Yunnan. RAPD was used to study 12 *Clematis* species of Yunnan (Pu et al., 2008). Selected ten RAPD primers were used in the random amplification and 89 polymorphic DNA bands were detected. The twelve species can be unambiguously identified. More importantly, the cluster tree largely reflected the phylogenetic relationship of the studied species. *C. napaulensis*, *C. montana*, and *C. chrysocoma* of subgenus *Cheiropsis*, the most primitive group, are basal to other taxa. *C. buchananiana*, *C. connata*, and *C. ranunculoides* of subgenus *Viorna* are basal to subgenus *Clematis*, which includes *C. puberula* var. *ganpiniana*, *C. metuoensis*, *C. meyeniana*, *C. gouriana*, and *C. peterae*. *Caulis clematidis armandii* (Chuan Mu Tong) is an important Chinese herbal medicine. The botanical origins of *Caulis clematidis armandii* are the dried stems of *C. armandii* and *C. montana*. Due to the similarity in morphology, it is hard to differentiate *C. armandii* and *C. montana* from other *Clematis* species. The following species are also marketed in China as Chuan Mu Tong (Ma and Zhang, 2002): *C. finetiana*, *C. kerriana*, *C. argentilucida*, *C. peterae*, *C. apiifolia* var. *obtusidentata*, *C. tangutica*, *C. meyeniana*, and *C. uncinata* of subgenus *Clematis*; *C. leschenaultiana*, *C. lasiandra*, *C. henryi*, *C. chrysocoma*, *C. pogonandra*, and *C. trullifera* of subgenus *Viorna*; and *C. macropetala* of subgenus *Atragene*. The distribution area of these species is much smaller than that of *C. montana* and *C. armandii*, and some endangered species should be protected (Chen and Pu, 2006). Moreover, although these species are similar to raw plants of Chuan Mu Tong in drug efficacy, they are sometimes used for specific purposes in some regions and their pharmaceutical ingredients have not been investigated thoroughly. Other plants (e.g., *Aristolochia* and *Akebia*) were also erroneously used as Chuan Mu Tong, which may cause dangerous toxicities (Ma and Zhang, 2002). Guo et al. (2010) established an efficient molecular method to distinguish *Caulis clematidis armandii* from its common adulterants. The dendrogram based on RAPD data showed that evolutionarily, ancient *C. montana* is basal to other taxa and *C. urophylla* and *C. lasiandra* of subgenus *Viorna* are basal to the seven species of subgenus *Clematis*, which is not incongruent with RAPD results of Pu et al. (2008). *C. armandii* is closer to *C. finetiana* than to *C. argentilucida*, *C. peterae* var. *trichocarpa*, and *C. apiifolia*, which well agrees the chemotaxonomic results of Ishtiaq et al. (2010a,b). Since *C. montana* and *C. armandii* are only distantly related, it is necessary to perform a detailed study to find the physical basis of their common drug efficacy. Two RAPD fragments specific for *C. armandii* and *C. montana* were sequenced and primers were designed for sequence-characterized amplified region (SCAR) markers (Guo et al., 2010). The two SCAR markers were specific to both *C. armandii* and *C. montana* and were not present in other species. This method can be used in quality control of Chuan Mu Tong. RAPD was also used to study the hybrid origin of progeny from crosses of *C. tubulosa* and *C. brevicaudata* (Yuan et al., 2010).

9.5.2 Phylogeny

Internal transcribed spacer (ITS) is a piece of nonfunctional RNA located between structural ribosomal RNAs (rRNA) of a common precursor transcript, which is especially useful for elucidating relationships among congeneric species and closely related genera (Hao et al., 2010). ITS sequences were PCR-amplified from leaf genomic DNA of eight *Clematis* species of Zhejiang, China (Jiang et al., 2011). The ITS region varied between 534 and 561 bp, with 50 variable sites and 22 parsimony informative sites. On the ITS-based phylogenetic tree, *C. patens* is closer to *C. tientaiensis* than to *C. finetiana* and *C. chinensis* in one group, while *C. henryi* of subgenus *Viorna* is closer to *C. lasiandra* (subgenus *Viorna*) than to *C. apiifolia* and *C. uncinata* of subgenus *Clematis* in the other group. ITS2 is a part of ITS, which is shorter and easier to amplify. The ITS2 region of *Radix et Rhizoma Clematidis* (Wei Ling Xian) and its adulterants varied between 220 and 230 bp (Zeng et al., 2011). On the ITS2-based phylogenetic tree, *C. chinensis* is closer to *C. hexapetala* than to *C. mandshurica*. These official species of Wei Ling Xian form one cluster, while *C. lasiandra* of subgenus *Viorna* is closer to *C. pogonandra* (subgenus *Viorna*) than to *C. uncinata* (subgenus *Clematis*), which does not contradict with ITS results (Jiang et al., 2011). With ITS2, it is also easy to differentiate Wei Ling Xian and plants of *Sarcandra*, *Podophyllum*, *Paeonia*, and *Smilax*, which were inadvertently or deliberately used as Wei Ling Xian in Chinese markets (Zeng et al., 2011). ITS2 was also used to identify Chuan Mu Tong, its adulterants (e.g., *Aristolochia* and *Akebia*), and closely related species (Liu et al., 2011). With ITS2, it is also easy to differentiate Chuan Mu Tong and other *Clematis* species. However, ITS2 cannot distinguish *C. aethusifolia* and *C. pogonandra* (subgenus *Viorna*), *C. leschenaultiana* and *C. siamensis* (subgenus *Viorna*), and *C. tangutica* and *C. tibetana* (subgenus *Clematis*).

Due to low resolving power, the actin I intron is not ideal for phylogenetic study of *Clematis* (Slomba et al., 2004). The nuclear (nr) ITS and five plastid DNA markers were used to infer the evolutionary relationship of 33 *Clematis* species and related taxa (Miikeda et al., 2006). The traditional subgenus *Viorna* and sect. *Atragene* (a subgenus in Wang and Li's system of 2005) were monophyletic, whereas the taxa of subgenus *Clematis*, as well as those of subgenus *Cheiropsis*, did not form a clade. Massive morphological diversification and very few nucleotide substitutions within *Clematis* indicate the recent radiation of the genus, especially in Qinghai-Tibetan Plateau (QTP) and Hengduan Mountains. Based on a sampling of 75 species, sequences of the nrITS, the plastid atpB-rbcL spacer, psbA-trnH-trnQ spacer, and rpoB-trnC spacer regions were analyzed using parsimony, maximum likelihood (ML), and Bayesian inference methods (Xie et al., 2011). Analyses of the combined data set by the three methods generated distinct trees. Ten major clades with various levels of support were detected (Figure 9.3a). The molecular results do not support previous infrageneric classifications based on morphology and suggest significant evolutionary convergence in floral and vegetative characters in *Clematis*. Several clades were resolved as regional geographic groups. *C. montana*, *C. chrysocoma*, and *C. fasciculiflora*, which are in southwest China and are members of subgenus *Cheiropsis*, form clade X and are basal to other clades when maximum parsimony method was used. *C. rehderiana*, *C. leschenaultiana*, *C. ranunculoides*, *C. siamensis*, and *C. lasiandra*, distributed in South, East, and Southeast Asia and members of subgenus *Viorna*, form clade VIII and are basal to subgenus *Clematis* and subgenus *Atragene*, which is quite different from topology inferred by ML method. The members of subgenus *Clematis*, *C. delavayi*, *C. peterae*, *C. heracleifolia*, *C. pinnata*, *C. brevicaudata*, *C. taiwaniana*, and *C. apiifolia*, are in clade VII, while *C. patens* of northeast Asia and *C. florida* of North America cluster in clade VI, implying disjunct distribution. The raw species of Wei Ling Xian and their relatives and *C. armandii* and its relatives with similar drug efficacy are in clade V. On an expanded ITS tree, all ten clades claimed by Xie et al. (2011) are recovered (Figure 9.3b). *C. songarica*, distributed in Central Asia and Mongolia, is basal to clade V. The genus *Naravelia* has nine species, and *N. laurifolia* is embedded in clade IX. More *Naravelia* species should be examined to determine whether this genus should be merged with *Clematis*.

Interestingly, *C. akebioides*, *C. tangutica*, *C. tibetana*, *C. orientalis*, and *C. serratifolia* of clade I are closer to clade II (subgenus *Atragene*) than to other members of subgenus *Clematis* (Figure 9.3). Moreover, relationships among some closely related species are not resolved, for example, *C. tangutica* and *C. tibetana*, *C. apiifolia* and *C. taiwaniana*, *C. brachiata* and *C. strigillosa*, and *C. uncinata* and *C. meyeniana*, suggesting that these taxa may have diverged recently. Bayesian dating suggests a relatively ancient origin of the genus in the Oligocene (25.99 million years ago (mya)) and a relatively recent species radiation in the Miocene (7.81 mya) following the uplift of QTP. Geologic and climatic changes in the late Tertiary to Quaternary (2.588 mya-present) might be important for the speciation of *Clematis*, especially in East Asia. Long-distance dispersal of the fruits by wind, water, and/or animals and strong environmental adaptability might explain the present worldwide distribution and high species diversity.

Yunnan, China, has the most *Clematis* species in the world and the Hengduan Mountains of northwestern Yunnan is regarded as the center of origin, differentiation, and endemism of the genus (Jiang et al., 2007). ITS and three plastid markers, same as those of Xie et al. (2011), were PCR-amplified to study the phylogenetic relationship of *Clematis* (Jiang, 2010). Seventy-three species (132 samples) cluster into 18 clades. *C. potaninii* and *C. fasciculiflora* of subgenus *Cheiropsis* are basal to other taxa. Clade XV containing *C. montana*, *C. gracilifolia*, *C. venusta*, and *C. chrysocoma* is closer to clade XIV of subgenus *Clematis*. Clade XIII containing *C. connata*, *C. rehderiana*, and *C. lasiandra* and clades XII and X containing other members of subgenus *Viorna* cluster together. Clade VIII represents subgenus *Atragene* and is closer to clade VII containing members of *Meclatis*, subgenus *Clematis*, which is congruent with the results of Xie et al. (2011). *C. hexapetala*, *C. terniflora*, *C. armandii*, *C. uncinata*, *C. crassifolia*, and *C. patens* cluster in clades IV and V. This study also suggests a recent species radiation. The studied species are mainly from southwestern China, especially the Hengduan Mountains of northwestern Yunnan. The major diversification of modern *Clematis* occurred in Himalaya-Hengduan Mountains region during periods of the uplift of QTP and the global geologic and ecological changes during the late Tertiary (Xie et al., 2011). Many plant groups had adaptive radiation, differentiation, and speciation after migrating to this region, for example, *Pedicularis* (Yang et al., 2003), *Aconitum* (Hao et al., 2013), *Corydalis*, *Primula*, *Silene*, and *Rhododendron* (Luo et al., 2005). Jiang (2010) found that relationships among some closely related species of the Hengduan Mountains are not resolved, for example, *C. tangutica*, *C. akebioides*, and *C. tibetana* of subgenus *Clematis* and *C. ranunculoides* and *C. yuanjiangensis* of subgenus *Viorna*, suggesting that these taxa might be the result of rapid radiation and speciation. *C. peterae* and *C. gouriana* of southwestern China, as well as *C. lasiandra* and *C. pseudootophora* of subgenus *Viorna*, might also come into being from species radiation. Morphology has diverged significantly in these pairs of species, but the molecular marker sequences have not yet diverged dramatically. It is motivating to study whether the chemical profile has also diverged or not, which could be helpful in choosing the source plant for the specific medicinal use. Moreover, natural hybridization may be common in *Clematis*, since artificial hybridization among closely related species or even among sections is easy and widely applied in horticulture (Johnson, 1997; Yuan et al., 2010). Many cultivars are obtained by hybridization and possess unique pharmaceutical components (Hashimoto et al., 2011; Sakaguchi et al., 2012). Due to the inconsistent position of *C. delavayi* in the chloroplast (cp) DNA and ITS trees, Miikeda et al. (2006) suggested a hybrid origin of this species in the early evolution of the genus. Based on the incongruent position in the cpDNA and ITS trees, Jiang (2010) inferred that *C. delavayi*, *C. peterae*, *C. parviloba* (subgenus *Clematis*), *C. connata*, *C. rehderiana*, and *C. ranunculoides* (subgenus *Viorna*) might have a hybrid origin. ITS and psbA-trnH-trnQ can be used as DNA barcodes in the authentication of *Clematis*.

Molecular phylogeography and species distribution modeling (SDM) were combined to predict the future responses of two *Clematis* species to climate change

(Zhang et al., 2013c), utilizing evidence of responses from the past. Genetic data for *C. sibirica* show a significant phylogeographic signal and demographic contraction during the glacial–interglacial cycles in the Pleistocene. This forest species would experience range reduction, though without genetic loss, in the face of future climate change. In contrast, SDMs, based on the existing potential distribution, predict that *C. songarica*, a steppe species, should maintain a consistently stable potential distribution under the Last Glacial Maximum (LGM) and the future climatic conditions. This species contains a high level of genetic differentiation among populations in cpDNA, likely benefiting from stable habitats over a long time period. Evidence from the molecular phylogeography of these two species, the forest species is more sensitive to past climate changes than the steppe species. SDMs predict that the forest species will face the challenge of potential range contraction in the future more than the steppe species. This provides a perspective on ecological management in arid eastern Central Asia, indicating that increased attention should be paid to montane forest species, due to their high sensitivity to disturbance.

9.6 Conclusions

Clematis is a botanical source for various pharmaceutically active components, which has been used in traditional Chinese medicine for thousands of years. Increasing interest in *Clematis* medicinal resources has led to additional discoveries of triterpenoid saponins, flavonoids, coumarins, alkaloids, and many other compounds in various *Clematis* species and to investigations on their chemotaxonomy, molecular phylogeny, and pharmacology. However, around 90% of all 355 species have not been explored in phytochemistry, molecular biology, and pharmacology, which is a gold mine of myriad medicinal compounds. Chemosystematics tries to classify and identify plants by detecting differences and similarities in their biochemical compositions, which is of great help in pharmaceutical resource discovery. Yet, the results of chemotaxonomy have to be cross-examined with those of molecular taxonomy and traditional morphology-based classification, in order to streamline the quality control and guarantee the authenticity of plant materials used in the clinical setting, research laboratory, and pharmaceutical industry. More importantly, chemical and biological research of *Clematis* medicinal resources should not be restricted in the above aspects, for example, the biosynthetic pathway of secondary metabolites has not been studied; the regulation of *Clematis* biological processes at genomic level, epigenomic level, transcriptional and posttranscriptional levels, and translational and posttranslational levels is totally unknown, although it is essential for the sustainable development and utilization of the *Clematis* medicinal resources. The biological and chemical studies of *Taxus* (Hao et al., 2011) and *Polygonum* (Hao et al., 2012) with the use of omics technologies provide a paradigm of the active integration of various state-of-the-art methodologies into the early stage of the drug research and development. Systems biology and omics techniques will play an increasingly significant role in future medical research involving bioactive compounds of *Clematis*.

References

Anonymous, 2013. Inhibition of key enzymes linked to obesity by preparations from Mediterranean dietary plants: effects on α-amylase and pancreatic lipase activities. Plant Foods Hum. Nutr. 68 (4), 340–346.

Chen, W.Y., Pu, C.X., 2006. Resource investigation of medicinal species of *Clematis* in Yunnan province. J. Yunnan Coll. Chin. Med. 29, 31–33.

Chen, J.H., Du, Z.Z., Shen, Y.M., Yang, Y.P., 2009. Aporphine alkaloids from *Clematis parviloba* and their antifungal activity. Arch. Pharm. Res. 32 (1), 3–5.

Chen, R.Z., Cui, L., Guo, Y.J., et al., 2011. In vivo study of four preparative extracts of *Clematis terniflora* DC. for antinociceptive activity and anti-inflammatory activity in rat model of carrageenan-induced chronic non-bacterial prostatitis. J. Ethnopharmacol. 134 (3), 1018–1023.

Cheng, L., Xia, T.S., Wang, Y.F., et al., 2014a. The anticancer effect and mechanism of α-hederin on breast cancer cells. Int. J. Oncol. 45 (2), 757–763.

Cheng, L., Xia, T.S., Wang, Y.F., et al., 2014b. The apoptotic effect of D Rhamnose β-hederin, a novel oleanane-type triterpenoid saponin on breast cancer cells. PLoS One 9 (6), e90848.

Choi, C.H., Kim, T.H., Sung, Y.K., et al., 2014. SKI306X inhibition of glycosaminoglycan degradation in human cartilage involves down-regulation of cytokine-induced catabolic genes. Korean J. Intern. Med. 29 (5), 647–655.

Dennis, W.M., Bierner, M.W., 1980. Distribution of flavonoids and their systematic significance in *Clematis* subsection *Viornae*. Biochem. Syst. Ecol. 8, 65–67.

Dilshara, M.G., Lee, K.T., Kim, H.J., et al., 2014a. Anti-inflammatory mechanism of α-viniferin regulates lipopolysaccharide-induced release of proinflammatory mediators in BV2 microglial cells. Cell. Immunol. 290 (1), 21–29.

Dilshara, M.G., Lee, K.T., Choi, Y.H., et al., 2014b. Potential chemoprevention of LPS-stimulated nitric oxide and prostaglandin E_2 production by α-L-rhamnopyranosyl-(1→6)-β-D-glucopyranosyl-3-indolecarbonate in BV2 microglial cells through suppression of the ROS/PI3K/Akt/NF-κB pathway. Neurochem. Int. 67, 39–45.

Dilshara, M.G., Lee, K.T., Jayasooriya, R.G., et al., 2014c. Downregulation of NO and PGE2 in LPS-stimulated BV2 microglial cells by trans-isoferulic acid via suppression of PI3K/Akt-dependent NF-κB and activation of Nrf2-mediated HO-1. Int. Immunopharmacol. 18 (1), 203–211.

Ding, Q., Yang, L.X., Yang, H.W., et al., 2009. Cytotoxic and antibacterial triterpenoids derivatives from *Clematis ganpiniana*. J. Ethnopharmacol. 126 (3), 382–385.

Dong, F.Y., Cui, G.H., Zhang, Y.H., et al., 2010. Clematomandshurica saponin E, a new triterpenoid saponin from *Clematis mandshurica*. J. Asian Nat. Prod. Res. 12 (12), 1061–1068.

Du, Z.Z., Yang, X.W., Han, H., et al., 2010. A new flavone C-glycoside from *Clematis rehderiana*. Molecules 15 (2), 672–679.

Fay, M.F., Dale, P.J., 1993. Condensed tannins in *Trifolium* species and their significance for taxonomy and plant breeding. Genet. Resour. Crop. Evol. 40 (1), 7–13.

Feng, L.F., Weng, M.Y., Tian, F.R., Chen, J.H., 2011. Comparison on total flavonoid content from aerial parts of 11 medicinal plants of *Clematis* L. Modern Chin. Med. 13, 14–18.

Fu, Q., Zan, K., Zhao, M., et al., 2010. Triterpene saponins from *Clematis chinensis* and their potential anti-inflammatory activity. J. Nat. Prod. 73 (7), 1234–1239.

Fu, Q., Zan, K., Zhao, M., et al., 2013. Three new triterpene saponins from *Clematis chinensis*. J. Asian Nat. Prod. Res. 15 (6), 610–618.

Gong, Y.X., Hua, H.M., Xu, Y.N., et al., 2013. Triterpene saponins from *Clematis mandshurica* and their antiproliferative activity. Planta Med. 79 (11), 987–994.

Grey-Wilson, C., 2000. *Clematis* the Genus. Timber Press, Portland, Oregon.
Guo, J.L., Ren, Y., Chen, L., et al., 2010. Authentication of *Caulis clematidis armandii* ("Chuanmutong") and differentiation of its common adulterants using RAPD and SCAR markers. J. Med. Plants Res. 4, 697–701.
Hai, W., Cheng, H., Zhao, M., et al., 2012. Two new cytotoxic triterpenoid saponins from the roots of *Clematis argentilucida*. Fitoterapia 83 (4), 759–764.
Han, W., Xiong, Y., Li, Y., et al., 2013. Anti-arthritic effects of clematichinenoside (AR-6) on PI3K/Akt signaling pathway and TNF-α associated with collagen-induced arthritis. Pharm. Biol. 51 (1), 13–22.
Hao, D.C., Chen, S.L., Xiao, P.G., Peng, Y., 2010. Authentication of medicinal plants by DNA-based markers and genomics. Chin. Herb. Med. 2 (4), 250–261.
Hao, D.C., Ge, G.B., Xiao, P.G., et al., 2011. The first insight into the tissue specific taxus transcriptome via Illumina second generation sequencing. PLoS One 6 (6), e21220.
Hao, D.C., Ma, P., Mu, J., Chen, S.L., Xiao, P.G., et al., 2012. De novo characterization of the root transcriptome of a traditional Chinese medicinal plant *Polygonum cuspidatum*. Sci. China Life. Sci. 55 (5), 452–466.
Hao, D.C., Gu, X.J., Xiao, P.G., et al., 2013. Recent advances in the chemical and biological studies of *Aconitum* pharmaceutical resources. J. Chin. Pharm. Sci. 22 (3), 209–221.
Hashimoto, M., Suzuki, T., Iwashina, T., 2011. New acylated anthocyanins and other flavonoids from the red flowers of *Clematis* cultivars. Nat. Prod. Commun. 6 (11), 1631–1636.
Hawaze, S., Deti, H., Suleman, S., 2012. In vitro antimicrobial activity and phytochemical screening of *Clematis* species indigenous to Ethiopia. Indian J. Pharm. Sci. 74 (1), 29–35.
He, Y.X., Li, L., Zhang, K., Liu, Z.R., 2011. Cytotoxic triterpene saponins from *Clematis mandshurica*. J. Asian Nat. Prod. Res. 13 (12), 1104–1109.
Hsieh, M.S., Wang, K.T., Tseng, S.H., et al., 2011. Using 18 F-FDG microPET imaging to measure the inhibitory effects of *Clematis chinensis* Osbeck on the pro-inflammatory and degradative mediators associated with inflammatory arthritis. J. Ethnopharmacol. 136 (3), 511–517.
Huang, W.W., Kong, D.Y., Yang, P.M., 2003. Studies on lignan constituents of *Clematis armandii* Franch. Chin. J. Nat. Med. 1, 199–203.
Ishtiaq, M.Ch., Mumtaz, A.S., Wang, Y., et al., 2010a. Proteins as biomarkers for taxonomic identification of Traditional Chinese Medicines (TCMs) from subsection *Rectae* genus *Clematis* from China. World Appl. Sci. J. 8, 62–70.
Ishtiaq, M.Ch., He, Q., Sun, F., Wang, Y., Xiao, P.G., et al., 2010b. Determination of taxonomic status of Chinese species of the genus *Clematis* by using high performance liquid chromatography-mass spectrometry (HPLC-MS) technique. Pak. J. Bot. 42, 691–702.
Jiang, N., 2010. Molecular Phylogeny and Reproductive Biology of *Clematis* (Ranunculaceae). PhD Dissertation, CAS, Kunming Institute of Botany.
Jiang, N., Guan, K.Y., Wang, Z.L., 2007. Species diversity and geographical distribution of *Clematis* (Ranunculaceae) in Yunnan. Acta Bot. Yunnan. 29 (2), 145–154.
Jiang, D., Shi, S.P., Cao, J.J., et al., 2008. Triterpene saponins from the fruits of *Akebia quinata*. Biochem. Syst. Ecol. 36, 138–141.
Jiang, M., Zhou, Y.Q., Li, R.R., 2011. ITS sequence analysis of eight medicinal plants in *Clematis* L. Chin. Trad. Herb. Drugs 42, 1802–1806.
Jin, F., Narkowicz, C., Jacobson, G.A., 2013. Protoanemonin content variation between *Clematis* spp.: leaf, stem and root. Nat. Prod. Commun. 8 (2), 211–212.
Johnson, M., 1997. Slaktet Klematis. M. Johnsons Plantskola AB, Södertälje.
Kim, H.K., Saifullah, S., Khan, S., et al., 2010. Metabolic classification of South American Ilex species by NMR-based metabolomics. Phytochemistry 71 (7), 773–784.

Lebreton, P., 1986. Flavonoids, taxonomic markers in the Ranunculaceae. Plant Med. Phytother. 20, 275–286.

Lee, C.W., Park, S.M., Kim, Y.S., et al., 2014. Biomolecular evidence of anti-inflammatory effects by *Clematis mandshurica* Ruprecht root extract in rodent cells. J. Ethnopharmacol. 155 (2), 1141–1155.

Li, Y., Wang, S.F., Zhao, Y.L., et al., 2009. Chemical constituents from *Clematis delavayi* var. *spinescens*. Molecules 14 (11), 4433–4439.

Li, W.T., Yang, B.X., Zhu, W., et al., 2013. A new indole alkaloidal glucoside from the aerial parts of *Clematis terniflora* DC. Nat. Prod. Res. 27 (24), 2333–2337.

Li, S.G., Huang, X.J., Li, M.M., et al., 2014. Triterpenoid saponins from the roots of *Clematis uncinata*. Chem. Pharm. Bull. 62 (1), 35–44.

Liu, L.F., Ma, X.L., Wang, Y.X., et al., 2009. Triterpenoid saponins from the roots of *Clematis chinensis* Osbeck. J. Asian Nat. Prod. Res. 11 (5), 389–396.

Liu, M.Z., Li, M.N., Yao, H., Liu, P., 2011. Molecular identification of *Clematidis Armandii Caulis* and its adulterants and closely related species by ITS2 sequence. Global Tradit. Chin. Med. 4, 446–450.

Lu, B., Zhang, B., Qi, W., et al., 2014. Conformational study reveals amino acid residues essential for hemagglutinating and anti-proliferative activities of *Clematis montana* lectin. Acta Biochim. Biophys. Sin. (Shanghai). pii: gmu085 [Epub ahead of print].

Luo, Y., Zhang, F.M., Yang, Q.E., 2005. Phylogeny of *Aconitum* subgenus *Aconitum* (Ranunculaceae) inferred from ITS sequences. Plant Syst. Evol. 252, 11–25.

Ma, H.M., Zhang, B.L., 2002. Comparison among families of Mutong. Chin. J. Chin. Mater. Med. 27, 412–418.

Miikeda, O., Kita, K., Handa, T., Yukawa, T., 2006. Phylogenetic relationships of *Clematis* (Ranunculaceae) based on chloroplast and nuclear DNA sequences. Bot. J. Linn. Soc. 152, 153–168.

Mostafa, M., Appidi, J.R., Yakubu, M.T., Afolayan, A.J., 2010. Anti-inflammatory, antinociceptive and antipyretic properties of the aqueous extract of *Clematis brachiata* leaf in male rats. Pharm. Biol. 48 (6), 682–689.

Peng, H., Lv, H., Wang, Y., et al., 2009. *Clematis montana* lectin, a novel mannose-binding lectin from traditional Chinese medicine with antiviral and apoptosis-inducing activities. Peptides 30 (10), 1805–1815.

Peng, C., Perera, P.K., Li, Y.M., et al., 2012. Anti-inflammatory effects of *Clematis chinensis* Osbeck extract (AR-6) may be associated with NF-κB, TNF-α, and COX-2 in collagen-induced arthritis in rat. Rheumatol. Int. 32 (10), 3119–3125.

Pu, C.X., Yang, Z.Y., Liu, X.L., 2008. RAPD analysis of 12 species of *Clematis*. J. Yunnan Univ. Traditional Chin. Med. 31, 37–41.

Sakaguchi, K., Kitajima, J., Iwashina, T., 2012. Kaempferol tri- and tetraglycosides from the flowers of *Clematis* cultivar. Nat. Prod. Commun. 7 (2), 173–174.

Sakaguchi, K., Kitajima, J., Iwashina, T., 2013. Kaempferol 3,7,4′-glycosides from the flowers of *Clematis* cultivars. Nat. Prod. Commun. 8 (8), 1093–1094.

Shi, S.P., Dong, C.X., Jiang, D., Tu, P.F., 2006. Macrocyclic glucosides from *Clematis mandshurica* and *Clematis hexapetala*. Biochem. Syst. Ecol. 35, 57–60.

Slomba, J., Garey, J.R., Essig, F.B., 2004. The actin I intron—a phylogenetically informative DNA region in *Clematis* (Ranunculaceae). Sida 21 (2), 879–886.

Sun, F., Yang, D.P., 2009. Advance in chemical constituents of genus *Clematis*. Chin. J. Chin. Mater. Med. 34, 2660–2668.

Sun, F., He, Q., Shi, P., Xiao, P.G., Cheng, Y.Y., 2007. Characterization and identification of triterpenoid saponins in crude extracts from *Clematis* spp. by high-performance liquid chromatography/electrospray ionization with multi-stage tandem mass spectrometry. Rapid Commun. Mass Spectrom. 21 (22), 3743–3750.

Sun, F., He, Q., Xiao, P.G., et al., 2008. Simultaneous quantification of five triterpenoid saponins in *Clematis* L. spp. by high-performance liquid chromatography with evaporative light scattering detection. Phytochem. Anal. 19, 40–45.

Sun, F., Cai, Z., Ishtiaq, M.Ch., Xiao, P.G., Cheng, Y.Y., 2010a. Distribution of the triterpenoid saponins and chemotaxonomy of the genus *Clematis* L. by high-performance liquid chromatography–mass spectrometry. Biochem. Syst. Ecol. 38, 1018–1025.

Sun, F., Cai, Z., Yang, D.P., Xiao, P.G., Cheng, Y.Y., 2010b. Determination of triterpenoid saponins in *Radix Clematidis* and related *Clematis* species by high performance liquid chromatography-mass spectrometry. Chin. J. Anal. Chem. 9, 1293–1298.

Sun, S.X., Li, Y.M., Fang, W.R., et al., 2010c. Effect and mechanism of AR-6 in experimental rheumatoid arthritis. Clin. Exp. Med. 10 (2), 113–121.

Tian, X., Feng, J., Tang, H., et al., 2013. New cytotoxic triterpenoid saponins from the whole plant of *Clematis lasiandra* Maxim. Fitoterapia 90, 233–239.

Wang, W.T., Li, L.Q., 2005. A new system of classification of the genus *Clematis* (Ranunculaceae). Acta Phytotax. Sin. 43 (5), 431–488.

Wang, F., Tang, Q.L., Ma, X.L., Liu, L.F., 2009a. Research development on the chemical constituents from *Clematis* spp. Chin. Wild Plant Resour. 28, 1–6.

Wang, X.P., Qian, H.B., Huang, J., et al., 2009b. Comparative study on *Clematis chinensis* and *Clematis uncinata* by HPLC in vitro and in vivo. Lishizhen Med. Materia Med. Res. 20, 2224–2225.

Wu, W., Xu, X., Dai, Y., Xia, L., 2010. Therapeutic effect of the saponin fraction from *Clematis chinensis* Osbeck roots on osteoarthritis induced by monosodium iodoacetate through protecting articular cartilage. Phytother. Res. 24 (4), 538–546.

Wu, W., Gao, X., Xu, X., et al., 2013. Saponin-rich fraction from *Clematis chinensis* Osbeck roots protects rabbit chondrocytes against nitric oxide-induced apoptosis via preventing mitochondria impairment and caspase-3 activation. Cytotechnology 65 (2), 287–295.

Xiao, P.G., Wang, F.P., Gao, F., et al., 2006. A pharmacophylogenetic study of *Aconitum* L. (Ranunculaceae) from China. Acta Phytotax. Sin. 44 (1), 1–46.

Xiao, P.G., Jiang, Y., Li, P., et al., 2007. The botanical origin and pharmacophylogenetic treatment of Chinese materia medica Beimu. Acta Phytotax. Sin. 45 (4), 473–487.

Xie, L., Wen, J., Li, L.Q., 2011. Phylogenetic analyses of *Clematis* (Ranunculaceae) based on sequences of nuclear ribosomal ITS and three plastid regions. Syst. Bot. 36 (4), 907–921.

Xiong, J., Bui, V.B., Liu, X.H., et al., 2014. Lignans from the stems of *Clematis armandii* ("Chuan-Mu-Tong") and their anti-neuroinflammatory activities. J. Ethnopharmacol. 153 (3), 737–743.

Yan, L.H., Xu, L.Z., Lin, J., et al., 2009. Triterpenoid saponins from the stems of *Clematis parviloba*. J. Asian Nat. Prod. Res. 11 (4), 332–338.

Yan, L.H., Xu, L.Z., Wang, Z.M., et al., 2010. Two new phenolic glycosides from the stems of *Clematis parviloba*. Yao Xue Xue Bao 45 (12), 1527–1532.

Yang, F.S., Wang, X.Q., Hong, D.Y., 2003. Unexpected high divergence in nrDNA ITS and extensive parallelism in floral morphology of *Pedicularis* (Orobanchaceae). Plant Syst. Evol. 240, 91–105.

Yang, A.M., Du, J., Miao, Z.H., Yuan, H.J., 2009. Study on the chemical constituents from *Clematis brevicaudata*. Zhong Yao Cai 32 (10), 1534–1537.

Yuan, T., Wang, L.Y., Roh, M.S., 2010. Confirmation of *Clematis* hybrids using molecular markers. Sci. Hortic. 125, 136–145.

Zeng, X., Li, L., Ye, N., et al., 2011. Molecular identification of radix et rhizome clematis and its adulterants based on ITS2 DNA barcode. Global Tradit. Chin. Med. 4, 264–269.

Zhang, H.X., Zhang, M.L., Sanderson, S.C., 2013a. Retreating or standing: responses of forest species and steppe species to climate change in arid Eastern Central Asia. PLoS One 8 (4), e61954.

Zhang, R., Fang, W., Han, D., et al., 2013b. Clematichinenoside attenuates myocardial infarction in ischemia/reperfusion injury both in vivo and in vitro. Planta Med. 79 (14), 1289–1297.

Zhang, W., Wang, X., Tang, H., et al., 2013c. Triterpenoid saponins from *Clematis tangutica* and their cardioprotective activities. Fitoterapia 84, 326–331.

Zhang, W., Yao, M.N., Tang, H.F., et al., 2013d. Triterpenoid saponins with anti-myocardial ischemia activity from the whole plants of *Clematis tangutica*. Planta Med. 79 (8), 673–679.

Zhang, Z., Wang, X., Zhao, M., Qian, K., 2014. Optimization of polysaccharides extraction from *Clematis huchouensis* Tamura and its antioxidant activity. Carbohydr. Polym. 111, 762–767.

Zhao, M., Tang, H.F., Qiu, F., et al., 2012. Triterpenoid saponins from *Clematis argentilucida*. Biochem. Syst. Ecol. 40, 49–52.

Zhao, M., Ma, N., Qiu, F., et al., 2014a. Triterpenoid saponins from the roots of *Clematis argentilucida* and their cytotoxic activity. Planta Med. 80 (11), 942–948.

Zhao, M., Ma, N., Qiu, F., et al., 2014b. Triterpenoid saponins from the roots of *Clematis argentilucida*. Fitoterapia 97, 234–240.

Potentilla and Rubus medicinal plants: potential non-*Camellia* tea resources

10

10.1 Introduction

Rosaceae (the rose family) comprises at least 2830 species in 95 genera. The genus *Potentilla* is a member of Rosaceae, which is mainly distributed in temperate, arctic, and alpine zones of the Northern Hemisphere. This genus has about 500 species (Dobeš and Paule, 2010), 90 of which are distributed in China. This genus has been known since ancient times in diverse ethnic populations for its therapeutic properties. Extracts of the aerial and/or underground parts have been used in traditional medicine for the treatment of inflammations, injuries, certain types of cancer, microbial infections, diarrhea, diabetes mellitus, and other disorders (Tomczyk and Latté, 2009). For example, *P. fragarioides* is used in traditional Chinese medicine as a haemostatic, while both European folk medicine and Tibet medicine use *P. anserina* to treat various diseases.

The genus *Rubus* of the family Rosaceae includes up to 1000 species, around 300 of which are found in China, especially southwest China (Dong, 2009). Raspberries are economically and medicinally important plants. Traditional Chinese medicine has a long history of using fruits and other parts of *Rubus* plants therapeutically. Chinese raspberries (e.g., *R. chingii*, Fu Pen Zi) are considered to have sweet and warm properties and are associated with the liver and kidney meridians. They tonify and stabilize the kidneys to preserve vital energy and treat liver and kidney deficiency. Chinese raspberries are used to prevent frequent urination, enuresis, premature ejaculation, and impotence. They are also applied to reduce sore lower back, improve eyesight, and prevent uterine, cervical, and colon cancer.

Herbal teas prepared from medicinal plants are popular remedies in many traditional cultures and are sources of many health-promoting phytochemicals for humans, who often use them in their everyday diet. Herbal teas can be made from non-*Camellia* plants, such as *Potentilla* and *Rubus*. For example, *P. arbuscula* var. *veitchii*, *P. fruticosa*, and *P. glabra* var. *mandshurica* are the raw plants of Yao Wang tea (named after famous medical doctor and herbalist Simiao Sun of the Tang dynasty), Jin Lao Mei tea, and Guan Yin tea, respectively. Jin Lao Mei tea is used as a substitute to green tea in mountainous regions in China. In some western countries, *Potentilla* and raspberry (*R. idaeus*) herb teas are used as a female tonic to alleviate menstrual cramps and discomfort. *R. caucasicus* leaves are raw materials for tea production in Russia (Melkadze et al., 2008). *R. suavissimus* is a shrub widely grown in

Medicinal plants: chemistry, biology and omics. http://dx.doi.org/10.1016/B978-0-08-100085-4.00010-4

southwestern China, leaves of which are usually used as an herbal tea (Chou et al., 2009). Flavonoids and other phenolic compounds, present in both green tea and non-*Camellia* tea, have significant antioxidant activities, and regular consumption may help prevent and/or moderate chronic diseases (Li et al., 2011). In this brief chapter, we summarize the phytochemical ingredients and their bioactivities and molecular pharmacognosy of *Potentilla* and *Rubus*, which have great potential in developing non-*Camellia* tea resources.

10.2 Chemical components

10.2.1 Flavonoids

Types and structures of flavonoids found in *Potentilla* and *Rubus* are summarized in Figures 10.1 and 10.2 and Table 10.1. A high-performance thin-layer chromatography (HPTLC)-densitometry method was developed for separation and quantitative determination of four flavonoids: quercetin 3-*O*-β-D-glucuropyranoside (miquelianin), quercetin 3-*O*-β-D-glucopyranoside (isoquercitrin), quercetin 3-*O*-β-D-galactopyranoside (hyperoside), and quercetin 3-*O*-β-D-(6″-α-l-rhamnosylo)-glucopyranoside (rutin) from aerial parts of *P. argentea*, *P. erecta*, *P. fruticosa* (Jin Lao Mei tea), *P. rupestris*, *P. nepalensis* var. "Miss Wilmott," and *P. thuringiaca* (Tomczyk et al., 2012). The method was relatively simple, specific, precise, and accurate and may be used for the quality control of simultaneous determination of quercetin derivatives in *Potentilla* extracts and also in chemotaxonomy. Aerial parts of four *Potentilla* species, *P. anserina*, *P. erecta*, *P. grandiflora*, and *P. nepalensis* var. "Miss Willmott", were extracted with methanol. Tiliroside (a glycosidic flavonoid in both *Potentilla* and *Rubus*) ranged from 20.3 ± 0.3 mg/g for *P. erecta* herbs to 197.7 ± 2.9 mg/g for *P. grandiflora* herbs (Tomczyk et al., 2010c).

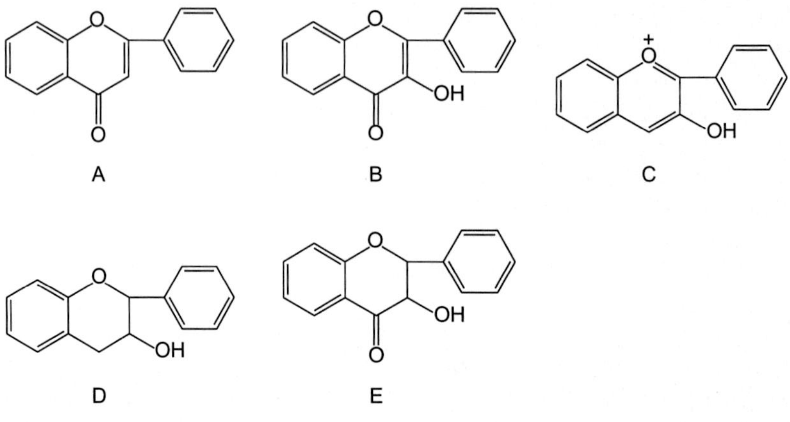

Figure 10.1 Types of flavonoids found in *Potentilla* and *Rubus*. A, Flavone type. B, Flavonol type. C, Anthocyanidin type. D, Flavan-3-ol type. E, Flavanonol type.

In *R. idaeus* (red raspberry), variation in flavonoid composition was primarily attributed to *R. idaeus* (red raspberry) genotype and associated with night temperature and hours exposed to temperatures over 29 °C (Bradish et al., 2012). The cultivar "Nantahala" had particularly high levels of cyanidin-3-sambubioside, indicative of its purple raspberry lineage. Quercetin-3-glucoside levels increased the most with elevated temperatures.

10.2.2 Tannins, anthocyanins, and phenolic acids

Hyperoside was the predominant phenolic compound in leaf extracts of three *Potentilla* species by RP-HPLC assay (Wang et al., 2013), with the content of 8.86 (*P. fruticosa*), 2.56 (*P. glabra*), and 2.68 mg/g (*P. parvifolia*), respectively. The highest content of total identified phenolic compounds (hyperoside, (+)-catechin, caffeic acid, ferulic acid, rutin, and ellagic acid) was observed in *P. parvifolia* (14.17 mg/g),

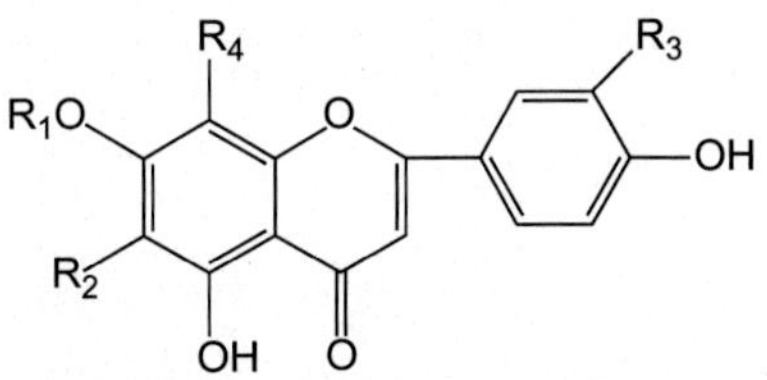

	R_1	R_2	R_3	R_4
1	Glc	H	H	H
2	Glc	H	OH	H
3	H	OGlc	H	H
4	H	OGlc	H	OXyl
5	H	Ara	H	Glc

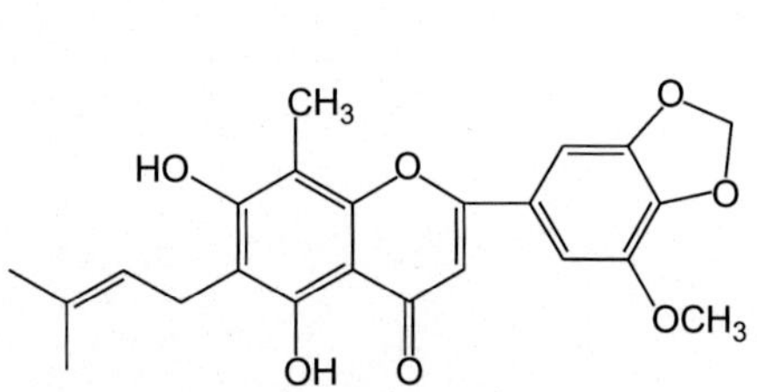

6

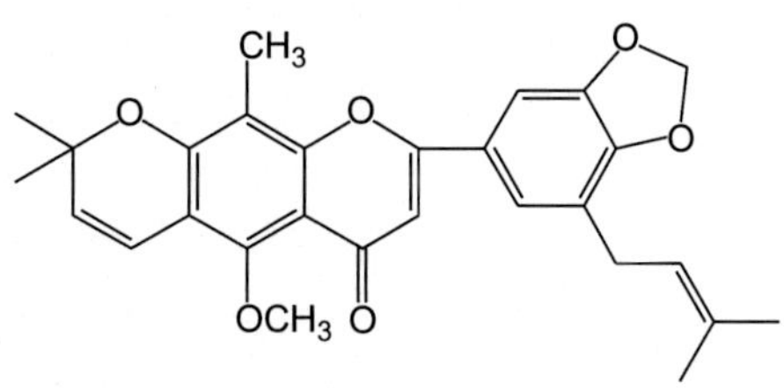

7

Figure 10.2 Chemical structures of flavonoids in *Rubus* and *Potentilla*. Refer to Table 10.1 for more details.

(Continued)

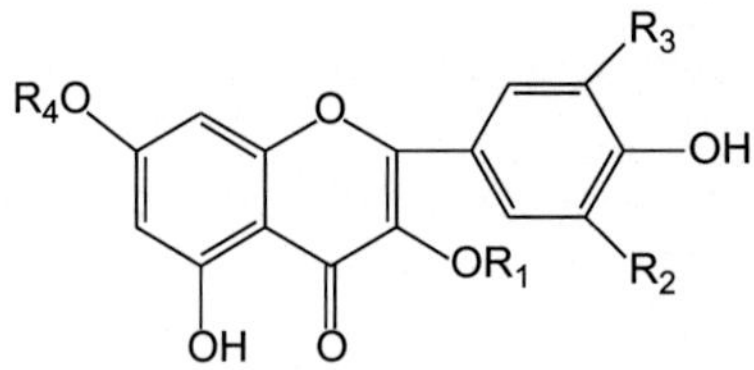

	R_1	R_2	R_3	R_4
8	Glc	H	OH	H
9	Glc	H	H	H
10	Glc	OH	H	H
11	Ara	H	OH	H
12	Rha	H	H	H
13	Glc	OH	OH	H
14	Glc	H	OMe	H
15	H	H	OH	H
16	H	H	H	H
17	*p*-coumaroyl-*O*-Glc	H	H	H
18	Gal	H	OH	H
19	Gal	H	H	H
20	Rut	H	OH	H
21	Rha(1→6)Glc	H	H	H
22	Rha	H	OH	H
23	GluA	H	OH	H
24	Gal	H	OH	Glc
25	Rha(1→6)Glc	H	H	H
26	Rha(1→6)Glc	H	OH	H
27	Ara(1→6)Glc	H	H	H
28	Ara(1→6)Glc	H	OH	H

Figure 10.2 Continued.

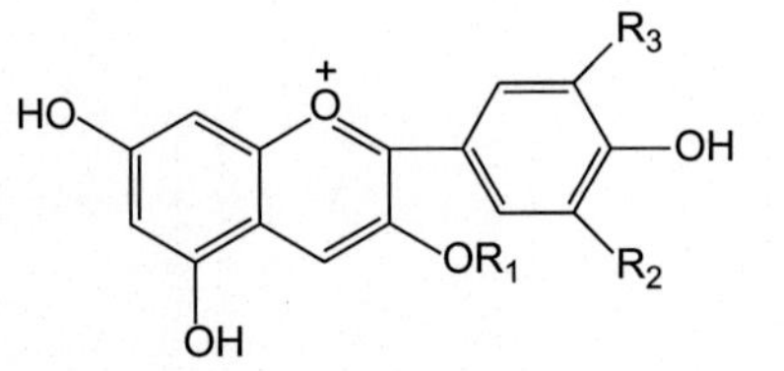

	R_1	R_2	R_3
29	Glc	H	OH
30	Rut	H	H
31	Glc	OH	OH

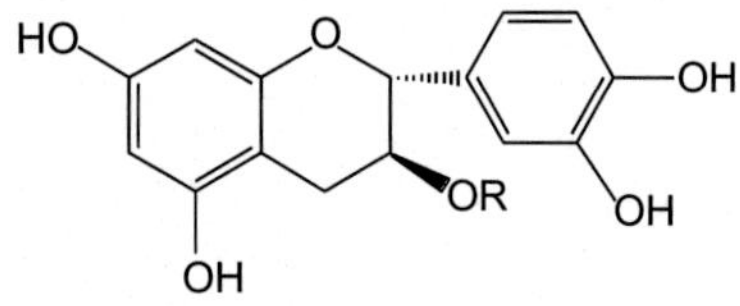

	R
32	H
33	Glc

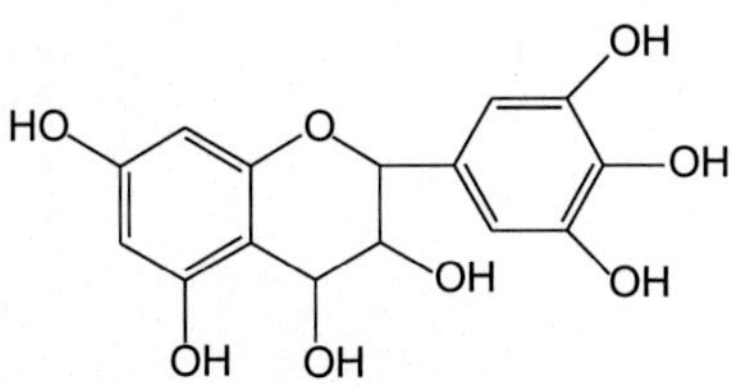

34

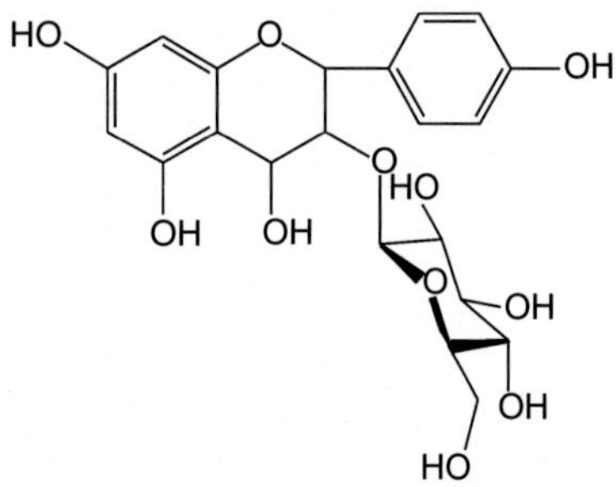

35

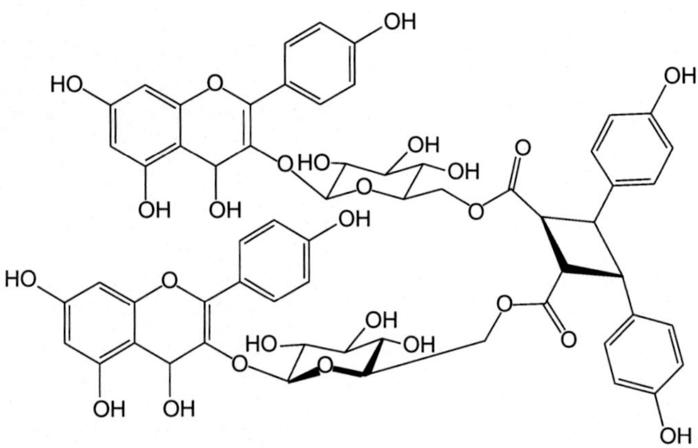

36

Figure 10.2 Continued.

Table 10.1 **Flavonoids found in *Rubus* and *Potentilla* medicinal plants**

	Compounds	Types	Species	References
1	Apigenin-7-*O*-glucoside	A	I_1, I_6, I_{10}, and II_5	Deng et al. (2001), Rezeng et al. (2014), Xue et al. (2005), and Tomczyk (2011)
2	Luteolin-7-*O*-glucoside	A	I_1, I_6, I_{10}, and II_5	Deng et al. (2001), Rezeng et al. (2014), Xue et al. (2005), and Tomczyk (2011)
3	Isovitexin	A	II_1	Xu (2008)
4	Apigenin-6-*O*-β-D-glucopyranosyl-8-*O*-β-D-xylopyranoside	A	II_1	Xu (2008)
5	Apigenin-6-C-arabinopyranosyl-8-C-glucopyranoside	A	II_5	Xue et al. (2005)
6	Fenzhimine A	A	I_2	Kang and Zheng (2007)
7	Fenzhimine B	A	I_2	Kang and Zheng (2007)
8	Quercetin-3-*O*-glucuronide	B	I_3, I_{14}, II_2, and II_4	Venskutonis et al. (2007), Paudel et al., 2013, Wang et al. (2009a,b), and Söhretoglu and Kırmızıbekmez (2011)
9	Kaempferol-3-*O*-glucuronide	B	I_3 and II_2	Venskutonis et al. (2007) and Wang et al. (2009a,b)
10	Isoquercitrin	B	I_4 and I_5	Byamukama et al. (2005)
11	Quercetin-3-*O*-α-D-arabinopyranoside	B	II_2 and I_{10}	Wang et al. (2009a,b) and Tomczyk (2011)
12	Kaempferol-3-*O*-α-L-rhamnoside	B	II_3	Gao (2007)
13	Myricetin-3-*O*-β-D-glucoside	B	II_1 and I_{14}	Kombal and Glasl (1995) and Paudel et al. (2013)
14	Isorhamnetin-3-β-D-glucuronide	B	II_1	Kombal and Glasl (1995)
15	Quercetin	B	I_6, I_{10}, II_2, and II_4	Rezeng et al. (2014), Wang et al. (2009a,b), Söhretoglu and Kırmızıbekmez (2011), and Tomczyk (2011)
16	Kaempferol	B	I_6, I_8, I_{10}, II_2, and II_4	Rezeng et al. (2014), Wei et al. (2012), Wang et al. (2009a,b), Söhretoglu and Kırmızıbekmez (2011), and Tomczyk (2011)
17	Tiliroside	B	I_7 and I_{10}	Ding, (2011) and Tomczyk (2011)
18	Nicotiflorin	B	I_7	Ding (2011)
19	Hyperoside	B	I_6	Rezeng et al. (2014)

20	Kaempferol-3-*O*-galactopyranoside	B	I_8	Wei et al. (2012)
21	Rutin	B	I_9, I_{14}, and II_2	Paudel et al. (2013) and Wang et al. (2009a,b)
22	Quercetin-3-*O*-α-L-rhamnpyranoside	B	II_2	Wang et al. (2009a,b)
23	Quercetin-3-*O*-β-D-glucuronide	B	II_2 and II_4	Wang et al. (2009a,b) and Söhretoglu and Kırmızıbekmez (2011)
24	Quercetin3-*O*-β-D-galactoside-7-*O*-β-D-glucoside	B	II_2	Wang et al. (2009a,b)
25	Kaempferol 3-*O*-α-rhamnopyranosyl-(1→6)-β-glucopyranoside	B	II_4	Söhretoglu and Kırmızıbekmez (2011)
26	Quercetin 3-*O*-α-rhamnopyranosyl-(1→6)-β-glucopyranoside	B	II_4	Söhretoglu and Kırmızıbekmez (2011)
27	Kaempferol 3-*O*-α-arabinopyranosyl-(1→6)-β-glucopyranoside	B	II_4	Söhretoglu and Kırmızıbekmez (2011)
28	Quercetin 3-*O*-α-arabinopyranosyl-(1→6)-β-glucopyranoside	B	II_4	Söhretoglu and Kırmızıbekmez (2011)
29	Cyanidin-3-*O*-glucoside	C	I_{11}	Zuo et al. (2008)
30	Pelargonidin-3-*O*-rutinoside	C	I_{12}	Tan et al. (2003)
31	Delphinidin-3-*O*-glucuronide	C	I_{12}	Tan et al. (2003)
32	(+)-Catechin	D	II_6	Zhang et al. (1990)
33	(+)-Catechin 3-*O*-β-D-glucopyranoside	D	II_6	Huang et al. (1990)
34	Herahydroxy dihydroflavanonol	E	I_{13}	Sun et al. (2004)
35	Dihydrokaempferol glucoside	E	I_{14}	Paudel et al. (2013)
36	Potentilin A	F	II_1	Xu et al. (2010)

Genus: I, *Rubus*, II, *Potentilla*.
I_1 *R. buergeri*; I_2 *R. biflorus Buch*; I_3 *R. idaeus*; I_4 *R. pinnatus*; I_5 *R. rigidus*; I_6 *R. amabilis*; I_7 *R. chingii*; I_8 *R. crataegifolius*; I_9 *R. suavissimus* S. Lee (Chinese sweet tea); I_{10} *R. rupestris* L.; I_{11} *R. obcordatus*; I_{12} *R. parvifolius*; I_{13} *R.raspberry*; I_{14} *R. occidentalis* L.
II_1 *P. anserina*; II_2 *P. discolor* Bunge; II_3 *P. chinensis*; II_4 *P. recta*; II_5 *P. multifida* L.; II_6 *P. viscose*.
Types of flavonoids: A flavone type; B flavonol type; C anthocyanidin type; D flavan-3-ol type; E flavanonol type; F diflavonol type.

followed by *P. fruticosa* (10.01 mg/g) and *P. glabra* (7.01 mg/g). *P. fruticosa* possessed the highest content of total phenolic (84.93 ± 0.50 mmol gallic acid equivalent/100 g) and total flavonoid (84.14 ± 0.03 mmol quercetin equivalent/100 g) contents, which were in good correlation with its significant DPPH IC50 (16.87 μg/mL), ABTS (2763.48 μmol Trolox equivalent/g), and FRAP (1398.70 μmol Trolox equivalent/g) capacities.

Plant phenolics include phenolic acids, flavonoids, and tannins. Tannins are usually divided into hydrolyzable tannins and condensed tannins. Like green tea, *Potentilla* and *Rubus* contain abundant tannins (Figures 10.3 and 10.4; Tables 10.2 and 10.3). Among the aerial parts of 10 common *Potentilla* plants, *P. fruticosa* has

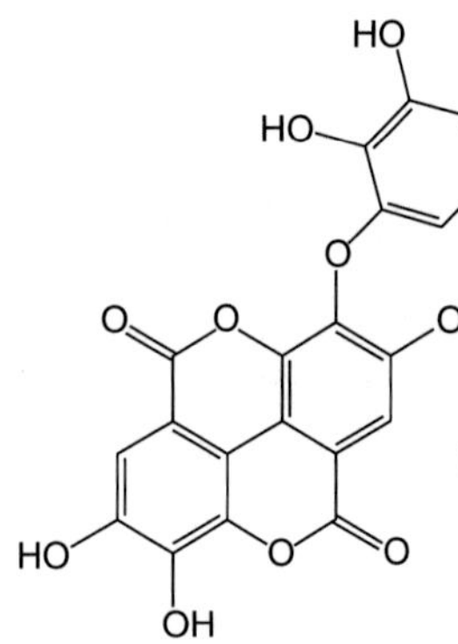

37

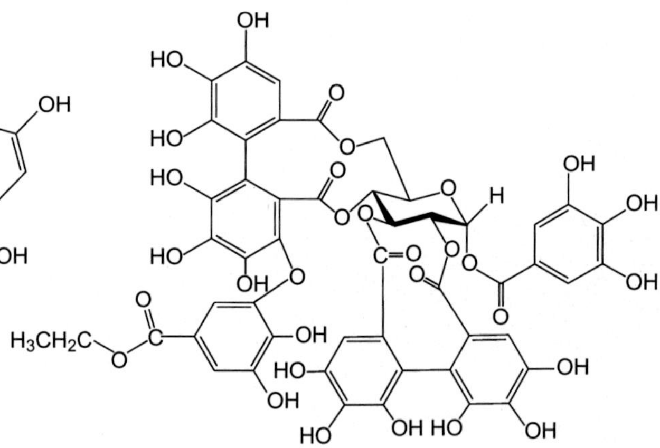

38

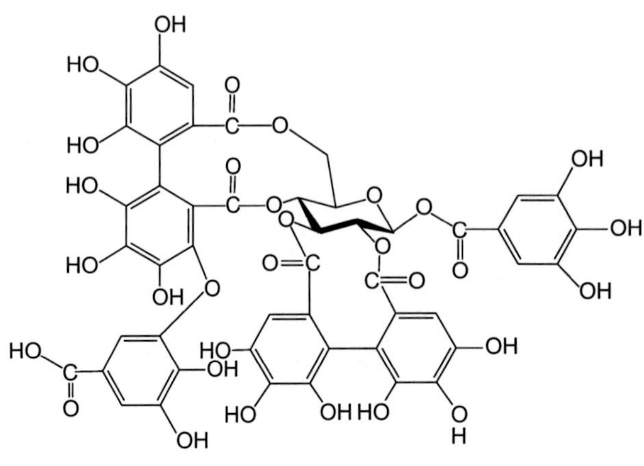

39

Figure 10.3 Tannins found in *Rubus*. Refer to Table 10.2 for more details.

(Continued)

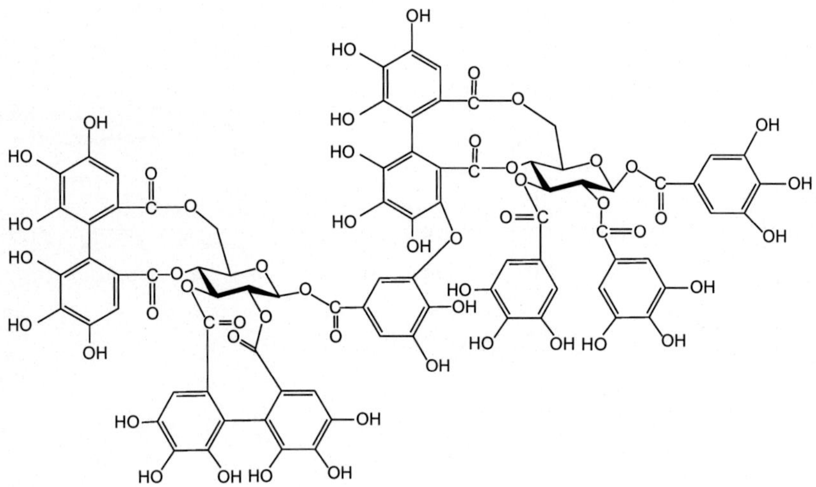

40

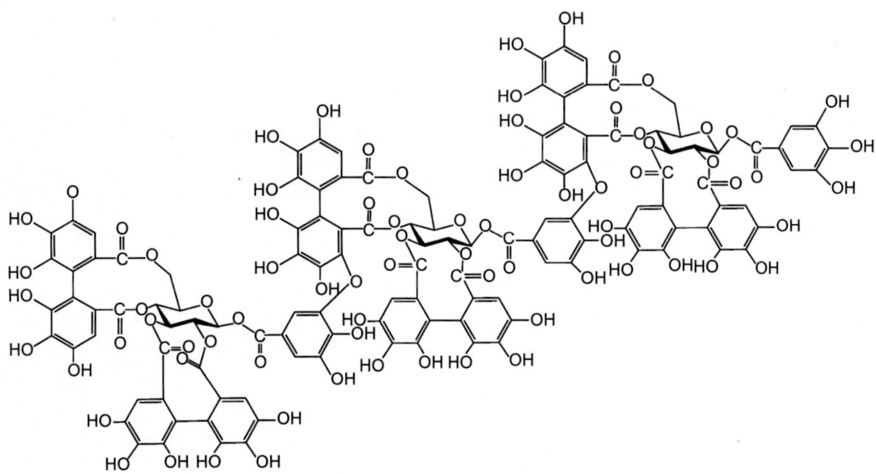

41

Figure 10.3 Continued.

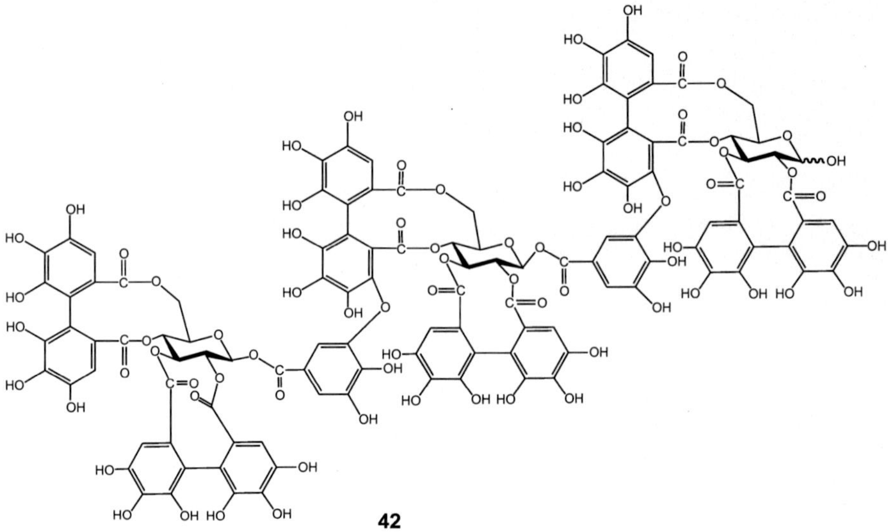

42

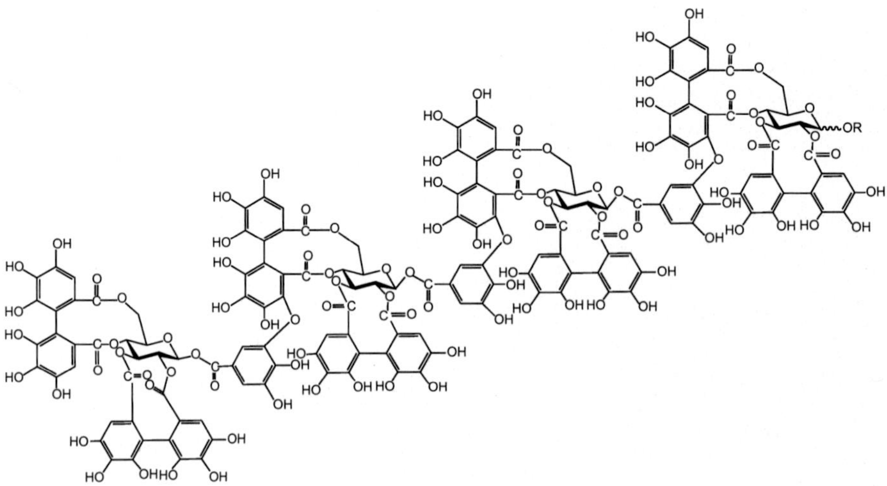

43 R=β galloyl

44 R=α, β H

Figure 10.3 Continued.

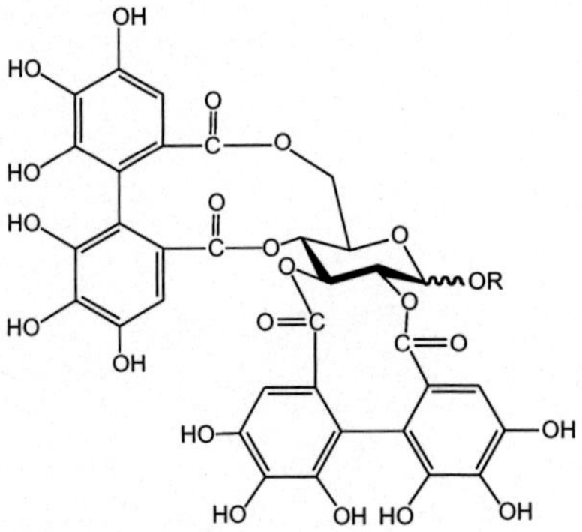

45 R=α, β H

46 R=β galloyl

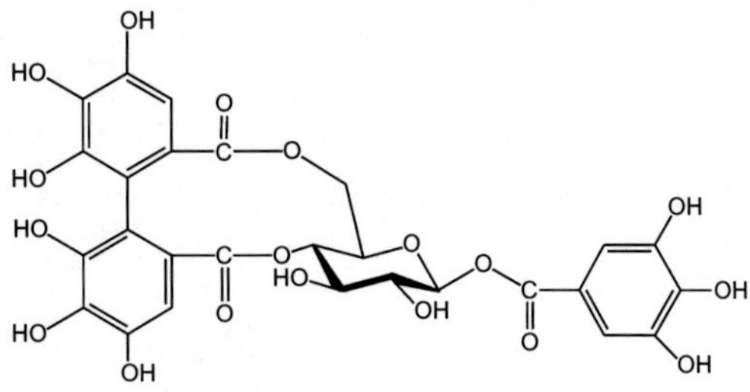

47

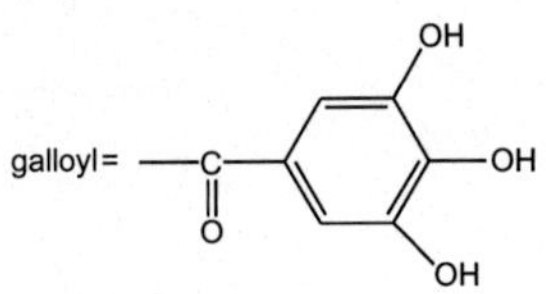

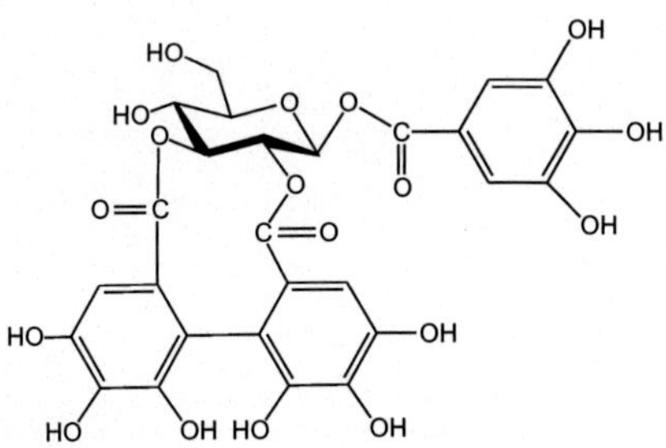

48

Figure 10.3 Continued.

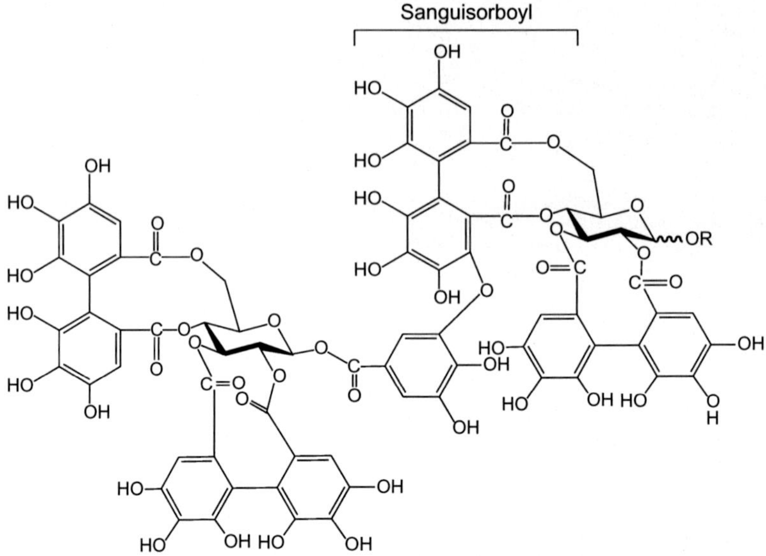

49 R=β galloyl

50 R=α galloyl

51 R=α, β H

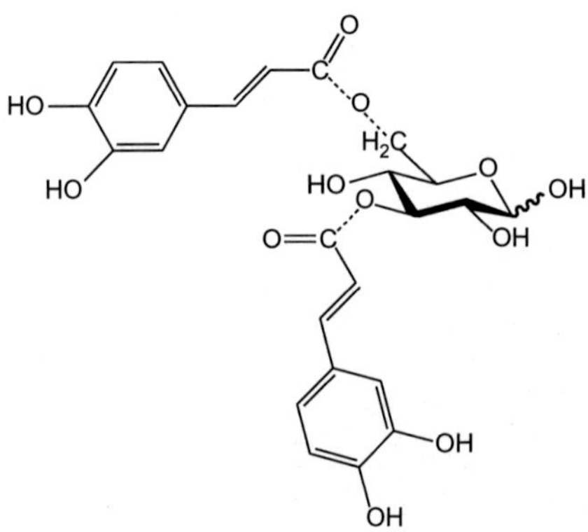

52

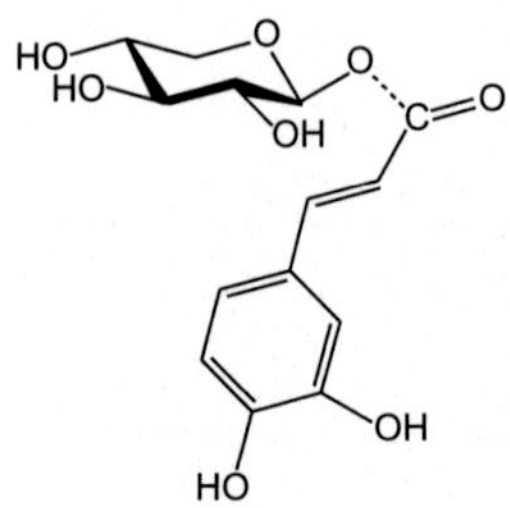

53

Figure 10.3 Continued.

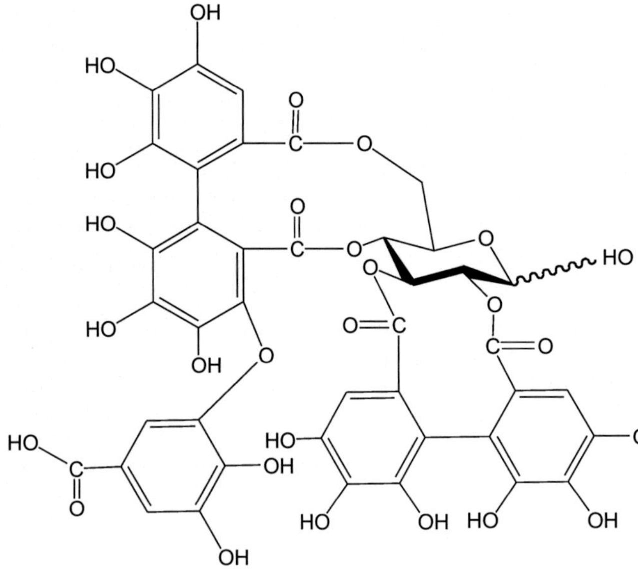

54

Figure 10.3 Continued.

relatively high concentrations of tannins (167.3 ± 2.0 mg/g dry weight), proanthocyanidins (condensed tannins; 4.6 ± 0.2 mg/g dry weight), phenolic acids (16.4 ± 0.8 mg/g dry weight), and flavonoids (quercetin 7.0 ± 1.1 mg/g dry weight) (Tomczyk et al., 2010b). Methyl brevifolincarboxylate ranged from 5.0 ± 0.6 mg/g for *P. erecta* herbs to 68.5 ± 3.4 mg/g for *P. nepalensis* flowers (Tomczyk et al., 2010c), and ellagic acid ranged from 24.0 ± 0.6 mg/g for *P. erecta* herbs to 216.2 ± 3.2 mg/g for *P. anserina* leaves.

Anthocyanins belong to a parent class of flavonoids synthesized via the phenylpropanoid pathway. A great variability in total anthocyanins was detected in Croatian wild blackberry (*R. discolor*) and raspberry (*R. idaeus*) genotypes (Dujmović Purgar et al., 2012). Blackberries are particularly rich in anthocyanins (2226–2367 mg/kg), compared to raspberries (279–582 mg/kg). Wild raspberry and blackberry had two- to fivefold more total phenolic contents compared to cultivated plants (Mikulic-Petkovsek et al., 2012). The anthocyanin content of *R. glaucus* was 45 mg/100 g FW (Garzón et al., 2009). The isolated anthocyanins were characterized as cyanidin 3-sambubioside, cyanidin 3-glucoside, cyanidin 3-xylorutinoside, cyanidin 3-rutinoside, pelargonidin 3-glucoside, and pelargonidin 3-rutinoside. The HPLC determined concentrations of anthocyanins in the fruits of *R. idaeus* varied from 76.22 to 277.06 mg/100 g dry weight (Sparzak et al., 2010). The content of sanguiin H-6 was in the range from 135.04 to 547.48 mg/100 g dry weight.

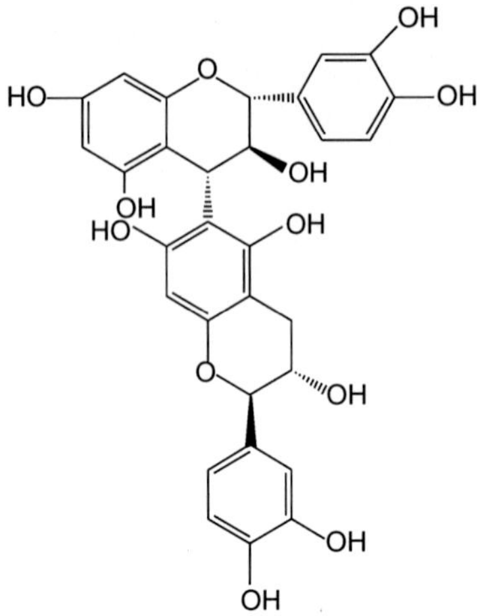

55

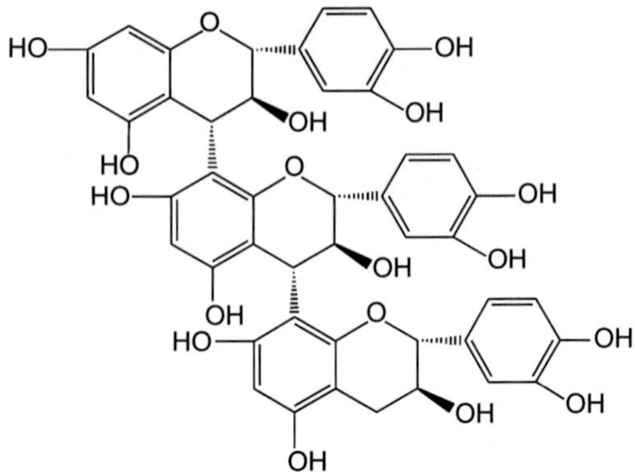

56

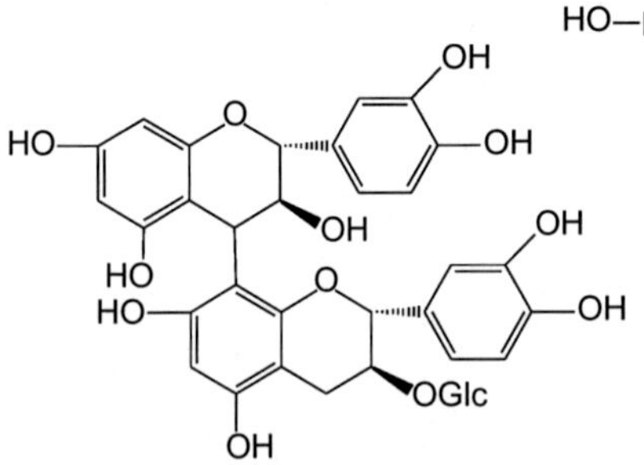

57

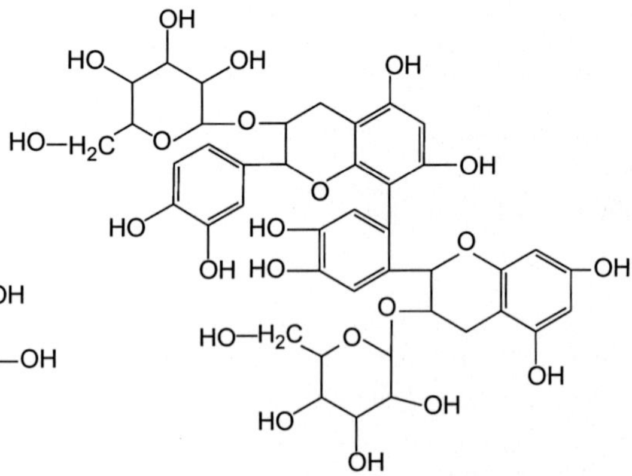

58

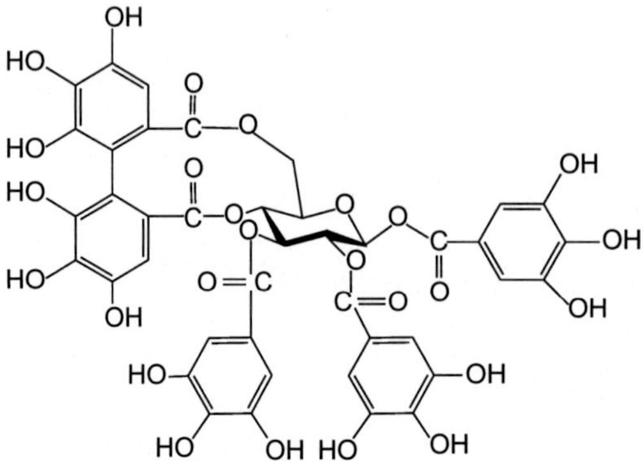

59

Figure 10.4 Tannins found in *Potentilla*. Refer to Table 10.3 for more details.

(Continued)

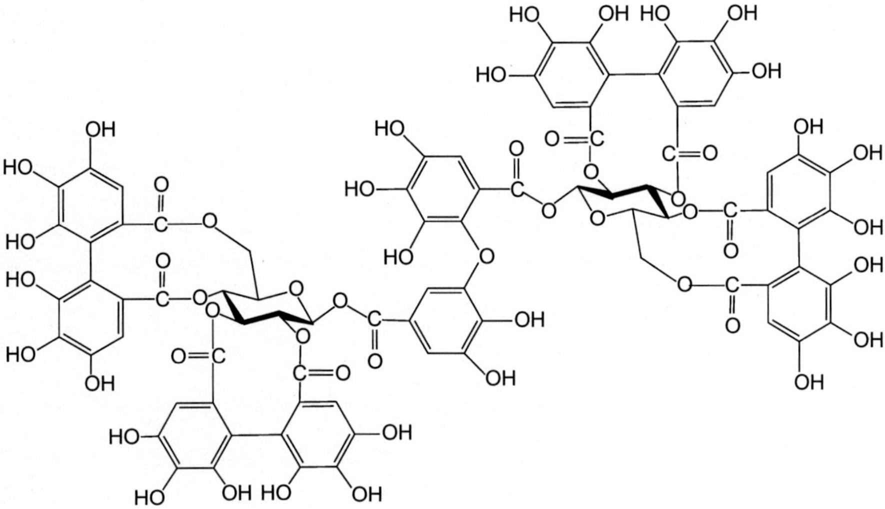

60

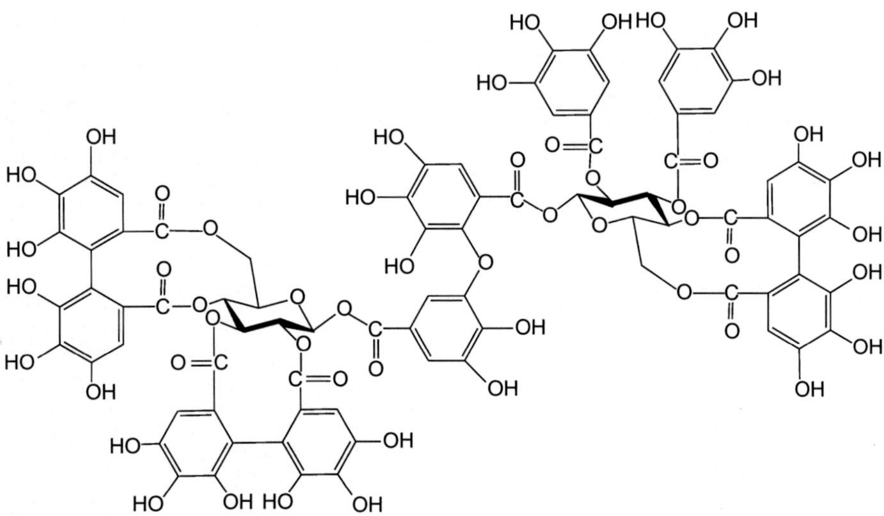

61

Figure 10.4 Continued.

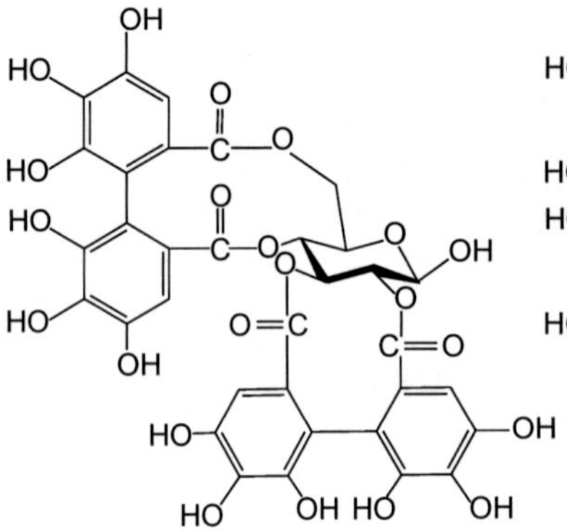

62

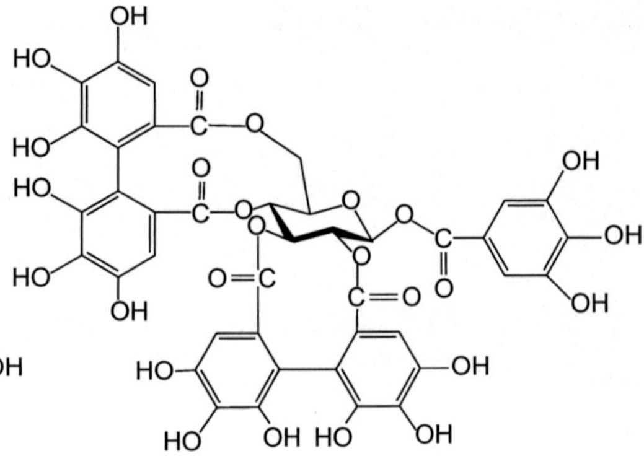

63

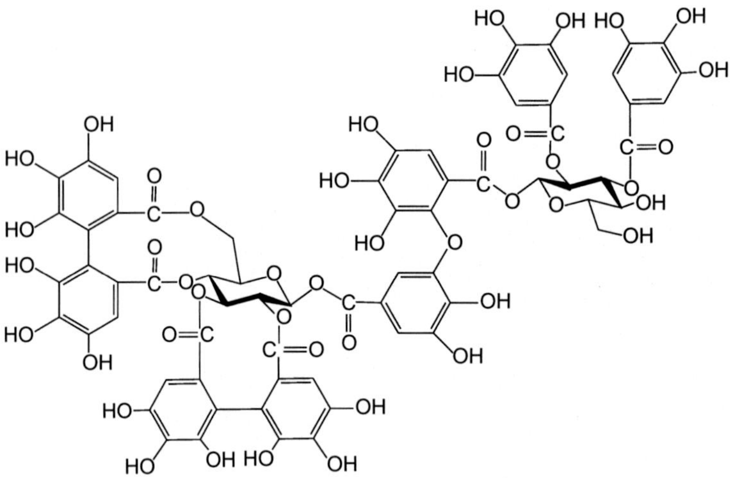

64

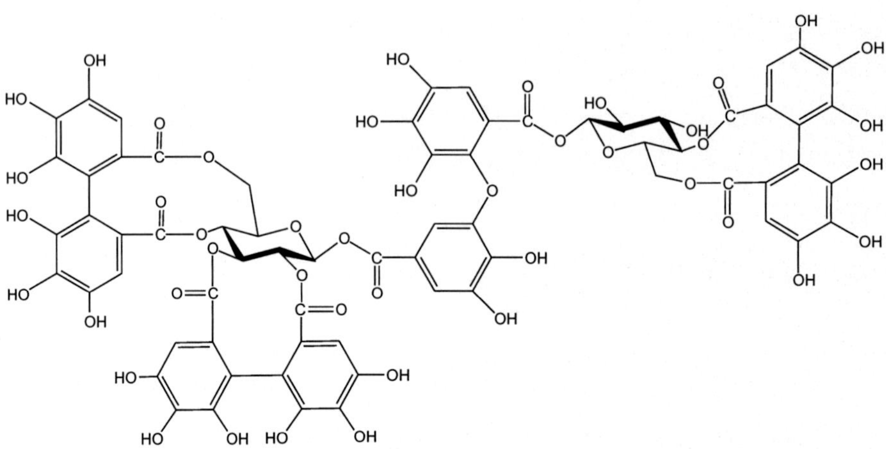

65

Figure 10.4 Continued.

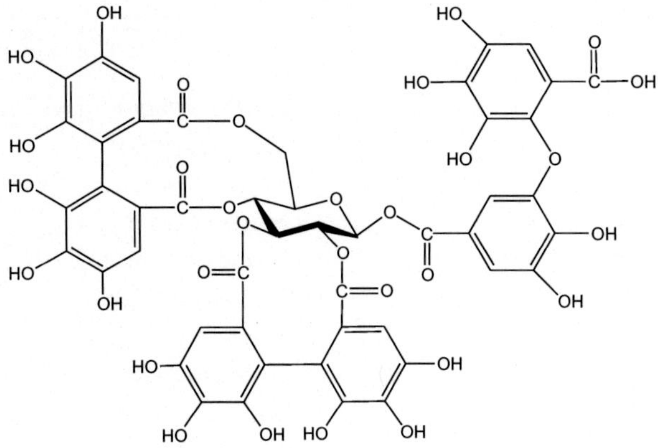

66

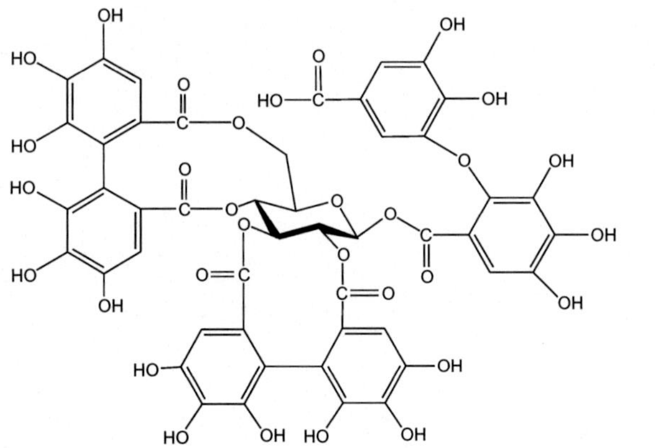

67

Figure 10.4 Continued.

10.2.3 Terpenoids

Terpenoids and other compounds are summarized in Table 10.4 and Figure 10.5.

The whole herb of *P. discolor* has long been used for the treatment of diarrhea, malaria, hemoptysis, and hematemesis in clinical applications. A simple and accurate HPLC-diode array detection (DAD)-electrospray ionization multistage mass spectrometry (ESI-MSn) method is developed for the identification of constituents in *P. discolor* and simultaneous quantification of its marker compounds (Yang et al., 2011). Twenty-three compounds including 13 flavonoids and 10 triterpenoids were identified or tentatively characterized. The method was successfully applied to the quantification of 11 marker compounds in 10 samples of *P. discolor* collected from different provinces

Table 10.2 Tannins found in *Rubus*

	Compound	Species	References
37	Rubuphenol	*Rubus aleaefolius*	Cui et al. (2002)
38	Sanguiin H-2 ethyl ester	*Rubus aleaefolius*	
39	Rubusuaviins A	Chinese sweet tea	Li et al. (2007)
40	Rubusuaviins B	Chinese sweet tea	
41	Rubusuaviins C	Chinese sweet tea	
42	Rubusuaviins D	Chinese sweet tea	
43	Rubusuaviins E	Chinese sweet tea	
44	Rubusuaviins F	Chinese sweet tea	
45	Pedunculagin	Chinese sweet tea	
46	1(β)-*O*-Galloyl pedunculagin	Chinese sweet tea	
47	Strictinin	Chinese sweet tea	
48	Sanguiin H-5	Chinese sweet tea	
49	Lambertianin A	Chinese sweet tea	
50	Sanguiin H-6	Chinese sweet tea	
51	1-Desgalloyl sanguiin H-6	Chinese sweet tea	
52	3,6-Di-*O*-caffeoyl-(α/β)-glucose	*Rubus sanctus*	Hussein et al. (2003)
53	1-*O*-Caffeoyl-β-xylose	*Rubus sanctus*	
54	2,3-*O*-Hexahydroxydiphenoyl-4,6-*O*-sanguisorboyl-(α/β)-glucose	*Rubus sanctus*	

Table 10.3 Tannins found in *Potentilla*

	Compound	Species	References
55	Procyanidin B6	*P .erecta*	Ren et al. (2010)
56	Procyanidin C2	*P. viscose*	Huang et al. (1990)
57	Procyanidin B3, 3′-*O*-β-D-glucopyranoside	*P. viscose*	
58	Potentillanin	*P. viscose* *P. kleiniana*	Ren et al. (2010)
59	Tellimagrandin	*P. discolor*	
60	Agrimoniin	*P. erecta* *P. discolor*	
61	Gemin A	*P. discolor*	
62	Pedunculagin	*P. erecta* *P. kleiniana*	
63	Casuarictin	*P. discolor*	
64	Laevigatin B	*P. erecta*	
65	Laevigatin F	*P. erecta*	
66	Agrimonic acid A	*P. kleiniana*	
67	Agrimonic acid B	*P. kleiniana*	

Table 10.4 Other compounds found in *Rubus* and *Potentilla* medicinal plants

	Compounds	Types	Species	References
68	Fulgic acid A	A	*P. fulgens*	Choudhary et al. (2013)
69	Fulgic acid B	A	*P. fulgens*	Choudhary et al. (2013)
70	Potentene A	A	*P. fulgens*	Jaitak et al. (2010)
71	Potentene B	A	*P. fulgens*	Jaitak et al. (2010)
72	3α,30-Dihydroxylup-20(29)-en-27-oic acid	A	*P. discolor* Bunge	Morikawa et al. (2014)
73	(20S)-3α,29-Dihydroxylupan-27-oic acid	A	*P. discolor* Bunge	Morikawa et al. (2014)
74	Potentillanoside A	A	*P. anserina*	Yang et al. (2008)
75	Potentillanoside B	A	*P. anserina*	Yang et al. (2008)
76	Potentillanoside C	A	*P. anserina*	Yang et al. (2008)
77	Potentillanoside D	A	*P. anserina*	Yang et al. (2008)
78	Potentillanoside E	A	*P. anserina*	Yang et al. (2008)
79	Potentillanoside F	A	*P. anserina*	Yang et al. (2008)
80	2α,3β,19α,23,30-Pentahydroxyurs-12-en-28-oic acid-28-*O*-β-D-glucopyranosyl ester	A	*P. multicaulis* Bunge	Jia et al. (2013)
81	Ent-kauran-3β,16β,17,19-tetrol	B	*R. corchorifolius* L. f	Zhang et al. (2007)
82	Ent-2-carbonyl-16β-hydroxy-kauran-17-β-D-glucopyranoside	B	*R. corchorifolius* L. f	Zhang et al. (2007)
83	2,6-Dimethyl-2,3-dihydro-4-oxo-4H-pyran-2-acetic acid	C	*P. anserina*	Xu et al. (2010)
84	3-*O*-β-Glucopyranosyl-3β,15β-dihydroxypregn-5-en-20-one	D	*R. phoenicolasius*	Liu et al. (2014)
85	*N*-Hydroxyl-hexahydroazepin-2,4-diones	E	*P. multicaulis* Bunge	Jia et al. (2013)

A, Triterpenoids; B, diterpenoids; C, monoterpenoids; D, pregnane glycosides; E, heterocyclic compounds.

of China and is also promising for chemotaxonomy and authentication of *Potentilla*. Monoterpene, sesquiterpene, and triterpene are found in *Potentilla*. Triterpenoids are mainly of oleane type, ursane type, and lupane type (Ren et al., 2010).

Monoterpene, diterpene, and triterpene are found in *Rubus* (Meng et al., 2011). According to the type of skeleton, diterpenes are classified into labdane type and kaurane type. Triterpenoids are mainly of oleane type, ursane type, and lupane type (Li et al., 2009a; Sohn et al., 2011).

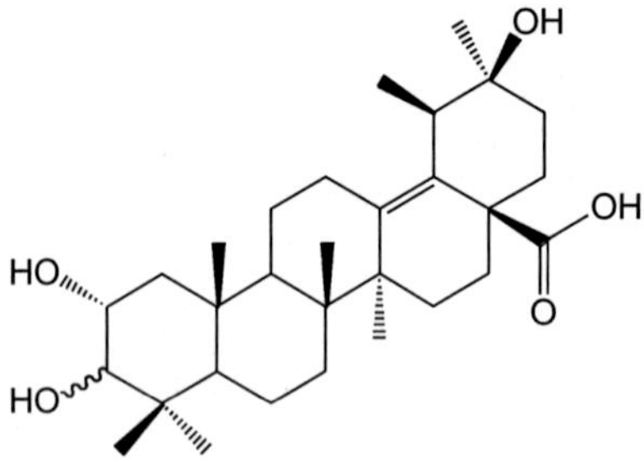

68	α
69	β

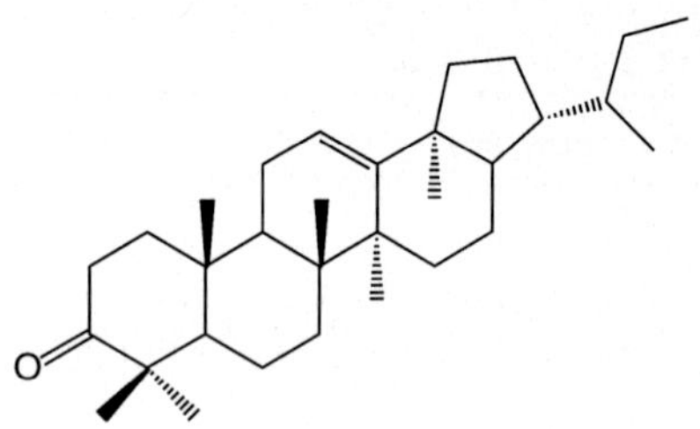

70

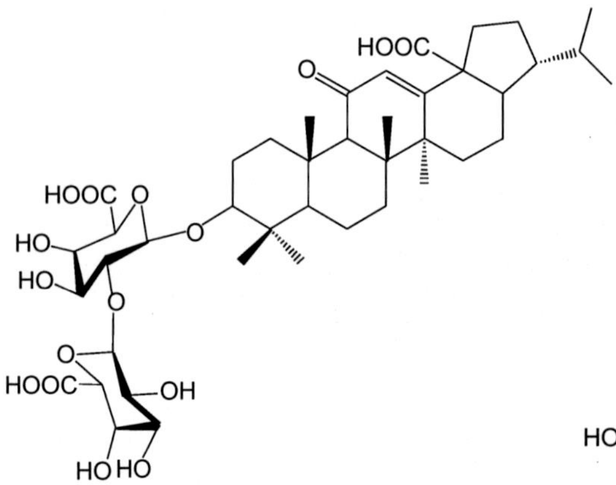

71

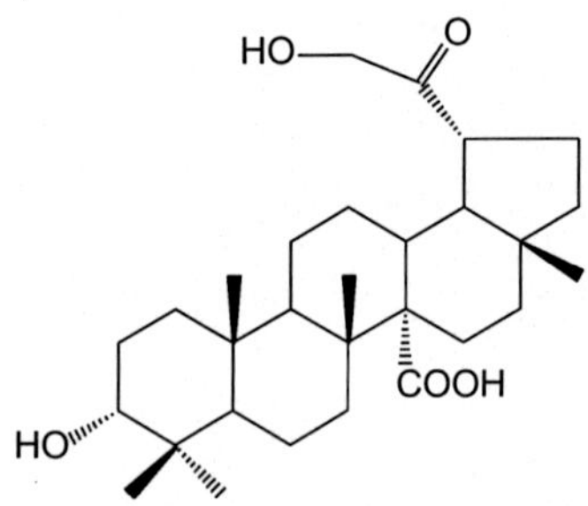

72

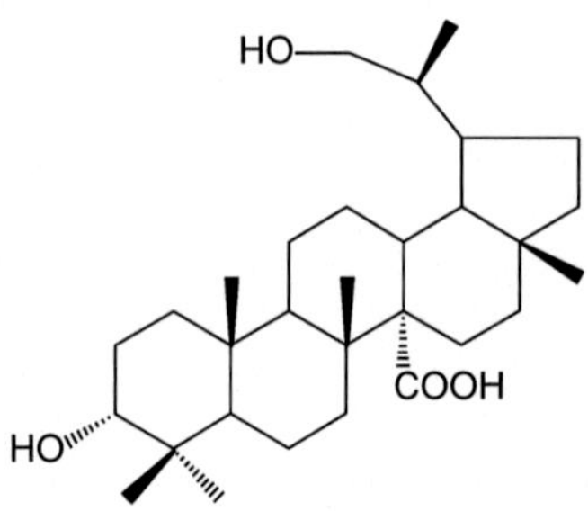

73

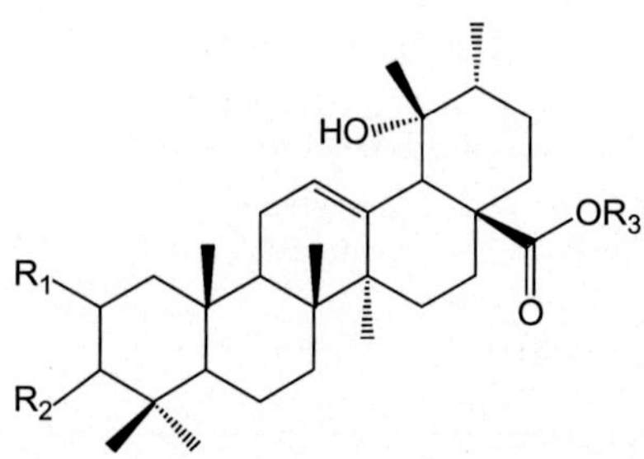

	R_1	R_2	R_3
74	α-OH	=O	Glc
75	=O	α-OH	Glc
76	α-OH	β-OH	6-O-Ac-Glc

Figure 10.5 Examples of other compounds found in *Potentilla* and *Rubus*.

(Continued)

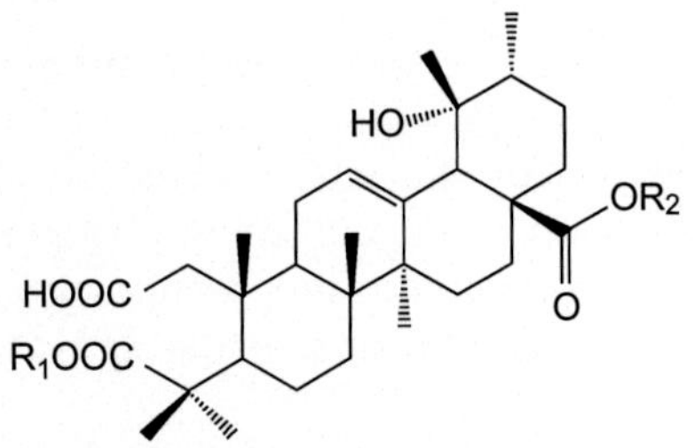

	R_1	R_2
77	H	Glc
78	CH_3	Glc

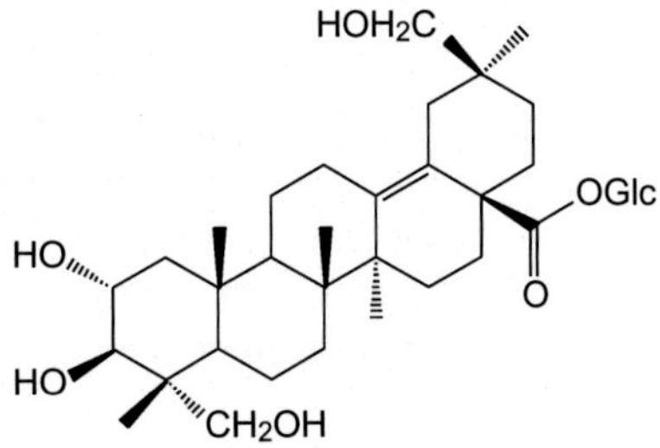

79

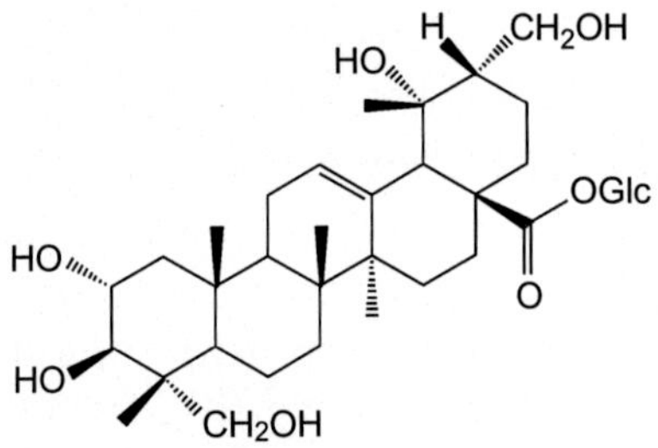

80

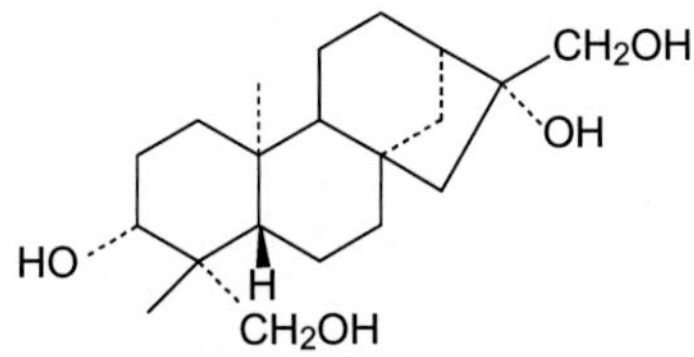

81

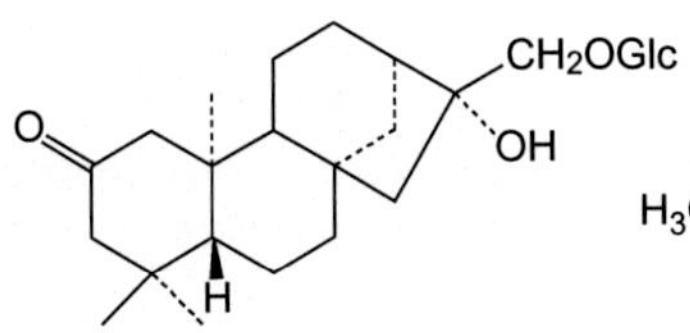

82

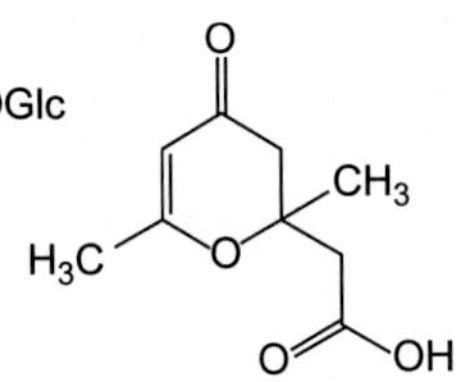

83

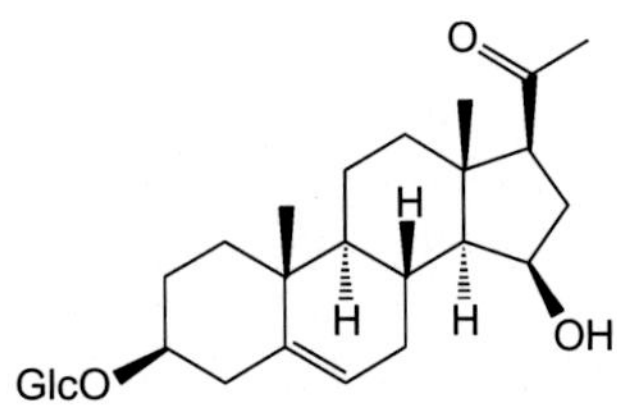

84

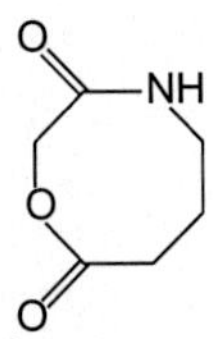

85

Figure 10.5 Continued.

10.2.4 Other compounds

Wild grown European blackberry (*R. fruticosus*) plants are widespread in different parts of northern countries. European blackberry plants are used for herbal medicinal purposes such as antimicrobial, anticancer, antidysentery, antidiabetic, antidiarrheal, and good antioxidant. Blackberry plant contains tannins, gallic acid, villosin, and iron (Verma et al., 2014); fruit contains vitamin C, niacin (nicotinic acid), pectin, sugars, and anthocyanins and also contains of berries albumin, citric acid, malic acid, and pectin.

Numerous research groups have reported zero detectable amounts of sugar alcohol in fully ripe *Rubus* fruit (Lee, 2015), with the exception of three out of 82 *Rubus* fruit samples (cloudberry 0.01 g/100 g, red raspberry 0.03 g/100 g, and blackberry 4.8 g/100 g; 73 other blackberry samples contained no detectable sorbitol).

10.3 Bioactivities

10.3.1 Antioxidant activity and cardioprotective effects

Afzelechin-4$\alpha \rightarrow 8''$-catechin, epiafzelechin, and rutin of *P. fulgens* exhibited significant 1,1-diphenyl-2-picrylhydrazyl (DPPH) radical scavenging activity (Jaitak et al., 2010). An increase in the expression of hypoxia-inducible factor (HIF)-1α and a decrease in the expression of insulin-like growth factor-1 receptor (IGF-1R) were observed in the cells subjected to oxidative stress. The antioxidative activity of tiliroside isolated from *P. argentea* may originate at the level of IGF-1R and HIF-1α signaling (Tomczyk et al., 2010a). *P. anserina* (Jue Ma) polysaccharide fraction reduced H2O2-induced oxidative damage in a dose-dependent manner and inhibited H_2O_2-induced apoptosis of murine splenic lymphocytes (Shuai et al., 2009).

Simultaneous overexpression of *P. astrisanguinea* SOD and RaAPX in transgenic *Arabidopsis thaliana* confers cold stress tolerance through increase in vascular lignifications (Shafi et al., 2014).

The alcohol extracts of *P. anserina* protect the myocardial tissue from apoptosis after acute myocardial ischemia and reperfusion injury in rats and inhibit the expressions of caspase-9 and caspase-3 mRNA and proteins (Qin et al., 2012a). The *n*-butanol extract of *P. anserina* has a notable protective effect on acute myocardial ischemic injury in mice (Li et al., 2009b). The n-butanol extract of Jue Ma root could attenuate the Ca^{2+} overload by elevating the Ca^{2+}-ATPase SERCA2 level, increasing the activity of myocardium in rats, and enhancing the capacity of sarcoplasmic reticulum Ca^{2+} reuptake (Li et al., 2012).

P-coumaric acid is the main phenolic acid responsible for the antioxidant and radical scavenging activity of lyophilized aqueous extracts of domesticated and wild ecotypes of raspberry fruits (Gülçin et al., 2011). A fermentation filtrate of *R. coreanus* showed antioxidant and aldose reductase-inhibitory effects and could improve eye injury and visual dysfunction of patients with dry eyes and diabetes (Kwon and Kim, 2011). Ellagitannins contribute significantly to the antioxidant capacity of cloudberries (*R. chamaemorus*) and red raspberries (*R. idaeus*) in lipoprotein and lipid emulsion environments, the latter being more relevant for food applications (Kähkönen et al., 2012). Short-term *R. coreanus* supplementation may offer health benefits by enhancing antioxidant capacity in a healthy Korean population (Lee et al., 2011a). The triterpenoid glycoside nigaichigoside F_1 and its aglycone 23-hydroxytormentic acid (23-HTA) isolated from the unripe fruits of *R. coreanus* ameliorate cisplatin-induced toxicity in renal epithelial LLC-PK_1 cells (Sohn et al., 2011). The protective effect of 23-HTA pretreatment on cisplatin-induced renal damage is associated with the attenuation of oxidative stress and the preservation of antioxidant enzymes. Lipid peroxidation is involved in calcium oxalate (CaOx) stone formation. *R. idaeus* has a remarkable prophylactic effect on CaOx stones in nephrolithic mice (Ghalayini et al., 2011). Coreanoside F1, a dimeric triterpenoid glycoside, has stronger antioxidant activity than its monomers (Nguelefack et al., 2011). Kaempferol-3-*O*-α-L-arabinofuranoside with 3-*O*-α-L-arabinofuranyl has much more potent antioxidant activity than kaempferol-3-*O*-β-D-glucopyranoside with 3-*O*-β-D-glucopyranosyl, and anthocyanin is more potent than its corresponding flavonol glycosides. Methyl (3-hydroxy-2-oxo-2,3-dihydroindol-3-yl)-acetate, vanillic acid, kaempferol, and tiliroside, isolated from the fruits of *R. chingii* (Fu Pen Zi), showed stronger DPPH free radical scavenging activity than that of ascorbic acid (Ding, 2011). The *R. ellipticus* fruits could be pharmaceutically exploited for antioxidant properties (Sharma and Kumar, 2011). Despite these studies, actually most *Rubus* species have not been explored for their constituents with antioxidant activities.

Polyphenolic compounds from *R. coreanum* inhibit the catecholamine secretion from the isolated perfused adrenal medulla of the spontaneously hypertensive rats evoked by the stimulation of cholinergic receptors as well as by direct membrane depolarization (Yu et al., 2009). The inhibitory effect of polyphenolic compounds is mediated by blocking the influx of calcium and sodium into the adrenal medullary chromaffin cells as well as by inhibition of Ca^{2+} release from the cytoplasmic calcium store, which might be involved in the increased nitric oxide (NO) production due to the activation of NO synthase. The constituents of *R. idaeus* seed extracts, mainly polyphenolic compounds, may prevent oxidative lymphocyte damage by reactive oxygen species and may also reduce the level of DNA damage (Godevac et al., 2009).

10.3.2 Effects on the nervous system

Umbelliferone, isolated from *P. evestita*, possessed strong antinociceptive and anti-inflammatory activities by inhibiting both peripherally and centrally acting pain mediators (Rauf et al., 2014).

The *n*-butanol extract of Jue Ma root could also protect the primary hippocampal neurons from hypoxic injury by attenuating mitochondrial cell death (Qin et al., 2012b). Blackberry ingestion attenuates brain degenerative processes with the benefits assigned to the polyphenolics (Tavares et al., 2013). Commercial blackberry extracts were ineffective, but both wild blackberries, *R. brigantinus* and *R. vagabundus*, presented neuroprotective effects.

R. coreanus exerts a memory-enhancing effect via the improvement of cholinergic function and the potentiated antioxidant activity in memory-impaired mice (Choi et al., 2014).

10.3.3 Anti-inflammatory and antiallergic activities

Tormentil (*P. recta*) extracts (TE) have antioxidative properties and are used as a complementary therapy for chronic inflammatory bowel disease. During ulcerative colitis therapy with 2400 mg TE per day, median clinical activity index and C-reactive protein decreased (Huber et al., 2007). TE appeared safe up to 3000 mg/d. Tannins from TE are not systemically absorbed.

Ellagitannins and other natural ellagic acid analogs are potent inhibitors of Janus kinase 2 (McGhie et al., 2012). *P. recta* is one of the numerous cinquefoil species growing in Poland. It is used in traditional medicine, for example, in the treatment of skin inflammation. The strong scavenging activity against H_2O_2 and the inhibition of the enzymatic activity of lipoxidase and, particularly, hyaluronidase observed for the tested *P. recta* subfractions and agrimoniin, a main ellagitannin, partly explain the beneficial effects of *P. recta* in treatment of skin inflammation (Bazylko et al., 2013).

A drink containing boysenberry (*R. loganbaccus* × *R. baileyanus*) juice concentrate with tannin compounds may have anti-inflammatory properties. Polyphenolic-enriched red raspberry extract showed anti-inflammatory effects in an antigen-induced arthritis rat model (Jean-Gilles et al., 2012). Ripe fruit of *R. coreanus* inhibits mast cell-mediated allergic inflammation (Kim et al., 2012a). Anthocyanins induce the activation of matrix metalloproteinase-2 and attenuate the activity of the proinflammatory molecules, such as tumor necrosis factor (TNF)-α and interleukin-1β (Kim et al., 2011). Mexican blackberry extracts inhibit proinflammatory responses and have antioxidant capacity (Cuevas-Rodríguez et al., 2010).

P. mooniana Wight. is a plant found in the Himalayan region where the root is traditionally used to treat stomach problems including gastric ulcer. Ethanol root extract from *P. mooniana* (EPM) significantly reduces the ulcer score against all the four tested gastric ulcer models (Laloo et al., 2014). In the PL model, EPM showed significant reduction in acid pepsin output and cell shedding. EPM (400 mg/kg, p.o.) when compared with ulcer control showed significant increase in gastric pH by 41.6% and decrease in H^+K^+-ATPase activity by 47.73%. EPM showed significant increase in mucus content by 58.60% and a decrease in the microvascular permeability of Evans blue by 85.00%, justifying its protective effects. EPM showed significant antioxidant activity and histopathologically possessed excellent cytoprotective effect. The gastroprotective effect of EPM is attributed mainly to the defensive mechanism owing to the presence of a good quantity of polyphenolic components.

Root extract of *P. alba* is used in the treatment of thyroid pathology in Russia (Turchaninova, 2014). The findings confirm visible reduction of size of the thyroid gland and normalization of its functions in children and adults with thyroid glands with normal or enhanced function treated by Alba per orem.

10.3.4 Antimicrobial and antiparasitic activities

The viscous extracts of *P. fruticosa* (shrubby cinquefoil, Jin Lu Mei in China) showed the antimicrobial activity against Gram-positive (*Staphylococcus aureus* and *Enterococcus faecalis*) and Gram-negative (*Escherichia coli*, *Klebsiella pneumoniae*, *Pseudomonas aeruginosa*, and *Proteus mirabilis*) bacteria, spore-forming bacteria (*Bacillus subtilis* and *Bacillus cereus*), and fungi (*Candida albicans*) (Jurkštienė et al., 2011). The aerial parts of *P. recta* have high polyphenolics (total phenol, phenolic acids, tannins, proanthocyanidins, and flavonoids) content and inhibit cariogenic *Streptococcus* spp. strains (Tomczyk et al., 2011). Preparations of 10 common *Potentilla* plants exhibited inhibitory effects on water-insoluble α-(1 $\rightarrow$ 3)-, α-(1 $\rightarrow$ 6)-linked glucan (mutan) and artificial dental plaque formation (Tomczyk et al., 2010b). The extract from *P. fruticosa* showed the highest antibiofilm activities. *Potentilla* species could be a useful supplement for pharmaceutical products as a new anticariogenic agent in a wide range of oral care products. Whether *Potentilla* tea drinks help prevent caries is worthy of studies. Aqueous extracts from aerial parts of nine *Potentilla* species showed strong antimicrobial activity against *Helicobacter pylori*, a pathogen of gastric disease (Tomczyk et al., 2008b). Among three species, *P. fruticosa* exhibited the strongest inhibition against Gram-positive bacteria, *Pseudomonas aeruginosa* and *Candida albicans* with MIC values of 0.78–6.25 mg/ml (Wang et al., 2013). *P. parvifolia* possessed antibacterial and antifungal activities against all the microorganisms tested, with EC50 and MIC values of 20.52–47.02 and 0.78–50 mg/ml, respectively.

The root of *P. reptans* inhibited growth of Gram-negative bacteria with the 75% ethanol extract having an MIC50 at 1 mg/ml against *Pseudomonas aeruginosa* and the decoction an MIC50 at 3.9 μg/ml against *E. coli* (Watkins et al., 2012).

P. fulgens, a medicinal plant of southwest China and northeast India, is a natural anthelmintic (Roy et al., 2010, 2012). Confirmatory TUNEL assay and DNA fragmentation assay of the extract-treated parasites confirmed the apoptotic nature of cell death of the cestode *Raillietina echinobothrida*, induced by *P. fulgens*, concluded to be responsible for paralysis and death of the parasite (Giri et al., 2013).

The inhibition of viral infection by black raspberry juice on murine norovirus-1 and feline calicivirus F9 probably occurs at the internalization of virions into the cell or the attachment of the viral surface protein to the cellular receptor (Oh et al., 2012). The polyphenol components in black raspberry (i.e., gallic acid and quercetin), however, did not show any activity against these viruses. Black raspberry might be used in the prevention of foodborne viral outbreaks. The extracted oil from the leaves of *R. parvifolius* effectively inhibited the growth of a wide range of Gram-positive and Gram-negative bacteria, including *S. aureus*, *S. epidermidis*, *Enterococcus faecalis*, *E. coli*, *Pseudomonas aeruginosa*, *Acinetobacter baumannii*, *Bacillus cloacae*, and *Klebsiella pneumonia* (Cai et al., 2012). Ellagic acid derivatives from *R. ulmifolius* inhibit *S. aureus* biofilm formation and improve response to antibiotics (Quave et al., 2012). Essential oil from the leaves of *R. corchorifolius* (Shan Mei) showed antibacterial activity (Zhou et al., 2009). All the isolated polyphenols as well as the *R. ulmifolius* leaf extract showed antibacterial activity against two *H. pylori* strains (Martini et al., 2009).

Two human pathogens *Corynebacterium diphtheriae* and *Moraxella catarrhalis* were most sensitive to raspberry extracts (Krauze-Baranowska et al., 2014). Contrary to the extracts, sanguiin H-6 and ellagic acid were only active against eight and nine bacterial strains, respectively. The determined MIC and MBC values of both compounds were several times lower than the tested extracts. The highest sensitivity of *C. diphtheriae* to extracts from both black and red raspberries may be due to its sensitivity to sanguiin H-6 and ellagic acid.

The flavan-3-ol (−)-epicatechin is the most important antiamoebic compound among the extracts from *R. coriifolius* (Soto et al., 2010). 3,5-Dihydroxybenzoic acid, gallic acid, and ethyl galactoside from the aerial parts of *R. niveus* are potent nematicides against root-knot nematode (Sultana et al., 2010).

10.3.5 Anticancer activity

The polyphenol methyl brevifolincarboxylate from *P. argentea* showed a higher level of cytotoxic potency against human breast carcinoma cell line MCF-7 and greater potency as a topoisomerase I/II inhibitor than tiliroside (Tomczyk et al., 2008a,b). The triterpenes corosolic acid and 2β,3β,19α-trihydroxyurs-12-en-28-oic acid of *P. chinensis* show strong anticancer activities (Liu et al., 2006). Apoptosis induction of the methanolic extract of *P. fulgens* root was achieved by suppression of survivin, an antiapoptosis protein (Radhika et al., 2012). Endogenous level of glutathione was depleted significantly in MCF-7 cells after treating with *P. fulgens* root, but not with kaempferol or ellagic acid. This apoptotic induction and lowering the level of glutathione are highly desirable mode for an anticancer agent.

Ethanol extracts of unripe *R. coreanus* suppress the growth of prostate cancer cells by antiproliferative and/or proapoptotic effects, and these effects are stronger in androgen-dependent cells (Kim et al., 2012a,b). SW480 colon adenocarcinoma cells are more susceptible to raspberry leaf extract in comparison with HEp2 laryngeal carcinoma cells (Durgo et al., 2012). Black raspberries (*R. occidentalis*) inhibit multiple stages of oral, esophageal, and colon cancer (Johnson et al., 2011). The water-soluble glucan RCP-1 of *R. crataegifolius* could increase both cytotoxic activity against B16 melanoma cells and the NO production of macrophages *in vitro* (Ni et al., 2009). Furthermore, *in vivo* tests indicated that RCP-1 could remarkably enhance T and B lymphocyte proliferations, augment the phagocytosis of macrophages, and increase the TNF-α levels in serum.

Total alkaloids of *R. alceifolius* inhibit tumor angiogenesis through suppression of the Notch signaling pathway in a mouse model of hepatocellular carcinoma (Zhao et al., 2014).

Unripe *R. coreanus* suppresses migration and invasion of human prostate cancer cells by reducing matrix metalloproteinase expression (Kim et al., 2014).

Aqueous extracts of *P. reptans* rhizome and aerial parts show concentration-dependent cytotoxic effect in the range of tested concentrations (Radovanovic et al., 2013). ICE50 value of *P. reptans* rhizome extract was 280.51 ± 1.16 μg/ml. IC50 value of *P. reptans* aerial parts' extract was 310.79 ± 1.22 μg/ml. The significant difference in cytotoxicity among tested concentrations was observed. Aqueous extracts of *P. reptans* rhizome and aerial parts demonstrated weak cytotoxic activity on 4T1 mouse breast cancer cell line, which is in correlation with current cytotoxicity data for aqueous herbal extracts. Rhizome extract of *P. reptans* has slightly higher antitumor activity than aerial parts' extract.

The total phenolic and total flavonoid contents of *P. fulgens* were 138.8 ± 1.6 mg gallic acid and 401.6 ± 4.6 mg quercetin per 1 g of the extract, respectively (Radhika et al., 2012). The extract resulted in increasing *in vivo* survivality of mice bearing EA cells and loss of cell viability in a dose-dependent manner in MCF-7 cells. This effect may be attributed to apoptotic cell death as confirmed by flow cytometric analysis and PARP1 proteolysis. Such induction of apoptosis was achieved by suppression of inhibitor of the apoptosis protein survivin. Endogenous level of glutathione was depleted significantly in MCF-7 cells after the treatment with root extract of *P. fulgens* only, but not with kaempferol or ellagic acid. This induction of apoptotic cell death and lowering the level of glutathione are highly desirable modes for an anticancer agent.

10.3.6 Hepatoprotective activity

The hepatoprotective activity of *Potentilla* is interrelated to its antioxidant, antiviral, and anti-inflammatory activities and is well documented (Tomczyk and Latté, 2009). *R. parvifolius* extract possessed prominent hepatoprotective effects against CCl4-induced hepatic injury in mice, which was at least partially ascribed to its strong antioxidant capacity (Gao et al., 2011). The presence of various phenolic compounds, including caffeic acid conjugates, ellagic acid glycosides, and flavonol glycosides, might be responsible for the hepatoprotective and antioxidant activities of *R. parvifolius*. *R. sanctus* aqueous alcoholic extract contains a complicated phenolic mixture (ellagitannins, flavonoids, and caffeic acid derivatives) and protects against CCl_4-induced toxicity in isolated rat hepatocytes (Badr et al., 2009). Total alkaloids of *R. alceaefolius* can protect liver in experimental nonalcoholic fatty liver disease, by clearing the oxygen free radical, reducing the product of lipid peroxidation, inhibiting the release of inflammatory cytokines, and reducing inflammatory response (Zheng et al., 2011). Alkaloids from *R. alceifolius* may protect the liver via decreasing cytochrome P450 (CYP) 2E1 enzymatic activity and decreasing its mRNA (Lin et al., 2011). They may also inhibit apoptosis of liver cells (Chen et al., 2010b). Total alkaloids of *R. alceaefolius* significantly reduce the alanine aminotransferase and aspartate aminotransferase levels, protect liver cells from injury, and inhibit the mRNA expressions of CYP2E1 and CYP3A1 in liver tissue (Hong et al., 2009).

Asiatic acid from *P. chinensis* attenuates ethanol-induced hepatic injury via suppression of oxidative stress and Kupffer cell activation (Wei et al., 2013). Tormentic acid (TA), isolated from *P. chinensis*, significantly prevented LPS/D-GalN-induced fulminant hepatic failure (Lin et al., 2014), as evidenced by the decrease in serum aminotransferase and total bilirubin activities and the attenuation of histopathologic changes. TA alleviated the proinflammatory cytokines including TNF-α and NO/iNOS by inhibiting nuclear factor-κB (NF-κB) activity. TA strongly inhibited lipid peroxidation, recruited the antioxidative defense system, and increased HO-1 activity. TA significantly attenuated increases in TUNEL-positive hepatocytes through decreasing the levels of cytochrome c, as well as caspase-3, caspase-8 and caspase-9, while augmenting the expression of Bcl-2. TA protects hepatocytes against LPS/D-GalN-induced injury by blocking NF-κB signaling pathway for anti-inflammatory response and attenuating hepatocellular apoptosis.

A methanol extract from the tuberous roots of *P. anserina* exhibited hepatoprotective effects against d-galactosamine (d-GalN)/lipopolysaccharide (LPS)-induced liver injuries in mice (Morikawa et al., 2014). Six triterpene 28-*O*-monoglucopyranosyl esters, potentillanosides A–F, were isolated from the extract along with 32 known compounds, including 15 triterpenes. Four ursane-type triterpene 28-*O*-monoglycosyl esters, potentillanoside A (IC50 46.7 μM), 28-*O*-β-D-glucopyranosyl pomolic acid (IC50 9.5 μM), rosamutin (IC50 = 35.5 μM), and kaji-ichigoside F1 (IC50 14.1 μM), inhibited d-GalN-induced cytotoxicity in primary cultured mouse hepatocytes. Among these four triterpenes, potentillanoside A, rosamutin, and kaji-ichigoside F1 exhibited *in vivo* hepatoprotective effects at doses of 50–100 mg/kg, p.o. The mode of action was ascribed to the reduction in cytotoxicity caused by d-GalN.

Total alkaloids of *R. alceaefolius* (TARAP) downregulate leptin and resistin and upregulate adiponectin expression in rats with nonalcoholic fatty liver disease (NAFLD) (Zheng et al., 2014). TARAP could improve treatment responses in patients with NAFLD.

10.3.7 Antidiabetic and antiobesity activities

P. discolor (Fan Bai Cao) is used as an ethnomedicine in treatments of diabetes mellitus (DM) in China for years. Four weeks' treatment of P. discolor decoction ameliorated the development of hyperlipidemia, lipid peroxidation, and hyperglycemia associated with hyperphagia and polydypsia in ob/db mice (Song et al., 2012). Total flavonoid extract and total triterpenoid extract of P. discolor showed antidiabetic and antioxidant effects on diabetic rats induced by high-fat diet (HFD) and streptozotocin (Zhang et al., 2010). Sulfated flavonoids from *P. discolor* improve the complications of DM by inhibition of the aldose reductase in both experimental animals and clinical trials. *Astragalus membranaceus* and *P. discolor* mixture improves the insulin resistance of mice with type 2 diabetes (Zhang et al., 2012). The mechanism may be related to increasing the mRNA expressions of phosphatidylinositol 3-kinase and peroxisome proliferator-activated receptor γ coactivator 1 in the liver. The synthesized tiliroside derivatives showed significant antidiabetic activities *in vitro* (Qin et al., 2011), while trans-tiliroside from *P. chinensis* is an active principle with antihyperglycemic, antihyperlipidemic, and antioxidant effects (Qiao et al., 2011). Tiliroside derivatives enhance GLUT4 translocation via AMP-activated protein kinase in muscle cells (Shi et al., 2011).

R. coreanus (Cha Tian Pao, Korean black raspberry) may rescue diabetic osteoporotic bone loss by simultaneous alteration of activation in osteoblasts and osteoclasts (Choi et al., 2012). These effects may be partially influenced by the upregulation of the endocannabinoid system. Dietary *R. coreanus* may be of use in improving the conditions of diabetic osteoporosis. In spite of chemical similarity between *Rubus* and *Potentilla*, the potential antidiabetic activity of *Rubus* has not been well studied.

The improved obesity phenotype by Chinese sweet leaf tea (*R. suavissimus*, Tian Cha) components may be associated with the attenuated expression of vascular endothelial growth factor in preadipocyte 3T3-L1 cells (Koh et al., 2011). A possible approach to treat obesity is to promote early adipogenesis in adipose tissue, thereby leading to replacement of enlarged adipocytes that secrete inflammatory factors with small adipocytes. Tian Cha extract enhances early adipogenesis by increasing the expression of adipogenic transcription factors C/EBPα and PPARγ (Ezure and Amano, 2011).

Rubi Fructus (*R. coreanus*) prevents the development of obesity in mice fed with a HFD and stimulates the differentiation of brown adipocytes through the modulation of thermogenic genes (Jeong et al., 2014).

10.3.8 Immunomodulatory effects

Polysaccharide from *P. anserina* showed immunomodulatory activity *in vitro* and *in vivo* (Chen et al., 2010a,b). Treatment with *P. indica* produced a time- and dose-dependent increase in cell viability and proliferation (Ang et al., 2014). Following 72 h of treatment with *P. indica* and *D. pentandra*, thymocyte proliferation was augmented by 18% and 41%, respectively, and splenocyte proliferation increased by 35% and 42%, respectively, when compared with untreated cells. These extracts may act as potential immunostimulants and, thus, represent an alternative source of immunomodulatory compounds for the treatment of human immune-mediated diseases.

10.3.9 Other effects

P. alba extracts significantly increased swimming endurance time and have anxiolytic-like action with a predominant locomotor component (Shikov et al., 2011). An ethanolic extract from *R. coreanus* showed fatigue-alleviating effect on mice (Lee et al., 2011b). The extract of *P. chinensis* (Chinese cinquefoil) flower inhibited neurotransmitter release from neuronal PC12 cells by approximately 90% at a concentration of 10 μg/ml (Jung et al., 2012). The extract paralyzed muscle as efficiently as botulinum neurotoxins, suggesting its potential versatility in cosmetics and therapeutics. Quercetin-4′-*O*-β-D-glucoside of *P. bifurca* showed strong tyrosinase inhibitory activities (Piao et al., 2009).

R. coreanus inhibits oxidized low-density lipoprotein (LDL) uptake by macrophages through regulation of c-Jun N-terminal kinase activation and prevents atherosclerosis (Bhandary et al., 2012). The blackberry nectar reduced the serum triglyceride level, total cholesterol, and LDL cholesterol in hypercholesterolemic hamsters, without influencing the HDL and blood glucose concentrations (Ferreira de Araujo et al., 2011). A decrease in the initiation of lipid peroxidation reactions in the blood, brain, and small intestine was also observed.

R. coreanus can potentially improve visual sensitivity and can be used to treat eye disease (Wahid et al., 2011). The unripe *R. coreanus* extract exerts a relaxing effect on penile cavernous tissue in part by activating the NO-cGMP system, and it may improve erectile dysfunction, which does not completely respond to sildenafil citrate (Zhao et al., 2011). Shoots of *Rubus* species have been used for wound healing, infected insect bites, and pimples in folk medicine for ages (Süntar et al., 2011). *R. coreanus* extracts stimulate cannabinoid receptors and control osteoporosis in aged male rats (Lim et al., 2014).

10.3.10 Toxicity and safety

P. erecta (tormentil) and its rhizome extracts have been known for a long time in traditional medicine as a remedy for the treatment of inflammations, wounds, and gastrointestinal disorders. Tormentil rhizomes have also been used as part of alcoholic beverages in Germany, Ukraine, and Russia. *P. erecta* rhizome extracts should be considered safe with respect to acute toxicity when applied to humans (Shushunov et al., 2009). The use of raspberry leaf in pregnancy is a traditional herbal therapy and is recommended by some midwives. Due to the lack of evidence for safety and efficacy, such recommendations are questionable (Holst et al., 2009). The safety issues of *Potentilla* and *Rubus* tea for long-term consumption should be investigated thoroughly.

Among the 128 *R. idaeus* samples, 66 (51.6%) were found to detect pesticide residues (Łozowicka et al., 2012): 14.1% contained one pesticide and around 38% multiple pesticide residues. The most frequently detected were pyrimethanil residues (36.0%). Twenty-seven (21.1%) raspberry samples exceeded the maximum residue limits. The estimated daily intakes ranged from 0.003% to 3.183% of the acceptable daily intake (ADI) for adults 0.008% and 9.7% for toddlers. The most critical case is procymidone; the acute risk was 180.9% of acute reference dose (ARfD) for toddlers and 83% of ARfD for adults, which is high.

10.4 Molecular pharmacognosy

10.4.1 Taxonomy, phylogeny, and evolution

10.4.1.1 Taxonomy and authentication

Leaves of nine kinds of plants from three subfamily of Rosaceae were used in FTIR analyses (Qiu et al., 2014). Genetic relationship was analyzed and species were identified through studying FTIR spectra. Leaves mainly contain large amounts of carbohydrates, proteins, lipids, nucleic acids, and other substances. The peaks of carbohydrates are mainly between 1440 and 775 cm^{-1}. The vibration peaks of the cellulose and lignin are between 1440 and 1337 cm^{-1}. The peaks between 1000 and 775 cm^{-1} are stretching vibration of ribose. The vibration peaks of protein are between 1620 and 1235 cm^{-1}. The peak at 1620 cm^{-1} is sensitive to C=O stretching vibration of protein amide I. The peak at 1523 cm^{-1} is assigned to N—H and C—N stretching vibration of protein amide II. Peaks of lipids mainly appeared between 2930 and 1380 cm^{-1}. The peak at 2922 cm^{-1} is CH_2 stretching vibration of fat. The peak at 1732 cm^{-1} is C=O stretching vibration of fatty acids. The mark peak of the nucleic acid appears in the region between 1250 and 1000 cm^{-1}. The peak at 1068 cm^{-1} is due to the symmetrical stretching vibration of PO^{2-} group of the phosphodiester-deoxyribose backbone, and the peak at 1246 cm^{-1} is associated to the asymmetrical stretch vibration of PO^{2-} group. The cluster model is established by smoothing, standardizing, the second derivative, principal component analysis (PCA), and hierarchical cluster analysis. It is in accordance with the traditional classification. *Prunus armeniaca* L. and *Prunus seudocerasus* Lindl. were clustered into Prunoideae. *P. fulgens* Wall., *Rosa chinensis* Jacd, and *Fragaria ananassa* Duchesne var. were clustered into Rosoideae. *Pyracantha fortuneana* Li, *Malus pumila* Mill. *Eriobotrya bengalensis* Hook. f., and *Malus hallianna* Koehne were clustered into Pomoideae. The correct rate of cluster at subfamily is 100%. The correct rate of cluster at genus is 55.56%. The correct rate of identification is 100% when unknown species were laid into the model of hierarchical cluster to identify.

10.4.1.2 Rubus

Internal transcribed spacer (ITS) sequences were used to generate the first phylogeny of *Rubus* (Alice and Campbell, 1999). 57 taxa including 20 species of subgenus *Rubus* (blackberries) were sampled and one to seven species from each of the remaining 11 subgenera, the monotypic and closely related *Dalibarda*. In *Rubus*, ITS sequences are most informative among subgenera, and variability is low between closely related species. Parsimony analysis indicates that *Rubus* and *Dalibarda* form a strongly supported clade, and *D. repens* may nest within *Rubus*. Of the subgenera with more than one species sampled, only subgenus *Orobatus* appears monophyletic. Three large clades are strongly supported: one contains all sampled species of nine of the 12 subgenera; another includes extreme Southern Hemisphere species of subgenera *Comaropsis*, *Dalibarda*, and *Lampobatus*; and a third clade consists of subgenus *Rubus* plus *R. alpinus* of subgenus *Lampobatus*. *R. ursinus* might be a hybrid between a close relative of *R. macraei* (subgenus *Idaeobatus*, raspberries) and an unidentified subgenus *Rubus* species. ITS sequences are generally consistent with biogeography and ploidy, but traditionally important morphological characters, such as stem armature and leaf type, appear to have limited phylogenetic value in *Rubus*. Figure 10.6a shows an extended ITS tree.

Hybridization has long been considered a source of taxonomic complexity in *Rubus*. Molecular data were used to examine natural hybridization between *R. caesius*, a facultatively agamospermous tetraploid of subgenus *Rubus* and *R. idaeus*, a sexual diploid of subgenus *Idaeobatus* (Alice et al., 2001). Both species are widespread in Europe, and hybridization between them is hypothesized to have given rise to numerous species, including *R. maximiformis* and *R. picticaulis*. ITS region of *R. caesius*, *R. idaeus*, five putative hybrids from Sweden, and one individual each of *R. maximiformis* and *R. picticaulis* from Germany was sequenced. *R. caesius* differs from *R. idaeus* at 20

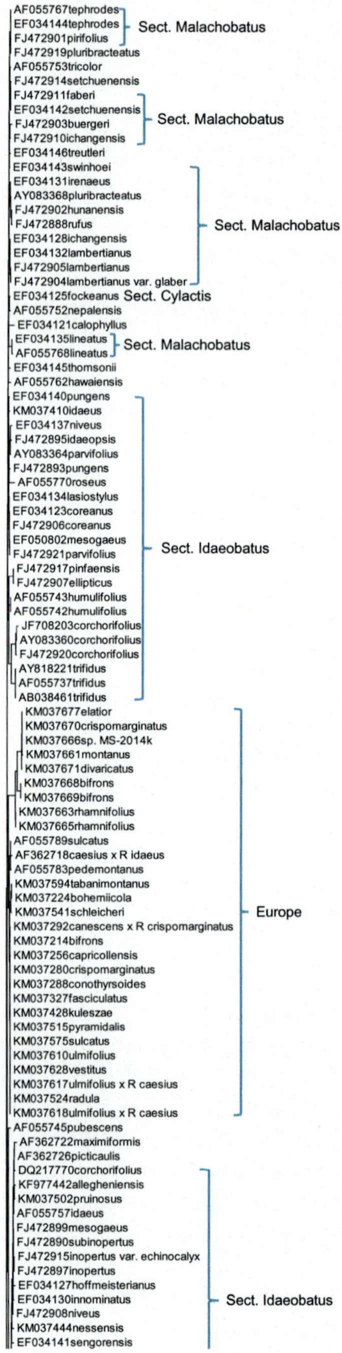

Figure 10.6 Phylogenetic relationship of *Rubus*. The evolutionary history was inferred by using the ML method based on the GTR+I+G model. Trees are drawn to scale, with branch lengths measured in the number of substitutions per site. The analysis involved 269 ITS (a) and 311 trnL-F (b) sequences. There were 981 (a) and 1271 (b) positions in the final datasets. Evolutionary analyses were conducted in MEGA6.

(Continued)

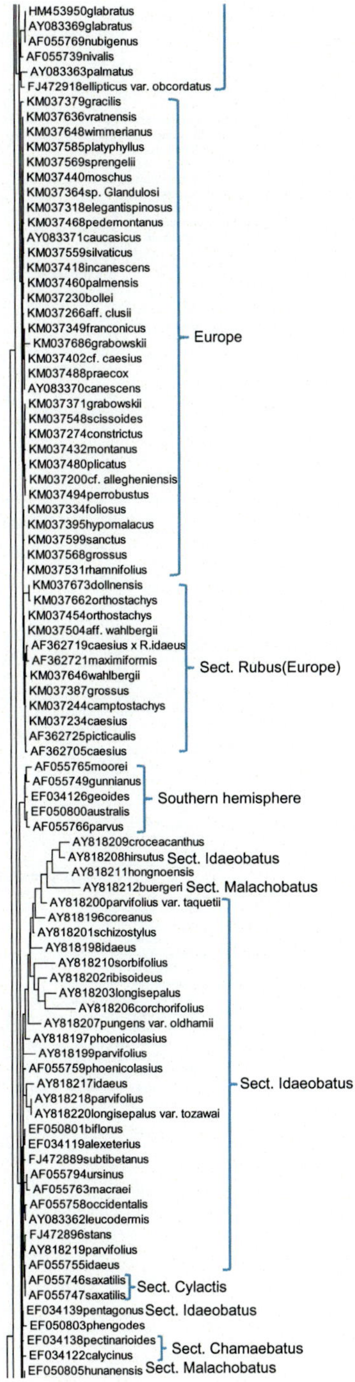

Figure 10.6 Continued.

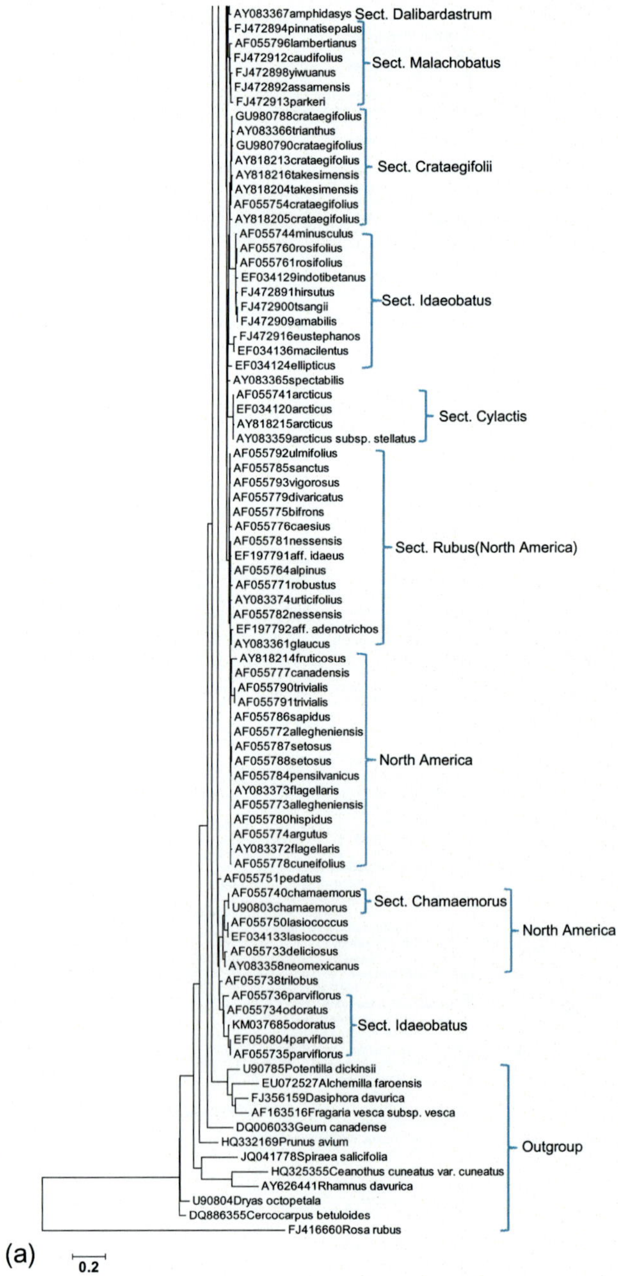

Figure 10.6 Continued.

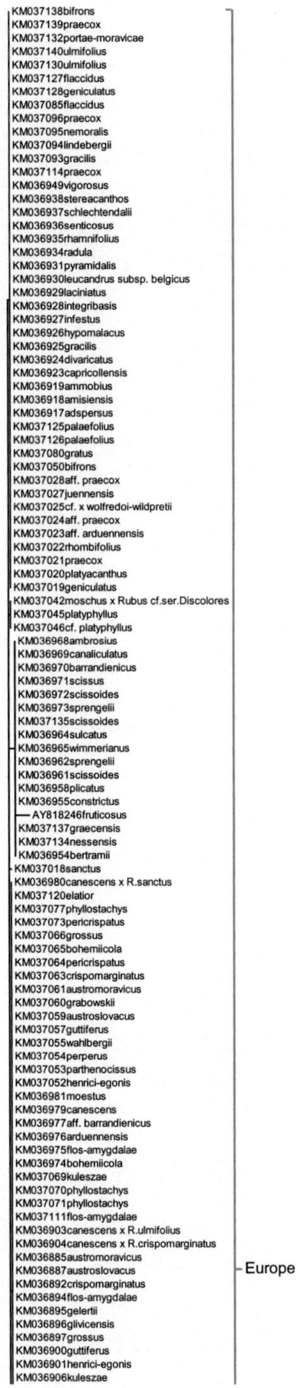

Figure 10.6 Continued.

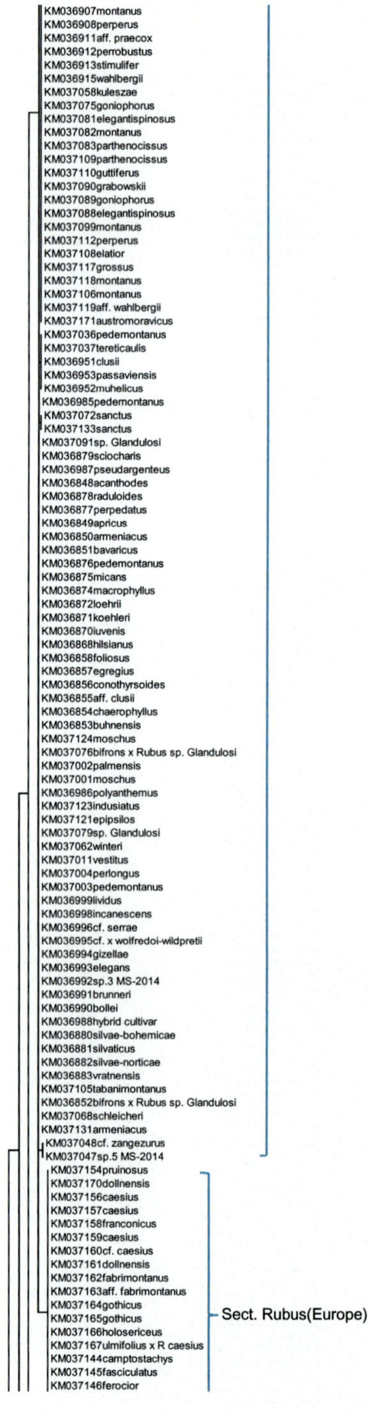

Figure 10.6 Continued.

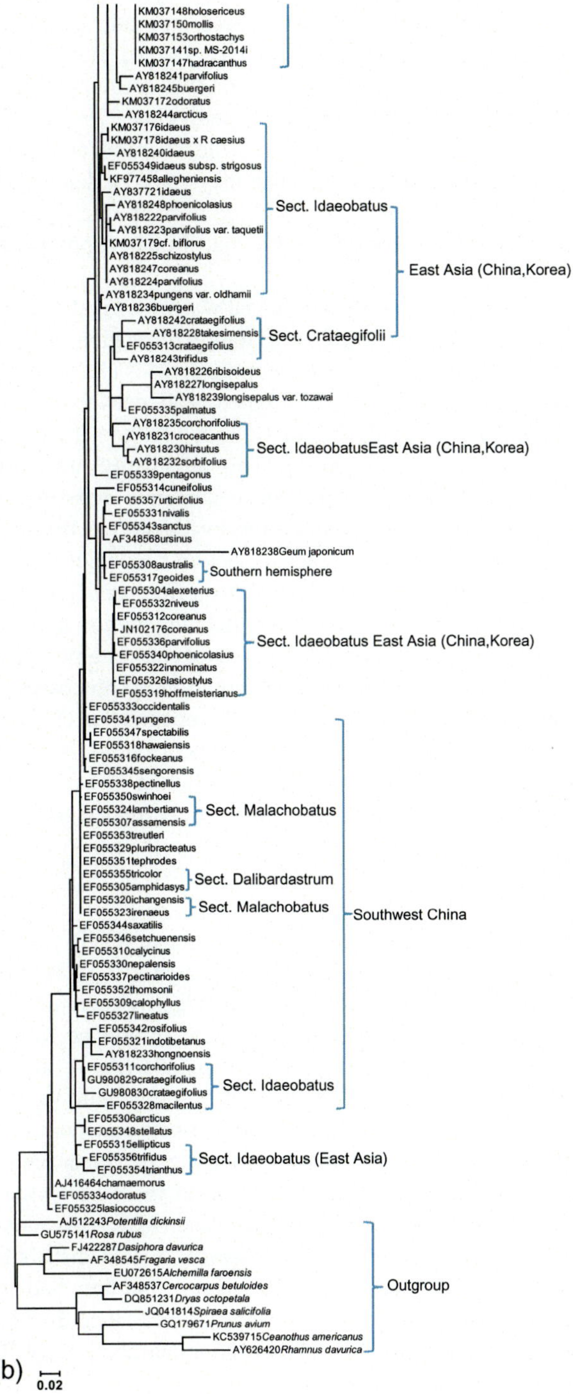

(b)

Figure 10.6 Continued.

nucleotide sites and one insertion/deletion event. The putative hybrids exhibit nucleotide polymorphism additivity at all or most of these sites indicating that gene flow does occur naturally between *R. caesius* and *R. idaeus*. Two to 21 ITS region clones were sequenced for each of five putative hybrids, and two restriction enzymes were used to distinguish the parental species at three sites. Individual 729 has only intact parental ITS sequences and may therefore be an F1, while the other four show chimeric (mixtures of parental) ITS sequences, suggesting that they may be later-generation hybrids. Figure 10.6b shows the phylogenetic tree inferred from cp trnL-F sequences. Conflicts between cp and ITS trees are easily observed, which support the hypothesis of hybridization.

The phylogeny of the genus *Rubus* in Korea was investigated using an ITS of the nuclear ribosomal (nr) DNA and a trnL-trnF intergenic region of the chloroplast (cp) DNA (Yang and Pak, 2006). Twenty-one ingroup species (1.2 kb for each) were analyzed using parsimony, resulting in 672 aligned sequences from ITS and 502 bases with trnL/F. Individual and combined analysis of ITS and trnL/F data proved that the genus *Rubus* is a monophyletic group. This phylogeny substantiated a previous sectional classification scheme rather than a subgenus classification. Results did not support the earlier sectional classification by Focke (section *Corchorifolii*) but support the sectional classification of Nakai: section *Crataegifolii* (*R. crataegifolius*, *R. takesimensis*, and *R. trifidus*) and section *Villosii* (*R. corchorifolius*). Most species, which are found in Korea and belong to subgenus *Idaeobatus*, appeared in two different groups in all data sets, suggesting that this subgenus is a polyphyletic group that has gone through at least two independent evolutionary processes. The taxa, when mapped onto the combined tree, showed that the occurrence of their morphological characters of simple and compound leaves was concurrent in Korean *Rubus*. ITS data were consistent overall with the geographic distribution of each species. The trnL/F data provided phylogenetic information within closely related species (Figure 10.6b).

10.4.1.3 Potentilla

Alpine species occurring in the Qinghai-Tibetan Plateau (QTP) responded to past climatic oscillations. The genetic distribution pattern of 101 individuals of *P. glabra* comprising both the interior QTP and the plateau edge was examined using variations of the cp trnT-L (Wang et al., 2009a). Phylogenetic and network analyses of 31 recovered haplotypes identified three tentative clades (A, B, and C). Analysis of molecular variance (amova) revealed that most of the genetic variability was found within populations (0.693), while differentiations between populations were obviously distinct (F_{st} 0.307). Two independent range expansions within clades A and B occurring at approximately 316 and 201 thousand years ago (kya) were recovered from the hierarchical mismatch analysis, and these two expansions were also confirmed by Fu's FS values and "g" tests. However, distant distributions of clade C and private haplotypes from clades A and B suggest that they had survived the Last Glacial Maximum (LGM) and previous glaciers in situ since their origins. Multiple refugia of a few cold-enduring species had been maintained in the QTP platform during LGM and/or previous glacial stages.

A reconstruction of phylogenetic relationships based on three cpDNA markers comprising 98 species of the genus *Potentilla* and 15 additional genera from the tribe Potentilleae (Rosaceae) supported the current generic concept of the subtribe Fragariinae and resolved major lineages within the subtribe Potentillinae (Dobeš and Paule, 2010). Resolved lineages corresponded to the combined genera *Argentina*, *Piletophyllum*, and *Tylosperma*, a European and Asian group of species approximately equivalent to the *Trichocarpae Herbaceae* sensu Th. Wolf, the series *Tormentillae* Th. Wolf, a North American clade comprising the genera *Horkelia*, *Horkeliella*, and *Ivesia*, the species *P. fragarioides* and *P. freyniana*, and a taxonomically highly diverse but molecularly little diverged core group of *Potentilla*, respectively. Age estimates of phylogenetic splits resolved in the Potentilleae using Bayesian inference suggested a diversification of the tribe in the Eocene and radiation of two major evolutionary lineages corresponding to the Fragariinae and Potentillinae at approximately comparable times. Ancestral area reconstructions based on the recent distribution ranges of species and collection sites of cpDNA haplotypes suggested an Asian origin for *Potentilla* s.str. and explained the arrival of this still informal taxon in Europe and particularly North America by multiple dispersal events. In combining the phylogenetic and geographic data with molecularly inferred time estimates and taxonomy, strongly contrasting evolutionary patterns were identified, including rapid speciation on a continental and worldwide scale accompanied by multiple intercontinental dispersals. Figure 10.7 shows the extended phylogenetic trees inferred from ITS and trnL-F sequences. Species such as *P. polyphylla*, *P. stenophylla*, *P. microphylla*, *P. peduncularis*, and *P. cuneata* are basal to most other species, implying that QTP and its flanking regions might be an origin center of *Potentilla*. Similar to *Rubus*, morphological convergence is not uncommon in *Potentilla*. The extended ITS tree (Figure 10.7a) is largely congruent for most major lineages with cp tree (Figure 10.7b; Dobeš and Paule, 2010; Töpel et al., 2011), suggesting that hybridization among distant lineages might not play a significant role in the history of *Potentilla* and related genera.

Several naturally occurring hybrids in *Potentilla* have been reported, which is supported by molecular evidence. An nr (ITS+ETS) and a cp (trnL-F+trnS-G) gene tree were compared to identify topological incongruences that may indicate hybridization events in the genus (Töpel et al., 2011). The monophyly and phylogenetic position of the proposed segregated genera *Argentina*, *Ivesia*, and *Horkelia* were tested. The systematic signal from the two morphological characters, style and anther shape, was investigated by ancestral state reconstruction to elucidate how well these characters concur with the results of the molecular phylogenies. Six major clades, Anserina, Alba, Fragarioides, Reptans, ivesioid, and Argentea, were identified within genus *Potentilla*. *Horkelia*, *Ivesia*, and *Horkeliella* (the ivesioid clade) form a monophyletic group nested within *Potentilla*. The origin of the proposed segregated genus *Argentina* (the Anserina clade) is uncertain but not in conflict with a new generic status of the group. Style morphology is an informative character that reflects the phylogenetic relationships within *Potentilla*. Five well-supported incongruences were found between the nr and the cp phylogenies, and three of these involved polyploid taxa. From the tree topology, it is unambiguous that ITS is of better resolving power than

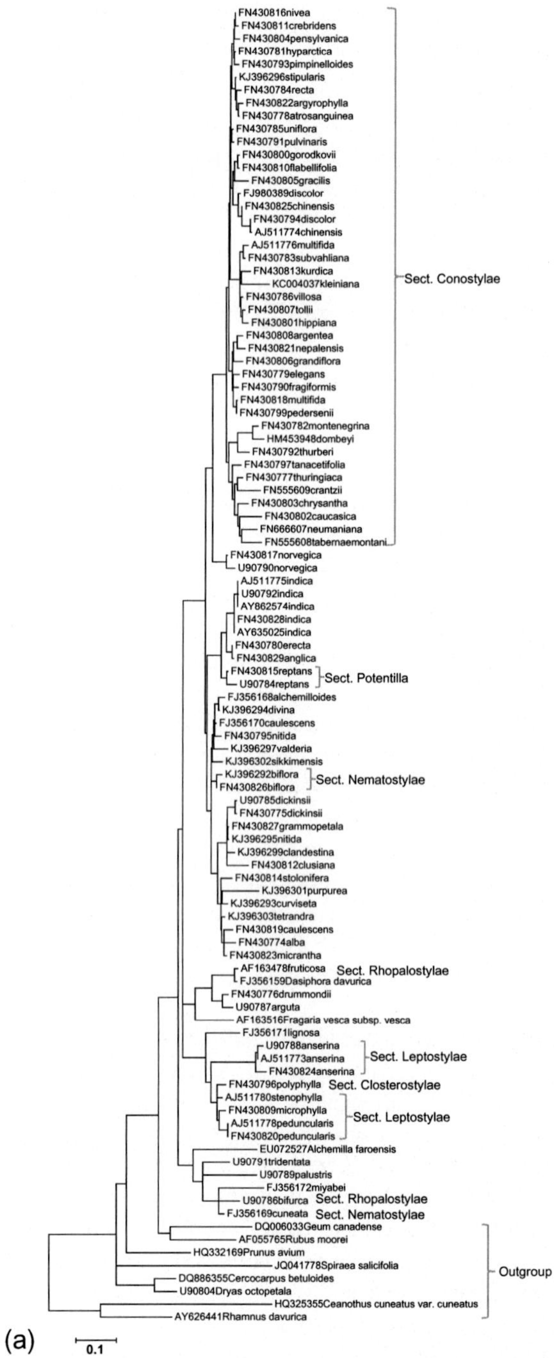

Figure 10.7 Phylogenetic relationship of *Potentilla*. The evolutionary history was inferred by using the maximum likelihood (ML) method based on the GTR+I+G model. Trees are drawn to scale, with branch lengths measured in the number of substitutions per site. The analysis involved 102 ITS (a) and 287 trnL-F (b) sequences. There were 896 (a) and 1462 (b) positions in the final datasets. Evolutionary analyses were conducted in MEGA6.

(Continued)

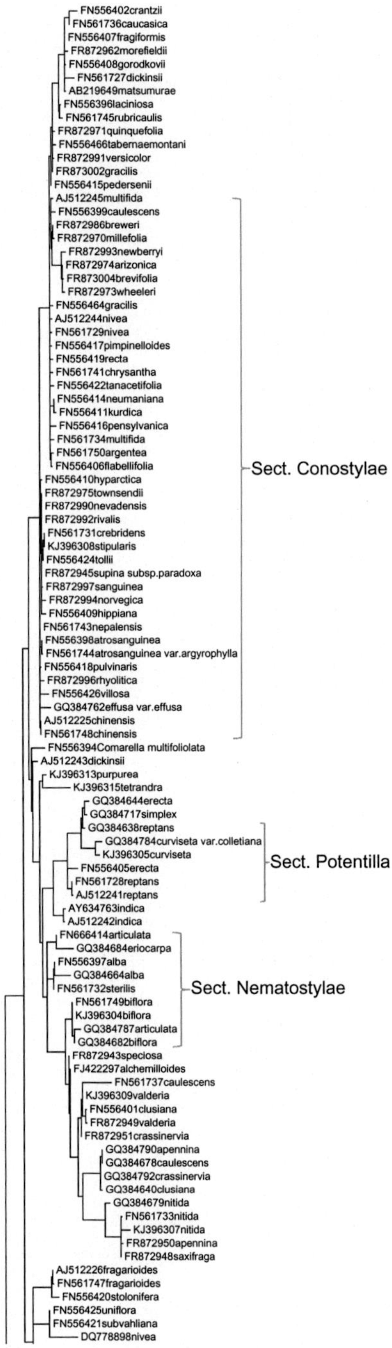

Figure 10.7 Continued.

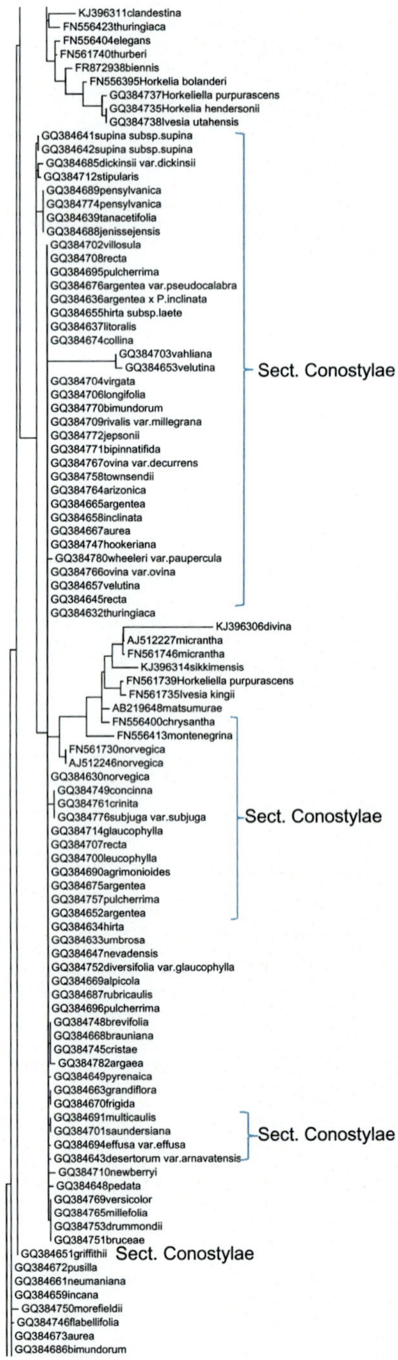

Figure 10.7 Continued.

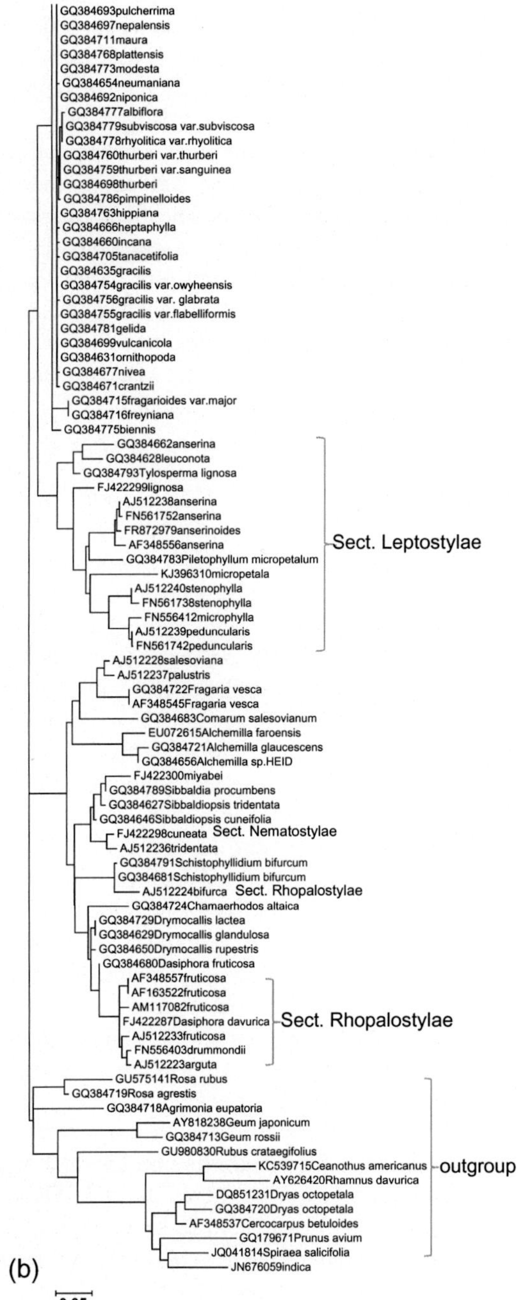

Figure 10.7 Continued.

the cp intergenic spacer sequences (Töpel et al., 2011; Dobeš and Paule, 2010). Further investigations, using low-copy molecular markers, are required to infer the phylogeny of these species and to test the hypothesis of hybrid origin.

Potential events of allopolyploidy may be indicated by incongruences between separate phylogenies based on plastid and nuclear gene sequences. Two plastid regions and two nr regions were sequenced for 34 ingroup taxa in Fragariinae (Rosaceae) and six outgroup taxa (Lundberg et al., 2009). Five well-supported incongruences were found that might indicate allopolyploidy events. The incongruences involved *Aphanes arvensis*, *P. miyabei*, *P. cuneata*, *Fragaria vesca/moschata*, and *Drymocallis* clade. Allopolyploidy may be hypothesized in the first four cases. Phylogenies were estimated using Bayesian inference and analyses were evaluated using convergence diagnostics. Two orphan *Potentilla* species, *P. miyabei* and *P. cuneata* (in Himalaya and Hengduan Mountains), are placed in Fragariinae. Due to unresolved topological incongruences, they are not reclassified in any genus.

10.4.2 *Potentilla*

Polyploidy and hybridization are prevalent in both *Potentilla* and *Rubus*. Combined analysis of AFLP, cpDNA sequences, and ploidy level variation revealed a hybrid origin of the *P. alpicola* populations from South Tyrol (Italy) resulting from crosses between *P. pusilla* and two cytotypes of *P. argentea* (Paule et al., 2012).

Microsatellites are valuable markers for the analysis of genetic diversity, linkage mapping, or genotyping. The 454 high-throughput pyrosequencing identified 1476 simple sequence repeat (SSR) containing sequences from *P. pusilla*, from which 72 polymorphic markers were developed (Dobeš and Scheffknecht, 2012). Fosmid genomic library and end sequencing are useful in molecular marker mining (Hao et al., 2011a) but have not been applied on *Potentilla*.

Second-generation sequencing has permitted detailed sequence characterization at the whole genome level of a growing number of nonmodel organisms, but the data produced have short read lengths and biased genome coverage leading to fragmented genome assemblies. The PacBio RS long-read sequencing platform offers the promise of increased read length and unbiased genome coverage and thus the potential to produce genome sequence data of a finished quality containing fewer gaps and longer contigs. The PacBio RS sequencing platform was used to sequence and *de novo* assemble the *P. micrantha* cp genome (Ferrarini et al., 2013). Following error correction, 28,638 PacBio RS reads were recovered with a mean read length of 1902 bp totalling 54,492,250 nucleotides and representing an average depth of coverage of $320\times$ the cp genome. The dataset covered the entire 154,959 bp of the cp genome in a single contig (100% coverage) compared to seven contigs (90.59% coverage) recovered from an Illumina data and revealed no bias in coverage of GC-rich regions. Post assembly, the data were largely concordant with the Illumina data generated and allowed 187 ambiguities in the Illumina data to be resolved. The additional read length permitted small differences in the two inverted repeat regions to be assigned unambiguously. PacBio data will be of immense utility for the development of genome sequence assemblies containing fewer unresolved gaps and ambiguities and a significantly smaller number of contigs than could be produced using short-read sequence data alone.

10.4.3 Rubus

10.4.3.1 Molecular marker

An F1 population developed from a cross between an advanced breeding selection of *R. occidentalis* and *R. idaeus* "Latham" was used to construct a new genetic map consisting of DNA sequence-based markers (Bushakra et al., 2012). The alignment of the orthologous markers suggests that the genomes of *Rubus* and *Fragaria* have a high degree of synteny, which decreases with phylogenetic distance. The genetic map gives unprecedented insights into the genome evolution of raspberry from the putative ancestral genome of the single ancestor common to Rosaceae. The cultivated *Rubus* hybrids were authenticated with random amplification of polymorphic DNA (RAPD) (Jia et al., 2008). SSR markers demonstrate genetic and molecular differentiation between thorny and thornless Andean blackberries (*R. glaucus*) (Marulanda et al., 2012). Southwest China is believed to be an origin center of extant *Rubus* species and possesses many endemic taxa (Dong, 2009), some of which have been used by local community for medicinal and food uses and comprise an invaluable non-*Camellia* tea resource. Various genomic and genetic tools should be combined to shed some light on those endemic *Rubus* species.

Morphological and molecular identification methods were developed for *R. arcticus* (Kostamo et al., 2013). The best morphological characteristics were the length-to-width ratio of the middle leaflet and leaf margin serration. A particular characteristic, fingertip touch, was shown by electron microscopy to be related to the density and quality of the leaf hairs. Red raspberry SSR marker no. 126 proved to be applicable for differentiation of the eight arctic bramble cultivars tested. These identification methods are critical to secure the maintenance and management of *R. arcticus*. The challenges faced and approaches taken are equally applicable to other species with similar biology.

Color and chemical composition of fruits of 10 red raspberry genotypes grown in Nordic climate during three harvest seasons were studied (Mazur et al., 2014). The main phenolic compounds in the fruits were ellagitannins and anthocyanins, contributing 57% and 42% to the quantified phenolic compounds, respectively. Cyanidin-3-sophoroside was the most abundant anthocyanin (61%). All quality parameters were significantly affected by genotypes, which could be categorized into three groups. "Veten" and "RU984 06038" were characterized by high concentrations of flavonoids, that is, anthocyanins and quercetin glycosides, and dark red color. "Octavia", "Glen Magna", "RU004 03067", "Glen Ample", and "RU974 07002" were characterized by light color, high titratable acids, and low flavonoid concentrations. "Malling Hestia", "RU024 01003", and "RU004 04095" had high content of dry matter, soluble solids, ascorbic acid, and ellagic acid containing compounds, in addition to high hue and chroma values. All quality parameters, except ascorbic acid and lambertianin C, varied significantly between harvest seasons. The lowest seasonal variation in fruit quality was observed in "RU024 01003" and "Glen Ample" and the highest "RU004 03067" and "Glen Magna".

UPLC, HPLC, and two *Rubus* linkage maps were used to conduct QTL analyses to explore the presence of associations between concentrations of five anthocyanins in fruit and genotype (Bushakra et al., 2013). Twenty-seven QTL were identified on the *Rubus* linkage maps, four of which are associated with molecular markers designed from transcription factors and three of which are associated with molecular markers designed from anthocyanin biosynthetic pathway candidate genes. While QTL for anthocyanin accumulation have been identified on six of seven *Rubus* linkage groups (RLG), the QTL on RLG2 and RLG7 may be very important for genetic control of cyanidin modification in *Rubus*.

10.4.3.2 Genomics and transcriptomics

Genotyping by sequencing (GBS) was used to produce highly saturated maps for a *R. idaeus* pseudo-testcross progeny (Ward et al., 2013). While low coverage and high variance in sequencing resulted in a large number of missing values for some individuals, a novel method of imputation based on maximum likelihood (ML) marker ordering from initial marker segregation overcame the challenge of missing values and made map construction computationally tractable. The two resulting parental maps contained 4521 and 2391 molecular markers spanning 462.7 and 376.6 cM, respectively, over seven linkage groups. Detection of precise genomic regions with segregation distortion was possible because of map saturation. Microsatellites (SSRs) linked these results to published maps for cross validation and map comparison. GBS together with genome-independent imputation provides a rapid method for genetic map construction in any pseudo-testcross progeny. Comparison of SSRs to published *R. idaeus* maps showed that the linkage maps constructed with GBS and the imputation method were robust and marker positioning was reliable. The high marker density allowed identification of genomic regions with segregation distortion in *R. idaeus*, which may help to identify deleterious alleles that are the basis of inbreeding depression in the species.

R. idaeus (red raspberry) is one of the important economical crops that possess numerous nutrients, micronutrients, and phytochemicals with essential health benefits to human. Due to the lack of *Rubus* genome sequences, over 70,000,000 86-nt Illumina high-throughput sequencing reads derived from *R. idaeus* roots were aligned to the *Fragaria vesca* genome (Ward et al., 2012). Alignment identified 16956 putatively expressed genes. On the other hand, *de novo* transcriptome assembly identified over 36,000 transcripts that had matches to plants. These methods generate resources for the studies of secondary metabolism and molecular marker mining. In the future, the tissue-specific transcriptome sequencing should be performed to study the gene expression difference between different tissues (Hao et al., 2011b), to correlate the genotype with metabolic phenotype, and thus to rationalize the use of *Rubus* leaves as tea drinks.

The large-scale transcriptome sequencing data, more than 51 million sequencing reads, containing abundant information on genes involved in the metabolic pathways in *R. idaeus* "Nova" fruits were generated using Illumina sequencing technology (Hyun et al., 2014a). The molecular mechanism underlying the ripening process and phytochemical biosynthesis in red raspberry is attributed to the changes in gene expression. 42,604 Unigenes with an average length of 812 bp were obtained by *de novo* assembly. At the protein level, Nova fruit transcriptome showed 77% and 68 % sequence similarities with *R. coreanus* and *Fragaria versa*, respectively, indicating the evolutionary relationship between them. Sixty-nine percent of assembled unigenes were annotated using public databases including NCBI nonredundant, Clusters of Orthologous Groups, and Gene Ontology database. To analyze the relationship between several novel transcripts and the amounts of metabolites such as γ-aminobutyric acid and anthocyanins, real-time PCR and target metabolite analysis were performed on two different ripening stages of Nova. The comprehensive transcriptome resource will be useful for understanding the ripening process and for breeding *R. idaeus* cultivars with improved fruit quality.

The Korean black raspberry (*R. coreanus*, KB) on ripening is usually consumed as fresh fruit, whereas the unripe KB has been widely used as a source of traditional herbal medicine. Such a stage-specific utilization of KB has been assumed due to the changing metabolite profile during fruit ripening process. To analyze biochemical changes during fruit ripening process at molecular level, the transcriptome of KB fruits was sequenced, assembled, and annotated (Hyun et al., 2014b). Over 4.86 Gb of normalized cDNA prepared from fruits was sequenced using Illumina HiSeq™ 2000 and assembled into 43,723 unigenes. Anthocyanins and proanthocyanidins varied greatly in two stages of fruits. Upregulation of F3′H1, DFR4, and LDOX1 resulted in the accumulation of cyanidin derivatives during the ripening process of KB, indicating the positive relationship between the expression of anthocyanin biosynthetic genes and the anthocyanin accumulation. The ability of *R. coreanus* Miquel chalcone flavanone isomerase 2 (RcMCHI2) gene to complement *Arabidopsis* transparent testa 5 mutant supported the feasibility of the transcriptome library to provide the gene resources for improving plant nutrition and pigmentation. Transcriptome library and metabolic profiling would help to define the gene-metabolite relationships in this nonmodel plant.

10.5 Proteomics

The alpine environment, which is characterized by sharp temperature shifts, high levels of ultraviolet radiation exposure, and low oxygen content, limits plant growth and distribution. *P. saundersiana* is widespread in the QTP. A comparative proteomics approach was used to investigate the mechanisms by which *P. saundersiana* withstands the alpine environment by examining plants located at five different altitudes (Ma et al., 2014). 118 Proteins spots with variable abundance were detected and functionally characterized. Proteins involved in antioxidant activity, primary metabolites, epigenetic regulation, and protein posttranslational modification play important roles in conferring tolerance to alpine environments. *P. saundersiana* modulated the root architecture and leaf phenotype to enhance adaptation to alpine environmental stress through mechanisms that involved hormone synthesis and signal transduction, particularly the cross talk between auxin and strigolactone (Ma et al., 2014). An increase of antioxidant proteins and primary metabolites as a response to the alpine environment in *P. saundersiana* was observed. Proteins associated with the epigenetic regulation of DNA stability and posttranslational protein degradation were involved in this process. *P. saundersiana* uses multiple strategies to adapt to the high-altitude environment of the Alpine region.

10.6 Metabolomics

P. discolor is an important crude materials in traditional Chinese medicine (TCM) for therapy of hyperglycemia and hyperlipidemia. A plasma metabonomic approach based on the combination of UPLC-Q-TOF with multivariate data analysis was applied to investigate the therapeutic effects of the extract of *P. discolor* (EPD) and corosolic acid (CA) (Li et al., 2014), the main bioactive compounds of *P. discolor*. Male C57BL/6 mice were fed with high-fat diet (HFD-fed group) for 8 weeks and then treated with EPD (EPD-treated group) or CA (CA-treated group) for another 8 weeks. After the experimental period, samples of plasma were collected and analyzed by ultra performance liquid chromatography/quadrupole time-of-flight mass spectrometry (UPLC-Q-TOF). The PCA and partial least squares discriminant analysis (PLS-DA) models were built to find biomarkers of type 2 diabetes mellitus (T2DM) and investigate the therapeutic effects of EPD and CA. 26 Metabolites, which are distributed in several metabolic pathways, were identified as potential biomarkers of T2DM. EPD and CA could reverse the pathological process of T2DM through regulating the disturbed pathway of metabolism. The metabonomic results are beneficial not only for the evaluation of the therapeutic effect of TCM but also for the elucidation of the underlying molecular mechanism.

Phenolic compounds are the most important class of secondary metabolites in raspberries and fulfill a broad range of biological functions in plants. Due to their presence in fruits, they are considered as important bioactive compounds in human nutrition and are closely related to fruit quality. A targeted UPLC–MS/MS method was used to screen various phenolic compounds in fruits of red and yellow raspberry cultivars (Carvalho et al., 2013). Fifty phenolic compounds were detected above the quantification limit. Besides the obvious lack of anthocyanins, all yellow fruits analyzed lack procyanidin B1. The presence of this dimer, along with B3 dimers, is described for the first time in raspberry fruits. Dihydrochalcone and stilbene derivatives and the quercetin metabolite, isorhamnetin with its glycosides, were identified in considerable concentrations in raspberries. Based on a PCA plot, the red cultivar "Heritage" and the yellow "Alpen Gold" could clearly be separated from the other tested cultivars due to their distinct metabolite profiles/concentrations. The obtained data will lead to a better understanding of the overall biosynthetic network of polyphenols in raspberry and will help to explain responsible factors for the different metabolite profiles in ongoing studies.

10.7 Conclusions

This chapter summarizes the major phytochemical constituents in the aerial and underground parts of *Potentilla* and *Rubus*, which are the physical basis of non-*Camellia* tea resources. *In vitro* and *in vivo* pharmacological studies of *Potentilla* species are reviewed and discussed, focusing on cardio- and neuroprotective, anticancer, antimicrobial, antihyperglycemic, anti-inflammatory, hepatoprotective, and antioxidative activities. Most pharmacological effects can be explained by the abundant flavonoids and tannins and to a lesser extent by triterpenes, present in all plant parts. However, future efforts should concentrate more on *in vitro* and *in vivo* studies and also on clinical trials in order to confirm traditional knowledge in the light of a rational disease prevention and health promotion. The efficacy of *Potentilla* parts/extracts in the long-term use as tea drinks should be further substantiated in clinical studies.

Rubus phenolic compounds have protective effects on age-related neurodegenerative diseases and bone loss *in vivo* and can inhibit LDL and liposomal oxidation *in vitro*. *Rubus* extracts have also exerted antimutagenic effects *in vitro* and *in vivo* by modifying cell signaling pathways and suppressing tumor promotion factors. However, the antiobesity, antidiabetic, antimicrobial, and anti-inflammatory properties of *Rubus* phenolic compounds need further investigation. Similarly, studies that elucidate the *in vivo* physiologically effective concentrations of *Rubus* phenolic compounds are necessary. The health-promoting effects of terpenoids should not be neglected. Targeted metabolite profiling is useful to identify compounds contributing to the antioxidant properties and health benefits and for tailored breeding for tea drinks and functional foods.

References

Alice, L.A., Campbell, C.S., 1999. Phylogeny of *Rubus* (rosaceae) based on nuclear ribosomal DNA internal transcribed spacer region sequences. Am. J. Bot. 86 (1), 81–97.

Alice, L.A., Eriksson, T., Eriksen, B., Campbell, C.S., 2001. Hybridization and gene flow between distantly related species of *Rubus* (Rosaceae): evidence from nuclear ribosomal DNA internal transcribed spacer region sequences. Syst. Bot. 26 (4), 769–778.

Ang, H.Y., Subramani, T., Yeap, S.K., et al., 2014. Immunomodulatory effects of *Potentilla indica* and *Dendrophthoe pentandra* on mice splenocytes and thymocytes. Exp. Ther. Med. 7 (6), 1733–1737.

Badr, A.M., El-Demerdash, E., Khalifa, A.E., et al., 2009. *Rubus sanctus* protects against carbon tetrachloride-induced toxicity in rat isolated hepatocytes: isolation and characterization of its galloylated flavonoids. J. Pharm. Pharmacol. 61 (11), 1511–1520.

Bazylko, A., Piwowarski, J.P., Filipek, A., et al., 2013. In vitro antioxidant and anti-inflammatory activities of extracts from *Potentilla recta* and its main ellagitannin, agrimoniin. J. Ethnopharmacol. 149 (1), 222–227.

Bhandary, B., Lee, G.H., So, B.O., et al., 2012. Rubus coreanus inhibits oxidized-LDL uptake by macrophages through regulation of JNK activation. Am. J. Chin. Med. 40 (5), 967–978.

Bradish, C.M., Perkins-Veazie, P., Fernandez, G.E., et al., 2012. Comparison of flavonoid composition of red raspberries (*Rubus idaeus* L.) grown in the southern United States. J. Agric. Food Chem. 60 (23), 5779–5786.

Bushakra, J.M., Stephens, M.J., Atmadjaja, A.N., et al., 2012. Construction of black (*Rubus occidentalis*) and red (*R. idaeus*) raspberry linkage maps and their comparison to the genomes of strawberry, apple, and peach. Theor. Appl. Genet. 125 (2), 311–327.

Bushakra, J.M., Krieger, C., Deng, D., et al., 2013. QTL involved in the modification of cyanidin compounds in black and red raspberry fruit. Theor. Appl. Genet. 126 (3), 847–865.

Byamukama, R., Kiremirea, B.T., Andersenb, Ø.M., Steigen, A., 2005. Anthocyanins from fruits of *Rubus pinnatus* and *Rubus rigidus*. J. Food Compos. Anal. 18, 599–605.

Cai, Y., Hu, X., Huang, M., et al., 2012. Characterization of the antibacterial activity and the chemical components of the volatile oil of the leaves of *Rubus parvifolius* L. Molecules 17 (7), 7758–7768.

Carvalho, E., Franceschi, P., Feller, A., et al., 2013. A targeted metabolomics approach to understand differences in flavonoid biosynthesis in red and yellow raspberries. Plant Physiol. Biochem. 72, 79–86.

Chen, J.R., Yang, Z.Q., Hu, T.J., et al., 2010a. Immunomodulatory activity in vitro and in vivo of polysaccharide from *Potentilla anserina*. Fitoterapia 81 (8), 1117–1124.

Chen, W., Hong, Z., Li, T., et al., 2010b. Ultrastructure and TUNEL staining on inhibition of *Rubus alceaefolius* total alkaloids for apoptosis of liver in rat models of acute hepatitis. China J.Chin. Mater Med. 35 (8), 1060–1063.

Choi, C., Lee, H., Lim, H., et al., 2012. Effect of *Rubus coreanus* extracts on diabetic osteoporosis by simultaneous regulation of osteoblasts and osteoclasts. Menopause 19 (9), 1043–1051.

Choi, M.R., Lee, M.Y., Hong, J.E., et al., 2014. *Rubus coreanus* Miquel ameliorates scopolamine-induced memory impairments in ICR mice. J. Med. Food 17 (10), 1049–1056.

Chou, G., Xu, S.J., Liu, D., et al., 2009. Quantitative and fingerprint analyses of Chinese sweet tea plant (*Rubus suavissimus* S. Lee). J. Agric. Food Chem. 57 (3), 1076–1083.

Choudhary, A., Mittal, A.K., Radhika, M., et al., 2013. Two new stereoisomeric antioxidant triterpenes from *Potentilla fulgens*. Fitoterapia 91, 290–297.

Cuevas-Rodríguez, E.O., Dia, V.P., Yousef, G.G., et al., 2010. Inhibition of pro-inflammatory responses and antioxidant capacity of Mexican blackberry (*Rubus* spp.) extracts. J. Agric. Food Chem. 58 (17), 9542–9548.

Cui, C.B., Zhao, Q.C., Cai, B., et al., 2002. Two new and four known polyphenolics obtained as new cell-cycle inhibitors from *Rubus aleaefolius* Poir. J. Asian Nat. Prod. Res. 4, 243–252.

Deng, Y., Peng, S.L., Zhang, X.R., et al., 2001. Triterpenoids from *Rubus buergeri*. Acta Bot. Sin. 43 (6), 644–646.

Ding, H.Y., 2011. Extracts and constituents of *Rubus chingii* with 1,1-Diphenyl-2-picrylhydrazyl (DPPH) free radical scavenging activity. Int. J. Mol. Sci. 12 (6), 3941–3949.

Dobeš, C., Paule, J., 2010. A comprehensive chloroplast DNA-based phylogeny of the genus *Potentilla* (Rosaceae): implications for its geographic origin, phylogeography and generic circumscription. Mol. Phylogenet. Evol. 56 (1), 156–175.

Dobeš, C., Scheffknecht, S., 2012. Isolation and characterization of microsatellite loci for the *Potentilla* core group (Rosaceae) using 454 sequencing. Mol. Ecol. Resour. 12 (4), 726–739.

Dong, X.L., 2009. Study on Molecular Phylogeny of the Genus *Rubus* from Southwest China and the Genetic Relationship Between Wild Excellent Germplasm and Introduced Cultivars. PhD Dissertation, Si Chuan Agricultural University.

Dujmović Purgar, D., Duralija, B., Voća, S., et al., 2012. A comparison of fruit chemical characteristics of two wild grown *Rubus* species from different locations of Croatia. Molecules 17 (9), 10390–10398.

Durgo, K., Belščak-Cvitanović, A., Stančić, A., et al., 2012. The bioactive potential of red raspberry (*Rubus idaeus* L.) leaves in exhibiting cytotoxic and cytoprotective activity on human laryngeal carcinoma and colon adenocarcinoma. J. Med. Food 15 (3), 258–268.

Ezure, T., Amano, S., 2011. *Rubus suavissimus* S. Lee extract increases early adipogenesis in 3T3-L1 preadipocytes. J. Nat. Med. 65 (2), 247–253.

Ferrarini, M., Moretto, M., Ward, J.A., et al., 2013. An evaluation of the PacBio RS platform for sequencing and de novo assembly of a chloroplast genome. BMC Genomics 14, 670.

Ferreira de Araujo, R.R., da Silva Santos, V., Rodrigues Machado, A., et al., 2011. Benefits of blackberry nectar (Rubus spp.) relative to hypercholesterolemia and lipid peroxidation. Nutr. Hosp. 26 (5), 984–990.

Gao, W., 2007. Studies on Chemical Constituents of *Potentilla chinensis* Ser. MSc Thesis, Shenyang Pharmaceutical University.

Gao, J., Sun, C.R., Yang, J.H., et al., 2011. Evaluation of the hepatoprotective and antioxidant activities of *Rubus parvifolius* L. J. Zhejiang Univ. Sci. B 12 (2), 135–142.

Garzón, G.A., Riedl, K.M., Schwartz, S.J., 2009. Determination of anthocyanins, total phenolic content, and antioxidant activity in Andes Berry (*Rubus glaucus* Benth). J. Food Sci. 74 (3), C227–C232.

Ghalayini, I.F., Al-Ghazo, M.A., Harfeil, M.N., 2011. Prophylaxis and therapeutic effects of raspberry (*Rubus idaeus*) on renal stone formation in Balb/c mice. Int. Braz J Urol 37 (2), 259–266.

Giri, B.R., Roy, B., Sinha Babu, S.P., 2013. Evidence of apoptosis in *Raillietina echinobothrida* induced by methanolic extracts of three traditional medicinal plants of Northeast India. Exp. Parasitol. 134 (4), 466–473.

Godevac, D., Tesević, V., Vajs, V., et al., 2009. Antioxidant properties of raspberry seed extracts on micronucleus distribution in peripheral blood lymphocytes. Food Chem. Toxicol. 47 (11), 2853–2859.

Gülçin, I., Topal, F., Çakmakçı, R., et al., 2011. Pomological features, nutritional quality, polyphenol content analysis, and antioxidant properties of domesticated and 3 wild ecotype forms of raspberries (*Rubus idaeus* L.). J. Food Sci. 76 (4), C585–C593.

Hao, D.C., Ge, G.B., Xiao, P.G., et al., 2011a. The first insight into the tissue specific taxus transcriptome via Illumina second generation sequencing. PLoS One 6 (6), e21220.

Hao, D.C., Yang, L., Xiao, P.G., 2011b. The first insight into the *Taxus* genome via fosmid library construction and end sequencing. Mol. Genet. Genomics 285 (3), 197–205.

Holst, L., Haavik, S., Nordeng, H., 2009. Raspberry leaf–should it be recommended to pregnant women? Complement. Ther. Clin. Pract. 15 (4), 204–208.

Hong, Z.F., Li, T.J., Zhao, J.Y., 2009. Effect of total alkaloids of *Rubus alceaefolius* poiron on gene expressions of CYP2E1 and CYP3A1 in rats with acute liver injury. Zhongguo Zhong Xi Yi Jie He Za Zhi 29 (8), 711–715.

Huang, Q., Zhang, B., Zhang, H.Y., et al., 1990. Studies on constituents of tannin from *Potentilla viscosa* J. don. Acta Chim. Sin. 48 (8), 825–829.

Huber, R., Ditfurth, A.V., Amann, F., et al., 2007. Tormentil for active ulcerative colitis: an open-label, dose-escalating study. J. Clin. Gastroenterol. 41 (9), 834–838.

Hussein, S.A., Ayoub, N.A., Nawwar, M.A., 2003. Caffeoyl sugar esters and an ellagitannin from *Rubus sanctus*. Phytochemistry 63, 905–911.

Hyun, T.K., Lee, S., Kumar, D., et al., 2014a. RNA-seq analysis of *Rubus idaeus* cv. Nova: transcriptome sequencing and de novo assembly for subsequent functional genomics approaches. Plant Cell Rep. 33 (10), 1617–1628.

Hyun, T.K., Lee, S., Rim, Y., et al., 2014b. De-novo RNA sequencing and metabolite profiling to identify genes involved in anthocyanin biosynthesis in Korean black raspberry (*Rubus coreanus* Miquel). PLoS One 9 (2), e88292.

Jaitak, V., Kaul, V.K., Himlata, et al., 2010. New hopane triterpenes and antioxidant constituents from Potentilla fulgens. Nat. Prod. Commun. 5 (10), 1561–1566.

Jean-Gilles, D., Li, L., Ma, H., et al., 2012. Anti-inflammatory effects of polyphenolic-enriched red raspberry extract in an antigen-induced arthritis rat model. J. Agric. Food Chem. 60 (23), 5755–5762.

Jeong, M.Y., Kim, H.L., Park, J., et al., 2014. *Rubi Fructus* (*Rubus coreanus*) activates the expression of thermogenic genes in vivo and in vitro. Int. J. Obes. 39 (3), 456–464. http://dx.doi.org/10.1038/ijo.2014.155.

Jia, J.B., Li, W.L., Wu, W.L., et al., 2008. RAPD analysis of *Rubus* spp. J. Plant Resour. Environ. 17 (3), 18–22.

Jia, L., Wang, J., Lv, C., et al., 2013. Two new compounds from *Potentilla multicaulis* Bunge. Nat. Prod. Res. 27 (15), 1361–1365.

Johnson, J.L., Bomser, J.A., Scheerens, J.C., Giusti, M.M., 2011. Effect of black raspberry (*Rubus occidentalis* L.) extract variation conditioned by cultivar, production site, and fruit maturity stage on colon cancer cell proliferation. J. Agric. Food Chem. 59 (5), 1638–1645.

Jung, C.H., Choi, J.K., Yang, Y., et al., 2012. A botulinum neurotoxin-like function of *Potentilla chinensis* extract that inhibits neuronal SNARE complex formation, membrane fusion, neuroexocytosis, and muscle contraction. Pharm. Biol. 50 (9), 1157–1167.

Jurkštienė, V., Pavilonis, A., Garšvienė, D., et al., 2011. Investigation of the antimicrobial activity of *Rhaponticum* (*Rhaponticum carthamoides* D.C. Iljin) and shrubby cinquefoil (*Potentilla fruticosa* L.). Medicina 47 (3), 174–179.

Kähkönen, M., Kylli, P., Ollilainen, V., et al., 2012. Antioxidant activity of isolated ellagitannins from red raspberries and cloudberries. J. Agric. Food Chem. 60 (5), 1167–1174.

Kang, S., Zheng, S., 2007. Two new flavones from *Rubus biflorus* Buch. Acta Pharm. Sin. 42 (12), 1288–1291.

Kim, S.J., Lee, H.J., Kim, B.S., et al., 2011. Antiulcer activity of anthocyanins from *Rubus coreanus* via association with regulation of the activity of matrix metalloproteinase-2. J. Agric. Food Chem. 59 (21), 11786–11793.

Kim, H.H., Choi, P.H., Yoo, J.S., et al., 2012a. Ripe fruit of *Rubus coreanus* inhibits mast cell-mediated allergic inflammation. Int. J. Mol. Med. 29 (2), 303–310.

Kim, Y., Kim, J., Lee, S.M., et al., 2012b. Chemopreventive effects of Rubus coreanus Miquel on prostate cancer. Biosci. Biotechnol. Biochem. 76 (4), 737–744.

Kim, Y., Lee, S.M., Kim, J.H., 2014. Unripe *Rubus coreanus* Miquel suppresses migration and invasion of human prostate cancer cells by reducing matrix metalloproteinase expression. Biosci. Biotechnol. Biochem. 78 (8), 1402–1411.

Koh, G.Y., McCutcheon, K., Zhang, F., et al., 2011. Improvement of obesity phenotype by Chinese sweet leaf tea (*Rubus suavissimus*) components in high-fat diet-induced obese rats. J. Agric. Food Chem. 59 (1), 98–104.

Kombal, R., Glasl, H., 1995. Flavan-3-ols and flavonoids from *Potentilla anserina*. Planta Med. 61 (5), 484–485.

Kostamo, K., Toljamo, A., Antonius, K., et al., 2013. Morphological and molecular identification to secure cultivar maintenance and management of self-sterile *Rubus arcticus*. Ann. Bot. 111 (4), 713–721.

Krauze-Baranowska, M., Majdan, M., Hałasa, R., et al., 2014. The antimicrobial activity of fruits from some cultivar varieties of *Rubus idaeus* and *Rubus occidentalis*. Food Funct. 5 (10), 2536–2541.

Kwon, S.C., Kim, Y.B., 2011. Antioxidative and aldose reductase-inhibitory effects of a fermentation filtrate of *Rubus coreanus*. Lab. Anim. Res. 27 (4), 365–368.

Laloo, D., Prasad, S.K., Sairam, K., Hemalatha, S., 2014. Gastroprotective activity of polyphenolic-rich extract of *Potentilla mooniana*. Pharm Biol. 52, 1532–1542.

Lee, J., 2015. Sorbitol, *Rubus* fruit, and misconception. Food Chem. 166, 616–622.

Lee, J.E., Park, E., Lee, J.E., et al., 2011a. Effects of a *Rubus coreanus* Miquel supplement on plasma antioxidant capacity in healthy Korean men. Nutr. Res. Pract. 5 (5), 429–434.

Lee, S., You, Y., Yoon, H.G., et al., 2011b. Fatigue-alleviating effect on mice of an ethanolic extract from *Rubus coreanus*. Biosci. Biotechnol. Biochem. 75 (2), 349–351.

Li, H.Z., Tanaka, T., Zhang, Y.J., et al., 2007. Rubusuaviins A-F, monomeric and oligomeric ellagitannins from Chinese Sweet Tea and their α-amylase inhibitory activity. Chem. Pharm. Bull. 55, 1325–1331.

Li, J.Y., Li, Y., Gong, H.Y., et al., 2009a. Protective effects of *n*-butanol extract of *Potentilla anserina* on acute myocardial ischemic injury in mice. Zhong Xi Yi Jie He Xue Bao 7 (1), 48–52.

Li, W., Fu, H., Bai, H., et al., 2009b. Triterpenoid saponins from *Rubus ellipticus var. obcordatus*. J. Nat. Prod. 72 (10), 1755–1760.

Li, L., Xu, L.J., Peng, Y., Shi, R.B., Xiao, P.G., 2011. Comparison of green tea and four other kind of teas. China J.Chin. Mater Med. 36 (1), 5–10.

Li, L.Z., Wang, L.J., Wang, Y., et al., 2012. Effect of *n*-butanol extract from *Potentilla anserina* on hypoxia-induced calcium overload and SERCA2 expression of rat cardiomyocytes. Chin. Herb. Med. 4 (2), 142–149.

Li, Y., Li, J.J., Wen, X.D., et al., 2014. Metabonomic analysis of the therapeutic effect of *Potentilla discolor* in the treatment of type 2 diabetes mellitus. Mol. Biosyst. 10 (11), 2898–2906.

Lim, H.K., Lee, H.R., Do, S.H., 2014. Stimulation of cannabinoid receptors by using *Rubus coreanus* extracts to control osteoporosis in aged male rats. Aging Male, 1–9.

Lin, J., Zhao, J., Li, T., et al., 2011. Hepatoprotection in a rat model of acute liver damage through inhibition of CY2E1 activity by total alkaloids extracted from *Rubus alceifolius* Poir. Int. J. Toxicol. 30 (2), 237–243.

Lin, X., Zhang, S., Huang, R., et al., 2014. Protective effect of tormentic acid from *Potentilla chinensis* against lipopolysaccharide/D-galactosamine induced fulminant hepatic failure in mice. Int. Immunopharmacol. 19 (2), 365–372.

Liu, P., Duan, H.Q., Pan, Q., et al., 2006. Triterpenes from herb of *Potentilla chinensis*. China J. Chin. Mater. Med. 31 (22), 1875–1879.

Liu, C., Liao, Z., Liu, S., et al., 2014. A new pregnane glycoside from *Rubus phoenicolasius* and its antiproliferative activity. Nat. Prod. Res. 28 (21), 1843–1846.

Łozowicka, B., Kaczyński, P., Jankowska, M., et al., 2012. Pesticide residues in raspberries (*Rubus idaeus* L.) and dietary risk assessment. Food Addit. Contam. Part B Surveill. 5 (3), 165–171.

Lundberg, M., Töpel, M., Eriksen, B., et al., 2009. Allopolyploidy in Fragariinae (Rosaceae): comparing four DNA sequence regions, with comments on classification. Mol. Phylogenet. Evol. 51 (2), 269–280.

Ma, L., Sun, X., Kong, X., et al., 2014. Physiological, biochemical and proteomics analysis reveals the adaptation strategies of the alpine plant *Potentilla saundersiana* at altitude gradient of the Northwestern Tibetan Plateau. J. Proteomics 112C, 63–82.

Martini, S., D'Addario, C., Colacevich, A., et al., 2009. Antimicrobial activity against *Helicobacter pylori* strains and antioxidant properties of blackberry leaves (*Rubus ulmifolius*) and isolated compounds. Int. J. Antimicrob. Agents 34 (1), 50–59.

Marulanda, M., López, A.M., Uribe, M., 2012. Molecular characterization of the Andean blackberry, *Rubus glaucus*, using SSR markers. Genet. Mol. Res. 11 (1), 322–331.

Mazur, S.P., Nes, A., Wold, A.B., et al., 2014. Quality and chemical composition of ten red raspberry (*Rubus idaeus* L.) genotypes during three harvest seasons. Food Chem. 160, 233–240.

McGhie, T.K., Martin, H., Lunken, R.C., 2012. The combination of analytical-scale HPLC separation with a TR-FRET assay to investigate JAK2 inhibitory compounds in a Boysenberry drink. Food Funct. 3 (11), 1170–1175.

Melkadze, R.G., Chichkovani, N.S., Kakhniashvili, E.Z., 2008. Characteristics of the composition of Caucasian blackberry (*Rubus caucasicus* L.) leaves as a raw material for tea production. Prikl. Biokhim. Mikrobiol. 44 (6), 714–719.

Meng, X.J., Liu, B., Zeng Cai Dan, R., et al., 2011. Progress of chemical constituents and pharmacology of genus *Rubus*. Nat. Prod. Res. Dev. 23, 767–775.

Mikulic-Petkovsek, M., Schmitzer, V., Slatnar, A., et al., 2012. Composition of sugars, organic acids, and total phenolics in 25 wild or cultivated berry species. J. Food Sci. 77 (10), C1064–C1070.

Morikawa, T., Ninomiya, K., Imura, K., et al., 2014. Hepatoprotective triterpenes from traditional Tibetan medicine *Potentilla anserina*. Phytochemistry 102, 169–181.

Nguelefack, T.B., Mbakam, F.H., Tapondjou, L.A., et al., 2011. A dimeric triterpenoid glycoside and flavonoid glycosides with free radical-scavenging activity isolated from Rubus rigidus var. camerunensis. Arch. Pharm. Res. 34 (4), 543–550.

Ni, W., Zhang, X., Bi, H., et al., 2009. Preparation of a glucan from the roots of *Rubus crataegifolius* Bge. and its immunological activity. Carbohydr. Res. 344 (18), 2512–2518.

Oh, M., Bae, S.Y., Lee, J.H., et al., 2012. Antiviral effects of black raspberry (*Rubus coreanus*) juice on foodborne viral surrogates. Foodborne Pathog. Dis. 9 (10), 915–921.

Paudel, L., Wyzgoski, F., Scheerens, J.C., et al., 2013. Nonanthocyanin secondary metabolites of black raspberry (*Rubus occidentalis* L.) fruits: identification by HPLC-DAD, NMR, HPLC-ESI-MS, and ESI-MS/MS analyses. J. Agric. Food Chem. 61 (49), 12032–12043.

Paule, J., Scherbantin, A., Dobeš, C., 2012. Implications of hybridisation and cytotypic differentiation in speciation assessed by AFLP and plastid haplotypes – a case study of *Potentilla alpicola* La Soie. BMC Evol. Biol. 12, 132.

Piao, X., Tian, Y., Mi, X., Cui, J., 2009. Tyrosinase inhibition of *Potentilla bifurca*. China J. Chin. Mater. Med. 34 (15), 1952–1954.

Qiao, W., Zhao, C., Qin, N., et al., 2011. Identification of trans-tiliroside as active principle with anti-hyperglycemic, anti-hyperlipidemic and antioxidant effects from *Potentilla chinensis*. J. Ethnopharmacol. 135 (2), 515–521.

Qin, N., Li, C.B., Jin, M.N., et al., 2011. Synthesis and biological activity of novel tiliroside derivants. Eur. J. Med. Chem. 46 (10), 5189–5195.

Qin, X., Lv, Q., Zhang, X., et al., 2012a. Study on protective effect of alcohol extract of *Potentilla Anserinea* against acute myocardial ischemia/reperfusion-induced myocardial apoptosis in rats. China J. Chin. Mater. Med. 37 (9), 1279–1284.

Qin, X.J., Li, L.Z., Lv, Q., et al., 2012b. Neuroprotection of *n*-butanol extract from roots of *Potentilla anserina* on hypoxic injury in primary hippocampal neurons. Chin. Herb. Med. 4 (3), 195–200.

Qiu, L., Li, X.Y., Liu, P., et al., 2014. Analysis of leave FTIR of nine kinds of plants from rosaceae with genetic relationship. Guang Pu Xue Yu Guang Pu Fen Xi 34 (2), 344–349.

Quave, C.L., Estévez-Carmona, M., Compadre, C.M., et al., 2012. Ellagic acid derivatives from *Rubus ulmifolius* inhibit *Staphylococcus aureus* biofilm formation and improve response to antibiotics. PLoS One 7 (1), e28737.

Radhika, M., Ghoshal, N., Chatterjee, A., 2012. Comparison of effectiveness in antitumor activity between flavonoids and polyphenols of the methanolic extract of roots of *Potentilla fulgens* in breast cancer cells. J. Complement. Integr. Med. 9, 24.

Radovanovic, A.M., Cupara, S.M., Popovic, S.L., et al., 2013. Cytotoxic effect of Potentilla reptans L. rhizome and aerial part extracts. Acta Pol. Pharm. 70 (5), 851–854.

Rauf, A., Khan, R., Khan, H., et al., 2014. In vivo antinociceptive and anti-inflammatory activities of umbelliferone isolated from *Potentilla evestita*. Nat. Prod. Res. 28 (17), 1371–1374.

Ren, Y.L., Qu, W., Liang, J.Y., 2010. The research progress of *Potentilla* plants. Strait. Pharm. J. 22 (5), 1–8.

Rezeng, C., Limao, C., Liu, B., 2014. Chemical constituents of Tibetan medicine kandrakari (I). China J. Chin. Mater. Med. 39 (7), 1168–1172.

Roy, B., Swargiary, A., Syiem, D., Tandon, V., 2010. *Potentilla fulgens* (Family Rosaceae), a medicinal plant of north-east India: a natural anthelmintic? J. Parasit. Dis. 34 (2), 83–88.

Roy, B., Dasgupta, S., Giri, B.R., 2012. Electron microscopic observations on the alterations of tegumental surface of *Raillietina echinobothrida* treated with root-peel extract of *Potentilla fulgens*. Microsc. Res. Tech. 75 (7), 1000–1005.

Shafi, A., Dogra, V., Gill, T., et al., 2014. Simultaneous over-expression of PaSOD and RaAPX in transgenic *Arabidopsis thaliana* confers cold stress tolerance through increase in vascular lignifications. PLoS One 9 (10), e110302.

Sharma, U.S., Kumar, A., 2011. In vitro antioxidant activity of *Rubus ellipticus* fruits. J. Adv. Pharm. Technol. Res. 2 (1), 47–50.

Shi, L., Qin, N., Hu, L., et al., 2011. Tiliroside-derivatives enhance GLUT4 translocation via AMPK in muscle cells. Diabetes Res. Clin. Pract. 92 (2), e41–e46.

Shikov, A.N., Lazukina, M.A., Pozharitskaya, O.N., et al., 2011. Pharmacological evaluation of Potentilla alba L. in mice: adaptogenic and central nervous system effects. Pharm. Biol. 49 (10), 1023–1028.

Shuai, X.H., Hu, T.J., Zhang, X., et al., 2009. Inhibitory action of *Potentilla anserina* polysaccharide fraction on H_2O_2-induced apoptosis of murine splenic lymphocytes. Yao Xue Xue Bao 44 (9), 987–993.

Shushunov, S., Balashov, L., Kravtsova, A., et al., 2009. Determination of acute toxicity of the aqueous extract of *Potentilla erecta* (Tormentil) rhizomes in rats and mice. J. Med. Food 12 (5), 1173–1176.

Sohn, S.I., Rim, H.K., Kim, Y.H., et al., 2011. The ameliorative effect of 23-hydroxytormentic acid isolated from *Rubus coreanus* on cisplatin-induced nephrotoxicity in rats. Biol. Pharm. Bull. 34 (9), 1508–1513.

Söhretoglu, D., Kırmızıbekmez, H., 2011. Polyphenols from *Potentilla recta*. Biochem. Syst. Ecol. 39, 132–134.

Song, C., Huang, L., Rong, L., et al., 2012. Anti-hyperglycemic effect of *Potentilla discolor* decoction on obese-diabetic (Ob-db) mice and its chemical composition. Fitoterapia 83 (8), 1474–1483.

Soto, J., Gómez, C., Calzada, F., Ramírez, M.E., 2010. Ultrastructural changes on *Entamoeba histolytica* HM1-IMSS caused by the flavan-3-ol, (−)-epicatechin. Planta Med. 76 (6), 611–612.

Sparzak, B., Merino-Arevalo, M., Vander Heyden, Y., et al., 2010. HPLC analysis of polyphenols in the fruits of *Rubus idaeus* L. (Rosaceae). Nat. Prod. Res. 24 (19), 1811–1822.

Sultana, N., Akhter, M., Khatoon, Z., 2010. Nematicidal natural products from the aerial parts of *Rubus niveus*. Nat. Prod. Res. 24 (5), 407–415.

Sun, C.Q., Shao, X.M., Wang, L.M., et al., 2004. The exploitation and utilization of raspberry (Rubus L.). Guihaia 24, 578–582.

Süntar, I., Koca, U., Keleş, H., Akkol, E.K., 2011. Wound healing activity of *Rubus sanctus* Schreber (Rosaceae): preclinical study in animal models. Evid. Based Complement. Alternat. Med. 2011, 816156.

Tan, M.X., Wang, H.S., Li, S., Chen, W., 2003. Studies on the chemical constituents from the Chinese traditional medicine *Rubus parvifolius*. Guihaia 23, 282–284.

Tavares, L., Figueira, I., McDougall, G.J., et al., 2013. Neuroprotective effects of digested polyphenols from wild blackberry species. Eur. J. Nutr. 52 (1), 225–236.

Tomczyk, M., 2011. Secondary metabolites from *Potentilla recta* L. and *Drymocallis rupestris* (L.) Soják (syn. *Potentilla rupestris* L.) (Rosaceae). Biochem. Syst. Ecol. 39, 893–896.

Tomczyk, M., Latté, K.P., 2009. Potentilla – a review of its phytochemical and pharmacological profile. J. Ethnopharmacol. 122 (2), 184–204.

Tomczyk, M., Drozdowska, D., Bielawska, A., et al., 2008a. Human DNA topoisomerase inhibitors from *Potentilla argentea* and their cytotoxic effect against MCF – 7. Pharmazie 63 (5), 389–393.

Tomczyk, M., Leszczyńska, K., Jakoniuk, P., 2008b. Antimicrobial activity of *Potentilla* species. Fitoterapia 79 (7–8), 592–594.

Tomczyk, M., Bazylko, A., Staszewska, A., 2010a. Determination of polyphenolics in extracts of Potentilla species by high-performance thin-layer chromatography photodensitometry method. Phytochem. Anal. 21 (2), 174–179.

Tomczyk, M., Pleszczyńska, M., Wiater, A., 2010b. Variation in total polyphenolics contents of aerial parts of *Potentilla* species and their anticariogenic activity. Molecules 15 (7), 4639–4651.

Tomczyk, M., Tumanov, A., Zaniewska, A., Surazynski, A., 2010c. The potential mechanism of tiliroside-dependent inhibition of t-butylhydroperoxide-induced oxidative stress in endometrial carcinoma cells. Planta Med. 76 (10), 963–968.

Tomczyk, M., Wiater, A., Pleszczyńska, M., 2011. In vitro anticariogenic effects of aerial parts of *Potentilla recta* and its phytochemical profile. Phytother. Res. 25 (3), 343–350.

Tomczyk, M., Bazylko, A., Bonarewicz, J., 2012. Method development and validation for optimized separation of quercetin derivatives in selected Potentilla species using high-performance thin-layer chromatography photodensitometry method. J. Pharm. Biomed. Anal. 61, 265–270.

Töpel, M., Lundberg, M., Eriksson, T., Eriksen, B., 2011. Molecular data and ploidal levels indicate several putative allopolyploidization events in the genus *Potentilla* (Rosaceae). PLoS Curr.. 3, RRN1237.

Turchaninova, L.I., 2014. Experience of using phytopreparation Alba (root extract of the *Potentilla alba*) in complex treatment of thyroid pathology in children and adolescents. Lik. Sprava 3–4, 125–129.

Venskutonis, P.R., Dvaranauskaite, A., Labokas, J., 2007. Radical scavenging activity and composition of raspberry (*Rubus idaeus*) leaves from different locations in Lithuania. Fitoterapia 78 (2), 162–165.

Verma, R., Gangrade, T., Punasiya, R., Ghulaxe, C., 2014. *Rubus fruticosus* (blackberry) use as an herbal medicine. Pharmacogn. Rev. 8 (16), 101–104.

Wahid, F., Jung, H., Khan, T., et al., 2011. Effects of *Rubus coreanus* extract on visual processes in bullfrog's eye. J. Ethnopharmacol. 138 (2), 333–339.

Wang, L.Y., Ikeda, H., Liu, T.L., et al., 2009a. Repeated range expansion and glacial endurance of *Potentilla glabra* (Rosaceae) in the Qinghai-Tibetan plateau. J. Integr. Plant Biol. 51 (7), 698–706.

Wang, Q., Xu, D.R., Shi, X.H., Qin, M.J., 2009b. Flavones from *Potentilla discolor* Bunge. Chin. J. Nat. Med. 7 (5), 361–364.

Wang, S.S., Wang, D.M., Pu, W.J., Li, D.W., 2013. Phytochemical profiles, antioxidant and antimicrobial activities of three *Potentilla* species. BMC Complement. Altern. Med. 13, 321.

Ward, J.A., Ponnala, L., Weber, C.A., 2012. Strategies for transcriptome analysis in nonmodel plants. Am. J. Bot. 99 (2), 267–276.

Ward, J.A., Bhangoo, J., Fernández-Fernández, F., et al., 2013. Saturated linkage map construction in *Rubus idaeus* using genotyping by sequencing and genome-independent imputation. BMC Genomics 14, 2.

Watkins, F., Pendry, B., Sanchez-Medina, A., Corcoran, O., 2012. Antimicrobial assays of three native British plants used in Anglo-Saxon medicine for wound healing formulations in 10th century England. J. Ethnopharmacol. 144 (2), 408–415.

Wei, Z., Sun, J., Li, P., et al., 2012. Antioxidant activity constituents from root of *Rubus crataegifolius*. China J. Chin. Mater. Med. 37 (23), 3591–3594.

Wei, J., Huang, Q., Huang, R., et al., 2013. Asiatic acid from *Potentilla chinensis* attenuate ethanol-induced hepatic injury via suppression of oxidative stress and Kupffer cell activation. Biol. Pharm. Bull. 36 (12), 1980–1989.

Xu, J.F., 2008. Studies on the Chemical Constituents and Bioactivities of *Mallotus apelta* & *Potentilla anserina*. PhD Dissertation, Chinese Academy of Medical Sciences.

Xu, J.F., Zheng, X., Liu, W., et al., 2010. Flavonol glycosides and monoterpenoids from *Potentilla anserina*. J. Asian Nat. Prod. Res. 12 (6), 529–534.

Xue, P., Liang, H., Wang, B., Zhao, Y., 2005. Chemical constituents from *Potentilla multifida* L. J. Chin. Pharm. Sci. 14 (2), 86–88.

Yang, J.Y., Pak, J.H., 2006. Phylogeny of korean rubus (rosaceae) based on its (nrDNA) and trnL/F intergenic region (cpDNA). J. Plant. Biol. 49 (1), 44–54.

Yang, J., Chen, X., Liu, X., et al., 2008. Structural determination of two new triterpenoids from *Potentilla discolor* Bunge by NMR techniques. Magn. Reson. Chem. 46 (8), 794–797.

Yang, J., Wen, X.D., Jia, B.X., et al., 2011. Quality evaluation of *Potentilla discolor* by high-performance liquid chromatography coupled with diode array detection and electrospray ionization tandem mass spectrometry. Phytochem. Anal. 22 (6), 547–554.

Yu, B.S., Na, D.M., Kang, M.Y., Lim, D.Y., 2009. Polyphenols of *Rubus coreanum* inhibit catecholamine secretion from the perfused adrenal medulla of SHRs. Korean J. Physiol. Pharmacol. 13 (6), 517–526.

Zhang, B., Huang, Q., Lv, L., et al., 1990. Tannins of Potentilla viscose. Chin. Sci. Bull. 35 (3), 239–243.

Zhang, M., Cao, Y., Du, F., et al., 2007. Isolation and identification of two new diterpenoid from *Rubus corchorifolius* L. f. Acta Pharm. Sin. 42 (11), 1155–1158.

Zhang, L., Yang, J., Chen, X.Q., et al., 2010. Antidiabetic and antioxidant effects of extracts from *Potentilla discolor* Bunge on diabetic rats induced by high fat diet and streptozotocin. J. Ethnopharmacol. 132 (2), 518–524.

Zhang, D.M., Lou, L.X., Wu, A.M., et al., 2012. Effects of Astragalus membranaceus and *Potentilla discolor* mixture on insulin resistance and its related mRNA expressions in KKAy mice with type 2 diabetes. Zhong Xi Yi Jie He Xue Bao 10 (7), 821–826.

Zhao, C., Kim, H.K., Kim, S.Z., et al., 2011. What is the role of unripe *Rubus coreanus* extract on penile erection? Phytother. Res. 25 (7), 1046–1053.

Zhao, J., Lin, W., Cao, Z., et al., 2014. Total alkaloids of *Rubus alceifolius* Poir inhibit tumor angiogenesis through suppression of the Notch signaling pathway in a mouse model of hepatocellular carcinoma. Mol. Med. Rep. 11, 357–361. http://dx.doi.org/10.3892/mmr.2014.2702.

Zheng, H., Zhao, J., Liu, Y., et al., 2011. Effect of total alkaloids of *Rubus alceiefolius* on oxidative stress in rats with non-alcoholic fatty liver disease. Zhongguo Zhong Yao Za Zhi 36 (17), 2383–2387.

Zheng, H., Zhao, J., Zheng, Y., et al., 2014. Protective effects and mechanisms of total alkaloids of *Rubus alceaefolius* Poir on non'alcoholic fatty liver disease in rats. Mol. Med. Rep. 10 (4), 1758–1764.

Zhou, S.D., Chen, X.X., Xiao, S.Y., et al., 2009. Studies on the chemical components and antibacterial activity of essential oil from the leaves of Rubus corchorifolius. Zhong Yao Cai 32 (10), 1547–1550.

Zuo, G.Y., Liu, S.L., Xu, G.L., et al., 2008. Triterpenoids from the roots of *Rubus obcordatus* (Rosaceae). Acta Bot. Yunnan. 30, 381–385.

Phytochemical and biological research of *Cannabis* pharmaceutical resources

11

11.1 Introduction

Cannabis, a representative genus of Cannabaceae family (Sytsma et al., 2002), has been a source of fiber, food, oil, medicine, and inebriant since prehistoric times. Marijuana (*Cannabis sativa* L.) is one of the most abused drugs due to its psychotropic effects. Interestingly, it is also used for medicinal purposes. Marijuana has been used for thousands of years as a treatment for medical conditions. In traditional Chinese medicine (TCM), hemp seed, which is rich in polyunsaturated fatty acids, has been used in atopic dermatitis, inflammation, and rheumatoid arthritis (RA) for many centuries (He and Shi, 2010; Jeong et al., 2014). Hemp Seed Pill (7.5 g b.i.d.) is safe and effective for alleviating functional constipation for subjects in excessive syndrome (Cheng et al., 2011). Ma Fen, the female inflorescence and unripe fruit (Wei et al., 2010), is used in TCM to dispel wind and relieve pain and spasm. *Cannabis* leaf is used by China's ethnic Uyghur group in to relieve pain and inflammation. *Cannabis* is also used extensively in Indian and Arabian folk medicine.

Over the last decades, the scientific interest in chemistry and pharmacology of cannabinoids has increased. Most attention has focused on Δ^9-tetrahydrocannabinol (9-THC) as it is the psychoactive constituent of *Cannabis sativa*. However, in recent years, the focus of interest in the second plant constituent with nonpsychotropic properties, cannabidiol (CBD), has been enhanced. This chapter summarizes the research progress of phytochemistry, pharmacology, and omics of *Cannabis* plants.

11.2 Cannabinoids

Cannabinoids are the most distinctive and specific class of compounds in *Cannabis*, which were only known to exist in the *Cannabis* plant. However, new cannabinoid-like chromane and chromene derivatives have been isolated from the Chinese medicinal plant *Rhododendron anthopogonoides* (Iwata and Kitanaka, 2011). Over 100 phytocannabinoids have been found in *Cannabis* (Russo, 2011). Most cannabinoids are closely related, which fall into 12 groups in terms of chemical structure (Figures 11.1 and 11.2, Table 11.1). A. 9-THC; B. Δ^8-tetrahydrocannabinol (Δ^8-THC); C. cannabichromene (CBC); D. CBD; E. cannabigerol (CBG); F. cannabinol (CBN); G. cannabidiol (CBD);

Medicinal plants: chemistry, biology and omics. http://dx.doi.org/10.1016/B978-0-08-100085-4.00011-6

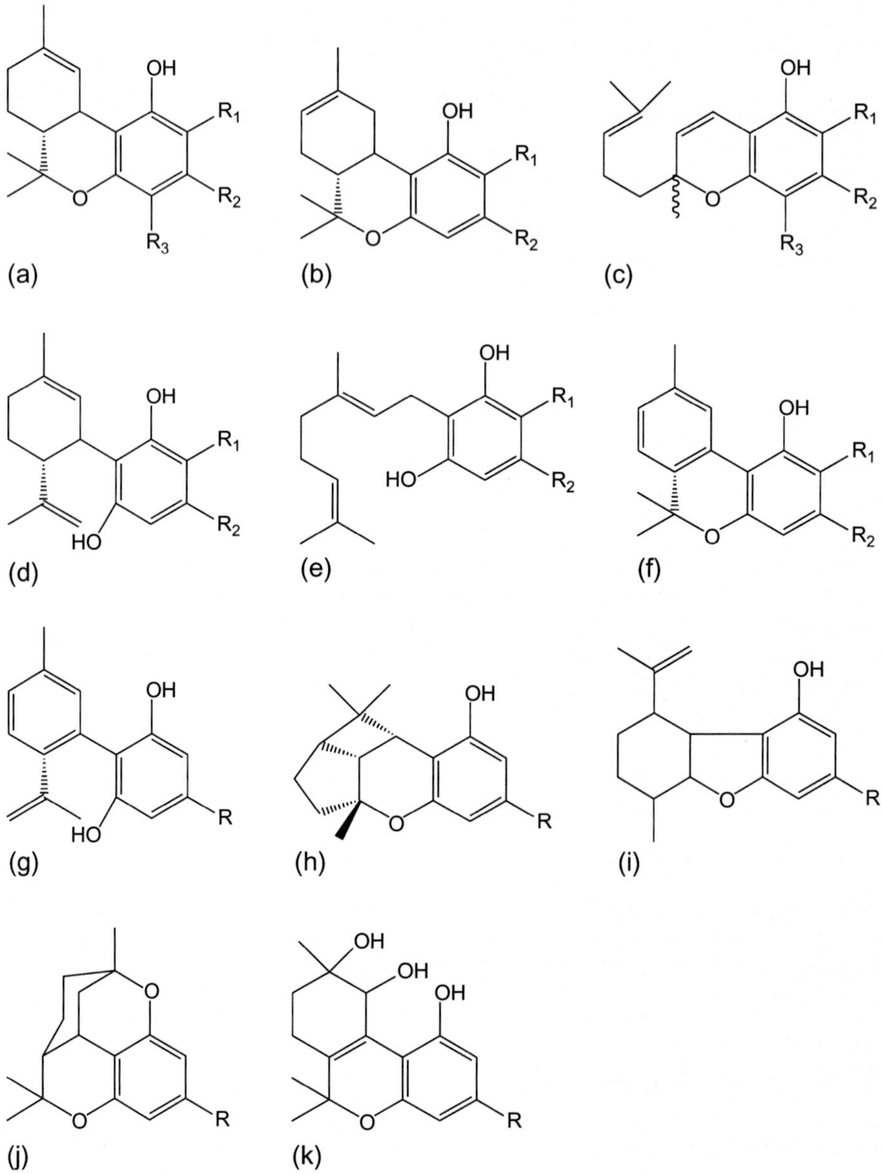

Figure 11.1 Types of cannabinoids in *Cannabis sativa* L.

H. cannabicyclol (CBL); I. cannabielsoin (CBE); J, cannabicitran type (CBCT); K. cannabitriol (CBT); L. other types. Many compounds differ by only a single chemical moiety and might be midpoints along biosynthetic pathways (Figure 11.3), that is, degradation products, precursors, or by-products. Compound 8-THC is easier and less expensive to prepare and is considered to be less psychotropic

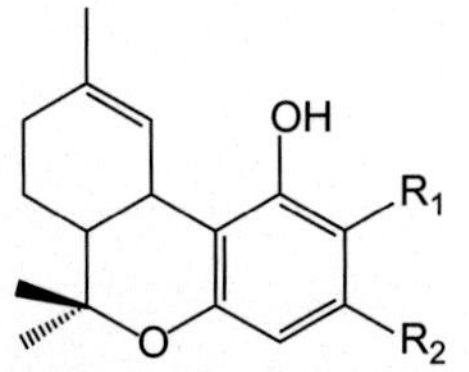

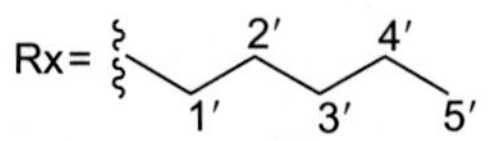

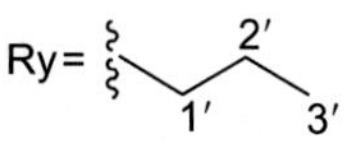

No.	R_1	R_2
1	COOH	Rx
2	H	Rx
42	H	CH_3
45	H	Ry

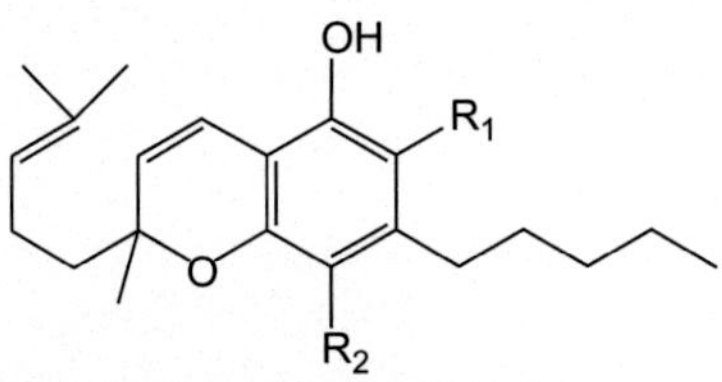

No.	R_1	R_2
3	COOH	H
4	H	H
11	H	OAc

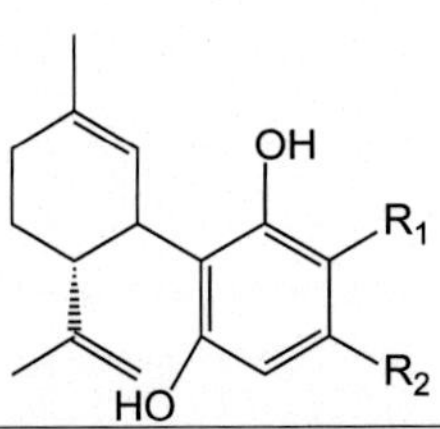

No.	R_1	R_2
5	COOH	Rx
6	H	Rx

Figure 11.2 Examples of various cannabinoids found in *Cannabis sativa* L.

(Continued)

43	H	Ry

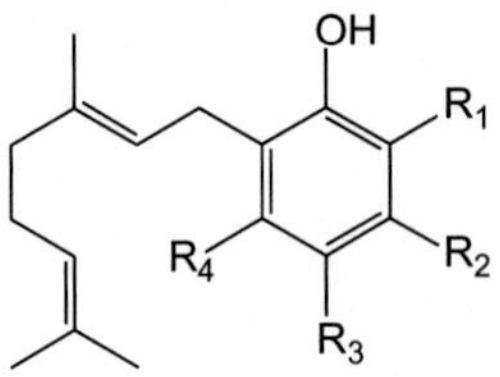

No.	R_1	R_2	R_3	R_4
7	COOH	Rx	H	OH
8	H	Rx	H	OH
15	H	Rx	OH	OAc
16	H	OH	OAc	Rx
24	COOH	Rx	H	OCH_3

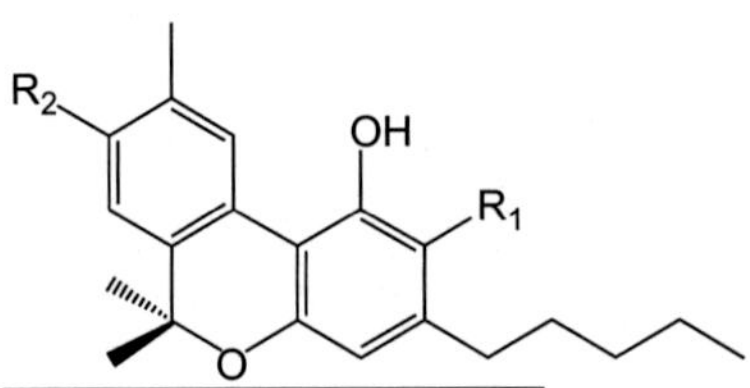

No.	R_1	R_2
9	H	H
17	H	OH
18	COOH	OH
41	COOH	H

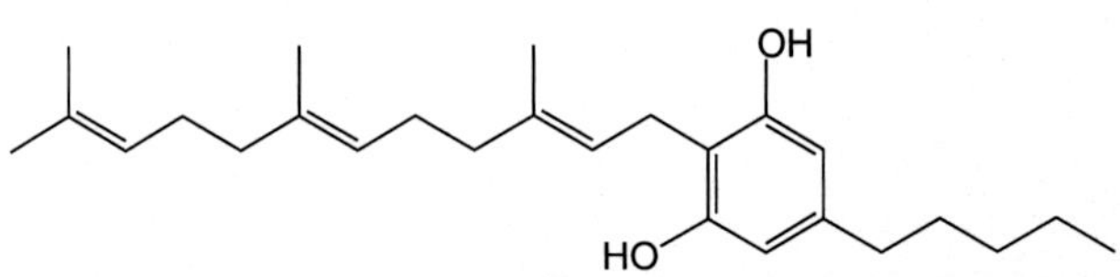

10

Figure 11.2 Continued.

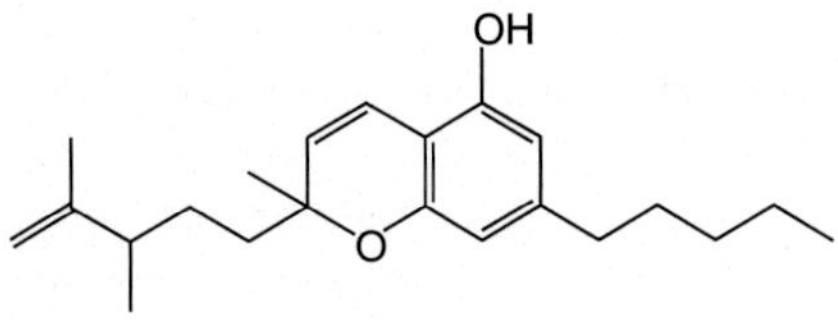

12

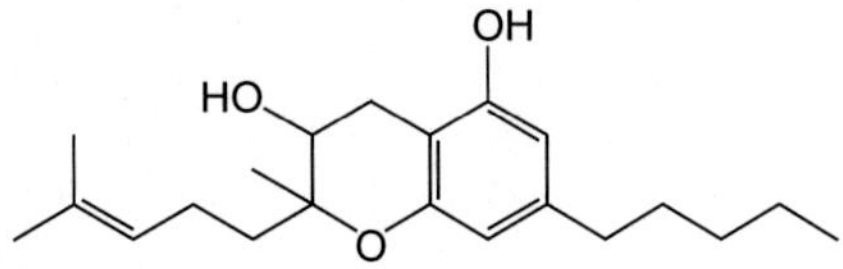

13

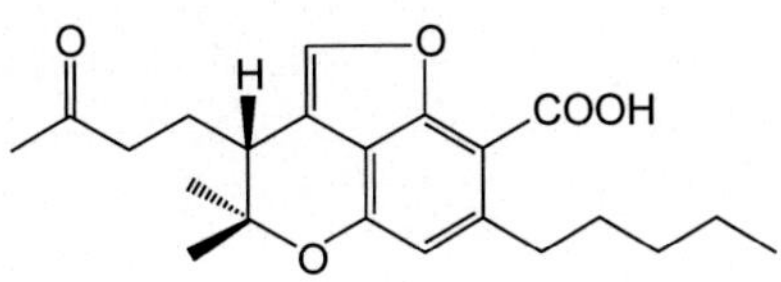

14

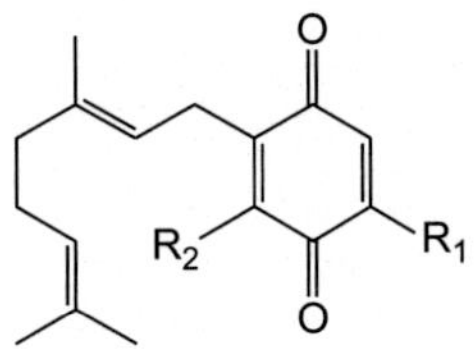

No.	R_1	R_2
19	OH	Rx
52	Rx	$OCOCH_3$

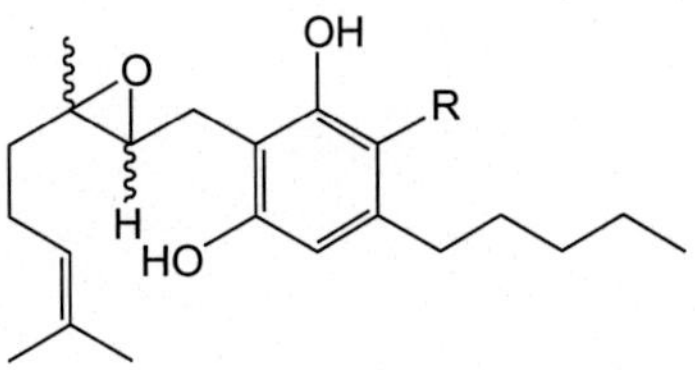

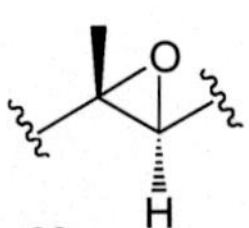

20a

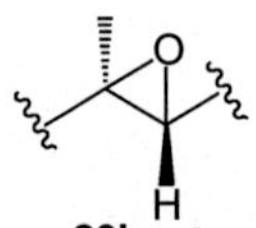

20b

R = COOH

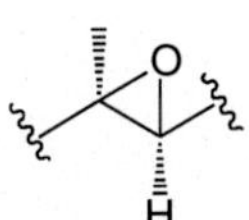

21a

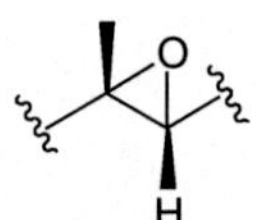

21b

R = COOH

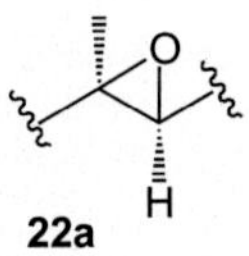

22a

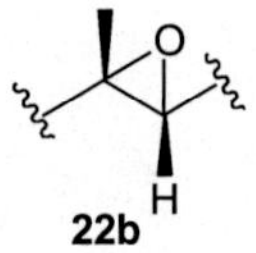

22b

R = H

Figure 11.2 Continued.

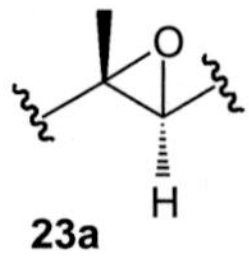

23a

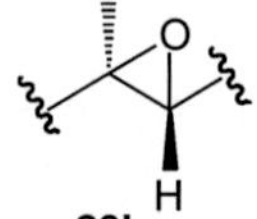

23b

R=H

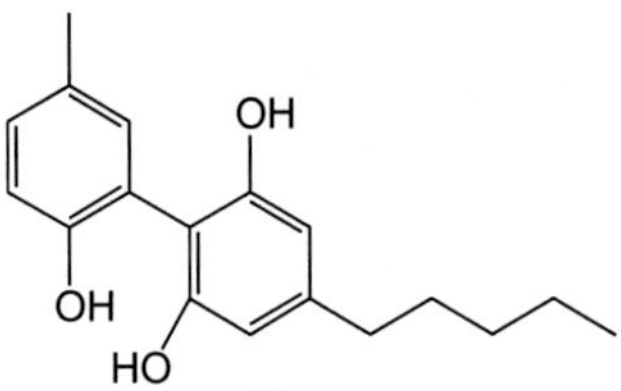

25

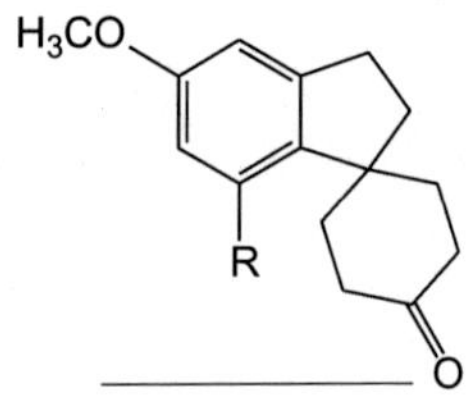

No.	R
26	OCH_3
27	OH

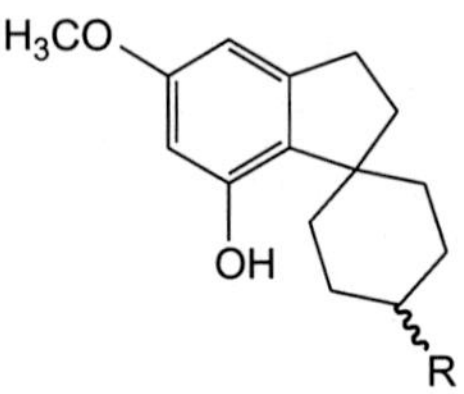

No.	R
28	β-OH
53	α-OH
54	β-OAc

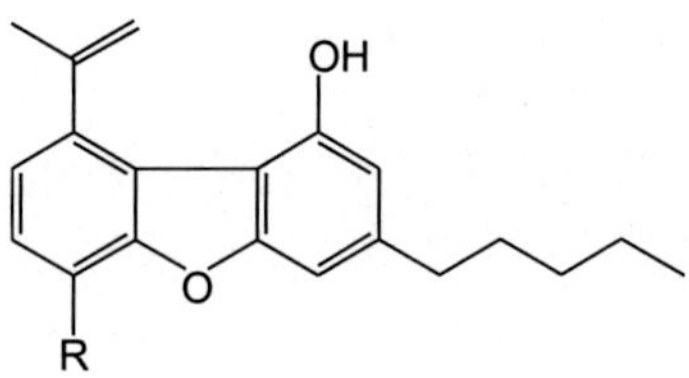

No.	R
29	H
47	OH

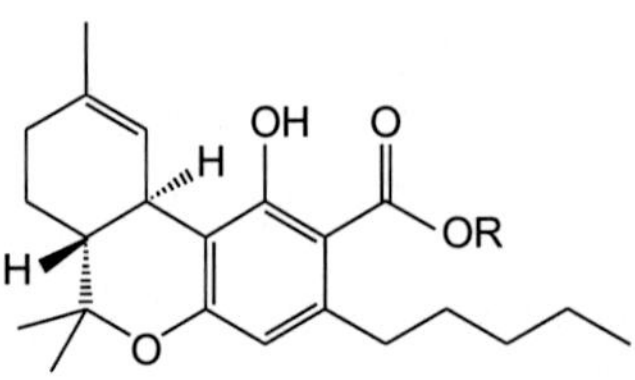

R =

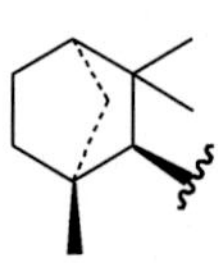

30

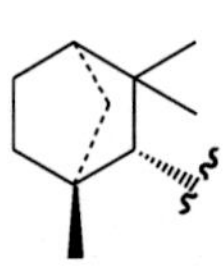

31

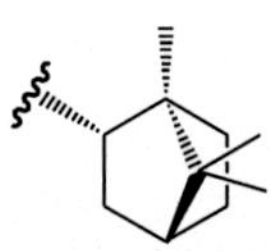

32

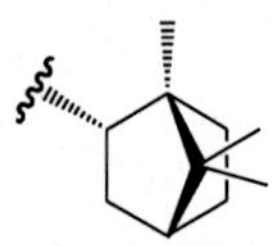

33

Figure 11.2 Continued.

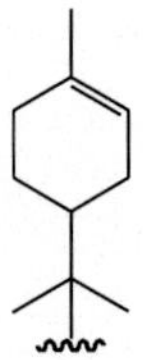
34

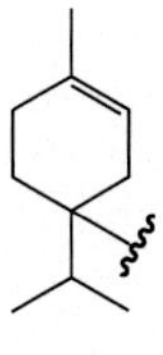
35

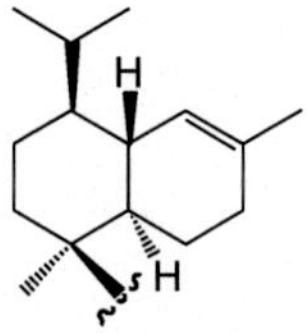

36

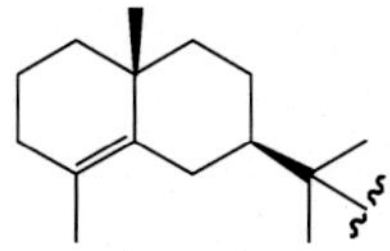
37

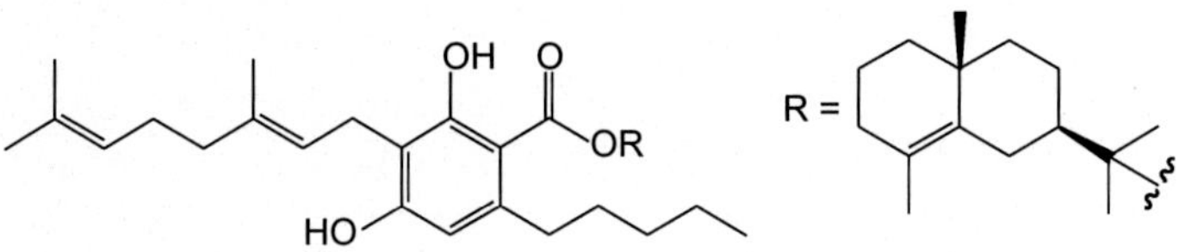

38

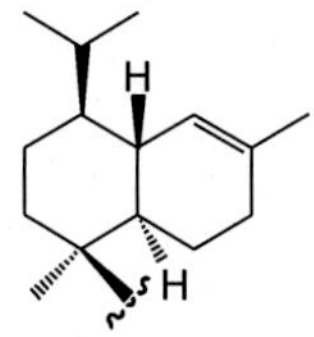

39

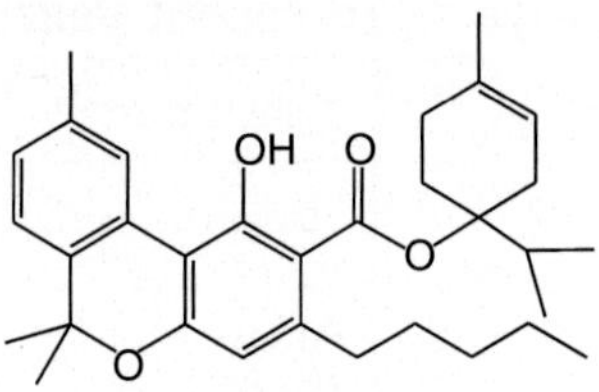

40

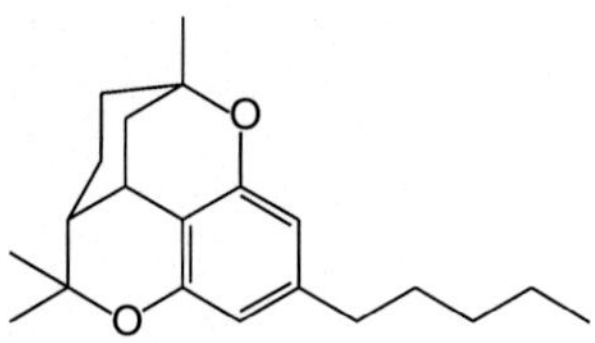

44

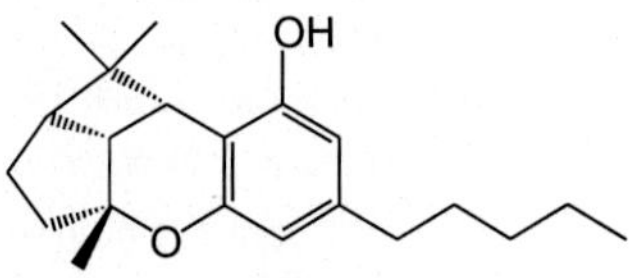

46

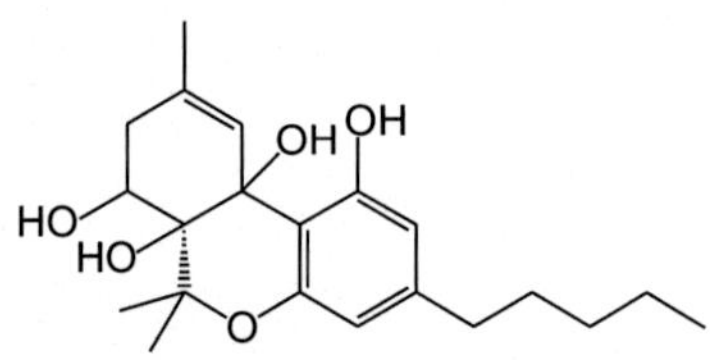

48

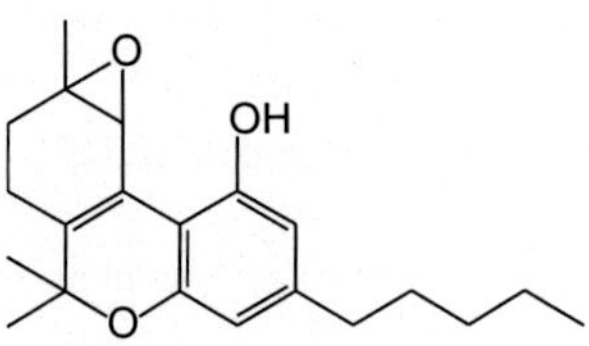

49

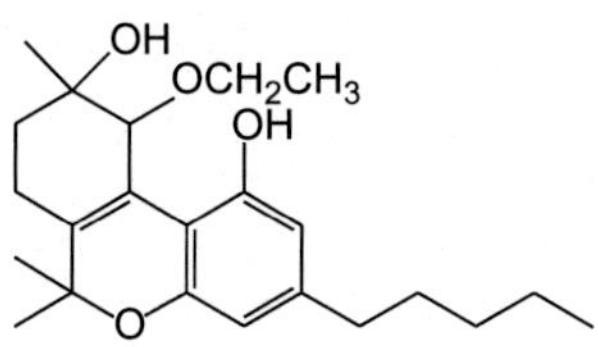

50

Figure 11.2 Continued.

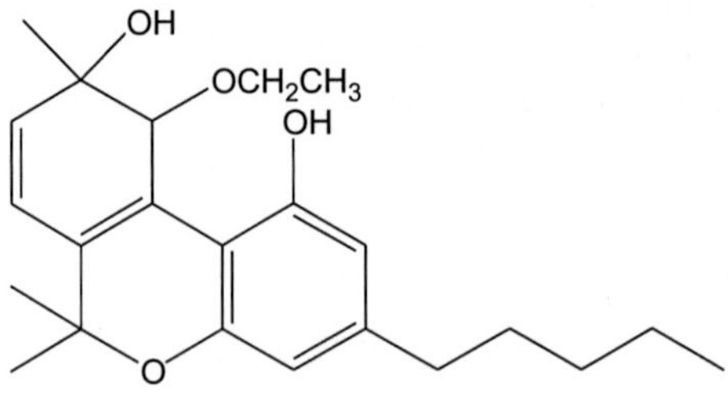

51

Figure 11.2 Continued.

Table 11.1 Cannabinoids found in *Cannabis sativa* L. in recent years

No.	Compounds	Types	Tissues	References
1	▽^{9}-Tetrahydrocannabinolic acid	A	Trichome Plant material	Happyana et al. (2013) and Ahmed et al. (2008)
2	▽^{9}-Tetrahydrocannabinol	A	Trichome Flowered aerial part Plant material Pollen grain	Happyana et al. (2013), Pollastro et al. (2011), Ahmed et al. (2008), and Ross et al. (2005)
3	Cannabichromenic acid	C	Trichome	Happyana et al. (2013)
4	Cannabichromene	C	Trichome Pollen grain	Happyana et al. (2013) and Ross et al. (2005)
5	Cannabidiolic acid	D	Trichome	Happyana et al. (2013)
6	Cannabidiol	D	Trichome Flowered aerial part Pollen grain	Happyana et al. (2013), Pollastro et al. (2011), and Ross et al. (2005)
7	Cannabigerolic acid	E	Trichome Plant material	Happyana et al. (2013), Radwan et al. (2008b), and Ahmed et al. (2008)

Table 11.1 **Continued**

No.	Compounds	Types	Tissues	References
8	Cannabigerol	E	Trichome Flowered aerial part Plant material Pollen grain	Happyana et al. (2013), Pollastro et al. (2011), Radwan et al. (2008b), and Ross et al. (2005)
9	Cannabinol	F	Trichome Pollen grain	Happyana et al. (2013) and Ross et al. (2005)
10	Sesquicannabigerol	L1	Flowered aerial part	Pollastro et al. (2011)
11	(±)-4-Acetoxycannabichromene	C	Plant material	Radwan et al. (2009)
12	(±)-3″-Hydroxy-$\bigtriangledown^{(4'',5'')}$-cannabichromene	C	Plant material	Radwan et al. (2009)
13	(−)-7*R*-hydroxycannabichromane	C	Plant material	Radwan et al. (2009)
14	(−)-7*R*-cannabicoumarononic acid A	L2	Plant material	Radwan et al. (2009)
15	5-Acetyl-4-hydroxycannabigerol	E	Plant material	Radwan et al. (2009)
16	4-Acetoxy-2-geranyl-5-hydroxy-3-*n*-pentylphenol	E	Plant material	Radwan et al. (2009)
17	8-Hydroxycannabinol	F	Plant material	Radwan et al. (2009)
18	8-Hydroxycannabinolic acid	F	Plant material	Radwan et al. (2009)
19	2-Geranyl-5-hydroxy-3-*n*-pentyl-1,4-benzoquinone	L3	Plant material	Radwan et al. (2009)
20	(±)-6,7-*Trans*-epoxycannabigerolic acid	E	Plant material	Radwan et al. (2008b)
21	(±)-6,7-*Cis*-epoxycannabigerolic acid	E	Plant material	Radwan et al. (2008b)
22	(±)-6,7-*Cis*-epoxycannabigerol	E	Plant material	Radwan et al. (2008b)
23	(±)-6,7-*Trans*-epoxycannabigerol	E	Plant material	Radwan et al. (2008b)
24	5′-Methoxycannabigerolic acid	E	Plant material	Radwan et al. (2008b)
25	5′-Methyl-4-pentylbiphenyl-2,2′,6-triol	G	Plant material	Radwan et al. (2008b)

Continued

Table 11.1 Continued

No.	Compounds	Types	Tissues	References
26	7-Methoxycannabispirone	L4	Plant material	Radwan et al. (2008b)
27	Cannabispirone	L4	Plant material	Radwan et al. (2008b)
28	β-Cannabispiranol	L4	Plant material	Radwan et al. (2008b)
29	Dehydrocannabifuran	I	Plant material	Radwan et al. (2008b)
30	β-Fenchyl ∇^9-tetrahydrocannabinolate	A	Plant material	Ahmed et al. (2008)
31	α-Fenchyl ∇^9-tetrahydrocannabinolate	A	Plant material	Ahmed et al. (2008)
32	Epi-bornyl ∇^9-tetrahydrocannabinolate	A	Plant material	Ahmed et al. (2008)
33	Bornyl ∇^9-tetrahydrocannabinolate	A	Plant material	Ahmed et al. (2008)
34	α-Terpenyl ∇^9-tetrahydrocannabinolate	A	Plant material	Ahmed et al. (2008)
35	4-Terpenyl ∇^9-tetrahydrocannabinolate	A	Plant material	Ahmed et al. (2008)
36	α-Cadinyl ∇^9-tetrahydrocannabinolate	A	Plant material	Ahmed et al. (2008)
37	γ-Eudesmyl ∇^9-tetrahydrocannabinolate	A	Plant material	Ahmed et al. (2008)
38	γ-Eudesmyl cannabigerolate	E	Plant material	Ahmed et al. (2008)
39	α-Cadinyl cannabigerolate	E	Plant material	Ahmed et al. (2008)
40	4-Terpenyl cannabinolate	F	Plant material	Ahmed et al. (2008)
41	Cannabinolic acid	F	Plant material	Ahmed et al. (2008)
42	Δ^9-Tetrahydrocannabiorcol	A	Pollen grain	Ross et al. (2005)
43	Cannabidivarin	D	Pollen grain	Ross et al. (2005)
44	Cannabicitran	J	Pollen grain	Ross et al. (2005)
45	Δ^9-Tetrahydrocannabivarin	A	Pollen grain	Ross et al. (2005)
46	Cannabicyclol	H	Pollen grain	Ross et al. (2005)
47	Cannabielsoin	I	Pollen grain	Ross et al. (2005)
48	6a,7,10a-Trihydroxytetrahydrocannabinol	A	Pollen grain	Ross et al. (2005)
49	9,10-Epoxycannabitriol	K	Pollen grain	Ross et al. (2005)
50	10-*O*-Ethylcannabitriol	K	Pollen grain	Ross et al. (2005)

Table 11.1 Continued

No.	Compounds	Types	Tissues	References
51	7,8-Dehydro-10-*O*-ethylcannabitriol	K	Pollen grain	Ross et al. (2005)
52	5-Acetoxy-6-geranyl-3-*n*-pentyl-1,4-benzoquinone	L3	Plant material	Radwan et al. (2008a)
53	α-Cannabispiranol	L4	Plant material	Radwan et al. (2008a)
54	β-Acetyl cannabispiranol	L4	Plant material	Radwan et al. (2008a)

A, Δ^9-THC type; B, Δ^8-THC type; C, CBC type; D, CBD type; E, CBG type; F, CBN type; G, CBND type; H, CBL type; I, CBE type; J, CBCT type; K, CBT type; L, other types (L1, sesqui-CBG type; L2, cannabicoumarononic acid type; L3, cannabinoid benzoquinone type; L4, cannabispiroindane type).

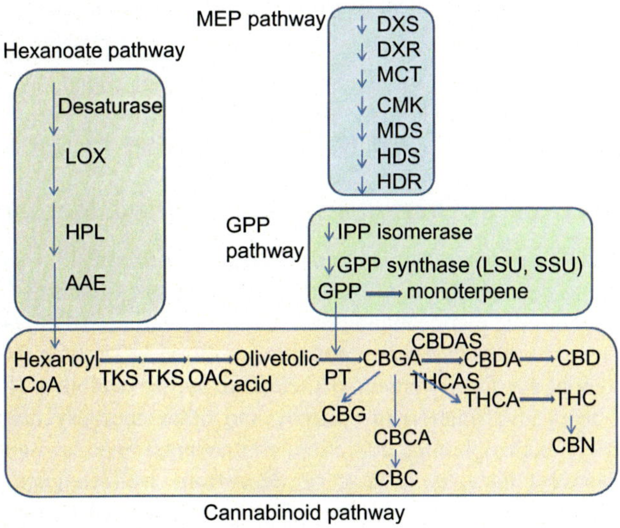

Figure 11.3 Schematic illustration of cannabinoid biosynthesis pathway, according to van Bakel et al. (2011). DXS, 1-deoxy-D-xylulose 5-phosphate synthase; DXR, 1-deoxy-D-xylulose 5-phosphate reductoisomerase; MCT, 2-C-methyl-D-erythritol 4-phosphate cytidylyltransferase; CMK, 4-diphosphocytidyl-2-C-methylerythritol kinase; MDS, 2-C-methyl-D-erythritol 2,4-cyclodiphosphate synthase; HDS, 4-hydroxy-3-methylbut-2-enyl diphosphate synthase; HDR, 4-hydroxy-3-methylbut-2-enyl diphosphate reductase; IPP, isopentenyl diphosphate; GPP, geranyl diphosphate; LSU, large subunit; SSU, small subunit; LOX, lipoxygenase; HPL, hydroperoxide lyase; AAE, acyl-activating enzyme; TKS, tetraketide synthase; OAC, olivetolic acid cyclase; PT, aromatic prenyltransferase; CBGA, cannabigerolic acid; CBDAS, cannabidiolic acid synthase; CBDA, cannabidiolic acid; CBD, cannabidiol; THCAS, Δ^9-tetrahydrocannabinolic acid synthase; THCA, Δ^9-tetrahydrocannabinolic acid; THC, Δ^9-tetrahydrocannabinol; CBCA, cannabichromenic acid; CBC, cannabichromene; CBG, cannabigerol.

than 9-THC (Hippalgaonkar et al., 2011). Additionally, 8-THC is chemically more stable, does not undergo oxidation to cannabinol, and has a much longer shelf life than 9-THC.

Cannabinoids were detected in extracts of 25–143 collected cells of capitate-sessile and capitate-stalked trichomes and separately in the gland (head) and the stem of the latter (Happyana et al., 2013). Δ^9-Tetrahydrocannabinolic acid (THCA), cannabidiolic acid (CBDA), and cannabigerolic acid (CBGA) were identified as the most abundant compounds in all analyzed samples while their decarboxylated derivatives, 9-THC, CBD, and CBG, codetected in all samples, were present at significantly lower levels. CBC and CBN were identified as minor compounds only in the samples of intact capitate-stalked trichomes and their heads harvested from 8-week-old plants.

The farnesyl prenylogue of CBG was identified on the basis of its spectroscopic properties (Pollastro et al., 2011). *C. sativa* could contain novel prenylogous versions of medicinally relevant cannabinoids, and their biological profiles could offer interesting opportunities for biomedical exploitation.

11.3 Pharmacology and therapeutic use of cannabinoids

11.3.1 Effects on the nervous system: multiple mechanisms

Compound 9-THC and other cannabinoids exert antidepressant-like actions and thus may contribute to the overall mood-elevating properties of *Cannabis* (El-Alfy et al., 2010). Nonpsychotropic plant cannabinoids, cannabidivarin (CBDV) and CBD, activate and desensitize transient receptor potential vanilloid 1 (TRPV1) channels *in vitro* and have potential for the treatment of neuronal hyperexcitability (Iannotti et al., 2014). CBD bears investigation in epilepsy and other neuropsychiatric disorders, including anxiety, schizophrenia, addiction, and neonatal hypoxic-ischemic encephalopathy. However, there is a lack of data from well-designed double-blind randomized, controlled studies on the efficacy of pure CBD for any disorder (Devinsky et al., 2014). CBD attenuates vomiting and nausea-like behavior via indirect agonism of 5-HT_{1A} somatodendritic autoreceptors in the dorsal raphe nucleus (Rock et al., 2012). Cannabinoids have therapeutic potential in schizophrenia (Kucerova et al., 2014). CBDA has therapeutic potential as a highly potent and selective treatment for anticipatory nausea without psychoactive or locomotor effects (Rock et al., 2014).

THC substantially increases the synthesis of pregnenolone in the brain via activation of cannabinoid receptor type 1 (CB1) (Vallée et al., 2014). Pregnenolone then, as a signaling-specific inhibitor of CB1, reduces several effects of THC. This negative feedback mediated by pregnenolone reveals a paracrine/autocrine loop protecting the brain from CB1 receptor overactivation that could open an unexpected approach for the treatment of *Cannabis* intoxication and addiction.

9-THC or CBD botanical extracts, as well as the combination of both natural cannabinoids, which are the components of an already approved cannabis-based medicine, preserved memory in AβPP/PS1 transgenic mice when chronically administered during the early symptomatic stage (Aso et al., 2014). THC could be a potential therapeutic option for Alzheimer's disease through multiple functions and pathways (Cao et al., 2014). Synaptic and cognitive impairments following repeated exposure to 9-THC are associated with the induction of cyclooxygenase-2 (COX-2), an inducible enzyme that converts arachidonic acid to prostanoids in the brain (Chen et al., 2013). Ablation of COX-2 eliminates 9-THC-impaired hippocampal long-term synaptic plasticity and working and fear memories. Importantly, the beneficial effects of decreasing β-amyloid plaques and neurodegeneration by 9-THC in Alzheimer's disease animals are retained in the presence of COX-2 inhibition. These results suggest that the applicability of medical marijuana would be broadened by concurrent inhibition of COX-2.

CBC raises the viability of adult neural stem progenitor cells while inhibiting their differentiation into astroglia (Shinjyo and Di Marzo, 2013), possibly through upregulation of ATP and adenosine signaling. CBDV significantly decreased PTZ-induced seizure severity and increased latency to the first sign of seizure (Amada et al., 2013). CBD treatment modulates cholesterol homeostasis in microglial cells (Rimmerman et al., 2011), and pretreatment with the cholesterol chelating agent MBCD reverses this effect without interfering with CBD's anti-inflammatory effects. 9-THC+CBD is a reasonable option for some patients with multiple sclerosis (Anonymous, 2014).

CBD and CBC stimulated descending pathways of antinociception and caused analgesia by interacting with several target proteins involved in nociceptive control (Maione et al., 2011). In a meaningful proportion of otherwise treatment-resistant patients, clinically important improvements in peripheral neuropathic pain and sleep quality are obtained with THC/CBD spray (Serpell et al., 2014), which was well tolerated, and no new safety concerns were identified.

11.3.2 Anti-inflammatory, antioxidant, and immunomodulatory activities

9-THC has been shown to mediate potent anti-inflammatory properties. Cannabinoid inhibits HIV-1 Tat-stimulated adhesion of human monocyte-like cells to extracellular matrix proteins (Raborn et al., 2014). The nonpsychotropic cannabinoid CBC inhibits nitric oxide (NO) production and exerts anti-inflammatory actions in activated macrophages. By negatively coupling with tonic CB1 cannabinoid signaling, CBC ameliorates experimental murine colitis (Romano et al., 2013). CBC selectively reduces inflammation-induced hypermotility *in vivo* in a manner that is not dependent on cannabinoid receptors or transient receptor potential ankyrin-1 (TRPA1) (Izzo et al., 2012). Cannabinoids can affect both the activity and the expression of transient receptor potential vanilloid (TRPV) types 1–4 (De Petrocellis et al., 2012), with various potential therapeutic applications, including in gastrointestinal tract inflammation.

CBD, CBG, cannabigevarin (CBGV), and Δ^9-tetrahydrocannabivarin (THCV) stimulated and desensitized human TRPV1 (De Petrocellis et al., 2011). CBC, CBD, and CBN were potent rat TRPA1 agonists and desensitizers, but THCV-botanical drug substance (BDS) was the most potent compound at this target. CBG-BDS and THCV-BDS were the most potent rat TRPM8 antagonists. All nonacid cannabinoids, except CBC and CBN, potently activated and desensitized rat TRPV2. CBDV and all the acids inhibited diacylglycerol lipase-α (DAGL-α). Some BDS, but not the pure compounds, inhibited monoacylglycerol lipase (MAGL). CBD was the only compound to inhibit fatty acid amide hydrolase (FAAH), whereas the BDS of CBC > CBG > CBGV inhibited *N*-acylethanolamine-hydrolyzing acid amidase (NAAA). CBC = CBG > CBD inhibited *N*-arachidonoyl ethanolamine (AEA) cellular uptake, as did the BDS of Δ^9-tetrahydrocannabivarin acid (THCVA), CBGV, CBDA, and THCA, but the latter extracts were more potent inhibitors. These results are relevant to the analgesic, anti-inflammatory, and anticancer effects of cannabinoids and *Cannabis* extracts.

CBD exerts sebostatic and anti-inflammatory effects on human sebocytes (Oláh et al., 2014) and might be a promising therapeutic agent for the treatment of acne vulgaris. CBD exposure during neuronal differentiation sensitizes cells against redox-active neurotoxins (Schönhofen et al., 2014).

The immunomodulatory activity of THC is mediated by epigenetic regulation (Yang et al., 2014). Histone modifications are associated with 9-THC-mediated alterations in antigen-specific T-cell responses.

9-THC, the psychoactive component of marijuana, is known to suppress the immune responses to bacterial, viral, and protozoan infections. Chronic 9-THC treatment decreased the efficacy of the memory immune response to *Candida* infection, which correlated with a decrease in IFN-γ that was only observed after the secondary *Candida* challenge (Blumstein et al., 2014). CBD and less so THC induce a cellular stress response in BV-2 microglial cells (Juknat et al., 2012, 2013), which underlies their high immunosuppressant and anti-inflammatory activities.

11.3.3 Antimicrobial activity

The mechanism of THC suppression of HIV-1 infection was traced to a reduction in cell surface HIV receptor (CD4, CCR5, and CXCR4) expression that diminished entry efficiency (Williams et al., 2014). 9-THC administration results in a generalized attenuation of viral load and tissue inflammation in simian immunodeficiency virus (SIV)-infected male rhesus macaques (Molina et al., 2014).

CBD, CBC, CBG, 9-THC, and CBN showed potent activity against a variety of methicillin-resistant *Staphylococcus aureus* (MRSA) strains of current clinical relevance (Appendino et al., 2008). CBG exhibited antimicrobial activity against *Mycobacterium intracellulare* (Radwan et al., 2008b).

The cannabinoid ester 4-terpenyl cannabinolate showed moderate antimicrobial activity against *Candida albicans* (Ahmed et al., 2008). 8-Hydroxycannabinol displayed significant antifungal activities (Radwan et al., 2009), while 5-acetyl-4-hydroxycannabigerol displayed strong antileishmanial activity. CBGA displayed moderate antileishmanial activity (Radwan et al., 2008b).

11.3.4 Anticancer activity

Cannabinoids inhibit angiogenic capacities of endothelial cells via the release of tissue inhibitor of matrix metalloproteinases-1 from lung cancer cells (Ramer et al., 2014). Cannabinoid-induced upregulation of ICAM-1 on lung cancer cells is responsible for increased cancer cell susceptibility to LAK cell-mediated cytolysis (Haustein et al., 2014). A standardized *C. sativa* extract with high content of CBD attenuates colon carcinogenesis and inhibits colorectal cancer cell proliferation via CB1 and CB2 receptor activation (Romano et al., 2014).

9-THC and 8-THC disturb the mitochondrial function and possess antitumor activities. They are potent inhibitors of Tu183 cellular respiration and are toxic to this highly malignant oral cancer (Whyte et al., 2010). Direct modulation of the outer mitochondrial membrane channel, voltage-dependent anion channel 1 (VDAC1), by CBD might be a novel mechanism for CBD-induced cell death (Rimmerman et al., 2013).

11.3.5 Other effects

Indian hemp has shown beneficial effects in various gastrointestinal conditions but it is not widely accepted due to high content of THC resulting in unwanted psychotropic effects. Since industrial hemp infuse rich in CBD reduces intestinal motility in healthy mice, CBD should be further evaluated for the treatment of intestinal hypermotility (Sabo et al., 2013).

Cannabinoids inhibit cholinergic contraction in human airways through prejunctional CB1, which may explain the acute bronchodilation produced by marijuana smoking (Grassin-Delyle et al., 2014).

Compound 9-THC produced dose-dependent increases in diuresis and decreases in colonic temperature, with slightly lower ED50 values for diuresis than for hypothermia (Paronis et al., 2013).

Smoking of marijuana and intravenous and oral administration of 9-THC and 8-THC reduce the intraocular pressure in animals and in humans (Hippalgaonkar et al., 2011).

CBD and CBG are transcriptional repressors that control cell proliferation and differentiation, which indicates that they have the potential to be lead compounds for the development of novel therapeutics for skin diseases (Pucci et al., 2013).

CBD reduced food intake, supporting the need for further mechanistic and behavioral work examining possible antiobesity effects (Farrimond et al., 2012). On the contrary, CBN has been shown to increase feeding.

11.4 Toxicity and safety

An acute challenge with 9-THC can induce psychotic symptoms including delusions (Nottage et al., 2014). THC triggers paranoid thoughts in vulnerable individuals (Freeman et al., 2014). The most likely mechanism of action causing paranoia was the generation of negative affect and anomalous experiences. Acute administration of THC impairs spatial but not associative memory function in zebrafish (Ruhl et al., 2014). 9-THC disrupts hippocampal neuroplasticity and neurogenesis in trained adolescent Sprague–Dawley rats (Steel et al., 2014).

Children exposed in utero to *Cannabis* present permanent neurobehavioral and cognitive impairments. 9-THC disrupts cortical development by inducing an SCG10/stathmin-2 degradation pathway (Tortoriello et al., 2014). Chronic exposure to THC during adolescence is responsible for heightened anxiety and increased vulnerability to drug relapse in adulthood (Stopponi et al., 2014).

9-THC both inhibits basal and bicarbonate-stimulated sperm motility *in vitro* and reduces male fertility *in vivo* (Morgan et al., 2012).

9-THC is subject to first pass hepatic metabolism primarily by hydroxylation to yield active and inactive oxygenated products. Cytochrome P450 2C9 (CYP2C9) and CYP3A4 are major enzymes involved in the 11-hydroxylation and the 8-(or 7-) hydroxylation, respectively, of the cannabinoids by human hepatic microsomes (Watanabe et al., 2007). CYP3A4 is a major enzyme responsible for the 7α- and 7β-hydroxylation of 8-THC and the 9α- and 10α-epoxidation of 9-THC. 9-THC increased the rate of phenytoin hydroxylation in human liver microsomes and expressed CYP2C9 enzyme (Bland et al., 2005). Similar increases in rate were observed with coincubation of 11-hydroxy-9-THC and 11-nor-9-carboxy-9-THC with phenytoin. These *in vitro* data suggest the potential for an interaction from the concomitant administration of 9-THC and phenytoin that could result in decreased phenytoin concentrations *in vivo*.

Cannabis is a possible risk factor for cardiovascular disease in young adults (Jouanjus et al., 2014). THC appeared to be safe and well tolerated by healthy older individuals (Ahmed et al., 2014). Data on safety and effectiveness of THC in frail older persons are urgently required, as this population could benefit from the therapeutic applications of THC.

11.5 Other compounds

Other constituents of the *Cannabis* plant are nitrogenous compounds (27 known), amino acids (18), proteins (3) (He and Shi, 2010), glycoproteins (6), enzymes (2), sugars and related compounds (34), hydrocarbons (50), simple alcohols (7), aldehydes (13), ketones (13), simple acids (21), fatty acids (22) (Jeong et al., 2014), simple esters (12), lactones (1), steroids (11) (Radwan et al., 2009; Figure 11.4 and Table 11.2), terpenes (120) (Russo, 2011; Happyana et al., 2013), noncannabinoid phenols (25) (Radwan et al., 2009; Figure 11.4), flavonoids (21) (Ross et al., 2005; Figure 11.4), vitamins (1) (vitamin A), pigments (2), and elements (9).

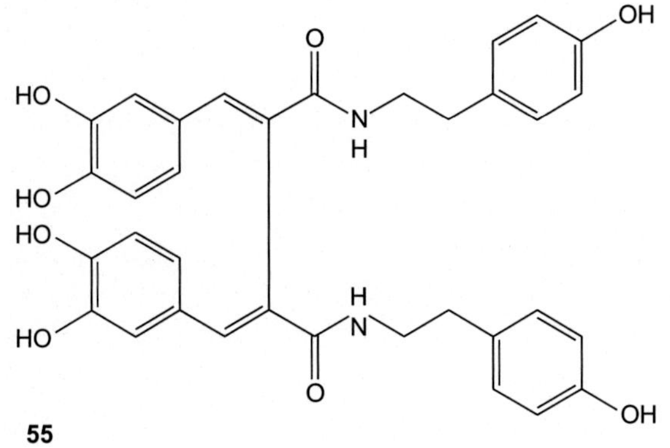

55

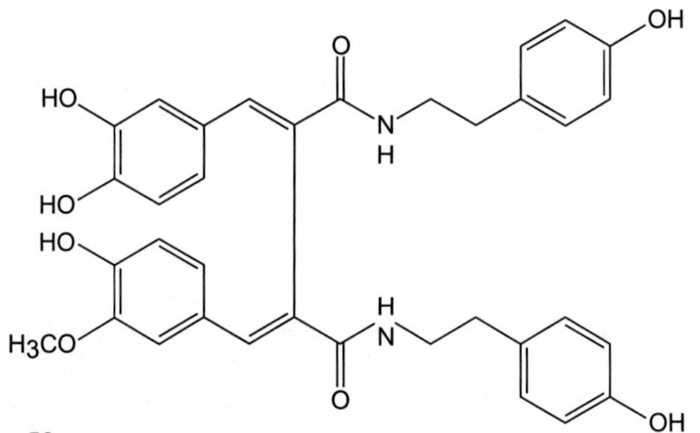

56

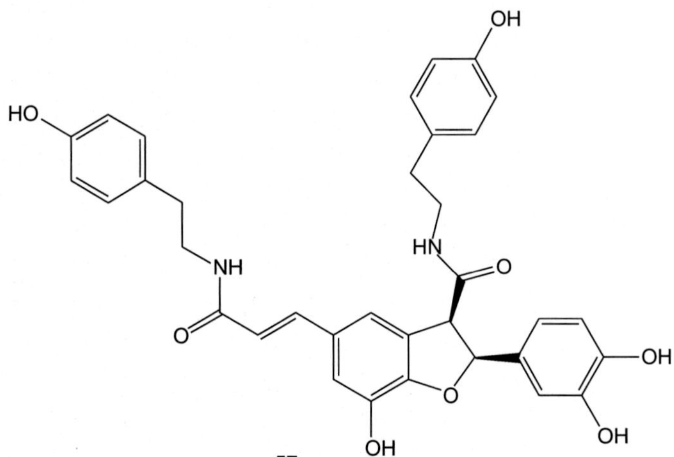

57

Figure 11.4 Examples of noncannabinoid compounds found in *Cannabis* plants.

(Continued)

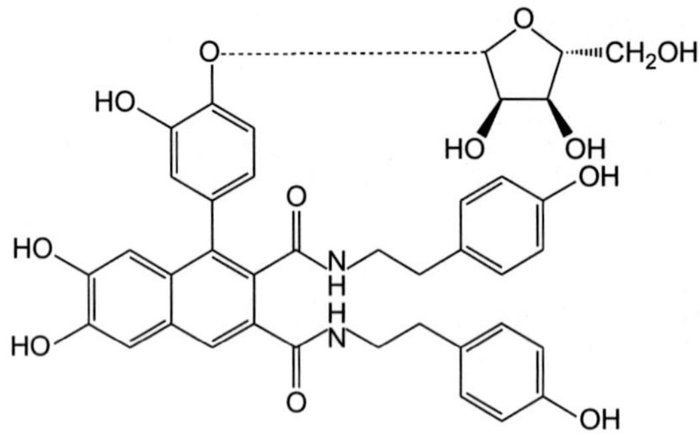

58

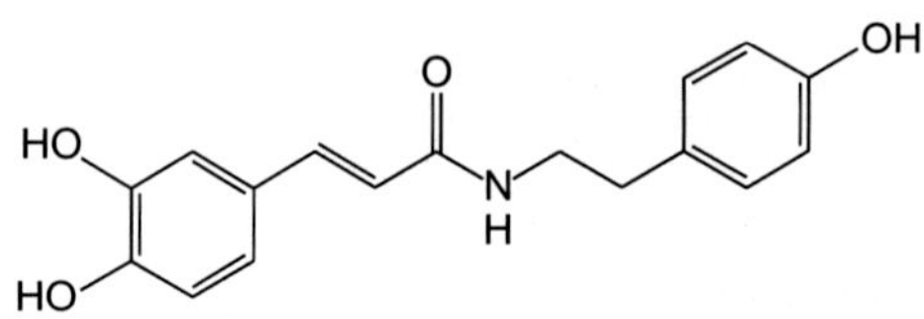

59

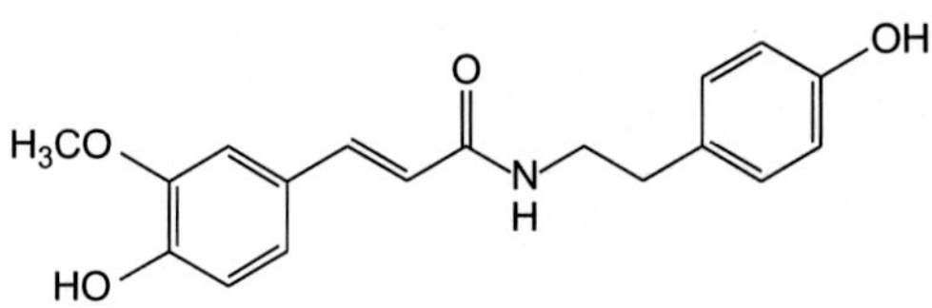

60

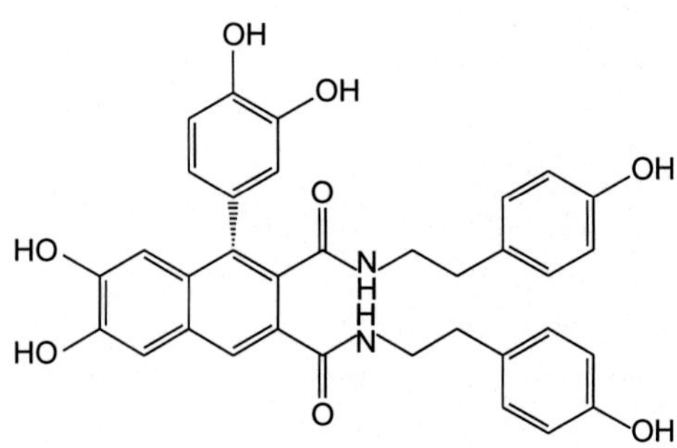

61

Figure 11.4 Continued.

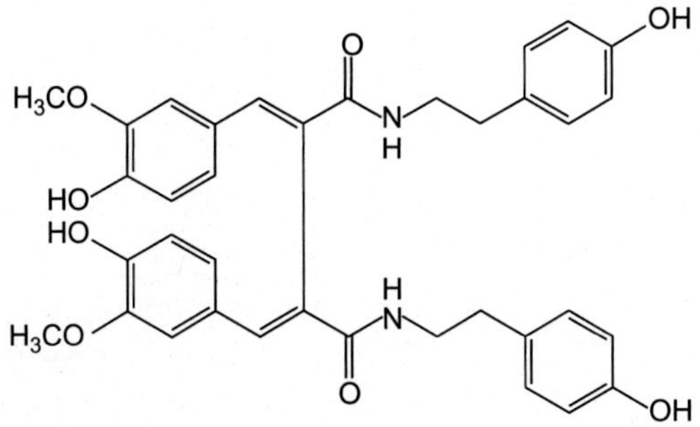

62

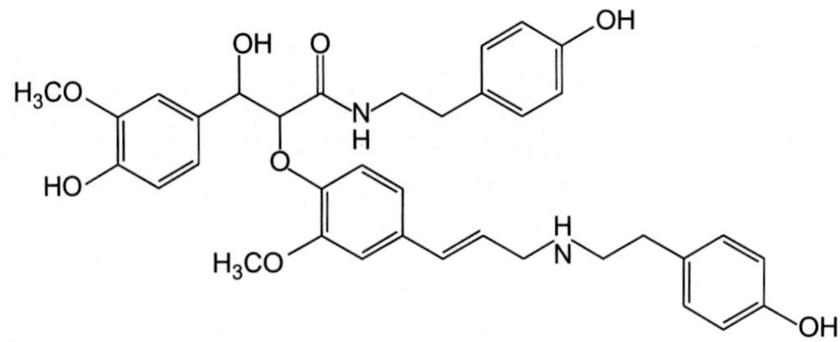

63

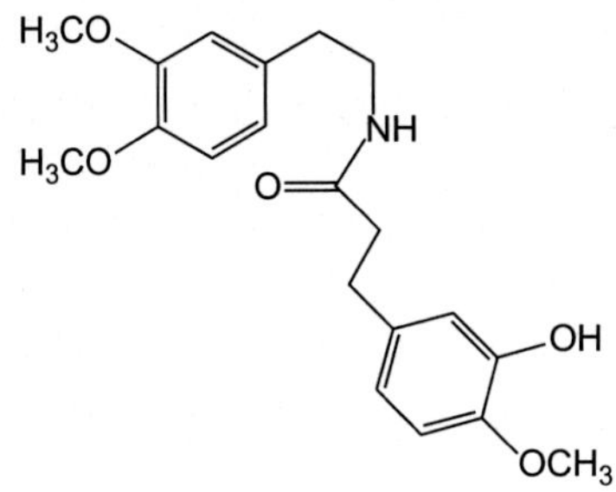

64

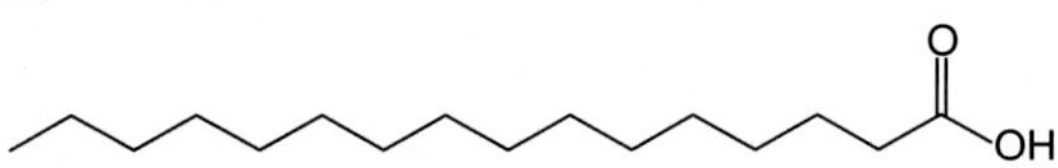

65

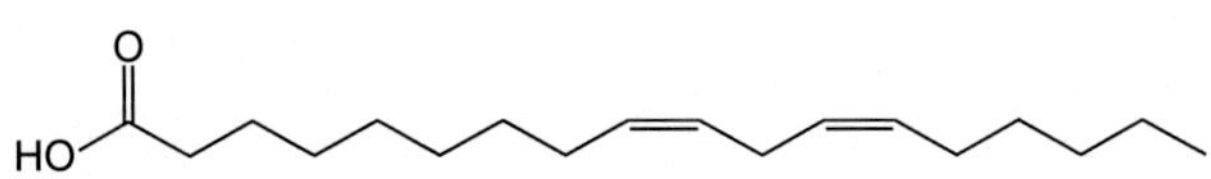

66

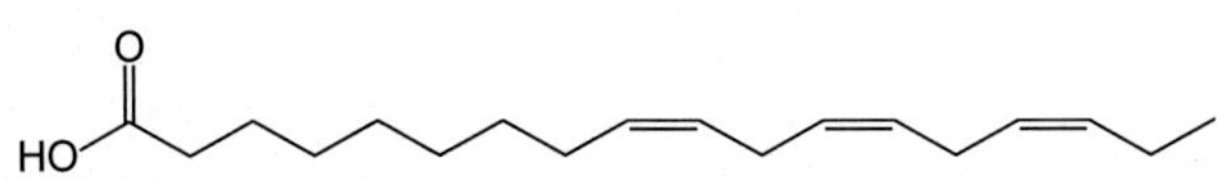

67

Figure 11.4 Continued.

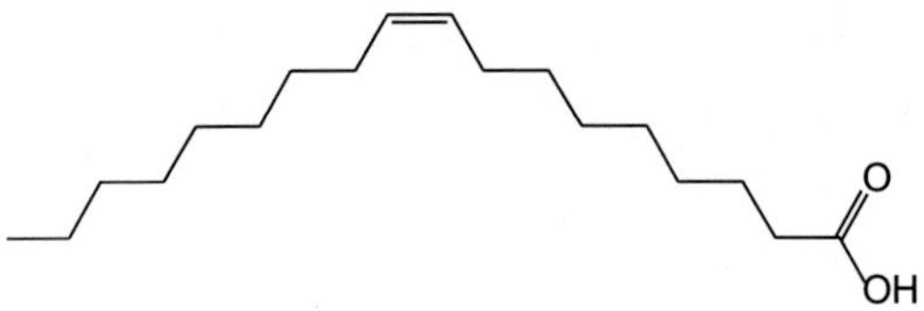

68

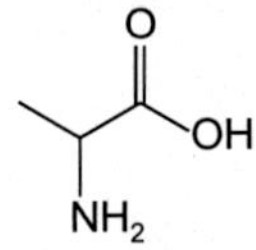

69

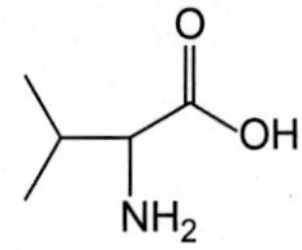

70

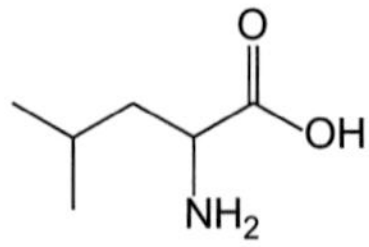

71

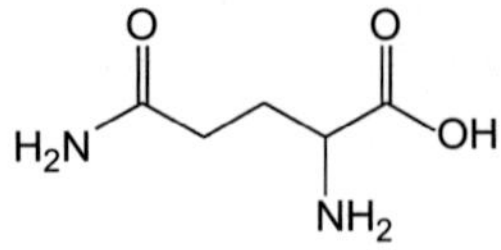

72

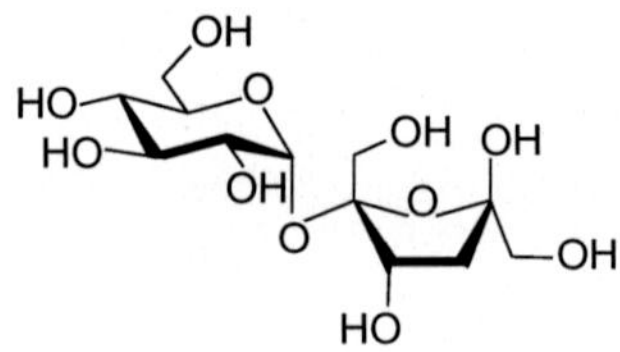

73

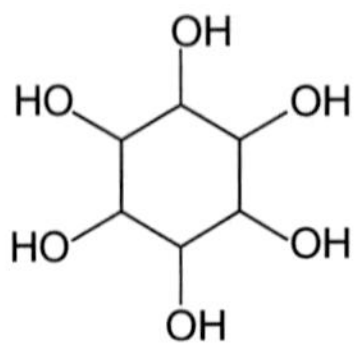

74

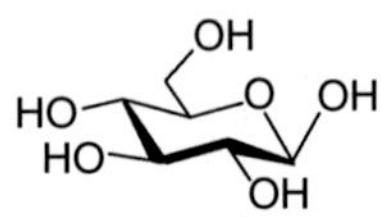

75

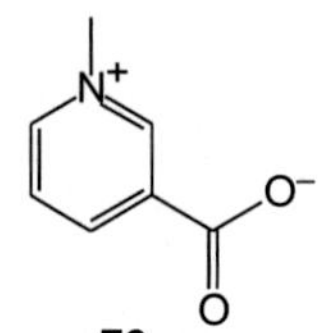

76

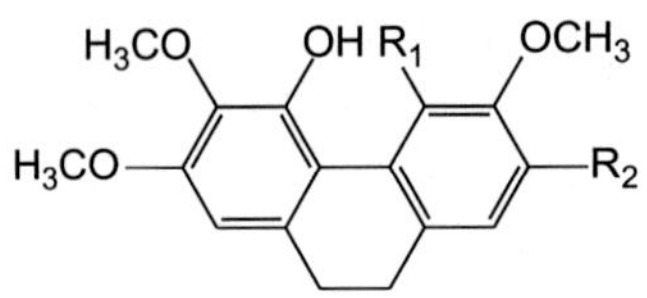

No.	R_1	R_2
77	OH	H
78	H	OCH_3

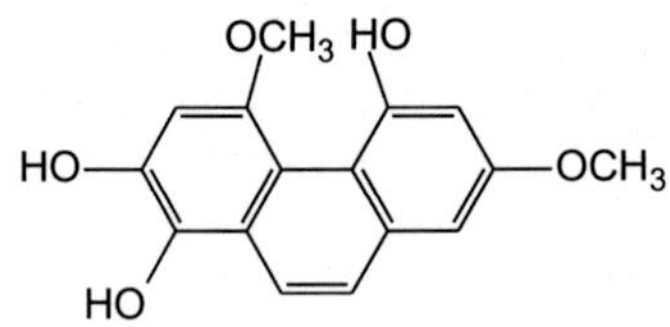

79

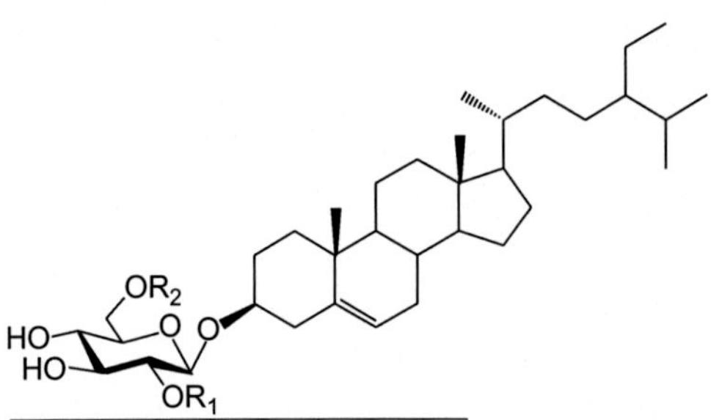

No.	R_1	R_2
80	$CO(CH_2)_{14}CH_3$	H
81	H	Ac

Figure 11.4 Continued.

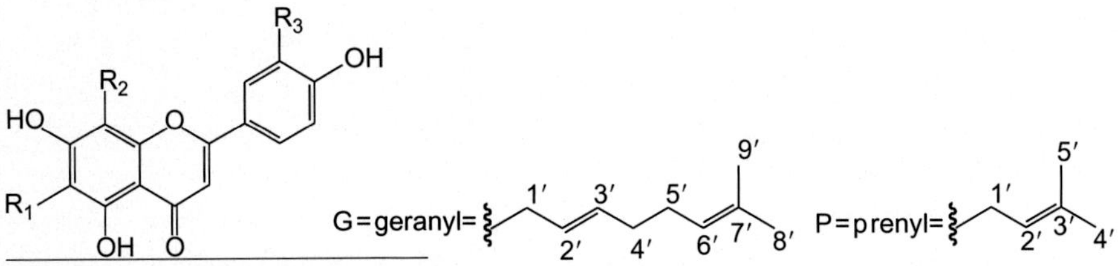

No.	R_1	R_2	R_3
82	H	G	OCH_3
83	H	H	OCH_3
84	P	H	H
85	G	H	OCH_3
86	P	H	OCH_3

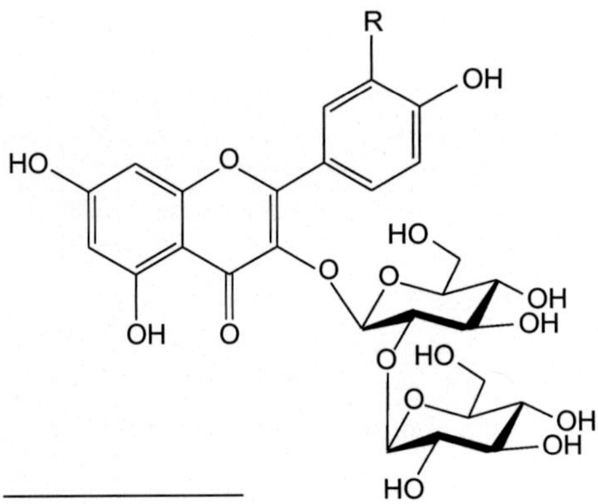

No.	R
87	H
88	OH

Figure 11.4 Continued.

Table 11.2 Other compounds found in *Cannabis sativa* L. in recent years

No.	Compounds	Types	Tissues	References
55	3, 3′-Demethyl-cannabisin G	I	Seed	Ross et al. (2005)
56	3-Demethyl-cannabisin G	I	Seed	Ross et al. (2005)
57	3, 3′-Demethyl-grossamide	I	Seed	Ross et al. (2005)
58	4′-*O*-β-D-ribofuranosyl cannabisin A	I	Seed	Ross et al. (2005)
59	*N*-Caffeoyltyramine	I	Seed	Ross et al. (2005)
60	*N*-Feruloyltyramine	I	Seed	Ross et al. (2005)
61	Cannabisin A	I	Seed	Ross et al. (2005)
62	Cannabisin G	I	Seed	Ross et al. (2005)
63	Cannabisin E	I	Seed	Ross et al. (2005)
64	*N*-(3,4-Dimethoxyphenethyl)-3-(3-hydroxy-4-methoxyphenyl) propanamide	I	Seed	Ross et al. (2005)
65	Palmitic acid	IIa	Seed	Ross et al. (2005)
66	Linoleic acid	IIb	Seed	Ross et al. (2005)
67	α-Linolenic acid	IIb	Seed	Ross et al. (2005)
68	Oleic acid	IIb	Seed	Ross et al. (2005)
69	Alanine	III	Seed	Ross et al. (2005)
70	Valine	III	Seed	Ross et al. (2005)
71	Leucine	III	Seed	Ross et al. (2005)
72	Glutamine	III	Seed	Ross et al. (2005)
73	Sucrose	IV	Seed	Ross et al. (2005)
74	Myoinositol	IV	Seed	Ross et al. (2005)
75	Glucose	IV	Seed	Ross et al. (2005)
76	Trigonelline	V	Seed	Ross et al. (2005)
77	4,5-Dihydroxy-2,3,6-trimethoxy-9,10-dihydrophenanthrene	VI	Plant material	Radwan et al. (2008a)
78	4-Hydroxy-2,3,6,7-tetramethoxy-9,10-dihydrophenanthrene	VI	Plant material	Radwan et al. (2008a)
79	4,7-Dimethoxy-1,2,5-trihydroxyphenanthrene	VI	Plant material	Radwan et al. (2008a)
80	β-Sitosteryl-3-*O*-β-D-glucopyranoside-2′-*O*-palmitate	VII	Plant material	Radwan et al. (2008a)
81	β-Sitosteryl-3-*O*-β-D-glucopyranoside-6′-*O*-acetate	VII	Plant material	Radwan et al. (2009)
82	Cannflavin C	VII	Plant material	Radwan et al. (2008a)
83	Chrysoeriol	VII	Plant material	Radwan et al. (2008a)
84	6-Prenylapigenin	VII	Plant material	Radwan et al. (2008a)
85	Cannflavin A	VII	Plant material	Radwan et al. (2008a)
86	Cannflavin B	VII	Plant material	Radwan et al. (2008b)
87	Kaempferol 3-*O*-sophoroside	VII	Pollen grain	Ahmed et al. (2008)
88	Quercetin 3-*O*-sophoroside	VII	Pollen grain	Ahmed et al. (2008)

I, lignanamide type; II, fatty acid type (IIa, saturated fatty acid type; IIb, unsaturated fatty acid type); III, amino acid type; IV, saccharide type; V, alkaloid type; VI, phenanthrene type; VII, sterol type; VII, flavonoid type.

11.5.1 Terpenoids

Terpenoids, such as limonene, myrcene, α-pinene, linalool, β-caryophyllene, caryophyllene oxide, nerolidol, and phytol, share a precursor with phytocannabinoids (Figure 11.3) and are flavor and fragrance components common in human diets that have been designated as generally recognized as safe (GRAS) by the US Food and Drug Administration and other regulatory agencies. They display unique therapeutic effects that may contribute meaningfully to the entourage effects of cannabis-based medicinal extracts. Phytocannabinoid–terpenoid interactions could produce synergy with respect to treatment of pain, inflammation, depression, anxiety, addiction, epilepsy, cancer, and fungal and bacterial infections (Russo, 2011).

11.5.2 Hemp seed oil

The oils of four Italy cultivars are rich in linoleic acid (ω-6) and α-linolenic acid (ω-3) (Porto et al., 2014). Felina 32 and Chamaeleon oils exhibited the highest amount of linoleic acid (59%) and α-linolenic acid (18%). Finola and Uso31 oils have the richest γ-linolenic acid (5–6%). All hemp seed oils presented high oxidation stability and an acceptable initial quality. These oils can be used to produce essential fatty acid (EFA) dietary supplements high in ω-6 and ω-3. The oil of Italy cannabinoid-free *C. sativa* was characterized by a high unsaturated/saturated fatty acid ratio and by an almost perfect balance of ω-3 and ω-6 fatty acids (Lesma et al., 2014), as requested for healthy foods. The alcoholic extracts, for which a high content of amino acids and phenolic compounds has been highlighted, could provide dietary supplements to help in preventing oxidative stress. Six known and four new lignanamides have been identified, confirming and assessing the general metabolic pattern in the seeds of these locally grown plants.

China's Yunma No. 1 and Bama Huoma, with more than 50% oil and 30% protein in dehulled seed, could be used as oil extraction material and protein source with respect to kernel yield (Chen et al., 2010). The concentration of unsaturated fatty acids exceeded 90%, higher than most conventional vegetable oils. Moreover, polyunsaturated fatty acids (PUFAs) ranged from 76.26% to 82.75% and were mainly composed of linoleic acid and α-linolenic acid with a ratio close to 3:1. γ-Tocopherol was found at an average concentration of 28.23 mg/100 g of oil. Hemp seed oil (HO) is a highly valuable vegetable oil. A yield (1.84–1.92%) of unsaponifiable matter was obtained from a Spain HO (Montserrat-de la Paz et al., 2014), and the most interesting compounds were β-sitosterol (1905.00 ± 59.27 mg/kg of oil), campesterol (505.69 ± 32.04 mg/kg), phytol (167.59 ± 1.81 mg/kg), cycloartenol (90.55 ± 3.44 mg/kg), and γ-tocopherol (73.38 ± 2.86 mg/100 g).

HO treatment induced lipid accumulation (Jeong et al., 2014), reactive oxygen species (ROS) production, C/EBP homologous protein (CHOP) expression, and apoptosis in MH7A human RA fibroblast-like synovial cells. CHOP functions as an antirheumatoid factor. Dietary HO caused significant changes in plasma fatty acid profiles and improved clinical symptoms of atopic dermatitis (Callaway et al., 2005). These improvements resulted from the balanced and abundant supply of PUFAs in HO.

Dietary hemp seed can provide significant cardioprotective effects during postischemic reperfusion, which might be due to its abundant PUFA (Al-Khalifa et al., 2007). Hemp seed meal and linoleic acid exert inhibitory effects on both Aβ42 cytotoxicity and cholesterol uptake and are potential candidates for the treatment of Alzheimer's disease and cardiovascular disease (Lee et al., 2011).

11.5.3 Others

The peptide fractions of hemp seed protein hydrolysate exhibited higher oxygen radical absorbance capacity and scavenging of 2,2-diphenyl-1-picrylhydrazyl, superoxide, and hydroxyl radicals (Girgih et al., 2013). Tyrosol, an antioxidant metabolite, was detected in *Cannabis* cell cultures (Pec et al., 2010). The tyrosol content increased after eliciting with jasmonic acid (JA).

Significant differences in isoperoxidase and isoesterase patterns were found between male and female plants (Truţa et al., 2002), in both the number and stain intensity of bands. For both esterase and peroxidase, the isozyme spectrum is richer for staminate plants. Also, some differences are obvious between the two sexes concerning catalase and peroxidase activities, as well as the level of soluble protein. The quantitative analysis of flavones, polyholozides, and polyphenols highlights differences depending not only on sex but also on the tested organ.

4-Acetoxy-2-geranyl-5-hydroxy-3-*n*-pentylphenol has antibacterial activities (Radwan et al., 2009). 6-Prenylapigenin displayed moderate anti-MRSA activity, cannflavin A exhibited strong antileishmanial activity, while 4,5-dihydroxy-2, 3, 6-trimethoxy-9, 10-dihydrophenanthrene, cannflavin C, and β-acetyl cannabispiranol displayed strong antioxidant activities in the DPPH assay (Radwan et al., 2008a). Cannflavin B displayed moderate antileishmanial activity (Radwan et al., 2008b).

11.6 Chemotaxonomy

Cannabinoids are important chemotaxonomic markers unique to *Cannabis*. The two cannabinoids usually produced in greatest abundance are CBD and 9-THC. *Cannabis* plants are categorized by their chemical phenotype or "chemotype," based on the overall amount of psychoactive THC produced and on the ratio of THC to CBD. Although overall cannabinoid production is influenced by environmental factors, the THC/CBD ratio is genetically determined and remains fixed throughout the life of a plant (de Meijer et al., 2003). Nondrug plants produce less THC and more CBD, while drug plants produce more THC and less CBD. When plants of these two chemotypes cross-pollinate, the plants in the first filial (F1) generation have an intermediate chemotype and produce similar amounts of CBD and THC. Female plants of this chemotype may produce enough THC to be utilized for drug production (Hillig and Mahlberg, 2004).

Alleles B(D) and B(T) encode alloenzymes that catalyze the conversion of CBG to CBD and THC, respectively. Mean THC levels and the frequency of B(T) were significantly higher in *C. indica*, a putative subspecies, than *C. sativa* (Hillig and Mahlberg, 2004). The proportion of high THC/CBD chemotype plants in most accessions assigned to *C. sativa* was <25% and in most accessions assigned to *C. indica* was >25%. Plants with relatively high levels of THCV and/or CBDV were common only in *C. indica*. This study supports a two-species concept of *Cannabis*.

Extracts of pistillate inflorescences of 162 greenhouse-grown plants of diverse origin were analyzed by gas chromatography (GC) (Hillig, 2004). Peak area ratios of 48 compounds were subjected to multivariate analysis, and the results were interpreted with respect to geographic origin and taxonomic affiliation. A canonical analysis in which the plants were preassigned to *C. sativa* or *C. indica* based on previous genetic, morphological, and chemotaxonomic studies resulted in 91% correct assignment of the plants to their preassigned species. A scatterplot on the first two principal component axes shows that accessions from Afghanistan assigned to the wide-leaflet drug biotype of *C. indica* group are apart from the other taxa. The essential oil of these plants usually had relatively high ratios of guaiol, isomers of eudesmol, and other unidentified compounds. Plants assigned to the narrow-leaflet drug biotype of *C. indica* tended to have relatively high ratios of *trans*-β-farnesene. Cultivars of the two drug biotypes may exhibit distinctive medicinal properties due to significant differences in terpenoid composition. The terpenoid composition of the essential oil of *Cannabis* is useful for chemotaxonomic discrimination.

In order to develop *Cannabis* plant material as a medicinal product, quality control and clear chemotaxonomic discrimination between varieties are a necessity. Chemical analysis of 11 *Cannabis* varieties, which grew under the same environmental conditions, used a GC flame ionization detection method that was validated for quantitative analysis of *Cannabis* monoterpenoids, sesquiterpenoids, and cannabinoids (Fischedick et al., 2010). Quantitative data were analyzed using principal component analysis (PCA) to determine which compounds are most important in discriminating *Cannabis* varieties. In total, 36 compounds were identified and quantified in the 11 varieties. Using PCA, each *Cannabis* variety could be chemically discriminated.

11.7 Molecular biology, phylogeny, and omics

11.7.1 Molecular markers

Three chemotype-associated randomly amplified polymorphic DNA (RAPD) markers were selected (de Meijer et al., 2003), one of which was transformed into a sequence-characterized amplified region (SCAR) marker and shows tight linkage to the chemotype and codominance. Within the dioecious populations, the presence of a high number of male-specific markers, presumably associated with the Y chromosome, was found by RAPD and amplified fragment length polymorphism (AFLP) analysis (Mandolino et al., 2002; Flachowsky et al., 2001). AFLP markers have practical utility for establishing conspiracy in the cultivation and distribution of marijuana (Datwyler and Weiblen, 2006), identifying geographic sources of seized drugs, and discriminating illegal, potent marijuana cultivars from hemp where the cultivation of industrial hemp is permitted.

Short tandem repeats (STRs) are codominant, can be standardized such that reproducibility between laboratories can be easily achieved, have a high discrimination power, and can be multiplexed. STR multiplex system can be used in genetic individualization of *C. sativa* (Mendoza et al., 2009). Microsatellite (SSR) markers have been developed for DNA typing and genetic relatedness analyses (Alghanim and Almirall, 2003). Inter-simple sequence repeat (ISSR) DNA fingerprinting was used to assess the genetic stability of synthetic seeds of *C. sativa* during *in vitro* multiplication and storage for six months at different growth conditions (Lata et al., 2011). All the ISSR profiles from *in vitro* stored plants were monomorphic and comparable to the mother plant, which confirms the genetic stability among the clones. GC analysis of six major cannabinoids (9-THC, THCV, CBD, CBC, CBG, and CBN) showed homogeneity in the regrown clones and the mother plant with insignificant differences in cannabinoid content, thus confirming the stability of plants derived from synthetic seeds following 6-month storage.

DNA-based methods have the potential to examine *Cannabis* material unsuitable for analysis using conventional means including seeds, pollen, and severely degraded material. DNA polymorphisms in the THCA synthase gene can be used to discriminate "drug-type" and "fiber-type" *Cannabis* (Kojoma et al., 2006; Rotherham and Harbison, 2011). An increased relative expression of THCA gene, determined by real-time quantitative reverse-transcription PCR (qRT-PCR), was found in samples with the highest percentage of THC coming from seized sample and adult plants (Cascini et al., 2013), while fewer gene expression was detected in samples from leaves of young plants. However, no correlation between THCA synthase gene copy number and the content of THC in the herbal *Cannabis* samples tested was found (Cascini et al., 2012). Gene quantification could be useful in forensics in order to complement chemical analysis for the identification and classification of seized *Cannabis* samples, thus distinguishing the drug-type from the fiber-type varieties.

11.7.2 Proteomics

Different tissues have significantly different levels of cannabinoids. Cannabinoids not only accumulate primarily in the glands but also can be found in flowers and leaves (Raharjo et al., 2004). Comparative proteomics of leaves, flowers, and glands of *C. sativa* was used to identify specific tissue-expressed proteins. Less than half of the proteins expressed in flowers were also expressed in the glands. Some selected gland protein spots were identified, for example, a polyketide synthase, an enzyme believed to be involved in cannabinoid biosynthesis. Proteomic profiling of hemp seed proteins from a nondrug type of industrial hemp, Cheungsam, identified 1102 protein spots (Park et al., 2012), 168 of which are unique. These proteins are involved in energy regulation (23%), metabolism (18%), stress response (16%), unclassified (12%), cytoskeleton (11%), binding function (5%), and protein synthesis (3%). These proteins might have important biological functions in germination, storage, or development of hemp seed. *Cannabis* is able to absorb and accumulate heavy metals in roots and shoots. Proteomic analysis found that copper stress induced the suppression of two root proteins and the downregulation of seven proteins, while five proteins were upregulated (Bona et al., 2007). Differences in root protein expression pattern were indicative of a plant adaptation to chronic stress and were associated with the reestablishment of the cellular and redox homeostasis in the root.

11.7.3 Genomics and phylogeny

The marijuana strain Purple Kush was sequenced using short-read approaches (van Bakel et al. 2011). The draft haploid genome sequence is 534 Mb and the transcriptome has 30,000 genes. Comparison of the transcriptome of Purple Kush with that of the hemp cultivar "Finola" revealed that many genes involved in cannabinoid and precursor pathways are more highly expressed in Purple Kush than in "Finola". The exclusive occurrence of THCA synthase in the Purple Kush transcriptome and its replacement by CBDA synthase in "Finola" may explain why THC is produced in marijuana but not in hemp. Resequencing the hemp cultivars "Finola" and "USO-31" showed little difference in gene copy numbers of cannabinoid pathway enzymes. However, single nucleotide variant analysis uncovered a relatively high level of variation among four *Cannabis* types and supported a separation of marijuana and hemp. *Cannabis* genome sequences will aid the development of therapeutic marijuana strains with tailored cannabinoid profiles and provide a basis for the breeding of hemp with improved agronomic characteristics.

To address the composition of the urticalean rosids and the relationships of the component families (Cannabaceae, Cecropiaceae, Celtidaceae, Moraceae, Ulmaceae, and Urticaceae), Sytsma et al. (2002) analyzed sequence variation for these families and various rosid outgroups using rbcL, trnL-F, and ndhF plastid regions. Urticalean rosids are derived from a lineage including Barbeyaceae, Dirachmaceae, Elaeagnaceae, and Rhamnaceae, with Rosaceae less closely related; thus, they are imbedded within Rosales. Ulmaceae are the sister to all remaining families. Of note is that Cannabaceae, which consists of *Humulus* and *Cannabis*, is a subclade of Celtidaceae; this expanded family should be called Cannabaceae, which consists of six genera (Figure 11.5). Monophyletic Moraceae is sister to Urticaceae. Support for these relationships comes from a number of morphological characters (floral sexuality, presence or absence of hypanthium, stamen type and dehiscence, pollen pore number, ovule position, and embryo alignment) and chromosome numbers. Most fruit types, in terms of ecological dispersal, are derived independently multiple times and are strongly correlated with habitat.

11.7.4 Metabolomics

NMR-based metabolomics is an analytic platform, which has been used to classify and analyze *C. sativa* L. cell suspension cultures and plants (Flores-Sanchez et al., 2012). Diverse groups of primary and secondary metabolites were identified by comparing NMR data with reference compounds and/or by structure elucidation using ^{1}H NMR, J-resolved, ^{1}H-^{1}H COSY, and ^{1}H-^{13}C HMBC spectroscopy. *Cannabis* cell cultures

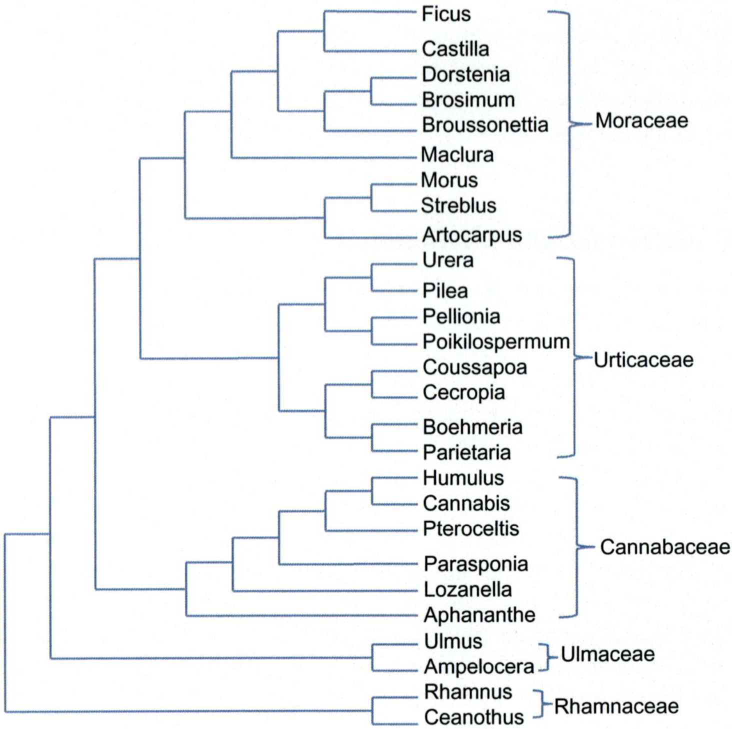

Figure 11.5 Phylogenetic relationship of *Cannabis* and related genera, according to Sytsma et al. (2002).

were treated with JA and pectin as elicitors to evaluate their effect on metabolism of two cell lines using NMR spectroscopy and multivariate data analysis (Pec et al., 2010). According to PCA and partial least square-discriminant analysis (PLS-DA), the chloroform extract of the pectin-treated cultures was more different than control and JA-treated cultures; but in the methanol/water extract, the metabolome of the JA-treated cells showed clear differences with control and pectin-treated cultures. No cannabinoids were found in either control or elicited *Cannabis* cell cultures (Flores-Sanchez et al., 2009). THCA synthase gene expression was monitored during a time course. Results suggest that other components in the signaling pathway might control the cannabinoid pathway. In ^{1}H NMR spectra, THCA and CBDA are important metabolites to differentiate 12 cultivars from each other (Choi et al., 2004). The discrimination of the cultivars could also be obtained from a water extract containing carbohydrates and amino acids. The level of sucrose, glucose, asparagine, and glutamic acid are found to be major discriminating metabolites of these cultivars. This method allows an efficient differentiation between *Cannabis* cultivars without any prepurification steps.

11.8 Conclusions

Untoward side effects of *Cannabis* limit its medical value. Research in chemistry, biology, and omics of *Cannabis* plants is definitely beneficial to synthetic drug design, which aims to maximize the potency of *Cannabis* and avoid its adverse effects. Innovative conventional plant breeding has yielded *Cannabis* chemotypes expressing high titers of each component for future study. The availability of the *C. sativa* genome enables the study of a multifunctional plant that occupies a unique role in human culture. Molecular breeding would play a vital role in obtaining *Cannabis* strain of high CBD (or other desired constituents) content and low (or no) THC content.

References

Ahmed, S.A., Ross, S.A., Slade, D., et al., 2008. Cannabinoid ester constituents from high-potency *Cannabis sativa*. J. Nat. Prod. 71 (4), 536–542.

Ahmed, A.I., van den Elsen, G.A., Colbers, A., et al., 2014. Safety and pharmacokinetics of oral delta-9-tetrahydrocannabinol in healthy older subjects: a randomized controlled trial. Eur. Neuropsychopharmacol. 24 (9), 1475–1482.

Alghanim, H.J., Almirall, J.R., 2003. Development of microsatellite markers in *Cannabis sativa* for DNA typing and genetic relatedness analyses. Anal. Bioanal. Chem. 376 (8), 1225–1233.

Al-Khalifa, A., Maddaford, T.G., Chahine, M.N., et al., 2007. Effect of dietary hempseed intake on cardiac ischemia-reperfusion injury. Am. J. Physiol. Regul. Integr. Comp. Physiol. 292 (3), R1198–R1203.

Amada, N., Yamasaki, Y., Williams, C.M., Whalley, B.J., 2013. Cannabidivarin (CBDV) suppresses pentylenetetrazole (PTZ)-induced increases in epilepsy-related gene expression. Peer J. 1, e214.

Anonymous, 2014. Delta-9-tetrahydrocannabinol + cannabidiol. A reasonable option for some patients with multiple sclerosis. Prescrire Int. 23 (150), 145–148.

Appendino, G., Gibbons, S., Giana, A., et al., 2008. Antibacterial cannabinoids from *Cannabis sativa*: a structure-activity study. J. Nat. Prod. 71 (8), 1427–1430.

Aso, E., Sánchez-Pla, A., Vegas-Lozano, E., et al., 2014. Cannabis-based medicine reduces multiple pathological processes in AβPP/PS1 mice. J. Alzheimers Dis. 43 (3), 977–991 (Epub ahead of print).

Bland, T.M., Haining, R.L., Tracy, T.S., Callery, P.S., 2005. CYP2C-catalyzed delta9-tetrahydrocannabinol metabolism: kinetics, pharmacogenetics and interaction with phenytoin. Biochem. Pharmacol. 70 (7), 1096–1103.

Blumstein, G.W., Parsa, A., Park, A.K., et al., 2014. Effect of delta-9-tetrahydrocannabinol on mouse resistance to systemic *Candida albicans* infection. PLoS One 9 (7), e103288.

Bona, E., Marsano, F., Cavaletto, M., Berta, G., 2007. Proteomic characterization of copper stress response in *Cannabis sativa* roots. Proteomics 7 (7), 1121–1130.

Callaway, J., Schwab, U., Harvima, I., et al., 2005. Efficacy of dietary hempseed oil in patients with atopic dermatitis. J. Dermatolog. Treat. 16 (2), 87–94.

Cao, C., Li, Y., Liu, H., et al., 2014. The potential therapeutic effects of THC on Alzheimer's disease. J. Alzheimers Dis. 42 (3), 973–984 (Epub ahead of print).

Cascini, F., Passerotti, S., Martello, S., 2012. A real-time PCR assay for the relative quantification of the tetrahydrocannabinolic acid (THCA) synthase gene in herbal *Cannabis* samples. Forensic Sci. Int. 217 (1–3), 134–138.

Cascini, F., Passerotti, S., Boschi, I., 2013. Analysis of THCA synthase gene expression in cannabis: a preliminary study by real-time quantitative PCR. Forensic Sci. Int. 231 (1–3), 208–212.

Chen, R., Zhang, J., Fan, N., et al., 2013. Δ^9-THC-caused synaptic and memory impairments are mediated through COX-2 signaling. Cell 155 (5), 1154–1165.

Chen, T., He, J., Zhang, J., et al., 2010. Analytical characterization of Hempseed (seed of *Cannabis sativa* L.) oil from eight regions in China. J. Diet. Suppl. 7 (2), 117–129.

Cheng, C.W., Bian, Z.X., Zhu, L.X., et al., 2011. Efficacy of a Chinese herbal proprietary medicine (Hemp Seed Pill) for functional constipation. Am. J. Gastroenterol. 106 (1), 120–129.

Choi, Y.H., Kim, H.K., Hazekamp, A., et al., 2004. Metabolomic differentiation of *Cannabis sativa* cultivars using 1H NMR spectroscopy and principal component analysis. J. Nat. Prod. 67 (6), 953–957.

Datwyler, S.L., Weiblen, G.D., 2006. Genetic variation in hemp and marijuana (*Cannabis sativa* L.) according to amplified fragment length polymorphisms. J. Forensic Sci. 51 (2), 371–375.

de Meijer, E.P., Bagatta, M., Carboni, A., et al., 2003. The inheritance of chemical phenotype in *Cannabis sativa* L. Genetics 163 (1), 335–346.

De Petrocellis, L., Ligresti, A., Moriello, A.S., et al., 2011. Effects of cannabinoids and cannabinoid-enriched *Cannabis* extracts on TRP channels and endocannabinoid metabolic enzymes. Br. J. Pharmacol. 163 (7), 1479–1494.

De Petrocellis, L., Orlando, P., Moriello, A.S., et al., 2012. Cannabinoid actions at TRPV channels: effects on TRPV3 and TRPV4 and their potential relevance to gastrointestinal inflammation. Acta Physiol. 204 (2), 255–266.

Devinsky, O., Cilio, M.R., Cross, H., et al., 2014. Cannabidiol: pharmacology and potential therapeutic role in epilepsy and other neuropsychiatric disorders. Epilepsia 55 (6), 791–802.

El-Alfy, A.T., Ivey, K., Robinson, K., et al., 2010. Antidepressant-like effect of delta9-tetrahydrocannabinol and other cannabinoids isolated from *Cannabis sativa* L. Pharmacol. Biochem. Behav. 95 (4), 434–442.

Farrimond, J.A., Whalley, B.J., Williams, C.M., 2012. Cannabinol and cannabidiol exert opposing effects on rat feeding patterns. Psychopharmacology 223 (1), 117–129.

Fischedick, J.T., Hazekamp, A., Erkelens, T., et al., 2010. Metabolic fingerprinting of *Cannabis sativa* L., cannabinoids and terpenoids for chemotaxonomic and drug standardization purposes. Phytochemistry 71 (17–18), 2058–2073.

Flachowsky, H., Schumann, E., Weber, W.E., Peil, A., 2001. Application of AFLP for the detection of sex-specific markers in hemp. Plant Breed. 120 (4), 305–309.

Flores-Sanchez, I.J., Pec, J., Fei, J., et al., 2009. Elicitation studies in cell suspension cultures of *Cannabis sativa* L. J. Biotechnol. 143 (2), 157–168.

Flores-Sanchez, I.J., Choi, Y.H., Verpoorte, R., 2012. Metabolite analysis of *Cannabis sativa* L. by NMR spectroscopy. Methods Mol. Biol. 815, 363–375.

Freeman, D., Dunn, G., Murray, R.M., et al., 2014. How *Cannabis* causes paranoia: using the intravenous administration of Δ^9-tetrahydrocannabinol (THC) to identify key cognitive mechanisms leading to paranoia. Schizophr. Bull. 41 (2), 391–399 (Epub ahead of print)

Girgih, A.T., Udenigwe, C.C., Aluko, R.E., 2013. Reverse-phase HPLC separation of hemp seed (*Cannabis sativa* L.) protein hydrolysate produced peptide fractions with enhanced antioxidant capacity. Plant Foods Hum. Nutr. 68 (1), 39–46.

Grassin-Delyle, S., Naline, E., Buenestado, A., et al., 2014. Cannabinoids inhibit cholinergic contraction in human airways through prejunctional CB1 receptors. Br. J. Pharmacol. 171 (11), 2767–2777.

Happyana, N., Agnolet, S., Muntendam, R., et al., 2013. Analysis of cannabinoids in laser-microdissected trichomes of medicinal *Cannabis sativa* using LCMS and cryogenic NMR. Phytochemistry 87, 51–59.

Haustein, M., Ramer, R., Linnebacher, M., et al., 2014. Cannabinoids increase lung cancer cell lysis by lymphokine-activated killer cells via upregulation of ICAM-1. Biochem. Pharmacol. 92 (2), 312–325. http://dx.doi.org/10.1016/j.bcp.2014.07.014.

He, H.B., Shi, M.Q., 2010. Research progress of chemical components and pharmacological activities of Huo Ma Ren. Chin. J. Ethnomed. Ethnopharm. 4, 56–57.

Hillig, K.W., 2004. A chemotaxonomic analysis of terpenoid variation in *Cannabis*. Biochem. Syst. Ecol. 32 (10), 875–891.

Hillig, K.W., Mahlberg, P.G., 2004. A chemotaxonomic analysis of cannabinoid variation in *Cannabis* (Cannabaceae). Am. J. Bot. 91 (6), 966–975.

Hippalgaonkar, K., Gul, W., ElSohly, M.A., et al., 2011. Enhanced solubility, stability, and transcorneal permeability of δ-8-tetrahydrocannabinol in the presence of cyclodextrins. AAPS PharmSciTech 12 (2), 723–731.

Iannotti, F.A., Hill, C.L., Leo, A., et al., 2014. Nonpsychotropic plant cannabinoids, cannabidivarin (CBDV) and cannabidiol (CBD), activate and desensitize transient receptor potential vanilloid 1 (TRPV1) channels in vitro: potential for the treatment of neuronal hyperexcitability. ACS Chem. Neurosci. 5 (11), 1131–1141 (Epub ahead of print).

Iwata, N., Kitanaka, S., 2011. New cannabinoid-like chromane and chromene derivatives from *Rhododendron anthopogonoides*. Chem. Pharm. Bull. 59 (11), 1409–1412.

Izzo, A.A., Capasso, R., Aviello, G., et al., 2012. Inhibitory effect of cannabichromene, a major non-psychotropic cannabinoid extracted from *Cannabis sativa*, on inflammation-induced hypermotility in mice. Br. J. Pharmacol. 166 (4), 1444–1460.

Jeong, M., Cho, J., Shin, J.I., et al., 2014. Hempseed oil induces reactive oxygen species- and C/EBP homologous protein-mediated apoptosis in MH7A human rheumatoid arthritis fibroblast-like synovial cells. J. Ethnopharmacol. 154 (3), 745–752.

Jouanjus, E., Lapeyre-Mestre, M., Micallef, J., et al., 2014. *Cannabis* use: signal of increasing risk of serious cardiovascular disorders. J. Am. Heart Assoc. 3 (2), e000638.

Juknat, A., Pietr, M., Kozela, E., et al., 2012. Differential transcriptional profiles mediated by exposure to the cannabinoids cannabidiol and Δ^9-tetrahydrocannabinol in BV-2 microglial cells. Br. J. Pharmacol. 165 (8), 2512–2528.

Juknat, A., Pietr, M., Kozela, E., et al., 2013. Microarray and pathway analysis reveal distinct mechanisms underlying cannabinoid-mediated modulation of LPS-induced activation of BV-2 microglial cells. PLoS One 8 (4), e61462.

Kojoma, M., Seki, H., Yoshida, S., Muranaka, T., 2006. DNA polymorphisms in the tetrahydrocannabinolic acid (THCA) synthase gene in "drug-type" and "fiber-type" *Cannabis sativa* L. Forensic Sci. Int. 159 (2–3), 132–140.

Kucerova, J., Tabiova, K., Drago, F., Micale, V., 2014. Therapeutic potential of cannabinoids in schizophrenia. Recent Pat. CNS Drug Discov. 9 (1), 13–25.

Lata, H., Chandra, S., Techen, N., et al., 2011. Molecular analysis of genetic fidelity in *Cannabis sativa* L. plants grown from synthetic (encapsulated) seeds following in vitro storage. Biotechnol. Lett. 33 (12), 2503–2508.

Lee, M.J., Park, S.H., Han, J.H., et al., 2011. The effects of hempseed meal intake and linoleic acid on Drosophila models of neurodegenerative diseases and hypercholesterolemia. Mol. Cells 31 (4), 337–342.

Lesma, G., Consonni, R., Gambaro, V., et al., 2014. Cannabinoid-free *Cannabis sativa* L. grown in the Po valley: evaluation of fatty acid profile, antioxidant capacity and metabolic content. Nat. Prod. Res. 28 (21), 1801–1807 (Epub ahead of print).

Maione, S., Piscitelli, F., Gatta, L., et al., 2011. Non-psychoactive cannabinoids modulate the descending pathway of antinociception in anaesthetized rats through several mechanisms of action. Br. J. Pharmacol. 162 (3), 584–596.

Mandolino, G., Carboni, A., Bagatta, M., et al., 2002. Occurrence and frequency of putatively Y chromosome linked DNA markers in *Cannabis sativa* L. Euphytica 126, 211–218.

Mendoza, M.A., Mills, D.K., Lata, H., et al., 2009. Genetic individualization of *Cannabis sativa* by a short tandem repeat multiplex system. Anal. Bioanal. Chem. 393 (2), 719–726.

Molina, P.E., Amedee, A.M., LeCapitaine, N.J., et al., 2014. Modulation of gut-specific mechanisms by chronic δ(9)-tetrahydrocannabinol administration in male rhesus macaques infected with simian immunodeficiency virus: a systems biology analysis. AIDS Res. Hum. Retrovir. 30 (6), 567–578.

Montserrat-de la Paz, S., Marín-Aguilar, F., García-Giménez, M.D., Fernández-Arche, M.A., 2014. Hemp (*Cannabis sativa* L.) seed oil: analytical and phytochemical characterization of the unsaponifiable fraction. J. Agric. Food Chem. (5), 1105–1110.

Morgan, D.J., Muller, C.H., Murataeva, N.A., et al., 2012. Δ^9-Tetrahydrocannabinol (Δ^9-THC) attenuates mouse sperm motility and male fecundity. Br. J. Pharmacol. 165 (8), 2575–2583.

Nottage, J.F., Stone, J., Murray, R.M., et al., 2014. Delta-9-tetrahydrocannabinol, neural oscillations above 20 Hz and induced acute psychosis. Psychopharmacology 232 (3), 519–528 (Epub ahead of print).

Oláh, A., Tóth, B.I., Borbíró, I., et al., 2014. Cannabidiol exerts sebostatic and antiinflammatory effects on human sebocytes. J. Clin. Invest. 124 (9), 3713–3724. http://dx.doi.org/10.1172/JCI64628.

Park, S.K., Seo, J.B., Lee, M.Y., 2012. Proteomic profiling of hempseed proteins from Cheungsam. Biochim. Biophys. Acta 1824 (2), 374–382.

Paronis, C.A., Thakur, G.A., Bajaj, S., et al., 2013. Diuretic effects of cannabinoids. J. Pharmacol. Exp. Ther. 344 (1), 8–14.

Pec, J., Flores-Sanchez, I.J., Choi, Y.H., Verpoorte, R., 2010. Metabolic analysis of elicited cell suspension cultures of *Cannabis sativa* L. by (1)H-NMR spectroscopy. Biotechnol. Lett. 32 (7), 935–941.

Pollastro, F., Taglialatela-Scafati, O., Allarà, M., et al., 2011. Bioactive prenylogous cannabinoid from fiber hemp (*Cannabis sativa*). J. Nat. Prod. 74 (9), 2019–2022.

Porto, C.D., Decorti, D., Natolino, A., 2014. Potential oil yield, fatty acid composition, and oxidation stability of the hempseed oil from four *Cannabis sativa* L. cultivars. J. Diet. Suppl. 12 (1), 1–10 (Epub ahead of print)

Pucci, M., Rapino, C., Di Francesco, A., et al., 2013. Epigenetic control of skin differentiation genes by phytocannabinoids. Br. J. Pharmacol. 170 (3), 581–591.

Raborn, E.S., Jamerson, M., Marciano-Cabral, F., Cabral, G.A., 2014. Cannabinoid inhibits HIV-1 Tat-stimulated adhesion of human monocyte-like cells to extracellular matrix proteins. Life Sci. 104 (1–2), 15–23.

Radwan, M.M., Elsohly, M.A., Slade, D., et al., 2008a. Non-cannabinoid constituents from a high potency *Cannabis sativa* variety. Phytochemistry 69 (14), 2627–2633.

Radwan, M.M., Ross, S.A., Slade, D., et al., 2008b. Isolation and characterization of new *Cannabis* constituents from a high potency variety. Planta Med. 74 (3), 267–272.

Radwan, M.M., Elsohly, M.A., Slade, D., et al., 2009. Biologically active cannabinoids from high-potency *Cannabis sativa*. J. Nat. Prod. 72 (5), 906–911.

Raharjo, T.J., Widjaja, I., Roytrakul, S., Verpoorte, R., 2004. Comparative proteomics of *Cannabis sativa* plant tissues. J. Biomol. Tech. 15 (2), 97–106.

Ramer, R., Fischer, S., Haustein, M., et al., 2014. Cannabinoids inhibit angiogenic capacities of endothelial cells via release of tissue inhibitor of matrix metalloproteinases-1 from lung cancer cells. Biochem. Pharmacol. 91 (2), 202–216.

Rimmerman, N., Juknat, A., Kozela, E., et al., 2011. The non-psychoactive plant cannabinoid, cannabidiol affects cholesterol metabolism-related genes in microglial cells. Cell. Mol. Neurobiol. 31 (6), 921–930.

Rimmerman, N., Ben-Hail, D., Porat, Z., et al., 2013. Direct modulation of the outer mitochondrial membrane channel, voltage-dependent anion channel 1 (VDAC1) by cannabidiol: a novel mechanism for cannabinoid-induced cell death. Cell Death Dis. 4, e949.

Rock, E.M., Bolognini, D., Limebeer, C.L., et al., 2012. Cannabidiol, a non-psychotropic component of cannabis, attenuates vomiting and nausea-like behaviour via indirect agonism of 5-HT(1A) somatodendritic autoreceptors in the dorsal raphe nucleus. Br. J. Pharmacol. 165 (8), 2620–2634.

Rock, E.M., Limebeer, C.L., Navaratnam, R., et al., 2014. A comparison of cannabidiolic acid with other treatments for anticipatory nausea using a rat model of contextually elicited conditioned gaping. Psychopharmacology 231 (16), 3207–3215.

Romano, B., Borrelli, F., Fasolino, I., et al., 2013. The cannabinoid TRPA1 agonist cannabichromene inhibits nitric oxide production in macrophages and ameliorates murine colitis. Br. J. Pharmacol. 169 (1), 213–229.

Romano, B., Borrelli, F., Pagano, E., et al., 2014. Inhibition of colon carcinogenesis by a standardized *Cannabis sativa* extract with high content of cannabidiol. Phytomedicine 21 (5), 631–639.

Ross, S.A., ElSohly, M.A., Sultana, G.N., et al., 2005. Flavonoid glycosides and cannabinoids from the pollen of *Cannabis sativa* L. Phytochem. Anal. 16 (1 (Jan–Feb)), 45–48.

Rotherham, D., Harbison, S.A., 2011. Differentiation of drug and non-drug *Cannabis* using a single nucleotide polymorphism (SNP) assay. Forensic Sci. Int. 207 (1–3), 193–197.

Ruhl, T., Prinz, N., Oellers, N., et al., 2014. Acute administration of THC impairs spatial but not associative memory function in zebrafish. Psychopharmacology 231 (19), 3829–3842 (Epub ahead of print).

Russo, E.B., 2011. Taming THC: potential cannabis synergy and phytocannabinoid-terpenoid entourage effects. Br. J. Pharmacol. 163 (7), 1344–1364.

Sabo, A., Horvat, O., Stilinovic, N., et al., 2013. Industrial hemp decreases intestinal motility stronger than Indian hemp in mice. Eur. Rev. Med. Pharmacol. Sci. 17 (4), 486–490.

Schönhofen, P., de Medeiros, L.M., Bristot, I.J., et al., 2014. Cannabidiol exposure during neuronal differentiation sensitizes cells against redox-active neurotoxins. Mol. Neurobiol. (Epub ahead of print).

Serpell, M., Ratcliffe, S., Hovorka, J., et al., 2014. A double-blind, randomized, placebo-controlled, parallel group study of THC/CBD spray in peripheral neuropathic pain treatment. Eur. J. Pain 18 (7), 999–1012.

Shinjyo, N., Di Marzo, V., 2013. The effect of cannabichromene on adult neural stem/progenitor cells. Neurochem. Int. 63 (5), 432–437.

Steel, R.W., Miller, J.H., Sim, D.A., Day, D.J., 2014. Delta-9-tetrahydrocannabinol disrupts hippocampal neuroplasticity and neurogenesis in trained, but not untrained adolescent Sprague–Dawley rats. Brain Res. 1548, 12–19.

Stopponi, S., Soverchia, L., Ubaldi, M., et al., 2014. Chronic THC during adolescence increases the vulnerability to stress-induced relapse to heroin seeking in adult rats. Eur. Neuropsychopharmacol. 24 (7), 1037–1045.

Sytsma, K.J., Morawetz, J., Pires, J.C., et al., 2002. Urticalean rosids: circumscription, rosid ancestry, and phylogenetics based on rbcL, trnL-F, and ndhF sequences. Am. J. Bot. 89 (9), 1531–1546.

Tortoriello, G., Morris, C.V., Alpar, A., et al., 2014. Miswiring the brain: Δ^9-tetrahydrocannabinol disrupts cortical development by inducing an SCG10/stathmin-2 degradation pathway. EMBO J. 33 (7), 668–685.

Truţa, E., Gille, E., Tóth, E., Maniu, M., 2002. Biochemical differences in *Cannabis sativa* L. depending on sexual phenotype. J. Appl. Genet. 43 (4), 451–462.

van Bakel, H., Stout, J.M., Cote, A.G., et al., 2011. The draft genome and transcriptome of *Cannabis sativa*. Genome Biol. 12 (10), R102.

Vallée, M., Vitiello, S., Bellocchio, L., et al., 2014. Pregnenolone can protect the brain from cannabis intoxication. Science 343 (6166), 94–98.

Watanabe, K., Yamaori, S., Funahashi, T., et al., 2007. Cytochrome P450 enzymes involved in the metabolism of tetrahydrocannabinols and cannabinol by human hepatic microsomes. Life Sci. 80 (15), 1415–1419.

Wei, Y., Wang, H., Guo, S., et al., 2010. Textual research of *Cannabis sativa* varieties and medicinal part. Zhongguo Zhong Yao Za Zhi 35 (13), 1773–1776.

Whyte, D.A., Al-Hammadi, S., Balhaj, G., et al., 2010. Cannabinoids inhibit cellular respiration of human oral cancer cells. Pharmacology 85 (6), 328–335.

Williams, J.C., Appelberg, S., Goldberger, B.A., et al., 2014. Δ(9)-Tetrahydrocannabinol treatment during human monocyte differentiation reduces macrophage susceptibility to HIV-1 infection. J. Neuroimmune Pharmacol. 9 (3), 369–379.

Yang, X., Hegde, V.L., Rao, R., et al., 2014. Histone modifications are associated with Δ^9-tetrahydrocannabinol-mediated alterations in antigen-specific T cell responses. J. Biol. Chem. 289 (27), 18707–18718.

Phytochemical and biological research of Polygoneae medicinal resources

12

12.1 Introduction

Polygonaceae plants have a worldwide distribution, most of which are in the temperate region of the Northern Hemisphere. There are 235 Polygonaceae species and 37 varieties in China. The use of medicinal species is listed in Table 12.1. In the tribe Polygoneae of the subfamily Polygonoideae, there are seven genera (http://frps.eflora.cn/frps/Trib.%20Polygoneae), namely, *Antenoron*, *Fagopyrum*, *Fallopia*, *Koenigia*, *Polygonum*, *Pteroxygonum*, and *Reynoutria*. *Polygonum* is the largest genus of the family Polygonaceae. This chapter summarizes the current knowledge of phytochemistry, bioactivity, phylogeny, and omics of Polygoneae medicinal plants.

12.2 Chemical components

12.2.1 Phenolics, flavonoids, and phenylpropanoids

Examples of Polygoneae compounds are shown in Figure 12.1 and Table 12.2. Gallic acid, rutin, coumaric acid, and quercetin were found in the leaves of *P. minus* (Vikram et al., 2014). Polyphenolic compounds, such as hydroxyanthraquinones, stilbenoids, and proanthocyanidins, are abundant in Polygoneae plants.

12.2.2 Essential oil, terpenoids, steroids, and benzoquinones

Nonane, heptane, octadecanal, β-caryophyllene, *trans*-α-bergamotene, β-farnesene, α-caryophyllene, *p*-benzoquinone, phenol, α-panasinsen, pentanoic acid, octane, heptane, undecane, 1,2 benzenedicarboxylic acid, and nonane were found in the roots of *P. minus* (Vikram et al., 2014). Many other components of essential oil are also present in *P. minus* (Vikram et al., 2014).

Twenty-one compounds were separated and determined from *P. cuspidatum* flower, which amounted to 99.29% of the total volatiles (Sun et al., 2012). (E)-2-hyexenoic acid methyl ester, 1-phenyl-L-pentanone, (E)-4-hexenoic acid methyl ester, 3-hexenoic acid methyl ester, 2-methyl-6-methylene-1, and 7-octadien-3-one were the main constituents of *P. cuspidatum* flower, which amounted to 63.23% of the total volatiles. Organic ester compounds amounted to 52.09% of the total volatiles. Hexenoic acid methyl ester compounds were highly abundant, corresponding to

Medicinal plants: chemistry, biology and omics. http://dx.doi.org/10.1016/B978-0-08-100085-4.00012-8

Table 12.1 Polygonaceae plants used in traditional Chinese medicine and folk medicine

Species	Chinese name	Habitat	Distribution	Therapeutic use	Note
Antenoron filiforme (*Polygonum filiforme*)	Jin Xian Cao, Ren Zi Cao (Zhe Jiang), Hua Xue Gui (Hu Bei), Xue Jing Cao (Hu Nan)	Altitude 500–1200 m, hillside, forest, bushes, ditch, creek	Central south China, Shan Xi, Shaan Xi, Shan Dong, Jiang Su, Zhe Jiang, Jiang Xi, Fu Jian, Tai Wan, Si Chuan, Gui Zhou, Yun Nan	Root tuber/whole plant: pungent, cool; blood-cooling, hemostasis, stasis-removing and pain-relieving; rheumatism, stomachache, hemoptysis, hematemesis, postpartum abdominal pain due to blood stasis, traumatic injury	*A.neofoliforme* (*Polygonum neofiliforme*) has the same therapeutic use; east China, Shan Xi, Shaan Xi, Gan Su, He Nan, Hu Bei, Hu Nan, Guang Dong, Si Chuan, Gui Zhou, Yun Nan
Calligonum leucocladum	Bai Pi Sha Guai Zao	Desert	North Xin Jiang	Whole plant: astringing to stop diarrhea	
Calligonum mongolicum	Sha Guai Zao	Desert	Inner Mongolia, Gan Su, Ning Xia, Xin Jiang	Root/twig with fruit: bitter, astringent, lukewarm; cloudy urine, chapped skin	
Fagopyrum dibotrys (*P. cymosum*)	Jin Qiao Mai, Qiao Mai Dang Gui (Hu Bei), Qiao Mai San Qi (Zhe Jiang)	Altitude 600–1000 m, ditch, roadside	Central south China, southwest China, Shaan Xi, Gan Su, Jiang Su, Zhe Jiang, Jiang Xi, Fu Jian	Root/rhizome: slightly pungent, astringent, cool; heat-clearing and detoxification, blood-activating and stasis-dissipating, wind-expelling and dampness-removing; sore throat, ulcerative carbuncle, scrofula, hepatitis, lung abscess, wind syndrome of head, stomachache, dysentery, leukorrheal disease	

Fagopyrum esculentum (*P. fagopyrum*)	Qiao Mai, Hua Mai (Hu Nan), Qiao Zi (Gui Zhou)		Cultivated all over China; semiwild, roadside	Stem/leaf: acid, cold; antihypertension, hemostasis; choking, carbuncle; seed: sweet, cool; lower Qi and comfort intestine, stagnation-relieving, detumescence; dyspepsia, diarrhea, carbuncle back, burn	
Fagopyrum gracilipes (*P. gracilipes*)	Xi Geng Qiao Mai	Altitude 300–2000 m, hillside, roadside, forest, riverside, crop field	Shaan Xi, Gan Su, Hu Bei, Si Chuan, Yun Nan	Whole plant: heat-clearing and detoxification, blood-activating and stasis-dissipating, spleen-invigorating and diuresis-promoting; seed: appetizer, comfort intestine	
Fagopyrum tataricum (*P. tataricum*)	Ku Qiao Mai, Qiao Ye Qi (Shaan Xi), Ku Qiao Tou (Gui Zhou)	Altitude 700–2000 m, hill, valley, riverside, in cultivation/semiwild	Northeast China, northwest China, southwest China	Root tuber/rhizome: sweet, bitter, flat; stomach-invigorating and comfort Qi, dehumidification and pain-relieving; stomachache, dyspepsia, dysentery, internal lesion caused by overexertion, pain in waist and lower extremities	

Continued

Table 12.1 Continued

Species	Chinese name	Habitat	Distribution	Therapeutic use	Note
Muehlenbeckia platycladum (*Coccoloba platyclada*; *Homalocladium platycladum*)	Zhu Jie Liao, Fei Tian Wu Gong (Guang Xi), Bai Zu Cao (Guang Zhou)	Native to south Pacific Solomon islands	Cultivated in garden and greenhouse throughout China	Stem/leaf: sweet, light, flat; stasis-removing, growth-promoting and itch-relieving, swell-relieving/pain killer; carbuncle, traumatic injury, snakebite, centipede bite	
Oxyria digyna (*Rumex digynus*)	Shen Ye Shan Liao, Suan Jiang Cao (Shaan Xi)	Altitude 2000–3600 m, mountain grassland, hillside, valley	Ji Lin, Shaan Xi, Gan Su, Qing Hai, Si Chuan, Yun Nan, Tibet	Whole plant: acid, cool; heat-clearing and diuresis-promoting; depression of hepatic Qi, hepatitis, scurvy	
P. alatum (*P. nepalense*; *Persicaria alata*)	Tou Zhuang Liao, Nepal Liao, Mao Er Yan Jing (Si Chuan)	Altitude 800–2700 m, hillside, valley, wetland, roadside, crop field	All over China	Whole plant: bitter, cold; heat-clearing and detoxification, astringe and secure the intestine; sore throat, bloodshot eyes, painful swollen gums, bloody dysentery, joint pain, stomachache	
P. alopecuroides	Hu Wei Liao	Riverside, water meadow	Ji Lin, Inner Mongolia	Rhizome: heat-clearing and detoxification, blood-cooling and hemostasis; used as *Paris polyphylla* (Cao He Che)in Inner Mongolia	

P. amphibium (*Persicaria amphibia*)	Liang Qi Liao, Xiao Huang Yao (Gui Zhou)	Low-altitude lake, shallow water, wetland	Northeast China, north China, Shaan Xi, Ning Xia, Shan Dong, north Jiang Su, An Hui, He Nan, Hu Bei, Si Chuan, Gui Zhou, Yun Nan	Whole plant: bitter, flat; heat-clearing and diuresis-promoting; dysentery; external use on furuncle	Used for hypertrophic spondylitis in Chang Wei, Shan Dong; *P.amphibium* var. *terrestre* has the same therapeutic efficacy and distribution area
P. amplexicaule (*Bistorta amplexicaule*)	Bao Jing Liao	Altitude >1000 m, mountain woodside, bushes, valley, ditch	Hu Bei, Si Chuan, Yun Nan	Rhizome: spasm-relieving, stasis-removing and hemostasis, myogenic/pain killer, antibacterial, anti-inflammatory	
P. amplexicaule var. *sinense*	Zhong Hua Bao Jing Liao, Xue San Qi (Hu Bei), Hong Hai Er (Si Chuan)	Altitude 1300–1500 m, wet valley, bushes	Shaan Xi, Hu Bei, Si Chuan, Yun Nan	Rhizome: slightly bitter, astringent, flat; slightly poisonous; heat-clearing and detoxification, astringing to stop diarrhea; blood-promoting and pain-relieving; dysentery, diarrhea, traumatic injury, bleeding	Not with the cockscomb and *Uncaria*
P. attenuatum	Mao Ye Er Liao	Mountain bushes	Ji Lin	Rhizome: heat-clearing and detoxification, blood-cooling, detumescence, and convulsion-relieving	

Continued

Table 12.1 Continued

Species	Chinese name	Habitat	Distribution	Therapeutic use	Note
P. aubertii (*Fallopia aubertii*)	Mu Teng Liao	Altitude 500–2000 m, hillside, ravine sunny shrubs, or in cultivation	southwest China, Inner Mongolia, Shaan Xi, Gan Su, Ning Xia	Root tuber: bitter, astringent, cool; heat-clearing and detoxification, regulate menstrual cycle and hemostasis; dysentery, dyspepsia, stomachache, irregular menstruation	
P. aviculare	Bian Xu, Niu Jin Cao (Shaan Xi), Tai Yang Cao (Zhe Jiang), Ban Jiu Tai (An Hui), Di Liao	Open field, wetland	All over China	Aboveground part: bitter, cool; disinhibit urine and free strangury, insecticidal, anti-itching; urinary infection, oliguria with reddish urine, urine leaching/acerbity/painful, eczema, pruritus vulvae disease, leukorrheal disease	China Pharmacopoeia 1990 version; *P. aviculare* var. *rigidum* has similar efficacy, distributed in Helan Mountains, Ning Xia; *P. aviculare* var. *vegetum*, distributed in Jiang Su, An Hui, Zhe Jiang, He Nan, Si Chuan, Gui Zhou
P. barbatum	Mao Liao, Si Ji Qing	Altitude <1000 m, roadside, ditch, woodland	Jiang Su, An Hui, Zhe Jiang, Fu Jian, Tai Wan, Hu Bei, Guang Dong, Guang Xi, Si Chuan, Gui Zhou, Yun Nan	Root: pungent, warm; astringent; whole plant: toxin-removing and myogenic, stranguria; seed: emetic, antidiarrheal	

P. barbatum var. *gracile* (*P. barbatum* subsp. *gracile*)	Xi Mao Ci Liao, Xiao Mao Liao, Liao Zi Cao (Si Chuan)	Altitude 500–1000 m, hill, ditch, roadside	Jiang Su, Shang Hai, Guang Xi, Si Chuan, Gui Zhou	Whole plant/root: slightly pungent, warm; cold dispelling and blood-activating; measles, deficiency-cold after disease, abdominal pain, take a chill after traumatic injury, Yin coldness, stale coldness	
P. bistorta (*Bistorta vulgaris*)	Quan Shen, Zi Shen (He Bei, Hu Nan, Gui Zhou), Dao Gen Cao (Xin Jiang), Xia Shen (Shan Dong), Ma Feng Qi (Guang Xi)	Hillside meadow, forest meadow	North China, northwest China, Ji Lin, Shan Dong, Jiang Su, An Hui, Zhe Jiang, He Nan, Hu Bei, Si Chuan, Gui Zhou; main producing area: He Bei, Shan Xi, Shan Dong, He Nan	Rhizome: bitter, astringent, cool; heat-clearing and detoxification, apocatastasis, hemostasis; bloody dysentery, heat diarrhea, cough with lung heat, scrofula, aphtha, hematemesis, epistaxis, hemorrhoidal bleeding, snakebite	China Pharmacopoeia 1990
P. bungeanum	Liu Ye Ci Liao	The edge of the field, roadside wetland	North China, Hei Long Jiang, Liao Ning, Shan Dong	Root: heat-clearing/ detoxification, diuresis	
P. caespitosum (*Persicaria caespitosa*)	Cong Zhi Liao, Tai Yang Cao (Si Chuan), Gu Zhi Lian (Gui Zhou)	Altitude 700–2000 m, hillside forest, roadside, ditch	Central south China, Shaan Xi, Gan Su, Jiang Su, Shang Hai, Jiang Xi, Fu Jian, Si Chuan, Gui Zhou, Yun Nan	Whole plant: heat-clearing/detoxification, blood-cooling and hemostasis; abdominal pain, diarrhea, dysentery	

Continued

Table 12.1 Continued

Species	Chinese name	Habitat	Distribution	Therapeutic use	Note
P. campanulatum	Zhong Hua Liao	Altitude 1200 m, hillside shrubs	Si Chuan, Gui Zhou, Yun Nan	Whole plant: blood-activating/stasis-removing, swell-reducing; innominate toxic swelling, dorsal furuncle, scrofula, leukorrheal disease	
P. capitatum (*Persicaria capitata*)	Tou Hua Liao, Yan Qiao Mai (Hu Bei), Shui Xiu Qiu (Gui Zhou), Shi Mang Cao (Guang Xi)	Altitude 250–500 m, hillside, roadside, rock seam	Jiang Xi, Hu Bei, Hu Nan, Guang Dong, Guang Xi, Si Chuan, Gui Zhou, Yun Nan	Whole plant: acid, cold; detoxification/stasis-removing, disinhibit urine and free strangury; dysentery, stone strangury, edema, rheumatism, traumatic injury, skin and external disease, eczema	
P. chinense (*Persicaria chinensis*)	Huo Tan Mu, Chi Di Li (Zhe Jiang), Bai Fan Cao (Guang Dong), Lao Shu Zhe (Guang Xi), Yun Yao (Si Chuan)	Altitude 1200–2200 m, hillside, roadside, creek, beside the house	Southwest China, An Hui, Zhe Jiang, Jiang Xi, Fu Jian, Tai Wan, Hu Bei, Hu Nan, Guang Dong, Guang Xi	Whole plant: slightly acid, cool; heat-clearing/detoxification, diuresis-promoting and anti-dyspepsia; diarrhea, dysentery, jaundice, wind-heat sore throat, heat dizziness, leukorrheal disease, scabies, carbuncle; root: acid, sweet, flat; Qi-promoting; Qi deficiency and dizziness, tinnitus and deafness, traumatic injury	*P. chinense f. ovalifolium* has similar efficacy, distributed in Gui Zhou and Yun Nan

P. chinense var. *hispidum*	Cu Mao Huo Tan Mu, Xiao Hong Ren (Si Mao, Yun Nan)	Valley, wetland, roadside	Gui Zhou, Yun Nan	Root tuber: acid, flat; collateral-dredging and blood-activating, hemostasis, detoxification; dysentery, diarrhea, irregular menses, metrorrhagia, postpartum bleeding	
P. cilinerve (*Pleuropterus ciliinervis*; *P. multiflorum* var. *cillinerve*)	Mao Mai Liao, Zhu Sha Qi (Shaan Xi), Zhu Sha Lian (Hu Bei)	Altitude 800–1800 m, hillside forest, ditch, roadside, chaotic crevice	Northeast China, Shaan Xi, Gan Su, Ning Xia, He Nan, Hu Bei, Si Chuan, Gui Zhou	Root tuber: sweet, slightly astringent, cool; slightly poisonous; heat-clearing/detoxification, pain-relieving, hemostasis, regulate menstrual cycle; tonsillitis, vomiting and diarrhea, ulcer, dysentery	
P. convolvulus (*P. dumctorum*)	Juan Jing Liao, Qiao Mai Man (Gui Zhou)	Altitude 600–1600 m, hillside grassland, crop field	Northeast China, north China, Shaan Xi, Gan Su, Xin Jiang, Shan Dong, north Jiang Su, He Nan, Gui Zhou	Whole plant: heat-clearing/detoxification, swell-reducing; root: strengthening the stomach, cough-relieving, analgesia, detoxification; tuberculosis hemoptysis, pertussis, peratodynia	
P. coriaceum	Ge Ye Liao, Ban She Lian (Guang Xi)	Roadside wetland, ditch	Guang Xi, Yun Nan, Tibet	Rhizome: bitter, cold; slightly poisonous; heat-clearing/detoxification; dysentery, oral ulceration, gingivitis, carbuncle, hemorrhoids, burn	

Continued

Table 12.1 Continued

Species	Chinese name	Habitat	Distribution	Therapeutic use	Note
P. cuspidatum (*Pleuropterus cuspidatus*)	Hu Zhang, Huo Xue Dan (Jiang Su), Suan Tong Gan (Hu Nan), Da Ye She Zong Guan (Guang Xi), Wo Gong Liao (Miao ethnic group)	Altitude <2500 m, hillside grassland, forest, ditch, roadside	East China, central south China, Liao Ning, Shaan Xi, Gan Su, Si Chuan, Gui Zhou, Yun Nan;in cultivation	Root/rhizome: slightly bitter, cool; wind-dispelling and diuresis-promoting, stasis-removing, analgesia, relieving cough and reducing sputum; joint pain, jaundice with damp-heat, amenorrhea, cough with abundance of phlegm, traumatic injury; leaf: slightly acidic, cool; wind-dispelling, blood-cooling, detoxification	China Pharmacopoeia 1990
P. cynanchoides	Niu Pi Xiao Liao, Mao Xue Teng (Gui Zhou)	Altitude 750–1200 m, riverbank, ravine	Shaan Xi, Gan Su, Hu Bei, Si Chuan, Gui Zhou, Yun Nan	Root: pungent, astringent, cool; astringing the lung to stop cough, spasm pain-relieving; tuberculosis hemoptysis, pertussis, peratodynia, overstrained cough, rheumatoid arthritis	
P. denticulatum	Xue Shao	Hillside sunny shrubs, winding from other plants	Gui Zhou	Root tuber: heat-clearing, detoxification, stasis-dissipating, hemostasis with astringents, relieving asthma; dysentery, tormina postpartum, traumatic swelling	

P. denta-alatum	Chi Chi Liao	Altitude 500–2000 m, hillside shrubs, riverbank, wasteland	Hei Long Jiang, Liao Ning, He Bei, Shaan Xi, Gan Su, He Nan	Whole plant: red eyes	
P. dissitiflorum (*Persicaria dissitiflorum*)	Xi Hua Liao, Hong Xiang Long Cao (Hu Bei)	Altitude 750–1500 m, forest wetland, riverbank	Northeast China, north China, east China, central south China, Shaan Xi, Gan Su	Whole plant: detoxification, diuresis; snakebite, stranguria	
P. divaricatum (*Persicaria divaricata*)	Cha Fen Liao, Suan Bu Liu (northeast China), Suan Jiang (northeast China, desert)	Hillside grassland	North China, Ji Lin, Liao Ning, Qing Hai, Shan Dong, Hu Bei, Si Chuan, Tibet	Root: acid, sweet, warm; dispelling cold, warming the kidney; cold hernia, scrotal sweating	
P. emodi (*Bistorta emodi*)	Hong Teng Liao	Hillside grassland	Si Chuan, Yun Nan	Whole plant: light, lukewarm; muscle-relaxing and blood-activating; traumatic injury, rheumatalgia	
P. excurrens (*Persicaria excurrens*)	Zhong Zhou Liao	Altitude <1000 m, hillside bushes, valley	Northeast China, north China, An Hui, Jiang Xi, Hu Bei	Whole plant: relieve pain, insecticide	

Continued

Table 12.1 Continued

Species	Chinese name	Habitat	Distribution	Therapeutic use	Note
P. flaccidum (*P.hydropiper* var. *flaccidum*; *Persicaria flaccida*)	La Liao, La Ma Liao (Jiang Xi), Liao Zi Cao (Si Chuan)	Waterfront, wetland	Northeast China, Jiang Xi, Fu Jian, Guang Dong, Si Chuan, Gui Zhou	Whole plant: pungent, warm; slightly poisonous; wind-dispelling and diuresis-promoting, stagnation-removing, stasis-dissipating, relieve pain, insecticide; dysentery, diarrhea, uterine bleeding, tonsillitis, rheumatalgia, traumatic swelling	
P. giraldii	Yi Liao	Altitude 600–1800 m, hillside, roadside, ditch, bushes	He Bei, Shan Xi, Shaan Xi, Gan Su, He Nan, Hu Bei, Si Chuan, Gui Zhou	Root tuber: acid, bitter, cool; heat-clearing, detoxification, hemostasis, pain-relieving; diarrhea, dysentery, pain in waist and lower extremities, hemafecia, uterine bleeding; external use for burn, sore and furuncle, dog bite	
P. griffithi	Chang Geng Liao	Altitude 3300–4400 m, hillside grassland, stone crevice	Cona and Zayü of Tibet	Rhizome: heat-clearing, detoxification	

P. heterophyllum	Yi Ye Liao	Roadside	Northeast China	Whole plant: heat-clearing, diuresis, insecticide	
P. hydropiper (*Persicaria hydropiper*)	Shui Liao, La Ren Cao (Zhe Jiang), Tong Gu Xiao (Guang Xi), Yao Liao (Si Chuan), Hu La Liao (northeast China)	Altitude <1500 m, hillside, roadside bushes, ditch, wetland	All over China	Whole plant: pungent, warm; slightly poisonous, wind-dispelling and diuresis-promoting, stagnation-reducing, stasis-dissipating, pain-relieving, insecticide; dysentery, diarrhea, indigestion, malnutrition, eczema, stubborn dermatitis, rheumatalgia, injuries from falls	*P. hydropiper* var. *angustifolia* and *P. hydropiper* var.*diffusa* of Hei Long Jiang have similar efficacy
P. hydropiper var. *hispidum*	Mao Shui Liao	Wetland	Fu Jian, Gui Zhou	Whole plant: bitter, slightly pungent, cool; dysentery, diarrhea, heatstroke, abdominal pain	
P. japonicum (*Persicaria japonia*)	Can Jian Cao, Liao Cao (Zhe Jiang), Shui Liao (Hu Bei), Hong Liao Zi (Gui Zhou)	Hill, plain roadside, bushes, ditch, wetland	Jiang Su, An Hui, Zhe Jiang, Jiang Xi, Fu Jian, Tai Wan, He Nan, Hu Bei, Guang Dong, Guang Xi, Si Chuan, Gui Zhou	Whole plant: pungent, warm; stasis-dissipating and blood-activating; waist and knee pain, measles, dysentery	
P. jucundum (*Persicaria jucunda*)	Yu Yue Liao	Altitude 200–700 m, hillside bushes, valley, roadside, ditch, wetland	Yangtze River basin, southeast China	Whole plant: diarrhea	*P. minus* (*Persicaria minor*) has the same efficacy, distributed in northeast, north, east, central south, and southwest China

Continued

Table 12.1 Continued

Species	Chinese name	Habitat	Distribution	Therapeutic use	Note
P. lapathifolium (*Persicaria lapathifolia*)	Suan Mo Ye Liao, Liao Diao Zi (northeast China), Han Tian Liao (Si Chuan), Da Ma Liao	Altitude <1900 m, hillside, roadside, creek	Northeast China, north China, Shaan Xi, Gan Su, Ning Xia, Shan Dong, Jiang Su, An Hui, Zhe Jiang, He Nan, Hu Bei, Guang Dong, Guang Xi, Si Chuan, Gui Zhou	Whole plant: pungent、bitter, cool; heat-clearing, detoxification, diuresis-promoting, relieve itching; dysentery, diarrhea; external use for eczema, scrofula	
P. lapathifolium var. *salicifolium*	Mian Mao Suan Mo Ye Liao, Hong La Liao (Hu Bei), La Liao Cao (Jiang Su), Liu Ye Liao (Gui Zhou)	Low mountain ditch, waterfront, wet lowland	All over the country	Whole plant: wind-dispelling and diuresis-promoting, heat-clearing, detoxification, hemostasis, help digestion	
P. lapathifolium var. *xanthophyllun*	Huang Ban Suan Mo Ye Liao, Liao Zi Cao (Si Chuan)	Altitude 500–1500 m, hillside grassland	Shaan Xi, Gan Su, Ning Xia, Hu Bei, Si Chuan	Whole plant: heat-clearing, diuresis	
P. lapidosum	Shi Sheng Liao	Hillside, hayfield	Northeast China, He Bei	Rhizome: bitter, slightly cold; slightly poisonous; heat-clearing, detoxification, blood-cooling, hemostasis	Used as *P. bistorta* in notrheast China
P. longisetum (*Persicaria Longiseta*)	Chang Zong Liao, La Liao (Hu Bei), Shui Hong Hua (Si Chuan)	By the fountain valley, damp place	All over the country	Whole plant: blood-activating and stasis-dissipating, apocatastasis, pain-relieving	Used as *P. flaccidum* in some regions

P. macranthum (*P. japonicum* var. *conspicuum*; *Persicaria conspicua*)	Da Hua Liao, Da Chang Hua Liao (Gui Zhou)	Altitude <700 m, hillside, roadside, bushes, wetland	Shan Xi, Shaan Xi, Jiang Su, An Hui, Zhe Jiang, Tai Wan, Hu Bei, Si Chuan, Gui Zhou, Yun Nan	Whole plant: rheumatalgia, injuries from falls, dysentery, diarrhea	
P. macrophyllum (*Bistorta macrophylla*; *P. sphaerostachyum*)	Yuan Sui Liao, Xie Zi Qi (Shaan Xi), Ran Bo (Tibetan)	Hilly area	Southwest China, Shaan Xi, Gan Su, Qing Hai, Hu Bei	Rhizome: bitter, astringent, cool; astringent, hemostasis, blood-activating, stop diarrhea; dysentery, hematemesis, bleeding from five sense organs, uterine bleeding, leukorrheal disease, injuries from falls	
P. mandshuriense	Er Ye Liao, Bei Chong Lou (Liao Ning)	Hillside, wet grassland	Northeast China, He Bei	Rhizome: bitter, cool; slightly poisonous; heat-clearing, detoxification, blood-cooling, hemostasis	Local standard of Liao Ning
P. microcephalum	Xiao Tou Liao	Altitude 1500–1800 m, forest, roadside bushes	Southwest China, He Nan, Hu Bei, Hu Nan	Whole plant: anti-inflammation, hemostasis	
P. molle (*Ampelygonum molle*; *Aconogonon molle*)	Juan Mao Liao	Waterfront, roadside wetland	Guang Xi	Whole plant: sores	

Continued

Table 12.1 Continued

Species	Chinese name	Habitat	Distribution	Therapeutic use	Note
P. muricatum (*P. strigosum* var. *muricatum*)	Fu Jing Liao	Low-altitude hillside, roadside bushes, wetland	Middle and lower reaches of the Yangtze River to the south	Whole plant: itch of skin, dysentery	
P. multiflorum (*Pleuropterus multiflorus*; *Reynoutria multiflora*; *Tiniaria multiflora*; *Fallopia multiflora*)	He Shou Wu, Jin Shou Wu (Zhe Jiang), Ye Shao (Hu Bei), Wo Pu Weng (Miao ethnic group), Ma Gan Shi	Altitude 500–2200 m, hillside, roadside, valley, waterfront, bushes	He Bei, Shan Xi, Shaan Xi, Gan Su, Xin Jiang, Shan Dong, Jiang Su, An Hui, Zhe Jiang, Jiang Xi, He Nan, Hu Bei, Guang Dong, Guang Xi, Si Chuan, Gui Zhou, Yun Nan; main producing area: He Nan, Hu Bei, Si Chuan	Root tuber: bitter, sweet, astringent, warm; detoxification, boils, relaxing bowel; scrofula, sore carbuncle, rubella itching, dry intestinal constipation, hyperlipemia; rattan: sweet, flat; nourishing the blood and tranquilization, wind-dispelling and collateral-dredging; insomnia and dreamful sleep, blood deficiency and body pain, rheumatalgia; external use for itch of skin	China Pharmacopoeia 1990
P. nodosum (*Persicaria nodosa*)	Jie Liao, Qu La Liao (Hu Bei)	Altitude 600–1400 m, ravine waterfront, roadside wetland	Northeast China, north China, northwest China, southwest China	Whole plant with root: insecticide, detoxification, stasis-dissipating and resuscitation-inducing; sores, toxic heat	Aboveground part is used as *P. orientale*

P. ochotense	Dao Gen Liao	Alpine tundra	Changbai Mountain, Ji Lin	Rhizome: blood-cooling, detoxification	Used as *Paris polyphylla* in Ji Lin
P. orientale (*Persicaria orientalis*)	Hong Liao, Jia Liao (Xin Jiang), Jiu Yao Cao (Zhe Jiang), Ba Zi Liao (Fu Jian), Dong Fang Liao	Low mountain, hills, plain, ditch, roadside, grassland, wetland, often grow in plots, or in cultivation	All over the country; main producing area: Jiang Su, Liao Ning	Fruit: saline, cool; blood-scattering and bulk breakdown, stagnation-reducing and pain-relieving; lump in the abdomen, cancer pain, indigestion, epigastric pain; whole plant: pungent, cool; slightly poisonous; wind-dispelling and diuresis-promoting, blood-activating and pain-relieving; rheumatalgia, malaria, hernia, smelly feet	China Pharmacopoeia 1990
P. pacificum (*Bistorta pacifica*)	Pacific Liao	High mountain bushes, forest, wet meadow	Hei Long Jiang	Rhizome: heat-clearing, detoxification, blood-cooling, hemostasis, astringent; bloody dysentery, hematemesis, burn, bleeding wound	
P. paleaceum (*Bistorta chinensis*)	Cao Xue Jie, Xia Zi Qi (Hu Bei), Zi Hua Gen (Yun Nan), Yi Kou Xue (Si Chuan)	Altitude 2000 m, alpine meadow	Qing Hai, Hu Bei, Si Chuan, Gui Zhou, Yun Nan	Rhizome: bitter, astringent, lukewarm; blood-activating and stasis-dissipating, pain-relieving, hemostasis; stomachache, indigestion, irregular menses, edema, injuries from falls	

Continued

Table 12.1 Continued

Species	Chinese name	Habitat	Distribution	Therapeutic use	Note
P. perfoliatum (*Persicaria perfoliata*; *Ampelygonum perfoliatum*; *Truellum perfoliatum*)	Gang Ban Gui, Shui Ma Ling (Shang Hai), Lan Lu Hu (Fu Jian), Da Meng Tui (Guang Xi), She Dao Tui (Si Chuan)	Altitude <1700 m, valley shrubs, ditch	All over the country	Whole plant: acid, cool; heat-clearing, detoxification, diuresis, apocatastasis; edema, jaundice, diarrhea, malaria, pertussis, eczema, scabies	
P. peregrinatoris (*Aconogonon peregrinatoris*)	Ni A Luo (Tibetan)	Altitude 3800–4900 m, hillside meadow, flood land	Tibet	Root: pungent, warm; antidysentery with astringent; dysentery, diarrhea	Tibetan medicine
P. persicaria (*Persicaria vulgaris*)	Tao Ye Liao, Shan La Liao (Hu Bei)	Altitude <1000 m, ditch, roadside wetland	Ji Lin, He Bei, Shaan Xi, Gan Su, Xin Jiang, Shan Dong, Jiang Xi, Fu Jian, He Nan, Hu Bei, Si Chuan, Gui Zhou, Tibet	Whole plant: pungent, warm; sweating dehumidification, help digestion; dysentery, diarrhea, snakebite	
P. pinetorum (*Persicaria pinetorum*)	Song Lin Liao	Altitude 2000–2500 m, coniferous and broadleaved mixed forest, wet valley, roadside	Shaan Xi, Gan Su, Hu Bei, Si Chuan	Whole plant: heat-clearing, detoxification, wind-dispelling and diuresis-promoting, reduce phlegm	

P. plebeium	Ye Hua Liao, Mi Zi Liao (Guang Xi), Xi Jian Liao (Gui Zhou)	Low mountain, hills, plain, hillside, roadside	Central south China, southwest China, He Bei, Shaan Xi, Jiang Su, An Hui, Jiang Xi, Fu Jian, Tai Wan	Whole plant: diuresis-promoting, stranguria, resolving turbidity, insecticide; malignant sore scabies, stranguria with turbid discharge, ascariasis	
P. polystachyum (*Aconogonon polystachyum*; *Persicaria polystachya*)	Duo Sui Liao, Zi Zi Sa Zeng (Tibetan)	Barren hills	Si Chuan, Tibet	Whole plant: pungent, cool; wind-dispelling and damp-removing, heat-clearing, detoxification, insecticide; dysentery, diarrhea, infantile indigestion, injuries from falls, rheumatalgia	Tibetan medicine; used as *P. flaccidum* in Tibet
P. rude (*Persicaria rudia*; *Aconogonon molle* var. *nide*)	Jiu Gu Niu	Ditch	Gui Zhou	Whole plant: pungent, slightly sweet, warm; collateral-dredging, analgesia; internal lesion caused by overexertion, irregular menses, arthralgia, and myalgia	
P. runcinatum (*Persicria runcinata*)	Chi Jing San, Fei E Qi (Hu Bei), She Tou Liao (Guang Xi), Xue Dang Gui (Yun Nan), Ku Cha Tou Cao	Altitude 1000–2500 m, hillside bushes, ditch wetland	Tai Wan, He Nan, Hu Bei, Hu Nan, Guang Xi, Si Chuan, Gui Zhou, Yun Nan	Whole plant: acid, bitter, cool; heat-clearing, detoxification, blood-activating, swell-reducing; dysentery, stomachache, leukorrheal disease, amenorrhea	*P. runcinatum* var. *corymbosum* of Gui Zhou has similar efficacy

Continued

Table 12.1 Continued

Species	Chinese name	Habitat	Distribution	Therapeutic use	Note
P. runcinatum var. *sinense*	Hua Chi Jing San, San Xue Dan (Hu Bei), Hua Liao Zi Cao (Si Chuan)	Altitude 1300–2600 m, forest, roadside bushes, ditch	Shaan Xi, Gan Su, He Nan, Hu Bei, Hu Nan, Guang Xi, Si Chuan, Yun Nan	Root/whole plant: bitter, astringent, cold; swell-reducing and detoxicating, blood-activating and muscle-relaxing; overstrained cough, irregular menses, rheumatalgia, injuries from falls	
P. sagittifolium	Da Jian Ye Liao, She Zi Cao (Guang Xi), She Dao Tui (Hu Bei)	Altitude 800–1500 m, mountain roadside, ditch wetland	Shaan Xi, Hu Bei, Hu Nan, Si Chuan, Gui Zhou	Whole plant: heat-clearing, detoxification, dysentery, furunculosis, itch of skin, snakebite	
P. senticosum (*Persicaria senticosa*)	Ci Liao, Mao Er Ci (Zhe Jiang)	Altitude <1000 m, hillside, roadside bushes, forest, ditch, rock seam	East China, centralsouth China, Ji Lin, Liao Ning, He Bei, Bei Jing, Gui Zhou	Whole plant: acid, slightly pungent, flat; detoxification, apocatastasis, diuresis-promoting, itch-relieving; paronychia, refractory carbuncle furuncle, hurt, eczema, itching and pain, internal and external hemorrhoids	

P. sibiricum (*Persicaria sibirica*; *Aconogonon sibiricum*)	Siberia Liao, Jian Dao Gu (Si Chuan)	Altitude <1600 m, saline meadow, saline lowland, gravel saline alkali soil	Hei Long Jiang, Liao Ning, He Bei, Bei Jing, Inner Mongolia, Gan Su, Ning Xia, Qing Hai, Shan Dong, north Jiang Su, He Nan, Si Chuan, Yun Nan, Tibet	Root: edema; whole plant: heat-clearing, detoxification, wind-dispelling, and damp-removing	
P. sieboldii (*Persicaaria sieboldii*)	Jian Ye Liao, Ju Ju Cao (Hu Bei), Zou You Cao (Si Chuan), Que Chi	Altitude <1800 m, hillside, ditch, wet bushes	Northeast China, He Bei, Shan Dong, Jiang Su, An Hui, Fu Jian, Hu Bei, Hu Nan, Si Chuan, Gui Zhou	Whole plant: acid, astringent, flat; wind-dispelling and damp-removing, heat-clearing, detoxification; rheumatalgia, snakebite	
P. suffultum (*Bistorta suffulta*)	Zhi Zhu Liao, Xue San Qi (Zhe Jiang), Suan Pan Qi (Hu Bei), Niu Zi Qi (Si Chuan), Hong San Qi (Shaan Xi)	Altitude 1200–2000 m, forest, wet bushes, creek, roadside	Southwest China, He Bei, Shan Xi, Shaan Xi, Gan Su, Ning Xia, Qing Hai, An Hui, Zhe Jiang, Jiang Xi, He Nan, Hu Bei, Hu Nan	Rhizome: bitter, astringent, cool; hemostasis with astringents, analgesia, myogenic; traumatic injury, overstrained hemoptysis, hemafecia, irregular menses	*P. suffultum* var. *pergracile* has similar efficacy, distributed in Shaan Xi, Gan Su, Hu Bei, Si Chuan, Yun Nan
P. taipaishanense	Tai Bai Liao	Altitude 2800–3200 m, mountain meadow	Taibai Mountain, Shaan Xi	Rhizome: bitter, astringent, cool; heat-clearing, detoxification, stasis-removing, hemostasis	Endemic in Shaan Xi

Continued

Table 12.1 Continued

Species	Chinese name	Habitat	Distribution	Therapeutic use	Note
P. thunbergii (*Persicaria thunbergii*)	Ji Ye Liao, Lu Ti Cao (Jiang Su), Shui Hu Die (Hu Bei), Shui Ma Liao	Altitude <1850 m, valley, forest, wet meadow	Northeast China, He Bei, Shaan Xi, Shan Dong, Jiang Su, An Hui, Zhe Jiang, Fu Jian, Tai Wan, Hu Bei, Hu Nan, Guang Xi, Si Chuan, Gui Zhou, Yun Nan	Rhizome/whole plant: acid, slightly pungent, flat; heat-clearing, detoxification, blood-cooling, hemostasis, wind-dispelling, analgesia, cough-relieving; acute filthy disease, snakebite, dysentery	
P. tinctorium (*Persicaria tinctoria*)	Liao Lan, Da Qing Ye (northeast and northern China), Qing Ban Shui La Liao (Zhe Jiang), Lan Dian, Lan Liao (northeast China)	Native to China, in cultivation/semiwild throughout the country	He Bei, Shan Xi	Leaf: bitter, cold; heat-clearing, detoxification, blood-cooling, removing ecchymoses; febrile fever, rash, dyspnea and cough due to lung heat, pharyngitis, mumps, erysipelas, carbuncle; processed (indigo naturalis): saline, cold; heat-clearing, detoxification, blood-cooling, arresting convulsion; febrile disease, hot blood and vomitting, chest pain, hemoptysis, aphtha, child's epilepsy	China Pharmacopoeia 1990

P. viscosum (*Persicaria viscosa*)	Nian Mao Liao	Low-altitude, roadside wetland, waterfront	Central south China, Ji Lin, Liao Ning, Shaan Xi, Jiang Su, Zhe Jiang, Jiang Xi, Fu Jian, Tai Wan, Si Chuan, Gui Zhou, Yun Nan	Rhizome: heat-clearing, detoxification, blood-cooling, hemostasis	
P. viviparum (*Bistorta vivipara*)	Zhu Ya Liao, Hong San Qi (Ning Xia), Hou Er Qi (Hu Bei), Ye Gao Liang (Si Chuan)	Altitude 1200–3000 m, alpine grassland, wetland, creek	North China, southwest China, Ji Lin, Shaan Xi, Gan Su, Qing Hai, Xin Jiang, He Nan	Rhizome: bitter, astringent, cool; heat-clearing, detoxification, stasis-removing, hemostasis; tonsillitis, sore throat, dysentery, diarrhea, leukorrheal disease, hemafecia	
Rheum altaicum	Altai Da Huang	Mountain rock seam, gravel	Xin Jiang	Root/rhizome: reducing excess heat, bowel-relaxing, break and remove stasis, swell-reducing	*R. compactum* has the same efficacy and distribution area
Rheum delavayi	Sha Qi, Bai Xiao Huang (Yun Nan)	Mountainous region	Yun Nan	Root: bitter, cool; heat-clearing, detoxification, hemostasis, myogenic; lung heat and cough, sore throat, hemafecia, traumatic bleeding, fracture, traumatic sprain, rheumatic arthralgia	

Continued

Table 12.1 **Continued**

Species	Chinese name	Habitat	Distribution	Therapeutic use	Note
Rheum emodi	Zang Bian Da Huang, Bai Niu Wei Qi	Altitude 3200–4300 m, hillside shrubs, boulders beach, cliff	Yun Nan, Tibet; cultivated in northwest China	Root: bitter, sweet, cold; heat-clearing, detoxification, hemostasis, myogenic; lung heat and cough, sore throat, hemafecia, carbuncle, traumatic bleeding	
Rheum forrestii	Niu Wei Qi, Hong Ma Ti Niao (Si Chuan), Xue San Qi (Yun Nan)	Altitude 3000 m, rock	Si Chuan, Yun Nan	Root: bitter, cool; muscle-relaxing and blood-activating, hemostasis, heat-clearing; sore throat, mumps, constipation, traumatic injury, and bleeding	
Rheum franzenbachii	Bo Ye Da Huang, Hua Bei Da Huang, Tang Da Huang	Hillside meadow	North China, Ji Lin, Liao Ning, Xin Jiang, Hu Bei; in cultivation	Rhizome: bitter, cold; purging heat, bowel-relaxing, break and remove stasis; heat constipation, jaundice with damp-heat, furuncle and carbuncle, traumatic stasis pain, aphthous ulcer, burn	*R.nanum* has similar efficacy, distributed in Inner Mongolia, Gan Su, Xin Jiang

Rheum nobile	Gao Shan Da Huang, Ta Huang (Tibet)	Altitude 4500–5200 m, mountainside, meadow	Tibet	Rhizome: bitter, cold; reducing excess heat, stagnation-breaking, stasis-removing; heat constipation, delirium, dyspeptic hysteresis, dysentery, abdominal pain, tenesmus, hygropyretic jaundice, edema, red eyes, headache, amenorrhea	
Rheum officinale	Da Huang, Nan Da Huang, Xi Da Huang, Chuan Jun, Jun (Tibetan)	Altitude 1300–1600 m, hilly area, rand, hillside	Southwest China, Shaan Xi, Gan Su, Qing Hai, He Nan, Hu Bei; main producing area: Gan Su, Qing Hai	Root/rhizome: bitter, cold; purgating heat and bowels, blood-cooling and detoxification, stasis-removing and collateral-dredging; heat constipation, stagnation of abdominal pain, uncomfortable diarrhea, hygropyretic jaundice, hot blood and vomitting, red eyes, sore throat, blood stasis/amenorrhea, traumatic injury	China Pharmacopoeia 1990; *R.tibeticum* of Qing Hai has similar efficacy; *R.wittrocki*, distributed in north and south Xin Jiang
Rheum palmatum	Zhang Ye Da Huang, Bei Da Huang, Jin Da Huang (Hu Bei)	Hilly area, forest, grass slope	Shaan Xi, Gan Su, Qing Hai, Hu Bei, Si Chuan, Tibet;in cultivation	Root/rhizome: same as *R. officinale*	China Pharmacopoeia 1990

Continued

Table 12.1 **Continued**

Species	Chinese name	Habitat	Distribution	Therapeutic use	Note
Rheum pumilum	Xiao Da Huang	Shady mountain slope, semishady slope, meadow	Qing Hai	Root: bitter, cold; purging stomach/bowel stagnation, reducing excess heat, blood stasis-discharging, subduing inflammation; stagnation, dyspepsia, abdominal pain, endoretention of heat, constipation	
Rheum tanguticum	Tang Gu Te Da Huang, Ji Zhua Da Huang	Mountain shrubs, woodside, damp place	Gan Su, Qing Hai, Hu Bei, Si Chuan, Tibet; cultivated in Shaan Xi	Root/rhizome: same as *R. officinale*	China Pharmacopoeia 1990; *R. tanguticum* has similar efficacy
Rumex acetosa	Suan Mo, Suan Jiang (northeast China), Shan Bo Cai (Hu Bei), Niu Er Da Huang (Hu Nan, Si Chuan, Gui Zhou)	Low-altitude to 3100 m, mountain, wet valley, woodside, meadow	North China, northwest China, Ji Lin, Liao Ning, Jiang Su, An Hui, Zhe Jiang, Jiang Xi, He Nan, Hu Bei, Hu Nan, Si Chuan, Gui Zhou, Yun Nan	Root/whole plant: acid, bitter, cold; blood-cooling, detoxification, bowel-relaxing, insecticide; heat dysentery, stranguria, hematemesis, malignant sore, scabies	
Rumex aquaticus	Shui Sheng Suan Mo	Altitude <2400 m, valley, ditch	Northeast China, North China, northwest China, Hu Bei	Root: dyspepsia, acute hepatitis, eczema, stubborn dermatitis	

Rumex chalepensis	Zhong Ya Suan Mo	Open field, riverside, wetland, roadside, wasteland, grassland	Northeast China, He Bei, Shaan Xi, Shan Dong, Jiang Su, Jiang Xi, He Nan, Hu Bei	Root: bitter, acid, cold; blood-cooling, hemostasis, heat-clearing, detoxification, insecticide; uterine bleeding, hematemesis, hemoptysis, nasal bleeding, stomach bleeding, hemafecia, hemokelidosis, constipation, edema	
Rumex crispus	Zhou Ye Suan Mo, Niu She Tou (Ning Xia), Yang Ti Gen (Ji Lin), Si Ji Cai Gen (Si Chuan)	Roadside, damp place, waterfront	Northeast China, He Bei, Inner Mongolia, Shaan Xi, Ning Xia, Qing Hai, Jiang Su, An Hui, Zhe Jiang, Fu Jian, Tai Wan, He Nan, Hu Bei, Guang Xi, Si Chuan, Gui Zhou, Yun Nan	Root/whole plant: bitter, acid, cold; heat-clearing, blood-cooling, bowel-relaxing, insecticide, resolve phlegm and relieve cough; acute hepatitis, cough and phlegm, hematemesis, metrorrhagia, constipation, dysentery, scabies	*R. crispus* var. *unicallosus* of Ning Xia has similar efficacy
Rumex dentatus	Chi Guo Suan Mo, Niu Er Da Huang (Si Chuan), Niu She Cao (Jiang Su)	Altitude 400–1200 m, fringe wetland	He Bei, Shan Xi, Shaan Xi, Gan Su, Qing Hai, Jiang Su, An Hui, Zhe Jiang, He Nan, Hu Bei, Si Chuan, Gui Zhou, Yun Nan	Root/whole plant: similar to *R. crispus*; leaf: breast swelling	

Continued

Table 12.1 Continued

Species	Chinese name	Habitat	Distribution	Therapeutic use	Note
Rumex gmelini	Mao Mai Suan Mo	Waterfront, wet meadow	Northeast China, He Bei, Inner Mongolia	Root: hemostasis, laxative	
Rumex hastatus	Chuan Dian Tu Da Huang	Hillside meadow	Si Chuan, Yun Nan	Whole plant: acid, astringent, slightly pungent, warm; inducing sweat and dispelling exogenous evils, moistening lung to arrest cough; common cold, cough, edema, asthma due to excessive phlegm	
Rumex japonicus	Yang Ti, Yang She Tou (Zhe Jiang), Niu Li Cai (Guang Xi), Niu She Da Huang	Mountains and plains, roadside damp place	East China,central south China, Gui Zhou	Root: bitter, acid, cold; slightly poisonous; blood-cooling and hemostasis, loosen stools, detoxification, insecticide; constipation, stranguria with turbid discharge, jaundice, hematemesis, hemorrhoidal hamorrhage, favus of the scalp; leaf: sweet, cold; hemorrhoidal hamorrhage, constipation, infantile malnutrition; fruit:bitter, astrigent, flat; mixed red and white dysentery	*R.nepalensis* has the same efficacy, distributed in southwest China, Shaan Xi, Gan Su, Qing Hai, Jiang Xi, He Nan, Hu Bei

Rumex maritimus	Chang Ci Suan Mo, Niu Shi Cao (Hu Bei), Xue Da Huang (Si Chuan), Zhou Ye Yang Ti (Guang Xi)	Roadside, ditch, damp place	Northeast China, north China, Yangtze River basin, Fu Jian, Guang Xi, Gui Zhou	Whole plant: slightly sweet, slightly bitter, cool; heat-clearing and blood-cooling, insecticide/detoxification; tuberculosis hemoptysis, carbuncle, favus of the scalp itch, itch of skin, traumatic injury, hemorrhoidal bleeding	
Rumex obtusifolius (*R. madaio*)	Dun Ye Suan Mo, Xue San Qi, Hua Xue Lian (Jiang Su)	Hillside, damp place, open forest	Yangtze River basin	Root: bitter, cold; clearing away heat and loosing the bowels, killing parasites to relieve itshing, hemostasis; lung abscess, hemoptysis, hepatitis, constipation; external use in traumatic injury, burn	
Rumex patientia	Ba Tian Suan Mo, Tu Da Huang (Shaan Xi), Niu Xi Xi (Jiang Xi)	Roadside, village wetland, ditch	Hei Long Jiang, Liao Ning, He Bei, Shan Xi, Shaan Xi, Gan Su, Qing Hai, south Jiang Su, Hu Bei, Gui Zhou	Root: bitter, acid, cold; slightly poisonous; blood-cooling and hemostasis, heat-clearing and detoxification, laxative, insecticide; dysentery, diarrhea, hepatitis, traumatic injury, constipation, ulcerative carbuncle, scabies	*R. patientia* var. *callosus* of northeast China has the same efficacy

Continued

Table 12.1 Continued

Species	Chinese name	Habitat	Distribution	Therapeutic use	Note
Rumex stenophyllus	Zhai Ye Suan Mo	Wet meadow, saline soil	Northeast China, north China, Xin Jiang	Root: bitter, acid, cold; blood-cooling and hemostasis, heat-clearing and detoxification, insecticidal; uterine bleeding, gastrorrhagia, hemafecia, hemokelidosis, edema	*R.stenophyllus* var. *ussuriensis* of Hei Long Jiang has the same efficacy

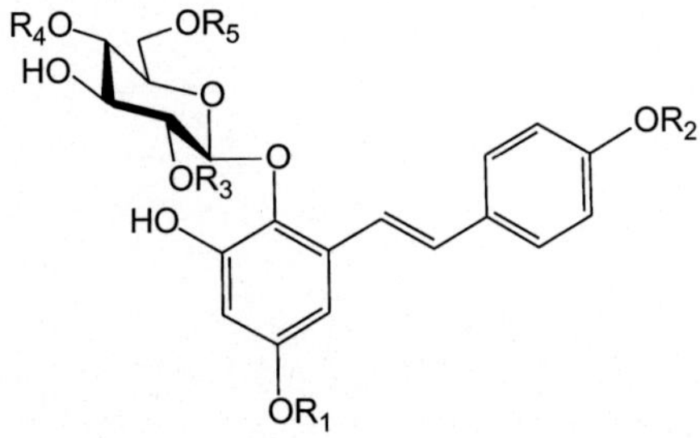

No.	R_1	R_2	R_3	R_4	R_5
1	H	H	H	H	H
2	H	H	H	H	Galloyl
3	H	H	Galloyl	H	H
4	H	H	H	α-D-Glc	H
5	H	H	H	H	α-D-Glc
6	H	α-D-Glc	H	H	H
7	α-D-Glc	H	H	H	H
8	H	H	β-D-Fru (f)	H	H

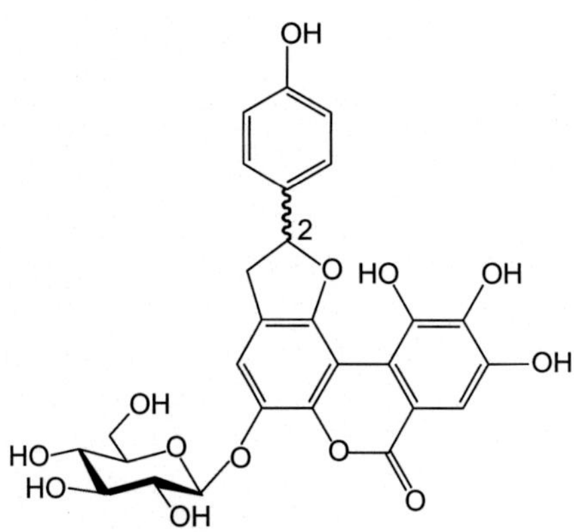

No.	2
9	S

Figure 12.1 Examples of compounds isolated from the *Polygonum* species. Refer to Table 12.2 for more details.

(Continued)

10	R

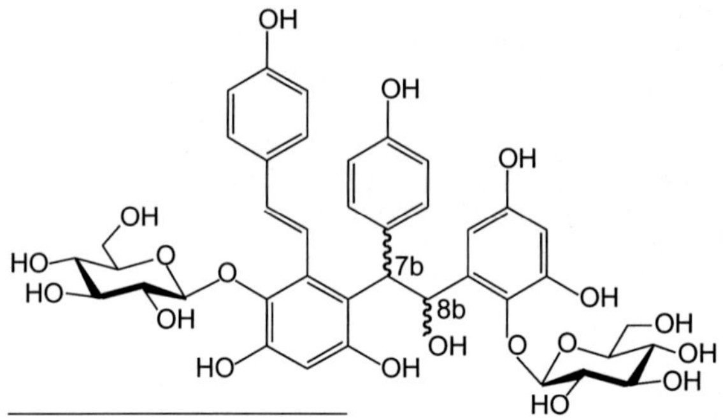

No.	7b	8b
11	R	S
12	S	R

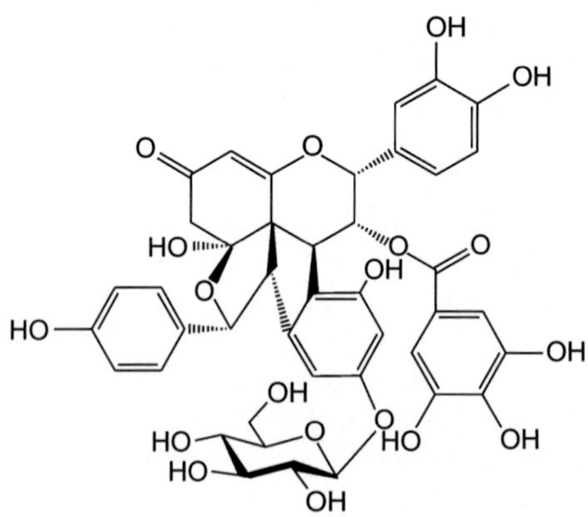

13

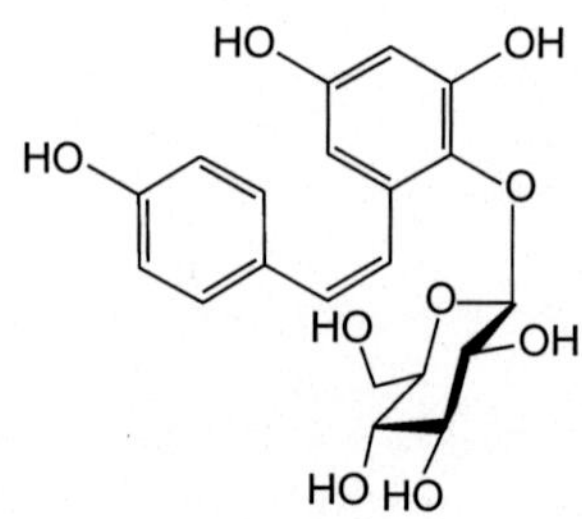

14

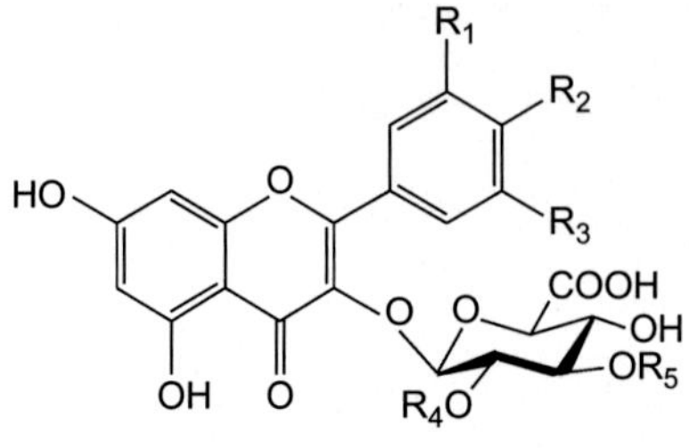

Figure 12.1 Continued.

No.	R_1	R_2	R_3	R_4	R_5
15	OH	OH	OH	H	H
16	OH	OCH_3	OH	H	H
17	H	OH	OH	H	H
18	H	OH	OCH_3	H	H
19	H	OCH_3	H	H	H
20	H	OH	H	H	H
21	H	OH	H	$COCH_3$	H
22	H	OH	OCH_3	$COCH_3$	H
23	H	OH	OH	$COCH_3$	H
24	H	OH	OH	H	$COCH_3$
25	H	OH	H	H	$COCH_3$
26	OH	OH	H	H	H
27	OH	OH	H	α-L-Rha	H
28	OH	OH	H	α-L-Rha	H

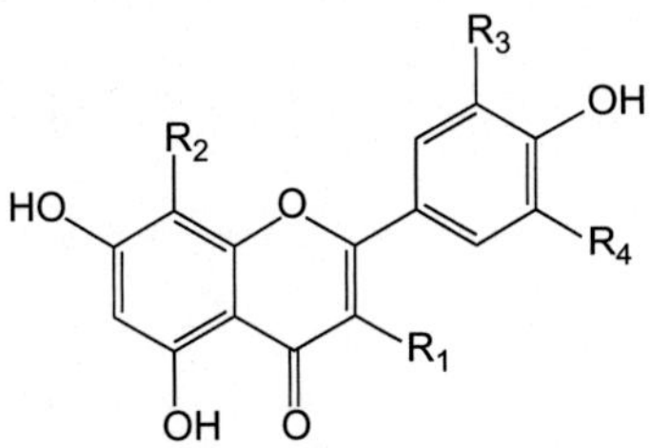

No.	R_1	R_2	R_3	R_4
29	OR_x	H	H	OCH_3
30	OR_y	H	H	OCH_3
31	OH	H	OH	H

Figure 12.1 Continued.

32	OH	H	OH	OH
33	*O*-β-D-Gal	H	OH	OH
34	*O*-α-L-Rha	H	OH	OH
35	*O*-β-D-Gal	H	OH	H
36	*O*-β-D-Glc(6-1)-α-L-Rha	H	OH	H
37	OH	H	*O*-β-D-Gal	H
38	OH	OCH_3	OH	H
39	H	H	H	H
40	H	H	OH	H
41	OH	H	H	H
42	*O*-(6″-galloyl)-β-D-Gal	H	OH	H

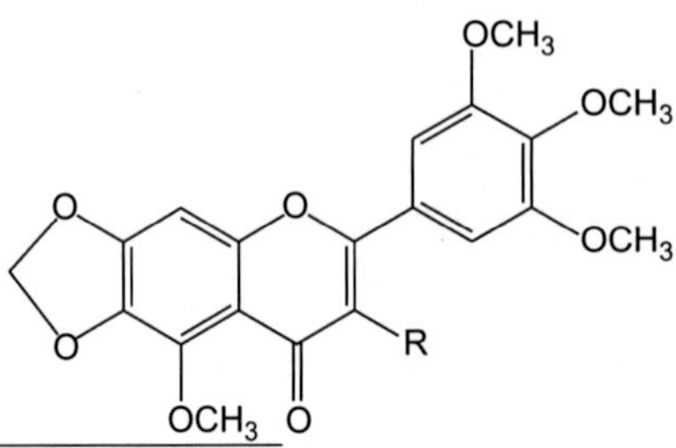

No.	R
43	H
44	OCH_3

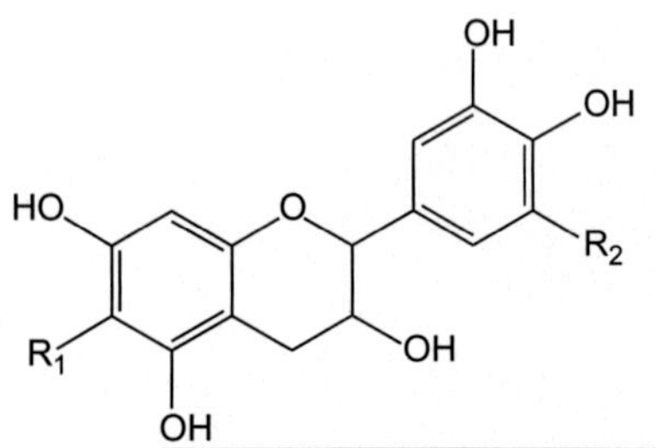

No.	R_1	R_2
45	H	H
46	H	OH
47	OH	H

Figure 12.1 Continued.

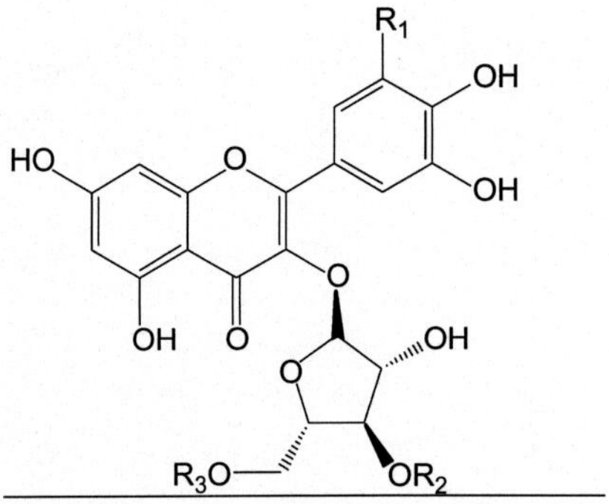

No.	R1	R2	R3
48	H	$COCH_3$	$COCH_3$
49	H	$COCH_3$	H
50	H	H	H
51	OH	$COCH_3$	$COCH_3$
52	OH	H	H

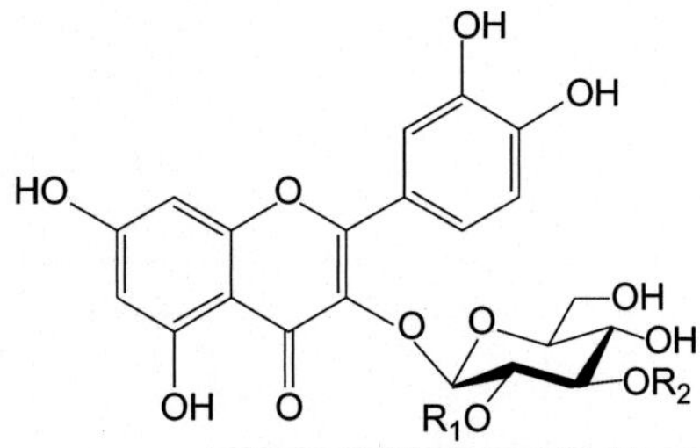

No.	R1	R2
53	Galloyl	H
54	H	Galloyl

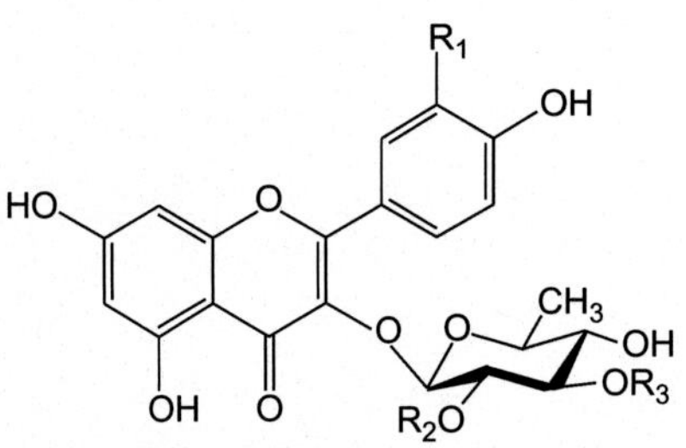

No.	R1	R2	R3
55	OH	Galloyl	H
56	OH	H	Galloyl
57	OH	H	H
58	H	H	H

Figure 12.1 Continued.

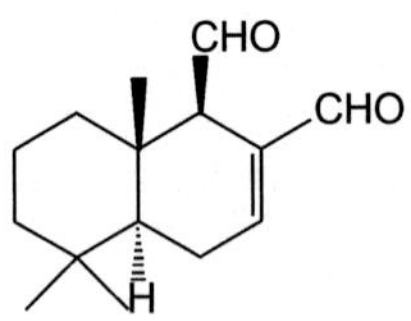
59

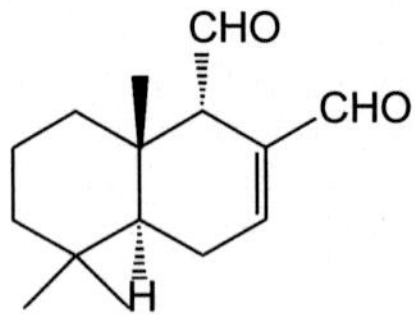
60

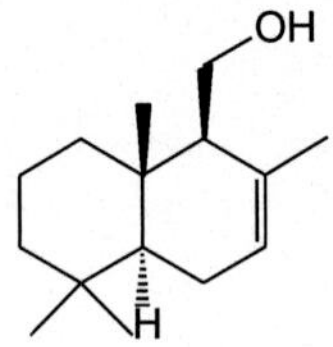
61

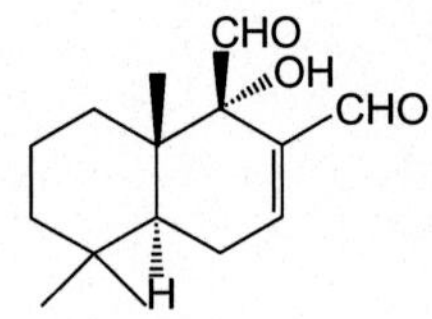
62

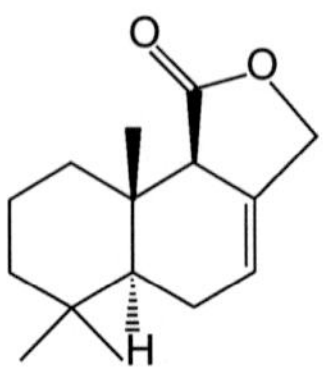
63

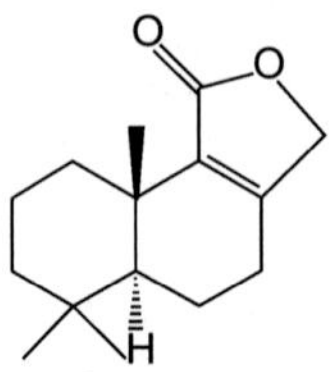
64

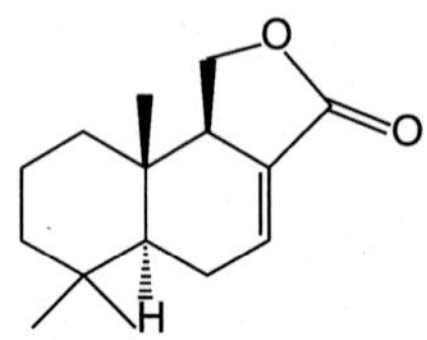
65

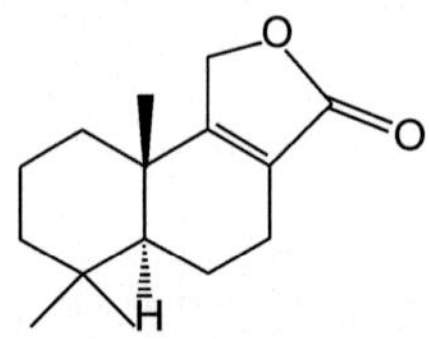
66

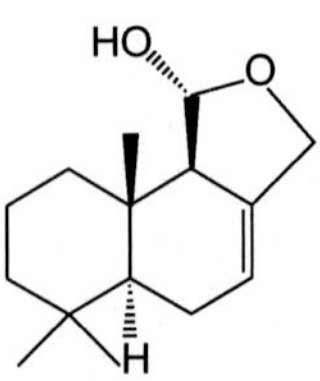
67

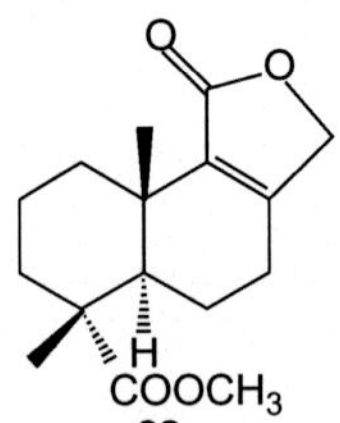
68

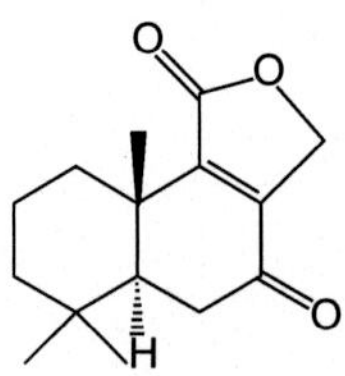
69

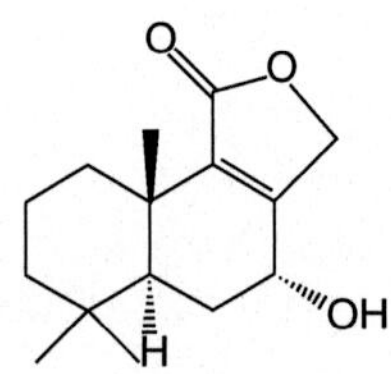
70

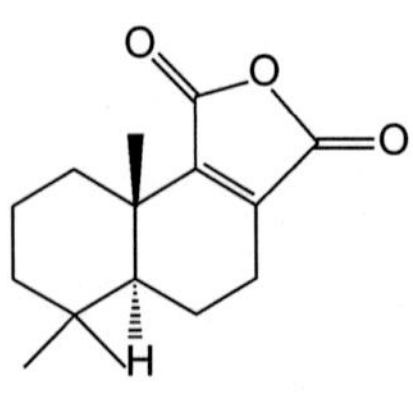
71

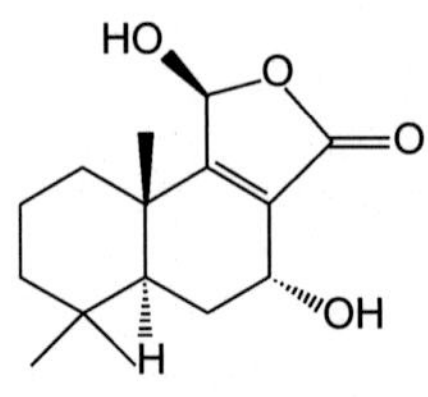
72

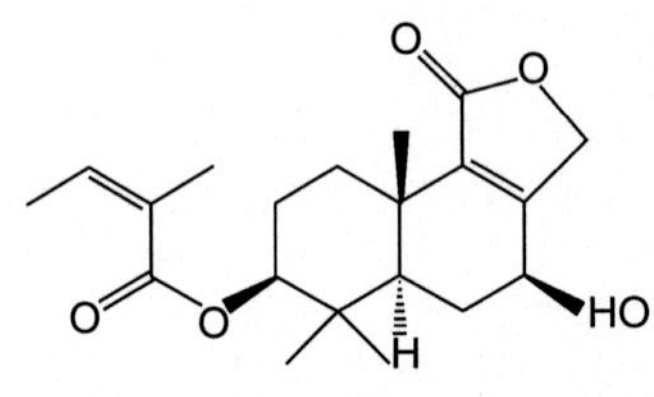
73

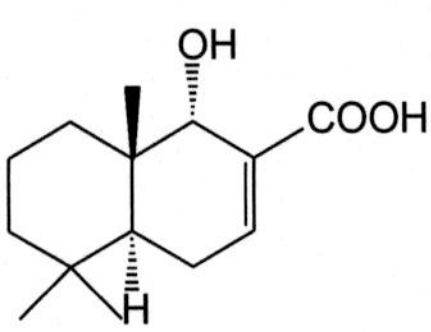
74

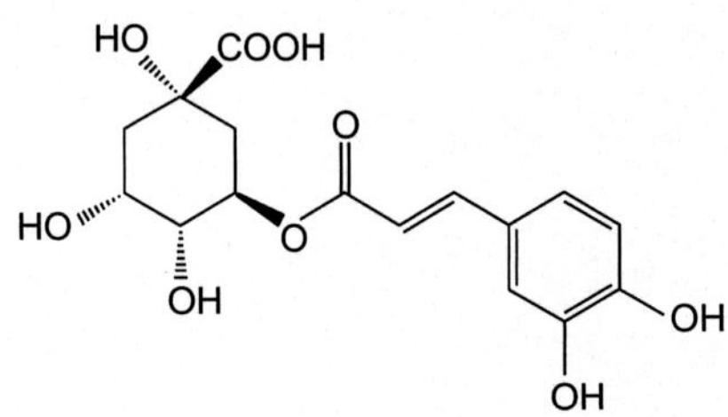
75

Figure 12.1 Continued.

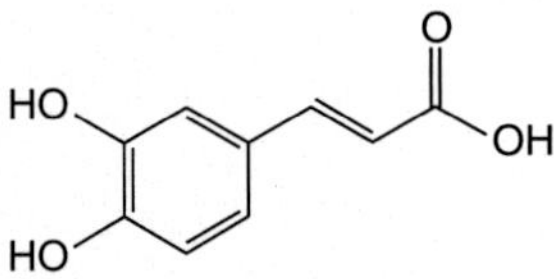
76

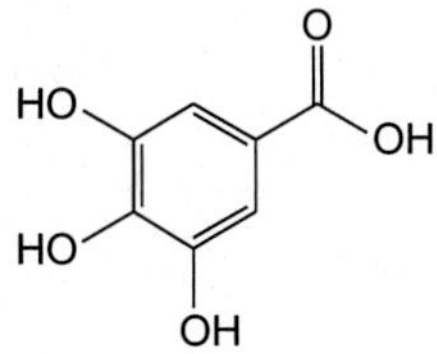
77

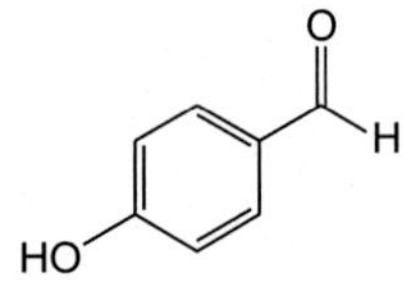
78

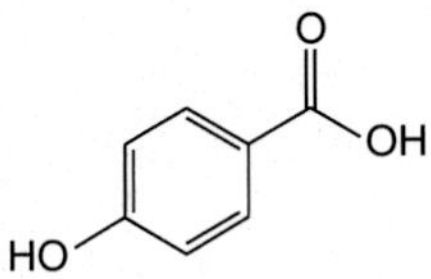
79

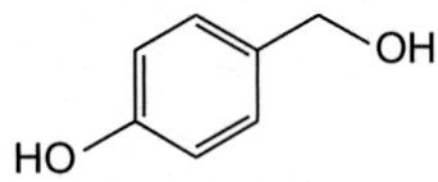
80

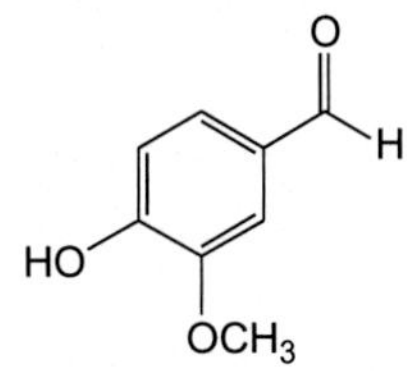
81

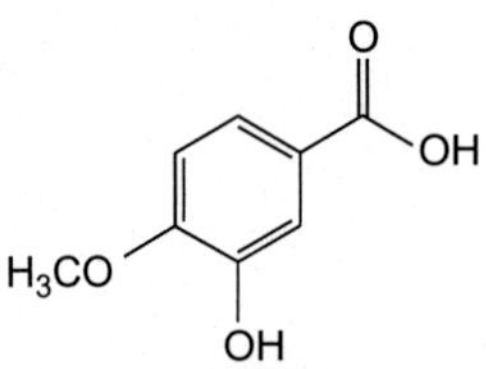
82

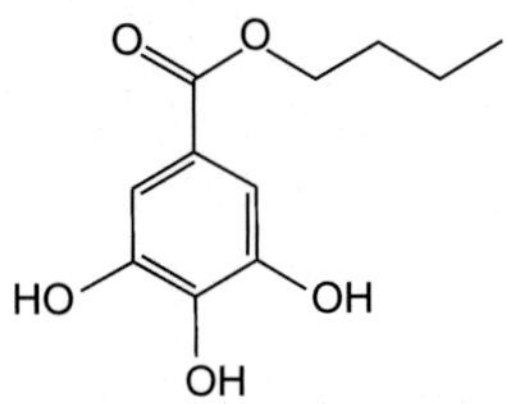
83

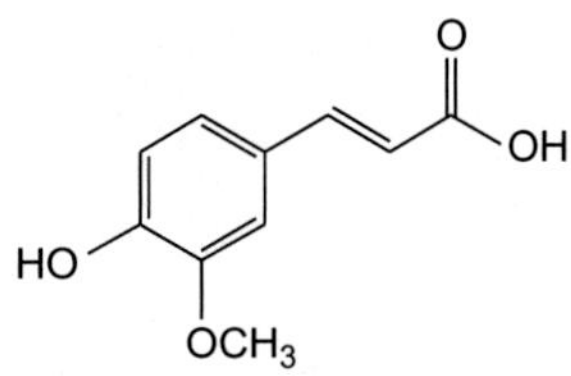
84

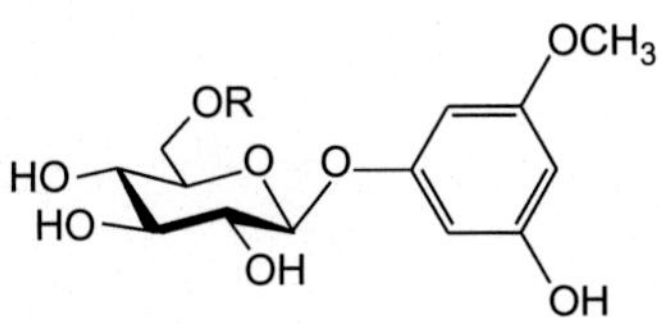
85

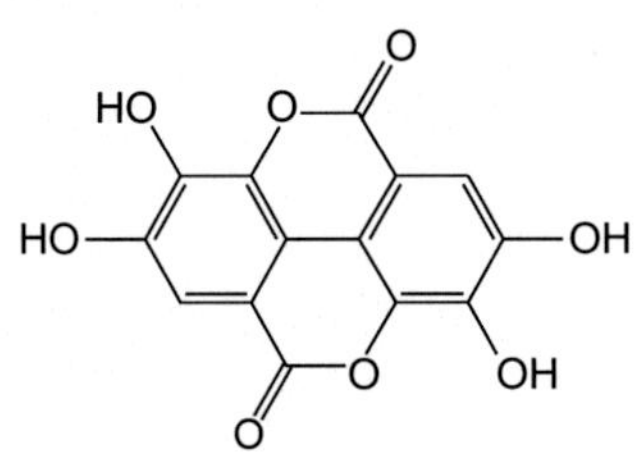
86

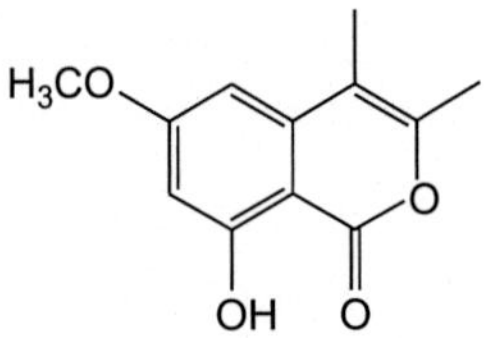
87

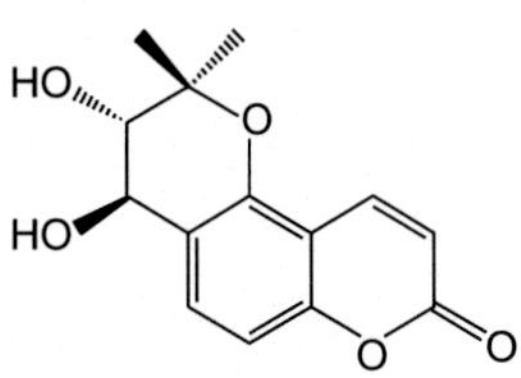
88

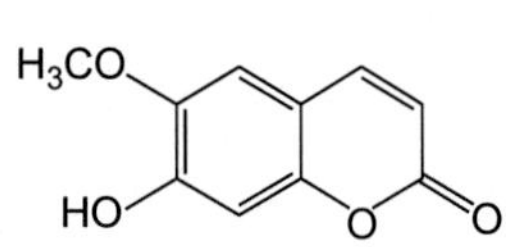
89

Figure 12.1 Continued.

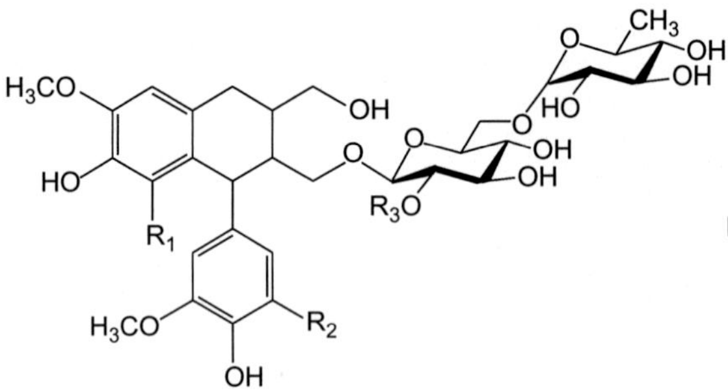

No.	R_1	R_2	R_3
90	OCH_3	OCH_3	H
91	H	H	α-L-Rha

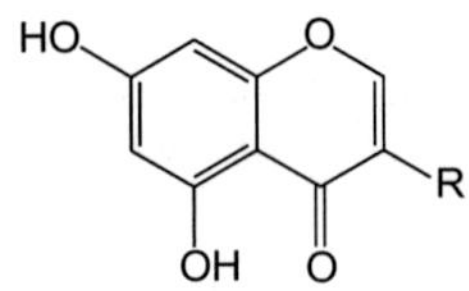

No.	R
92	OH
93	H

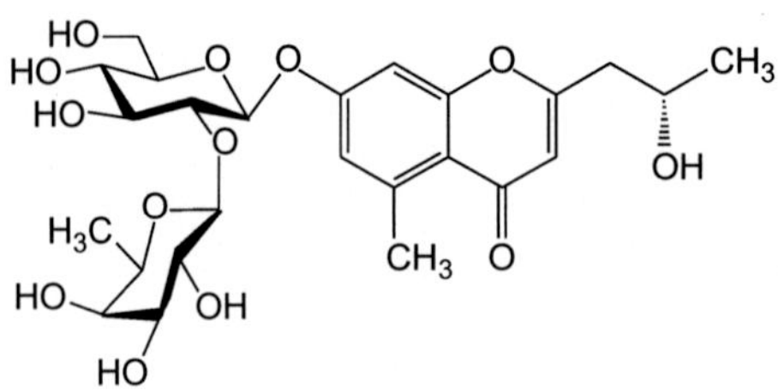

94

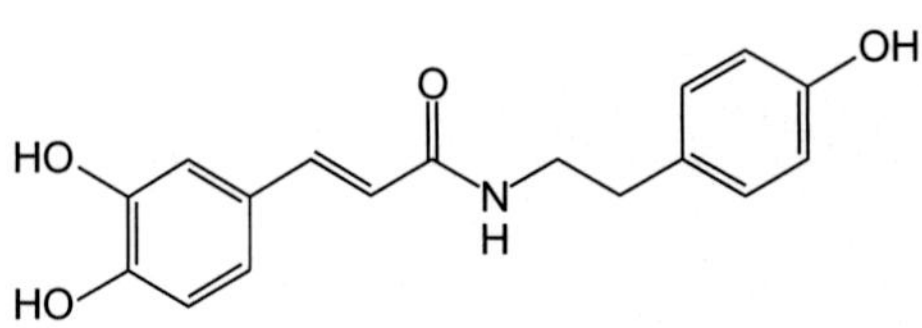

95

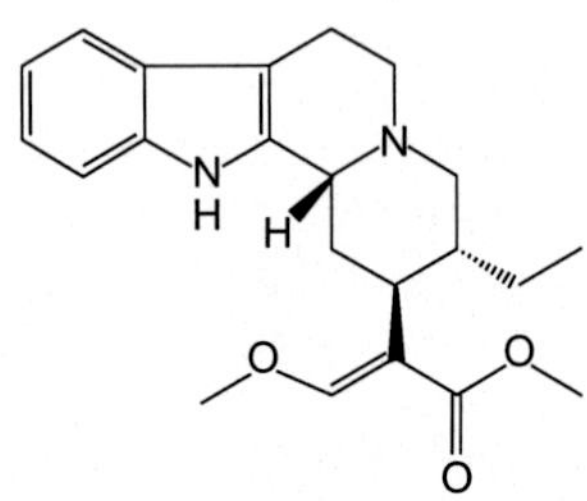

96

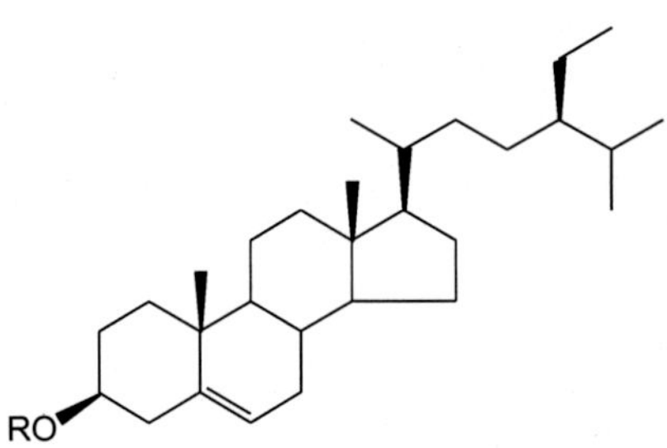

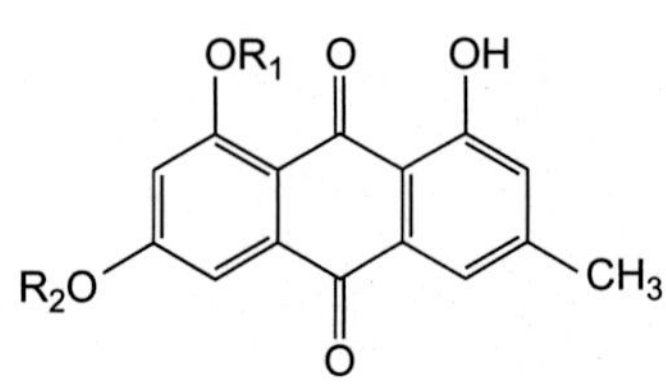

Figure 12.1 Continued.

No.	R
97	Ac
98	H
99	β-D-Glc

No.	R_1	R_2
100	H	CH_3
101	H	H
102	β-D-Glc	H
103	β-D-Glc	CH_3

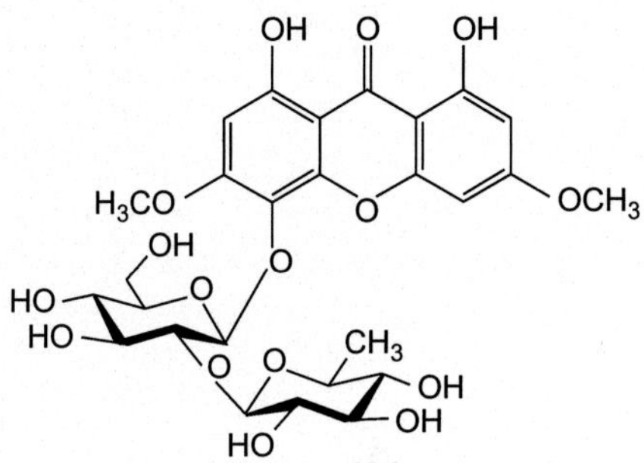

104

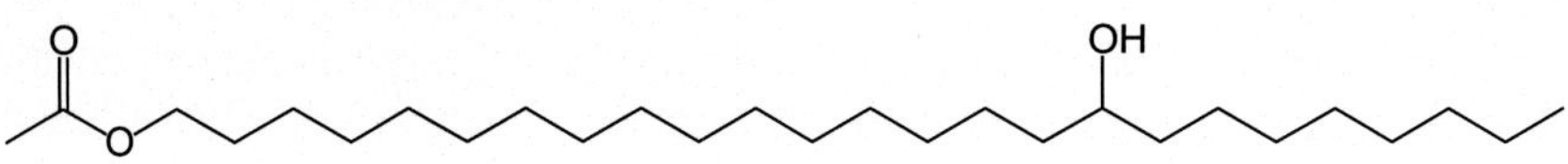

105

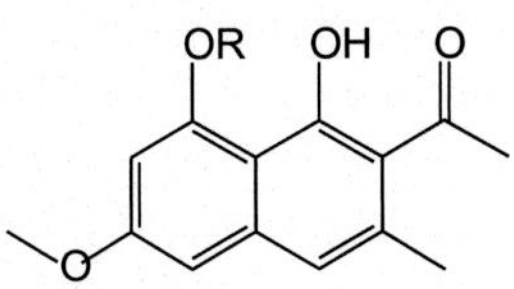

No.	R
106	β-D-Glc

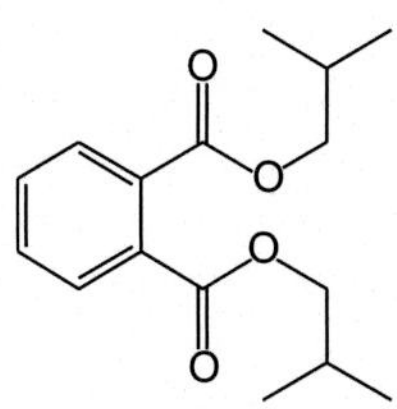

107

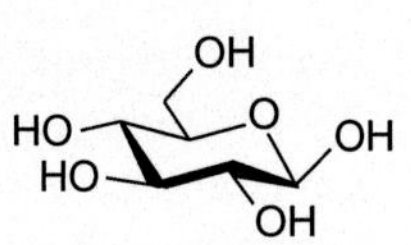

Glc: β-D-glucopyranosyl

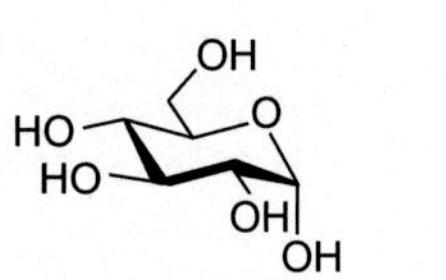

Glc: α-D-glucopyranosyl

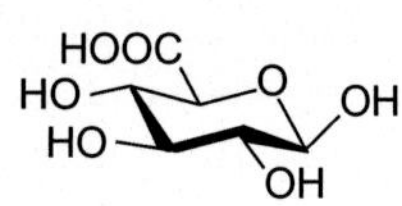

GlcA: β-D-glucuropyranosyl

Figure 12.1 Continued.

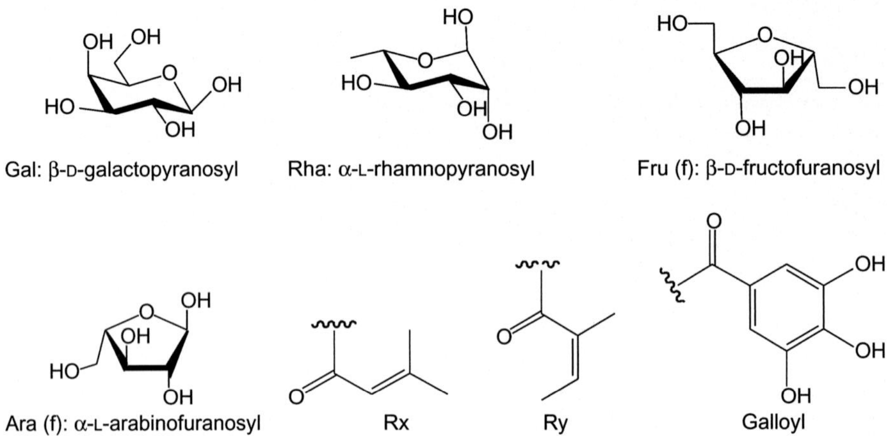

Figure 12.1 Continued.

85.66% of the total organic ester compounds. 1-Phenyl-1-pentanone and acetophenone were 16.30% and 4.33% of the total volatiles, respectively.

The liverwort (*Porella vernicosa*) complex produces a very hot tasting polygodial, a drimane-type sesquiterpene dialdehyde. The same compound has been isolated from two ferns, *Thelypteris hispidula* and *Blechnum fluviatile*, as well as from the higher plants *Polygonum hydropiper* and *P. hydropiper f. purpurascens* (Polygonaceae); *Cinnamosma*, *Caspicodendron*, *Canella*, and *Warburgia* species (Canellaceae); and *Pseudowintera colorata*, *Tasmannia lanceolata*, *Drimys*, and *Zygogynum* species (Winteraceae) (Asakawa et al., 2012). In addition, the liverworts and higher plants that elaborate polygodial and its related pungent drimane dials contain a small amount of α-tocopherol, γ-tocopherol, or δ-tocotrienol. The drimane-type sesquiterpenoids and tocopherols might be useful chemotaxonomic markers in some lower and higher plants, including Polygonaceae.

12.2.3 Other compounds

2-Furanmethanol, 2 (5H)-furanone, 2-hydroxy-2-cyclopenten-1-one, 2, 4-dihydroxy-2,5-dimethyl-3(2H)-furan-3-one, 2H-pyran-2,6 (3H)-dione, 2-hydroxy-gamma-butyrolactone, 2,5-dimethyl-4-hydroxy-3 (2H)-furanone, 2,5-furandicarboxaldehyde, 2,3-dihydro-3,5-dihydroxy-6-methyl 4H-pyran-4-one, (S)-(+)-2′, 3′-dideoxyribonolactone, 5-(hydroxymethyl)-furancarboxaldehyde, 3-deoxy-D-mannoic lactone, etc. were found in the callus of *P. minus* (Vikram et al., 2014).

Table 12.2 Various types of compounds found in *Polygonum* medicinal plants in recent years

No.	Compounds	Types	Species	Tissues	References
1	2,3,5,4′-Tetrahydroxystilbene-2-*O*-β-D-glucopyranoside	A	a	Tubers, roots	Cheung et al. (2014), Kim et al. (2008a,b), and Xu et al. (2006)
2	(E)-2,3,4′,5-Tetrahydroxystilbene-2-β-D-(6″-galloyl)-glucopyranoside	A	a	Roots	Kim et al. (2008a,b)
3	(E)-2,3,4′,5-Tetrahydroxystilbene-2-β-D-(2″-galloyl)-glucoside	A	a	Roots	Kim et al. (2008a,b)
4	(E)-2,3,5,4′-Tetrahydroxystilbene-2-*O*-(4″-*O*-α-D-glucopyranosyl)-β-D-glucopyranoside	A	a	Roots	Li et al. (2013a)
5	(E)-2,3,5,4′-Tetrahydroxystilbene-2-*O*-(6″-*O*-β-D-glucopyranosyl)-β-D-glucopyranoside	A	a	Roots	Li et al. (2013a)
6	(E)-2,3,5,4′-Tetrahydroxystilbene-2-*O*-β-D-glucopyranosyl-4′-*O*-α-D-glucopyranoside	A	a	Roots	Li et al. (2013a)
7	(E)-2,3,5,4′-Tetrahydroxystilbene-2-*O*-β-D-glucopyranosyl-5-*O*-α-D-glucopyranoside	A	a	Roots	Li et al. (2013a)
8	(E)-2,3,5,4′-Tetrahydroxystilbene-2-*O*-(2″-*O*-β-D-fructofuranosyl)-β-D-glucopyranoside	A	a	Roots	Li et al. (2013a)
9	Polygonumoside A	A	a	Roots	Yan et al. (2014)
10	Polygonumoside B	A	a	Roots	Yan et al. (2014)
11	Polygonumoside C	A	a	Roots	Yan et al. (2014)
12	Polygonumoside D	A	a	Roots	Yan et al. (2014)
13	Polyflavanostilbene A	A	b	Rhizomes	Li et al. (2013b)
14	(Z)-2,3,4′,5-Tetrahydroxystilbene-2-β-D-glucoside	A	a	Roots	Kim et al. (2008a,b) and Xu et al. (2006)
15	Myricetin 3-*O*-β-D-glucuronide	B	c	Herbs	Granica et al. (2013)
16	Mearsetin 3-*O*-β-D-glucuronide	B	c	Herbs	Granica et al. (2013)

Continued

Table 12.2 Continued

No.	Compounds	Types	Species	Tissues	References
17	Quercetin 3-*O*-β-D-glucuronide	B	c	Herbs	Granica et al. (2013)
18	Isorhamnetin 3-*O*-β-D-glucuronide	B	c	Herbs	Granica et al. (2013)
19	Kaempferide 3-*O*-β-D-glucuronide	B	c	Herbs	Granica et al. (2013)
20	Kaempferol 3-*O*-β-D-glucuronide	B	c	Herbs	Granica et al. (2013)
21	Kaempferol 3-*O*-β-(2″-*O*-acetyl-β-D-glucuronide)	B	c	Herbs	Granica et al. (2013)
22	Isorhamnetin 3-*O*-β-(2″-*O*-acetyl-β-D-glucuronide)	B	c	Herbs	Granica et al. (2013)
23	Quercetin 3-*O*-β-(2″-*O*-acetyl-β-D-glucuronide)	B	c	Herbs	Granica et al. (2013)
24	Quercetin 3-*O*-β-(3″-*O*-acetyl-β-D-glucuronide)	B	c	Herbs	Granica et al. (2013)
25	Kaempferol 3-*O*-β-(3″-*O*-acetyl-β-D-glucuronide)	B	c	Herbs	Granica et al. (2013)
26	Quercetin-3-*O*-β-glucuronide	B	d	Herbs	Smolarz et al. (2008)
27	Quercetin-3-*O*-α-rhamnosyl-(1→2)-β-glucuronide	B	d	Herbs	Smolarz et al. (2008)
28	Kaempferol-3-*O*-α-rhamnosyl-(1→2)-β-glucuronide	B	d	Herbs	Smolarz et al. (2008)
29	3-*O*-senecioyl-isorhamnetin	B	e	Herbs	Lajter et al. (2013a)
30	3-*O*-angeloyl-isorhamnetin	B	e	Herbs	Lajter et al. (2013a)
31	Quercetin	B	f, i, h, b	Aerial parts (f), herbs (i,h), roots (b)	Moradi-Afrapoli et al. (2012a), Lin et al. (2009) and Zhang et al. (2012c)
32	Myricetin	B	f	Aerial parts	Moradi-Afrapoli et al. (2012a)
33	Myricetin 3-*O*-β-D-galactopyranoside	B	f	Aerial parts	Moradi-Afrapoli et al. (2012a)
34	Myricetin 3-*O*-α-L-rhamnopyranoside (myricitrin)	B	f, h	Aerial parts (f), herbs (h)	Moradi-Afrapoli et al. (2012a) and Zhang et al. (2012c)
35	Quercetin 3-*O*-β-D-galactopyranoside	B	f, g	Aerial parts	Moradi-Afrapoli et al. (2012a) and Tantry et al. (2012)
36	Rutin	B	g, h,b	Aerial parts (g), herbs (h), leaves (b)	Tantry et al. (2012) and Zhang et al. (2012c)

37	Quercetin-3′-*O*-β-D-galactopyranoside	B	i	Herbs	Lin et al. (2009)
38	8-Methoxyquercetin	B	i	Herbs	Lin et al. (2009)
39	Apigenin	B	i,b	Herbs (i), roots (b)	Lin et al. (2009)
40	Luteolin	B	i	Herbs	Lin et al. (2009)
41	Kaempferol	B	h	Herbs	Zhang et al. (2012c)
42	Quercetin 3-*O*-(6″-galloyl)-β-D-galactoside	B	j	Whole plant	Datta et al. (2000)
43	5, 3′,4′,5′-Tetramethoxy-6,7-methylenedioxyflavone	B	e	Herbs	Lajter et al. (2013a)
44	3,5,3′,4′,5′-Pentamethoxy-6,7-methylenedioxyflavone	B	e	Herbs	Lajter et al. (2013a)
45	(+) Catechin	B	f, g, b	Aerial parts (f, g), roots (b)	Moradi-Afrapoli et al. (2012a) and Tantry et al. (2012)
46	(−) Gallocatechin	B	f	Aerial parts	Moradi-Afrapoli et al. (2012a)
47	Amplexicine	B	g	Aerial parts	Tantry et al. (2012)
48	Quercetin 3-*O*-α-L-(3″,5″-diacetyl-arabinofuranoside)	B	f	Aerial parts	Moradi-Afrapoli et al. (2012a)
49	Quercetin 3-*O*-α-L-(3″-acetyl-arabinofuranoside)	B	f	Aerial parts	Moradi-Afrapoli et al. (2012a)
50	Quercetin 3-*O*-α-L-arabinofuranoside (avicularin)	B	f	Aerial parts	Moradi-Afrapoli et al. (2012a)
51	Myricetin 3-*O*-α-L-(3″,5″-diacetyl-arabinofuranoside)	B	f	Aerial parts	Moradi-Afrapoli et al. (2012a)
52	Myricetin 3-*O*-α-L-arabinofuranoside	B	f	Aerial parts	Moradi-Afrapoli et al. (2012a)
53	Quercetin-3-*O*-(2″-*O*-galloyl)-β-D-glucopyranoside	B	h	Herbs	Zhang et al. (2012c)
54	Quercetin-3-*O*-(3″-*O*-galloyl)-β-D-glucopyranoside	B	h	Herbs	Zhang et al. (2012c)
55	Quercetin-3-*O*-(2″-*O*-galloyl)-α-L-rhamnopyranoside	B	h	Herbs	Zhang et al. (2012c)
56	Quercetin-3-*O*-(3″-*O*-galloyl)-α-L-rhamnopyranoside	B	h	Herbs	Zhang et al. (2012c)
57	Quercitrin	B	h, b	Herbs (h), roots (b)	Zhang et al. (2012c)
58	Kaempferol-3-*O*-α-L-rhamnopyranoside	B	h	Herbs	Zhang et al. (2012c)
59	Polygodial	C	k, l, m, n	Whole plant (k, l), leaves, stem (m, n)	Asakawa et al. (2012)

Continued

Table 12.2 Continued

No.	Compounds	Types	Species	Tissues	References
60	Isopolygodial	C	k, l, m, n	Whole plant (k,l), leaves (m,n), stem (m)	Asakawa et al. (2012)
61	Drimenol	C	k1, k3, l, m, n	Whole plant (k1, k3, l), leaves (m, n)	Asakawa et al. (2012)
62	Warburganal	C	k3	Leaves	Asakawa et al. (2012)
63	Drimenin	C	k1, k3, l	Whole plant (k1, l), leaves (k3)	Asakawa et al. (2012)
64	Isodrimenin	C	k1, k3, l	Whole plant (k1, l), leaves (k3)	Asakawa et al. (2012)
65	Cinnamolide	C	k3, m, n	Leaves (k3, m, n), stem (m)	Asakawa et al. (2012)
66	Confertifolin	C	k3, l	Whole plant (l), leaves (k3)	Asakawa et al. (2012)
67	Isodrimeninol	C	k3	Leaves	Asakawa et al. (2012)
68	Poygonumate	C	k2	Whole plant	Asakawa et al. (2012)
69	7-Ketocoisodrimenin	C	k2	Whole plant	Asakawa et al. (2012)
70	Futronolide	C	k2	Whole plant	Asakawa et al. (2012)
71	Winterin	C	k2	Whole plant	Asakawa et al. (2012)
72	Dendocarbin L	C	k2	Whole plant	Asakawa et al. (2012)
73	7β-Hydroxyiso-angeloyloxy-7-epi-futronolide	C	k2	Whole plant	Asakawa et al. (2012)
74	Changweikagnic acid A	C	k2	Whole plant	Asakawa et al. (2012)
75	Chlorogenic acid	D	g, b	Aerial parts (g), roots (b)	Tantry et al. (2012)
76	Caffeic acid	D	g	Aerial parts	Tantry et al. (2012)

77	Gallic acid	D	g, b	Aerial parts (g), roots (b)	Tantry et al. (2012)
78	*p*-Hydroxybenzaldehyde	D	i	Herbs	Lin et al. (2009)
79	*p*-Hydroxybenzoic acid	D	g	Rhizomes	Xiang et al. (2012)
80	*p*-Hydroxybenzoic ethanol	D	g	Rhizomes	Xiang et al. (2012)
81	Vanillin	D	g	Rhizomes	Xiang et al. (2012)
82	Isovanillic acid	D	g	Rhizomes	Xiang et al. (2012)
83	3,4,5-Trihydroxy-benzoic acid-butyl ester	D	g	Rhizomes	Xiang et al. (2012)
84	4-Hydroxy-3-methoxycinnamic acid	D	g	Rhizomes	Xiang et al. (2012)
85	1-*O*-β-D-(6′-*O*-galloyl)-glucopyranosyl-3-methoxy-5-hydroxybenzene	D	h	Herbs	Zhang et al. (2012c)
86	Ellagic acid	D	h	Herbs	Zhang et al. (2012c)
87	Polygonolide	E	k3	Leaves	Asakawa et al. (2012)
88	5, 6-Dihydropyranobenzopyrone	E	g	Aerial parts	Tantry et al. (2012)
89	Scopletin	E	g	Aerial parts, rhizomes	Tantry et al. (2012) and Xiang et al. (2012)
90	(+)-Lyoniresinol-3a-*O*-[α-L-rhamnopyranosyl-(1‴→6″)]-β-D-glucopyranoside	F	o	Aerial parts	Abd El-Kader et al. (2013a,b)
91	(+)-Isolariciresinol-3a-*O*-[α-L-rhamnopyranosyl-(1‴→2″)-α-L-rhamnopyranosyl-(1⁗→6″)]-β-D-glucopyranoside	F	o	Aerial parts	Abd El-Kader et al. (2013a,b)
92	3,5,7-Trihydroxychromone	G	i	Herbs	Lin et al. (2009)
93	5,7-Dihydroxy-4H-chromen-4-one	G	h	Herbs	Zhang et al. (2012c)
94	(S)-2-(2′-hydroxypropyl)-5-methyl-7-hydroxychromone-7-*O*-α-L-fucopyranosyl(1→2)-β-D-glucopyranoside	G	a	Roots	Zhao et al. (2014)

Continued

Table 12.2 Continued

No.	Compounds	Types	Species	Tissues	References
95	N-*trans*-caffeoyl-tyramine	H	f	Aerial parts	Moradi-Afrapoli et al. (2012b)
96	Hirsutine	H	h	Herbs	Zhang et al. (2012c)
97	β-Sitosterol-3-*O*-acetate	I	o	Aerial parts	Abd El-Kader et al. (2013a,b)
98	β-Sitosterol	I	i,a	Herbs (i), roots (a)	Lin et al. (2009) and Xu et al. (2006)
99	β-Sitosterol-3-*O*-β-D-glucoside	I	a	Roots	Xu et al. (2006)
100	Physcion	J	a	Roots	Kim et al. (2008a,b) and Xu et al. (2006)
101	Emodin	J	a	Roots	Kim et al. (2008a,b)
102	Emodin-8-*O*-β-D-glucoside	J	a	Roots	Kim et al. (2008a,b)
103	Physcion-8-*O*-β-D-glucoside	J	a	Roots	Kim et al. (2008a,b) and Xu et al. (2006)
104	1,8-Dihydroxy-3,6-dimethoxy-xanthone-5-*O*-[α-L-rhamnopyranosyl-(1″ → 2′)]-β-D-glucopyranoside	K	o	Aerial parts	Abd El-Kader et al. (2013a,b)
105	17-Hydroxypentacosanyl acetate	L	o	Aerial parts	Abd El-Kader et al. (2013a,b)
106	Torachrysone-8-*O*-β-D-glucoside	M	a	Roots	Kim et al. (2008a,b)
107	Diisobutyl phthalate	N	g	Rhizomes	Xiang et al. (2012)

A, stilbene type (No. 1–14); B, flavonoid type (No. 15–58); C, drimane sesquiterpenoid type (No. 59–74); D, phenol type (No. 75–86); E, coumarin and isocoumarin types (No. 87–89); F, lignan type (No. 90–91); G, chromone type (No. 92–94); H, alkaloid type (No. 95–96); I, sterol type (No. 97–99); J, anthraquinone type (No. 100–103); K, xanthone type (No. 104); L, fatty alcohol acetate type (No. 105); M, naphthalene type (No. 106); N, ester type (No. 107).

a, *P. multiflorum* Thunb.; b, *P. cuspidatum* Sieb. et Zucc; c, *P. aviculare* L.; d, *P. amphibium* L.; e, *P. persicaria* L. (syn. *Persicaria maculosa* Gray); f, *P. hyrcanicum* Rech. f.; g, *P. amplexicaule* D. Don; h, *P. capitatum* Burch. –Ham. ex D. Don; i, *P. jucundum* Meisn.; j, *P. viscosum* Burch. –Ham. ex D. Don; k, *P. hydropiper* (k1, French *P. hydropiper*; k2, Bangladesh *P. hydropiper*; k3 Japanese *P. hydropiper*); l, *P. hydropiper* f. *purpurascens*; m, *P. minus* Huds.; n, *P. punctatum* var. *punctatum*; o, *P. bellardii*.

12.3 Bioactivities and therapeutic uses

12.3.1 Effects on cardiovascular system

Resveratrol (3,5,4′-trihydroxy-*trans*-stilbene) is a natural polyphenolic compound that exists in *P. cuspidatum*, grapes, peanuts, and berries, as well as their manufactured products. Resveratrol is a pharmacologically active compound that interacts with multiple targets in a variety of cardiovascular disease models to exert protective effects or induce a reduction in cardiovascular risk parameters (Tang et al., 2014). Endothelial hyperpermeability induced by hyperglycemia is the initial step in the development of atherosclerosis, one of the most serious cardiovascular complications in diabetes. Resveratrol ameliorates high-glucose-induced hyperpermeability mediated by caveolae via VEGF/KDR pathway (Tian et al., 2013). Polydatin, a resveratrol glucoside extracted from *P. cuspidatum*, attenuates cardiac hypertrophy through modulation of cardiac Ca^{2+} handling and calcineurin-NFAT signaling pathway (Ding et al., 2014).

2,3,4′,5-Tetrahydroxystilbene-2-*O*-β-D-glucoside (TSG), extracted from the root of *P. multiflorum* (He Shou Wu), had inhibitory effects on angiotensin II-induced proliferation of vascular smooth muscle cells (Xu et al., 2014a). Its antiproliferative effect might be associated with the downregulation of intracellular reactive oxygen species (ROS), followed by the suppression of the Src-MEK1/2-ERK1/2 signaling pathway, hence blocking cell cycle progression. TSG could prevent cardiac remodeling induced by pressure overload in rats (Xu et al., 2014b), which may be related to a decreasing angiotensin II level, an antioxidant effect, suppression of transforming growth factor-β1 expression, and inhibition of extracellular signal-regulated kinase 1/2 and p38 mitogen-activated protein kinase activation. Proteomic analysis was used to investigate the molecular events occurring in the atherosclerotic rats after TSG treatment (Yao et al., 2013). TSG treatment suppresses atherosclerosis by altering the expression of different proteins. Calreticulin, vimentin, HSP 70, lipocortin 1, and Apo A-I are key proteins that may be novel molecular targets responsible for atherogenesis suppression induced by TSG treatment.

Dysregulated tonic tension and calcium sensitization in blood vessels have frequently been observed in many cardiovascular diseases. Emodin of *P. multiflorum* inhibits tonic tension through suppressing PKCδ-mediated inhibition of myosin phosphatase in rat isolated thoracic aorta (Lim et al., 2014).

P. viviparum (PV) is a perennial herb and widely distributed in high-elevation mountain regions, such as the Tibetan Plateau. In Tibetan traditional medicine, PV is usually used to boost the blood circulation to dissipate blood stasis. PV induces vasorelaxation in the rat thoracic aorta via activation of nitric oxide (NO) synthase in endothelial cells (Chang et al., 2014).

2,3,4′,5-Tetrahydroxystilbene-2-*O*-β-D-glucoside (TSG) of *P. multiflorum* could raise the content of CYP (cytochrome p450) 7A and then promote the lipolysis of cholesterol (Wang et al., 2014a). TSG also showed the best LDL-reducing effect. Emodin could inhibit HMG-CoA reductase and DGAT1, which were key enzymes in the synthesis of TC and TG. Physcion increased the content of HTGL and then could boost the lipolysis of triglyceride. Physcion showed the best VLDL-reducing effect. The lipid regulation activity might be due to an overall synergy of TSG, emodin, and physcion. TSG attenuates human platelet aggregation, secretion, and spreading *in vitro* (Xiang et al., 2014).

Antiatherosclerotic effects of *P. aviculare* L. ethanol extract in ApoE knockout mice fed a Western diet are mediated via the MAPK pathway (Haeng Park et al., 2014).

Extracts of *P. persicaria* inhibit G protein-activated inwardly rectifying K^+ channels (Lajter et al., 2013a). The electrophysiologically active agents, not the new flavonoids from the chloroform extract, are among the minor compounds of HPLC eluates.

12.3.2 Anti-inflammatory activity

Resveratroloside and catechin-(4α→8)-catechin, the newly found constituents in the invasive variety, have similar NO inhibition potency as that of piceid (the major constituent of *P. cuspidatum*) (Fan et al., 2013b), but the newly found major constituent, that is, piceatannol glucoside, showed no apparent effect. The total content of resveratrol measured in the root extracts of the Swiss sample was about 2.5 times less than that of the Chinese one. When the invasive variety of *P. cuspidatum* is used in traditional medicine, the chemical difference should be kept in mind.

P. tinctorium, a traditional medicine used in China and Korea, improves 2, 4-dinitrofluorobenzene-induced atopic dermatitis-like lesional skin (Han et al., 2014). *P. tinctorium* inhibited IL-32 and TSLP production through the attenuation of caspase-1 activation in an animal model of allergic rhinitis (Jeong et al., 2014).

Flavonol glucuronides at physiologically achievable concentrations within the range of 0.5–10 μM significantly inhibited the production of ROS as well as elastase release in human neutrophils model and should be considered as responsible for the anti-inflammatory activity of the *P. aviculare* (Granica et al., 2013).

P. chinense aqueous leaf extract has gastroprotective activity on ethanol-induced hemorrhagic mucosal lesions in rats (Ismail et al., 2012). *P. viviparum* inhibits the lipopolysaccharide-induced inflammatory response in RAW264.7 macrophages through heme oxygenase-1 induction and activation of the Nrf2 pathway (Cheng et al., 2013).

12.3.3 Antimicrobial, antiparasitic, and insecticidal activities

Guizhou Miao medicine *P. capitatum* has antibacterial activity and can be used in bladder infection (Hu et al., 2014). Methanolic crude extract of aerial parts of *P. maritimum* showed a high antibacterial activity against gram-positive bacterial strains: *Bacillus cereus*, *Bacillus subtilis*, and *Staphylococcus aureus* with a highest MIC of 120 μg/ml (El-Haci et al., 2013). *P. orientale* extracts had protective effect against *Clavibacter michiganense* subsp. *sepedonicum*, the causal agent of bacterial ring rot of potato (Cai et al., 2013).

Ethanol extracts of *P. cuspidatum*, resveratrol, (+)-catechin, and emodin 8-*O*-β-D-glucopyranoside inhibit HIV-1-induced syncytium formation (Zhang et al., 2013b). An ethyl acetate subfraction separated from *P. cuspidatum* root and its major component, emodin, inhibited Epstein-Barr virus lytic cycle (Yiu et al., 2014). *In vitro* and *in vivo* studies suggested the inhibitory effects of emodin isolated from *P. cuspidatum* on Coxsackievirus B_4 (Liu et al., 2013). Resveratrol, (E)-3, 5, 12-trihydroxystilbene-3-*O*-beta-D-glucopyranoside-2′-(3″, 4″, 5″-trihydroxybenzoate), and catechin-3-*O*-gallate, isolated from *P. cuspidatum*, had inhibitory effect against neuraminidase (NA) activity (Chen et al., 2012), with IC50 values of 129.8, 44.8, and 21.3 μmol/l, respectively. (E)-3, 5, 12-trihydroxystilbene-3-*O*-beta-D-glucopyranoside-2′-(3″, 4″, 5″-trihydroxybenzoate) and catechin-3-*O*-gallate had significant inhibitory effect against H1N influenza virus (EC50 = 5.9, 0.9 μmol/l, respectively), with very low cytotoxicity to the host cells; their therapeutic selective index (SI) in MDCK cells ranged from 56 to 269. Emodin blocks the SARS coronavirus spike protein and angiotensin-converting enzyme 2 interaction (Ho et al., 2007).

P. chinense had antidiarrheal activity (Xiao et al., 2013). Ellagic acid and corilagin are two components contributing to this effect. Cinnamoylphenethyl amides from *P. hyrcanicum* possess antitrypanosomal activity (Moradi-Afrapoli et al., 2012b).

The essential oil of *P. hydropiper* exhibited LC50 values of 194.63 and 199.65 and confertifolin exhibited LC50 values of 2.02 and 3.16 against the second and fourth instar larvae of *Aedes albopictus* (dengue vector mosquito), respectively (Maheswaran and Ignacimuthu, 2014). The ovicidal activity of 100% on 0–6 h old eggs, repellent activity of 320.6 min, oviposition deterrent activity of 98.51%, and adulticidal activity of 100% at 10 ppm concentration of confertifolin were recorded.

12.3.4 Anticancer activity

Trans-piceid (T-Pc) is abundant in *P. cuspidatum* and in grapes and grape products such as wine. Piceid presents antiproliferative effects in intestinal epithelial Caco-2 cells, which is unrelated to resveratrol release (Storniolo et al., 2014). 2-Methoxystypandrone of *P. cuspidatum* inhibits signal transducer and activator of transcription 3 and nuclear factor-κB signaling by inhibiting Janus kinase 2 and IκB kinase (Kuang et al., 2014). Resveratrol inhibits invasion and metastasis of colorectal cancer cells via MALAT1-mediated Wnt/β-catenin signal pathway (Ji et al., 2013). Constitutively activated STAT3 plays a pivotal role in oncogenesis and metastasis in many human cancers, and STAT3 has been validated as a novel anticancer drug target. 2-Methoxystypandrone of *P. cuspidatum* demonstrated a potent inhibitory effect on STAT3 activation and significantly inhibited cell proliferation of human breast cancer cells (Liu et al., 2012b), especially those with constitutively activated STAT3 (IC50 2.7–3.1 μM). The SAR analysis of quinone analogs suggested that the phenolic and carbonyl groups are the key structures contributing to their inhibitory activities against the STAT3 signaling.

Aqueous and organic extracts of 27 selected species from five genera (*Fallopia*, *Oxyria*, *Persicaria*, *Polygonum*, and *Rumex*) of the family Polygonaceae occurring in the Carpathian Basin were screened *in vitro* for antiproliferative activity against HeLa (cervix epithelial adenocarcinoma), A431 (skin epidermoid carcinoma), and MCF7 (breast epithelial adenocarcinoma) cells (Lajter et al., 2013b), using the 3-(4,5-dimethylthiazol-2-yl)-2,5-diphenyltetrazolium bromide assay. A total of 196 *n*-hexane, chloroform, 50% methanol, and water extracts of different plant parts were investigated. Extracts of *P. hydropiper*, *Rumex acetosa*, *R. alpinus*, *R. aquaticus*,

R. scutatus, and *R. thyrsiflorus* at 10 or 30 μg/ml demonstrated substantial cell growth inhibitory activity (at least 50% inhibition of cell proliferation) against one or more cell lines. *R. acetosa* and *R. thyrsiflorus* proved to be the most active and are considered worthy of activity-guided phytochemical investigations.

3,5-Dihydroxybenzyl alcohol of *Reynoutria japonica* (*P. cuspidatum*) showed more potent inhibitory activity against topoisomerase I (IC50 4 μM) than camptothecin (Hwangbo et al., 2012), the positive control (IC50 18 μM). 3,5-Dihydroxybenzyl alcohol, citreorosein, *cis*-resveratrol, *trans*-resveratrol, and *trans*-resveratrol-5-*O*-β-D-glucopyranoside showed stronger inhibitory activities toward DNA topoisomerase II (IC50 0.54, 14, 15, 0.77, and 3 μM, respectively) than the positive control, etoposide (IC50 44 μM). Emodin and citreorosein displayed weak cytotoxicities against human lung cancer (A549), ovarian cancer (SK-OV-3), human liver hepatoblastoma (HepG2), and colon adenocarcinoma (HT-29) cell lines.

Flavonoid glucuronides of *P. amphibium* had antileukemic activity (Smolarz et al., 2008). A potential antitumor ellagitannin of *P. capitatum*, davidiin, inhibited hepatocellular tumor growth by targeting EZH2 (Wang et al., 2014b). The *n*-butanol extract of *P. bellardii* has the highest cytotoxicity in HeLa, MCF-7, and HepG-2 cells, with IC50 values of 15.26, 50.66, and 30.09 μg/ml, respectively (Abd El-Kader et al., 2013b). Myricetin-3-*O*-(5″-acetyl-α-arabinofuranoside) exhibited a marked cytotoxicity in HeLa (IC50 75.04 μg/ml) and HepG-2 (41.03 μg/ml) cells.

(−)-Loliolide of *P. aviculare* exerted inhibitory activity against cellular senescence in human dermal fibroblasts (Yang et al., 2014a). Quercetin-3-*O*-β-D-glucuronide isolated from *P. aviculare* inhibits cellular senescence in human primary cells (Yang et al., 2014b).

Apoptosis induction *by P. minus* is related to antioxidant capacity, alterations in expression of apoptotic-related genes, and S-phase cell cycle arrest in HepG2 cell line (Mohd Ghazali et al., 2014). Resveratrol-4-*O*-D-(2′-galloyl)-glucopyranoside isolated from *P. cuspidatum* exhibits anti-hepatocellular carcinoma viability by inducing apoptosis via the JNK and ERK pathway (Xie et al., 2014). Polydatin inhibits growth of lung cancer cells by inducing apoptosis and causing cell cycle arrest (Zhang et al., 2014).

8-Methyltryptanthrin of *P. tinctorium* induced differentiation of P19CL6 embryonal carcinoma cells into spontaneously beating cardiomyocyte-like cells (Seya et al., 2014).

12.3.5 Antioxidant activity

Polydatin from *P. cuspidatum* exhibited protective effects against carbon tetrachloride-induced liver injury in mice (Zhang et al., 2012a). *P. cuspidatum* contained many more contributing antioxidants other than resveratrol (Kurita et al., 2014).

Tetrahydroxystilbene glucoside of *P. multiflorum* exhibited protective effect against hydrogen peroxide-induced dysfunction and oxidative stress in osteoblastic MC3T3-E1 cells (Zhang et al., 2012b). Compared with *P. multiflorum* polysaccharide PMP-1, PMP-2 exhibited a much stronger antioxidant capacity against free radicals, lipid oxidation, and protein glycation (Lv et al., 2014). The IC_{50} values of PMP-2 were 0.47, 0.6, and 0.93 mg/ml for superoxide anion scavenging, hydroxyl radical scavenging, and hydroxyl peroxide scavenging, respectively. The inhibitory ability of PMP-2 on lipid oxidation marked in the rat liver, followed by the heart and kidney. PMP-2 also showed satisfactory suppression of AGEs formation.

The EtOAc fraction of *P. aviculare* displayed the highest content of flavonoids (sum, 208.9 mg/g) with the strongest peroxynitrite scavenging activity (IC50, 2.68 μg/ml) (Nugroho et al., 2014). The activities of the eight compounds (myricitrin, isoquercitrin, avicularin, quercitrin, myricetin, desmanthin-1, quercetin, and kaempferol) were comparable to that of the positive control (l-penicillamine; IC50: 1.03 μg/ml). The folkloric medicinal uses of *P. aviculare* are mainly attributed to flavonoids, such as particularly highly contained myricetin, myricitrin, and desmanthin-1.

Xanthone and lignan glycosides from the aerial parts of *P. bellardii* showed significant antioxidant potential by DPPH(·) scavenging activity test (Abd El-Kader et al., 2013a). 5, 6-Dihydropyranobenzopyrone, amplexicine, catechin, rutin, quercetin-3-*O*-β-D-galactopyranoside, chlorogenic acid, galloyl glucose, caffeic acid, gallic acid, and scopoletin, isolated from *P. amplexicaule*, exhibited considerable antioxidant activity in a DPPH radical scavenging assay (Tantry et al., 2012).

12.3.6 Effects on the nervous system

Resveratrol reverses the effects of chronic unpredictable mild stress on behavior, serum corticosterone levels, and BDNF expression in rats (Liu et al., 2014). Resveratrol attenuates oxidative damage and ameliorates cognitive impairment in the brain of senescence-accelerated mice (Liu et al., 2012a).

P. minus possesses antioxidant and anticholinesterase activity and demonstrated enhanced cognition *in vivo* (George et al., 2014). The data suggest neuroprotective properties of the extract. Hexane extracts of *P. multiflorum* improve tissue and functional outcome following focal cerebral ischemia in mice (Lee et al., 2014). Tetrahydroxystilbene glucoside of *P. multiflorum* attenuates neuroinflammation through the inhibition of microglia activation (Zhang et al., 2013a). 8-Hydroxycalamenene of *Reynoutria elliptica* (*P. ellipticum*) attenuated the cell death of transformed RGC-5 cells (Jo et al., 2013).

Neurons rely on the release and subsequent cleavage of GSH to cysteinylglycine (Cys-Gly) by astrocytes in order to maintain optimal intracellular GSH levels. In neurodegenerative diseases characterized by oxidative stress, neurons need an optimal GSH supply to defend themselves against free radicals released from activated microglia and astroglia. The rate of GSH synthesis is controlled largely by the activity of γ-glutamyl cysteine ligase. Expression of γ-glutamyl cysteine ligase and of the Xc-system, which facilitates cystine uptake, is regulated by the redox-sensitive transcription factor, nuclear factor erythroid-2-related factor 2 (Nrf2). Resveratrol and *P. multiflorum* were all identified as potent Nrf2 activators and "GSH and Cys-Gly boosters" (Steele et al., 2013).

The antidepressant-like effects of resveratrol in the FST (forced swim test) and TST (tail suspension test) are mediated, at least in part, by modulating hypothalamic-pituitary-adrenal axis, BDNF expression, and ERK phosphorylation in the brain region of mice (Wang et al., 2013a).

12.3.7 Other activities

The antimelanogenic activity of 2,3,5,4′-tetrahydroxystilbene-2-*O*-β-D-glucopyranoside, isolated from *P. multiflorum*, is likely mediated through a noncompetitive inhibition on tyrosinase, downregulation of the expression of melanogenic proteins, and reduction of tyrosinase/TRP-1 complex formation (Cheung et al., 2014).

Supplementation for 12 weeks with *P. minus* and the proprietary *Eurycoma longifolia* extract, Physta, was well tolerated and more effective than placebo in enhancing sexual performance in healthy volunteers (Udani et al., 2014). *P. cuspidatum* inhibits pancreatic lipase activity and adipogenesis via attenuation of lipid accumulation (Kim et al., 2013), thus having antiobesity effect.

Constituents from the root of *P. multiflorum* have promotion effect on hair growth (Sun et al., 2013).

The semialcoholic extract of *P. senegalensis* possesses α-glucosidase inhibitory activity and antioxidant potency (Bothon et al., 2013). Proanthocyanidins, isolated from stems of *P. multiflorum*, strongly inhibit α-amylase with an acarbose equivalence (AE) value of 1954.7 μmol AE/g and inhibit α-glucosidase with an AE value of 211.1 μmol AE/g (Wang et al., 2013b). They have potential as functional ingredients in reducing postprandial hyperglycemia. Phenolic constituents from aerial parts of *P. hyrcanicum* had *in vitro* α-glucosidase inhibitory activity (Moradi-Afrapoli et al., 2012a).

12.3.8 Toxicity and safety

Adverse reactions induced by *P. multiflorum* are common, with patients developing drug-induced liver injury (DILI) and even liver failure (Ma et al., 2014). Although the mechanism is unknown, many studies have suggested that an idiosyncratic reaction occurs, which is related to genetic polymorphisms of CYP1A2. The influence of stilbene glucoside on the pharmacokinetics of emodin may be attributed to the inhibition of UGT1A8 mRNA expression (Ma et al., 2013). P. cuspidatum markedly increased the systemic exposure and brain concentration of carbamazepine (CBZ) and CBZE through inhibiting the activities of CYP 3A and MRP 2 (Chi et al., 2012).

12.4 Molecular biology and genomics

12.4.1 Molecular marker

Molecular markers can help elucidate how neutral evolutionary forces and introduction history contribute to genetic variation in invaders. Genetic diversity, population structure, and colonization patterns in the invasive *P. cespitosum*, a highly selfing, tetraploid Asian annual introduced to North America, were examined using nine diploidized polymorphic microsatellite markers (SSR; Matesanz et al., 2014). Low heterozygosity was found in all 16 populations, consistent with the selfing mating system of *P. cespitosum*. Despite the high selfing levels, substantial genetic variation within and among *P. cespitosum* populations was revealed, based on the percentage of polymorphic loci, allelic richness, and expected heterozygosity. Inferences from individual assignment tests and pairwise FST values indicated high among-population differentiation, which indicates that the effects of gene flow are limited relative to those of genetic drift, probably due to the high selfing rates and the limited seed dispersal ability of *P. cespitosum*. Population structure appears to be the result of the random movement of propagules across the introduced range, possibly associated with human dispersal. The high population differentiation, genetic diversity, and fine-scale genetic structure (populations founded by individuals from different genetic sources) in the introduced range suggest that multiple introductions to this region may have occurred. High genetic diversity may further contribute to the invasive success of *P. cespitosum* in its introduced range.

Among 50 RAPD primers, only C29 primer had two specific bands, which could distinguish *P. capitatum* from *P. nepalense* (Zhou et al., 2013). Four pairs of specific primers were designed based on the two sequences of RAPD marker bands, and only one pair primer (Z1-2) was successfully converted into SCAR marker, which could be used as an effective SCAR mark to identify Z300 DNA for *P. capitatum*.

AFLP was used to identify the knotgrass (*P. aviculare*) population at the crime site as the most likely origin of the botanical evidence (Koopman et al., 2012). The genetic diversity of 48 representational populations of *P. capitatum* including 240 individuals was investigated by ISSR marker (Zhou et al., 2010). SRAP (sequence-related amplified polymorphism) was used to detect the polymorphisms of *Radix Polygoni Multiflori* in Chongqing (Cheng et al., 2013). SRAP might be superior to RAPD in genetic diversity studies.

Analysis of nuclear internal transcribed spacer (nrITS) sequences reveals polymorphism among five *P. barbata* populations belonging to five geographic locations in India (Choudhary et al., 2011). UPGMA analysis based on the ITS datasets shows that the sampled populations are grouped according to their geographic locations and are supposed to evolve under reproductive isolation, which most likely are due to the long-distance distribution and population fragmentation.

DNA bar codes are used to discriminate the Polygonaceae in China Pharmacopoeia and their adulterants (Song et al., 2009). The amplification efficiency of six candidate DNA bar codes (rbcL, trnH-psbA, ndhJ, rpoB, rpoC1, and accD) was 100%, while the efficiency of YCF5 and nrITS was 56% and 44%, respectively. The interspecific divergence was highest for the trnH-psbA (20.05%), followed by the nrITS (14.01%) across all species pairs, while intraspecific variation both within populations and between populations was absent. The trnH-psbA can not only distinguish 10 species of Polygonaceae in China Pharmacopoeia but also recognize eight other species of Polygonaceae including their adulterants.

P. multiflorum (*Fallopia multiflora*) is often confused and substituted with the roots of *Fallopia multiflora* var. *ciliinervis*, *Pteroxygonum giraldii*, *Cynanchum auriculatum*, and *Stephania cepharantha*. The nrITS regions of six *Fallopia* species were sequenced and analyzed (Zheng et al., 2009). The diagnostic primers PMITS28 and PMITS545, which amplified an expected 517 bp ITS fragment from *F. multiflora*, were designed. No amplified product was observed when other species was used. Based on

18S rRNA gene sequence analysis, *P. multiflorum* could be easily distinguished from adulterants and other herbs with similar medicinal components (Yan et al., 2008). All *P. multiflorum* samples were divided into four clades based on matK sequence. Permutations of matK were related to the geographic distributions of the samples. These markers can be used to authenticate the botanical origin of *P. multiflorum*. The best bar code of *P. multiflorum* is psbA-trnH, with significant interpopulation variability (Sun et al., 2013). The combination of loci gave better performance for distinguishing populations than a single locus. It could be good to use matK + rbcL + psbA-trnH + ITS2 or psbA-trnH alone for *P. multiflorum* in geoherbalism identification.

12.4.2 Transcriptome

cDNA-amplified fragment length polymorphism (cDNA-AFLP) transcript profiling was applied to generate the expression profiles of *P. minus* in response to salicylic acid (SA) and methyl jasmonate (MeJA) elicitations (Ee et al., 2013). Two sets of genes were induced by SA and MeJA, respectively. Stress-related genes were proved to lead to the expression of genes involved in secondary metabolite biosynthetic pathways. A total of 98 transcript-derived fragments (TDFs) were upregulated, including 46 from SA-treated and 52 from MeJA-treated samples. The cDNA-AFLP transcripts generated using 64 different Mse1/Taq1 primer combinations showed that treatments with SA and MeJA induced genes mostly involved in scavenging ROS, including zeaxanthin epoxidase, cytosolic ascorbate peroxidase 1, and peroxidase. Of these stress-related genes, 15% of other annotated TDFs are involved mainly in secondary metabolic processes, and two genes encoding (+)-delta cadinene synthase and cinnamoyl-CoA reductase were highlighted.

The leaf of an aromatic plant *P. minus* is widely used as a food additive and in the perfume industry. The leaf also accumulates secondary metabolites such as flavonoid. A standard cDNA library of *P. minus* leaves was constructed, and two normalized full-length enriched cDNA libraries were constructed from stem and root in order to create a gene resource for the biosynthesis of secondary metabolites, especially flavonoid biosynthesis (Roslan et al., 2012). Large-scale sequencing of *P. minus* cDNA libraries identified 4196 expressed sequence tags (ESTs). From the three cDNA libraries, 11 ESTs encoding seven genes were mapped to the flavonoid biosynthetic pathway. Three flavonoid biosynthetic pathway-related ESTs, namely, chalcone synthase, CHS (JG745304); flavonol synthase, FLS (JG705819; Figure 12.2); and leucoanthocyanidin dioxygenase, LDOX (JG745247), were selected for further examination by quantitative RT-PCR (qRT-PCR) in different *P. minus* organs. Expression was detected in leaf, stem, and root. Based on transcriptome data, gene expression studies can be initiated to better understand the underlying physiological processes.

Various active components have been extracted from the root of *P. cuspidatum*. However, the genetic basis for their activity is little known. 25,600,002 short reads (2.3 Gb) of *P. cuspidatum* root transcriptome were obtained via Illumina HiSeq 2000 sequencing (Hao et al., 2012). A total of 86,418 unigenes were assembled de novo and annotated. 12, 18, 60, and 54 unigenes were, respectively, mapped to the mevalonic acid (MVA), methyl-D-erythritol 4-phosphate (MEP), shikimate (Figure 12.3), and resveratrol biosynthesis pathways, suggesting that they are

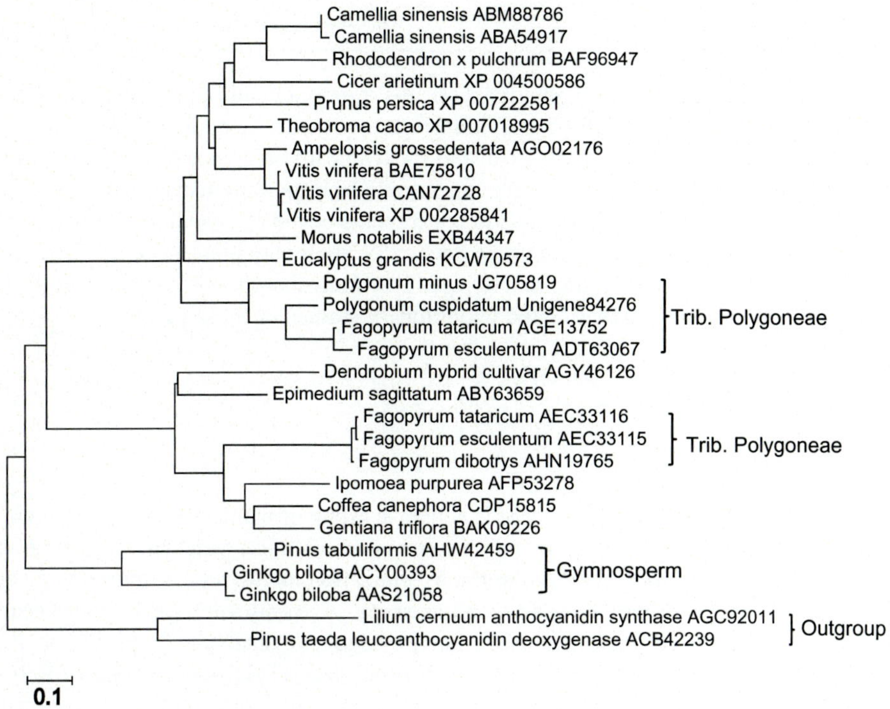

Figure 12.2 Phylogenetic relationship of plant flavonol synthases. The evolutionary history was inferred by using the ML method and the JTT+I (13.26% sites)+G model. The tree is drawn to scale, with branch lengths measured in the number of substitutions per site. The analysis involved 29 amino acid sequences, including a *P. cuspidatum* sequence predicted from transcriptome Unigene 84276 (Hao et al., 2012) and a *P. minus* sequence predicted from EST (Roslan et al., 2012). There were a total of 365 positions in the final dataset. Evolutionary analyses were conducted in MEGA6.

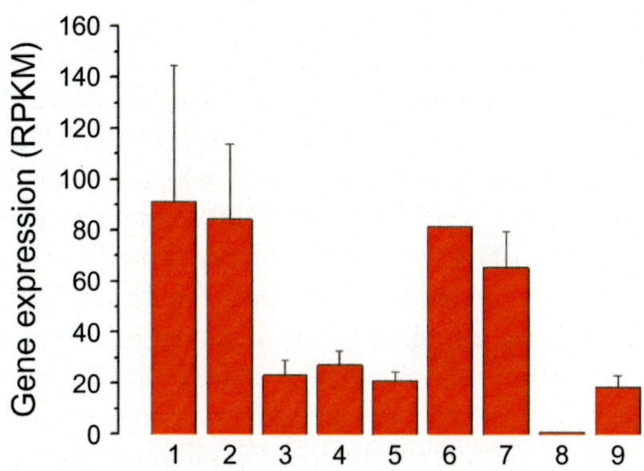

Figure 12.3 Expression levels (RPKM value) of the *P. cuspidatum* unigene in shikimate metabolic pathway (Hao et al., 2012), which is responsible for biosynthesis of anthraquinone and stilbene. 1, 3-deoxy-D-arabino-heptulosonate-7-phosphate (DAHP) synthase; 2, 3-dehydroquinate synthase; 3, shikimate 5-dehydrogenase; 4, shikimate: NADP oxidoreductase; 5, shikimate kinase; 6, EPSP synthase; 7, chorismate synthase; 8, isochorismate synthase; 9, chorismate mutase. Bars represent the standard error of the average.

involved in the biosynthesis of pharmaceutically important anthraquinone and resveratrol. Eighteen potential UDP-glycosyltransferase unigenes were identified as the candidates most likely to be involved in the biosynthesis of glycosides of secondary metabolites. Identification of relevant genes could be important in eventually increasing the yields of the medicinally useful constituents of the *P. cuspidatum* root. From the previously published transcriptome data of 19 nonmodel plant taxa, 1127 shared orthologs were identified and characterized. This information will be very useful for future functional, phylogenetic, and evolutionary studies of these plants.

12.5 Phylogeny

The buckwheat family Polygonaceae is a diverse group of plants and is a good model for investigating biogeography, breeding systems, coevolution with symbionts such as ants and fungi, functional trait evolution, hybridization, invasiveness, morphological plasticity, pollen morphology, and wood anatomy. Age estimates for Polygonaceae were obtained by calibrating a Bayesian phylogenetic analysis (Schuster et al., 2013), using a relaxed molecular clock with fossil data. Eighty-one species of Polygonaceae were analyzed with MrBayes to infer species relationships. One nuclear (nrITS) and three chloroplast (cp) markers (the trnL-F spacer region and matK and ndhF genes) were used. Seven calibration points including fossil pollen and a leaf fossil of *Muehlenbeckia* (a Southern Hemisphere group) were used to infer node ages. Results of the Beast analyses indicate an age of 110.9/118.7 million years (My) with an uncertainty interval of 90.7–125.0 My for the stem age of Polygonaceae. This age is older than previously thought (approximately 65.5–70.6 My). The estimated divergence time for *Muehlenbeckia* is 41.0/41.6 (39.6–47.8) My and its crown clade is 20.5/22.3 (14.2–33.5) My. Because the breakup of Gondwana occurred from 95 to 30 My ago, diversification of *Muehlenbeckia* is best explained by oceanic long-distance and maybe stepping-stone dispersal rather than vicariance.

Interspecific hybridization and the following polyploidization play a major role in plant diversification, but quantifying the contribution of this mechanism to diversification within taxonomically complex clades remains difficult. Incongruence among gene trees can provide critical insights, especially when combined with data on chromosome numbers, morphology, and geography. Molecular phylogenetic studies using three cpDNA regions and nrITS sequences were performed to explore the hybrid speciation in *Persicaria* (*Polygonum*, Polygonaceae; Figure 12.4) (Kim and Donoghue, 2008), with an emphasis on sampling within section *Eupersicaria*. There are major conflicts between the combined cpDNA tree and the nrITS tree; a variety of incongruence tests rejected stochastic error as the cause of incongruence in most cases. Both the tree incongruence results and information on chromosome numbers suggest that the origin of 10 polyploid species involved interspecific hybridization. The recognition of several previously named species that have been treated as belonging within other species was supported. Repeated allotetraploidy (as distinct from radiation at the

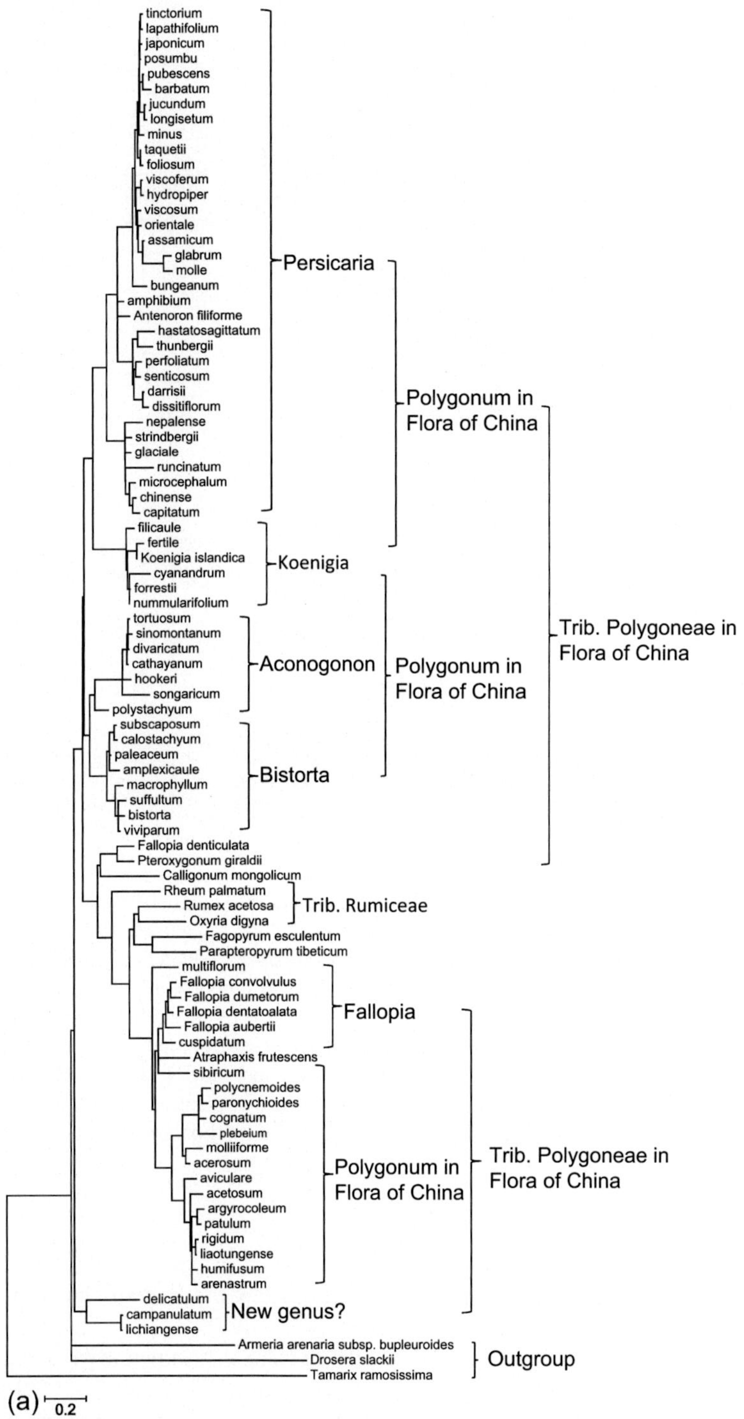

Figure 12.4 Phylogenetic relationship of *Polygonum* and related groups. A, ITS tree inferred from maximum likelihood (ML) method based on GTR+I (10.90% sites)+G model.

(Continued)

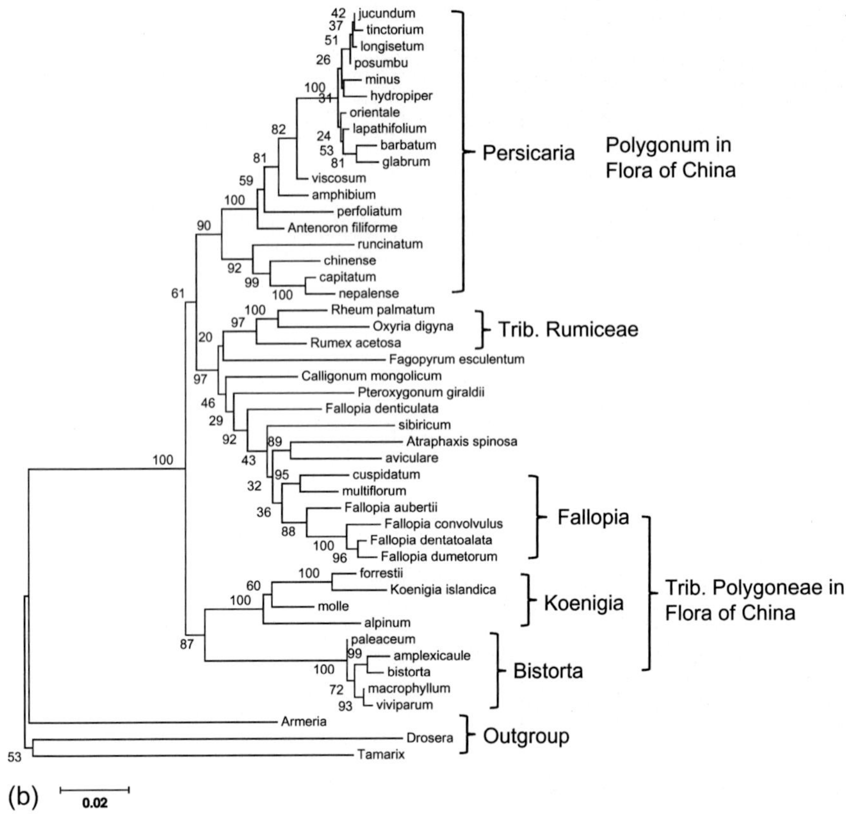

Figure 12.4 Continued. The tree is drawn to scale, with branch lengths measured in the number of substitutions per site. The analysis involved 91 nucleotide sequences. There were a total of 1006 positions in the final dataset. Evolutionary analyses were conducted in MEGA6 (Tamura et al., 2013). B, matK+rbcL+trnL-F tree inferred from neighbor joining method. The tree is drawn to scale, with branch lengths in the same units as those of the evolutionary distances used to infer the phylogenetic tree. The evolutionary distances were computed using the maximum composite likelihood method and are in the units of the number of base substitutions per site. The analysis involved 46 nucleotide sequences. All ambiguous positions were removed for each sequence pair. There were a total of 4645 positions in the final dataset.

tetraploid level) might be the key mechanism governing the diversification of this taxonomically challenging group.

Multiple instances of allopolyploid speciation were also shown in *Persicaria* using a low-copy nuclear gene region (LEAFY second intron) (Kim et al., 2008b), which belong to *Polygonum* in Flora of China. Fifteen species seem to be allopolyploids, which is higher than the number found in previous comparisons of cp DNA and nrITS phylogenies (Kim and Donoghue, 2008). This underestimation of the extent of allopolyploidy is due in at least three cases to homogenization of nrITS toward the maternal lineage. The diploid species, *P. lapathifolia*, has been involved in at least six cases of allopolyploid speciation. Of the diploids, this species is the most widespread geographically and ecologically and also bears more numerous and conspicuous flowers, illustrating ecological factors that may influence hybridization frequency. With a few exceptions, the allopolyploid species also are widespread, plastic, ecological generalists. Hybridization events fostered by human introductions may be fueling the production of new species that have the potential to become aggressive weeds.

ITS sequences from 44 Indian *Polygonum* taxa were examined to investigate relationships among various sections proposed previously (Choudhary et al., 2012). The maximum parsimony trees obtained from ITS sequences suggested eight major groups of the Indian *Polygonum* spp. The relationships among different sections were largely congruent with those inferred from morphological characters as described by Hooker. The treatment of the *Persicaria* suggested by Haraldson on the basis of anatomical characters proved to be nearly in line with that based on ITS data. A high resolution of phylogeny of the Himalayan *Polygonum* (e.g., *P. microcephalum*, *P. assamicum*, *P. recumbens*, and *P. effusum*) was provided and merger of the section *Amblygonon* in the section *Persicaria* was supported. Molecular differences were detected among *Persicaria barbata* collected from different geographic locations of India, although these were not differentiated at the morphological level.

To examine the phylogenetic relationships of *Koenigia* (Polygonaceae), 43 samples representing all species of *Koenigia* and closely related taxa (*Aconogonon*, *Bistorta*, and *Persicaria*) were sequenced for nr ITS and four cp regions (trnL-F, atpB-rbcL, rbcL, and rpl32-trnL(UAG); Fan et al., 2013a,b). Trib. Polygoneae and trib. Rumiceae are recovered on both cp and ITS trees (Figure 12.4), while trib. Atraphaxideae is not. The placement of *P. bistorta* is uncertain due to conflict between cp and ITS trees (Figure 12.4). It was proposed that the genus *Koenigia* be circumscribed to include five species, that is, perennials (*K. forrestii* and *K. nummularifolia*) and annuals (*K. islandica*, *K. pilosa*, and *K. nepalensis*). However, on both cp and ITS trees, *K. nepalensis* (*P. nepalense*) is more closely related to *P. capitatum* and *P. chinense* (Figure 12.4), instead of *Koenigia*. *K. islandica* and *K. fertilis* (*P. fertile*, Figure 12.4a), both of which are from the Himalayan region, can be merged into a single species. *P. delicatulum* (*K. delicatula*), *P. campanulatum*, and *P. lichiangense* occupy an isolated position at the base of the Polygonaceae (Figure 12.4a), which might be reassigned to a new genus.

The tetraploid *P. minor* (*P. minus*), which is sister to the *P. hydropiperoides* complex in the nrITS tree (Kim and Donoghue, 2008), may have arisen through hybridization between an unknown diploid lineage or possibly a tetraploid in the *P. hydropiperoides* complex and the diploid *P. hydropiper*. It is not entirely clear that *P. hydropiper* served as the maternal parent, since the relationship between *P. hydropiper* and *P. minor* is weakly supported in the cpDNA tree (Figure 12.4b). Morphologically, *P. minor* is more similar to the diploid *P. foliosa* (*P. foliosum*), which should also be considered as a possible diploid maternal lineage for *P. minor* (Kim and Donoghue, 2008).

The pollen apertures likely evolved in parallel in the *Aconogonon-Koenigia-Bistorta* clade and *Persicaria* clade (Fan et al., 2013a,b), and tricolpate pollen is probably the ancestral one. Quincuncial aestivation likely evolved during the early evolution of *Koenigia* and its close relatives. The uplift of the Himalayas has played a vital role in promoting species diversification of *Koenigia*. *K. islandica* expanded its range during Pleistocene glacial cycles by tracking changes in newly available habitats.

12.6 Conclusion

There is little study addressing chemotaxonomy and metabolomics of Polygonum and related genera, although such studies can provide vital information for sustainable utilization of Polygonaceae medicinal resources. The family Polygonaceae has been supported as monophyletic, but the circumscription of some subgroups, especially *Polygonum* and related taxa, has been controversial. The use of population genetic approaches, sampling from multiple populations, would shed light on this issue. More extensive sampling inside and outside the Himalayas and Hengduan Mountains is indispensable to explore the phylogenetic hypothesis and facilitate species diversity conservation.

References

Abd El-Kader, A.M., Ahmed, A.S., Nafady, A.M., Ibraheim, Z.Z., 2013a. Xanthone and lignan glycosides from the aerial parts of *Polygonum bellardii* all growing in Egypt. Pharmacogn. Mag. 9 (34), 135–143.

Abd El-Kader, A.M., El-Readi, M.Z., Ahmed, A.S., et al., 2013b. Polyphenols from aerial parts of *Polygonum bellardii* and their biological activities. Pharm. Biol. 51 (8), 1026–1034.

Asakawa, Y., Ludwiczuk, A., Harinantenaina, L., et al., 2012. Distribution of drimane sesquiterpenoids and tocopherols in liverworts, ferns and higher plants: Polygonaceae, Canellaceae and Winteraceae species. Nat. Prod. Commun. 7 (6), 685–692.

Bothon, F.T., Debiton, E., Avlessi, F., et al., 2013. In vitro biological effects of two anti-diabetic medicinal plants used in Benin as folk medicine. BMC Complement. Altern. Med. 13, 51.

Cai, J., Xie, S., Feng, J., et al., 2013. Protective effect of *Polygonum orientale* L. extracts against *Clavibater michiganense* subsp. *sepedonicum*, the causal agent of bacterial ring rot of potato. PLoS One 8 (7), e68480.

Chang, M.L., Chang, J.S., Yu, W.Y., et al., 2014. *Polygonum viviparum* L. induces vasorelaxation in the rat thoracic aorta via activation of nitric oxide synthase in endothelial cells. BMC Complement. Altern. Med. 14, 150.

Chen, K.T., Zhou, W.L., Liu, J.W., et al., 2012. Active neuraminidase constituents of *Polygonum cuspidatum* against influenza A(H1N1) influenza virus. Zhongguo Zhong Yao Za Zhi 37 (20), 3068–3073.

Cheng, H.W., Lee, K.C., Cheah, K.P., et al., 2013. *Polygonum viviparum* L. inhibits the lipopolysaccharide-induced inflammatory response in RAW264.7 macrophages through haem oxygenase-1 induction and activation of the Nrf2 pathway. J. Sci. Food Agric. 93 (3), 491–497.

Cheung, F.W., Leung, A.W., Liu, W.K., Che, C.T., 2014. Tyrosinase inhibitory activity of a glucosylated hydroxystilbene in mouse melan-a melanocytes. J. Nat. Prod. 77 (6), 1270–1274.

Chi, Y.C., Lin, S.P., Hou, Y.C., 2012. A new herb-drug interaction of *Polygonum cuspidatum*, a resveratrol-rich nutraceutical, with carbamazepine in rats. Toxicol. Appl. Pharmacol. 263 (3), 315–322.

Choudhary, R.K., Ajmal Ali, M., Lee, J., 2011. Studies on genetic diversity among populations of *Persicaria barbata* (L.) H. Hara from India based on internal transcribed spacer sequences of nuclear ribosomal DNA. Saudi J. Biol. Sci. 18 (2), 123–127.

Choudhary, R.K., Park, S.H., Lee, J., 2012. Phylogeny and systematics of Indian *Polygonum* sensu lato in the subfamily Polygonoideae based on ITS sequences of nuclear ribosomal DNA. Genet. Mol. Res. 11 (4), 4370–4382.

Datta, B.K., Datta, S.K., Sarker, S.D., 2000. Quercetin 3-*O*-(6″-galloyl)-beta-D-galactoside from *Polygonum viscosum* (Polygonaceae). Biochem. Syst. Ecol. 28 (8), 805–807.

Ding, W., Dong, M., Deng, J., et al., 2014. Polydatin attenuates cardiac hypertrophy through modulation of cardiac Ca^{2+} handling and calcineurin-NFAT signaling pathway. Am. J. Physiol. Heart Circ. Physiol. 307 (5), H792–H802.

Ee, S.F., Oh, J.M., Mohd Noor, N., et al., 2013. Transcriptome profiling of genes induced by salicylic acid and methyl jasmonate in *Polygonum minus*. Mol. Biol. Rep. 40 (3), 2231–2241.

El-Haci, I.A., Bekkara, F.A., Mazari, W., et al., 2013. Screening of biological activities of *Polygonum maritimum* L. from Algerian coast. Asian Pac. J. Trop. Biomed. 3 (8), 611–616.

Fan, D.M., Chen, J.H., Meng, Y., et al., 2013a. Molecular phylogeny of *Koenigia* L. (Polygonaceae: Persicarieae): implications for classification, character evolution and biogeography. Mol. Phylogenet. Evol. 69 (3), 1093–1100.

Fan, P., Zhang, T., Hostettmann, K., 2013b. Anti-inflammatory activity of the invasive neophyte *Polygonum cuspidatum* Sieb. and Zucc. (Polygonaceae) and the chemical comparison of the invasive and native varieties with regard to resveratrol. J. Tradit. Complement. Med. 3 (3), 182–187.

George, A., Ng, C.P., O'Callaghan, M., et al., 2014. In vitro and ex-vivo cellular antioxidant protection and cognitive enhancing effects of an extract of *Polygonum minus* Huds (Lineminus™) demonstrated in a Barnes Maze animal model for memory and learning. BMC Complement. Altern. Med. 14, 161.

Granica, S., Czerwińska, M.E., Zyżyńska-Granica, B., Kiss, A.K., 2013. Antioxidant and anti-inflammatory flavonol glucuronides from *Polygonum aviculare* L. Fitoterapia 91, 180–188.

Haeng Park, S., Sung, Y.Y., Jin Nho, K., Kyoung Kim, H., 2014. Anti-atherosclerotic effects of *Polygonum aviculare* L. ethanol extract in ApoE knock-out mice fed a Western diet mediated via the MAPK pathway. J. Ethnopharmacol. 151 (3), 1109–1115.

Han, N.R., Kang, S.W., Moon, P.D., et al., 2014. Genuine traditional Korean medicine, Naju Jjok (Chung-Dae, *Polygonum tinctorium*) improves 2,4-dinitrofluorobenzene-induced atopic dermatitis-like lesional skin. Phytomedicine 21 (4), 453–460.

Hao, D.C., Ma, P., Mu, J., Chen, S., Xiao, P.G., et al., 2012. De novo characterization of the root transcriptome of a traditional Chinese medicinal plant *Polygonum cuspidatum*. Sci. China Life Sci. 55 (5), 452–466.

Ho, T.Y., Wu, S.L., Chen, J.C., et al., 2007. Emodin blocks the SARS coronavirus spike protein and angiotensin-converting enzyme 2 interaction. Antiviral Res. 74 (2), 92–101.

Hu, N., Tang, S.H., Cheng, L., 2014. Prescription parsing of miao medicine *Polygonum capitatum* and kelintong capsule. Zhongguo Zhong Yao Za Zhi 39 (7), 1318–1320.

Hwangbo, K., Zheng, M.S., Kim, Y.J., et al., 2012. Inhibition of DNA topoisomerases I and II of compounds from *Reynoutria japonica*. Arch. Pharm. Res. 35 (9), 1583–1589.
Ismail, I.F., Golbabapour, S., Hassandarvish, P., et al., 2012. Gastroprotective activity of *Polygonum chinense* aqueous leaf extract on ethanol-induced hemorrhagic mucosal lesions in rats. Evid. Based Complement. Alternat. Med. 2012, 404012.
Jeong, H.J., Oh, H.A., Lee, B.J., Kim, H.M., 2014. Inhibition of IL-32 and TSLP production through the attenuation of caspase-1 activation in an animal model of allergic rhinitis by Naju Jjok (*Polygonum tinctorium*). Int. J. Mol. Med. 33 (1), 142–150.
Ji, Q., Liu, X., Fu, X., et al., 2013. Resveratrol inhibits invasion and metastasis of colorectal cancer cells via MALAT1 mediated Wnt/β-catenin signal pathway. PLoS One 8 (11), e78700.
Jo, H., Lee, H.J., Kim, C.Y., et al., 2013. 8-Hydroxycalamenene isolated from the rhizomes of *Reynoutria elliptica* exerts neuroprotective effects both in vitro and in vivo. Food Chem. Toxicol. 51, 231–241.
Kim, S.T., Donoghue, M.J., 2008. Incongruence between cpDNA and nrITS trees indicates extensive hybridization within *Eupersicaria* (Polygonaceae). Am. J. Bot. 95 (9), 1122–1135.
Kim, H.K., Choi, Y.H., Choi, J.S., et al., 2008a. A new stilbene glucoside gallate from the roots of *Polygonum multiflorum*. Arch. Pharm. Res. 31 (10), 1225–1229.
Kim, S.T., Sultan, S.E., Donoghue, M.J., 2008b. Allopolyploid speciation in *Persicaria* (Polygonaceae): insights from a low-copy nuclear region. Proc. Natl. Acad. Sci. U. S. A. 105 (34), 12370–12375.
Kim, Y.S., Lee, Y.M., Kim, J.H., Kim, J.S., 2013. *Polygonum cuspidatum* inhibits pancreatic lipase activity and adipogenesis via attenuation of lipid accumulation. BMC Complement. Altern. Med. 13, 282.
Koopman, W.J., Kuiper, I., Klein-Geltink, D.J., et al., 2012. Botanical DNA evidence in criminal cases: knotgrass (*Polygonum aviculare* L.) as a model species. Forensic Sci. Int. Genet. 6 (3), 366–374.
Kuang, S., Qi, C., Liu, J., et al., 2014. 2-Methoxystypandrone inhibits signal transducer and activator of transcription 3 and nuclear factor-κB signaling by inhibiting Janus kinase 2 and IκB kinase. Cancer Sci. 105 (4), 473–480.
Kurita, S., Kashiwagi, T., Ebisu, T., et al., 2014. Content of resveratrol and glycoside and its contribution to the antioxidative capacity of *Polygonum cuspidatum* (Itadori) harvested in Kochi. Biosci. Biotechnol. Biochem. 78 (3), 499–502.
Lajter, I., Vasas, A., Orvos, P., et al., 2013a. Inhibition of G protein-activated inwardly rectifying K^+ channels by extracts of *Polygonum persicaria* and isolation of new flavonoids from the chloroform extract of the herb. Planta Med. 79 (18), 1736–1741.
Lajter, I., Zupkó, I., Molnár, J., et al., 2013b. Antiproliferative activity of polygonaceae species from the Carpathian Basin against human cancer cell lines. Phytother. Res. 27 (1), 77–85.
Lee, S.V., Choi, K.H., Choi, Y.W., et al., 2014. Hexane extracts of *Polygonum multiflorum* improve tissue and functional outcome following focal cerebral ischemia in mice. Mol. Med. Rep. 9 (4), 1415–1421.
Li, F., Zhan, Z., Liu, F., et al., 2013a. Polyflavanostilbene A, a new flavanol-fused stilbene glycoside from *Polygonum cuspidatum*. Org. Lett. 15 (3), 674–677.
Li, S.G., Chen, L.L., Huang, X.J., et al., 2013b. Five new stilbene glycosides from the roots of *Polygonum multiflorum*. J. Asian Nat. Prod. Res. 15 (11), 1145–1151.
Lim, K.M., Kwon, J.H., Kim, K., et al., 2014. Emodin inhibits tonic tension through suppressing PKCδ-mediated inhibition of myosin phosphatase in rat isolated thoracic aorta. Br. J. Pharmacol. 171 (18), 4300–4310.

Lin, Y., Zhang, C., Zhang, M., 2009. Chemical constituents in herbs of *Polygonum jucundum*. Zhongguo Zhong Yao Za Zhi 34 (13), 1690–1691.

Liu, G.S., Zhang, Z.S., Yang, B., He, W., 2012a. Resveratrol attenuates oxidative damage and ameliorates cognitive impairment in the brain of senescence-accelerated mice. Life Sci. 91 (17–18), 872–877.

Liu, J., Zhang, Q., Chen, K., et al., 2012b. Small-molecule STAT3 signaling pathway modulators from *Polygonum cuspidatum*. Planta Med. 78 (14), 1568–1570.

Liu, Z., Wei, F., Chen, L.J., et al., 2013. In vitro and in vivo studies of the inhibitory effects of emodin isolated from *Polygonum cuspidatum* on Coxsackievirus B_4. Molecules 18 (10), 11842–11858.

Liu, D., Xie, K., Yang, X., et al., 2014. Resveratrol reverses the effects of chronic unpredictable mild stress on behavior, serum corticosterone levels and BDNF expression in rats. Behav. Brain Res. 264, 9–16.

Lv, L., Cheng, Y., Zheng, T., et al., 2014. Purification, antioxidant activity and antiglycation of polysaccharides from *Polygonum multiflorum* Thunb. Carbohydr. Polym. 99, 765–773.

Ma, J., Zheng, L., Deng, T., et al., 2013. Stilbene glucoside inhibits the glucuronidation of emodin in rats through the down-regulation of UDP-glucuronosyltransferases 1A8: application to a drug-drug interaction study in *Radix Polygoni Multiflori*. J. Ethnopharmacol. 147 (2), 335–340.

Ma, K.F., Zhang, X.G., Jia, H.Y., 2014. CYP1A2 polymorphism in Chinese patients with acute liver injury induced by *Polygonum multiflorum*. Genet. Mol. Res. 13 (3), 5637–5643.

Maheswaran, R., Ignacimuthu, S., 2014. Effect of *Polygonum hydropiper* L. against dengue vector mosquito *Aedes albopictus* L. Parasitol. Res. 113 (9), 3143–3150.

Matesanz, S., Theiss, K.E., Holsinger, K.E., Sultan, S.E., 2014. Genetic diversity and population structure in *Polygonum cespitosum*: insights to an ongoing plant invasion. PLoS One 9 (4), e93217.

Mohd Ghazali, M.A., Al-Naqeb, G., Krishnan Selvarajan, K., et al., 2014. Apoptosis induction by *Polygonum minus* is related to antioxidant capacity, alterations in expression of apoptotic-related genes, and S-phase cell cycle arrest in HepG2 cell line. Biomed. Res. Int. 2014, 539607.

Moradi-Afrapoli, F., Asghari, B., Saeidnia, S., et al., 2012a. In vitro α-glucosidase inhibitory activity of phenolic constituents from aerial parts of *Polygonum hyrcanicum*. Daru 20 (1), 37.

Moradi-Afrapoli, F., Yassa, N., Zimmermann, S., et al., 2012b. Cinnamoylphenethyl amides from *Polygonum hyrcanicum* possess anti-trypanosomal activity. Nat. Prod. Commun. 7 (6), 753–755.

Nugroho, A., Kim, E.J., Choi, J.S., Park, H.J., 2014. Simultaneous quantification and peroxynitrite-scavenging activities of flavonoids in *Polygonum aviculare* L. herb. J. Pharm. Biomed. Anal. 89, 93–98.

Roslan, N.D., Yusop, J.M., Baharum, S.N., et al., 2012. Flavonoid biosynthesis genes putatively identified in the aromatic plant *Polygonum minus* via expressed sequences tag (EST) analysis. Int. J. Mol. Sci. 13 (3), 2692–2706.

Schuster, T.M., Setaro, S.D., Kron, K.A., 2013. Age estimates for the buckwheat family Polygonaceae based on sequence data calibrated by fossils and with a focus on the amphi-Pacific *Muehlenbeckia*. PLoS One 8 (4), e61261.

Seya, K., Yamaya, A., Kamachi, S., et al., 2014. 8-Methyltryptanthrin-induced differentiation of P19CL6 embryonal carcinoma cells into spontaneously beating cardiomyocyte-like cells. J. Nat. Prod. 77 (6), 1413–1419.

Smolarz, H.D., Budzianowski, J., Bogucka-Kocka, A., et al., 2008. Flavonoid glucuronides with anti-leukaemic activity from *Polygonum amphibium* L. Phytochem. Anal. 19 (6), 506–513.

Song, J., Yao, H., Li, Y., et al., 2009. Authentication of the family Polygonaceae in Chinese pharmacopoeia by DNA barcoding technique. J. Ethnopharmacol. 124 (3), 434–439.

Steele, M.L., Fuller, S., Patel, M., et al., 2013. Effect of Nrf2 activators on release of glutathione, cysteinylglycine and homocysteine by human U373 astroglial cells. Redox Biol. 1 (1), 441–445.

Storniolo, C.E., Quifer-Rada, P., Lamuela-Raventos, R.M., Moreno, J.J., 2014. Piceid presents antiproliferative effects in intestinal epithelial Caco-2 cells, effects unrelated to resveratrol release. Food Funct. 5 (9), 2137–2144.

Sun, Y.S., Yang, S.F., Wang, J.H., et al., 2012. Analysis of head volatile constituents of *Polyonum cuspidatum* flower by HS-GC/MS. Zhong Yao Cai 35 (2), 241–243.

Sun, Y.N., Cui, L., Li, W., et al., 2013. Promotion effect of constituents from the root of *Polygonum multiflorum* on hair growth. Bioorg. Med. Chem. Lett. 23 (17), 4801–4805.

Tamura, K., Stecher, G., Peterson, D., et al., 2013. MEGA6: Molecular Evolutionary Genetics Analysis version 6.0. Mol. Biol. Evol. 30, 2725–2729.

Tang, P.C., Ng, Y.F., Ho, S., et al., 2014. Resveratrol and cardiovascular health – promising therapeutic or hopeless illusion? Pharmacol. Res. 90, 88–115. http://dx.doi.org/10.1016/j.phrs.2014.08.001.

Tantry, M.A., Radwan, M.M., Akbar, S., Khan, I.A., 2012. 5, 6-Dihydropyranobenzopyrone: a previously undetermined antioxidant isolated from *Polygonum amplexicaule*. Chin. J. Nat. Med. 10 (1), 28–31.

Tian, C., Zhang, R., Ye, X., et al., 2013. Resveratrol ameliorates high-glucose-induced hyperpermeability mediated by caveolae via VEGF/KDR pathway. Genes Nutr. 8 (2), 231–239.

Udani, J.K., George, A.A., Musthapa, M., et al., 2014. Effects of a proprietary freeze-dried water extract of *Eurycoma longifolia* (Physta) and *Polygonum minus* on sexual performance and well-being in men: a randomized, double-blind, placebo-controlled study. Evid. Based Complement. Alternat. Med. 2014, 179529.

Vikram, P., Chiruvella, K.K., Ripain, I.H., Arifullah, M., 2014. A recent review on phytochemical constituents and medicinal properties of kesum (*Polygonum minus* Huds.). Asian Pac. J. Trop. Biomed. 4 (6), 430–435.

Wang, H., Song, L., Feng, S., et al., 2013a. Characterization of proanthocyanidins in stems of *Polygonum multiflorum* Thunb as strong starch hydrolase inhibitors. Molecules 18 (2), 2255–2265.

Wang, Z., Gu, J., Wang, X., et al., 2013b. Antidepressant-like activity of resveratrol treatment in the forced swim test and tail suspension test in mice: the HPA axis, BDNF expression and phosphorylation of ERK. Pharmacol. Biochem. Behav. 112, 104–110.

Wang, W., He, Y., Lin, P., et al., 2014a. In vitro effects of active components of *Polygonum Multiflorum Radix* on enzymes involved in the lipid metabolism. J. Ethnopharmacol. 153 (3), 763–770.

Wang, Y., Ma, J., Chow, S.C., et al., 2014b. A potential antitumor ellagitannin, davidiin, inhibited hepatocellular tumor growth by targeting EZH2. Tumour Biol. 35 (1), 205–212.

Xiang, M.X., Hu, Y.J., Yan, Y.J., 2012. Study on chemical composition of ethylacetate fraction from *Polygonum amplexicaule var. sinense*. Zhong Yao Cai 35 (10), 1610–1614.

Xiang, K., Liu, G., Zhou, Y.J., et al., 2014. 2,3,5,4′-tetrahydroxystilbene-2-*O*-β-D-glucoside (THSG) attenuates human platelet aggregation, secretion and spreading in vitro. Thromb. Res. 133 (2), 211–217.

Xiao, H.T., Tsang, S.W., Qin, H.Y., et al., 2013. A bioactivity-guided study on the anti-diarrheal activity of *Polygonum chinense* Linn. J. Ethnopharmacol. 149 (2), 499–505.

Xie, Q., Yang, Y., Wang, Z., et al., 2014. Resveratrol-4-*O*-D-(2′-galloyl)-glucopyranoside isolated from *Polygonum cuspidatum* exhibits anti-hepatocellular carcinoma viability by inducing apoptosis via the JNK and ERK pathway. Molecules 19 (2), 1592–1602.

Xu, M.L., Zheng, M.S., Lee, Y.K., et al., 2006. A new stilbene glucoside from the roots of *Polygonum multiflorum* Thunb. Arch. Pharm. Res. 29 (11), 946–951.

Xu, X.L., Huang, Y.J., Ling, D.Y., Zhang, W., 2014a. Inhibitory effects of 2,3,4′,5-tetrahydroxystilbene-2-*O*-β-D-glucoside on angiotensin II-induced proliferation of vascular smooth muscle cells. Chin. J. Integr. Med. 21 (3), 204–210.

Xu, X.L., Zhu, Q.Y., Zhao, C., et al., 2014b. The effect of 2, 3, 4′, 5-tetrahydroxystilbene-2-*O*-β-D-glucoside on pressure overload-induced cardiac remodeling in rats and its possible mechanism. Planta Med. 80 (2–3), 130–138.

Yan, P., Pang, Q.H., Jiao, X.W., et al., 2008. Genetic variation and identification of cultivated *Fallopia multiflora* and its wild relatives by using chloroplast matK and 18S rRNA gene sequences. Planta Med. 74 (12), 1504–1509.

Yan, S.L., Su, Y.F., Chen, L., et al., 2014. Polygonumosides A-D, stilbene derivatives from processed roots of *Polygonum multiflorum*. J. Nat. Prod. 77 (2), 397–401.

Yang, H.H., Hwangbo, K., Zheng, M.S., et al., 2014a. Inhibitory effects of (−)-loliolide on cellular senescence in human dermal fibroblasts. Arch. Pharm. Res. http://dx.doi.org/10.1007/s12272-014-0435-0.

Yang, H.H., Hwangbo, K., Zheng, M.S., et al., 2014b. Quercetin-3-*O*-β-D-glucuronide isolated from *Polygonum aviculare* inhibits cellular senescence in human primary cells. Arch. Pharm. Res. 37 (9), 1219–1233.

Yao, W., Fan, W., Huang, C., et al., 2013. Proteomic analysis for anti-atherosclerotic effect of tetrahydroxystilbene glucoside in rats. Biomed. Pharmacother. 67 (2), 140–145.

Yiu, C.Y., Chen, S.Y., Yang, T.H., et al., 2014. Inhibition of Epstein-Barr virus lytic cycle by an ethyl acetate subfraction separated from *Polygonum cuspidatum* root and its major component, emodin. Molecules 19 (1), 1258–1272.

Zhang, H., Yu, C.H., Jiang, Y.P., et al., 2012a. Protective effects of polydatin from *Polygonum cuspidatum* against carbon tetrachloride-induced liver injury in mice. PLoS One 7 (9), e46574.

Zhang, J.K., Yang, L., Meng, G.L., et al., 2012b. Protective effect of tetrahydroxystilbene glucoside against hydrogen peroxide-induced dysfunction and oxidative stress in osteoblastic MC3T3-E1 cells. Eur. J. Pharmacol. 689 (1–3), 31–37.

Zhang, L.J., Wang, Y.L., Wang, Z., et al., 2012c. Study on the chemical constituents of the active fraction of *Polygonum capitatum*. Zhong Yao Cai 35 (9), 1425–1428.

Zhang, F., Wang, Y.Y., Yang, J., et al., 2013a. Tetrahydroxystilbene glucoside attenuates neuroinflammation through the inhibition of microglia activation. Oxid. Med. Cell. Longev. 2013, 680545.

Zhang, H., Li, C., Kwok, S.T., et al., 2013b. A review of the pharmacological effects of the dried root of *Polygonum cuspidatum* (Hu Zhang) and its constituents. Evid. Based Complement. Alternat. Med. 2013, 208349.

Zhang, Y., Zhuang, Z., Meng, Q., et al., 2014. Polydatin inhibits growth of lung cancer cells by inducing apoptosis and causing cell cycle arrest. Oncol. Lett. 7 (1), 295–301.

Zhao, H.N., Chen, L.L., Huang, X.J., et al., 2014. A new chromone glycoside from roots of *Polygonum multiflorum*. Zhongguo Zhong Yao Za Zhi 39 (8), 1441–1444.

Zheng, C.J., Zhao, S.J., Zhao, Z.H., Guo, J., 2009. Molecular authentication of the traditional medicinal plant *Fallopia multiflora*. Planta Med. 75 (8), 870–872.

Zhou, T., Jin, Y., Wu, Y., et al., 2010. Genetic diversity of *Polygonum capitatum* from Guizhou populations by ISSR markers. Zhongguo Zhong Yao Za Zhi 35 (12), 1511–1517.

Zhou, T., Xie, Y., Zhang, L.Y., et al., 2013. Study on sequence characterized amplified region (SCAR) markers of *Polygonum capitatum*. Zhongguo Zhong Yao Za Zhi 38 (16), 2577–2580.

Phytochemistry and biology of *Ilex* pharmaceutical resources

13

13.1 Introduction

Ilex, whose common name is holly, is the only living genus of almost 600 species in the family Aquifoliaceae (Manen et al., 2010). The species are evergreen and deciduous trees, shrubs, and climbers from the tropics to temperate regions worldwide. Various *Ilex* species are utilized by worldwide ethnic groups to make non-*Camellia* tea drinks for daily consumption and health promotion (Li et al., 2011a). Mate tea from *I. paraguariensis* originated from the southern part of South America (Brazil, Argentina, Uruguay, and Paraguay) and is now a popular health-promoting drink in western countries (Bracesco et al., 2011). Large-leaved Kudingcha (bitter spikeleaf tea) is made from *I. kudingcha* and *I. latifolia* and has been drunk as a functional food in southern China for about 2000 years (Li et al., 2011a; Table 13.1). Various compounds have been isolated from *Ilex* plants since the 1980s and have been characterized in chemistry and pharmacology (Liu and Ding, 2008; Li et al., 2011a; Table 13.2). Studies in chemotaxonomy and molecular phylogeny are performed to facilitate further conservation and exploitation of *Ilex* pharmaceutical and food resources (Manen et al., 2010; Kim et al., 2010). In this chapter, we summarize the recent progress in the phytochemical and biological research of *Ilex* plants.

13.2 Chemical components and biological activities

13.2.1 Terpenoids and triterpenoid saponins

Most triterpenoid compounds in medicinal plants are saponin glycosides, which are the attachment of various sugar molecules to the triterpene unit. These sugars can be easily cleaved off in the gut by bacteria, allowing the aglycone (triterpene) to be absorbed. Terpenoid molecules could insert into cell membranes and modify the composition, influence membrane fluidity, and potentially affect signaling by many ligands and cofactors (Lindsey et al., 2003). Both large-leaved Kudingcha and mate tea contain abundant saponins (Tables 13.2 and 13.3; Li et al., 2011a, 2012b; de Souza et al., 2011). The newly discovered saponins and terpenoids are summarized in Figures 13.1 and 13.2 and Tables 13.4 and 13.5.

Medicinal plants: chemistry, biology and omics. http://dx.doi.org/10.1016/B978-0-08-100085-4.00013-X

Table 13.1 Traditional use of green tea and four kinds of non-*Camellia* tea

	Original region	Use history	Main application region	Traditional use
Camellia sinensis (green tea)	South of the Yangtze River, China	Around 3000 years	Worldwide, raw materials for drink, food, and cosmetic and healthcare products	Sweet and bitter taste, cool in property; refresh oneself; eliminate worries and thirst; reduce phlegm; help in digestion, diuresis, and detoxification; treat headaches, blurred vision, somnolence, GI upset, thirst, dyspepsia, phlegm stagnation, dysentery
Ilex paraguariensis (mate tea)	South America	300–400 years	Drink and healthcare products in America, Europe, and Asia	Hepatic protection; help in digestion, arthritis, inflammation, hypertension, and hypercholesterolemia; antiobesity; antirheumatism
I. latifolia, I. kudingcha (large-leaved Kudingcha)	South of the Yangtze River, China	Around 2000 years	Tea drink in China and Southeast Asia	Sweet and bitter taste, cold; liver meridian, lung meridian, and stomach meridian; heat clearing; relieve summer heat; smooth the liver; dispel wind; improve eyesight; help produce saliva; treat sunstroke and high fever, headache, toothache, red eye, aphtha, polydipsia, diarrhea, dysentery, and mastitis; relieve swelling and pain, astringency, and hemostasis
Ligustrum robustum (*L. purpurascens* (small-leaved Kudingcha))	Southwest China	Since ancient times	Tea drink and healthcare products in China and Southeast Asia	Slightly bitter taste, sweet, slightly cold; heat clearing; detoxicating; eliminate worries; relieve swelling, pain, scald, and furuncle; antirheumatism; arthralgia therapy
Litsea coreana (eagle tea)	Guizhou and Sichuan of China	Since ancient times	Southwest China	Detoxification, detumescence, improve eyesight, help produce saliva, defend thirst, prevent sunstroke

Table 13.2 Comparing the polyphenol constitutents of green tea, mate tea, and large-leaved Kudingcha

	Total polyphenol	Caffeine	Total flavonoid	Total saponin
Green tea	30%	1–4%	>0.4%	NR
Mate tea	More abundant than green tea 178.32 mg/g spray-dried extract (Berté et al., 2011)	0.7–2%	>1%	0.35 g/L
Large-leaved Kudingcha	2–6%	None or trace (0.228–0.436 μg/g)	9–12%	>20%
Small-leaved Kudingcha	20–30%	–	>5%	–
Eagle tea	>30%	–	>10%	>1%

NR, not reported.
Modified from Li et al. (2011a).

Table 13.3 Major components of non-*Camellia* tea and green tea

Component	Green tea	Mate tea	Large-leaved Kudingcha
Caffeine	+	+	
Theobromine	+	+	
Theophylline	+	+	
Chlorogenic acid		+	+
Caffeic acid	+	+	
Quinic acid	+	+	
Caffeoyl derivative		+	+
Coumaric acid	+		
Feruloylquinic acid		+	
Gallic acid	+		
Caffeoylshikimic acid		+	+
Gallocatechin gallate	+		
Catechin (C)	+		+
Epigallocatechin (EGC)	+		
Epigallocatechin gallate (EGCG)	+		
Epicatechin (EC)	+		+
Catechin gallate (CG)	+		+
Epicatechin gallate (ECG)	+		+
Gallocatechin gallate (GCG)	+		
Gallocatechin (GC)	+		
Kaempferol	+	+	+
Myricetin	+		+
Quercetin	+	+	+
Rutin	+	+	+

Modified from Li et al. (2011a).

A, oleanolic acid type

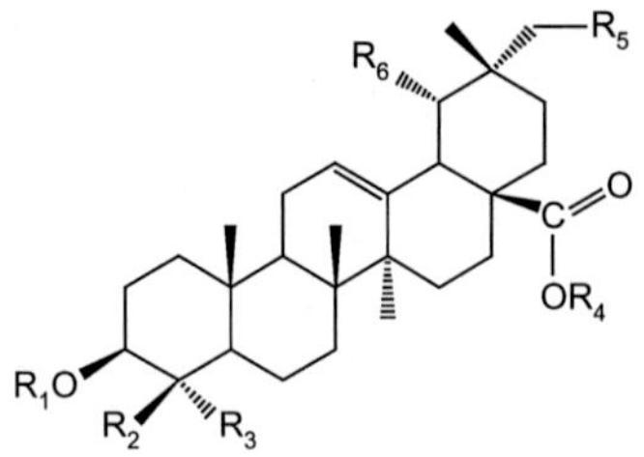

	R_1	R_2	R_3	R_4	R_5	R_6
1	Glc(1→4)-Glc(1→3)-Rha(1→2)-Ara	CH_3	CH_2OH	H	H	H
2	Glc(1→4)-GluA	CH_3	CH_3	Glc	H	OH
3	Glc(1→4)-GluA	CH_3	CH_3	Glc	H	H
4	Xyl	CH_2OH	CH_3	H	H	OH
5	Glc(1→2)-Glc	CH_3	CH_2OH	H	H	OH
6	Rha(1→2)-[Glc(1→3)]-Ara	CH_3	CH_3	Glc(1→6)-Glc	H	H
7	H	COOH	CH_3	Glc	H	H
8	H	COOH	CH_3	Glc	H	OH
9	H	COOH	CH_3	Glc	OH	H
10	Xyl	CH_3	CH_2OH	Glc	H	OH
11	Xyl	CH_3	CH_3	Glc	H	OH
12	Xyl	CH_3	CH_3	Glc	OH	H
13	Xyl	CH_3	CH_2OH	Glc	H	OH
14	H	CH_2OH	CH_3	Glc	H	OH
15	Glc(1→4)-GluA	CH_3	CH_3	Glc	H	OH
16	Glc(1→4)-Glc(1→2)-GluA	CH_3	CH_3	Glc	H	OH
17	Glc(1→4)-Glc(1→2)-GluA	CH_3	CH_3	Glc	H	H
18	Ara(1→2)-GluA	CH_3	CH_3	Glc	H	H
19	H	CH_3	COOH	Glc	H	OH
20	2-*O*-Ac-Xyl	CH_3	CH_3	H	H	OH
21	GluA	CH_3	CH_3	H	H	OH
22	GluA	$COCH_3$	CH_3	Glc	H	OH
23	GluA	$COCH_3$	CH_3	H	H	OH

Figure 13.1 Triterpenoids and saponins of *Ilex*. Refer to Table 13.4 for more details.

(Continued)

B, ursolic acid type

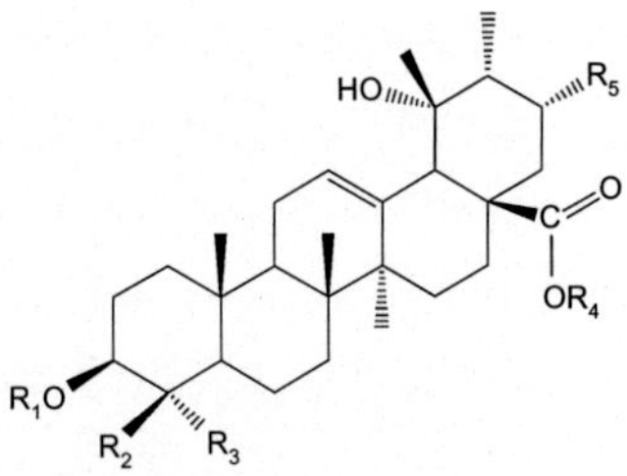

24	R_1 = GlurA	$R_2 = CH_3$	$R_3 = CH_3$	R_4 = H	R_5 = H
25	R_1 = 3-*O*-sulfo-GlurA	$R_2 = CH_3$	$R_3 = CH_3$	R_4 = H	R_5 = H
26	R_1 = GlurA	$R_2 = CH_3$	$R_3 = CH_3$	R_4 = Glc	R_5 = H
27	R_1 = 3-*O*-sulfo-GlurA	$R_2 = CH_3$	$R_3 = CH_3$	R_4 = Glc	R_5 = H
28	R_1 = Xyl	$R_2 = CH_3$	R_3 = CHO	R_4 = Glc	R_5 = H
29	R_1 = H	$R_2 = CH_3$	R_3 = CHO	R_4 = Glc(1→2)Xyl	R_5 = H
30	R_1 = H	$R_2 = CH_3$	$R_3 = CH_2OH$	R_4 = H	R_5 = OGlc
31	R_1 = 6-*O*-butyl ester-GlurA	$R_2 = CH_3$	$R_3 = CH_2OH$	R_4 = Glc	R_5 = H
32	R_1 = Glc	$R_2 = CH_3$	$R_3 = CH_2OH$	R_4 = H	R_5 = OGlc
33	R_1 = Glc(1→2)-Ara	$R_2 = CH_3$	$R_3 = CH_2OH$	R_4 = Rha(1→2)Glc	R_5 = H
34	R_1 = Glc(1→3)-Ara	$R_2 = CH_3$	$R_3 = CH_2OH$	R_4 = Glc	R_5 = H
35	R_1 = Glc	$R_2 = CH_3$	$R_3 = CH_2OH$	R_4 = Glc	R_5 = H
36	R_1 = Ara(1→2)-Glc(1→2)-Xyl	$R_2 = CH_3$	$R_3 = CH_3$	R_4 = H	R_5 = H
37	R_1 = Xyl	$R_2 = CH_3$	$R_3 = CH_3$	R_4 = Glc	R_5 = H
38	R_1 = Xyl	$R_2 = CH_3$	$R_3 = CH_2OH$	R_4 = Glc	R_5 = H
39	R_1 = Glc(1→3)-(2′-*O*-acetyl)Ara	$R_2 = CH_3$	$R_3 = CH_3$	R_4 = Glc	R_5 = H
41	R_1 = Glc(1→4)-GlcA	$R_2 = CH_3$	$R_3 = CH_3$	R_4 = Glc	R_5 = H
42	R_1 = Glc(1→4)-Glc(1→2)-GluA	$R_2 = CH_3$	$R_3 = CH_3$	R_4 = Glc	R_5 = H
43	R_1 = 6-*O*-methyl ester-GluA	$R_2 = CH_3$	$R_3 = CH_2OH$	R_4 = H	R_5 = H
44	R_1 = 6-*O*-methyl ester-GluA	$R_2 = CH_3$	$R_3 = CH_2OH$	R_4 = Glc	R_5 = H
45	R_1 = 2-*O*-Ac-Xyl	$R_2 = CH_3$	$R_3 = CH_3$	R_4 = H	R_5 = H

Figure 13.1 Continued.

46	R_1 = GluA	$R_2 = CH_3$	$R_3 = CH_3$	R_4 = H	R_5 = H
47	$R_1 = S_1$	$R_2 = CH_3$	$R_3 = CH_3$	R_4 = Glc	R_5 = H
48	R_1 = Glc	R_2 = CHO	$R_3 = CH_3$	R_4 = Glc	R_5 = H
49	R_1 = Glc	R_2 = CHO	$R_3 = CH_3$	R_4 = H	R_5 = H
50	R_1 = 3-*O*-sulfo-Xyl	$R_2 = CH_3$	$R_3 = CH_3$	R_4 = Glc	R_5 = H
51	R_1 = 6-*O*-butyl ester-GluA	$R_2 = CH_3$	$R_3 = CH_3$	R_4 = H	R_5 = H
52	R_1 = GluA	$R_2 = CH_3$	$R_3 = CH_3$	R_4 = H	R_5 = H
53	R_1 = 6-*O*-ethyl ester-GluA	$R_2 = CH_3$	$R_3 = CH_3$	R_4 = H	R_5 = H
54	R_1 = Ara(1→2)-Glc	$R_2 = CH_3$	$R_3 = CH_3$	R_4 = H	R_5 = H
55	R_1 = Ara(1→2)-Glc	$R_2 = CH_3$	$R_3 = CH_3$	R_4 = Glc	R_5 = H
56	R_1 = Ara(1→2)-Glc	$R_2 = CH_3$	$R_3 = CH_3$	R_4 = Glc	R_5 = H
57	R_1 = Rha(1→2)-Glc(1→2)-Xyl	R_2 = CHO	$R_3 = CH_3$	R_4 = Glc	R_5 = H
58	R_1 = Xyl	R_2 = CHO	$R_3 = CH_3$	R_4 = Glc	R_5 = H

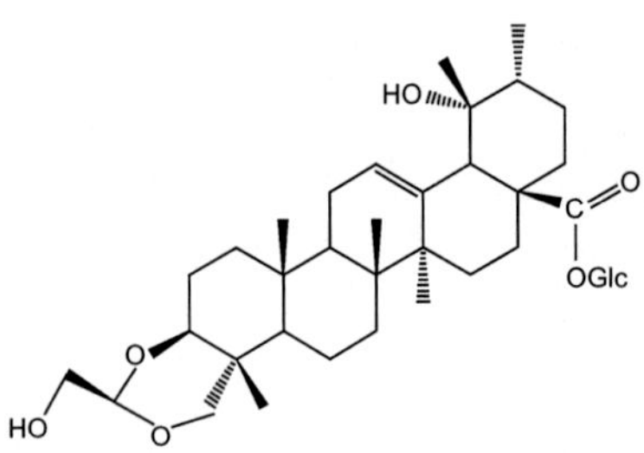

40

C, 3β-hydroxy-12(13),18(19)-diene-20-one-kudinone type

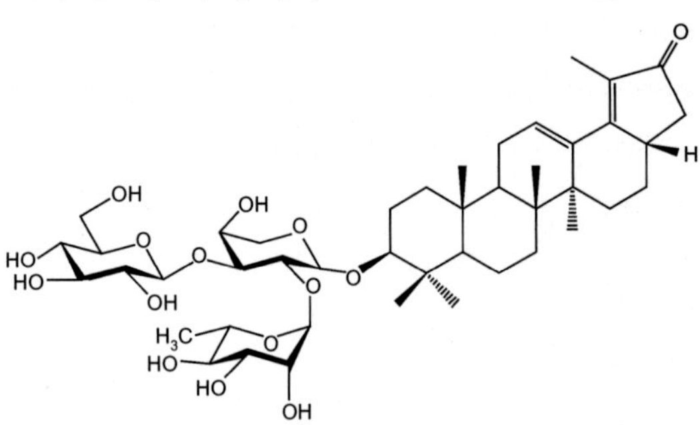

59

Figure 13.1 Continued.

D, kudinchalactone type

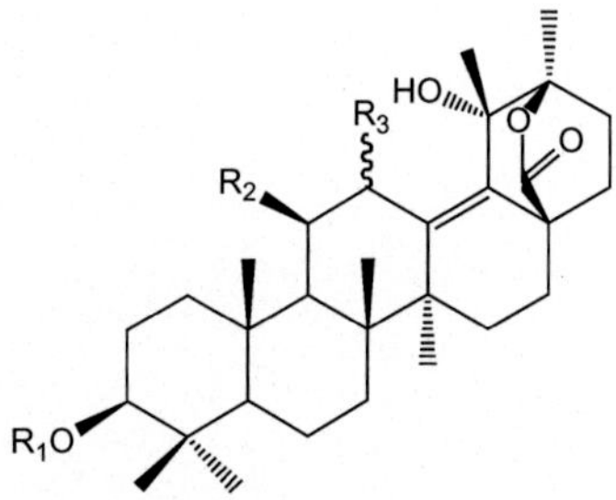

60	R_1 = Rha(1→2)-[Glc(1→3)]-Ara	R_2 = H	R_3 = α-OCH_2CH_3
61	R_1 = Rha(1→2)-Ara	R_2 = H	R_3 = α-OCH_2CH_3
62	R_1 = H	R_2 = H	R_3 = β-OH
63	R_1 = Rha(1→2)-[Glc(1→3)]-Ara	R_2 = H	R_3 = β-OH
64	R_1 = Rha(1→2)-[Glc(1→3)]-Ara	R_2 = H	R_3 = α-OCH_3
65	R_1 = Rha(1→2)-[Glc(1→3)]-Ara	R_2 = H	R_3 = β-OCH_2CH_3
66	R_1 = Glc(1→3)-Ara	R_2 = H	R_3 = β-OH
67	R_1 = Rha(1→2)-Ara	R_2 = OH	R_3 = H
69	R_1 = Rha(1→2)-Glc	R_2 = H	R_3 = β-OH
69	R_1 = Rha(1→2)-Ara	R_2 = H	R_3 = β-OH
70	R_1 = Rha(1→2)-Glc	R_2 = OH	R_3 = H

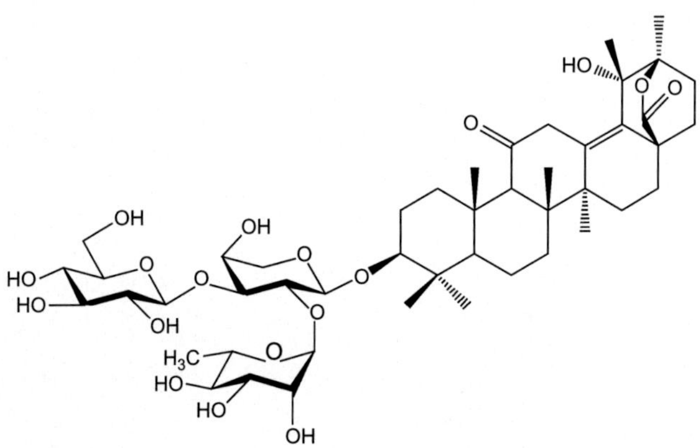

71

Figure 13.1 Continued.

E, kudinlactone type

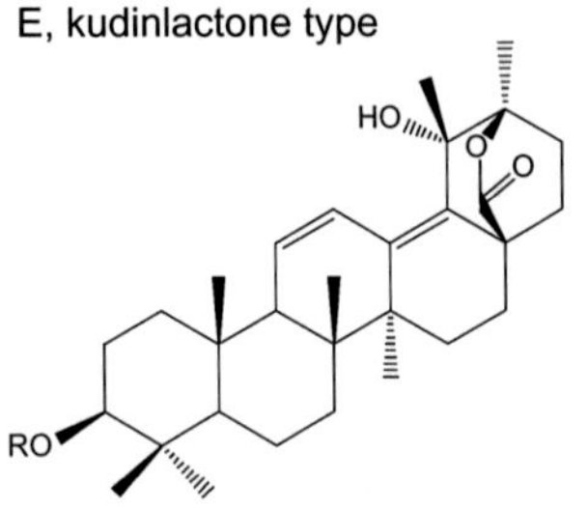

72	R = Ara(f)(1→2)-Ara
73	R = Rha(1→2)-Ara
74	R = Rha(1→2)-Glc
75	R = Glc(1→2)-Glc
76	R = Glc(1→2)-Ara
77	R = Glc(1→3)-Ara

F, 3β-hydroxy-urs-12,18-diene-28-oic acid type

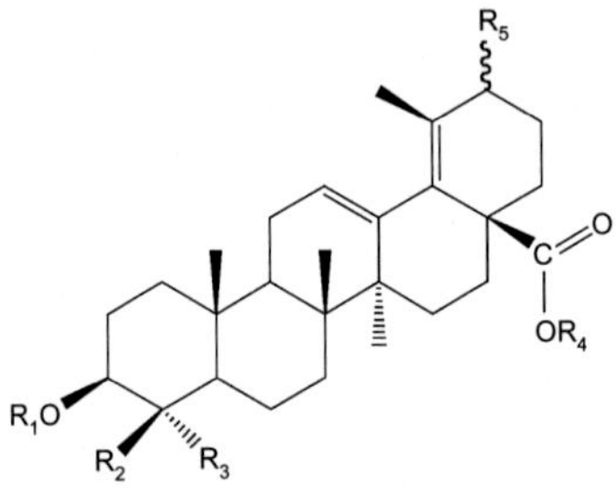

78	R_1 = GlurA	$R_2 = CH_3$	$R_3 = CH_3$	R_4 = Glc	$R_5 = \alpha\text{-}CH_3$
79	R_1 = GlurA	$R_2 = CH_3$	$R_3 = CH_3$	R_4 = Glc	$R_5 = \beta\text{-}CH_3$
80	R_1 = Xyl	$R_2 = CH_3$	$R_4 = CH_3$	R_4 = Glc	$R_5 = \beta\text{-}CH_3$
81	R_1 = H	$R_2 = CH_3$	$R_3 = CH_3$	R_4 = Glc	$R_5 = \beta\text{-}CH_3$
82	R_1 = Glc(1→2)-Xyl	$R_2 = CH_3$	$R_3 = CH_2OH$	R_4 = Glc	$R_5 = \alpha\text{-}CH_3$
83	R_1 = Rha(1→2)-Glc(1→2)-Ara	R_2 = COOH	$R_3 = CH_3$	R_4 = Glc	$R_5 = \alpha\text{-}CH_3$
84	R_1 = Xyl	$R_2 = CH_3$	$R_3 = CH_3$	R_4 = H	$R_5 = \alpha\text{-}CH_3$
85	R_1 = Glc(1→2)-Xyl	$R_2 = CH_3$	$R_3 = CH_3$	R_4 = H	$R_5 = \beta\text{-}CH_3$
86	R_1 = Glc(1→2)-Ara	R_2 = COOH	$R_3 = CH_3$	R_4 = Glc	$R_5 = \alpha\text{-}CH_3$
87	R_1 = Glc(1→2)-Ara	R_2 = COOH	$R_3 = CH_3$	R_4 = H	$R_5 = \alpha\text{-}CH_3$

Figure 13.1 Continued.

88	R_1 = Glc(1→3)-Ara	R_2 = COOH	R_3 = CH_3	R_4 = Glc	R_5 = α-CH_3
89	R_1 = Glc(1→3)-Ara	R_2 = COOH	R_3 = CH_3	R_4 = H	R_5 = α-CH_3

G, 3β-hydroxy-urs-12,19-diene-28-oic acid type

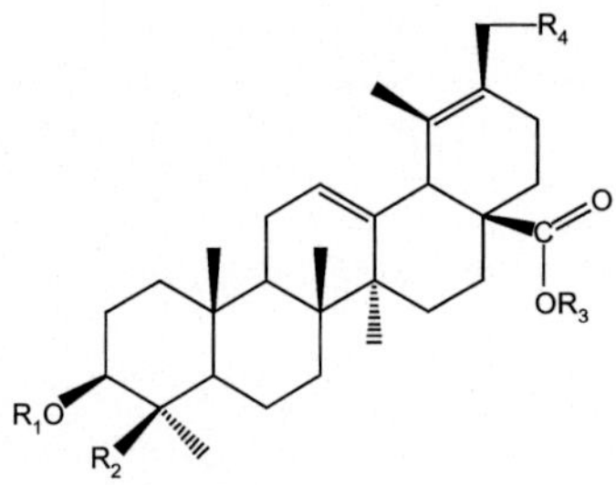

90	R_1 = GlurA	R_2 = CH_3	R_3 = Glc	R_4 = H
92	R_1 = Xyl	R_2 = CH_3	R_3 = Glc	R_4 = H
93	R_1 = H	R_2 = COOH	R_3 = Glc	R_4 = OH
94	R_1 = Glc(1→3)-Ara	R_2 = COOH	R_3 = Glc	R_4 = OH
95	R_1 = 3-*O*-sulfo-Xyl	R_2 = CH_3	R_3 = Glc	R_4 = H
96	R_1 = 3-*O*-sulfo-Xyl	R_2 = CH_3	R_3 = H	R_4 = H

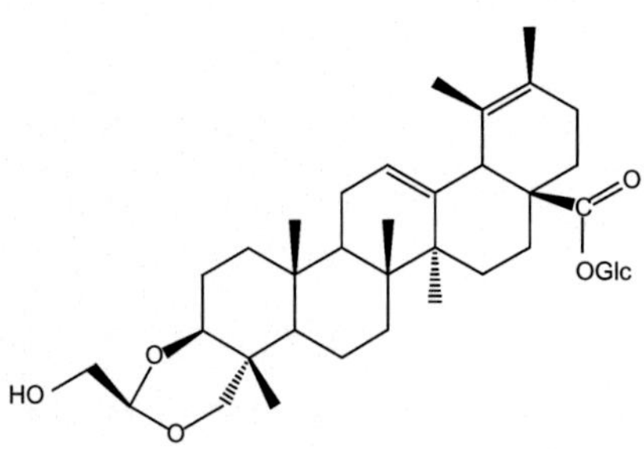

91

H, 3β-hydroxy-urs-12,19(29)-diene-28-oic acid type

Figure 13.1 Continued.

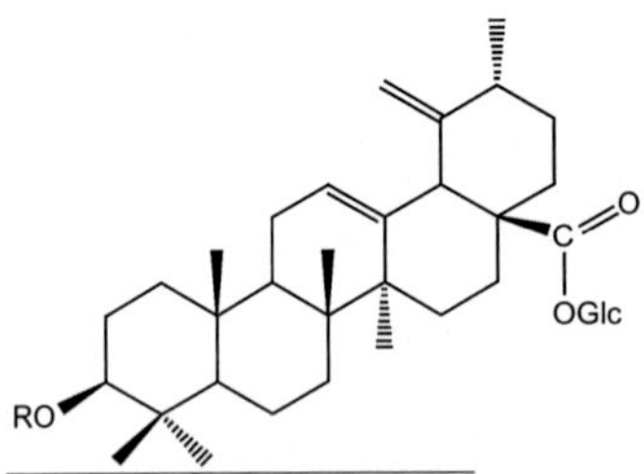

97	R = GlurA
98	R = Xyl
99	R = 3-*O*-sulfo-Xyl

I, ilexgenin B type

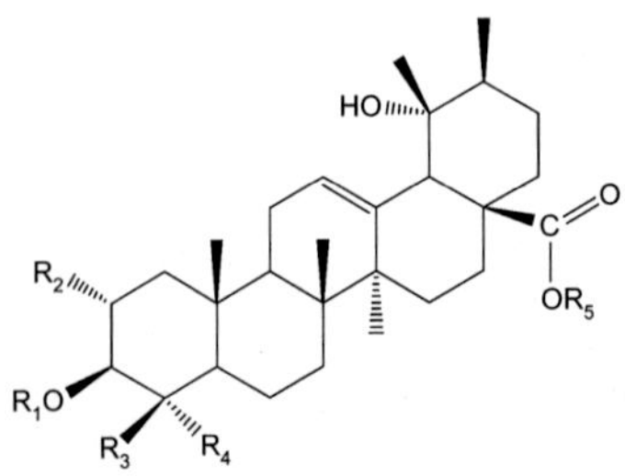

100	R_1 = H	R_2 = OH	R_3 = CH_3	R_4 = CH_2OH	R_5 = Glc
102	R_1 = H	R_2 = H	R_3 = COOH	R_4 = CH_3	R_5 = Glc
103	R_1 = H	R_2 = H	R_3 = COOH	R_4 = CH_3	R_5 = Glc
104	R_1 = H	R_2 = H	R_3 = CH_2OH	R_4 = COOH	R_5 = Glc
105	R_1 = Ara	R_2 = H	R_3 = CH_3	R_4 = CH_3	R_5 = H
106	R_1 = Glc(1→3)-Ara	R_2 = H	R_3 = CH_3	R_4 = CH_3	R_5 = H
107	R_1 = Ara	R_2 = H	R_3 = CH_3	R_4 = CH_2OH	R_5 = Glc
108	R_1 = Glc(1→2)-Ara	R_2 = H	R_3 = CH_3	R_4 = CH_3	R_5 = Glc
109	R_1 = Xyl	R_2 = H	R_3 = CH_2OH	R_4 = CH_3	R_5 = Glc
110	R_1 = 3-*O*-sulfo-Ara	R_2 = H	R_3 = CH_3	R_4 = CH_3	R_5 = Glc
111	R_1 = 2-*O*-sulfo-Xyl	R_2 = H	R_3 = CH_3	R_4 = CH_3	R_5 = Glc
112	R_1 = S_2	R_2 = H	R_3 = CH_3	R_4 = CH_3	R_5 = Glc
113	R_1 = S_3	R_2 = H	R_3 = CH_3	R_4 = CH_3	R_5 = Glc

Figure 13.1 Continued.

114	R_1 = 3-*O*-sulfo-Ara	R_2 = H	$R_3 = CH_3$	$R_4 = CH_3$	R_5 = H
115	R_1 = Rha(1→2)-Glc(1→2)-Xyl	R_2 = H	$R_3 = CH_3$	$R_4 = CH_3$	R_5 = Glc
116	R_1 = Glc(1→2)-Xyl	R_2 = H	$R_3 = CH_3$	$R_4 = CH_3$	R_5 = Glc
117	R_1 = Ara	R_2 = H	$R_3 = CH_3$	$R_4 = CH_2OH$	R_5 = Glc

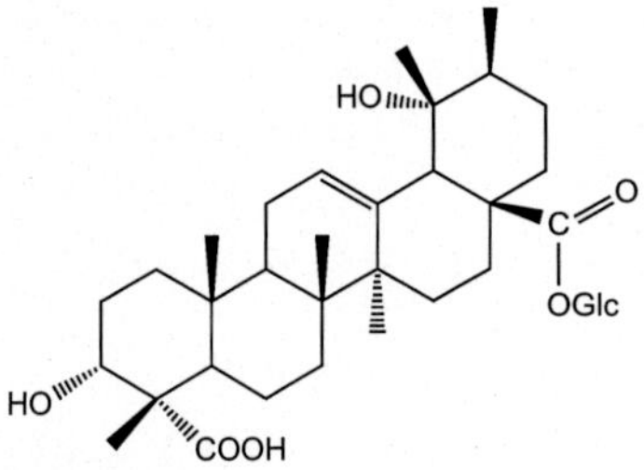

101

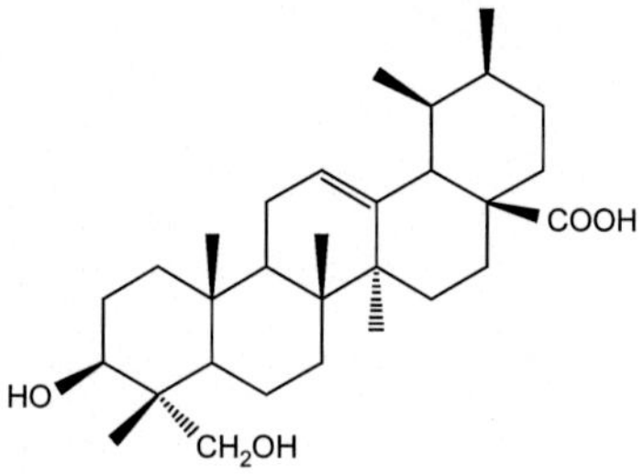

118

J, heterobetulinic acid type

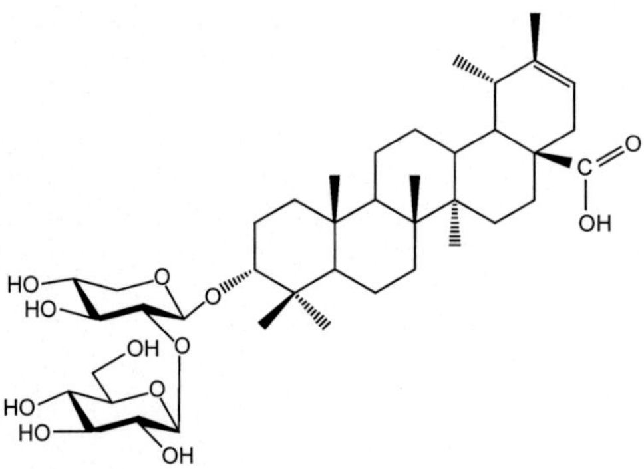

119

K, 3β, 19α-dihydroxy-urs-12-en-28,20-lactone type

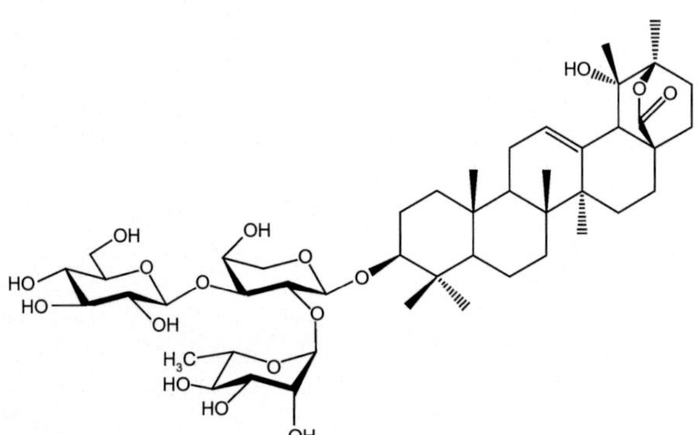

120

L, ulmoidol type

Figure 13.1 Continued.

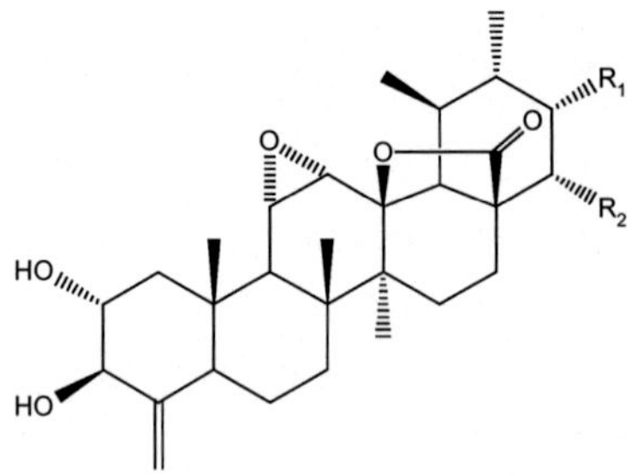

121	R_1 = OGlc	R_2 = H
122	R_1 = H	R_2 = OGlc—Glc

M, 3β,19α,23-trihydroxy-urs-12,20(30)-dien-28-oic acid type

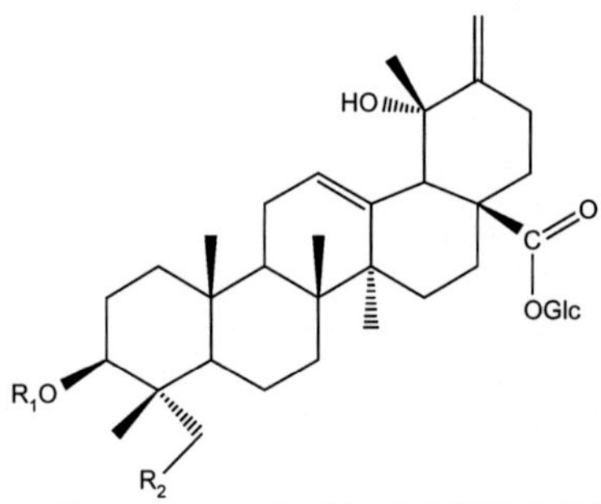

123	R_1 = H	R_2 = OH
124	R_1 = Xyl	R_2 = H

N, seven-membered ring triterpene

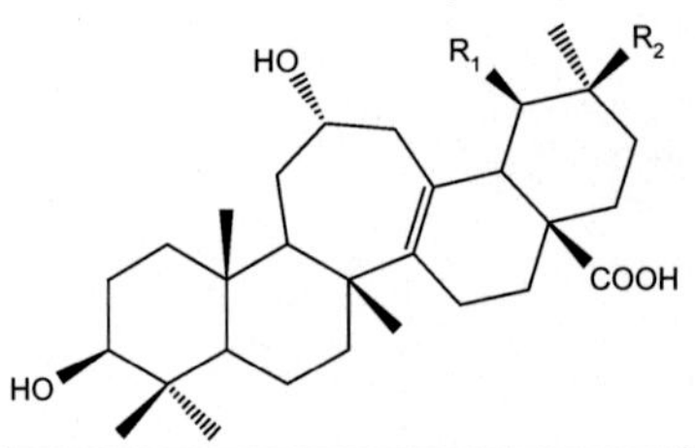

125	R_1 = CH_3	R_2 = H
126	R_1 = H	R_2 = CH_3

Figure 13.1 Continued.

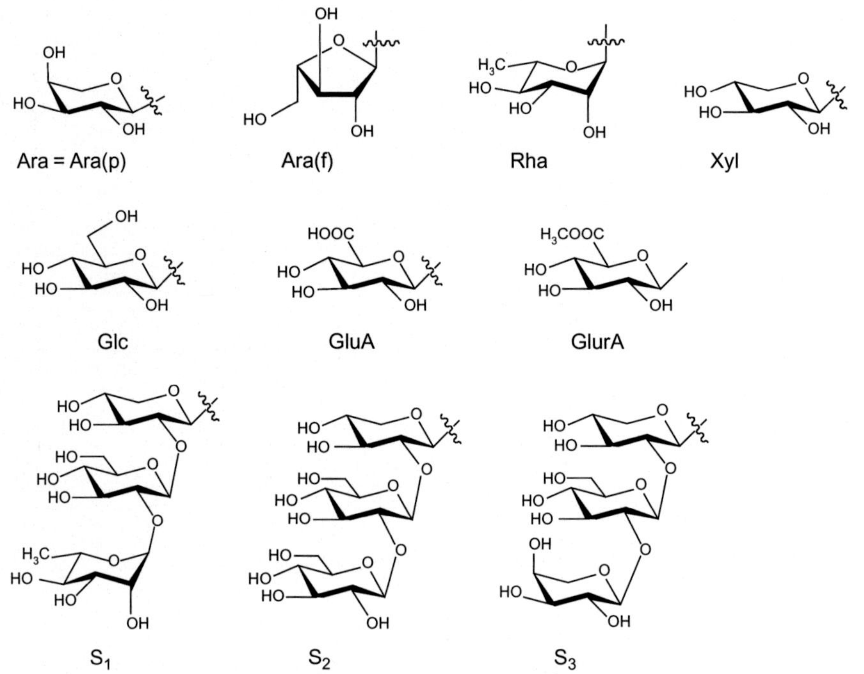

Figure 13.1 Continued.

13.2.1.1 *Anti-inflammatory activity*

Triterpenoid saponins from *I. mamillata*, endemic in Yunnan and Guangxi, China, showed inhibitory activities in anti-inflammatory assay (Yan-Yun et al., 2012). The saponin fraction from the ethanolic extracts of the root of *I. pubescens* exhibited potent anti-inflammatory effects on carrageenan-induced paw edema in rats (Wang et al., 2008a). A purified saponin fraction derived from the root of *I. pubescens* showed anti-inflammatory and analgesic activities (Wang et al., 2008b). The molecular mechanisms might be associated with inhibition of the elevated expression of cyclooxygenase (COX)-2 protein and the overproduction of the proinflammatory cytokines and augmentation of the anti-inflammatory cytokines IL-4 and IL-10 in the carrageenan-injected paw tissues of rats. Saponins in mate tea and quercetin synergistically inhibit iNOS and COX-2 in lipopolysaccharide (LPS)-induced macrophages through NF-κB pathways (Puangpraphant and de Mejia, 2009).

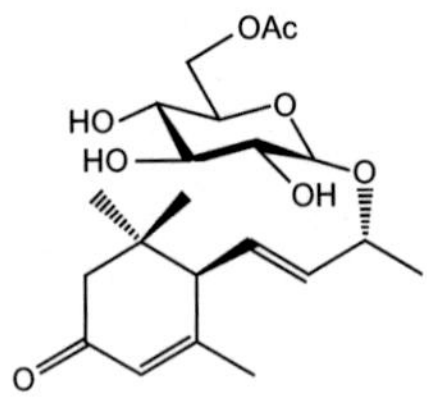

132

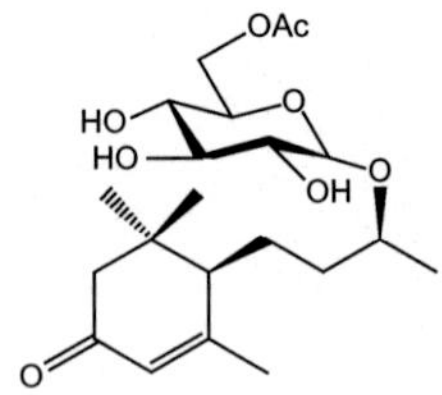

133

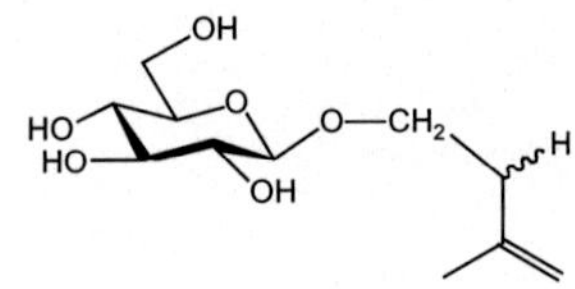

134

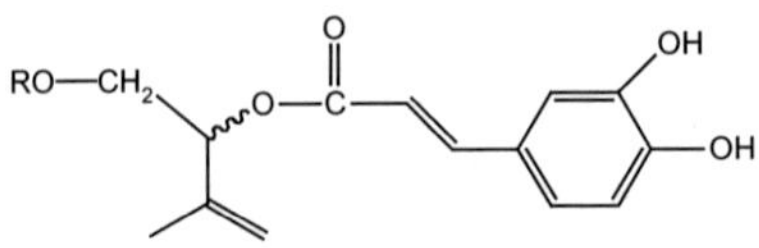

135	R = H
136	R = Glc

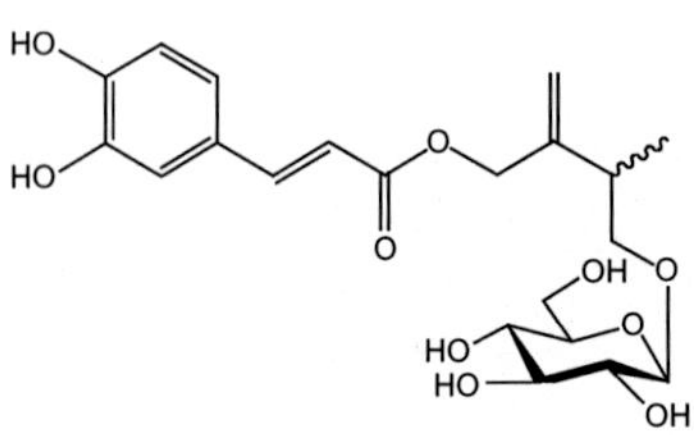

137

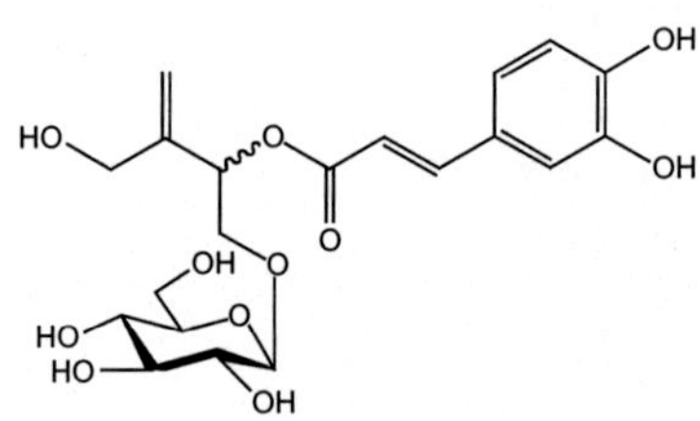

138

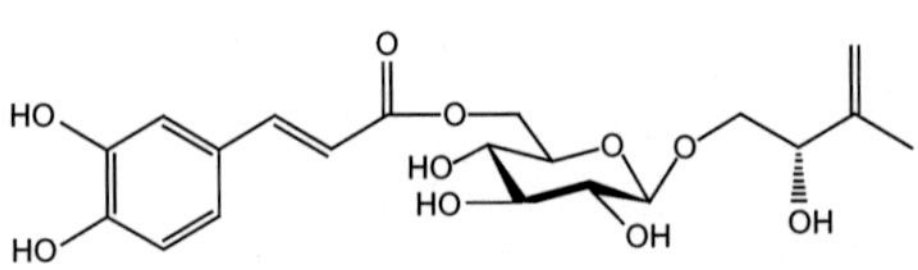

139

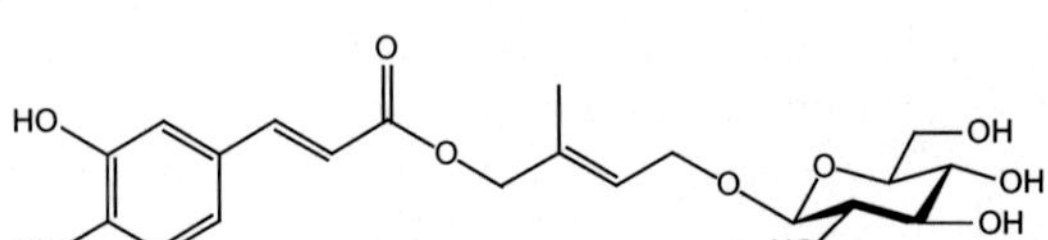

140

Figure 13.2 Terpenoids of *Ilex*. Refer to Table 13.5 for more details.

Table 13.4 Triterpenes and triterpenoid saponins found in *Ilex* medicinal plants in recent years

	Compound	Aglycone	Species	Tissue	References
1	3-*O*-β-D-glucopyranosyl-(1-4)-β-D-glucopyranosyl-(1-3)-α-L-rhamnopyranosyl-(1-2)-α-L-arabinopyranosyl-hederagenin	A	a	Leaf	Song et al. (2012)
2	3-*O*-β-D-glucopyranosyl-(1-4)-β-D-glucuronopyranosyl siaresinolic acid-28-*O*-β-D-glucopyranosyl ester	A	a	Leaf	Song et al. (2012)
3	3-*O*-β-D-glucopyranosyl-(1-4)-β-D-glucuronopyranosyl oleanolic acid-28-*O*-β-D-glucopyranosyl ester	A	a	Leaf	Song et al. (2012)
4	Ilexsaponin D	A	b	Root	Li et al. 2012a
5	3β,19α,23-Trihydroxy-olean-12-en-28-oic acid-3-*O*-β-D-glucopyranosyl-(1→2)-*O*-b-D-glucopyranoside	A	e	Leaf	Xie et al. (2009)
6	Mateglycosides A	A	f	Leaf	Sugimoto et al. (2009)
7	Ilexhainanoside C	A	g	Leaf	Chen et al. (2009a,b)
8	Ilexhainanoside D	A	g	Leaf	Chen et al. (2009a,b)
9	Ilexhainanoside E	A	g	Leaf	Chen et al. (2009a,b)
10	Ilexsaponin C	A	b	Root	Feng et al. (2008)
11	Oblonganoside K	A	h	Leaf	Wu et al. (2007b)
12	Oblonganoside L	A	h	Leaf	Wu et al. (2007b)
13	Oblonganoside M	A	h	Leaf	Wu et al. (2007b)
14	Spathodic acid 28-*O*-β-D-glucopyranoside	A	i	Stem	Zhang et al. (2005)
15	Hylonoside III	A	l	Leaf	Ouyang and Wu, (2003)
16	Hylonoside IV	A	l	Leaf	Ouyang and Wu, (2003)
17	Hylonoside V	A	l	Leaf	Ouyang and Wu, (2003)
18	3β-*O*-[α-L-arabinopyranosyl-(1→2)-β-D-glucuronic acid]-oleanolic acid 28-*O*-β-D-glucuronopyranosyl ester	A	o	Aerial part	Li et al. (2014b)

Continued

Table 13.4 Continued

	Compound	Aglycone	Species	Tissue	References
19	Ilexasprellanoside D	A	c	Root	Lei et al. (2014)
20	Ilexasprellanoside E	A	c	Root	Lei et al. (2014)
21	Ilexasprellanoside F	A	c	Root	Lei et al. (2014)
22	Ilexpublesnin L	A	b	Root	Zhou et al. (2013a)
23	Ilexpublesnin M	A	b	Root	Zhou et al. (2013a)
24	Ilexasoside A	B	c	Root	Wang et al. (2012a)
25	Ilexasoside B	B	c	Root	Wang et al. (2012a)
26	Ilexasoside C	B	c	Root	Wang et al. (2012a)
27	Ilexasoside D	B	c	Root	Wang et al. (2012a)
28	Ilexpublesnin A	B	b	Root	Zhang et al. (2010)
29	Ilexpublesnin B	B	b	Root	Zhang et al. (2010)
30	3β,21α,23-Trihydroxy-urs-12-en-28-oic acid 21-*O*-β-D-glucopyranoside	B	e	Leaf	Xie et al. (2009)
31	28-*O*-β-D-glucopyranosyl-3β,23-dihydroxy-urs-12-en-28-oic acid 3-*O*-β-D-glucuronopyranoside 6-*O*-butyl ester	B	e	Leaf	Xie et al. (2009)
32	3-*O*-β-D-glucopyranosyl-3β,21α,23-trihydroxy-urs-12-en-28-oic acid 21-*O*-β-D-glucopyranoside	B	e	Leaf	Xie et al. (2009)
33	3-*O*-β-D-glucopyranosyl-(1→2)-*O*-α-L-arabinopyranosyl-3β,19α,23-trihydroxy-urs-12-en-28-oic acid 28-*O*-α-L-rhamnopyranosyl-(1→2)-*O*-β-D-glucopyranoside	B	e	Leaf	Xie et al. (2009)
34	Mateglycosides B	B	f	Leaf	Sugimoto et al. (2009)
35	Mateglycosides C	B	f	Leaf	Sugimoto et al. (2009)
36	Ilexsaponin B	B	b	Root	Feng et al. (2008)
37	Oblonganoside H	B	h	Leaf	Wu et al. (2007b)
38	Oblonganoside I	B	h	Leaf	Wu et al. (2007b)
39	Oblonganoside J	B	h	Leaf	Wu et al. (2007b)
40	Affinoside I	B	j	Leaf	Taketa et al. (2004)

Table 13.4 Continued

	Compound	Aglycone	Species	Tissue	References
41	Hylonoside I	B	l	Leaf	Ouyang, (2003)
42	Hylonoside II	B	l	Leaf	Ouyang, (2003)
43	19α,23-Dihydroxyurs-12-en-28-oic acid 3β-*O*-β-D-glucuronopyranoside-6-*O*-methyl ester	B	o	Aerial part	Li et al. (2014b)
44	19α,23-Dihydroxyurs-12-en-28-oic acid 3β-*O*-[β-D-glucuronopyranoside-6-*O*-methyl ester]-28-*O*-β-D-glucuronopyranosyl ester	B	o	Aerial part	Li et al. (2014b)
45	Ilexasprellanoside B	B	c	Root	Lei et al. (2014)
46	Ilexasprellanoside C	B	c	Root	Lei et al. (2014)
47	Ilexpublesnin E	B	b	Root	Zhou et al. (2013a)
48	Ilexpublesnin C	B	b	Root	Zhou et al. (2013a)
49	Ilexpublesnin D	B	b	Root	Zhou et al. (2013a)
50	Ilexoside P	B	b	Root	Zhou et al. (2013b)
51	Ilexpublesnin A	B	b	Root	Zhou et al. (2013b)
52	Asprellanosides A	B	c	Root	Zhou et al. (2012)
53	(3β)-19-Hydroxy-28-oxours-12-en-3-yl β-D-glucopyranosiduronic acid *n*-butyl ester	B	c	Root	Zhao et al. (2013)
54	19α-Hydroxyurs-12-en-28-oic acid 3β-*O*-β-D-glucuronopyranoside	B	o	Root	Wang et al. (2014a)
55	19α-Hydroxyurs-12-en-28-oic acid 3β-*O*-β-D-glucuronopyranoside-6-*O*-ethyl ester	B	o	Root	Wang et al. (2014a)
56	19α-Hydroxyurs-12-en-28-oic acid 3β-*O*-α-L-arabinopyranosyl-(1→2)-β-D-glucuronopyranoside	B	o	Root	Wang et al. (2014a)
57	3β-*O*-[α-L-arabinopyranosyl-(1→2)-β-D-glucuronopyranosyl]-19α-hydroxyurs-12-en-28-oic acid 28-*O*-β-D-glucopyranosyl ester	B	o	Root	Wang et al. (2014a)

Continued

Table 13.4 **Continued**

	Compound	Aglycone	Species	Tissue	References
58	3β-*O*-[α-L-arabinopyranosyl-(1→2)-β-D-glucuronopyranoside-6-*O*-methyl ester]-19α-hydroxyurs-12-en-28-oic acid 28-*O*-β-D-glucopyranosyl ester	B	o	Root	Wang et al. (2014a)
59	Ilekudinchoside E	C	a	Leaf	Zuo et al. (2012a)
60	Ilekudinchoside F	D	a	Leaf	Zuo et al. (2012b)
61	Ilekudinchoside G	D	a	Leaf	Zuo et al. (2012b)
62	Kudinchalactone A	D	a	Leaf	Zuo et al. (2011)
63	Ilekudinchoside A	D	a	Leaf	Zuo et al. (2011)
64	Ilekudinchoside B	D	a	Leaf	Zuo et al. (2011)
65	Ilekudinoside U	D	a	Leaf	Tang et al. (2009)
66	Ilekudinoside L	D	a	Leaf	Tang et al. (2005)
67	Ilekudinoside M	D	a	Leaf	Tang et al. (2005)
68	Ilekudinoside Q	D	a	Leaf	Tang et al. (2005)
69	Ilekudinoside R	D	a	Leaf	Tang et al. (2005)
70	Ilekudinoside S	D	a	Leaf	Tang et al. (2005)
71	Ilekudinoside V	D	a	Leaf	Tang et al. (2009)
72	Ilekudinchoside C	E	a	Leaf	Zuo et al. (2011)
73	Ilekudinchoside D	E	a	Leaf	Zuo et al. (2011)
74	Ilekudinoside K	E	a	Leaf	Tang et al. (2005)
75	Ilekudinoside N	E	a	Leaf	Tang et al. (2005)
76	Ilekudinoside O	E	a	Leaf	Tang et al. (2005)
77	Ilekudinoside P	E	a	Leaf	Tang et al. (2005)
78	Ilexasoside E	F	c	Root	Wang et al. (2012a)
79	Ilexasoside F	F	c	Root	Wang et al. (2012a)
80	Pubescenoside C	F	b	Root	Wang et al. (2008a)
81	Pubescenoside D	F	b	Root	Wang et al. (2008a)
82	Oblonganoside C	F	h	Leaf	Wu et al. (2007a)
83	Ilexhainanoside A	F	g	Leaf	Zhou et al. (2007)
84	Ilexasprellanoside A	F	c	Root	Lei et al. (2014)
85	Ilexpublesnin F	F	b	Root	Zhou et al. (2013a)
86	Ilexsaponin G	F	b	Root	Li et al. (2014a)
87	Ilexsaponin H	F	b	Root	Li et al. (2014a)
88	Ilexsaponin D	F	b	Root	Li et al. (2012b)
89	Ilexsaponin E	F	b	Root	Li et al. (2012b)

Table 13.4 Continued

	Compound	Aglycone	Species	Tissue	References
90	Ilexasoside G	G	c	Root	Wang et al. (2012a)
91	Oblonganoside A	G	h	Leaf	Wu et al. (2007a)
92	Oblonganoside B	G	h	Leaf	Wu et al. (2007a)
93	Ilexhainanoside B	G	g	Leaf	Zhou et al. (2007)
94	Ilexsaponin F	G	b	Root	Li et al. (2012b)
95	Asprellanoside D	G	c	Root	Zhang et al. (2013)
96	Asprellanoside E	G	c	Root	Zhang et al. (2013)
97	Ilexasoside H	H	c	Root	Wang et al. (2012a)
98	Oblonganoside D	H	h	Leaf	Wu et al. (2007a)
99	Asprellanoside C	H	c	Root	Zhang et al. (2013)
100	(20*S*)-niga-ichigoside F1	I	i	Stem	Zhang et al. (2005)
101	Buxifolioside I	I	k	Leaf	Taketa et al. (2004)
102	Buxifolioside II	I	k	Leaf	Taketa et al. (2004)
103	28-*O*-β-D-glucopyranosyl ester of 20(*S*)-ilexgenin A	I	m	Leaf	Pires et al. (2002)
104	28-*O*-β-D-glucopyranosyl ester of 20(*S*)-3β,19α,24-trihydroxyurs-12-ene-23, 28-dioic acid	I	m	Leaf	Pires et al. (2002)
105	Ilemaminoside A	I	d	Leaf	Yan-Yun et al. (2012)
106	Ilemaminoside B	I	d	Leaf	Yan-Yun et al. (2012)
107	3β-*O*-α-L-arabinopyranosyl-19α,23-dihydroxy-20α-urs-12-en-28-oic acid 28-*O*-β-D-glucuronopyranosyl ester	I	o	Aerial part	Li et al. (2014b)
108	3β-*O*-β-D-glucuronopyranosyl-(1→2)-α-L-arabinopyranosyl-19-hydroxy-20α-urs-12-en-28-oic acid 28-*O*-β-D-glucuronopyranosyl ester	I	o	Aerial part	Li et al. (2014b)
109	Ilexpublesnin G	I	b	Root	Zhou et al. (2013a)
110	Ilexpublesnin H	I	b	Root	Zhou et al. (2013a)
111	Ilexpublesnin I	I	b	Root	Zhou et al. (2013a)
112	Ilexpublesnin J	I	b	Root	Zhou et al. (2013a)
113	Ilexpublesnin K	I	b	Root	Zhou et al. (2013a)
114	Asprellanosides B	I	c	Root	Zhou et al. (2012)
115	Ilexoside O	I	b	Root	Zhou et al. (2013b)

Continued

Table 13.4 Continued

	Compound	Aglycone	Species	Tissue	References
116	Ilexsaponin B_3	I	b	Root	Zhou et al. (2013b)
117	3β,19α,23-Trihydroxy-20α (H)-urs-12-en-28-oic acid 3β-*O*-α-L-arabinopyranoside	I	o	Root	Wang et al. (2014b)
118	3β, 23-Dihydroxy-20α(H)-urs-12-en-28-oic acid	I	o	Root	Wang et al. (2014b)
119	Heterobetulinic acid	J	b	Root	Wu et al. (2012)
120	Ilekudinoside T	K	a	Leaf	Tang et al. (2009)
121	Ilexpernoside A	L	e	Leaf	Xie et al. (2007)
122	Ilexpernoside B	L	e	Leaf	Xie et al. (2007)
123	Oblonganoside E	M	h	Leaf	Wu et al. (2007a)
124	Oblonganoside F	M	h	Leaf	Wu et al. (2007a)
125	Ilelic acid A	N_1	p	Leaf	Wang et al. (2012b)
126	Ilelic acid B	N_2	p	Leaf	Wang et al. (2012b)

Aglycone: A, Oleanolic acid type; B, ursolic acid type; C, 3β-hydroxy-12(13),18(19)-diene-20-one-kudinone type; D, kudinchalactone type; E, kudinlactone type; F, 3β-hydroxy-urs-12,18-diene-28-oic acid type; G, 3β-hydroxy-urs-12,19-diene-28-oic acid type; H, 3β-hydroxy-urs-12,19(29)-diene-28-oic acid type; I, ilexgenin B type; J, heterobetulinic acid type; K, 3β,19α-dihydroxy-urs-12-en-28,20-lactone type; L, ulmoidol type; M, 3β,19α,23-trihydroxy-urs-12,20(30)-dien-28-oic acid type; N, seven-membered ring triterpene (N_1: ilesane type; N_2: dubosane type).
a, *I. kudingcha*; b, *I. pubescens*; c, *I. asprella*; d, *I. mamillata*; e, *I. pernyi*; f, *I. paraguariensis*; g, *I. hainanensis*; h, *I. oblonga*; i, *I. litseaefolia*; j, *I. affinis*; k, *I. buxifolia*; l, *I. hylonoma*; m, *I. psammophila*; n, *I. centrochinensis*; o, *I. cornuta*; p, *I. latifolia*; q, *I. rotunda*; r, *I. glabra*.
Refer to Figure 13.1 for more details.

Table 13.5 Terpenoids found in *Ilex* in recent years

Taxon	Type	Terpenoid	Tissue	References
f	A	Matenosides A **132**	Leaf	Xu et al. (2010)
f	A	Matenosides B **133**	Leaf	Xu et al. (2010)
h	B	Oblongaroside A **134**	Leaf	Ouyang et al. (2007)
h	B	Oblongar ester A **135**	Leaf	Ouyang et al. (2007)
h	B	Oblongaroside B **136**	Leaf	Ouyang et al. (2007)
b	B	Pubescenoside A **137**	Root	Jiang et al. (2005)
b	B	Pubescenoside B **138**	Root	Jiang et al. (2005)
q	B	Rotundarpenoside A **139**	Leaf	Kim et al. (2012a,b)
q	B	Rotundarpenoside B **140**	Leaf	Kim et al. (2012a,b)

A, Megastigmane type; B, hemiterpene type. Refer to Figure 13.2 for more details.

Rotundarpene, a hemiterpene glycoside of *I. rotunda*, prevents TRAIL-induced apoptosis in human keratinocytes by suppressing the caspase-8 and Bid pathways and the mitochondrial pathway (Lee et al., 2014). Rotundarpene inhibits toll-like receptor 2 activation-induced production of inflammatory mediators in keratinocytes by suppressing the Akt and NF-κB pathways (Kim et al., 2014).

Two acetylated megastigmane glycosides of *I. paraguariensis* exhibited human neutrophil elastase (HNE) inhibitory activity (Xu et al., 2010).

13.2.1.2 Modulation of lipid metabolism and antiobesity activity

Triterpenoid saponins from the leaves of *I. kudingcha* inhibit aggregated LDL-induced lipid deposition in macrophages (Zheng et al., 2009a). The total *I. kudingcha* saponins may have a significant therapy application of hypercholesterolemia and atherosclerosis, as they improve abnormal hemorheological parameters in ApoE$^{-/-}$ mice (Zheng et al., 2009b). Several triterpenoid saponins and monoterpene oligoglycosides were found to exhibit potent inhibitory activities, and these compounds seem to be responsible for the antiobesity activity of mate tea (Sugimoto et al., 2009). Lupeol and betulin isolated from *I. macropoda* inhibit the activity of human acyl-CoA: cholesterol acyltransferases 1 and 2 in a dose-dependent manner (Im et al., 2006).

The fat weight and both liver and adipose glucose oxidation were reduced significantly by mate saponin fraction (MSF) (35%, 90%, and 60%, respectively), while LAE (crude extracts from mate leaves) and FHE (unripe fruit extract) were less active (Resende et al., 2012). A significant lowering of the blood triglyceride level was observed in rats treated with MSF and LAE. All creatinine, urea, and transaminase plasma levels remained unaffected no matter what mate preparation was considered. The glucose blood level was increased after treatment with FHE, which did not correlate with the content of either methylxanthines, polyphenols, or saponins.

13.2.1.3 Antimicrobial activity

Monodesmosidic triterpenoid saponins, matesaponin-1 (a bidesmosidic one), caffeic and chlorogenic acids, and rutin, extracted from *I. paraguariensis*, could synergistically inhibit herpes simplex virus types 1 and 2 replication (Lückemeyer et al., 2012). Four triterpenoid saponins of the ursane type isolated from the leaves of *I. oblonga* showed appreciable inhibitory activity against tobacco mosaic virus replication (Wu et al., 2007a,b). Asprellanoside A and oblonganoside H, isolated from the roots of *I. asprella*, exhibited anti-HSV-1 activity with TIC values of 0.14 and 0.18 mM, respectively (Zhou et al., 2012).

Four triterpenoid saponins of *I. kudingcha* showed antibacterial activities against *Staphylococcus aureus* and methicillin-resistant *S. aureus* (Zuo et al., 2011).

13.2.1.4 Antiparasitic activity

Ursolic acid (a pentacyclic triterpene acid) derivatives inhibit the hydrogen peroxide- and glutathione-mediated degradation of hemin, providing an additional mechanism of action for antimalarial activity (Mullié et al., 2010). Triterpenoid glycosides isolated from the leaves of *I. affinis* and *I. buxifolia*, two adulterant species of *I. paraguariensis*, showed *in vitro* antitrypanosomal activity (Taketa et al., 2004).

13.2.1.5 Effects on cardiovascular system

The triterpene chikusetsusaponin IVa (1) prolongs the recalcification time, prothrombin time, activated partial thromboplastin time, and thrombin time of normal human plasma in a dose-dependent manner (Dahmer et al., 2012), (2) inhibits the amidolytic activity of thrombin and factor Xa upon synthetic substrates S2238 and S2222, (3) inhibits thrombin-induced fibrinogen clotting (50% inhibition concentration, 199.4 ± 9.1 μM), and (4) inhibits thrombin- and collagen-induced platelet aggregation. Chikusetsusaponin IVa preferentially inhibits thrombin in a competitive manner (Ki 219.6 μM). When administered intravenously to rats, chikusetsusaponin IVa inhibited thrombus formation in a stasis model of venous thrombosis, and it did not induce a significant bleeding effect. Chikusetsusaponin IVa also prolonged the *ex vivo* activated partial thromboplastin time.

Hemiterpene glucosides from *I. pubescens* showed antiplatelet aggregation activities (Jiang et al., 2005). Triterpenoid saponins from the roots of *I. cornuta* exhibited protective effects against H_2O_2-induced cardiomyocyte injury (Wang et al., 2014a). Triterpenoid saponins from the aerial parts of *I. cornuta* showed protective effects against H_2O_2-induced myocardial cell injury (Li et al., 2014b).

13.2.1.6 Cytotoxic effects

Multiple triterpenoids/saponins, isolated from various *Ilex* species, displayed cytotoxic activity toward various human cancer cell lines (e.g., Wang et al., 2012b, 2014b; Li et al., 2014a).

13.2.1.7 Other effects

Nonhemolytic saponins purified from *I. dumosa*, *I. argentina*, and *I. paraguariensis* showed an alum-type adjuvant effect, that is, immunostimulating potential (Silveira et al., 2011).

13.2.2 Polyphenols, flavonoids, and phenolic glycosides

Phenolics are broadly distributed in plants and are the most abundant secondary metabolites. Plant polyphenols have drawn increasing attention due to their potent antioxidant properties and their significant effects in the prevention of multiple oxidative stress-associated diseases such as cancer. The common polyphenol components

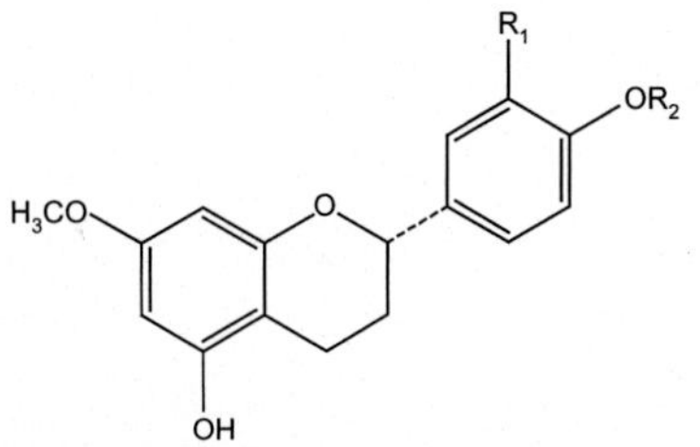

127	$R_1 = OH$	$R_2 = H$
128	$R_1 = OCH_3$	$R_2 = CH_3$

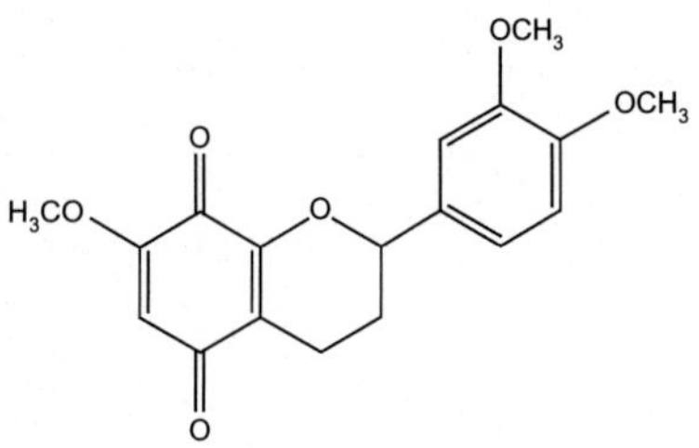

129

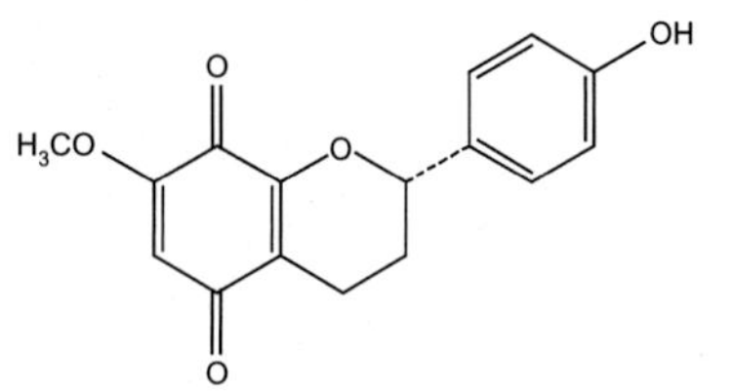

130

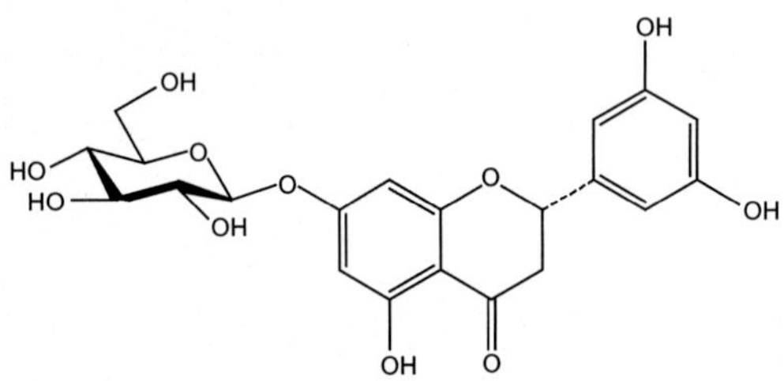

131

Figure 13.3 Flavonoids of *Ilex*. Refer to Table 13.6 for more details.

of *Ilex* are listed in Table 13.3. The newly discovered flavonoids and phenolic glycosides are summarized in Figures 13.3 and 13.4 and Tables 13.6 and 13.7.

Using TLC densitometry, the following bioactive compounds of mate tea were identified and quantified: chlorogenic acid (2.1 mg/g), caffeic acid (1.5 mg/g), rutin (5.2 mg/g), quercetin (2.2 mg/g), and kaempferol (4.5 mg/g) (Bojić et al., 2013).

13.2.2.1 Modulation of lipid metabolism

The polyphenol extract from the dried leaves of *I. paraguariensis* was the most effective in the inhibition of triglyceride accumulation in 3T3-L1 adipocytes, and rutin (100 µg/ml) likely accounted for a large portion of this activity (Gosmann et al.,

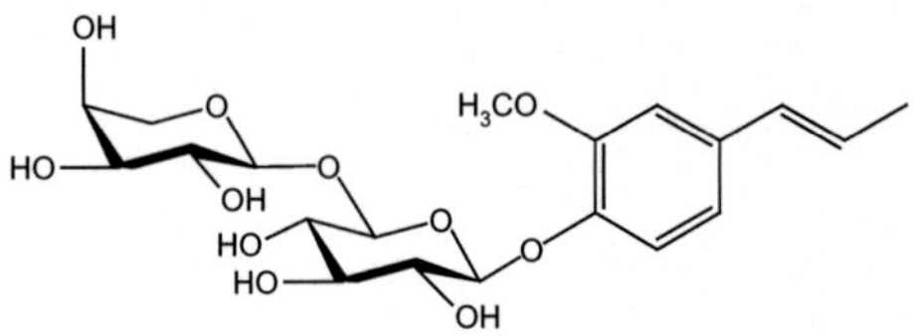

141

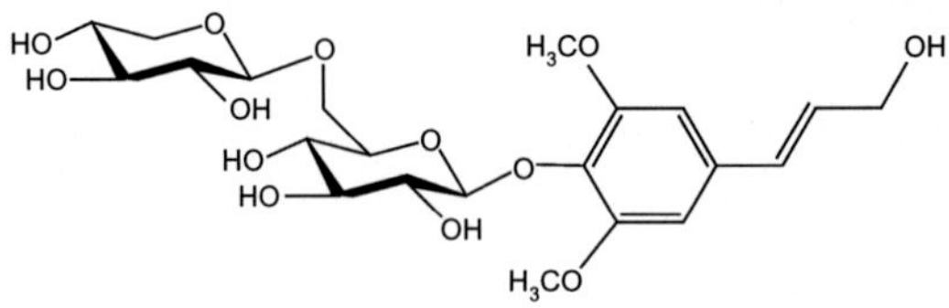

142

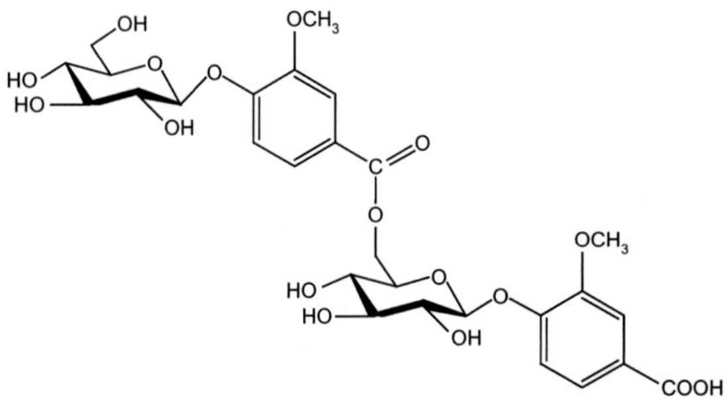

143

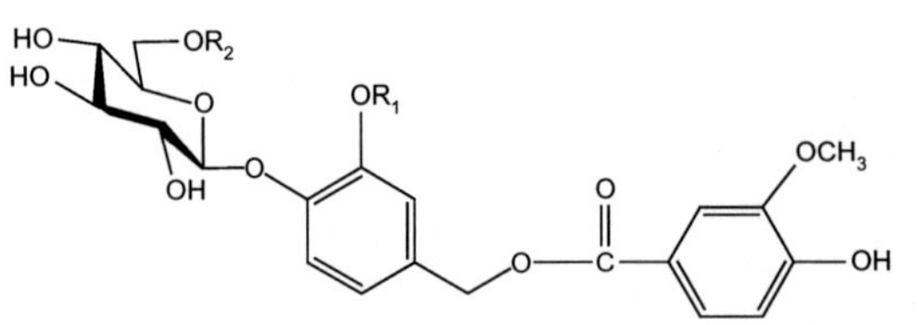

144	$R_1 = H$	$R_2 = COCH_3$
145	$R_1 = CH_3$	$R_2 = H$
146	$R_1 = H$	R_2 = caffeoyl
147	$R_1 = H$	R_2 = vanilloyl

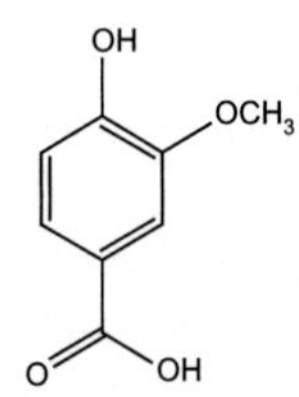

148

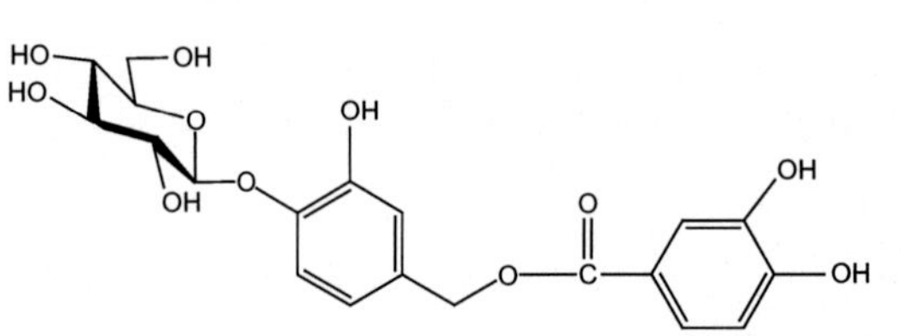

149

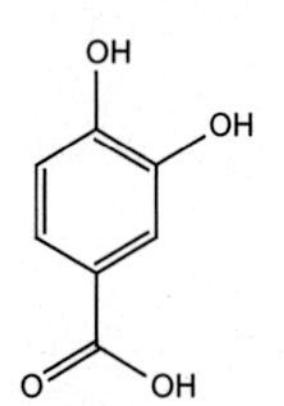

150

Figure 13.4 Phenolic glycosides of *Ilex*. Refer to Table 13.7 for more details.

(Continued)

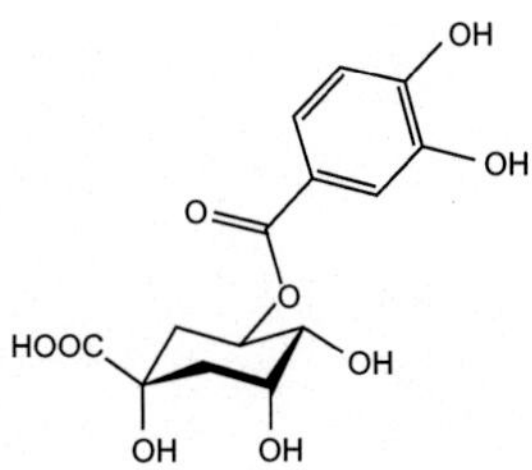

151

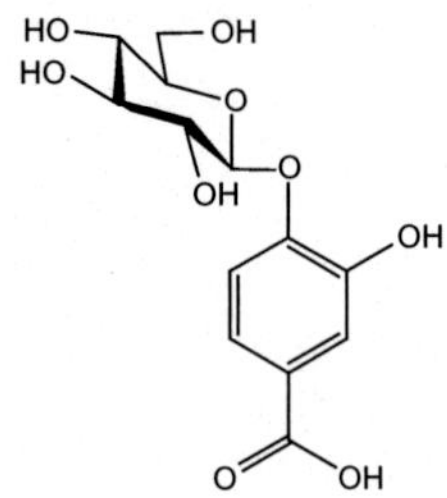

152

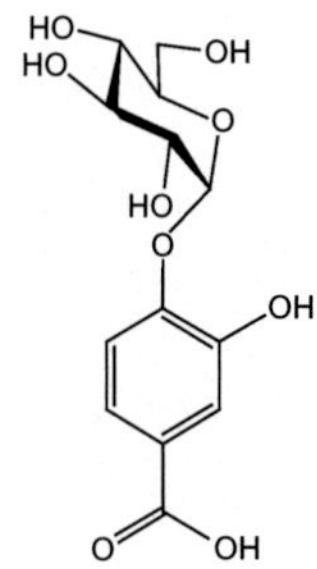

153

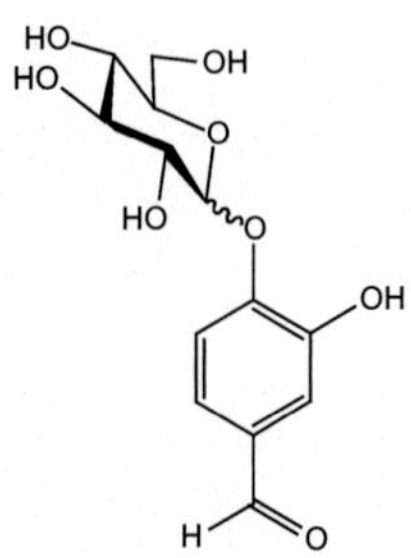

154

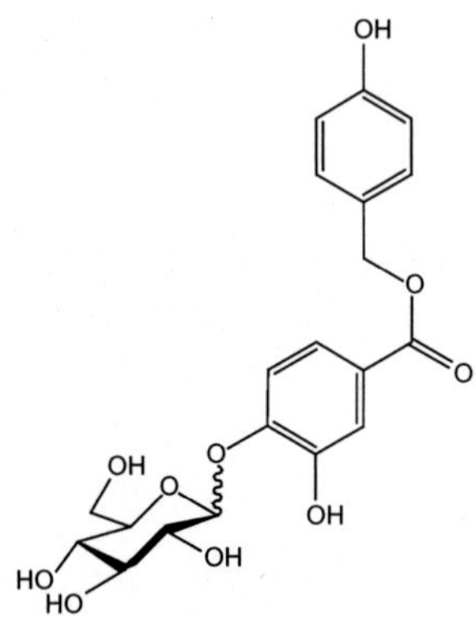

155

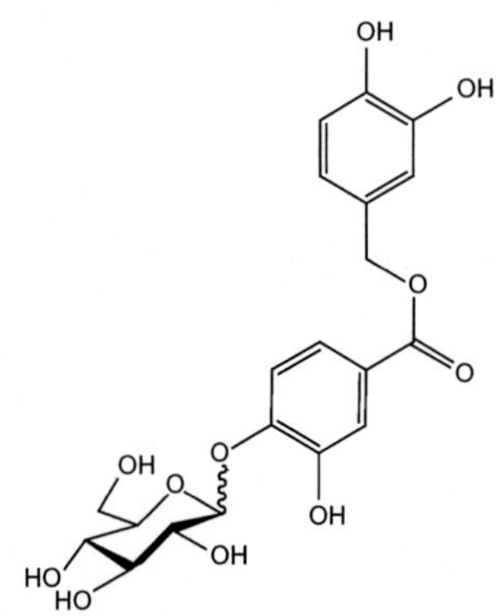

156

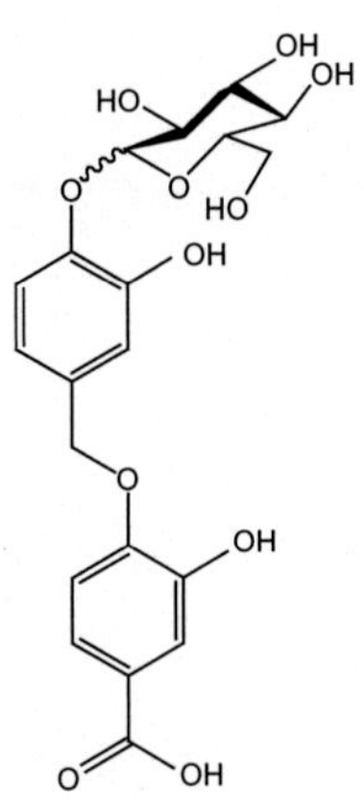

157

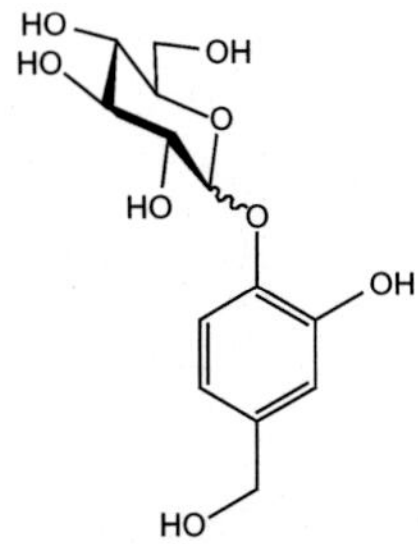

158

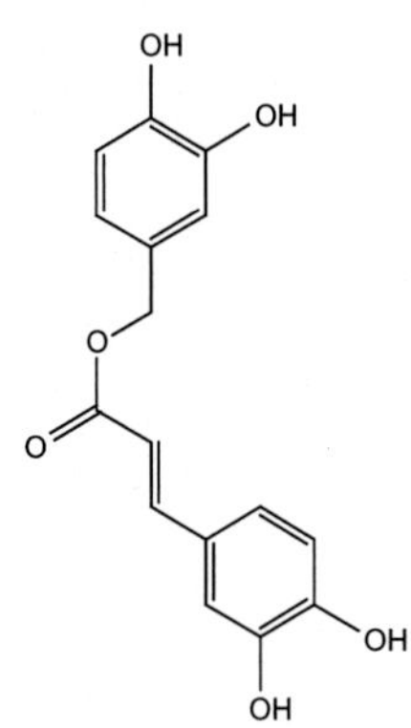

159

Figure 13.4 Continued.

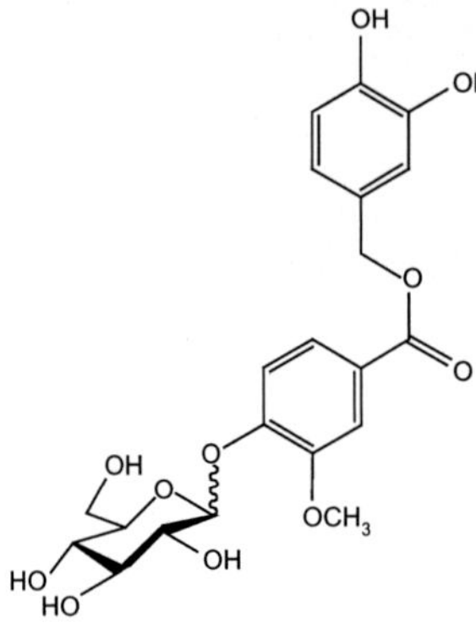

160

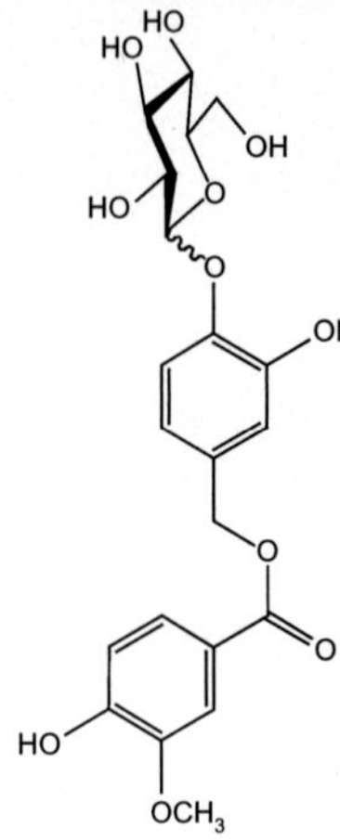

161

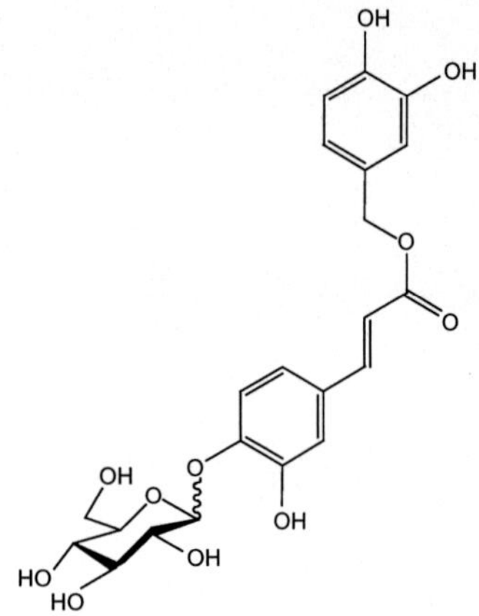

162

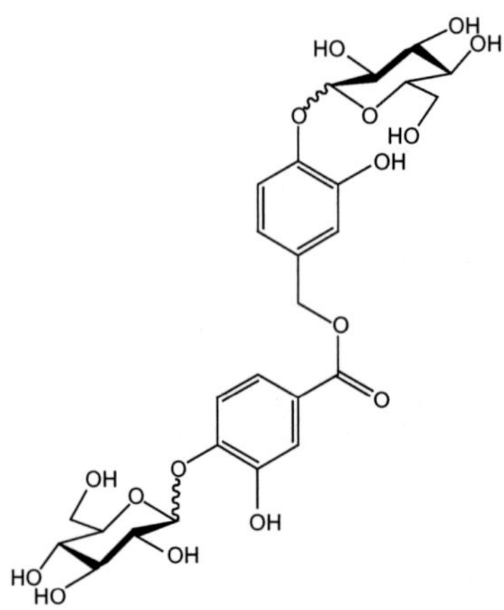

163

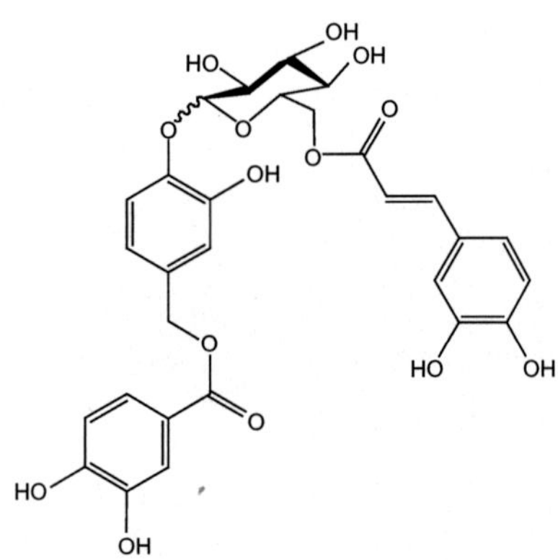

164

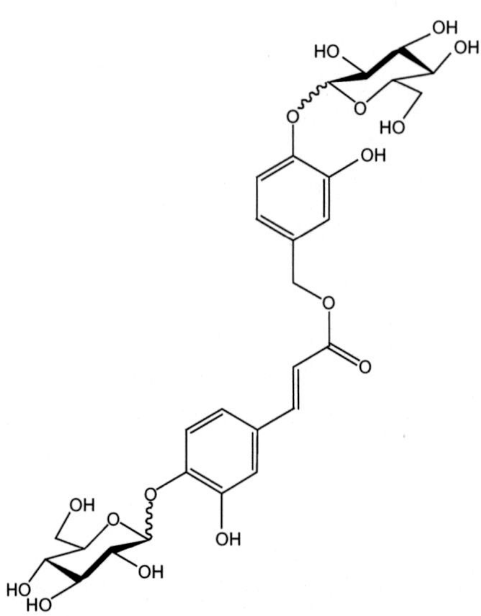

165

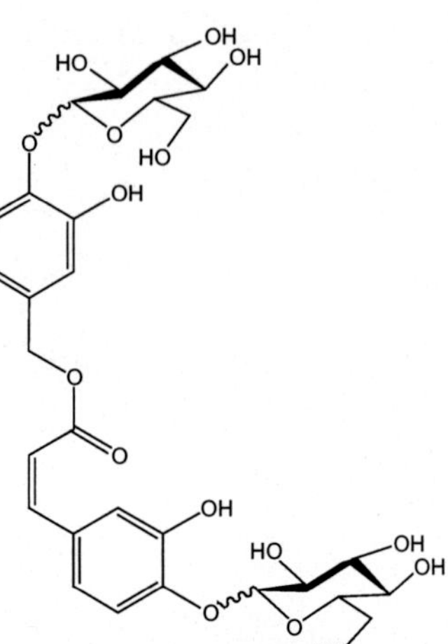

166

Figure 13.4 Continued.

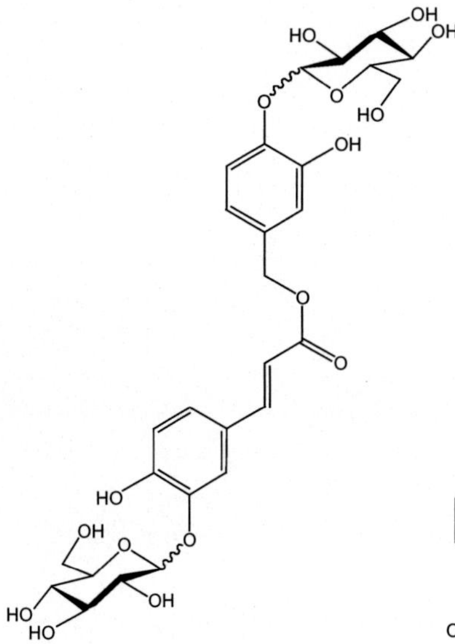

167

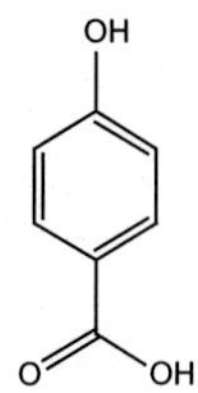

168

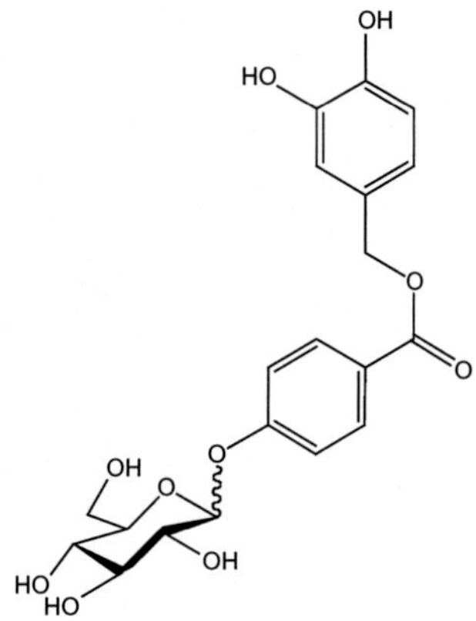

169

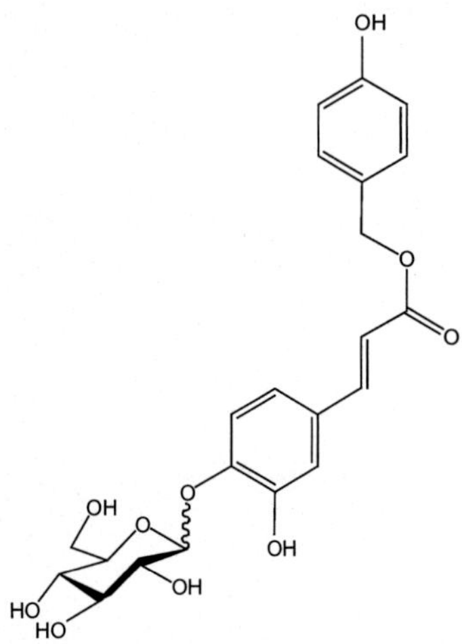

170

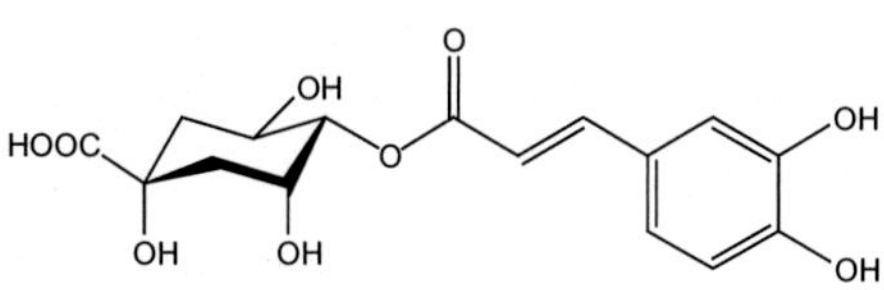

171

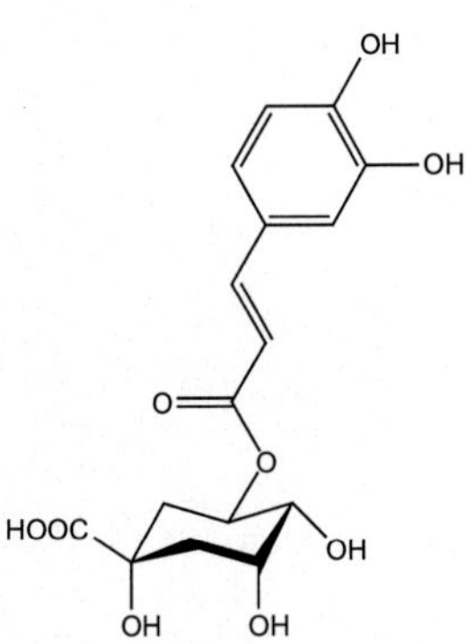

172

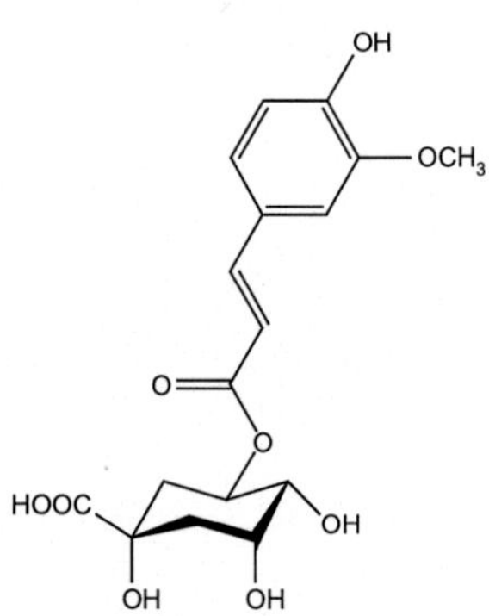

173

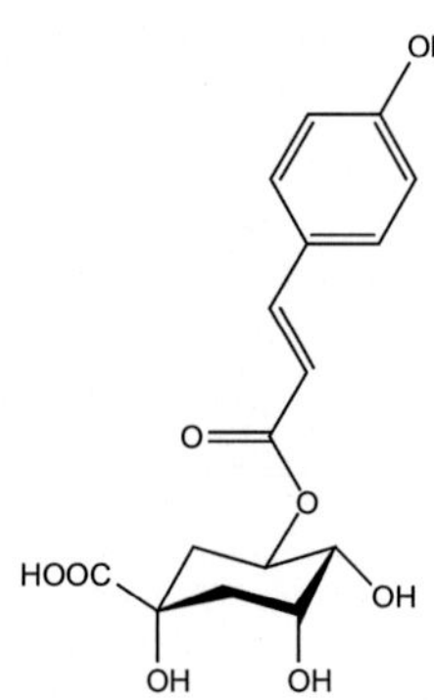

174

Figure 13.4 Continued.

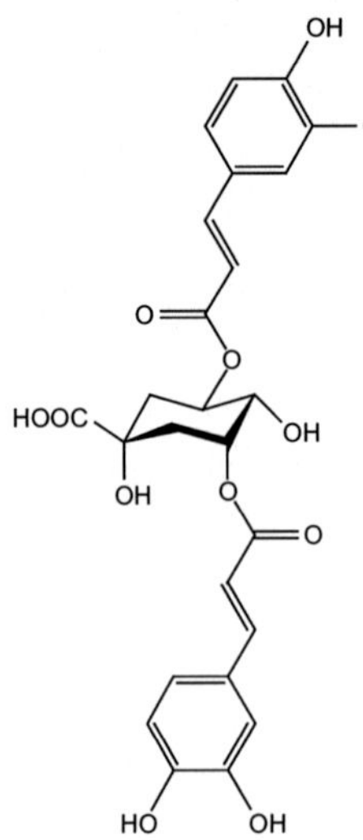

175

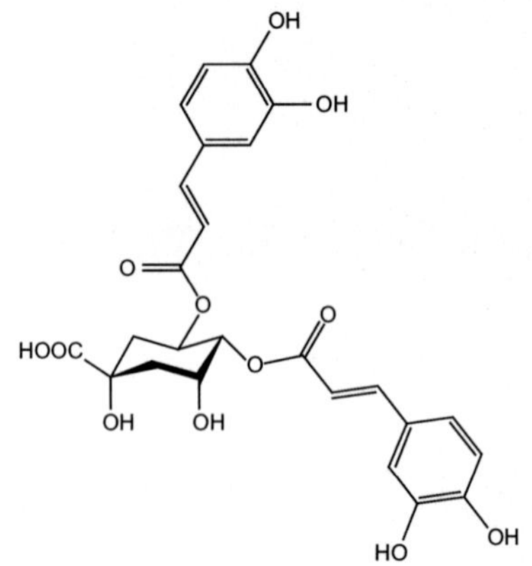

176

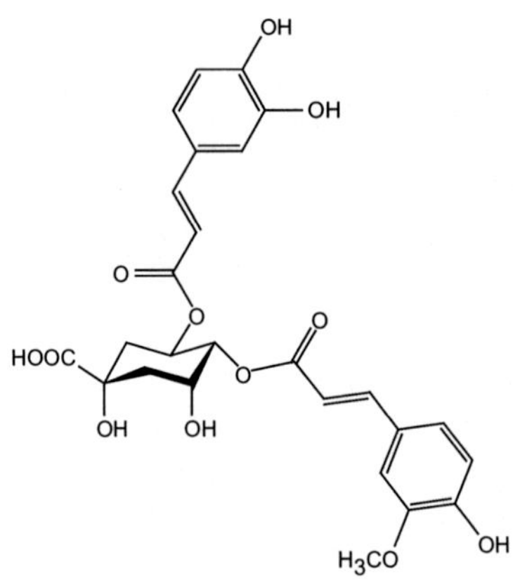

177

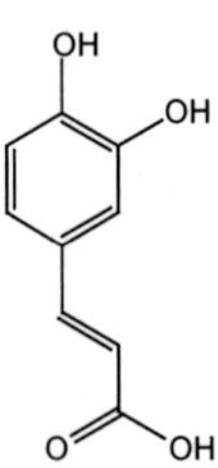

178

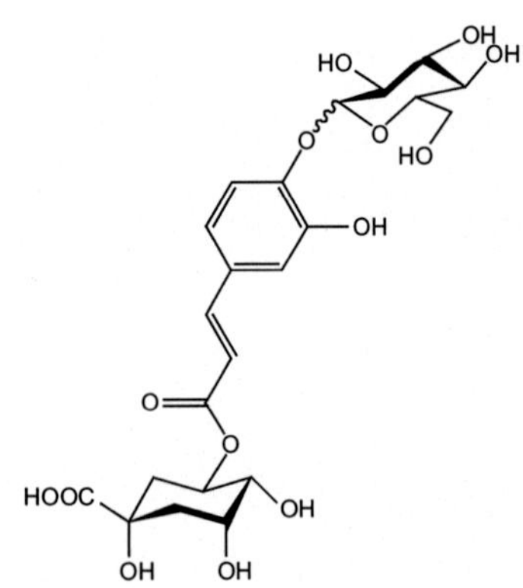

179

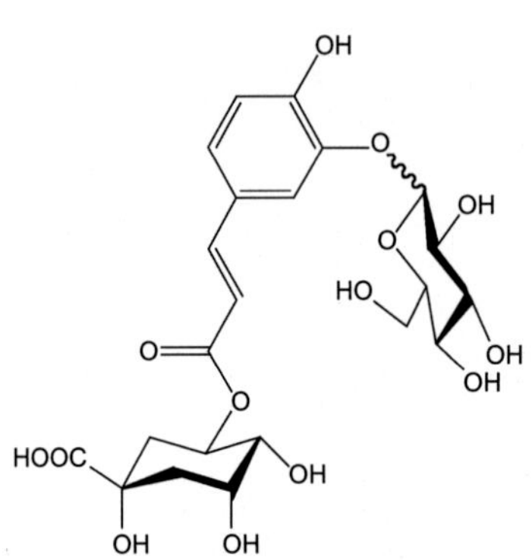

180

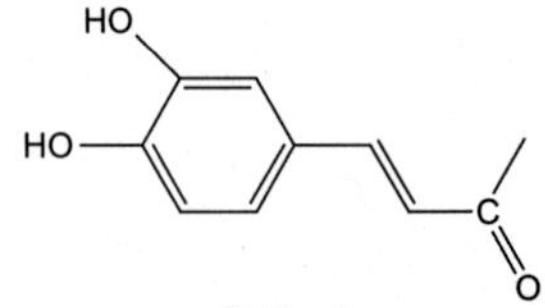

Caffeoyl

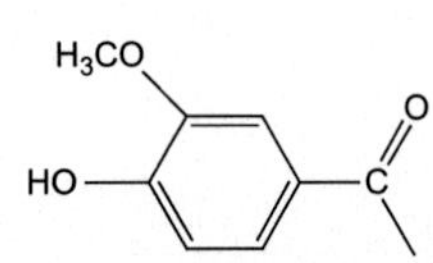

Vanilloyl

Figure 13.4 Continued.

2012). Additionally, mate extracts had a modulatory effect on the expression of genes related to adipogenesis such as PPARγ2, leptin, TNF-α, and C/EBPα. Chlorogenic acid protects paraoxonase 1 activity in high-density lipoprotein from inactivation caused by physiological concentrations of hypochlorite (Gugliucci and Bastos, 2009).

13.2.2.2 Antioxidant activity

High-performance liquid chromatography (HPLC) analysis showed that 4,5-dicaffeoylquinic acid is the major component of the phenolic fraction of mate powder (Vieira et al., 2010). Among the extracts prepared with different solvents, the 80% methanol extract showed the highest total polyphenol content (11.51/100 g) and antioxidant activity. Organic extracts containing phenolic antioxidants might be used as natural antioxidants by the food industry, replacing the synthetic phenolic additives used nowadays (Bastos et al., 2007). Phenolic (e.g., chlorogenic acid) and xanthine compounds (e.g., caffeine) of *I. paraguariensis* are involved in the prevention of hydrogen peroxide-induced red blood cell lysis (Peralta et al., 2013). Caffeoylquinic acid-derived free radicals identified during antioxidant reactions of bitter tea (*I. latifolia* and *I. kudingcha*) have sufficient stability for biological activity, which contributes appreciably to the antioxidant chemistry of large-leaved Kudingcha (Pirker and Goodman, 2010). Major phenolics in *I. kudingcha* and *I. cornuta* were mono- and dicaffeoylquinic acids, whereas those in *Ligustrum robustum*, that is, the popular small-leaved Kudingcha of southwest China, were phenylethanoid and monoterpenoid glycosides (Zhu et al., 2009). *Ilex* exhibited significantly stronger antioxidant capacities than *Ligustrum*. Huazhongilexone-7-*O*-β-D-glucopyranoside, a novel flavanone from *I. hainanensis*, showed antioxidant activity (Chen et al., 2009b). Two antioxidant phenylacetic acid derivatives, 2,4-dihydroxyphenylacetic acid and 2,4-dihydroxyphenylacetic acid methyl ester, were isolated from the seeds of *I. aquifolium* (Nahar et al., 2005).

13.2.2.3 Antidiabetic activity

The leaves of *I. kudingcha* are used as an ethnomedicine in the treatment of diabetes mellitus and obesity in China. An active components group, consisting of three dicaffeoylquinic acids and three novel triterpenoid saponins, significantly reduced the elevated levels of serum glucose and lipids in type II diabetic mice (Song et al., 2012). Litseaefoloside C, a phenolic glycoside of *I. litseaefolia*, showed inhibitory activities *in vitro* for α-glucosidase and lipase (Zhang et al., 2005). Dicaffeoylquinic acids and matesaponins dramatically increased satiety marker glucagon-like peptide 1 levels and induced anorexic effects in mice (Hussein et al., 2011a). Caffeic and chlorogenic acids in *I. paraguariensis* extracts are the main inhibitors of advanced glycation end-product generation by methylglyoxal in model proteins (Gugliucci et al., 2009).

13.2.2.4 Anti-inflammatory and anticancer activities

The 3,4,5-tricaffeoylquinic acid inhibits the tumor necrosis factor-α-stimulated production of inflammatory mediators in keratinocytes via suppression of Akt and NF-κB pathways (Lee et al., 2011). Protocatechuic acid (PCA), isolated from the leaves of *I. chinensis*, suppresses ovalbumin-induced airway inflammation in a mouse allergic asthma model (Wei et al., 2013). The protective effect of PCA might be mediated by the inhibition of the extracellular signal-regulated protein kinase (ERK), p38 mitogen-activated protein kinase (MAPK) phosphorylation, and NF-κB activation.

Dicaffeoylquinic acids in mate tea inhibit NF-κB nucleus translocation in macrophages and induce apoptosis by activating caspases-8 and caspases-3 in human colon cancer cells (Puangpraphant et al., 2011). Flavonol-rich fractions of *I. vomitoria* leaves induce microRNA-146a and have anti-inflammatory and chemopreventive effects in intestinal myofibroblast CCD-18Co cells (Noratto et al., 2011). Epigallocatechin gallate showed greater cytotoxicity than quercetin and gallic acid against HepG2 cells (Ramirez-Mares et al., 2004).

13.2.2.5 Other effects

Rutin and caffeoylshikimic acid inhibit the extracellular glucosyltransferase activity from *Streptococcus mutans* and prevent the development of dental caries (Battagim et al., 2012). Biologically, elastase activity significantly increased with age, which results in a reduced skin elasticity and in the appearance of wrinkles or stretch marks. Dicaffeoylquinic acid derivatives and flavonoids of *I. paraguariensis* exhibited potent HNE inhibitory activity (Xu et al., 2009).

13.2.3 Purine alkaloids

The results obtained with TLC densitometry for caffeine (5.4 mg/g) and theobromine (2.7 mg/g) in mate tea show no statistical difference to the content of total xanthines (7.6 mg/g) obtained by UV-vis spectrophotometry (Bojić et al., 2013).

Stimulant properties of mate tea are attributed to methylxanthines, such as caffeine (Vieira et al., 2010). Despite its antioxidant capacity and well-known health benefits, mate tea has been shown to possess some genotoxic and mutagenic activities and to increase the incidence of some types of cancer (Wnuk et al., 2009). Caffeine is one of the most abundant compounds found in the dry mass of mate (Table 13.2), and it was found that caffeine manifests a more potent cyto- and genotoxic effect that may account, at least in part, for the detrimental effects observed for mate extract (Wnuk et al., 2009). In contrast, large-leaved Kudingcha contains only trace amounts of caffeine, while polyphenol and flavonoids are abundant (Tables 13.2 and 13.3).

Table 13.6 Flavonoids found in *Ilex* in recent years

Taxon	Type	Flavonoid	Tissue	References
n	A	(2*S*)-5,3′,4′-Trihydroxy-7-methoxyflavan **127**	Leaf	Li et al. (2011a)
n	A	(2*S*)-5-Hydroxy-7,3′,4′-trimethoxyflavan **128**	Leaf	Li et al. (2011a)
n	B	(2*R*)-7,3′,4′-Trimethoxy-5,8-quinoflavan **129**	Leaf	Li et al. (2011b)
n	B	(2*S*)-7-Methoxy-4′-hydroxy-5,8-quinoflavan **130**	Leaf	Li et al. (2011b)
g	C	Huazhongilexone-7-*O*-β-D-glucopyranoside **131**	Leaf	Chen et al. (2009a,b)

A, Flavanol type; B, 5,8-quinoflavan type; C, flavanone type.

Table 13.7 Phenolic glycosides found in *Ilex* in recent years

Taxon	Type	Phenolic glycosides	Tissue	References
e	A	*Trans*-isoeugenyl-α-L-rabinopynosyl (1→6)-β-D-glucopyranoside **141**	Leaf	Xie et al. (2008)
b	B	Ilexpubside B **142**	Root	Yang et al. (2006)
b	C	Ilexpubside A **143**	Root	Yang et al. (2006)
i	C	Litseaefoloside A **144**	Stem	Zhang et al. (2005)
i	C	Litseaefoloside B **145**	Stem	Zhang et al. (2005)
i	C	Litseaefoloside C **146**	Stem	Zhang et al. (2005)
i	C	Litseaefoloside D **147**	Stem	Zhang et al. (2005)
r	C	Vanillic acid **148**	Leaf	Jaiswal et al. (2014)
h	D	Oblongaroside C **149**	Leaf	Ouyang et al. (2007)
r	D	3,4-Dihydroxybenzoic acid **150**	Leaf	Jaiswal et al. (2014)
r	D	5-*O*-(3′,4′-dihydroxy)benzoylquinic acid **151**	Leaf	Jaiswal et al. (2014)
r	D	3-Hydroxy-4-*O*-β-glucosylbenzoic acid **152**	Leaf	Jaiswal et al. (2014)
r	D	3-Hydroxy-4-*O*-α-glucosylbenzoic acid **153**	Leaf	Jaiswal et al. (2014)
r	D	3-Hydroxy-4-*O*-glucosyl-benzaldehyde **154**	Leaf	Jaiswal et al. (2014)
r	D	4-Hydroxy-7-*O*-(3′-hydroxy-4′-*O*-glucosylbenzoyl) benzyl alcohol **155**	Leaf	Jaiswal et al. (2014)
r	D	3,4-Dihydroxy-7-*O*-(3′-hydroxy-4′-*O*-glucosylbenzoyl) benzyl alcohol **156**	Leaf	Jaiswal et al. (2014)
r	D	3-Didroxy-4-*O*-(3′-hydroxy-4′-*O*-glucosylbenzyl) benzoic acid **157**	Leaf	Jaiswal et al. (2014)
r	E	3-Hydroxy-4-*O*-glucosylbenzyl alcohol **158**	Leaf	Jaiswal et al. (2014)

Continued

Table 13.7 Continued

Taxon	Type	Phenolic glycosides	Tissue	References
r	E	3,4-Dihydroxy-7-*O*-caffeoylbenzyl alcohol **159**	Leaf	Jaiswal et al. (2014)
r	E	3,4-Dihydroxy-7-*O*-(4′-*O*-glucosylvanilloyl) benzyl alcohol **160**	Leaf	Jaiswal et al. (2014)
r	E	3-Hydroxy-4-*O*-glucosyl-7-*O*-vanilloylbenzyl alcohol **161**	Leaf	Jaiswal et al. (2014)
r	E	3,4-Dihydroxy-7-*O*-(3′-hydroxy-4′-*O*-glucosylcaffeoyl) benzyl alcohol **162**	Leaf	Jaiswal et al. (2014)
r	E	3-Hydroxy-4-*O*-glucosyl-7-*O*-(3′-hydroxy-4′-*O*-glucosylbenzoyl)benzyl alcohol **163**	Leaf	Jaiswal et al. (2014)
r	E	3-Hydroxy-4-*O*-(6″-*O*-glucosylcaffeoyl)-7-*O*-(3′,4′-dihydroxybenzoyl) benzyl alcohol **164**	Leaf	Jaiswal et al. (2014)
r	E	3-Hydroxy-4-*O*-glucosyl-7-*O*-(3′-hydroxy-4′-*O*-*trans*-glucosylcaffeoyl) benzyl alcohol **165**	Leaf	Jaiswal et al. (2014)
r	E	3-Hydroxy-4-*O*-glucosyl-7-*O*-(3′-hydroxy-4′-*O*-*cis*-glucosylcaffeoyl) benzyl alcohol **166**	Leaf	Jaiswal et al. (2014)
r	E	3-Hydroxy-4-*O*-glucosyl-7-*O*-(4′-hydroxy-3′-*O*-glucosylcaffeoyl)benzyl alcohol **167**	Leaf	Jaiswal et al. (2014)
r	F	4-Hydroxybenzoic acid **168**	Leaf	Jaiswal et al. (2014)
r	F	3,4-Dihydroxy-7-*O*-(4′-*O*-glucosylbenzoyl)benzyl alcohol **169**	Leaf	Jaiswal et al. (2014)
r	F	4-Hydroxy-7-*O*-(3′-hydroxy-4′-*O*-glucosyl caffeoyl)benzyl alcohol **170**	Leaf	Jaiswal et al. (2014)
r	G	4-*O*-caffeoylquinic acid **171**	Leaf	Jaiswal et al. (2014)
r	G	5-*O*-caffeoylquinic acid **172**	Leaf	Jaiswal et al. (2014)
r	G	5-*O*-feruloylquinic acid **173**	Leaf	Jaiswal et al. (2014)
r	G	5-*O*-p-coumaroylquinic acid **174**	Leaf	Jaiswal et al. (2014)
r	G	3,5-Di-*O*-caffeoylquinic acid **175**	Leaf	Jaiswal et al. (2014)
r	G	4,5-Di-*O*-caffeoylquinic acid **176**	Leaf	Jaiswal et al. (2014)
r	G	4-*O*-feruloyl-5-*O*-caffeoylquinic acid **177**	Leaf	Jaiswal et al. (2014)
r	G	Caffeic acid **178**	Leaf	Jaiswal et al. (2014)
r	G	5-*O*-(4′-*O*-glucosylcaffeoyl)quinic acid **179**	Leaf	Jaiswal et al. (2014)
r	G	5-*O*-(3′-*O*-glucosylcaffeoyl)quinic acid **180**	Leaf	Jaiswal et al. (2014)

A, Phenylpropanoid type; B, syringic acid type; C, vanillic acid type; D, 3,4-dihydroxybenzoyl type; E, 3,4-dihydroxybenzyl type; F, 4-dihydroxybenzoyl type; G, chlorogenic acid type.

13.2.4 Polysaccharides

Polysaccharides (SP), isolated from mate leaves, consist of type I arabinogalactan (AG1) containing a (1→4)-linked β-Galp chain, with substituents of arabinosyl units at *O*-6 (Maria-Ferreira et al., 2013). This AG1 seems to be attached to rhamnosyl units from an RG1, via 1→4 linkage. Oral administration of SP1, SP9, SP10, SP11, and SP12 inhibited the gastric lesions induced by ethanol in rats.

Crude ILPS (SP from the leaves of *I. latifolia*) significantly prevented the increase of serum alanine aminotransferase and aspartate aminotransferase levels (Fan et al., 2014), reduced the formation of malondialdehyde, and enhanced the activities of superoxide dismutase and glutathione peroxidase in carbon tetrachloride-induced liver injury mice. The results suggested that ILPS should be a potent natural polymer with antioxidant and hepatoprotective activities.

Via oral administration, rhamnogalacturonan from *I. paraguariensis* (SPI) prevented the late mortality of infected mice by a rate of 60% at 10 mg/kg (Dartora et al., 2013), in comparison with untreated mice. Dexamethasone, the positive control, was slightly less effective, with an overall survival rate of 16.7% of mice at the end of the observation period. SPI affected neutrophil influx, avoiding its accumulation in the lungs and significantly decreased the tissue expression of iNOS and COX-2. Mate is a potential nutraceutical, and its SP is a promising adjuvant for sepsis treatment, being consumed as tealike beverages with no related adverse effects.

13.3 Pharmacology of *Ilex* tea and extracts

13.3.1 Effects on lipid metabolism and cardiovascular disease

The ingestion of mate tea, independent of the dietary intervention, increased plasma and blood antioxidant protection in patients with dyslipidemia (Boaventura et al., 2012). Consumption of mate improves serum lipid parameters in healthy dyslipidemic subjects and provides an additional LDL cholesterol reduction in individuals on statin therapy (de Morais et al., 2009), which may reduce the risk of cardiovascular diseases. Mate tea inhibits *in vitro* pancreatic lipase activity and has a hypolipidemic effect on high-fat diet-induced obese mice (Martins et al., 2010). The gross mate tea reduces fat more efficiently but produces a greater increase in blood glucose when compared with commercial mate tea and water groups (Silva et al., 2011). Dietary supplementation with *I. latifolia* significantly decreased the plasma total cholesterol levels and circulating immune complexes and increased high-density lipoprotein cholesterol in rats fed with a high-cholesterol diet (Luo et al., 2013). *I. latifolia*, similar to *C. sinensis* (green tea), could be used as food supplements to protect against the development of hypercholesterolemia.

Twenty-two metabolites, which are distributed in several metabolic pathways, were identified as potential biomarkers of nonalcoholic fatty liver disease (NAFLD) (Li et al., 2013). Taking these biomarkers as screening indexes, the extract of

I. hainanensis could reverse the pathological process of NAFLD through regulating the disturbed pathway of metabolism.

The *I. asprella* root decoction has some protective effects on regulating the related genes of lipid metabolism caused by chronic stress and hyperlipidemic fatty liver in rats (Hu et al., 2012).

13.3.2 Antidiabetic and antiobesity effects

Mate extract has potent antiobesity activity *in vivo*. Additionally, mate extract had a modulatory effect on the expression of several genes related to obesity (Arçari et al., 2009). *I. paraguariensis* extracts inhibit advanced glycation end-product formation more efficiently than green tea (Lunceford and Gugliucci, 2005). Mate extract has the ability to decrease the differentiation of preadipocytes and to reduce the accumulation of lipids in adipocytes, both of which contribute to a lower growth rate of adipose tissue, lower body weight gain, and obesity (Kang et al., 2012). Mate tea consumption improved the glycemic control and lipid profile of type II diabetes mellitus (T2DM) subjects, and mate tea consumption combined with nutritional intervention was highly effective in decreasing serum lipid parameters of prediabetes individuals, which may reduce their risk of developing coronary diseases (Klein et al., 2011). Oral administration of mate (100 mg/kg) for seven weeks induced significant decreases in body weight, body mass index, and food intake in obese diabetic mice (Hussein et al., 2011b). Anti-inflammatory effects of mate extract ameliorate insulin resistance in mice with high-fat diet-induced obesity (Arçari et al., 2011). Mate aqueous extract decreases intestinal SGLT1 (a glucose transporter) gene expression but does not affect other biochemical parameters in alloxan-diabetic Wistar rats (Oliveira et al., 2008).

Yerba mate (*I. paraguariensis*) modulates the NF-κB pathway and Akt expression in the liver of rats fed on a high-fat diet (Fujii et al., 2014). Ethanol extract of kuding tea (EK) has both preventive and therapeutic roles in metabolic disorders in mice induced with high-fat diets (Fan et al., 2012). The effects might be mediated through the antagonism of liver X receptor (LXR) β transactivity. Kuding tea is a useful dietary therapy and a potential source for the development of novel antiobesity and lipid-lowering drugs.

13.3.3 Antioxidant and anti-inflammatory effects

Ethanol extracts of *I. centrochinensis*, an endemic species in southern China, possess significant anti-inflammatory and free radical scavenging activities, whereas those of *I. ficoidea*, another Chinese species, exhibited a negligible anti-acute inflammatory activity and a moderate anti-chronic inflammatory activity (Li et al., 2011b). Mate tea possesses potent antioxidant effects against hydroxyl and superoxide radicals in both chemical and cell culture systems, as well as DNA protective properties (Leonard et al., 2010). The daily rectal application of enemas containing aqueous extracts of *I. paraguariensis* decreases oxidative tissue damage in the colon without fecal stream regardless of the time of irrigation (Cunha et al., 2011). After the supplementation period with mate tea, lipid peroxidation was acutely lowered, an effect that was maintained after prolonged administration. Total antioxidant status and the level of antioxidant enzyme gene expression were demonstrated after prolonged consumption (Matsumoto et al., 2009a). Regular consumption of mate tea may increase the antioxidant defense of the body by multiple mechanisms. Ingestion of mate tea increased LDL resistance toward *ex vivo* copper oxidation (Matsumoto et al., 2009b). An Argentinean *Ilex* species, *I. brevicuspis*, has choleretic, intestinal propulsion, and antioxidant activities, which may lead to the potential development of a new tea product and/or phytopharmaceutical products, without central nervous system stimulant activity (Filip and Ferraro, 2003).

Paraoxonase 2 is an intracellular antioxidant enzyme that can be modulated by polyphenols. Yerba mate (*I. paraguariensis*) enhances the gene modulation and activity of paraoxonase 2 (Fernandes et al., 2012).

I. paraguariensis extract increases life span and protects against the toxic effects caused by paraquat in *Caenorhabditis elegans* (Lima et al., 2014).

The extract of *I. asprella* (IA) at 500 mg/kg could effectively diminish the mortality rate and ameliorate lung edema and inflammation (Dai et al., 2014). The administration of IA extract significantly depressed the expression of IL-6, TNF-α, and MCP-1 and significantly increased the expression of IL-10 and IFN-γ in serum. *I. kudingcha* methanol extract has anti-inflammatory effects on 3% dextran sulfate sodium-induced ulcerative colitis in mice by reducing the colonic levels and inhibiting the mRNA expression of proinflammatory cytokines (Song et al., 2013).

Rats exposed to ultraviolet radiation presented DNA damage in blood and increased protein carbonylation and lipid peroxidation in skin. Oral and topic treatments with mate tea prevented lipid peroxidation (Barg et al., 2014), and both treatments with mate tea also prevented DNA damage. Topic treatment with mate tea prevented increases in protein carbonylation.

13.3.4 Antimicrobial effects

Aqueous extracts of mate are a natural antimicrobial against *Escherichia coli* O157:H7 and *Staphylococcus aureus* (Burris et al., 2011, 2012). Yerba mate had an antimicrobial activity against foodborne pathogens and enhanced the growth of probiotic lactic acid bacteria *in vitro* (Gonzalez-Gil et al., 2014), but *in vivo*, yerba mate did not decrease *Salmonella enteritidis* colonization. *Staphylococcus aureus*, *Listeria monocytogenes*, and *Salmonella enteritidis* were inhibited by yerba mate ethanolic extract (Prado Martin et al., 2013), which showed the lowest MIC/MBC (0.78/0.78 mg/ml), the highest phenolic content (193.9 g.GAE/kg), and the presence of chlorogenic acid derivatives, especially 3-*O*-caffeoylquinic and caffeic acid. This extract was able to inhibit microbial growth at pH 7 and 8.

The aqueous extract of *I. paraguariensis* (1000 mg/ml) possesses inhibitory activity against the fungus *Malassezia furfur* (Filip et al., 2010). IA extract reduced the lung influenza viral load and improved pulmonary ventilation (Dai et al., 2014).

13.3.5 Effects on nervous system

Acute administration of the hydroalcoholic extract of *I. paraguariensis* differentially modulates short- and long-term learning and memory in rats probably through its antagonist's action on adenosine receptors, which partly substantiate the traditional use of mate tea for improvement of cognition (Prediger et al., 2008).

I. pubescens extracts decreased significantly the number of escape failures relative to the control and showed antidepressant effects (Xu et al., 2004). The antidepressant-like effect of *I. paraguariensis* was not accompanied by inhibitory effect on monoamine oxidase activity (Reis Ede et al., 2014). There were no significant alterations on lipid peroxidation, thiol content, and vitamin C levels among the groups. Aqueous extract of *I. paraguariensis* decreases the time of immobility in rats suggesting an antidepressant-like effect.

The ethanol extract of *I. latifolia* ameliorated ischemic injury induced by middle cerebral artery occlusion/reperfusion in rats, and this neuroprotective effect might be associated with its antiapoptotic effect, resulting from antioxidative and anti-inflammatory actions (Kim et al., 2011).

Organic and conventional yerba mate commercial samples did not change the behavior (locomotion, exploration, or anxiety) of the treated animals. In both organic and conventional infusions, the presence of the polyphenols rutin and chlorogenic acid and their acyl derivatives were detected (Branco Cdos et al., 2013), which could be associated with the anticonvulsant and neuroprotective effects. Mate may provide new perspectives for the development of therapeutic approaches with natural compounds in the pharmaceutical area, both to reduce the convulsions' frequency and to minimize the neuronal damage associated with recurrent seizures.

I. latifolia and its active compounds prevented glutamate-induced neuronal cell damage by inhibiting the increase of Ca^{2+}(i), generation of ROS, and resultantly apoptotic pathway (Kim et al., 2012a). The neuroprotective effects of *I. latifolia* on ischemia-induced brain damage might be associated with the antiexcitatory and antioxidative actions and could be attributable to the active compounds, caffeoylquinic acids.

13.3.6 Anticancer activity

Mate tea, ardisia tea (*Ardisia compressa*), and green tea are cytotoxic to the HepG2 cells, with mate tea demonstrating dominant cytotoxicity (Ramirez-Mares et al., 2004). The cytotoxic activity and the inhibition of topoisomerase II may contribute to the overall chemopreventive activity of mate tea and ardisia tea extracts. Mate and ardisia teas may be used as chemopreventive agents.

Kuding tea (*I. kudingcha*) significantly induced apoptosis in TCA8113 cancer cells by upregulating Bax, caspase-3, and caspase-9 expressions and downregulating Bcl-2 expression (Zhu et al., 2014). Kuding tea also exerted an antimetastatic effect on cancer cells, which was demonstrated by the decreased expression of matrix metalloproteinases (MMPs) and the increased expression of tissue inhibitors of metalloproteinases (TIMPs) and confirmed by the inhibition of the metastasis of U14 squamous cell carcinoma cells in imprinting control region (ICR) mice.

13.3.7 Other effects

Mate consumption is associated with higher bone mineral density in postmenopausal women (Conforti et al., 2012).

13.4 Toxicity and safety

Contamination in yerba mate from the Argentinean market, expressed as the sum of 16 analyzed PAHs, ranged between 224.6 and 4449.5 μg/kg on dry mass (Garcia Londoño et al., 2014). The contamination expressed as PAH4 (sum of benzo(a)pyrene, chrysene, benzo(a)anthracene, and benzo(b)fluoranthene) varied between 8.3 and 512.4 μg/kg. The correlation coefficient for the PAH2 (sum of benzo(a)pyrene and chrysene) and PAH4 groups was 0.99; for PAH2 and PAH8 (sum of benzo(a)pyrene, chrysene, benzo(a)anthracene, benzo(b)fluoranthene, benzo(k)fluoranthene, benzo(g,h,i)perylene, dibenzo(a,h)anthracene, and indeno(1,2,3-cd)pyrene), 0.97; and for PAH4 and PAH8, 0.98.

Drinking mate, common in southern South America, may increase the risk of esophageal squamous cell carcinoma (ESCC). Very high total and carcinogenic PAH concentrations in yerba mate might contribute to the high incidence of ESCC in southern South America (Golozar et al., 2012). The large PAH content variation by brand, batch, and processing method suggests that it may be possible to reduce the content of carcinogenic PAHs in commercial yerba mate, making it a healthier beverage.

The acute exposure to mate tea did not induce an increase in frequency of micronuclei (FMic) in the exfoliated buccal cells of healthy nondrinker and nonsmoker subjects and may not have genotoxic effects (Bortoluzzi et al., 2014). More human studies are needed before a conclusion can be made on the oral carcinogenic risk of mate tea to humans.

The contents of cadmium in 10 batches of *Cortex Ilex Rotunda* ranged from 3.4 to 10.4 mg/kg (Yan et al., 2012), while contents of cadmium in *I. asprella* and *I. pubescens* were more than 0.3 mg/kg.

13.5 Chemotaxonomy

I. paraguariensis is one of the most commercialized plants in South America, which grows naturally or in cultivation in Argentina, Uruguay, Brazil, and Paraguay, where other *Ilex* species, such as *I. brevicuspis*, *I. buxifolia*, *I. affinis*, and *I. theezans*, are also distributed. HPLC was used to distinguish the adulterant species from *I. paraguariensis* (Filip and Ferraro, 2003). Novel triterpenoid glycosides detected in *I. buxifolia* and *I. affinis* confirm the structural specificity of the *I. paraguariensis* triterpenoids and reinforce the proposal to detect mate adulteration by triterpenoid analysis (Taketa et al., 2004). Arbutin-2′-sulphonyl could be a discriminating metabolite of *I. theezans*

(Andrade et al., 2004). However, the metabolic similarities between species, which can be used for chemotaxonomy of the species, were not clear yet.

The metabolomic analysis of 11 *Ilex* species, *I. argentina*, *I. brasiliensis*, *I. brevicuspis*, *I. dumosa* var. *dumosa*, *I. dumosa* var. *guaranina*, *I. integerrima*, *I. microdonta*, *I. paraguariensis* var. *paraguariensis*, *I. pseudobuxus*, *I. taubertiana*, and *I. theezans*, was carried out by nuclear magnetic resonance (NMR) spectroscopy and multivariate data analysis (Choi et al., 2005). *Ilex* samples can be discriminated by the principal component analysis (PCA) and classification of the ^{1}H NMR spectra based on the phytochemicals present in the organic and aqueous fractions. Notably, in the classification of metabolites obtained from aqueous fractions, no species overlapped. The major metabolites that contribute to the discrimination are arbutin, caffeine, phenylpropanoids (e.g., caffeic acid, chlorogenic acid, and dicaffeoylquinic acid), and theobromine. For the first time, arbutin is reported as an ingredient of *Ilex* species and is found to be a biomarker for *I. argentina*, *I. brasiliensis*, *I. brevicuspis*, *I. integerrima*, *I. microdonta*, *I. pseudobuxus*, *I. taubertiana*, and *I. theezans*. There is a clear discrimination between *I. paraguariensis* and other species. The method based on the determination of multiple metabolites illustrates the feasibility of chemotaxonomic analysis of *Ilex* species.

Numerous metabolites including saponins and phenylpropanoids have been reported from *Ilex* species. Based on the above study, 1-D- and 2-D-NMR-based metabolomics were used to classify 11 South American *Ilex* species (Kim et al., 2010). ^{1}H NMR combined with PCA, partial least square discriminant analysis (PLS-DA), and hierarchical cluster analysis (HCA) showed a clear separation between species and resulted in four groups based on metabolomic similarities. The signal congestion of ^{1}H NMR spectra was overcome by the implementation of 2-D J-resolved spectroscopy and heteronuclear single-quantum coherence. Species included in group A (*I. paraguariensis*) were characterized by a higher amount of xanthines and phenolics including phenylpropanoids and flavonoids; group B (*I. dumosa* var. *dumosa* and *I. dumosa* var. *guaranina*), oleanane-type saponins; group C (*I. brasiliensis*, *I. integerrima*, *I. pseudobuxus*, and *I. theezans*), arbutin and dicaffeoylquinic acids; and group D (*I. argentina*, *I. brevicuspis*, *I. microdonta*, and *I. taubertiana*), the highest level of ursane-type saponins. The *Ilex* species used in this study were grown in the same conditions after collecting the seeds from diverse places in South America. Thus, it is expected that the metabolomic differences observed could not be due to environmental factors such as climate, soil condition, and biotic stress, but rather to the inherent character of each species, that is, their genetic composition. The above findings agree with the molecular analysis of amplified fragment length polymorphism (AFLP; Gottleib et al., 2005). For example, AFLP results showed a very close relationship between *I. brasiliensis*, *I. integerrima*, and *I. theezans*, which was congruent to the metabolomic analysis. However, the above studies did not include East Asia species, especially the source plants of Kudingcha.

The band differences of the esterase isoenzymes of 25 Kudingcha germplasm materials in Aquifoliaceae, which involved five species, *I. kudingcha*, *I. latifolia*, *I. cornuta*, *I. pentagona*, and *I. huoshanensis*, were analyzed by polyacrylamide gel electrophoresis (Zhang et al., 2008). The similarity coefficients of the test samples were calculated for cluster dendrogram. There were 18 bands of esterase isoenzymes in all samples. Based on the band differences, the 25 germplasm materials could be distinguished on the species level. The five Kudingcha species could be unmistakably distinguished. *I. latifolia* is closer to *I. pentagona* than to *I. kudingcha*, and *I. cornuta* is further to *I. latifolia* than to *I. kudingcha*. Different germplasm materials from the same species were clustered in line with their origins, which is congruent to the morphology. The analysis results of the esterase isoenzyme are useful references for the classification of Kudingcha species and could be used to determine the origin of Kudingcha species.

The large-leaved Kudingcha from the genus *Ilex*, a traditional Chinese tea, contains several distinctive triterpenoid saponins that can be useful in classification and quality control (Li et al., 2012a). The reversed-phase ultra performance liquid chromatography coupled with evaporative light scattering detection (UPLC-ELSD) is used to simultaneously determine five triterpenoid saponins kudinoside L (1), kudinoside C (2), kudinoside A (3), kudinoside F (4), and kudinoside D (5) in several species of the large-leaved Kudingcha and mate tea (*I. paraguariensis*). UPLC separation took only 13 min with low detection and quantification limits and can be used for the quality control of large-leaved Kudingcha. The different *Ilex* species showed differences in the distribution of the five triterpenoids. *I. kudingcha*, the bona fide species of large-leaved Kudingcha, contains the maximum amount of triterpenoid saponins. The chemical profile of *I. kaushue* is almost identical to that of *I. kudingcha*, thus supporting the view that these two species are synonyms. The kudinosides (1–5) are abundant in *I. kudingcha* of Hainan and Guangxi, China. Compounds 2 and 3 are present in *I. latifolia*, while 1, 4, and 5 are not. Notably, none of the kudinosides were detected in *I. pentagona*, *I. cornuta*, and *I. paraguariensis*. Therefore, *I. pentagona* and *I. cornuta* should be considered as adulterants and removed from the large-leaved Kudingcha for tea quality control and utilization. Actually, *I. cornuta* is the raw plant of Gougucha, another health-promoting tea consumed in southwest China. *I. kudingcha* and *I. paraguariensis* are chemically dissimilar, although they both have a bitter taste and share similar pharmaceutical effects.

13.6 Molecular taxonomy, molecular phylogeny, and genomics

13.6.1 Phylogeny

The evolutionary history and diversification of the genus *Ilex*, based on 108 species, are deduced by analyzing two nuclear (ITS and nepGS) and three plastid (rbcL, trnL-F, and atpB-rbcL) sequences (Manen et al., 2010). Nuclear and plastid trees are highly dissimilar, and the nuclear tree is more in agreement with current taxonomic classifications (Figure 13.5a and b). Many Chinese species, including *I. cornuta*, *I. zhejiangensis*, *I. integra*, *I. hylonoma*, *I. latifolia*, *I. pernyi*, and *I. ficoidea*, are in the *Aquifolium* clade of subgenus *Ilex*. *I. latifolia*, *I. pernyi*, and *I. ficoidea* are closer to *I. hylonoma* than to *I. cornuta*, an adulterant of large-leaved Kudingcha. *I. pentagona*, a proxy of Ku Ding Cha, is closer to *I. pernyi* than to *I. hainanensis*

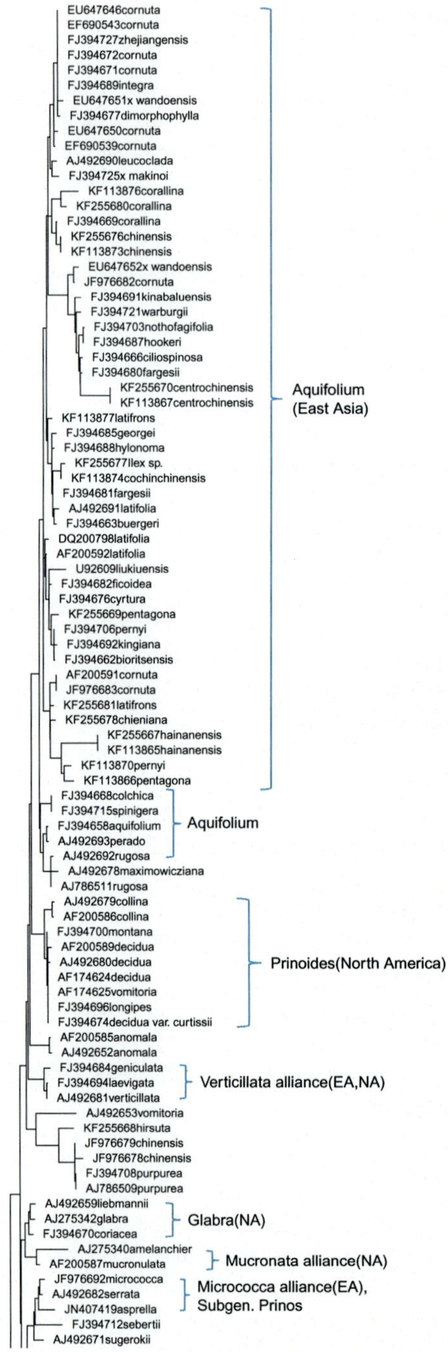

Figure 13.5 Phylogenetic relationship of *Ilex*. The evolutionary history was inferred by using the maximum likelihood method based on the GTR + I + G model. Trees are drawn to scale, with branch lengths measured in the number of substitutions per site. The analysis involved (a) 272 ITS sequences and

(Continued)

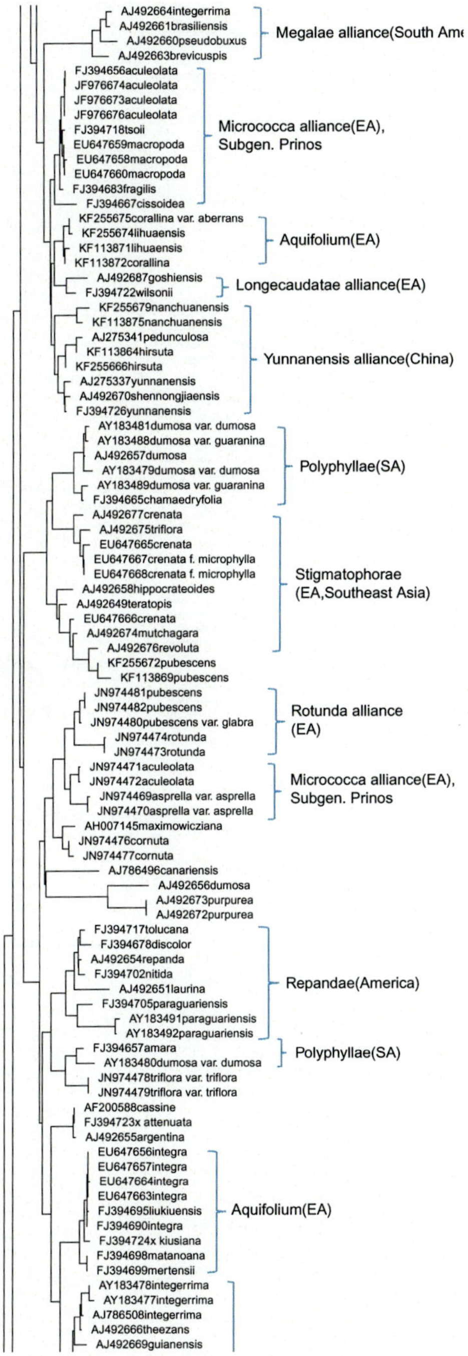

Figure13.5 Continued.

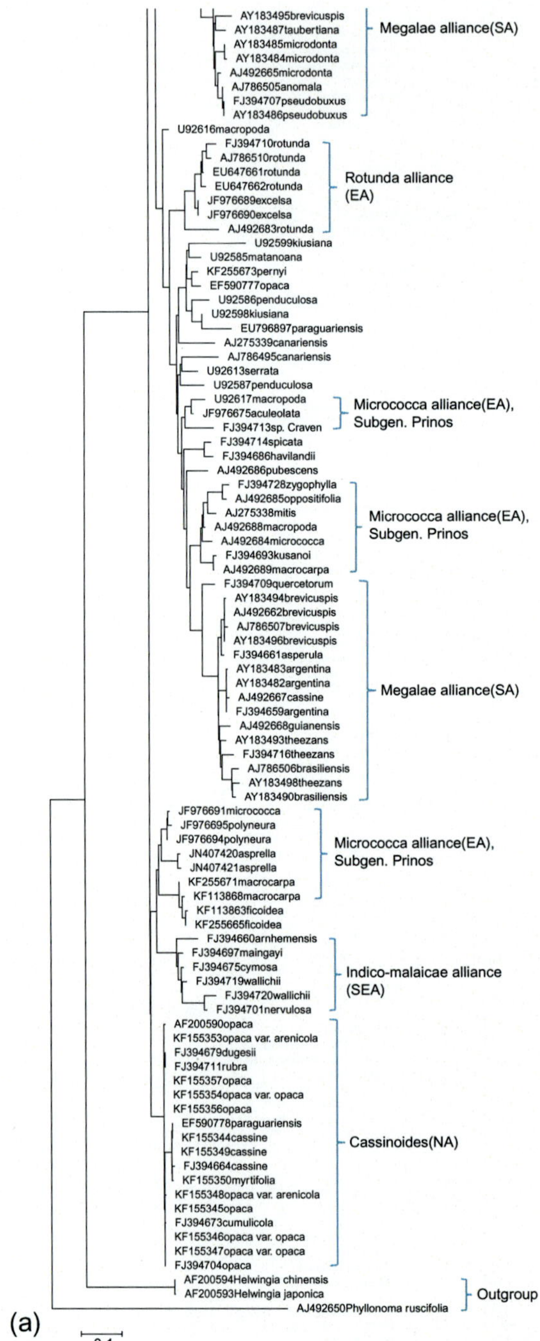

Figure13.5 Continued.

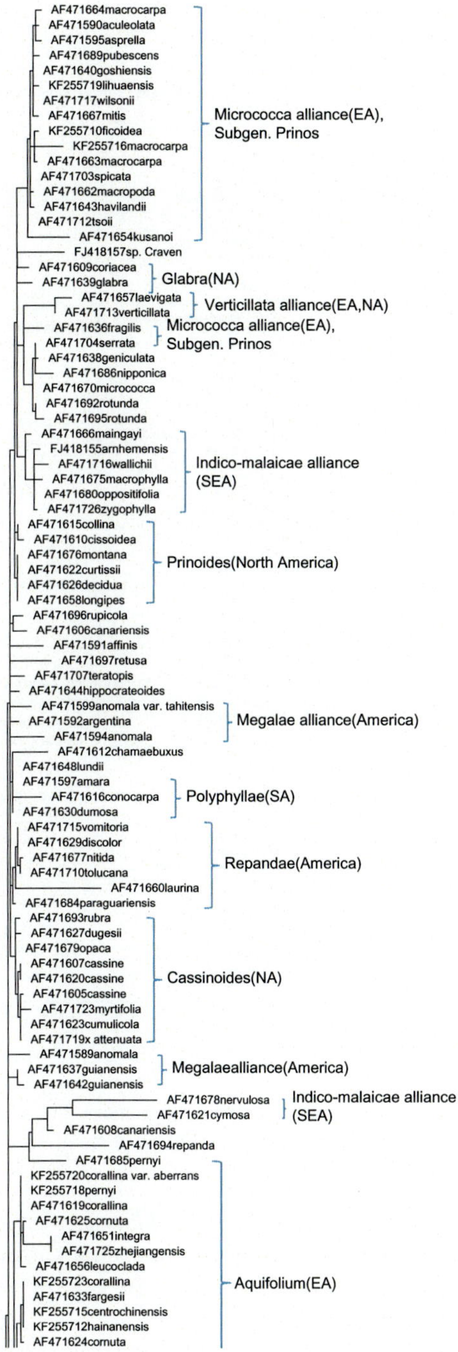

Figure13.5 Continued.

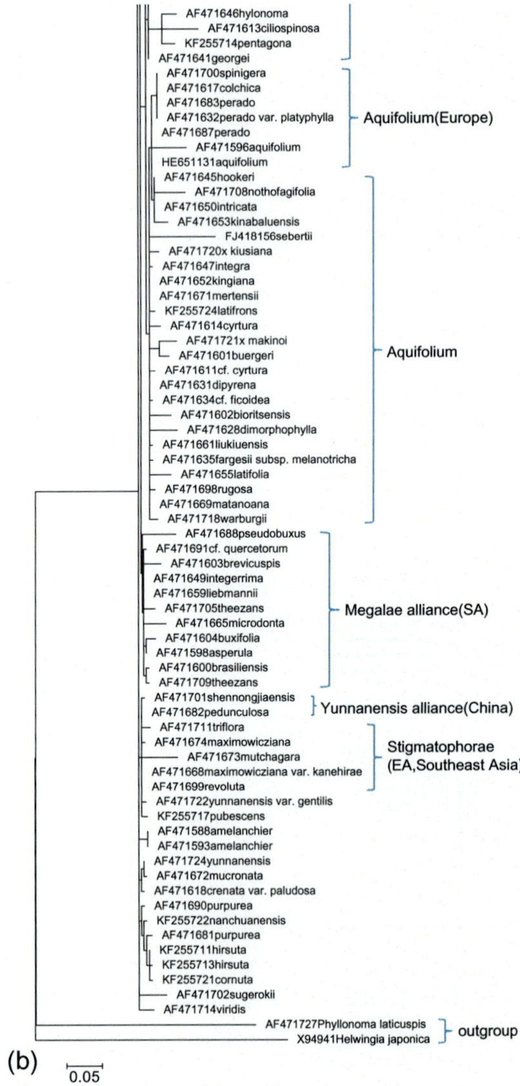

Figure13.5 Continued. (b) 159 atpB-rbcL spacer sequences. There were (a) 1088 and (b) 1135 positions in the final data set, respectively. Evolutionary analyses were conducted in MEGA6.

(Figure 13.5a). *I. fargesii* is closer to *I. centrochinensis* than to other Chinese species, and their distribution region is overlapping in central China. *I. chinensis*, the most common *Ilex* in China, is closer to *I. corallina*. Subgenus *Prinos* is embedded in subgenus *Ilex* on both nuclear and plastid trees.

I. pubescens of *rotunda* alliance is closer to *yunnanensis* alliance than to *aquifolium*. The *Repandae* clade includes *I. paraguariensis* and *I. vomitoria*, which is basal to the above taxa. *I. latifolia* was found to be closer to *I. liukiuensis* than to *I. cornuta* (Gottleib et al., 2005). However, these studies did not include *I. kudingcha* and *I. pentagona*. We sequenced 11 chloroplast noncoding regions for five *Ilex* species (Hao et al., 2012c and unpublished data). *I. latifolia*, *I. kudingcha*, *I. cornuta*, and *I. pentagona* intermingled in a cluster and cannot discriminate from each other in the chloroplast phylogenetic tree. *I. paraguariensis* is basal to these four China species. The significant discrepancy between nuclear and chloroplast phylogenies might be due to intraspecific polymorphism, hybridization, and introgressions (Manen et al., 2010). The most recent common ancestor (MRCA) of extant species is dated from the Miocene, while the fossil records suggest that the *Ilex* stem lineage originated in the Late Cretaceous, implying extensive lineage extinctions between the Cretaceous and Miocene and explaining the difficulties in clarifying the relationships between *Ilex* and its closest relatives (Manen et al., 2010). The MRCA ancestral area is in the North America and/or East Asia. Several bidirectional North America/East Asia and North America/South America dispersal events are suggested to explain current geographic and phylogenetic patterns. Hybridization and introgression events between distantly related lineages are inferred, indicating weak reproductive barriers between *Ilex* species. This characteristic might be useful in selecting cultivars with desired pharmaceutical substances.

13.6.2 Molecular marker

Intersimple sequence repeat (ISSR), random amplified polymorphic DNA (RAPD), and AFLP markers have been used to guide interspecific hybridization and germplasm improvement (Qian et al., 2008; Zhou et al., 2009; Zhang et al., 2011). Population studies on *I. paraguariensis* (Gauer and Cavalli-Molina, 2000), *I. aquifolium* (Rendell and Ennos, 2003), *I. leucoclada* (Torimaru et al., 2003), *I. canariensis* (Manen, 2004), and *I. cornuta* (Son et al., 2007) reveal very high population diversity and polymorphism. More studies are needed at population or species level to ensure a better characterization of *Ilex* species.

Using microsatellites cloned from an *I. paraguariensis* shotgun genomic library, 25 pairs of primers were designed and synthesized (Pereira et al., 2013). Levels of polymorphism were evaluated in 24 individuals from two populations. Twenty loci were polymorphic, and an average of 4.8 and 4.5 alleles per locus were detected in the two populations, respectively. The mean observed heterozygosity was lower than the expected heterozygosity (0.54 vs. 0.60), indicating a departure from the Hardy–Weinberg equilibrium and suggesting endogamy in both populations. The reported set of markers is highly informative and constitutes a powerful tool for the genetic characterization studies in *I. paraguariensis*.

13.6.3 Genomics

Representational difference analysis (RDA) was adopted to screen *I. paraguariensis* genomes (Gottlieb and Poggio, 2010). The occurrence of sex-related genomic differences was studied to develop an early gender detection molecular method that could help reduce energy inputs during the mate tea processing and that could be helpful in breeding programs. Fragments isolated in RDA assays fall into three categories. The first category is specific to *I. paraguariensis*. The second category comprises sequences identified as organellar or ribosomal plant DNA. The third category consists of clones representing conserved domains of retrotransposons (RNaseH, integrases, and/or chromodomains), including Ty3/gypsy retrotransposons and Ty1/copia retroelements, which are associated with sex determination regions of the Solanaceae, Caricaceae, and Salicaceae. These sequences could be used in phylogenetic analysis and species authentication. RDA and other powerful methods, such as fosmid genomic library (Hao et al., 2011a) and the second-generation high-throughput sequencing (Hao et al., 2011b, 2012a,b), could be of great help in gaining deep insight into the *Ilex* genome structure and function and gene expression regulation and pharmacophylogeny.

I. asprella, which contains abundant α-amyrin-type triterpenoid saponins, is an anti-influenza herbal drug widely used in south China. The transcriptome of the *I. asprella* root was deciphered using RNA-Seq (Zheng et al., 2014), which provided a data set for functional gene mining. An Illumina HiSeq™ 2000 generated 55,028,452 clean reads. De novo assembly of these reads generated 51,865 unigenes, in which 39,269 unigenes were annotated (75.71% yield). According to the structures of the triterpenoid saponins of *I. asprella*, a putative biosynthetic pathway downstream of 2,3-oxidosqualene was proposed, and candidate unigenes in the transcriptome data that were potentially involved in the pathway were screened using homology-based BLAST and phylogenetic analysis. Further amplification and functional analysis of these putative unigenes will provide insight into the biosynthesis of *Ilex* triterpenoid saponins.

13.7 Conclusions

Mate tea and large-leaved Kudingcha are widely consumed nonalcoholic beverages in South America and East Asia, respectively, which are gaining rapid popularity into the world market, either as tea itself or as an element in formulated foods or dietary supplements. The local people have used them for centuries as a social and medicinal beverage. Mate is hypocholesterolemic, hepatoprotective, a central nervous system stimulant, and a diuretic and is beneficial to cardiovascular system and obesity management. Mate has a high antioxidant capacity and is associated with both the prevention and the cause of some types of cancers. Kudingcha and the related Chinese *Ilex* regulate lipid metabolism, antagonize cardiovascular diseases, and exert antioxidant and anti-inflammatory effects. Terpenoid/saponin and polyphenol/flavonoid extracted from *Ilex* plants have distinct pharmacological effects. Novel pharmacological effects of different species and various constituents are being revealed. Both mate and Kudingcha have gained public attention outside of their traditional use region, and research on non-*Camellia* tea has been expanding. This review presents the usage, chemistry, biological activities, health effects, and some taxonomic considerations of mate tea and Kudingcha. *I. paraguariensis*, *I. kudingcha*, and other *Ilex* plants have great potential as the source of biological compounds for the nutraceutical and pharmaceutical industry. To date, the regulation of *Ilex* biological processes at the genomic level, epigenomic level, transcriptional and posttranscriptional levels, and translational and posttranslational levels is totally unknown, although such knowledge is essential for the sustainable development and utilization of the *Ilex* medicinal and food resources. The integration of systems biology and omics techniques would dramatically change the whole scenario of *Ilex* research and development.

References

Andrade, F.D., Piacente, S., Pizza, C., Vilegas, W., 2004. Arbutin-2′-sulphonyl from the infusion of *Ilex theezans* leaves. Fitoterapia 75, 782–784.

Arçari, D.P., Bartchewsky, W., dos Santos, T.W., et al., 2009. Antiobesity effects of yerba maté extract (*Ilex paraguariensis*) in high-fat diet-induced obese mice. Obesity 17, 2127–2133.

Arçari, D.P., Bartchewsky Jr., W., dos Santos, T.W., et al., 2011. Anti-inflammatory effects of yerba maté extract (*Ilex paraguariensis*) ameliorate insulin resistance in mice with high fat diet-induced obesity. Mol. Cell. Endocrinol. 335, 110–115.

Barg, M., Rezin, G.T., Leffa, D.D., et al., 2014. Evaluation of the protective effect of *Ilex paraguariensis* and *Camellia sinensis* extracts on the prevention of oxidative damage caused by ultraviolet radiation. Environ. Toxicol. Pharmacol. 37 (1), 195–201.

Bastos, D.H., Saldanha, L.A., Catharino, R.R., et al., 2007. Phenolic antioxidants identified by ESI-MS from Yerba maté (*Ilex paraguariensis*) and green tea (*Camellia sinensis*) extracts. Molecules 12, 423–432.

Battagim, J., Souza, V.T., Miyasaka, N.R., et al., 2012. Comparative study of the effect of green and roasted water extracts of mate (*Ilex paraguariensis*) on glucosyltransferase activity of *Streptococcus mutans*. J. Enzyme Inhib. Med. Chem. 27, 232–240.

Berté, K.A., Beux, M.R., Spada, P.K., et al., 2011. Chemical composition and antioxidant activity of yerba-mate (*Ilex paraguariensis* A.St.-Hil., Aquifoliaceae) extract as obtained by spray drying. J. Agric. Food Chem. 59, 5523–5527.

Boaventura, B.C., Di Pietro, P.F., Stefanuto, A., et al., 2012. Association of mate tea (*Ilex paraguariensis*) intake and dietary intervention and effects on oxidative stress biomarkers of dyslipidemic subjects. Nutrition 28, 657–664.

Bojić, M., Simon Haas, V., Sarić, D., Maleš, Z., 2013. Determination of flavonoids, phenolic acids, and xanthines in mate tea (*Ilex paraguariensis* St.-Hil.). J. Anal. Methods Chem. 2013, 658596. http://dx.doi.org/10.1155/2013/658596.

Bortoluzzi, M.C., Campagnoli, E.B., Milan, J.R., et al., 2014. Frequency of micronucleus in oral epithelial cells after exposure to mate-tea in healthy humans. Med. Oral Patol. Oral Cir. Bucal 19 (4), e345–e349.

Bracesco, N., Sanchez, A.G., Contreras, V., et al., 2011. Recent advances on *Ilex paraguariensis* research: minireview. J. Ethnopharmacol. 136, 378–384.

Branco Cdos, S., Scola, G., Rodrigues, A.D., et al., 2013. Anticonvulsant, neuroprotective and behavioral effects of organic and conventional yerba mate (*Ilex paraguariensis* St. Hil.) on pentylenetetrazol-induced seizures in Wistar rats. Brain Res. Bull. 92, 60–68.

Burris, K.P., Davidson, P.M., Stewart Jr., C.N., Harte, F.M., 2011. Antimicrobial activity of Yerba Mate (*Ilex paraguariensis*) aqueous extracts against *Escherichia coli* O157:H7 and *Staphylococcus aureus*. J. Food Sci. 76, M456–M462.

Burris, K.P., Davidson, P.M., Stewart Jr., C.N., et al., 2012. Aqueous extracts of yerba mate (*Ilex paraguariensis*) as a natural antimicrobial against *Escherichia coli* O157:H7 in a microbiological medium and pH 6.0 apple juice. J. Food Prot. 75, 753–757.

Chen, X.Q., Yang, J., Liu, X.X., et al., 2009a. Complete assignments of ^{1}H and ^{13}C NMR spectral data for three new triterpenoid saponins from *Ilex hainanensis* Merr. Magn. Reson. Chem. 47, 169–173.

Chen, X.Q., Zan, K., Yang, J., et al., 2009b. A novel flavanone from *Ilex hainanensis* Merr. Nat. Prod. Res. 23, 442–447.

Choi, Y.H., Sertic, S., Kim, H.K., et al., 2005. Classification of *Ilex* species based on metabolomic fingerprinting using nuclear magnetic resonance and multivariate data analysis. J. Agric. Food Chem. 53, 1237–1245.

Conforti, A.S., Gallo, M.E., Saraví, F.D., 2012. Yerba Mate (*Ilex paraguariensis*) consumption is associated with higher bone mineral density in postmenopausal women. Bone 50, 9–13.

Cunha, F.L., Silva, C.M., Almeida, M.G., et al., 2011. Reduction in oxidative stress levels in the colonic mucosa without fecal stream after the application of enemas containing aqueous *Ilex paraguariensis* extract. Acta Cir. Bras. 26, 289–296.

Dahmer, T., Berger, M., Barlette, A.G., et al., 2012. Antithrombotic effect of chikusetsusaponin IVa isolated from *Ilex paraguariensis* (Maté). J. Med. Food 15 (12), 1073–1080.

Dai, W.P., Li, G., Li, X., et al., 2014. The roots of *Ilex asprella* extract lessens acute respiratory distress syndrome in mice induced by influenza virus. J. Ethnopharmacol. 155 (3), 1575–1582.

Dartora, N., de Souza, L.M., Paiva, S.M., et al., 2013. Rhamnogalacturonan from *Ilex paraguariensis*: a potential adjuvant in sepsis treatment. Carbohydr. Polym. 92 (2), 1776–1782.

de Morais, E.C., Stefanuto, A., Klein, G.A., et al., 2009. Consumption of yerba mate (*Ilex paraguariensis*) improves serum lipid parameters in healthy dyslipidemic subjects and provides an additional LDL-cholesterol reduction in individuals on statin therapy. J. Agric. Food Chem. 57, 8316–8324.

de Souza, L.M., Dartora, N., Scoparo, C.T., et al., 2011. Comprehensive analysis of maté (*Ilex paraguariensis*) compounds: development of chemical strategies for matesaponin analysis by mass spectrometry. J. Chromatogr. A 1218, 7307–7315.

Fan, S., Zhang, Y., Hu, N., et al., 2012. Extract of Kuding tea prevents high-fat diet-induced metabolic disorders in C57BL/6 mice via liver X receptor (LXR) β antagonism. PLoS One 7 (12), e51007.

Fan, J., Wu, Z., Zhao, T., et al., 2014. Characterization, antioxidant and hepatoprotective activities of polysaccharides from *Ilex latifolia* Thunb. Carbohydr. Polym. 101, 990–997.

Feng, F., Zhu, M.X., Xie, N., et al., 2008. Two new triterpenoid saponins from the root of *Ilex pubescens*. J. Asian Nat. Prod. Res. 10, 71–75.

Fernandes, E.S., Machado Mde, O., Becker, A.M., et al., 2012. Yerba mate (*Ilex paraguariensis*) enhances the gene modulation and activity of paraoxonase-2: in vitro and in vivo studies. Nutrition 28 (11–12), 1157–1164.

Filip, R., Ferraro, G.E., 2003. Researching on new species of "Mate": *Ilex brevicuspis*: phytochemical and pharmacology study. Eur. J. Nutr. 42, 50–54.

Filip, R., Davicino, R., Anesini, C., 2010. Antifungal activity of the aqueous extract of *Ilex paraguariensis* against *Malassezia furfur*. Phytother. Res. 24, 715–719.

Fujii, T.M., Jacob, P.S., Yamada, M., et al., 2014. Yerba Mate (*Ilex paraguariensis*) modulates NF-kappaB pathway and AKT expression in the liver of rats fed on a high-fat diet. Int. J. Food Sci. Nutr. 65 (8), 967–976 (Epub ahead of print).

Garcia Londoño, V.A., Reynoso, M., Resnik, S., 2014. Polycyclic aromatic hydrocarbons (PAHs) in yerba mate (*Ilex paraguariensis*) from the Argentinean market. Food Addit. Contam. Part B Surveill. 7 (4), 247–253 (Epub ahead of print).

Gauer, L., Cavalli-Molina, S., 2000. Genetic variation in natural populations of mate (*Ilex paraguariensis* A. St.-Hil., Aquifoliaceae) using RAPD markers. Heredity 6, 647–656.

Golozar, A., Fagundes, R.B., Etemadi, A., et al., 2012. Significant variation in the concentration of carcinogenic polycyclic aromatic hydrocarbons in yerba maté samples by brand, batch, and processing method. Environ. Sci. Technol. 46 (24), 13488–13493.

Gonzalez-Gil, F., Diaz-Sanchez, S., Pendleton, S., et al., 2014. Yerba mate enhances probiotic bacteria growth in vitro but as a feed additive does not reduce *Salmonella Enteritidis* colonization in vivo. Poult. Sci. 93 (2), 434–440.

Gosmann, G., Barlette, A.G., Dhamer, T., et al., 2012. Phenolic compounds from Maté (*Ilex paraguariensis*) inhibit adipogenesis in 3T3-L1 preadipocytes. Plant Foods Hum. Nutr. 67, 156–161.

Gottleib, A.M., Giberti, G.C., Poggio, L., 2005. Molecular analyses of the genus *Ilex* (Aquifoliaceae) in southern South America, evidence from AFLP and ITS sequence data. Am. J. Bot. 92, 352–369.

Gottlieb, A.M., Poggio, L., 2010. Genomic screening in dioecious "yerba mate" tree (*Ilex paraguariensis* A. St. Hill., Aquifoliaceae) through representational difference analysis. Genetica 138, 567–578.

Gugliucci, A., Bastos, D.H., 2009. Chlorogenic acid protects paraoxonase 1 activity in high density lipoprotein from inactivation caused by physiological concentrations of hypochlorite. Fitoterapia 80, 138–142.

Gugliucci, A., Bastos, D.H., Schulze, J., Souza, M.F., 2009. Caffeic and chlorogenic acids in *Ilex paraguariensis* extracts are the main inhibitors of AGE generation by methylglyoxal in model proteins. Fitoterapia 80, 339–344.

Hao, D.C., Yang, L., Xiao, P.G., 2011a. The first insight into the *Taxus* genome via fosmid library construction and end sequencing. Mol. Genet. Genomics 285, 197–205.

Hao, D.C., Ge, G.B., Xiao, P.G., et al., 2011b. The first insight into the tissue specific taxus transcriptome via Illumina second generation sequencing. PLoS One 6 (6), e21220.

Hao, D.C., Chen, S.L., Xiao, P.G., 2012a. Application of high-throughput sequencing in the medicinal plant transcriptome studies. Drug Dev. Res. 73 (8), 487–498.

Hao, D.C., Ma, P., Mu, J., Chen, S.L., Xiao, P.G., et al., 2012b. De novo characterization of the root transcriptome of a traditional Chinese medicinal plant *Polygonum cuspidatum*. Sci. China Life Sci. 55, 452–466.

Hao, D.C., Xiao, P.G., Peng, Y., et al., 2012c. Evaluation of the chloroplast barcoding markers by mean and smallest interspecific distances. Pak. J. Bot. 44, 1271–1274.

Hu, X.Y., Shu, X.C., Guo, Y., Ma, Y., 2012. Effect of an *Ilex asprella* root decoction on the related genes of lipid metabolism from chronic stress and hyperlipidemic fatty liver in rats. Chin. Med. J. (Engl.) 125 (19), 3539–3542.

Hussein, G.M., Matsuda, H., Nakamura, S., et al., 2011a. Mate tea (*Ilex paraguariensis*) promotes satiety and body weight lowering in mice: involvement of glucagon-like peptide-1. Biol. Pharm. Bull. 34, 1849–1855.

Hussein, G.M., Matsuda, H., Nakamura, S., et al., 2011b. Protective and ameliorative effects of maté (*Ilex paraguariensis*) on metabolic syndrome in TSOD mice. Phytomedicine 19, 88–97.

Im, K.R., Jeong, T.S., Kwon, B.M., et al., 2006. Acyl-CoA: cholesterol acyltransferase inhibitors from *Ilex macropoda*. Arch. Pharm. Res. 29, 191–194.

Jaiswal, R., Halabi, E.A., Karar, M.G., Kuhnert, N., 2014. Identification and characterisation of the phenolics of *Ilex glabra* L. Gray (Aquifoliaceae) leaves by liquid chromatography tandem mass spectrometry. Phytochemistry 106, 141–155.

Jiang, Z.H., Wang, J.R., Li, M., et al., 2005. Hemiterpene glucosides with anti-platelet aggregation activities from *Ilex pubescens*. J. Nat. Prod. 68, 397–399.

Kang, Y.R., Lee, H.Y., Kim, J.H., et al., 2012. Anti-obesity and anti-diabetic effects of Yerba Mate (*Ilex paraguariensis*) in C57BL/6J mice fed a high-fat diet. Lab. Anim. Res. 28, 23–29.

Kim, H.K., Saifullah, Khan, S., et al., 2010. Metabolic classification of South American *Ilex* species by NMR-based metabolomics. Phytochemistry 71, 773–784.

Kim, J.Y., Jeong, H.Y., Lee, H.K., et al., 2011. Protective effect of *Ilex latifolia*, a major component of "kudingcha", against transient focal ischemia-induced neuronal damage in rats. J. Ethnopharmacol. 133, 558–564.

Kim, J.Y., Lee, H.K., Hwang, B.Y., et al., 2012a. Neuroprotection of *Ilex latifolia* and caffeoylquinic acid derivatives against excitotoxic and hypoxic damage of cultured rat cortical neurons. Arch. Pharm. Res. 35 (6), 1115–1122.

Kim, M.H., Park, K.H., Oh, M.H., et al., 2012b. Two new hemiterpene glycosides from the leaves of *Ilex rotunda*. Thunb. Arch. Pharm. Res. 35 (10), 1779–1784.

Kim, Y.J., Jung, E.B., Lee, M.S., et al., 2014. Rotundarpene inhibits toll-like receptor 2 activation-induced production of inflammatory mediators in keratinocytes by suppressing the Akt and NF-κB pathways. Int. Immunopharmacol. 18 (2), 325–332.

Klein, G.A., Stefanuto, A., Boaventura, B.C., et al., 2011. Mate tea (*Ilex paraguariensis*) improves glycemic and lipid profiles of type 2 diabetes and pre-diabetes individuals: a pilot study. J. Am. Coll. Nutr. 30, 320–332.

Lee, C.S., Lee, S.A., Kim, Y.J., et al., 2011. 3,4,5-Tricaffeoylquinic acid inhibits tumor necrosis factor-α-stimulated production of inflammatory mediators in keratinocytes via suppression of Akt- and NF-κB-pathways. Int. Immunopharmacol. 11, 1715–1723.

Lee, D.H., Nam, Y.J., Kim, Y.J., et al., 2014. Rotundarpene prevents TRAIL-induced apoptosis in human keratinocytes by suppressing the caspase-8- and Bid-pathways and the mitochondrial pathway. Naunyn Schmiedebergs Arch. Pharmacol. 387 (12), 1209–1219 (Epub ahead of print).

Lei, Y., Shi, S., Song, Y., et al., 2014. Triterpene saponins from the roots of *Ilex asprella*. Chem. Biodivers. 11 (5), 767–775.

Leonard, S.S., Hogans, V.J., Coppes-Petricorena, Z., et al., 2010. Analysis of free-radical scavenging of Yerba Mate (*Ilex paraguariensis*) using electron spin resonance and radical-induced DNA damage. J. Food Sci. 75, C14–C20.

Li, L., Xu, L.J., Peng, Y., Shi, R.B., Xiao, P.G., 2011a. Comparison of green tea and four other kind of teas. China J. Chin. Mater. Med. 36, 5–10.

Li, L., Zhang, Y., Zhang, P., et al., 2011b. Appraisal of anti-inflammatory and free radical scavenging activities of ethanol extract of *Ilex ficoidea* Hemsl and *Ilex centrochinensis* S.Y. Hu. Environ. Toxicol. Pharmacol. 32, 122–127.

Li, L., He, Y.X., Gou, M., Dai, C., 2012a. Three new triterpenoid saponins from *Ilex pubescens*. J. Asian Nat. Prod. Res. 14 (12), 1169–1174.

Li, L., Peng, Y., Ma, G., He, C., Feng, Y., Lei, Q., Xiao, P.G., 2012b. Quantitative analysis of five kudinosides in the large-leaved Kudingcha and related species from the genus *Ilex* by UPLC-ELSD. Phytochem. Anal. 23 (6), 677–683.

Li, J.J., Yang, J., Cui, W.X., et al., 2013. Analysis of therapeutic effect of Ilex hainanensis Merr. extract on nonalcoholic fatty liver disease through urine metabolite profiling by ultraperformance liquid chromatography/quadrupole time of flight mass spectrometry. Evid. Based Complement. Alternat. Med. 2013, 451975.

Li, L., Feng, L.S., He, Y.X., 2014a. Cytotoxic triterpene saponins from *Ilex pubescens*. J. Asian Nat. Prod. Res. 16 (8), 830–835.

Li, S., Zhao, J., Liu, Y., et al., 2014b. New triterpenoid saponins from *Ilex cornuta* and their protective effects against H_2O_2-induced myocardial cell injury. J. Agric. Food Chem. 62 (2), 488–496.

Lima, M.E., Colpo, A.C., Salgueiro, W.G., et al., 2014. *Ilex paraguariensis* extract increases lifespan and protects against the toxic effects caused by paraquat in *Caenorhabditis elegans*. Int. J. Environ. Res. Public Health 11 (10), 10091–10104.

Lindsey, K., Pullen, M.L., Topping, J.F., 2003. Importance of plant sterols in pattern formation and hormone signalling. Trends Plant Sci. 8, 521–525.

Liu, J., Ding, P., 2008. Research progress of *Ilex* medicinal plant resources, chemical constituents and pharmacology. J. Guangzhou Univ. Trad. Chin. Med. 25, 277–280.

Lückemeyer, D.D., Müller, V.D., Moritz, M.I., et al., 2012. Effects of *Ilex paraguariensis* A. St. Hil. (yerba mate) on herpes simplex virus types 1 and 2 replication. Phytother. Res. 26, 535–540.

Lunceford, N., Gugliucci, A., 2005. *Ilex paraguariensis* extracts inhibit AGE formation more efficiently than green tea. Fitoterapia 76, 419–427.

Luo, X.Y., Li, N.N., Liang, Y.R., 2013. Effects of *Ilex latifolia* and *Camellia sinensis* on cholesterol and circulating immune complexes in rats fed with a high-cholesterol diet. Phytother. Res. 27 (1), 62–65.

Manen, J.F., 2004. Are both sympatric species *Ilex perado* and *Ilex canariensis* secretly hybridizing? Indication from nuclear markers collected in Tenerife. BMC Evol. Biol. 4, 46.

Manen, J.F., Barriera, G., Loizeau, P.A., Naciri, Y., 2010. The history of extant *Ilex* species (Aquifoliaceae): evidence of hybridization within a Miocene radiation. Mol. Phylogenet. Evol. 57, 961–977.

Maria-Ferreira, D., Dartora, N., da Silva, L.M., et al., 2013. Chemical and biological characterization of polysaccharides isolated from *Ilex paraguariensis* A. St.-Hil. Int. J. Biol. Macromol. 59, 125–133.

Martins, F., Noso, T.M., Porto, V.B., et al., 2010. Maté tea inhibits in vitro pancreatic lipase activity and has hypolipidemic effect on high-fat diet-induced obese mice. Obesity 18, 42–47.

Matsumoto, R.L., Bastos, D.H., Mendonça, S., et al., 2009a. Effects of mate tea (*Ilex paraguariensis*) ingestion on mRNA expression of antioxidant enzymes, lipid peroxidation, and total antioxidant status in healthy young women. J. Agric. Food Chem. 57, 1775–1780.

Matsumoto, R.L., Mendonça, S., de Oliveira, D.M., et al., 2009b. Effects of maté tea intake on ex vivo LDL peroxidation induced by three different pathways. Nutrients 1, 18–29.

Mullié, C., Jonet, A., Dassonville-Klimpt, A., et al., 2010. Inhibitory effect of ursolic acid derivatives on hydrogen peroxide- and glutathione-mediated degradation of hemin: a possible additional mechanism of action for antimalarial activity. Exp. Parasitol. 125, 202–207.

Nahar, L., Russell, W.R., Middleton, M., et al., 2005. Antioxidant phenylacetic acid derivatives from the seeds of *Ilex aquifolium*. Acta Pharm. 55, 187–193.

Noratto, G.D., Kim, Y., Talcott, S.T., Mertens-Talcott, S.U., 2011. Flavonol-rich fractions of yaupon holly leaves (Ilex vomitoria, Aquifoliaceae) induce microRNA-146a and have anti-inflammatory and chemopreventive effects in intestinal myofibroblast CCD-18Co cells. Fitoterapia 82, 557–569.

Oliveira, D.M., Freitas, H.S., Souza, M.F., et al., 2008. Yerba Maté (*Ilex paraguariensis*) aqueous extract decreases intestinal SGLT1 gene expression but does not affect other biochemical parameters in alloxan-diabetic Wistar rats. J. Agric. Food Chem. 56, 10527–10532.

Ouyang, M.A., 2003. Glycosides from the leaves of *Ilex hylonoma*. Nat. Prod. Res. 17, 183–188.

Ouyang, M.A., Wu, C.L., 2003. Three new saponins from the leaves of *Ilex hylonoma*. J. Asian Nat. Prod. Res. 5, 89–94.

Ouyang, M.A., Wang, C.Z., Wang, S.B., 2007. Water-soluble constituents from the leaves of *Ilex oblonga*. J. Asian Nat. Prod. Res. 9, 399–405.

Peralta, I.N., Cogoi, L., Filip, R., Anesini, C., 2013. Prevention of hydrogen peroxide-induced red blood cells lysis by *Ilex paraguariensis* aqueous extract: participation of phenolic and xanthine compounds. Phytother. Res. 27 (2), 192–198.

Pereira, M.F., Ciampi, A.Y., Inglis, P.W., et al., 2013. Shotgun sequencing for microsatellite identification in *Ilex paraguariensis* (Aquifoliaceae). Appl. Plant Sci. 1 (3), 1200245.

Pires, V.S., Gosmann, G., Guillaume, D., Schenkel, E.P., 2002. Triterpenes and saponins from *Ilex psammophila*. Nat. Prod. Lett. 16, 401–406.

Pirker, K.F., Goodman, B.A., 2010. Caffeoylquinic acid derived free radicals identified during antioxidant reactions of bitter tea (*Ilex latifolia* and *Ilex kudingcha*). Food Funct. 1, 262–268.

Prado Martin, J.G., Porto, E., de Alencar, S.M., et al., 2013. Antimicrobial activity of yerba mate (*Ilex paraguariensis* St. Hil.) against food pathogens. Rev. Argent. Microbiol. 45 (2), 93–98.

Prediger, R.D., Fernandes, M.S., Rial, D., et al., 2008. Effects of acute administration of the hydroalcoholic extract of mate tea leaves (*Ilex paraguariensis*) in animal models of learning and memory. J. Ethnopharmacol. 120, 465–473.

Puangpraphant, S., de Mejia, E.G., 2009. Saponins in yerba mate tea (*Ilex paraguariensis* A. St.-Hil) and quercetin synergistically inhibit iNOS and COX-2 in lipopolysaccharide-induced macrophages through NFkappaB pathways. J. Agric. Food Chem. 57, 8873–8883.

Puangpraphant, S., Berhow, M.A., Vermillion, K., et al., 2011. Dicaffeoylquinic acids in Yerba mate (*Ilex paraguariensis* St. Hilaire) inhibit NF-κB nucleus translocation in macrophages and induce apoptosis by activating caspases-8 and -3 in human colon cancer cells. Mol. Nutr. Food Res. 55, 1509–1522.

Qian, Y.S., Wang, H.Z., Shi, N.N., et al., 2008. Studies of genetic diversity among ten species of *Ilex* based on RAPD and AFLPs. J. Mol. Cell Biol. 41, 35–43.

Ramirez-Mares, M.V., Chandra, S., de Mejia, E.G., 2004. In vitro chemopreventive activity of *Camellia sinensis*, *Ilex paraguariensis* and *Ardisia compressa* tea extracts and selected polyphenols. Mutat. Res. 554, 53–65.

Reis Ede, M., Schreiner Neto, F.W., Cattani, V.B., et al., 2014. Antidepressant-like effect of *Ilex paraguariensis* in rats. Biomed. Res. Int. 2014, 958209.

Rendell, S., Ennos, R.A., 2003. Chloroplast DNA diversity of the dioecious European tree *Ilex aquifolium* L. (English holly). Mol. Ecol. 12, 2681–2688.

Resende, P.E., Verza, S.G., Kaiser, S., et al., 2012. The activity of mate saponins (*Ilex paraguariensis*) in intra-abdominal and epididymal fat, and glucose oxidation in male Wistar rats. J. Ethnopharmacol. 144 (3), 735–740.

Silva, R.D., Bueno, A.L., Gallon, C.W., et al., 2011. The effect of aqueous extract of gross and commercial yerba mate (*Ilex paraguariensis*) on intra-abdominal and epididymal fat and glucose levels in male Wistar rats. Fitoterapia 82, 818–826.

Silveira, F., Rossi, S., Fernández, C., et al., 2011. Alum-type adjuvant effect of non-haemolytic saponins purified from *Ilex* and *Passiflora* spp. Phytother. Res. 25, 1783–1788.

Son, S.W., Kim, J.H., Kim, Y.S., Park, S.J., 2007. ITS sequences variations in populations of *Ilex cornuta* (Aquifoliaceae). Korean J. Plant Taxon. 37, 131–141.

Song, C., Xie, C., Zhou, Z., et al., 2012. Antidiabetic effect of an active components group from *Ilex kudingcha* and its chemical composition. Evid. Based Complement. Alternat. Med. 2012, 423690.

Song, J.L., Qian, Y., Li, G.J., Zhao, X., 2013. Anti-inflammatory effects of kudingcha methanol extract (*Ilex kudingcha* C.J. Tseng) in dextran sulfate sodium-induced ulcerative colitis. Mol. Med. Rep. 8 (4), 1256–1262.

Sugimoto, S., Nakamura, S., Yamamoto, S., et al., 2009. Brazilian natural medicines. III. Structures of triterpene oligoglycosides and lipase inhibitors from mate, leaves of Ilex paraguariensis. Chem. Pharm. Bull. 57, 257–261.

Taketa, A.T., Gnoatto, S.C., Gosmann, G., et al., 2004. Triterpenoids from Brazilian *Ilex* species and their in vitro antitrypanosomal activity. J. Nat. Prod. 67, 1697–1700.

Tang, L., Jiang, Y., Chang, H.T., et al., 2005. Triterpene saponins from the leaves of *Ilex kudingcha*. J. Nat. Prod. 68, 1169–1174.

Tang, L., Jiang, Y., Tian, X.M., et al., 2009. Triterpene saponins from the leaves of *Ilex kudingcha*. J. Asian Nat. Prod. Res. 11, 554–561.

Torimaru, T., Tomaru, N., Nishimura, N., Yamamoto, S., 2003. Clonal diversity and genetic differentiation in *Ilex leucoclada* M. patches in an old-growth beech forest. Mol. Ecol. 12, 809–818.

Vieira, M.A., Maraschin, M., Pagliosa, C.M., et al., 2010. Phenolic acids and methylxanthines composition and antioxidant properties of mate (*Ilex paraguariensis*) residue. J. Food Sci. 75, C280–C285.

Wang, J.R., Zhou, H., Jiang, Z.H., Liu, L., 2008a. Two new triterpene saponins from the anti-inflammatory saponin fraction of *Ilex pubescens* root. Chem. Biodivers. 5, 1369–1376.

Wang, J.R., Zhou, H., Jiang, Z.H., et al., 2008b. In vivo anti-inflammatory and analgesic activities of a purified saponin fraction derived from the root of *Ilex pubescens*. Biol. Pharm. Bull. 31, 643–650.

Wang, C.Q., Wang, L., Fan, C.L., et al., 2012a. Ilelic acids A and B, two unusual triterpenes with a seven-membered ring from *Ilex latifolia*. Org. Lett. 14 (16), 4102–4105.

Wang, L., Cai, Y., Zhang, X.Q., et al., 2012b. New triterpenoid glycosides from the roots of *Ilex asprella*. Carbohydr. Res. 349, 39–43.

Wang, W., Zhao, J., Li, S., et al., 2014a. Five new triterpenoidal saponins from the roots of *Ilex cornuta* and their protective effects against H_2O_2-induced cardiomyocytes injury. Fitoterapia 99C, 40–47.

Wang, W.L., Zhou, X., Liu, Y.L., et al., 2014b. Two new 20α(H)-ursane-type triterpenoids from *Ilex cornuta* and their cytotoxic activities. J. Asian Nat. Prod. Res. 16 (2), 175–180.

Wei, M., Chu, X., Guan, M., et al., 2013. Protocatechuic acid suppresses ovalbumin-induced airway inflammation in a mouse allergic asthma model. Int. Immunopharmacol. 15 (4), 780–788.

Wnuk, M., Lewinska, A., Oklejewicz, B., et al., 2009. Evaluation of the cyto- and genotoxic activity of yerba mate (*Ilex paraguariensis*) in human lymphocytes in vitro. Mutat. Res. 679, 18–23.

Wu, Z.J., Ouyang, M.A., Wang, C.Z., Zhang, Z.K., 2007a. Six new triterpenoid saponins from the leaves of *Ilex oblonga* and their inhibitory activities against TMV replication. Chem. Pharm. Bull. 55, 422–427.

Wu, Z.J., Ouyang, M.A., Wang, C.Z., et al., 2007b. Anti-tobacco mosaic virus (TMV) triterpenoid saponins from the leaves of *Ilex oblonga*. J. Agric. Food Chem. 55, 1712–1717.

Wu, T., Zhang, Q.W., Zhang, X.Q., et al., 2012. Two new compounds from the roots of *Ilex pubescens*. Nat. Prod. Res. 26, 1408–1412.

Xie, G.B., Lei, L.D., Tu, P.F., 2007. Complete assignments of ^1H and ^{13}C NMR spectroscopic data for two new triterpenoid saponins from *Ilex pernyi*. Magn. Reson. Chem. 45, 997–1000.

Xie, G.B., Niu, F., Wang, X.J., et al., 2008. Chemical constituents from the leaves of *Ilex pernyi*. Yao Xue Xue Bao 43, 60–62.

Xie, G.B., Zhou, S.X., Lu, Y.N., et al., 2009. Triterpenoid glycosides from the leaves of *Ilex pernyi*. Chem. Pharm. Bull. 57, 520–524.

Xu, C., Luo, L., Tan, R.X., 2004. Antidepressant effect of three traditional Chinese medicines in the learned helplessness model. J. Ethnopharmacol. 91, 345–349.

Xu, G.H., Kim, Y.H., Choo, S.J., et al., 2009. Chemical constituents from the leaves of *Ilex paraguariensis* inhibit human neutrophil elastase. Arch. Pharm. Res. 32, 1215–1220.

Xu, G.H., Kim, Y.H., Choo, S.J., et al., 2010. Two acetylated megastigmane glycosides from the leaves of *Ilex paraguariensis*. Arch. Pharm. Res. 33, 369–373.

Yan, X.H., Hou, H.C., Zhang, L.W., Wang, Z., 2012. Analysis of cadmium in *Cortex Ilicis Rotundae*. Zhong Yao Cai 35 (4), 531–533.

Yang, X., Ding, Y., Sun, Z.H., Zhang, D.M., 2006. Studies on chemical constituents from *Ilex pubescens*. J. Asian Nat. Prod. Res. 8, 505–510.

Yan-Yun, C., Liang, Z., Ning, L., et al., 2012. Triterpenoid saponins from *Ilex mamillata* C.Y. Wu ex C.J. Tseng. Nat. Prod. Res. 26 (21), 1991–1995.

Zhang, A.L., Ye, Q., Li, B.G., et al., 2005. Phenolic and triterpene glycosides from the stems of *Ilex litseaefolia*. J. Nat. Prod. 68, 1531–1535.

Zhang, G.H., Zheng, D.J., Liu, G.M., Li, L.Y., 2008. Study on esterase isozymes of five *Kuding-cha* species in Aquifoliaceae. Chin. Agri. Sci. Bull. 24, 50–56.

Zhang, C.X., Lin, C.Z., Xiong, T.Q., et al., 2010. New triterpene saponins from the root of *Ilex pubescens*. Fitoterapia 81, 788–792.

Zhang, J.H., Gao, Y.Z., Zhang, B., Wang, Z.J., 2011. Genetic diversity of *Ilex* L. tree species. Acta Bot. Boreal-Occident Sin. 31, 504–510.

Zhang, Z., Fu, Q., Zheng, K., 2013. Three new triterpene glycosides from *Ilex asprella*. J. Asian Nat. Prod. Res. 15 (5), 453–458.

Zhao, Z., Lin, C., Zhu, C., He, W., 2013. A new triterpenoid glycoside from the roots of *Ilex asprella*. Chin. J. Nat. Med. 11 (4), 415–418.

Zheng, J., Tang, L., Xian, X.D., et al., 2009a. Inhibitory effect of triterpenoid saponins from the leaves of *Ilex kudingcha* on aggregated LDL-induced lipid deposition in macrophages. Planta Med. 75, 1410–1414.

Zheng, J., Wang, X., Li, H., et al., 2009b. Improving abnormal hemorheological parameters in $ApoE^{-/-}$ mice by *Ilex kudingcha* total saponins. Clin. Hemorheol. Microcirc. 42, 29–36.

Zheng, X., Xu, H., Ma, X., et al., 2014. Triterpenoid saponin biosynthetic pathway profiling and candidate gene mining of the *Ilex asprella* root using RNA-Seq. Int. J. Mol. Sci. 15 (4), 5970–5987.

Zhou, S., Yang, J., Liu, F., Tu, P., 2007. Complete assignments of ^1H and ^{13}C NMR spectroscopic data for two new triterpenoid saponins from *Ilex hainanensis*. Magn. Reson. Chem. 45, 179–181.

Zhou, X.J., Zhang, D.M., Luo, Y.L., et al., 2009. Inter-simple sequence repeats (ISSR) marker analysis of *Ilex* plants species and its application. J. Henan Agri. Univ. 43, 196–200.

Zhou, M., Xu, M., Ma, X.X., et al., 2012. Antiviral triterpenoid saponins from the roots of *Ilex asprella*. Planta Med. 78 (15), 1702–1705.

Zhou, Y., Chai, X., Zeng, K., et al., 2013a. Ilexpublesnins C-M, eleven new triterpene saponins from the roots of *Ilex pubescens*. Planta Med. 79 (1), 70–77.

Zhou, Z., Feng, Z., Fu, C., Zeng, L., 2013b. A new triterpene saponin from the roots of *Ilex pubescens*. Nat. Prod. Res. 27 (15), 1343–1347.

Zhu, F., Cai, Y.Z., Sun, M., et al., 2009. Comparison of major phenolic constituents and in vitro antioxidant activity of diverse Kudingcha genotypes from *Ilex kudingcha*, *Ilex cornuta*, and *Ligustrum robustum*. J. Agric. Food Chem. 57, 6082–6089.

Zhu, K., Li, G., Sun, P., et al., 2014. In vitro and in vivo anti-cancer activities of Kuding tea (*Ilex kudingcha* C.J. Tseng) against oral cancer. Exp. Ther. Med. 7 (3), 709–715.

Zuo, W.J., Dai, H.F., Chen, J., et al., 2011. Triterpenes and triterpenoid saponins from the leaves of *Ilex kudingcha*. Planta Med. 77, 1835–1840.

Zuo, W., Wang, Q., Li, W., et al., 2012a. Structure elucidation and NMR assignments of an unusual triterpene saponin derivative from *Ilex kudingcha*. Magn. Reson. Chem. 50, 325–328.

Zuo, W.J., Dai, H.F., Zeng, Y.B., et al., 2012b. Two new triterpenoid saponins from the leaves of *Ilex kudingcha*. J. Asian Nat. Prod. Res. 14, 308–313.

Phytochemical and biological research of *Salvia* medicinal resources

14

14.1 Introduction

Salvia is the largest genus of the family Lamiaceae having about 800 species throughout the world. Most of the plants of this genus are well known for their biologically active constituents, specifically those with antitumor activity. Several species of *Salvia* are used in folk medicines, for instance, *Salvia cavaleriei* is used for the treatment of dysentery, haemoptysis, boils, and fall injuries; *Salvia desoleria* for the treatment of menstrual, digestive, and central nervous system diseases; and *Salvia bucharica* as a traditional medicine for the treatment of liver disorders. The plants of this genus are rich in essential oils, and among their constituents, 1,8-cineol and guaiane-type mono- and sesquiterpenes are very common.

In China, over 40 *Salvia* species have been used as medicinal plants for various diseases (Table 14.1), some for thousands of years. The biological activities of *Salvia* medicinal plants used in traditional Chinese medicine (TCM) have been a research hot spot. Ten field surveys of *Salvia* medicinal plants were performed in 17 provinces including detailed information of their local names, growth environment, distribution, and therapeutic effects (Li et al., 2013a). 35 *Salvia* species have been used in TCM in different regions of China, including 20 species used as danshen to treat heart diseases and 15 species used to treat many other conditions including gynecologic diseases, muscular or skeletal problems, hepatitis, urological diseases, and mouth and eye ailments. Some *Salvia* species possess significant pharmacological activity in the context of ethnopharmacological knowledge, especially in the treatment of heart disease. This chapter summarizes the research progress on phytochemistry, bioactivity, phylogeny, and omics of the important *Salvia* medicinal plants.

14.2 Chemical components

Seven types of compounds are present in *Salvia* plants (Wu et al., 2012), namely, sesquiterpenoids (Figure 14.1), diterpenoids (Figure 14.2), sesterterpenoids (Figure 14.3), triterpenoids (Figure 14.4), steroids, polyphenols (Figure 14.5), and others (Figure 14.6).

14.2.1 Diterpenoids, essential oil, and other terpenoids

Diterpenes (Table 14.2) are one class of natural compounds that possess a core skeleton of 20 carbons. Their formation can be rationalized by considering the different types of cyclization of geranylgeranyl diphosphate (GGPP). Five types of diterpenoids

Medicinal plants: chemistry, biology and omics. http://dx.doi.org/10.1016/B978-0-08-100085-4.00014-1

Table 14.1 *Salvia* medicinal plants used in China

Species	Habitat	Therapeutic use	Diterpenoids	Caffeic acid derivative (group)
***Subgenus* Salvia**				
Section *Eurysphace*				
S. bulleyana	Northwest Yunnan, altitude 2100–3400 m, hillside, roadside	Root: used as danshen	Cryptotanshinone (S3), dihydrotanshinone I (S4), tanshinone I (S4), tanshinone IIA (S3)	Caffeic acid (A), danshensu (A), rosmarinic acid (B), salviaflaside (B), salvianolic acid B (D), salvianolic acid K (C)
S. brachyloma	Southwest Sichuan, northwest Yunnan, altitude 3200–3800 m, grass slope, woodside, grassland, forest	Root: used as danshen; flower: cough, hepatitis		
S. campanulata	Northwest Yunnan, altitude 2800–3800 m, ditch, forest, hillside, roadside, hassock	Root: used as danshen	Cryptotanshinone (S3), dihydrotanshinone I (S4), tanshinone I (S4), tanshinone IIA (S3)	Caffeic acid (A), danshensu (A), rosmarinic acid (B), salvianolic acid B (D)
S. castanea	Southwest Sichuan, northwest Yunnan, altitude 2500–2800 m, woodside, grassland	Root: used as danshen	Cryptotanshinone (S3), dihydrotanshinone I (S4), tanshinone I (S4), tanshinone IIA (S3)	Caffeic acid (A), danshensu (A), rosmarinic acid (B), salvianolic acid B (D)
S. digitaloides	Southwest Sichuan, northwest Yunnan, altitude 2500–3400 m, forest, shady wetland, grass slope	Root: used as danshen	16,17-Dinorpisferal A (S4), cryptotanshinone (S3), dihydroneotanshinlactone (S4), dihydrotanshinone I (S4), tanshinone I (S4), tanshinone IIA (S3)	Caffeic acid (A), danshensu (A), rosmarinic acid (A), salvianolic acid B (A)

S. evansiana	Southwest Sichuan, northwest Yunnan, altitude 3400–4200 m, mountain, forest, grass slope	Root: used as danshen	Cryptotanshinone (S3), dihydrotanshinone I (S4), tanshinone I (S4), tanshinone IIA (S3)	Caffeic acid (A), danshensu (A), lithospermic acid (C), rosmarinic acid (B)
S. flava	Southwest Sichuan, northwest Yunnan, altitude 2500–4000 m, mountain, forest, grass slope	Root: used as danshen	Cryptotanshinone (S3), dihydrotanshinone I (S4), tanshinone I (S4), tanshinone IIA (S3)	Caffeic acid (A), danshensu (A), rosmarinic acid (B), salviaflaside (B), salviaflaside methyl ester (B), salvianolic acid A (C), salvianolic acid B (D), salvianolic acid J (C), salvianolic acid K (C)
S. kiaometiensis	Northeast Yunnan, central south Sichuan, altitude 2500–3200 m, grass slope, roadside	Root: used as danshen		
S. maximowicziana	Southern Shaanxi, southern Gansu, western Henan, western Hubei, Sichuan, northeast Yunnan, Tibet, altitude 1500–3450 m, roadside, grass slope, woodside, hillside, mountain top, forest	Root: substitute for *Gentiana macrophylla* in west Sichuan; rheumatism, heat-clearing, pain; spastic muscle, joint pain, infantile fever	Cryptotanshinone (S3), dihydrotanshinone I (S4), tanshinone I (S4), tanshinone IIA (S3)	Caffeic acid (A), danshensu (A), lithospermic acid (C), rosmarinic acid (B), salvianolic acid B (D)

Continued

Table 14.1 **Continued**

Species	Habitat	Therapeutic use	Diterpenoids	Caffeic acid derivative (group)
S. przewalskii	Western Gansu, western Sichuan, northwest Yunnan, Tibet, Qinghai, altitude 2100–4050 m, woodside, roadside, ditch	Root: substitute for *Gentiana macrophylla* in southwest Sichuan; used as danshen in northwest Yunnan	12-Methoxy-carnosic acid (S2), 1,2-didehydrotanshinone (S4), 8,11,13-icetantriene-10,11,12-triol (S15), barbatusol (S15), carnosol (S2), cryptotanshinone (S3), danshenspiroketallactone (S9), danshenxinkun B (S4), danshenxinkun C (S4), dihydrotanshinone I (S4), epi-danshenspiroketallactone (S9), ganxintriol A (S4), ganxintriol B (S4), pisiferanol (S15), pisiferin (S15), przewaquinone A (S3), przewaquinone B (S4), przewaquinone C (S4), przewaquinone D (S4), przewaquinone F (S4), przewalskin (S2), przewalskin A (S14), przewalskin B (S16), przewalskin C (S15), przewalskin D (S15), rosmadial (S2), salvicanol (S15), sugiol (S2), tanshindiol A (S4), tanshinone I (S4), tanshinone IIA (S3)	9″-Monomethyl lithospermate B (D), 9‴-monomethyl lithospermate B (D), caffeic acid (A), danshensu (A),dimethyl lithospermate B (D), lithospermic acid (C), przewalskinic acid A(C), rosmarinic acid (B), salviaflaside (B), salvianolic acid B (D), salvianolic acid K (C)
S. roborowskii	Western Gansu, Qinghai, western/southwestern	Flower: hepatic heat, strengthening teeth,	Cryptotanshinone (S3), dihydrotanshinone I (S4),	Caffeic acid (A), danshensu (A), rosmarinic

	Sichuan, northwest Yunnan, Tibet, altitude 2500–3900 m, hillside, grassland, ditch, in the shade, mountainside	toothache, abdominal pain; whole plant: liver-clearing and vision-improving, hepatitis; fruit: vision-improving	tanshinone I (S4), tanshinone IIA (S3)	acid (B), salvianolic acid B (D)
S. umbratica	Hebei, Shanxi, Shaanxi, Gansu, Ningxia, Anhui, Sichuan, altitude 600–2000 m, hillside, valley, roadside	Whole plant/seed: heat-clearing and blood-cooling, hemostasis, blood-activating; irregular menstruation	Cryptotanshinone (S3), dihydrotanshinone I (S4), tanshinone I (S4), tanshinone IIA (S3)	Caffeic acid (A), danshensu (A), rosmarinic acid (B), salvianolic acid B (D)
Section *Eusphace*				
S. officinalis	Native to Europe, cultivated in China	Leaf: laryngopharyngitis, hypertension, aid digestion	Cryptotanshinone (S3), dihydrotanshinone I (S4), tanshinone I (S4), tanshinone IIA (S3)	Caffeic acid (A), danshensu (A),methyl salvianolate I(C), rosmarinic acid (B), sagecoumarin (C), salvianolic acid I (C), salvianolic acid K (C), salvianolic acid L (D), sagerinic acid (D)
***Subgenus* Sclare**				
Section *Drymosphace*				
S. bowleyana	Zhejiang, Anhui, Jiangxi, Fujian, Hunan, Guangdong, Guangxi,	Root: used as danshen	7-Carbonyl-12-hydroxy-dehydroabietane (S2), cryptotanshinone (S3),	Caffeic acid (A), danshensu (A), lithospermic acid (C),

Continued

Table 14.1 Continued

Species	Habitat	Therapeutic use	Diterpenoids	Caffeic acid derivative (group)
	altitude 0–1000 m, hill, valley, roadside, forest, waterfront		dihydrotanshinone I (S4), dihydroisotanshinone (S2), tanshinone I (S4), tanshinone IIA (S3), tanshinlactone (S4)	methyl rosmarinate (B), rosmarinic acid (B), salviaflaside (B), salvianolic acid B (D), salvianolic acid C (C), salvianolic acid K (C)
S. cavaleriei	Hubei, Guangdong, Guangxi, Sichuan, Guizhou, altitude 530–1300 m, hillside, forest, ditch	Whole plant: hematemesis, hemoptysis, dysentery with bloody stool, metrorrhagia, traumatic bleeding, hypermenorrhea, gastralgia; traumatic injury, furuncle	Dihydrotanshinone I (S4), tanshinone I (S4), cryptotanshinone (S3), tanshinone IIA (S3)	Rosmarinic acid (B), caffeic acid (A),danshensu (A), isosalvianolic acid C (C), salvianolic acid A (C), salvianolic acid B (D), salvianolic acid C (C), salvianolic acid H (C), salvianolic acid I (C), lithospermic acid (C)
S. dabieshanensis	Anhui, Hubei, Dabie Mountains, altitude 600–1100 m, hillside, woodside	Root: used as danshen	Dihydrotanshinone I (S4), tanshinone I (S4), cryptotanshinone (S3), tanshinone IIA (S3)	Caffeic acid (A), danshensu (A), lithospermic acid (C), rosmarinic acid (B), salvianolic acid B (D)
S. honania	Southern Henan, plain, hillside, facing the sun, wetland	Root: used as danshen		
S. meiliensis	Western Anhui, altitude 400–1300 m, hillside	Root: used as danshen	Dihydrotanshinone I (S4), tanshinone I (S4), cryptotanshinone (S3), tanshinone IIA (S3)	Caffeic acid (A), danshensu (A), lithospermic acid (C), rosmarinic acid (B), salviaflaside (B),

				salvianolic acid B (D), salvianolic acid K (C)
S. miltiorrhiza	Hebei, Shanxi, Anhui, Jiangxi, Zhejiang, Jiangsu, Sichuan, altitude 120–1300 m, hillside, forest, hassock, bushwood, creek, valley, in cultivation	Root: removing blood stasis and relieving pain, activating blood to promote menstruation, relieving restlessness; irregular menstruation, amenorrhea, menalgia, mass in the abdomen, stabbing pain in the chest and abdomen, pyretic arthralgia, ulcer and sores, sleeplessness, swelling of the liver and spleen, angina pectoris	1,2,5,6-Tetrahydro-tanshinone I (S4), 1,2-didehydrotanshinone II (S3), 1,2-dihydro-1,6-dimethylfuro[3,2-c]-naphtha [2,1-e] oxepine-10,12-dione (S3), 3-hydroxytanshinone IIA (S3), 4-methylenemiltirone (S4), 6,7,8,9-tetrahydro-1,6,6-trimethylfuro[3,2-c]naphtha-[2,1-e]oxepine-10,12-dione (S3), 7β-hydroxy-8,13-abietadiene-11,12-dione (S3), acetyl danshenxinkun A (S4), arucadiol (S3), abietatriene (S2), cryptoacetalide (S10), cryptotanshinone (S3), danshenol A (S5), danshenol B (S6), danshenspiroketallactone (S3), danshenxinkun A (S4), danshenxinkun C (S4), dihydrotanshinone I (S4), epicryptoacetalide (S10), epi-danshenspiroketallactone (S9), formytanshinone I (S4), hydroxytanshinone (S3),	2-(3-Methoxy-4-hydroxyphenyl)-5-(3-hydroxypropyl)-7-methoxybenzofuran-3-carbaldehyde (B), 9″-methyl lithospermate (C), ammonium potassium salvianolic acid B (D), caffeic acid (A), danshensu (A), dimethyl lithospermate (C), ethyl salvianolic acid B (D), lithospermic acid (C), lithospermic acid B(D), magnesium salvianolic acid B (D), rosmarinic acid (B), methyl rosmarinate (B), salviaflaside (B), salvianolic acid A (C), salvianolic acid B (D), salvianolic acid C (C), salvianolic acid D (B), salvianolic acid E (D), salvianolic acid F (C),

Continued

Table 14.1 **Continued**

Species	Habitat	Therapeutic use	Diterpenoids	Caffeic acid derivative (group)
			isocryptotanshinone (S3), isotanshinone I (S4), isotanshinone IIB (S3), ketoisocryptotanshinone (S3), methylenedihydrotanshinquinone (S7), miltionone I (S3), miltionone II (S3), miltipolone (S12), miltirone (S3), neocryptotanshinone (S3), neocryptotanshinone II (S8), nortanshinone (S7), oleoyl danshenxinkun A (S4), oleoyl neocryptotanshinone (S3), przewaquinone A (S3), przewalskin (S2), salvinone (S3), salviol (S2), salviolone (S13), secodialdialdehyde (S2), sugiol (S2), tanshinaldehyde (S3), tanshindiol A (S4), tanshindiol B (S4), tanshindiol C (S4), tanshinketolactone (S3), tanshinlactone (S4), tanshinone I (S4), tanshinone IIA (S3), tanshinone IIB (S3), tanshinone VI (S4), trijuganone A (S4),	salvianolic acid G (C), salvianolic acid K (C)

			trijuganone B (S4), trijuganone C (S3)	
S. paramiltiorrhiza	Anhui, Hubei, altitude 300–1000 m, hillside, riverside	Root: used as danshen	Cryptotanshinone (S3), dihydrotanshinone I (S4), tanshinone I (S4), tanshinone IIA (S3)	Caffeic acid (A), danshensu (A), lithospermic acid (C), rosmarinic acid (B), salviaflaside (B), salvianolic acid B (D), salvianolic acid K (C)
S. plectranthoides	Shaanxi, Hubei, Guizhou, Yunnan, Sichuan, Guangxi, altitude 400–2500 m, hillside, valley, open forest, creek	Whole plant: anemofrigid cold, bellyache; root: irregular menstruation, common cold with cough	Cryptotanshinone (S3), dihydrotanshinone I (S4), tanshinone I (S4), tanshinone II (S3)	Caffeic acid (A), danshensu (A), lithospermic acid (C), rosmarinic acid (B), salvianolic acid B (D)
S. prionitis	Jiangsu, Anhui, Zhejiang, Jiangxi, Hubei, Hunan, Guangdong, Guangxi, altitude 100–800 m, hillside, hassock	Whole plant: heat-clearing, detoxification; acute amygdalitis, laryngopharyngitis, bronchitis, enteritis, dysentery	1,2-Didehydromiltirone (S4), 3-ketosapriparaquinone (S1), 3-hydroxysalvilenone (S11), 4-hydroxysapriparaquinone (S1), 4-hydroxysaprorthoquinone (S1), 7,8-secoparaferruginone (S2), 8,11,13-dehydroabietane (S2), arucadiol (S3), cryptotanshinone (S3), danshenxinkun B (S4), danshenxinkun C (S4), dehydromiltirone (S3),	Caffeic acid (A), danshensu (A), lithospermic acid (C), rosmarinic acid (B), methyl rosmarinate(B), salvianolic acid A (C), salvianolic acid B (D), salvianolic acid C (C)

Continued

Table 14.1 **Continued**

Species	Habitat	Therapeutic use	Diterpenoids	Caffeic acid derivative (group)
			dihydroisotanshinone I (S4), ferruginol (S2), hongencaotone (S2), miltirone (S3), prineoparaquinone (S1), prioketolactone (S4), royleanone (S2), salvinolone (S3), salvilenone (S11), sapriparaquinone (S1), saprirearine (S11), saprothoquinone (S1), sclareapinone (S1), sugiol (S2), taxodione (S2), tanshinone I (S4), tanshinone IIA (S3)	
S. sinica	Zhejiang, Anhui, altitude 300–1000 m, hillside, creek, ditch	Root: used as danshen	Cryptotanshinone (S3), dihydrotanshinone I (S4), tanshinone I (S4), tanshinone IIA (S3)	Caffeic acid (A), danshensu (A), lithospermic acid (C), rosmarinic acid (B), salviaflaside (B), salvianolic acid B (D), salvianolic acid K (C)
S. trijuga	Sichuan, Yunnan, Tibet, altitude 1900–3900 m, hillside, valley, ditch, bushwood, forest, grassland	Root: used as danshen	Cryptotanshinone (S3), dihydrotanshinone I (S4), ferruginol (S2), sugiol (S2), tanshinone I (S3), tanshinone IIA (S3), tanshinone acid methyl ester (S3), trijuganone A (S4), trijuganone B (S4), trijuganone C (S3)	Caffeic acid (A), danshensu (A), lithospermic acid (C), rosmarinic acid (B), salviaflaside (B), salvianolic acid B (D), salvianolic acid K (C)

S. vasta	Hubei, altitude 1400 m, hillside, woodside, roadside, facing the sun	Root: used as danshen	Cryptotanshinone (S3), dihydrotanshinone I (S4), tanshinone I (S4), tanshinone IIA (S3)	Caffeic acid (A), danshensu (A), lithospermic acid (C), rosmarinic acid (B), salvianolic acid B (D)
S. yunnanensis	Sichuan, Guizhou, Yunnan, altitude 1800–2900 m, hillside, grassland, woodside, open forest, dryland	Root: used as danshen	1,2-Dihydrotanshinone (S4), 5,6-dehydrosugiol (S2), cryptotanshinone (S3), dihydrotanshinone I (S4), danshenol A (S5), danshenol C (S6), przewalskin (S2), tanshinone I (S4), tanshinone IIA (S3), tanshinlactone (S4), yunnanin A (S2)	Caffeic acid (A), danshensu (A), ethyl salvianolate A(C), lithospermic acid (C), methyl salvianolate A(C), rosmarinic acid (B), salvianolic acid B (D), yunnaneic acid A (D), yunnaneic acid B (D), yunnaneic acid C (B), yunnaneic acid D (B), yunnaneic acid E (B), yunnaneic acid F (B), yunnaneic acid G (D), yunnaneic acid H (D)
Section *Plethiosphace*				
S. deserta	Northern Xinjiang, altitude 270–1850 m, open country, sandland, grassland, forest	Whole plant: heat-clearing, detoxification, cough expectorant, detumescence, diuresis; common cold, fever	6,7-Dehydroroyleanon (S2), 7-*O*-acetylhorminone (S2), 7-*O*-methylhorminone (S2), ferruginol (S2), horminon (S2), royleanon	Caffeic acid (A), danshensu (A), rosmarinic acid (B), lithospermic acid (C), salviaflaside (B),

Continued

Table 14.1 **Continued**

Species	**Habitat**	**Therapeutic use**	**Diterpenoids**	**Caffeic acid derivative (group)**
			(S2), taxochinon (S2), taxodione (S2)	salvianolic acid B (D), salvianolic acid K (C)
Section *Notiosphace*				
S. plebeia	Throughout China, altitude 60–2800 m, hillside, open country, grassland, roadside, ditch	Whole plant: heat-clearing, detoxification, blood-cooling, diuresis; sore throat, bronchitis, nephritis, edema, furuncle, mastitis, hemorrhoid	Epoxysalviacoccin (S17), salviacoccin (S17)	Caffeic acid (A), danshensu (A), rosmarinic acid (B), salvianolic acid B (D)
Subgenus *Jungia*				
S. splendens	Native to America, cultivated in China	Whole plant: heat-clearing, swell-dispersing, detoxification, blood-cooling; snake bite	Olearin (S17), salviaclerodan A (S17), salviarin (S17), salvisplendin A (S17), salvisplendin B (S17), salvisplendin C (S17), salvisplendin D (S17), splendidin (S17), splendidin A (S17), splendidin B (S17), splendidin C (S17), splenolide A (S17), splenolide B (S17)	Caffeic acid (A), rosmarinic acid (B)
S. coccinea	Native to America, cultivated in China	Whole plant: blood-cooling, hemostasis, heat-clearing and diuresis-promoting; metrorrhagia, hyperpyrexia, bellyache	Salviacoccon (S17)	

Subgenus *Allagospadonopsis*				
S. chinensis	East China, South Central China, Sichuan, Yunnan, altitude 120–500 m, hillside, plain, forest, grassland, roadside, open country	Whole plant: heat-clearing, detoxification, blood-activating, regulating qi to alleviate pain; hepatitis, abdominal flank pain, hygrothermal leucorrhea, acute mastitis, achalasia of cardia, sputum, asthma, furuncle		Caffeic acid (A), danshensu (A), isosalvianolic acid C(C), lithospermic acid (B), rosmarinic acid (B), salvianolic acid B (D), salvianolic acid D (C), syringaresinol (B)
S. japonica	East China, Hubei, Guangdong, Guangxi, Guizhou, Taiwan, altitude 200–1200 m, hillside, waterfront, forest	Whole plant: heat-clearing, detoxification, blood-activating and stasis-removing, swell-relieving, pain killer; traumatic injury, rheumatism, edema, leukorrheal disease, menalgia, postpartum bleeding, hepatitis, facial paralysis, neuralgia, acute mastitis, furuncle		Caffeic acid (A), danshensu (A), lithospermic acid (C), rosmarinic acid (B)
S. kiangsiensis	Jiangxi, Fujian, Hunan, valley, forest	Whole plant: blood-cooling, hemostasis, blood-activating and swell-relieving, heat-		Caffeic acid (A), danshensu (A), lithospermic acid (C),

Continued

Table 14.1 Continued

Species	Habitat	Therapeutic use	Diterpenoids	Caffeic acid derivative (group)
		clearing, detoxification; hematemesis, hematochezia, irregular menstruation, traumatic injury, acute mastitis, furuncle		rosmarinic acid (B), salvianolic acid B (D)
S. liguliloba	Anhui, Zhejiang, altitude 800 m, hillside, woodside, roadside, foothill, crop field, grassland	Whole plant: blood-activating and collateral-dredging; arthralgia, irregular menstruation		
S. substolonifera	Zhejiang, Jiangxi, Fujian, Hubei, Hunan, Guangxi, Sichuan, Guizhou, altitude 40–950 m, forest, waterfront	Whole plant: lung heat-clearing, hemostasis; wind-heat cough, sputum, asthma		
S. scapiformis	Jiangxi, Fujian, Taiwan, Hunan, Guangdong, Guangxi, hillside, grassland, valley, forest, hilltop	Whole plant: strengthening tendons and bones; as tonic; lung disease, dizziness		Caffeic acid (A), danshensu (A), rosmarinic acid (A)

A, Caffeic acid monomers; B, caffeic acid dimmers; C, caffeic acid trimmers; D, caffeic acid tetramer; E, higher caffeic acid oligomers.

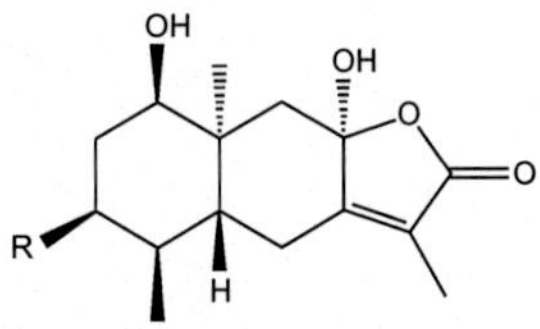

No.	R
1	OAc
2	OH

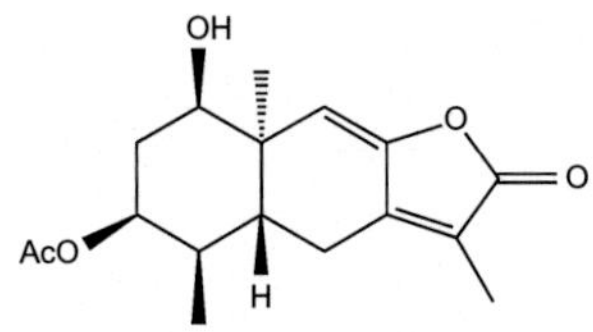

3

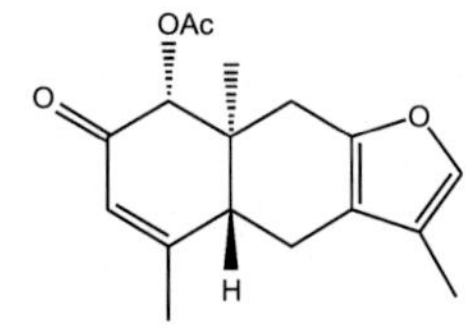

4

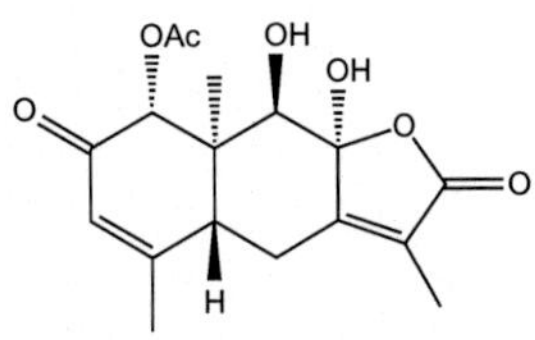

5

Figure 14.1 Sesquiterpenoids isolated from *Salvia*. Refer to Table 14.2 for more details.

are present in *Salvia*, namely, abietane diterpenoids, clerodane diterpenoids, pimarane diterpenoids, labdane diterpenoids, and others.

Six types of sesquiterpenoids (15C) are found in *Salvia*, namely, aliphatic sesquiterpene, germacranes, carotanes, caryophyllanes, guaianes, and others. Both C-23 and C-25 sesterterpenoids are present in *Salvia*. Five types of triterpenoids (30C) are found in *Salvia*, namely, ursanes, oleananes, lupanes, dammaranes, and others.

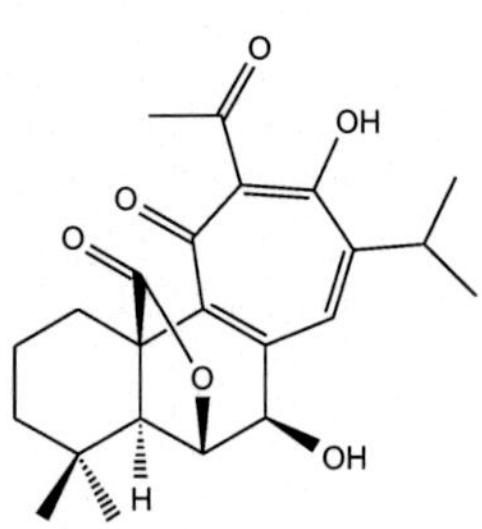

6

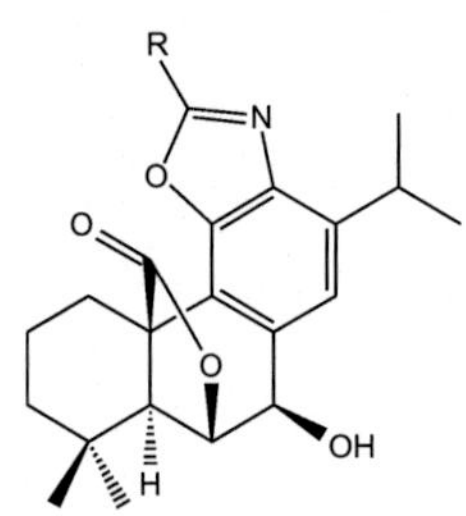

No.	R
7	H
8	CH_3

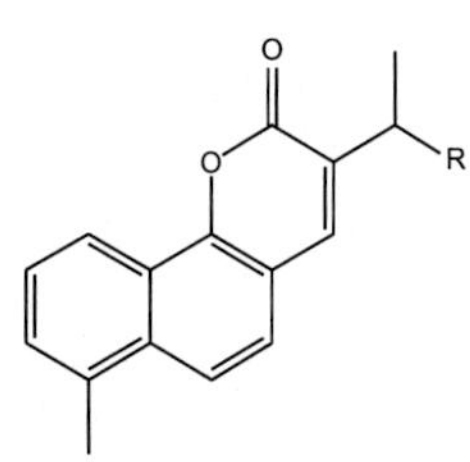

No.	R
9	CH_3
10	CH_2OH

Figure 14.2 Diterpenoids isolated from *Salvia*. Refer to Table 14.2 for more details.

(Continued)

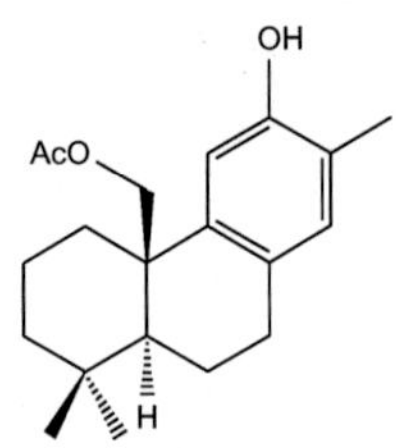

11

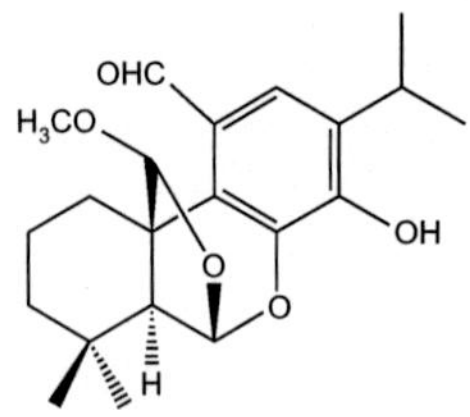

12

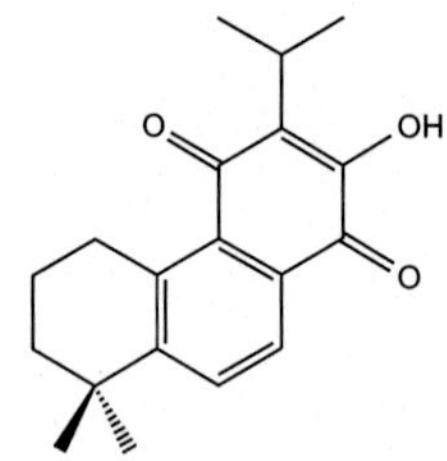

13

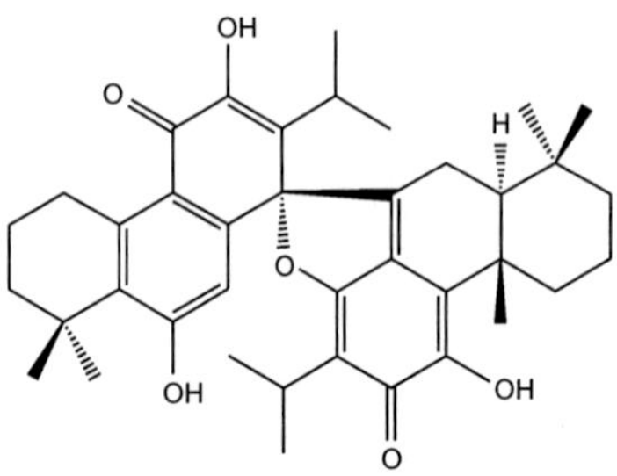

14

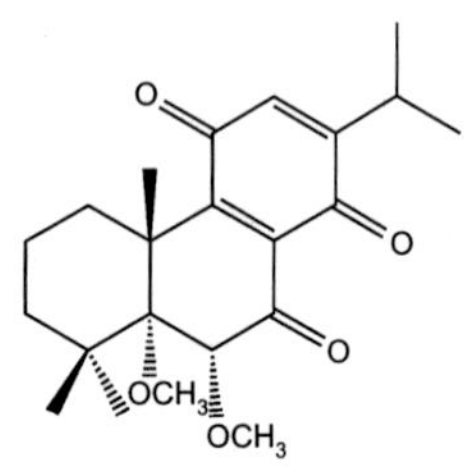

15

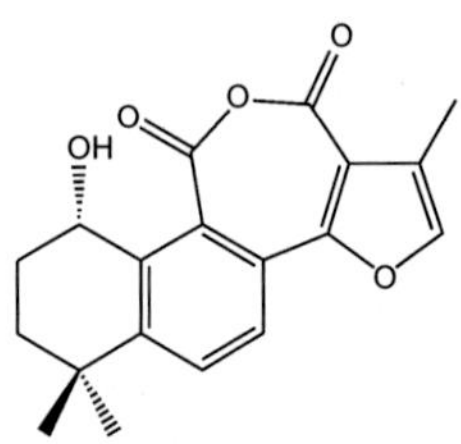

16

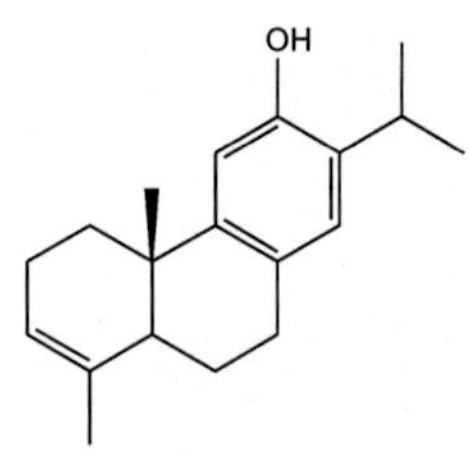

17

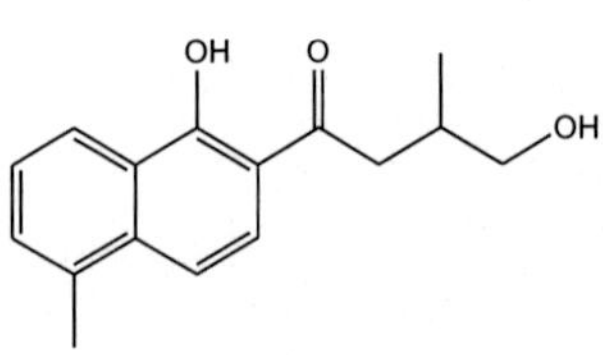

18

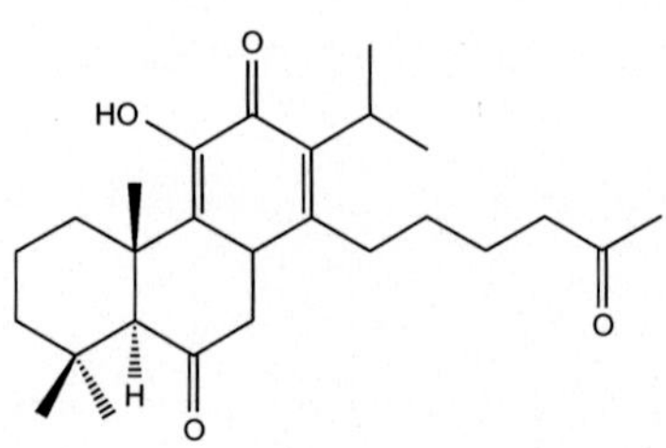

19

Figure 14.2 Continued.

The essential-oil contents of *S. tomentosa* ranged from 1.1% to 3.3% (v/w, based on the dry weight of the plant material) (Hanlidou et al., 2014). The populations from Thassos of Greece had high contents of α-pinene (18.0 ± 2.9%), 1,8-cineole (14.7 ± 3.0%), *cis*-thujone (14.0 ± 6.9%), and borneol (12.8 ± 2.2%) and smaller amounts of camphene, camphor, and β-pinene, whereas the populations from Thrace

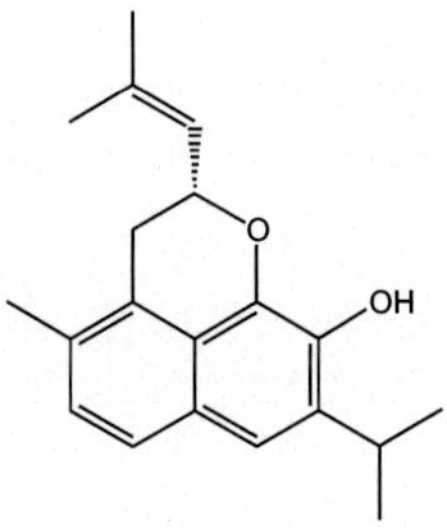

20

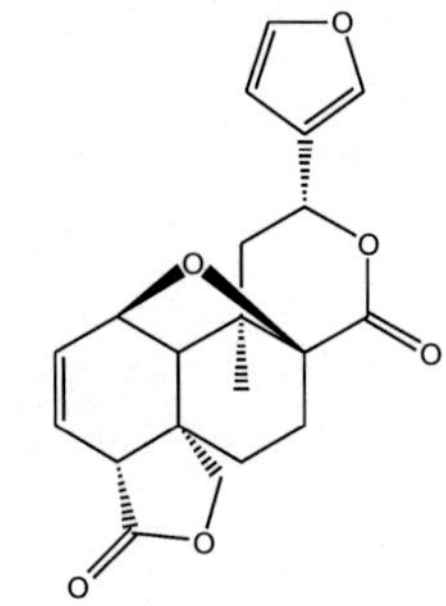

21

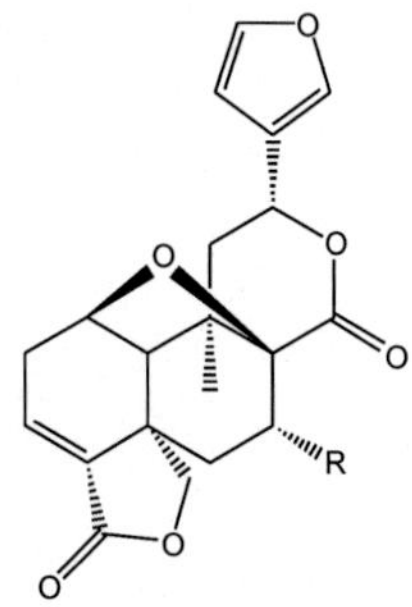

No.	R
22	H
23	OH_3

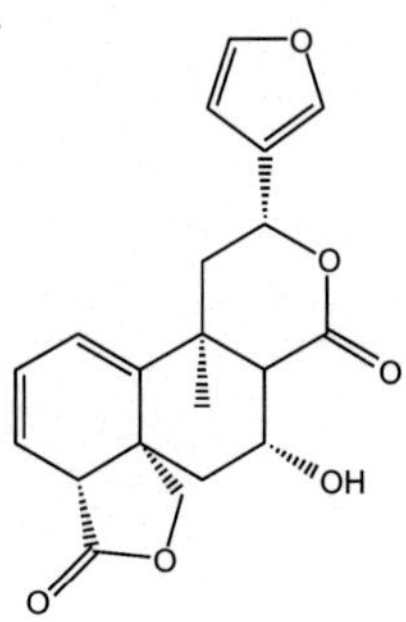

24

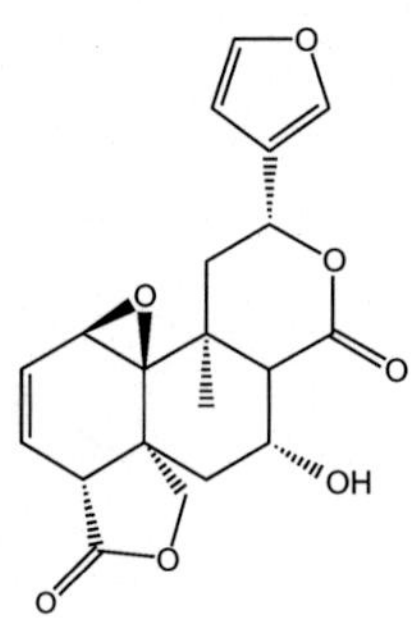

25

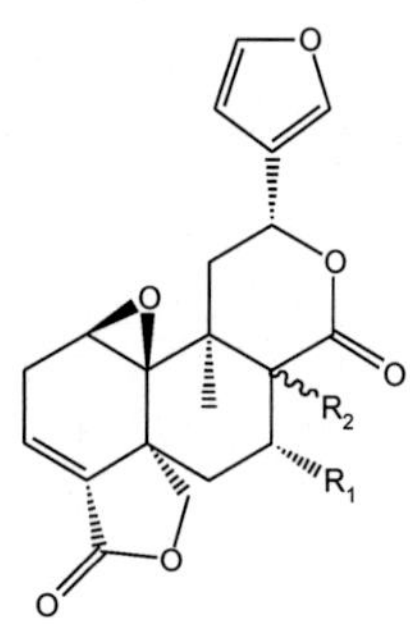

No.	R_1	R_2
26	OH	β-H
27	H	α-OH
28	H	α-H

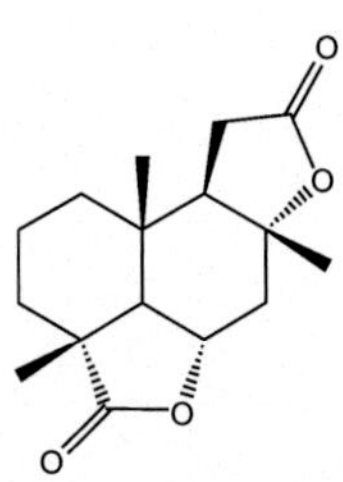

29

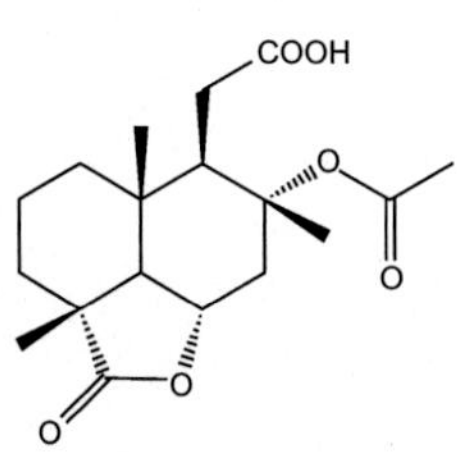

30

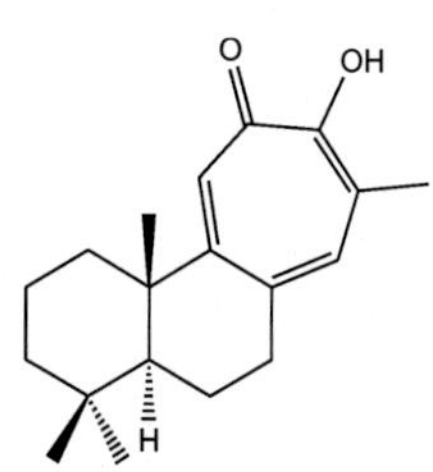

31

Figure 14.2 Continued.

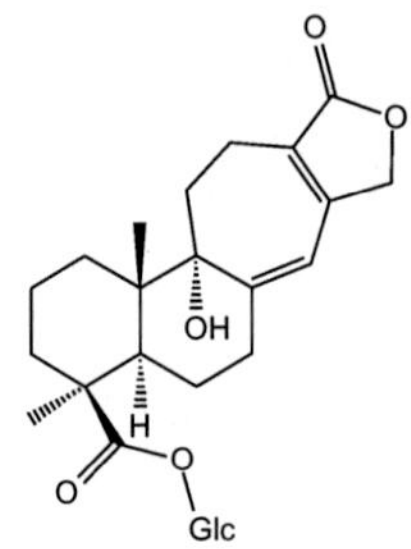

32

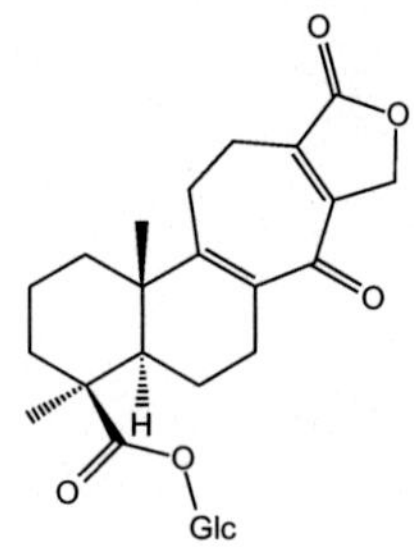

33

Figure 14.2 Continued.

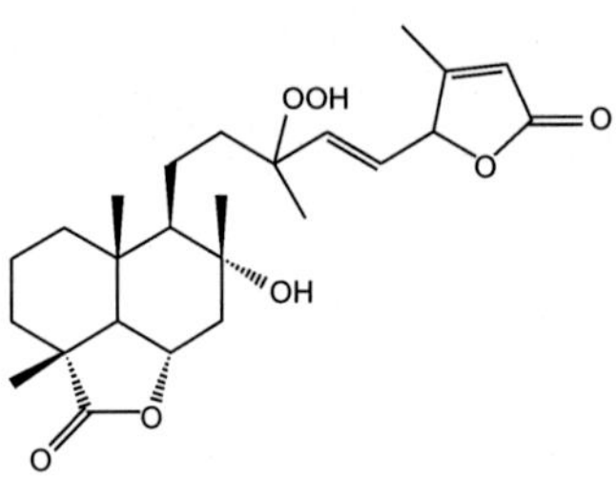

34

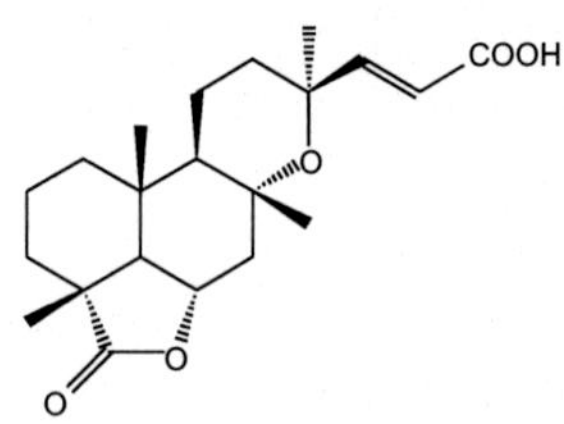

35

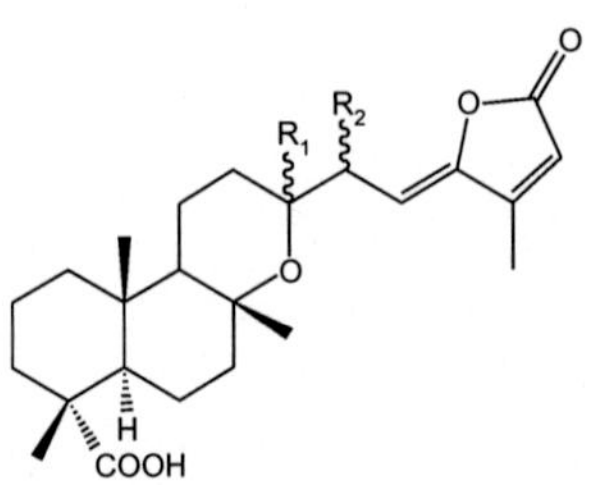

No.	R_1	R_2
36	α-CH_3	β-OH
37	β-CH_3	β-OH
38	α-CH_3	α-OH
39	β-CH_3	β-H

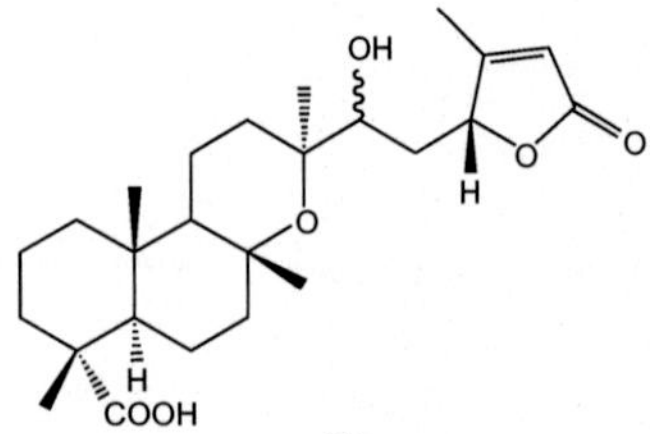

40

Figure 14.3 Sesterterpenoids isolated from *Salvia*. Refer to Table 14.2 for more details.

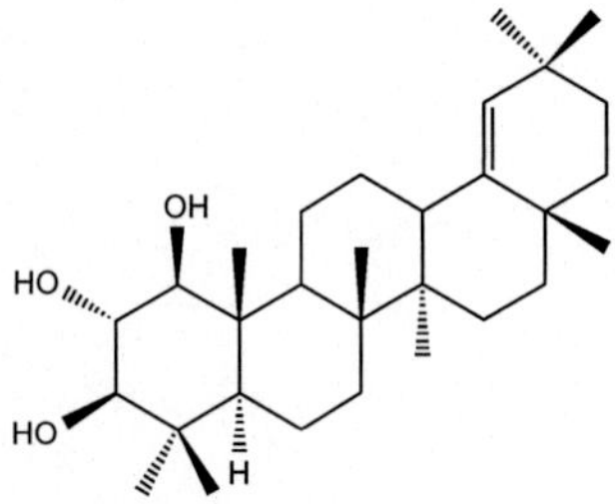

41

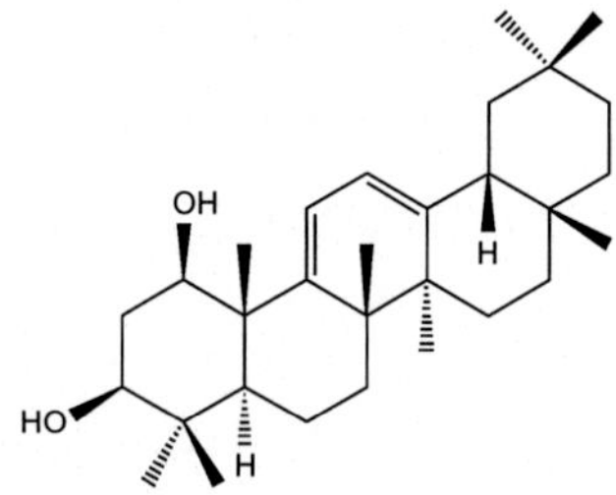

42

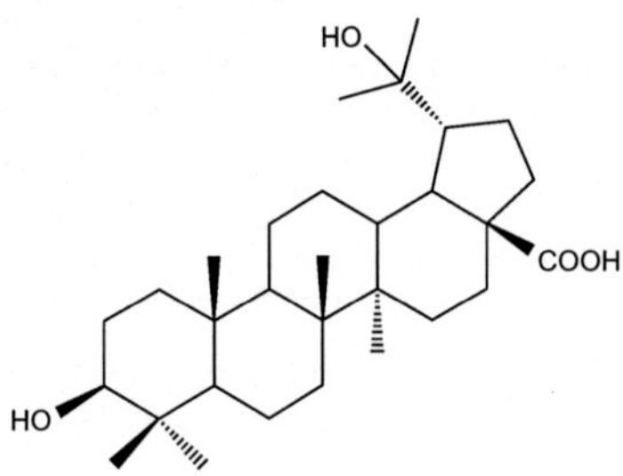

43

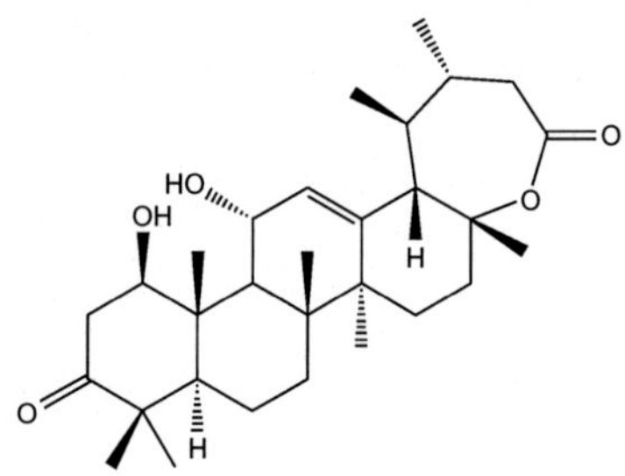

44

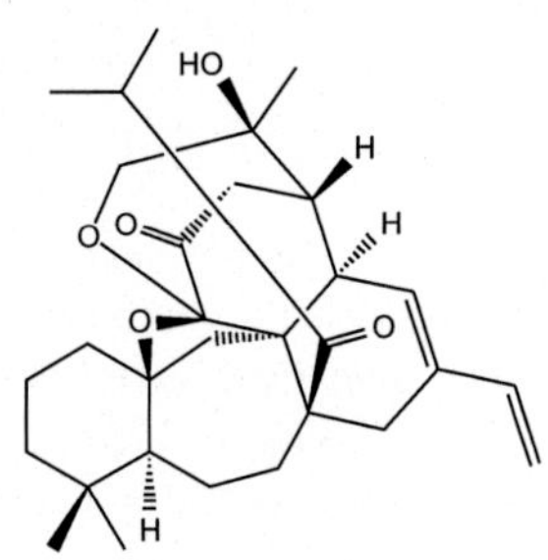

44

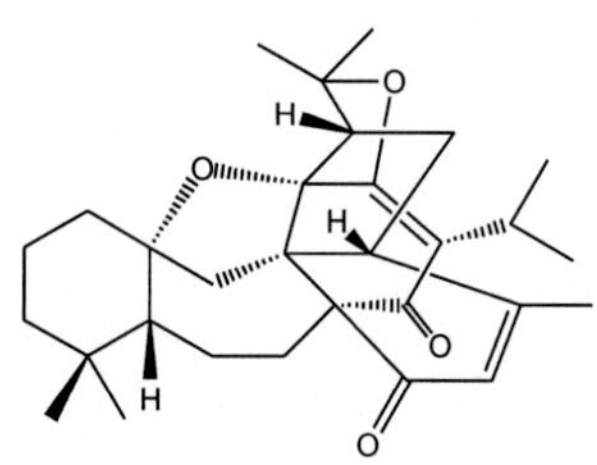

46

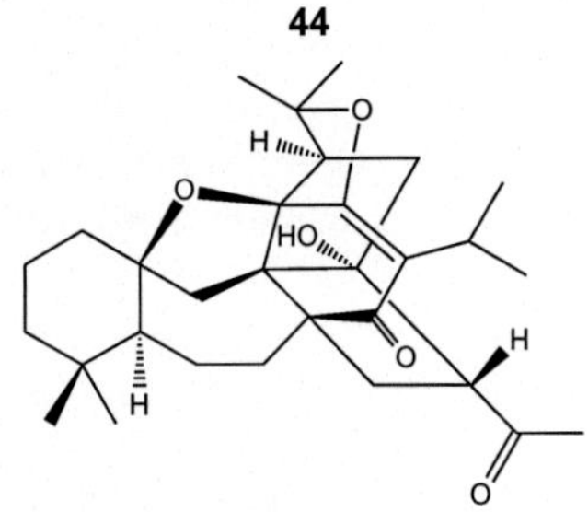

47

Figure 14.4 Triterpenoids isolated from *Salvia*. Refer to Table 14.2 for more details.

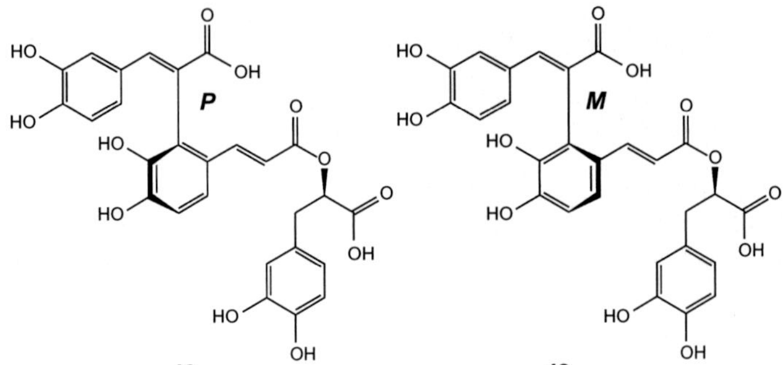

48 49

Figure 14.5 Phenolic acids isolated from *Salvia*. Refer to Table 14.2 for more details.

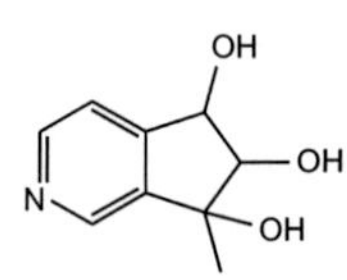

50

Figure 14.6 Alkaloid isolated from *Salvia*. Refer to Table 14.2 for more details.

showed high α-pinene (16.7±4.0%), β-pinene (22.8±4.5%), camphor (18.3±4.3%), and camphene (10.3±2.4%) contents, much lower 1,8-cineole and borneol amounts, while *cis*-thujone was completely lacking. Multivariate statistical analyses indicate that the observed essential-oil variation was related to geographic and environmental factors. 33 compounds were identified from aerial parts of *S. sclarea*, representing the 85.0% of the total oil (Yuce et al., 2014). The most abundant components (>5%) were caryophyllene oxide (24.1%), sclareol (11.5%), spathulenol (11.4%), 1H-naphtho [2,1-b]pyran (8.6%), and β-caryophyllene (5.1%). 54 Components were detected in the essential oil of *S. chionantha* (Tel et al., 2010). Germacrene D (25.03%), β-caryophyllene (8.71%), spathulenol (5.86%), and α-humulene (4.82%) were identified as the major compounds.

14.2.2 Phenolic compounds

Caffeic acid (3,4-dihydroxycinnamic acid), one of the most common phenolic acids, frequently occurs in fruits, grains, and dietary supplements for human consumption as simple esters with quinic acid or saccharides and are also found in traditional Chinese herbs. Caffeic acid derivatives occur as major water-soluble components of *Salvia miltiorrhiza* (Sm; danshen), including caffeic acid monomers and a wide variety of

Table 14.2 Various types of compounds found in *Salvia* L. plants in recent years

No.	Compounds	Types	Subtype	Species	Tissues	References
1	Plebeiolide A	A	Eudesmane	a	Basal leaves	Dai et al. (2014)
2	Plebeiolide B	A	Eudesmane	a	Basal leaves	Dai et al. (2014)
3	Plebeiolide C	A	Eudesmane	a	Basal leaves	Dai et al. (2014)
4	Plebeiafuran	A	Eudesmane	a	Basal leaves	Dai et al. (2014)
5	1α-Acetoxy-8α,9β-dihydroxy-2-oxo-eudesman-3,7(11)-dien-8,12-olide	A	Eudesmane	a	Whole herbs	Cao et al. (2013)
6	Salyunnanin A	B_1	6,20-Epoxyabietane	b	Roots	Wu et al. (2014d)
7	Salyunnanin B	B_1	Abietane alkaloid	b	Roots	Wu et al. (2014d)
8	Salyunnanin C	B_1	Abietane alkaloid	b	Roots	Wu et al. (2014d)
9	Salyunnanin D	B_1	Norabietane	b	Roots	Wu et al. (2014d)
10	Salyunnanin E	B_1	Norabietane	b	Roots	Wu et al. (2014d)
11	Salyunnanin F	B_1	Abietane-8,11,13-triene	b	Roots	Wu et al. (2014d)
12	Salvialerial	B_1	6,20-Epoxyabietane	c	Whole herbs	Choudhary et al. (2012)
13	Salvialerione	B_1	Abietane-11,14-dione	c	Whole herbs	Choudhary et al. (2012)
14	Salvialeriafone	B_1	Abietane dimer	c	Whole herbs	Choudhary et al. (2012)
15	Xantoquinone	B_1	Abietane-7,11,14-trione	d	Aerial parts	Habibi et al. (2013)
16	Castanol A	B_1	14,17-Epoxyabietane	e	Whole herbs	Pan et al. (2012)

Continued

Table 14.2 **Continued**

No.	Compounds	Types	Subtype	Species	Tissues	References
17	Castanol B	B_1	Norabietane	e	Whole herbs	Pan et al. (2012)
18	Castanol C	B_1	Norabietane	e	Whole herbs	Pan et al. (2012)
19	7-(2-Oxohexyl)-11-hydroxy-6,12-dioxo-7,9(11),13-abietatriene [=7-(2-oxohexyl)-taxodione]	B_1	Abietane-6,12-dione	f	Hairy roots	Kuźma et al. (2012b)
20	Sahandol	B_1	4,5-Seco-abietane	g	Roots	Ebrahimi et al. (2013)
21	Tehuanin A	B_2	Clerodane-17,12:18,19-diolide	h	Aerial parts	Bautista et al. (2012)
22	Tehuanin B	B_2	Clerodane-17,12:18,19-diolide	h	Aerial parts	Bautista et al. (2012)
23	Tehuanin C	B_2	Clerodane-17,12:18,19-diolide	h	Aerial parts	Bautista et al. (2012)
24	Tehuanin D	B_2	Clerodane-17,12:18,19-diolide	h	Aerial parts	Bautista et al. (2012)
25	Tehuanin E	B_2	Clerodane-17,12:18,19-diolide	h	Aerial parts	Bautista et al. (2012)
26	Tehuanin F	B_2	Clerodane-17,12:18,19-diolide	h	Aerial parts	Bautista et al. (2012)
27	Tehuanin G	B_2	Clerodane-17,12:18,19-diolide	h	Aerial parts	Bautista et al. (2012)
28	Tehuanin H	B_2	Clerodane-17,12:18,19-diolide	h	Aerial parts	Bautista et al. (2012)
29	Norambreinolide-18,6α-olide	B_3	Norlabdane	g	Aerial parts	Moghaddam et al. (2010)

30	8α-Acetoxy-13,14,15,16-tetranorlabdan-12-oic acid-18,6α-olide	B_3	Norlabdane	g	Aerial parts	Moghaddam et al. (2010)
31	Przewalskin Y-1	B_4	Other diterpenoids	i	Whole herbs	Yang et al. (2011)
32	Salviadigitoside A	B_4	Other diterpenoids	j	Roots	Wu et al. (2014b)
33	Salviatalin A-19-*O*-β-glucoside	B_4	Other diterpenoids	j	Roots	Wu et al. (2014b)
34	8α-Hydroxy-13-hydroperoxylabd-14,17-dien-19,16;23,6α-diolide	C	C-25	g	Aerial parts	Moghaddam et al. (2010)
35	17,18,19,20-Tetranor-13-epi-manoyloxide-14-en-16-oic acid-23, 6α-olide	C	C-25	g	Aerial parts	Moghaddam et al. (2010)
36	(4R,5R,8R,9R,10S,13S,14S)-14-hydroxymanoyloxide-15,17-dien-15 (Z)-16,19-olide	C	C-25	k	Aerial parts	Ebrahimi et al. (2014)
37	(4R,5R,8R,9R,10S,13R,14S)-14-hydroxymanoyloxide-15,17-dien-15 (Z)-16,19-olide	C	C-25	k	Aerial parts	Ebrahimi et al. (2014)
38	(4R,5R,8R,9R,10S,13S,14R)-14-hydroxymanoyloxide-15,17-dien-15 (Z)-16,19-olide	C	C-25	k	Aerial parts	Ebrahimi et al. (2014)
39	(4R,5R,8R,9R,10S,13R)-manoyloxide-15,17-dien-15(Z)-16,19-olide	C	C-25	k	Aerial parts	Ebrahimi et al. (2014)
40	(4R,5R,8R,9R,10S,13S,16R)-14-hydroxymanoyloxide-17-en-16,19-olide	C	C-25	k	Aerial parts	Ebrahimi et al. (2014)

Continued

Table 14.2 Continued

No.	Compounds	Types	Subtype	Species	Tissues	References
41	Olean-18-ene-1β,2α,3β-triol	D_1	Oleanane	l	Aerial parts	Habibi et al. (2012)
42	1β,3β-Dihydroxyolean-9(11),12-dienyl	D_1	Oleanane	d	Aerial parts	Gandomkar et al. (2012)
43	3β,20-Dihydroxylupane-28-oic acid	D_2	Lupane	m	Aerial parts	Farimani et al. (2012a)
44	Urmiensolide	D_3	Other triterpenoids	n	Whole herbs	Moridi Farimani et al. (2013)
45	Salvadione C	D_3	Other triterpenoids	o	Aerial parts	Moridi Farimani et al. (2011)
46	Perovskone B	D_3	Other triterpenoids	o	Aerial parts	Moridi Farimani et al. (2011)
47		D_3	Other triterpenoids	o	Aerial parts	Farimani et al. (2012b)
48	Salvianolic acid T	E	Trimeric caffeic acids derivative	p	Roots	Li et al. (2014a)
49	Salvianolic acid U	E	Trimeric caffeic acids derivative	p	Roots	Li et al. (2014a)
50	Salviadiginine A	F	Cyclopenta[c]pyridine alkaloid	j	Roots	Wu et al. (2014b)

A, Sesquiterpenoids; B, diterpenoids (B_1, abietane type; B_2, clerodane type; B_3, labdane type; and B_4, other diterpenoids); C, sesterterpenoids; D, triterpenoids (D_1, oleanane type; D_2, lupane type; and D_3, other triterpenoids); E, phenolic acids; F, other compounds.

a, *S. plebeian*; b, *S. yunnanensis*; c, *S. leriaefolia*; d, *S. xanthocheila* Boiss; e, *S. castanea* Diels f. *pubescens* Stib.; f, *S. austriaca*; g, *S. sahendica*; h, *S. herbacea*; i, *S. przewalskii* Maxim; j, *S. digitaloids*; k, *S. mirzayanii*; l, *S. atropatana* Bunge; m, *S. eremophila*; n, *S. urmiensis*; o, *S. hydrangea*; p, *S. miltiorrhiza*.

oligomers (Jiang et al., 2005; Figure 14.5). This class of compounds is worthy of further studies that may lead to new drug discovery.

Rosmarinic acid and salvianolic acids occurred as the major caffeic acid derivatives (polyphenolic acids) in Chinese *Salvia* species. The caffeic acid is the building block of a variety of the plant metabolites from the simple monomers to multiple condensation products to give rise to a variety of caffeic acid derivatives. According to the number of caffeic acid unit in the structures of caffeic acid derivatives, these compounds could be divided into caffeic acid monomers (A), dimers (B), trimers (C), tetramers (D), and higher caffeic acid oligomers (E) (Li et al., 2013b). For instance, rosmarinic acid, a caffeic acid dimer (an ester of caffeic acid and 3,4-dihydroxy-phenyl lactic acid), is common and abundant in Salvia species, and salvianolic acid B, a caffeic acid tetramer, is the major phenolic compound responsible for the high biological activities of Sm and the related species. Some caffeic acid derivatives occur as potassium, ammonium, or magnesium salts, such as magnesium lithospermate B in Sm, and some caffeic acid derivatives occur as phenolic glycosides, such as salviaflaside (rosmarinic acid 3-glucoside), a common compound in *Salvia* species.

14.3 Pharmacology and therapeutic use

14.3.1 Effects on the nervous system

Salvianolic acid B (Sal B) increases the level of antioxidant substances and decreases free radical production (Jiang et al., 2014b). It also improves antiapoptotic Bcl-2 expression and reduces proapoptotic Bax expression. SalB may exert the neuroprotective effect through mitochondria-dependent pathway on hippocampal CA1 neurons in mice with cerebral ischemia and reperfusion (I/R) injury. Tanshinone I had anti-inflammatory effect in neuroprotection against cerebral I/R injury in the gerbil hippocampus (Park et al., 2014).

Tanshinone (Tan) IIA reduces the risk of Alzheimer's disease (AD) by inhibiting iNOS, matrix metalloproteinase (MMP)-2, and NF-κBp65 transcription and translation in the temporal lobes of rat AD models (Jiang et al., 2014a).

Acetylcholinesterase and butyrylcholinesterase enzymes play a role in the pathogenesis of AD. The essential oil of *S. chionantha* showed moderate acetylcholinesterase ($56.7 \pm 1.9\%$) inhibitory activity (Tel et al., 2010).

Sm extract had neuroprotective effect against Aβ25-35-induced apoptosis in SH-SY5Y cells (Yu et al., 2014a). Sm injection exhibited protective effects against learning and memory impairments in streptozotocin-induced diabetic rats (Cai et al., 2014). *S. officinalis* and *S. lavandulaefolia* have a long-standing use as traditional herbal remedies that can enhance memory and improve cognitive functions. Unfortunately, promising beneficial effects are debased by methodological issues, the use of different herbal preparations (extracts, essential oil, and raw material), and the lack of details on herbal products used (Miroddi et al., 2014). Sage promising effects need further higher methodological standard clinical trials.

The active ingredient in *S. divinorum* is the neoclerodane diterpene, salvinorin A, which has been shown to be a kappa agonist both *in vitro* and *in vivo* (Prisinzano, 2005). Salvinorin A differs from other commonly abused compounds since although it affects accumbal dopamine transmission, it is unable, at least at the tested doses, to sustain stable intravenous self-administration behavior (Serra et al., 2014).

14.3.2 Effects on the cardiovascular system

Taxodione, a diterpenoid from the roots of *S. chorassanica*, had protective effect in serum/glucose deprivation-induced ischemic injury due to its putative role of antiapoptosis (Shafaei-Bajestani et al., 2014). Tan IIA, a major component of Sm, inhibits platelet activation via ERK-2 signaling pathway (Maione et al., 2014). Tan IIA promotes nonamyloidogenic processing of amyloid precursor protein in platelets via estrogen receptor signaling to phosphatidylinositol 3-kinase/Akt (Shi et al., 2014). Salvianolic acid A of Sm, a matrix metalloproteinase-9 inhibitor, attenuates aortic aneurysm formation in apolipoprotein E-deficient mice (Zhang et al., 2014b).

Tanshinol inhibits β-adrenergic receptor-mediated cardiac fibrosis by ROS/p38 MAPK axis (Lu et al., 2014). Magnesium lithospermate B of Sm has vasodilation effect and mediates sGC/cGMP/PKG translocation in experimental vasospasm (Chang et al., 2014).

Sm aqueous extract had protective effect on myocardium oxidative injury in I/R rats (Ge et al., 2014a). Compound danshen tablet (CDT), an herbal preparation consisting of Sm, notoginseng (radix and rhizome of *Panax notoginseng*), and *Borneolum syntheticum*, is widely used to improve coronary heart disease, cardiac angina, and atherosclerosis in clinics in Asia, especially in China. CDT protects cardiomyocytes against myocardial I/R injury and inhibits apoptosis in rats by activating Akt-eNOS signaling pathway (Ren-An et al., 2014).

14.3.3 Anti-inflammatory and antioxidant effects

Tan IIA possesses significant anti-inflammatory, antioxidant, and antiadipogenic effects in primary orbital fibroblasts (Rhiu et al., 2014). These results provide the basis for further study of the potential use of Tan IIA to treat Graves' orbitopathy. Anti-inflammation via inhibition of NF-κB pathways in hepatic stellate cells (HSCs) is one therapeutic approach to hepatic fibrosis. Tan IIA decreased lipopolysaccharide (LPS)-induced HSC activation (Liu and Huang, 2014). Tan IIA induces apoptosis in fibroblast-like synoviocytes in rheumatoid arthritis via blockade of the cell cycle in the G2/M phase and a mitochondrial pathway (Jie et al., 2014).

Eudesmane-type sesquiterpenes of *S. plebeian* leaf, plebeiolide C and plebeiafuran (Figure 14.1), inhibit NO production in LPS-induced RAW264.7 cells (Dai et al., 2014).

Sm compounds protect the liver from acute injury by the regulation of p38 and NF-κB signaling in Kupffer cells (Yue et al., 2014). *S. plebeia* suppresses atopic dermatitis-like skin lesions (Choi et al., 2014).

Lithospermic acid B of Sm, an antioxidant, had a preventive effect on the development of diabetic retinopathy in spontaneously obese diabetic rats (Jin et al., 2014).

S. hispanica (chia) seeds have high antioxidant capacity and could be a novel isoflavone source that can be incorporated in the human diet (Martínez-Cruz and Paredes-López, 2014).

S. fruticosa reduces intrinsic cellular and H_2O_2-induced DNA oxidation in HEK 293 cells (Hani and Bayachou, 2014). *S. nemorosa*, *S. atropatana*, *S. santolinifolia*, and *S. eremophila* were the most active radical scavengers with higher total phenol contents (Firuzi et al., 2013), while *S. limbata*, *S. xanthocheila*, and *S. aegyptiaca* were the weakest ones. *S. glutinosa* displayed the highest DPPH radical-scavenging activity with an IC50 of 3.2 μg/ml (Loizzo et al., 2014). These species are characterized by the highest total phenol and flavonoid contents.

Wrinkle formation is a striking feature of intrinsic and photoinduced skin aging, which are both associated with oxidative stress and inflammatory response. Magnesium lithospermate B of Sm exerts antiwrinkle effect through inhibition of MMPs via NF-κB signaling (Jung et al., 2014).

Polysaccharides of Sm had antioxidant and antitumor activities (Jiang et al., 2014c).

14.3.4 Anticancer and immunomodulatory activity

Tan IIA induces apoptosis in human oral cancer KB cells through a mitochondria-dependent pathway (Tseng et al., 2014). Tan IIA induces cytochrome c-mediated caspase cascade apoptosis in A549 human lung cancer cells via the JNK pathway (Zhang et al., 2014c). Tan IIA blocks epithelial–mesenchymal transition through HIF-1α downregulation and reverses hypoxia-induced chemotherapy resistance in breast cancer cells (Fu et al., 2014).

Cryptotanshinone of Sm is a powerful stimulator of suicidal erythrocyte death or eryptosis, which is effective mainly by stimulation of Ca^{2+} entry (Bissinger et al., 2014). Several abietane-type diterpenoids of *S. yunnanensis* exhibited substantial cytotoxicity with IC50 of 0.86–10.1 μM (Wu et al., 2014b). Salvialeriafone, a diterpene–norditerpene conjugate, salvicanaric acid, dehydroroyleanone, and cariocal, isolated from *S. leriaefolia*, exhibited *in vitro* antiproliferative activity against the human cervical cancer cell line (HeLa) (Choudhary et al., 2012), while dehydroroyleanone showed cytotoxicity against the human prostate cancer cell line (PC3).

Toxic effects of manool—a diterpene from *S. officinalis*—against Chinese hamster lung fibroblasts (V79) were observed at concentrations of up to 8.0 μg/ml (Nicolella et al., 2014). The micronucleus test was used to evaluate the genotoxicity and antigenotoxicity of manool in V79 and HepG2 cells at concentrations of 0.5–6.0 and 0.5–8.0 μg/ml, respectively. For evaluation of antigenotoxicity, the concentrations of manool were combined with methyl methanesulfonate (MMS, 44 μg/ml). The frequency of micronuclei was significantly increased in cultures of both cell lines treated with the highest concentration tested, demonstrating a genotoxic effect. Manool exhibited a protective effect against chromosome damage induced by MMS in HepG2 cells, but not in V79 cells. Some manool metabolite may be responsible for the antigenotoxic effect in HepG2 cells.

Cryptotanshinone and dihydrotanshinone of Sm reverse P-glycoprotein (P-gp) mediated multidrug resistance in colon cancer cells (Hu et al., 2014). Neotanshinlactone of Sm had potent selective antibreast cancer activity (Wang et al., 2007a).

A lupane triterpenoid of *S. eremophila*, 3β, 20-dihydroxylupane-28-oic acid, had anticancer property (Farimani et al., 2012a).

All of the extracts of *S. eremophila* and *S. santolinifolia* were active at IC50 values of 10.5–75.2 μg extract/ml (Firuzi et al., 2013), while the methanol and dichloromethane extracts of *S. limbata*, *S. hypoleuca*, and *S. aethiopis* showed considerable cytotoxic activity against the tested cell lines. *S. ceratophylla* exhibited the strongest antiproliferative activity against C32 cells with an IC50 value of 20.8 μg/ml (Loizzo et al., 2014), while *S. glutinosa* exhibited an IC50 value of 29.5 μg/ml against ACHN cell line.

A neutral polysaccharide fraction (SMPA), prepared from the roots of Sm, improves immune function in gastric cancer rats (Wang et al., 2014b).

14.3.5 Antimicrobial and antiparasitic activities

The antibacterial action of cryptotanshinone (CT) may be associated with its action as active oxygen radical generator (Feng et al., 2009); *Staphylococcus aureus* undergoes an oxygen-limiting state upon exposure to CT. CT and antibiotics had synergistic effect against clinical methicillin- and vancomycin-resistant *Staphylococcus aureus* (Cha et al., 2014). 7-(2-Oxohexyl)-taxodione, a novel taxodione derivative of *S. austriaca*, has antimicrobial and antibiofilm properties (Kuźma et al., 2012a). A sesquiterpene lactone of *S. nubicola*, nubiol, was moderately active against *Pseudomonas aeruginosa* (Ali et al., 2007). *S. macrochlamys* essential oil had good antimycobacterial activity against *Mycobacterium intracellulare* (Tabanca et al., 2006).

The methanol extracts of *S. santolinifolia*, *S. eremophila*, *S. sclarea*, and *S. limbata* inhibited the growth of all tested bacterial strains (Firuzi et al., 2013). *S. officinalis*, compared to six other medicinal plants, had the greatest inhibitory effect on three cariogenic bacteria (Kermanshah et al., 2014).

Antifungal activity of the essential oils from *S. macrochlamys* and *S. recognita* was nonselective at inhibiting growth and development of reproductive stroma of the plant pathogens *Colletotrichum acutatum*, *Colletotrichum fragariae*, and *Colletotrichum gloeosporioides* (Tabanca et al., 2006). Chemotypes of *S. africana-lutea* in controlling fungal growth and consequently mycotoxin production were found (Nkomo et al., 2014). Products for use in the agricultural sector may be developed from such chemotypes.

The IC50 values of abietane diterpenoids of *S. sahendica* ranged from 0.8 μM to over 8.8 μM against *P. falciparum* and from 1.8 μM to over 32.3 μM against *T. brucei* rhodesiense (Ebrahimi et al., 2013). Tanshinone-type diterpenoids of Sm ranged from 4.1 μM to over 30 μM against *P. falciparum* K1 strain with selectivity indexes (SI) from 0.3 to 1.9 (Slusarczyk et al., 2011). IC50 against *T. brucei* rhodesiense STIB 900 were from 0.5 μM to over 30 μM, and methylenetanshinquinone showed the greatest selective activity with an SI of 24.

A triterpenoid of *S. urmiensis*, urmiensolide, showed *in vitro* antitrypanosoal activity with an IC50 5.6 μM against the *Trypanosoma brucei* rhodesiense STIB 900 strain (Moridi Farimani et al., 2013). Salvadione C and perovskone B of *S. hydrangea*, triterpenoids with rare carbon skeletons, showed *in vitro* antiplasmodial activity (Moridi Farimani et al., 2011), with IC50 1.43 and 0.18 μM, respectively. IC50 values against

T. brucei rhodesiense were 4.33 and 15.92 μM, respectively. *S. officinalis* leaves could control cutaneous leishmaniasis (Nikmehr et al., 2014).

Hydrangenone of *S. hydrangea*, a new heptacyclic isoprenoid with a 6/7/6/5/5 membered carbon ring skeleton, showed *in vitro* antiplasmodial activity, with an IC50 1.4 μM against *P. falciparum* (Farimani et al., 2012b).

The methanol extract of *S. connivens* showed antidiarrheal effect on the animal models (Pérez Gutiérrez et al., 2014). Phytochemical screening revealed the presence of alkaloids, tannins, and saponins, which may be responsible for this effect. The extract did not cause any mortality, and no visible signs of toxicity or differences in food and water uptake were seen.

14.3.6 Effects on the liver

PF2401-SF, a standardized fraction of Sm, had antifibrotic effect in thioacetamide-induced experimental rat liver fibrosis (Parajuli et al., 2014). Standardized Sm extract suppresses hepatic stellate cell activation and attenuates steatohepatitis induced by a methionine-choline-deficient diet in mice (Lee et al., 2014).

Sal B inhibited PPARγ expression and attenuated weight gain in mice with high-fat diet-induced obesity (Wang et al., 2014c). Extracts from *Salvia*-Nelumbinis naturalis alleviate hepatosteatosis via improving hepatic insulin sensitivity (Zhang et al., 2014a).

Sal B protects against acute ethanol-induced liver injury through SIRT1-mediated deacetylation of p53 in rats (Li et al., 2014b).

Danhong injection (DHI) is a Chinese medical product extracted from *Radix et Rhizoma Salviae Miltiorrhizae* and *Flos carthami* (*Carthamus tinctorius*). DHI treatment was accompanied by significantly increased lipolysis in the liver and decreased fatty acid synthesis (Chen et al., 2014b).

14.3.7 Effects on the bone

Sal B promotes osteogenesis of human mesenchymal stem cells through activating ERK signaling pathway (Xu et al., 2014).

In TCM, the antiosteoporotic effect of Sm is ascribed to its action on liver and blood stasis as main therapeutic targets defining osteoporosis (Guo et al., 2014). 36 clinical trials were identified, which used Sm in combination with other herbs and components to treat postmenopausal, senile, and secondary osteoporosis. The trials were characterized by high efficacy (>80%) and low toxicity problems. However, various limitations such as small patient samples, short treatment duration, frequent lack of detailed numerical data, and no clear endpoints must be taken into account. More than 100 compounds have been isolated from this plant and tested in various animal models and biochemical assays. Compounds display antiresorptive and bone formation-stimulating features targeting different pathways in the bone remodeling cycle. Pathways affected include the activation of osteoblasts, the modulation of osteoclastogenesis, and the inhibition of collagen degradation by cathepsin K.

14.3.8 Other effects

Sal A prevents pulmonary fibrosis by inducing fibroblast cell cycle arrest and promoting apoptosis (Pan et al., 2014). Tanshinone IIA was a potent postoperative adhesion preventer by enhancing fibrinolytic activity and decreasing cyclooxygenase-2 activity (Wang et al., 2014a). Compound *Astragalus* and Sm extract suppresses rabbits' hypertrophic scars by modulating the TGF-β/Smad signal (Wu et al., 2014a).

Dietary fatty acids have been extensively evaluated for nutritional and cosmetic benefits. A novel combination of *S. hispanica* seed, rich in ω3 (linolenic acid) and ω6 (linoleic acid) PUFAs, and *Punica granatum* (pomegranate) fruit extracts had inhibitory effect on melanin production (Diwakar et al., 2014).

Sm had reducing effect on alcohol self-administration in Sardinian alcohol-preferring rats (Maccioni et al., 2014). Cryptotanshinone might be involved in the regulation of Tor1, Sch9, Gcn2, and Sod2, the highly conserved longevity proteins modulated by nutrients from yeast to humans (Wu et al., 2014c).

S. haematodes root extract enhances anabolic activity, testicular function, and sexual behavioral performance in a dose-dependant manner (Bansode et al., 2014).

Protocatechualdehyde (PCA), a phenolic acid compound of Sm, prevents methylglyoxal-induced mitochondrial dysfunction and advanced glycation end products (AGEs)-RAGE axis activation in human lens epithelial cells (Wang et al., 2014e).

Embryonic stem cells (ESCs) and induced pluripotent stem cells (iPSCs) are considered the most powerful in terms of differentiating into three-germ-layer cells. Sal B can be used as a replacement of the expensive LIF for maintaining ESC pluripotency while increasing cell proliferation (Liu et al., 2014).

14.3.9 Safety

Multiple components in Sm preparations had cumulative organic anion transporter-mediated drug–drug interaction potential (Wang et al., 2014d). Danshen–Gegen formula (DGF), *Radix S. miltiorrhiza* (danshen), and *Radix Puerariae lobatae* (Gegen) for the treatment of cardiovascular disease could significantly induce the liver phase I metabolism of warfarin (Zhang et al., 2014d) and, to a less extent, potentially increase the intestinal absorption and decrease the plasma protein binding of warfarin.

14.4 Molecular phylogeny and genomics

14.4.1 S. miltiorrhiza

454 Pyrosequencing was used to produce a substantial expressed sequence tag (EST) dataset from the root of Sm (Li et al., 2010). A total of 46,722 ESTs with an average read length of 414 bp were generated. 454 ESTs were combined with the Sm ESTs from NCBI GenBank to assemble into 18,235 unigenes, of which 454 sequencing identified 13,980 novel unigenes. 73% of these unigenes (13,308) were annotated using BLAST searches (*E*-value $\leq$ 1e-5) against the Swiss-Prot, Kyoto Encyclopedia of Genes and

Genomes (KEGG), the Arabidopsis Information Resource (TAIR), and nonredundant (nr) and nucleotide (nt) databases. Twenty-seven unigenes (encoding 15 enzymes) were found to be involved in tanshinone biosynthesis and 29 unigenes (encoding 11 enzymes) involved in phenolic acid biosynthesis. Seventy putative genes were found to encode cytochromes P450 (CYP) and 577 putative transcription factor (TF) genes.

56,774 Unigenes (average length 467 bp) were obtained by Illumina deep sequencing over the entire growing cycle of Sm (Wenping et al., 2011). Unigenes (34,340; 60.49%) were annotated and 2545 unigenes were assigned to specific pathways. Unigenes (1539) were identified as part of five major secondary-metabolism pathways, covering almost all nodes in the phenylpropanoid and terpenoid pathways. Using BLAST search against AGRIS, 1341 unigenes were found homologous to 686 *Arabidopsis* TF genes. Real-time PCR was used to verify the spatiotemporal expression patterns of several novel transcripts related to the biosynthesis of active ingredients.

454 GS-FLX pyrosequencing reads of Sm were assembled and clustered into 64,139 unigenes (29,883 isotigs and 34,256 singletons) (Yang et al., 2013). NCBI nr protein databases and Swiss-Prot database searches anchored 32,096 unigenes (50%) with functional annotations. Further assignments with Gene Ontology (GO) terms and KEGG biochemical pathways identified 168 unigenes referring to the terpenoid backbone biosynthesis (including 144 MEP and MVA pathway genes and 24 terpene synthases). Comparative analysis of the transcriptomes identified 2863 unigenes that were highly expressed in roots, including those enzymes of early steps of tanshinone biosynthetic pathway, such as copalyl diphosphate synthase (SmCPS), kaurene synthase-like (SmKSL), and CYP76AH1. Other differentially expressed unigenes predicted to be related to tanshinone biosynthesis fall into CYPs, dehydrogenases and reductases, and regulatory factors. The above large unigene datasets provide valuable resources for further investigation of the radix development and biosynthesis of tanshinones and other medicinal compounds.

Methyl jasmonate (MeJA) is an effective elicitor to enhance the production of secondary metabolites. The transcriptional profiles of Sm leaves at 12 h (T12) after MeJA elicitation and mock-treated leaves (T0) were generated using the Illumina deep RNA sequencing (RNA-Seq) strategy (Luo et al., 2014). 37,647 Unique sequences were obtained from 21 million reads, 25,641 (71.53%) of which were annotated. A total of 5287 unique sequences were expressed differentially between T0 and T12, which covered almost all known genes involved in tanshinone and phenolic acid biosynthesis. Many TFs (e.g., MYB, bHLH, and WRKY) and genes involved in plant hormone biosynthesis and signal transduction were expressed differentially in response to the MeJA induction. Importantly, three and four candidate CYPs that could be involved in the tanshinone and phenolic acid biosynthesis, respectively, were selected from the RNA-Seq data based on coexpressed pattern analysis with SmCPS1/SmKSL1 and SmRAS, which are the key genes responsible for biosynthesis.

Metabolomics and transcriptomics were combined to investigate the inducible biosynthesis of Sm tanshinones (Gao et al., 2014). Untargeted metabolomics investigation of elicited hairy root cultures indicated that tanshinone production was a dominant component of the metabolic response, increasing at later time points. A transcriptomic approach was applied to define not only a comprehensive transcriptome (composed of 20,972 nonredundant genes) but also its response to induction, revealing 6358 genes that exhibited differential expression, with significant enrichment for the upregulation of genes involved in stress, stimulus, and immune response processes. There is a slower but more sustained increase in transcript levels of known genes of tanshinone biosynthesis. Among the coregulated genes were 70 TFs and 8 CYPs, providing targets for future investigation. There is a biphasic response of Sm terpenoid metabolism to elicitation, with early induction of sesquiterpenoid and triterpenoid biosynthesis, followed by later and more sustained production of the diterpenoid tanshinones. The above investigations of induced gene expression profiles shed light on the molecular mechanisms of the induced bioactive compound biosynthesis and regulation in *Salvia* plants.

Genome-wide characterization and comparative analysis of R2R3-MYB TFs, which regulate secondary metabolite biosynthesis, show the complexity of MYB-associated regulatory networks in Sm (Li and Lu, 2014).

A cDNA-AFLP analysis with 128 primer pairs revealed that 2300 transcript-derived fragments (TDFs) were differentially expressed among the four samples, that is, the red roots from Sm and *S. castanea* and two tanshinone-free roots from Sm (Yang et al., 2012). About 323 TDFs were sequenced, of which 78 were annotated with known functions through BLASTX search, and 14 annotated TDFs were assigned into secondary metabolic pathways through searching the KEGG database. The quantitative real-time (qRT) PCR analysis indicated that the expression of nine TDFs was positively correlated with the accumulation of phenolic compounds and tanshinones. These TDFs additionally showed coordinated transcriptional response with 6 previously identified genes involved in the biosynthesis of tanshinones and phenolic compounds in Sm hairy roots treated with yeast extract.

14.4.2 Chloroplast genome

The chloroplast (cp) genome of Sm is 151,328 bp in length and exhibits a typical quadripartite structure of the large single-copy (LSC, 82,695 bp) and small single-copy (SSC, 17,555 bp) regions (Qian et al., 2013), separated by a pair of inverted repeats (IRs, 25,539 bp). It contains 114 unique genes, including 80 protein-coding genes, 30 tRNAs, and four rRNAs. The genome structure, gene order, GC content, and codon usage are similar to the typical angiosperm cp genomes. Four forward, three inverted, and seven tandem repeats were detected in the Sm cp genome. Simple sequence repeat (SSR) analysis among the 30 asterid cp genomes revealed that most SSRs are AT-rich, which contribute to the overall AT richness of these cp genomes. Fewer SSRs are distributed in the protein-coding sequences compared to the noncoding regions. Entire cp genome comparison of Sm and three other Lamiales cp genomes showed a high degree of sequence similarity and a relatively high divergence of intergenic spacers. The 10 most divergent and 10 most conserved genes, as well as their length variation, will be supportive for phylogenetic studies in asterids. Both regional and functional constraints affect gene sequence evolution. Phylogenetic analysis of cp genomes demonstrated a sister relationship between Sm and *Sesamum indicum*.

Cp genomic DNA and RNA of Sm were subjected to strand-specific RNA-Seq and single-molecule real-time (SMRT) sequencing analyses (Chen et al., 2014a,b). Mapping the RNA-Seq reads to the genome assembly allowed to determine the relative expression levels of 80 protein-coding genes. 19 Polycistronic transcription units and 136 putative antisense and intergenic noncoding RNA (ncRNA) genes were identified. The comparison of the abundance of protein-coding transcripts (cRNA) with and without overlapping antisense ncRNAs (asRNA) suggests that the presence of asRNA is associated with increased cRNA abundance ($p < 0.05$). 2687 Potential DNA modification sites and two potential DNA modification motifs were predicted. The two motifs include a TATA boxlike motif (CPGDMM1, "TATANNNATNA") and an unknown motif (CPGDMM2, "WNYANTGAW"). Specifically, 35 of the 97 CPGDMM1 motifs (36.1%) and 91 of the 369 CPGDMM2 motifs (24.7%) were found to be significantly modified. Analysis of genes downstream of the CPGDMM1 motif revealed the significantly increased abundance of ncRNA genes that are less than 400 bp away from the significantly modified CPGDMM1 motif, implying a complex interplay among DNA modifications and ncRNA and cRNA expressions in cp genome.

14.4.3 *Other* Salvia *species*

In family Lamiaceae, sepals are fused into a tube that is densely covered by hairs for mechanical defense and contains secondary metabolites for chemical defense against insects and abiotic stresses. Because of its large size and its impressive metabolic activity (both in terms of quantity and diversity of secondary metabolites), Legrand et al. (2010) used *S. sclarea* (clary sage) calyx as a model system to generate the first high-throughput sequencing of the transcriptome of an angiosperm calyx. Massive parallel 454 pyrosequencing technology was used to a normalized cDNA extract, and potential candidate genes for all steps of secondary metabolite pathways (phenylpropanoids and terpenoids) were unveiled. It is efficient in predicting the expression of large numbers of TFs and a novel class of gene transcription regulatory elements, miRNAs, without the need to generate a separate miRNA library. 18 Conserved miRNAs were predicted, 15 of which were present in most studied plant species, while the others were only shared with limited or discrete plant lineages. A separate data mining of the same clary sage EST library suggested the presence of 19 potential target genes to the 18 predicted conserved miRNAs. These coded for only 6 TFs or F-box proteins, 11 metabolism or abiotic stress response-related proteins, and 2 products with no known predicted function.

RNA from *S. sclarea* leaf samples taken 0, 10, and 26 h after MeJA treatment was subjected to Illumina HiSeq 2000 single-end sequencing, which generated 35,757,567 reads with a total of 1752,120,783 base pairs (1.75 Gb). In total, 29.03% (0 h), 24.76% (10 h), and 23.93% (26 h) of reads were mapped to *Salvia* unigene datasets. Of these, 12.44%, 9.88%, and 9.19% were perfect matches, respectively. Unique matches accounted for 8,604,567 reads. At least 99.24% of reads were clean and the sequencing reached saturation in all samples. The sequencing randomness is shown in Figure 14.7. The distribution of gene coverage at three time points is shown in

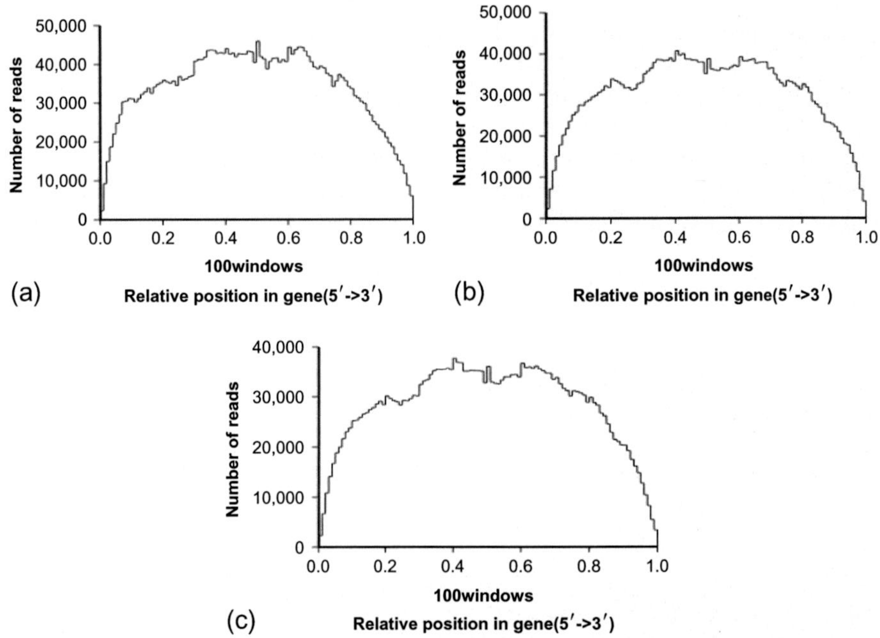

Figure 14.7 Assessment of *S. sclarea* transcriptome sequencing randomness. (a) 0 h; (b) 10 h; (c) 26 h. Seeds were germinated in the glasshouse and 8-week-old plantlets were used for MeJA treatment. MeJA (7.5 mM) was sprayed onto the leaves and poured into root-containing soil. Plantlets were treated for 0, 10, 26, and 73 h. The leaf RNA was extracted for sequencing library construction. The cDNA library of *S. sclarea* leaf was subject to Illumina HiSeq 2000 sequencing.

Figure 14.8. In total, 45,334 unigenes were expressed at all three time points, while a small portion of unigenes was only found in one or two time points. The most abundant unigenes at the three time points are distinct.

The results of differential gene expression obtained via RNA-Seq were confirmed by qRT-PCR, which was also used to characterize the temporal gene expression change after 0–73 h of MeJA treatment in Sm leaf and root (Figure 14.9a and b). In roots, the MeJA-induced response was possibly as extensive as that in leaves but qualitatively/quantitatively different from that in the aboveground parts. MeJA response is tissue-specific, and the differential response between leaf and root was very early observed. For instance, at 10 h of 0.75 mM MeJA, eight Sm leaf genes, HPPD, HPPR (hydroxyphenylpyruvate reductase), PAL1 and PAL2, C4H (cinnamate-4-hydroxylase), 4CL (4-coumarate/CoA ligase) 2, RAS (rosmarinic acid synthase), and DFR, were downregulated (Figure 14.9b), while four root genes, PAL2, C4H, 4CL2, and QGOMT, were suppressed (Figure 14.9a). The present results suggest that both tissues have a different JA signaling cascade that provides plants with a mechanism to tailor their responses specifically to the tissue that is affected.

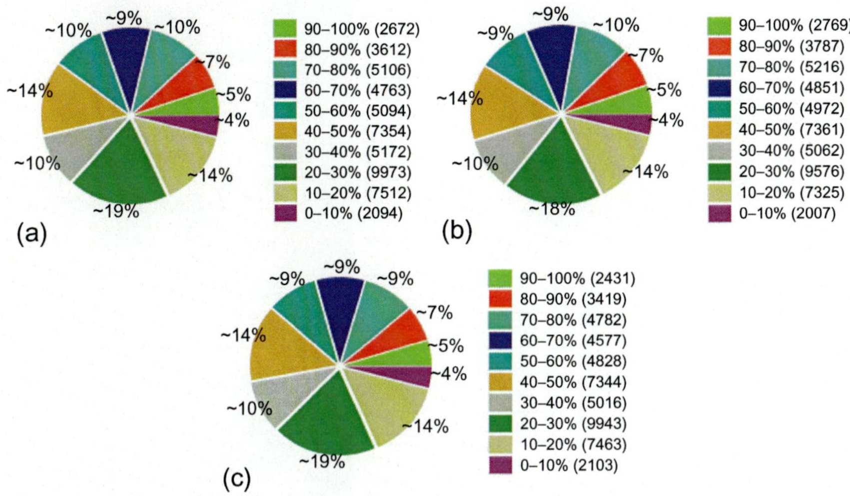

Figure 14.8 Gene coverage of three time points. (a) 0 h; (b) 10 h; (c) 26 h. Clean reads were mapped to reference genes using SOAPaligner/SOAP2 (Li et al., 2009). No more than two mismatches were allowed in the alignment. One unigene dataset of *S. sclarea* (Legrand et al., 2010) and two unigene datasets of *S. miltiorrhiza* (Li et al., 2010; Wenping et al., 2011) were used as reference datasets. Gene coverage is the percentage of a reference gene that is covered by reads. This value is the ratio of the number of bases in a gene covered by unique mapping reads to the number of total bases in that gene.

Greek sage (*S. fruticosa*, syn. *S. triloba*) is appreciated for its essential oil, which is used as an aromatic spice and active against a wide range of microorganisms and viruses. The essential oil is dominated by terpenoids and flavonoids, which are produced and stored in glandular trichomes on the plant surface. Chatzopoulou et al. (2010) gave insights into the metabolic activities of *S. fruticosa* trichomes on a transcriptome level. A total of 2304 clones were sequenced from a cDNA library from leaves' trichomes. Exclusion of sequences shorter than 100 bp resulted in 1615 high-quality ESTs with a mean length of 592 bp. Cluster analysis indicated the presence of 197 contigs (908 clones) and 707 singletons, generating a total of 904 unique sequences. Of them, 628 (69.5%) had significant hits in the nonredundant (nr) protein database and were annotated. A total of 517 (82.3%) sequences were functionally classified using the gene ontologies (GOs) and established pathway associations to 220 (24.3%) sequences in KEGG. 52 (5.8%) of the unique ESTs revealed a GO biological term with relation to terpenoid (78 ESTs), phenylpropanoid (43 ESTs), flavonoid (18 ESTs), or alkaloid (10 ESTs) biosynthesis or to CYPs (26 ESTs). Expression analysis of a selected set of genes known to be involved in the pathways of secondary metabolite synthesis showed higher expression levels in trichomes, validating the tissue specificity of the analyzed glandular trichome library.

De novo transcriptome sequencing on two different branching types of *S. splendens* (strain 35 and Cailinghong) was performed using the Illumina paired-end sequencing

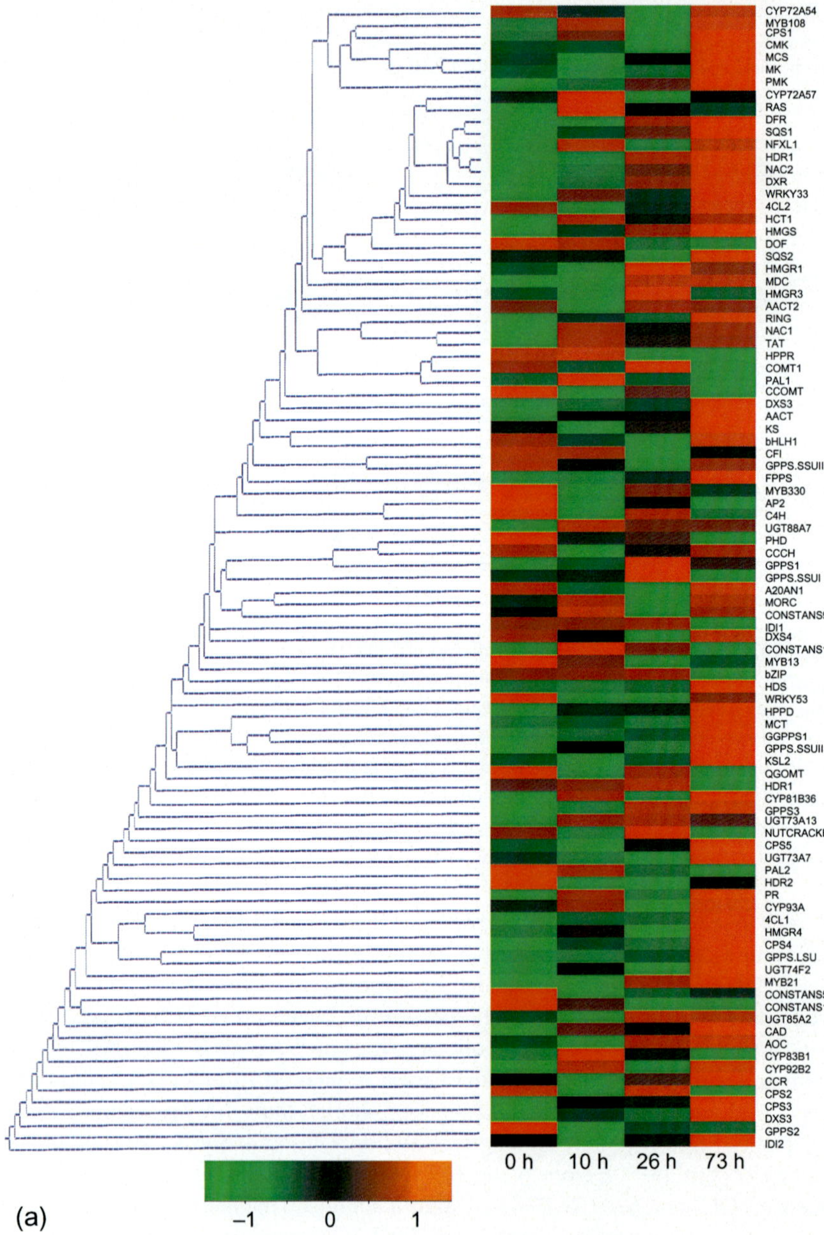

Figure 14.9 Fuzzy cluster analysis and heat map showing the temporal coexpression pattern of Sm (a) root (b) leaf with 0.75 mM MeJA. DNase-treated RNA was used to synthesize first-strand cDNA by using SuperScript II reverse transcriptase (200 units/μl; Invitrogen). SYBR® Green JumpStart™ Taq ReadyMix™ (Sigma-Aldrich) was used in all qRT-PCR reactions, which were performed in three technical replicates in the CFX96 real-time PCR detection system (Bio-Rad) using 0.2 ml nonskirted low-profile 96-well PCR plates (Thermo Scientific) and PCR sealers Microseal B film (Bio-Rad). qRT-PCRs were performed to study the time course of secondary metabolism gene expression. CFX Manager software (Bio-Rad) was used for data analysis. Fuzzy cluster analysis (Huang et al., 2010) was used to construct the phylogenetic tree based on the fold change of gene expression at four time points.

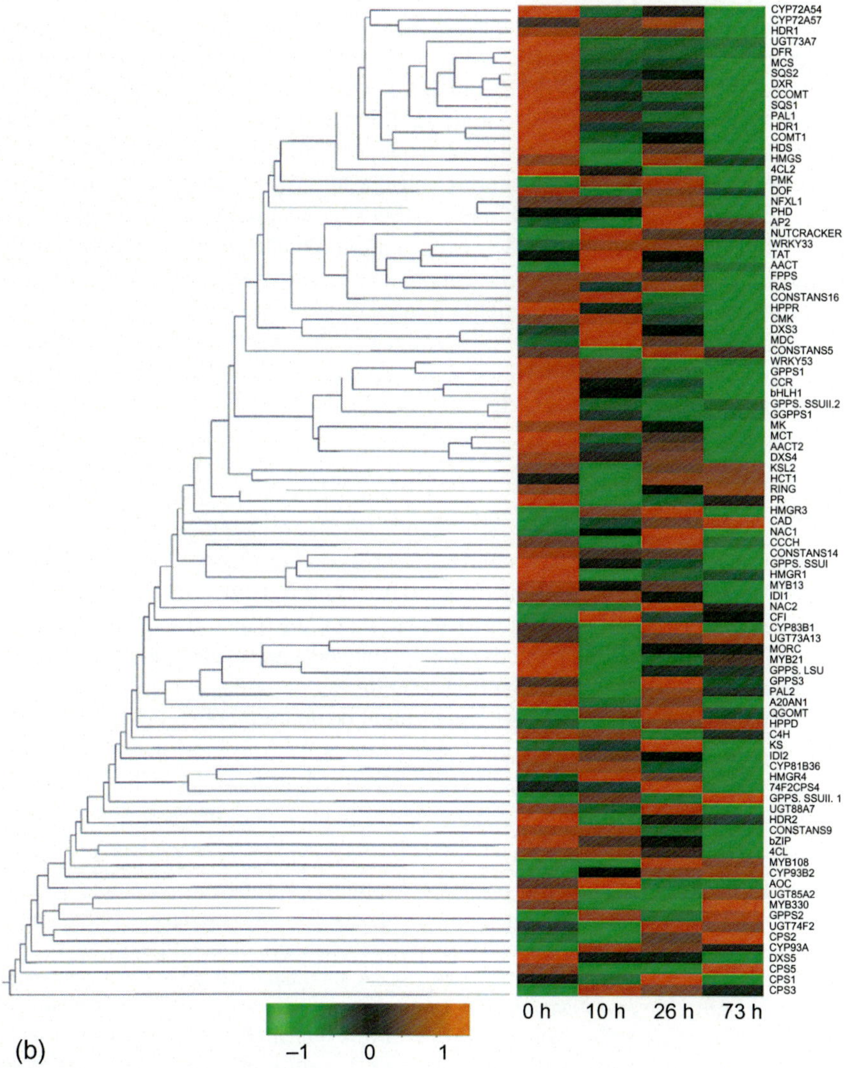

Figure 14.9 Continued. The expression data were standardized by shifting transformation and ranging transformation, and the fuzzy similarity matrix was set up by the max–min method. The phylogenetic tree was obtained using the MATLAB software. Red, upregulation in gene expression; green, downregulation in gene expression. ANR, anthocyanidin reductase; ANS, anthocyanidin synthase; AOC, allene oxide cyclase; AP2, AP2 domain transcription factor RAP2; bZIP, G-box binding factor bZIP transcription factor; CAD, cinnamyl alcohol dehydrogenase; CCOMT, caffeoyl-CoA *O*-methyltransferase; CCR, cinnamoyl-CoA reductase; CFI(CHI), chalcone-flavanone isomerase-like protein; C4H, cinnamate 4-hydroxylase; 4CL, 4-coumarate-CoA ligase; CMK, 4-diphosphocytidyl-2-C-methylerythritol kinase; COMT, caffeic acid *O*-methyltransferase; CONSTANS, zinc finger protein CONSTANS-LIKE; CPS, copalyl diphosphate synthase; CYP, cytochrome P450; DFR, dihydroflavonol-4-reductase; DOF, Dof zinc finger protein; DXR, 1-deoxy-D-xylulose

(Continued)

technology (Ge et al., 2014a,b). For strain 35, 16,488,829 reads were generated and assembled into 38,498 unigenes, with a mean length 779 bp. For Cailinghong, 16,464,713 reads were generated and assembled into 34,302 unigenes, with a mean length 812 bp. A total of 49,310 unigenes were identified, among them 33,925 (68.80%) were annotated in the nonredundant NCBI database, 25,371 (51.45%) were annotated in the Swiss-Prot database, while 24,888 (50.47%) and 9896 (20.07%) unigenes were assigned to GO categories and clusters of orthologous groups, respectively. 134 Differently expressed unigenes between strain 35 and Cailinghong were mapped to 79 KEGG pathways. Moreover, some candidate genes potentially involved in branching were identified. 2453 Simple sequence repeats (SSRs) detected from transcriptome dataset might contribute to marker-assisted selection.

S. officinalis SSR markers have potential use in conservation-genetic studies of narrow endemic *S. brachyodon* (Radosavljević et al., 2012) and can be cross amplified in other related species (Radosavljević et al., 2011).

14.4.4 Phylogeny

The tribe Mentheae and all five subtribes have a Mediterranean origin and have dispersed to the New World multiple times (Drew and Sytsma, 2012). In the subtribe Salviinae, *Salvia* clade II is closer to clade III (including *S. przewalskii* of China) than to clade I. Worldwide cooling trends probably played a large role in the diversification and present-day distribution of the tribe Mentheae.

Salvia, the largest genus of Lamiaceae, might be nonmonophyletic. On the cp rbcL + trnL-F phylogenetic tree, *Rosmarinus* and *Perovskia* are sister to an Old World clade of *Salvia* (Walker et al., 2004), and the section *Audibertia* is sister to subgenus *Calosphace* (clade II) or the monotypic Asian genus *Dorystaechas*. The Old World

Figure 14.9 Continued. 5-phosphate reductoisomerase; DXS, 1-deoxy-D-xylulose 5-phosphate synthase; F3H, flavanone 3-hydroxylase; F4′ST, flavonol 4′-sulfotransferase; FPPS, farnesyl diphosphate synthase; GGPPS, geranylgeranyl diphosphate synthase; GPPS, geranyl diphosphate synthase; HCT, hydroxycinnamoyl-CoA shikimate/quinate hydroxycinnamoyl transferase; HDR, 4-hydroxy-3-methylbut-2-enyl diphosphate reductase; HDS, 4-hydroxy-3-methylbut-2-enyl diphosphate synthase; HMGR, hydroxymethylglutaryl-CoA reductase; HMGS, hydroxymethylglutaryl-CoA synthase; HPPD, hydroxyphenylpyruvate dioxygenase; IDI, isopentenyl diphosphate isomerase; IDS, isoprenyl diphosphate synthase; I2′H, isoflavone 2′-hydroxylase; KS, kaurene synthase; LSU, large subunit; MADS, MADS-box transcription factor; MCT, 2-C-methyl-D-erythritol 4-phosphate cytidylyltransferase; MDH, menthol dehydrogenase; MDS(MCS), 2-C-methyl-D-erythritol 2,4-cyclodiphosphate synthase; MK, mevalonate kinase; MORC, MORC family CW-type zinc finger protein; MYBpre, MYB transcription factor precursor; NUTCRACKER, zinc finger protein NUTCRACKER; PAL, phenylalanine ammonia lyase; PHD, PHD finger protein; PMK, 5-phosphomevalonate kinase; PR, (+)-pulegone reductase; PS, pinene synthase; QGOMT, quercetin 3-*O*-glucoside-6″-*O*-malonyltransferase; RING, RING finger and CHY zinc finger domain-containing protein; Sesq, sesquiterpene synthase; SQS, squalene synthase; SSU, small subunit; TPS, terpene synthase; UFGT, UDP-glucose: flavonoid 7-*O*-glucosyltransferase; UGT, UDP-glycosyltransferase.

taxa *S. officinalis* and *S. sclarea* are in clade I. On the trnL-F+psbA-trnH+nuclear internal transcribed spacer (ITS) tree, *Salvia* is clearly polyphyletic, with five other genera intercalated within it (Walker and Sytsma, 2007). In *Salvia* clade III, *S. przewalskii* (purple danshen), *S. cynica*, and *S. roborowskii* are sister to each other, which are closer to *S. digitaloides* than to *S. glutinosa*. Sm is basal to these five Chinese species. The character that defines *Salvia* within the Mentheae (the significantly elongate connective) has independently originated in each of the three major *Salvia* lineages. The independent origin of the defining character for *Salvia* is supported by the molecular phylogeny that each of the major clades of *Salvia* is associated with a genus that does not express the significantly elongate connective, and by the distinct staminal morphology in each clade of *Salvia*. *Salvia* is polyphyletic in that it is defined by a convergent character.

More marker sequences of Chinese species are retrieved from NCBI GenBank to reconstruct the phylogenetic relationship (Figure 14.10a and b). The three *Salvia* clades proposed by Walker and Sytsma (2007) are present on the ITS tree (Figure 14.10a).

The complete sequences of *Salvia* ITS plus 5.8S rDNA are about 612–617 bp (Wang et al., 2007b). A phylogenetic tree generated by neighbor-joining method partly supported the morphological classification within *Salvia*, but incompatible results were also obtained in the treatment of phylogenetic positions of some species such as *S. trijuga*, *S. flava* var. *flava*, and *S. flava* var. *megalentha*. Such incongruence is also conspicuous when much more species are included in ITS analysis (Figure 14.10a), for example, *S. trijuga*, *S. glutinosa*, and *S. yunnanensis*. The *Salvia* ITS regions showed considerable variation between subgenera in contrast to the conservative 5.8S rDNA sequences. The Chinese *Salvia* species might have a different origin from the European species. Polytomies found in all single-marker trees, as well as extremely short internal branch lengths (Figure 14.10a and b), indicate that a rapid radiation might occur during the evolution of *Salvia*. Similar to *Isodon* (Lamiaceae; Yu et al., 2014a,b), the Qinghai-Tibet Plateau (QTP) and adjacent region are the most likely ancestral areas of Chinese *Salvia*, with the Hengduan Mountains region (southeastern edge of the QTP) currently possessing the greatest diversity of Chinese *Salvia* species (Table 14.1). Many species of this area, such as *S. brachyloma*, *S. bulleyana*, *S. campanulata*, *S. castanea*, *S. digitaloides*, *S. evansiana*, *S. flava*, *S. heterochroa*, *S. kiaometiensis*, and *S. przewalskii*, are used as danshen in local pharmacopeia (Li et al., 2013a). Bursts of speciation triggered by geologic and/or climatic changes during the uplift of QTP are a common phenomenon and of significant importance in generating the current biodiversity in this region. Similar scenarios were reported within Andean regions (Yu et al., 2014a,b).

The phylogenetic positions of subgenera and sections inferred by ITS analysis were comparable to that of traditional classification (Figure 14.10a), while the phylogeny within sections is still confusing due to limited information in genomic sequence. ITS analysis not only supports the rationality of using species from *Drymosphace* section, for example, *S. bowleyana*, *S. dabieshanensis*, *S. honania*, *S. sinica*, *S. trijuga*, and *S. yunnanensis*, as alternative drug resources of danshen but also reveals significant genetic differences between alpine danshen species such as *S. przewalskii* and traditional danshen plant (i.e., Sm). ITS1 was superior to the cp marker for discriminating

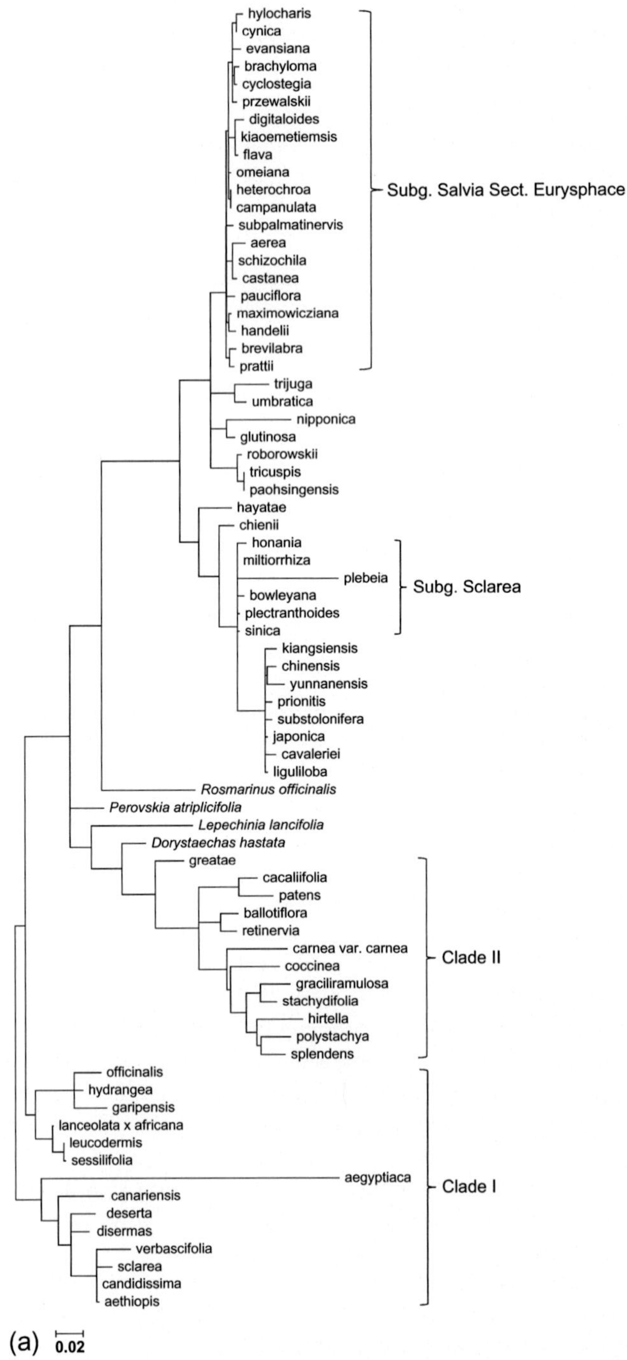

Figure 14.10 The evolutionary history was inferred by using the maximum likelihood method based on the GTR+I+G model. The tree with the highest log likelihood is shown and is drawn to scale, with branch lengths measured in the number of substitutions per site. (a) 74 ITS sequences; (b) 29 trnL-trnF+ITS+matK sequences. Evolutionary analyses were conducted in MEGA6.

(Continued)

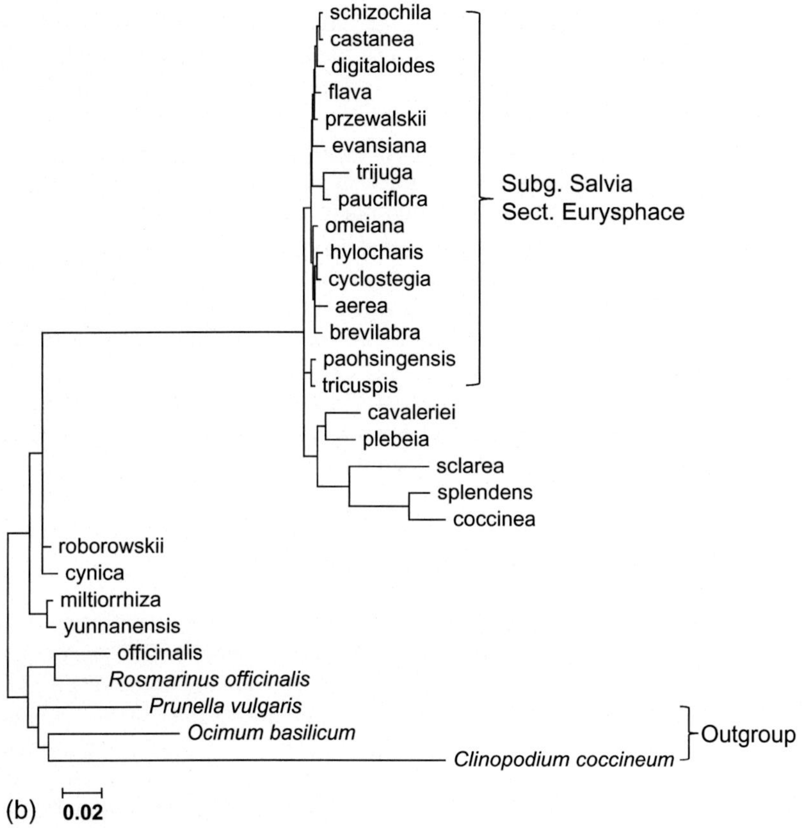

Figure 14.10 Continued.

between *Salvia* species, with an accuracy of 81.48% (Wang et al., 2013). The three combinations did not increase species discrimination.

On the rbcL+trnL-F+ITS tree, all Japanese and two Taiwanese *Salvia* species are included in a clade with other Asian *Salvia* species (Takano and Okada, 2011), and Japanese *Salvia* species were distributed among three subclades: (1) *S. plebeia* (subgenus *Sclarea*), (2) species belonging to subgenus *Salvia*, and (3) species belonging to subgenus *Allagospadonopsis*. This phylogenetic tree reveals a close relationship between *S. nipponica* and *S. glabrescens*. No support for monophyly of *S. lutescens* and its varieties was obtained in cpDNA, nrDNA, and cpDNA+nrDNA trees, and *S. pygmaea* var. *simplicior* may be more closely related to *S. japonica* (Shu Wei Cao) than to other varieties of *S. pygmaea*.

Phylogenetic reconstruction of Africa *Salvia* was based on ITS, external transcribed spacer (ETS), and plastid rpl32-trnL. Sequence data were generated for 41 of the 62 African taxa (66%) (Will and Claßen-Bockhoff, 2014). *Salvia* in Africa is nonmonophyletic. Each of the five major regions in Africa, except Madagascar, was colonized at least twice, and floristic links between North African, southwest Asian, and European species are strongly supported. The large radiation in sub-Saharan Africa (23 species) can be traced back to dispersal from North Africa via East Africa to the Cape Region. Adaptation to bird pollination in southern Africa and Madagascar reflects parallel evolution. The phenotypic diversity in African *Salvia* is associated with repeated introductions to the continent. Many important evolutionary processes, such as colonization, adaptation, parallelism, and character transformation, are reflected in this comparatively small group, which help to understand the evolution of *Salvia* and other large genera.

S. divinorum is a potent hallucinogenic plant that is classified within subgenus *Calosphace* and section *Dusenostachys* and hypothesized to be an interspecific hybrid. It is of ethnobotanical significance due to its employment in traditional healing ceremonies by the Mazatecs of Oaxaca and due to its unique pharmacology—a highly selective, nonnitrogenous, κ-opioid receptor agonist. On the ITS + trnL-F + psbA-trnH tree of 52 species representing the major lineages of subgenus *Calosphace*-and six accessions of *S. divinorum*, *S. divinorum* is not within *Dusenostachys* and is not a hybrid (Jenks et al., 2011). The closest known relative of this psychoactive Mexican sage is *S. venulosa*, a rare endemic of Colombia.

14.5 Chemotaxonomy and metabolomics

HPLC/DAD was used for studying the distribution of tanshinones and phenolic acids in some Chinese *Salvia* species (Li et al., 2013b). The distributions of tanshinones are distinct in the subgenus level. Tanshinones occurred frequently in the species of subgenus *Salvia* and subgenus *Sclarea*, while they were absent in subgenus *Jungia* and subgenus *Allagospadonopsis*. Section *Eurysphace* (subgenus *Salvia*) were abundant in abietane diterpenoids such as tanshinones, and tanshinones occurred in almost all the investigated species of this group. 20-Norabietanes and 19,20-dinorabietanes occurred frequently in the species of subgenus *Salvia* and subgenus *Sclarea*, while they were absent in subgenus *Jungia* and subgenus *Allagospadonopsis*. The chemotaxonomic data support the circumscription of section *Drymosphace* (subgenus *Sclarea*) proposed by Bentham. Walker et al. (2004), based on the phylogenetic tree inferred from rbcL + trnL-F, suggested that all members of clade III (Chinese species) are in Bentham's section *Drymosphace*. Clerodane diterpenes such as Tan IIA and CT were isolated from some New World species (*S. mellifera*, *S. apiana*, *S. axillaris*, and *S. columbariae*), which could be considered as chemical evidence of a probable evolutionary link between Asian and American *Salvia* species.

Polyphenolics are also abundant in *Salvia*, especially in section *Drymosphace* and section *Eurysphace*. However, the intersectional variation of polyphenolics is significant (Li et al., 2013b), which may be due to both genetic and environmental factors. The Chinese *Salvia* species were very rich in caffeic acid derivatives, while most species from Europe and New World produced few caffeic acid derivatives, except for rosmarinic acid. Tanshinones and phenolic acids were abundant in different danshen species, which justifies their use in TCM.

The LC/DAD/ESI-MS^n analysis of the *Salvia* extracts led to the separation and identification of their constituents. Identification was based on accurate mass measurements of the pseudomolecular $[M-H]^-$ ions and their fragmentation pattern. No obvious change of peak intensity was found within 24 h treatment of 7.5 mM MeJA. Plantlets were then treated with MeJA for 72 h. Data obtained from the ESI-MS^n analysis of the extracts are summarized in Table 14.3.

The compounds identified in *S. officinalis* extracts belonged to four classes of phytochemicals: phenolic acids (caffeic acid and rosmarinic acid), flavonoids (cirsimaritin, salvigenin, and genkwanin), diterpenes (rosmadial, carnosol, carnosic acid, and methoxy carnosate), and triterpenes (oleanolic acid and ursolic acid). Overall, 16 constituents were identified. Increased amounts of caffeic acid (peak one), rosmarinic acid (peak three), and salvianolic acid F derivative (peak eight) were detected after MeJA treatment in UV chromatograms at 330 nm (Figure 14.11a and Table 14.3) and 280 nm (Figure 14.11b). Peaks 4 and 10, representing unknown compounds, also increased after MeJA elicitation. Together, metabolic phenotype agrees with the gene expression data in that MeJA stimulates the phenylpropanoid pathway and terpenoid pathway (Figure 14.12a and b) and then results in the accumulation of the phenolic compounds and terpenoids. The lack of major changes of some secondary metabolites in these pathways might be explained by the yet unknown posttranslational factors and/or other regulatory factors.

Among the 40 identified constituents of indigenous Dalmatian sage (*S. officinalis* L.) populations in Montenegro, the most abundant were *cis*-thujone (16.98–40.35%), camphor (12.75–35.37%), 1,8-cineol (6.40–12.06%), *trans*-thujone (1.5–10.35%), camphene (2.26–9.97%), borneol (0.97–8.81%), viridiflorol (3.46–7.8%), limonene (1.8–6.47%), α-pinene (1.59–5.46%), and α-humulene (1.77–5.02%) (Stešević et al., 2014). The composition of the essential oils under study did not meet the ISO 9909 requirements, while the oils of populations P02-P04, P09, and P10 complied with the German Drug Codex. A few of the main essential-oil constituents appeared to be highly intercorrelated. Strong positive correlations were observed between α-pinene and camphene, camphene and camphor, and *cis*-thujone and *trans*-thujone. Strong negative correlations were evidenced between *cis*-thujone and α-pinene, *cis*-thujone and camphene, *cis*-thujone and camphor, and *trans*-thujone and camphene. Multivariate analyses allowed the grouping of the populations into three distinct chemotypes, that is, chemotype A, rich in total thujones; chemotype B, with intermediate contents of thujones, α-pinene, camphene, and camphor, and high borneol contents; and chemotype C, rich in camphor, camphene, and α-pinene. The chemotypes did not significantly differ in the total essential-oil content and the *cis*-thujone/*trans*-thujone ratio.

A chemotaxonomic survey of the triterpenoids of *Salvia* species shows that triterpenoids from the exudate of *S. argentea* var. *aurasiaca* provide some features, such as hydroxylation at C-15 and hydroxylation at C-7 (Lakhal et al., 2014). On this basis, *S. argentea L.* var. *aurasiaca* is quite different from *S. argentea*.

Table 14.3 Peak assignments of *S. officinalis* extract before and after MeJA treatment

Peak	Retention time (min)	$[M–H]^-$	MS^2	Compound
1	8.3	179 (100%)	153 (100%)	Caffeic acid
2	10.3	431 (100%), 545 (68%),	249 (100%)	Apigenin-7-*O*-glucoside
3	11.5	359(100%)	161 (100%), 179 (30%)	Rosmarinic acid
4	12.1	563 (100%)	387 (100%)	Unknown
5	14.3	285 (100%)	283 (100%)	Luteolin
6	14.9	327 (100%), 425 (83%)	229 (100%)	Salvigenin
7	15.7	491 (100%), 605 (96%)	311 (100%), 267 (28%), 179 (13%)	Salvianolic acid C isomer
8	16.3	427 (100%), 269 (59%), 313 (53%)	313 (100%)	Salvianolic acid F derivative
9	16.7	329 (100%)	313 (100%)	Salvianolic acid F derivative
10	17.3	505 (100%)		Unknown
11	19.2	313 (100%), 459 (86%), 345 (65%)	298 (100%)	Cirsimaritin
12	19.7	625 (100%)	311 (100%)	Unknown
13	21.3	283 (100%)	–	Genkwanin
14	24.0	343 (100%)	299 (100%)	Rosmadial
15	25.7	329 (100%), 443 (61%), 659 (20%)	285 (100%)	Carnosol
16	27.0	633 (100%)	589 (100%), 615 (79%)	Unknown
17	29.9	331 (100%)	287 (100%)	Carnosic acid
18	33.0	345 (100%)	301 (100%), 286 (28%)	Methyl carnosate
19	33.5	615 (100%), 325 (56%)	571 (100%), 495 (92%)	Unknown
20	44.8	339 (100%)	163 (100%)	Unknown
21	45.9	455 (100%)	407 (100%), 451 (35%)	Oleanolic acid
22	46.7	455 (100%)	453 (100%)	Ursolic acid
23	56.0	591(100%)	515 (100%), 559 (61%)	Unknown

The pseudomolecular $[M–H]^-$ ion of each peak represents the molecular weight of each component of the extract minus one, because we used negative ionization mode. As many natural compounds have the same molecular weight, we use the fragmentation pattern of the pseudomolecular $[M–H]^-$ ion (fourth column of the table), by giving energy to generate smaller ions. The fragments are useful in the identification of each compound. The parent $[M–H]^-$, for every class of compounds, follows a special fragmentation pathway. For example, *O*-flavonoid glucosides commonly lose a hexose moiety (162 amu) giving a fragment ion at m/z equal to the deprotonated molecular ion of the respective flavonoid aglycone. The intensity of the molecular ions expressed as % ratio is given in the parentheses. This intensity is also used in order to discriminate compounds with the same molecular weight, like isomers. All these data are compared with literature data in order to identify the constituents of the leaf extract.

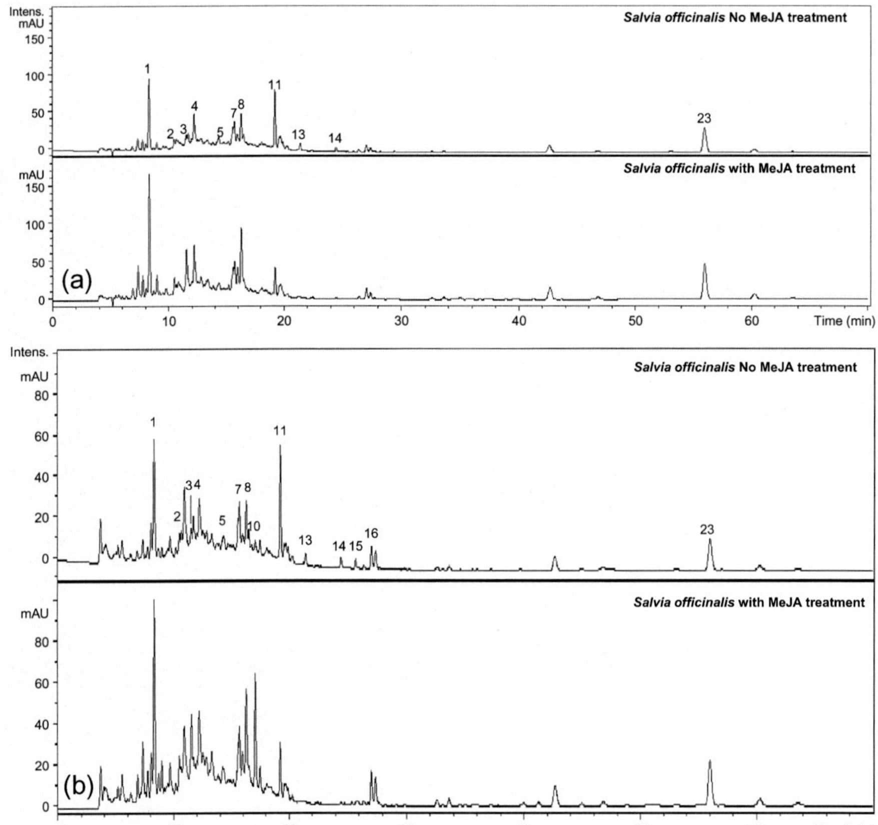

Figure 14.11 UV chromatogram at (a) 330 nm (b) 280 nm of *S. officinalis* (above) and of *S. officinalis* with MeJA treatment (below). Fresh plant material was dried with liquid nitrogen and pulverized into a fine powder. The ground leaves and roots were subsequently extracted with two solvents of increasing polarity, hexane (extraction volume 200 ml) and ethyl acetate (extraction volume 200 ml), in a Soxhlet apparatus for ~6 h for each solvent. The ethyl acetate extracts were concentrated in a rotary evaporator and kept into sealed flasks. All LC-MS^n experiments were performed on a quadrupole ion trap mass analyzer (Agilent Technologies, model MSD Trap SL). Peak 1, caffeic acid; 2, apigenin-7-*O*-glucoside; 3, rosmarinic acid; 5, luteolin; 6, salvigenin; 7, salvianolic acid C isomer; 8, 9, salvianolic acid F derivative; 11, cirsimaritin; 13, genkwanin; 14, rosmadial; 15, carnosol; 17, carnosic acid; 18, methyl carnosate; 21, oleanolic acid; 22, ursolic acid.

The red roots of *S. castanea*, used as danshen in Hengduan Mountains of China, produced high contents of rosmarinic acid (21.77 mg/g) and tan IIA (12.60 mg/g) but low content of Sal B (1.45 mg/g) (Yang et al., 2012). The red roots of Sm produced high content of Sal B (18.69 mg/g), while tanshinone accumulation was much less than that in *S. castanea*. Tanshinones were not detected in the two tanshinone-free Sm samples, which produced high contents of phenolic compounds.

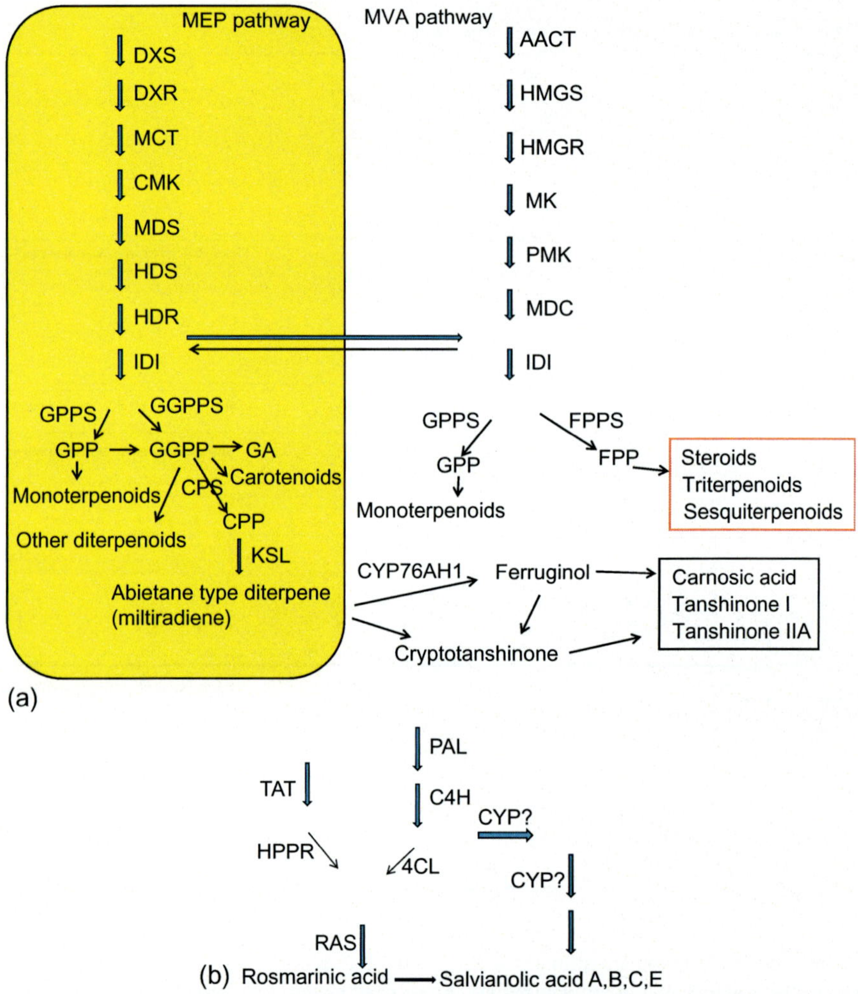

Figure 14.12 Biosynthetic pathway of terpenoids (a) and phenolic acids (b) in Sm. 4CL, 4-coumarate: CoA ligase; AACT, acetyl-CoA C-acetyltransferase; C4H, cinnamate 4-hydroxylase; CDP-ME, 4-(cytidine 5′-diphospho)-2-C-methyl-D-erythritol; CDP-ME2P, 2-phospho-4-(cytidine 5′-diphospho)-2-C-methyl-D-erythritol; CMK, 4-(cytidine 5′-diphospho)-2-C-methyl-D-erythritol kinase; CPS, copalyl diphosphate synthase; DMAPP, dimethylallyl diphosphate; DXP, 1-deoxy-D-xylulose 5-phosphate; DXR, 1-deoxy-D-xylulose 5-phosphate reductoisomerase; DXS, 1-deoxy-D-xylulose 5-phosphate synthase; FPP, farnesyl diphosphate; FPPS, farnesyl diphosphate synthase; G3P, glyceraldehyde 3-phosphate; GA, gibberellin; GGPP, geranylgeranyl diphosphate; GGPPS, geranylgeranyl diphosphate synthase; GPP, geranyl diphosphate; GPPS, geranyl diphosphate synthase; HDR, 4-hydroxy-3-methylbut-2-enyl diphosphate reductase; HDS, 4-hydroxy-3-methylbut-2-enyl diphosphate synthase; HMBPP, 4-hydroxy-3-methylbut-2-enyl diphosphate; HMG-CoA, 3-hydroxy-3-methylglutaryl-CoA; HMGR, hydroxymethylglutaryl-CoA reductase; HMGS, hydroxymethylglutaryl-CoA synthase; HPPR, hydroxyphenylpyruvate reductase; IDI, isopentenyl diphosphate isomerase; IDS, isoprenyl diphosphate synthase; IPP, isopentenyl diphosphate; KS, kaurene synthase; KSL, kaurene synthase-like; MCT, 2-C-methyl-D-erythritol

Figure 14.12 Continued. 4-phosphate cytidylyltransferase; MDC, mevalonate pyrophosphate decarboxylase; MDS, 2-C-methyl-D-erythritol 2,4-cyclodiphosphate synthase; MEcPP, 2-C-methyl-D-erythritol 2,4-cyclodiphosphate; MEP, 2-C-methyl-D-erythritol 4-phosphate; MK, mevalonate kinase; MVA, mevalonate; PAL, phenylalanine ammonia lyase; PMK, 5-phosphomevalonate kinase; RAS, rosmarinic acid synthase; TAT, tyrosine aminotransferase; TPS, terpene synthase.

14.6 Conclusions

Salvia, the largest genus of the family Labiatae, is economically important because it has exhibited various biological and pharmacological activities including antitumor, antiallergic, antioxidant, antimicrobial, and antiplatelet aggregation activities. The diterpenoid constituents of this genus are of great interest due to their diverse structures and significant biological activities. This group of plants has been characterized from a phytochemical standpoint by the production of monoterpenoids, mainly diterpenoids with an abietane or clerodane skeleton, triterpenoids, phenolic acids, and flavonoids. Many diterpenoids isolated from *Salvia* species have shown antioxidant, antifeedant, antibacterial, antimutagenic, anti-inflammatory, and antiplatelet aggregation activities or cytotoxic properties. Species diversity of this genus is becoming a gold mine for sustainable development of traditional and modern medicine industry.

References

Ali, M.S., Ibrahim, S.A., Ahmed, S., Lobkovsky, E., 2007. Guaiane sesquiterpene lactones from *Salvia nubicola* (Lamiaceae). Chem. Biodivers. 4 (1), 98–104.

Bansode, F.W., Rajendran, S.M., Singh, R.K., 2014. Dose-dependent effects of ethanol extract of *Salvia haematodes* Wall roots on reproductive function and copulatory behaviour in male rats. Andrologia 47, 266–275. http://dx.doi.org/10.1111/and.12255.

Bautista, E., Maldonado, E., Ortega, A., 2012. Neo-clerodane diterpenes from *Salvia herbacea*. J. Nat. Prod. 75 (5), 951–958.

Bissinger, R., Lupescu, A., Zelenak, C., et al., 2014. Stimulation of eryptosis by cryptotanshinone. Cell. Physiol. Biochem. 34 (2), 432–442.

Cai, H., Lian, L., Wang, Y., et al., 2014. Protective effects of *Salvia miltiorrhiza* injection against learning and memory impairments in streptozotocin-induced diabetic rats. Exp. Ther. Med. 8 (4), 1127–1130.

Cao, S.Y., Ke, Z.L., Xi, L.M., 2013. A new sesquiterpene lactone from *Salvia plebeia*. J. Asian Nat. Prod. Res. 15 (4), 404–407.

Cha, J.D., Lee, J.H., Choi, K.M., et al., 2014. Synergistic effect between cryptotanshinone and antibiotics against clinic methicillin and vancomycin-resistant *Staphylococcus aureus*. Evid. Based Complement. Alternat. Med. 2014, 450572.

Chang, C.Z., Wu, S.C., Kwan, A.L., 2014. Magnesium lithospermate B, an active extract of *Salvia miltiorrhiza*, mediates sGC/cGMP/PKG translocation in experimental vasospasm. Biomed Res. Int. 2014, 272101.

Chatzopoulou, F.M., Makris, A.M., Argiriou, A., et al., 2010. EST analysis and annotation of transcripts derived from a trichome-specific cDNA library from *Salvia fruticosa*. Plant Cell Rep. 29 (5), 523–534.

Chen, H., Zhang, J., Yuan, G., Liu, C., 2014a. Complex interplay among DNA modification, noncoding RNA expression and protein-coding RNA expression in *Salvia miltiorrhiza* chloroplast genome. PLoS One 9 (6), e99314.

Chen, J., Deng, J., Zhang, Y., et al., 2014b. Lipid-lowering effects of Danhong injection on hyperlipidemia rats. J. Ethnopharmacol. 154 (2), 437–442.

Choi, J.K., Oh, H.M., Lee, S., et al., 2014. *Salvia plebeia* suppresses atopic dermatitis-like skin lesions. Am. J. Chin. Med. 42 (4), 967–985.

Choudhary, M.I., Hussain, A., Ali, Z., et al., 2012. Diterpenoids including a novel dimeric conjugate from *Salvia leriaefolia*. Planta Med. 78 (3), 269–275.

Dai, Y., Liu, L., Xie, G., et al., 2014. Four new eudesmane-type sesquiterpenes from the basal leaves of *Salvia plebeia* R. Br. Fitoterapia 94, 142–147.

Diwakar, G., Rana, J., Saito, L., et al., 2014. Inhibitory effect of a novel combination of *Salvia hispanica* (chia) seed and *Punica granatum* (pomegranate) fruit extracts on melanin production. Fitoterapia 97, 164–171.

Drew, B.T., Sytsma, K.J., 2012. Phylogenetics, biogeography, and staminal evolution in the tribe Mentheae (Lamiaceae). Am. J. Bot. 99 (5), 933–953.

Ebrahimi, S.N., Zimmermann, S., Zaugg, J., et al., 2013. Abietane diterpenoids from Salvia sahendica–antiprotozoal activity and determination of their absolute configurations. Planta Med. 79 (2), 150–156.

Ebrahimi, S.N., Farimani, M.M., Mirzania, F., et al., 2014. Manoyloxide Sesterterpenoids from *Salvia mirzayanii*. J. Nat. Prod. 77 (4), 848–854.

Farimani, M.M., Moghaddam, F.M., Esmaeili, M.A., Amin, G., 2012a. A lupane triterpenoid and other constituents of *Salvia eremophila*. Nat. Prod. Res. 26 (21), 2045–2049.

Farimani, M.M., Taheri, S., Ebrahimi, S.N., et al., 2012b. Hydrangenone, a new isoprenoid with an unprecedented skeleton from *Salvia hydrangea*. Org. Lett. 14 (1), 166–169.

Feng, H., Xiang, H., Zhang, J., et al., 2009. Genome-wide transcriptional profiling of the response of *Staphylococcus aureus* to cryptotanshinone. J. Biomed. Biotechnol. 2009, 617509.

Firuzi, O., Miri, R., Asadollahi, M., et al., 2013. Cytotoxic, antioxidant and antimicrobial activities and phenolic contents of eleven salvia species from Iran. Iran. J. Pharm. Res. 12 (4), 801–810.

Fu, P., Du, F., Chen, W., et al., 2014. Tanshinone IIA blocks epithelial-mesenchymal transition through HIF-1α downregulation, reversing hypoxia-induced chemotherapy resistance in breast cancer cell lines. Oncol. Rep. 31 (6), 2561–2568.

Gandomkar, S., Yousefi, M., Habibi, Z., As'habi, M.A., 2012. A new triterpene from *Salvia xanthocheila* Boiss. Nat. Prod. Res. 26 (7), 648–653.

Gao, W., Sun, H.X., Xiao, H., et al., 2014. Combining metabolomics and transcriptomics to characterize tanshinone biosynthesis in *Salvia miltiorrhiza*. BMC Genomics 15, 73.

Ge, G., Zhang, Q., Ma, J., et al., 2014a. Protective effect of *Salvia miltiorrhiza* aqueous extract on myocardium oxidative injury in ischemic-reperfusion rats. Gene 546 (1), 97–103.

Ge, X., Chen, H., Wang, H., et al., 2014b. De novo assembly and annotation of *Salvia splendens* transcriptome using the Illumina platform. PLoS One 9 (3), e87693.

Guo, Y., Li, Y., Xue, L., et al., 2014. *Salvia miltiorrhiza*: an ancient Chinese herbal medicine as a source for anti-osteoporotic drugs. J. Ethnopharmacol. 155, 1401–1416. http://dx.doi.org/10.1016/j.jep.2014.07.058.

Habibi, Z., Cheraghi, Z., Ghasemi, S., Yousefi, M., 2012. A new highly hydroxylated triterpene from *Salvia atropatana* Bunge. Nat. Prod. Res. 26 (20), 1910–1913.

Habibi, Z., Gandomkar, S., Yousefi, M., Ghasemi, S., 2013. A new abietane diterpenoid from *Salvia xanthocheila* Boiss. Nat. Prod. Res. 27 (3), 266–269.

Hani, S.B., Bayachou, M., 2014. *Salvia fruticosa* reduces intrinsic cellular and H_2O_2-induced DNA oxidation in HEK 293 cells; assessment using flow cytometry. Asian Pac. J. Trop. Biomed. 4 (5), 399–403.

Hanlidou, E., Karousou, R., Lazari, D., 2014. Essential-oil diversity of Salvia tomentosa Mill. in Greece. Chem. Biodivers. 11 (8), 1205–1215.

Hu, T., To, K.K., Wang, L., et al., 2014. Reversal of P-glycoprotein (P-gp) mediated multidrug resistance in colon cancer cells by cryptotanshinone and dihydrotanshinone of *Salvia miltiorrhiza*. Phytomedicine 21 (11), 1264–1272.

Huang, W., Zhang, J., Wang, Y., Huang, D., 2010. A simple method to analyze the similarity of biological sequences based on the fuzzy theory. J. Theor. Biol. 265 (3), 323–328.

Jenks, A.A., Walker, J.B., Kim, S.C., 2011. Evolution and origins of the Mazatec hallucinogenic sage, *Salvia divinorum* (Lamiaceae): a molecular phylogenetic approach. J. Plant Res. 124 (5), 593–600.

Jiang, R.W., Lau, K.M., Hon, P.M., et al., 2005. Chemistry and biological activities of caffeic acid derivatives from *Salvia miltiorrhiza*. Curr. Med. Chem. 12 (2), 237–246.

Jiang, P., Li, C., Xiang, Z., Jiao, B., 2014a. Tanshinone IIA reduces the risk of Alzheimer's disease by inhibiting iNOS, MMP'2 and NF'κBp65 transcription and translation in the temporal lobes of rat models of Alzheimer's disease. Mol. Med. Rep. 10 (2), 689–694.

Jiang, Y.F., Liu, Z.Q., Cui, W., et al., 2014b. Antioxidant effect of salvianolic acid B on hippocampal CA1 neurons in mice with cerebral ischemia and reperfusion injury. Chin. J. Integr. Med. http://dx.doi.org/10.1007/s11655-014-1791-1.

Jiang, Y.Y., Wang, L., Zhang, L., et al., 2014c. Characterization, antioxidant and antitumor activities of polysaccharides from Salvia miltiorrhiza Bunge. Int. J. Biol. Macromol. 70, 92–99.

Jie, L., Du, H., Huang, Q., et al., 2014. Tanshinone IIA induces apoptosis in fibroblast-like synoviocytes in rheumatoid arthritis via blockade of the cell cycle in the G2/M phase and a mitochondrial pathway. Biol. Pharm. Bull. 37 (8), 1366–1372.

Jin, C.J., Yu, S.H., Wang, X.M., et al., 2014. The effect of lithospermic acid, an antioxidant, on development of diabetic retinopathy in spontaneously obese diabetic rats. PLoS One 9 (6), e98232.

Jung, Y.R., Kim, D.H., Kim, S.R., et al., 2014. Anti-wrinkle effect of magnesium lithospermate B from *Salvia miltiorrhiza* BUNGE: inhibition of MMPs via NF-kB signaling. PLoS One 9 (8), e102689.

Kermanshah, H., Kamangar, S.S., Arami, S., et al., 2014. The effect of hydro alcoholic extract of seven plants on cariogenic bacteria – an in vitro evaluation. Oral Health Dent. Manag. 13 (2), 395–401.

Kuźma, L., Wysokińska, H., Różalski, M., et al., 2012a. Antimicrobial and anti-biofilm properties of new taxodione derivative from hairy roots of *Salvia austriaca*. Phytomedicine 19 (14), 1285–1287.

Kuźma, L., Wysokińska, H., Różalski, M., et al., 2012b. An unusual taxodione derivative from hairy roots of *Salvia austriaca*. Fitoterapia 83 (4), 770–773.

Lakhal, H., Kabouche, A., Alabdul Magid, A., et al., 2014. Triterpenoids from Salvia argentea var. aurasiaca (Pomel) Batt. & Trab. and their chemotaxonomic significance. Phytochemistry 102, 145–151.

Lee, H.S., Son, W.C., Ryu, J.E., et al., 2014. Standardized *Salvia miltiorrhiza* extract suppresses hepatic stellate cell activation and attenuates steatohepatitis induced by a methionine-choline deficient diet in mice. Molecules 19 (6), 8189–8211.

Legrand, S., Valot, N., Nicolé, F., et al., 2010. One-step identification of conserved miRNAs, their targets, potential transcription factors and effector genes of complete secondary metabolism pathways after 454 pyrosequencing of calyx cDNAs from the Labiate *Salvia sclarea* L. Gene 450 (1–2), 55–62.

Li, C., Lu, S., 2014. Genome-wide characterization and comparative analysis of R2R3-MYB transcription factors shows the complexity of MYB-associated regulatory networks in *Salvia miltiorrhiza*. BMC Genomics 15, 277.

Li, R., Yu, C., Li, Y., et al., 2009. SOAP2: an improved ultrafast tool for short read alignment. Bioinformatics 25 (15), 1966–1967.

Li, Y., Sun, C., Luo, H.M., et al., 2010. Transcriptome characterization for *Salvia miltiorrhiza* using 454 GS FLX. Yao Xue Xue Bao 45 (4), 524–529.

Li, M.H., Li, Q.Q., Liu, Y.Z., Cui, Z.H., Zhang, N., Huang, L.Q., Xiao, P.G., 2013a. Pharmacophylogenetic study on plants of genus *Salvia* L. from China. Chin. Herb. Med. 5 (3), 164–181.

Li, M.H., Li, Q.Q., Zhang, C.H., Zhang, N., Cui, Z.H., Huang, L.Q., Xiao, P.G., 2013b. An ethnopharmacological investigation of medicinal *Salvia* plants (Lamiaceae) in China. Acta Pharm. Sin. B 3 (4), 273–280.

Li, M., Lu, Y., Hu, Y., et al., 2014a. Salvianolic acid B protects against acute ethanol-induced liver injury through SIRT1-mediated deacetylation of p53 in rats. Toxicol. Lett. 228 (2), 67–74.

Li, W., Zhou, S.P., Jin, Y.P., et al., 2014b. Salvianolic acids T and U: a pair of atropisomeric trimeric caffeic acids derivatives from root of *Salvia miltiorrhiza*. Fitoterapia 98, 248–253.

Liu, Y.W., Huang, Y.T., 2014. Inhibitory effect of tanshinone IIA on rat hepatic stellate cells. PLoS One 9 (7), e103229.

Liu, C.H., Shyu, W.C., Fu, R.H., et al., 2014. Salvianolic acid B maintained stem cell pluripotency and increased proliferation rate by activating Jak2-Stat3 combined with EGFR-Erk1/2 pathways. Cell Transplant. 23 (4–5), 657–668.

Loizzo, M.R., Abouali, M., Salehi, P., et al., 2014. In vitro antioxidant and antiproliferative activities of nine *Salvia* species. Nat. Prod. Res. 28, 2278–2285.

Lu, H., Tian, A., Wu, J., et al., 2014. Danshensu inhibits β-adrenergic receptors-mediated cardiac fibrosis by ROS/p38 MAPK axis. Biol. Pharm. Bull. 37 (6), 961–967.

Luo, H., Zhu, Y., Song, J., et al., 2014. Transcriptional data mining of *Salvia miltiorrhiza* in response to methyl jasmonate to examine the mechanism of bioactive compound biosynthesis and regulation. Physiol. Plant. 152, 241–255. http://dx.doi.org/10.1111/ppl.12193.

Maccioni, P., Vargiolu, D., Falchi, M., et al., 2014. Reducing effect of the Chinese medicinal herb, *Salvia miltiorrhiza*, on alcohol self-administration in Sardinian alcohol-preferring rats. Alcohol 48, 587–593. http://dx.doi.org/10.1016/j.alcohol.2014.06.002.

Maione, F., De Feo, V., Caiazzo, E., et al., 2014. Tanshinone IIA, a major component of *Salvia milthorriza* Bunge, inhibits platelet activation via Erk-2 signaling pathway. J. Ethnopharmacol. 155 (2), 1236–1242.

Martínez-Cruz, O., Paredes-López, O., 2014. Phytochemical profile and nutraceutical potential of chia seeds (*Salvia hispanica* L.) by ultra high performance liquid chromatography. J. Chromatogr. A 1346, 43–48.

Miroddi, M., Navarra, M., Quattropani, M.C., et al., 2014. Systematic review of clinical trials assessing pharmacological properties of *Salvia* species on memory, cognitive impairment and Alzheimer's disease. CNS Neurosci. Ther. 20 (6), 485–495.

Moghaddam, F.M., Farimani, M.M., Seirafi, M., et al., 2010. Sesterterpenoids and other constituents of *Salvia sahendica*. J. Nat. Prod. 73 (9), 1601–1605.

Moridi Farimani, M., Bahadori, M.B., Taheri, S., et al., 2011. Triterpenoids with rare carbon skeletons from *Salvia hydrangea*: antiprotozoal activity and absolute configurations. J. Nat. Prod. 74 (10), 2200–2205.

Moridi Farimani, M., Nejad Ebrahimi, S., Salehi, P., et al., 2013. Antitrypanosomal triterpenoid with an ε-lactone E-ring from *Salvia urmiensis*. J. Nat. Prod. 76 (9), 1806–1809.

Nicolella, H.D., Oliveira, P.F., Munari, C.C., et al., 2014. Differential effect of manool – a diterpene from *Salvia officinalis*, on genotoxicity induced by methyl methanesulfonate in V79 and HepG2 cells. Food Chem. Toxicol. 72C, 8–12.

Nikmehr, B., Ghaznavi, H., Rahbar, A., et al., 2014. In vitro anti-leishmanial activity of methanolic extracts of *Calendula officinalis* flowers, *Datura stramonium* seeds, and *Salvia officinalis* leaves. Chin. J. Nat. Med. 12 (6), 423–427.

Nkomo, M.M., Katerere, D.D., Vismer, H.H., et al., 2014. Fusarium inhibition by wild populations of the medicinal plant *Salvia africana-lutea* L. linked to metabolomic profiling. BMC Complement. Altern. Med. 14, 99.

Pan, Z.H., Li, Y., Wu, X.D., et al., 2012. Norditerpenoids from *Salvia castanea* Diels f. pubescens. Fitoterapia 83 (6), 1072–1075.

Pan, Y., Fu, H., Kong, Q., et al., 2014. Prevention of pulmonary fibrosis with salvianolic acid A by inducing fibroblast cell cycle arrest and promoting apoptosis. J. Ethnopharmacol. 155, 1589–1596. http://dx.doi.org/10.1016/j.jep.2014.07.049.

Parajuli, D.R., Zhao, Y.Z., Jin, H., et al., 2014. Anti-fibrotic effect of PF2401-SF, a standardized fraction of *Salvia miltiorrhiza*, in thioacetamide-induced experimental rats liver fibrosis. Arch. Pharm. Res. 38 (4), 549–555.

Park, J.H., Ok, P., Cho, J.H., et al., 2014. Anti-inflammatory effect of tanshinone I in neuroprotection against cerebral ischemia-reperfusion injury in the gerbil hippocampus. Neurochem. Res. 39 (7), 1300–1312.

Pérez Gutiérrez, S., Zavala Mendoza, D., Soto Peredo, C., et al., 2014. Evaluation of the antidiarrheal activity of *Salvia connivens*. Pharm. Biol. 52, 1467–1470.

Prisinzano, T.E., 2005. Psychopharmacology of the hallucinogenic sage *Salvia divinorum*. Life Sci. 78 (5), 527–531.

Qian, J., Song, J., Gao, H., et al., 2013. The complete chloroplast genome sequence of the medicinal plant *Salvia miltiorrhiza*. PLoS One 8 (2), e57607.

Radosavljević, I., Jakse, J., Javornik, B., et al., 2011. New microsatellite markers for *Salvia officinalis* (Lamiaceae) and cross-amplification in closely related species. Am. J. Bot. 98 (11), e316–e318.

Radosavljević, I., Satovic, Z., Jakse, J., et al., 2012. Development of new microsatellite markers for Salvia officinalis L. and its potential use in conservation-genetic studies of narrow endemic Salvia brachyodon Vandas. Int. J. Mol. Sci. 13 (9), 12082–12093.

Ren-An, Q., Juan, L., Chuyuan, L., et al., 2014. Study of the protective mechanisms of compound Danshen tablet (Fufang Danshen Pian) against myocardial ischemia/reperfusion injury via the Akt-eNOS signaling pathway in rats. J. Ethnopharmacol. 156, 190–198. http://dx.doi.org/10.1016/j.jep.2014.08.023.

Rhiu, S., Chae, M.K., Lee, E.J., et al., 2014. Effect of tanshinone IIA in an in vitro model of Graves' orbitopathy. Invest. Ophthalmol. Vis. Sci. 55, 5900–5910. http://dx.doi.org/10.1167/iovs. 14-14008.

Serra, V., Fattore, L., Scherma, M., et al., 2014. Behavioural and neurochemical assessment of salvinorin A abuse potential in the rat. Psychopharmacology (Berl) 232, 91–100.

Shafaei-Bajestani, N., Emami, S.A., Asili, J., Tayarani-Najaran, Z., 2014. Anti-apoptotic effect of taxodione on serum/glucose deprivation-induced PC12 cells death. Cell. Mol. Neurobiol. 34, 1103–1109.

Shi, C., Zhu, X., Wang, J., Long, D., 2014. Tanshinone IIA promotes non-amyloidogenic processing of amyloid precursor protein in platelets via estrogen receptor signaling to phosphatidylinositol 3-kinase/Akt. Biomed. Rep. 2 (4), 500–504.

Slusarczyk, S., Zimmermann, S., Kaiser, M., et al., 2011. Antiplasmodial and antitrypanosomal activity of tanshinone-type diterpenoids from *Salvia miltiorrhiza*. Planta Med. 77 (14), 1594–1596.

Stešević, D., Ristić, M., Nikolić, V., et al., 2014. Chemotype diversity of indigenous Dalmatian sage (*Salvia officinalis* L.) populations in Montenegro. Chem. Biodivers. 11 (1), 101–114.

Tabanca, N., Demirci, B., Baser, K.H., et al., 2006. Chemical composition and antifungal activity of *Salvia macrochlamys* and *Salvia recognita* essential oils. J. Agric. Food Chem. 54 (18), 6593–6597.

Takano, A., Okada, H., 2011. Phylogenetic relationships among subgenera, species, and varieties of Japanese *Salvia* L. (Lamiaceae). J. Plant Res. 124 (2), 245–252.

Tel, G., Oztürk, M., Duru, M.E., et al., 2010. Chemical composition of the essential oil and hexane extract of *Salvia chionantha* and their antioxidant and anticholinesterase activities. Food Chem. Toxicol. 48 (11), 3189–3193.

Tseng, P.Y., Lu, W.C., Hsieh, M.J., et al., 2014. Tanshinone IIA induces apoptosis in human oral cancer KB cells through a mitochondria-dependent pathway. Biomed. Res. Int. 2014, 540516.

Walker, J.B., Sytsma, K.J., 2007. Staminal evolution in the genus *Salvia* (Lamiaceae): molecular phylogenetic evidence for multiple origins of the staminal lever. Ann. Bot. 100 (2), 375–391.

Walker, J.B., Sytsma, K.J., Treutlein, J., Wink, M., 2004. *Salvia* (Lamiaceae) is not monophyletic: implications for the systematics, radiation, and ecological specializations of *Salvia* and tribe Mentheae. Am. J. Bot. 91 (7), 1115–1125.

Wang, X., Morris-Natschke, S.L., Lee, K.H., 2007a. New developments in the chemistry and biology of the bioactive constituents of Tanshen. Med. Res. Rev. 27 (1), 133–148.

Wang, Y., Li, D.H., Zhang, Y.T., 2007b. Analysis of ITS sequences of some medicinal plants and their related species in *Salvia*. Yao Xue Xue Bao 42 (12), 1309–1313.

Wang, M., Zhao, H.X., Wang, L., et al., 2013. Potential use of DNA barcoding for the identification of *Salvia* based on cpDNA and nrDNA sequences. Gene 528 (2), 206–215.

Wang, C., Li, X., Meng, X., et al., 2014a. Prevention of experimental postoperative peritoneal adhesions through the intraperitoneal administration of tanshinone IIA. Planta Med. 80 (12), 969–973.

Wang, L., Venitz, J., Sweet, D.H., 2014b. Cumulative organic anion transporter-mediated drug-drug interaction potential of multiple components in *Salvia Miltiorrhiza* (Danshen) preparations. Pharm. Res. 31, 3503–3514.

Wang, N., Yang, J., Lu, J., et al., 2014c. A polysaccharide from *Salvia miltiorrhiza* Bunge improves immune function in gastric cancer rats. Carbohydr. Polym. 111, 47–55.

Wang, P., Xu, S., Li, W., et al., 2014d. Salvianolic acid B inhibited PPARγ expression and attenuated weight gain in mice with high-fat diet-induced obesity. Cell. Physiol. Biochem. 34 (2), 288–298.

Wang, Y.H., Han, Y.P., Yu, H.T., et al., 2014e. Protocatechualdehyde prevents methylglyoxal-induced mitochondrial dysfunction and AGEs-RAGE axis activation in human lens epithelial cells. Eur. J. Pharmacol. 738, 374–383.

Wenping, H., Yuan, Z., Jie, S., et al., 2011. De novo transcriptome sequencing in *Salvia miltiorrhiza* to identify genes involved in the biosynthesis of active ingredients. Genomics 98 (4), 272–279.

Will, M., Claßen-Bockhoff, R., 2014. Why Africa matters: evolution of Old World *Salvia* (Lamiaceae) in Africa. Ann. Bot. 114 (1), 61–83.

Wu, Y.B., Ni, Z.Y., Shi, Q.W., et al., 2012. Constituents from *Salvia* species and their biological activities. Chem. Rev. 112 (11), 5967–6026.

Wu, C., Jiang, J., Boye, A., et al., 2014a. Compound *Astragalus* and *Salvia miltiorrhiza* extract suppresses rabbits' hypertrophic scars by modulating the TGF-β/Smad signal. Dermatology 229, 363–368.

Wu, C.Y., Liao, Y., Yang, Z.G., et al., 2014b. Cytotoxic diterpenoids from *Salvia yunnanensis*. Phytochemistry 106, 171–177.

Wu, S.J., Huang, C.H., Chan, Y.Y., et al., 2014c. Two diterpenoids and a cyclopenta[c]pyridine derivative from roots of *Salvia digitaloids*. Int. J. Mol. Sci. 15 (7), 11566–11577.

Wu, Z., Song, L., Liu, S.Q., Huang, D., 2014d. Tanshinones extend chronological lifespan in budding yeast *Saccharomyces cerevisiae*. Appl. Microbiol. Biotechnol. 98, 8617–8628.

Xu, D., Xu, L., Zhou, C., et al., 2014. Salvianolic acid B promotes osteogenesis of human mesenchymal stem cells through activating ERK signaling pathway. Int. J. Biochem. Cell Biol. 51, 1–9.

Yang, L.X., Li, X.C., Liu, C., et al., 2011. Chemical constituents from *Salvia przewalskii* Maxim. Yao Xue Xue Bao 46 (7), 818–821.

Yang, D., Ma, P., Liang, X., et al., 2012. Metabolic profiles and cDNA-AFLP analysis of *Salvia miltiorrhiza* and *Salvia castanea* Diel f. tomentosa Stib. PLoS One 7 (1), e29678.

Yang, L., Ding, G., Lin, H., et al., 2013. Transcriptome analysis of medicinal plant *Salvia miltiorrhiza* and identification of genes related to tanshinone biosynthesis. PLoS One 8 (11), e80464.

Yu, H., Yao, L., Zhou, H., et al., 2014a. Neuroprotection against Aβ25-35-induced apoptosis by *Salvia miltiorrhiza* extract in SH-SY5Y cells. Neurochem. Int. 75, 89–95.

Yu, X.Q., Maki, M., Drew, B.T., et al., 2014b. Phylogeny and historical biogeography of *Isodon* (Lamiaceae): rapid radiation in south-west China and Miocene overland dispersal into Africa. Mol. Phylogenet. Evol. 77, 183–194.

Yuce, E., Yildirim, N., Yildirim, N.C., et al., 2014. Essential oil composition, antioxidant and antifungal activities of *Salvia sclarea* L. from Munzur Valley in Tunceli, Turkey. Cell. Mol. Biol. 60 (2), 1–5.

Yue, S., Hu, B., Wang, Z., et al., 2014. Salvia miltiorrhiza compounds protect the liver from acute injury by regulation of p38 and NFκB signaling in Kupffer cells. Pharm. Biol. 52, 1278–1285.

Zhang, J., Wang, J., Jiang, J.Y., et al., 2014a. Tanshinone IIA induces cytochrome c-mediated caspase cascade apoptosis in A549 human lung cancer cells via the JNK pathway. Int. J. Oncol. 45 (2), 683–690.

Zhang, L., Xu, J., Song, H., et al., 2014b. Extracts from *Salvia*-Nelumbinis naturalis alleviate hepatosteatosis via improving hepatic insulin sensitivity. J. Transl. Med. 12 (1), 236.

Zhang, T., Xu, J., Li, D., et al., 2014c. Salvianolic acid A, a matrix metalloproteinase-9 inhibitor of *Salvia miltiorrhiza*, attenuates aortic aneurysm formation in apolipoprotein E-deficient mice. Phytomedicine 21 (10), 1137–1145.

Zhang, Z., Ge, B., Zhou, L., et al., 2014d. Induction of liver cytochrome P450s by Danshen-Gegen formula is the leading cause for its pharmacokinetic interactions with warfarin. J. Ethnopharmacol. 154 (3), 672–686.

Phytochemical and biological research of *Panax* medicinal resources

15

15.1 Introduction

Panax of the Araliaceae family (Umbelliales) has some commonly used medicinal plants. Medical application of *P. ginseng* (Ren Shen) was first found in "Shen Nong's Herbal Classic" around AD 200 (Chen et al., 2008). *P. quinquefolium* (Xi Yang Shen, American ginseng) was first introduced in "Essential of Materia Medica" in 1694 in China, which accelerated its diffusion. The shortage of Asian ginseng resources resulted in the urgent need for the introduction of American ginseng. *P. ginseng* is mainly used to maintain the homeostasis of the body, and the pharmacological efficacy of ginseng identified by modern science includes improved brain function, pain-relieving effects, preventive effects against tumors as well as antitumor activity, enhanced immune function, antidiabetic effects, enhanced liver function, adjusted blood pressure, antifatigue and antistress effects, improved climacteric disorder and sexual functions, and antioxidative and antiaging effects (Choi, 2008). In this chapter, we briefly summarize the recent research progress in chemistry, pharmacology, and omics of *Panax* medicinal resources.

15.2 Chemical components

15.2.1 Saponin

At least 322 saponins were reported from 11 *Panax* species (this study and Yang et al., 2014a), which can be classified into (1) protopanaxadiol (PPD) (Figure 15.1), (2) protopanaxatriol (PPT), (3) octillol, (4) oleanolic acid (OA), (5) C17 side-chain varied, and (6) miscellaneous subtypes (Figure 15.2 and Table 15.1), according to structural differences in sapogenins. Glucose (Glc), glucuronic acid (GlcA), rhamnose (Rha), xylose (Xyl), and arabinose (Ara) are predominant monosaccharides composing the *Panax* saponins, of which an Ara occurs in either the pyran (p-Ara) or furan (f-Ara) configuration (Figure 15.1). Many compounds differ by only a single chemical moiety and might be midpoints along biosynthetic pathways (Figure 15.3).

The most important bioactive components contained in *P. ginseng* and *P. quinquefolium* are ginseng saponins (GS). The number of ginsenoside types contained in ginseng (38 ginsenosides) is substantially more than that of ginsenoside types contained in American ginseng (19) (Choi, 2008). The contents of ginsenoside Rb1, Re, and Rd in *P. quinquefolium* are higher than those in *P. ginseng* (Chen et al., 2008).

Medicinal plants: chemistry, biology and omics. http://dx.doi.org/10.1016/B978-0-08-100085-4.00015-3

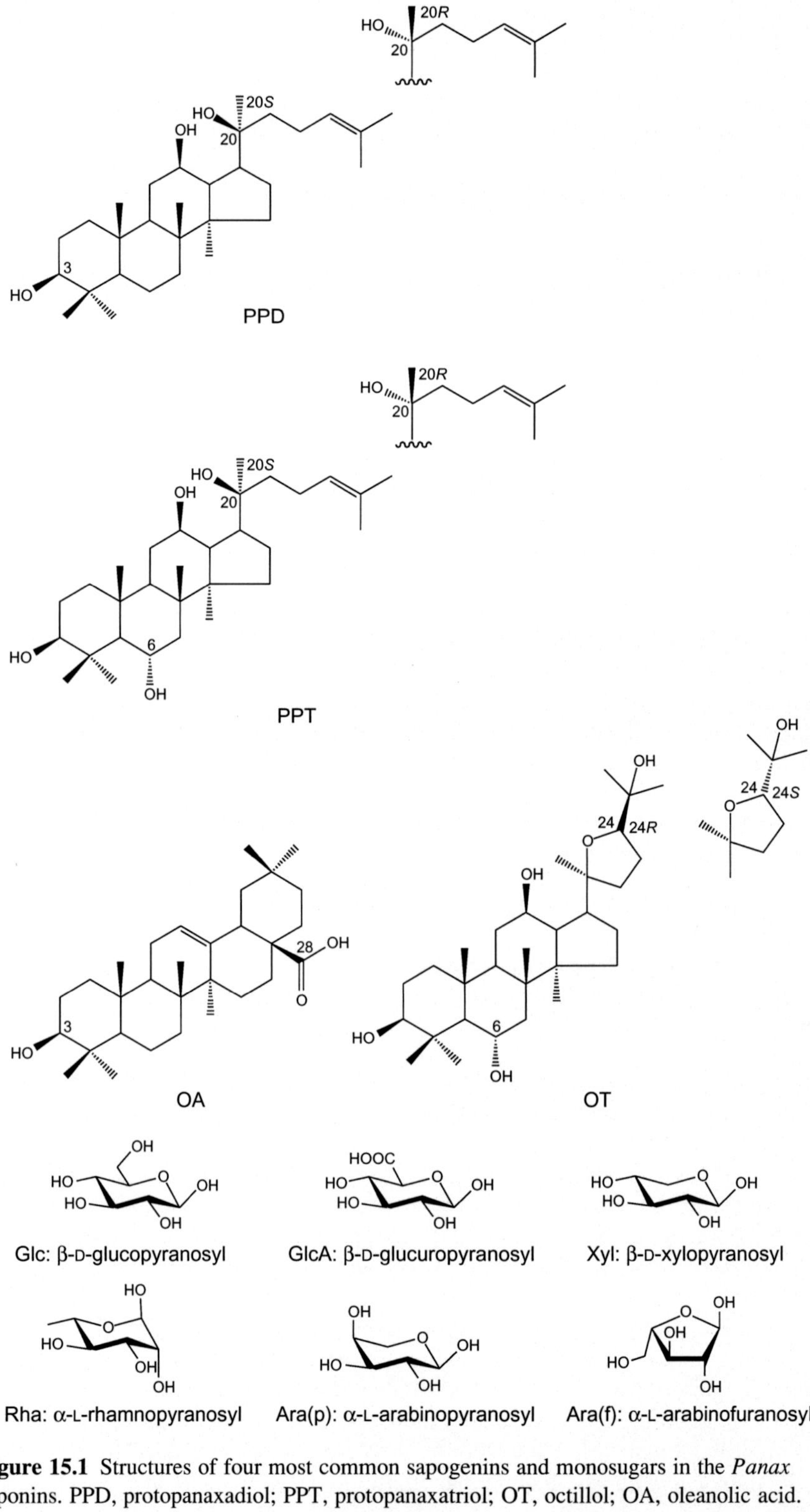

Figure 15.1 Structures of four most common sapogenins and monosugars in the *Panax* saponins. PPD, protopanaxadiol; PPT, protopanaxatriol; OT, octillol; OA, oleanolic acid.

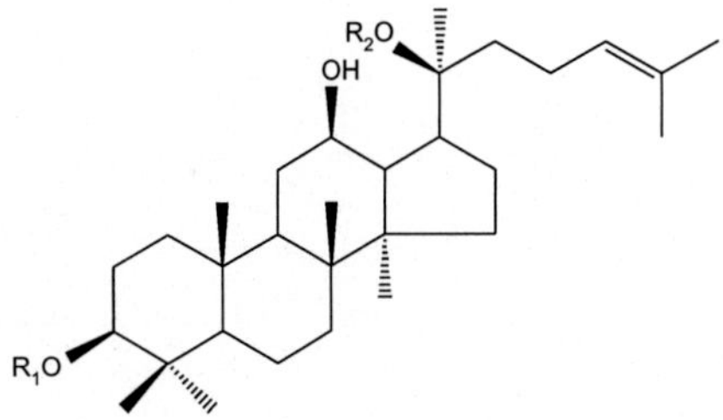

No.	R1	R2
1	Glc(2-1)Glc(2-1)Xyl	H
15	Glc(2-1)Glc-6-Ac	Glc(6-1)Glc
16	Glc(2-1)Glc-6-Ac	Glc(6-1)Ara(p)
17	Glc(2-1)Glc-6-Ac	Glc(6-1)Ara(f)
18	Glc(2-1)Glc-6-Ac	Glc(6-1)Glc
19	Glc	Glc(6-1)Xyl
20	Glc	Glc(6-1)Ara(f)
31	Glc(6–1)Xyl	Glc(6-1)Glc

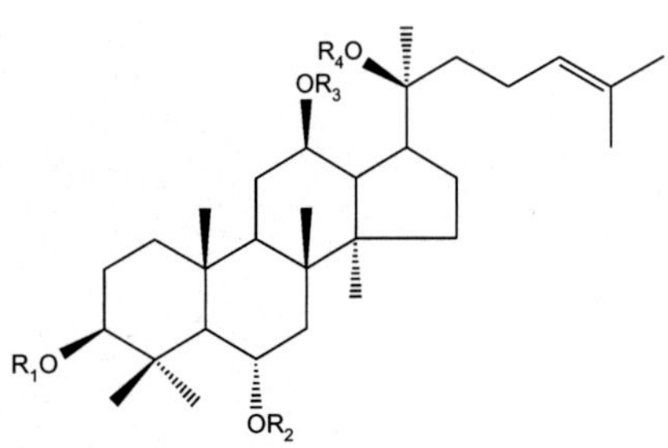

No.	R1	R2	R3	R4
2	H	Glc(6-1)Ac	H	Glc
3	H	Glc(2-1)Rha	H	Glc(6-1)Glc
27	H	H	H	Glc(6-1)Ara(f)(5–1)Xyl
28	Glc(2-1)Glc	H	Glc	H
29	Glc(2-1)Glc	H	H	Glc(6-1)Ara(f)
30	Glc(2-1)Glc	H	H	Glc(6-1)Ara(p)(4-1)Xyl

Figure 15.2 Examples of saponins isolated from the *Panax* species. Ac, acetyl.

(Continued)

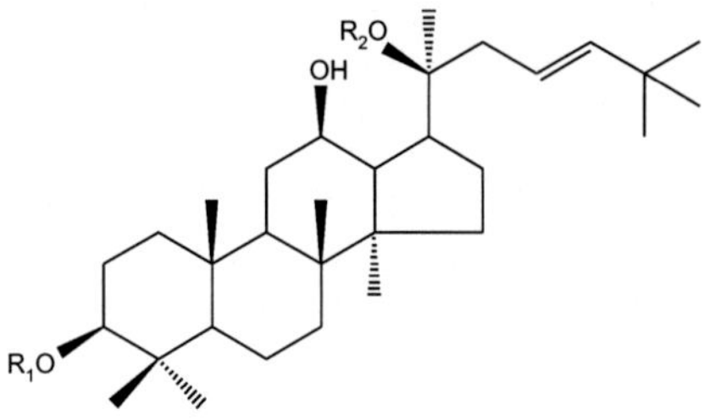

No.	R_1	R_2
4	Glc(2–1)Glc	Glc(2-1)Xyl

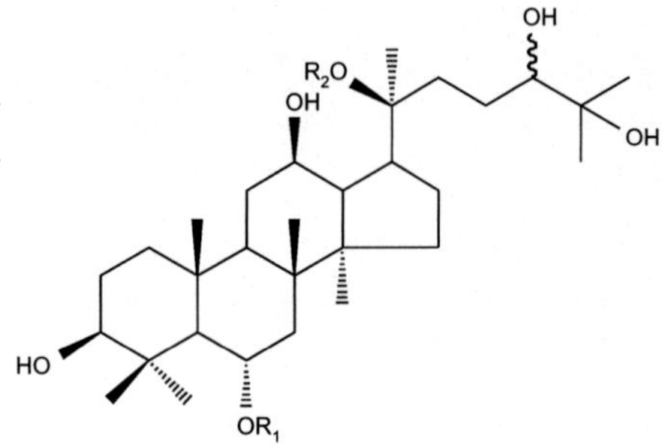

No.	R_1	R_2
5	Glc	Glc

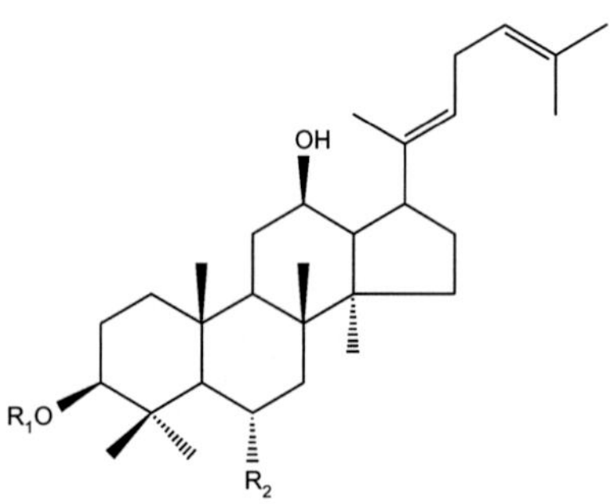

No.	R_1	R_2
6	Glc(2-1)Glc	H
7	H	OGlc

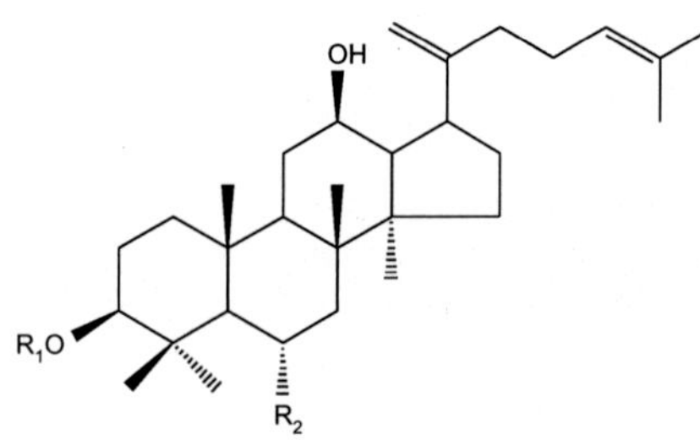

No.	R_1	R_2
8	Glc(2-1)Glc	H
9	H	OGlc

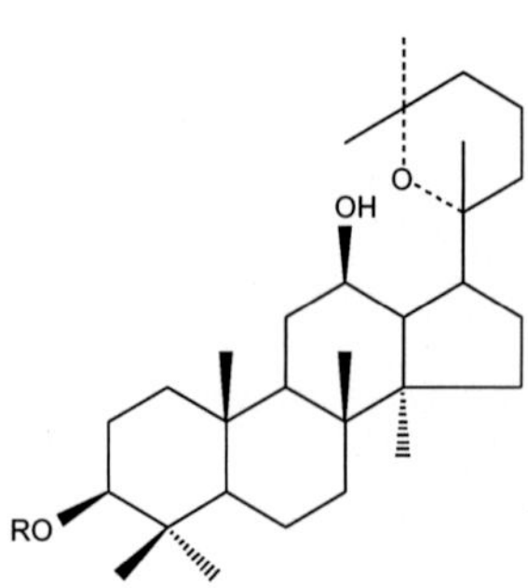

No.	R
10	Glc

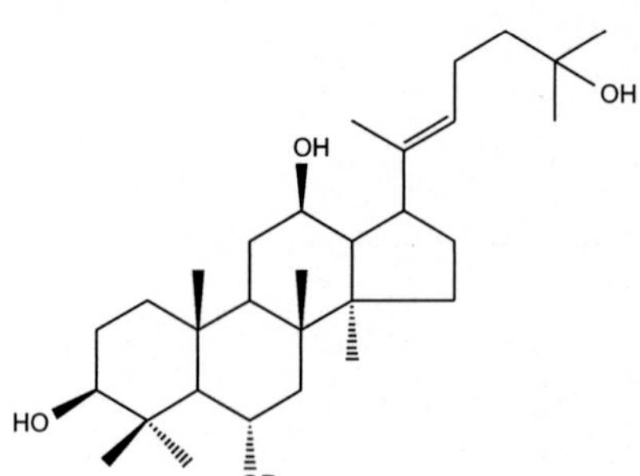

No.	R
11	Glc

Figure 15.2 Continued.

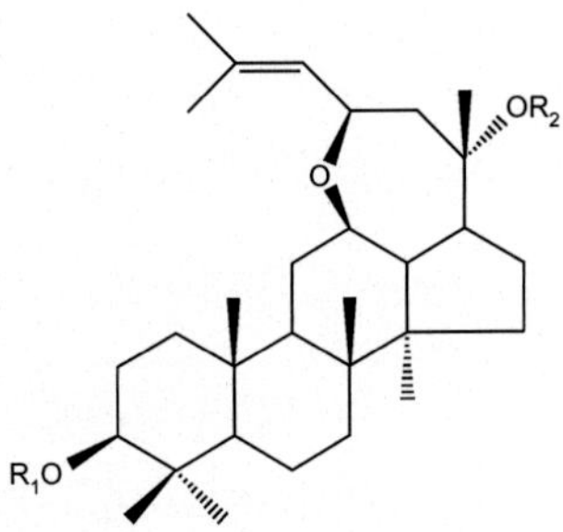

No.	R1	R2
12	Glc	Glc(6-1)Ara(f)
13	H	Glc(6-1)Ara(f)

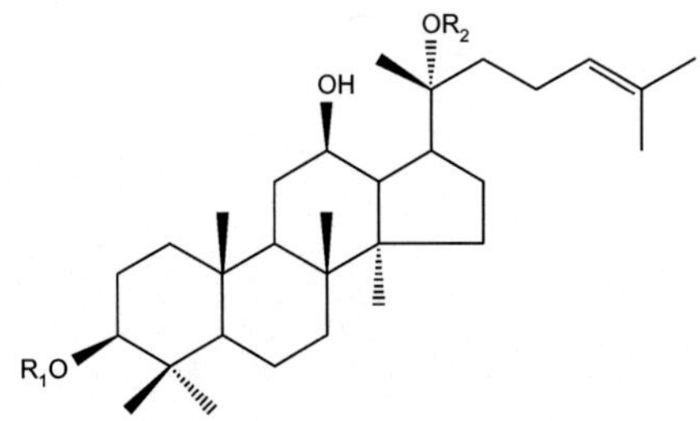

No.	R1	R2
14	Glc(2-1)Glc(2–1)Xyl	Glc(6-1)Ara(p)

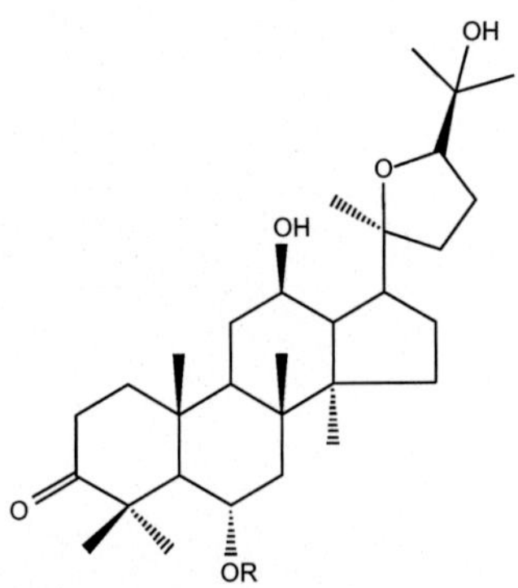

No.	R
21	Glc
22	H

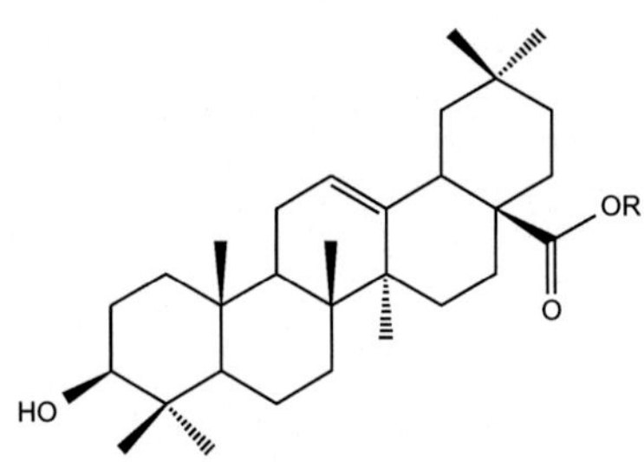

No.	R
23	Glc

Figure 15.2 Continued.

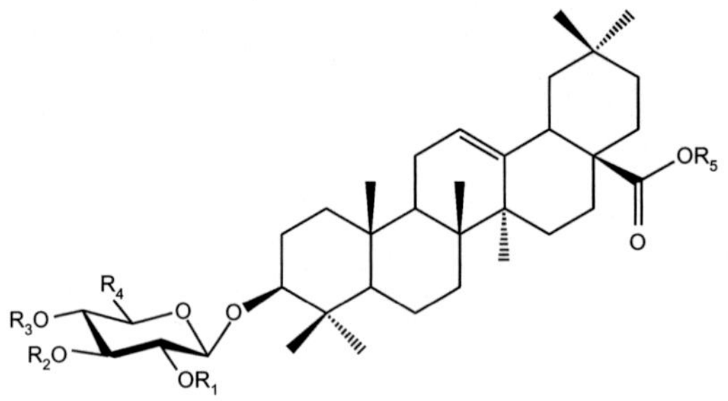

No.	R_1	R_2	R_3	R_4	R_5
24	Xyl	Ara(p)	H	COOH	Glc
25	Xyl	Gal	H	COOH	H
26	H	Glc	Ara(f)	$COOCH_3$	H
36	Ara(p)	H	H	$COOCH_3$	H
37	H	Glc(6–1)Xyl	H	$COOCH_3$	H
38	Xyl	Ara(p)	H	$COOCH_3$	Glc

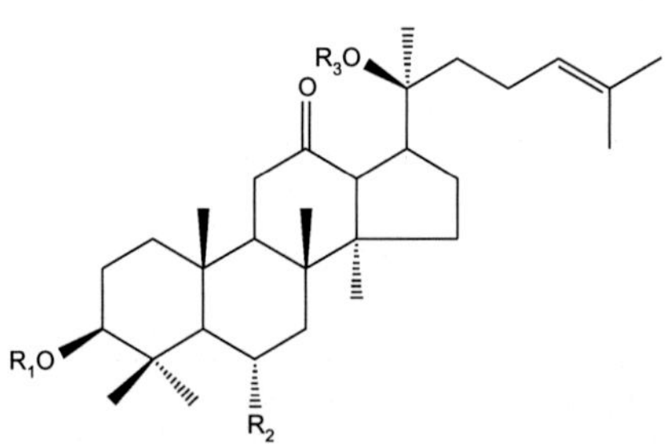

No.	R_1	R_2	R_3
32	H	OH	Glc(6–1)Ara(p)
33	[6-Xyl]Glc(2–1)Glc	H	Glc(6–1)Glc
34	[6-Xyl]Glc(2–1)Glc	H	Glc(6–1)Ara(p)
35	Glc(6–1)Xyl	H	Glc

Figure 15.2 Continued.

Table 15.1 Various types of saponins isolated from the *Panax* species

No.	Compounds	Types	Species	Tissues	References
1	Notoginsenoside ST-4[a]	A	I	Roots	Liu et al. (2014d)
2	Yesanchinoside D[a]	B	I	Roots	Liu et al. (2014d)
3	Yesanchinoside E	B	I	Roots	Liu et al. (2014d)
4	Yesanchinoside H	E	I	Roots	Liu et al. (2014d)
5	Notoginsenoside J	E	I	Roots	Liu et al. (2014d)
6	Ginsenoside Rg5[a]	E	I	Roots	Liu et al. (2014d)
7	Ginsenoside Rh4[a]	E	I	Roots	Liu et al. (2014d)
8	Ginsenoside Rk1[a]	E	I	Roots	Liu et al. (2014d)
9	Ginsenoside Rk3[a]	E	I	Roots	Liu et al. (2014d)
10	Notoginsenoside R7	E	I	Roots	Liu et al. (2014d)
11	Sanchinoside B1[a]	E	I	Roots	Liu et al. (2014d)
12	Notoginsenoside-LX	E	I	Leaves	Li et al. (2014b)
13	Notoginsenoside-LY	E	I	Leaves	Li et al. (2014b)
14	Notoginsenoside-FZ	A	I	Leaves	Li et al. (2014b)
15	6″-*O*-acetylginsenoside Rb1	A	II	Roots	Samimi et al. (2014)
16	Ginsenoside Rs1	A	III, VII	Leaves	He et al. (2014) and Yang et al. (2014a)
17	Ginsenoside Rs2	A	III, VII	Leaves	He et al. (2014) and Yang et al. (2014a)
18	Quinquenoside R1	A	III, II	Leaves and roots	He et al. (2014) and Yang et al. (2014a)
19	Gypenosiden IX	A	III, I	Leaves, fruits, and pedicels	He et al. (2014) and Yang et al. (2014a)
20	Notoginsenoside Fe	A	III, I	Leaves, leaves, and seeds	He et al. (2014) and Yang et al. (2014a)
21	Pseudoginsenoside RT6	F	II	Leaves and stems	Liu et al. (2013)
22	Pseudoginsengenin R1	F	II	Leaves and stems	Liu et al. (2013)
23	Oleanolic acid 28-*O*-β-D-glucopyranosyl ester	D	IV	Rhizomes	Liang et al. (2013)
24	Hemslosides Ma2	D	IV	Rhizomes	Liang et al. (2013)
25	Elatoside A	D	IV	Rhizomes	Liang et al. (2013)
26	Stipuleanoside R1 methyl ester	D	IV	Rhizomes	Liang et al. (2013)

Continued

Table 15.1 Continued

No.	Compounds	Types	Species	Tissues	References
27	Chikusetsusaponin LM3	B	V	Leaves	Yoshizaki et al. (2013a)
28	Chikusetsusaponin LM4	B	V	Leaves	Yoshizaki et al. (2013a)
29	Chikusetsusaponin LM5	B	V	Leaves	Yoshizaki et al. (2013a)
30	Chikusetsusaponin LM6	B	V	Leaves	Yoshizaki et al. (2013a)
31	Chikusetsusaponin VII	A	V	Taproots	Yoshizaki et al. (2013b)
32	Chikusetsusaponin FT1	F	V	Fruits	Yoshizaki et al. (2012)
33	Chikusetsusaponin FT2	F	V	Fruits	Yoshizaki et al. (2012)
34	Chikusetsusaponin FT3	F	V	Fruits	Yoshizaki et al. (2012)
35	Chikusetsusaponin FT4	F	V	Fruits	Yoshizaki et al. (2012)
36	Bifinosides A	D	VI	Roots	Nguyen et al. (2011)
37	Bifinosides B	D	VI	Roots	Nguyen et al. (2011)
38	Bifinosides C	D	VI	Roots	Nguyen et al. (2011)

A, PPD type (protopanaxadiol); B, PPT type (protopanaxatriol); C, OT type (octillol); D, OA type (oleanolic acid); E, C17 side-chain varied type; F, other types.
I, *P. notoginseng* (Burk) F.H. Chen; II, *P. quinquefolius* L.; III, *P. japonicus* C.A. Mey. var. *major* (Burk) C.Y. Wu & K.M. Feng; IV, *P. stipuleanatus* Tsai & K.M. Feng; V, *P. japonicus* C.A. Meyer; VI, *P. bipinnatifidus* Seem.; VII, *P. ginseng* C.A. Meyer
[a]Saponin isolated from steam-treated notoginseng.

In *P. ginseng*, the contents of Rg1, Rb2, and Rc are higher than those in *P. quinquefolium*. *P. ginseng* had a higher ratio of Rg1:Rb1 than *P. quinquefolium*. After steaming for several hours, the total GS will decrease. However, some ginsenosides (Rg2, 20R-Rg2, Rg3, Rh1, and Rh2) increase, while others (Rb1, Rb2, Rb3, Rc, Rd, Re, and Rg1) decrease. However, variation of saponin composition, especially in *P. quinquefolium*, is high.

P. ginseng and *P. quinquefolium* produce saponins that are found in *P. notoginseng* (San Qi), such as ginsenosides Rb1 and Rg1. Additionally, *P. notoginseng* provides unique notoginsenoside R1 and contains 50–300% higher content of Rb1, Rg1, and total saponins than the two other species, suggesting that it is a more desirable herbal source of certain saponins (Liu et al., 2014d).

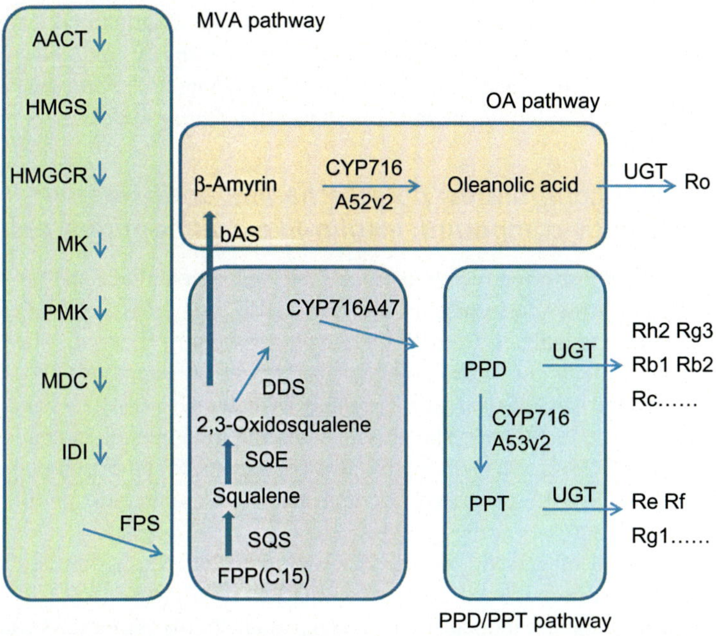

Figure 15.3 Schematic illustration of ginsenoside biosynthetic pathway. HMG-CoA, 3-hydroxy-3-methylglutaryl-coenzyme A; IPP, isopentenyl pyrophosphate; DMAPP, dimethylallyl pyrophosphate; FPP, farnesyl diphosphate; OA, oleanolic acid; PPD, protopanaxadiol; PPT, protopanaxatriol; HMGCR, 3-hydroxy-3-methylglutaryl-CoA reductase; SQS, squalene synthase; SQE, squalene epoxidase; bAS, β-amyrin synthase; OAS, oleanolic acid synthase; CPR, cytochrome P450 reductase; DDS, dammarenediol-II synthase; PPDS, protopanaxadiol synthase; and PPTS, protopanaxatriol synthase.

Dammarane ginsenosides (PPD and PPT) are transformed into compounds such as the ginsenosides Rg3, Rg5, and Rk1 by steaming and heating and are converted into metabolites such as compound K, Rh1, and PPT by intestinal microflora (Kim, 2012). These metabolites are nonpolar, pharmacologically active, and easily absorbed via the intestinal mucosa. However, the activities metabolizing these components into bioactive compounds differ considerably among individuals because different individuals possess distinctive indigenous strains of intestinal bacteria. To overcome this difference, ginsengs fermented with enzymes or microbes have been developed.

Ginsenoside Rs2, quinquenoside R1, ginsenoside Rs1, notoginsenoside Fe, ginsenoside Rd2, and gypenosiden IX were isolated from *P. japonicus* var. *major* (He et al., 2014; Table 15.1 and Figure 15.2). Large amounts of ginsenosides malonyl-Rb1, malonyl-Rc, malonyl-Rb2, and malonyl-Rd were found in a suspension culture of *P. japonicus* var. *repens* cells (Kochkin et al., 2013). Twenty triterpenoid saponins were isolated from rhizomes, taproots, and lateral roots of *P. japonicus* (Yoshizaki et al., 2013b).

Ten oleanane-type saponins, including three new compounds, bifinosides A–C, were isolated from the roots of *P. bipinnatifidus* (Nguyen et al., 2011). Dammarane-type glycosides are predominant in *P. ginseng*, *P. quinquefolium*, and *P. vietnamensis*, while oleanane-type saponins dominate *P. stipuleanatus*. These results could be implicated in chemotaxonomy of the *Panax* species.

15.2.2 Other compounds: polysaccharide, antioxidative aromatic compound, insulin-like acid peptides, etc.

Ginseng contains more phenolic compounds, acid polysaccharides, and polyethylene compounds than American ginseng and San Qi (Choi, 2008). 5,7-Dihydroxy-8-methoxyflavone was isolated from *P. japonicus* var. *major* (He et al., 2014).

The sugar contents of water-soluble ginseng oligosaccharide (WGOS)-root, WGOS-flower, and WGOS-leaf in *P. ginseng* were 95.87%, 87.07%, and 83.09%, respectively (Jiao et al., 2014a). The ginsenosides and total phenol content decreased in the order of WGOS-L > WGOS-F > WGOS-R. WGOS-R comprised only Glc, and WGOS-F and WGOS-L comprised Glc and Rha in a molar ratio of 6:1 and 7:1, respectively.

The essential oil of ginseng leaves (EOGLs), collected via hydrodistillation and analyzed by GC/MS, contained a complex mixture of aliphatic (69.0%), terpenoid (21.5%), and aromatic compounds (2.4%) (Jiang et al., 2014). Among 54 components identified, the major ones were palmitic acid (36.1%), beta-farnesene (15.4%), linoleic acid (9.8%), and phytol (5.6%).

15.3 Pharmacology and therapeutic use

15.3.1 Effects on the nervous system

P. ginseng and *P. quinquefolium* are general tonics and adaptogens. Rg1 and Rb1 enhance central nervous system (CNS) activities (Chen et al., 2008), but the effect of the latter is weaker. Thus, for the higher contents of Rg1, *P. ginseng* is a stimulant, whereas the Rb1 contents of *P. quinquefolium* are mainly calming to the CNS. Network pharmacology is especially useful to reveal underlying multicompound, multitarget, multipathway mode of action of traditional Chinese medicinal plants (Hao and Xiao, 2014). Rg1 prevents cognitive impairment and hippocampus senescence in a rat model of D-galactose-induced aging (Zhu et al., 2014a).

Ginseng total saponins (GTSs) have the antidepressant effect in the corticosterone-induced mouse depression model (Chen et al., 2014a). Enhancing hippocampal inhibitory phosphorylation of glycogen synthase kinase (GSK)-3β may be one of the underlying mechanisms. GTSs improve the neurorestoration of rat after traumatic brain injury (Hu et al., 2014). Rd regulate mitochondrial permeability transition pore (MPTP) formation and cytochrome c release through protein kinase-dependent mechanism involving the activation of intramitochondrial AKT and ERK pathways (Zhou et al., 2014a). Phosphorylated tau was found to be regulated after cerebral ischemia and linked to high risk for the development of poststroke dementia. Rd attenuates tau protein phosphorylation via the PI3K/AKT/GSK-3β pathway after transient forebrain ischemia (Zhang et al., 2014c).

The spontaneous recovery of function after injury in the adult CNS is limited due to several proteins, such as Nogo-A, that have repulsive or inhibitory effects on growing nerve ending. *P. notoginseng* saponins (PNSs) promote stroke recovery by influencing expression of Nogo-A, NgR, and p75NGF, *in vitro* and *in vivo* (Liu et al., 2014c). Notoginsenoside R1 attenuates amyloid-β-induced damage in neurons by inhibiting reactive oxygen species (ROS) and modulating mitogen-activated protein kinase (MAPK) activation (Ma et al., 2014a). Total saponin in leaves of *P. notoginseng* has antioxidative effects, which may be neuroprotective in neurological disorders (Zhou et al., 2014b). Gypenoside XVII (GP-17) is a novel phytoestrogen isolated from *Gynostemma pentaphyllum* or *P. notoginseng*, which has neuroprotective effects. It attenuates Aβ25-35-induced parallel autophagic and apoptotic cell death through the estrogen receptor-dependent activation of Nrf2/ARE pathways (Meng et al., 2014).

Multiple sclerosis (MS) is a common disabling autoimmune disease without an effective treatment in young adults. Rd ameliorates experimental autoimmune encephalomyelitis in C57BL/6 mice (Zhu et al., 2014b).

Chikusetsu saponin V of *P. japonicus* attenuates MPP^+-induced neurotoxicity in SH-SY5Y cells via regulation of Sirt1/Mn-SOD and GRP78/caspase-12 pathways (Yuan et al., 2014).

15.3.2 Antidiabetic and cardiovascular effects

Ginsenoside Rb3 protects cardiomyocytes against ischemia–reperfusion injury via the inhibition of JNK-mediated NF-κB pathway (Ma et al., 2014b). Through the activation of the PI3K/AKT pathway and MPTP inhibition, *P. quinquefolium* saponin might protect the heart against ischemia injury and apoptosis of myocardial cells (Guo et al., 2015). Saponin extract from *P. japonicus* has protective effect on myocardial infarction, which is involved in NF-κB, Sirt1, and MAPK signaling pathways and the inhibition of inflammation (Wei et al., 2014). Rg3-enriched Korean red ginseng extract acutely lowers central and peripheral arterial pressures in healthy adults (Jovanovski et al., 2014a). Ginsenoside PPT (ginseol K-g1) has blood pressure-lowering effect (Rhee et al., 2014). Modulation of endothelial function by *P. ginseng* ginsenosides in healthy individuals has been validated by a randomized controlled trial (Jovanovski et al., 2014b).

PNSs, including notoginsenoside R1, ginsenoside Rb1, and ginsenoside Rg1, have protective effects on cardiovascular diseases (Yang et al., 2014b). Notoginsenoside R1 attenuates hypoxia and hypercapnia-induced vasoconstriction in isolated rat pulmonary arterial rings by reducing the expression of ERK (Xu et al., 2014a). PNSs exert tissue-specific regulatory effects on angiogenesis in part through modulating the expression of miR-18a (Yang et al., 2014e), which could be responsible for its bidirectional effect on complex disease conditions where paradoxical angiogenesis is implicated. PNS and related bioactive component might be a rational therapy for complex disease conditions including comanifestation of cancer and ischemic cardiovascular disease.

Re, Rg1, panaxan A, and panaxan B from *P. ginseng* are good for diabetes (Chen et al., 2008). 20*S*-ginsenoside Rh2 from *P. ginseng* is an aldose reductase inhibitor (Fatmawati et al., 2014). *P. ginseng* might be an important herbal medicine in preventing diabetic complications. Korean white ginseng (*P. ginseng*) has a beneficial effect on vascular and glycemic health in type II diabetes, which is confirmed by a randomized, double blind, placebo-controlled, multiple-crossover, and acute dose escalation trial (Shishtar et al., 2014). Hyperlipidemia and insulin resistance were attenuated effectively in response to GTS treatment (Liu et al., 2014a), which may be associated with the increased secretion of glucagon-like peptide (GLP)-1. Preoperative Rg3 administration enhanced islet function before islet transplantation and attenuated both cytokine-induced damage associated with nitric oxide (NO) production and apoptosis (Kim et al., 2014c).

Four ginsenosides of *P. notoginseng* significantly inhibit the enzyme activity of protein tyrosine phosphatase (PTP) 1B in a dose-dependent manner (Li et al., 2014b), with inhibitory concentration 50 (IC50) values of 29.08, 21.27, 28.12, and 26.59 μM, respectively. *P. notoginseng* leaves might have potential as a new therapeutic agent for the treatment of diabetes.

15.3.3 Antioxidant and anti-inflammatory effects

Red ginseng is a steamed and dried *P. ginseng*, which has been used as alternative medicine for thousands of years. Red ginseng extract ameliorates autoimmune arthritis via regulation of STAT3 pathway, Th17/Treg balance, and osteoclastogenesis in mice and human (Jhun et al., 2014). Ursan-3α,11β-diol-3-*O*-α-D-glucopyranosyl-(6′ → 1″)-α-D-glucopyranosyl-(6″ → 1‴)-α-D-glucopyranosyl-(6‴ → 1⁗)-α-D-glucopyranoside and lanost-5,24-dien-3β-ol-3-*O*-β-D-glucopyranosyl-(6′ → 1″)-β-D-glucopyranosyl-(6″ → 1‴)-β-D-glucopyranoside have potential anti-inflammatory activities (Chung et al., 2014). Two sterols of *P. ginseng* seeds have inhibitory effects on inflammation-related factors in HepG2 cells (Kim et al., 2013). Anti-inflammatory effect of Rb1 contributes to the recovery of gastrointestinal motility in the rat model of postoperative ileus (Tan et al., 2014). Rb1 protected the liver and lung in septic shock rats by downregulating the expression of TLR4 and inhibiting the production of tumor necrosis factor (TNF)-α (Wu et al., 2014c). Anti-inflammatory effects of PNSs ameliorate acute lung injury induced by oleic acid and lipopolysaccharide (LPS) in rats (Chen et al., 2014b). Oleanane triterpenoids from *P. stipuleanatus* inhibit NF-κB (Liang et al., 2013). Ginsenoside Ro and chikuseksusaponin IVa might be the anti-inflammatory constituents of kou zi qi (rhizomes of *P. japonicus* var. *major*) (Zou et al., 2004). MR2, the major saponin of *P. vietnamensis*, may protect the hepatocytes from apoptosis via an inhibition of TNF-α production by activated macrophages and a direct inhibition of apoptosis induced by TNF-α (Tran et al., 2002).

PNS is protective to rat hemorrhagic shock model by antioxidative stress and anti-inflammatory pathways (Liu et al., 2014b), and intercellular adhesion molecule (ICAM)-1 may play an important role. Panaxatriol saponins may activate endogenous cytoprotective mechanism against oxygen-Glc deprivation–reperfusion-induced oxidative injury via the activation of PI3K/AKT and Nrf2 signaling pathway (Huang et al., 2014).

In vitro antioxidant tests showed that WGOS-R exhibited higher antioxidant activity than WGOS-F and WGOS-L (Jiao et al., 2014a). *In vivo* tests showed that WGOS-R significantly enhanced activities of superoxide dismutase (SOD), catalase (CAT), glutathione peroxidase (GSH-Px), and total antioxidant capability in the serum and liver and decreased malondialdehyde (MDA) level in the serum and liver. Phenolic compound-rich extracts from white ginseng (*P. ginseng*) have hypolipidemic and antioxidant properties in cholesterol-fed rabbits (Lee et al., 2013a). Polysaccharides from *P. japonicus* have antioxidant activities (Yang et al., 2014f). EOGLs exhibited weak DPPH radical scavenging (12.0 ± 0.4 mg/mL) and ABTS radical scavenging activities (Jiang et al., 2014).

15.3.4 Anticancer and immunomodulatory effects

One molecular mechanism underlying the cancer chemopreventive and/or chemotherapeutic activity of ginsenosides and their intestinal metabolites is the modulation of cell signaling pathways associated with oxidative stress, inflammation, cell proliferation, apoptosis, angiogenesis, and the metastasis of cancer cells (Choi et al., 2013).

Re and Rg1 enhance angiogenesis, whereas Rb1, Rg3, and Rh2 inhibit it (Chen et al., 2008). Rh2, an antitumor agent, can be obtained from Rb1 by steaming. The content of Re in *P. quinquefolium* is more than three times higher than that in *P. ginseng*. Re has vasorelaxing, antioxidant, antihyperlipidemic, and angiogenic effects. Thus, for the CNS "hot," wound healing and hypoglycemic effects, *P. ginseng* is better than *P. quinquefolium*. For anticancer effects, *P. quinquefolium* is better. Rg1 induces apoptosis through the inhibition of the EpoR-mediated JAK2/STAT5 signaling pathway in the TF-1/Epo human leukemia cell line (Li et al., 2014a). Autophagy often serves as cell survival mechanism in cancer cells exposed to chemotherapeutic agents. 20*S*-ginsenoside Rg3 is a novel inhibitor of autophagy and sensitizes hepatocellular carcinoma to doxorubicin (Kim et al., 2014a). Stereospecific effects of Rg3 inhibit TGF-β1-induced epithelial–mesenchymal transition and suppress lung cancer migration, invasion, and anoikis resistance (Kim et al., 2014d). Targeting matrix metalloproteinase (MMP)-2 by microRNA (miR)-29b is a mechanism by which American ginseng suppresses the migration of colon cancer cells (Poudyal et al., 2013).

ERK1/2 and p38 MAPK pathways are involved in the proapoptotic effect of notoginsenoside Ft1, a natural compound exclusively found in *P. notoginseng*, on human neuroblastoma SH-SY5Y cells (Gao et al., 2014). PNS inhibits breast cancer metastasis (Wang et al., 2014c).

The antiproliferative activity of *P. vietnamensis* is closely related to the contents of ginsenoside Rg3, Rg5, and Rk1 (Le et al., 2014). Oleanane-type triterpenoids from *P. stipuleanatus* have anticancer activities (Liang et al., 2010).

P. ginseng extract exerted a protective effect in a rat model of cachexia and suggests that it may be a therapeutic promising tool for supportive care in oncology (Lobina et al., 2014).

Three dammarane triterpenes significantly increased interleukin (IL)-12 expression in LPS-activated mouse peritoneal macrophage at a concentration of 100 ng/mL (Tran et al., 2014). 27-Demethyl-(*E*, *E*)-20(22), 23-dien-3β, 6α, 12β-trihydroxydammar-25-one strongly increased the Th1 response-mediated cytokine IL-2 and decreased Th2 response-mediated cytokines IL-4 and IL-6 expression at 100 ng/mL on ConA-activated splenocytes. Rapeseed oil and GS work synergistically to enhance Th1 and Th2 immune responses induced by the foot-and-mouth disease vaccine (Zhang et al., 2014a). 6″-*O*-acetylginsenoside Rb1, isolated from American ginseng, demonstrated an immunosuppressive activity towards inhibiting the production of NO and TNF-α in LPS-induced macrophage cells in a dose-dependent manner using murine macrophages (Samimi et al., 2014). 20*S*-dihydroprotopanaxatriol modulates functional activation of monocytes and macrophages (Kim and Cho, 2013).

The rhamnogalacturonan I-rich ginseng pectin fragment, RG-I-4, potently inhibited galectin-3-mediated hemagglutination, cancer cell adhesion and homotypic aggregation, and binding of galectin-3 to T cells (Gao et al., 2013). WGOS inhibited tumor growth *in vivo* and significantly increased relative spleen and thymus weight, serum TNF-α level, spleen lymphocyte proliferation, natural killer cell activity, phagocytic function, and NO production by macrophage in H22-bearing mice (Jiao et al., 2014b). However, no direct cytotoxicity was detected. Therefore, the antitumor activity of WGOS may be related to their immunomodulatory effects.

EOGLs exhibited obvious cytotoxic activities against different cancer cell lines, including Hela, A549, ZR-75-1, HT-29, SGC7901, and B16 cells (Jiang et al., 2014). Furthermore, annexin V-FITC/PI staining assay indicated that EOGLs can induce late apoptosis of ZR-75-1 cells and the percentage of apoptotic cells increased in a concentration-dependent manner.

15.3.5 Antimicrobial activity

Red ginseng (*P. ginseng*) and ginsenosides Rg1 and Rb1 could decrease hepatitis A virus (HAV) titers (Lee et al., 2013b). Ginseng extract is active against respiratory syncytial virus infection (Lee et al., 2014c). It also has the potential beneficial effects on preventing influenza A virus infections via multiple immunomodulatory functions (Lee et al., 2014d). Evidence for the favorable effects of *radix ginseng* extracts on viral, bacterial, fungal, and parasitic infections and the possible underlying mechanisms has been summarized (Wu et al., 2014b). EOGLs showed antibacterial activity against the Gram-positive bacteria, *Staphylococcus aureus* and *Bacillus subtilis*, and the Gram-negative bacterium, *Escherichia coli* (Jiang et al., 2014).

15.3.6 Other effects

P. ginseng has been used to treat chronic obstructive pulmonary disease (COPD) (Wu et al., 2014a).

Two new dammarane triterpenes, dammar-20(22), 24-diene-3β, 6α, 12β-triol and 20*S*-ginsenoside Rg3, showed potential as SIRT1 activators (Yang et al., 2014d), as analyzed by *in vitro* enzyme-based SIRT1-NAD/NADH and SIRT1-p53 luciferase cell-based assays. They also increase the level of NAD^+/NADH ratio in HEK293 cells.

They may be developed as SIRT1 activators for antiaging and treatment of age-associated diseases.

Ginsenoside Rh1 induces mouse osteoblast growth and differentiation through the bone morphogenetic protein 2/runt-related gene 2 signaling pathway (Siddiqi et al., 2014).

P. ginseng extract treatments increased energy expenditures by modulating PKA-dependent lipid mobilization in adipose tissue and prevented mice from diet-induced obesity (Cho et al., 2014). Ginseng may effectively treat female obesity and related disorders in part by the inhibition of angiogenesis (Lee et al., 2014a). Treatment with *P. ginseng* antagonizes the estrogen decline in ovariectomized mice (Xu et al., 2014b). Ginseng has beneficial effects on women's health due to its estrogen-like function (Yang et al., 2014c). Ginseng is efficient for protection for female quality of life and bisphenol A (BPA) exposure and BPA-related oxidative stress.

Ginseng seeds inhibit melanin biosynthesis and show great potential as a skin-depigmenting substance (Lee et al., 2014b). Rb1, the most predominant ginsenoside isolated from the roots of ginseng, has melanogenesis inhibitory effect (Wang et al., 2015).

P. ginseng induces the expression of cation channel of sperm (CatSper) genes and sperm hyperactivation (Park et al., 2014a).

Polysaccharides and low-molecular-weight compounds of *P. japonicas* have hematopoietic effect on blood deficiency model mice (Zhang et al., 2014b).

The effects of ginseng are not fully explained by ginsenosides. Another class of active ingredients, gintonin, is a complex of glycosylated ginseng proteins containing lysophosphatidic acids (LPAs), the intracellular lipid mitogenic mediator (Im and Nah, 2013). Gintonin specifically and potently activates the G protein-coupled receptors (GPCRs) for LPA. The actions of ginseng are also linked to LPA and its GPCRs.

15.3.7 Safety

American ginseng extract is safe in long-term use in type II diabetic patients (Mucalo et al., 2014). Neither lopinavir nor ritonavir steady-state pharmacokinetics were altered by two weeks of *P. ginseng* administration in healthy volunteers (Calderón et al., 2014). Thus, a clinically significant interaction between *P. ginseng* and LPV-r is unlikely to occur in HIV-infected patients who choose to take these agents concurrently. It is also unlikely that *P. ginseng* will interact with other ritonavir-boosted protease inhibitor combinations, although confirmatory data are necessary.

15.4 Molecular phylogeny and genomics

15.4.1 P. ginseng

Complete genome information for *Panax* species remains unavailable due to their large genome size. Two 454 pyrosequencing runs generated 2,423,076 reads from *P. ginseng* roots, stems, leaves, and flowers (Li et al., 2013a). In the separately assembled database, 45,849, 6172, 4041, and 3273 unigenes were specifically found in the roots, stems, leaves, and flowers, respectively. In the jointly assembled database, 178,145 unigenes were observed, including 86,609 contigs and 91,536 singletons, 105,522 of which were identified for the first time. 223 Unigenes are involved in ginsenoside backbone biosynthesis (Figure 15.3). 326 Potential cytochrome P450 (CYP450; Figure 15.4) and 129 potential UDP-glycosyltransferase (UGT) sequences were predicted, some of which may encode enzymes responsible for ginsenoside backbone modification. A BLAST search identified 14 potential microRNAs, which were estimated to target 100 protein-coding genes, including transcription factors (TFs), transporters, and DNA binding proteins. A total of 13,044 simple sequence repeats (SSRs) were identified from the 178,145 unigenes. The novel unigenes identified enlarge the available *P. ginseng* gene pool and will facilitate gene discovery. The identification of microRNAs and the prediction of targets provide information on gene transcriptional regulation in *P. ginseng*. SSRs are useful genetic makers for molecular breeding and genetic applications in *Panax* species.

P. ginseng has been cultivated for centuries, and nine commercial cultivars have been registered in Korea. These nine elite cultivars are grown in less than 10% of ginseng fields, and an authentication system is needed for each cultivar since their values are higher than those of local landraces. Nineteen microsatellite (SSR) markers were developed using expressed gene sequences to establish an authentication system for elite cultivars (Kim et al., 2012). Cultivars "Chunpoong," "Sunpoong," "Gumpoong," "Sunun," and "Sunone" can be identified by one cultivar-specific allele, gm47n-a, gm47n-c, gm104-a, gm184-a (or gm129-a), and gm175-c, respectively. "Yunpoong" can be identified by the coappearance of gm47n-b and gm129-c. "Sunhyang" can be distinguished from the other eight cultivars by the coappearance of gm47n-b, gm129-b, and gm175-a. The two other cultivars "Gopoong" and "Cheongsun" can be identified by their specific combinations of five marker alleles. This marker set can identify the cultivars among 70 ginseng individuals and can be used to select true F1 hybrid plants between two cultivars. The homogeneity of each cultivar and phylogenetic relationships among cultivars can be analyzed using these markers. This marker system will be useful to the seed industry and for breeding of ginseng. Multiloci expressed sequence tag (EST)–SSR bands are also useful to reveal the recent genome duplication of *P. ginseng*, which resulted in two highly similar paralogous regions in the ginseng genome (Kim et al., 2014b).

A comprehensive transcriptome from adventitious roots of *P. ginseng*, which were treated with methyl jasmonate (MeJA) for different time lengths (0, 2, 6, 12, and 24 h), was obtained by 454 pyrosequencing (Subramaniyam et al., 2014). Reference transcriptome 39,304,529 (40 Mb) was obtained from 5,724,987,880 bases (5.7 Gb) of 22 libraries by *de novo* assembly, and 35,266 (58.5%) transcripts were annotated with Gene Ontology (GO) and Kyoto Encyclopedia of Genes and Genomes (KEGG). The digital gene expression patterns were obtained from *in vitro*-grown adventitious root sequences, which mapped to reference. 3813 (6.3%) Unique transcripts had significantly differential expression ($\geq$2-fold up- or downregulations). Candidates of ginsenoside pathway genes were predicted from observed expression patterns. Among them, 30 TFs, 20 CYPs, and 11 glycosyl transferases were predicted as ginsenoside candidates.

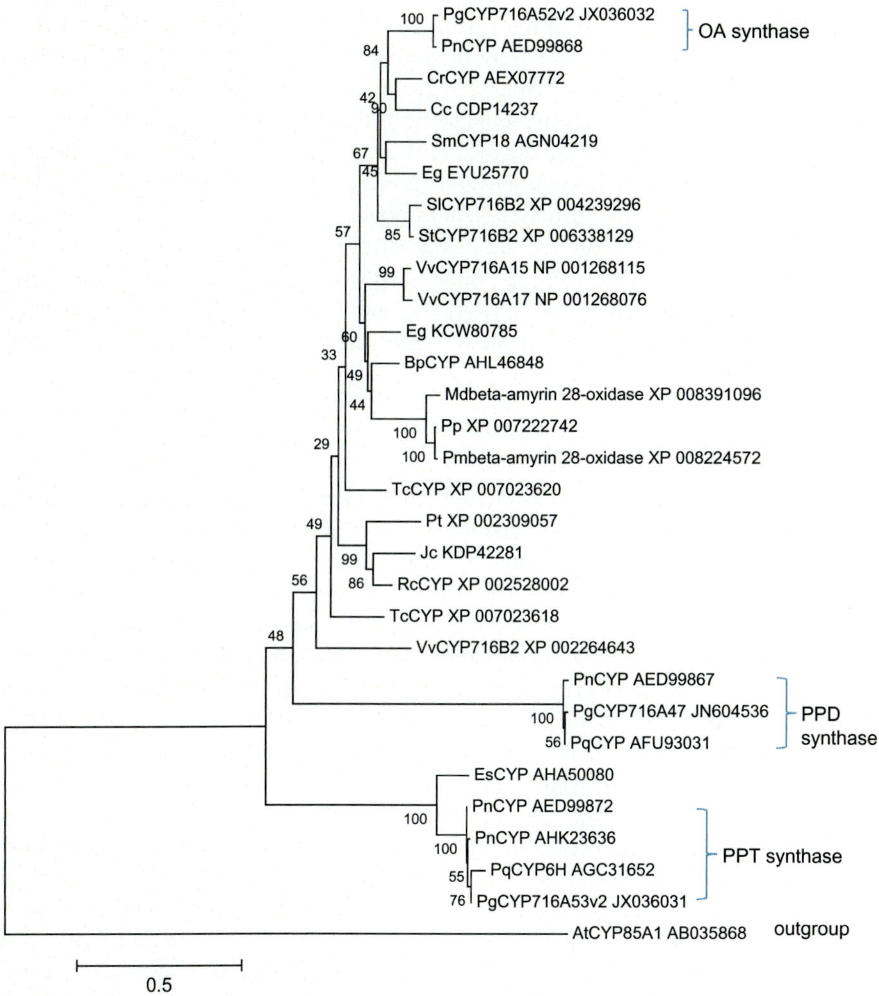

Figure 15.4 Phylogenetic tree constructed from the deduced amino acid sequences of *Panax* CYPs and related plant CYPs. The evolutionary history was inferred using the neighbor-joining (NJ) method. Bootstrap values are shown on the branches. The tree is drawn to scale, with branch lengths in the same units as those of the evolutionary distances used to infer the phylogenetic tree. The evolutionary distances were computed using the JTT matrix-based method and are in the units of the number of amino acid substitutions per site. Evolutionary analyses were conducted in MEGA6 (http://megasoftware.net/). Pg, *Panax ginseng*; Pq, *Panax quinquefolius*; Pn, *Panax notoginseng*; Es, *Eleutherococcus senticosus*; Sm, *Salvia miltiorrhiza*; Tc, *Theobroma cacao*; Sl, *Solanum lycopersicum*; Bp, *Betula platyphylla*; Jc, *Jatropha curcas*; Rc, *Ricinus communis*; Vv, *Vitis vinifera*; Pt, *Populus trichocarpa*; At, *Arabidopsis thaliana*; Eg, *Erythranthe guttata*; Cc, *Coffea canephora*; Cr, *Catharanthus roseus*; St, *Solanum tuberosum*; Pp, *Prunus persica*; Pm, *Prunus mume*; Md, *Malus domestica*; Eg, *Eucalyptus grandis*.

A full-length mlnc (long noncoding) RNA, MAR (multiple-function-associated mlncRNA), was cloned in *P. ginseng* (Wang et al., 2014a). It generated the most abundant siRNAs. The MAR siRNAs were predominantly 24 nt, and some of them were distributed in a phased pattern. A total of 228 targets were predicted for 71 MAR siRNAs. High-throughput degradome sequencing validated 68 predicted targets involved in diverse metabolic pathways, suggesting the significance of MAR in *P. ginseng*. MAR was detected in all tissues analyzed and responded to MeJA treatment, which sheds light on the function of mlncRNAs in plants.

AFLP markers are useful for the characterization of the genetic structure and genetic relationships of the ginseng populations (Reunova et al., 2010). AFLP, random amplified polymorphic DNA (RAPD), and ISSR markers are recommended for use in large-scale population genetic studies of *P. ginseng* to develop measures of its conservation.

15.4.2 *P. quinquefolius*

A one-quarter 454 sequencing of American ginseng root transcriptome produced 209,747 high-quality reads with an average sequence length of 427 bp (Sun et al., 2010). *De novo* assembly generated 31,088 unique sequences comprising 16,592 contigs and 14,496 singletons. 93.1% of the reads were assembled into contigs with an eightfold coverage. A total of 21,684 (69.8%) unigenes were annotated by a BLAST similarity search against four public databases, and 4097 unigenes were assigned to specific metabolic pathways by KEGG. All known enzymes involved in ginsenoside backbone synthesis, starting from acetyl-CoA via the isoprenoid pathway (Figure 15.3), were found. A total of 150 CYP and 235 glycosyltransferase unigenes were found, some of which encode enzymes responsible for the conversion of the ginsenoside backbone into the various ginsenosides. One CYP (Figure 15.4) and four UGTs were selected as the candidates most likely to be involved in ginsenoside biosynthesis through a MeJA inducibility experiment and tissue-specific expression pattern analysis based on a real-time PCR assay. The transcriptome dataset provides an important resource for molecular genetics and functional genomics of *Panax* species. 6678 ESTs from the flower, leaf, and root cDNA libraries of American ginseng were generated via 454 pyrosequencing (Wu et al., 2010). Assembly of ESTs resulted in 3349 unigenes including 534 contigs and 2815 singletons. By analyzing the predominant transcripts within specific tissues, a tissue-specific gene expression pattern was revealed. Functional classification assigned unigenes to broad ranges of GO categories (biological processes, cellular components, and molecular functions). Based on BLASTX search, 24 unigenes representing candidates related to ginsenoside biosynthesis were identified. Cloning and characterization of 3-hydroxy-3-methylglutaryl-coenzyme A reductase (HMGCR; Figure 15.3), the rate-limiting enzyme in mevalonic acid (MVA) pathway, demonstrated that it belonged to the plant HMGCR family and was highly expressed in leaves. Putative TFs were detected in 63 unigenes, including zinc finger, WRKY, homeobox, and MADS-box family proteins. Five hundred and eighty-eight SSR motifs were identified, of which dimer was most abundant.

Root RNA samples from seven development stages of American ginseng were subjected to 454 sequencing (Wu et al., 2013). *De novo* assembly resulted in 41,623 transcripts. Annotation information was available for 34,801 (84%) transcripts. In addition, 3955 genes were assigned to metabolic pathways using KEGG. All known enzymes involved in the ginsenoside backbone biosynthesis were found. Coexpression analysis was used to identify a number of candidate sequences involved in the latter stages of ginsenoside biosynthesis pathway. Transcript profiles suggest ginsenoside biosynthesis occurs at several stages of development. Expression profiles across seasonal development suggest a primary dammarane-type ginsenoside biosynthesis occurs just prior to plant senescence, with secondary ginsenoside production occurring throughout development.

Roots, flowers, and processed products of *P. ginseng* and American ginseng can be difficult to differentiate from each other, leading to illegal trade in which one species is sold as the other. Convenient and reliable chloroplast genome-derived DNA markers can be developed for authentication of *P. ginseng* and American ginseng (Jung et al., 2014). One codominant marker could reproducibly identify both species and intentional mixtures of the two species. A set of species-specific dominant DNA markers were developed. Each marker could detect 1% cross contamination with other species by low-resolution agarose gel electrophoresis or quantitative PCR. These markers were successfully applied to evaluate the original species from various processed ginseng products purchased from markets in Korea and China.

Single nucleotide polymorphisms (SNPs) are widely employed in the studies of population genetics, molecular breeding, and conservation genetics. SNPs were developed based on screening the library of single-copy nuclear genes (Li et al., 2013c). 160 and 171 SNPs were identified from *P. quinquefolium* and *P. ginseng*, respectively, suggesting that both species possessed a high level of nucleotide diversity. The number of haplotype per locus ranged between one and 12 for *P. ginseng* and between one and nine for *P. quinquefolium*, respectively. The nucleotide diversity of total sites (π_T) varied between 0.000 and 0.023 for *P. ginseng* and 0.000 and 0.035 for *P. quinquefolium*, respectively.

15.4.3 *P. notoginseng*

454 Pyrosequencing of *P. notoginseng* root resulted in 188,185 reads with an average length of 410 bp (Luo et al., 2011). A total of 70.2% of assembled 30,852 unigenes were annotated by BLAST search. The KEGG assignment discovered 41 unigenes representing 11 genes involved in triterpene saponin backbone biosynthesis. The transcript encoding dammarenediol synthase (DDS), which is the first committed enzyme in the biosynthetic pathway of major triterpene saponins (Figure 15.3), is highly expressed in the root of 4-year-old *P. notoginseng*. The candidate CYP (Pn02132 and Pn00158; Figure 15.4) and UGT (Pn00082) genes most likely to be involved in hydroxylation or glycosylation of aglycones for triterpene saponin biosynthesis were inferred from 174 CYPs and 242 glycosyltransferases by phylogenetic analysis, respectively. On the CYP tree (Figure 15.4), there are two main clades: One clade contains PPT synthase of three *Panax* species; in the other clade, the subclade containing

PPD synthase is basal to that of OA synthase, implying that OA biosynthesis pathway is evolutionarily younger than dammarane biosynthesis pathway.

Putative TFs were detected in 906 unigenes, including Myb, homeobox, WRKY, and basic helix-loop-helix (bHLH) (Luo et al., 2011). Moreover, 2772 SSRs were identified from 2361 unigenes, of which dinucleotide motifs were most abundant. In contrast, only 12 polymorphic microsatellite loci were isolated from the microsatellite-enriched genomic library of *P. notoginseng* (Liu et al., 2011).

Cleaved amplified polymorphic sequence (CAPS) marker system using mitochondrial consensus primers was applied for molecular identification of *P. ginseng* cultivars (Lee et al., 2012). A total of 34 primers were tested in six *P. ginseng* cultivars, *P. quinquefolius*, and *P. notoginseng*. Four primers (mt7, mt11, mt13, and mt18) generated codominant polymorphic banding patterns discriminating *P. ginseng* from the other two species. The majority of the cleaved PCR products also yielded additional latent polymorphisms between *P. ginseng* and two other *Panax* species. Specific latent CAPS polymorphisms for cultivar Gopoong and Chunpoong were detected from internal region amplified with mt9 primer by treating HinfI and Tsp509I endonucleases, respectively. The length of amplified region of *P. ginseng* was 2179 bp, and those of *P. quinquefolius* and *P. notoginseng* were 2178 and 2185 bp, respectively. BLAST search revealed that the amplified region was a mitochondrial cytochrome oxidase subunit 2 (COX-2) gene intron II region. Nineteen SNPs including each specific SNP for Gopoong and Chunpoong and three insertion and deletion (InDel) polymorphisms were detected by sequence alignment. The CAPS markers serve as a practical and reliable tool for identification, purity maintenance, and selection of candidate lines and cultivars.

Authentication of *P. notoginseng* by 5*S*-rRNA spacer domain and RAPD analysis was successful (Cui et al., 2003). SNPs were detected in 377 sequences from four *Panax* species to recognize patterns of intragenomic variation in internal transcribed spacer 2 (ITS2) (Chen et al., 2013). Intraspecific variations were analyzed based on three commonly used barcoding markers (ITS2, matK, and psbA-trnH). Intraspecific genetic distances in *P. ginseng* and *P. quinquefolius* were quite low (0–0.002), and the multicopy ITS2 could be considered a single locus in the genomes of these species. Five stable SNPs were detected in ITS2 region to identify the *Panax* species. Double peaks could be clearly examined at SNP positions for the mixed powder of *P. ginseng* and *P. quinquefolius*, and the height of the peaks could indicate the mixed ratio roughly. SNP-based molecular barcodes could be developed for the identification of the closely related *Panax* species.

15.4.4 Phylogeny

It is not easy to resolve the internal relationships of major clades within the core ginseng plant family, perhaps due to rapid radiation of these lineages. Phylogenomics based on chloroplast (cp) genomes is useful in resolving complex relationships. Five Araliaceae cp genomes were explored using Illumina HiSeq 2000 sequencing technology (Li et al., 2013b). The five cp genomes are 156,333–156,459 bp in length including a pair of inverted repeats (IRs; 25,551–26,108 bp) separated by the large single-copy (SC; 86,028–86,566 bp) and small SC (18,021–19,117 bp) regions. Each cp genome contains the same 114 unique genes consisting of 30 tRNA genes, four rRNA genes, and 80 protein-coding genes. Gene size, content, and order, AT content, and IR/SC boundary structure are similar among all Araliaceae cp genomes. A total of 140 repeats were identified in the five cp genomes with palindromic repeat as the most common type. Phylogenomic analyses using parsimony, likelihood, and Bayesian inference based on the complete cp genomes strongly supported the monophyly of the Asian Palmate group and the *Aralia–Panax* group (Figure 15.5a). Twenty-six DNA markers with the percentage of variable sites higher than 5% were identified,

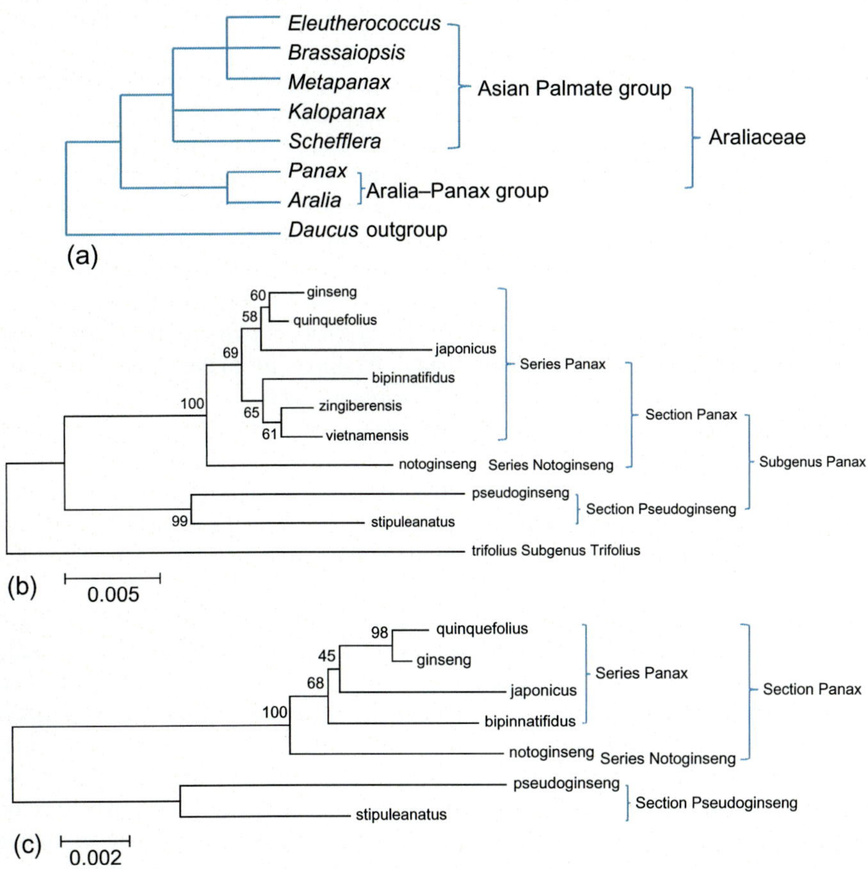

Figure 15.5 (a) Phylogenetic relationship of *Panax* and related genera (modified from Li et al., 2013b). (b) Phylogenetic relationship of 10 *Panax* species inferred by ITS+matK sequences and NJ method. Bootstrap values are shown on the branches. The evolutionary distances were computed using the maximum composite likelihood method and are in the units of the number of base substitutions per site. Evolutionary analyses were conducted in MEGA6 (http://megasoftware.net/). (c) Phylogenetic relationship of seven *Panax* species inferred by psbA-trnH+trnK (including matK)+ITS sequences and NJ method.

which may be useful for phylogenetic studies of Araliaceae. The cp genomes of Araliaceae are highly conserved in all aspects of genome features. The large-scale phylogenomic data based on the complete cp DNA sequences are shown to be effective for the phylogenetic reconstruction of Araliaceae.

Sequences of the cp trnC-trnD region and the ITS regions of nuclear rDNA were used to reconstruct phylogenetic relationships of *Panax* species (Lee and Wen, 2004). The trnC-trnD phylogeny is congruent with the ITS phylogeny for the diploid taxa of *Panax*. This DNA region has phylogenetic utility at the lower taxonomic level for other flowering plants. The trnC-trnD region includes the trnC-petN intergenic spacer, the petN gene, the petN-psbM intergenic spacer, the psbM gene, and the psbM-trnD intergenic spacer. The petN and psbM genes are small, 90 and 104–114 bp across angiosperms, respectively, and have conserved sequences. The entire trnC-trnD region can be amplified by universal primers for many flowering plant groups. In *Panax*, the trnC-trnD region provides a similar number of informative phylogenetic characters as the ITS regions and a slightly higher number of informative characters than the cp ndhF gene.

Cp trnK and nuclear 18*S* rRNA gene sequences of 13 *Panax* taxa were used to construct phylogenetic relationship and to assist taxonomic delimitation within this genus (Zhu et al., 2003). The length of trnK gene sequence varied from 2537 to 2573 bp among taxa, while matK sequences, embedded in the intron of trnK gene, were 1512 bp in all taxa. Species-specific trnK/matK sequence provided much insight into phylogeny and taxonomy of this genus (Figure 15.5b). In ITS + matK phylogenetic reconstruction, series *Panax* (subgenus *Panax*) consists of two groups. In one group, *P. ginseng* is closer to *P. quinquefolius* than to *P. japonicas*; in the other, *P. zingiberensis* is closer to *P. vietnamensis* than to *P. bipinnatifidus*.

18*S* rRNA gene sequences were 1808 or 1809 bp in length (Zhu et al., 2003). Only nine types of 18*S* rRNA sequences were observed among 13 taxa. Parsimony and neighbor-joining analyses of the combined datasets of trnK-18*S* rRNA gene sequences yielded a well-resolved phylogeny within genus *Panax*, which had three main clades. *P. pseudoginseng* and *P. stipuleanatus* formed a sister group located at a basal position in the phylogenetic tree (also in Figure 15.5b and c), which suggested that they are relatively primitive. Monophyly of *P. ginseng*, *P. japonicus* (Japan), and *P. quinquefolius* was well supported (northern clade; also in Figure 15.5b and c). The remaining taxa distributed in southern parts of Asia formed a relatively large clade (southern clade). The taxonomic debated taxa traditionally treated as subspecies or varieties of *P. japonicus* or *P. pseudoginseng* showed various nucleotide sequences, but all fell into one cluster. These taxa might differentiate from a common ancestor and are in a period of high variation, which is revealed not only on morphological appearance but also on molecular divergence. Comparing trnK, but not 18*S* rRNA sequences, among 13 *Panax* taxa is useful for the identification of ginseng drugs.

Eleven candidate DNA barcoding loci were screened to establish an accurate and effective *Panax* species identification system, for both commercial and conservation purposes (Zuo et al., 2011). Considerable differences were found in the performance of the potential barcoding regions. The sequencing of atpF-atpH was unsuccessful due to poly-*N* structures. The RbcL, RpoB, and RpoC1 regions were found to be mostly invariable, with only four to eight variable sites. MatK, PsbK-I, PsbM-TrnD, Rps16, and NAD1 could identify four to six out of eight considerably divergent species but only one to five out of 19 clusters within the *P. bipinnatifidus* species group. PsbA-trnH and ITS were most variable, in both species and cluster identifications. The combination of psbA-trnH and ITS is sufficient for identifying all the species and clusters in the genus.

15.5 Metabolomics

Reverse-phase high-performance liquid chromatography (RP-HPLC) method was developed to compare triterpene saponins of ginseng drugs (Zhu et al., 2004). Eleven ginsenosides, which represent four types of typical sapogenins (Figure 15.1), were chosen as quantitative standards to characterize the chemical profile of each ginseng drug and probe the relationship between genetic varieties and chemical constituent pattern. The ginsenoside compositions in Ginseng drugs of different origins had considerable variability. Total saponin contents varied by 10-fold from the highest drug to the lowest one. Chikusetsu-ninjin derived from *P. japonicus* (Japan) has the highest content (192.80–296.18 mg/g) and *P. ginseng* the lowest (5.78–15.63 mg/g). Two main groups (I and II) suggested by phytochemical data were clearly observed: group I mainly containing dammarane saponins consisted of *P. ginseng*, *P. quinquefolius*, *P. notoginseng*, *P. vietnamensis*, and *P. vietnamensis* var. *fuscidiscus*, and group II containing abundant OA saponins consisted of *P. japonicus* (Japan), *P. zingiberensis*, *P. japonicus* (China), *P. japonicus* var. *angustifolius*, *P. japonicus* var. *major*, *P. japonicus* var. *bipinnatifidus*, and *P. stipuleanatus*. The ratios of dammarane saponins to OA saponins were >1.9 and <0.25 for groups I and II, respectively. The drug samples derived from the same botanical origin had similar constituent patterns. Similarities of chemical constitution were seen among the closely related taxa (Figure 15.5), but *P. japonicus* (Japan) and *P. zingiberensis* were exceptions.

Although both rhizome and root of *P. notoginseng* are officially utilized as notoginseng in *Chinese Pharmacopoeia*, individual parts of the root were differentially used in practice. Quantitative comparison and metabolite profiling of different portions derived from the whole root, as well as commercial unassorted samples, showed an overall higher content of saponins in rhizome, followed by main root, branch root, and fibrous root (Wang et al., 2014b). Ginsenoside Rb2 was proposed as a potential marker with content of 0.5 mg/g as a threshold value for the discrimination of rhizome and other parts. Multivariate analysis of metabolite profile further suggested 32 saponins as potential markers for the discrimination of different parts of notoginseng, which is important for the rational exploitation of individual parts.

In total, 86 components were identified from leaf samples of *P. ginseng* and *P. quinquefolius* by ultrahigh performance liquid chromatography (UPLC) quadrupole/time-of-flight mass spectrometry (QTOF-MS/MS) (Mao et al., 2014), in which nine ginsenosides could be regarded as the characteristic chemical markers for the discrimination of *P. ginseng* from *P. quinquefolius*. UPLC–QTOF-MS is also useful in the discrimination of processed ginseng species (Park et al., 2014b). Metabolomics approach is a powerful tool for rapid finding of characteristic markers for the quality control of *Panax* leaves.

Quantitative HPLC coupled with PCA was used in the detection of adulteration of notoginseng root extract with other *Panax* species (Wang et al., 2009). HPLC and multivariate analysis were also used in adulteration and cultivation region identification of American ginseng (Yu et al., 2014). The adulteration of ginseng of different ages is a serious problem. It is crucial to develop objective criteria or standard protocols for differentiating ginseng root samples according to their cultivation age. GC/MS combined with multivariate statistical analysis with variable selection was used to obtain metabolic profiling, and an optimal partial least squares discriminant analysis (PLS-DA) model was developed for the differentiation of ginseng according to cultivation age (Park et al., 2013). Relative levels of various metabolites, such as amino acids, alcohols, fatty acids, organic acids, and sugars, were measured for various cultivation ages. Increasing cultivation age resulted in the production of more panaxynol and panaxydol, which are active polyacetylene compounds in ginseng. Optimized PLS-DA models for the prediction of ginseng age were obtained by selecting variables based on a variable importance in the projection cut-off value of 1.3. Proline, glucaric acid, mannose, gluconic acid, GlcA, myoinositol, panaxydol, and panaxynol are suggested as key and relevant compounds with which to differentiate the age of ginseng samples.

15.6 Conclusions

Wild ginseng is a scarce and rare commodity. Wild American ginseng is near extinction due to multiple factors such as climate change, habitat fragmentation, and the supply to the Asian traditional medicine market. Field cultivation of the ginseng plant is a time-consuming and labor-intensive process. Sustainable biotechnological production of ginsenosides should be highlighted. Microbial cell factories and transgenic plants hold great promise in large-scale production of cost-effective ginseng drugs. Genomic, proteomic, and metabolomics studies of *Panax* plants provide essential information for sustainable utilization of *Panax* medicinal resources.

References

Calderón, M.M., Chairez, C.L., Gordon, L.A., et al., 2014. Influence of *Panax ginseng* on the steady state pharmacokinetic profile of lopinavir-ritonavir in healthy volunteers. Pharmacotherapy 34 (11), 1151–1158. http://dx.doi.org/10.1002/phar.1473.

Chen, C.F., Chiou, W.F., Zhang, J.T., 2008. Comparison of the pharmacological effects of *Panax ginseng* and *Panax quinquefolium*. Acta Pharmacol. Sin. 29 (9), 1103–1108.

Chen, L., Dai, J., Wang, Z., et al., 2014a. The antidepressant effects of ginseng total saponins in male C57BL/6N mice by enhancing hippocampal inhibitory phosphorylation of GSK-3β. Phytother. Res. 28 (7), 1102–1106.

Chen, X., Liao, B., Song, J., et al., 2013. A fast SNP identification and analysis of intraspecific variation in the medicinal *Panax* species based on DNA barcoding. Gene 530 (1), 39–43.

Chen, Y.Q., Rong, L., Qiao, J.O., 2014b. Anti-inflammatory effects of *Panax notoginseng* saponins ameliorate acute lung injury induced by oleic acid and lipopolysaccharide in rats. Mol. Med. Rep. 10 (3), 1400–1408.

Cho, H.M., Kang, Y.H., Yoo, H., et al., 2014. *Panax* red ginseng extract regulates energy expenditures by modulating PKA dependent lipid mobilization in adipose tissue. Biochem. Biophys. Res. Commun. 447 (4), 644–648.

Choi, J.S., Chun, K.S., Kundu, J., Kundu, J.K., 2013. Biochemical basis of cancer chemoprevention and/or chemotherapy with ginsenosides. Int. J. Mol. Med. 32 (6), 1227–1238.

Choi, K.T., 2008. Botanical characteristics, pharmacological effects and medicinal components of Korean *Panax ginseng* C A Meyer. Acta Pharmacol. Sin. 29 (9), 1109–1118.

Chung, I.M., Kim, Y.O., Ali, M., et al., 2014. Triterpene glycosides from red ginseng marc and their anti-inflammatory activities. Bioorg. Med. Chem. Lett. 24 (17), 4203–4208.

Cui, X.M., Lo, C.K., Yip, K.L., et al., 2003. Authentication of *Panax notoginseng* by 5S-rRNA spacer domain and random amplified polymorphic DNA (RAPD) analysis. Planta Med. 69 (6), 584–586.

Fatmawati, S., Ersam, T., Yu, H., et al., 2014. 20(S)-Ginsenoside Rh2 as aldose reductase inhibitor from *Panax ginseng*. Bioorg. Med. Chem. Lett. 24 (18), 4407–4409.

Gao, B., Shi, H.L., Li, X., et al., 2014. p38 MAPK and ERK1/2 pathways are involved in the pro-apoptotic effect of notoginsenoside Ft1 on human neuroblastoma SH-SY5Y cells. Life Sci. 108 (2), 63–70.

Gao, X., Zhi, Y., Sun, L., et al., 2013. The inhibitory effects of a rhamnogalacturonan I (RG-I) domain from ginseng pectin on galectin-3 and its structure-activity relationship. J. Biol. Chem. 288 (47), 33953–33965.

Guo, C.Y., Ma, X.J., Wang, J.S., et al., 2015. Effects of *Panax quinquefolium* saponin on phosphatidylinositol 3-kinase/serine threonine kinase pathway of neonatal rat myocardial cells subjected to hypoxia. Chin. J. Integr. Med. 21 (5), 384–388.

Hao, D.C., Xiao, P.G., 2014. Network pharmacology: a rosetta stone for traditional chinese medicine. Drug Dev. Res. 75 (5), 299–312.

He, R., Liu, Q., Liu, Y.H., et al., 2014. Chemical constituents of leaves of *Panax japonicus* var. *major*. Zhongguo Zhong Yao Za Zhi 39 (9), 1635–1638.

Hu, B.Y., Liu, X.J., Qiang, R., et al., 2014. Treatment with ginseng total saponins improves the neurorestoration of rat after traumatic brain injury. J. Ethnopharmacol. 155 (2), 1243–1255.

Huang, Y., Yu, J., Wan, F., et al., 2014. Panaxatriol saponins attenuated oxygen-glucose deprivation injury in PC12 cells via activation of PI3K/Akt and Nrf2 signaling pathway. Oxid. Med. Cell. Longev. 2014, 978034.

Im, D.S., Nah, S.Y., 2013. Yin and Yang of ginseng pharmacology: ginsenosides vs gintonin. Acta Pharmacol. Sin. 34 (11), 1367–1373.

Jhun, J., Lee, J., Byun, J.K., et al., 2014. Red ginseng extract ameliorates autoimmune arthritis via regulation of STAT3 pathway, Th17/Treg balance, and osteoclastogenesis in mice and human. Mediat. Inflamm. 2014, 351856.

Jiang, R., Sun, L., Wang, Y., et al., 2014. Chemical composition, and cytotoxic, antioxidant and antibacterial activities of the essential oil from ginseng leaves. Nat. Prod. Commun. 9 (6), 865–868.

Jiao, L., Li, B., Wang, M., et al., 2014a. Antioxidant activities of the oligosaccharides from the roots, flowers and leaves of *Panax ginseng* C.A. Meyer. Carbohydr. Polym. 106, 293–298.

Jiao, L., Zhang, X., Li, B., et al., 2014b. Anti-tumour and immunomodulatory activities of oligosaccharides isolated from *Panax ginseng* C.A. Meyer. Int. J. Biol. Macromol. 65, 229–233.

Jovanovski, E., Bateman, E.A., Bhardwaj, J., et al., 2014a. Effect of Rg3-enriched Korean red ginseng (*Panax ginseng*) on arterial stiffness and blood pressure in healthy individuals: a randomized controlled trial. J. Am. Soc. Hypertens. 8 (8), 537–541.

Jovanovski, E., Peeva, V., Sievenpiper, J.L., et al., 2014b. Modulation of endothelial function by Korean red ginseng (*Panax ginseng* C.A. Meyer) and its components in healthy individuals: a randomized controlled trial. Cardiovasc. Ther. 32 (4), 163–169.

Jung, J., Kim, K.H., Yang, K., et al., 2014. Practical application of DNA markers for high-throughput authentication of *Panax ginseng* and *Panax quinquefolius* from commercial ginseng products. J. Ginseng Res. 38 (2), 123–129.

Kim, D.G., Jung, K.H., Lee, D.G., et al., 2014a. 20(S)-Ginsenoside Rg3 is a novel inhibitor of autophagy and sensitizes hepatocellular carcinoma to doxorubicin. Oncotarget 5 (12), 4438–4451.

Kim, D.H., 2012. Chemical diversity of *Panax ginseng*, *Panax quinquifolium*, and *Panax notoginseng*. J. Ginseng Res. 36 (1), 1–15.

Kim, J.A., Son, J.H., Song, S.B., et al., 2013. Sterols isolated from seeds of *Panax ginseng* and their antiinflammatory activities. Pharmacogn. Mag. 9 (34), 182–185.

Kim, M.Y., Cho, J.Y., 2013. 20S-dihydroprotopanaxatriol modulates functional activation of monocytes and macrophages. J. Ginseng Res. 37 (3), 300–307.

Kim, N.H., Choi, H.I., Ahn, I.O., Yang, T.J., 2012. EST-SSR marker sets for practical authentication of all nine registered ginseng cultivars in Korea. J. Ginseng Res. 36 (3), 298–307.

Kim, N.H., Choi, H.I., Kim, K.H., et al., 2014b. Evidence of genome duplication revealed by sequence analysis of multi-loci expressed sequence tag-simple sequence repeat bands in *Panax ginseng* Meyer. J. Ginseng Res. 38 (2), 130–135.

Kim, S.S., Jang, H.J., Oh, M.Y., et al., 2014c. Ginsenoside Rg3 enhances islet cell function and attenuates apoptosis in mouse islets. Transplant. Proc. 46 (4), 1150–1155.

Kim, Y.J., Choi, W.I., Jeon, B.N., et al., 2014d. Stereospecific effects of ginsenoside 20-Rg3 inhibits TGF-β1-induced epithelial-mesenchymal transition and suppresses lung cancer migration, invasion and anoikis resistance. Toxicology 322, 23–33.

Kochkin, D.V., Kachala, V.V., Shashkov, A.S., et al., 2013. Malonyl-ginsenoside content of a cell-suspension culture of *Panax japonicus* var. *repens*. Phytochemistry 93, 18–26.

Le, T.H., Lee, S.Y., Kim, T.R., et al., 2014. Processed Vietnamese ginseng: preliminary results in chemistry and biological activity. J. Ginseng Res. 38 (2), 154–159.

Lee, H., Kim, M., Shik Shin, S., Yoon, M., 2014a. Ginseng treatment reverses obesity and related disorders by inhibiting angiogenesis in female db/db mice. J. Ethnopharmacol. 155 (2), 1342–1352.

Lee, J.S., Hwang, H.S., Ko, E.J., et al., 2014b. Immunomodulatory activity of red ginseng against influenza A virus infection. Nutrients 6 (2), 517–529.

Lee, J.S., Ko, E.J., Hwang, H.S., et al., 2014c. Antiviral activity of ginseng extract against respiratory syncytial virus infection. Int. J. Mol. Med. 34 (1), 183–190.

Lee, J.W., Bang, K.H., Kim, Y.C., et al., 2012. CAPS markers using mitochondrial consensus primers for molecular identification of *Panax* species and Korean ginseng cultivars (*Panax ginseng* C. A. Meyer). Mol. Biol. Rep. 39 (1), 729–736.

Lee, L.S., Cho, C.W., Hong, H.D., et al., 2013a. Hypolipidemic and antioxidant properties of phenolic compound-rich extracts from white ginseng (*Panax ginseng*) in cholesterol-fed rabbits. Molecules 18 (10), 12548–12560.

Lee, M.H., Lee, B.H., Lee, S., Choi, C., 2013b. Reduction of hepatitis A virus on FRhK-4 cells treated with Korean red ginseng extract and ginsenosides. J. Food Sci. 78 (9), M1412–M1415.

Lee, Y., Kim, K.T., Kim, S.S., et al., 2014d. Inhibitory effects of ginseng seed on melanin biosynthesis. Pharmacogn. Mag. 10 (Suppl. 2), S272–S275.

Lee, C., Wen, J., 2004. Phylogeny of *Panax* using chloroplast trnC-trnD intergenic region and the utility of trnC-trnD in interspecific studies of plants. Mol. Phylogenet. Evol. 31 (3), 894–903.

Li, C., Zhu, Y., Guo, X., et al., 2013a. Transcriptome analysis reveals ginsenosides biosynthetic genes, microRNAs and simple sequence repeats in *Panax ginseng* C. A. Meyer. BMC Genomics 14, 245.

Li, D., Cao, J., Bi, X., et al., 2014a. New dammarane-type triterpenoids from the leaves of *Panax notoginseng* and their protein tyrosine phosphatase 1B inhibitory activity. J. Ginseng Res. 38 (1), 28–33.

Li, J., Wei, Q., Zuo, G.W., et al., 2014b. Ginsenoside Rg1 induces apoptosis through inhibition of the EpoR-mediated JAK2/STAT5 signalling pathway in the TF-1/Epo human leukemia cell line. Asian Pac. J. Cancer Prev. 15 (6), 2453–2459.

Li, M.R., Wang, X.F., Zhang, C., et al., 2013b. A simple strategy for development of single nucleotide polymorphisms from non-model species and its application in *Panax*. Int. J. Mol. Sci. 14 (12), 24581–24591.

Li, R., Ma, P.F., Wen, J., Yi, T.S., 2013c. Complete sequencing of five araliaceae chloroplast genomes and the phylogenetic implications. PLoS One 8 (10), e78568.

Liang, C., Ding, Y., Nguyen, H.T., et al., 2010. Oleanane-type triterpenoids from *Panax stipuleanatus* and their anticancer activities. Bioorg. Med. Chem. Lett. 20 (23), 7110–7115.

Liang, C., Ding, Y., Song, S.B., et al., 2013. Oleanane-triterpenoids from *Panax stipuleanatus* inhibit NF-κB. J. Ginseng Res. 37 (1), 74–79.

Liu, C., Hu, M.Y., Zhang, M., et al., 2014a. Association of GLP-1 secretion with anti-hyperlipidemic effect of ginsenosides in high-fat diet fed rats. Metabolism. 63 (10), 1342–1351.

Liu, H., Xia, T., Zuo, Y.J., et al., 2011. Development and characterization of microsatellite markers for *Panax notoginseng* (Araliaceae), a Chinese traditional herb. Am. J. Bot. 98 (10), e274–e276.

Liu, H.Z., Liu, Z.L., Zhao, S.P., et al., 2014b. Protective mechanism of *P. notoginseng* saponins on rat hemorrhagic shock model in recovery stage. Cell Biochem. Biophys. 70 (3), 1719–1724.

Liu, J., Wang, Y., Qiu, L., et al., 2014c. Saponins of *Panax notoginseng*: chemistry, cellular targets and therapeutic opportunities in cardiovascular diseases. Expert Opin. Investig. Drugs 23 (4), 523–539.

Liu, J.P., Tian, X., Liu, H.Y., et al., 2013. Two novel dammarane-type compounds from the leaves and stems of *Panax quinquefolium* L. J. Asian Nat. Prod. Res. 15 (9), 974–978.

Liu, L., Zhu, L., Zou, Y., et al., 2014d. *Panax notoginseng* saponins promotes stroke recovery by influencing expression of Nogo-A, NgR and p75NGF, in vitro and in vivo. Biol. Pharm. Bull. 37 (4), 560–568.

Lobina, C., Carai, M.A., Loi, B., et al., 2014. Protective effect of *Panax ginseng* in cisplatin-induced cachexia in rats. Future Oncol. 10 (7), 1203–1214.

Luo, H., Sun, C., Sun, Y., et al., 2011. Analysis of the transcriptome of Panax notoginseng root uncovers putative triterpene saponin-biosynthetic genes and genetic markers. BMC Genomics 12 (Suppl. 5), S5.

Ma, B., Meng, X., Wang, J., et al., 2014a. Notoginsenoside R1 attenuates amyloid-β-induced damage in neurons by inhibiting reactive oxygen species and modulating MAPK activation. Int. Immunopharmacol. 22 (1), 151–159.

Ma, L., Liu, H., Xie, Z., et al., 2014b. Ginsenoside Rb3 protects cardiomyocytes against ischemia-reperfusion injury via the inhibition of JNK-mediated NF-κB pathway: a mouse cardiomyocyte model. PLoS One 9 (8), e103628.

Mao, Q., Bai, M., Xu, J.D., et al., 2014. Discrimination of leaves of *Panax ginseng* and *P. quinquefolius* by ultra-high performance liquid chromatography quadrupole/time-of-flight mass spectrometry based metabolomics approach. J. Pharm. Biomed. Anal. 97, 129–140.

Meng, X., Wang, M., Sun, G., et al., 2014. Attenuation of Aβ25-35-induced parallel autophagic and apoptotic cell death by gypenoside XVII through the estrogen receptor-dependent activation of Nrf2/ARE pathways. Toxicol. Appl. Pharmacol. 279 (1), 63–75.

Mucalo, I., Jovanovski, E., Vuksan, V., et al., 2014. American ginseng extract (*Panax quinquefolius* L.) is safe in long-term use in type 2 diabetic patients. Evid. Based Complement. Alternat. Med. 2014, 969168.

Nguyen, H.T., Tran, H.Q., Nguyen, T.T., et al., 2011. Oleanolic triterpene saponins from the roots of *Panax bipinnatifidus*. Chem. Pharm. Bull. 59 (11), 1417–1420.

Park, E.H., Kim, D.R., Kim, H.Y., et al., 2014a. *Panax ginseng* induces the expression of CatSper genes and sperm hyperactivation. Asian J. Androl. 16 (6), 845–851.

Park, H.E., Lee, S.Y., Hyun, S.H., et al., 2013. Gas chromatography/mass spectrometry-based metabolic profiling and differentiation of ginseng roots according to cultivation age using variable selection. J. AOAC Int. 96 (6), 1266–1272.

Park, H.W., In, G., Kim, J.H., et al., 2014b. Metabolomic approach for discrimination of processed ginseng genus (*Panax ginseng* and *Panax quinquefolius*) using UPLC-QTOF MS. J. Ginseng Res. 38 (1), 59–65.

Poudyal, D., Cui, X., Le, P.M., et al., 2013. A key role of microRNA-29b for the suppression of colon cancer cell migration by American ginseng. PLoS One 8 (10), e75034.

Reunova, G.D., Kats, I.L., Muzarok, T.I., Zhuravlev, I.N., 2010. Polymorphism of RAPD, ISSR and AFLP markers of the *Panax ginseng* C. A. Meyer (Araliaceae) genome. Genetika 46 (8), 1057–1066.

Rhee, M.Y., Cho, B., Kim, K.I., et al., 2014. Blood pressure lowering effect of Korea ginseng derived ginseol K-g1. Am. J. Chin. Med. 42 (3), 605–618.

Samimi, R., Xu, W.Z., Lui, E.M., Charpentier, P.A., 2014. Isolation and immunosuppressive effects of 6″-O-acetylginsenoside Rb1 extracted from North American ginseng. Planta Med. 80 (6), 509–516.

Shishtar, E., Jovanovski, E., Jenkins, A., Vuksan, V., 2014. Effects of Korean white ginseng (*Panax Ginseng* C.A. Meyer) on vascular and glycemic health in type 2 Diabetes: results of a randomized, double blind, placebo-controlled, multiple-crossover, acute dose escalation trial. Clin. Nutr. Res. 3 (2), 89–97.

Siddiqi, M.H., Siddiqi, M.Z., Ahn, S., et al., 2014. Ginsenoside Rh1 induces mouse osteoblast growth and differentiation through the bone morphogenetic protein 2/runt-related gene 2 signalling pathway. J. Pharm. Pharmacol. 66 (12), 1763–1773.

Subramaniyam, S., Mathiyalagan, R., Natarajan, S., et al., 2014. Transcript expression profiling for adventitious roots of *Panax ginseng* Meyer. Gene 546 (1), 89–96.

Sun, C., Li, Y., Wu, Q., et al., 2010. De novo sequencing and analysis of the American ginseng root transcriptome using a GS FLX Titanium platform to discover putative genes involved in ginsenoside biosynthesis. BMC Genomics 11, 262.

Tan, S., Yu, W., Lin, Z., et al., 2014. Anti-inflammatory effect of ginsenoside Rb1 contributes to the recovery of gastrointestinal motility in the rat model of postoperative ileus. Biol. Pharm. Bull. 37 (11), 1788–1794.

Tran, Q.L., Adnyana, I.K., Tezuka, Y., et al., 2002. Hepatoprotective effect of majonoside R2, the major saponin from Vietnamese ginseng (*Panax vietnamensis*). Planta Med. 68 (5), 402–406.

Tran, T.L., Kim, Y.R., Yang, J.L., et al., 2014. Dammarane triterpenes from the leaves of *Panax ginseng* enhance cellular immunity. Bioorg. Med. Chem. 22 (1), 499–504.

Wang, C.Z., Ni, M., Sun, S., et al., 2009. Detection of adulteration of notoginseng root extract with other panax species by quantitative HPLC coupled with PCA. J. Agric. Food Chem. 57 (6), 2363–2367.

Wang, J.R., Yau, L.F., Gao, W., et al., 2014a. Quantitative comparison and metabolite profiling of saponins in different parts of the root of *Panax notoginseng*. J. Agric. Food Chem. 62 (36), 9024–9034.

Wang, L., Lu, A.P., Yu, Z.L., et al., 2014b. The melanogenesis-inhibitory effect and the percutaneous formulation of ginsenoside Rb1. AAPS PharmSciTech. 15 (5), 1252–1262.

Wang, P., Cui, J., Du, X., et al., 2014c. *Panax notoginseng* saponins (PNS) inhibits breast cancer metastasis. J. Ethnopharmacol. 154 (3), 663–671.

Wang, M., Wu, B., Chen, C., Lu, S., 2015. Identification of mRNA-like non-coding RNAs and validation of a mighty one named MAR in *Panax ginseng*. J. Integr. Plant Biol. 57 (3), 256–270.

Wei, N., Zhang, C., He, H., et al., 2014. Protective effect of saponins extract from *Panax japonicus* on myocardial infarction: involvement of NF-κB, Sirt1 and mitogen-activated protein kinase signalling pathways and inhibition of inflammation. J. Pharm. Pharmacol. 66 (11), 1641–1651.

Wu, D., Austin, R.S., Zhou, S., Brown, D., 2013. The root transcriptome for North American ginseng assembled and profiled across seasonal development. BMC Genomics 14, 564.

Wu, H., Høiby, N., Yang, L., et al., 2014a. Effects of radix ginseng on microbial infections: a narrative review. J. Tradit. Chin. Med. 34 (2), 227–233.

Wu, L., Zhang, A.L., Di, Y.M., et al., 2014b. *Panax ginseng* therapy for chronic obstructive pulmonary disease: a clinical trial protocol and pilot study. Chin. Med. 9, 20.

Wu, L.L., Jia, B.H., Sun, J., et al., 2014c. Protective effects of ginsenoside Rb1 on septic rats and its mechanism. Biomed. Environ. Sci. 27 (4), 300–303.

Wu, Q., Song, J., Sun, Y., et al., 2010. Transcript profiles of *Panax quinquefolius* from flower, leaf and root bring new insights into genes related to ginsenosides biosynthesis and transcriptional regulation. Physiol. Plant. 138 (2), 134–149.

Xu, Y., Ding, J., Ma, X.P., et al., 2014a. Treatment with *Panax ginseng* antagonizes the estrogen decline in ovariectomized mice. Int. J. Mol. Sci. 15 (5), 7827–7840.

Xu, Y., Lin, L., Tang, L., et al., 2014b. Notoginsenoside R1 attenuates hypoxia and hypercapnia-induced vasoconstriction in isolated rat pulmonary arterial rings by reducing the expression of ERK. Am. J. Chin. Med. 42 (4), 799–816.

Yang, J.L., Ha, T.K., Dhodary, B., et al., 2014a. Dammarane triterpenes as potential SIRT1 activators from the leaves of *Panax ginseng*. J. Nat. Prod. 77 (7), 1615–1623.

Yang, M., Lee, H.S., Hwang, M.W., Jin, M., 2014b. Effects of Korean red ginseng (*Panax ginseng* Meyer) on bisphenol A exposure and gynecologic complaints: single blind, randomized clinical trial of efficacy and safety. BMC Complement. Altern. Med. 14, 265.

Yang, Q., Wang, X., Cui, J., et al., 2014c. Bidirectional regulation of angiogenesis and miR-18a expression by PNS in the mouse model of tumor complicated by myocardial ischemia. BMC Complement. Altern. Med. 14, 183.

Yang, W.Z., Hu, Y., Wu, W.Y., et al., 2014d. Saponins in the genus *Panax* L. (Araliaceae): a systematic review of their chemical diversity. Phytochemistry 106C, 7–24.

Yang, X., Wang, R., Zhang, S., et al., 2014e. Polysaccharides from *Panax japonicus* C.A. Meyer and their antioxidant activities. Carbohydr. Polym. 101, 386–391.

Yang, X., Xiong, X., Wang, H., Wang, J., 2014f. Protective effects of panax notoginseng saponins on cardiovascular diseases: a comprehensive overview of experimental studies. Evid. Based Complement. Alternat. Med. 2014, 204840.

Yoshizaki, K., Devkota, H.P., Fujino, H., Yahara, S., 2013a. Saponins composition of rhizomes, taproots, and lateral roots of Satsuma-ninjin (*Panax japonicus*). Chem. Pharm. Bull. 61 (3), 344–350.

Yoshizaki, K., Devkota, H.P., Yahara, S., 2013b. Four new triterpenoid saponins from the leaves of *Panax japonicus* grown in southern Miyazaki prefecture. Chem. Pharm. Bull. 61 (3), 273–278.

Yoshizaki, K., Murakami, M., Fujino, H., et al., 2012. New triterpenoid saponins from fruit specimens of *Panax japonicus* collected in Toyama prefecture and Hokkaido (2). Chem. Pharm. Bull. 60 (6), 728–735.

Yu, C., Wang, C.Z., Zhou, C.J., et al., 2014. Adulteration and cultivation region identification of American ginseng using HPLC coupled with multivariate analysis. J. Pharm. Biomed. Anal. 99, 8–15.

Yuan, D., Wan, J.Z., Deng, L.L., et al., 2014. Chikusetsu saponin V attenuates MPP^+-induced neurotoxicity in SH-SY5Y cells via regulation of Sirt1/Mn-SOD and GRP78/caspase-12 pathways. Int. J. Mol. Sci. 15 (8), 13209–13222.

Zhang, C., Wang, Y., Wang, M., et al., 2014a. Rapeseed oil and ginseng saponins work synergistically to enhance Th1 and Th2 immune responses induced by the foot-and-mouth disease vaccine. Clin. Vaccine Immunol. 21 (8), 1113–1119.

Zhang, H., Wang, H.F., Liu, Y., et al., 2014b. The haematopoietic effect of *Panax japonicus* on blood deficiency model mice. J. Ethnopharmacol. 154 (3), 818–824.

Zhang, X., Shi, M., Ye, R., et al., 2014c. Ginsenoside Rd attenuates tau protein phosphorylation via the PI3K/AKT/GSK-3β pathway after transient forebrain ischemia. Neurochem. Res. 39 (7), 1363–1373.

Zhou, J.S., Wang, J.F., He, B.R., et al., 2014a. Ginsenoside Rd attenuates mitochondrial permeability transition and cytochrome C release in isolated spinal cord mitochondria: involvement of kinase-mediated pathways. Int. J. Mol. Sci. 15 (6), 9859–9877.

Zhou, N., Tang, Y., Keep, R.F., et al., 2014b. Antioxidative effects of *Panax notoginseng* saponins in brain cells. Phytomedicine 21 (10), 1189–1195.

Zhu, D., Liu, M., Yang, Y., et al., 2014a. Ginsenoside Rd ameliorates experimental autoimmune encephalomyelitis in C57BL/6 mice. J. Neurosci. Res. 92 (9), 1217–1226.

Zhu, J., Mu, X., Zeng, J., et al., 2014b. Ginsenoside Rg1 prevents cognitive impairment and hippocampus senescence in a rat model of D-galactose-induced aging. PLoS One 9 (6), e101291.

Zhu, S., Fushimi, H., Cai, S., Komatsu, K., 2003. Phylogenetic relationship in the genus *Panax*: inferred from chloroplast trnK gene and nuclear 18S rRNA gene sequences. Planta Med. 69 (7), 647–653.

Zhu, S., Zou, K., Fushimi, H., et al., 2004. Comparative study on triterpene saponins of Ginseng drugs. Planta Med. 70 (7), 666–677.

Zou, K., Liu, Z.X., Zhu, S., et al., 2004. Research of ginsenosides in kou zi qi using HPLC-MS-MS. Yao Xue Xue Bao 39 (5), 385–388.

Zuo, Y., Chen, Z., Kondo, K., et al., 2011. DNA barcoding of *Panax* species. Planta Med. 77 (2), 182–187.

Index

Note: Page numbers followed by *f* indicate figures and *t* indicate tables.

D

E

F

G

N

O

P

Q

R

Printed in the United States
By Bookmasters